Robert Dundas Thomson

Dictionary of Chemistry

with Its Applications

to Mineralogy, Physiology, and the Arts

Elibron Classics
www.elibron.com

Elibron Classics series.

ISBN 0-543-78336-7 (paperback)
ISBN 0-543-78335-9 (hardcover)

This Elibron Classics Replica Edition is an unabridged facsimile
of the edition published byRichard Griffin and Company, London and Glasgc

THOMSON'S

CYCLOPÆDIA

OF

CHEMISTRY.

In active preparation,

UNIFORM WITH THIS VOLUME,

CYCLOPÆDIA OF THE PHYSICAL SCIENCES,

COMPRISING

ACOUSTICS.	MAGNETISM.
ASTRONOMY.	MATHEMATICS, PHILOSOPHY OF.
DYNAMICS.	METEOROLOGY.
ELECTRICITY.	OPTICS.
HEAT.	PNEUMATICS.
HYDRODYNAMICS.	STATICS.

With Numerous Illustrations.

BY

JOHN P. NICHOL, LL.D.

PROFESSOR OF PRACTICAL ASTRONOMY IN THE UNIVERSITY OF GLASGOW.

DICTIONARY

OF

CHEMISTRY,

WITH ITS APPLICATIONS

TO

MINERALOGY, PHYSIOLOGY, AND THE ARTS.

BY

ROBERT DUNDAS THOMSON, M.D., F.R.S.L. & E., F.C.S.,

FELLOW OF THE ROYAL MEDICAL AND CHIRURGICAL SOCIETY; HONORARY MEMBER OF THE PHILOSOPHICAL SOCIETY OF GLASGOW; PRESIDENT OF THE BRITISH METEOROLOGICAL SOCIETY, ETC.; AND PROFESSOR OF CHEMISTRY IN ST THOMAS'S HOSPITAL COLLEGE, LONDON.

With Numerous Illustrations.

NEW EDITION.

LONDON AND GLASGOW:

RICHARD GRIFFIN AND COMPANY,

Publishers to the University of Glasgow.

PREFACE.

THE present volume was projected two years ago, but has been much interrupted in its progress. It is intended to serve as an introduction and familiar work of reference to the Science of Chemistry, a study in which, from its rapid advance, new names, and facts are accumulating with amazing rapidity. Had the object been to write an exhaustive treatise, the task would have been a comparatively light one, as materials had, to a great extent, been collected; but as the work was destined to be comprised within a very limited compass, the labour of selection, and therefore of an extensive criticism understood rather than expressed, was necessarily involved in the undertaking. The opportunities which the author has possessed for many years—generously afforded to him by the liberal and intelligent manufacturers of Glasgow and Lancashire—of personally studying practical chemistry on a great scale, induced him to commence a series of researches on the composition of various chemical substances, from the first stage of the process of manufacture, through its varied phases, to the finished condition of the product. He was most energetically assisted in this inquiry by his ardent and indefatigable pupils, many of whom are now filling important positions. The results of several of these investigations have been included in the present volume. And although the author's removal to London has interfered with the prosecution of this branch of his subject to an extent which was not contemplated when the volume was originally projected, he takes the opportunity of referring to the articles on the manufacture of COPPER, KELP, LEAD, CARBONATE OF SODA, POTTERY, as illustrations of a considerable amount of research in this department of practical science. On a future occasion it is anticipated that the results of further investigations will be added, aided by the liberal arrangements of the authorities of the noble Institution with which the author has the honour to be connected, and by the extended field of study presented by a chemical practice in the metropolis.

A familiar acquaintance with the study of ores and minerals for many years, together with the advantages of an extensive private museum, rendered historical by its having been founded by the late Dr. Thomas Thomson, have enabled the author to supply a copious index of mineralogical synonymes, in addition to many new analyses, published for the first time. The preservation of the study of minerals, that is of elements and salts found in nature, as the essential basis of the science of Chemistry, seems so important, that attempts to restrict mineralogical arrangements to mere crystallographic systems, are much to be regretted, since that valuable auxiliary in the discrimination of physical forms is as thoroughly applicable to artificial salts as to those found in nature. Following out this view, many species have been restored, in this volume, which crystallographers have removed from the mineral list, either from their similarity in per centage composition to previously described compounds, or from their amorphous condition, without, perhaps, viewing the subject in the true chemical aspect, which usually presents us with a prospect of the most important bodies in nature, not in a definite shape; but as "clods" and "glebe" without crystalline characters, and therefore in a condition admirably suited to produce that endless variety of form in nature which "even custom cannot stale," and only renders still more worthy of our admiration and regard.

To make the work available as a practical manual, it would be advisable to study it in a systematic manner. The elementary bodies, beginning with oxygen, may be taken up in the order in which they are arranged, under the article ATOMIC WEIGHT, the reading of the context being illustrated by the simple experiments detailed under the various heads, and the theory of the processes by which the elements and their compounds are isolated being thoroughly followed out by the construction of diagrams and formulæ; without the precision of which Chemistry is deprived of its mathematical exactness, which renders it so well calculated to furnish the student with the most happy opportunities for correct inductive reasoning. For the sake of acquiring a knowledge of the language and of the phenomena of the science with as much rapidity as possible, it will be found advantageous to practice the system of testing arranged under the tables headed ANALYSIS QUALITATIVE.

The author can confidently recommend such an experimental introduction to the science after a long experience, being fully satisfied that Chemistry ought to be taught not only by lectures, but by direct manipulation from the very outset of the study, as was recommended in the preface to his "School Chemistry;" and he is convinced that the time has arrived when examining boards must find it incumbent, particularly for the physician, to render practice in the chemical laboratory as imperative as attendance in the dissecting room for the student of anatomy.

In the preparation of the work, the author is desirous of acknowledging his obligations to many friends who have afforded him numerous facilities for acquiring information. To John Wood, Esq., Chairman of the Board of Inland Revenue, he is indebted for many reports connected with the department of Government over which he so ably presides, and for permission to engrave the beautiful drawings by Dr. Joseph Dalton Hooker, illustrative of the anatomy of various seeds and roots used as adulterating agents. From the following pupils he has received much assistance, generally acknowledged throughout the work:—Messrs. George William Brown, John Brown, Andrew Buchanan, Roderick Alison Couper, James King, Robert Kirkwood, James Napier, Robert Thomson Pattison, W. V. Simons, James Taylor, Hugh Brown Tennent, David Walker, John Wilson. His obligations to other friends are recognised in the body of the volume. In addition to many original engravings, the publishers have placed at the author's disposal a series of woodcuts purchased abroad, which have been freely used, and it is hoped will be found to render the descriptions much more intelligible. In conclusion, it may be observed that it is intended to publish periodical additions to the present work, so as to preserve it on a level with the rapid progress of the science. The Supplement at the end of the volume comprehends many substances which have been described since a portion of the work was printed.

SOME ABBREVIATIONS USED IN THIS VOLUME.

F.P.,	Fusing point.
B.P.,	Boiling point.
S.G.,	Specific gravity.
B.B.,	Before the blowpipe.
H,	Hardness.
C,	Carbon.
H,	Hydrogen.
N,	Nitrogen.
O,	Oxygen.
SiO_2,	Silica.
CO_2,	Carbonic acid.
Al_2O_3,	Alumina.
CaO,	Lime.
MgO,	Magnesia.
NaO,	Soda.
KO,	Potash.
MnO,	Protoxide of manganese.
SH,	Sulphuretted hydrogen.
SO_3,	Sulphuric acid.
HCl,	Hydrochloric acid.
NO_5,	Nitric acid.
NH_3,	Ammonia.

For other Symbols, see ATOMIC WEIGHT.

NOTE.—For the sake of precision in describing bodies, the numerical characters are usually stated first. For example, under IRON we read, "*Metallic iron*, Fe 3·5; 28. Spec. grav. 7·645," &c. Fe is the symbol (Ferrum) for iron, 3·5 is the atomic weight by the oxygen scale; 28 the same number multiplied by 8, or the atomic weight of iron, by the hydrogen scale.

CYCLOPÆDIA

OF

CHEMISTRY AND MINERALOGY.

A. The letter *a* commencing a word has often the effect of deprivation, from the Greek privative α or αν: thus, *hydrous*, with water; *anhydrous*, without water; ā ā, in a prescription, denotes *ana ana* of each.

Abichite. Syn. of Terhydrous hexarseniate of copper.

Abietine. A crystalline resin in Strasburg turpentine.

Abrazite. See ZEAGONITE.

Absinthine. $C_{16}H_{11}O_5$. Yellow resinous fluid, becoming crystalline; very bitter, soluble in alcohol and acetic acid; soluble in sulphuric acid, with a red and then blue colour; obtained by alcohol and ether from wormwood root, *Artemisia absinthium*.

Absolute Alcohol. Spirit deprived of water, having a specific gravity of ·794, and consisting of (C_4H_5O, HO) carbon, hydrogen, oxygen, and water. See ALCOHOL.

Absorbent. A body which possesses the property of sucking another into its pores or interstices: thus, a sponge is an absorbent of water.

Absorption. The act by which one substance encloses another in its pores; paper filtration consists of the absorption of water by paper, and its replacement by new additions of water. Charcoal absorbs gases, as ammonia, chlorohydric acid, in great quantity, see CHARCOAL. Water absorbs these gases likewise with great avidity. See WATER.

Acadialite. An amber-coloured variety of chabasite from Nova Scotia.

Acanticonite. A subspecies of epidote from Norway.

Acerdese, or hydrous sesquioxide of manganese. See VARVICITE.

Acetochloryle, or chloracetyle, (C_4Cl_3) the hypothetic base of *chloral*.

Acescent. Substances which have a tendency to become acid are so termed, which is delayed by bitters in solution. Beer is acescent, and is less liable to acescency when much hops are present.

Acetal. A colourless ethereal fluid with a peculiar odour, boiling at 203°·3 or 221°, of spec. grav. ·823; of vapour 4·082; soluble in 6 parts of water, and miscible in all proportions in alcohol; converted into resin of aldehyde by potash; obtained by placing alcohol in the bottom of a jar, and suspending in the jar a number of watch-glasses, containing platinum black moistened with water. In the course of some months acetal, aldehyde, acetic acid, and acetic ether are formed. These are neutralized with chalk and distilled: the distilled fluid is treated repeatedly with chloride of calcium till that salt is no longer moistened; it is then distilled, and as soon as the boiling point rises to 203° it is acetal. It consists of $C_8H_9O_3$, or it may be viewed as 3 atoms oxide of ethyle and 1 atom acetic acid $= C_{16}H_{18}O_6$, or $C_{12}H_{14}O_4$ (Stas.)

Acetanilide. $C_{16}H_9NO_2$, F.P. 233°½. Brilliant colourless plates, soluble in boiling water, alcohol, and ether, distilling without alteration; formed by dropping anhydrous acetic acid into aniline, or acetic chloride into aniline.

Acetates are compounds of acetic acid with 1 or more atoms of basic oxides. They are decomposed by a red heat; the compounds with silver and copper leaving the metal and yielding strong acetic acid. Acetates of alkaline earths yield acetone and a carbonate. They are all soluble in water; the least soluble are the acetates of tungsten, molybdenum, silver, and mercury. The stronger acids displace acetic acid from its combinations. The acetate of alumina is used as a mordant in calico printing; acetate of lead, in printing, and for the formation of chrome yellow; acetate of copper or verdigris, for various purposes.—Acetates are distinguished by their yielding the odour of vinegar when sulphuric acid is poured on them, and by their solutions giving a deep yellow, with a colourless solution of sesquichloride of iron.

Acetate of Ammonia, or spirit of Mindererus, or liquor of acetate of ammonia, difficultly crystalline, (oblique rhombs?) obtained by saturating carbonate of ammonia with acetic acid. It is used as a diaphoretic in medicine, and is said to act as an antidote for a fit of inebriety. It may be made of nearly definite strength by saturating 6 parts of liquor ammoniæ (spec. grav. ·960) with strong acetic acid, and making the mixture up to 24 parts with water.

The *Binacetate* is procured in deliquescent needles by distilling a mixture of equal parts of salammoniac and acetate of potash, or lime; or by adding an additional quantity of acetic acid to the acetate and evaporating: it melts at 169° and sublimes at 250° undecomposed.

Acetate of Potash, in deliquescent plates, destitute of crystalline water, is formed by saturating carbonate of potash with acetic acid, and evaporating; it is decomposed by ignition into acetone, acetic acid, oil, water, inflammable gas, carbonic acid, charcoal, and carbonate of potash.

Binacetate of Potash, in flat pearly prisms, or needles and plates, with 1 atom of water, is formed by adding acetic acid to the acetate, and evaporating. (T. Thomson.)

Acetate of Soda, in oblique rhombic prisms, with 6 atoms of water, is formed by saturating carbonate of soda with acetic acid, and evaporating; it tastes bitter, and fuses under 212°, splitting, on cooling, in all directions; it dissolves in 3·9 water, at 43°; in 1 part of boiling water, 5 alcohol; effloresces in dry air. It is formed on the large scale in the production of pure acetic or pyroligneous acid, by the distillation of wood. The raw pyroligneous acid is distilled and saturated with chalk, and lastly with milk of lime, as it is said that chalk does not entirely remove its acid reaction, (125 to 128 chalk, to 1000 pyroligneous acid.) The concentrated solution is precipitated by sulphate of soda, as long as a precipitate falls. A double salt of gypsum and sulphate of soda forms, acetate of soda remaining in solution. When 125 chalk are required, 800 crystalline sulphate of soda will be necessary; in cooling, some carbonate of soda is added, to remove the last traces of lime; the solution is filtered, evaporated in an iron boiler to the spec. grav. 1·356, (27 Twaddel,) and then crystallized in iron coolers. To purify the crystals from attached black oily matter they are kept melted in an iron boiler until the oil is volatilized, digested in cold water, the clear liquor drawn off, and crystallized.

Acetate of Alumina, Red liquor. An uncrystalline compound, used extensively as a mordant for vegetable colours, parting with its acid when cloth impregnated with it is exposed to the air, and depositing half its base when heated. It is prepared by precipitating acetate of lead by sulphate of alumina or alum, in nearly equal parts, neutralizing any excess of acid by chalk or carbonate of soda, or by adding 3 lbs. alum to every gallon of acetate of lime at 13° Twad. Thus made, it contains some double salt of alumina in solution. Its commercial red colour is sometimes due to peach or Brazil wood.

Acetate of Lime, Pyrolignate of lime. Silky needles when pure; is prepared on the large scale in the manufacture of acetic acid from wood in an amorphous brown mass. In this state I have found it composed of acetic acid 48·22, water 5·62, lime 35·, carbonate of lime 8·17, tary matter 2·99. It is used for preparing acetate of alumina, by decomposing alum or sulphate of alumina.

Acetate of protoxide of Iron, iron liquor, pyrolignate of iron, tincture of iron. Whitish-green needles, very soluble in water, usually in the state of solution as used in printing; obtained by dissolving iron in acetic acid, or precipitating 198 acetate of lead with 138 protosulphate of iron. Sulphuretted hydrogen causes a precipitate in its solution. When dissolved in alcohol it constitutes the tinctura ferri acetatis of the pharmacopœia. In dyeing it is employed to form blacks. It is used to preserve timber, by introducing a solution into a wound at the bottom of the tree before it is cut down. On the large scale, finely divided iron is heated in a cast iron boiler to 150°, with pyroligneous acid, spec. grav. 1035, (7° Twaddel) until the liquid, in some days, acquires the density of 1090, (18 Twaddel.)

Peracetate of Iron, sesquiacetate. A brownish-red fluid, obtained by dissolving sesquioxide of iron in acetic acid, or by precipitating equal parts of iron, alum, and acetate of lead, or by mixing 198 dry sesquisulphate of iron and 570 acetate of lead. By heat a portion of the base is deposited. As used in dyeing and printing, these acetates of iron are prepared by means of impure pyroligneous acid; they are used in dyeing wool, and to form on cotton, blues, by means of yellow prussiate of potash. When the ground is to be of a uniform shade, it is necessary to have the peracetate of iron free from protoacetate. The ethereal acetate of iron, used in medicine, is made by adding to 9 parts of the aqueous solution of the acetate 1 part acetic naphtha and 2 parts rectified spirit.

Acetates of Copper. There are four combinations of acetic acid and oxide of copper. The *neutral acetate*, (CuO, $C_4H_3O_3$, HO, and also $5\,HO$,) obtained by dissolving vert-de-gris in acetic acid, or by precipitating acetate of lead by sulphate of copper, filtering and crystallizing. The crystals are bluish-green oblique rhombic prisms, efflorescent, poisonous, scarcely soluble in alcohol; afford by heat acetone, strong acetic acid, gas, and metallic copper. Sugar when boiled with it causes the deposition of red oxide or dinoxide of copper. The crystals contain 51 per cent. of acetic acid. It is employed as a mordant for black in wool, as a resist-paste in the blue vat dyes, the dinoxide of copper being formed, and the indigo deoxidized, before coming in contact with the cotton, which prevents their union.

Subsesquiacetate of Copper. $3\,CuO$, $2\,C_4H_3O_3$ $6\,HO$. Pale blue minute crystals, procured by digesting verdigris in water, and evaporating the soluble portion.

Disacetate of Copper, vert-de-gris. $2\,CuO$, $C_4H_3O_3$ $6\,HO$, or CuO $C_4H_3O_3$ CuO HO $5\,HO$. Firm concrete mass of a greenish-blue colour, separated by cold water into soluble subsesquiacetate and insoluble crystalline trisacetate, prepared in France at Montpellier and Grenoble, by stratifying thin plates of copper with grape lees moistened with wine; a pulverulent layer of verdigris forms, which is scraped off; the alcohol contained in the wine lees is converted into acetic acid, which unites with the oxide of copper, formed by the union of the oxygen of the air with the copper. In this country copper plates are wrapped in cloths moistened with pyroligneous acid. It is used as a pigment, and to form the neutral acetate; it enters as an ingredient of green basilicon ointment; it dissolves, when pure, in acetic and dilute sulphuric acids, and leaves, by ignition, metallic copper mixed with charcoal.

Arsenio-acetate of Copper, Schweinfurt green, Vienna green. CuO, $C_4H_3O_3 + 3$ CuO AsO_3. Rich green coloured powder, insoluble in water, soluble in ammonia; decomposed by acids and caustic alkalies; prepared by forming a thin paste with 10 parts of verdigris in water of 130°, passing through a hair sieve to separate copper and other impurities. This paste is added to a boiling hot solution of 8 parts of arsenious acid in 100 water, when the fine green paint separates. It is poisonous, and used as a fine paint.

Acetate of Lead, sugar of lead, saccharum saturni. PbO, $C_4H_3O_3$ 3 HO. Contains 59 per cent. oxide of lead and 26·84 acetic acid. White rhomboidal prisms or needles, terminated by dihedral summits, sweet and astringent, poisonous, acid reaction; effloresces slightly in dry air, losing acid; soluble in $1\frac{1}{2}$ water and 8 alcohol; most of the lead is precipitated by a current of carbonic acid gas. Sugar of lead is prepared in a state of purity, by dissolving litharge or lead in pure acetic acid. For dyeing purposes, scales of metallic lead or granulated lead may be digested in hot pyroligneous acid and the solution evaporated. It is commonly used in the form of a brown mass. The vessels employed may be of lead or earthenware, and the crystallizing vessels of the same nature or of wood. It is used as a mordant for chromic acid, to prepare orange and yellow, and also to form, by double decomposition, chrome yellow paint, iron, and red liquor, &c.

Trisacetate of Lead, Goulard's lotion. 3 PbO, $C_4H_3O_3$. Fine needles, very poisonous, with an alkaline reaction, insoluble in alcohol; a thick solution, as used in medicine, absorbing carbonic acid from the air. It may be made by boiling a solution of $23\frac{3}{4}$ parts of sugar of lead in 120 water, with 28 parts of litharge till the latter is dissolved, or by adding to a cold saturated solution of sugar of lead one-fifth of its volume of caustic ammonia, and allowing it to stand; 1 part of the salt dissolved in 24 parts of water constitutes Goulard's lotion or water. It is used in organic chemistry to precipitate solutions of gum, extractive, and albuminous substances. As it precipitates gum and not sugar, these two bodies can be readily separated by this salt. It is believed that in the formation of white lead this salt is first produced.

Hexacetate of Lead. 6 PbO, $C_4H_3O_3$, 3 HO. White crystalline precipitate, somewhat soluble in boiling water, and separating in feathery crystals; loses in a vacuum all its water; white lead frequently contains this salt. It is formed by adding an excess of caustic ammonia to sugar of lead in solution.

Acetic Acid. *Vinegar, Spirit of Verdigris, Wood Vinegar, Pyroligneous Acid, Acetous Acid. Essig,* Gr., *Vinaigre,* Fr.

History.—Vinegar is mentioned by Moses, and appears to have been employed by mankind from the earliest ages. Pliny states that a fluid (cedrium) was obtained by fire from the *tæda* (pinus sylvestris) which was used in Egypt to preserve dead bodies. This may probably be an allusion to pyroligneous acid. A description of the old French process for vinegar is published in the Phil. Trans. for 1670, and also in the Elemens de Chimie of the Dijon Academie, vol. 3, p. 5. Berthollet, in 1785, held that radical vinegar, or acetic acid, as it was called, distilled from *verdigris*, differed from distilled vinegar or acetous acid, but Adet, in 1797, concluded that they were identical. Chaptal, in 1798, sided with Berthollet, and Dabit of Nantes held with the latter, that acetic acid contained more oxygen, Darracq at last proved that the two only differed in their concentration, and abolished the term *acetous* acid. Dr. Higgins first attempted to determine the composition of vinegar, (On Acetous Acid, p. 26,) and was afterwards followed by Gay Lussac and Thenard, who analyzed the acetate of barytes with chlorate of potash.

Theory of Acetification.—When we keep wine or beer carefully corked up from access of air they have no tendency to become sour, but if the cork is removed from the bottles, so as to admit oxygen, they speedily turn acid. Pure spirit, however, when exposed to the air never becomes sour. Wine and beer, therefore, must contain some matter in solution which assists the action of the air upon the alcohol contained in these fluids. This is the albuminous matter in solution which acts as a ferment. Mr. E. Davy has observed, that when alcohol in vapour is brought in contact with platinum black, the latter becomes red hot, and acetic acid is formed. It is probable that the ferment is a cellular plant, which absorbs, like the porous metal, oxygen from the air in such a condition as to produce union with the hydrogen of the alcohol. The view at present entertained, is that alcohol is first, by the action of 2 atoms oxygen, converted into aldehyde, (C_4H_5O, HO $+ O_2 =$ C_4H_3 O HO, aldehyde and 2 HO,) then as a second stage, by the further addition of 2 atoms oxygen, acetic acid is produced, (C_4H_3O, HO $+$ $O_2 = C_4H_3O_3HO =$ hydrous acetic acid.) The various processes for preparing acetic acid are as follows:—

1. *Quick Vinegar Method.*—This, usually called the German process, was discovered independently, and patented by Mr. Ham of Bristol soon after 1820, and is followed by Messrs. Hill

and Evans of Worcester. In Germany this process is considered to have been first practised by Wagenmann and Schützenbach. A cask of 6 feet in height, similar to that used for wine vinegar, is filled with beech wood shavings which have previously had hot water poured over them, and been soaked in vinegar. The cask is covered with a top, and is perforated with a funnel, which leads into a space formed by a false top, perforated with numerous holes. Alcohol being poured into the funnel trickles through the false bottom into the shavings, toward the bottom of the cask, which tapers downwards. The air from the cask escapes by small glass tubes which enter into the space between the tops. About a foot from the bottom of the cask there is a circle of 8 holes, which enter downwards obliquely, for the admission of air. A glass syphon, inserted in the bottom of the cask, conducts the fluid which accumulates into an exterior vessel. It is found advantageous to mix with the alcohol to be acetified some vinegar, so as to cover the shavings with the vinegar plant, (mother of vinegar.) The temperature at first rises to 80° then to 100°, where it remains, if the process advances properly. In a cask of 6 feet high and 3½ diameter, a charge consists of 16 gallons of alcohol of 60 per cent. diluted with 60 gallons water; 30 gallons of vinegar are daily thus prepared. The liquid which passes through the first cask is transferred to a second similar cask, and to a third if necessary, until it is thoroughly acetified. The presence of various substances prevents the conversion of the alcohol into vinegar; among these may be enumerated aromatics, fluid oils, and the most minute traces of pyroligneous acid. Messrs. Hill, Evans, & Co. of Worcester, were among the first to adopt and carry out Mr. Ham's ingenious ideas. Malt vinegar being the kind manufactured at their works, the antecedent operations of mashing the grain in water, at a temperature of about 154°, and fermenting the sweet infusion thus obtained in large vats, by the aid of yeast, so as to convert the saccharine matter into alcohol, are necessarily the same as those of ordinary brewing as seen practised in distilleries. The spirituous liquid thus furnished is suitable, without further preparation, for the oxidizing, which appears as a species of aëration, the spirituous liquid being showered upon the surface of a mass of faggots of birch twigs, occupying the upper part of a large vat, and, after trickling to the bottom, being returned again and again to the top by the constant action of a pump. The oxygen of the air is continually absorbed by the circulating fluid; and the air, which is admitted by a small aperture in the vat below the level of the faggots, passes away by the openings in the cover of the vat, more or less exhausted of its active element. The process comes to a termination when all the alcohol is oxidized; and this is learned from the progressive rise in the proportion of acid in the liquid being found to cease. The product, when drawn off, is already finished vinegar, but is always kept in store for some time to clarify, or, as it is said, to mature it, before being sent into the market. This mode of oxidizing the alcohol, which is rapid and effective, appears to have the incidental advantage of changing and rendering insoluble certain glutinous and albuminous matters in the fermented wort, which are apt, if not got rid of at this stage of the process, to occasion after-muddiness in the vinegar, and to prevent its keeping. It was generally considered necessary in the vinegar trade, at a former period, to add a small portion of sulphuric acid to vinegar in order to counteract this tendency of the liquid to decomposition, and to preserve it from turbidity. This addition of sulphuric acid was permitted to the extent of 1 gallon of sulphuric acid to 1000 gallons of vinegar, by an excise regulation, and had, therefore, a legal sanction. But sulphuric acid is now known to be unnecessary in properly-prepared vinegars, although still added by some manufacturers, for the purpose of increasing the strength of their vinegar, or, in some instances, merely from habit and the indisposition to disturb the routine of an old-established practice. The presence of sulphuric acid in vinegar should be looked upon as the mark of inferior quality, for it is only where the mode of manufacture is defective that the addition appears to be at all necessary.

2. *Malt Vinegar.*—In this country vinegar for domestic use is made from malt by a process similar to that employed for producing alcohol. The malt prepared from barley is mashed in the tun with water at the temperature 168° to 170°. This liquor is replaced by fresh water, the heat of which may be raised still higher. A third repetition completes the process. The liquid is then fermented at 70° by means of yeast; 100 gallons of wort requiring about 3½ gallons of yeast. After the fermentation ceases the liquor is placed in casks, which are laid on their sides in an apartment at the temperature of summer heat. As the alcohol produced by the fermenting process is converted into vinegar entirely by the action of the oxygen of the air, it is necessary to leave the bungs out of the casks, and not to fill them completely, in order that the whole upper surface of the wort may be in contact with air. From two to three months are required to complete the acetification of the contents of the casks. Another method often employed for the production of vinegar from malt is to place the wort in upright casks, which are supplied with a double bottom, the upper one

drilled with holes, on which the refuse raisins from British wines, termed *rape*, are deposited. The liquor is then fermented and drawn off into another cask, where it is allowed to acidify. In consequence of the length of time required for this process, it is obvious that the quick vinegar method is applicable to it. It is calculated that a cwt. of good malt will afford about 60 to 65 gallons of vinegar. In Germany potatoes are used for procuring the alcohol, and to obtain the preceding quantity of vinegar, it is necessary to employ about 30 gallons of potatoes.

3. *Wine Vinegar.*—In France vinegar is usually made from wine, by two methods. One plan consists in placing the wine to be acidified in contact with vinegar already formed, and to allow the action to proceed with access of air, at the requisite temperature. This is the oldest method, and it is practised with great success at Orleans. The apartment where the process is conducted is usually one where the temperature may be economically kept up, the air being renewed through apertures, which may be closed at will. By means of a stove the heat is raised to 86°. The vessels employed are oak casks that have frequently been previously in use for holding wine, and have a capacity of from 40 to 50 gallons. These casks are placed horizontally upon supports, and arranged in three or four rows, in tiers upon each other, for the sake of economy of space, and for uniformity of temperature. They are filled up to one-third with good vinegar; 2¼ gallons of wine to be acidified are then added, and after eight days a similar quantity is added. The same addition is made twice successively, and in eight days after the last addition the whole liquor is acidified. The 9 gallons which have been added are then withdrawn, and the process repeated as before. When the wine is turbid, it is necessary to filter it through beech chips, and the same operation should be performed if the resulting vinegar is similarly circumstanced. Weak and new wines acidify most easily; strong and old wines with greater difficulty. The weak wines are sometimes strengthened by adding sugar, molasses, and yeast. By the method described 30 days are required to convert the alcohol completely into vinegar. But by admitting the oxygen more freely, vinegar has been made in three days by a modification of the preceding process. In this case casks are used 6½ feet in height and 3¼ feet in diameter. The top of the cask is replaced by a cover, fitting as tight as possible. At the distance of 6 or 8 inches below the cover is fitted a false top, pierced with minute holes, about a tenth or two of an inch in diameter. A piece of thread, 6 inches long, like a wick, is suspended from each of the holes, and partially closes it, being tied by a knot above. The wine is introduced between the cover and false top, passes down by the threads, and falls drop by drop into the interior of the cask, which is filled with beech chips. Perforations in the cask below enable the air to enter. It passes through the false bottom by means of tubes, which rise above the surface of the wine, and makes its escape from the cask by an opening in the cover, through which the wine is introduced. This process is precisely similar to the quick vinegar method already described.

4. *Cider Vinegar* is made in America from refuse cider, which is placed in a cask and mixed with vinegar and the mother of vinegar plant, which acts as a ferment. The fermentation is also assisted by adding a piece of dough or lean muscle of an animal.

5. *Sugar Vinegar.*—Various proportions of the ingredients are recommended in the production of vinegar from sugar. It has been recommended to employ 20 lbs. sugar, 15 to 17 gallons water, and 24 pints of bruised gooseberries mixed with yeast and raisins. The saccharine liquor attenuates and becomes acidified in the course of a summer. The colour may be removed by distillation. A pleasant vinegar is made for the table by dissolving 3 lbs. of sugar in a gallon of water, placing the solution in an open vessel, loosely covered with a cloth, and adding yeast to produce fermentation. The mother of vinegar, or vinegar plant, (Mycoderma aceti,) answers the purpose of yeast. In both cases the same plant is formed or propagated. Alcohol is first formed, which speedily changes into vinegar by access of atmospheric oxygen.

6. *Pyroligneous Acid* or *Wood Vinegar. Holzessig*, Gr.—In this country wood vinegar is usually prepared from hardwoods; but by the patent still of Mr. Halliday, any refuse of dyewoods can be subjected to dry distillation, and is thus saved from waste. The common stills consist of cylinders of iron similar to those used for gas-making, placed horizontally or sometimes vertically, connected with a metallic worm for condensing the products. The iron cylinders or retorts vary in size, being often 6 feet long, and from 3 to 4 feet in diameter. A retort of 6 feet by 4 is capable of holding a charge of 800 lbs. of wood. The wood is cut into convenient lengths; heat is then applied to the cast-iron retort for about 12 hours, care being taken that it never attains a red heat; the fireman judges of the temperature by ejecting saliva on the door of the furnace, when, if it evaporates with a hissing sound, he allows the fire to be extinguished. With some manufacturers, however, the distillation is hurried on by smart firing, when the operation terminates in 8 hours. But no doubt by this process the destruction of a portion of the valuable products occurs. The products of the distillation are conveyed by a series of refrigeratory tubes, which are surrounded with cold water, into a cistern or receiver. Here two layers are formed; the lower stratum consisting of tar, creasote, oils saturated with acetic acid, with some trace of pyroxilic spirit, while acetic acid and pyroxilic spirit, with tar, oils, creasote, and mesite float above. The gaseous products

escape, and are again passed through the furnace, the tube which conveys them previously dipping into water to prevent explosion. When ignited in the furnace, these gases save the use of fuel. The tar or lower stratum is drawn off by means of a tap at the bottom of the cistern, and it is washed with water to remove acetic acid. The tar contains the following liquids:—creasote, eupion, capnomore, picamar, mesite; and the following solids:—pittacal, cedriret, paraffine, pyroxanthine, asphalt. The acid or upper layer is first purified by rectification, that is, distillation in a copper still. Pyroxilic spirit being most volatile, comes over first. It is removed and distilled over lime.

Purification as Acetate of Soda.— In some works, for the sake of economy, the acid is placed in a cast-iron or copper boiler, a quantity of sulphate of soda is added, proportional to that of the acid, the whole is heated, and the salt being dissolved, the liquor is evaporated to the spec. grav. 1·114. Some chalk is then added to the liquor, when an effervescence takes place, and a precipitate of sulphate of lime falls. To render the liquor neutral, it is necessary to add milk of lime. During the neutralization a quantity of tar rises, which is skimmed off, the liquor evaporated to 1·12 and allowed to rest. The evaporation is carried on to 1·23, when the liquor is run into wooden coolers, where it is allowed to crystallize. In three or four days the crystals of acetate of soda are formed in rhomboidal prisms. The mother liquors are crystallized, and when no more crystals can be obtained, the liquor when evaporated and ignited, yields carbonate of soda. The salt thus obtained is very impure, from the tar with which it is contaminated. This is removed by carefully heating the acetate for 24 hours in an iron furnace, shallow but broad, and capable of containing about 8 cwt. When the salt is sufficiently heated it becomes pyrophorus, and burns like tinder when it accidentally catches fire. The salt after cooling is dissolved in water and again crystallized. To obtain the acid from the pure salt, it is mixed with sulphuric acid diluted with half its weight of water; 100 parts of pounded salt require 36 of strong sulphuric acid. The reaction is allowed to continue for some time, and the sulphate of soda deposits in the form of a powder or granular crystals. When thus prepared the acetic acid contains a small portion of sulphate of soda, and therefore precipitates chloride of barium. To obtain the acid perfectly pure, the acetate must be distilled with sulphuric acid, and the product, in order to free it from sulphuric and sulphurous acids, must be distilled with peroxide of lead, mixed with peroxide of manganese.

German purification as Acetate of Lime.—The impure acid may be deprived of a considerable part of its tar, &c. by filtration. A vessel similar to the alkaline tub of the soap boiler, with a double perforated bottom, is filled to the height of an inch with straw, and the latter is covered with a layer of sacking: over this is laid 6 inches of moist wood sawings, which are pressed down with a bat of wood; above this gravel is strewed to the depth of 4 inches, to prevent the sawings from being displaced and swimming about. This constitutes the filter upon which the impure acid is placed. The tar remains on the sand, and the acid passes to the bottom, and is let off by a stopcock. The acid is now placed in a cast-iron boiler, heated and saturated with milk of lime previously passed through a hair sieve. An excess of lime is considered necessary. The liquor is boiled down to one-half, and allowed to stand for two days; it is then drawn off, evaporated in a flat vessel, accurately neutralized, and reduced to the consistence of thick turpentine when warm, and of such a nature when cool that it should crumble down without adhering to the fingers. When it is beginning to thicken it should be carefully stirred with an iron spatula. The acetate is then transferred to an iron plate, and spread out to the depth of 2 or 3 inches, for the purpose of being dried; where it should not remain long, however, as it is apt to absorb moisture. The salt when cool is placed in the drying furnace, which is a simple round furnace from 7 to 8 feet long and 4½ to 5 broad, built of brick, floored by an iron plate. The salt is first spread out to the depth of 2 inches, and after it has become somewhat dry, the depth is increased to 4 or 5 inches. The heat should be continued for 24 hours; and towards the end of the process may be pushed as high as 257°, the mass being frequently stirred. No smoke should be permitted to arise, a sign of decomposition; nor should sparks be allowed to gain access to it, otherwise it will catch fire and burn like tinder. To separate the acid, 20 lbs. of the acetate of lime are introduced into an iron retort, and stirred, with 5 lbs. of water. After 10 hours, 20 lbs. of oil of vitriol, diluted with 5 lbs. of water, are poured over it, and the porcelain or tin cover of the retort cemented on with lime and fine sand. The first tenth of acid which comes over contains sulphurous acid, which may be removed by distilling it over black oxide of manganese and a little wood charcoal. The spec. grav. of the first distillation is 1045 to 1050; after being twice distilled, it ranges from 1042 to 1049. When carbonate of potash is added to this acid a yellow tinge is produced, indicating organic matter. This may be removed, previous to distillation, by adding a solution of nutgalls, or oak bark, and allowing the acid to clear. The colour may also be removed by means of animal charcoal, freed previously from bone earth by digestion in muriatic acid. (*R. D. Thomson's Records of Gen. Science*, ii. 133.) Schwartz's method is similar to the preceding. Pasch recommends the removal of the colour by means of animal charcoal, freed previously from bone earth; while Stoltze places the distilled acid in contact with $\frac{1}{16}$

of its weight of *black* oxide of manganese, for six hours, at the temperature of 248° F. He then adds wood charcoal to the amount of 8 times the weight of the manganese, and allows the mixture to digest at the same temperature for 12 hours. The acid is then distilled. The manganese may be replaced by sulphuric acid of equal weight, or by a mixture of manganese, common salt, and sulphuric acid.

A. P. Halliday's patent process for the manufacture of Pyroligneous Acid, from sawdust, spent dyewoods, tan, &c.—I am indebted to my late friend Mr. Halliday of Manchester for the following figure of his ingenious invention:—

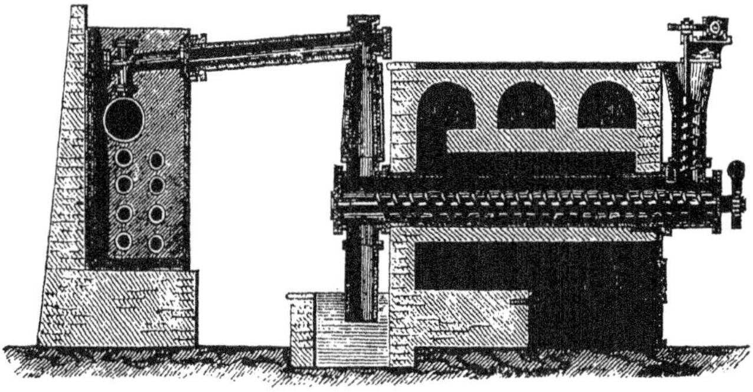

In the ordinary process of manufacturing pyroligneous acid, branches or billets of oak or any other wood are introduced into air-tight cast-iron cylinders, and subjected to destructive distillation, by the application of heat. It is also known that *sawdust*, *wood turnings*, *wood chips*, *spent dyewoods*, *spent tan*, *turf*, *&c.* are capable of yielding pyroligneous acid; but, owing to the minutely divided state in which these substances exist, the ordinary method of effecting their destructive distillation becomes difficult, if not impossible, in consequence of that portion in immediate contact with the retort becoming completely charred, whilst the non-conducting property of the charcoal prevents the heat from penetrating into the interior. The first part of this invention consists in effecting the destructive distillation of sawdust, spent dyewoods, &c. in order to obtain pyroligneous acid therefrom, by causing such substances to pass in *continuous motion* through heated retorts, by means of suitable machinery or apparatus. The sawdust, spent dyewoods, &c. from which pyroligneous acid is to be distilled, are introduced into a hopper, in which vertical screws or worms revolve, conveying the materials (regulating, at the same time, the supply) to the retorts, placed in a horizontal position, and heated by means of a furnace; and in which retorts revolving screws or worms keep the material in constant agitation, moving it, at the same time, forward, until the whole is completely carbonized, and all the pyroligneous acid evolved. The charcoal thus formed, falls through pipes dipping into a vessel of water, or into an air-tight vessel, and from which it may be withdrawn through a door at the side. The pyroligneous acid is condensed, in the usual way, in pipes of iron or copper, surrounded with or immersed in water,—the other products obtained by the destructive distillation of the material being employed in the ordinary manner. The superiority of this retort is shown by the circumstance, that 8 retorts, 12 inches in diameter and 6 feet long, turn out as much pyroligneous acid in 24 hours as 16 retorts 3 feet in diameter and 6 feet long. The material is, of course, comparatively costless, having been previously thrown away. The product from 20 tons of pine sawdust is, pyroligneous acid, 2258 gals. 8° Twaddel; tar, 218 gals. Oak sawdust, 20 tons, gives 2400 gals. pyroligneous acid, 7° Twaddel, and 80 gals. tar. Spent logwood gives 2310 gals. 6° Twaddel, and 90 gals. tar.

French Method.—In France large plate-iron cylinders are employed, cased with iron to give them greater strength, and placed vertically on the fire. Each of them is capable of holding a large quantity of wood, which, by some manufacturers, is previously dried by the heat of the flue. At the upper part of the vessel there is a small cone, placed laterally, which closes with a cover, and serves as the neck of the retort. By means of three rings and three chains, the cylinder may be elevated or lowered at pleasure. The operation is commenced by means of hot charcoal placed beneath the cylinder, but the remainder of the distillation is produced by the heat of the combustible gases disengaged from the wood. The quantity of fuel required amounts to an eighth of the wood submitted to distillation. Communicating with the neck is a series of zig-zag tubes, surrounded by other cylinders, which constitute the condenser. In the space between the two series of cylinders, a sufficient quantity of water is kept circulating, by entering below and passing out above in a heated state. The condenser communicates with the first receiver placed underground. The liquid products pass from thence into a larger receiver, through a tube so disposed as to intercept any communication with the interior of the apparatus. The gas, which is disengaged without condensation, is brought back by tubes under the ash-pit of the furnace, and made to serve as fuel. Towards the end of the operation the cylinder is heated red hot. The distillation is completed in eight hours. If the heat has been applied slowly, the charcoal in the cylinder is greater, and the acid less in quantity than when the firing has been rapid. It is believed that the harder kind of woods, or those which contain much ligneous matter in their cells, yield the greatest amount of acid, although cellulin also affords acid. Along with the acid much empyreumatic oil and tar come over; a portion of these pass into the aqueous liquid and colour it brown; another portion separates by remaining at rest, which is pumped off. The acid liquor is also pumped off, and first purified by distillation in a copper alembic. It is in this distillation that the pyroxilic spirit separates. In consequence of its volatility, it passes over first, and must be managed with care, otherwise it will be lost. The spirit which comes over first is removed, and rectified with lime. To purify the acid in some manufactories, it is usual to saturate it with carbonate of soda. In others, for the sake of economy, the acid is placed in a cast-iron or copper boiler; a quantity of sulphate of soda is added, proportional to that of the acid; the whole is heated, and the salt being dissolved, the liquor is evaporated to the spec. grav. 1·114 (15° Beaume). Some chalk is then dissolved in the liquor, when an effervescence takes place, and a precipitation of the sulphate of lime occurs. Generally the liquor remains always acid, even with an excess of chalk. (Is this only when hot?) The saturation is therefore completed with milk of lime. During the neutralization, a quantity of tar rises to the surface, which is to be skimmed off. The liquor is evaporated to 16° Beaume (spec. grav. 1·12). It is then allowed to rest, and treated as above described. It is calculated that 1000 lbs. of distilled pyroligneous acid require 127½ lbs. carbonate of lime or chalk, and 2½ lbs. caustic lime for neutralization. This solution has a spec. grav. of 1090 (12° B.), and contains 197·95 lbs. of dry acetate of lime. It requires for decomposition 188⅓ lbs. of anhydrous sulphate of soda, or 426 lbs. of the same salt in crystals. But as the glauber salt does not precipitate the acetate of lime completely, it is necessary to add a little carbonate of soda. According to Stoltze, all kinds of wood, when they have been deprived of resins, and well washed with caustic alkalies, and dried, give nearly the same products by distillation, as exhibited in the following table. (Gründliche anleitung, die rohe holzäure zu benutzen, Halle, 1820, and Berthier's Traité i. 243.)

Wood.	Acetic Acid.	Oil.	Charcoal.	Gas.
White Birch (Bouleau),	45·00	8·60	24·40	20·00
Beech	44·00	9·55	24·60	22·85
Oak	43·00	9·06	26·20	21·74
Ash	46·80	8·80	22·10	22·30
White Poplar ..	45·80	8·05	23·40	22·85
Wild Prune.....	43·70	10·35	21·60	24·35
Juniper Tree....	45·80	10·73	22·70	20·77
Fir (Sapin)	41·20	13·70	21·20	23·90
Pine (Pin)......	42·40	11·80	21·50	24·30

Acetic Acid Anhydrous. $C_4H_3O_3$, spec. grav. 1·073, of vapour 3·47, B.P. 279°·5. Colourless fluid, very mobile, powerful refractor of light, with a strong smell; falls to the bottom of water in oily drops, and only dissolves after some time; fuming sulphuric acid disengages from it carbonic acid, and forms a conjugate compound, which unites with lead; potassium and zinc form anhydrous acetate and a new acid. Anhydrous acetic acid is obtained by heating together acetate of potash and chloride of benzoyle; there are first formed acetate of benzoyle and chloride of potassium, and at an elevated temperature, acetate of benzoyle $2(C_9H_4O_3)$ doubles into anhydrous acetic acid $(C_4H_3O_3)$ and anhydrous benzoic acid, $(C_{14}H_5O_3.)$ It may be also made by dropping oxichloride of phosphorus or protochloride of phosphorus on fused acetate of potash, and rectifying the product.

Acetic Acid, Hydrate of. *Glacial Acetic Acid.* $C_4H_3O_3\,HO = 7{\cdot}5$, or 60: acetic acid = 85·01 water = 14·99, spec. grav. of vapour, 2·764 = ⅔ vapour of acetic acid and ⅔ steam Spec. grav. of the liquid crystals, 1·06296. Colourless plates and tables at a temperature under 40°; obtained by distilling with 9¾ sulphuric acid, 3 acetate of soda, twice fused, and crystallized after each fusion, or by distilling dry acetate of lead with sulphuric acid. The following table exhibits the spec. grav. of the crystals in solution, and combined with various quantities of water, (T. Thomson):—

Acid.	Water.	At 60 deg.
1 atom,	1 atom,	1·06296
1 "	2 "	1·07060
1 "	3 "	1·07084
1 "	4 "	1·07132
1 "	5 "	1·06820
1 "	6 "	1·06708
1 "	7 "	1·06349
1 "	8 "	1·05974
1 "	9 "	1·05794
1 "	10 "	1·05439

The specific gravity appears to be a maximum when the liquid is a compound of 1 atom acid and 4 atoms water, or of 58·2 dry acid and 41·8 water. In the following table, by Mollerat, the composition of acid and water is exhibited:—

Hydrous Acid.	Water.	Spec. Grav.
110 parts,	0·0 parts,	1·0630
...	10· "	1·0742
...	22·5 "	1·0770
...	32·5 "	1·0791
...	43· "	1·0763
...	55· "	1·0742
...	66·5 "	1·0728
...	97·5 "	1·0658
...	108·5 "	1·0637
...	118·2 "	1·0630

The following table gives the strength of acetic acid variously diluted (Mohr):—

Hydrous Acid in 100 *parts.*

100	1·0635	...	67 1·0690	...	34	1·0450	
97	1·0680	...	64 1·0680	...	31	1·0410	
94	1·0706	...	61 1·0670	...	28	1·0380	
91	1·0721	...	58 1·0660	...	25	1·0340	
88	1·0730	...	55 1·0640	...	22	1·0310	
85	1·0730	...	52 1·0620	...	19	1·0260	
82	1·0730	...	49 1·0590	...	16	1·0230	
79	1·0735	...	46 1·0550	...	13	1·0180	
76	1·0730	...	43 1·0530	...	10	1·0150	
73	1·0720	...	40 1·0513	...	7	1·0107	
70	1·0700	...	37 1·0480	...	4	1·0050	

Acid consisting of 1 atom acid and 1 water, and acid consisting of 1 atom acid and 7 atoms water, have the same spec. grav. 1·0630. (T. Thomson.)

Impurities.—We often observe in weak vinegar the occurrence of the *Vibrio aceti* (vinegar eel), the *Musca cellaris*, or vinegar fly, and the mother of vinegar (Mycoderma aceti).

Aceto-Sulphuric Acid. $C_4H_2O_2$ 2 SO_3. Colourless silky prisms, fusing at 143½°, decomposing at 320°, and smelling of caramel. The crystallized acid contains 2 atoms basic water, and 3 atoms additional water; obtained by acting on strong acetic acid with vapour of anhydrous sulphuric acid.

Acetic Ether. *Acetate of Ethyle.* C_4H_5O, $C_4H_3O_3$ 11 or 88: sp. grav., vapour 3041. Colourless fluid; odour pleasant. B.P. 165° sp. grav. ·89; soluble in 7 water, miscible with alcohol and ether in all proportions; when boiled with a spirituous solution of alkali alcohol is formed and an alkaline acetate; prepared by distilling 16 parts sugar of lead 4½ absolute alcohol, 6 sulphuric acid, or 10 acetate of soda 6 alcohol (·840), and 15 sulphuric acid; the product is agitated with a strong solution of carbonate of soda, the liquor poured off, agitated with chloride of calcium, and distilled off from this salt.

Aceto-Nitrile. C_4H_3N. A fluid boiling at 170°, identical with cyanide of methyle, obtained by acting on acetate of ammonia, with anhydrous phosphoric acid.

Acetometry. The process of measuring the value of acetic acid or quantity of dry acid in a sample. One method consists in weighing a piece of marble, immersing it in the acetic acid or vinegar, and reweighing it after it has saturated the acid. Every 6·25 grains which have disappeared are equivalent to 6·375 of acetic acid. A preferable mode is to weigh out 100 grains of the sample, and also 50 grains of crystals of bicarbonate of potash pounded. The powder is added gradually to the acid until reddened litmus becomes slightly blue when immersed in the solution. The residual bicarbonate is again weighed. Every 12·53 grs. used represent 6·375 grs. of acid in the 100 grains of sample. From the table which is given above, it is obvious that no correct information is obtained by taking the specific gravity of vinegar.

Acetone. *Pyroacetic Spirit, Mesitic Alcohol, Naphtha.* C_3H_3O. Sp. grav. ·7921; of vapour 2022, B.P. 100°. Clear, colourless liquid, smell peculiar, taste pungent; miscible in all proportions with water, ether, and alcohol, and may be separated from water by caustic potash, chloride of calcium, or salts insoluble in acetone. Bleaching powder changes it into carbonic acid and chloroform. It has been considered as the hydrate of a hypothetic body, mesityle, C_6H_5, acetone being C_6H_5O, HO, or the hydrate of the oxide of this radical—but this is doubtful. It is obtained by passing acetic acid through a red hot tube, or by distilling acetates of lead or lime. Acetate of lime (CaO $C_4H_3O_3$) becomes carbonate of lime (CaO CO_2), and acetone (C_3 H_3O). It has been recommended in consumption. By fuming sulphuric acid it is changed into *mesitilole* (C_6H_4, or C_{18} H_{12}), and by perchloride of phosphorus into chloride of mesityle, or terchloro mesitilole (C_{18} H_9Cl_3).

Acetous Acid. A synonyme of aldehydic acid.

Acetyle, *Acetule.* C_4H_3, the hypothetic radical of acetic acid.

Acetyle Mercaptan. $C_4H_4S_2$, or aldehyde, in which sulphur replaces oxygen. Dazzling white needles, subliming at 113°; obtained by adding sulphohydric acid to aldehyde, and to the oil thus formed adding sulphuric acid, and reprecipitating by water.

Acetyle Urea. $C_4H_6N_2O_2$. Crystals by adding ammonia to cyanate of methyle.

Acetylic Acid, or acetic acid.

Acetylous Acid, or aldehydic acid.

Achirite. Dioptase or subsesquisilicate of copper.

Achmite. Reddish-brown right oblique prisms, or 6-sided prism terminated by a 4-sided pyramid. Sp. grav. 3.398; hardness 4. B.B. fuses into a black bead. SiO_3 52·016, FeO 28·080, NaO 13·33, MnO 3·487, CaO ·876, MgO ·504, Al_2O_3 ·685. Found at Eger, in S. Norway. NaO SiO_3, Fe_2O_3 2 SiO_3.

Achroite. A species of tourmaline from Elba, in which it is supposed that boracic and carbonic acids replace silica. (Hermann.)

Achromatic. (α and χρωμα,) destitute of

colour. Achromatic glasses refract light without producing coloured images.

Acicular Bismuth Ore, or needle ore of bismuth.

Acidity. Sourness; the effect produced on the palate by certain acids, although not universally characteristic of all acids. The cause of the sour taste is not yet known; nor does it certainly proceed from the presence of any one element unless it be hydrogen.

Acids. (*Acidus*, sour). Acids were formerly characterized by their sour taste, their reddening effect on vegetable blue colours, and their neutralizing influence on alkalies. It is not easy to give, in the present state of chemistry, an accurate definition of an acid, but generally the term is applied to such bodies as replace each other in their union with bases. The following tables, although by no means complete, contain the principal of the inorganic and organic acids:—

Inorganic Acids.

Oxygen Acids.

Hypochlorous or protochlorous,	ClO
Chlorous or terchlorus,	ClO_3
Hypochloric or quaterchlorous,	ClO_4
Chloric or quintochloric,	ClO_5
Perchloric or septachloric,	ClO_7
Chlorochloric,	Cl_3O_{15}
Chloroperchloric,	Cl_3O_{17}
Iodous,	?
Hypoiodic (quatriodous,)	IO_4
Iodic,	IO_5
Periodic, septiodic,	IO_7
Hypobromous,	BrO ?
Bromic,	BrO_5
Nitrous, ternitrous,	NO_3
Hyponitric, quaternitrous,	NO_4
Nitric,	NO_5
Carbonic,	CO_2
Chlorocarbonic,	$COCl$
Boracic,	BO_3
Silicic,	SiO_3
Hypophosphorous,	PO
Phosphorous,	PO_3
Phosphoric,	PO_5
Sulphurous,	SO_2
Sulphuric,	SO_3
Chlorosulphuric,	SO_2Cl
Hyposulphurous,	SO_2S
Hyposulphuric,	SO_2SO_3
Sulpho-hyposulphuric,	SO_2SO_3S
Bisulpho-hyposulphuric,	$SO_2SO_3S_2$
Tersulpho-hyposulphuric,	$SO_2SO_3S_3$
Sulphazous,	$3\,SO_2NO_3\ 3\,HO$
Sulphazic,	$4\,SO_2NO_3\ 3\,HO$
Sulphazotic,	$5\,SO_2NO_3\ 3\,HO$
Metasulphazic,	$7\,SO_2\ 2\,NO_3\ 6\,HO$
Sulphazotous,	$2SO_2NO_3\ 2HO\ 2SO_3$
Sulphazidic,	$S_2O\ NO_3\ 3\,HO$
Sulphazilic,	$4\,SO_2NO_3HO$
Metasulphazilic,	$4\,SO_2NO_3\ 3\,HO\ 2\,SO_3$
Sulphammonic,	$8\,SO_2NO_3\ 3\,HO$
Metasulphammonic,	$4\,SO_2S_2O_2NO_3\ 3\,HO$
Sulphamidic,	$2\,S_2O_2NO_3\ 3\,HO$
Chlorohyposulphuric,	S_2O_5Cl
Nitrosulphuric,	SO_2NO_2
Selenious,	SeO_2
Selenic,	SeO_3
Arsenious,	AsO_3
Arsenic,	AsO_5
Antimonious,	SbO_4
Antimonic,	SbO_5
Chromic,	CrO_3
Chlorochromic,	CrO_2Cl
Vanadic,	VO_3
Uranic,	U_2O_3
Molybdic,	MoO_3
Tungstic,	TnO_3
Tellurous,	TeO_2
Telluric,	TeO_3
Titanic,	TiO_2
Columbic,	CmO_3
Pelopic,	PeO ?
Niobic,	NiO ?
Ferric,	FeO_3
Manganic,	MnO_3
Permanganic,	MnO_7
Cupric,	CuO_3 ?
Stannic,	SnO_2
Auric,	Au_2O_3
Osmous,	OsO_2
Osmic,	OsO_4

Hydrogen Acids.

Chlorohydric,	HCl
Iodohydric,	HI
Bromohydric,	HBr
Fluohydric,	HF
Sulphohydric,	SH
Seleniohydric,	SeH
Arseniohydric,	AsH_3
Tellurohydric,	TeH

Organic Acids.

Carbonic Oxide Series.

Carbonic,	$CO + O$
Chlorocarbonic,	$CO + Cl$
Oxalic,	$CO + CO_2$
Oxamic,	$CO + CO_2 + C_2O_3NH_2$
Crocoic,	C_5O_4HO
Rhodizoic,	$C_7O_7\ 3\,HO$
Mellitic,	C_4O_3HO
Euchroic,	$C_{12}O_6N\ 2\,HO$

Cyanogen Series.

Cyanic,	CyO
Fulminic,	Cy_2O_2
Cyanuric,	Cy_3O_3
Mellanuric,	$C_6H_4N_4O_4$
Cyanohydric,	CyH
Sulphocyanohydric,	CyS_2H
Ferrocyanohydric,	Cy_3FeH_2

Ferridcyanohydric,	$Cy_6Fe_2H_3$
Nitroprussic,	$Cy_5Fe_2NOH_2$
Cobaltocyanohydric,	$Cy_6Co_2H_3$
Chromocyanohydric	$Cy_6Cr_2H_3$
Platinocyanohydric,	Cy_2PtH_2
Iridiocyanohydric,	Cy_3IrH_2
Manganocyano-hydric,	$Cy_6Mn_2H_3$
Mellonohydric,	C_6N_4H
Sulphomellono-hydric,	$C_6N_4H_3S_4H$
Cyameluric,	$C_{12}N_7H_3O_3$ 3 HO
Uric,	$C_{10}N_4H_4O_6$
Alloxanic,	$C_8N_2H_2O_8$
Leucoturic,	$C_6N_2H_3O_6$
Hydurilic,	$C_{12}N_3H_5O_{11}$
Mesoxalic,	C_3O_4 2 HO
Mycomelic,	$C_8N_4H_5O_5$
Parabanic,	$C_6N_2O_4$ 2 HO
Oxaluric,	$C_6N_2H_3O_7HO$
Alloxanosulphurous	?
Thionuric,	$C_8N_3H_7S_2O_{14}$
Uramilic,	$C_{16}N_5H_{10}O_{15}$?
Dialuric,	$C_8N_2H_4O_8$
Alituric,	$C_6N_2H_3O_4$
Dilituric,	$C_8N_3H_3O_{10}$
Lantanuric,	$C_6N_2H_5O_7$
Hidantoic,	$C_8N_4H_9O_9$
Uroxanic,	$C_{10}N_3H_9O_{14}$
Hyperuric or Guanic,	$C_{10}N_4H_5O_9$?

Acids of Decaying Woody Matter.

Humous,	$C_{40}H_{14}O_{12}$
Humic,	$C_{40}H_{15}O_{15}NH_3$
Ulmic,	$C_{40}H_{12}O_{12}$?
Geic,	$C_{40}H_{12}O_{12}$?
Crenic,	$C_{40}H_{16}NO_{12}$?
Apocrenic,	$C_{28}H_{14}N_6O_6$?
Mudesous,	$C_{12}H_5O_8 + 3$ HO ?
Mudesic,	$C_{12}H_5O_{10} +$ HO ?

Acids produced by Distillation of Wood and of Coal, &c.

Acetic,	$C_4H_3O_3 + 3$ HO
Carbolic or phenic,	$C_{12}H_5O +$ HO
Sulpho-phenic,	$C_{12}H_5OHO + 2\ SO_3$
Chloro-phenesic,	$C_{12}H_3Cl_2OHO$
Chloro-phenisic,	$C_{12}H_2Cl_3OHO$
Chloro-phenusic,	$C_{12}Cl_5OHO$
Bromo-phenisic,	$C_{12}H_2Br_3OHO$
Nitro-phenesic,	$C_{12}H_3\ 2\ NO_4OHO$
Nitro-phenisic,	$C_{12}H_2\ 3\ NO_4OHO$
Hypo-sulpho-naphthalic,	$C_{20}H_8S_2O_5 +$ HO ?
Hypo-sulpho-naphthic,	$C_{11}H_4S_2O_5 +$ HO ?
Hypo-sulpho-glutic,	?
Naphthionic,	$C_{20}NS_2H_8O_5HO +$ HO
Thio-naphthamic,	$C_{20}NS_2H_8O_5HO$
Nitro-naphthesic,	$C_{32}H_9N_3O_8$?
Nitro-naphthaléisic,	$C_{24}H_8N_3O_{12}$?
Nitro-naphthisic,	?

ACI

Chloronaphthisic,	$C_{20}H_5ClO_6$
Chloroxénaph-thalesic	$C_{20}HCl_5O_6$
Naphthalic or phthalic,	$C_{16}H_4O_6$ 2 HO
Nitrophthalic,	$C_{16}H_3NO_{10}$ 2 HO
Chlorophthalisic,	$C_{16}HCl_3O_6$
Ampelic,	$C_{14}H_6O_6$?
Retinic,	$C_{21}H_9O_3$
Sulphomesitylic,	$C_6H_5OHO\ 2\ SO_3$
Sulphethylic,	$C_4H_5OSO_3$

Methyle and Formyle Series.

Sulphomethylic,	$C_2H_3OHO\ 2\ SO_3$
Methyloxamic,	$C_6NH_4O_5HO$
Formic,	C_2HO_3HO
Cacodylic,	$C_4H_6AsO_3HO$

Acids existing in, or derived from, the Cells and Juices of Plants.

Sulpho-amidic,	?
Pectic,	$C_{32}H_{20}O_{28}$ 2 HO
Pectosic,	$C_{32}H_{20}O_{28}$ 3 HO
Meta-pectic,	$C_8H_5O_7$ 2 HO
Parapectic,	$C_{24}H_{15}O_{21}$ 2 HO
Sulphosaccharic,	?
Sacchulmic,	$C_{35}H_{10}O_{15}$
Glucic,	$C_{24}H_{15}O_{15}$ 6 HO
Melassic acid,	$C_{24}H_{12}O_{10}$?
Saccharic,	$C_{12}H_8O_{14}$ 2 HO ?
Saccho-lactic or mucic,	$C_{12}H_8O_{14}$ 2 HO
Pyro-mucic,	$C_{10}H_3O_5 +$ HO
Lactic,	$C_6H_5O_5 +$ HO
Butyric,	$C_8H_7O_3HO$

Ethyle Series.

Sulphovinic,	$C_4H_5OHO,\ 2\ SO_3$
Phosphovinic,	$C_4H_5O\ 2\ HOPO_5$
Oxalovinic,	$C_4H_5OHO\ 2\ (C_2O_3)$
Xanthic,	$C_4H_5OHO\ 2\ (CS_2)$
Allophanic,	$C_4H_3O_5N_2HO$
Ethionic,	$C_4H_5O\ 2\ SO_3$
Isethionic,	$C_4H_5O_2HOS_2O_5$
Methionic,	$C_2H_3S_2O_7$
Althionic,	$C_4H_5OHO\ 2\ SO_3$
Sulphethamic,	$C_{16}NH_{28}O_4\ 4\ SO_3$
Carbamic,	CO_2CONH_2

Acetyle Series.

Acetylous,	$C_4H_3O_2HO$
Acetylic, acetic,	$C_4H_3O_3HO$
Sulphacetic,	$C_4H_4O_2\ 2\ SO_3$
Chloracetic,	$C_4Cl_3O_3HO$
Sulphacetylic,	$C_4H_4\ 2\ SO_3$

Series of Propyle, Butyryle, &c.

Propylic,	$C_6H_5O_3HO$
Nitropropylic,	$C_6H_4NO_4HO$
Butyric,	$C_8H_7O_3HO$
Nitrobutyric,	$C_8H_6NO_7HO$
Sulphamylic,	$C_{10}H_{11}OHO\ 2\ SO_3$
Hyposulphamylic,	$C_{10}H_{11}S_2O_5HO$

Valerianic,	$C_{10}H_9O_3HO$
Chlorovalerisic,	$C_{10}H_7Cl_3O_4$
Chlorovalerosic,	$C_{10}H_5Cl_4O_3$

Benzoyle Series.

Benzoic,	$C_{14}H_5O_3HO$
Bromobenzoic,	$C_{28}H_9BrO_8$ 2 HO
Formobenzoilic,	$C_2HO_3 + C_{14}H_6O_2 + HO$
Hippuric,	$C_{18}NH_8O_5HO$
Benzoglycolic,	$C_{18}H_7O_7HO$
Glycolic,	$C_4H_3O_5HO$
Hyposulphobenzoic,	$C_{14}H_4O_3 + S_2O_5 + 2\ HO$
Nitrobenzoic,	$C_{14}H_4NO_7HO$
Hyposulphobenzidic	$C_{12}H_5S_2O_5O$
Benzilic,	$C_{28}H_{11}O_5 + HO$
Amygdalic,	$C_{40}H_{26}O_{24} + HO$

Salicyle Series.

Salicylous,	$C_{14}H_5O_4H$
Melanic,	$C_{10}H_4O_5$
Salicylic,	$C_{14}H_5O_4OHO$
Chlorosalicylic,	$C_{14}H_5O_4Cl$
Bromosalicylic,	$C_{14}H_5O_4Br$
Iodosalicylic,	$C_{14}H_5O_4I$
Nitrosalicylic,	$C_{14}H_5NO_{10}$
Nitrophloretic,	$C_{30}H_{12}NO_{15}$
Anisic,	$C_{16}H_7O_5HO$
Bromanisic,	$C_{16}H_7BrO_4$?
Nitranisic,	$C_{16}H_7NO_4O_6$
Dinitranisic,	$C_{16}H_6$ 2 NO_4O_6

Cinnamyle Series.

Cinnamic,	$C_{18}H_7O_3 + HO$
Nitrocinnamic,	$C_{18}H_6NO_4O_3HO$
Sulphocinnamic,	$C_{18}H_6O_2$ 2 $SO_3 + 2\ HO$
Myriospermic,	?
Toluylic,	$C_{16}H_7O_3HO$
Sulphotoluylic,	$C_{14}H_7S_2O_5$
Pyroguaiacic,	$C_{14}H_7O_4 + H$
Guaiacic,	$C_{12}H_7O_5 + HO$

Acids of Colouring Matters secreted in Plants, &c.

Carminic,	$C_{28}H_{14}O_{16}$
Nitrococcusic,	$C_{16}H_5O_{18}N_3$
Alizaric,	$C_{14}H_5O_7$ or $C_{16}H_4O_6$ 2 HO
Lecanoric,	$C_{18}H_8O_8HO$
Parellic,	$C_{21}H_8O_{10}HO$
Erythric,	$C_{20}H_{10}O_9HO$
Erythrelesic,	?
Alpha-orsellic,	$C_{32}H_{15}O_{13}HO$
Beta-orsellic,	$C_{34}H_{17}O_{14}HO$
Alpha-orsellesic,	$C_{16}H_8O_7HO$
Evernic,	$C_{34}H_{15}O_{13}HO$
Evernesic,	$C_{18}H_9O_7HO$
Usnic,	$C_{38}H_{17}O_{14}$
Parietic, or chrysophanic,	$C_{10}H_8O_3$
Erythroleic,	$C_{26}H_{22}O_8$
Sulphopurpuric,	$2(C_{16}H_5NO_2)$ 2 SO_3 ?
Sulpho-indigotic,	$C_{16}H_4NO_2S_2O_5$?
Hyposulpho-indigotic,	?

ACI

Isatic,	$C_{16}H_6NO_5HO$
Chlorisatic,	$C_{16}H_5ClNO_5HO$
Bichlorisatic,	$C_{16}H_4Cl_2NO_5HO$
Bromisatic,	$C_{16}H_5BrNO_5HO$
Bibromisatic,	$C_{16}H_4Br_2NO_5HO$
Indic,	$C_{16}H_6NO_2HO$?
Hydrindic,	$C_{64}H_{21}N_4O_9HO$?
Nitrindic,	$C_{16}H_4NO_4ONO_2HO$?
Chlorohydrindic,	$C_{32}H_{11}Cl_2N_2O_6$
Sulphisatanous,	$C_{16}H_5NO_4$ 2 $SO_2 + HO$
Chloranilic,	$C_{12}Cl_2O_6$ 2 HO
Imasatic,	$C_{16}H_5NO_3NHHO$
Anilic,	$C_{14}H_4NO_9HO$
Picric,	$C_{12}H_2$ 3 NO_4OHO
Picramic,	$C_{12}H_4O_2NO_4NHO$
Chrysanilic,	?
Anthranilic,	$C_{14}H_6NO_3HO$
Chrysammic,	$C_{15}HN_2O_{12}HO$
Chrysolepic,	$C_{12}H_2N_3O_{13}HO$
Styphnic,	$C_{12}H_2$ 3 NO_4O_3HO ?

Acids secreted by Plants and their Derivatives.

Citric,	$C_{12}H_5O_{11}$ 3 HO
Aconitic,	C_4HO_3HO
Itacic,	$C_5H_2O_3HO$
Citracic,	$C_{10}H_4O_6$ 2 HO
Mesacic,	$C_{10}H_4O_6$ 2 HO
Nitrocitracic,	$C_{10}H_3NO_4O_6$ 2 HO ?
Citracanilic,	$C_{22}H_{10}NO_5HO$
Tartaric,	$C_8H_4O_{10}$ 2 HO
Tartralic,	$C_{12}H_6O_5$ 2 HO
Tartrelic,	$C_{16}H_8O_{20}$ 2 HO
Anhydrous tartaric,	$C_4H_2O_5$
Pyrotartaric,	$C_5H_3O_3HO$
Pyro-racemic,	$C_6H_3O_5HO$
Racemic,	$C_4H_2O_5HO$
Malic,	$C_8H_4O_8$ 2 HO
Malamidic,	$C_8NH_7O_8$
Maleic,	$C_8H_2O_6$ 2 HO
Paramaleic,	C_4HO_3HO
Tannic,	$C_{18}H_5O_9$ 3 HO
Gallic,	C_7HO_3 2 HO
Pyrogallic,	C_2HO ?
Metagallic,	$C_6H_2O_2$?
Ellagic,	$C_7H_3O_5$
Catechuic,	$C_{18}H_9O_8$?
Pyromorintannic,	$C_{12}H_6O_4$?
Aspertannic,	$C_{14}H_9O_9$?
Rufimoric,	$C_{14}H_7O_8$?
Tanningenic,	$C_{15}H_6O_6$?
Japonic,	$C_{12}H_4O_4HO$?
Rubinic,	$C_{18}H_6O_4$?
Caffeo-tannic,	$C_{14}H_8O_7$
Viridic,	$C_{14}H_7O_8$
Boheic,	$C_{14}H_6O_8$
Opianic,	$C_{20}H_9O_9HO$
Xanthopenic,	?
Opianosulphurous,	$C_{20}H_9O_{11}S_2HO$
Sulphopianic,	$C_{20}H_9O_7S_2HO$
Hemipic,	$C_{20}H_8O_{10}$ 2 HO
Humopic,	?
Apophyllic,	$C_{16}NH_7O_8$
Narcotic,	?

Amalic,	$C_{12}N_2H_7O_8$
Meconic,	$C_{14}HO_{11}$ 3 HO
Ethylomeconic,	$C_{14}HO_{11}C_4H_5O$ 2 HO
Mecon-amidic,	$C_{84}H_{24}N_7O_{13}$ 9 HO + 6 Aq
Mecono-ethylo-meconic,	2 $(C_{14}HO_{11})C_4H_5O$ 5 HO
Bi-ethylo-meconic,	$C_{14}HO_{11}$ 2 $(C_4H_5O)HO$
Comenic,	$C_{12}H_2O_8$ 2 HO
Paracomenic,	$C_{12}H_2O_8$ 2 HO
Chlorocomenic,	2 $HOC_{12}HClO_8$
Bromocomenic,	2 $HOC_{12}HBrO_8$
Ethylocomenic,	$HOC_4H_5O,C_{12}H_2O_8$
Comenamic,	$C_{12}H_4NO_7HO$
Pyromeconic,	$C_{10}H_3O_5HO$
Bromopyromeconic,	$C_{10}H_3BrO_6$
Iodopyromeconic,	$C_{10}H_3IO_6$?
Kinic,	$C_7H_4O_4$ 2 HO ?
Chelidonic,	$C_{14}H_5O_{13}$?

Acids of Oils, Fats, &c.

Caproic,	$C_{12}H_{11}O_3HO$
Nitrovalerianic,	$C_{10}H_8NO_4O_3HO$
Œnanthylic,	$C_{14}H_{13}O_3HO$
Caprylic,	$C_{16}H_{15}O_3HO$
Pelargonic,	$C_{18}H_{17}O_3HO$
Capric,	$C_{20}H_{19}O_3HO$
Margaritic,	$C_{22}H_{21}O_3HO$
Laurostearic,	$C_{24}H_{23}O_3HO$
Cocinic,	$C_{26}H_{25}O_3HO$
Myristic,	$C_{28}H_{27}O_3HO$
Benic,	$C_{30}H_{29}O_3HO$
Sulphocetylic,	$C_{32}H_{33}OHO$ 2 SO_3
Cetylic or ethalic,	$C_{32}H_{31}O_3HO$
Palmitic,	$C_{32}H_{31}O_3$
Margaric,	$C_{34}H_{33}O_3HO$
Stearic,	$C_{34}H_{33}O_3HO$
Suberic,	$C_8H_6O_3HO$
Succinic,	$C_4H_2O_3HO$
Stearophanic,	$C_{35}H_{34}O_3HO$
Bassic,	$C_{36}H_{35}O_3HO$
Balenic,	$C_{38}H_{37}O_3HO$
Behenic,	$C_{42}H_{41}O_3HO$
Cerotic,	$C_{54}H_{53}O_3HO$
Melissic,	$C_{60}H_{59}O_3HO$
Camphoric,	$C_{10}H_7O_3HO$
Sulphocamphoric,	$C_9H_7O_3SO_2HO$ + 2 Aq
Sulphocamphic,	$C_{20}H_{13}S_2O_5HO$
Œnanthic,	$C_{14}H_{13}O_2HO$
Roccellic,	$C_{17}H_{15}O_3HO$?
Oleic,	$C_{36}H_{33}O_3HO$
Sebacic,	$C_{10}H_8O_3HO$
Elaidic,	$C_{36}H_{33}OHO$
Azelaic,	$C_{10}H_8O_4HO$
Pimelic,	$C_7H_5O_3HO$
Adipic,	$C_6H_4O_3HO$
Lipic,	$C_5H_3O_4HO$
Azoleic,	$C_{13}H_{13}O_4$?
Ricinic,	$C_{35}H_{31}O_5$?
Palmic,	$C_{34}H_{32}O_5HO$
Acrylic,	$C_6H_3O_3HO$
Sulpholeic, ?	Sulphomargaric, ?
Metamargaric,	$C_{70}H_{67}O_6$ 3 HO ?
Hydromargaritic,	$C_{70}H_{71}O_{10}$ 2 HO ?

ACI

Hydromargaric,	$C_{70}H_{70}O_3HO$?
Metoleic,	$C_{70}H_{63}O_8HO$?
Hydroleic,	$C_{70}H_{63}O_8$ 2 HO ?
Ambreic,	$C_{13}H_{10}NO_6$
Caincic,	$C_{16}H_{13}O_7$
Crameric,	$C_{10}H_8O_5$?
Caffeic,	$C_{16}H_7O_6$ 2 HO ?
Chiococcic,	$C_{12}H_9O_3$
Ipecacuanic,	$C_{14}H_9O_7$

Boletic.	Solanic.
Fungic.	Coneic.
Tanacetic.	Aceric.
Lactucic.	Myroxilic.
Atropic.	Quinovic.
Cocognidic.	Menispermic.

Butyroleic,	$C_{34}H_{30}O_4HO$
Crotonic, ?	Sabadillic, ?

Acids from essential Oils, and from the excretions of Plants.

Coumaric,	$C_{18}H_7O_5HO$
Eugenic,	$C_{20}H_{12}O_4HO$?
Picranisic,	$C_{12}H_3N_3O_{14}$
Cuminic,	$C_{20}H_{11}O_3HO$
Cyminic,	?
Terebinic,	$C_{14}H_9O_7HO$
Pinic acid,	$C_{40}H_{30}O_4$
Sylvic,	$C_{20}H_{15}O_2$
Colopholic,	?
Dammaric,	$C_{40}H_{31}O_7$
Pimaric,	$C_{48}H_{30}O_4HO$
Pyromaric,	?
Azomaric,	$C_{20}H_{11}NO_8$
Camphamic,	?
Campholic,	$C_{20}H_{17}O_3HO$
Polychromic,	?
Aloetic,	$C_{14}H_2N_2O_{10}HO$?
Aloeretinic,	$C_{14}H_3NO_{12}$

Acids of the Bile, Brain, &c.

Cholic,	$C_{52}NH_{44}O_{11}HO$
Paracholic,	$C_{52}NH_{45}O_{12}$
Cholonic,	$C_{52}NH_{42}O_{11}$
Cholalic,	$C_{40}H_{40}O_{10}$
Choloidic,	$C_{48}H_{39}O_9$
Choleic,	$C_{52}NS_2H_{44}O_{13}HO$
Hyocholic,	$C_{54}NH_{43}O_{10}$
Hyocholalic,	$C_{50}H_{40}O_8$
Hyocholeic,	$C_{54}NS_2H_{45}O_{12}$
Nitrocholic,	$C_2N_4HO_9HO$
Choloidanic,	$C_{16}H_{12}O_7$
Cholesteric,	$C_8H_4O_4HO$
Lithofellic,	$C_{40}H_{35}O_7HO$
Nitrolithofellic,	$C_{40}NH_{35}O_{12}$?
Pyrolithofellic,	$C_{40}H_{34}O_6$
Cerebric,	?
Oleophosphoric,	?
Inosic,	$C_{10}N_2H_6O_{10}HO$
Taurylic,	$C_{14}H_8O_2$?
Damaluric,	$C_{14}H_{12}O_4$
Damolic,	$C_{26}H_{24}O_4$

Acidifiable. Capable of being converted into an acid; thus nitrogen, hydrogen, carbon, boron, &c. are acidifiable bases, because acids are formed by their union with oxygen, chlorine, &c.

Acidimetry. The process for measuring the strength of acids. See under various acids.

Acidulous Waters are such as contain free acids in solution.

Aconitic Acid. *Equisetic.* C_4HO_3HO. Crystalline mass soluble in alcohol, ether, and water; melts without change, but decomposes by a higher heat, yielding an oil, itacic acid. It is obtained from aconitum napellus, and equisetum fluviatile by solution of the juice of these plants in water, uniting the acid with oxide of lead, decomposing the lead salt by sulphohydric acid, and crystallizing out of ether. It may also be obtained by heating citric acid till it ceases to yield inflammable vapours; acetone, carbonic oxide, and acid being evolved. What remains is dissolved in alcohol, and heated with chlorohydric acid gas, by which *aconitic ether*, an oily fluid, is formed. Caustic potash forms with it aconitate of potash. Aconitate of lead is formed from this salt, which is decomposed as above. When heated it yields itacic and citracic acids.

Aconitine. White granules of unknown composition, obtained from the root and leaves of A. napellus and other species, by digesting the plant in alcohol, evaporating to an extract, adding water, filtering, evaporating to a syrup, adding sulphuric acid, when sulphate of aconitine is procured. This salt is precipitated by ammonia, and the filtered aconitine redissolved in dilute sulphuric acid, decolorized by animal charcoal, filtered and precipitated by ammonia, and crystallized out of alcohol; $\frac{1}{50}$ of a grain is poisonous; it contracts the pupil of the eye.

Acorn. The fruit of the oak; is sometimes used to adulterate coffee. The ash consists of silica 1·01 PO_5 11·15 CO_2 13·69 SO_3 4·79 Cl 2·51 Fe_2O_3 ·54 MgO 4·32 CaO 6·01 NaO ·63 KO 54·93. The seed is composed of thin sided cells filled with starch. The skin is a cellular membrane containing bundles of spiral vessels.

Cells. [*Hooker.*] Skin.

Acrospire. The maltster's name for the rudiments of the future stem or plumule of barley, when germinating in the process of malting.

Acroleine. *Hydrous oxide of acryle.* C_6H_3O, HO B.P 125° C = 64·55 H = 7·08 O = 28·37 spec. grav. of vapour = 1·897. Volatile oily fluid, with a peculiar odour, acting on the eyes with great force, uniting with atmospheric oxygen and becoming acid. The odour is distinguished when the wick of a tallow candle is allowed to glow after being blown out. It is obtained by distilling fats, but in greatest purity by distilling glycerine, the sweet principle of oil, with phosphoric acid; the products are distilled over oxide of lead, and chloride of calcium, in an atmosphere of carbonic acid. When acroleine or hydrate of oxide of acryle makes its appearance, we infer the presence of glycerine in the distilled substance.

Acrusite. A black coated crystal of carbonate of lead.

Acryle. C_6H_3. The hypothetic radical of acroleine.

Acrylic Acid. $C_6H_3O_3HO$. Obtained in union with silver as fine white crystals, by pouring acroleine upon oxide of silver in a tubulated retort, until the action ceases; by distilling over a water bath, foreign oils are removed, and acrylate of silver may be taken up by water from the residue and crystallized. The acryle series consists of

Acryle, hypothetic, ...	C_6H_3
Oxide of acryle, ...	C_6H_3O
Hydrous oxide, or acroleine,	C_6H_3OHO
Acrylous acid, ...	$C_6H_3O_2HO$?
Acrylic acid, ...	$C_6H_3O_3HO$

Actinolite. *Actinote.* (ακτιν, ray of the sun.) *Calamite, byssolite, strahlstein.* A variety of Hornblende, distinguished by green radiated oblique 6-sided prisms, spec. grav. 3·175 to 3·482; of different shades of colours—leek, olive, and emerald green. Constituents according to Bonsdorff, SiO_3 59·75, MgO 21·10, CaO 14·25, FeO 3·95, MnO ·31, FH ·76. B.B. melts into a glass coloured by iron. When of the lustre of glass it is called *vitreous*, and of resin, *common* actinolite; occurs in talc rocks.

Adamant. A synonyme of the diamond.

Adamantine Spar. A synonyme of the corundum.

Adept. An old term applied to proficients in alchemy.

Adhesion. (*Adhaerio*, I stick to.) A species of attraction by which the surfaces of bodies, either different or of the same nature, adhere to each other. A weak adhesive power is exemplified when a knife is dipped in water and thus wetted. An example of strong adhesion is afforded when the same instrument is immersed in glue or starch paste. Plane surfaces of glass by means of pressure may be made to adhere to each other, and cut lead surfaces also adhere under pressure with great force. The force of adhesion is measured by the weight required to separate two surfaces. The force of the adhesion of gold to mercury is 446 grains; of silver 429, tin 418, lead 397, bismuth 372, platinum 282, zinc 204, copper 242, antimony 126, iron 115, cobalt 8. (Brooke Taylor.) The force of adhesion seems in the order of their chemical affinities between solids and fluids, inversely as the temperature, and in direct ratio to the squares of the surfaces. Adhesion differs from cohesion, in that the last term applies to the force which retains the particles of the same mass, as of iron or lead together.

Adhesive Slate. A yellow or gray slaty mass, infusible, adhering strongly to the tongue, occurring in the Paris gypsum beds. Spec. grav. 2·08; it contains SiO_3 86·5, Al_2O_3 7, MgO 1·5, CaO 1·25, FeO 2·50.

Adinole. Compact felspar, orthose, petrosilex, or fusible hornstone.

Adipic Acid. $C_6H_4O_3HO$. Fusing point 266° (Laurent) 293° (Bromeis.) Crystals brownish, radiated tubercles, sparingly soluble in cold, very soluble in boiling water, easily soluble in alcohol and ether; distils over without decomposition; obtained by boiling oleic acid with nitric acid, after the separation by evaporation of the suberic and pimelic acids. It consists of C 49·78, H 6·77, O 43·45. It is distinguished from pimelic acid by its not precipitating salts of lead and copper. Sesquichloride of iron yields with it a brick precipitate.

Adipocere. (*Adeps*, fat; *cera*, wax.) Or *Margarate of Ammonia* principally. A fatty or waxy substance, first found by Fourcroy in the Cimitiere des Saints Innocens in Paris, as a product of the decomposition of the human remains. It has been frequently confounded with cholesterin; it fuses about 100°; it is soluble in alcohol and ammonia. It has been reported to have been found in bogs, (bog butter?) but it requires more careful examination.

Adipocere. A synonyme of hatchettine, or rock fat.

Adit, *Galerie.* The horizontal entrance to a vein in a mine, as in fig. at A.

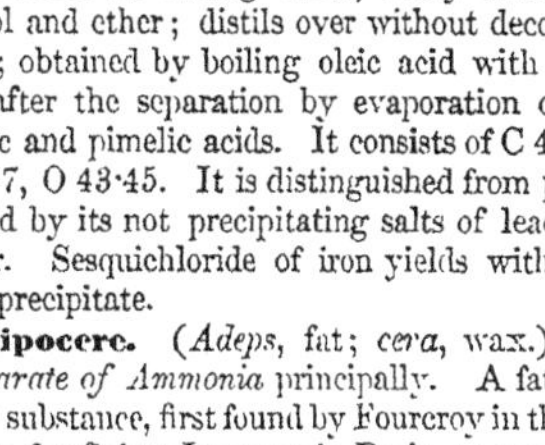
A

Adopter, Adapter. An elongated vessel applied to the beak of a retort, and conducting into the receiver.

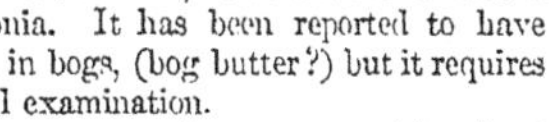

Adularia, (from Mount Adula.) *Moonstone.* That form of felspar which is very translucent, passing into blue, green, or gray, with a pearly lustre.

Aedelforsite, (from Aedelfors, Sweden.) A red zeolite, consisting of SiO_3 60·28, Al_2O_3 15·42, CaO 8·18, FeO 4·46, HO 11·07 MgO, and MnO ·42 (Retzius.) Its formula is Al_2O_3, 3 SiO_3, CaO, SiO_3, 4 HO, or stilbite, without 2 atoms of water.

Aedelite. A variety of mesotype.

Aegyrine. Consists of SiO_3 46·571, CaO 5·913, MgO 5·878, FeO 24·384, MnO 2·068, NaO 7·79, KO 2·961, Al_2O_3 3·418, TiO_2 2·017.

Aeolipyle. (Aeölipylae, Vitruvius.) Αιολου πυλαι—the doors of Æolus; originally applied to machines for discovering the cause of the winds. The term is now given to hollow spherical vessels terminating in a curved tube. Water or spirit being introduced into the interior, and heat applied, steam issues from the point of the tube, and may be used as a blowpipe when adjusted to a lamp, or as a bellows for blowing the fire. A similar instrument was used by Solomon de Caus in 1615, to show the pressure of steam, "causing water, by the assistance of heat, to mount above its level."

Aequinolite. A mineral allied to the obsidian of Mexico, or soda felspar.

Aerated Waters. Waters impregnated with carbonic acid.

Aerial Acid. An old synonyme of carbonic acid.

Aeriform Bodies. Substances existing in the rarest form, or as gases, as distinguished from solids and liquids.

Aerolite. (αηρ, air; λιθος, stone.) *Meteorolite. Meteorite.* Stones which are believed to fall from the atmosphere or moon, although many which have been so classed are obviously derived from the earth. They generally contain nickel and chromium. The analysis, by Mr. Faraday, of a very large one found in the Cape Colony, affords a view of the general nature of their constituents—silica, 28·9; protoxide of iron, 33·22; magnesia, 19·2; alumina, 5·22; lime, 1·61; oxide of nickel, ·82; sesquioxide of chromium, ·7; sulphur, 4·24; water, 6·5. Their spec. grav. is 3·352 to 4·281. The view generally entertained is, that these meteorites are mixtures of various minerals, of which nickel ore and chrome iron always appear to form a part. The compound minerals usually present, it is considered, are native iron, magnetic iron ore, which, after being reduced to powder, is removed by the magnet; chrome iron ore, bisulphuret of iron, olivine. With the exception of the speculation of Laplace—that they may be derived by volcanic action from the moon—no feasible explanation has been given of the source of these stones. It may be worth inquiry whether they may not be connected with electrical discharges, or volcanic forces acting on the ores existing on our earth. Great care would require to be exercised in the selection of such reputed stones, and in the examination of the localities connected with their fall, especially when we call to mind the numerous instances of powders deposited on ships at great distances from land, and of inorganic matter falling in Britain simultaneously with the occurrence of violent volcanic eruptions at Mount Hecla in Iceland. See Howard, Phil. Trans., 1802. Shepherd, Catalogue of American Meteorites. Silliman's Journ., 2d series. Chladni, Ann. Chem. v. 31. Powell's List of Meteors, Brit. Ass. Rept.

Aerology. (αηρ, air; and λογος, a discourse.) The science of the air, its properties, &c.

Aerometer. (αηρ, air; μετρον, a measure.) An instrument by which the necessary corrections are made in experimenting with aerial fluids, to ascertain the mean bulk of the gases; invented by Dr. Marshall Hall. (Quart. Journ. of Science, 5.)

Aeronautics. The mode of navigating or sailing through the air.

Aerosite. Antimonial sulphuret of silver, or dark red silver ore—1½ Ag, S_3, Sb. Spec. grav. 5·85; red six-sided prisms.

Aerostation. (αηρ, air; and ιστημι—I weigh.) Properly the science of weights suspended in the air, but now applied to the navigation of the air by balloons. It seems to have first occurred to Dr. Black, in 1767 or 1768, to suggest the idea of aerostation, and he put the idea in practice soon afterwards. "There is an anecdote of Dr. Black," says Dr. T. Thomson, (Hist. of Chem. i. 328) "which I was told by the late Mr. Benjamin Bell of Edinburgh, author of a well-known system of surgery, and he assured me that he had it from the late Sir George Clark of Pennycuik, who was a witness of the circumstance related. Soon after the appearance of Mr. Cavendish's paper on hydrogen gas, in which he made an approximation to the specific gravity of that body, showing that it was at least 10 times lighter than common air, Dr. Black invited a party of his friends to supper, informing them that he had a curiosity to show them. Dr. Hutton, Mr. Clerk of Eldin, and Sir George Clark of Pennycuik, were of the number. When the company invited had assembled, he took them into a room; he had the allantois of a calf filled with hydrogen; upon setting it at liberty, it immediately ascended and adhered to the ceiling. The phenomenon was easily accounted for; it was taken for granted that a small black thread had been attached to the allantois, that this thread passed through the ceiling, and that some one in the apartment above, by pulling the thread, elevated it to the ceiling, and kept it in this position. This explanation was so probable, that it was acceded to by the whole company, though, like many other plausible theories, it turned out wholly unfounded; for when the allantois was brought down, no thread whatever was found attached to it. Dr. Black explained the cause of its ascent to his admiring friends; but such was his carelessness of his own reputation, and of the information of the public, that he never gave the least account of this curious experiment even to his class; and more than 12 years elapsed before this obvious property of hydrogen was applied to the elevation of air balloons by M. Charles, in Paris." The first ascent in the atmosphere was made by means of a fire balloon, in 1782, by Stephen and John Montgolfier, who, by burning paper under a silk bag, so rarified the air, that the bag ascended, with some animals attached. On the 15th Oct., 1783 Pilatre de Rozier ascended in a fire balloon. On the 1st Dec., 1783, Messrs. Robert and Charles ascended from Paris in a balloon filled with hydrogen. Hydrogen continued to be used for the inflation of balloons until the employment of coal gas for illuminating purposes brought it also into action for acrostation, and superseded the use of the more expensive hydrogen. The first balloon ascent in England was made by Lunardi, an Italian, on the 15th Sept., 1784; and the second by Mr. Blanchard and Mr. Shelden, on the 16th October following. Little alteration has been made, apparently, in the original construction of balloons. They are composed of silk, and may be varnished with caoutchouc, to render them air-tight. A net is spread over the balloon, and is fastened to a hoop which passes round the balloon. From the hoop a car of wicker work is suspended, in which the aeronaut takes his seat, supplied with sand-bags for ballast, and bags of air for the purpose of breathing, should the balloon ascend with great rapidity, and the coal gas escape so as to surround the aerial voyager. Without the supply of fresh air, he would be in danger of being asphyxiated. M. Garnerin first descended, on the 21st Sept., 1802, from a balloon, at a considerable elevation, (a few hundred feet) by means of a parachute or umbrella, without any considerable injury. In his progress towards the earth, it was observed that the whole apparatus began to vibrate violently, so that the basket with the aeronaut was frequently in the same plane with the parachute itself. Another descent of a similar kind recorded, is that of Mr. Cocking, who ascended with Mr. Green, from the Surrey Gardens, on the 25th July, 1837. His parachute was made to descend in the form of an inverted umbrella—an unfortuate idea, as it very speedily collapsed by the pressure of the air. This martyr in the parachute cause was desirous of descending from an elevation of 8,000 feet; but in consequence of the great weight attached to the balloon, it could not be made to ascend above 5,000 feet. It was at this height, then, that the parachute, detached by a knife from the balloon, began its descent, along with the unfortunate victim. For three seconds the machine continued to move steadily, though with great velocity. The tin tube which constituted the extending hoop now broke, the parachute collapsed and descended like a closed umbrella, with the staff upwards. Little or no more resistance was now presented to its descent than by a heavy stone. Mr. Cocking fell near Blackheath, and was killed on the spot, many of his ribs and vertebrae being broken. Recent descents have been successfully made by Mad. Poitivin, in an umbrella-shaped parachute, perforated at the top, so as to allow the air to stream from below equally upwards; the result being, that the oscillating motion is obviated, and considerable steadiness of the apparatus attained. The great desideratum at the present time is, a method of navigating the air according to the will, instead of being at the mercy of the wind. I was once witness of a very interesting experiment by the veteran Green, at the Polytechnic Institution, in which he was most successful, in directing a model balloon to any part of a room filled with still air. A small windmill placed in front of the balloon, and motion given to it by a weight and rolled up cord, produced rarification of the air, and consequent horizontal motion of the machine. An upward movement was gained in the same way, by changing the position of the tiny windmill. The inventor hoped by such means to conduct his balloon to heights where he would meet with steady atmospheric currents blowing in different directions.

Aerugo. (Literally, rust of brass.) Carbonate of copper.

Aeschynite. (αἰσχύνη, modesty.) Black rhomboidal pyramids. Hardness, 5 to 6; spec. grav. 5·08-5·55; infusible, per se, and on charcoal before the blowpipe; on the outer edges forming a black slag; forms a dark yellow bead with borax, and a colourless clear bead with salt of phosphorus; with soda, a rust yellow mass without fusing. It consists of, titanic acid 56, zirconia 20, peroxide of cerium 15, lime 3·8, peroxide of iron 2·6, peroxide of tin 0·5. (Hartwall.) The specimen was from Miask, in Ural. By another analysis, the composition has been found columbic acid 33·39, titanic acid 11·94, zirconia 17·52, protoxide of iron 17·6, yttria 9·35, oxide of lanthanum 4·76, protoxide of cerium 2·48, lime 2·40, water 1·5, fluorine trace.

Aethamin Sulphuric Acid. See ETHAMIN SULPHURIC ACID.

Aethal, or **Ethal.** *Cetylic Alcohol.* The alcohol of spermaceti.

Aethionic. See ETHIONIC ACID.

Aetites. *Lapis Collymus.* Eaglestone; a hollow geode of oxide of iron, used formerly in medicine.

Affinity. (*Ad* and *finis*), a term first introduced into chemistry by Dr. Hooke, to distinguish chemical actions from gravitation, cohesion, and adhesion. When we pour a few drops of sulphuric acid into lime water, we observe a white turbidity produced, and on standing a white precipitate subsides. On examination we find it to consist of lime and sulphuric acid. Hence it is said that the lime has been removed from the solution by the sulphuric acid in consequence of the chemical affinity existing between the acid and the lime, which gave rise to the formation of the sulphate of lime. The early chemists supposed, that as in such an experiment the lime lost its alkaline reaction, that of restoring the colour of reddened litmus, and the acid the power of reddening litmus, the acid and the lime were both destroyed, and the resulting sulphate of lime was formed out of their ruins. This view was refuted in 1674 by Mayow, who showed that sal ammoniac, formed by adding muriatic acid to ammonia, contained both of these substances, as might be proved by adding potash when the ammonia was evolved, and by adding sulphuric acid when the muriatic acid was set free. Geoffroy, in 1718, generalized Mayow's view, and drew up tables of affinities, in which he showed that metals are separated from acids by the absorbent earths; the absorbent earths from acids by volatile alkalies; while the latter are separated by the fixed alkalies. In 1751, Gellert published new tables; and in 1758, Limbourg gave others. In consequence of the publication of these tables, two theories of affinity were broached. Some adopted the view of Geoffroy, that every body A had a particular degree or intensity of affinity by which it was united to another body X, and that whenever a third body, B, having a greater affinity for X than A had, is presented to the compound, A is displaced, and B unites in its stead with the body X. 2. Others were of opinion that no such difference in the intensity of affinity existed; that B might be capable of displacing A, and separating it from X; while, at the same time, A, in its turn, might displace B from X—or that B might be able to remove A, and incapable of disengaging C, though A were capable of removing C. In 1775, Bergman decided in favour of the first of these hypotheses, and carried chemists generally with him. According to him, the affinity of each of the bodies A B C D for X differs in intensity in such a manner that the intensity of the affinity of each may be expressed in numbers. He considered also that affinity was *elective*, in consequence of which, if A have a greater affinity for X than B has, if we present A to B X, X separates altogether from B and unites to A. Thus barytes has a stronger affinity for sulphuric acid than potash has; hence, when barytes is added to a solution of sulphate of potash, the sulphuric acid leaves the potash, and forms sulphate of barytes. In 1803, Berthollet opposed Bergman's view. He considered affinity as an attraction existing between the bodies which combine, similar probably to that which exists between the heavenly bodies. If this be so, the affinity must increase with the mass of the acting body. Hence, though barytes has a stronger affinity for sulphuric acid than potash, yet if we present a great quantity of potash to a small quantity of sulphate of barytes, the potash will separate a portion of the acid. In certain cases some countenance is lent to this view. For example, if we add to a blue solution of sulphate of copper a large quantity of colourless chlorohydric acid, the blue solution becomes green from the formation of chloride of copper, although the affinity of sulphuric acid is the greater for copper. Berthollet, therefore, considered affinity a species of attraction, depending on mass as in other cases of gravitation, and he explained chemical decompositions upon the view that the change which occurs is owing either to the insolubility or the elasticity of the ingredient which separates. Nitrate of barytes and sulphate of soda decompose each other, because sulphate of barytes is insoluble in water, while nitrate of soda is soluble. The two theories, therefore, may be simply stated as follows:—Bergman affirmed that affinity is *elective;* that the body which has the stronger affinity displaces that which possesses the weaker; and the strength of affinity may be measured by decomposition. According to Berthollet, affinity is not elective; the strength of affinity is not an absolute quantity, but increases with the mass of the attracting body. Hence this view leads to the conclusion, that bodies are capable of uniting in any proportion whatever indefinitely. The theory of union in definite proportions, first broached by Richter in 1792-1802, and established by Dalton in 1804, without any knowledge of Richter's views, has refuted the ideas of Berthollet, and has confirmed, in a modified manner, the theory of Bergman. Modern chemistry would tend to the inference, that instead of affinity the expression replacement or substitution may be employed. Thus, when barytes is added to sulphate of soda, barytes replaces soda, and sulphate of barytes is the result. Richter observed that when two neutral salts, which mutually decompose each other, are mixed together, the two newly formed salts still retain the same neutral state as the two original ones from which they were formed. He made a series of experiments, from which he drew up tables exhibiting the weight of each base capable of saturating 100 parts by weight of each acid, and the weight of each acid capable of saturating 100 parts of each base. He observed that the numbers in each table formed a series which have the same ratio to each other in all the tables. If, for example, in the table representing the muriates, the quantity of potash required to saturate 100 parts of muriatic acid were three times as great as the quantity of alumina requisite to produce the same effect, the same thing would hold in the sulphates, nitrates, and in all the other series of salts. Thus Richter had the merit of showing that the saturating power of acids and bases may be represented by numbers attached to them. Mr. Dalton, in 1804, communicated his views on the *atomic theory* to Dr. Thomas Thomson, who first published it to the world. According to him, the ultimate particles of all bodies consist of atoms incapable of further division; they are spherical, and of different weights. It is these atoms which unite together, and we can determine the atomic weight of a body by ascertaining how much of it will combine with another body. Thus, when 200 parts by weight of oxygen are exposed to the action of a burning glass, in a vessel containing a piece of charcoal, the oxygen is entirely converted into carbonic acid, which has increased in weight by 75 parts of carbon or charcoal. If we pass this carbonic acid through red hot charcoal, the 200 parts oxygen are found united to 150 carbon or charcoal, or 100 oxygen to 75 charcoal, to form carbonic oxide. We have thus two compounds of oxygen and carbon, carbonic acid CO_2, and carbonic oxide CO, in which the relation of oxygen is 2 to 1 with the same weight of carbon in each. The more recent view, which attributes chemical decomposition to electrical conditions, has not been confirmed by the later researches in organic chemistry, since it has been found that hydrogen, the type of the electro-positive metals, may be replaced by chlorine, the type of electronegative bodies, atom for atom, even where the amount of hydrogen is great or small.

Agalmatolite. *Bildstein, Figurestone, Koreite, Lardite, Pagodite.* White, gray, green, yellow, red, or brown mass, with a somewhat soapy feel. Sp. grav. 2·895; hardness, 2; silica, 49·816; alumina, 29·596; lime, 6; protoxide of iron, 1·5; potash, 6·8; water, 5·5. Infusible before blowpipe, but becomes white; fracture splintery. Brought from China in the form of images and ornaments, and from Nagyag.

Agaric Mineral. *Rock milk.* A white, light powder, soiling the fingers, and floating on water for some time. Found in calcareous rocks in Switzerland. It is a carbonate of lime.

Agaphite. A synonyme of turquoise.

Agate. A variety of quartz, found in amygdaloid rocks in rounded masses. When polished, it presents a variety of colours, depending upon alternations of layers of different tints; it has probably been deposited from solution in water.

Agedoite. A synonyme of asparagine, found in liquorice root.

Agent. A substance which produces a chemical action. A *reagent* is a term applied to such bodies as by their reaction occasion changes of colour or precipitation. In this sense it is equivalent to *test.*

Aggregate. (*Ad,* to; *grex,* a flock.) To gather together the smaller particles of substances into a denser mass.

Aggregation. The advantage of the aggregation of particles is very observable in filtering precipitates. Sulphate of barytes and oxalate of lime, when first formed, are very fine powders, which pass through the pores of a filter; but when they have stood for some hours exposed to heat, the finer particles aggregate into larger molecules, and are retained on the filter.

Agriculture. (*Ager,* a field; *cultura,* culture,) may be defined to be the proper treatment of the soil so as to enable plants sown in it to derive a ready and proper supply of food. The basis, therefore, of the study is the discrimination of the true nature of the food of plants. In former times, although many applications at present used in agriculture were employed, their mode of operation was not properly appreciated. It is only of later years that natural agriculture has been studied. At one time it was supposed that plants derived their food alone from water and the air, and that the only use of the soil was to fix them in their place. The first agriculturist who seems to have taken the proper view of the subject was Jethro Tull of Shelborne, in the county of Berks, who published a most valuable work on agriculture in the year 1733. "Cattle," he says, "feed on vegetables that grow upon the earth's external surface; but vegetables themselves first receive from within the earth the nourishment they give to animals;" and he denominates "the superficies of the pores, cavities, or interstices of the divided parts of the earth" the "pasture of plants." He appears, however, to have undervalued the importance of water and air; and it was not till the time of Kirwan, in 1796, that the various sources of the food of plants began to obtain their proper position in the science. "The most fertile soil," says he, "is that which contains the greatest quantity of the food of those vegetables that nourish men and useful animals, and administers it to them with due economy. The first essential requisite, therefore, of a fruitful soil is, that it contain a sufficient quantity of the three or four simple earths (lime, magnesia, alumina, and silica), and of the soluble carbonaceous principle. The other requisites are, that the proportion of each and general texture of the soil be such as to enable it to admit and retain as much water as is necessary to vegetation, and no more:" and he states, that "manures are applied to supply either the defective ingredients of the soil, or improve its texture, or correct its vices." Afterwards Davy advocated the same view. "The great use of the soil," says he, "is to afford support to the plant, to enable it to fix its roots, and to derive nourishment by its tubes slowly and gradually from the soluble and dissolved substances mixed with the earths," while the greater part of the weight of plants is derived from the carbonic acid and oxygen of the air, and from water. Liebig, still more recently, has followed up the same view, and has added most materially to our knowledge of the growth of vegetables. The theory or mode of explaining the nutrition of plants at present is, that the carbon which they contain is absorbed by the leaves and roots in the form of carbonic acid; the carbon is deposited in the system of the plant; the hydrogen is derived from water; the nitrogen from the ammonia and nitrogen of the air; the oxygen from water, &c.; and the ash which all vegetables leave when they are burned, is that portion of the food which is obtained from the soil. One important object to be held in view in agriculture is, to restore to the land the ash or salts which have been removed by the crop; hence the importance of manures, which not only supply salts, but likewise ammonia and carbonic acid, by the decay of the vegetable matter. The soil which contains the inorganic food of plants, may be considered to have been originally rock, and to be in a condition of gradual disintegration, by which the soluble food of plants is slowly formed through the action of the atmosphere. Hence the importance of tilling the earth, in order that new portions may arrive at the surface, to be exposed to the influence of air and water. The object of fallowing appears to be, to allow the upturned soil to enjoy for some time the disintegrating energies of the atmosphere, and to enable the agriculturalist to plough and pulverize the soil. See MANURES, SOILS.

Agrostemmine. Yellowish white fusible plates, difficultly soluble in water, more readily in alcohol; obtained by exhausting the seed of Agrostemma githago (corn cockle) with diluted spirit containing acetic acid, distilling off the alcohol, adding magnesia, and treating the precipitate with alcohol.

Agustite. A synonyme of the emerald.

Air. *Atmospheric and Common Air.* The term airs was originally applied to different species of gases before their characteristic distinctions were detected. Dr. Hales was the first person who began the investigations which led to this discrimination. The word *air* is now limited to the atmosphere, q. v.

Akcethine. Yellow, tasteless, rhombohedrons, F. P. 302°, soluble in alcohol, ether, and acetone; soda evolves from it ammonia; formed by the action of ammonia and sulphur on acetone.

Alabandine. A synonyme of sulphide of manganese.

Alabaster. *Alabuss*, (Scot.) *sulphate of lime, gypsum, or plaster of Paris.* A white substance used for the purpose of making casts and moulds, or when in fine white masses, as at Florence, it is sculptured into figures and vases. There are two varieties, gypseous and calcareous.

Alabastrites. A synonyme of alabaster. The older mineralogists by this term understood calcsinter.

Alalite. A synonyme of white augite, from the place of its occurrence; four-sided, doubly oblique prisms, spec. grac. 3·297, silica 54·4, CaO 15·7, MgO 22·57, Fe_2O_3 2·5, Al_2O_3 2·83 B.B. fuses into a glass.

Alanine. $C_6\ NH_7O_4$ a crystalline body isomeric with sarcosine, urethane, and lactamide, converted by hyponitrous acid into lactic acid and N; homologous with glycocoll; formed by acting on aldehydammonia with cyanohydric acid, and an excess of chlorohydric acid.

Alantine. A German synonyme for inulin.

Albene. A provincial name of a calctuff occurring at Erding, in Bavaria.

Albine. A synonyme of apophyllite.

Albite. *Soda felspar.* Occurs crystallized in white doubly oblique rhombs. Spec. grav. 2·6. Fuses like felspar before the blowpipe, but yields

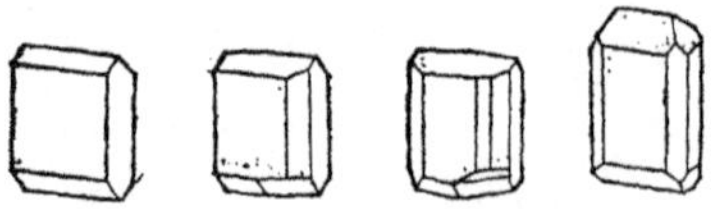

a yellow tinge to the flame. Its composition is SiO_3 68·, Al_2O_3 22·, MgO ·4, potash ·7, soda 7·8. It enters as a constituent into greenstones, which therefore differ from granites by containing more soda than potash—although albite also occurs in granite sometimes, along with felspar.

Album Græcum. The white fæces of dogs, consisting chiefly of bone earth were so named formerly, and used as a medicine.

Albumen. (*Album Ovi*) *Albumin*, *Oonin.* The type of albumen is the white of egg, a transparent glairy body enclosed in the cells of a membrane, which when broken by whipping with a spoon, or twigs, yields a fluid soluble in water. Albumen also exists in the serum of blood, and in lymphatic fluids. When exposed to the vacuum of an air-pump, there is left a brittle translucent substance, which, when pulverized, is a white powder with a saltish taste. A similar substance is procured by heating under a temperature of 120°, until all water is evaporated, when it is soluble in cold water. When heated to 140° it becomes white, opaque, and coagulated, and is then no longer soluble in water. It may be freed from the greater portion of its salts and impurity by digestion in water, ether, and alcohol. Its solutions are precipitated by chlorohydric, sulphuric, nitric, pyrophosphoric, and metaphosphoric acids. The precipitates contain a portion of these acids, and are insoluble in dilute acids, but, when freed from the excess of acid, dissolve in pure water. Acetic and phosphoric acids produce in dilute solutions no precipitate; but the strong acids form a gelatinous precipitate, easily soluble in water. When acetic acid is added in small quantity to a solution to overcome its alkaline character, flocks of pure albumen fall. Albumen is precipitated from its solutions by alcohol. Carbonate of soda, when added to a solution of albumen, prevents its coagulation by heat, even at a boiling temperature. Pure albumen dissolves easily in acetic and dilute phosphoric acids. The temperature at which albumen coagulates depends on its state of dilution; a solution containing 8 per cent. coagulates at 140°, but when more dilute it requires a temperature of even 167°; and when very dilute the solution may even require boiling, or evaporation, before coagulation occurs. Coagulated albumen is soluble in dilute caustic alkali. When procured by boiling serum of blood, it is found to contain nearly 90 per cent. of water, drying up when heated into a hard, pale yellow translucent mass, resembling gum arabic, becoming soft when digested in water; insoluble in alcohol, ether, and volatile oils. By exposure to the action of a long continued heat in a Papin's digester, it dissolves, and forms a yellow solution. Albumen consists of C 53·32 H 7·29 N 15·7 S 1·3 O 22·39. Its formula is S_2 N_{27} C_{216} H_{169} O_{68}. When acetic acid is added to a solution of albumen, and then a solution of yellow prussiate of potash, a copious white precipitate falls. Albumen of silk, when similarly treated, gives a green precipitate insoluble in water.

Albuminates. Protosulphate of iron and sulphate of copper precipitate albumen, but in excess redissolve it. The salts of tin, lead, bismuth, silver, and mercury, precipitate albumen white. Hence the best antidote for corrosive sublimate is white of egg.

Alburnum. The colourless part of wood situated between the bark and the duramen, or heart-wood.

Alcarazas. Porous earthenware vessels used in Spain for cooling water and wine by the evaporation through their sides, similar to water jugs used in India, and now in this country.

Alcargene. See Kacodylic Acid.

Alcarsine. A synonyme of oxide of cacodyle, (alcohol and arsenic.)

Alchemy, Alkemy. The knowledge of the substance or composition of bodies. So named from the Arabic substantive *Kiyamon*, that is, the substance or constitution of anything—from the root *Kama*, (Mr. Palmer.) This science, the incipient stage of chemistry, seems to be of Egyptian origin; it first signified natural philosophy; afterwards it was applied to the art of working metals; in the third century it was limited to the art of making gold and silver. From Egypt it passed to Greece, and thence to Arabia, and ultimately into the west of Europe, the followers of the science being termed alchemists. They conceived that the ingredients which enter into the constitution of gold are present in all metals mixed with impure bodies, but susceptible by proper treatment of being produced in a perfect state. The object of their studies was, therefore, to discover the method or agent for effecting this change, or in other words, of converting the inferior metals into gold. This desideratum they designated the philosopher's stone. Alchemy then came to signify the art of making the philosopher's stone. It was the search after this wonderful agent which gave rise to numerous experiments, and caused the discovery of an immense number of facts which constitute the basis of the present important science of chemistry. True alchemy must not be confounded with the quackery which ran parallel with it, any more than the science of medicine of the present day is to be mistaken for the empirical remedies which meet the eye in the ephemeral periodicals of our time.

Alcoate, Alcoholate. Weak compounds,

which alcohol forms with certain salts, are termed alcoates. Chloride of zinc combines with 1 atom of alcohol. Their existence has been called in question, but apparently without reason.

Alcohol. C_4H_5O, HO 6·75; 54. This term was originally applied to sulphuret of antimony, and bodies reduced to a fine state of division were said to be alcoholized. But in more recent times it is employed to designate the chemical compound which constitutes the essence of fermented liquors. It is said to have been first accurately described by Arnold de Villa Nova, who was born about the middle of the 13th century; although this is doubtful; but the method of distilling it from fermented liquors, and strengthening it by removing the water by means of carbonate of potash, was known to Raymond Lully about the middle of the 13th century. Alcohol is an ingredient of all fermented liquors. It is a colourless, mobile, combustible fluid, with a peculiar taste, a non-conductor of electricity, spec. grav. ·7947 at 59°; boiling point 173°; spec. grav. of its vapour, 1·6133. It consists of $\frac{1}{2}$ vol. ether vapour, 1·29044, and $\frac{1}{2}$ vol. vapour of water $=$ ·31005 $=$ 1.60049. Its formula is $C_4H_5O + HO$, and it contains C $=$ 52·658 H 12·896 O 34·446. Brandy and whisky contain about 50 per cent. of alcohol; spirit of wine 55 to 60 per cent. of alcohol; highly rectified spirit contains 80 to 85 per cent.; and alcohol 90 to 95 per cent. of alcohol. The common brandies contain usually an essential volatile oil, and frequently acetic acid. The acid may be removed by distilling with carbonate of soda, which retains the acetic acid, and the peculiar oil (fusel oil) may be destroyed by a small portion of bleaching powder or manganate of potash, and a few drops of sulphuric acid. Fat oils, when shaken with alcohol, likewise remove the volatile oils; but the best separater is wood charcoal. When brandy or whisky is distilled, the first spirit in the receiver contains 75 to 80 per cent. alcohol; the second portion is between 60 and 70 per cent.; and lastly, weak spirit and water. When spirit, containing 60 per cent. alcohol, with a spec. grav. of ·8952, is distilled, the first product has a spec. grav. of ·840 to ·833, and contains from 85 to 90 per cent. alcohol. Alcohol cannot be rendered anhydrous or destitute of water by mere distillation. Some substance must be mixed with the alcohol which has a stronger affinity with the water than with the alcohol. Slaked lime is such a body, and to the same class belongs chloride of calcium fused in a strong heat, on an iron plate or crucible. One part of this salt with 2 parts of alcohol is shaken in a bottle for several days. The liquid is poured off into a retort containing half as much chloride of calcium as the spirit weighs, and distillation performed by means of a sand heat; the first ounce is set aside. By two distillations in this way, anhydrous alcohol may be obtained. Strong alcohol may also be procured by tying up spirit in a bladder and allowing it to hang in the air; the particles of water pass more rapidly through the coats of the bladder than the alcohol. Alcohol may also be deprived of water by placing it in a vacuum over caustic or quick lime. The lime absorbs the water, while the vacuum is filled with the vapour of alcohol, and fluid alcohol remains of great strength in the vessel in which the spirit was deposited. To remove water from spirit also, carbonate and tartrate of potash, and caustic potash, may be employed; likewise Glauber's salt, sulphate of lime, carbonate of soda, &c. The best table of the strength of alcohol is that of Dr. Steel, (R. D. Thomson's Records of Gen. Science, vol. i. 257. 1835.)

Alcohol per cent.	Sp. Gr. at 60 deg. Fah.	Cor. for ea. deg.	Alcohol per cent.	Sp. Gr. at 60 deg. Fah.	Cor. for ea. deg.
100·	·79460	48	78·5	·85136	47
99·5	·79568	48	78·	·85248	47
99·	·79716	48	77·5	·85369	47
98·5	·79845	48	77·	·85491	47
98·	·79973	48	76·5	·85612	47
97·5	·80102	48	76·	·85733	47
97·	·80231	48	75·5	·85854	47
96·5	·80361	48	75·	·85975	47
96·	·80490	48	74·5	·86095	47
95·5	·80620	48	74·	·86216	47
95·	·80750	48	73·5	·86336	47
94·5	·80880	48	73·	·86456	47
94·	·81010	48	72·5	·86576	46
93·5	·81146	48	72·	·86696	46
93·	·81271	48	71·5	·86815	46
92·5	·81398	48	71·	·86935	46
92·	·81524	48	70·5	·87053	46
91·5	·81660	48	70·	·87172	46
91·	·81797	48	69·5	·87291	46
90·5	·81953	48	69·	·87411	46
90·	·82108	48	68·5	·87529	46
89·5	·82286	48	68·	·87648	46
89·	·82465	48	67·5	·87767	46
88·5	·82594	48	67·	·87886	46
88·	·82724	47	66·5	·88006	46
87·5	·82857	47	66·	·88123	46
87·	·82982	47	65·5	·88238	46
86·5	·83112	47	65·	·88354	46
86·	·83242	47	64·5	·88473	46
85·5	·83371	47	64·	·88593	46
85·	·83499	47	63·5	·88709	45
84·5	·83627	47	63·	·88826	45
84·	·83754	47	62·5	·88973	45
83·5	·83881	47	62·	·89121	45
83·	·84008	47	61·5	·89208	45
82·5	·84133	47	61·	·89295	45
82·	·84259	47	60·5	·89411	45
81·5	·84385	47	60·	·89528	45
81·	·84509	47	59·5	·89643	45
80·5	·84633	47	59·	·89759	45
80·	·84757	47	58·5	·89873	45
79·5	·84840	47	58·	·89988	45
79·	·84924	47	57·5	·90101	45

Alcohol per cent.	Sp. Gr. at 60 deg. Fah.	Cor. for ea. deg.	Alcohol per cent.	Sp. Gr. at 60 deg. Fah.	Cor. for ea. deg.
57·	·90215	45	28·5	·95979	30
56·5	·90328	45	28·	·96057	30
56·	·90442	45	27·5	·96132	29
55·5	·90554	45	27·	·96207	29
55·	·90666	45	26·5	·96279	28
54·5	·90778	45	26·	·96351	28
54·	·90891	44	25·5	·96421	27
53·5	·91005	44	25·	·96491	26
53·	·91119	44	24·5	·96558	25
52·5	·91232	44	24·	·96626	24
52·	·91345	44	23·5	·96691	23
51·5	·91487	44	23·	·96757	23
51·	·91570	43	22·5	·96821	22
50·5	·91680	43	22·	·96886	22
50·	·91791	43	21·5	·96949	21
49·5	·91896	43	21·	·97012	21
49·	·92001	43	20·5	·97074	21
48·5	·92115	43	20·	·97136	20
48·	·92229	43	19·5	·97198	20
47·5	·92337	43	19·	·97260	20
47·	·92446	43	18·5	·97321	19
46·5	·92554	42	18·	·97382	19
46·	·92663	42	17·5	·97443	18
45·5	·92770	42	17·	·97504	18
45·	·92877	42	16·5	·97564	17
44·5	·92986	42	16·	·97625	17
44·	·93095	41	15·5	·97686	16
43·5	·93200	41	15·	·97748	16
43·	·93306	41	14·5	·97806	15
42·5	·93408	41	14·	·97865	15
42·	·93511	41	13·5	·97930	14
41·5	·93612	40	13·	·97995	14
41·	·93714	40	12·5	·98059	13
40·5	·93815	40	12·	·98124	12
40·	·93916	40	11·5	·98188	12
39·5	·94015	40	11·	·98253	12
39·	·94115	40	10·5	·98320	12
38·5	·94213	39	10·	·98387	12
38·	·94311	39	9·5	·98456	11
37·5	·94406	39	9·	·98525	11
37·	·94502	39	8·5	·98595	10
36·5	·94595	38	8·	·98666	10
36·	·94689	38	7·5	·98739	10
35·5	·94782	37	7·	·98812	10
35·	·94876	37	6·5	·98887	10
34·5	·94967	36	6·	·98963	10
34·	·95059	36	5·5	·99041	9
33·5	·95148	35	5·	·99120	9
33·	·95238	35	4·5	·99201	9
32·5	·95325	34	4·	·99282	9
32·	·95412	34	3·5	·99373	9
31·5	·95496	33	3·	·99464	9
31·	·95581	33	2·5	·99547	9
30·5	·95662	32	2·	·99630	8
30·	·95743	32	1·5	·99721	8
29·5	·95822	31	1·	·99813	7
29·	·95902	31	0·5	·99906	7

Alcohol has a strong affinity for water, and mixes with it in all proportions, the spec. grav. of the resulting mixture increasing with the amount of water. When the water is added at first heat is evolved, showing that a chemical union has taken place; but the density of the resulting fluid is less than the mean of the two fluids; consequently there is a condensation or mutual penetration. The spec. grav. therefore of such mixtures cannot be calculated, but must be determined by experiment. This point is one of great importance, particularly in reference to excise questions. The following table by Marozeau, was determined by Gay Lussac's hydrometer:—

Vol. of Alcohol per cent.		Vol. of Alcohol per cent.	
0	1·000	10	·987
1	·999	20	·976
2	·997	30	·966
3	·996	40	·953
4	·994	50	·936
5	·993	60	·915
6	·992	70	·891
7	·990	80	·865
8	·989	90	·835
9	·988	100	·795

The following table by Gröning affords a proximate idea of the strength of alcohol, according to the boiling point of the fluid:—

Boiling Point.	Alcohol by Vol. per cent. In the Boiling Fluid.	Alcohol by Vol. per cent. In the Distilled Fluid.	Boiling Point.	Alcohol by Vol. per cent. In the Boiling Fluid.	Alcohol by Vol. per cent. In the Distilled Fluid.
77°·25 C.	...92	93	87°·50 C.	...20	71
77·50	...90	92	88·75	...18	68
77·81	...85	91·5	90·00	...15	66
78·15	...80	90·5	91·25	...12	61
78·75	...75	90·	92·50	...10	55
79·50	...70	89·	93·75	... 7	50
80·00	...65	87·	95·	... 5	42
81·25	...50	85	96·25	... 3	36
82·50	...40	82	97·50	... 2	28
83·75	...35	80	98·75	... 1	13
85·00	...20	78	100·	... 0	0
86·25	...25	76			

The tables used in this country for the determination of the strength of spirit from its spec. grav. are those of Gilpin, (Phil. Trans. 1794.) The alcohol employed was of spec. grav. ·825 at 60° F., and was composed of 100 parts alcohol, of spec. grav. ·814, and 4·5 water. Alcohol of ·825 consists of 89 alcohol and 11 water.

Real Specific Gravities at the different Temperatures.

Heat.	Pure Alcohol.	100 Alcohol 5 Water.	100 Alcohol 10 Water.	100 Alcohol 15 Water.	100 Alcohol 20 Water.	100 Alcohol 25 Water.	100 Alcohol 30 Water.	100 Alcohol 35 Water.	100 Alcohol 40 Water.	100 Alcohol 45 Water.
30°	·83896	·84995	·85957	·86825	·87585	·88282	·88921	·89511	·90054	·90558
35	·83672	·84769	·85729	·86587	·87357	·88059	·88701	·89294	·89839,	·90345
40	·83445	·84539	·85507	·86361	·87134	·87838	·88481	·89073	·89617	·90127
45	·83214	·84310	·85277	·86131	·86105	·87613	·88255	·88849	·89396	·89909
50	·82977	·84076	·85042	·85902	·86676	·87384	·88030	·88626	·89174	·89684
55	·82736	·83834	·84802	·85664	·86441	·87150	·87796	·88393	·88945	·89458
60	·82500	·83599	·84568	·85430	·86208	·86918	·87569	·88169	·88720	·89232
65	·82262	·83362	·84334	·85193	·85976	·86686	·87337	·87938	·88490	·89006
70	·82023	·83124	·84092	·84951	·85736	·86451	·87105	·87705	·88254	·88773
75	·81780	·82878	·83851	·84710	·85496	·86212	·86864	·87466	·88018	·88538
80	·81530	·82631	·83603	·84467	·85248	·85966	·86622	·87228	·87776	·88301
85	·81291	·82396	·83371	·84243	·85036	·85757	·86411	·87021	·87590	·88120
90	·81044	·82150	·83126	·84001	·84797	·85518	·86172	·86787	·87360	·87889
95	·80794	·81900	·82877	·83753	·84550	·85272	·85928	·86542	·87114	·87654
100	·80548	·81657	·82639	·83513	·84308	·85031	·85688	·86302	·86879	·87421

Heat.	100 Alcohol 50 Water.	100 Alcohol 55 Water.	100 Alcohol 60 Water.	100 Alcohol 65 Water.	100 Alcohol 70 Water.	100 Alcohol 75 Water.	100 Alcohol 80 Water.	100 Alcohol 85 Water.	100 Alcohol 90 Water.	100 Alcohol 95 Water.
30°	·91023	·91449	·91847	·92217	·92563	·92889	·93191	·93474	·93741	·93991
35	·90811	·91241	·91640	·92009	·92355	·92680	·92986	·93274	·93541	·93790
40	·90596	·91026	·91428	·91799	·92151	·92476	·92783	·93072	·93341	·93592
45	·90380	·90812	·91211	·91584	·91937	·92264	·92570	·92859	·93131	·93382
50	·90160	·90596	·90997	·91370	·91723	·92051	·92358	·92647	·92919	·93177
55	·89933	·90367	·90768	·91144	·91502	·91837	·92145	·92436	·92707	·92963
60	·89707	·90144	·90549	·90927	·91287	·91622	·91933	·92225	·92499	·92758
65	·89479	·89920	·90328	·90707	·91066	·91400	·91715	·92010	·92283	·92546
70	·89252	·89695	·90104	·90484	·90847	·91181	·91493	·91793	·92069	·92333
75	·89018	·89464	·89872	·90252	·90617	·90952	·91270	·91569	·91849	·92111
80	·88781	·89225	·89639	·90021	·90385	·90723	·91046	·91340	·91622	·91891
85	·88605	·89043	·89460	·89843	·90209	·90558	·90882	·91186	·91465	·91729
90	·88376	·88817	·89230	·89617	·89988	·90342	·90668	·90967	·91248	·91511
95	·88146	·88588	·89003	·89390	·89763	·90119	·90443	·90747	·91029	·91290
100	·87915	·88357	·88769	·89158	·89536	·89889	·90215	·90522	·90305	·91066

Heat.	100 Alcohol 100 Water.	95 Alcohol 100 Water.	90 Alcohol 100 Water.	85 Alcohol 100 Water.	80 Alcohol 100 Water.	75 Alcohol 100 Water.	70 Alcohol 100 Water.	65 Alcohol 100 Water.	60 Alcohol 100 Water.	55 Alcohol 100 Water.
30°	·94222	·94447	·94675	·94920	·95173	·95429	·95681	·95944	·96209	·96470
35	·94025	·94249	·94484	·94734	·94988	·95246	·95502	·95772	·96048	·96315
40	·93827	·94058	·94295	·94547	·94802	·95060	·95328	·95602	·95879	·96159
45	·93621	·93860	·94096	·94348	·94605	·94871	·95143	·95423	·95705	·95993
50	·93419	·93658	·93897	·94149	·94414	·94683	·94958	·95243	·95534	·95831
55	·93208	·93452	·93696	·93948	·94213	·94486	·94767	·95057	·95357	·95662
60	·93002	·93247	·93493	·93749	·94018	·94296	·94579	·94876	·95181	·95493
65	·92794	·93040	·93285	·93546	·93822	·94099	·94388	·94689	·95000	·95318
70	·92580	·92828	·93076	·93337	·93616	·93898	·94193	·94500	·94813	·95139
75	·92364	·92613	·92865	·93132	·93413	·93695	·93989	·94301	·94623	·94957
80	·92142	·92393	·92646	·92917	·93201	·93488	·93785	·94102	·94431	·94768
85	·91969									
90	·91751									
95	·91531									
100	·91310									

Heat.	50 Alcohol 100 Water.	45 Alcohol 100 Water.	40 Alcohol 100 Water.	35 Alcohol 100 Water.	30 Alcohol 100 Water.	25 Alcohol 100 Water.	20 Alcohol 100 Water.	15 Alcohol 100 Water.	10 Alcohol 100 Water.	5 Alcohol 100 Water.
30°	·96719	·96967	·97200	·97418	·97635	·97860	·98108	·98412	·98813	·99334
35	·96579	·96840	·97086	·97319	·97556	·97801	·98076	·98397	·98804	·99344
40	·96434	·96706	·96967	·97220	·97472	·97737	·98033	·98373	·98795	·99345
45	·96280	·96563	·96840	·97110	·97384	·97666	·97980	·98338	·98774	·99338
50	·96126	·96420	·96708	·96995	·97284	·97589	·97920	·98293	·98745	·99316
55	·95966	·96272	·96575	·96877	·97181	·97500	·97847	·98239	·98702	·99284
60	·95804	·96122	·96437	·96752	·97074	·97410	·97771	·98176	·98654	·99244
65	·95635	·95962	·96288	·96620	·96959	·97309	·97688	·98106	·98594	·99194
70	·95469	·95802	·96143	·96484	·96836	·97203	·97596	·98028	·98527	·99134
75	·95292	·95638	·95987	·96344	·96708	·97086	·97495	·97943	·98454	·99066
80	·95111	·95467	·95826	·96192	·96568	·96963	·97385	·97845	·98367	·98991

Alcohol, when inflamed in the open air, burns with a bluish-white flame, the products of combustion being carbonic acid and water. When the amount of oxygen is deficient, lamp black is deposited. When the vapour of alcohol is mixed with oxygen in the proper proportion, and fired, it detonates with a loud explosion. If the vapour be passed with atmospheric air or oxygen through a red hot tube, the products are carbonic acid, acetic acid, aldehydic acid and water, with a peculiar body smarting the eyes. When platinum black is moistened with spirit, similar products result with a rise of temperature. If alcoholic vapour be passed through a glass tube at a low red heat there form aldehyde, olefiant gas, carburetted hydrogen and water. When distilled with bleaching powder, chloroform is produced; and by the continuous action of chlorine, chloral is formed. Sulphuric acid removes water from alcohol, and sets ether free. The various acids form ethers with the basis of alcohol. Various substances are soluble in alcohol. All salts of inorganic acids insoluble or difficultly soluble in water, are insoluble in alcohol.

Alcoometry, Alcoholometry. The process of estimating the amount of alcohol contained in a fluid mixture. The original method of effecting this object was a very rude one. A measured quantity of spirits was poured upon a portion of gunpowder in a plate and ignited. If at the end of the combustion the gunpowder was dry, it took fire and exploded; but if the spirit was weak, the water which it contained melted the powder and prevented it from igniting; in the former case it was said to be *above*, in the latter case *below proof*. The term proof spirit is now applied to a spirit consisting nearly of equal weights of alcohol and water. By different estimates proof spirit has a spec. grav. of ·91921 ·92003 ·91957. The density at 62° is ·918633. When the fluid containing the alcohol has in it nothing besides water, the determination of the specific gravity by means of a hydrometer or specific gravity bottle is sufficient, and from the preceding tables the strength may be calculated; but when sugar or other ingredients, as in beer, are dissolved in the liquid to be tested, it is necessary to have recourse to other methods. The fluid containing the alcohol may be distilled in a retort or tube apparatus, so arranged that no escape of alcoholic vapour can occur. A convenient apparatus of simple construction will be found to be two flasks connected by a Liebig's condensing tube, or a tube 3 or 4 feet long, passing through corks. One of these flasks acts as a retort, and the other as a receiver, being immersed in a large vessel of cold water to condense the vapour as it passes over.

The alcoholic fluid being placed in the retort flask, heat is applied by means of a gas flame through the medium of a small sand bath. The distillation is continued until one-half of the fluid is distilled over, or until the liquid dropping over yields a green colour and the odour of aldehyde, when treated with bicromate of potash and sulphuric acid. The specific gravity of the distilled fluid is then taken, from which, by the preceding tables, the amount of absolute alcohol contained in the whole amount of original fluid distilled may be calculated; and hence the per centage composition of the alcoholic fluid. *Example.*—Retort flask = 634· grs., beer = 2294·6, total = 2928·6 grs. Receiver flask = 880 grs., distilled spirit = 987·4 grs., spec. grav. of spirit = ·9848 = 9·98 per cent. of absolute alcohol. Then 100 : 9·98 : : 987·4 : 98·54 = total absolute alcohol in 2294·6 grs. of beer. Then 2294·6 : 98·54 : : 100 : 4·29 per cent. of absolute alcohol in the beer. Another method of quicker application from which the original gravity of the alcoholic fluid can be calculated, but probably not so precise, has been, it is said, long used by German brewers. Whether it is originally a German discovery is doubtful;

but certain it is, that it has been known to the excise of this country for at least 20 years. 1. The specific gravity of the beer or alcoholic fluid is ascertained by means of the hydrometer. 2. A measured quantity of the fluid is boiled in an open vessel till one-half has evaporated, or until all the alcohol is dissipated. 3. The remaining fluid or *extract fluid*, thus deprived of its alcohol, and consisting of water, albuminous matter, and sugar, or extract as it is termed, in solution, is made up to its original bulk by the addition of distilled water. 4. The specific gravity of this prepared fluid is taken. 5. By deducting the specific gravity of the original fluid, *a*, from the specific gravity of the extract fluid, *b*, after the addition of the water, the spirit indication, *c*, of so many degrees is obtained, which when added to the specific gravity of the extract fluid, *b*, yields the specific gravity of the original fluid before it was fermented. A beer having a spec. grav. of 1035·1, when boiled down and made up to its original bulk with water, gives a spec. grav. of 1044·7. Now 1044·7 — 1035·1 = 9·6 degrees of spirit indication, and by the following table 9·6 degrees are equivalent to 43·7 degrees of gravity lost. Then 1044·7 + 43 = 1087·7 = the original specific gravity of the beer or alcoholic fluid. The following tables, for which I am indebted to the kind attention of John Wood, Esq., the very able chairman of the Board of Inland Revenue, are those which are used by the excise. It is proper to mention that the first table is the result of the labours of Messrs. Dobson and Phillips, and the second table was made long afterwards, to test its accuracy, at the request of the excise board, by a chemical commission consisting of Messrs. Graham, Hofmann, and Redwood. Simple comparison of the respective numbers exhibits, in a strong point of view, the value of the experiments in the first table, and of the ability of the gentlemen in the service of the excise.

TABLE 1.

Spirit Indication.	TENTHS.									
	0	1	2	3	4	5	6	7	8	9
1	3·1	3·5	3·7	4·0	4·3	4·6	5·0	5·3	5·6	5·9
2	6·2	6·6	7·1	7·5	7·9	8·3	8·8	9·2	9·6	10·1
3	10·5	10·9	11·4	11·8	12·2	12·6	13·1	13·5	13·9	14·4
4	14·8	15·2	15·7	16·1	16·5	16·9	17·4	17·8	18·2	18·7
5	19·1	19·5	20·0	20·4	20·9	21·3	21·7	22·2	22·6	23·1
6	23·5	23·9	24·4	24·8	25·3	25·7	26·2	26·6	27·1	27·5
7	28·0	28·5	29·0	29·5	30·0	30·5	31·0	31·5	32·0	32·5
8	33·0	33·5	34·0	34·5	35·1	35·6	36·1	36·6	37·1	37·7
9	38·2	38·7	39·2	39·8	40·3	40·8	41·3	41·9	42·4	42·9
10	43·4	43·9	44·4	45·0	45·5	46·0	46·5	47·0	47·6	48·1
11	48·6	49·1	49·7	50·2	50·7	51·2	51·8	52·3	52·8	53·4
12	53·9	54·4	55·0	55·5	56·1	56·6	57·1	57·7	58·2	58·8
13	59·3	59·8	60·4	60·9	61·5	62·0	62·6	63·1	63·7	64·2
14	64·8	65·4	65·9	66·5	67·1	67·6	68·2	68·8	69·4	69·9
15	70·5									

TABLE 2.

Deg. of Spirit Indication.	·0	·1	·2	·3	·4	·5	·6	·7	·8	·9
0	..	·3	·6	·9	1·2	1·5	1·8	2·1	2·4	2·7
1	3·0	3·3	3·7	4·1	4·4	4·8	5·1	5·5	5·9	6·2
2	6·6	7·0	7·4	7·8	8·2	8·6	9·0	9·4	9·8	10·2
3	10·7	11·1	11·5	12·0	12·4	12·9	13·3	13·8	14·2	14·7
4	15·1	15·5	16·0	16·4	16·8	17·3	17·7	18·2	18·6	19·1
5	19·5	19·9	20·4	20·9	21·3	21·8	22·2	22·7	23·1	23·6
6	24·1	24·6	25·0	25·5	26·0	26·4	26·9	27·4	27·8	28·3
7	28·8	29·2	29·7	30·2	30·7	31·2	31·7	32·2	32·7	33·2
8	33·7	34·3	34·8	35·4	35·9	36·5	37·0	37·5	38·0	38·6
9	39·1	39·7	40·2	40·7	41·2	41·7	42·2	42·7	43·2	43·7
10	44·2	44·7	45·1	45·6	46·0	46·5	47·0	47·5	48·0	48·5
11	49·0	49·6	50·1	50·6	51·2	51·7	52·2	52·7	53·3	53·8
12	54·3	54·9	55·4	55·9	56·4	56·9	57·4	57·9	58·4	58·9
13	59·4	60·0	60·5	61·1	61·6	62·2	62·7	63·3	63·8	64·3
14	64·8	65·4	65·9	66·5	67·1	67·6	68·2	68·7	69·3	69·9
15	70·5									

It is obvious that the two methods may be conjoined. In the second case, by conducting the process by distillation, the alcohol may be collected and estimated by the first method.

Alcornine. A neutral crystalline body, obtained by alcohol from the bark of *Alcornea latifolia*.

Aldehyde. (*Alcohol dehydrogenatus.*) *Hydrous oxide of acetyle.* $C_4H_4O_2$ or C_4H_3OHO. A colourless, limpid, volatile fluid, with a peculiar ethereal odour; irrespirable, as it produces suffocation. Spec. grav. ·790, boiling point 71¼°. Spec. grav. of its vapour 1·512=2 vols. carbon vapour ·84279, 2 vols. hydrogen=·13760 ½ vol. oxygen =·55280=1·53319. It is miscible with water, alcohol, and ether, in all proportions. By the action of air and sun light it passes into acetic acid, and very rapidly if platinum black be present. It dissolves sulphur and iodine; is converted by chlorine and bromine into chloral and bromal; by nitric acid into acetic and carbonic acids; converted by caustic potash into a brown resin. When heated with nitrate of silver, with the addition of so much solution of barytes as to precipitate all the oxide of silver, metallic silver coats the glass tube like a mirror, and acetate of barytes remains in solution. Aldehyde is derived from ether and alcohol. When ether vapour is passed through a red hot tube, aldehyde, carburetted hydrogen, and olefiant gas are evolved. Aldehyde is made in largest quantity by distilling 2 alcohol of 80 per cent. with 3 black oxide of manganese, 2 water, and 3 sulphuric acid, C_4H_5OHO, 2 $MnO2SO_3$ = $C_4H_3OHO + 2\ HO\ 2\ (MnOSO_3)$, or bichromate of potash may be used as the oxidizing agent; it occurs also in the oxidation of casein, and several organic substances, and is produced by the action of many chlorides, as chloride of antimony, chloride of zinc, &c. upon spirit. Pure aldehyde is prepared by passing dry ammoniacal gas through aldehyde, when crystals of aldehyde ammonia separate ($C_4H_4O_2NH_3$) as a white crystalline compound; 2 parts of these crystals are dissolved in 3 parts water, and distilled with 3 parts sulphuric acid diluted with 5 parts water; 1½ parts are distilled off. When aldehyde is kept at 32° it changes into prismatic icy crystals, *elaldehyde*, and when kept at the ordinary temperature it becomes needles, or 4-sided prisms. *Chlorine* acts upon aldehyde, and forms chlor-aldehyde $C_4H_4Cl_2$.

Aldehydes, *Metaldehyde.* Aldehyde having a peculiar relation to acetic acid, from which it differs only in the amount of oxygen, it is believed that aldehyde is an intermediate stage in the production of acetic acid from alcohol. A class of such compounds exists connected with other organic acids, as the *aldehyde* of butyric acid, the *aldehyde* of metacetonic acid, the aldehyde of valerianic acid, ($C_{10}H_{10}O_2$,) the composition of valerianic acid being $C_{10}H_{10}O_4$.

Aldehydic Acid. *Lampic Acid.* $C_4H_4O_3$. An acid not isolated; believed to be formed when aldehyde is boiled with nitrate of silver, and also in the lamp without flame.

Aldehydene. The hypothetic radical of the aldehyde series, C_4H_3.

Aldide. A name given to a class of which aldehyde is the type.

Ale. A well-known amber-coloured fermented liquor of considerable strength, but varying in the amount of alcohol which it contains, according to the option of the particular ale brewer. Edinburgh ale contains apparently more saccharine matter unfermented than other ales, and is strong. Burton ale has more hops added, and is hence termed pale bitter ale. The Scotch ale is said to be brewed during the cold months of the year, only one mash of half-an-hour's duration being made, and the heat of the liquor raised to 180°; it is then drained off into the wort copper. The malt is then deprived of all its soluble matter by sparging, or dashing over hot water of 180° for eight or ten times successively, the liquor draining through by apertures placed at the sides of the mash tun, and armed with stopcocks. One gallon of yeast is added to 240 gallons of wort, and the fermentation begun at 50°, and continued for two or three weeks: four lbs. of hops are used to the quarter of malt. The following table by my pupil, Mr. John Wright Currie, of the Perth Brewery, gives the strength of various ales brewed in that establishment. The experiments were made with great care in my laboratory.

Scottish Ales.

	Absol. Alcohol per cent.	Extract per cent.	Water per cent.
Export Ale,...	7·96	3·75	88·29
India Ale,.....	8·97	2·75	89·28
No. 3 Ale,....	7·055	5·7	87·25
No. 4 Ale,....	7·855	6·675	85·47

Scottish Ale Brewing.—To brew 20 barrels of ale, 80 bushels of malt and 80 lbs. of hops are required. Three or four barrels of water at 180° are let down into the mash tun, and at

a, Mash tun. *b*, Under back.

the same time the sluice of the malt bung is opened, and the malt and remainder of the liquor at 175° run down together, and stirred. The mashing requires three hours, when the sparger (sprinkler) is fixed to the head of the tun. The sparger is a copper cylinder five or six inches in diameter, closed at both ends, and nearly so to within a foot of the centre, which is open, with a cross division against which a run of liquor by a spout from the copper strikes and sends it round the tun. An iron bar is fixed across the latter, on which the sparger is placed on a pivot. Its two arms extend the width of the tun, the inferior side of these being pierced with small holes similar to the mouth of a watering-pan, from which as it revolves, the liquor escapes and sprinkles the mash. The water in the boiler being tempered to the heat required for sparging, (185°,) the taps of the mash tun are slacked, and the worts permitted to flow out slowly, the sparger being set in motion—this operation being merely a continuation of the mashing. I may add, that having examined most of the waters used by the Edinburgh ale brewers, I have found them all very hard waters, containing a large quantity of carbonate of lime. If any virtue is to be attributed to the water, it may be presumed that the carbonate may act by neutralizing any acid as soon as it is formed. The 30 barrels of wort, 72 lbs. saccharine extract per barrel, are boiled for half-an-hour, and 40 lbs. hops added; another half-hour's boiling takes place, the remaining hops are added, and the wort boiled for another half-hour. Worts of the gravity of 50 lbs. extract per barrel strengthen 5 lbs. per barrel in one hour's boiling, and worts of 100 lbs. in one and a half-hour strengthen 15 lbs. So that the brewer can easily judge of the amount of boiling required by the use of the saccharometer. The worts, after a quarter of an hour, are run into the hop back, and then spread on the coolers, where they remain twelve hours; during this time they lose one-eighth of their bulk by evaporation. The worts being cooled to 53°, one barrel of wort is run into the gyle, and 6 gallons of yeast added, and thoroughly mixed; the remaining wort is then added. In Scotland, the temperature for commencing the fermentation is about 52½°, in England 62½°. In twenty-four hours the first stage ends, the surface being characterized on the edges by a white circle, and irregular patches of white breaking through, and soon being covered with froth. The head of froth is beat down, and the process continued for twenty-four hours more. In eight days the heat has increased 10°. The brewer judges of the period to stop the fermentation by the saccharometer, which indicates the amount of sugar which has been converted into alcohol. The next process is cleansing. In Edinburgh the ale is run finished from the gyle into the casks in which it is sold. In Alloa and Stirling it is run into butts, from which it is rocked into casks, a pint of fillings or prepared wort being put at the same time into each.

English Ale Brewing.—To make 20 barrels of ale, as in the preceding case, 24 barrels of water at 170° are added to the materials in the mash tun, allowed to extract for three hours, and run off; 17 barrels are drawn; a second and third mash are made at 185°, 9 barrels of water being used for each. After all these are removed, 8 barrels of water at 170° are run in for small beer. The 33 barrels of wort are boiled one hour; then 40 lbs. hops are added, the boiling continued for half-an-hour; the remaining hops added, and the boiling continued for half-an-hour, or two hours altogether. The worts are then run into the coolers. The fermentation is made

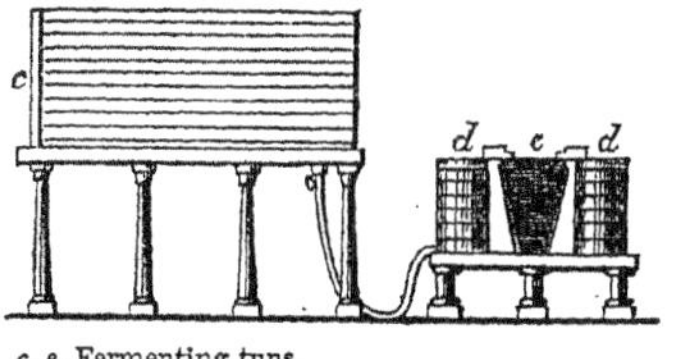

c, e, Fermenting tuns.
d, d, Troughs to receive the yeast thrown off.

at 60° to 65°. 9 gallons of yeast are added. In twelve hours a head has gathered; in twenty to twenty-four hours the head is formed, the heat being 70°. The great difference between the Scottish and English systems of brewing is that the first is conducted 10° under the English; it is therefore a slow process, while the English is a rapid one, and ales of different descriptions are produced. The cleansing is performed by mixing the yeast and ale together in the gyle, turning into barrels placed in close troughs, and continuing the fermentation until the yeast forms and separates from the ale, which is managed by keeping the barrels repeatedly filled up until fermentation ceases, and the process is finished.

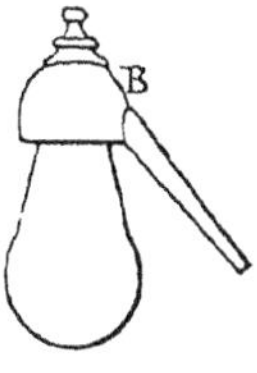

Alembic. (*Alembicus.*) A term applied to a peculiar head or capital of a still. (B.) See Still. In France, the term alembic or alambic is used to designate a glass still, consisting of a retort and head

Alembroth Salt. *Salt of Wisdom. Salt of Art.* A salt known to the alchemists; rhomboidal crystals obtained by mixing 2 atoms corrosive sublimate, (34,) 1 atom sal ammoniac, ($6\frac{3}{4}$,) and 8 atoms water, (9.) At 140° this mixture is fluid, but on cooling, sal alembroth crystallizes. Its composition is chloride of mercury and chloride of ammonium, with 1 atom water, ($NH_4Cl\ HgCl\ HO$,) but the water varies.

Alexandrite. A synonyme of phenakite.

Algaroth Powder. *English Powder. Oxychloride of Antimony.* Discovered by Victor Algarothi, a Veronese physician. White flocky precipitate, or soft powder, obtained by boiling black sulphuret of antimony, ($Sb\ S_3$,) with chlorohydric acid, and pouring the solution into water; terchloride of antimony is thus converted into an oxychloride of antimony. When it is allowed to stand in water it becomes crystalline, and then consists nearly of 11·32 chlorine, 76·6 antimony, and 12·08 oxygen, or 9 SbO_3, 2 $Sb\ Cl_3$. It is used in the preparation of tartar emetic, and is said to have an emetic action when swallowed.

Algæ. Seaweeds are plants which yield certain chemical products. The carragheen gives bassorin when boiled, cooling in the form of a jelly; the fucus saccharinus is covered with manna, and the fronds of these plants, when distilled with water acidulated with sulphuric acid, give an oily base fucusol. Their ash is *Kelp.*

Algerite. (After Mr. F. Alger, of Boston.) Honey-yellow prisms in limestone, at Franklin, New Jersey. Spec. grav. 2·78, hardness 3 to 3·5, fuses B.B. into a white blebby glass. Consists of silica 49·96, Al_2O_3 24·41, Fe_2O_3 1·48, MgO 5·18, $CaOCO_2$ 4·21, KO 9·97, HO 5·06. Form. 3 (Al_2O_3 2 SiO_3); (MgO KO) 3 SiO_3, 3 HO.

Alica. A grain used by the ancients for making tisanes, perhaps triticum spelta.

Aliment. *Alimentary Bodies,* (from *alere,* to nourish.) By this term is to be understood all those substances which answer as food for animals, and support their system under the continual wasting which it is undergoing. They are of two classes. 1. Nitrogenous, azotized, nutritive, or plastic substances, consisting of carbon, hydrogen, nitrogen, oxygen, and sulphur, which are deposited from the blood in the solid form of muscle and organized tissue. These include fibrine, albumen, caseine, bodies common to both vegetable and animal food, gluten of vegetable origin, and gelatin peculiar to animals. 2. Non-nitrogenous, non-azotized, calorifiant, or respiratory aliment, including starch, sugar, gum, fat. Neither class is sufficient by itself to retain the system in a healthy condition; but the true theory of alimentation depends on the proper proportioning of the two classes to the wants or waste of the system. The best type to study this ratio by is milk, the natural food of the human species, and of mammalia in the early stage of their existence. In milk, we have caseine, or curd, and albumen, representing the nitrogenous, or muscular part of animals; sugar, fat, or butter, supplying the system with the means of keeping up the supply of animal heat through the medium of the respiratory or breathing apparatus, and the salts which form also an essential constituent of the blood and solid parts of the body. The law that there is a settled ratio between these constituents of food was pointed out by the author in 1846, (Experimental Researches on Food, 8vo, 1846, p. 167,) and it has since been ably substantiated and illustrated by Liebig, Knapp, and Fresenius. "It is the elevated mission of science," says Liebig, "to show why man and animals require such an admixture in the constituents of their food for the support of the vital functions, and what the influences are which determine in accordance with the law of instinct changes in this admixture." It is obvious that although nature has ordained the animal system so that it is capable of sustaining a great amount of strain from bad management, yet, that there is a point beyond which discretion should restrain us from proceeding in utter disregard of nature's laws. Hitherto, especially in the higher classes of society, men have been obliged to submit their health, and even existence, to the tender mercies of the individual whom they may have hired to cook their victuals, obliged, should the results be prejudicial, to call in the aid of a physician, instead of using their reason and applying it to the discrimination of proper aliment. Man is elevated "in regard to an important function which he possesses in common with the lower animals above the level of those beings which are destitute of reason, and supplies him, in the regulation of those bodily wants which are essential to his existence and prosperity, with a protection which the lower animals do not require, because in them the commands of the instinctive will are not opposed or overpowered by the allurements of sense, nor by a perverted and resisting will." The following table gives the proximate composition of vegetable food in the moist state, from my own experiments:—

	Nitrogen.	Albuminous Matter.	Water per cent.
Beans,	4·050	25·361	10·60
Linseed,	3·780	23·620	14·20
Oatmeal,	2·488	15·612	5·30
Semolina,	2·050	12·811	13·32
Canada Flour,	1·860	11·625	11·80
Salep,	1·855	11·512	...
Barley,	1·810	11·310	9·46
Maize,	1·750	10·930	...
Essex Flour,	1·690	10·558	13·10
Lothian Flour,	1·55	9·742	12·86
Malt,	1·35	8·437	5·05
Rice,	1·430	8·930	...
Sago,	·534	3·337	13·32
Arrowroot,	·515	3·218	12·76
Tapioca,	·502	3·137	11·82
Ryegrass,	·460	2·875	75·
Starch,	·349	2·185	13·40
Potatoes,	·358	2·237	78·48
Swedish Turnips,	·212	1·325	84·55
Red Mangel Wurzel,	...	...	91·57
White Do.	...	...	87·63
Beet Root,	...	...	89·14
Salsify,	...	...	77·42

The following table, beginning with milk, affords a view of the relative position which various articles of food hold in respect to that important type of the aliment of mammalia, and is from the author's own experiments:—

Approximate relation of nitrogenous or nutritive to non-nitrogenous, or calorifiant matter.

	Nut. Calorif.
Cow's Milk, Food for a growing animal,	1— 2
Human Milk,	1— 4
Beans,	1— 2½
Oatmeal,	1— 5
Semolina, Barley,	1— 7
English Wheat Flour, Food for an animal at rest,	1— 8
Potatoes,	1— 9
Rice,	1—10
Turnips,	1—11
Arrowroot, Tapioca, Sago,	1—26
Starch,	1—40

From this table we infer that the food destined for the animal in a state of exercise should range between wheat and milk, flour varying in its degree of dilution with calorifiant matter, according to the nature and extent of the demands upon the system. The animal system is thus viewed as in an analogous condition to a field from which different crops extract various amounts of matter from the soil, which must be ascertained by experiment. An animal at rest consumes more calorifiant food in relation to the nutritive constituents than an animal in full exercise. The food therefore employed by a person of sedentary habits should contain more calorifiant, and less nutritive matter than one whose occupations cause him to take more exercise. The food of animals, and the manure of plants, we thus see afford somewhat of a parallelism. Milk may therefore be used with a certain amount of farinaceous matter, such as the class of meals and flours, with probable advantage, but the dilution should not exceed the prescribed limits. It is thus that we may explain the fact of beans, oats, oatmeal, and barley-meal being used so extensively in the feeding of horses. These articles of food, however, do not suffice alone; calorifiant matter in the form of hay should also be administered. From this table, likewise, we infer that as nature has provided milk for the support of the infant mammalia, the constitution of their food should always be formed after this type. Hence we learn that milk, in some form or other, being the true food of children, the use of arrowroot, or any of the members of the starch class, where the relation of the nutritive to the calorifiant matter is as 1 to 26, instead of being as 1 to 4, by an animal placed in the circumstances of a human infant, is opposed to the principles unfolded by the preceding table. Due attention to these principles should be attended to in the preparation of all our various forms of food.

Alixia Camphor. White capillary crystals, occurring on the interior of the bark of *Alixia aromatica*, subliming at 150°; soluble in hot, insoluble in cold water; soluble in alcohol, ether, oil of turpentine, acetic acid, carbonates of potash, and ammonia; coloured yellow by nitric acid.

Alizaric Acid. Identical with phthalic acid.

Alizarine. (*Ali-Zari*, oriental name for madder.) Orange-yellow needles, volatile with elevation of temperature; insoluble in cold, little soluble in hot water; soluble in alcohol and ether, forming a dark yellow solution; forming a red solution with strong sulphuric acid, and yellow flakes falling on the addition of water; soluble in caustic potash, with a red colour; precipitated yellow by acids. It consists of C 68·18, H 4·55, O 27·27, $C_{20}H_8O_8$. (Rochleder.) It seems to be identical with chrysophanic acid, or parietin, obtained from the parmelia parietina lichen, and also with *morindon*, a derivative from the *Morinda citrifolia*. It may be obtained by exhausting madder by boiling water, and precipitating by acetate of lead. The violet precipitate thus obtained is washed with water, diffused through water, and decomposed by a current of sulphohydric acid. Sulphide of lead, mixed with fat, alizarin, and purpurin fall. This precipitate is filtered and boiled with alcohol, and yields a dark brownish-yellow solution of fat, alizarin, and purpurin. Water allows the somewhat impure alizarin to fall. Fat is removed from it by cold ether. By solution of the remaining alizarin in boiling ether and spontaneous evaporation, alizarin is obtained in orange plates.

Alkahest. An universal solvent of the alchemists. Van Helmont distinguished carbonate of potash by this title.

Alkali. (*Al*, the article; and *kali*, Arabic for salt ash,) a term now generally restricted to a class of bases—potash, soda, lithia, and ammonia—which are soluble in alcohol and water, neutralize acids, change blue colours to green, render reddened litmus blue, yellow turmeric paper brown. Potash was formerly called the vegetable alkali, because it is derived from the ash of trees; soda the mineral alkali, because its source is common salt, a product of the mineral kingdom; and ammonia was designated the volatile alkali, from its great tendency to assume the gaseous form at common temperatures. Certain earthy bodies having like characters, to a certain extent, of alkalies, in so far that they are somewhat soluble in water, turn blue colours green, and neutralize acids, have been termed alkaline earths. These are lime, barytes, and strontian. A numerous class of organic bases exists, which form salts with acids, and having certain affinities with the alkalies, has been called alkaloids (alkali, *ειδος*, form).

Alkaligene. (Alkali and *γενάω*, I produce.) Alkali producer; a hypothetic term.

Alkalimeter. (Alkali and *μετρον*, a measure.) An instrument for measuring the amount of alkali in a salt mixture, particularly in commercial samples of soda-ash, and carbonates of soda and potash. The principle depends on using a known quantity of acid, and ascertaining when the alkali present has been neutralized. Alkalimeters are usually graduated glass tubes, or narrow cylinders, of varied form.

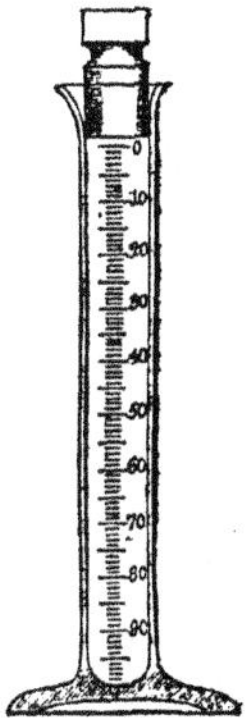

Alkalimetry, consists in the mode of using the alkalimeter.

Alkalimetry with a Solid Acid.—I have long used a very convenient system of alkalimetry. It consists in dissolving 50 grains of the alkaline sample in 2 ounces of water, in a flask, or stoppered phial, weighing out, on a watch-glass, 60 grains of well crystallized oxalic acid, reduced to a fine powder. Small portions of the acid powder are then to be added at a time, by means of a spatula or knife, to the alkaline solution, shaking the liquid between each addition, or stirring it with a glass rod, heating and testing it with litmus paper, until the latter becomes slightly reddened, while the liquid is hot. The residual oxalic acid is then weighed. Suppose it to be 20 grains. It is obvious that to saturate the alkali in 50 grains of the sample, 40 grs. of oxalic acid have been required. Now, 7·875 grains of oxalic acid are capable of saturating or removing the alkaline reaction of nearly 4 grains of soda (3·897 grs.), or 6 grains of potash (more precisely 5·888 grs.) We say, then, that the relation between the atomic weight of oxalic acid and of the soda or potash is the same as between the quantity of oxalic acid consumed in the experiment, and the soda or potash neutralized by it, and contained in the 50 grains of the sample, or as 7·875 : 3·897 for soda, and 5·888 for potash : : 40 : x. By multiplying the second term of the proportion by 40, and dividing by the 7·875, we get x equal to the amount of alkali contained in the sample.

If we wish to ascertain the amount of carbonate of alkali present, all that we require is to substitute for the number for the alkali the atomic weight of the carbonate of the alkali: and then as the quantity of sample used was 50 grs., by multiplying the product x by 2, we obtain the per centage amount of alkali. Whatever process be adopted in alkalimetry, this plan will be found advantageous as a rapid check, as no preparation of test acid is required.

Alkalimetry with a Fluid Acid.—By adding 9 ounces of distilled water to the strongest oil of vitriol of commerce, an acid is obtained, having the spec. grav. 1068 at 60°. An alkalimeter, containing 400 graduations, if filled up with the test acid, will contain in each division 5 grains of absolute acid. After adding the test acid gradually to a solution of 100 grains of the alkaline sample in 5 or 6 ounces of water, placed in a tumbler or beaker until the fluid slightly reddens litmus paper, we read off the number of divisions of the test acid which have been used. Multiply this number by 48, and divide by 100; or, more precisely, as every division of the acid contains 5 grains, or the atomic weight of acid, it will saturate the atomic weight of potash (5·888 grs.) or of soda (3·897 grs.) we have the proportion 1 : 5·888 or 3·897 : : x = number of divisions : y amount of alkali in the sample.

Alkalescent. The act of the evolution of alkaline characters.

Alkaloids. (Alkali and εἶδος, form.) Organic bases approximating to alkalies in character. See BASES ORGANIC for a table containing the principal alkaloids.

Alkana. Oriental alkanet, from the root of the Henna plant (*Lawsonia inermis*).

Alkanet. The root of the *Anchusa tinctoria*, a species of bugloss, containing anchusin, a red colouring matter, used principally for colouring oils and ointments.

Alkarsine, or oxide of cacodyle.

Allagite. A synonyme of manganese spar or carbonate of manganese, which often is mixed with silicate of manganese.

Allanite. A mineral brought from Allick in Greenland by Giesecké, and analyzed by Dr. Thos. Thomson in 1803, who named it after Mr. Allan. It is usually black, massive, brittle, but when crystallized is a right oblique 4-sided prism. Sp. grav. 4·001. Hardness 6. Fuses B.B. into a scoria or magnetic glass. Gelatinizes with nitric and muriatic acids. It consists of silica 35·15, Al_2 O_3 16·23, FeO 15·55; oxide of cerium 13·34; oxide of lanthanum 5·80 MnO 0·98; lime 12·02, MgO 0·78; water ·50.

Allantoic Fluid. The urine of the fœtus in certain mammalia, placed in a bladder between the allantois and the amnium. It contains allantoine.

Allantoine, Allantoic Acid, Amniotic Acid. Discovered by Vauquelin and Buniva, and analyzed by Liebig. Yellowish or colourless 4-sided prisms, neutral, without taste and smell, losing nothing at 212°; soluble in 400 parts cold, in 40 hot water; soluble in alcohol. Its formula is C_4 H_3 N_2 O_3. It contains the elements of oxalic acid and ammonia, and is changed into these bodies by alkalies. It may be obtained from the allantoic fluid by evaporation, to a quarter of its volume; the crystals, which separate on cooling, are decolorized by animal charcoal. It may also be formed by boiling 1 uric acid and 20 water with brown oxide of lead added, as long as the colour of the fluid changes.

Allanturic Acid. C_{10} H_7 N_4 O_9, or uric acid and 3 atoms water; a white matter formed in the first stage of the preparation of allantoine, and also by treating uric acid with nitric acid, and precipitating by ammonia.

Allituric Acid. Crystalline powder, disengaging ammonia with potash; soluble in 18 parts cold water; obtained by boiling alloxantine with chlorohydric acid. Form. C_6 N_2 H_3 O_4.

Allochroite, Grossularite, or Green Garnet, or even Colophonite.

Allogonite. A synonyme of Herderite.

Allomorphite. A synonyme of a variety of sulphate of barytes, containing 2 per cent. sulphate of lime.

Allophane. (ἄλλος, high; and φαίνω, I shine.) Light sky-blue green, brown, or yellow, massive, occurring in cavities in marl, Thuringia, Schneeberg, &c. Sp. grav. 1·870; loses colour before the blowpipe, the flame being tinged green; infusible except by oxygen hydrogen blowpipe, when it becomes a white opaque bead; brittle; semitransparent. Consists of silica 21·92, Al_2 O_3 32·20, HO 41·30, $CuOCO_2$ 3.06, FeO ·28, $CaOCO_2$ ·70, $CaOSO_3$ .52; but its constitution varies.

Allophanic Acid. An acid not yet isolated, obtained in union with oxide of ethyle, when the vapour of hydrous cyanic acid is mixed with ether and alcohol. C_4 N_2 H_3 O_5 would be its composition in the anhydrous state.

Allotropy. (ἄλλος, another; τρόπος, habit.) A body having different physical forms, is said to be allotropic. Carbon, for example, occurring as coke, plumbago, charcoal, diamond, in different crystalline forms, is a body subject to allotropy.

Alloxan, Erythric Acid. Colourless light prismatic crystals, efflorescent, losing 27 p. cent. of water, and becoming anhydrous. Very soluble in water. Odour disagreeable; reddens vegetable colours, and stains the skin purple; formed by adding cautiously uric acid to nitric acid, of spec. grav. 1·35. When crystals make their appearance, the addition of uric acid is stopped for the time, and the solution allowed to cool, when crystals of alloxan separate. These are thrown on a filter, upon a stopper of asbestus, and washed once with ice-cold water. They are then purified by recrystallization. When oxidized, it changes into uric and carbonic acid. Form. $C_8 H_4 N_2 O_{10}$. It is formed from uric acid, by the addition of 2 atoms oxygen, and 4 water, when it becomes equivalent to alloxan and urea. Alloxan yields an abundance of compounds: by alkalies, *alloxanic* acid; by ammonia, *mycomelic* acid; by NO_5, *parabanic acid;* with sulphohydret of ammonia, *dialurate* of *ammonia* is formed; with sulphate of ammonia, *thionurate* of *ammonia*.

Alloxanic Acid. $C_8 N_2 H_2 O_8$, 2 HO. Isomeric with alloxan; radiated needles, obtained by heating alloxan in barytes water. Alloxanate of barytes separates in crystals on cooling, which is decomposed by sulphuric acid: it is bibasic.

Alloxano-Sulphurous Acid. Only known in combination with potash in a crystalline form; a salt obtained by adding a strong solution of sulphurous acid to a cold saturated solution of alloxan till the odour of the acid becomes permanent. It is probably 1 at alloxan with 2 SO_2. By boiling a solution of alloxanic acid to dryness, and heating pretty strongly in a capsule, *leucoturic* acid remains, while water takes up *difluan*. By heat, alloxanate of barytes changes into *mesoxalic* acid.

Alloxantine. $C_8 N_2 H_5 O_{10}$. White hard crystals, sparingly soluble in cold, more soluble in hot water, yielding a violet colour with barytes water, reducing nitrate of silver, and yielding a black powder of silver, with formation of alloxan or oxaluric acid. It is obtained by passing sulphohydric acid through a solution of alloxan, or through the acid mother liquor from that substance. It contains an atom more hydrogen than alloxan, which it takes from the sulphohydric acid SH, sulphur being deposited. By HCl *dilituric* and *alituric* acids are derived from it; likewise *hidantoic* acid, and, with the aid of alloxan, *murexide*.

Alloy. Mixtures or compounds of various metals are called alloys (λυγοω, I bind). The various coins used as a circulating medium are alloys of the more precious metals. Thus the gold coinage is alloyed with copper or silver, and silver coin is alloyed with copper. The composition of many alloys will be given under the various species of bronze, coin, bell metal, brass, German silver, Britannia metal, pewter, pinchbeck, cannon metal, similor, packfong, tombac, tutenag, solder, prince's metal, Muntz's metal, type metal, &c. In uniting the various metals, sometimes the alloy has a density greater, sometimes less, than the means of the metals separated, showing that there has been either a contraction or an expansion in the union. The following alloys have a greater density than the mean density of the metals of which they consist:—Gold, with the following metals: zinc, tin, bismuth, antimony, cobalt. Silver, with the following metals: zinc, lead, tin, bismuth, antimony. Copper, with zinc, or tin, or palladium, or bismuth, or antimony. Lead, with bismuth, or zinc, or antimony. Platinum, with molybdenum. Palladium, with bismuth. The following alloys have a specific gravity less than the mean of the metals of which they consist:—Gold, with silver, or iron, or lead, or copper, or iridium, or nickel. Silver, with copper. Copper, with lead or platinum. Iron, with bismuth, or zinc, or antimony, or lead. Tin, with lead, or zinc, or palladium, or antimony. Nickel, with arsenic. Zinc, with antimony. The electrical conducting power of the alloys is inferior to that of the metals of which they consist. When the fusing point of the constituent metals is very different, they may be separated by *eliquation* or fusion. Tin, which fuses at 442°, may be separated from copper, whose fusing point is 1996°. Brass, which consists of copper and zinc, when melted for casting, and exposed to the air, exhibits a beautiful white light, and gives out fumes of oxide of zinc. In general, alloys are less easily oxidated than the separate metals.

Allspice. An aromatic mild spice, the fruit of the Eugenia pimento, a W. Indian tree.

Alluaudite. *Dufrenite*, *Craurite*. Protophosphate of iron, from Liege, consisting of 28·42 PO_5, 57·6 FeO, and 12·15 HO.

Alluvium. *Alluo*, I wash upon; or *ăllurio*, an inundation; *Alluvion*. Earth, sand, gravel, stones, and other transported matter, which have been washed away and thrown down by rivers, floods, or other causes, upon land not permanently submerged beneath the waters of lakes or seas.

Allyle. $C_6 H_5$. The hypothetic radical of oil of garlic, or sulphuret of allyle, C_6H_5S = All S; and of oil of mustard, or sulphocyanide of allyle.

Almagra. A purplish red ochre, formerly used as a pigment, and in medicine; the *sil atticum* of the ancients.

Almandine Garnet. A synonyme of fine red transparent garnet.

Almandine Ruby. A synonyme of spinelle.

Almonds, *Fixed Oil of.* Obtained by expression from the fruit of the *Amygdalus communis;* pleasant taste, without smell. Spec. grav. ·918. At 14° deposits solid matter. Both sweet and bitter almonds, varieties from the same plant, contain the same fixed oil.

Almonds, *Volatile Oil of Bitter.* Procured by distilling bitter almonds with water. When the fixed oil of bitter almonds is obtained by expression without the application of heat, it contains no trace of the volatile oil, showing that the volatile oil does not exist in the bitter almonds, but it results from an action between the amygdaline and emulsine which they contain. If the amygdaline be removed by means of alcohol, no volatile oil can be formed. If the amygdaline be separated, and brought in contact with the emulsion of sweet almonds, volatile oil results. The volatile oil, when purified, is hydret or hydride of benzoyle ($C_{12} H_6 O_2$). 100 parts of amygdaline form 47 of crude oil. It is obvious, therefore, that emulsine acts as a ferment, and that the volatile oil is one of the products of this action; no change is effected in the emulsine, but the amygdaline $C_{40}\,NH_{27}\,O_{22}$ is converted into 1 atom $HCy = C_2\,NH$; 2 hydride of benzoyle $= C_{28}\,H_{12}\,O_4$; $\frac{1}{2}$ sugar $C_6\,H_5\,O_5$; 2 formic acid $= C_4H_2O_6$; 7 water H_7O_7. Oil of bitter almonds is used in France for perfuming soap.

Aloes. The reddish-brown evaporated juice of the leaves of the *Aloe spicata, vulgaris*, or *socotrina.* The leaves are cut off close to the stem, then cut in pieces, and the juice allowed to run out. This fluid remains standing for 24 hours, when extractive matter is deposited. The clear liquor is then decanted, and allowed to concrete in the sun, or evaporated in copper boilers, as in the West Indies. The bitter principle is taken up by cold water, and hence the best mode of preparing aloetic pills from the commercial article is to form them from a cold water extract. It is an active purgative, and is best given in union with soap, which renders it more soluble. From this insolubility it has been supposed to act specifically on the rectum. Socotrine aloes consists of pure aloes 85, sulphate of potash 2, sulphate of lime 2, gallic acid ·25, albumen 8—traces of carbonates of potash and lime, and phosphate of lime.

Aloetamide. $C_{14}H_5N_3O_{10}$. Dark red body formed by ammonia on aloetic acid.

Aloetic Acid. Yellow or brown, of a bitter taste, formed by heating 1 part aloes with 8 nitric acid, of sp. grav. 1·25, and adding water; aloetic acid falls; it is to be dissolved in water, and purified: its composition is $C_{14}H_2N_2O_{10}HO$?

Alocretinic Acid. A product of the action of nitric acid on aloes, $C_{14}H_3NO_{12}$.

Aloine. A supposed alkali in aloes, forming a brown solution, changing reddened litmus, and forming a crystalline salt with SO_3 (Meissner). Small granular crystals were obtained by Smith from aloes, by pounding aloes with sand, macerating in cold water, and evaporating in vacuo. 2 to 4 grs. of these crystals are said to be more active than 10 to 15 of the common extract. As it has not been obtained from socotrine aloes, it is doubtful whether it be the principle of aloes. These crystals consist of $C_{34}\,H_{18}\,O_{14}\,HO$.

Alphaorsellesic Acid. $C_{16}\,H_9\,O_8$. Obtained by boiling alphaorsellate of barytes with water. When further boiled it yields colourless orcine and carbonic acid.

Alphaorsellic Acid. $C_{32}\,H_{16}\,O_{14}$. Crystals obtained from the *Roccella tinctoria* of South America, nearly insoluble in cold water; soluble in alcohol. Gives a red with bleaching liquor.

Alpha Resins. A mode of nomenclature sometimes used when several resins are mixed together, as alpha resin of colophony, alpha resin of benzoine, &c.

Alstonite. A synonyme of baryto-calcite of Brooke, from the town of Alston in Cumberland, near which it is found.

Altaite. Telluride of lead.

Altheine. A synonyme of *Asparagine.*

Althionic Acid. When sulphuric acid is added in excess to alcohol, olefiant gas is evolved, and in the residue is an acid, the salts of which are composed exactly as the sulphovinates, the crystalline form being, however, different. It is possible these may be mixtures of isethionates and sulphovinates.

Aludel. A clay vessel with the form of a pear, with two necks connected together, and used at Almaden in Spain for subliming mercury. It has long been replaced in Idria by condensing chambers, in consequence of loss sustained.

Aluchi Resin. Obtained from a tree in Madagascar, called timpi (Bomare), or from *Wintera aromatica.* It is very soluble in alcohol and ether, but not in soda.

Alum. A term now applied to a class of 8-hedral salts, formed on the same model. Thus we have potash alum, ($KO\,SO_3\,Al_2O_3\,3\,SO_3$, $24HO$) a compound of 1 atom sulphate of potash, 1 atom tersulphate of alumina, and 24 atoms water. Two others, also, may be formed by replacing the potash by soda or ammonia. These salts are, therefore, isomorphous. The potash may likewise be replaced by mananese and magnesia, as in manganese alum and by protoxide of iron. The composition of these salts will then be—

$KOSO_3$	$Al_2O_3\,3\,SO_3\,24\,HO$	...Potash alum.
$NaOSO_3$	$Al_2O_3\,3\,SO_3\,24\,HO$	...Soda alum.
$LOSO_3$	$Al_2O_3\,3\,SO_3\,24\,HO$	...Lithia alum.
NH_4OSO_3	$Al_2O_3\,3\,SO_3\,24\,HO$	Ammonia alum.
$\left.\begin{matrix}MnO\\MgO\end{matrix}\right\}$ SO_3	$Al_2O_3\,3\,SO_3\,24\,HO$	Manganese magnesia alum.
FeO SO_3	$Al_2O_3\,3\,SO_3\,24\,HO$	Protoiron alum.
KO SO_3	$Cr_2O_3\,3\,SO_3\,24\,HO$	Chrome alum.

In addition to these, the sesquioxide of aluminum may be replaced by other sesquioxides, as sesquioxide of iron, sesquioxide of manganese, yielding sesqui-iron alum, sesquimanganese alum. These are all formed on the same type as in the last formula.

Potash Alum.—Crystallizes in colourless 8-

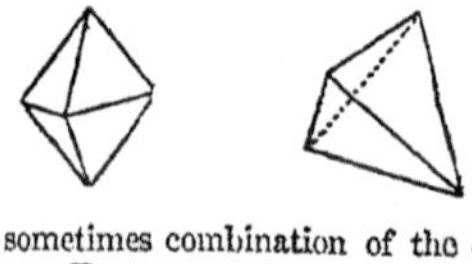

hedrons, sometimes combination of the cube and 8-hedron. Taste, astringent and sweetish. Reddens litmus; spec. gra. 1·724, of saturated solution 1·045; soluble in 13·3 parts water, at 56°, in 8·2 at 70°, in 2·2 at 100°, in 0·06 at 190°; loses 18 atoms water at 142°; the alumina parts with its sulphuric acid at a white heat, alumina and sulphate of potash being left. It contains, per cent. $KO = 9{\cdot}86$, Al_2O_3 11·09, SO_3 32·85, HO 46·2. The isomorphism of the oxides of iron, with potash and alumina, causes the presence of these oxides in alum. Their presence may be distinguished by yellow prussiate of potash, which gives with them a blue colour. The iron is best removed by boiling and repeated crystallization. Roman alum is so called because it comes from Italy; sometimes, though rarely, in prisms. Alum is used extensively as a mordant for vegetable colours in dyeing, and calico printing. It is used, from its affinity for colours, to prepare lakes. It is also mixed, in small proportion, with bread, to render it white, and to raise it. Ammonia alum is, however, preferred for this purpose, as it is nearly decomposed by the temperature of firing. A saturated solution of 2 nitre, 1 common salt, 1 alum, in a boiling state, is used by workers in gold to produce fine colours on gold. The solution dissolves copper, silver, and gold. As a mordant, it is usually employed in the state of acetate.

Magnesia Alum, Pickeringite.—$MgOSO_3$, Al_2O_3 3 SO_3, 24 HO. White fibrous masses with silky lustre.

Manganese Alum.—MnO SO_3, Al_2 O_3 3 SO_3 24 HO. Asbestus-like filaments: first examined by Stromeyer, from near Algoa Bay.

Iron Alum, Feather Alum.—FeO SO_3 Al_2 O_3 3 SO_3 24 HO. Silky, styptic fibres, found in alum shale.

Soda Alum.—8-hedrons, or fibrous masses. Sp. grav. 1·88. Found native at Mendoza, S. America. SO_3 37·7, Al_2 O_3 12·4, NaO 7·5, HO 42·4 (T. Thomson). B.B. gives a yellow flame.

Ammonia Alum.—8-hedrons, and fibrous. Sp. grav. 1·56. When boiled with caustic potash or soda yields ammonia, and also a green flame, B.B. When sublimed, gives sulphate of ammonia. Occurs at Tschermig in Bohemia. The following table will be found useful in discriminating whether an alum be an ammonia or potash alum, a point of considerable importance in the arts, as the value of the salt depends on the amount of alumina present. Ammonia, however, may sometimes interfere with the shade of colour, although that species of alum contains most of this earth:—

Ammonia.. 2·125	3·74	...Potash	5·899	10·
Alumina... 6·5	11·450...	„	6·5	10·92
4 Sul. acid. 20·	35·240...	„	20·	33·61
25 Water.. 28·125	49·570...	24W.	27·	45·47
56·75	100		59·399	100

The quantitative determination of the alumina or sulphuric acid will show to which species a sample may belong, according as the result may approximate to either column, as the two species are often mixed together.

Manufacture of Alum.—Alum is prepared in this country by two different methods. 1. The simplest mode is to dissolve alumina or pure clay in sulphuric acid, as by either method the first step consists in the production of a tersulphate of alumina. The object of importance in this process is to obtain a form of alumina as free as possible from iron, and likewise if lime is absent entirely, much sulphuric acid will be saved. The porcelain clay of the S. of England answers well for the purpose; and the shale of the coal strata of Scotland and Lancashire is advantageously used for the same object. The clay or shale is first calcined; it is found preferable in practice not to reduce the shale to powder, but to use it in small pieces, an inch or so in diameter, as this prevents the formation of a quantity of sludge. It is then heated in a wooden vessel with sulphuric acid, by passing steam through the mixture. Care must be taken, on the application of the heat, to prevent the formation of subsulphates, by which a great loss of acid will be sustained. The character of these subsalts is, that they are insoluble in boiling solutions, and redissolved on cooling. On the evaporation of the liquors to a convenient bulk, a salt of potash or ammonia is added, either sulphate or muriate, as they are sold in commerce, and the alum, on further concentration and cooling, crystallizes out in fine 8-hedral crystals. To purify them from iron, and other foreign bodies, a second solution and crystallization is required. The alum earth of the Freienwalde contains a small quantity of tersulphate of alumina, with much protosulphate of iron and sulphate of potash. When these two salts are exposed to the air, a difficultly soluble salt of sulphate of iron with potash is formed, setting free sulphuric acid, which unites with alumina. At Solfatara, in Italy, an impure alum is obtained by dissolving the efflorescence of lavas, evaporating, and recrystallizing. Alumstone contains the basic sulphate of potash and alumina, which it is difficult to dissolve. It is broken in small pieces, and calcined at Civita Vecchia in reverberatory furnaces, at Tolfa, in furnaces like lime-kilns, until sulphurous acid begins to be perceptibly evolved. It is then moistened with water, and exposed, when it acquires a soft consistence. It is then boiled in lead vessels, the liquor evaporated at a gentle heat, and then allowed to crystallize in wooden vessels. It partly crystallizes in the state of cubic alum. It is termed Roman alum, of a reddish hue, from an admixture with some sesquioxide of iron, but it is the purest alum in commerce. When dissolved in water, and heated under 104°, it affords cubic alum; but when the temperature is raised above 110°, a basic alum falls, and octahedral alum remains in solution, which crystallizes in 8-hedrons, if the liquor be decanted and crystallized, but in cubes if it remain in contact with the precipitated cubic alum. It seems to be a basic alum, and is hence preferred in dyeing, &c. from its containing a larger amount of alumina. 2.

From Alum Slate.—The greatest amount of alum in this country is made from alum slate, a dark-coloured slaty rock often found in connection with coal measures. This substance contains alumina, protoxide of iron, a trace of potash, and dispersed through it bisulphuret of iron, or iron pyrites. When exposed to moist air for a great length of time, the sulphur is oxidized, and is converted into sulphuric acid. A hair-like salt results, consisting of protosulphate of iron and tersulphate of alumina. When this is dissolved, and the water evaporated, sesquioxide of iron falls, and tersulphate of alumina remains in solution. If a salt of potash or ammonia is now added, crystals of alum are obtained. The old mode of máking alum in Scotland depended on the decomposition of the alum slate in the exhausted coal mines of Hurlet and Campsie. The demand for alum rendered it necessary to accelerate the process of oxidization as at present pursued in alum works in this country. The following description of the process, from a visit to an alum work, is by my assistant, Mr. Robert Kirkwood:—"The alum slate lies in two layers or strata; beneath all is a stratum of coal, above this lies the gentle slate; above this the diamond slate, which is covered by a stratum of limestone. The alum slate has a grayish or bluish-black appearance. It is a siliceous clay, containing a considerable portion of coaly matter, and bisulphuret of iron, either in small crystals, or in a state of fine division. When exposed to the air, it soon exfoliates, owing to the bisulphuret of iron absorbing oxygen with avidity, thus forming protosulphate of iron, while the excess of sulphuric acid unites with the alumina. This natural process of decomposition, and subsequent formation of new compounds, being too slow for the manufacturer, he takes advantage of the coaly matter in the slate, to cause, by slow combustion, the more rapid oxidation of the sulphur; his object being to convert the sulphur into sulphuric acid with as little loss of sulphur as possible. If combustion were too rapid, some of the sulphur would be driven off in a volatile form, or the bisulphuret and earthy matter might form a slag, which, being insoluble, would be of no value in the subsequent stage of lixiviation. To prevent too rapid combustion, ridges or heaps are piled up of alternate layers of fresh slate, and of spent or already exhausted ore. Sometimes the ridge is ignited at different parts along its base; sometimes it is not ignited at all, but becomes decomposed by a slow spontaneous combustion. One of these ridges requires 12, 18, or 20 months for its perfect calcination. This may be said to constitute the first stage in the preparation of alum, and comprehends the oxidation of the sulphur and iron. The next stage is the lixiviation and evaporation of the sulphates of iron and alumina. The calcined ore is wheeled into large square troughs or pits in the ground, named steeps, and the washings of a former steep are run upon it, and the whole allowed to remain at rest for some time. It is then run off into a cistern to settle; the ore in the steep being repeatedly digested until it is quite exhausted. The last washings are run into a separate cistern, and instead of being evaporated (which would be unprofitable, owing to their weakness), they are kept for the purpose of being run upon fresh portions of calcined ore. The exhausted ore is either removed to the waste heap, or is spread in layers alternately with the fresh ore upon the ridges, for the purpose of moderating the combustion. The washings are allowed to settle, and then run into a boiler for evaporation. The boilers are large, water-tight cisterns, made of stone, and arched over with brick, sometimes about 60 feet in length, 7 feet in breadth, and 4½ feet in depth. At one end is a large furnace, and at the opposite end a chimney, and by this means a current is formed, which, sweeping the flame and heated gases over the surface of the liquid, causes speedy evaporation. The loss by evaporation is frequently supplied by fresh liquor, and after several days the contents of the boiler are allowed to rest. This constitutes the second stage. The third stage consists in the addition of the potash or ammonia salt. This is not done at random. Before the liquor is removed from the boiler, it is tested to see how much potash is required for a given quantity. Having ascertained this, the boiler is guaged, and this is done by introducing a guaging rod at an orifice in the roof of the boiler. The liquor is then run into coolers, or large stone troughs, and the potash or ammonia salt added. After allowing the mother liquor to drain off into another cistern, the 'first alum' is shovelled out, and placed on a sloping platform, to allow the mother liquor to drain off. The object of this stage is to form a double salt, which shall crystallize, sulphate of alumina not being capable of crystallizing. Solution, evaporation, and recrystallization of the 'first alum' constitute the fourth stage. The 'first alum' is washed to remove all the mother liquor thoroughly, and is then introduced into cisterns covered with wooden lids, where it is dissolved in alum washings, and is evaporated by means of steam. When sufficiently evaporated, the lids are put on, the steam shut off, and the liquor allowed to settle for some hours. It is then run into another series of alum coolers, or sunk stone cisterns, like the former, and yields, after several days, tolerably white crystals, called 'second alum.' In the fifth stage, these crystals are drained to free them from the adhering mother liquor, and they are subsequently washed with water. It is then introduced into a large and closely-covered lead cistern, named the roaching pan, where it is entirely, or almost entirely, dissolved by means of steam. On cooling, it is run off into the roaching casks, which are large wooden vessels. After standing for a sufficient length of time in the casks, the casks (which are so constructed as to admit of this) are taken to pieces; the mother liquor is allowed to drain off by means of holes bored in the centre of the crystallized alum. When it has become sufficiently dry and hard, it is broken into masses, and is then fit for the market."

Alumina. *Alumine, argil, sesquioxide of aluminum.* Al_2O_3, 6·4234, 51·3872, or 6·5, 52. White powder, very infusible, but fusible by the oxyhydrogen blowpipe more easily than silica; prepared by precipitating alum by caustic ammonia, washing and igniting; or by igniting ammoniacal alum; spec. grav. 4·2.

Trishydrate of Alumina, $3\ Al_2O_3$, HO, or diaspore, is found native. It may be purified from iron by solution in acid, and boiling with caustic potash or soda; it dissolves, and may be precipitated by adding acid and caustic ammonia.

Terhydrate or *Gibbsite*, $Al_2O_3\ 3\,HO$, is found in America.

There is also a native *Bihydrate*.

Corundum or anhydrous alumina, with a specific gravity of 3·95, occurs native in granite.

Aluminum. Al = 1·7117, or 1·75, or 13·693. A gray powder; is prepared by placing alternate layers of potassium or sodium and chloride of aluminum in a platinum crucible, and fastening down the cover with platinum wire. A gentle heat, gradually raised, is sufficient to produce the decomposition. When placed in cold water, chloride of potassium dissolves out, and aluminum remains. When heated in the air, it absorbs oxygen, and is converted into alumina.

Characters of Salts of Alumina.—1. Most of the salts of alumina are soluble in water, but few of them are capable of crystallizing. 2. They are usually sweet and astringent, resembling salts of yttria and glucina. 3. They are not precipitated by oxalate of ammonia and tartaric acid, by which they are distinguished from salts of yttria. 4. Precipitated by *caustic ammonia*, in concentrated solutions, not in very dilute solutions; slightly soluble in excess, but not so when salts of ammonia are present; precipitated by alkaline carbonates. 5. Precipitated and redissolved by excess of *caustic soda* and *potash*. 6. *Sulphohydric acid* produces no precipitate; but *sulphurets* form a white precipitate of alumina, sulphohydric acid being evolved. 7. *Yellow prussiate of potash*, no immediate precipitate, but after some time. 8. *Phosphate of ammonia* gives a white precipitate. 9. *Iodide of potassium*, white becoming yellow. 10. *Nitrate of cobalt*, with a mineral or salt containing alumina, before the blowpipe, gives a fine blue, called *Thenard's blue.*

Sesquichloride of Aluminum. — A yellow waxy crystalline body, boiling at 360°, and volatilizing, fuming in the air; obtained by passing dry chlorine over a mixture of charcoal and alumina.

Tersulphate of Alumina.—A white mass, or crystalline when native, with a sweetish inky taste, acid reaction, soluble in 2 parts water, scarcely so in alcohol. It consists of water, 48·50; tersulphate of alumina, 51·5; and its formula is $Al_2O_3\ 3\ SO_3\ 18\,HO$. It is the salt formed in the first stage of the production of alum, and is prepared separately, by a patent process, at Newcastle, by acting on calcined porcelain clay, with sulphuric acid. Native sulphate or alunogen, hair salt or feather alum, contains 35·68 SO_3 14·98 Al_2O_3 49·34 HO. The composition is, therefore, the same as the artificial salt.

Alum Earth. An earthy slaty substance, in the brown coal series of Germany, France, &c. probably brown coal in an impure state, containing pyrites and clay.

Aluminite, *Hallite.* Earthy kidney-shaped masses, occurring at Halle, &c. containing 21·5 SO_3, 31 Al_2O_3 45 HO, impurities 2·. It adheres to the tongue, gives off water, and SO_2 when ignited.

Alum Meal. A name sometimes given to the granular alum which occurs in the manufacture of that salt.

Alumo Calcite. A name which has been given to an impure quartz in the Erzgebirge. Spec. grav. 2·174, SiO_3 86·6, CaO 6·25, Al_2O_3 2·23, HO 4·.

Alum Pyrophorus, or Homberg's pyrophorus.

Alum Slate. A bituminous slate of a dark colour, consisting of two species, the diamond and gentle slate, the former distinguished by having visible crystals of pyrites, the latter with this substance more equally diffused, with a rhomboidal structure. The mean of analyses of Hurlet alum slate gave me—silica, 50·92; alumina, 13·32; sulphur, 9·51.

Alumstone, *Alunite.* White rhombohedrons, spec. grav. 2·694, consisting of 9·9 KO 42·2 Al_2O_3 33·1 SO_3 14·8 HO, and also in an amorphous form, as Tolfa rock—a crystalline rock produced by the action of sulphurous acid from volcanoes on trachytes. It consists of Al_2O_3 19· KO 4 SO_3 16·5 HO 3· SiO_3 56·5. When made from this rock, alum was called *rock alum.*

Alunogen. A synonyme of tersulphate of alumina.

Amadou. A tinder prepared from the Boletus igniarius, a fungus occurring on trees. It is boiled with water, and beat in a mortar or with a hammer, to soften the matter of which it consists. Since the introdution of lucifers it has gone much out of use.

Amalgam. A combination of mercury with any other metal. All metals seem to possess such a tendency in a greater or less degree. Some of them are solid, others fluid; the former being often crystalline, and the fluid, solutions of a solid amalgam in mercury. The latter is proved by straining through chamois leather, when the solid amalgam remains on the filter, and mercury passes through, containing but a trace of the alloy. Amalgams are readily formed with potassium and sodium, either by heat or simply by bringing them in contact. The product is a white crystalline solid, when it contains 1½ per cent. potassium. Amalgams of this description are used for the demonstration of ammonium.

Silver Amalgam.—When 3 parts of a saturated solution of nitrate of silver, with 2 parts of a saturated solution of protonitrate of mercury, are mixed with an amalgam of 7 parts mercury and 1 silver, in 36 hours the whole solution is pervaded with a beautiful crystalline amalgam, resembling in form a tree—*arbor Dianae*, or tree of Diana—consisting of 65 per cent. mercury and 35 silver. The amorphous amalgam may be formed by adding mercury to silver which has been precipitated by copper in the form of a powder. A native amalgam occurs in 12-hedrons, spec. grav. 13·7 to 14·1 of the same composition as the above.

Gold Amalgam is made by heating together mercury with gold grains or foil; or by bringing mercury and gold leaf together in the cold state; or by precipitating a solution of sesquichloride of gold with excess of mercury. When filtered through leather, the excess of mercury passes, and the remaining amalgam consists of 2 gold and 1 mercury. When 1 part gold is heated with 6 mercury, and the mass cooled, an amalgam in 4-sided prisms is formed. When an amalgam of gold is heated, the mercury is dissipated, and the gold remains. This amalgam is employed in the old process of gilding, and also in the separation of the precious metals.

Tin Amalgam is obtained in cubic crystals, consisting of 3 parts mercury and 1 tin, by bringing melted tin in contact with mercury, and even in the cold state. This is used in silvering looking-glasses. The glass is laid on a horizontal stone table, and covered with tin foil; mercury is then poured on it, and the excess of mercury pressed out by covering the whole with leather and exposing it to pressure. In twenty-four hours the table is inclined at an angle of 5 to 10 degrees, so as to allow the excess of mercury to run off. In some weeks the amalgam becomes very hard. To silver glass balls an amalgam is used of 1 tin, 1 lead, 2 bismuth, and 10 mercury.

The Electric Machine Amalgam is composed of equal parts of tin and zinc, and 3 of mercury, mixed with some lard, to give it the proper consistence. The zinc and tin are melted in a crucible; the melted mixture poured into thrice its weight of mercury, in a wooden box made for the purpose; the box well shaken till the metals are cold. The amalgam is then reduced to a very fine powder, and mixed with lard.

Amalgamation. The technical term employed for the process of extracting gold and silver from their matrix by means of mercury. When gold or silver exists in a native state as an ore, all that is necessary is to triturate 1 part of the ore deprived of rocky matter with 2 parts of mercury, in a mortar of greater or less dimensions. The excess of mercury is poured off, the amalgam strained through leather, and distilled in a retort, the mercury being condensed in water.

American Method.—To amalgamate sulphuret of silver, chloride of silver, and such poor ores, in America, the earths are triturated with water in a wide shallow cylinder, in a similar manner to clay at a pottery, and then dried; from 1 to 5 per cent. of common salt is then added to it, according to its purity, and the whole triturated. Sulphate of copper, in the proportion of ½ lb. to 1 lb. to the hundredweight, (magistral) is next added, and the whole triturated; 6 times as much quicksilver is now added as the ore contains, which is added at three separate times, and trituration performed. The amalgam is washed and distilled. In this operation it appears that, in the first instance, chloride of copper and sulphate of soda are formed when the common salt and sulphate of copper are added. The latter is speedily changed into chloride of copper and sulphuret of copper, with excess of sulphur, for which a corresponding quantity of sulphuret of silver is changed into chloride of silver. The chloride of copper, dissolved by the common salt, completes this change, while it passes into protosulphuret of copper. The chloride of silver dissolves in excess of common salt, and is reduced by the mercury, calomel being formed.

European Method.—The triturated earth, which must contain so much pyrites as to leave 30 per cent. sulphuret of iron when melted in a crucible, is mixed with 10 per cent. common salt. The mixture is roasted in a furnace, by which process the pyrites loses half its sulphur; a considerable part is converted into sulphurous acid, and another portion into sulphuric acid, which acts on the common salt, and expels chlorine, while it becomes sulphurous acid. The chlorine forms chloride of silver. The amalgamation then is performed with the matter taken from the furnace in wooden vessels.

Amalic Acid. $C_{12}\,N_2\,H_7\,O_8$. A weak acid obtained by acting on caffeine with chlorine; it is analogous to alloxantine; when oxidized it yields cholestrophane, homologous with parabanic acid; and when it acts on ammonia, it forms caffeomurexide, homologous with murexide.

Amandine. A synonyme of emulsine.

Amanitine. A poisonous principle in some mushrooms—unexamined.

Amarine. *Benzoline.* $C_{42}H_{18}N_2$. White needles; a base from hydrobenzamide by boiling with potash.

Amarone. $C_{32}H_{10}N$.

Amaphenase. A synonyme of aniline.

Amarythrine. *Picroerythrine, erythrine bitter.* $C_{34}\,H_{23}\,O_{20}$. Obtained by boiling erythric acid with water.

Amasatine. $C_{16}H_5NO_3\ NH_2\ HO$. A yellow body formed in the action of ammonia on isatine.

Amausite. Petrosilex; a variety of felspar.

Amazon Stone. A synonyme of green felspar, and also of nephrite.

Amber. A yellow fossil resin, found on the shores of the Baltic, in connection with beds of wood coal, especially in the Prussian territories, where it is cast up by the sea, and often includes, buried in its substance, fossil insects. It is also found in Greenland, France, Switzerland, &c. Spec. grav. 1·065 to 1·075. Electrical when rubbed; fusing point 550°; insoluble in water; slightly soluble in alcohol and ether; soluble in sulphuric acid and nitric acid. Potash extracts succinic acid from it. Besides, amber contains a volatile oil and two resins soluble in alcohol and ether. When in fine powder, and digested with ether, 10 per cent. of the amber is dissolved, consisting of succinic acid, balsam, and volatile oil. If the ether solution be mixed with water and distilled, a resin swims on the water in the retort, and when heated, a volatile oil passes with the water, while the water contains succinic acid. The resin consists of two, which may be separated by cold alcohol of ·840, which dissolves one, and leaves the other. The greater portion of the amber 90 per cent., is insoluble in ether, and is bitumen of amber.

Oil of Amber, $C_{20}H_{16}$, (Döpping) is obtained by distilling amber, and, therefore, the products are principally those of the bitumen. The first product, when distilled with charcoal, becomes rectified oil of amber; 1 rectified oil dissolved in 24 alcohol of ·830, and 96 caustic ammonia of ·916, forms *Eau de Luce*. When amber is treated with fuming nitric acid, a resin is formed (*artificial musk*) which is soluble in an excess of nitric acid, $C_{15}H_8NO_7$, but it is probably a mixture of resins. Amber is used for making varnishes, and for ornamental purposes, tobacco pipe mouthpieces, &c.

Ambergris. *Ambreine.* An ash-gray light soft substance which floats in water; spec. grav. ·78 to ·92; smell agreeable; taste insipid; soluble in alcohol of ·827, and depositing crystals. A concretion apparently formed in the stomach or intestines of the *Physeter macrocephalus* or spermaceti whale.

Ambreine. Colourless needles, or white brilliant insipid solid, with an agreeable smell, which may be removed by keeping the ambreine long in a fused state; obtained by digesting ambergris in hot alcohol; composed of C 83·37 H 13·32 O 3·31; fusing point about 86°; distils without alteration.

Ambreic Acid. Yellow or white plates or masses with a peculiar odour; reddens litmus. Solid at 212°, forms salts with bases; obtained by heating ambreine with nitric acid, evaporating to dryness, washing with cold water, boiling with carbonate of lead, washing with cold water to remove any traces of lead, and crystallizing out of alcohol. It is used as a perfume in the form of tincture. Composition uncertain.

Amblygonite. (αμϐλυς, blunt; γονη, an angle). Greenish-white rhombic prisms, lustre vitreous, semitransparent; H 6, spec. grav. 3, B.B. easily fusible, becoming a white enamel. Occurs at Chursdorf, near Penig, in Saxony, in granite; consists of PO_5 35·69, Al_2O_3 25·69, LO 9·11. *Form.* 2 LO, PO_5, 4 Al_2O_3 3 PO_5.

Amethyst. (αμεθυστος, sober—from being used against drunkenness). Violet-coloured quartz or rock crystal, used as a precious stone, owing its colour, probably, to manganese. The oriental amethyst is the violet variety of corundum or spinelle.

Amianthus. *Amianthoide, Asbestus.* A species of amphibole; spec. grav. 1·551, melts before the blowpipe into a bead. White with a slight shade of green, lustre silky, composed of fine soft flexible threads consisting of silica 43·5 MgO 40·, FeO 2·08, Al_2O_3 ·4, HO 13·8 It occurs in Sardinia, and is used to make incombustible cloth. Common asbestus seems to be rather a variety of pyroxene.

Amiatite. Hyalite or resinous quartz, contains 80 to 92 per cent. of silica, with from 7 to 10 per cent. water.

Amide, Amidogen. The term amide is given to a class of substances which contain ammonia deprived of an atom of hydrogen, (NH_2.) Hence, when they are treated with water, soda, or potash, ammonia is disengaged, the third atom of hydrogen derived from water again uniting, so as to form the necessary quantity in ammonia. That such a substance as amide or amidogen exists, may be proved by passing dry ammonia over heated potassium or sodium. A green body is formed, and hydrogen is evolved, which is replaced by the metal, $NH_2H + Na$, becoming NH_2Na and H. Some of the most familiar of these amides are potassamide NH_2K, sodamide NH_2Na, oxamide $NH_2C_2O_2$, benzamide $NH_2C_{14}H_5O_2$, sulphamide NH_2SO_2, carbamide NH_2CO, platosamine NH_2Pt. Some of these amides are formed by heating organic salts of ammonia, by which they lose water, and leave an amide. Thus oxalate of ammonia, when distilled, gives water and oxamide ($NH_3C_2O_3 = NH_2C_2O_2$ and HO). It is probable that many animal substances, as albumen, belong to this class.

Amidochrysammic Acid. $C_{15}H_4N_3O_{13}$. Olive-green needles from chrysaminamide, by HCl.

Amidine. A yellow or white substance when hydrous, is the name applied by Guerin Varry to the soluble part of starch obtained by placing 1 part of potatoe starch in 100 water, boiling for a quarter of an hour, and then pouring it into a tall glass to allow the insoluble matter to subside. Pour off the supernatant liquor, and boil it down to the consistence of syrup; pass is through a cloth, evaporate below 212°; the filtrations are to be continued as long as any precipitate appears. The insoluble portion, or outer covering of the starch granules, is called *amylin* according to this view, and the enclosed soluble portion *amidin*. Varry called the soluble part *amidin*, and the insoluble *amidine*; but as these words have the same sound in English, Dr. T. Thomson employed for the latter term *amylin*. According to Raspail, starch granules are vesicular, enclosing a soluble portion; but this idea is replaced, by considering starch to consist of layers of a homogeneous body, which is insoluble in cold water, and swells up, without dissolving, into a paste when treated with hot water.

Amidon. The continental name for starch.

Amidosulphuric Acid. Formed by the action of oil of vitriol on starch.

Amiduline. A modified starch.

Amilene. *Amylene.* $C_{10}H_{10}$. A colourless fluid with a peculiar odour, floating on water, boiling at 320°, spec. grav. of vapour 5·061; obtained by distilling oil of potatoes or grain with anhydrous phosphoric acid.

Ammelide. $C_{12}N_9H_9O_6$. A white powder, obtained by dissolving melamine or ammeline in strong oil of vitriol or melam in nitric acid; mixing the solution with 2 vols. water, and then with 4 of alcohol.

Ammeline. $C_6N_5H_5O_2$. White powder, a base obtained by super-saturating with acetic acid the alkaline liquid which has deposited *melamine*.

Ammoline. An oily base, occurring in Dippel's animal oil, smelling of horse-radish; separated from several other oils, as animine, olanine, by fractional distillation.

Ammonialdehyde. A compound of aldehyde and ammonia.

Ammonia. *Spirit of hartshorn, Volatile alkali.* NH_3 2·125 or 17. 1½ vol. H = ·1032, ½ vol. N = ·4880 = ·5912. The experimental spec. grav. by Dr. Thomson is ·5931, 100 cubic inches weighing 18·303 grains. Colourless gas, odour peculiar, taste caustic, blistering the tongue and mucous membrane of the mouth, irrespirable except in a very dilute state. Ammonia restores reddened litmus to a blue; makes moistened turmeric paper brown. With a pressure of 6½ atmospheres it becomes a fluid, or at a temperature of -40°. The fluid has a density of ·76. When a taper is dipped in the gas it is extinguished with a greenish-yellow flame, water being formed at the expense of the hydrogen of the ammonia, and nitrogen being set free. When electric sparks are passed through ammonia, the gas is resolved into hydrogen and nitrogen. One vol. water dissolves 670 vols. ammonia. Ammonia is obtained in the gaseous state by heating equal parts of slaked lime and sulphate or muriate of ammonia, water being added to form a paste, and received over mercury, in bottles filled with mercury, and inverted in a mercurial trough. In this process the lime replaces the ammonia, which being volatile is removed by the heat, and a lime salt remains in the retort ($CaO, NH_3\ SO_3\ HO$ becomes $CaO\ SO_3$ and NH_3 and HO). Ammonia is produced by the distillation and decomposition of substances containing nitrogen, and hence its origin in the products of the distillation of coal. It is a product of volcanoes (?) and exists in minute quantities in the atmosphere. It is formed in the decomposition of prussic acid, and of many cyanogen compounds; during the solution of tin and iron in nitric acid; in the reduction of platinum out of its solution in aqua regia, as platinum black by caustic alkali, in spirit; in the decomposition of sulphohydric acid by nitric acid; by the action of iron filings on air and water; occurs along with boracic acid in the hot springs which supply that acid in Tuscany and Tibet.

Liquid Ammonia, Caustic Ammonia, Spirits of Hartshorn.—Ammonia is absorbed most abundantly by water. If a small portion of water be admitted to a jar full of ammonia, over mercury, the gas is absorbed almost explosively. A current of ammonia when passed through 6 cubic inches of water, is absorbed to the extent of 780 times the volume of the water, while the bulk of the water increases to 10 cubic inches; hence 1 cubic inch of water absorbs, according to this experiment, 468 of ammonia. The solution is colourless, precipitates many metallic solutions, and has the characters of the gas; its specific gravity is less than that of water; the boiling point depends on the amount of water, falling according to the quantity of ammonia contained in the solution. The specific gravity of mixtures of ammonia and water is a mean of the two ingredients. A strong solution of ammonia when cooled to —36° or —40° solidifies in the form of silky needles. When heated it parts with its ammonia below 212°, and thus the gas may be conveniently made from the liquid ammonia of commerce. To prepare it, 1 part of sal ammoniac or sulphate of ammonia is to be introduced into a retort, and upon it is poured, by means of a funnel, 1 part of slaked lime, converted into the consistence of a thin paste with water. The beak of the retort is then connected with a series of Wolfe's bottles, q. v., containing water which is to be impregnated with the gas to any required strength. By determining the specific gravity of the solution, the amount of ammonia contained in it may be ascertained. The following table is one of the best:—

Spec. Grav.	Gas per ct.	Spec. Grav.	Gas per ct.
·8720	32·5	·9435	14·53
·8875	29·25	·9476	13·46
·9000	26·00	·9513	12·40
·9054	25·37	·9545	11·56
·9166	22·07	·9573	10·82
·9255	19·54	·9599	10·17
·9326	17·52	·9619	9·60
·9385	15·88	·9692	8·50

Uses.—In medicine, caustic ammonia is sometimes used as a blistering agent, or as a rubefacient, mixed with oil (1 part ammonia, and 2 parts oil), for sore throats, under the name of volatile oliniment. It is used as an antidote to prussic acid, by inhalation; as smelling salts, when poured on a spunge, in a bottle, and kept close, or by mixing 1 part of sal ammoniac with 2 parts of lime, moistened with water, in a close-stoppered bottle. For the manufacture of alum, ammonia is largely employed; it is used as a manure in union with acids. In the laboratory it is an indispensable agent for the separation of metallic oxides, &c. *Estimation.*—Ammonia is estimated by precipitating the muriate by bichloride of platinum in a minimum of water, evaporating to dryness at 212° in a water bath, treating the residue with 1 part of ether and 2 of alcohol, throwing on a small weighed filter, drying at 212°, and weighing. The composition of this salt is $NH_3HCl + PtCl_2$, 27·891 grs. of which contain 2·125 grs. of ammonia, and 12·33 metallic platinum. The salt, when burned, leaves platinum, from which the ammonia may likewise be calculated.

Salts of Ammonia.—Ammonia possesses the characters of an alkali or base, and forms a series of interesting and important salts. The oxygen acid salts contain at least 1 atom of water; some of them are volatile, while others are decomposed at a red heat, as phosphate of ammonia. Most of them, when boiled, become acid by loss of ammonia. *Characters.*—1. Salts of ammonia are generally soluble in water. 2. When caustic potash or soda is added to a salt of ammonia, the odour of ammonia is evolved, and moistened turmeric paper is rendered brown. 3. The addition of a salt of magnesia and phosphate of soda to a salt of ammonia is followed by a crystalline precipitation. 4. Ammoniacal salts, with the exception of the phosphate and borate, are expelled by heat. 5. They are precipitated yellow by bichloride of platinum.

Chlorohydride of Ammonia, Sal Ammoniac, Chloride of Ammonium, Muriate of Ammonia.—NH_3HCl or NH_4Cl. 8-hedrons cubes or 24-hedrons, or in feathery crystals, or in cubes, when sublimed. Sp. grav. 1·45 to 1·53. Pungent, saline taste; fixed at the usual temperatures, but is volatilized by heat, and then has the effect of decomposing many salts. It contains 31·98 per cent. ammonia. It is formed on the large scale by saturating ammoniacal liquor of the gas works with muriatic acid, evaporating in lead vessels, running the liquor into wooden coolers, drying the salt in open stoves, and then subliming in subliming pots, which are formed of iron, with a large dome. After some days the sublimation is complete.

Bisulphohydride of Ammonia, Hydrosulphuret of Ammonia.—$NH_3\ 2SH$. Colourless foliated crystals, with an alkaline reaction, and a strong smell of ammonia; usually, however, only known in the state of solution, and extensively employed as one of the most valuable chemical reagents for the detection of metals, &c. It is prepared by passing a current of sulphohydric acid through caustic ammonia, until a salt of magnesia is no longer precipitated by it. The addition of some flowers of sulphur ensures the presence of a sufficient amount of that body to produce the necessary reactions, particularly with tin. It must be kept in a closely-stoppered bottle, full of the fluid, otherwise air will gradually change it into hyposulphite of ammonia. It requires, therefore, to be frequently renewed.

Sulphate of Ammonia, (Glauber, 1640 ?) *Secret Sal Ammoniac, Vitriolated Ammoniac.*—*Muscagnin* (when native) $NH_3\ SO_3\ HO$ 8·28, 4-sided prism with square base; spec. grav. 2·; sharp bitter *taste*; soluble in 2 parts of cold, and 1 of boiling water; melts at 284°; decomposes at 536°. Deliquesces slowly in air. *Formed* by saturating carbonate of ammonia with sulphuric acid, and manufactured on the large scale for agricultural and other purposes, as for alum making, by saturating gas liquor or urine with oil of vitriol. There are two modes of making sulphate of ammonia from the crude ammoniacal liquor from the gas works: one consists in simply saturating the liquor with sulphuric acid, and evaporating the solution in either cast iron boilers, by a fire, or in a leaden vessel, by steam, till it has a spec. grav. of 1·3, when it is run out into a vessel lined with lead, and crystallized. The other is to distil the ammoniacal liquor in a close boiler, and to conduct the products into a dilute solution of sulphuric acid, or acid of a spec. grav. of 1·7; in the latter case, the sulphate of ammonia precipitates, or, as the workmen call it, "drops" from the saturated solution, and is "fished" out by a copper ladle, and is called "fished salt." The latter process has been further improved by certain patents, on the principle of Coffey's still, whereby cast iron columns, 30 feet in height, and 4 feet in diameter, and filled with plates 4 inches apart, and perforated with holes to admit of passage both ways, have been to a small extent adopted. The crude gas liquor, from a cistern, flows in at the top, while high pressure steam enters at the bottom. The steam, on its upward passage, carries with it the carbonate and sulphuret of ammonia, and escapes at the top by an exit pipe into a vessel containing sulphuric acid, diluted with twice its volume of water. The distillation is continued till the sulphuric acid is saturated, when it is syphoned into a vessel lined with lead, called the "evaporating basin," and evaporated by steam till the sulphuret of ammonia drops, and is then fished out as before noticed, and allowed to drain till it is ready for being stored, or put into casks, and constitutes commercial sulphate of ammonia. It not unfrequently happens that some makers mix it either with sulphate of soda, or chloride of sodium, for the purpose of adulterating it. There exists also a *bihydrate*, $NH_3SO_3\ 2HO$, which loses its additional water at a gentle heat.

Bisulphate.—$NH_3\ 2SO_3\ 2HO$ (?) Thin rhomboidal scales; acid and bitter taste. Obtained by adding to the preceding sulphuric acid, and crystallizing. Anhydrous sulphuric acid unites with ammonia, and forms an *anhydrous sulphate* (NH_3 SO_3), or *sulphamide* (SO_2 NH_2 HO), a white amorphous powder, dissolving in 9 parts of water, and giving out, when heated, sulphurous acid, ammonia, and sulphate of ammonia, leaving sulphate of ammonia. Chloride of barium precipitates only half the sulphuric acid present, and bichloride of platinum only half the ammonia, while lime, when rubbed with it, yields no smell of ammonia. When the *anhydrous sulphate* is dissolved and evaporated, *anhydrous parasulphate* crystallizes in square prisms, having the same composition. A *deliquescent sulphate* remains in the mother liquor after the crystallization of the parasulphate, which must be precipitated in vacuo. It appears to consist of the common sulphate, with the anhydrous sulphate.

Nitrate.—NH_3 NO_5 HO 10, 80. *Nitrum semivolatile, nitrum flammans.* 6-sided colourless prisms, terminated by 6-sided pyramids. Obtained by neutralizing carbonate of ammonia by nitric acid, and evaporating. Spec. grav. 1·707. (Kopp.) Taste acrid bitter; deliquescent. Begins to fuse at 133°, and is quite fluid at 226°. At 302° gives out white fumes; at 460° gives out protoxide of nitrogen, and is entirely decomposed at 480° into NO and HO. (Davy.) Dissolves in 2 parts cold, ½ part boiling water. (Fourcroy.) Produces great cold in dissolving. Most metals are oxidized when heated with the nitrate, as lead, zinc, antimony, bismuth, nickel, copper, and silver—but not arsenic, iron, mercury, and tin. Nitrate of ammonia and salammoniac in fusion dissolve gold, platinum, rhodium, and iridium, especially with addition of a small quantity of chlorate or nitrate of potash. (Emmet.) In contact with spongy platinum the nitrate decomposes at 320°. The nitrate is used in the preparation of protoxide of nitrogen, and for burning away the last traces of carbon in organic analyses.

Sesquicarbonate of Ammonia.—2 NH_3 3 CO_2 2 HO, 14·75. *Sal volatile, commercial carbonate of ammonia, volatile salts of hartshorn.* Hard white cakes, formed by distilling 1½ part carbonate of lime or chalk, with 4 parts of sulphate or muriate of ammonia, in a clay or iron retort. The product at first is coloured, but may be purified by redistillation. The first products of the distillation are water and ammonia; the sesquicarbonate then passes over and condenses in white crystalline crusts; when this salt is treated with a small portion of cold water, a bicarbonate of ammonia, in granular crystals, remains, while protocarbonate is dissolved. (Scanlan.) Hence this may be viewed as its composition (NH_3 CO_2 HO + NH_3 2 CO_2 HO = 2 NH_3 3 CO_2 2 HO). The commercial salt often is contaminated with hyposulphite of ammonia, sulphate of ammonia, and sal ammoniac, lime, and chloride of calcium.

Bicarbonate.—$NH_3 2CO_2 2HO$. 6-sided prisms, having no smell, and less taste than the sesquicarbonate; soluble in 8 parts cold water. Obtained by passing a current of carbonic acid through a solution of the sesquicarbonate, and crystallizing in vacuo; or by exposing the commercial salt in powder to an atmosphere of carbonic acid; or by washing the same salt in powder with alcohol of sp. grav. ·821. This salt is isomorphous with bicarbonate of potash; it volatilizes slowly in the air, giving out a slight odour of ammonia. The bicarbonate may be procured with 2½ atoms of water, by digesting the sesquicarbonate with boiling water, sufficient to dissolve it in a strong flask, kept close: on cooling, the salt is left. It is obtained with 3 atoms of water, when the salt 4 NH_3 5O_2 4 HO is distilled. When the terhydrated bicarbonate is further distilled, it yields the salt 4 NH_3 7 CO_2 12 HO. (Rose.)

Carbamide.—NH_2 CO, (Regnault.) Isomeric with urea. Obtained by passing dry ammonia into flasks full of chlorocarbonic acid (CO Cl); a mixture of salammoniac and carbamide results, which are difficult to separate completely. Carbamide is soluble in weak alcohol and in water; it does not precipitate lime or barytes water. Strong acids convert it into ammonia. It forms no precipitate with nitric acid as urea does.

Diphosphate. Neutral phosphate.—2 NH_3PO_5 3HO 16·5. Large colourless transparent crystals of oblique rhombic prisms. Taste saline and cooling, with impression of bitterness. Alkaline reaction; soluble in 4 parts cold water; insoluble in alcohol. Effloresces in the air, and loses ammonia. Obtained by adding caustic ammonia to the solution of the superphosphate of lime, obtained from bones, filtering off the phosphate of lime, evaporating and crystallizing. When heated it fuses, loses its ammonia, and is converted into the monophosphate of water (HO PO_5), or pyrophosphoric acid. By boiling, it is changed also into the *monophosphate*, formerly termed the *acid* or *biphosphate* NH_3 PO_5 3 HO 14·5. 8-hedrons, with square bases, having commonly the base prolonged into a square prism. Obtained by adding to the common phosphate phosphoric acid until litmus is reddened, or by adding the acid in excess to liquid ammonia till the solution does not precipitate chloride of barium. Crystals transparent, undergoing no change in the air. Taste acid, cooling, saline. By heat, monophosphate of water is left. Precipitates nitrate of silver white. The phosphates of ammonia have been recommended to impregnate cloth, to render it less combustible. Cloth, immersed in a solution of these salts, if exposed to ignition, would have the ammonia expelled, and the hydrous acid deposited on the fibre, which would be carbonized, but not inflamed.

Neutral Arseniate.—2 NH_3 HO, AsO_5 2 HO, 22·125. Oblique rhomboidal prisms exactly resembling phosphate of ammonia. Obtained by adding ammonia to a concentrated solution of arsenic acid till a precipitate appears. This precipitate is dissolved by aid of heat, and the liquid is set aside to crystallize. Reaction to litmus alkaline, but when exposed to the air half of the ammonia is lost, and its action is acid. When heated, it evolves ammonia, water, and lastly arsenic sublimes, and nitrogen. *Binarseniate, Acid arseniate,* NH_3 2 HO AsO_5 HO 20·. 8-hedrons, with square bases (2d system), the base being drawn out generally, as with biphosphate of ammonia, into a 4-sided square prism. Reaction, acid; obtained by adding an atom of arsenic acid to the neutral salt. *Arsenite of Ammonia* can only be obtained in solution; which, when evaporated, leaves 8-hedrons of arsenious acid.

Antimonite (NH_3 SbO_4 HO?), by digesting the hydrous acid in ammonia. It only exists in solution, and deposits by heat a white biantimonite.

Oxalate of Ammonia.—NH_3 C_2 O_3 2 HO 8·875; 71; needles and right rhombic prisms, the faces of which are inclined at angles of 104°·7; the longitudinal edges of the prisms are often replaced by tangent planes, and the angle at the base by one or two small faces. Bitter and unpleasant. Sp. grav. 1·582. 100 water dissolve 4½; insoluble in alcohol. By heat it loses 2 atoms water. At 428° the salt is converted into carbonic oxide and carbonate of ammonia. When more strongly heated it yields a trace of carbon, carbonic oxide, carbonic acid, ammonia, water, oxamide (NH_2 C_2 O_2), and also oxamic acid. When the solution is heated for some time it loses ammonia and becomes acid. This salt is a very important reagent in the detection and estimation of lime, which it precipitates entirely from hot neutral or alkaline solutions; the precipitate consists of CaO C_2O_3 2 HO, when dried at 212°; and by ignition loses carbonic oxide, and is converted into carbonate of lime (CaO CO_2), every 6¼ grs. of which contain 3½ of lime. It is prepared by saturating a cold solution of oxalic acid with carbonate of ammonia.

Binoxalate of Ammonia.—NH_3 2 C_2O_3 2 HO. 8-hedrons, or rectangular flat 4-sided prisms; reaction acid. Obtained by adding an additional atom of oxalic acid to a solution of oxalate of ammonia, or by adding a mineral acid. When heated, it yields oxamic acid (NH_2 C_4 O_5).

For *Acetate of Ammonia*, see ACETIC ACID.

Theories of Ammonia.—1. *Ammonia Theory.* By this view ammonia is considered as a gas, and combines directly with acids. With SO_3 it forms NH_3 SO_3 + HO. With Cl it forms sal ammoniac, NH_3 Cl H. 2. *Ammonium Theory.* If we rub in a mortar a small portion of sodium, with some mercury (filtered through leather or paper) until an amalgam is formed, and then drop it into a strong solution of sal ammoniac in a tube, the amalgam swells and rises up in the tube to an extraordinary height. This increase of volume is supposed to be owing to the union of ammonium (NH_4) with the mercury. Na + Hg, and NH_4 + Cl becoming Na Cl and Hg NH_4. According to this view, the ammoniacal salts are no longer compounds of the gas with an acid, but of acids with the aid of a metal; sulphate of ammonia being NH_4OSO_3. 3. *Amide Theory.* According to this theory ammonia is an amide of hydrogen (H + NH_2). This is proved by passing dry ammonia over potassium in a tube at a red heat; a green substance is formed called potassamide (KNH_2), and hydrogen is liberated. Sulphate of ammonia is, upon this view, HNH_2, HO, SO_3.

Ammonide, *Ammoniuret.* A basic double salt of ammonia.

Ammoniac Gum. A gum resin; the dried juice of the *Dorema Armeniacum*, from Persia. Spec. grav. 1·207; it contains 70 per cent. of resin, which fuses at 130°, and is very soluble in alcohol. This gum is used in medicine as an antispasmodic.

Ammonium, *Ammonyle, Hydramide.* NH_4. The hypothetic radical of ammoniacal salts, but which has not been isolated. It is only known in union with mercury, as the ammoniacal amalgam. This amalgam may be formed by acting on mercury placed in a depression of a cake of sal ammoniac with a very powerful galvanic battery, or by adding to a saturated solution of sal ammoniac, in a tube, an amalgam of mercury with sodium, containing a tenth of sodium.

Amoibite. *Nickel glance.* An ore of nickel, consisting of S 14, As 45·34, Ni 37·34, Fe 2·50, Pb 0·82, Co trace.

Amomum. *Granum paradisi, grains of paradise.* The fruit of a species of ginger used in India with betel to promote digestion.

Amniotic Acid, or allantoine.

Amorphous. (Privative α, destitute of μορφη, shape.) *Formless;* the opposite of the crystallized or definite form of a body. Charcoal is the amorphous, and diamond the crystalline states of carbon. In these different states substances seem to possess even differences in other characters besides appearance, their fusing point varying.

Ampelic Acid. White flakes, without taste and smell, nearly insoluble in cold, very soluble in hot water, and in hot alcohol and ether; fuses at 500°, and sublimes at a higher temperature; soluble in sulphuric acid, and reprecipitated by water; combines with alkalies, forming very soluble salts. Obtained in the rectification of that portion of the oils from alum slate (ampelite) which distil at 302°. A similar acid, consisting of C_{14} H_6 O_6, was obtained from coal oils in that portion boiling between 266° and 320°.

Ampeline. Colourless oil, without smell, soluble in water, alcohol, and ether. Obtained by distilling sulphuric acid and the oil from alum slate, which boils between 392° and 536°; it seems isomeric with salicylic acid, and in some characters resembles creasote.

Ampelite. A generic name given to slates, the aluminous ampelite being alum slate, and the graphic, common writing slate.

Amphibole. A class of minerals, consisting of many varieties, as actinolite, asbestus, hornblende, tremolite, pargasite. It was named from *αμφιβολος*, ambiguous, by Hauy, from its being confounded at first with tourmaline. White amphibole is *tremolite;* compact amphibole, *corneine;* black amphibole is *hornblende;* green amphibole is *actinote.* Other synonymes are *amianthinite, amianthoide, asbestinite, foliated augite, byssolite, corinthine, green diallage, grammatite, blue hypersthene, keraphyllite, keratophyllite, lotalite, pargasite, smaragdite, asbestus, amianthus, raphilite.* Oblique rhombic prisms, with angles of 130° 15, 124° 30′. Spec. grav. 2·9257

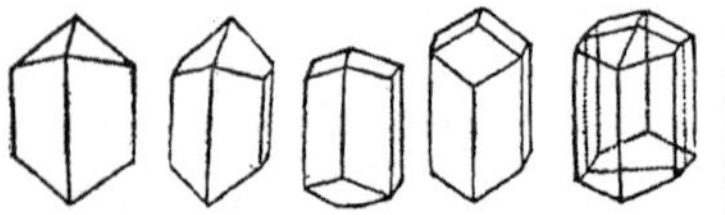

to 3·48. H. 4.75. Very frangible, lustre vitreous. B.B. fuses into a semitransparent glass, with borax and soda into a transparent glass, opaline glass with salt of phosphorus. This mineral affords an example of isomorphism in a striking degree. We may take hornblende as an average of its composition. SiO_3 42·24, MgO 13·74, CaO 12·24, Al_2O_3 13·92, FeO 16·26, MnO ·33. *Formula.*—4RO,$3SiO_3$. Found in primary rocks.

Amphibolites. Trap rocks, with a basis of amphibole.

Amphide Salts, are those which are formed of a binary acid and a binary base, as sulphate of soda (NaO SO_3), to distinguish them from haloid or binary salts, as common salt (NaCl). These terms have never been applied in this country.

Amphigene, or *Leucite.*

Amphodelite. A mineral, from Lojo, in Finland, having nearly the crystalline form of felspar, a spec. grav. of 2·763, and consisting of silica 45·8, Al_2 O_3 35·45, CaO 10·15, MgO 5·05, FeO 1·7. Water and loss 1·85.

Amygdalic Acid. C_{40} H_{26} O_{24} HO. Soft crystalline mass when prepared in the cold state, gummy when hot. Obtained by boiling amygdaline with dilute solutions of caustic alkalies; ammonia is evolved, and amygdalate formed. When barytes is used, the barytes salt is decomposed by sulphuric acid. Insoluble in alcohol, soluble in water. This acid may be viewed as a compound of oil of bitter almonds, anhydrous formic acid, and sugar.

Amygdaline. C_{40} NH_{27} O_{22}. Pearly scales without water, when crystallized from alcohol; colourless prisms when crystallized from water; its solution in water is slightly bitter; very soluble in water and alcohol, insoluble in ether; distilled with nitric acid, it is converted into prussic acid, oil of bitter almonds, formic, and benzoic acids; it is procured from bitter almonds, by pounding and pressing them between heated plates of iron, to free them from the fat oil. The residue is then boiled twice with alcohol of ·810, the solution filtered through a clean cloth, and the cake pressed. On standing some days, fine crystals separate, but the greater portion remains in solution. The liquid is distilled to the consistence of syrup. This is fermented with yeast to remove sugar, water being added; it is then filtered, evaporated, and again treated with alcohol of ·810; the amygdaline falls. It is purified from fat oil by frequent pressure, if necessary, and crystallization from alcohol. Amygdaline, when brought in contact with emulsine, is decomposed into oil of bitter almonds. This oil does not exist in the almonds, the two principles being prevented from acting on each other by the fat oil contained in the almonds.

Amygdaloid. A species of trap rock, so named from its containing almond-shaped cavities filled with crystals. It consists of silica 50· Al_2 O_3 18·5, CO 3· FeO 16·75, HO 5· NaO 4· (Kennedy).

Amyle. Ayl. C_{10} H_{11}. An oil boiling at 311°. Obtained by acting on iodide of amyle by zinc.

Amyle, Hydrous Oxide of. *Amylic alcohol, oil of grain, fusel oil, oil of whisky, oil of potatoes.* C_{10} H_{11} OHO. A colourless oil, boiling at 269½°, and crystallizing at -4°. Peculiar penetrating odour, affecting the chest; burns with a white flame, and is poisonous. Spec. grav. ·812, of vapour 3·147. This oil comes over in the distillation of the fermented infusion of barley, oats, and potatoes, mixed with alcohol and water, and is purified by washing it with water, and distilling over chloride of calcium. The valerate of amyle, formed by distilling this oil with sulphuric acid and bichromate of potash, and the acetate with the same materials, and the addition of an acetate, is employed to flavour confections and British brandy, under the name of oil of pears, &c. The *oxide of amyle,* or amylic ether, is obtained by acting on chloride of amyle with a solution of potash in alcohol, while the chloride is formed by distilling grain oil with chloride of phosphorus. The sulphate of amyle and water, or sulphamylic acid, is produced by boiling the oil with sulphuric acid.

Amylene. *Valerene.* $C_{10}H_{10}$. See AMILENE.

Amylamine. C_{10} H_{13} N or NH_2 Ayl. *Amide of amyle,* a fluid; boiling point 203°. Spec. grav. ·7503. A base with the smell of ammonia. Obtained by treating cyanate or cyanurate of amyle, or amyle-urea, with potash.

Amylate. A compound of starch with a base.

Amyle-Urea. C_{12} H_{14} NO_2. Obtained by acting on sulphate of amylamine with cyanate of potash.

Amylic Acid. *Valerianic Acid, Delphinic Acid.* $C_{10}H_9O_3HO$. A colourless oil. Sp. grav. ·937; of vapour 3·66; boiling point 270°. Obtained by treating oil of grain at 400° with a mixture of caustic potash or soda, and hydrate of potash, and when valerol is oxidized; it occurs in the oxidation of some animal substances; it is obtained from the valerian, and exists in the berries of the guelder rose.

Amyline, *Amylum,* or *Amidine.* See AMIDIN, a constituent of starch.

Amylonic Acid, probably formic acid, obtained by distilling peroxide of manganese with muriatic acid.

Amylo-urethane. *Carbonate of Amyle.* $C_{12}H_{13}NO_4$. An oil by the action of phosgene gas on amylic alcohol.

Amyle-diethyle-amine. $C_{18}\ H_{21}$ N or N, Ae_2, Ayl. A base obtained by heating amyl-triethyl-ammonium.

Amyle-triethyl-ammonium. N, Ae_3, Ayl. A base formed upon the model of ammonium.

Amyle-phenyl-amine. Ayl Ph NH, by bromide of amyle on aniline.

Amyrine. A crystalline resin from arbol-a-brea by hot alcohol.

Anacardic Acid. $C_{44}\ H_{32}\ O_7$. *White* crystalline inodorous mass, of a feebly aromatic, and afterwards burning taste; fuses at 79°; at 212 has a peculiar odour, and is decomposed at 392° into liquid products; stains paper, liquefies in the air, smelling of rancid oil; its alcoholic and ethereal solutions are acid; is obtained from Anacardium occidentale, along with cardole, both of which exist in the balsam. The alcoholic solution of the mixture is digested with freshly-precipitated protoxide of lead, which takes up the acid and leaves the cardole. The lead compound is decomposed by sulphohydride of ammonia, and the liquor filtered. The anacardate of ammonia passing through is decomposed by sulphuric acid; the resulting acid is washed with water and dissolved in alcohol; the colouring matter thrown down by the cautious addition of trisacetate of lead; the clear liquor boiled with carbonate of barytes. The anacardic acid is then precipitated with an alcoholic solution of protoacetate of lead, and the lead salt decomposed by sulphuric acid. It is a bibasic acid.

Anacardium. The cashew nut, with a milky juice.

Anaethene. $C_{30}\ H_{24}$. From oil of Fennel.

Analcime. *Cubizite, Sarcolite.* (α privative, and αλκη, strength, from its weak electric power.) Cubic crystals, or 24-hedrons, like the leucite crystal. Colour white, passing into gray. Lustre vitreous; brittle, sometimes nearly transparent. H 6·25. Spec. grav. 2·278 to 2·068; melts on charcoal B.B. without swelling; decomposed by HCl; yielding a jelly. Consists of SiO_3 55·60_3, $Al_2\ O_3$ 23·, NaO 14·65, HO 7·9. Occurs in the Kilpatrick Hills; Giants' Causeway.

Analysis. From ανα from, and λυω I loose. The separation of the constituents of a compound body. The determination of the nature of the component parts of a body is termed testing, or *qualitative* analysis, and the appreciation of the exact weights of these constituents is denominated *quantitative* analysis.

Analysis Qualitative. *Testing Tables.*

Mode of Using the Tables.—To discover the nature of the acid, a portion of the substance to be tested is dissolved in a small quantity of distilled water. A part of the solution is poured into a test-glass, three drops of a solution of nitrate of silver are added, and the first column examined to observe if the precipitate corresponds with any of those described. Another portion is placed in a second test-glass, and three drops of a solution of chloride of barium added. If the acid be a common one, it will be found in the table. If it is of rarer occurrence we must look for it in the body of the work. For the bases we proceed in a similar manner. The reagents required in these experiments are described in the body of the work. The principal are hydrochloric or chlorohydric acid, sulphuric acid, nitric acid, acetic acid, caustic ammonia, caustic soda and potash, chloride of barium, nitrate of barytes, oxalate of ammonia, sulphohydric acid, sulphohydride of ammonia, bichloride of platinum, nitrate of silver, phosphate of soda, solution of sulphate of lime, ferrocyanide of potassium (yellow prussiate of potash), ferridcyanide of potassium (red prussiate of potash).

Testing Analyses.—By means of the tables we should not only be able to distinguish the presence of single bases when dissolved, but also to discriminate between a series of bases, supposing them to be mingled together. The following examples point out the method of working the tables:—

When more than one salt exists in a mixture, we can succeed in separating more distinctly the different acids, by treating it with successive portions of waters; the more soluble salts being removed first, and treating each portion by reagents.

Example of Two Acids in Solution.—Add to a portion of the solution 3 drops of nitrate of silver; a white curdy precipitate falls, which does not disappear when nitric acid (NO_5) is added, even after boiling. It dissolves in ammonia (NH_3). It is, therefore, *chlorohydric* acid, the precipitate formed being chloride of silver. Add to a second portion 3 drops of solution of chloride of barium (BaCl); a white powder falls, insoluble in nitric acid (NO_5), which is sulphate of barytes (BaO SO_3), indicating the presence of sulphuric acid (SO_3) in the solution. The solution then contains the two acids, chlorohydric acid (HCl) and sulphuric acid (SO_3). It is necessary that the action in these cases should be clearly understood; and, therefore, diagrams should be constantly used by the beginner, so as to follow exactly every change which has occurred. If the solution reddens litmus, the acid may be presumed to be uncombined; but if it has no action, the acid may be inferred to be in union with a base, and the change may be thus exhibited:—

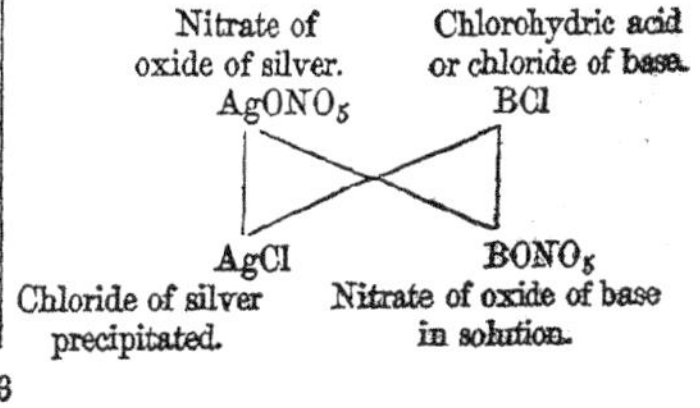

Chloride of barium.	Sulphate of oxide of base.
BaCl	$BOSO_3$

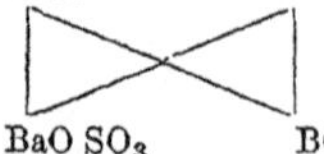

BaO SO_3	BCl
Sulphate of barytes precipitated.	Chloride of base in solution.

Example of Two Bases in Solution.—Add 2 drops of caustic ammonia (NH_3) slowly; a brownish precipitate falls, with a white shade. The addition of a few drops more of the ammonia (NH_3) removes the white tint; a reddish-brown precipitate remains undissolved. Throw the whole on a filter. To the filtered liquid add chlorohydric acid (HCl) till the solution renders litmus paper slightly red. When a drop of caustic soda (NaO) is added to a portion of it, a white flocky precipitate appears. An excess of soda redissolves the precipitate. Sulphohydric acid (SH), when passed through another portion, gives a white precipitate; and sulphohydride of ammonia (NH_3SH) has a similar effect, not redissolved in excess. *The precipitate dissolved by ammonia* (NH_3) *is, therefore, oxide of zinc.* The reddish-brown precipitate on the filter is dissolved off by dilute chlorohydric acid (HCl). On the addition of soda (NaO) to a portion of the solution, a reddish-brown precipitate falls. Sulphohydric acid (SH) drops sulphur when passed through another portion of the solution. Sulphohydride of ammonia (NH_3SH) renders another portion black, while yellow prussiate of potash yields with another portion Prussian blue. *The precipitate undissolved by the ammonia* (NH_3) *is, therefore, sesquioxide of iron.*

Process when the Precipitate is White.—Supposing the precipitate to be of an unmixed white colour, we observe from the first table that it may be one or more of twelve bodies, and besides, when we have added an excess of ammonia (NH_3), that there may be a base held in solution by that reagent. The first step then is to filter the liquid; the precipitate remains on the filter, and the ammoniacal liquor passes through. We saturate the latter with a few drops of acid, and then add ammonia drop by drop, and watch carefully if any precipitation takes place before an excess is added. If a precipitate appears and is again dissolved, the substance belongs to Class III. If no precipitate appears, some of the substances included under Class II. may be present, which are therefore to be tested for by the appropriate reagents. Dilute chlorohydric acid is now to be poured on the filter, when the precipitate will pass through in solution. Caustic soda is to be added in excess to the solution of the precipitates, and the liquid filtered. The filtered liquor is to be saturated with chlorohydric acid (HCl), and ammonia (NH_3) added. If a precipitate appears, it must be one or more of the following substances:

Alumina,	Lead protoxide,
Glucina,	Tin protoxide,

as these are the only white hydrous oxides which are soluble in caustic alkalies. To distinguish these, we find by Table III. that lead and tin are precipitated by sulphohydric acid (SH), while alumina and glucina are not affected. We therefore dissolve the precipitate in chlorohydric acid (HCl), and pass a current of sulphohydric acid (SH) through the solution. If a black precipitate appears, it may consist of both lead and tin. The precipitate is thrown on a filter, and washed with sulphohydride of ammonia. If tin is present it will be dissolved, and pass through the filter, while the lead will remain as a black precipitate. To detect the tin the filtered liquor is boiled with chlorohydric acid (HCl) till sulphohydric acid (SH) is driven off, and the distinguishing tests for tin employed. The sulphide of lead on the filter is dissolved in chlorohydric acid (HCl), and tested with sulphuric acid (SO_3) and chromate of potash. If alumina and glucina are present, they will remain dissolved in the liquid through which the SH has been passed. The gas is to be boiled away, and carbonate of ammonia added in excess. If alumina is present it will remain undissolved, while the glucina will be dissolved by the alkali. The liquid is therefore to be filtered, the solution which passes through neutralized by chlorohydric acid (HCl), and ammonia (NH_3) added. If glucina is present it will fall. The substances which may remain on the filter, which are insoluble in caustic soda, are the following:—

Magnesia,	Zirconia,
Yttria,	Titaniate,
Cerium oxides,	Bismuth,
Lantana,	Mercury oxide.

The first six are not precipitated by sulphohydric acid (SH), while the two last are precipitated by that gas, as appears by Table III. By transmitting a stream of sulphohydric acid (SH) through the solution, therefore, these two metals will be precipitated. After filtration, by adding sulphohydride of ammonia (NH_3SH), according to Table IV., any antimony detained will be dissolved, and may be precipitated from its solution by boiling the liquor and passing sulphohydric acid (SH) through it. An orange precipitate will fall. The bismuth and mercury may be separated by evaporating to dryness, dissolving the residue in nitric acid (NO_5), and adding water, the white trisnitrate of bismuth will precipitate as a white powder, while after filtration the mercury may be thrown down as yellow hydrate with caustic soda, or as iodide, by means of iodide of potassium, in the form of a beautiful red powder. The precipitate on the filter is dissolved in chlorohydric acid (HCl), the solution neutralized, and sulphohydride of ammonia (NH_3SH) added. Magnesia, if present, will remain in solution, while the titanic acid and earths will be thrown down in the form of bulky white precipitates. After filtration the magnesia may be precipitated by phosphate of soda and carbonate of ammonia. The earths remaining on the filter are now to be distinguished. The oxides of cerium, lantanum, zirconium, and yttrium, are first separated by means of a saturated solution of sulphate of potash; a double salt is formed, insoluble in the solution. The double salt of zirconia is nearly insoluble in water and acids, while those of the other two earths can be dissolved in much water. Zirconia, after ignition, is insoluble in acids, sulphuric acid excepted, while yttria is soluble in the same circumstances. Titanic acid is easily distinguished by its being precipitated by boiling, and when precipitated, by its insolubility in chlorohydric acid.

Proceeding when the Precipitate is Coloured.—By acting on the precipitate with caustic soda we remove sesquioxide of chromium and oxide of iridium, which are both soluble in that reagent. These may be separated from each other by sulphohydride of ammonia (NH_3SH), which throws down iridium, and redissolves it, but not sesquioxide of chromium. The remaining oxides may then be divided into those which are precipitated by sulphohydric acid (SH), and those which are not acted on by that gas.

Precipitated by SH.	*Not precipitated.*
Antimony oxide, ...Orange.	The earths.
Mercury oxides, ...Black.	Uranium oxides.
Bismuth oxide,......Black.	Manganese oxides.
Molybdenum oxides, Brown.	Vanadium oxides.
Lead oxide,..........Black.	Iron protoxide.
Osmium oxide, ...Yellow.	Iron peroxide (S falls.)
Tin oxide, ...Brown.	
Platinum oxide, ...Brown.	
Rhodium oxide, ...Brown.	
Iridium oxide, ...Brown.	
Palladium oxide, ...Black.	

The precipitate is to be washed by decantation, and then treated with an excess of sulphohydride of ammonia (NH_3SH). The following will be the result:—

Soluble in NH_3SH.	*Insoluble in NH_3SH.*
Antimony oxide.	Mercury oxides.
Molybdenum oxide.	Bismuth oxide.
Tin protoxide.	Lead —
Platinum —	Rhodium —
Iridium —	Palladium —
Osmium —	

These are then to be tested for by the distinguishing tests.

TESTS FOR ACIDS.—*Acids of Common Occurrence.*

		NITRATE OF SILVER. AgONO5.	CHLORIDE OF BARIUM. BaCl.
		Insoluble in cold Nitric Acid.	
Sulphuric, ...	SO_3	O.	*White* powder, insol. in NO_5 ($BaO\ SO_3$).
Nitric, ...	NO_5	O.	O.
Chlorohydric, (Muriatic,)	HCl	*White* curdy, (AgCl.)	O.
Iodohydric, ...	HI	*Whitish-yellow*, insol. in NH_3 insol. in cold NO_5 (AgI.)	O.
Cyanohydric, (Prussic,) ...	HCy	*White* curdy, sol. in boiling NO_5 (Ag Cy.) Smell of prussic acid.	O.
		Soluble in Nitric Acid.	
Carbonic, ...	CO_2	*White* curdy, sol. with effervescence in NO_5 ($AgO\ CO_2$).	*White*, sol. with effervescence in NO_5 ($BaO\ CO_2$).
Pyrophosphoric, ... (Bibasic phosphoric acid,)	PO_5	*White*, sol. without effervescence in NO_5 ($2\,AgO\ PO_5$).	*White*, sol. without effervescence in NO_5 ($2\,BaO\ PO_5$).
Oxalic, ...	C_2O_3	*White*, sol. in NO_5.	O. free; *White*, neut.
Antimonic, ...	SbO_5	*White* do.	*White*.
Sulphurous, ...	SO_2	*White*.	*White*, sol. in HCl.
Hyposulphurous, ...	SO_2S	*White*, then *black*.	O.
Hyposulphuric, ...	$\left\{\begin{matrix}SO_2\\SO_3\end{matrix}\right.$	O.	O.
Acetic, ...	$\left\{\begin{matrix}C_4H_3\\O_3\end{matrix}\right.$	O.	O.
Tartaric, ...	$\left\{\begin{matrix}C_8H_4\\O_{10}\end{matrix}\right.$	*White*.	*White*.
Phosphoric, (Tribasic,)......	PO_5	*Yellow*, sol. without effervescence in NO_5 ($3\,AgO\ PO_5$).	*White*, sol. without effervescence in NO_5 ($3\,BaO\ PO_5$).
Arsenious, ...	AsO_3	*Yellow* with NH_3 ($2\,AgO\ AsO_3$).	*White*, with NH_3 in time.
Arsenic, ...	AsO_5	*Brown*.	*White*, sol. in NO_5.
Chromic, ...	CrO_3	*Brown*.	*Yellow*, sol. in NO_5.
Sulphohydric, ...	SH	*Black*, sol. in NO_5, with evol. of SH.	O.

Distinguishing Tests for Acids.

Nitric acid is distinguished by dropping in a solution of copperas and adding sulphuric acid (SO_3); if the nitric acid (NO_5) is combined, a dark characteristic colour is produced. With SO_3 and heat, the odour of NO_5.

Oxalic acid precipitates a solution of sulphate of lime.

Antimonic and *Antimonious acids*, an orange precipitate by sulphohydric acid, SH.

Iodohydric acid. The precipitate with nitrate of silver ($AgO\ NO_5$) is scarcely soluble in ammonia (NH_3), while most silver salts are quite soluble. Starch and commercial nitric acid give a blue. Yellow with nitrate of lead ($PbO\ NO_5$).

Cyanohydric acid. Add a few drops of solution of decomposing sulphate of iron, then 2 drops of ammonia (NH_3), then chlorohydric acid (HCl) till the precipitate appears like to Prussian blue. For other tests, see HYDROCYANIC ACID.

Chromic acid becomes green by sulphohydric acid (SH), yellow by nitrate of lead ($PbO\ NO_5$).

Arsenic and *Arsenious acids* give a yellow precipitate by sulphohydric acid (SH).

Acetic acid. Smell of vinegar when heated with sulphuric acid (SO_3).

Phosphoric acid, when ignited, gives *white* with nitrate of silver ($AgO\ NO_5$).

Sulphurous acid, with sulphuric acid (SO_3) and heat, smell of burning sulphur.

Hyposulphurous acid with sulphuric acid (SO_3), sulphur (S) falls, and sulphurous acid (SO_2) evolved.

Hyposulphuric acid, with chlorohydric acid (HCl), sulphurous acid (SO_2) evolved, and sulphuric acid (SO_3) in solution.

Tartaric acid. *White* with chloride of calcium (CaCl) lime water (CaO) and potash (KO).

Nature of the Actions in the preceding and following Tables.

In the Acid Table the precipitates formed are combinations of the acid examined, with the base added, a new insoluble salt being formed. In the 1st Table of Bases all the precipitates produced are oxides, except those of mercury and peroxide of platinum, palladium, and gold. When NH_3 is added to calomel (Hg_2Cl) a blackish powder falls, 2 (Hg_2Cl) and 2 NH_3 becoming Hg_2Cl, Hg_2NH_2, and NH_3HCl. When NH_3 is added to corrosive sublimate (HgCl) *white precipitate* falls ($HgCl\ HgNH_2$). When NH_3 is poured on bichloride of platinum, the yellow ammonia chloride of platinum is formed ($NH_3\ HCl\ PtCl_2$). Ammonia forms with chloride of gold fulminating gold. In the 2d Table all the precipitates with soda are oxides, except with antimony, when a salt is formed. Potash forms with $PtCl_2$ a compound salt ($KCl\ PtCl_2$). In the 3d Table SH forms sulphides, except with chromic acid, when the green oxide appears. In the 4th Table the earths fall as oxides, $Al_2\ O_3$ 3 SO_3 and 3 (NH_3SH) becoming $Al_2\ O_3$ and 3 (NH_3SO_3) with SH evolved, and chromium as green oxide. Similar diagrams should be employed as in the case of acids.

Note.—*When bases are dissolved in organic acids, the effect of reagents is often interfered with, and the expected precipitate does not fall. A few drops of chlorohydric acid* (HCl) *are then required.*

TABLE I.—TESTS FOR BASES.

Caustic Ammonia (NH_3).

FIRST CLASS.

Precipitated and not Redissolved.

Earths.	Magnesia,	WHITE.
	Alumina,	
	Glucina,	
	Yttria,	
	Cerium oxides,	
	Zirconia,	
Metals.	Antimony oxide,	
	Titanic acid, with oxide in HCl.	
	Lead oxide, except organized salts,	
	Tin protoxide,	
	Bismuth oxide,	
	Mercury oxide,	
	Mercury suboxide,	BLACK.
	Chromium oxide,	GREEN.
	Uranium oxide,	BROWN.
	——— *peroxide*,	YELLOW.
	Molybdenum protoxide,	BROWN.
	——— *peroxide*,	YELLOW.
	Manganese protoxide,	WHITE, then BROWN.
	——— sesquioxide,	BROWN.
	Osmium oxide,	
	Vanadium peroxide,	
	Iron protoxide,	GREEN.
	—— sesquioxide,	BROWNISH-RED.
	Platinum protoxide, With Cl.	GREEN.
	——— peroxide, With Cl.	YELLOW.
	Rhodium peroxide,	
	Iridium oxide, After boiling.	BLUE.

SECOND CLASS.

Not Precipitated.

Alkalies,	Ammonia.
	Potash.
	Soda.
	Lithia.
Alkaline Earths,	Lime.
	Barytes.
	Strontian.
Arsenious acid.	
Arsenic acid.	

THIRD CLASS.

Precipitated and Redissolved in Excess.

Substance	Colour
Antimonious acid, } With HCl.	
Antimonic acid,	
Tellurium oxide, with HCl.	
Zinc oxide,	WHITE.
Cadmium oxide,	
Tin peroxide,	
Nickel oxide,	GREEN.
Cobalt oxide,	BLUE.
Copper oxide,	
Silver oxide, (neut.)	BROWN.
Palladium protoxide, With HCl.	REDDISH-WHITE.
Gold peroxide,	YELLOW.

Note.—The substances in italics are of rare occurrence.

TABLE II.—TESTS FOR BASES.

Caustic Soda or Potash (NaO or KO).

FIRST CLASS.

Precipitated and not Redissolved.

Group	Substance	Colour
Alkaline Earths,	Lime,	
	Barytes,	
	Strontian,	
Earths,	Magnesia,	
	Yttria,	
	Cerium oxides,	WHITE.
	Zirconia,	
Metals.	*Titanic acid, with oxide in HCl*,	
	Bismuth oxide,	
	Cadmium oxide,	
	Manganese protoxide, then brown,	
	Iron protoxide,	WHITISH-GREEN.
	—— sesquioxide,	REDDISH-BROWN.
	Uranium protoxide,	BROWN.
	—— *peroxide*,	
	Platinum peroxide, with KO, **O.** with NaO.	
	Rhodium peroxide, by boiling,	YELLOW.
	Palladium protoxide,	
	Mercury oxide,	
	—— suboxide,	BLACK.
	Molybdenum protoxide,	BROWNISH-BLACK.
	—— *deutoxide*,	
	Nickel oxide,	GREEN.
	Cobalt oxide,	BLUE.
	Copper oxide,	
	Osmium oxide,	BLACK.
	Gold oxide, Au. on standing.	
	Silver oxide,	BROWN.

SECOND CLASS.

Not Precipitated.

Alkalies, { Ammonia, *Lithia.* Potash. Soda.

Arsenious acid.

Arsenic acid.

Platinum protoxide.

ANA

THIRD CLASS.

Precipitated, and in Excess Dissolved.

Substance	Colour
Antimony oxide,	
Antimonious acid, } with KO	
Antimonic acid,	
Tellurium oxide,	
Columbic acid,	
Alumina,	WHITE.
Glucina,	
Zinc oxide,	
Tin protoxide,	
—— peroxide,	
Lead protoxide,	
Iridium peroxide,	BROWN.
Chromium oxide,	GREEN.
Vanadium oxide,	DIRTY WHITE.

TABLE III.—TESTS FOR BASES.

Sulphohydric Acid (SH).

FIRST CLASS.

Precipitated.

Substance	Colour
Tellurium oxide,	
Molybdenum oxides,	
Tin protoxide,	
Copper oxides,	BROWN.
Platinum oxides,	
Rhodium oxide,	
Iridium oxide,	
Antimony oxide,	
Antimonious acid,	ORANGE.
Antimonic acid,	
Arsenious acid (acid,)	
Arsenic acid —	
Cadmium oxide, —	
Tin peroxide, — and neut.	YELLOW.
Osmium oxide,	
Selenious acid, with HCl.	
Selenic acid,	
Lead oxide,	
Bismuth oxide,	
Mercury oxides,	
Silver oxide,	BLACK.
Palladium oxide,	
Gold oxide,	
Osmic acid,	
Chromic acid,	GREEN.
Zinc oxide, (neut.)	WHITE.

SECOND CLASS.

Not Precipitated.

Alkalies.

Alkaline Earths, { Lime. Barytes. Strontian.

Earths, { Magnesia. Alumina. *Glucina. Yttria. Cerium oxides. Zirconia.*

Metals,	Iron protoxide.
	—— sesquioxide, sulphur falls.
	Manganese oxides.
	Nickel oxide.
	Cobalt oxide.
	Chromium oxide.
	Vanadium oxide.
	Uranium oxide.
	Tungstic acid.
	Columbic acid.
	Titanic acid.

Copper in dilute solutions is *brown*, in concentrated solutions *black*.

TABLE IV.—TESTS FOR BASES.
Sulphohydride of Ammonia, $NH_3 SH$.

First Class.
Precipitated and not Redissolved.

Titanic acid, Alumina, *Glucina*, *Yttria*, *Cerium oxides*, *Zirconia*, Zinc oxide,	White.
Uranium oxide, Iron oxides, Nickel oxide, Cobalt oxide, Copper oxide, Bismuth oxide, Lead oxide, Mercury oxide, Silver oxide, *Palladium oxide*,	Black.
Tungstic acid, *Rhodium oxide*,	Brown.
Cadmium oxide,	Yellow.
Chromium oxide,	Green.
Manganese oxides,	Flesh-Red.

Second Class.
Not Precipitated.

Alkalies.

Alkaline Earths,	Magnesia. Lime. Barytes. Strontian.

Columbic acid.

Third Class.
Precipitated, and Redissolved in Excess.

Antimony oxide, Antimonious acid, Antimonic acid,	Orange.
Selenic acid, Arsenious acid, Arsenic acid, *Osmium oxide*, *Molybdenum oxide*, Tin peroxide, Tin protoxide, Gold oxide, *Iridium oxide*, *Tellurium oxide*,	Yellow. / Brown.
Vanadium oxide, Platinum oxides,	Brownish-Black.

Magnesia possesses some of the characters of *earths* and *alkaline earths* in common.

DISTINGUISHING TESTS FOR BASES.

First Class.
Precipitated by NH_3.

Magnesia,	*White* by NH_3, **O.** by NH_3SH. *White* by phosphate of soda and NH_3CO_2. Soluble in NH_3HCl.
Alumina,	Soluble in NaO and KO. Insoluble in alkal-carbonates. *White* by NH_3SH.
Glucina,	Soluble in NaO and KO. Soluble in carbonate of NH_3.
Yttria,	*White* by oxalic acid. Soluble in carbonate of NH_3. Insoluble in NaO.
Cerium oxides,	*White* by $KO\ SO_3$, nearly insoluble in HO. *Brownish-red* when ignited.
Zirconia,	Insoluble in HCl after ignition. Soluble in alkal-carbonates. Insoluble in NaO.
Antimony ox.	*Orange* by SH. Chloride decomposed by HO.
Titaniate, in HCl.	*Blue* colour produced by metallic zinc.
Lead oxide,	*Black* by NH_3SH. Soluble in NaO or KO. *White* by SO_3. *Yellow* by chromate of KO.
Tin protoxide,	*Brown* by NH_3SH; dissolved by an excess. Precipitates gold *purple*.
Mercurysubox.	*White* by HCl. *Black* by NH_3, NaO, and SH.
Mercury oxide,	*White* by NH_3. *Yellow* with KO or NaO. *Black* by SnCl. *Whitish-black* with a drop of NH_3SH. *Red* by KI.
Bismuth oxide,	Salts decomposed by HO. *Black* by NH_3SH. Distinguished from PbO by no precipitate with SO_3.
Chromium ox.	*Green* by NH_3. Soluble in KO.
Uranium protoxide,	*Brownish-black* by NH_3. *Black* by NH_3SH. Becomes peroxide by NO_5.
——*peroxide*,	*Yellow* by NH_3. *Brown* by NH_3SH, insoluble in excess.
Molybdenum oxides,	*Brown*, not soluble in NH_3 and NaO. *Yellow*, soluble in NH_3SH. *Brown* phosph. of NaO.
Tungstic acid,	Blue by Zn and HCl.

Manganese ox.	*White* then *brown* by NH_3 and NaO. **O.** by SH. Soluble in salammoniac. *Brown* by red prussiate. *Flesh-red* by NH_3SH.
Osmium oxide,	*Brown* by NH_3 and NaO. *Yellow* by SH. *Black* by Zn.
Iron protoxide,	*Blue* by red prussiate of potash in acid solutions.
— sesquioxide,	*Blue* by yellow prussiate of potash in acid solutions.
Gold oxide,	*Blue* by sulphate of iron. *Purple* by SnCl.
Platinum peroxide,	*Yellow* by KCl and NH_3HCl.
Rhodium oxide,	*Rose* colour by HCl, and evaporation. *Black* by Zn.
Iridium oxide,	*Blue* by NH_3. A fine *blue* solution with KO.

SECOND CLASS.
Not Precipitated by NH_3.

Ammonia,	Alkalies.	Smell of NH_3 when boiled with NaO, and vapour browns turmeric. *Yellow* by $PtCl_2$. Tinges flame *green.*
Potash,	Alkalies.	*Yellow* by $PtCl_2$, and on igniting the precipitate KCl taken up by HO. Tinges flame *violet.* Precipd. by tartaric acid.
Soda,	Alkalies.	**O.** by $PtCl_2$. *White* by antimoniate of potash. Tinges flame *yellow.*
Lithia,	Alkalies.	*White* by phosphate of soda and ammonia. Tinges flame *purple.*
Lime,	Alkaline Earths.	**O.** by sulphate of lime. *White* by oxalic acid. Tinges flame *red.*
Barytes,	Alkaline Earths.	Immediate *white* precipitate by sulphate of lime. **O.** by oxalic acid. Tinges flame *green.*
Strontian,	Alkaline Earths.	No immediate *white* precipitate by sulphate of lime, but after some time. Tinges flame *crimson.*
Arsenious acid,		*Yellow* by SH. Precipitated on copper black by boiling with HCl. Soluble in NH_3SH; solution precipitated *yellow* by HCl. or heat.

THIRD CLASS.
Precipitated and Redissolved by NH_3.

OXIDES.	
Tellurium,	*White* dissolved by NH_3 and NaO. *Brown* redissolved by NH_3SH.
Zinc,	*White* dissolved by NH_3 and NaO, and *white* by NH_3SH.
Cadmium,	*White* dissolved by NH_3. *White* and not dissolved by NaO. *Yellow* by NH_3SH and SH.
Tin peroxide,	*Yellow* by NH_3SH, soluble in excess. Zn precipitates Sn.
Nickel,	*Bluish-green* by NH_3, and dissolved with a violet blue colour. *Yellow-green* by red prussiate. *Bluish-green* by NaO, from ammoniacal sols.
Cobalt,	*Blue,* and dissolved by NH_3, with a brown shade. *Blue* undissolved by KO. *Brown* by red prussiate.
Copper,	*Blue,* and dissolved by NH_3, with a rich blue. *Blue* undissolved by KO becoming *black* on boiling. *Brown* by yellow prussiate. Turns a knife dipped into it *yellow.*
Silver,	*Brown,* and dissolved by NH_3. *White curdy* by HCl, insoluble in NO_5. *Yellow-white* by KI.
Palladium,	*Flesh-red,* dissolved by NH_3, *yellow* by KO; *yellowish-white* by HgCy, precipitate soluble in HCl.

TEST-CHARACTERS OF GASES.

I. *Soluble in Water.*

1. Chlorine and chlorohydric acid precipitate $AgONO_5$, *White curdy,* insol. in nitric acid, NO_5.
2. Carbonic acid precipitates lime water; precipitate redissolved by an acid with effervescence.
3. Cyanogen, with $AgONO_5$, *White curdy,* soluble by boiling in nitric acid, NO_5.
4. Ammonia. Odor. and precipitates bichloride of platinum, $PtCl_2$, *yellow.*
5. Protoxide of nitrogen, rekindles a glowing splinter.
6. Sulphurous acid. Smell of burning sulphur. Precipitated by BaCl; precipitate sol. in NO_5.
7. Sulphohydric acid, or sulphuretted hydrogen, blackens lead salts.

II. *Insoluble in Water.*

1. Oxygen rekindles a glowing splinter.
2. Hydrogen explodes when mixed with half its bulk of O, and fired.
3. Binoxide of nitrogen forms orange fumes with oxygen.
4. Nitrogen extinguishes a taper without precipitating lime water.
5. Carbonic oxide burns with a blue flame.
6. Carburetted hydrogen absorbed by terchloride of antimony $SbCl_3$.
7. Olefiant gas absorbed by chlorine.
8. Phosphuretted hydrogen takes fire in contact with air.
9. Arsenietted hydrogen burns with a blue flame, depositing metallic arsenic.

Analysis Quantitative.

Inorganic.—When the nature of a substance or compound has been determined by testing or qualitative analysis, the next object is to ascertain the quantities of each of the ingredients which are present. This is effected by quantitative, *analysis* (ανα, from; λυω, I loose). When the compound substance is of an explosive nature, or from other circumstances, it may be advisable to judge of its composition by bringing together its ingredients, and observing the proportions which unite, we are said to determine the constitution of the body by *synthesis* (συνθεσις, formation of a body from its elements). In proceeding to analyze a substance, it is necessary to obtain it either partially or wholly in a state of solution. Many salts are easily soluble in water, or by the addition of an acid; but the greater number of the ores and minerals, as they occur in nature, require special treatment before they are in a condition to be subjected to chemical analysis.

Analysis of Silicates soluble in Chlorohydric Acid.—A considerable number of minerals which have silica present as an acid, and contain from 10 to 20 per cent. of water, are decomposed by chlorohydric acid, the silica being replaced by that acid, which then unites with the bases and forms soluble salts. *Estimation of Water.*—To determine the amount of water present in the mineral, 25 grains of it in small portions, about the size of a pea, are weighed out in a platinum crucible, heated to redness, and weighed after cooling. The mineral should be previously dried at 212° in a water bath, unless it is found to lose water at this temperature. After ignition the silicates soluble in acid lose this property. *Estimation of Silica.*—To determine the amount of silica, and other constituents, a fresh portion of mineral is used; it is first pounded in a diamond or steel mortar, which is a steel box fitted

with a piston. Fragments of the mineral are placed in the mortar, the piston is introduced, and struck with repeated blows of a hammer on an anvil. A sheet of white glazed paper being spread on a table, the steel mortar with its contents are placed on the paper. About a grain of the rough powder is then transferred by means of a platinum spatula, knife point, or fragment of glass, to an agate mortar, and pounded with the agate pestle until it feels no longer gritty under the pestle. When the pounding of this portion is finished, another similar quantity is pounded; the fine powder being placed in a small weighing tube or watch-glass, covered by another inverted watch-glass, to prevent moisture being absorbed from the air by the powder. The quantity of powder to be analyzed is then weighed out in a glass tube or watch-glass, and transferred to a Berlin porcelain basin or platinum capsule, on a sand bath. Some distilled water is now added, and the powder formed into a paste by stirring with a glass or porcelain rod. When it is thoroughly diffused through the water, and no lumps exist, strong chlorohydric acid is added, and the agitation with the rod continued. In a short time the silica separates in a hydrous or gelatinous form (SiO_3 HO), and then care must be taken that none of the powder escapes decomposition by being enclosed in the jelly, by constant agitation and rubbing it against the bottom of the basin. The jelly is to be diluted with a small portion of water, and the whole evaporated to dryness with constant stirring, to prevent sputtering. The residue is well pulverized with the broad end of a glass rod, and heated on a sand bath until the smell of chlorohydric acid is no longer evolved. The object in applying the heat is to remove the water in combination with the silica, which renders it somewhat soluble. The powder in the

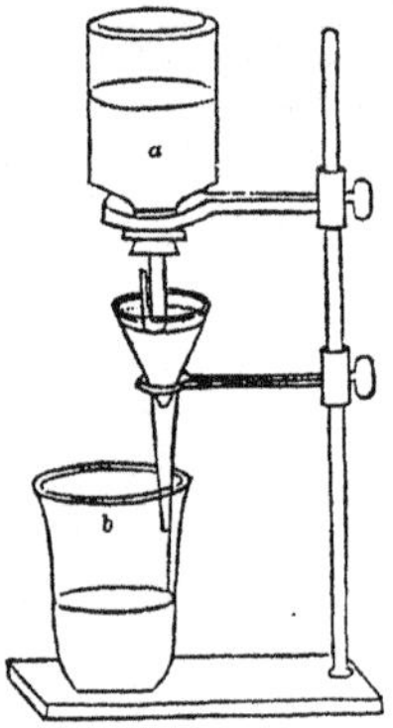

Filtering Bottle.

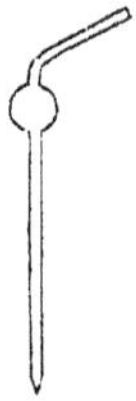

Sucker or Pipette

basin is now allowed to cool, and being moistened with strong hydrochloric acid, to supply acid to any of the bases which may have lost it by the heat, it is allowed to stand for 20 minutes. Water is then digested on it, and the whole filtered. (See FILTRATION.) The silica remains on the filter. Hot water is now to be added to the filter, by means of a sucker or washing bottle, until a drop of the solution, passing through, caught in a test-glass, gives no precipitate with nitrate of silver. Sometimes this operation is very tedious, if the silica has not been properly

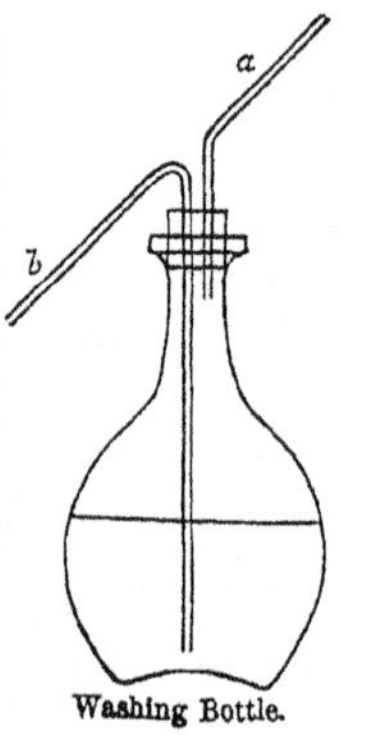

Washing Bottle.

evaporated to dryness, as in the gelatinous state it fills up the pores of the paper.

Separation of Silica from Minerals insoluble in Chlorohydric Acid.—When the silicate is found to be insoluble in acid, it is necessary to unite the silica with an alkaline base, an alkaline silicate being then formed which is decomposable by a mineral acid. Some minerals, undecomposable by acids before ignition, gelatinize after being heated in contact with acid (garnet, idocrase, epidote, and axinite). The silicate is to be pounded as already described, but if it is harder than agate, it is necessary to weigh the agate mortar before and after use, and deduct from the silica procured in the analysis the amount abraded from the mortar. The fine powder is mixed in a platinum crucible with 3½ to 4 times its weight of powdered anhydrous carbonate of soda (dried by heating the soda crystals on a sand bath). The crucible being covered with its lid, is placed in a *cylindrical* clay crucible, containing a little magnesia in the bottom, and introduced into a furnace, where it is kept at a bright red heat from half-an-hour to an hour. The fused mass frequently is separated by contraction from the sides of the crucible, and can be easily transferred to a porcelain basin; but when it adheres to the crucible, the crucible is placed in the porcelain basin. Some water and chlorohydric acid are added to the mass, and the whole digested until it is all obtained in the basin; when it is evaporated to dryness, and filtered, &c. as described in the previous section. After the silica has been properly washed, the funnel, filter, and silica are placed in a hot air stove, or oven, until the water evaporates; the silica and filter are then placed in a platinum crucible, and ignited over a gas or spirit flame, and weighed on cooling, covered with its lid to prevent absorption of moisture from the air. To ascertain that it contains no sand or undecomposed mineral, it is boiled with a solution of carbonate of soda in a flask or basin. If pure, it should dissolve completely. The insoluble matter, if undecomposed mineral, must be fused again with carbonate of soda, and the silica separated as described; if it is sand, it will be recognized by its grittiness between the teeth. When the silicate contains an alkali, instead of carbonate of soda or potash, we may use carbonate of barytes, or nitrate of barytes, a base which can be readily separated; but in this case a more elevated temperature is required for fusion.

Separation of Alumina and Iron.—After the separation of the silica as detailed, the liquid which passes through the filter is placed in a basin, and concentrated by evaporation on the sand bath. It contains all the constituents of the mineral, except the silica. After being reduced to the volume of about a pint, and made sufficiently acid, caustic ammonia is added to the concentrated fluid, when alumina and sesquioxide of iron fall, in gelatinous white or red flakes; if no iron is present, the whole of the precipitate will be white. It is necessary that the fluid should contain ammoniacal salts, which are formed by the addition of ammonia to an acid solution, otherwise magnesia will be thrown down, and a part of the alumina will be dissolved. If the precipitate is greenish, it is a proof that the protoxide of iron is present; but as it is soluble in ammonia, it must be converted by the addition of some nitric acid into sesquioxide which is insoluble. When alumina is alone, the best precipitant is sulphide of ammonium. The precipitate of alumina and sesquiox. of iron are thrown on a filter, and washed with boiling water, until a drop of the fluid, passing through, gives no precipitate with nitrate of silver. They are placed with the filter and funnel in a hot air stove, and after being dry are ignited in a platinum crucible, and weighed. To separate the iron from the alumina, the precipitate is transferred to a flask, and boiled with strong nitric and chlorohydric acids. Sometimes a skeleton of silica remains after the solution of the iron and alumina, which must be filtered off, washed, and weighed. The acid solution being concentrated, is boiled in a basin with an excess of strong caustic soda or potash. The alumina dissolves, while the sesquioxide of iron remains undissolved. When the quantity of iron is considerable, a portion of alumina remains undissolved. In this case it is often necessary to repeat the boiling with the caustic alkali two or three times. The whole is then diluted with hot water, filtered, and washed with boiling water, and tested with nitrate of silver. The sesquioxide of iron is dried in a stove, ignited, and weighed. It often contains potash or soda attached to it. It is best, therefore, to dissolve it off the filter in chlorohydric acid, and precipitate by carbonate of ammonia, filter, wash, dry, ignite, and weigh. The alkaline fluid containing the alumina is now supersaturated with chlorohydric acid, until it turns litmus paper slightly red, and is precipitated with carbonate of ammonia or sulphide of ammonium.

Separation of Lime.—The original liquid from which the alumina and iron have been separated is to be concentrated on the sand bath. The heat may render it acid, as ammoniacal salts become acid by heat. A slight excess of caustic ammonia is added, and care must be taken that a sufficient quantity of ammoniacal salts is present to retain the magnesia in solution. If a precipitate falls, it is probably magnesia. Acid must be added, and an excess of caustic ammonia. If it does not dissolve, it must be the alumina, and must be filtered off. Oxalate of ammonia is added in solution, when oxalate of lime falls. This is allowed to stand in a hot place for some hours. The supernatant liquor, while hot, is poured on the filter, hot water thrown on the precipitate, and allowed to clear, again poured off, and the process repeated. This precaution prevents the liquor from passing through in a milky state. The oxalate of lime is filtered, dried, ignited, and weighed. $CaO, CO_2 CO$. It remains as a gray carbonate, having lost carbonic oxide, which burns with a blue flame. It may be rendered white by ignition with access of air; but sometimes loses carbonic acid. To prevent error after weighing, it is moistened with a strong solution of carbonate of ammonia, ignited, and weighed. Every 6·25 grs. of carbonate of lime contain 3·5 of lime.

Separation of Magnesia.—If no alkalies are present in solution with the magnesia, the magnesia may be separated by adding carbonate of potash, evaporating to dryness, and heating to dispel ammoniacal salts, filtering, washing, igniting, and weighing the carbonate of magnesia, of which every 5·25 grains contain 2·5 magnesia; or the magnesia may be precipitated (Wollaston) by adding phosphate of soda to an ammoniacal solution, allowing the crystalline precipitate to stand for at least twelve hours, and then filtering and washing with water, containing one-tenth of caustic ammonia, as the precipitate of ammonia-phosphate of magnesia is soluble in water, but not in water containing caustic ammonia. When ignited (NH_3 $2MgO$ PO_5 becomes $2MgO$ PO_5), every 14 grs. contain 5 of magnesia.

Separation of Magnesia from Alkalies.—If, however, it is required to separate magnesia, and afterwards one or more of the alkalies, they are to be converted into sulphates by evaporating to dryness, and igniting in a platinum capsule. The fused residue is treated with sulphuric acid, and ignited. If there is an excess of acid, it may be removed by neutralizing it by ammonia, when it is carried off as sulphate of ammonia. The alkaline and magnesian sulphates are then weighed, dissolved in water, precipitated with pure acetate of barytes, and filtered. The filtered fluid is evaporated and ignited. Alkaline and earthy carbonates remain; but the magnesian salt is partially or wholly decarbonated. By digestion in water, the alkaline carbonates are alone dissolved, and are to be separated from each other by bichloride of platinum, (see SODA,) while carbonates of barytes and magnesia remain. These earthy carbonates are dissolved in chlorohydric acid, and the barytes precipitated by sulphuric acid. The filtered liquor is evaporated and ignited,, and the dried mass of sulphate of magnesia weighed. Every 7·5 grs. contain 2·5 of magnesia.

Analysis of Oxides.—Oxides of Iron. Iron is always estimated in analyses in the state of sesquioxide (Fe_2 O_3), although it occurs in minerals in the condition of protoxide (FeO), sesquioxide, protosesquioxide, (Fe_3 O_4 or FeO, Fe_2 O_3), or magnetic oxide. When the quantity of protoxide is considerable, it is necessary to boil the solution with some nitric acid, or to pass chlorine through it. When the oxide is sesquioxidated, that is, when the protoxide is entirely converted into sesquioxide, or peroxide, as it is sometimes called, it is precipitated by caustic ammonia, which throws it completely down. Sometimes it is necessary to throw down iron with sulphide of ammonium, in order to separate it from other bodies. In this case the black sulphide of iron is thrown on a filter, and covered with a glass plate, to prevent oxidation. After being washed, it is dissolved in chlorohydric acid, and filtered. Nitric acid is then added, and the sesquioxide of iron precipitated by means of caustic ammonia. Every 5· grs. of sesquioxide are equal to 4·5 protoxide of iron.

Oxides of Manganese.—This metal is always estimated in the state of brown oxide (Mn_3 O_4), a state of oxidization corresponding with the magnetic oxide of iron, into which all the oxides of manganese are converted by ignition. When the protoxide is in solution, it is precipitated by carbonate of soda, the liquor being heated after the addition of the reagent. The carbonate of manganese falls, but by ignition is converted into brown oxide; every 14·340 grs. of which are equivalent to 3·447 manganese, or 4·447 protoxide of manganese.

Nickel, Cobalt, and Copper, Oxides of.—These are separated in the condition of protoxides, by precipitation with an alkaline carbonate, or caustic alkali.

Silver Oxide.—Silver is usually precipitated by chlorohydric acid, and estimated in the form of chloride of silver; every 17·9433 grains being equivalent to 13·5 of silver.

Oxide of Zinc.—The best precipitant of zinc is carbonate of soda, in great excess. If ammoniacal salts are present, they must first be removed by boiling with an alkaline carbonate, and evaporating to dryness, so as to dissipate all ammonia, then digesting in water, filtering, igniting, and weighing. The oxide of zinc remains. Every 5·0659 grs. are equivalent to 4·0659 zinc.

Oxide of Lead is separated from its solutions by oxalate of ammonia, in a neutral or alkaline solution. When separated from the filter, and ignited in a porcelain crucible, there remains oxide of lead, every 14 grains of which are equal to 13 metallic lead.

Tin, Binoxide of.—*Tin* is best estimated in the state of binoxide, to which condition it is easily converted by the action of nitric acid. In the analysis of a metallic alloy containing tin, the alloy, when heated with nitric acid, leaves the binoxide of tin undissolved, in the form of a white or yellow powder, every 9·25 grs. of which are equal to 7·25 metallic tin.

Teroxide of Bismuth is separated from its solutions by carbonate of ammonia in excess. At first the oxide is dissolved, but after standing for some hours in a warm place, it is wholly separated. The precipitate is separated from the filter as much as possible, and ignited in a porcelain crucible. Every 29 grs. are equal to 26 grs. metallic bismuth.

Protosesquioxide of Uranium is separated from its solutions by caustic ammonia.

Oxide of Mercury.—Mercury is usually estimated in the metallic state. The mercurial compound is dissolved in an excess of chlorohydric acid, and boiled with a strong solution of protochloride of tin in a flask; the flask is well closed, and the whole allowed to cool. The clear liquor is poured off from the globules, and the metal washed and dried with blotting paper, but not filtered.

Gold is also procured in the metallic state, in absence of nitric acid, by adding to a hot solution of gold a solution of sulphated protoxide of iron (copperas). It may also be reduced by oxalic acid, which should be digested for twenty-four to forty-eight hours.

Platinum is separated from its solutions by concentrating its acid solution, adding to it a strong solution of sal ammoniac, and then as much strong alcohol as is necessary to separate completely the double yellow salt of ammonium-bichloride of platinum (Wollaston). It is washed with alcohol, containing some sal ammoniac, and is then burned at first in a covered crucible to prevent the platinum from being partially carried away with the sal ammoniac. The spongy platinum which remains is then weighed.

Palladium is separated by iodide of potassium. See IODINE. For RHODIUM, IRIDIUM, &c., see these metals.

Analysis of Sulphides (sulphurets).—The sulphide is pulverized as described under silicates, placed in a narrow-mouthed flask, and digested in fuming nitric acid, or in aqua regia (1 nitric, 2 chlorohydric acids). The greater portion of the sulphur is converted into sulphuric acid, and precipitated by chloride of barium. If any sulphur remains undissolved, it is thrown on a weighed filter and washed, dried at 212°, and weighed.

Arsenides are estimated by treatment with aqua regia, which converts them into arsenic acid. A current of sulphohydric acid is passed through the solution, which is fully saturated with gas, and the liquor placed in a very warm place till odour of sulphohydric acid gas is destroyed. The precipitate of tersulphide of arsenic, with excess of sulphur, is thrown on a weighed filter, washed, dried at 212°, and weighed. As much of the precipitate is removed from the filter as possible, and weighed. It is digested in aqua regia, and the sulphuric acid precipitated by chloride of barium. Every 14·567 of the sulphate of barytes is equivalent to 2 grains of sulphur, which is to be deducted from the tersulphide of arsenic. The remainder is the arsenic. As some of the precipitate adhered to the filter, the quantity of arsenic contained in it must be calculated by simple proportion from the amount obtained from the analysis of the larger portion.

Selenides.—Selenium may be estimated by passing chlorine over the selenide placed in powder, in a bulbed tube, and receiving the chloride of selenium in a flask. Water is added, and a stream of chlorine being passed through the solution, selenic acid is formed, which is precipitated as selenate of barytes by the addition of chloride of barium. For the mode of estimating *Iodides, Bromides, Chlorides, Fluorides,* see these substances.

Analysis of Waters. See WATERS.

Analysis of Organic Substances. See ORGANIC ANALYSIS.

Analysis of Gases. See GASES.

Anatase. (ανατείνω, I extend). *Oisanite* (from Oisans), *octahedrite,* from the length of the 8-hedra, *blue schorl, octahedral schorl.* $Ti\,O_2$. 8-hedron, with a square base, the summit being sometimes replaced by a small square parallel to the base of the pyramid. Colour, when pure, indigo blue, but more usually steel-gray or olive-brown; yellowish-green or blue by transmitted light. Lustre splendent and adamantine. Sp. grav. 3·857 to 3·826, H 5·5 to 6· B.B. infusible; dissolves in borax, yielding, with reducing flame, an amethyst bead. Scarcely fusible with microcosmic salt; but forms a blue bead in reducing flame. It may be fused with carbonate of potash and soda, but is not affected by acids. It consists only of titanic acid when in its pure state, but generally has present from ¼ to 1 per cent. oxide of iron, and is identical in composition with *Arkansite.* It is found in granite and slate in Cornwall, Dartmoor, Dauphine, Tyrol, &c.

Anauxite. (ἀναύξητος, without augmentation.) Greenish-white grains, in plates with translucent edges; found in decomposing volcanic rocks at Bilin. Spec. grav. 2·265. Contains 55·7 silica, 11·05 water; the other constituents being alumina, magnesia, and protoxide of iron.

Anchosine. A mineral containing silica 52·52; $Al_2\,O_3$ 30·98, CaO 3·82, FeO ·80, KO 6·38, HO 1·60. (Kobell. Jahr. fur Min. 1834.)

Anchusine, Anchusic Acid. C 71·33, H 7·00, O 21·67. A dark red resinous body, permanent in light, softening at 140°, and subliming, without change, at a higher temperature, in the form of violet fumes, like iodine; insoluble in water, soluble in alcohol and ether, with a red colour; in turpentine and fat oils. The red alcoholic and ethereal solutions are changed by light, and the latter, by boiling, into blue and green colours. Sulphuric acid dissolves it of a red colour. It is the red colouring matter of the Anchusa tinctoria, or alkanet, obtained by acting on the root with boiling absolute alcohol or ether.

Ancramite. A synonyme of carbonate of zinc.

Andalusite (from Andalusia,) *Disilicate of Alumina, Micaphyllite, Stanzaite, Adamantine spar, Apyritic felspar.* Sp. grav. 3·314 to 3·13. H 7·5. Right rhombic prisms, sometimes of great size. Colour flesh-red, pearl gray, brownish-red; fracture uneven; lustre glassy; translucent on edges. It consists of silica 35·30, $Al_2\,O_3$ 60·19, FeO 1·32, MgO 1, HO. 2·03. Infusible B.B., but becomes covered with white spots; fuses with great difficulty with borax and microcosmic salt; not attacked by acids. Occurs in Andalusia, Saxony, Tyrol, near Aberdeen, Glen Malor in Wicklow.

Andesine. A mineral, allied to felspar, grayish or greenish, consisting of silica 59·6, Al_2O_3 24·28, CaO 5·77, NaO 6·53, KO 1·08, MgO 1·08, Fe_2O_3 1·58. Spec. grav. 2·736 to 2·651.

Andreasbergolite, Andreolite. A synonyme of harmotome.

Andronia. A substance described by Winterl in wood coal.

Anemic Acid. A yellowish-white substance from the infusion of the *Anemone nemorosa*, by exposure to light; without taste and smell, but slightly soluble in cold water; reaction acid; insoluble in alcohol and ether.

Anemonic Acid. $C_7 H_4 O_3 HO$. Brown brittle mass, soluble in water, scarcely so in alcohol and ether; reaction acid; yielding uncrystalline salts. Obtained by boiling anemonine with barytes water; removing excess of barytes by carbonic acid; precipitating the filtered liquor with acetate of lead, decomposing the precipitate by sulphohydric acid, and evaporating the filtered liquid to dryness.

Anemonine. $C_7 H_3 O_4$ (Loewig), or $C_{15} H_6 O_6$ (Fehling). White crystalline needles or plates, heavier than water at the usual temperature, without smell, at first without taste, but afterwards having a permanent burning impression; yields by heat pungent vapour, affecting the eyes and nose. But slightly soluble in water; very soluble in alcohol and oil of turpentine. Obtained as an oil, which, with air, changes into anemonine and anemonic acid, by distilling the root of the Anemone nemorosa, pulsatilla, and pratensis with water, and setting the distilled fluid aside.

Anemometer. An instrument for measuring the strength of the wind.

Angelica Balsam. A blackish-brown resinous syrupy matter, with a bitter taste, obtained from the root of the *Angelica archangelica*, by extracting, with absolute alcohol, evaporating for the extract, washing it with water, and treating with ether.

Angelic Acid, $C_{10} H_7 O_3$, and

Angelicine, from the root of the angelica.

Anglarite. A synonyme of phosphate of iron, or vivianite.

Anglesite. A synonyme of sulphate of lead.

Angosturine. *Cusparine, Galipeine.* See CUSPARINE.

Anhydrite. *Anhydrous Sulphate of Lime, Muriacite, Karstenite, Cube Spar.* CaO SO_3 8·5 or 68. White right rectangular prisms or fibres. Sp. grav. 2·899 to 2.94. H 2·75 to 3·5. Usually associated with common salt. It is found in the salt mines of Austria and Tyrol, at Bex, in Switzerland, Lockport, New York, &c.

Anhydrous. (α priv destitute of; ὕδωρ, water.) A substance without any water united to it, or mechanically mixed with it, is said to be anhydrous: gypsum, for example, when ignited, loses water, and becomes anhydrous.

Anil, Nil. The native name of indigo.

Anilamine Urea. $C_{14} H_8 N_2 O_2$. *Carbanilamide*, or *anilo-urea.* Crystalline body, soluble in water, alcohol, and ether. Fuses at 161½°, when anhydrous at 212°; combines with acids. Obtained by acting on nitro-benzamide with sulphohydride of ammonia at a boiling temperature.

Anilic Acid. *Indigotic Acid, Nitrosalicylic Acid.* $C_{14}H_4NO_9HO$. Light yellowish-white prisms, soluble in 1000 parts water, fusible and volatile; it unites readily with bases, and forms well crystallized salts. It is obtained by adding pounded indigo, in small portions at a time, to nitric acid, of spec. grav. 1·28, diluted with an equal volume of water, gentle heat being applied; along with it, carbazotic acid remains in solution. On the addition of acetate of lead, anilate of lead is formed, which separates in crystals. Sulphuric acid will decompose the anilate of lead. Anilic acid is also formed by the action of nitric acid on salicine or salicylic acid.

Aniline. *Crystalline, Cyanol, Benzidam, Phenylamine, Phenamine, Phenamide.* $C_{12} H_7 N$. A colourless oily fluid of sp. grav. 1·020. A powerful base. Taste burning aromatic; smell vinous; boiling point 320°; does not freeze at -4°; evaporates at all temperatures, and becomes brown in air; it does not in solution affect vegetable colours; dissolves phosphorus, sulphur, camphor, and rosin; coagulates albumen, and forms crystalline salts with acids. It is formed by heating isatine with potash, the action being $C_{16}NH_5O_4$ 4 KO HO = $C_{12} H_7 N$, 4 KO CO_2, H_2; by distilling anthranilic acid, by distilling nitrobenzoin, or salicylamide, or nitrotoluole, with lime; by distilling oil of coal tar, first heating the oils with chlorohydric acid, which unites with the basic oils. These form the lowest layer, which is to be neutralized by milk of lime, or an alkali, and the oil which separates is to be distilled. A white fluid condenses in the first instance, containing aniline and leucol. To purify this mixture, the oils are treated successively with chlorohydric acid and an alkali, and the product distilled repeatedly until an oil is obtained, whose boiling point is 360°. It is lastly purified by heating it with oxalic acid, and decomposing the oxalate by potash. It is obtained from nitrobenzide by dissolving the latter in alcohol, saturating the solution with ammonia, and passing a current of sulphuretted hydrogen through the deep red mixture. The clear liquor, after standing for some time, becomes clouded, and sulphur separates in crystals. The current of gas is again to be renewed, until sulphur ceases to separate on standing. Hydrochloric acid is then added in excess to the solution, and the alcohol being removed by evaporation, a clear, slightly coloured fluid, *aniline*, remains, possessing the characters of an organic base. It is obviously produced by the separation of all the oxygen from nitrobenzide, by the action of the hydrogen of the sulphuretted hydrogen, in the form of water, the sulphur being dropped, and the place of the oxygen supplied by hydrogen. Aniline is readily detected by its producing, in minute quantities, with solutions of chloride of lime, a fine violet-blue, resembling ammoniacal oxide of copper. To detect the presence of benzole, we heat a drop of the tested fluid with fuming nitric acid, until the mixture becomes yellowish; the nitrobenzide thus formed is separated by water, and dissolved in ether. This solution is mixed with some weak alcohol and sulphuric acid; a fragment of granulated zinc is then added; a solution of sulphate of aniline is thus formed, to which caustic potash is added; the aniline, which separates, is dissolved in ether, and mixed with a solution of chloride of lime. Aniline has a remarkable analogy with ammonia, and from this and other circumstances it has been proposed to consider ammonia as an amide of hydrogen, Ad H (NH_2 H), and aniline as an amide of phenyl. Ad Ph (C_{12} H_5). Aniline has a compound oxanilide corresponding to the oxamide of ammonia, and carbanilide, parallel to the carbamide of ammonia. Cyanate of aniline passes into anil-urea, as cyanate of ammonia does into urea. In aniline we have an example of an organic body (phenyle), which replaces the third atom of hydrogen of ammonia. Its natural formula will, according to this view, therefore be C_{12} H_5 NH_2, and aniline would be styled phenylamine (C_{12} H_7 N). It unites directly with hydrogen acids, like ammonia. Cyanogen, when acting on an alcoholic solution of aniline, unites with the latter, and forms a new base, cyaniline (C_{12} H_7 N, NC_2), which unites with acids and yields salts. Acted on by chloride of cyanogen, it forms *melaniline*, a new base, where 1 atom of aniline is coupled with another atom of aniline, in which 1 atom of hydrogen is replaced by cyanogen 2 C_{12} H_7 N, NC_2 Cl = C_{26} H_{13} N_3, H Cl). The hydrogen of aniline may be replaced by chlorine, bromine, and iodine, in more than one proportion, and hence arises a series of bases, probably of greater extent than has yet been pointed out. Those described are chloraniline C_{12}, H_6, Cl, N, dichloraniline, trichloraniline, bromaniline, dibromaniline, tribromaniline, chlorodibromaniline, nitraniline, nitrodibromaniline. Again, the hydrogen is capable of being replaced by organic radicals, as in Ethyl-aniline NH, E, Ph, or, as it may be termed, Ethyl-phenylamine.

Anilo-cyanic Acid. $C_{14}NH_5O_2$. Volatile liquid, with a pungent odour, boiling at 354°; bearing to aniline the same relation as that of cyanic acid to ammonia; it is obtained by heating melanoximide.

Anilo-mellone. *Phenyle-mellone.* C_{18}, H_4 N_4. A fluid obtained by heating melaniline. It consists of mellone (C_6 N_4), coupled with phenyle (C_{12} H_4), less hydrogen.

Animal Chemistry. Minerals differ from plants in the circumstance that minerals grow without possessing any organization, or any vessels by which they can be supplied with nourishment from within. They increase in size by depositions of solid matter to their circumference. Plants, again, differ from animals by their possessing no nervous system, to which animals owe their powers of sensation, and their varied supply of intellectual power, from the humble bee, which collects its food and stores it with mathematical precision, to the dog and elephant, which, in many points, are superior in sagacity to man himself. Much mystery has been thrown over the organic world, and particularly over the animal creation, by a kind of silent agreement entered into by many cultivators of science, that the processes to which we are indebted for our subsistence are too minute, and are guided by what have been termed such vital dynamical laws, that they must ever elude human inquiry. Now, this is a mode of raising up artificial difficulties which are not to be found in the extensive field of nature; and it may be with truth affirmed, that, as far as the production and reproduction of plants and animals are concerned, we are not in a condition to predicate where our knowledge must stop. The true method of philosophy is to push inquiry as far as possible, not by metaphysical and hypothetical means, but by the sure process of induction, and not by raising artificial barriers long ere we have reached the destined goal. It has been well said by Liebig, that the term dynamic is about equal to that of specific in medicine. Everything is *specific* which we cannot explain, and *dynamic* is the explanation of all that we do not understand. There is no greater mystery in the growth of a plant or animal from seed than there is in the production of a crystal in a saturated solution of a salt. In the manufacture of sulphate of iron, a saturated solution of this salt is introduced into a large evaporating cooler. A piece of wood is dipped into the solution, and in the course of a short time it is surrounded with a circle of beautiful green rhombic prisms, which never existed before. We call this crystallization; but this is no explanation, it is merely a word; and we are in no respect nearer our object than we are in explaining the deposition of solid fibres from the blood of animals. The products of the animal and vegetable world constitute organic chemistry. Scientific men have had much difficulty in defining the characters of vegetables and animals. Vegetables being confined to one spot, and animals being moveable beings, some have considered that animals were distinctively locomotive. But it so happens that many of the lesser tribes of animals are incapable of locomotion; and hence this definition is untenable. Others observing that plants are destitute of sensation, have proposed to ascribe this attribute alone to animals, and define them as nervous beings.

But in opposition to this view, we find many inferior animals apparently destitute of sensation, and only supplied with a degree of irritability even inferior to that of the sensitive plant, cultivated so frequently in our botanic gardens; and hence this definition also fails us. We believe that the true distinction between plants and animals will be detected more readily by discovering the nature of the matter by means of which they increase in bulk, or, in other words, by the nature of their food; the term food being a word applied to express the matter which enables plants and young animals to increase in size, and full-grown animals to preserve their forms unimpaired. It is to chemistry, then, we are to look for an answer to the questions, What is an animal? What is a vegetable? To one accustomed to view only the larger kinds of animated beings, it might seem an easy task to give a reply to these questions. But when we know that nature is simple in her works, that in her we find no sudden leaps from great to small, that the whole animated world consists of a chain formed of a series of beings, passing down in regular gradation from the most perfect to the most imperfect state, the lowest plant being closely allied to the lowest form of animal, it will at once be obvious that to say where plants begin and animals end, cannot be a problem of easy solution. To apply, however, the test which we have suggested. Let us begin with plants. We find a plant cultivated among the Chinese, and introduced among ourselves, termed the *air-plant*, which, by being merely suspended in the air, increases in bulk and weight without even the application of water. This is one of the most simple forms of vegetable life, as the plant has nothing to feed on save the air, which, however, contains all the elements necessary for its growth, oxygen, vapour of water, carbonic acid, and nitrogen. But all these are gaseous bodies or vapours, while the air-plant is a solid—hence we infer that this plant is capable of reducing gases to the solid state, and of thus increasing in bulk and weight. According to the present views of persons best qualified to judge, it appears that all plants are endowed with similar properties, and that they mainly subsist by feeding on the gases which surround them, by converting these gases, by means of the organs with which they are supplied, into the solid forms of the vegetable kingdom, so endless in figure, but yet so lovely, that the greatest familiarity renders them only objects of superior admiration. When we turn to the animal world, we find that the individuals of which it consists are incapable of condensing gases; in fact, the least educated person knows that animals cannot exist upon air, but that they require to imbibe solid matter similar to that of which they consist. Man lives upon animal food, and those kinds of grain which contain matter nearly allied to it. The question, Why has grass—perhaps the most abundant vegetable in nature—never constituted a portion of human food? may not strike one as being in its answer fraught with important information: and yet the only reason which can be given for the fact that it has never been an article of human food, except perhaps among the lowest portions of the human family, is because it contains such a small portion of matter similar in its nature to the constituents of man's frame, that the quantity required would be too voluminous for the digestive capacity of the stomach and other organs. An animal may, therefore, be defined to be a being which subsists by appropriating to itself food similar to the matter of which its own body is composed. Hence we see the necessity for its locomotion, while a plant, finding its nourishment in the constituents of the air which surrounds it, has its food brought to it by the usual laws of nature. We believe, then, that such will be found the only legitimate mode of separating animals from plants. It is possible that, among the inferior tribes of animals, where an approximation is made to the vegetable kingdom, there may be individuals partaking of a semi-vegetable and animal nature, partly living on air, and partly on solids; although it does not follow that such an occurrence is necessary; yet, from the simplicity and gradation which we find subsisting throughout nature, we might expect to discover some such union of the two kingdoms or some equally simple transition from one set of beings to the other. Accordingly, we find that cellulose, hitherto considered as a characteristic of vegetables, exists in some of the lower animals. This circumstance occurs in some of the ascidiæ. A sack in these animals forms a congregation of large spherical cells, similar to the tissue of the cacti, and some of the fruits produced in the garden. The substance of this tissue is insoluble in water, alcohol, ether, acids, and alkalies; is free from nitrogen, and contains per cent. carbon 45·38 H = 6·47 O = 48·15. It is, therefore, identical in composition with vegetable cellulin. Schmidt has examined among the zoophytes, a class of animals which, in salt springs, are developed underground in the form of a gelatinous white mass, to which Ehrenberg has given the name of *Frustulia salina*. This animal was treated with solvents, then by acids and dilute alkalies. When analyzed, it was found to consist of C 46·19, H 6·06, O 47·85. This result agrees exactly with analyses which have been made of the basic tissue of some of the lichens, of which the composition was found to be C 46·08, H 6·67. Plants, then, it might be considered, are supplied with a more complicated apparatus than animals; for they are capable of generating solids from gases, or of producing from these elements compound bodies, which never existed before; while animals merely take from plants solid food ready prepared for them, and add it without any change to their own substance. Animals, then, it would appear, are made up of plants; and hence we infer that plants must have existed before animals. So strikingly is it the fact, that animals are derived from plants, that it is usual to judge of the fertility of a pasture field by the quantity and richness of the milk, the essence, as it were, supplied by the cattle which feed upon it. Plants, however, are scarcely susceptible of remaining so long without food as animals. Deprive a plant of moisture and air, and it speedily withers and decays. Man is enabled, perhaps, to endure hunger for a longer period than any other class of animals. Mr. Catlin has stated, that when he visited the Mandans, on the Missouri river, some years ago, he found their numbers amounting to 2000. But soon after he left them, the small-pox, introduced by the N. W. Fur Company, and the assaults of their enemies, contributed to effect the total annihilation of the tribe. One of the chiefs, described by Mr. C. as a man of noble spirit, when he found his friends and relatives dying around him, and that single-handed he could no longer defend himself from the enemies of his tribe, resolved upon terminating his existence by starving himself to death; and although strongly urged by the fur traders to desist from his intention, he, in the most determined manner, adhered to his purpose, and died on the ninth day. Since plants, then, are less capable of subsisting without food than animals, it would appear that the food of plants is less permanent in its effects than that of animals. But animals have the power of laying up a kind of food, which renders them capable of existing for much longer periods than that mentioned without the use of aliment. An instance, which occurred to myself, strikingly illustrated this fact. A poor idiot, who had been bed-rid for years, had received a sudden fright by a peal of thunder, which caused him to refuse nearly all sustenance; and so far did he carry his prejudice, that he would allow no water to cross his lips which was not clear and crystal from the spring. In this condition he remained for seventy-four days, when tired nature at last yielded, and gave up the struggle. The explanation was simple, but it was as beautiful as it was simple. The poor creature, by want of exercise, and by previously possessing a strong appetite, had become enormously fat. This deposit acted as so much fuel, and enabled respiration to proceed as usual; but as his frame was acquiring no addition, and every part of it was dwindling, or more properly burning, away, as soon as the fat was removed, the individual became thin and exhausted, and death inevitably closed the scene. The power of the human body, displayed in this fact, I believe to be peculiar, and not to be shared by the vegetable kingdom. These observations tend to delineate the characteristic features of the two living kingdoms of nature, and to show that there is a point at which vegetable unites with animal chemistry, inasmuch as plants supply food for animals. Hence the consideration of the materials of which animals are composed naturally includes the study of the plants, or at least of those parts of vegetables upon which they principally subsist. The chemistry of animal substances may, therefore, be legitimately viewed as including the nature of the substances derived from animals, and of the food from which they are primarily obtained. For the details of animal chemistry, see BLOOD, CHYLE, DIGESTION, RESPIRATION, &c. The ultimate constituents of animals are as follows:— Oxygen, chlorine, fluorine; hydrogen, nitrogen, carbon, silica, phosphorus, sulphur; potassium, sodium, calcium, magnesia; iron, manganese. The proximate constituents obtained from animals are:—*Animal amides?* Fibrin, albumen, casein, globulin, collin, chondrin, gelatin, hematosin, spermatin, salivin? pepsin? pancreatin (?) *Colouring matters:* Carmin, sericin, cancrin, peristerin, anserin. *Oily oxides, saponifiable:* Stearine, oleine, hircine? *Oily oxides not saponifiable:* Castorin, ambrein, cholesterin, serolin, cantharidin. *Animal bases:* Urea, odorin, animin, creatin, creatinin, sarcosin, alanin, ammolin, fuscin, crystallin, aposepedin, taurin, chitin, ammonia. *Acids*, see ACIDS.

Animal Heat. The heat generated in animals by the action of the oxygen of the air, through the medium of the lungs and blood vessels, upon the solids in the circulation, is so designated. As I consider Dr. Crawford to have been the discoverer of the theory of animal heat, and to have developed his views in a more remarkable manner than he usually obtains credit for, I shall here give a sketch of his theory. His great work was his "Experiments and Observations on Animal Heat, and the Inflammation of Combustible Bodies, being an attempt to resolve these phenomena into a general law of Nature," of which the first edition was published in 1779, and the second in 1788. In his preface to the second edition, he states that the experiments which gave rise to his explanation of animal heat and combustion were made at Glasgow in the summer of the year 1777, and that they were communicated in the autumn of that year to Dr. Reid, Mr. Wilson, and Dr. Irvine. In the beginning of the ensuing session they were made known to many of the professors and students in the University of Edinburgh, and in the course of the winter they were explained by their author in the Royal Medical Society. In this work, Dr. Crawford commences by stating the theory of specific heat, as elucidated by Drs. Black and Irvine, and proceeds to determine the specific heats of a variety of bodies, the general result of which was, that flesh, milk, and vegetables contain less absolute heat than water, and water less than arterial blood. Arterial blood, therefore, he concludes, contains a greater quantity of absolute heat than the principles of which it is composed. Thence he was led to suspect that, from the remarkable accumulation of heat in this fluid, it must absorb heat from the air in the process of respiration; and this suspicion was much confirmed by the following considerations: 1. Animals supplied with lungs, which respire air in large quantities, preserve a temperature considerably above that of the surrounding atmosphere. But animals which are not furnished with organs of respiration, are nearly of the same temperature as the medium which surrounds them. 2. Among hot animals, those are warmest which have the largest breathing organs, and which consequently respire the greatest quantity of air in proportion to their bulk: thus the respiratory organs of birds are more extensive than those of other animals, and birds possess the greatest amount of animal heat. 3. In the same animal, the degree of heat is in some measure proportional to the quantity of air inspired in a given time. Thus we find that animal heat is increased by exercise, and by whatever accelerates respiration. Dr. Crawford then proceeds to support these positions by experiments. He proves, that by placing an animal in a given quantity of atmospherical air, over lime water, the air will diminish nearly *one-fifth* in bulk, and the liquid will become turbid, and the quantity of calcareous earth precipitated will be in proportion to the diminution of the air. The correctness of this observation deserves remark, as some experimenters shortly afterwards found the quantity of oxygen in air to be upwards of *one-fourth* of the bulk of atmospheric air. Other experiments induced him to infer that the arterial blood assumed a livid hue in the capillaries during its passage into the venous system, and that it again acquired its florid hue in the lungs. The cause of the venous colour he attributed to the absorption of inflammable air; a portion of this air he considered combined with the pure air (oxygen) of the atmosphere, and formed aqueous vapour, while another portion, with the same gas, produced carbonic acid, or fixed air. The remaining propositions, which he endeavoured to substantiate by an elaborate series of experiments, are as follow:—1. The quantity of absolute heat contained in pure air is diminished by the change which it undergoes in the lungs of animals, and the quantity of heat in any kind of air that is fit for respiration is nearly proportional to its power in supporting life. This alone gave rise to a large series of experiments on the specific heat of gases, which are remarkable for the variety of modes which he invented for the purpose of determining his data. 2. The blood which passes from the lungs to the heart by the pulmonary vein contains more absolute heat than that which passes from the heart to the lungs by the pulmonary artery. 3. The comparative quantities of heat in bodies supposed to contain phlogiston, are increased by the changes which they undergo in the processes of calcination and combustion. His exposition of this proposition shows, that he considered that in the combination of pure air (oxygen) with the inflammable principle (hydrogen and carbon), a portion of their absolute heat will be separated, and fly off in the state of sensible heat. 4. When an animal is placed in a warm medium, the colour of the venous blood approaches more nearly to that of the arterial than when it is placed in a cold medium; the quantity of respirable air which it phlogisticates in a given time, in the former instance, is less than that which it phlogisticates during an equal space of time in the latter; and the quantity of heat produced when a given portion of pure air is altered by the respiration of an animal, is nearly equal to that which is produced when the same quantity of air is altered by the burning of wax or charcoal. The experiments which illustrate this, or rather upon which this proposition is founded, are highly important. He placed a guinea-pig in an inverted glass jar, containing nearly five pints of common air, and surrounded it with water of the temperature $55\frac{1}{2}$°, that of the atmosphere being 61°. In forty-two minutes the animal was removed, and the air tested by nitrous air, "and the measures of the test were found to be $2{\cdot}7\frac{1}{2}$ nearly, or $\frac{1}{4}$ as pure as common air." Another animal, of the same kind, was surrounded with water of the temperature 104°, and "the test of nitrous air gave 2·5, or it was nearly half as pure as common air." "Hence," says Crawford, "it appears that the degree of purity in the air necessary to the support of life is greater when an animal is placed in a warm than in a cold medium, the phlogistication (consumption of oxygen) of the air in the first or cold air experiment being twice as great as in the second or hot air trial." These trials were repeated in various ways, and always with uniform results. The summary of these experiments is briefly the fact, that as cold condenses air, and heat expands it, a larger amount of oxygen will be consumed by respiration in a given time in a cold than in a warm country. This is an interesting confirmation of the view of Liebig in his late work on animal chemistry, who holds that in different climates the quantity of oxygen introduced into the system by respiration varies according to the temperature of the external air. It may be called a confirmation rather than an anticipation, because we believe that Crawford's work is almost unknown to most of the present continental chemists, much less certainly than it deserves. Berzelius, aware of the experiments and views of Crawford on animal heat, yet affirms, that he was acquainted with no fact which supported such a view. The illustrations presented by Liebig of the immense supplies of food required by the Esquimaux, contrasted with its limited imbibition by the natives of tropical climates, is a sufficient answer to such an objection, and a flattering proof of the sagacity of Crawford. Dr. Crawford made a comparative set of experiments on the effects produced upon air by combustion and respiration, in which he observed some interesting facts. For example, he found that the diminution produced in the volume of pure air, by the combustion of wax, was to the entire diminution after the fixed air was absorbed, as 1 to 3 nearly, while, when charcoal was substituted, the diminution was as 1 to 7 nearly. Hence he concluded that when wax was burned, the pure air was partly converted into vapour, and partly into fixed air; but with charcoal it was entirely changed into fixed air. This induced him to determine if the process of respiration was analogous to the combustion of wax and tallow, or to that of charcoal; that is to say, if the expired air of animals contained also aqueous vapours. The result of his experiment was in the affirmative. In reference to the comparative quantities of heat produced, he found that with equal quantities of air, the amount of heat communicated to 31 lbs. 7 ounces of water was as follows:—

100 ounce measures of pure air by the combustion of wax, gave...		21 degrees.
"	charcoal......	19·3 "
"	respiration ..	17·3 "

These results differ considerably from each other; they are attributable to errors of experiment, as was inferred by himself; but he concludes that they approach so nearly to each other as to prove that the heat in these processes arises principally, if not entirely, from the conversion of pure air into fixed air or into water. He had concluded, from his experiments, that the specific heat of pure air (oxygen) is greater than that of fixed air (carbonic acid) and aqueous vapour, or, in more particular terms, that the same heat which would raise pure air one degree, would raise fixed air and aqueous vapour three degrees. The oxygen of the air, therefore, in combining with the inflammable principle (carbon and hydrogen) of the lungs, to form carbonic acid and water, gave out a certain quantity of heat, which was absorbed by the blood, in the change from the venous to the arterial state, the capacity for heat of venous to arterial blood being as 11½ to 10. This ingenious explanation was founded on the theory of the source of heat in combustion and chemical combination, entertained by Dr. Irvine. It can scarcely be said to be disproved at the present day, as no accurate idea of the source of heat in these cases has yet been formed. It is obvious that Crawford endeavoured to present explanations of phenomena of which we are still unable to afford a more plausible one. His fault consisted in his restricting the action of oxygen in the production of heat to the lungs, and in the adoption of what appears rather a hypothesis than a true explanation. Under the article *Aliment* it has been shown that the food used by animals in order that they may be retained in a state of health, should consist of nutritious and calorifiant, or heat-producing food. But there are some animals which apparently make use of no calorifiant matter. How, then, it may be inquired, is their heat kept up? The American savage, for example, consumes animal food alone, and beasts of prey are nourished after a similar fashion, without having an opportunity of partaking of the fruits of the field; for as the first do not sow, it is impossible for them to reap. Yet their animal heat is sustained; not, however, it must be observed, without great exertion, and a restless and wandering existence, impressed on them, in a great degree, by the nature of the food upon which they subsist. All the food which they eat, with the exception of fat (which is, however, always present to a certain extent in muscular tissue), enters into the solid constitution of their frames; and it is only when the muscular substance of the body is passing into gelatinous and other soluble states, that it is capable of evolving heat. This degradation of muscular tissue must proceed even in an animal in a perfect state of rest, but it is accelerated by motion in proportion to the degree and violence of the exercise. To keep up the animal heat, more particularly in winter, when it is tending to be conducted away by the colder surrounding media with rapidity, the savage is compelled to cover himself with warm skins, and to take a great amount of exercise. These expedients serve, likewise, as antagonists to the action of the atmosphere, and assist in forming and retaining heat, which is but tardily extricated. The function, then, which the calorifiant food fulfils in the animal economy is at once obvious from this view. It is to save the waste of the muscular tissue, and to admit of the formation of a sufficient amount of heat, by a more direct process of combustion. It is not a little remarkable that this physiological condition should constitute an essential element in civilization, by enabling the body to fulfil its functions without violent exertion, and to admit of unrestrained exercise of the mental powers.

Animalcules. Small animals, seldom discernible except by means of magnifying glasses, are so popularly termed. They occur in all

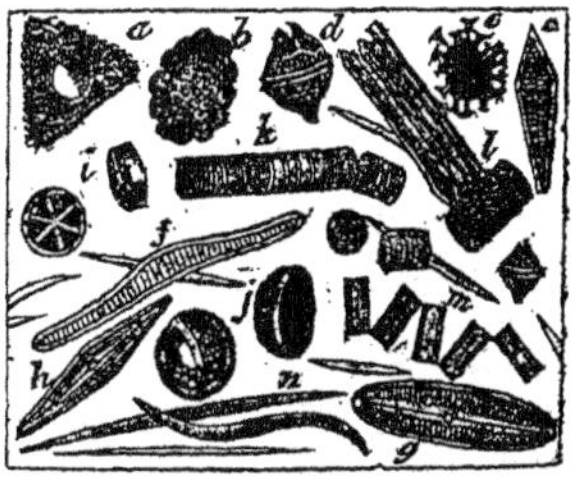

a Desmidium apiculosum.	*h* Navicula viridis.
b Euastrum verrucosum.	*i* Actinocyclus senarius.
c Xantidium ramosum.	*j* Pixidula prisca.
d Peridinium pyrophorum.	*k* Gaillonella distans.
e Gomphonema lanceolata.	*l* Synedra ulna.
f Hemanthidium arcus.	*m* Bacillaria vulgaris.
g Pinnularia dactylus.	*n* Sponge spiculæ.

waters exposed to the atmosphere, as rivers, town wells, &c. belonging to the genera naviculæ, exilariæ, bacillariæ, gaillonellæ, &c. Their occurrence in many solutions likewise renders them especial objects of attention to the chemist. An example is furnished in fossil and other deposits, especially in the tripoli slate of Bilin.

Animine. An oily base, with a peculiar smell, soluble in 20 parts cold water, less soluble in hot water, the cold solution becoming milky by heat; changes reddened litmus to a violet-blue; very soluble in alcohol, ether, and oils; obtained from rectified oil of Dippel by saturating with ammonia, and distilling; the first liquid is odorin; the next portions are odorine and animine; when the mixture is agitated with water, the odorine dissolves; by supersaturating with sulphuric acid, evaporating and distilling the residue with a base, animine remains as an oil. *Muriate of animine* forms double salts with copper, gold, and platinum: little soluble in water.

Anime Resin, a resin much resembling copal, obtained from the *Hymenæa courbarii* of Cayenne; it is distinguished from copal by its ready solubility in alcohol; it is used for varnishes and for scenting pastilles. There are two kinds, the American and Oriental; the latter consisting apparently of two resins. When digested in cold alcohol, a portion remains undissolved, which crystallizes from hot alcohol (*subresin*).

Anion. (ανα, upwards; and ιω, I go.) When a substance is decomposed by the galvanic battery, the elements into which it is resolved are termed *ions;* the element going to the *anode* is an electro-negative body, or *an-ion*, while the element which proceeds to the *cathode* is an electro-positive, or *cat-ion* (κατα and ιὼν). When water is decomposed, oxygen is attracted by the positive plus, or zinc pole, or anode, and is therefore the *anion*, or electro-negative body; while hydrogen, going to the cathode minus, or negative pole, is the *cation*.

Anise, Oil of. A transparent colourless oil, with the odour of anise seeds. Sp. grav. ·9857 to ·977. Solidifies at 50°, and becomes fluid at 62°; soluble in all proportions in alcohol (·806); obtained from the seeds of the *Pimpinella anisum*.

Anilonitriles. Neutral salts of ammonia are converted, by losing 2 atoms water, into *amides;* by losing 4 atoms, into *nitriles*. To carry out the exact parallelism of ammonia and aniline, attempts have been made to obtain anilo nitriles, but without success, and their existence seems impossible.

Anisamide. An amide obtained by acting on chloride of anisyle with ammonia.

Anisanilide. $C_{28}H_{13}NO_4$. White brilliant subliming needles, formed by the action of aniline on chloride of anisyle.

Anisic Acid. *Draconic Acid.* $C_{16}H_7O_5HO$. Colourless brilliant needles, somewhat soluble in cold, easily soluble in hot water, very soluble in alcohol and ether; forming crystalline salts with alkalies, oxides of lead, and silver; formed by boiling anise or estragon stearoptene with nitric acid (spec. grav. 1·2), washing the crystals, when separated, with cold water, and subliming them.

Anisidine. $C_{14}H_9NO_2$, or toluidine less 2 atoms oxygen. An oil formed by treating nitranisole with an alcoholic solution of sulphohydride of ammonia, evaporating the alcoholic solution at a gentle heat to $\frac{1}{4}$, and adding chlorohydric acid, and filtering. By distilling the muriate with potash, anisidine passes over.

Anisoine. $C_{20}H_{12}O_2$. White, solid, without odour, insoluble in water, little soluble in alcohol, soluble and crystallizable in ether; melts above 212°; formed by treating anise stearoptene with $1\frac{1}{2}$ its volume of sulphuric acid; a resin results, which is freed from acid by boiling water, and distilled, when it is converted into anisoine, a solid which distils over with a volatile oil.

Anisole. *Carbolate of Methyle.* $C_{14}H_8O_2$. A colourless aromatic fluid, insoluble in water, soluble in alcohol and ether; boiling at 305°; formed by distilling anisic acid with an excess of lime or barytes. It is anisic acid, with 2 atoms carbonic acid removed. It forms chloranisole, &c.

Anisyle. $C_{16}H_7O_4$. The hypothetic radical of anisic acid, the hypothetic hydride of which would be $C_{16}H_7O_4H$.

Anise Stearoptene. *Anise Camphor.* $C_{10}H_6O$. Crystalline plates, obtained by exposing anise oil to 32°, and pressing in blotting paper to separate it from the fluid elaoptene, and crystallizing it from 90 per cent. alcohol. Sp. grav. at $53\frac{1}{2}$°, 1·044; melted at 77°, ·9849; at 200, ·9256; soluble in 4 alcohol (·806) at 50°; in 0·6 alcohol at 59°; it unites with chlorohydric gas; fusing point 61°; boiling point, 428°; not changed by air, but when kept fluid it is no longer crystallizable. It is identical with the stearopten of fennel and star anise.

Ankerite. *Pyrotamous lime haloid.* $CaOCO_2$ 51·1, $MgOCO_2$ 25·7, $FeOCO_2$ 20·, $MnOCO_2$ 3·. Rhomboids with angles of 106° ·12; differing, therefore, from calc. spar. Spec. grav. 3·080, H 3·5 to 4. White, with tints of gray, red, and brown; foliated, slightly translucent, brittle, lustre pearly. B.B. becomes black magnetic; occurs in Salzburg upon beds in mica slate, and also in Stiria. It is often considered a variety of dolomite or brown spar.

Annealing. (Saxon, œlan, to heat.) *Tempering; recuire* (F.); *anlassen* (G.) When glass and metals, particularly iron or steel, have been heated to a red heat, they are very brittle. Glass drops, for example, made in the form of what are termed Prince Rupert drops, are so brittle, that when touched sharply with a stone, or when a portion is broken off, they fly into a thousand fragments: glass requires, therefore, to be annealed; a process consisting in placing it in a furnace for many hours, and gradually drawing it to a cooler part of the oven. Malleable iron, when it is to be subjected to pressure, requires annealing. Thus, boiler plates, which are drawn out by rollers, are placed for some time in an annealing furnace. Tempering of steel is an analogous process, and consists in heating the metal at various temperatures. Under these circumstances, steel acquires a peculiar tint by oxidization, according to the heat applied. When heated in oil to 430°, the temperature required for razors and scalpels, steel assumes a very pale yellowish tinge; at 460°, the colour is a pale straw tint, and the instrument has the usual temper of penknives, razors, and other fine-edged tools. As the heat rises, the colour becomes successively yellow, brown, red, and purple, to 580°, when the metal possesses an uniform deep blue tinge, as exemplified in watch springs. The blue weakens to a water colour, which is the last step that can be distinguished before the metal becomes red hot. The explanation of the process of annealing will depend upon the theory of heat which may be adopted. According to the material theory, annealing may restore the heat which has been lost in the preparation of the iron. By the immaterial theory, the particles of iron and glass are placed by the high temperatures of preparation in a peculiar condition, opposed to the attraction of cohesion; the cohesive force being restored by the modified application of heat in the annealing or tempering process. The following table (Parkes) gives the metallic mixtures of tin and lead, for baths, for tempering various instruments. The quantity of lead varies, but that of tin is always 4:—

	Lead.	Tin.	Temp.
Lancets, ...	7	4	420°
Razors, ...	8	4	442°
Penknives, ...	8½	4	450°
Do. larger, ...	10	4	470°
Scissors, shears, ...	14	4	490°
Axes, plane irons, ...	19	4	509°
Table knives, ...	30	4	530°
Watch springs, swords,.	48	4	550°
Large springs, augers,...	100	4	558°

Annotto. *Annatto. Arnotto. Onoto. Roucou* (F); *Orlean* (G.) Brown cakes, being the pulp of the seeds of the *Bixa orellana*, a shrub of S. America, but now cultivated in the W. and E. Indies. It is prepared by triturating gently the grains of the bixa under water, so as to rub off the superficial colouring matter without dissolving mucilage; the colouring matter settles, and is called *achiote;* it is then dried in cakes for sale. It has no taste, but a smell of urine, which is said to be added to it; soluble in water, slightly in alcohol, ether with orange colour, in caustic potash with a red colour, also in oil of turpentine; sulphuric acid makes it indigo blue; nitric acid makes it green; it contains a yellow and red-colouring matter. Annotto is sometimes employed for dyeing silk of an orange colour, by immersing the goods in a solution of the dye in potash, or soda, and brightening by means of alum, vinegar, or lime juice. It is extensively used for dyeing cheese and butter.

Anode. (ἀνα, upwards; and ὁδος, a way—the way which the sun rises.) That surface at which the electric current enters, being the *negative* extremity of the decomposing body by the agency of a galvanic battery, and is where oxygen, chlorine, acids, &c. are evolved. The term is founded on the view that in any case of electric decomposition the decomposing body is considered as placed so that the current passing through it shall be in the same direction and parallel to that supposed to exist in the earth from east to west, then the surfaces at which the electricity is passing into and out of the substance would have an invariable reference; that towards the east is the *anode*, and that towards the west the *cathode*.

Anorthite. (ανορθος, without right angles.) *Christianite.* White doubly-oblique prisms; fracture conchoidal, lustre on cleavage planes pearly, otherwise vitreous. H 6, spec. grav. 2·762 to 2·656, B.B. fuses on the edges with great difficulty; with borax and microcosmic salt fuses into a transparent bead, in the latter case leaving a skeleton of silica; with carbonate of soda forms an enamel. It consists of silica 44·49 Al_2O_3 34·46, CaO 15·68, MgO 5·26, Fe_2O_3 ·74, sometimes 2 per cent. of potash. It is closely allied to felspar; occurs at Mount Vesuvius in limestone along with greenish pyroxene.

Anoxydic Bodies, are those whose carbon, when they are charred, yields nothing to solvents, as in blood.

Anserine. $C_{10}H_9O_3$. Yellow oil liquid at common temperatures, but becoming a tallow at 45½°; it is the pigment of the feet and bill of the goose, pigeon's foot, and in the craw fish.

Antalogene. A synonyme of iodine, from its positive relation to chlorine (halogene.)

Anthiarine. The peculiar poison of the upas anthiar, said to consist of $C_{14}H_{15}O_5$. It exists in the resin to the extent of 3·56 per cent.

Anthocyane. The blue colour of flowers soluble in alcohol.

Anthophyllite, Anhydrous. (ανθος, a flower; and φυλλον, a leaf.) *Strelite, Karstine.* Yellow-gray crystals or crystalline fibres often radiating, cleavable parallel to a right rhombic prism with angles of 125° 30′ and 54° 30′; fracture uneven, lustre pearly, inclining to metallic; translucent. H 5 to 5½, spec. grav. 2·94 to 3·1558. B.B. infusible; forms a glass with borax coloured by iron; consists of silica 57·12, FeO 13·52, MgO 25·92, CaO 1·32, HO 1·36. It occurs at Kongsberg in Norway in mica slate; in Greenland and Mecklenburg.

Anthophyllite, Hydrous. H 2·5, spec. grav. 2·911. Greenish-yellow diverging fibres, consisting of a series of plates or very imperfect

ANT

crystals diverging from various centres (scopiform) of which the fibres are separable, but with greater difficulty than asbestus; lustre silvery; opaque; sectile; feel soft; it consists of silica 54·98, MgO 18·376, Fe_2O_3 9·832, MnO 1·20, KO 6·804, Al_2O_3 1·56, HO 11·448. New York; Girvan, Ayrshire.

Anthosiderite, or hydrous tersilicate of iron, (Fe_2O_3 3 SiO_3 HO) from Brazil, consisting of SiO_3 61·36, Al_2O_3 34·66, HO 3·98.

Anthoxanthine. The yellow colour of flowers.

Anthracene. *Paranaphthaline.* $C_{30}H_{12}$. Colourless crystalline grains, fusing at 356°, subliming at 592°, spec. grav. of vapour 6·741, obtained in the rectification of coal tar; it separates by application of cold in grains, not in plates, insoluble in water, slightly in alcohol and ether, very soluble in oil of turpentine. Nitric acid causes oxygen to replace hydrogen, and forms nitrites. Anthracene is distinguished from pyrene by placing them on a bit of glass and heating—anthracene gives a crystalline, pyrene no crystalline fume.

Anthracenese, Binitrite of. $C_{30}H_{10}O_2$, 2 NO_3. Orange-yellow needles or powder without smell, insoluble in water, slightly soluble in boiling alcohol, more easily soluble in ether, soluble in sulphuric acid, and reprecipitated by water.

Anthracenise, Ternitrite of. $C_{30}H_9O_3$, 3 NO_3 3 HO. An orange-red resin, very fusible and soluble in ether. Obtained by the action of excess of nitric acid on anthracene.

Anthracenise, Nitrite of. $C_{30}H_8O_4NO_3$. Long yellow needles or prisms with two acute terminations, somewhat soluble in hot alcohol and ether. Obtained by the action of nitric acid on anthracene.

Anthracenose, Nitrite of. $C_{30}H_8O_4NO_3$ HO. Nearly colourless or yellowish needles, insoluble in water. Obtained by acting on the mother liquors of the preceding bodies by nitric acid. (Laurent, Ann. Chim. 72, 421.)

Anthracenuse. $C_{30}H_7O_5$. White needles, becoming yellow until washed with ether, free from resin, prepared by converting anthracene into nitrite of anthracenose, and gently heating in a basin covered with a capsule, when anthracenuse sublimes.

Anthracite. (ανθραξ, carbo, charcoal.) *Glance coal, Columnar coal, Kilkenny coal, Blind coal, Culm.* Spec. grav. 1·52 to 1·55 (Pennsylvania) 1·75 (Rhode Island) 1·4354 (Kilkenny) 1·4 (Meissner) 1·482 (Schönfeld) H 2·. Regular 6-sided prisms, and acute 8-hedrons (?); distinguished from common coal by its higher specific gravity, its semi-metallic lustre, and by its burning without emitting smoke. Colour, black; lustre splendent and semi-metallic; sometimes beautifully iridescent, as at Lehigh. It occurs in primary rocks, but is most abundant in altered or transition rocks, as at Lehigh, in beds of 12 to 15 feet, up to 30 feet in thickness, alternating with clay slate, mica slate, and sandstone, also in Rhode Island; it is found in Wales in the coal measures. The following table gives the composition of some anthracites:—

	Sable.	Swansea.	Isere.	Vizille.	Pembroke.
C...	87·22	90·58	94·	94·09	94·1
H...	2·49	3·60	1·49	1·85	2·39
N ..	2·31	0·29	0·58	2·85	·87
O...	1·08	3·81	...	...	1·34
Ash,	6·90	1·72	4·	1·90	1·3

It is now used in the hot blast process for iron.

Anthracolite. *Anthraconite. Swinestone, Stinkstone.* A foliated or compact bituminous fœtid variety of limestone, occurring at Andreasberg, in the Harz, Christiania, Andrarum, Neudorf, and Dalmatia.

Anthrakometer. (ανθραξ, carbon; μετρον, measure.) An instrument used for measuring the carbonic acid of the air. See ATMOSPHERE.

Anthranilic Acid. C_{14} H_6 NO_3 HO. Yellowish translucent regular plates with a fine lustre; sublimes like benzoic acid by greater heat; when distilled yields carbonic acid and *aniline*; slightly soluble in water; very soluble in alcohol and ether, the solutions being acid, and having the taste of benzoic acid. Obtained by boiling indigo with caustic potash of spec. grav. 1·35, and diluting with water; before the indigo is entirely dissolved black oxide of manganese is cautiously added, until a portion of the solution taken out no longer deposits indigo. After dilution it is saturated with sulphuric acid, and evaporated to dryness, the anthranilate of potash taken up with alcohol and decomposed by acetic acid; it is further purified by uniting it with lime, and decomposing.

Anthropic Acid. (ανθρωπος, a man.) An acid described as being found in human fat, but which seems to be an impure mixture.

Anthrazothionic Acid. A synonyme of hydrosulphocyanic acid.

Antichlore. (αντι, against, chlorine.) The commercial name of sulphite of soda, when used to remove the last traces of chlorine from paper bleached with chloride of lime.

Antiedrite. A synonyme of Edingtonite.

Antigorite. A variety of diallage, belonging to the regular system and optically binaxial. Specific gravity 2·622; H 2·5. Greenish-black dichromatic slate; fuses B.B. on the edges to a brown slag; becomes white by ignition, soluble in chlorohydric acid leaving silica. As found in Antigorio Valley, in Piedmont, it consists of silica 46·2, MgO 34·79, FeO 12·86, Al_2O_3 1·98, HO 3·70.

Antimiasmatic, or *Disinfectant.* A substance employed to destroy the cause of infectious disease.

Antimonphyllite, or Teroxide of Antimony. SbO_3. Occurring native in veins traversing primary or graywacké rocks. Spec. grav. 5·566; H 2·5 to 3; 6-sided white rhombic prisms, with a shade of yellow; found in Bohemia, Saxony,

Dauphiny, Baden, Nassau, Hungary. Volatilizes B.B. in a white vapour, and fuses when simply placed in the flame of a candle.

Antimony. The conditions in which antimony is found in nature are as follows:—1. Native antimony, in obtuse rhomboids, spec. grav. 6·72. 2. Arsenide of antimony, $SbAs_3$, spec. grav. 6·13; reddish-gray. 3. Teroxide of antimony, white antimony. 4. Sesquisulphide of antimony, black antimony of the shops, SbS_3. 5. Red antimony, SbO_{32} SbS_3, in capillary crystals. Besides, antimony occurs associated in greater or smaller proportions with many metals, as in antimonial copper, antimonial nickel, antimonial silver.

Characters.—Sb 16·128. Colour, silver-white, with a good deal of lustre; crystal, acute rhombohedron; spec. grav. 6·702 (Brisson), 6·86 (Bergman), 6·712 (Hatchett), 6·715 (Marchand and Scheerer), 6·700 (Karsten). Fusing point 810° (Dr. Cromwell Mortimer, Phil. Trans. 1747, 672), 809½° (Dalton), 955° (Morveau). Texture fibrous or foliated; brittle and easily pulverized. When heated white hot, by the blowpipe, and thrown on the table, it burns and smokes, yielding an oxide. Bismuth and tin resemble it in this respect; spec. heat ·05077.

Preparation.—1. Antimony may be reduced from the sulphide by first freeing it from its matrix by fusion in a pot (R) supplied with an aperture at the bottom through which the fused mass flows into a vessel of the same kind (P) placed beneath it. The fused sulphide is then roasted in

a reverberatory furnace to convert it into oxide. This oxide is reduced to the metallic state by mixing 6½ parts with 1 charcoal powder moistened with a solution of carbonate of soda, and heated in a crucible till the deoxidation is completed. 2. When 10 tersulphide, 4¼ iron filings, 1 anhydrous sulphate of soda, and ¼ charcoal are ignited in a crucible, 6 to 6·4 parts metallic antimony are obtained (Liebig, Handwörterbuch.) The melted mass is poured out of the crucible and cooled. It contains some iron, and is the regulus antimonii martialis. It may be further purified by fusing 2 or 3 times with 2, 1½, and 1 part respectively of anhydrous carbonate of soda. 3. The sulphide of antimony when fused with cyanide of potassium yields metallic antimony. 4. Yellow prussiate of potash yields a similar result. (R.D.T.) To free antimony from sulphur, arsenic, iron, and copper, 8 parts antimony in rough powder, 1 part anhydrous carbonate of soda, ½ part black sulphide are fused in a clay crucible for an hour; on cooling, the metal is again fused with ½ part carbonate of soda; 7½ parts of pure antimony are thus obtained, (Liebig's Ann. 19, 22.) When pure antimony is deflagrated with nitre it should leave a white residue. When treated with nitric acid the solution should not be precipitated brown or blue by ferrocyanide of potassium, otherwise it contains copper or iron.

Teroxide. Flowers of Antimony. Protoxide.—SbO_3 19, 153. Spec. grav. 5·56 (Mohs.) Right prisms when native, in needles or regular 8-hedrons when sublimed; or by dissolving teroxide in caustic soda, 8-hedrons separate on cooling in close vessels. Colour white, lustre silky like asbestus, becomes yellow by heat and white on cooling; fuses at a red heat and volatilizes, the fumes condensing into needles. Specific gravity, 5·778. It distils in close vessels when heated, and does not decompose. Ignited in air it glows like tinder, and becomes antimonious acid. It acts as an oxide in tartar emetic, but it performs the part of an acid to soda and potash.

Preparation.—Dissolve antimony in nitrochlorohydric acid, and dilute the solution with water; a white precipitate falls (*algaroth powder*) consisting of teroxide of antimony, with some portion of terchloride ($5SbO_3, SbCl_3$, or $Sb_2\ ClO_2$). Wash the precipitate with water and boil it with a solution of an alkaline carbonate to remove chlorine. Wash again and dry it (Proust. Journ. de Phys. 55, 328). The sulphide may also be used to supply the terchloride which may be decomposed and purified as above.

Glass of Antimony consists principally of teroxide. It is procured by roasting the tersulphide until as much as possible of the sulphur is volatilized. It contains a portion of sulphur. This glass when pulverized is sometimes used in the preparation of tartar emetic by boiling it with cream of tartar. It has a brownish-yellow colour, some transparency, a glassy fracture, tasteless. *Glass of Antimony* consists of 8 teroxide, 1 tersulphide. *Crocus* is 8 teroxide, 2 tersulphide. When the amount of sulphur is considerable it is termed *Liver of Antimony*, consisting of 8 teroxide, 4 sulphide.

Teroxide Salts.—They are precipitated by caustic ammonia, soda, and potash, redissolved by the last two. Sulphohydric acid gives an orange precipitate in an acid solution. Sulphohydride of ammonia gives an orange precipitate soluble in excess. They are generally emetics; lose acid by ignition when it is volatile; colourless or yellow. When boiled with copper and chlorohydric acid, antimony is deposited from its solutions.

Sulphate of Antimony.—When antimony is heated in strong sulphuric acid, sulphurous acid is evolved, sulphur sublimes, and a white neutral salt results; soluble in dilute sulphuric acid; on the addition of water, a supersulphate dissolves, while a white subsulphate is left, composed of 17·99 SO_3 82·01 SbO_3.

Preparation of Tartar Emetic.—The process consists of three stages. 1. The conversion of tersulphide into terchloride of antimony by boiling with chlorohydric acid. SbS_3, 3 HCl become $SbCl_3$, 3 SH evolved. The black sulphide is introduced into a flask, is boiled with the acid until the black powder disappears, and the odour of sulphohydric acid is no longer evolved. The solution when cool is then filtered through paper or through calico, when it is ready for the second stage. 2. The second stage consists in decomposing the terchloride into teroxide by pouring it into water. ($SbCl_3$ and 3 HO become SbO_3 and 3 HCl.) The precipitate which falls in flocks is the powder of algaroth (consisting of 5 SbO_3, $SbCl_3$.) When thrown on a filter and washed free from acid with cold water it is a white powder, but when it stands in contact with water for some days it is converted into needles. When dried at 212° it is ready for the third operation. 3d stage. Union of teroxide of antimony with the bibasic tartrate of potash, or cream of tartar. This salt consists of tartaric acid united to 2 atoms of base, one of which is potash, the other water. When 19 grains of teroxide of antimony are boiled with 23·625 of cream of tartar the teroxide dissolves, and on evaporation tartar emetic crystallizes out in tetrahedrons, belonging to 8-hedrons with a rhombic base.

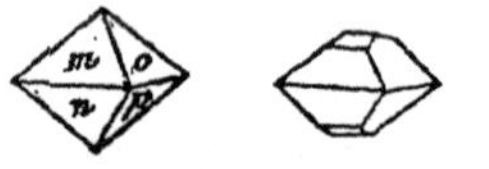

(SbO_3 and KOHO $C_8H_4O_{10}$ become $KOSbO_3$ $C_8H_4O_{10}$ 2 HO.) Instead of the teroxide, prepared as above, we may employ glass, liver, or crocus of antimony, which are impure forms of oxide. The crystals when first formed are clear, but they gradually become opaque by losing 1 atom of water (?) A series of emetics may be formed by replacing the basic water in cream of tartar by a metallic oxide. When tartar emetic is dried at 212° it loses 2 atoms of water of crystallization, and becomes anhydrous. When heated to 428° it loses the elements of 2 atoms of water, and becomes KO SbO_3 $C_8H_2O_8$, while the other emetics undergo a similar change. See EMETICS.

Hydrous Teroxide.—SbO_3 2 HO 21·375, 171. Dissolve freshly precipitated tersulphide in caustic soda, boil; add a solution of sulphate of copper, till a portion of the solution, mixed with acids, gives a pure white. It is filtered from the sulphide of copper—filtered and washed (Fresenius).

Antimonious Acid.—SbO_4 or (SbO_3 SbO_5) 20·; 161. Specific gravity, 6·695. Fine white powder, becoming yellow when heated, not decomposed by ignition but volatilized before the blowpipe; acid reaction; is reduced to the metallic state by ignition with potassium or sodium, likewise by charcoal, cyanide of potassium, ferrocyanide of potassium. *Process.*—This acid is obtained by oxidizing antimony by nitric acid, evaporating to dryness, and igniting the antimonic acid until it is changed into antimonious acid.

Antimonites are colourless, and are decomposed by nitric acid; some of these salts glow when ignited, and are then very insoluble in acids; but this is not the case with the alkaline salts.

Antimonic Acid.—SbO_5 21; 169. Spec. grav. 6·525 (Boullay). Straw-yellow powder, tasteless and insoluble in water, becoming of a deeper yellow when repeatedly heated; reaction acid. When ignited oxygen is given out and antimonious acid remains. When fused with some metallic sulphides, it is changed into teroxide, sulphurous acid being evolved. *Process.*—Dissolve antimony in aqua regia, and evaporate until all the nitric acid is removed—a red heat will expel an atom of oxygen; it is also formed by igniting the metal with nitre or oxide of mercury, or by dropping quintochloride of antimony into water. It decomposes the alkaline carbonates by fusion, but not in solution; soluble in caustic potash, from which acids precipitate the following:—

Hydrous Antimonic Acid.—When antimony is fused with 4 parts of nitre and washed with water, antimoniate of potash dissolves, and biantimoniate remains; from both of which the hydrate is obtained by means of nitric acid. The hydrate contains an atom of water, and is a fine white powder. It is sparingly soluble in water, and then precipitated by sulphohydric acid. It is slightly soluble in chlorohydric and tartaric acids.

Antimoniates.—Antimonic acid replaces carbonic acid, when fused with alkaline carbonates. It unites with 1, 1½, and 2 atoms of base, and also 1 base unites with 2 acid; colourless salts; decomposed by feebler acids; the acid then falls as a hydrate, which is soluble in chlorohydric and tartaric acids.

Bimeta-antimoniate of Potash. Bibasic Antimoniate.—KO SbO_5 7 HO. *Granular Antimoniate of Potash.* White, crystalline. Very soluble in cold water, more soluble in hot. When boiled with water it is changed into neutral antimoniate. It is this salt which is a valuable reagent for detecting soda. *Process.*—Fuse 1 part of metallic antimony with 4 nitre, digest in warm water to remove nitrite and nitrate of potash, when neutral antimoniate of potash remains. This is boiled for about an hour until it is all dissolved. It is filtered to separate the biantimoniate which remains undissolved. The filtered liquid is evaporated in a silver or platinum capsule to the consistence of syrup, till the gummy antimoniate begins to deposit. A few pieces of caustic potash are now added, to change the antimoniate into the meta or bibasic antimoniate, and the evaporation is continued until a portion, placed on a plate of glass, begins to crystallize. The capsule is withdrawn from the heat; a crystalline deposit soon occurs—a mixture of *meta-antimoniate and of bimeta-antimoniate of potash.* The alkaline liquor is poured off, and the salt dried in blotting paper, or on a porous tile. *To use this reagent.*—Place about 70 grains in a test-glass, pour upon it about 150 grains of cold water, in order to dissolve the excess of potash which the salt may contain, and to decompose the neutral meta-antimoniate into the acid salt, which is only slightly soluble in cold water. The liquid is poured off, and the bimeta-antimoniate is washed repeatedly. The last washings should be made as rapidly as possible, to dissolve as little as possible of the bimeta-antimoniate of potash. When we are satisfied that the excess of potash has been removed, we leave the acid salt in contact with water for a few seconds; we filter the liquor, which is always somewhat turbid, and preserve it for detecting minute quantities of soda, ($\frac{1}{300}$ even.) If the bimeta-antimoniate of soda does not fall immediately it is because the solution is alkaline, but it falls in the course of a short period.

Terchloride of Antimony. Butter of Antimony.—Sb_4Cl_3 29·5. This compound may be prepared by various methods:—1. By distilling 3 parts of corrosive sublimate with 1 part of antimony in fine powder, or 7 parts of sublimate with 3 parts of glass of antimony. 2. When 1 part of tersulphide of antimony is dissolved in 5 parts of fuming hydrochloric acid, with the assistance of a gentle heat, a solution of the antimony is obtained. This is to be allowed to clear by standing, and is then evaporated in a porcelain basin, until a drop, taken up by a glass rod, deposits crystals, when placed on a cold plate of glass. The whole is transferred into a small retort and distilled. When a drop becomes solid on a piece of metal, pure chloride is present. The evaporation may be conducted entirely in the porcelain vessel, with the loss, however, of a small portion by volatilization. Instead of the sulphide the metal may be employed. 3. The older methods consisted in distilling 6 parts of common salt, 5 parts of oxide of antimony, 7 parts sulphuric acid, and 3½ water. Butter of antimony is a grayish solid, crystallizing in octahedrons. It fuses at 212°, and possesses the consistence of an oil; it boils at 356°. By exposure to the air, it first smokes, attracts moisture, and gradually deposits the powder of algaroth. When heated with nitric acid it is converted into antimonic acid. The anhydrous chloride absorbs ammoniacal gas; the fluid chloride is decomposed by it. It is soluble in a warm solution of common salt, and on cooling it separates in large crystals.

Tersulphide of Antimony. Crude Antimony.—SbS_3. Occurs native, but is sold under the form of crude antimony of a conical shape. To procure it in this form, the ore is pounded, fused in a crucible, with a hole in the bottom, which is placed over another of the same size. *Kermes* is a brownish-red oxysulphide, possessing some appearance of a crystalline structure. It leaves a red stain on paper; destitute of taste and smell. In cold water it is quite insoluble, but in boiling water it gives out SH; scarcely soluble in caustic ammonia. It is a good deal employed on the continent, and has powerful emetic properties. It must not be administered along with acids, as they decompose it. It is used also as a paint.

Quintosulphide or Golden Sulphuret. Persulphuret. Sulphur auratum.—Sb S_5. Was known to Basil Valentine in the 15th century. It may be obtained by boiling equal parts of sulphur and crude antimony in caustic potash, or soda, till a clear solution is obtained. When sulphuric acid is added to the solution the golden sulphuret falls, which is to be filtered and washed with cold water. It can also be obtained by dissolving the crystallized soda sulphuret of antimony in 3 times its weight of water, filtering and precipitating with sulphuric acid. Any soluble sulphide of antimony will serve for its precipitation. The hydrate of this sulphide is obtained with a fine orange colour when a current of SH is passed through a solution of the oxide of antimony. It is an orange powder, becoming brown, almost destitute of taste and smell, insoluble in cold water; caustic ammonia, potash, and soda, dissolve it. When ignited in close crucibles sulphur is evolved, and the black sulphuret remains. It acts less powerfully than the preceding as a medicine.

The Washed White Oxide of Antimony is prepared in various ways. A simple method is to mix 4 parts of tersulphide of antimony with 10 parts of nitre, and add them gradually to a red hot crucible. The fused mass consists of sulphate, nitrate, antimonite, and antimoniate of potash. By washing, the sulphate, nitrate, and subantimoniate of potash are removed, while the insoluble acid antimoniate remains. It is seldom now used in medicine, but still stands in the Prussian Pharmacopoeia.

The Hepar Antimonii, or Liver of Antimony, is a mixture of a compound consisting of tersulphide of antimony united to sulphide of potassium, with oxide of antimony and antimonite of potash. The processes for its preparation are very numerous; the products are various.

The Crocus of Antimony is a dark yellow or rust-coloured powder consisting of one part oxide of antimony, 2 parts sulphide of antimony, and variable quantities of potash and oxide of antimony. It is prepared by deflagrating equal parts of sulphide of antimony and nitre, and washing the residue till the washings are tasteless.

Red Antimony.—$SbO_3 + 2\ SbS_3$. Found in Saxony, in veins; occurs in capillary crystals, their primitive form being a right square prism. Spec. grav. 4·09—4·6. Colour cherry-red, fracture foliated.

Separation of Arsenic and Antimony.—When the metals occur together in the form of a metallic alloy, the arsenic may be volatilized by passing a current of hydrogen gas over the alloy, with the application of heat. This should be moderate, otherwise the antimony will also be driven over. By weighing the alloy and tube containing it before and after the experiment, the loss measures the amount of arsenic; or a simpler method is to dissolve the alloy in nitric acid and add water; the teroxide of antimony falls and the arsenious acid remains in solution. It is necessary to add a few drops of caustic ammonia to precipitate a small portion of antimony which still remains in solution.

Estimation of Antimony.—The antimony is precipitated from its solutions by sulphuretted hydrogen; the resulting compound is the tersulphide ($Sb\ S_3$), which is thrown on a weighed filter, washed, and dried at 212°. When weighed, every 22 grains indicate 16 of antimony, or 14 of the teroxide. This may be further examined by converting the sulphur into SO_3 by NO_5.

Antimonyle. (SbO_2). A theoretical mode of viewing the base of teroxide of antimony, which would be an oxide of antimonyle ($SbO_2\ O$).

Antiphlogistic Theory. (αντι, opposed to, and *phlogiston.*) The phlogistic theory of Stahl considered oxides of metals as simple bodies, and the metals as compounds of the oxide with a hypothetical substance, phlogiston. Lavoisier started the antiphlogistic theory, now in use, which considers the metals as simple, and the oxides compounds of metals and oxygen. In medicine, antiphlogistic remedies (αντι, against, and φλογιζω, I burn) are those which are used to prevent inflammation.

Antiseptic. (αντι, against, and σηπτος, putrid.) Substances which prevent putrefaction or decomposition, as the mineral acids, charcoal, chloride of lime, chloride of soda, chlorine, creasote, yeast.

Antrimolite, (from county Antrim.) Stalactitic masses surrounding a nucleus of calcareous spar, adhering to the summit of cavities in an amygdaloidal rock at Bengane, four miles east from the Giants' Causeway, in Antrim. H 3·75; spec. grav. 2·0964. Colour chalk-white; texture fine silky, fibrous, radiated; opaque, dull. B.B. gives out water, reddening vegetable blues, and containing chlorohydric acid; does not froth, but softens into an enamel. With microcosmic salt it becomes a transparent colourless glass; gelatinizes in muriatic acid. It consists of silica 43·47, Al_2O_3 30·26, CaO 7·5, KO 4·10, FeO ·19, Cl ·098, HO 15·32. It belongs to the zeolite class, allied to mesolite.

Apatelite. A persulphate of iron, consisting of SO_3 42·9, $Fe_2\ O_3$ 53·3, HO 3·96.

Apatite. *Phosphate of Lime, Asparagus Stone, Moroxite, Phosphorite, Chrysolite* of the jewellers, &c. See PHOSPHATE OF LIME.

Apatoide. Small yellow half translucent grains, found in meteorites from Alais and Weston. H 5·5; B.B. fuses on charcoal, and becomes black; dissolves in borax, forming a greenish-yellow glass: contains no phosphoric acid.

Aphanese. *Triangular prismatic arseniate of copper.* 6 CuO AsO_5 PO_5 3 HO; contains AsO_5 27·08, PO_5 1·5, CuO 62·8, $Fe_2\ O_3$ ·49, HO 7·57. It appears the same as olivenite.

Aphanite. Sp. grav. 2·968. Dark green rock, having a felspar basis, found in granite at St. Bresson. Consists of silica 46·83, $Al_2\ O_3$ and $Fe_2\ O_3$ 30·33, CaO 9·55, MgO 6·86, NaO 3·57, KO ·81.

Apherese. A synonyme of tetraphosphate of copper.

Aphlogistic Lamp. *Flameless lamp* of Davy. If very thin platinum wire be formed into a coil by winding it round a thin glass rod, and be then adjusted so as to enclose the cotton wick of a lamp, containing ether or pyroxylic spirit; if the wick be ignited and blown out, the wire will continue red hot as long as the ether lasts. The products of this action are acetic, formic, and aldehydic (lampic), acids possessing a very acrid odour. It is an example of low combustion, and is the same as that exhibited when a platinum crucible is heated over gas; when the gas is extinguished, the crucible remains red hot, if the gas be allowed to stream upon it.

Aphrite. *Pearly carbonate of lime, Slate spar.* Found in the primary mountains, in layers, at Konigsberg, Cornwall, &c.

Aphrizite. *Common schorl, black tourmaline.*

Aphrodite. (αφρος, scum.) Allied to magnesite, or meerschaum; found at Faberg. Consisting of silica 51·6, MgO 33·7, HO 12·3, $Al_2\ O_3$ ·2, MnO 1·6, FeO ·6 (Belin). It is hydrous silicate of magnesia 4 (3 MgO, 2 SiO_3,) 3 HO).

Aphronitrum. The ancient name for the efflorescence on walls, consisting of sulphate of magnesia, sulphate and carbonate of soda.

Aphthalose and **Aphthitalite,** or sulphate of potash.

Aphthonite. (αφθονος.) So called from its richness in silver, found in Wermland, at Wermskog; resembles gray copper. Spec. grav. 4·87. Steel-gray, massive. H = calc. spar; brittle. B.B. fuses. It contains Ag 3·09, Cu 32·91, Zn 6·4, Fe 1·31, Sb 24·77, S 30·04, Pb ·04, Co ·49, As trace. Gangue 1·29.

Apiine. $C_{24}\ H_{14}\ O_{13}$? An uncrystallizable alkaloid from common parsley (*Apium petroselinum*), by boiling parsley with water, filtering and evaporating, when it becomes a jelly, like pectic acid; on exposure to the air it dries without change. With protosulphate of iron it gives a blood-red colour, which is therefore a delicate test for it; soluble in boiling alcohol; not altered by alkalies; acids prevent it from becoming a jelly, and some sugar is formed. (Braconnot.)

Apirine. White, tasteless insoluble powder, like starch, said to be an alkaloid, and to exist in the kernel of the *Cocos lapidea*. (Bizio.)

Aplome. *Ferrocalcareous garnet;* usually referred to green garnet; but, according to Scheerer, belonging to the same class with melanite and rothoffite.

Apocrenic Acid. $C_{14}\ H_7\ O_3\ N1\frac{1}{2}$, or C 62·57, H 4·8, O 17·63, N 15· (?) The result of the action of the air on crenic acid; obtained from natural waters; a brown extract, little soluble in water, but easily soluble when mixed with crenic acid; more soluble in alcohol; precipitated from its aqueous solution by salammoniac; forms amorphous salts.

Apophyllic Acid. $C_{16}\ H_7\ NO_8$. Right square prisms, resembling those of apophyllite; obtained by acting on cotarnine with nitric acid; when distilled it yields an oil, quinoline isomeric, with aniline.

Apophyllite. (απο, from; φυλλον, a leaf.) *Albine*, *Fishy-stone*, *Ichthyophthalmite*, *Tesselite.* Right square prisms, terminating in a 4-sided pyramid. White pearly lustre; brittle. H 4·25. Spec. grav. 2·335 to 2·359. B.B. exfoliates, and fuses into a blebby glass; decomposed by acids. It consists of SiO_3 51·, CaO 26·23, KO 5·88, HO 16·5, F trace. *Form.*—KO 2 SiO_3, 8 (CaO SiO_3), 16HO. Found in trap rocks in Skye, Bannat, Auvergne, Bohemia.

Aporetine. Resin ψ of rhubarb. $C_{16}\ H_8\ O_7$ = C 58·89, H 4·35, O 36·76.

Aposepidine. *Oxide of Caseine*, *Caseic acid.* (Prout.) Crystalline shining scales, insoluble in cold, soluble in 15 parts hot water; little soluble in alcohol and ether; easily soluble in alkalies. Obtained by allowing gluten to remain moistened till it decomposes; diluting with water, filtering, and evaporating. A crystalline mass remains, with the characters of decaying cheese. Mixed with alcohol, the greater part dissolves, a white powder separates, which is purified by washing with water, solution in water, heating with animal charcoal, and crystallization or precipitation of the concentrated solution with alcohol.

Apotheme. (ἀπὸ, from θεμα, a deposit), *Brown extractive.* Brown mass, slightly soluble in water, giving a yellowish colour to the solution, but again deposited by evaporation; more soluble in boiling water, the excess being deposited on cooling; still more soluble in hot alcohol. Caustic, and carbonated alkalies dissolve it, from which it is precipitated by acids; reddens then litmus; obtained by evaporating the infusion or juice of a plant to a thick consistence, and then adding water; apotheme remains undissolved.

Apparatus. (*parare*, to prepare.) The vessels arranged to exhibit the phenomena of chemistry.

Apples. The fruit of the pyrus malus. Their composition is similar to that of the pear or pyrus communis, which is as follows, when fresh:—Chlorophylle, ·08; sugar, 6·45; gum, 3·17; woody matter, 3·8; albumen, ·08; malic acid, ·11; lime, ·03; water, 86·28. (Berard.) The constituents of peaches and apricots are analogous.

Apyre. (α πυρ, permanent in fire.) A peculiar alkali, found in the urine and calculi by Brugnatelli; but, according to Döbereiner, it is phosphate of magnesia and phosphate of lime.

Apyrite, or **Apirite.** *Rubellite*, *Sieberite*, *Red tourmaline.*

Aqua. The Latin term for water; its formulæ in chemistry are *Aq.* and HO, the latter being often used for basic, and the former for crystalline water. A body containing water is said to be *hydrous*, and the compound with water is called a *hydrate.*

Aqua Binelli. So called from its proposer of Piedmont, and was at one time much lauded on the continent as a styptic. It is probably a preparation of animal or tar oils, and has been supposed to owe any virtue it possesses to the presence of creasote or carbolic acid.

Aquafortis. (Strong water.) The commercial name of nitric acid.

Aqua Marine. A sea-green variety of emerald or beryl, much used in brooches.

Aqua Regis or **Regia.** (Royal water.) *Nitro-chlorohydric acid*, *nitro muriatic* acid. A mixture of 1 part nitric and 3 to 4 parts chlorohydric acid.

Aqua Reginæ. (Queen's water.) A mixture of concentrated sulphuric acid, nitric acid or oil of vitriol, with $\frac{1}{8}$ to $\frac{1}{10}$ of its weight of nitre. (Keir.)

Aqua Vitæ. A name applied to whisky.

Aquila Alba, *Mitigata*, *Cœlestis*, *Mercurii*, old names for dichloride of mercury, or calomel.

Arabine. $C_{12}\ H_{11}\ O_{11}$. That portion of gum Arabic soluble in cold water, is so called; colourless and tasteless, without smell; between 282° and 392°, softens, and may be drawn into threads; becomes acid in a moist atmosphere; insoluble in alcohol, and precipitated from its aqueous solution in flakes; converted by boiling with sulphuric acid into sugar, but does not ferment. With nitric acid it is converted into mucic or saclactic acid (100 Arabine and 400 acid give 17 mucic acid), precipitated by sesquichloride of iron. It consists of C 42·11, H 6·43, O 51·46. A cold aqueous solution, containing more than 17¾ Arabine in 100 water, will not filter, nor will it filter at 212°, if it contains more than 23·54 per cent. This solution is known by the name of mucilage, and is employed in pharmacy to make cough mixtures, and in calico printing to thicken colours and mordants. Arabine is largely contained in gum Arabic, gum Senegal, and it is likewise present in mucilage of lint seed.

Arabic Gum. Small rounded drops or tears, either colourless or with a tinge of yellow, obtained from Acacia tortilis, A. sejal, A. Ehrenbergi, vera, Arabica, Senegal, and brought from Levant, Senegal, India. Spec. grav. 1·355; fracture vitreous; hard, and easily pounded; dried in vacuo at 257°; loses 17·6 water, and leaves 3 per cent. ash, consisting of carbonates of potash and lime, phosphate of lime, chloride of potassium, &c.

Araeoxene. *Vanadiate of Lead and Zinc*, with some arsenic and phosphoric acids. Red radiated crystals, containing 48·7 PbO and 16·32 ZnO. Found at Dahn in the Palatinate.

Arbol-a-brea. A greenish-gray resin, from the Canarium album of Manilla. 61 per cent. is easily soluble in alcohol.

Arbor Dianæ. *Tree of Diana.* Metallic silver deposited by the influence of mercury in the form of a tree. The experiment is made by dissolving 15 grains of nitrate of silver in half a wine glassful of water, adding a few globules of mercury, and allowing the glass to stand at rest for some hours.

Arbor Saturni. *Saturn's tree*, is formed by hanging a plate of zinc in a solution of acetate of lead.

Arbutine. $C_{32}H_{22}O_{19}$. Colourless bitter needles, soluble in water, alcohol, and ether; reaction neutral; not precipitated by salts of iron and lead; by the action of emulsine it is converted into *Arctuvine* ($C_{20}H_{10}O_7$) and sugar ($C_{12}H_{12}O_{12}$); obtained from the leaves of the Arctostaphylos uva ursi by the same process as for bitter principles.

Archil. See CUDBEAR, LITMUS, LICHENS, ORCHIL.

Arctizite. A synonyme of meionite, scapolite, or wernerite.

Arctuvine. $C_{20}H_{10}O_7$. Colourless 4-sided prisms, obtained by treating the brown residue formed by the action of emulsine on arbutine with ether and alcohol.

Arctuveine. $C_{20}H_{10}N_3O_{20}$. A black substance produced by the action of ammonia on arctuvine.

Arendalite. *Acanticonite.* The large crystals of epidote, found at Arendal in Norway, have been so termed.

Areometer. (*αραιος*, light; *μετρον*, a measure.) The continental term for the British *hydrometer*, which see.

Arfvedsonite. *Peritomous augite spar*, *Ferruginous hornblende.* Oblique 4-sided prisms (?) the angles of the prism being 123° 55′ and 56° 5′, while those of amphibole are 124° 30′ and 55° 30′. Spec. grav. 3·369, H 4·5. Lustre resinous; opaque; colour pure black. The prism splits parallel to its longitudinal faces, but no cleavage can be observed parallel to the base; we are therefore ignorant whether the prism be right or oblique. It consists of SiO_3 50·508, Al_2, O_3 2·488, $Fe_2 O_3$ 35·144, $Mn_2 O_3$ 8·920, CaO 1·56, HO ·96. B.B. fuses per se into a black globule; with borax and salt of phosphorus into a glass coloured by iron. Brought from Kargardluarduk by Giesecké.

Argand Lamp. Invented by Argand in 1789, with a circular wick to admit a double current of air; in common use.

Argal or **Argols.** The commercial name of impure cream of tartar, or acid tartrate of potash, as it is taken in the form of an incrustation from the interior of wine casks. It has usually a red tinge, due to the colouring matter of the grape.

Argentane. *German silver*, *Packfong*, *British plate.* An alloy consisting of copper, nickel, and zinc, in various proportions, in imitation of the Chinese silver or packfong, which has been found to contain Cu 43·8, Ni 15·6, Zn 40·6 (Engström, 1776), and Cu 40·4, Ni 31·6, Zn 25·4, Fe 2·6 (Fyfe, 1821), with a spec. grav. of 8·432. Each maker seems to have his own receipt for imitating this preparation; the maillechort of Paris, with a spec. grav. of 7·18, consists of Cu 65·, Ni 16·8, Zn 13·, Fe 3·4. I have found the composition of some used in Glasgow Cu 68·69, Ni 24·28, Zn 9·89. It is now extensively employed as a basis for plating with silver; forks and spoons, and articles of domestic use, being made of this material, and electro-plated. Before being dipped in the plating trough, they are usually immersed in nitric acid.

Argentarium. An alloy, consisting of equal parts of lead and tin.

Argentine. *Slate spar*, in thin plates, in Norway, Glentilt, &c. A variety of calcareous spar, with a pearly lustre: also applied to oxide of antimony.

Argentvive. (Living silver.) The alchemical name of mercury.

Argil. The old name of alumina or clay.

Argillaceous. Containing clay or alumina; applied to slates.

Argillite. A synonyme of clay slate.

Argyrithrose. *Argento sulphide of antimony*, *Red silver ore.*

Argyrose. *Sulphide of silver ore.*

Aricine. *Cusco cinchonine.* $C_{20} H_{24} N O_3$. Obtained from arica bark, in the same way as cinchonine, from Peruvian bark; it is closely allied to this alkaloid, as it differs from it by containing 2 atoms more oxygen, and 1 more than in quinine; it crystallizes in white brilliant translucent needles, with a slightly bitter taste. The salts of aricine are very bitter: soluble in water and alcohol, insoluble in ether.

Arkansite. A variety of brookite or titanic acid, found in Arkansas, United States.

Arkose. The felspathic sandstone of Poivin, in the Vosges, formed by the assimilation of hyalite and felspar, resting on granite.

Armenian Bole. See BOLE.

Armenite, or blue carbonate of copper.

Armenium. Probably a yellow or orange ochre.

Arnicine. An acrid resin, soluble in alcohol and ether. Similar to cytisin; derived from the flowers of arnica montana.

Arnotto. See ANNOTTO.

Aroma. (αρωμα, spice.) The cause of the peculiar odour of vegetables, due to volatile oils.

Arquerite. An ore from Arqueros, in Chili, consisting of silver, 86·5, mercury, 13·5.

Arrack. (*Alrak.*) A peculiar flavoured spirit, obtained by the fermentation of rice, in Bengal.

Arragonite. *Igloite, Flosferri, Needle-spar, Arragon spar.* CaO CO_2. Spec. grav. 2·90 to 2·9467, H 3·75. Right rhomboidal prisms, with angles of 116° 10; three of the crystals being frequently grouped together so as to form a 6-sided prism; the two acute angles of the prism are sometimes replaced by two faces, which conceal the base, and convert the crystal into an elongated 8-hedron; sometimes the edges are replaced by tangent planes, so that the crystal terminates in a long 6-sided pyramid, the extremity of which is not a point, but a ridge; colour white, yellow, gray, green; lustre glassy; translucent or transparent; refracts doubly, but not so much as calcareous spar. Arragonite differs from calcareous spar in its hardness, slight deviation in its angles, in containing a trace of water, and in possessing, according to Stromeyer, some carbonate of strontian. CaO CO_2 98·, SrO CO_2 1·014, Fe_2 O_3 ·14, HO ·21. It is found in Arragon in Spain, Leadhills, Bohemia.

Arrow Root. The starch of the tubers of the root of *Maranta arundinacea* and *M. Indica.* It is prepared in the West Indies and in the Pacific, by bruising the tubers in a mortar, mixing with water, and passing through a cotton filter, or hair sieve; the fibrous matter remains on the filter, and the starch passes through, which is allowed to settle in a tub, the water poured off, and the arrow root dried in the sun. Notwithstanding the process which is employed—the same as that for the preparation of starch—I find that arrow root contains a small quantity of albuminous matter; the result of analysis being—starch, 84·29; albuminous matter, 3·21; water, 12·50, where the albumen is to the starch as 1 to 25, instead of as 1 to 5, as it ought to be in nutritious food. From this reason, and from the circumstance that the soluble salts required for the blood have been removed by washing, arrow root is by no means to be recommended as an article of food by itself, but should be placed on a parallel precisely with wheat, or potato starch. It is distinguished under the microscope from potato starch by the smaller size of its grains. See STARCH. When dried at 212°, it consists, according to Prout, of C 42·8, HO 57·2.

Arsa. *Koumiss.* A spirit prepared by the Calmucks from mare's milk.

Arsenic. *Metallic Arsenic.* As 9·375, or 75. Spec. grav. 5·763 (Lavoisier), 5·672 (Herapath), 5·7 (Berzelius), 5·959 (Guibourt), 5·621 (Karsten). Colour, bluish-white, not unlike that of steel, with a good deal of brilliancy. Crystallizes in acute rhombohedrons. No smell when cold, but when heated it emits a strong odour of garlic, whether due to the metal or arsenious acid, is doubtful. It is soft and brittle, being easily pounded. Its melting point is not known, as when heated to 365° (Bergman) above 561° (Mitchell) it sublimes. It occurs in two modified forms; as a crystalline gray metal, when it is condensed on a cold surface, and as a white metal with a strongly metallic lustre. *Preparation.*—Metallic arsenic may be prepared by triturating arsenious acid, or white arsenic of commerce, with black flux (an intimate mixture of charcoal and carbonate of potash, obtained by calcining bitartrate of potash) and subliming in a tube (Brandt). It may also be obtained by igniting arsenical pyrites in close tubes; arsenic sublimes and sulphuret of iron remains, or by heating 5 parts of glassy acid with 1 of charcoal powder, in a close vessel.

Arsenious Acid. White Arsenic. Flowers of Arsenic.—AsO_3 12·375; 99. 1. 8-hedral crystals. 2. Right rhombic prisms. 3. Glassy masses. 4. And as a white powder. Spec. grav. of the clear glass, 3·699 (Wollaston), 3·698 (Le Royer and Dumas), 3·7385 (Guibourt), 3·7026 (Karsten); of the opaque glass, 3·695 (Guibourt), 3·529 (Taylor); of the powder, 3·706 (Bergman), 3·729 (Herapath), 3·7202 (Karsten). It is therefore a dimorphous body, and exists likewise in two allotropic states, the vitreous and opaque forms. Spec. grav. of vapour, 13·85. Arsenious acid is volatile under a red heat (424° ?). The vapour possesses no odour, except when the acid is thrown on hot charcoal; in this case a portion is reduced to the metallic state, and is converted into the vapour of the metal, which thus appears to possess the smell of garlic. When arsenious acid is sublimed and allowed to condense on a hot surface, the resulting product is the transparent, vitreous, or glassy acid; but when the acid is received in a space where there is a free circulation of air, it is deposited in the form of 8-hedrons or 4-hedrons. When a hot solution of potash is saturated with arsenious acid, and allowed to stand, rhombic crystals of acid separate (Pasteur). When the clear vitreous acid is exposed to the air it becomes opaque. The same change is effected at 212° (Regnault) when placed under chlorohydric acid, but not under water (Christison) or alcohol, or in a vacuum (H. Rose). The opaque acid is changed into the vitreous by boiling with soda. When 1¼ part of the vitreous acid is dissolved in 2 parts water, containing 6 chlorohydric acid, and set aside, 8-hedrons are deposited, and a spark is evolved at the deposition of each crystal (Rose, Pogg. Ann. 35, 48.) When the solution is agitated, the number of crystals, and the intensity of the light, increase. When slowly cooled the sparks continue to be emitted for forty-eight hours. The opaque and crystallized acid when similarlytreated give out no light. A similar result is observed with other isomeric modifications. The vitreous and opaque acids are soluble in alcohol, and are both soluble in water, the solution reddening litmus paper, but the opaque acid is slower in its action. 1 glassy acid dissolves in 9·33 boiling water (Guibourt), 21 (Taylor); 1 part opaque acid in 7·75 boiling water (Klaproth). If arsenious acid be boiled with water till the latter is saturated, on standing 1 part of acid is found to be dissolved in 16 water, at the temperature of 60° (Bucholz), in 33 (Klaproth); in 33½ if the acid had been opaque before solution; in 55 if it was transparent (Guibourt). 100 parts of cold water digested on arsenious acid dissolve ·25 parts of acid (Klaproth, An. Phil. 4, 132). Chlorohydric acid dissolves arsenious acid more readily than pure water; the acid when concentrated forming a terchloride of arsenic, which is volatile. Nitric acid and aqua regia convert it into arsenic acid (AsO_5). It is reduced by substances which have a strong affinity for oxygen, as hydrogen, charcoal, and chlorine, &c. *Sources.*—Arsenious acid is found native, crystallized in 8-hedrons and capillary crystals, at Andreasberg, Joachimsthal, and St. Marie mines. I have a stalactitic specimen of a reddish colour internally, white externally, from Java.

Preparation.—Arsenious acid may be formed by igniting compounds containing arsenic, or by passing oxygen from a bladder or gas-holder over the ignited metal.

Practical Application.—In medicine arsenious acid is used as an antiperiodic in intermittent fevers. The usual form of exhibition is Fowler's solution, or improperly termed solution of arsenite of potash, in the London Pharmacopoeia. This preparation is made by boiling 80 grains arsenious acid, 80 grains carbonate of potash, with 10 ounces of water, and adding 5 fluid drachms of compound tincture of lavender, and as much water as to make up the whole to a pint or 20 ounces. The dose is from 4 to 30 minims. It is mixed in minute quantities with the stearic acid in what are termed stearine candles to prevent crystallization on cooling in the mould. It enters as an ingredient into some of the preparations for preserving skins of animals, mixed with spices. It is employed in calico printing, and as an ingredient in glass to convert the protoxide into the sesquioxide of iron, which imparts less colour than the protoxide in enamel glass. It is also used as a steep for grain, and as a preservative against insects.

Arsenites.—Arsenious acid acts as a feeble base and likewise as a weak acid, being displaced by carbonic acid. Caustic ammonia when added to a solution of arsenious acid and evaporated spontaneously, allows the arsenious acid to crystallize out in 8-hedrons. Arsenious acid dissolves in caustic soda, and potash, without neutralizing them. With lime, barytes, and strontian, white precipitates fall in the form of flocks, which are soluble in ammonia and acids. The arsenites are usually bibasic; diarsenite of silver is yellow, of copper, green. When an acid is added to a concentrated solution of an arsenite, arsenious acid separates. They are easily decomposed; by heat the acid, in some cases, is volatilized; the alkaline arsenites give off arsenic and leave arseniates; arsenite of lead is not decomposed by ignition; alkaline arsenites in solution give a white precipitate with lime water. The arsenites insoluble in water, dissolve in chlorohydric acid, and several of them in ammoniacal salts.

Decompositions.—Arsenious acid may be deprived of its oxygen by the voltaic pile, arsenic appearing at the negative, and oxygen at the positive pole; by hydrogen, carbonic oxide, charcoal, phosphorus, sulphur, potassium, sodium, and zinc, below a red heat; a solution of phosphorous or hypophosphorous acids boiled down deoxidizes it. Arsenic is removed partially from solutions of arsenious acid by zinc, cadmium, tin, antimony, bismuth, copper, more rapidly when chlorohydric acid is present (Fischer, Pogg. Ann. 9, 260.) Copper and chlorohydric acid at common temperatures act but slowly in the reduction of arsenious acid, but, on boiling, the metal is deposited in solutions containing 1 $As\,O_3$ in 100,000 of water. Without the presence of chlorohydric acid there is no action. Sulphohydric acid colours water yellow when 1 acid is dissolved in 10,000 water.

Poisoning.—It is used in medicine for the cure of ague; when taken in over-doses it acts as a violent poison, producing burning pain in the stomach, vomiting, convulsions, weak pulse, delirium, and death. *Antidote.*—The best *antidote* is to dissolve in hot water a quantity of copperas (sulphate of iron), or, in its absence, to dissolve a piece of iron in nitric acid. When the solution of copperas is formed, nitric acid is added until the fluid, which becomes at first dark-coloured, turns of a clear yellow, heat being applied after every addition of acid. On the addition of caustic ammonia or carbonate of ammonia to this solution, the peroxide of iron falls. This precipitate is thrown on a calico or paper filter, and rapidly washed with boiling-hot water, until the liquid passing affords no precipitate with chloride of barium. The moist peroxide ($Fe_2\,O_3\,HO$) is administered to the poisoned person by taking it off the filter with a knife, and diffusing it through water in a tumbler. Magnesia has also been used.

Mode of detecting Arsenious Acid. Solid

Contents.—When death has been produced by arsenious acid it frequently happens that in consequence of the comparative insolubility of this substance, particles of it can be detected among the contents of the stomach, and especially upon its mucous coats. These should be carefully picked up by means of a spatula, or subjected to examination by the dry and liquid tests. 1. *Conversion into Tersulphide.*—A portion of the powder is placed on the inside of the cover of a porcelain crucible and a few drops of bisulphohydride of ammonia are poured on it, heat being applied by means of a sand bath; the powder, if it be arsenious acid, dissolves, and oatmeal or other organic matters remain. On evaporating the solution to dryness, a lemon-yellow spot remains, (with a very gentle heat, if the temperature be urged, the spot is orange) consisting of orpiment or tersulphide of arsenic, ($As\ O_3$ 3 SH, becoming $As\ S_3$ and 3HO.) When a drachm of water is poured on this spot with 1 drop of strong caustic ammonia the whole disappears. If any tersulphide of antimony be present it is not dissolved. On filtration the tersulphide of arsenic passes through and reappearing on the addition of chlorohydric acid to the solution, may then be thrown on a filter, while the tersulphide of antimony remains as an orange precipitate on the filter. If slips of white filtering paper be immersed in the solution of the tersulphide and be dried before the fire or touched with chlorohydric acid, they immediately become of a lemon-yellow colour, which disappears on immersion in caustic ammonia. Spots of metallic arsenic when exposed to vapour of iodine at 60° become brownish-yellow, and lemon-yellow on exposure to the air. Antimonial spots in the same circumstances become dark brown and then orange. 2. *Reduction Test.*—A portion of the powder may be mixed with equal parts of charcoal and carbonate of potash previously ignited, or with black flux (ignited cream of

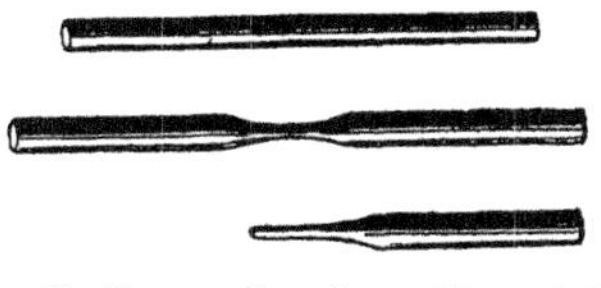

tartar—See Potash Salts) or with a mixture of cyanide of potassium and carbonate of potash, both dry. The mixture is to be introduced into a tube terminating in a hermetically sealed point or bulb, which may be readily formed by means of a spirit lamp, if made of flint glass, or with the aid of a blowpipe, if of hard glass. The tube may be of the diameter of a quill, and is to be heated in the middle, and turned constantly in the flame. When it softens it is drawn out, as in the second figure, and ultimately into two distinct tubes, as in the third figure. The arsenious acid, mixed with dry black flux, or other mixture, is placed in the narrow end of the tube; heat being applied to

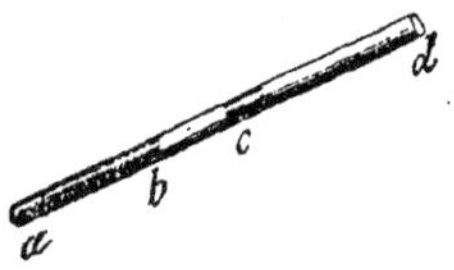

the mixture, metallic arsenic sublimes and forms a brilliant metallic ring above the mixture; this may be converted into arsenious acid by heat.

3. *Marsh's Test.*—The *fluid* contents of the stomach are diluted with distilled water, heated and filtered. The filtered liquor, if it possesses a high colour, is to be mixed with pure animal charcoal, and boiled or heated until the colour disappears, and then filtered. Take a small flask; fit to it a perforated cork, into which is inserted a short quill tube about an inch in length, terminated by a capillary opening. Place in the flask an ounce of water, a few bits of pure zinc, and then pure sulphuric acid: fix the cork and tube, and having waited till the common air is expelled, light the hydrogen escaping from the tube. It burns with a pale yellow flame. Bring horizontally in contact with the flame the bottom of a white porcelain basin, and observe if any stain is left upon it. If, after five minutes, no stain is produced, we may conclude that the zinc and acid are free from arsenic. Add now a few drops of solution of arsenious acid; the colour of the flame becomes blue, and a metallic stain or spot is left on the porcelain. If this is arsenic, it will dissolve if 2 drops of sulphohydride of ammonia are poured on it and heated; a lemon-yellow spot is left. If antimony is present it leaves an orange stain. The only metal with which arsenic can be confounded in this experiment is antimony. In this experiment arsenious acid and hydrogen become arsenietted hydrogen and water ($As\ O_3$ and 6 H become $As\ H_3$ and 3 HO); while on being inflamed arsenietted hydrogen and the oxygen of the air produce arsenic and water ($As\ H_3$ and 3 O become As and 3 HO). 4. *E. Davy's or Reinsch's Test.*—Place a clean piece of copper foil, about the size of a shilling, in a test-tube; add to it the solution of arsenious acid, and about 4 drops of hydrochloric acid, and boil. The copper becomes black by the deposition of arsenic. Transfer the copper to a clean dry test-tube; apply heat to the tube; arsenious acid sublimes in octahedrons into the upper part of the tube, and is prevented from escaping by placing the thumb on the mouth of the tube. In this case 3 Cu and 3 H Cl and $As\ O_3$ become 3 (Cu O H Cl) and As. 5. *Hume's Test.*—To a solution of arsenious acid add a few drops of a solution of

nitrate of silver. No change occurs. Dip a glass rod in caustic ammonia, and touch the surface of the solution with the moist rod, when instantly yellow diarsenite of silver ($2\,AgO\,AsO_3$) falls, which cannot be distinguished from the phosphate of silver. This test is therefore only of advantage in conjunction with the preceding tests. 6. Add to a solution of arsenious acid a few drops of a weak solution of sulphate of copper, so as not to colour the fluid; then touch the solution with a rod moistened in caustic ammonia, as in the preceding case. Green diarsenite of copper falls ($2\,CuO,\,AsO_3$). This also is a dubious test, as phosphoric acid affords a similar result. 7. Add to a solution of arsenious acid some sulphate of copper and caustic soda, and boil; dinoxide or brown oxide of copper falls ($AsO_3\ 4\,(CuO\,SO_3)\ 4\,NaO = AsO_5\ 2\,Cu_2O\ 4\,NaO\,SO_3$). The solution then contains arsenic acid, which gives a brown or brick-red with nitrate of silver. 8. Add to the solid acid some drops of acetic acid in a test-tube with some drops of caustic soda, evaporate and ignite. The disagreeable odour of the oxide of cacodyle will be perceived. The addition of protochloride of tin converts this into chloride of cacodyle, which has also a peculiar smell (Bunsen). 9. Add in a porcelain capsule a few drops of nitric acid to arsenious acid, and evaporate to dryness; dissolve in water; on the addition of nitrate of silver and ammonia, a brick-red precipitate falls.

Arsenic Acid.—AsO_5 14·375, 115. Spec. grav. 3·391 (Bergman), 3·729 (Herapath), 3·734 (Karsten). Discovered by Scheele in 1775. *Characters.*—Crystalline in the hydrous state; when dry, colourless, transparent, and vitreous powder or mass; or opaque after keeping, or when slightly heated; fuses at a low red heat. It is very insoluble in water, and hence has little taste till it has remained some time in contact with the palate, when it is acid, very poisonous, dissolves in 6 parts cold, 2 hot water. When evaporated, the solution becomes syrupy, and deposits crystals of hydrous arsenic acid. The solution is slowly precipitated by sulphohydric acid, and is precipitated white by barytes, strontian, and lime water; in contact with phosphuretted hydrogen, black phosphide of arsenic is formed. Sulphohydric acid precipitates quintosulphide of arsenic slowly; hyposulphite of soda precipitates the same yellow salt rapidly when heated. A solution of sulphurous acid mixed with arsenic acid, speedily deposits 8-hedrons of arsenious acid (Wöhler). Arsenic acid is decomposed when heated above its fusing point into arsenious acid and oxygen; it is deoxidized by many simple bodies, as hydrogen, carbon, iron, phosphorus, sulphur, &c. *Process.*—Arsenic acid was formed by Scheele by dissolving and evaporating to dryness 3 arsenious acid, 7 chlorohydric acid, 5 nitric acid. Bucholz recommended 2 chlorohydric acid, (spec. grav. 1·2) 8 arsenious acid, 24 nitric acid, (1·25) to be mixed in a crucible, evaporated, and exposed to a slight red heat. A simple method is to dissolve arsenic in nitric acid, and evaporate the solution to dryness, and dissolve up any arsenious acid by water, and again evaporate (Thomson). The acid is very insoluble in water, but may be dissolved by long digesting or boiling with water. *Arseniates.*—Arsenic acid forms more permanent compounds with bases than the arsenious acid. It usually is tribasic, but sometimes also dibasic and monobasic. The salts, with the exception of the alkaline arseniates, are insoluble, but those which are insoluble dissolve in chlorohydric acid. The tri and dibasic salts are alkaline. Most of them can be ignited without decomposition. When treated with boracic acid and charcoal in a glass tube, they yield a sublimate of arsenic. When potash or soda is added to the earthy arseniates, the acid is in a great measure removed. Nitrate of silver yields a *brick-red* precipitate, with soluble arseniates, which is not soluble in ammoniacal salts. The soluble arseniates yield a white precipitate of triarseniate of lead, with acetate of lead, or if acetate of lead is poured into a solution in nitric acid, of an insoluble arseniate, a white precipitate falls, which requires much nitric acid to dissolve it. The arseniates and phosphates are isomorphous (Mitscherlich). When carbonate of soda is added to a solution of arsenic acid, triarseniate of soda is formed, having 2 atoms of basic soda, and 1 of water, with 14 water of crystallization, ($2\,NaO\ HO\ AsO_5\ 14\,HO$.) When completely saturated with soda, the formula of the salts is $3\,NaO\ AsO_5\ 24\,HO$, isomorphous with trisphosphate of soda.

Arseniohydric Acid. Arseniettted Hydrogen. Arseniuretted Hydrogen.

	Vols.	Spec. grav.	Atoms.		Per cent.
As	$\frac{1}{4}$	2·5999	1	9·375	96·15
H	$1\frac{1}{2}$	0·1038	3	·375	3·85
		2·7037		9·750	

Experimental specific gravity 2.695 (Dumas). *Characters.*—Colourless gas; disagreeable odour; 1 part is absorbed by 5 parts of water; absorbed by oil of turpentine; in small quantity by fixed oils; extinguishes flame and destroys animal life; burns with a blue flame, and deposits arsenic on substances brought in contact with the flame; at —40° becomes a colourless fluid, but does not solidify at —166°. Neutral reaction to test paper. It is easily decomposed by heat into hydrogen and arsenic. When mixed with oxygen in presence of a taper, it explodes, water and arsenious acid being formed; it is not altered by air, nitrogen, and hydrogen; protoxide of nitrogen produces no alteration in it. Chlorine gas evolves arsenic, and forms chlorohydric acid. Nitric acid decomposes it into water, arsenic, and some arsenic acid. Heated tin absorbs the arsenic, and sets hydrogen free, but does not alter the volume. Hence it is easily analyzed in a graduated tube retort. It is decomposed by sulphuric, hyponitric, and nitric acids, by iodine slowly, by chlorine water, by heated sulphur, phosphorus, potassium, zinc, tin, caustic soda or potash, barytes, sulphate of copper, corrosive sublimate and calomel, nitrate of silver, bichloride of platinum. *Process.*—This gas may be prepared by dissolving an alloy of arsenic and tin, or of zinc and arsenic, in chlorohydric acid, or by heating arsenic in an alkaline lye (Gehlen). An alloy of potassium and arsenic may be prepared by igniting a mixture of 2 tersulphide of antimony, 2 bitartate of potash, and 1 arsenious acid, which is then treated with water. This gas is produced when an oxide of arsenic is present in a vessel from which hydrogen is evolved, but in this case it is contaminated with hydrogen gas, (AsO_3 6 Zn 3 HO 6 SO_3 = AsH_3 6 (ZnO SO_3). Iron is said not to evolve As H_3 with an acid and arsenious acid (Buchner), but with arsenic acid (Scheele, Gmelin). *Analyses.*—Metallic tin, when heated in a bulbed tube in contact with this gas, absorbs arsenic, and leaves hydrogen without any change of volume.

Bisulphide. Realgar, Red Suphuret, Sandarach.—As S_2 13·375, 107 As = 70·09 S = 29·91 per cent. Spec. grav. 3·3384 (Hauy), 3·5444 (Karsten), red oblique prisms; often translucent; forming, when pounded, an orange powder; occurs native in Germany, Spain, &c.; burns with a blue flame in presence of oxygen, being converted into sulphurous and arsenious acids; nitric acid converts it into sulphur and arsenic acid. When its vapour is passed through a red hot tube, it is changed into sulphohydric acid, sulphide of arsenic, and arsenious acid. When boiled with potash or soda, it yields disulphide.

Tersulphide, Tersulphuret, Sesquisulphuret, Orpiment, Auripigmentum.—As S_3 15·375, 123 spec. grav. 3·4522 (Hauy 4, 235), 3·480 (Mohs), 3·459 (Karsten). Right prismatic system, or in thin plates when native, possessing some flexibility, lustre pearly, translucent, lemon-yellow. When artificially prepared it is likewise lemon-yellow, but when heated becomes reddish-brown or orange. Boiling point 1292° (Mitscherlich), is volatilized by heat without decomposing; slightly decomposed by boiling with water; with difficulty in boiling chlorohydric acid; passed over ignited lime yields sulphide of calcium and arsenic; when heated with cyanide of potassium, sulphocyanide of potassium is formed and arsenic sublimed; forms calomel when mixed with corrosive sublimate, arsenious acid and chlorohydric acid; soluble to a slight extent when water is digested on it. When sulphohydric acid is passed through a solution of arsenious acid the liquid becomes yellow, but no precipitate falls unless the solution be saturated with arsenious acid, or unless chlorohydric, sulphuric, or nitric, oxalic, acetic, or tartaric acids be added. The mineral acids act best; but even carbonic acid has some influence. From its solubility in water arsenic cannot be removed entirely from its solutions by sulphohydric acid. When heated in the air, it burns into sulphurous and arsenious acids. It is much less poisonous than arsenious acid. It is very soluble in weak solutions of caustic ammonia, and caustic soda or potash. *Process.*—Occurs native, and is formed artificially by passing sulphohydric acid through arsenious acid dissolved in water containing chlorohydric acid, or by fusing bisulphide with sulphur (Thenard). *Application.*—Orpiment is sometimes used to dissolve indigo in the cold vat. It has been used also in calico printing, but it does not resist soap from its being soluble in alkalies. The orpiment is dissolved in ammonia, the cloth dipped in the solution; on exposure to the air the ammonia evaporates and leaves the orpiment. It is the basis of the colour called king's yellow.

Arsenical Ores. The native ores of arsenic are arsenic, As; arsenious acid, AsO_3; arsenical glance, As_6 S; realgar, As S_2; orpiment, As S_3; arsenical iron, Fe As; arsenical manganese, Mn_2 As; arsenical nickel, Ni As (Kupfer nickel); arsenical cobalt, Co As and Co_2 As_3; arsenical pyrites, Fe S_2, Fe As; glance cobalt, Co S_2, Co As; nickel glance, Ni S_2, Ni As; red silver ore, 3 AgS, As S_3. Besides these, arsenic acid is found in nature associated with lead, copper, &c. and from its isomorphous character, generally along with phosphoric acid.

Arsenide. Arseniuret, arseniet, a compound of arsenic with a base.

Arseniosiderite. Brownish-yellow concretions arranged in a fibrous manner. Spec. grav. 3·52, H 1. B.B. fuses with the reactions of iron and arsenic; consists of AsO_5 34·26, Fe_2O_3 41·31, Mn O_2 1·29, CaO 8·43, Si O_3 4·04, Ko ·76, HO 8·75; found in a manganese mine near Maçon. Form. 4 Fe_2 O_3, AsO_5 2 CaO, AsO_5, 6 HO.

Arsenicite. Arseniate of lime.

Arsenic-sinter. Arseniate of iron.

Artificial Bases. Organic radicals forming salts, which can be formed in the laboratory, as furfurine from meal, &c.

Arthanitine, *Cyclamine.* White needles, without smell; neutral; sharp taste; insoluble in ether, soluble in 500 cold water; converted by nitric acid into oxalic acid; obtained by digesting the fresh roots of the *Cyclamen Europæum* in cold alcohol; evaporating the solution, heating with ether to remove wax, then with water to take up extract, and crystallizing out of hot alcohol.

Asafœtida. C_{40}, H_{26}, O_{10}. Specific grav. 1·327. A fœtid gum resin, having the smell of garlic, being the exuded and concreted juice of the *Ferula asafœtida*, a plant cultivated in Persia and Khorassan; soluble in alcohol; water takes up $\frac{1}{4}$; ether divides it into two resins, one more soluble in ether than the other; asafœtida is used in medicine as an antispasmodic in asthma, chlorosis, in doses of from 5 to 20 grains.

Asarine. $C_{10} H_{11} O_4$. Oblique rhombic prisms, as 4-sided colourless plates and prisms, with a taste and smell of camphor; little soluble in water; very soluble in alcohol; melts at 104°; boils at 536°, rising to 572°, when it decomposes without subliming. Soluble in sulphuric acid, and reprecipitated by water; changed into a resin by nitric acid; obtained by distilling 1 of the dry root of *Asarum Europæum* (asarabacca), with 8 water, until 3 have passed over; it collects in crystals in the neck of the retort.

Asarite. Perhaps the stearoptene of asarum oil, in small silky needles, fusing at 158°, effervescing with nitric acid, which converts it into oxalic acid.

Asarone. The same as asarine.

Asbestus, *Amianthus.* The mineral so called from Sardinia, of which incombustible cloth was made, is a variety of amphibole; it is white, with a shade of green, and consists of fine soft flexible fibres, easily separated from each other, and admitting of being spun into threads. B.B. melts into a white bead. Spec. grav. 1·551; constituents: silica, 55·908; MgO, 27·068; CaO, 14.632; $Al_2 O_3$, 1·820; FeO, 6·528. What is commonly called asbestus is more nearly allied to pyroxene.

Asboline. A yellow bitter oil, not volatile, lighter than water, somewhat soluble in water, very soluble in alcohol; insoluble in turpentine; soluble in nitric acid, with a reddish-yellow colour; obtained by boiling soot with water, evaporating the solution, dissolving the extract in water, precipitating pitch by the addition of chlorohydric acid, and purifying by treatment with water, alcohol, and ether.

Asclepine. A peculiar yellow emetic principle, obtained from the root of asclepias vincetoxicum; soluble in water, alcohol, and ether; obtained by precipitating the decoction by acetate of lead, filtering, precipitating excess of lead by sulphohydric acid, evaporating; extracting the residue with water, which leaves resin.

Asclepione. $C_{40} H_{34} O_6$. White crystalline masses, tasteless and odourless, insoluble in water and alcohol; soluble in ether; fuses at 219°. It is closely allied to lactucone; obtained from the concentrated juice of the Asclepias Syriaca by ether.

Ashes of Plants. See various Plants.

Asparagine, *Asparamide, Malamide, Althein Agedoile.* $C_8 N_2 H_{10} O_8$. Right rhombic 8-hedrons or 6-sided prisms. Spec. grav. 1·519. Soluble in 58 water; insoluble in alcohol, ether, and oils; converted by acids, bases, and heat into aspartic acid and ammonia, and also by fermentation; with nitrous acid it gives malic acid. It is obtained from the root of the althæa officinalis, from asparagus and liquorice root, &c. also from vetches or pease, which are grown in the dark by evaporating the expressed juice; formed from neutral malate of ammonia. ($C_8 H_4 O_8$, $2 NH_3 HO = 2HO\ C_8 N_2 H_{10} O_8$.)

Asparagolite, *Asparagus stone*, or *phosphate of lime*, having a green colour.

Asparolite, a variety of iolite or dichroite.

Aspartic Acid, *Asparagic acid, Malamidic acid.* $C_8 NH_7 O_8$. Pearly scales or leaves; no smell; sour taste; soluble in 128 cold water; soluble in spirit; insoluble in absolute alcohol. Obtained by boiling asparagine with barytes water until ammonia ceases to be evolved; precipitating the barytes by sulphuric acid, filtering and evaporating, or by boiling asparagine with oxide of lead, and precipitating by sulphohydric acid. The aspartates are bibasic.

Asphalt, *Native pitch, Dead Sea bitumen, Trinidad bitumen.* A black substance with a resinous fracture; fusible; sectile; fracture conchoidal. Spec. grav. 1·073 to 1·16 and 1·205; fuses at 212°. Absolute alcohol takes up 5 per cent. of a yellow resin, and ether 70 per cent. of the insoluble part, which is also soluble in naphtha. The portion insoluble in ether is called asphaltene, is soluble in turpentine and naphtha, and becomes soft at 572°. This portion has the formula $C_{20}H_{16}O_3$. Asphalt is found on the shores of the Dead Sea, in a lake in Trinidad, and in Albania, where the celebrated Greek fire existed. I have examined the deposit from the Chimæra of Lycia, and found it to be lamp black or soot; hence it is probable that the two phenomena may be of a similar nature. Pitcoal yields asphalt of analogous character. Asphalt, as procured in gas works, is extensively employed for pavements, being mixed with sand, chalk, and sometimes with pounded iron slag, to give it consistence. It is also valuable for forming the flooring of houses, stables, &c., where it is desirable to preserve warmth, in consequence of its bad conducting power. It may also be used for lining the spaces beneath a ground flooring, to preserve the wood dry. When distilled, it yields a naphthous oil, and at a high temperature is converted into a porous coke.

Asphaltene. See ASPHALT.

Asphaltite. A magnesian limestone, impregnated with asphalte (Karsten).

Aspirator. An apparatus filled with a fluid, and connected with other vessels or tubes, so that when the fluid contained in it is allowed to flow out, the air in these other vessels is drawn over. It is a valuable instrument in the examination of the atmosphere. Instead of the aspirator, the vessels containing the air to be drawn over may be connected with a water pipe; the water of which, when made to run, will aspirate the required air.

Brunner's Aspirator. A and B two cylinders of tin, connected by a bar *a b*, through which passes the axis *d c*; *g h i* K are tubes connecting the cylinders, supplied with stop-cocks. One of these is a common stop-cock, the other is gouged out as in fig. 2. The cylinder A is filled with water, the stop-cocks being shut, and is connected with the vessel from which air is to be drawn by the tube K; a corresponding tube existing in B, but on the opposite side, both supplied with stop-cocks; the common stop-cock is turned so as to

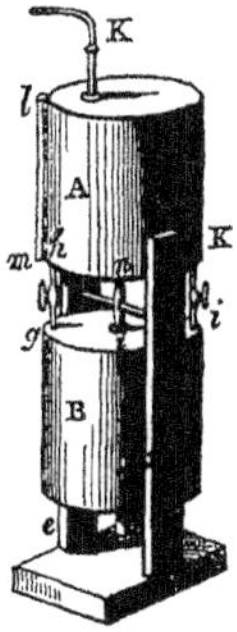

2

allow the water to flow into B, and the other tap so that the air in B may pass into the atmosphere. The water flowing into B, as indicated in the tube *l m*, is replaced by air through K. When A is empty, K is closed, and B placed uppermost, and the process continued. There is a side hook on the standard at the bottom to retain the lower cylinder when the upper is full of water.

Assay. A process by which the important metals are separated quantitatively when mixed with others of inferior value, or from their ores; it is particularly applicable to gold and silver. See these metals.

Assamarc. $C_{20}H_{11}O_{11}$. A reddish-yellow translucent syrup; neutral; soluble in water; reduces nitrate of silver; obtained by roasting various bodies; it exists in the tar from sugar, which is heated with water, and the aqueous residue treated with alcohol and ether; converted into a brown substance, which seems a mixture, and not ulmic acid, as has been supposed.

Astrakanite. $MgOSO_3NaOSO_34HO$.

Astringents (*astringere*, to constringe). Substances which, when used in medicine, have the property of drawing together the organic tissues. When employed externally, they are called styptics. The principal astringents are sulphuric acid, alum, catechu, oak bark, sulphate, and chloride of iron.

Acatamite, *Greensand of Peru.* Hydrous *hexmuriate* of copper. 6CuOHCl 6HO. Green right rhombic prisms. Spec. grav. 4·43, H 2·5. B.B. gives out fumes of chlorohydric acid, and melts into a globule of copper, giving blue and green colours to the flame: found in the desert of Atacama, between Chili and Peru.

Atelestite. A yellow mineral, resembling sphene in crystalline structure; found at Schneeberg in Saxony.

Athamantine. $C_{24}H_{15}O_7$. A fatty crystalline substance, from the root of Athamanta oreoselinum; it consists of oreoselone ($C_{14}H_5O_3$), a substance resembling in its functions glycerine, united to valerianic acid, $C_{10}H_{10}O_4$.

Athanor, Acanor. A peculiar furnace of the old chemists.

Atheriastite, Wernerite. A green scapolite from Arendal.

Atmometer. An instrument for measuring the amount of water evaporated from a moist surface in a certain time, invented by Leslie; now out of use.

Atmosphere. *Common air, atmospheric air*, is a colourless mixture of gases surrounding the surface of the earth. From barometric observations it has been inferred that a stratum of air reaching from the earth's surface to the height of 4 miles, at all times contains above one-half of the quantity of matter in the whole atmosphere, and by extending the laws resulting from these data it is concluded that a stratum of 12 or 13 miles high contains $\frac{29}{30}$ths of the whole; or if a barometer standing at 30 inches were elevated to that height the mercury would fall 29 inches or would stand at 1 inch. The pressure of the air, on the surface of the earth, is capable of supporting a column of mercury of nearly 30 inches, or more precisely of 29·92 inches, and a column of water of about 33 feet. From this it may be calculated that the pressure on every square inch on the surface of the globe when the barometer is at 29·92 is about 15 lbs. But as the pressure diminishes with the density of the atmosphere, as measured by the barometer, it is obvious that if the pressure on the human body be at 30 inches, (considering a man's body to have a surface of 15 square feet,) 13½ tons, it must very much diminish when the barometer subsides, so that we may have changes in pressure exceeding a ton upon the human body. This must undoubtedly exercise a considerable influence upon human health. If the human body has a mean surface of 2000 square inches, the pressure at 15 lbs. on the square inch will, therefore, be on the human body 30,000 lbs. The powerful pressure of the air is exhibited by the Magdeburg hemispheres, two hollow hemispheres of metal, which, when exhausted, cannot be separated, although when they contained air they were loosely united. If a jar be placed over the pump, and having its upper end tied over with bladder, on exhausting the jar, the bladder will be drawn inwards, and ruptured. This experiment should be made without a mercurial guage on the pump, otherwise it will be fractured. Caoutchouc will resist fracture, but be impelled inwards. The rarefaction of the air as we ascend in the atmosphere so rapidly increases that at the height of 500 miles it may be shown that a cubic inch of air would expand so as to occupy a sphere equal in diameter to the orbit of Saturn. The atmosphere is so named from its constituting a sphere of vapour or breathing (ατμος and σφαιρα). It embraces in its lower surface an extent of 2 millions of square miles. Its upper limit is probably not more distant than 50 miles. Its influence in sustaining a column of mercury may be shown by adapting a piston to a long tube open at both ends, introducing one end into mercury, and allowing the piston to touch the mercury. On drawing it up the mercury will be observed to follow the piston as far nearly as 30 inches. If the experiment be made with water instead of mercury, the water will reach an elevation of 33·12 feet (Daniel).

Weight of Atmospheric Air, according to Regnault.

1 litre of dry air = 61·0271 cubic inches.

Weight of a litre of air at 32° and 29·92 inches	= 1·293187 grammes.	
(1 gramme = 15·438 grains)	= 19·9642209 grains.	
100 *cubic inches air* at 32° and 29·92 inches	= 32·713697 "	
100 *cubic inches air* at 32° and 30 "	= 32·8011667 "	
1 *cubic foot air* at 32° and 30 "	= 566·8041605 "	
*100 cubic inches dry air at 60° and 30 "	= 30·948824 "	(Regnault.
Do. do.	= 31·0117 "	(Prout.)

The density of air at various altitudes is shown in the following table:—

Miles.	Vols.	Density.	Barometer.
0	1	1	30 inches.
2·705	2	$\frac{1}{2}$	15 "
5·41	4	$\frac{1}{4}$	7·5 "
8·115	8	$\frac{1}{8}$	3·75 "
10·82	16	$\frac{1}{16}$	1·875 "
13·525	32	$\frac{1}{32}$	0·9375 "
16·23	64	$\frac{1}{64}$	0·46875 "

When the height increases in an arithmetical ratio the volume increases in a geometrical ratio. It is upon the principle involved in this table that we are enabled to determine the altitude of any situation above the level of the sea. There are several corrections to be made for temperature, &c. but we may obtain an approximation to the height of a place by taking the logarithms of the upper and lower pressure, thus:—

	Inches.		
Lower station,	29·82	Logarith. =	1·4750
Upper station,	29·309	Do.	1·4670
		Fathoms =	80
		Feet =	480

In the same manner the temperature of the atmosphere diminishes as we ascend in the atmosphere, from which we can also calculate the elevation. But a simpler method of taking the height by means of the thermometer depends on the diminution of atmospheric pressure as ascertained by the temperature at which water boils; thus affording us the equivalent for the diminution of atmospheric pressure. It has been found that water boils 1° lower for every 549½ feet of ascent, and is equal to ·589 inch of barometric depression. At the summit of the Col. d'Erin the boiling point of water was found to be 191°·93 or 20°·07 below 212°, the external thermometer being 34°. Therefore 549·5 × 20°·07 = 11,028 feet, and supposing the mean temperature of the column 54°, the height will be 11,586 feet above the sea. The correction for the temperature of the air is made in the same manner as for barometric observations. (Forbes, Edin. Trans. 15, 409.)

Analysis of Air.—The atmosphere consists of nitrogen, oxygen, carbonic acid, and vapour of water, dispersed through each other. The mode in which this transposition takes place may be easily illustrated by filling a flask with carbonic acid, and another with common air, and uniting them by means of two corks and a connecting tube, or simply by applying them to each other. Although the carbonic acid is very heavy it will speedily, although it exists in the lower flask, make its way into the upper one, and lime water when poured into each flask will become milky. *a*, That water is contained in the air is proved by bringing a glass filled with ice into a warm room or atmosphere; dew is speedily deposited on the exterior of the glass. *b*, Lime water when exposed to the air becomes milky from the carbonic acid existing in the air. *c*, The oxygen may be removed from air by burning phosphorus in air over water in an inverted jar. *d*, When it is all removed the residual gas is incapable of supporting flame and life. Such are the tests for the constituents of atmospheric air. The mode of determining the quantities of each of these ingredients requires important precautions.

1. *Dumas's Method* consists in having a series of tubes connected, through which air is drawn by placing at their extremity a large flask, *a*, as an aspirator, from which the air has been pumped. An organic analysis furnace, *b*, contains a hard glass tube filled with copper turnings which are heated red hot by a fire of charcoal surrounding the whole of the tube. *c*, Are U tubes filled with pumice moistened with sulphuric acid. *d*, Is a Liebig's potash apparatus

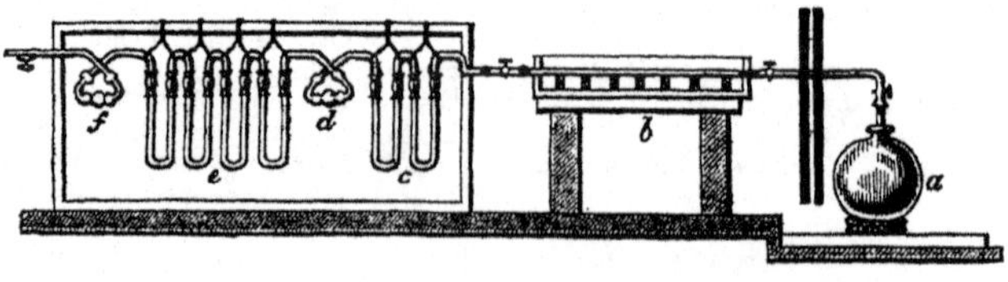

* The expansion of air for every degree from 32° to 212 is calculated at $\frac{1}{491}$ part.

filled with caustic potash to absorb carbonic acid. *e*, Are U tubes filled with pumice and sulphuric acid; and *f* is a potash apparatus. When the stop-cock of the vacuum flask is partially turned, the external air bubbles through the potash apparatus *f*; in passing through *e* tubes it is deprived of moisture, and in *d* the last traces of carbonic acid are removed; and this may be further secured by placing in a portion of *e* solid pieces of caustic potash. In *c* water is removed, and in *b* the oxygen is absorbed by the copper, which is converted into oxide of copper, while the nitrogen enters the flask *a*, and fills up the vacuum. By weighing the flask when empty, and when filled with nitrogen the weight of nitrogen is obtained, and by weighing the tube containing the copper, before and after the experiment, the corresponding weight of oxygen is procured. The other tubes are similarly treated. By means of this apparatus Dumas and Regnault have ascertained by a mean of 6 experiments that the atmosphere consists of

	By Weight.			By Vol.	
Oxygen,	23·01	23·10	20·833	20·81	20·88
Nitrogen,	76·99	76·90	79·166	79·19	79·12
	100·	100·	100·	100·	100·
	Dumas.	Regnault.	Cavendish,	Dumas.	Regnlt.

Mr. Dalton viewed the atmosphere as a mechanical mixture of gases and vapour of water, each of which exerted a certain pressure in bearing up the mercurial column as follows:—

	By Bulk.	Weight.	Pressure in in. of Mercury.
Nitrogen, ...	77·5	75·55	23·36
Oxygen, ...	21·	23·32	6·18
Vapour of water,	1·42	1·03	0·44
Carbonic acid,.....	0·08	0·10	0·02
	100·	100·	30·00

The composition by weight is deduced directly by weighing the gases as described, and from these numbers the vol. is determined by dividing the weights by the respective specific gravities of the gases. These must be considered very close approximations to the true composition of air in relation to the more abundant constituents. But there may still be reasonable doubts as to the true specific gravity of oxygen gas. By very accurate experiments it has been agreed that the atomic weight of oxygen is 8 times greater than that of hydrogen, and as 2 volumes are equivalent to 1 of oxygen when they combine, the specific gravity of oxygen should be 16 times greater than that of hydrogen. But $·06926 \times 16 = 1·10816$ specific gravity of oxygen, instead of 1·1056, the experimental number of Regnault; and if the number deduced for the specific gravity of oxygen by Regnault be correct, then $1·10563 \div 16 = ·069101$ instead of ·06926.

2. *Brunner's Method.*—Instead of a vacuum flask, in this process aspiration is used, or a vessel being filled with water, and put in communication with the air tubes, the water is made to flow out slowly by a stop-cock beneath into a graduated vessel. The air first passes through a chloride of calcium tube by which it is dried, then through caustic potash to absorb the carbonic acid, then through chloride of calcium to take up water; and, lastly, over heated phosphorus in a narrow part of the tube. The water which runs from the water vessel if caught in a graduated jar measures the amount of nitrogen in a given bulk of air.

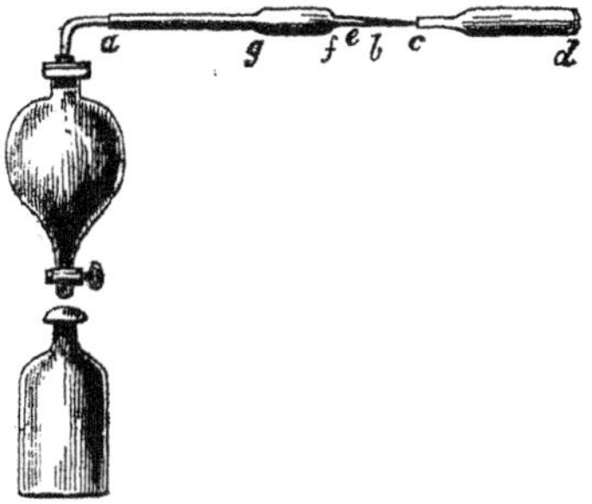

e, *f*, filled with carded cotton.
f, empty space.
b, *c*, a piece of phosphorus.
c, *d*, tube, the wide part contains slaked lime to absorb CO_2, the narrow part asbestus moistened with SO_3.

The aspirator is filled with olive oil, by which means the tension of vapour is obviated.

3. Instead of aspiration, the propulsion of the air by replacement by water may be adopted.

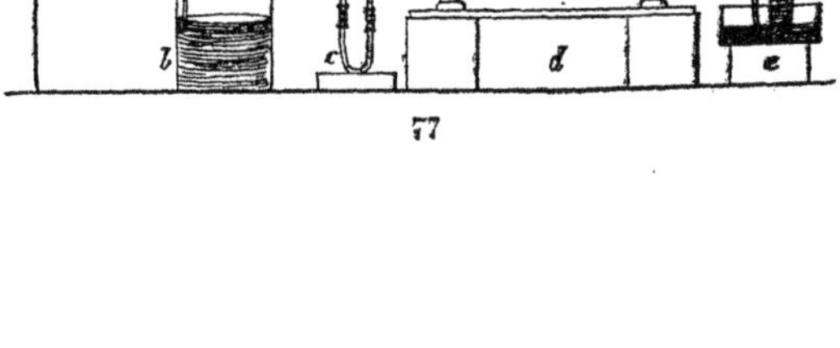

a, Vessel of water to drop into *b*, a Wolfe's bottle; the air is thus propelled through *c*, a tube filled with pounded pumice soaked with sulphuric acid to dry the gas; *d*, a glass tube filled with copper turnings kept at a red heat by charcoal in an organic analysis furnace; *e*, a graduated tube for collecting the nitrogen.

4. *Eudiometrical Method.*—The oxygen may be determined in a given bulk of air by introducing into it a piece of phosphorus on the extremity of a platinum wire. The air is contained in a graduated tube (eudiometer) standing over mercury. The air is measured by depressing the tube in the mercury, which should be of a proper depth, till the mercury in the exterior vessel stands on the same level with that in the tube. The number of degrees on the tube is then read off. Should the trough or glass containing the mercury be too shallow, the distance between the height of the mercury in the tube, and that in the trough must be measured and deducted from the height of the barometer. The tube is then fixed and the phosphorus introduced. It is deposited in a warm room, or in the rays of the sun, and left for twenty-four hours. At the same time the heights of the barometer and thermometer are marked, and the volume calculated at 60° F. and 30 inches, (Thomson's Heat and Electricity, p. 8) and correction must also be made for the tension of the vapour of water. The usual formula is volume of dry gas $=x$, p pressure $=30$ inches and $f=$ $\cdot524$, $\frac{p\,v}{p-f}=$. The formula of Regnault is $V\,\frac{1}{1+0\cdot003665\,t}\cdot\frac{H-f}{30}$ to determine its volume at 32° and at 30 inches—where V is the observed volume, *t*, the exterior temperature, *h*, the height of the barometer, *f*, the tension of the vapour at the temperature, *t*. After the phosphorus has ceased to act the wire is withdrawn and the gas again measured. The correction is made by the following formula $V'\,\frac{1}{1+\cdot003665\,t'}\cdot\frac{H'-f'}{30}$. The difference between the first and second calculation gives the amount of oxygen which has disappeared.

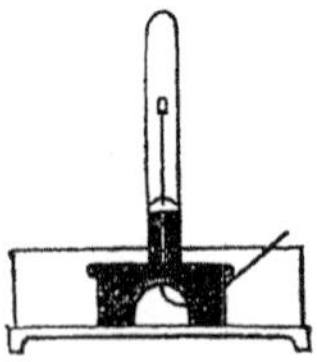

5. *Volta's Eudiometric Method* consists in measuring in an eudiometer a certain volume of air, and calculating its volume free from vapour, then mixing it with hydrogen and passing an electric spark through it by means of a Leyden jar, and measuring the volume after explosion; by dividing the volume which has disappeared by 3, the quotient is the oxygen contained in the mixture. This experiment will be repeated frequently with different proportions of hydrogen so as to secure the combustion of all the oxygen.

6. *Priestley's Method* was to mix with 100 volumes of air 80 of binoxide of nitrogen; the bulk of the gases will be reduced to 79·2, indicating that 20·8 of oxygen have disappeared, and 79·2 of nitrogen remain. But this is not a correct method, as it affords fallacious results. There are various other methods of extracting the oxygen from common air, which have been noticed under nitrogen. (For Liebig's method, see *Pyrogallic Acid.*) The watery vapour and carbonic acid in the atmosphere vary with its condition. The presence of carbonic acid in air was first shown by Dr. Black. Dalton found 6·8 volumes of carbonic acid in 10,000; Thenard, 3·91; Saussure, 4·15; the largest quantity being 5·01 per 10,000. In 100 volumes of air, the constituents may be arranged as follows:—

Nitrogen,	79·14
Oxygen,	20·82
Carbonic acid,	·04
Ammonia,	trace
Carburetted hydrogen,	trace
	100·

It has been calculated that 2¼ lbs. of charcoal will render irrespirable 25 cubic yards of air. The amount of ammonia found in the atmosphere varies greatly in 1,000,000 parts; the quantities found have been—

	NH3	Carb. NH3
Graeger, in Germ., in the country,	0.338	·938
Kemp, Isle of Man,	3·88	10·37
Fresenius, Wiesbaden, mean of day and night,	·133	·379
R.D. Thomson, Glasgow, dur. mist	·060	·170

Boussingault has recently repeated these experiments, and has found the quantity of carbonic acid to vary from 4 to 6 10,000th by volume, or from 6 to 9 10,000th by weight. But as a general statement, the proportions scarcely vary in respect to the nitrogen and oxygen at whatever height we collect the common air. At Brussels, Moscow, and Santa Fe de Bogota, the oxygen and nitrogen have been found nearly constant. At the latter the oxygen has been found rather less at night than during the day, and the carbonic acid from March to July was 3 to 4 volumes per 10,000, while in August and September it amounted to 47 volumes. At Bogota, the composition of the air during the rainy and fair seasons is as follows, in 10,000 volumes, as ascertained by the process of Regnault and Reiset:—

	Rainy Season.	Dry Season.
Carbonic acid,	3·822	4·573
Oxygen,	2099·542	2102·195
Nitrogen,	7896·636	7893·232

From the most recent experiments of Regnault, made on air from different latitudes, it appears that the general mean of oxygen in the air at Paris is 20·96, and in other places varies from 20·9 to 21.; but in certain cases, most frequent in hot countries, the proportion falls to 20·3. Chatin has found iodine in the atmospheric air of Paris, and in rain and snow water. In 4000 litres of air, or 244108·36 cubic inches = 141·2 cubic feet, he has found ·002 milligramme of iodine = ·03 grains. If a man consume 8000 litres, or 282½ cubic feet of air per day, he will introduce into his lungs ·06 grains of iodine daily. He has found less iodine in the air of alpine regions. Barral has likewise found nitric acid in the rain water of Paris, received even in a platinum gauge, and indications have been obtained in this country.

Impure Atmospheres.—When we are acquainted with the air in its pure state, we are prepared to examine it in its vitiated condition. Such effects may be produced by the products of combustion being detained in close apartments, without having free space to escape, or by the respiration of animals in crowded rooms, or by the results of vegetable and animal decay. According to Leblanc, when a piece of charcoal was burned in a close vessel, until it rendered the air so noxious as to cause a dog to fall down insensible in ten minutes, and to die in twenty-five minutes, although a candle was not extinguished in the same air, the composition of the atmosphere was as given in column 1; column 2 gives the composition of the air in the mine of Poullaouen, at 200 feet deep; No. 3 the air in the adit of the same mine; No. 4 the air in the mine of Huelgoat:—

	1	2	3	4
Oxygen, ...	19·19	20·4	15·70	17·6
Nitrogen, ...	75·62	79·6	80·4	82·
Carbonic acid, ...	4·61		3·90	0·4
Carbonic oxide, ..	0·54			
Carburetted hydrogen, ..	0·04			
	100·	100·	100·	100·

In some parts of Huelgoat, the oxygen did not exceed 10 per cent. and then a lamp would not burn. When the carbonic acid exceeded 3 per cent. a candle would not burn; the respiration then becomes difficult. The abundance of pyrites in these mines may probably have some influence on the amount of oxygen. Pure carbonic acid being introduced into an enclosed portion of air, it was found that animals might remain for upwards of half-an-hour in air containing 30 per cent. of carbonic acid (where a candle would not burn), under considerable suffering, but undergoing restoration gradually to a healthy state when exposed to the free air. This is an important observation, because it proves that carbonic oxide is the deadly poison engendered by burning stoves without proper ventilation; 1 per cent. of carbonic oxide having been found to prove instantly fatal to animals. According to Peclet, a man requires from 6 to 10 cubic metres of air per hour for respiration. Reckoning the metre at 3·28 feet, this will amount to from 211·725 to 352·875 cubic feet. The amount of carbonic acid in this air should, by a proper system of ventilation, never be allowed to accumulate. The air expired contains 4½ per cent. of carbonic acid. From this source alone, therefore, air would speedily be contaminated, and rendered dangerous, since air, containing 5 to 6 per cent. of carbonic acid, extinguishes a candle. When the oxygen falls to 15 per cent. in mines, the air is incapable of supporting respiration.

Atom. (ἀ, not; τέμνω, I cut; ἄτομος, an indivisible particle.) Previous to the Christian era, two theories of the nature of matter attracted the attention of philosophers; one of these was, that matter is infinitely divisible; the other, that it consists of small ultimate indivisible particles or atoms, which cannot be further cut or diminished in magnitude. Although, according to our present knowledge, it is highly probable that matter may be infinitely divisible, nay, that even it may be penetrable, and that the atoms may be mathematical points, surrounded with spheres of attraction and repulsion, after the view of Boscovich, still, with our limited views, it is convenient to adopt the hypothesis of the existence of atoms. The strongest arguments in favour of the limited division of matter are those supplied by Drs. Wollaston and Faraday. According to Wollaston, if matter be infinitely divisible, an atmosphere should extend through all space, and should be attracted by the planets according to their respective masses; but the fact is otherwise, for the sun itself has no sensible atmosphere. Dr. Faraday found that evaporation has a limit, as when mercury is placed in the bottom of a clean dry bottle, and a piece of gold leaf applied to the under part of the stopper, and the whole left for some months at a temperature between 60° and 80°, the gold leaf is whitened by amalgamation; but in the winter of 1824-25, the experiment failed; showing that the elasticity of the vapour of mercury, at that cold temperature, was less than the force of gravity upon it, and that consequently the mercury at that time was fixed. It is certain, too, that by experiment matter can be demonstrated to be capable of extremely minute subdivision. Dr. T. Thomson has shown that the size of a particle of lead cannot amount to so much as the $\frac{1}{888,492,000,000,000}$ part of a cubic inch

Atomic Theory, or the mode of explaining the way in which the elements unite together. The first light thrown on this subject was by the analysis of salts, that is, by the determination of the weight of the substances entering into the composition of salts. In 1777 Wenzel ascertained the important fact, that if two neutral salts are mixed together in solution, the resulting decomposition does not alter their state of neutrality. Thus when nitrate of lead and sulphate of soda are mixed, sulphate of lead falls, and nitrate of soda remains in solution, while the supernatant liquor is still neutral, and in certain proportions the decomposition is complete. This is the basis of the atomic theory, but it attracted no attention at the time. In 1782 Bergman also published analyses of salts. In 1789 Higgins published some ingenious views on combination. Among others occur the following:—"Water is composed of a single ultimate particle of oxygen, and an ultimate particle of hydrogen, and its atoms are incapable of uniting to a third particle of either of its constituents." In 1792-1802 Richter extended the analyses of Wenzel, and endeavoured to ascertain the capacity of saturation of each acid and base. He published a table with the names of chemical substances, and numbers attached to show their relative saturating powers. In 1803 Dalton communicated some of his atomic views to the Manchester Society, but they were not published or noticed. In August, 1804, he explained them privately to Dr. T. Thomson, who published them with his own deductions, in the third edition of his System of Chemistry, in 1807. I extract the following from Dr. Thomson's journal of the time, which contains the first notice of this important theory:—"26th August, 1804. Sunday.—Called on Mr. Henry and found him. Dined with his father, and drank tea in Mr. Henry's, junior, in company with Mr. Dalton. He (Mr. Dalton) has been lately occupied with experiments on the carburetted hydrogen. He finds 3 species. 1. Olefiant gas, composed of an atom of hydrogen and an atom of carbon. 2. Gas of marshes, composed of 2 atoms of hydrogen and 1 of carbon. 3. Oxide of carbon, composed of an atom of carbon and 1 of oxygen. He has suggested the following ingenious method of ascertaining the constituents of bodies:—

○ Oxygen.
⦶ Hydrogen.
⊕ Carbon.
⭘ Sulphur.
⊞ Phosphorus.
△ Azote.
○ ⦶ Water.
⦶ △ Ammonia.
⦶ ⊕ Olefiant gas.
○ △ Nitrous gas.
○ △ ○ Nitric acid.
○ ⭘ ○ Sulphuric acid.
○ ⊞ ○ Phosphoric acid.
⦶ ⊕ ⦶ Carburetted hydrogen.
○ ⦶ ⊕ Ether."

These were the main facts which Dr. Thomson noted to assist his memory, and they set him to think on the subject. He afterwards states that Dalton viewed the ultimate particles of matter as atoms incapable of further division. These atoms are all spheres, and are each of them possessed of particular weights, which may be denoted by numbers. The next step was supplied by Dr. Thomson in 1807, in a paper on oxalic acid, published in the Phil. Trans. He there shows that oxalic acid unites with potash and with strontian in two proportions, as a simple oxalate, and as a binoxalate, the acid in the binoxalate being twice the quantity united to the same amount of base which exists in the simple oxalate. These examples were further extended in the same year by Dr. Wollaston, after perusing Dr. Thomson's paper. In 1808 Berzelius took up the subject and confirmed the views of Dalton and Thomson. The general laws of the atomic theory, as at present received, are, 1. That in chemical compounds the elements of which they are composed are uniformly disposed throughout the mass. Thus water is composed of oxygen and hydrogen, in the proportion of 1 part oxygen to $\frac{1}{8}$ part hydrogen; and however small a portion of the water be examined, it will be found to consist of the same substances, in precisely the same proportions. We term this a law, because it is impossible at present to account for it otherwise than as an original impress of nature. For the sake of simplicity it has been agreed to represent the ultimate particle or *atom* of oxygen as weighing 1, when compared with the *atom* of hydrogen, as weighing $\frac{1}{8}$ (·125), or we may say that the one weight is *equivalent* to the other weight, and hence the term *equivalent* is used often instead of *atomic weight*, to denote the number attached to each element, expressive of its combining weight. 2. When one body unites with another body in more than one proportion, the relation between the two bodies, in the new compound, is a simple one. Thus water is composed of 1 atom of oxygen weighing 1, and 1 atom of hydrogen weighing ·125 = 1·125 the atomic weight or equivalent of water. But these two elements form another compound, binoxide of hydrogen, in which 2 atoms oxygen, weighing 2, unite with 1 atom of hydrogen, weighing ·125 = 2·125, the atomic weight or equivalent of binoxide of hydrogen. The combinations of nitrogen and oxygen afford a good example of the extension of this law. Protoxide of nitrogen consists of 1 atom nitrogen weighing 1·75 + 1 of oxygen weighing 1 = NO 2·75. Binoxide of nitrogen consists of 1 atom nitrogen weighing 1·75 + 2 atoms oxygen = NO_2 = 3·75. Ternitrous acid of 1 atom nitrogen = 1·75 + 3 atoms oxygen = NO_3 4·75. Quaternitrous acid of 1 atom nitrogen = 1·75 and 4 atoms oxygen = NO_4 5·75. Nitric acid of 1 atom nitrogen = 1·75 + 5 atoms oxygen = NO_5 6·75. This series, it is obvious, is in the simple relation of 1 nitrogen to 1, 2, 3, 4, and 5 oxygen.

Atomic Volume. *Diameter of an atom.* By this term is indicated the relative space which an atom occupies, obtained by dividing the atomic weight by the specific gravity of a substance. That this size of atoms is different, was pointed out by Dalton in 1808, and he gave the following formula for calculating the diameter of an atom, and at the same time published a table of numerical values of the diameters of various gases:—"The diameter of a particle of oxygen in its elastic state is to that of one of hydrogen as ·794 to 1. For the diameter of an elastic particle is $\sqrt[3]{}$ (weight of one atom ÷ specific gravity of the fluid). Whence denoting the weight of an atom of hydrogen by 1, and the specific gravity of hydrogenous gas also by 1, the weight of an atom of oxygen will be 7, and the specific gravity of hydrogenous gas 14; we have then $\sqrt[3]{\frac{7}{14}} : 1$ or $\sqrt[3]{\frac{1}{2}} : 1$ or ·794 : 1 :: diameter of an atom of oxygen: diameter of one of hydrogen." In this way Dalton affixed the following numbers to various gases:—Hydrogen 1·000, oxygen ·794, nitrogen ·747, muriatic acid 1·12, ammonia ·909, chlorine ·981, nitrous gas ·980, nitrous oxide ·947, carbonic oxide 1·020, carbonic acid 1·, sulphurous acid ·95, olefiant gas ·81, carburetted hydrogen 1·, sulphuretted hydrogen 1·, phosphuretted hydrogen 1·, fluosilicic acid 1·15. In addition to this, he supplies a column representing the number of atoms in a given volume, and he gives diagrams of 16 atoms of different elastic fluids, drawn in the centres of squares of different magnitudes, so as to be proportionate to the diameters of the atoms. In 1828 Dumas made some remarks on the same subject, and observed that the atoms of isomorphous bodies have the same volume.

	Density.	Wt. of Atom.	Vol. of Atom.
Copper,	8·89	395·7	44·4
Manganese,	8·01	355·7	44·4
Nickel,	8·38	369·7	44·1
Cobalt,	8.50	369·	43·4
Iron,	7·80	339·2	43·5
Zinc,	7·19	403·2	56·0

In 1831 Dr. Thomas Thomson gave a similar table without being aware of what Dumas had done, as his remarks were written before the French author's work appeared. Indeed it is unnecessary to say more, since the subsequent views are all anticipated by Dalton. It may, however, be added that Kopp and Schrœder have extended the same inquiries to salts and organic compounds, and have obtained some interesting results, in reference to their physical properties; while Playfair and Joule have shown that the volume occupied by compounds is always a multiple by a whole number of the volume of an atom of water. See SPECIFIC VOLUME.

Atomic Weight. *Combining Weight, Proportion, Chemical Equivalent, Equivalent.* These terms, which are synonymous, denote the number which has been found by experiment to express the relation by weight in which one element or compound unites with another. The value of an atomic or equivalent weight may be determined either by a direct or an indirect experiment—the former being the preferable mode. To ascertain how much sulphuric acid will unite with potash, we may proceed by weighing out 4·9 grains of potassium; by burning this in dry oxygen gas it will become potash, and weigh 5·9 grains. This potash may then be dissolved in water, the solution slightly acidulated with sulphuric acid, the solution evaporated to dryness and ignited; the residue, consisting of sulphate of potash, will be found to weigh 10·9 grains. The relative atomic, or equivalent weights, therefore, of the various bodies operated with, we infer to be from this experiment:—Potassium, 4·9; oxygen, 1; sulphuric acid, 5; since 5 is the excess of weight which the potash has received by the addition of the sulphuric acid. These may be termed the combining weights; they will always bear the same relation to each other if they are multiplied or divided by the same terms. If we represent hydrogen by 1, then oxygen will be 8, as its *combining* weight has been found to be 8 times greater. Each of these numbers will, therefore, be increased 8 times. Potassium will be 39·2; oxygen, 8·; sulphuric acid, 40·. The following table gives the result of the atomic weights at present adopted:—Column 2 gives the atomic or combining weights or equivalents, according to the view that all the weights are multiples of hydrogen by a whole number, oxygen being 1; col. 3 is obtained from col. 1, by multiplying by 8, hydrogen being 1; col. 4 represents the experimental numbers, oxygen being 100; col. 5 is obtained by multiplying the numbers in the previous column by 8; col. 6 contains the specific gravities of the various elements. The substances in italics are least common.

TABLE OF ATOMIC WEIGHTS.

1. SALTS FORMING RADICALS, or SUBSTITUTIONAL BODIES.	1	2	3	4	5	6
		Atomic Weight.				Spec. Grav.
	Symbols.	O=1	H=1	O=100	H=1	
Oxygen,	O	1	8	100	8	1·1056
Chlorine,	Cl	4·5	36·	443·75	35·5	2·44
Iodine,	I	15·75	126·	1588·75	127·1	4·948
Bromine,	Br	10·	80	1000·	80·	2·96
Fluorine,	Fl	2·375	19	23·75	19·	?
2. ACID RADICALS.						
Non-Metallic.						
Hydrogen,	H	·125	1	12·5	1	·06926
Nitrogen,	N	1·75	14	175·	14	·9713

ATO				ATO		
Carbon, ...	C	·75	6	75·	6	3·5
Boron, ...	B	1·375	11	136·25	10·9	1·183
Silicon, ...	Si	2·5	20	266·25	21·3	1·837?
Phosphorus, ...	P	4·	32	387·5	31	1·833
Sulphur, ...	S	2·	16	200	16	2·087
Selenium, ...	Se	5·	40	493·75	39·5	4·32
Metallic.						
Arsenic, ...	As	9·375	75	937·5	75·	5·7
Antimony, (Stibium,)	Sb	16·	128·	1612·5	129·	6·7
Chromium, ...	Cr	3·5	28·	333·75	26·7	5·9
Tellurium, ...	Te	8·	64·	802·5	64·2	6·305
Vanadium, ...	V	8·5	68	857·5	68·6	?
Uranium, ...	U	7·5	60	750·	60·	8·425
Molybdenum, ...	Mo	5·75	46·	575·	46·	8·6
Tungsten, (Wolfram,)	Tn or W	11·75	94·	1187·5	95·	17·6
Titanium, ...	Ti	3·	24·	312·5	25·	?
Columbium, (Tantalum,)	Cm or Ta	23·	184·	2300·	184·	?
Niobium, ...	Nb	?				
Pelopium, ...	Pe	?				
3. Basic Radicals.						
Alkalies, Metals of.						
Potassium, (Kalium,)	K	5·	40·	490·	39·2	·865
Sodium, (Natrium,)	Na	3·	24·	287·5	23	·953
Lithium, ...	L	·75	6	81·25	6·5	?
Alkaline Earths, Metals of.						
Calcium, (Lime,)	Ca	3·5	20	250·	20·	4 to 8?
Barium, ...	Ba	8·5	68·	856·25	68·5	4·?
Strontium, ...	Sr	5·5	44	547·5	43·8	?
Magnesium, ...	Mg	1·5	12·	152·5	12·2	2·24
Earths, Metals of.						
Aluminum, ...	Al	1·75	14	171·25	13·7	2.58
Glucinum, (Beryllium,)	G or Be	·5875	4·7	58·75	4·7	?
Zirconium, ...	Zr	2·755	22·	280·	22·4	?
Thorium, ...	Th	7·5	60	745·	59·6	?
Didymium, ...	D	6·	48·	600·	48·	?
Lantanum, ...	La	5·875	47·	587·5	47·	?
Cerium, ...	Ce	5·875	47	587·5	47	?
Yttrium, ...	Y	?				?
Terbium, ...	Tb	?				?
Erbium, ...	E	?				?
Norium, ...	No	?				?
Difficultly Fusible Metals.						
Iron, ...	Fe	3·5	28·	'350	28	7·79
Manganese, ...	Mn	3·5	28	345·	27·6	7·25
Nickel, ...	Ni	3·75	30·	370·	29·6	8·33
Cobalt, ...	Co	3·75	30	368·75	29·5	8·5
Copper, (Cuprum,)	Cu	4·	32	396·25	31·7	8·95
Silver, (Argentum,)	Ag	13·5	108	1351·25	108·1	10·498
Easily Fusible Metals.						
Zinc, ...	Zn	4·0625	32·5	407·5	32·6	6·864
Cadmium, ...	Cd	7·	56	700·	56·	8·64
Lead, (Plumbum,)	Pb	13·	104	1296·25	103·7	11·33
Tin, (Stannum,)	Sn	7·25	58	725	58·	7·283
Bismuth, ...	Bi	26·75	214	266·25	213	9·776
Mercury, (Hydrargyrum,)	Hg	12·5	100	1250·	100	13·596
Noble Metals.						
Gold, (Aurum,)	Au	12·25	98	2462·5	197	19·3
Platinum, ...	Pt	12·25	98	1233·75	98·7	22·
Palladium, ...	Pd	6·75	54	662·5	53·3	11·9
Rhodium, ...	R	6·5	52	652·5	52·2	10·65
Osmium, ...	Os	12·5	100	1245·	99·6	10·
Iridium, ...	Ir	12·375	99	1237·5	99·	18·68
Ruthenium, ...	Ru	6·5	52	652·5	52·2	?

Atramentum. The ancient name for lamp black.

Atropic Acid. Long volatile crystals, in union with atropine, in the root of the atropa belladonna, obtained as atropate of ammonia in the liquid from which atropine is precipitated by ammonia. The ammonia is replaced by caustic potash, decolourized by animal charcoal, and the atropate of potash decomposed by sulphuric acid.

Atropine. C_{34} H_{23} NO_6? Colourless white silky prisms, without smell, with a bitterish taste; soluble in 500 cold, and 30 boiling water; in 8 parts cold alcohol; 63 cold and 32 hot ether; soluble in turpentine and almond oils, and in caustic ammonia; strongly alkaline in its action. Atropine is obtained by exhausting pulverized belladonna root with cold water; the solution fermented, in three days filtered and boiled; evaporated to a thin extract; caustic ammonia added, and alcohol digested on it. The alcohol takes up atropate of atropine and other substances. The alcohol is distilled off; the residue treated with caustic ammonia and spirit. After agitation and standing, ether is added, the liquor is poured off and distilled. Caustic ammonia digested on the residue causes the atropine to separate as a yellowish-brown precipitate. The addition of sulphuric acid converts it into sulphate of atropine, and the addition of ammonia precipitates the atropine. The various salts may be obtained by dissolving atropine in the respective acids; the sulphate and muriate alone crystallize. Extract of belladonna is much used to dilate the pupils in affections of the eyes, and the salts may be employed probably for the same purpose.

Attenuation. A term used in brewing and distillation, applies to the thinning or weakening of saccharine worts during fermentation, by the conversion of the sugar into alcohol and carbonic acid. It is usual to speak of so many degrees of attenuation, indicating the decrease of specific gravity by the fermentation.

Attraction. See AFFINITY.

Attar, or **Otto of Roses.** *Oil of Roses.* C_{23} H_{23} O_3. Nearly colourless. Spec. grav. ·872; congeals below 80°; melts at 84°; at 57° 1000 alcohol dissolve 7½ oil, and at 72° 33 oil; constitutes a delicious perfume. It is present in very small quantity in rose petals, from which it is distilled along with water. It is much adulterated with sandal-wood and fixed oils, but may be generally distinguished by its great volatility. It is principally prepared in Bengal.

Augite. See PYROXENE.

Aurade. Tasteless needles: a camphor separable from orange flower oil by means of alcohol.

Aurantiine. See HESPERIDINE.

Auric Acid. See SESQUIOXIDE OF GOLD.

Aurichalcite. *Hydrous Cupro-carbonate of Zinc.* CuO 28·192, ZnO 45·839, CO_2 16·056, HO 9·950. A mineral found in the Altai mountains, in small broad columnar pieces, imbedded in limestone and brown hematite. Its colour is verdigris-green. Its formula is equivalent to blue malachite, with 1 atom oxide of zinc replacing 1 atom oxide of copper.

Auripigmentum. The ancient name for orpiment or tersulphide of arsenic.

Aurum Problematicum, or native tellurium.

Automalite. See GAHNITE.

Avenine. The caseine of the oat-seed (avena sativa).

Aventurine Quartz. A variety of quartz of a brownish-red colour, due to spangles of mica.

Axe-Stone, *jade, nephrite.* A species of serpentine.

Axinite (from the resemblance of its crystals to an axe). *Thumerstein, Thumite, Yanolite. Violet schorl.* Spec. grav. 3·271, H = 7, Clove-brown, of various shades, in doubly-oblique prisms, having angles of 134° 40′, 115° 17′, and 135° 10′. Streak white; lustre vitreous; commonly splendent; very fragile; transparent, often translucent. It occurs in Dauphiné, Thum, Kongsberg in Norway, and Botallack in Cornwall. B.B. melts into a green glass, which becomes black in the oxidizing flame; heated with fluor spar and bisulphate of potash, it colours the flame green. A specimen from the Pyrenees I found to consist of silica 44·03, Al_2 O_3 16·31, CaO 20·22, FeO 10·11, BO_3 trace, MnO 9·14.

Axis. In crystallography, the hypothetic lines connecting the opposite sides of a crystal.

Azaderine. An alkaloid found in the root of the malea azadirachta, which is useful in fever, and forms a crystalline salt with sulphuric acid.

Azelaic Acid. C_{10} H_9 O_5. A crystalline powder, more fusible than suberic acid; obtained by evaporating the nitric acid solution, which has been acting on oleic acid to the fourth of its volume; a soft mass, solidifies; which, pressed and washed with water, yields suberic and azelaic acids. The azelaic acid is dissolved up by ether.

Azobenzide. C_{12} H_5 N. Red crystals; fusing at 149°; boiling at 379°; formed by distilling an alcoholic solution of nitro-benzide with dry hydrate of potash.

Azobenzoide. C_{42} $H_{16\frac{1}{2}}$ $N_{2\frac{1}{2}}$? White powder, fusing by care into a crystalline mass; obtained by acting on oil of bitter almonds with caustic ammonia during several weeks.

Azobenzoidine. C_{28} H_{11} $N_{\frac{5}{3}}$. Colourless, transparent, oblique crystals; nearly insoluble in alcohol, and little soluble in ether. Obtained by acting for some time on distilled bitter almond oil with ammonia.

Azobenzoilide. C_{28} H_{11} N, N_2. White powder or plates, obtained by heating oil of bitter almonds with ammonia, and treating the brown viscid mass formed with alcohol and ether. This compound remains.

Azobenzoyle. $C_{42}\ H_{15}\ N_2$. White tasteless powder; soluble in 100 boiling alcohol; insoluble in water; obtained by adding to crude oil of bitter almonds an equal volume of caustic ammonia, and allowing it to stand for a month; a yellow resinous mass results. Boiling ether takes up hydrobenzamide, and leaves azobenzoyle and some azotide of benzoyle, which is separated by boiling alcohol.

Azoleic Acid. $C_{13}\ H_{13}\ O_4$. An oily fluid, insoluble in water; soluble in boiling nitric acid, and reprecipitated by water. Obtained from the oily substance which swims after the treatment of olive oil with nitric acid, by boiling with alcohol and sulphuric acid; azoleate of ethyle is formed, which is decomposed by an alcoholic solution of potash, and the addition of chlorohydric acid.

Azoerythrine. $C_{22}\ H_{19}\ O_{22}\ N$. A substance said to exist in archil, roccella tinctoria.

Azolitmine. $C_{18}\ H_{10}\ NO_{10}$. The principal constituent of litmus; insoluble in water and alcohol.

Azomaric Acid. $C_{20}\ H_{11}\ NO_8$. An acid formed by the action of nitric acid on pimaric acid.

Azote. (α, depriver of; ζωη, life.) The French name for nitrogen.

Azotic and **Azotous Acids.** Synonymes of nitric and nitrous acids.

Azoxybenzide. $C_{12}\ H_5\ NO$. Obtained, along with azobenzide, when nitrobenzide is distilled with an alcoholic solution of potash.

Azotide. A compound of nitrogen with bases; also called nitride, nitruret, nitret.

Azotized Bodies. Substances containing nitrogen.

Azulmic Acid, *Azulmine.* $C_5\ N_2\ H$? Black substance proceeding from the spontaneous decomposition of cyanogen in water. According to others it is composed of 2 cyanogen + 1 HO.

Azure Stone, or *Lazulite,* or Lapis lazuli.

Azure Spar. *Blue Spar.* A variety of lazulite.

B

Babingtonite. Spec. grav. 3·4; H 5·5. Black 8-sided prisms, from a doubly-oblique prism, the angles being 92° 34, 88° and 112° 30′. I found a specimen from Arendal to consist of SiO_3 47·46, $Al_2\ O_3$ 6·48, FeO 16·81, MnO 10·16, CaO 11·74, MgO 2·21, HO 1.24. Formula, from Arppe's analysis 3(CaO SiO_3), 3FeO2 SiO_3. Arppe found in it SiO_3 54·4, $Al_2\ O_3$ ·3 FeO 21·3, MnO 1·3, CaO 19·6, MgO 2·2, HO ·9. B.B. fuses into a black enamel. It is closely allied to hornblende.

Bablah. The brown fruit or seed of the Mimosa Arabica, or M. cinerea, from India and Senegal. It contains gallic and tannic acids, and is used in calico printing to give different shades of brown with an alum mordant.

Bagrationite. Allied to gadolinite. Spec. grav. 4·115, H=6·5. Found in the Ural mountains.

Baierine. *Bavarian tantalite, Columbite.*

Baikalite. A dark green variety of pyroxene, found on lake Baikal in Siberia.

Balance. The instrument by means of which the accurate weight of substances is determined in chemical analyses. The construction of a balance being the work of a mechanic, it is only necessary here simply to allude to those points which are requisite for the chemist to be acquainted with, in order to direct him in his purchase, and to enable him to adjust his instrument so as prevent it from becoming disordered from misuse. The chemist requires a very delicate balance for minute weights, and likewise another for heavy weights, sufficiently delicate, however, that it shall turn with a few tenths of a grain when loaded with some pounds. One of the best balances for delicate weighing is that contrived by Captain Kater and Mr. Robinson, and usually known as Robinson's balance. The accompanying cut will give an idea of its construction.

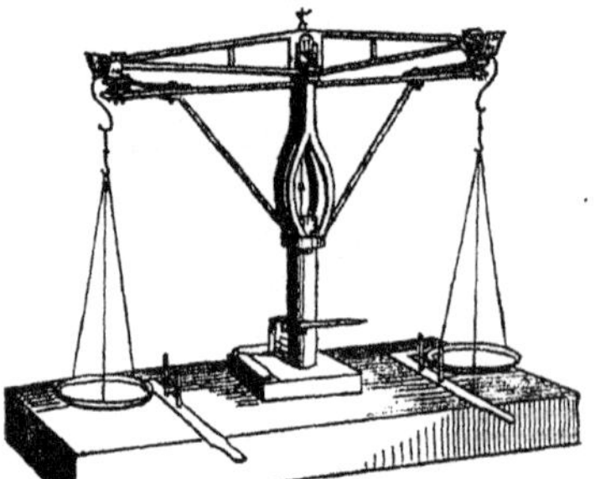

The beam of the balance, which is a frame of bell metal, in the form of a rhomb, is only 10 inches long. The fulcrum is an equilateral triangular prism of steel, an inch in length; the edge on which the beam vibrates being formed to an angle of 120°, that no injury may result from the weight with which it may be loaded. The knife edge of the fulcrum rests on an agate plane throughout its whole length. The supports for the scales are knife edges, on which the agate planes hook on—a plan suggested by Cavendish—and these are supplied with two pressing screws, so as to make them parallel to the central knife edge. Each end of the beam is sprung obliquely upwards and towards the middle, so as to form a spring through which a pushing screw passes, which serves to vary the point of suspension from the fulcrum, and, at the same time, by its oblique action, to raise or depress it, so as to furnish a means of bringing the points of suspension and the fulcrum into a right line. A piece of wire, 4 inches long, on which a screw is cut, points from the centre of the beam downwards. This tapers so as to form an index, and a small brass ball moves on the screw to enable the centre of gravity to be changed at will. On the lower half of the pillar on which the agate plane is fixed for the support of the fulcrum, a tube slides up and down, by means of a lever, which passes to the outside of the case. From the top of this tube arms proceed obliquely towards the end of the balance, which support a horizontal piece, carrying at each extremity two sets of Y's, one a little above the other. The upper Y's are destined to receive the agate planes to which the scales are attached, and thus to relieve the knife edges from their pressure; the lower to receive the knife edges themselves, which form the points of suspension of the pans, consequently the lower Y's, when in action, sustain the whole beam. When the lever is freed from a notch in which it is lodged, a spring is allowed to act upon the tube already mentioned, and to elevate it. The upper Y's first meet the agate planes carrying the scale pans, and free them from the knife edges. The lower Y's then come into action, and raise the whole beam, elevating the central knife edge above the agate plane. This is the usual state of the balance when not in use. When it is to be brought into action, the reverse of what has been described takes place. On pressing down the lever, the central knife edge first meets the agate plane, and afterwards the two agate planes, carrying the scale pans, are deposited on them, supporting knife edges. To adjust this balance, Captain Kater, who used it in arranging the imperial standard weights, gives the following directions:—1. To bring the points of suspension and the fulcrum into a right line, make the vibrations of the balance very slow, by moving the weight which influences the centre of gravity, and bring the beam into a horizontal position, by means of small pieces of paper thrown into the scales. Then load the scales with nearly the greatest weight the beam is fitted to carry. If the vibrations are performed in the same time as before, no further adjustment is necessary; but if the beam vibrates quicker, or if it oversets, cause it to vibrate in the same time as at first, by moving the adjusting weights, and note the distance through which the weight has passed. Move the weight then in the contrary direction, through double this distance, and then produce the former slow motion, by means of the screw acting vertically on the points of suspension. Repeat this operation till the adjustment is perfect. 2. To make the arms of the beam of an equal length, put weights in the scales as before; bring the beam as nearly as possible to a horizontal position, and note the division at which the index stands; unhook the scales, and transfer them with their weights to the other ends of the beam, when, if the index points to the same division, the arms are of equal length; but if not, bring the index to the division which had been noted, by placing small weights on one or the other scale. Take away half these weights, and bring the index again to the observed division by the adjusting screw, which acts horizontally on the points of support. If the scale pans are known to be of the same weight, it will not be necessary to change the scales, but merely to transfer the weights from one scale pan to the other. The beam is enclosed in a glass case; and, to prevent rusting, a basin containing quicklime, carbonate of potash, or sulphuric acid should be inserted within the case to absorb moisture. The weights commonly sold along with such balances have the larger ones formed of brass, the smaller weights of platinum. They usually consist of 400, 300, 200, 100 separate grain weights in brass = 1000, and 40, 30, 20, 10, grains in platinum. The smaller grain weights all in platinum are 4, 3, 2, 1 grs. The tenths are ·4 ($\frac{4}{10}$ths), ·3, ·2, ·1. The hundredths are made of platinum wire or foil, ·04, ·03, ·02, ·01. The thousandths of a grain, of a still finer wire, consist of ·004, ·003, ·002, ·001. Robinson's balance, when in good order, will turn with the $\frac{1}{1000}$ (·001) of a grain, when loaded with 2000 grains in each pan, a delicacy indicating the two-millionth part of the body weighed; but such minuteness is never required; indeed, a chemist, who understands the value of numbers, will never present more than two or three decimals. If the arms of the beam be of different lengths, counterpoising weights, when reversed in the pans, will not equilibriate; hence it is of advantage always to put the weights in the same scale. The weights should frequently be examined to ascertain that they agree with one another; and if they do, it does not much signify whether the grain weight corresponds with the imperial standard, although it is proper that it should be so. The weights should always be lifted with a brass or steel forceps. The substance to be weighed should be placed by means of a spatula on a watch glass, or glass tube, or glazed paper. When hot substances are weighed. they should be allowed to remain till they cool, and have ceased to depress the scale. When long tubes are to be weighed, the lower extremity is placed in the pan, and the superior end supported by a loop of hair passing round the upper part of the wires or threads of the scales.

Balduin's Phosphorus. Nitrate of lime, when evaporated, loses its water of crystallization, and becomes luminous in the dark, as discovered by Balduin in 1675.

Balenic Acid. C_{38} H_{38} O_4. A fatty acid, fusing at 164°; obtained from a natural fatty oil.

Balloon. (*Ballon*, Fr.) The silk apparatus for retaining coal gas, so as to cause an ascent in the atmosphere. The best material for their construction is thin silk, covered with a solution of India rubber in naphtha.

Balas Ruby, or rose-red ruby.

Baltimorite. Allied to serpentine; found near Baltimore, and in Perthshire. It is grayish-green, consisting of longitudinal fibres, somewhat resembling asbestus; lustre silky; it consists of SiO_3 40·95, MgO 34·70, FeO 10·05, Al_2O_3 1·5, HO 12·6.

Balsams are aromatic semi-resinous fluids, often containing benzoic acid. In France, the presence of this acid is necessary to form a balsam, but not so in this country. On distillation they yield an oil, benzoic acid, when present, and leave a resin. The balsams in use are B. of copaiba, B. of Mecca, of Peru, of Tolu. The balsams containing benzoic acid are soluble in alcohol, resin being precipitated on the addition of water, forming a milky fluid. Benzoine, storax, and liquid amber possess similar properties.

Balsam of Sulphur. *Sulphuretted oil.* A dark-coloured adhesive mixture used in medicine, prepared by heating a pint of olive oil with 2 ounces of washed sulphur until they unite. When considered formerly a balsam it was used in catarrhs, but is only now employed sometimes to wash foul ulcers.

Bamlite. A grayish-white radiated splintery mineral, from Bamle in Norway. Spec. grav. 2·984, H 6·, composition SiO_3 56·9, Al_2O_3 40·75, Fe_2O_3 1·04, CaO 1·04, Fl trace. *Form.* 2 Al_2O_3 3 SiO_3.

Banded Jasper is formed of silica sometimes united with some alumina in which the appearance of a banded or ribbon-like deposition is presented; probably from solution in water.

Barbotine. A vegetable product from the Levant and India, consisting of wax, bitter extract, earthy and gummy matter.

Bardiglione. A blue variety of anhydrous sulphate of lime used for ornamental purposes.

Baregine. *Glairine, Zoogene, Plombierine.* A gelatinous organic deposit at the hot wells of Bareges, and other places, destitute apparently of organization, and probably therefore produced by the action of water on plants. Sometimes the deposits at wells are species of *nostoc.*

Barilla. The name given to commercial alkalies, formerly imported from Alicant, Malaga, Carthagena, &c. as derived from the combustion of sea plants. The term *British barilla* is applied to the soda ball, or crude soda, as it occurs as one of the stages in the production of carbonate of soda from common salt in British soda manufactories, and also to kelp.

Barium. Ba 8·5, 68; 68·6. A silver-white metal, inferior in lustre to cast iron. Spec. grav. 4· (?) Sinks in oil of vitriol; melts below redness, and is not volatile at a red heat; ductile and malleable; decomposes water rapidly. 1. Obtained by exposing hydrate, carbonate, chloride, or nitrate made into a pasty cup with water, and filled with mercury, in a platinum vessel, to the action of a battery of 500 plates. 2. By passing the vapour of potassium over red hot barytes, or chloride of barium (Sir H. Davy, Phil. Trans. 1808). 3. By exposing barytes on charcoal to the oxygen-hydrogen blowpipe (An. Phil. 17, 419). Davy first obtained indications of this metal in October, 1807, and in March, 1808, formed an alloy of it with iron. The process of electrifying mercury in contact with barytes was pointed out to him in May, 1808, by Berzelius and Pontin, and in June he obtained the metal.

Barytes. *Baryta. Oxide of Barium.*—BaO 9·5, 9·575, 76·6. A grayish-white porous body. Spec.grav. 4·; 4·732. *Taste* harsh, and more caustic than lime, acting on the stomach as a violent poison; turns vegetable blues green; a non-conductor of electricity; decomposes animal bodies like the fixed alkalies, though with less energy. Before the blowpipe on charcoal it fuses, and penetrates the charcoal, an effect probably occasioned by the persence of water, as it has since been found, when anhydrous, to fuse only by the strongest heat of a forge (Abich. Gay Lussac; Rech. Phys.-Chim. 1, 170), and by the oxyhydrogen blowpipe (Clarke). When sulphuric or chlorohydric acid is poured on barytes, it becomes red hot (Barry, Ann. Phil. 2d ser. 2, 77). When water is poured on it, barytes is slaked like quicklime with the evolution of heat; the mass becomes white and swells; if more water is added, so as to dilute it completely, the barytes crystallizes on cooling as a hydrate (BaOHO), and it then absorbs carbonic acid from the air, and must, therefore, be preserved in closely-stoppered bottles (Mem. de l'Institut. 2, 59; Fourcroy, 2, 198). The hydrate loses no water by ignition (Bucholz and Gehlen). Anhydrous barytes dissolves in 35 water at 55⅓°, in 7½ at 116½°, in 5·6 at 158° (Osann); while th ehydrate dissolves in 20 cold, and 2 boiling water. *Process.*—1. Barytes may be easily prepared by dissolving the native carbonate of Anglesark in Lancashire in nitric acid, crystallizing, and exposing the nitrate to a strong red heat, gradually applied, in a clay crucible, covered after the escape of the greater portion of the nitric acid. 2. The carbonate may be formed from the sulphate by mixing it in fine powder with one-fourth its weight of charcoal, and igniting the mixture for two hours in a clay crucible ($BaOSO_3$ and 2C become BaS and $2CO_2$ evolved); sulphide of barium is formed and carbonic acid evolved. The sulphide is dissolved in water, and precipitated by an alkaline carbonate. The carbonate of barytes being washed, is treated with nitric acid, and ignited (Scheele). 3. Barytes may be obtained from the carbonate by mixing the carbonate with one-tenth its weight of lamp black or charcoal, made into a paste with gum or oil, and heating to whiteness in a black lead crucible (Hope, Pelletier).

Protohydrate. — BaOHO 10·625. White powder or fused white mass, forming an oily liquid at a low red heat.

Novohydrate, or crystalline hydrate.—BaO9HO. Spec. grav. 2·188 (Filhol). Clear, colourless, 4-sided or flat 6-sided prisms, with 4-sided pyramids (Hope, Edin. Trans. 4, 36). At 212° they undergo the watery fusion, and their water is reduced to 2 atoms or 20 per cent.; the original water being about 53 per cent.; they are soluble in 17½ water at 60°, and in any quantity of boiling water, as the crystals themselves fuse at that heat (Hope). *Process.*—The crystalline hydrate is obtained by dissolving anhydrous barytes from nitrate in water, and evaporating in vacuo.

Binoxide, Deutoxide, Peroxide. — BaO_2 10·5 (Thenard, Rec. Phys.-Chim. 1, 169). Grayish powder, more fusible than the equioxide; when placed in powder in water, it becomes white and hydrous, without giving out heat; it does not decompose at a low red heat, but when the heat is increased, or in boiling water, it gives out oxygen. With carbonic acid it affords oxygen and carbonate of barytes. Carbon, boron, phosphorus, sulphur, and the metals, remove oxygen from it when heated, and likewise cold sulphuretted hydrogen decomposes it. It is precipitated in pearly scales with 6 atoms water (Liebig and Wöhler, Poggen. Ann. 26, 172), by adding a solution of binoxide of hydrogen to barytes water; the hydrate is decomposed at 212° into oxygen and solution of barytes. *Process.*—The binoxide is prepared by passing dry oxygen gas slowly over anhydrous barytes (carefully prepared from the nitrate) in a hard glass tube (Gay Lussac and Thenard), or by sprinkling chlorate of potash in small portions over barytes at a low red heat, in a platinum crucible, and washing out the chloride of potassium with water, the hydrated binoxide remaining (Liebig and Wöhler).

Salts of Barytes.—1. They are generally white or colourless, and crystalline. 2. The soluble salts of barytes give a white precipitate by sulphuric acid and alkaline sulphates, insoluble in nitric acid. 3. Ammonia does not precipitate barytes, while soda and potash do. 4. Yellow prussiate gives no precipitate. 5. The soluble salts of barytes are poisonous. 6. Before the blowpipe, salts of barytes communicate a green tinge to the flame. 7. Sulphate of lime in solution precipitates barytes salts immediately. 8. Chromate of potash precipitates alkaline and neutral solutions of salts of barium yellow; insoluble in alkalies and acetic acid; while no precipitate occurs with strontian and lime salts.

Chloride Hydrous.—BaCl2HO 15·25; 122· Ba 55·73, Cl 29·5, HO 14·77. Spec. grav. 3.049 (Karsten). Commonly colourless tables, referable to the right prismatic system. *Taste,* pungent and disagreeable; poisonous; not altered in the air, decrepitating on being heated, but not liquefying; by ignition it fuses, but does not decompose. When the anhydrous salt is exposed to the air, it takes up two atoms of crystalline water, or 14¾ per cent.; 100 water at 61¼° dissolve 39·6 salt, with rise of temperature, when anhydrous (Brandes and Melm, Repert. 14, 97); at 60°, 43·5 (Gay Lussac); at 32°, 32·62; and 0·2711 for every 1·8 degree above 32° (ib.); at 222°, 78 parts (ib.); at 212°, 72 (B. and M.) 100 parts of boiling absolute alcohol, (·7964) dissolve 0·25 parts of the salt (Bucholz, Beitrage, 3, 24); not when cold; 100 parts of alcohol (93·3 per cent.), at 57° 2 dissolve ·0012 salt; at a boiling temperature, ·0020 (Fresenius, *Quantitative Analysis*). Alcohol of ·808 sensibly dissolves it (Thomson); a saturated, or even a strong solution of chloride of barium is precipitated by chlorohydric and nitric acids, and hence, in testing for sulphuric acid with this salt, this fact is to be noted. The *anhydrous chloride* (BaCl), which results from igniting the hydrate, is a white mass, with a specific gravity of 3·7037 (Karsten), 3·8 (Richter), 3·86 to 4·156 (Boullay); when heated in contact with steam, it gives out chlorohydric acid, and becomes alkaline (Kraus, Poggen. Ann. 43, 140). It is decomposed in solution by soda or potash nitre (Karsten). It is not decomposed in the cold by the vapour of anhydrous sulphuric acid (Rose, Pogg. Ann. 38, 120). *Process.*—It may be prepared by dissolving the native or artificial carbonate of barytes in chlorohydric acid, and crystallizing, or by boiling the solution of sulphuret of barium (from the decomposed sulphate with charcoal) with the same acid, until the sulphuretted hydrogen ceases to be evolved; filtering, and crystallizing. Chlorides of strontium and calcium may be removed from it by means of alcohol, which scarcely dissolves chloride of barium. Small portions of these salts may be detected by burning the alcohol, which gives a red with them, and a green when in contact with barytes salts. *Use.*—It is principally employed for testing and precipitating sulphuric acid in solutions; every 14½ parts of the resulting sulphate of barytes are equivalent to 5 sulphuric acid.

Sulphate. Heavy Spar. Barytine. Hepatite. Baroselenite. Cawk. Bolognian Spar. Litheospore. —$BaOSO_3$ 14·51 14·575, 116·6; BaO 65·52, SO_3 34·48. Spec. grav. 4·446 (Mohs). *Crystal.*—Primary form, a right rhombic prism, the faces of which are inclined at angles of

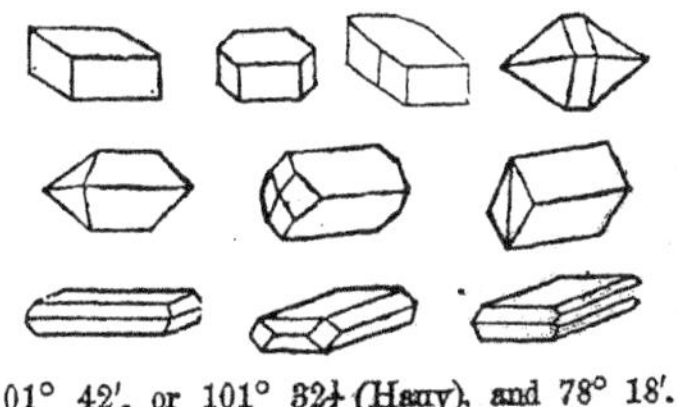

101° 42′, or 101° 32¼ (Hauy), and 78° 18′. It occurs in nature amorphous, and artificially

as a white powder. *Solubility.*—1 part is soluble in 43,000 parts cold water (Kirwan, Mineral. 1, 136); oil of vitriol ($HO\ SO_3$) dissolves it by boiling, but it is again precipitated on the addition of water (Withering); an acid sulphate being formed, which is deposited in needles, but decomposed by water. It becomes red hot when first mixed with the acid (Barry, Ann. Phil. 18, 77), and likewise with anhydrous acid (Kuhlmann). Oil of vitriol (1·848) does not unite with barytes at the usual temperature; but if the barytes, when moistened with oil of vitriol, be touched with a hot iron, the combination extends throughout, and moist air has a similar effect (ib). Alcohol, ether, and pyroxylic spirit, if added to the acid, prevent its action on barytes. Anhydrous sulphuric acid, in vapour, over barytes, is absorbed, and renders it red hot (Bussy, Journ. Pharm. 10, 370). When heated, it decrepitates from the conversion of hygroscopic water between its plates into steam. Before the blowpipe it fuses into an opaque white globule—a temperature of 35° Wedgewood being required (Saussure, Journ. Phys. 45, 15). When made into a paste with flour and water, and ignited, it phosphoresces in the dark, from the sulphide formed by heat probably again uniting with oxygen, and becoming sulphate. An Italian shoemaker, named Vincenzo Casciarolo, first observed that the Bologna stone (found at the foot of mount Paterno), a variety of heavy spar, when ignited became luminous in the dark (Lemery). When boiled with carbonate of potash or soda, sulphate of barytes is partially decomposed ($KO\ CO_2$ and $BaO\ SO_3$, become $KO\ SO_3$ and $BaO\ CO_2$); in the cold state the result is the opposite (Klaproth, Beit. 2, 70, and 73; Dulong, Ann. Chim. 82, 273). When fused with the alkaline carbonates, it is decomposed; fused with chloride of calcium, it yields chloride of barium and sulphate of lime; but in solution the reverse results. When barytes salts are precipitated with sulphuric acid, or sulphates, the salts contained in the liquid are frequently taken down; as nitrates of soda and barytes, but not chloride of barium. Ignition is frequently required in the case of nitrates, and then washing to remove it (Mitscherlich, Pogg. Ann. 55, 214); chloride of magnesium, and oxides of iron, cobalt, copper, &c. fall with the sulphate of barytes, when the sulphates of these bases are mixed with chloride of barium (Berzelius, Ann. Chim. 14, 374). *Sources.*—Sulphate of barytes is found along with galena in the graywacke formations, and likewise in the coal series, and in the old red sandstone conglomerate, as at Arran. It is used to mix with white lead in painting, but is of no value, as it is transparent, and thus injures the white lead. To render it fit for this purpose it is ground by millstones, and then by stones and water, into a fine powder. The powder is digested in sulphuric acid in iron pots, with the application of heat to remove iron. The sulphuric acid is washed out by water and decantation, and the powder dried into cakes like magnesia over a flue. It is frequently mixed with ochre, chrome yellow, &c. according to the colour required.

Nitrate.—$BaONO_5$ 16·25, 16·325, 130·6, BaO 58·47, NO_5 41·53. Spec. grav. 2·9149 (Hassenfratz, Ann. Chim. 28, 12); 3·1848 (Karsten). *Crystal.*—Regular 8-hedrons and cubes, often adhering to each other in the form of stars; the summits of the pyramids are often truncated. It sometimes crystallizes in 3-sided tables, with truncated angles, or in brilliant plates. *Taste* hot, acrid, and austere. *Solubility.*—100 water at 32° dissolve 5 parts, diminishing the temperature; at 58° 15·8 parts; at 120° 17 parts; and at 215° 35·2 parts (Gay Lussac). Insoluble in alcohol and strong nitric acid; its solubility in water is diminished by the presence of nitric acid; and carbonate of barytes is not dissolved by strong nitric acid (Mitscherlich, Pogg. Ann. 18, 159). It decrepitates by heat, and fuses; by ignition it decomposes, yielding oxygen, nitrogen, quaternitrous fumes, and leaving barytes. It detonates less violently with combustibles than most of the other nitrates. *Process.*—This salt may be prepared by dissolving native or artificial carbonate in nitric acid, and crystallizing, or by adding nitric acid to a dilute solution of the sulphuret prepared by decomposing the sulphate by charcoal. Nitrate of soda likewise partially decomposes the sulphide (Mohr, Ann. Pharm. 25, 290.)

Carbonate. Witherite. Barolite.—$BaOCO_2$ 12·25, 12·325, 98·6, BaO 77·55, CO_2 22·45. Specific gravity, native, 4·331; artificial, 3·763 (Thomson); 4·301 (Mohr); 4·3019 (Karsten). *Crystal.*—Right rhombic prisms, with angles of 118·30, and 61° 30′ (Phillips). The usual crystal is a 6-sided prism, terminated by a 6-sided pyramid. *Taste,* none: yet it is poisonous. *Soluble* in 4304 parts cold, and 2304 hot water (Fourcroy, An. Chim. 4, 64); in 14,137 cold, and 15,421 boiling water, and in 141,000 water containing ammonia or carbonate of ammonia (Fresenius, Quant. Analys.) Turns red litmus blue; fuses before the oxyhydrogen blowpipe; decomposed by a strong heat in a black lead crucible, or in the form of a paste with charcoal (Hope). In a blast furnace it loses its carbonic acid (Abich. Pogg. Ann. 23, 314), and at a lower temperature if vapour be present (Priestley). When mixed with sulphate of potash in water, partial double decomposition occurs, but at a higher temperature the reverse occurs (Kölreuter, Gossmann). It dissolves readily in cold solutions of muriate, nitrate, or succinate of ammonia (Vogel, Erdman. Journ. 7, 458). *Source.*—It was first examined by Bergman. Dr. Withering found it native in 1783, and hence it was called Witherite. It occurs in veins along with lead ore in various parts of England, as at Anglesark, near Chorley, in Lancashire, and in Cumberland. It is found at Neuberg, in Styria, and in Hungary, Sicily, Siberia, &c. It is formed artificially

(1) by exposing barytes water to the open air; (2) by decomposing the sulphate by charcoal into the sulphide, and precipitating the yellow solution of the sulphide by carbonate of ammonia, or soda, and washing; or (3) it may be formed by fusing 14½ parts of powdered native sulphate of barytes with 7 parts of anhydrous carbonate of soda, digesting the fused mass in cold water, and washing out the sulphate of soda.

Acetate of Barytes.—$BaOC_4H_3O_33HO$. Oblique rhombic prisms, or 8-sided very soluble prisms, obtained by dissolving carbonate of barytes in acetic acid, and evaporating. This salt is used in analysis when it is necessary to prevent the presence of nitric and chlorohydric acids, in a solution, or to convert a soluble sulphate into an acetate, thence into a carbonate or chloride, as in the estimation of alkaline sulphates.

Barm (Bier-rahm, beer-cream). A synonyme of yeast.

Barolite. Carbonate of barytes.

Baroselenite. Sulphate of barytes.

Barsovite. A mineral resembling scapolite, found at Barsowskoj in the Ural, but distinguishable by its action with the blowpipe and with acids. Spec. grav. 2·74, H 5·5. Colour snow white; massive, sometimes several cubic feet in size; lustre pearly; translucent on the edges; B.B. fuses with difficulty. With borax, fuses into a transparent glass; with muriatic acid it forms a jelly. *Composition.* SiO_3 49·08, Al_2O_3 32·76, CaO 18·16. *Form.* 3 $CaO2SiO_3$, 3 (Al_2O_3 SiO_3).

Baryto Calcite, of *Brooke* and *Children*, *Alstonite. Calcareo-carbonate of Barytes.* $BaOCO_2CaOCO_2$. White, with a shade of gray, yellow, or green oblique rhombic prisms, with angles of 106° 54′ and 102° 54′. Spec. grav. 3·66 to 3·636 H 4·; lustre vitreous, inclining to resinous. B.B. does not fuse, but melts with borax into a transparent glass; it gives indications of some iron and manganese. It contains CaO CO_2 31·65, $BaOCO_2$ 62·20, with some impurities, but when quite pure it has been found to consist of $CaOCO_2$ 31·89, $BaOCO_2$ 66·2, silica ·27. It occurs only at Alston Moor.

Baryto Calcite *of Bromley Hill. Bromlite. Bicalcareo-carbonate of Barytes. Dimorphous baryto-calcite.* Usually occurring in the form of two 6-sided pyramids, applied base to base; the fracture, however, indicating that these pyramids are formed of 3 quadrangular pyramids with a rhombic base, penetrating so as to confound their axes; belonging, therefore, to the right prismatic system, or doubly-oblique rhombic system. Colour snow-white; translucent; lustre vitreous. Spec. grav. 3·718, H 2·25. When pure, it consists of the same ingredients as the preceding. $BaOCO_2$ 65·31, CaO CO_2 32·9, $SrOCO_2$ 1.1, SiO_3 ·20, MnO ·16; its formula being Ba OCO_2 $CaOCO_2$. It is, therefore, dimorphous baryto calcite. It is found at Fallowfield and Bromley Hill, near Alston Moor, in Cumberland.

Baryto Fluorspar. Specific gravity 3·75. $BaOSO_3$ 51·5, Ca Fl 48·5, = $BaOSO_3$, 3 Ca Fl. Constitutes a bed an inch thick in limestone in Derbyshire.

Baryto Strontianite, *Stromnite.* $SrOCO_2$ 68·6, BaO SO_3 27·5, $CaOCO_2$ 2·6, FeO ·1. Spec. grav. 3·7, H 3·5. Colour grayish-white; lustre shining; fracture glistening and resinous; radiated, or in small plates; brittle. B.B. infusible. Found at Stromness.

Baryto Sulphate of Strontian. *Radiated Celestine. Baryto Celestine.* $BaOSO_3$ 35·19, $SrOSO_3$ 63·20, $FeOSO_3$ 1·24, HO ·72. Spec. grav. 3·921, H 2·75. Colour white, with a blue shade; texture laminated, or in imperfect crystals. B.B. becomes of a dazzling white, but does not easily fuse. Found in Drummond Island, Lake Erie, and at Kingston.

Basalt. Spec. grav. 2·87 to 3·. A grayish or bluish-black rock with a splintery fracture, consisting of a black base of basanite or flinty slate, with crystals of augite and olivine interspersed. It has also been described as an intimate mixture of augite, labradorite, and magnetic iron. When of a columnar form, as at the Giants' Causeway and Staffa, the form of the pillar is a pentagonal or 5-sided prism. Before the blow-

pipe it melts into a green glass, and is attracted by the magnet. It frequently contains augite, olivine, and hornblende, mica, hyacinth, obsidian, and titaniate of iron. It is frequently confounded with fine-grained greenstone, which, however, consists of albite and hornblende in about equal proportions. That which occurs near Edinburgh was analyzed by Dr. Kennedy in 1800, and found to consist of SiO_3 48·, Al_2O_3 16, CaO 9, NaO 4, FeO 16, HCl 1·, HO 5·; and from Staffa, SiO_3 44·5, Al_2O_3 16·75, CaO 9·5, FeO 20·2, MgO 2·25, MnO. 12, NaO 2·6, HO 2. Ebelmen found basalt from Polignac to contain SiO_3 53·, Al_2O_3 18·4 CaO 6·8, FeO 9·5, MgO 3·5, KO 2·7, NaO 3·1, HO, and organic matter 3·7. From the action of acids on basalt, it has been represented as composed of zeolite 39·81, augite 55·5, FeO 4·61 (Lowe); but this is certainly not constant. Basalt, when fused and cooled rapidly, presents the appearance of bottle glass, but when slowly cooled it resumes its crystalline aspect. It has, therefore, been known for upwards of fifty years that trap rocks contained water, although the fact has recently been announced as a new discovery. The presence of water distinguishes the trap rocks from lava or recent volcanic rocks, which usually contain a mere trace of hygroscopic water. According to C. Gmelin, basalt is resolvable by acids into a decomposable and a nondecomposable part.

Basaltic Hornblende. Hornblende occurring in basalt.

Basaltine, or pyroxene.

Basanite. *Lydian Stone. Touchstone. Flinty Slate.* Spec. grav. 2·585 to 2·644. H 7· Gray, bluish, yellow, brown, and red. When grayish-black called lydian stone; the other colours belong to flinty slate; fracture of flinty slate is slaty; that of basanite even; lustre glimmering; composed principally of silica, its colour being due, according to Humboldt, to carbon; it occurs in beds in trap rocks, &c. It is considered by some to be the basis of basalt. It occurs at Niederminnig and Mayen on the Rhine, where it is used for building and for millstones.

Basanomelanite. Titaniferous iron.

Basicirine. Hydrofluoride of cerium.

Base. A term usually applied to those bodies which are capable of uniting with acids to form salts, and are replaced by other bases. For example, ammonia (NH_3), when neutralized by sulphuric acid ($HOSO_3$), constitutes the base of the *salt* sulphate of ammonia (NH_3SO_3HO). When we add to this salt caustic potash (KO), the smell of ammonia is evident; in other words, the base ammonia is replaced by the base potash, and instead of the salt sulphate of ammonia, we have now formed sulphate of potash ($KOSO_3$). The term base, in this case, is used instead of the older term *alkali*, a name which only applies to certain bases. But each of the bases and acids possesses likewise a base. Instead, however, of employing this expression in such instances, it is now usual to speak of the *radical* or root of a base or acid. Hence we have *basic* and *acid radicals*. In ammonia, nitrogen is the radical of that base, or *basic radical;* and in sulphuric acids, sulphur is the radical of that acid, or *acid radical*. Some bodies are both base and acid formers, hence they have been sometimes termed *amphigene bodies* (*both formers*). These are oxygen, sulphur, selenium, and tellurium; and the salts formed by them are thence called amphide salts. A *basic* or *subsalt* is a compound consisting of an acid and a base, in which the base preponderates over the acid. Thus, corrosive sublimate consists of 1 atom chlorine and 1 atom mercury (HgCl), and is an equisalt, or neutral salt, while calomel consists of 2 atoms mercury, and 1 atom chlorine (Hg_2Cl), and is a *subsalt*, or *basic salt*, or *disalt*. Bases are divisible into *inorganic* and *organic bases*.

The inorganic bases consist of *metallic* bases and *oxymetallic* bases. Thus, in chloride of potassium (KCl), potassium is the metallic base; while in sulphate of potash (KO,SO_3) the base is an oxymetallic one, viz. potash.

Bases Organic.—A class of organic substances, many of them existing ready formed in plants, which, like inorganic oxymetallic bases, unite with acids, form salts, and are capable of being replaced by other bases, particularly the inorganic bases. For example, disulphate of quinine ($\overset{+}{Qu_2}SO_3$), so familiar as a tonic medicine when dissolved in water, with the aid of a little sulphuric acid, so as to render it an equal or neutral salt, has all its quinine removed from the sulphuric acid, and precipitated in the form of white flocks, when caustic ammonia is added to the solution ($\overset{+}{Qu}SO_3$ and NH_3) becoming NH_3SO_3 and $\overset{+}{Qu}$). From the circumstance that many organic bases have recently been formed artificially in the laboratory, they have excited a good deal of attention, and perhaps younger chemists have been too much carried away with the novelty of the subject to the neglect of more important, though less striking departments of the science. The fact that the alkaloids or organic bases contain nitrogen in small quantity, had early attracted the attention of chemists. Berzelius supposed that they might be conjugate compounds of ammonia, with an organic body: methylamine (C_2NH_5), for example, he would have considered a compound of an organic body, C_2H_2 with NH_3, or ammonia. Liebig, on the other hand, conceived that these bases were all formed on the type or model of ammonia (NH_3), in which the third atom of hydrogen was capable of being replaced by an organic substance; that atom being removable under various circumstances; whence it may be viewed as an amide of hydrogen (H,NH_2). By this view, methylamine will not be C_2H_2,NH_3, but NH_2, C_2H_3, or ammonia, in which the third atom of hydrogen is replaced by methyle, a gas which can be isolated by another process. This idea has been fully confirmed by further researches; and it has been shown that an infinite series of alkaloids or organic bases may be artificially formed upon the models of ammonium (NH_4), ammonia (NH_3), amide (NH_2), imide (NH), and nitryle, Nx_3, where the whole of the hydrogen is replaced by organic substances. See TABLE. In the artificial preparation of many alkaloids the general principles are as follows:— 1. Certain of them are produced by the action of ammonia on an organic body, frequently an oil: oil of mustard with ammonia becomes *thiosinnamine;* others of this class are urea, furfurine, fucusine, amarine, melamine, ammeline, lophine, aniline. 2. Alkaloids formed by reduction by sulphuretted hydrogen: aniline, toluidine, naphthalidine. 3. By distillation with potash: quinoleine, aniline, coniine, nicotine. 4. By dry distillation of organic bodies: aniline, picoline, petinine, nicotine, lophine. 5. By decomposition of a nitrogenous acid, aniline. 6. Removal of sulphur from a sulphuretted alkaloid as sinnamine from thiosinnamine. 7. Removal of sulphur from an essential oil, as sinapoline from oil of mustard. 8. Alteration of natural alkaloids by oxidation, as narcogenine, from narcotine. It is necessary to distinguish between organic bases, or alkaloids, and organic radicals; the alkaloids, in fact, being built up principally of organic radicals. The distinguishing character of these two classes of organic bodies consists in the circumstance that the organic bases, like metallic oxides, may be replaced by metallic oxides. Thus, disulphate of quinine and caustic ammonia ($\overset{+}{Qu}_2SO_3$ and NH_3) form a white solid quinine, which falls, and sulphate of ammonia ($\overset{+}{Qu}$ and NH_3SO_3); but no such decomposition takes place with the compounds of organic radicals with acids. Thus, with chloride of ethyle, ammonia does not remove the chlorine, neither does nitrate of silver, when added to it, form chloride of silver.

Natural Alkaloids from the Vegetable Kingdom, and Bases of Cinchona Bark.

Quinine,	$C_{20}H_{12}NO_2$
Cinchonine,	$C_{38}H_{22}N_2O_2$
Aricine,	$C_{20}H_{12}NO_3$?
Quinoidine,	Pitayine,
Chinovine,	Blanquinine,
Cinchovatine,	$C_{46}H_{27}N_2O_8$

Bases of the Papaveraceæ and Derivatives.

Morphine,	$C_{34}H_{19}NO_6$
Codeine,	$C_{36}H_{21}NO_6$
Nitro-codeine,	$C_{36}H_{20}NO_4NO_6$
Bromo-codeine,	$C_{36}H_{20}BrNO_6$
Chloro-codeine,	$C_{36}H_{20}ClNO_6$
Dicyano-codeine,	$C_{40}H_{21}N_3O_6$
Papaverine,	$C_{40}H_{21}NO_8$
Opianine,	$C_{66}H_{36}NO_{23}$
Thebaine,	$C_{38}H_{21}NO_6$
Pseudomorphine,	$C_{27}H_{18}NO_{14}$
Narceine,	$C_{46}H_{29}NO_{18}$
Narcotine,	$C_{46}H_{25}NO_{14}$
Methyl-narcotine,	$C_{44}NH_{23}O_{14}$
Ethyl-narcotine,	$C_{46}NH_{25}O_{14}$
Propyl-narcotine,	$C_{48}NH_{27}O_{14}$
Cotarnine,	$C_{26}H_{12}NO_5$
Narcogenine,	$C_{36}H_{19}NO_{10}$
Chelidonine,	$C_{40}H_{20}N_3O_6$
Chelerythrine,	?
Glaucine,	?
Glaucopicrine,	?

Bases of the Solanaceæ and other Vegetable Families.

Hyosciamine, ?	Stramonine, ?
Daturine, ?	Sabadilline,
Atropine,	$C_{34}H_{23}NO_6$?
Solanine,	$C_{84}H_{68}NO_{28}$?
Veratrine,	$C_{24}H_{26}NO_6$
Colchicine, ?	Aconitine, ?
Delphine,	$C_{27}H_{19}NO_2$?
Staphisine,	$C_{32}H_{23}NO_4$?
Emetine,	$C_{37}H_{27}NO_{10}$?
Chiococcine, ?	Violine, ?
Strychnine,	$C_{42}H_{22}N_2O_4$
Chlorostrychnine	$C_{44}H_{23}ClN_2O_4$?
Brucine,	$C_{46}H_{26}N_2O_8$
Bromobrucine,	$C_{46}H_{25}BrN_2O_8$
Jervine,	$C_{60}H_{45}N_2O_5$
Curarine,	Daphnine,
Corydaline,	$C_{50}H_{30}NO_{20}$?
Carapine,	?
Cusparine,	?
Bebeerine,	$C_{38}H_{21}NO_6$?
Sanguinarine,	Azadirine,
Capsicine,	Crotonine,
Buxine,	
Apyrine,	
Cynapine,	
Cissampeline, Pelosin,	$C_{36}H_{21}NO_6$
Pelluteine,	$C_{42}H_{21}NO_7$
Oxyacanthine,	
Berberine,	$C_{42}H_{18}NO_9$
Surinamine,	Jamaicine,
Piperine,	$C_{70}H_{37}N_2O_{10}$
Piperidine,	$C_{10}H_{11}N$
Amylpiperidine,	$C_{10}H_{11}N, C_{10}H_{11}$
Ethyl-piperidine,	$C_{10}H_{11}N, C_4H_5$
Methyl-piperidine,	$C_{10}H_{11}N, C_2H_3$
Menispermine,	$C_{18}H_{12}NO_2$
Paramenispermine,	$C_{18}H_{12}NO_2$
Harmaline,	$C_{27}H_{14}N_2O_2$
Harmine,	$C_{27}H_{12}N_2O_2$
Hydrocyanharmaline,	$C_{29}H_{15}N_3O_2$
Nitroharmalidine,	$C_{27}H_{13}N_3O_6$
Caffeine,	$C_{16}H_{10}N_4O_4$
Theobromine,	$C_{18}H_{10}N_6O_4$?
Castine,	Cicutine,
Chœrophylline,	Esenbeckine,
Digitaline,	
Apiine,	$C_{24}H_{14}O_{13}$?
Eupatorine,	Euphorbine,
Convolvuline,	Pereirine,
Asparagine,	$C_8H_{10}N_2O_8$
Santonine,	$C_{30}H_{18}O_6$
Peucedanine,	$C_{24}H_{12}O_6$?
Nitropeucedanine,	$C_{24}H_{11}NO_4O_6$
Columbine,	$C_{42}H_{22}O_{14}$?
Anisidine,	$C_{14}H_9NO_2$

Volatile Oily Bases.

Nicotine,	$C_{10}H_8N$
Coniine,	$C_{16}H_{16}N$?

Alkaloids found in the Animal Organism.

Urea,	$C_2N_2H_4O_2$
Creatinine,	$C_8N_3H_7O_2$

Volatile Oily Artificial Alkaloids.

Aniline,	$C_{12}H_7N$
Chloraniline,	$C_{12}H_6ClN$
Dichloraniline,	$C_{12}H_5Cl_2N$

Trichloraniline,	$C_{12}H_4Cl_3N$
Bromaniline,	$C_{12}H_6BrN$
Dibromaniline,	$C_{12}H_5Br_2N$
Tribromaniline,	$C_{12}H_4Br_3N$
Iodaniline,	$C_{12}H_6IN$
Nitraniline,	$C_{12}H_6NO_4N$
Ethyl-nitraniline,	$C_{12}H_4NO_4, HEN$
Diethyl-chloraniline,	$C_{12}H_4ClNE_2$
Cyaniline,	$C_{14}H_7N_2$
Melaniline,	$C_{26}H_{13}N_3$
Dichloromelaniline,	$C_{26}H_{11}Cl_2N_3$
Dibromomelaniline,	$C_{26}H_{11}Br_2N_3$
Di-iodomelaniline,	$C_{26}H_{11}I_2N_3$
Dinitromelaniline,	$C_{26}H_{11}2NO_4N_3$
Dicyanomelaniline,	$C_{26}H_{13}2C_2N$
Odorine or Picoline,	$C_{12}H_7N$
Toluidine,	$C_{14}H_9N$
Cyano-toluidine,	$C_{14}H_9NCy$
Lutidine,	$C_{14}H_9N$
Xylidine,	$C_{16}H_{11}N$
Cyano-cumidine,	$C_{18}H_{13}NCy$
Nitro-cumidine,	$C_{18}H_{12}NO_4N$
Cacotheline,	$C_{42}H_{22}N_4O_{20}$
Leucoline,	$C_{18}H_8N$
Nitromesidine,	$C_{18}H_{10}2NO_4$

Artificial Alkaloids from Oil of Mustard, &c.

Thiosinnamine,	$C_8H_8N_2S_2$
Sinnamine,	$C_8H_6N_2$
Sinapine,	$C_{32}H_{26}NO_{12}$
Sincaline,	$C_{10}H_{14}NO_2HO$
Sinapoline,	$C_{14}H_{12}N_2O_2$
Thialdine,	$C_{12}H_{13}NS_2$
Selenaldine,	$C_{12}H_{13}NSe_2$
Melamine,	$C_6H_6N_6$
Ammeline,	$C_6H_5N_5O_2$
Amarine,	$C_{42}H_{18}N_2$
Lophine,	$C_{46}H_{16}N_2$
Picrine,	$C_{42}H_{15}NO_4$
Furfurine,	$C_{30}H_{12}N_2O_6$
Fucusine,	$C_{30}H_{12}N_2O_6$

Artificial Alkaloids from Animal Products.

Glycocollin,	$C_4H_5NO_4$
Alanine,	$C_6H_7NO_4$
Leucine,	$C_{12}H_{13}NO_4$
Sarcosine,	$C_6H_7NO_4$
Guanine,	$C_{10}H_5N_5O_2$

From Naphthaline.

Semi-naphthalidine,	$C_{10}H_5N$
Naphthalidine,	$C_{20}H_9N$

1. *With the type of Ammonium*, (NH_4).

Ammonium,	N, H_4OHO
Tetramethyl-ammonium,	N, Me_4OHO
Tetr-ethyl-ammonium,	N, E_4OHO
Triethyl-amyl-ammonium,	$N, E_3AylOHO$
Tetr-amyl-ammonium,	N, Ayl_4OHO
Triethyl-methyl-ammonium,	N, E_3MeOHO
Diethyl-methyl-amyl-ammonium,	N, E_2Me, Ayl, O, HO
Triethyl-phenyl-ammonium,	N, E_3PhOHO
Methyl-ethyl-amyl-phenyl-ammonium,	N, E, Me, Ayl, Ph, O, HO

Artificial Alkaloids referred to Ammonia, (Amine).

2. *With a base of Amide*, (NH_2).

Ammonia,	NH_2H
Methyl-amine,	NH_2, C_2H_3
Formyl-amine,	NH_2C_2H
Ethyl-amine,	NH_2, C_4H_5
Propyl-amine,	NH_2, C_6H_7
Butyl-amine,	NH_2, C_8H_9
Amyl-amine,	$NH_2, C_{10}H_{11}$
Caprol-amine,	$NH_2, C_{12}H_{13}$
Phenyl-amine, aniline,	$NH_2, C_{12}H_5$
Toluyl-amine, toluidine,	$NH_2, C_{14}H_7$
Xylamine, xylidine,	$NH_2, C_{16}H_9$
Cumidine,	$NH_2, C_{18}H_{11}$
Cymidine,	$NH_2, C_{20}H_{13}$
Sinethylamine,	$C_{12}H_{10}N_2$
Thio-sin-ethylamine,	$C_{12}H_{12}N_2O_2$

3. *With a base of Imide*, (NH).

Ammonia,	NH, H_2
Di-methyl-amine,	$NH, 2(C_2H_3)$
Di-ethyl-amine,	$NH, 2(C_4H_5)$
Di-amyl-amine,	$NH, 2(C_{10}H_{11})$
Di-phenyl-amine,	$NH, 2(C_{12}H_5)$
Methyl-phenyl-amine, (methyl-aniline,)	$NH2C_2H_3, C_{12}H_5$
Ethyl-phenyl-amine, (ethyl-aniline,)	$NHC_4H_5, C_{12}H_5$
Amyl-phenyl-amine, amyl-aniline,	$NHC_{10}H_{11}, C_{12}H_5$

4. *With a base of Nitryle*, (Nx_3).

Ammonia,	NH_3
Triethyl-amine,	NE_3
Trimethyl-amine,	NMe_3
Triamyl-amine,	$NAyl_3$
Triphenyl-amine,	NPh_3
Diethyl-amyl-amine,	NE_2Ayl
Diethyl-phenyl-amine,	NE_2Ph
Methyl-ethyl-amyl-amine,	NMe, E, Ayl
Methyl-ethyl-phenyl-amine,	NMe, E, Ph
Diamyl-phenyl-amine,	$NAyl_2Ph$
Ethyl-amyl-phenyl-amine,	$NE, Ayl Ph$

Basil, Oil of. An aromatic ethereal oil, distilled from the root of the *Ocymum basilicum*, a plant of the family of *Labiatæ*. It deposits a stearoptene in 4-sided pyramids ($C_{20}H_{22}O_6$); little soluble in cold, very soluble in hot water; again deposited on cooling; partly soluble in cold alcohol; the solution reddening vegetable blues. They are soluble in boiling alcohol, ether, nitric and acetic acids. Sulphuric acid turns it red.

Bassic Acid. $C_{36}H_{36}O_4$. Fusing point, 159°. Crystalline fatty acid in the fat of Bassia latifolia and Coculus indicus. In 1842, I extracted an acid from the Shea butter of Africa, a fat from a species of Bassia, and termed it bassic acid; described and showed it in my lectures, but afterwards inferred that it was somewhat impure margaric acid. I think it highly probable that the present acid is of the same nature.

Bassora Gum. Spec. grav. 1·3591. Colour yellowish-white. A gum obtained from different species of acacia (A. leucophlæa?). Besides bassorin, it contains chlorophyllin, wax, acetate of potash, chloride of calcium, and bimalate of lime, but consists essentially of—water, 21·89; ash, 5·6; arabin, 11·2; bassorin, 61·31.

Bassorine. $C_{12}H_{11}O_{11}$. A solid; colourless, semitransparent, insipid, inodorous, and amorphous; it is tough, and not easily pulverized; insoluble in water, but swells up, and becomes like a jelly; insoluble in alcohol, and does not ferment. It is isomeric with starch, gum, and sugar; 100 parts heated with 1000 nitric acid form 22·61 mucic acid, and some oxalic acid; forms a crystalline sugar with sulphuric acid. It is obtained by exhausting gum bassora, or tragacanth, with cold water; bassorin remains in a gelatinous form. It exists in gum bassora, gum tragacanth (Astragalus tragacantha), and in gum kuteera (Sterculio urens).

Bastœsite. Fluohydride of cerium.

Bath, in chemistry, is a term applied to apparatus employed for the purpose of communicating a graduated temperature to bodies. Various fluids and solids are employed as the means through which the heat is applied.

Water Bath, Bain Marie, Bain d'Eau, Wasserbad, Marienbad. — In this bath, the substance to be heated can never have its temperature elevated above 212°, or a steam heat. It is a form of bath very generally in requisition, and it is used in a great variety of forms in the laboratory. One of the simplest shapes is a porcelain or salt ware cup, into which a smaller one fits, the intermediate space being filled with water, to which heat is applied below; the steam making its way at the points of junction, or by an aperture left for the purpose. A saucepan may be converted into a temporary steam chamber, so as to heat flasks, a piece of tin plate perforated answering as a cover (*b*). A common kettle, filled with water, has a tin box fitted through its lid, for the reception of the flasks, which become thus surrounded with

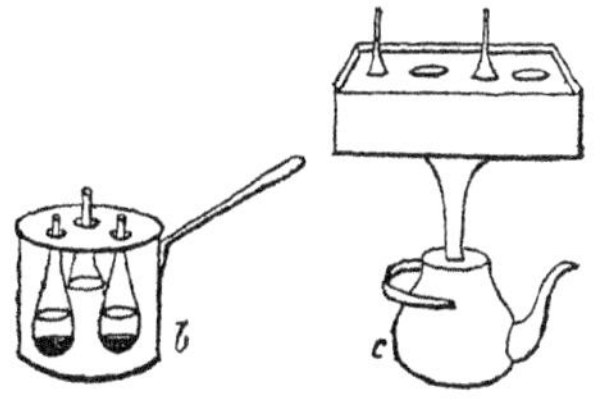

Water Bath.

steam (*c*). For the purpose of drying organic substances, or others which require a long sustained steady temperature, we employ a double copper box, which, if made with hard solder, will answer also as an oil bath, where a higher temperature is required to be sustained than soft solder can resist (*d*). It has one aperture, which merely enters the external box by which steam escapes. By the other, a thermometer may be adjusted through a perforated cork, and may have its bulb immersed in the steam.

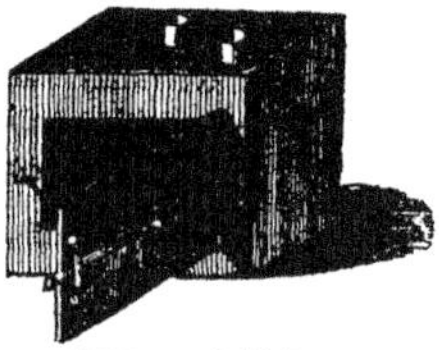

d, Water and Oil Bath.

Saline baths are used when a somewhat higher temperature than a steam heat is desirable. Salts, when dissolved in water, elevate the boiling point of the fluid, although the steam has exactly the usual temperature of 212°. The following saturated solutions boil at the annexed temperatures:—

Sulphate of soda,	213°·26
Alum,	213·8
Acetate of lead,	215·6
Chlorate of potash,	219·5
Chloride of barium,	220·0
Borax,	220·0
Chloride of potassium,	226·9
Chloride of sodium,	227·3
Sal ammoniac,	237·9
Nitrate of potash,	240·6
Chloride of strontium,	244·2
Nitrate of soda,	249·8
Chlorate of potash,	275·
Nitrate of potash,	303·8
Chloride of calcium,	355·
Nitrate of ammonia,	361·8
Chloride of zinc,	572·
Oil of vitriol,	636·8

Steam and Dry Air Bath.—When an organic body requires to be dried in a dry atmosphere, especially if it has a tendency to absorb moisture when exposed to the air, it is necessary to place it in a tube and draw dry air over it, in order to remove the moisture which emanates from it. For this purpose, the apparatus (*e*) is used. The body is placed in a bent tube, which

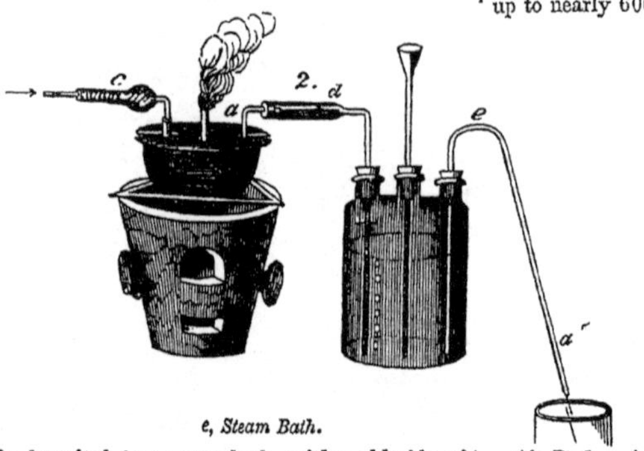

e, Steam Bath.

is deposited in a water bath, with a chloride of calcium tube on either side. The aspirating bottle, filled with water, causes air dried by the first chloride of calcium to pass over it, until it is thoroughly dry.

Steam and Vacuum Bath.—When the atmospheric air acts on the substance, and when it is desirable to exhaust the moist air rapidly, an air-pump is used as in fig. *f.* A, Air-pump, fixed by a screw into a table; B, chloride of

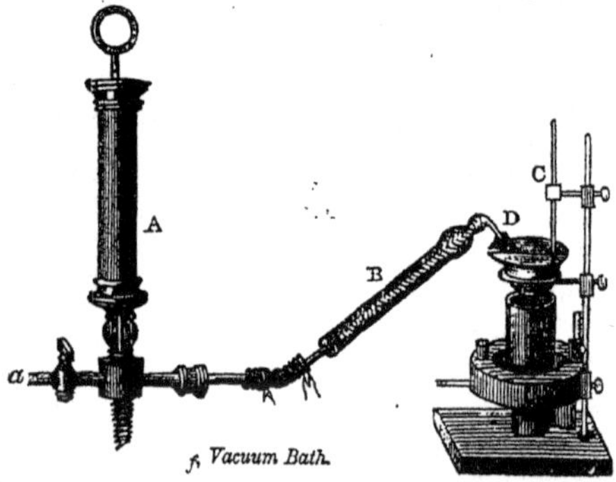

f, Vacuum Bath.

calcium tube; D, tube containing the substance to be dried, hermetically sealed at C, and placed in a water bath.

Metallic Baths.—When the temperature requires to be higher than can be attained by the preceding media, a mixture of easily fusible metals is used. Mercury may be employed between 300° and 400°; above this, unwholesome fumes are evolved. D'Arcet's fusible alloy is useful, consisting of 2 bismuth, 5 lead, 3 tin, with a fusing point of 212°, the temperature rising with the application of heat.

Oil Baths.—Oil answers very well for high temperatures to be long sustained. The fluid may be introduced into a simple bath, or into the copper bath, (fig. *d*,) and the heat regulated by a thermometer. The temperature may be carried up to nearly 600° by an oil bath.

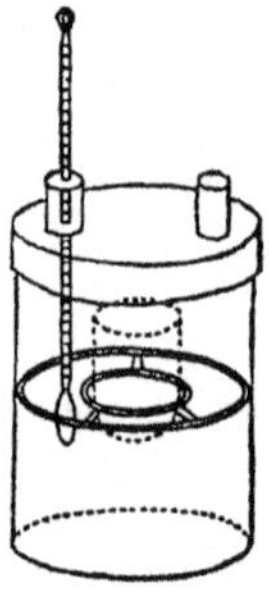

g, Air Bath.

Air Bath.—An air bath, *g*, may be formed of tin plate, and may consist of a cylindrical chamber, with a valve at the side, and a tall tube above, so as to produce a current of hot air over the surface of the body to be dried. A thermometer, adjusted to an aperture on the upper surface, enables the heat to be regulated, which may be applied by means of gas, or a spirit lamp.

Sand Bath, h.—For the evaporation of fluids, in glass vessels and in porcelain basins heat may

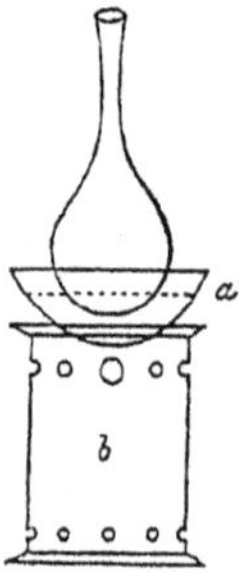

h, Sand Bath.

be applied through the medium of sand placed on an iron plate, or fire tiles. For distillation, a pot of sand is often employed for the purpose of communicating an equable temperature, and preventing fracture of glass vessels, or destruction of the substances by too elevated heat. A cast or wrought iron tray, sprinkled over with sand, and placed over a gas flame or spirit lamp, or over a fire, furnace, or stove, properly regulated so as to moderate the heat, will be found a very useful form of sand bath, where more expensive apparatus cannot be obtained.

Batrachite. According to some, a variety of peridote or of olivine. Spec. grav. 3·038, H. 5·, SiO_3 37·69, CaO 35·45, MgO 21·79, FeO 2·99,

HO 1·27. Massive, with traces of the cleavage of a rhombic prism of 115°; colour light greenish-gray; lustre resinous or vitreous. B.B. infusible; soluble in salt of phosphorus: it is found at Bizoni, in S. Tyrol.

Baudisserite. Carbonate of magnesia.

Baulite. Small crystals of the oblique prismatic system. Spec. grav. 2·656, H 5·5 to 6. B.B. fusible in thin splinters into a colourless glass; not attacked by muriatic acid; allied to felspar. SiO_2 80·23, Al_2O_3 11·71, CaO 1·46, NaO 2·26, KO 4·92. Form. RO 2 SiO_3, Al_2O_3, 6 SiO_3. Found in the volcanic mountain Baula.

Bavalite. Carbono-phosphate of iron, chamoisite, Berthierine.

Bdellium. A gum resin from the Levant (amyris?), in reddish angular masses, with a disagreeable smell, nauseous taste, having some analogy to turpentine; becomes soft between the teeth; burns like resin. Spec. grav. 1·371; alcohol dissolves $\frac{3}{8}$ of its weight. The resin consists of $C_{40}H_{31}O_5$. The gum resin consists of resin 59, gum 9·2, mucilage 30·6, vol. oil 1·2. A variety, in oval tears, comes from Senegal (from Hendolotia Africana).

Bean. The seed of *Vicia faba*, &c. The garden bean is sometimes used to adulterate coffee. It consists of very thick cells, filled with large starch grains, which become blue with iodine The seed skin is formed of hexagonal cellular tissue. I have found the composition of the ash of horse beans—silica 13·12, PO_5

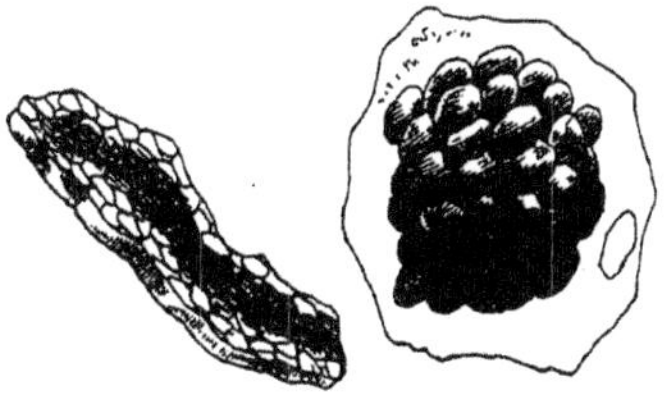

[*Hooker.*]

35·26, SO_3 1·29, Cl 1·75, CaO 5·18, Fe_2O_3 1·8, MgO 9·03, KO 23·15, NaO 9·42.

Bearberry. The leaves of Uva ursi are used in tanning, from their containing tannin.

Beaumontite. Spec. grav. 2·24. Honey-yellow crystals, similar to heulandite, of which it is supposed to be a variety. SiO_3 64·2, Al_2O_3 14·1, CaO 4·8, MgO 1·7, FeO 1·2, HO 13·4, NaO and loss ·6.

Bebeerine, $C_{38}H_{21}O_6$.? A white bitter powder, obtained by the same process as quinine from the bebeeru or red heart bark of Demerara. The sulphate is used in medicine as a tonic and febrifuge. It occurs in the form of shining scales, and is extensively manufactured by Messrs. Inerarity of Glasgow.

Bee Poison. The nature of this substance is unknown.

Beer, from the German *Bier*. The term is often applied to the class of fermented liquors, without discrimination of strength; but it is perhaps proper to restrict it to the weaker kinds of malt liquors, usually known as *small* and *table* beers. Malt liquors brewed in families are termed either ales or beers, and to such description of fluids this article will be confined. The worts of small and table beers of the commercial brewer are usually obtained from the malt from which previously the worts for ale or porter have been extracted. The operations in making beer are, 1. Malting, see MALT; 2. mashing; 3. boiling and cooling the wort; 4. fermenting the wort, see FERMENTATION. In home brewing, various qualities of beer are produced. 1. The best description of beer is made upon a considerable scale; for example, 9 hogsheads of ale from 9 bushels of malt to each hogshead, with 1 lb. hops to each bushel of malt. *Mashing.* — 12 hogsheads of water, at 170° of temperature, are poured on 81 bushels of malt, in a wooden vessel or mash tun. This is stirred for a quarter of an hour, and allowed to extract for three hours, when the wort is drawn off; 7 hogsheads of wort are thus obtained, weighing 136 lbs. saccharum, per barrel of 36 gallons. 6 hogsheads of water, at 180°, are then added for the second mash, for two hours, which yield 6 hogsheads of wort of 80 lbs. per barrel. 5 hogsheads of water, at 170°, are then added for table beer, which, after infusing one hour, are drawn, at the weight of 30 lbs. per barrel. *Boiling.*—The 13 hogsheads of worts, weighing 103 lbs. saccharum, are now boiled in the copper for an hour; 40 lbs. of hops are added, and the boiling continued for half-an-hour. The remaining 41 lbs. hops are then added, and the boiling repeated for half-an-hour; the fire withdrawn, and the worts allowed to remain for half-an-hour in the boiler. They are then filtered through the hop back, and cooled down in a cooler, by evaporation, to 60°. Half a barrel of wort is then run into the gyle tun, with which 6 gallons of yeast are mixed, and the rest of the worts then run on. In forty-eight hours the fermentation will be complete. 1 lb. salt and 3 lbs. flour are now thrown into the gyle, and the whole thoroughly mixed. *Cleansing* consists in separating the beer or ale from the yeast, and may be performed by placing the worts in pipes or butts set on their bilge on stillions. The pipes must be filled up every hour for the first six hours, then every two hours until the worts come to form yeast, then every four hours until cleansed. In six weeks after cleansing, the beer has reached the stage of first fining, and may be bunged down and used. But it is usual to hop down the new ale for keeping, by putting 6 lbs. of hops into each pipe of new ale, which is sufficient to keep it over summer. The bungs are left out for the first two or three months, and the cask being kept full, the hops remain at the surface, and prevent the beer from becoming acid by the action of the air. Where much ale is kept, it is necessary, before the access of the summer, to put in 1½ lb. hops into each pipe, which are kept slack bunged, with a vent plug in each cask. 2. In brewing on a smaller scale, as for example, 2 barrels or 72 gallons—9 bushels of malt, with 12 lbs. of hops, are used. The malt is ground small and mashed with 72 gallons of water, at 160°. The mash tun is covered with sacking for three hours; 40 gallons of wort are then drawn and mixed with the 12 lbs. of hops, while 60 gallons of water, at 170°, are run into the mash tun for a second mash, and drawn off in two hours; the worts are heated for two hours, cooled to 65°, strained through a flannel bag into the gyle tun, 1¼ gallon of yeast is then added and allowed to ferment for twenty-four to thirty hours, and cleansed as already described. 3. On a still smaller scale beer may be economically made. A wine pipe or rum puncheon sawn in two makes both a mash and gyle tun. An iron boiler of the capacity of 36 to 54 gallons, will answer for boiling the worts; and a cooler, with a few casks, can easily be obtained for the rest of the process. The proportions used may be 3 bushels malt, 2½ to 3 lbs. hops, and half a gallon of yeast, to make 36 gallons of beer. It is necessary that all the vessels should be thoroughly cleansed, and the presence of metals in contact with the beer should be avoided as much as possible. When litmus paper dipped into the fermenting worts indicates the formation of much acid, by rise of temperature, it may be overcome by the addition of a small portion of carbonate of soda. 100 gallons of pure water are boiled, and cooled until the temperature reaches 180°; 32 gallons are run into the mash tun; 3 bushels of ground malt are added, a bushel at a time, and mixed. The tun is then covered for three hours, and drawn off to the extent of 22 gallons, the remainder being absorbed by the malt; 34 gallons water, at 180°, are then run in for the second mash, and stirred and covered up for two hours, and run off; of the remaining water in the boiler, 14 gallons are used for a third mashing—the product after fermentation being *small beer*, and kept separate. The 54 gallons of worts are now boiled for half-an-hour, the hops added, and the boiling continued for an hour, the fire drawn, and in half-an-hour the whole filtered through the hop-drainer into the coolers, and cooled down to 65°. Half a gallon of yeast is added to the worts in the gyle tun, and fermented for thirty or thirty-six hours. When the head of yeast begins to sink, the worts are placed in casks and allowed to work over into a tub or trough, and are kept filled up every two hours until the yeast separates, and the fermentation ceases. The worts from 54 gallons become reduced, by waste, &c. to 36 gallons. (Dr. T. Thomson on Brewing.)

Beetroot. (*Beta Vulgaris.*) *Betterave, Runkelrübe.* This plant is cultivated extensively for the sugar contained in its root. In the white beet the amount per cent. varies from 5·8 to 9·2, and in the red beet from 7·5 to 10 per cent. The quantity of beetroot sugar manufactured on the continent of Europe is rapidly increasing. In Germany, in 1848, the quantity of sugar raised was 26,000 tons; in 1849, 34,000; in 1850, 40,000; and in 1851, 43,000. In Russia, in 1851, it was 35,000 tons. In Austria, in 1848, it was 8,000 tons; in 1851, 15,000 tons. In Belgium, in 1848-49, 4,500; in 1850, 7,000; in 1851, 10,000. In France, in 1851, the quantity raised was 60,000 tons; while the cane sugar imported into England, in 1851, was 6,636,214 cwts. or 331,810 tons. According to Pelouze, the manufacturers only express 70 per cent. of juice, and leave 30 of moist residue; but in the laboratory there is only 12 per cent. of residue.

Behenic Acid. $C_{44}H_{44}O_4$. Crystalline fat acid, fusing at 170°, found in the oil of Ben, or Moringa oleifera—the basis of Macassar oil.

Belladonna, or *Atropa belladonna*—deadly nightshade. See ATROPINE.

Bell Metal is a grayish-white alloy, hard, sonorous, and elastic. It usually consists of 75 copper and 25 tin, or of 78 copper and 22 tin; so that the proportion, in some measure, varies; less tin is used for church bells than for clock bells; and in small bells, as in those of watches, a little zinc is added to the alloy. The greater portion of the tin may be separated by melting the alloy and pouring water on the surface—oxide of tin is formed on the surface. Cymbals and gongs contain 81 copper and 19 tin.

Ben, Oil of. A whitish-yellow, thickish oil, from the fruit of the *Hyperanthera moringa*, in Egypt, Ceylon, &c.; used in perfumery.

Benic Acid. $C_{30}H_{29}O_3HO$. ? A fatty acid, fusing at 125°, found in certain kinds of oil of behen or ben.

Benzamide. $C_{14}NH_7O_2$, or $C_{14}H_5O_2$, NH_2. Needles, or pearly volatile scales, obtained by acting on chloride of benzoyle with ammonia, or by boiling hippuric acid with brown oxide of lead.

Benzanilide. Crystals by aniline and anhydrous benzoic acid.

Benzamile, C_{14} H_5 NO.

Benzene. *Benzine, benzole, phene.* See BENZOLE.

Benzidam, or *Aniline.*

Benzidine, $C_{12}H_8O_7$.

Benzilam, $C_{28}H_9N$.

Benzile. *Benzoyle.* $C_{28}H_{10}O_4$. Yellow 6-sided prisms, fusing at 197½°, insoluble in water; soluble in alcohol and ether; subliming without decomposing; obtained by acting on fused benzoine with chlorine, or by heating benzoine with nitric acid. The cold mass resulting is boiled in alcohol, and benzile deposited on cooling. This, from its composition, was supposed to be the radical benzoyle; but its formula is believed to be the double of that base.

Benzilic Acid. $C_{28}H_{11}O_5HO$. Colourless brilliant crystals, fusing at 248°, decomposed at a higher temperature into benzoic acid and purple vapours; formed by dissolving benzile or benzoine in a hot solution of potash or alcohol, and boiling till a violet colour is dissipated; benzilate of potash is thus formed, from which the base is separated by an excess of chlorohydric acid.

Benzilimide. $C_{28}H_{11}NO_2$.

Benzimic Acid. White needles, sparingly soluble in water and alcohol; formed in the process for amarine, when that substance is formed by the action of ammonia on pure oil of bitter almonds; composition unknown.

Benzimide. $C_{28}NH_{11}O_4$. Crystals deposited in the crude oil of bitter almonds; decomposed by acids into benzoic acid and ammonia. It is formed also by adding an alcoholic solution of potash to a mixture of hydride of benzoyle, or oil of bitter almonds and strong prussic acid.

Benzine. See BENZOLE.

Benzoene. See TOLUOLE.

Benzoglycolic Acid. $C_{18}H_8O_8$. Crystals little soluble in water; soluble in alcohol and ether; decomposed by boiling with water, and by heat; yields, when boiled with acids, glycolic acid; formed by acting on hippuric acid with nitrous acid (NO_3).

Benzoglycolamidic Acid, or Hippuric Acid. This name is derived from the view that hippuric acid is the amide of the preceding acid.

Benzohydramide. $C_{42}H_{18}N_2$. 6-sided prisms, isomeric with *hydrobenzamide*, obtained along with that body and *azobenzoyle* and *nitrobenzoyle*, by treating crude oil of bitter almonds with ammonia, and acting on the resinous mass which falls with boiling ether; on cooling, benzohydramide falls in crystals, with azobenzoyle; boiling alcohol takes up the former and leaves the latter; the benzohydramide is then purified by repeated solution in alcohol.

Benzohydrol. $C_{28}H_{15}O_5$. Colourless plates. The stearoptene of oil of cassia.

Benzoic Acid. $C_{14}H_5O_3$. HO or Bz OHO. Light white crystals in prisms or needles or pearly scales; usually with an aromatic smell, but when pure quite odourless; soluble in 200 cold, 25 hot water; melts at 249° and boils at 462¼°; but sublimes in a current of air with a gentle heat, and gives out vapours even at the temperature of the atmosphere. Spec. grav. of vapour about 4·271. It may be obtained by placing gum benzoin, pulverized with sand, in a pan or evaporating basin, pasting a piece of filtering paper over its mouth as a lid—and over all a paper cap so as to receive the vapours. On applying heat carefully by a sand bath to the vessel, the acid vapours pass through the pores of the filtering paper—are deposited on its upper surface within the paper cap in the form of fine light crystals, all oily matter being retained by the paper. 2. Gum benzoin when boiled with slaked lime or carbonate of soda yields benzoate of lime or soda, which may be decomposed by muriatic acid. 3. Dissolve gum benzoin in alcohol—precipitate the resin by muriatic acid; on distilling, benzoate of ethyle or benzoic ether passes over; which is changed into benzoate of potash by heating with caustic potash; when decomposed by muriatic acid, benzoic acid crystallizes. By distilling the residue with water, more benzoic acid is obtained in the retort. 4. Boil hippuric acid with muriatic; benzoic acid results. 5. Benzoic acid is obtained along with hippuric acid by evaporating cow's urine; ether dissolves up the benzoic acid, which may be purified by the preceding methods; by union with a base, or by dissolving it in 4 or 5 times its weight of sulphuric acid diluted with 6 times its weight of water; boil and add animal charcoal to decolourize; on cooling, benzoic acid separates in crystals. *Use.*—Benzoic acid is used in the form of *benzoate of ammonia* (a salt in feathery crystals, obtained by saturating caustic ammonia or its carbonate with benzoic acid), to precipitate sesquioxide of iron from neutral solutions, and thus separate iron from manganese, which remains in solution. The sesquibenzoate of iron is a buff-coloured precipitate, easily distinguishable from sesquioxide of iron. Benzoic acid constitutes the first of a series of homologous acids related to certain oils and bases also homologous.

Benzoic acid,	$C_{14}H_6O_4$	Xylole, ...	$C_{16}H_{10}$
Toluylic acid,	$C_{16}H_8O_4$	Cumole, ...	$C_{18}H_{12}$
?	$C_{18}H_{10}O_4$	Aniline, ...	$C_{12}H_7N$
Cuminic acid,	$C_{20}H_{12}O_4$	Toluidine, ...	$C_{14}H_9N$
Benzole, ...	$C_{12}H_6$	Xylidine, ...	$C_{16}H_{11}N$
Toluole, ...	$C_{14}H_8$	Cumidine, ...	$C_{18}H_{13}N$

Benzoic Acid, Anhydrous. $C_{14}H_5O_3$. F.P. 107½°, distils at 590°, without change. Oblique prisms, soluble in alcohol and ether; neutral; ammonia converts it into benzamide; with aniline, forms benzanilide; obtained by heating chloride of benzoyle with benzoate of soda, or with oxalate of potash.

Benzoin Gum. A solid brittle substance, sometimes in yellowish-white tears, and sometimes as a brown substance resembling rosin, obtained by incision from the *Styrax benzoe* in Sumatra, &c. each tree yielding 3 or 4 lbs. Spec. grav. 1·092; boiling water takes up some benzoic acid; soluble in ether and alcohol; reprecipitated as a white powder by water, muriatic, sulphuric, and acetic acids, but not by alkalies; soluble in acetic acid, sulphuric acid, fixed alkalies, and by ammonia sparingly. It contains 3 or 4 resins, having different formulæ, to the extent of 83 per cent. and benzoic acid from 12 to 19½ per cent. When distilled it yields benzoic acid, carbon, olefiant gas, and phenole. It is used for supplying benzoic acid.

Benzoinam. $C_{56}N_2H_{24}O_2$. White needles, consisting of the union of 2 atoms benzoine, ammonia being taken up and water separated; soluble in acids; precipitated by alkalies having the characters of a weak base; formed when a mixture of ammonia, alcohol, and benzoine are allowed to remain in a stoppered bottle for several months, along with hydrobenzoinamide.

Benzoinamide. *Hydrobenzoinamide.* $C_{42}H_{18}N_2$. White tasteless needles or powder, sublimable, isomeric or polymeric with hydrobenzamide, prepared by allowing benzoine to remain in contact with ammonia for several months.

Benzoine. $C_{28}H_{12}O_4$. Small sublimable crystals, insoluble in water, soluble in alcohol; formed by acting on crude oil of bitter almonds, which contains cyanohydric acid, with a solution of potash, or sulphide, or cyanide of potassium in alcohol; fused with hydrate of potash, it yields benzoate of potash; when passed through a red hot glass tube it becomes oil of bitter almonds. F.P. 248°.

Benzolactic Acid. $C_{20}H_{10}O_8$. A member of a homologous series, of which lactic acid ($C_6H_6O_6$) is the first.

Benzole. $C_{12}H_6$. *Benzene, Phene, Bicarburet of Hydrogen* (Faraday), *Benzine* (Mitscherlich). 1. This oil was discovered by Mr. Faraday among the oils condensed from oil gas, at a pressure of 30 atmospheres. The crude oils obtained from oil gas were distilled at a heat of 187°, and the distilled oil exposed to a temperature of 0°, when it solidified into a crystalline mass. This when pressed between sheets of blotting paper was freed from oily impurity. 2. Mitscherlich afterwards obtained this substance by distilling benzoic acid with hydrate of lime. When 1 part of crystallized benzoic acid is intimately mixed with 3 parts of slaked lime and the powder introduced into a glass or porcelain retort and exposed to a high temperature, benzole passes into the receiver. It may be further purified by distillation with water or over slaked and unslaked lime. 3. It may also be procured by passing the vapour of benzoic acid through a red hot gun barrel, according to D'Arcet. 4. It exists in coal naphtha, and may be separated by fractional distillation. *Properties.*—When freed from benzoic acid by potash and redistillation, and from water by chloride of calcium, benzole is a clear, colourless fluid with an ethereal odour, lighter than water. Spec. grav. = 0·85; spec. grav. of its vapour = 2·738. It solidifies at 32°, becomes fluid at 44½, and boils at 186°. It dissolves in alcohol and ether, but not in water. Its production from benzoic acid is easily explained.

$$C_{14}H_6O_4 + 2\,CaO = C_{12}H_6 + 2\,CaO,\ CO_2.$$

Benzoic acid and 2 lime. Benzole + 2 Carb. of lime. So that the lime simply acts by removing carbonic acid from the benzoic acid. *Benzone.*—If after the distillation of all the benzole the residue is distilled separately benzone passes over, holding naphthaline in solution, which crystallizes out at —5°, and leaves pure benzone (Peligot). This oil is decomposed by SO_3 or Cl, but not by NO_5 or NaO. Its composition is—

$$\begin{array}{l} C_{13}H_5O \text{ Benzone} \\ \underline{C\ \ H\ O_3} = CO_2 + HO \\ C_{14}H_6O_4 \text{ Benzoic acid.} \end{array}$$

Benzole and Chlorine. Chlorobenzole. Chloride of Benzole. — $C_{12}H_6Cl_6$, or $C_{12}H_3Cl_3 + 3\ HCl$ (Mitscherlich and Peligot). When benzole and chlorine are exposed in a large glass flask to the action of the sun's rays, the sides of the glass in a few minutes become covered with white crystals, which consist of benzole and chlorine united together, without change. They melt at 270°, are insoluble in water, but soluble in alcohol and ether, from which they may be obtained, particularly from the latter, in fine crystals. From the circumstance that when *chlorobenzole* is distilled at 555°—its boiling point—hydrochloric acid is given out and chlorobenzide formed, it is possible that chlorobenzole is a hydrochlorate of chlorobenzide.

Chlorobenzide.—(Mitscherlich) $C_{12}H_3Cl_3$. An oily, colourless fluid, boiling at 410°; obtained by the distillation of chlorobenzole. It may also be procured by distilling chlorobenzole with slaked lime.

Benzole and Bromine.—Similar compounds to the preceding are formed with bromine.

Benzole and Nitric Acid. Nitrobenzide.—$C_{12}H_5NO_4$ (Mitscherlich). When benzole is added gradually to fuming nitric acid the compound produced dissolves completely in the acid, but on cooling collects on the surface, and on dilution with water, precipitates in the form of a yellow oil, which may be freed from NO_5 by washing with water and distillation. When pure it has a faint yellow colour, very sweet taste, and the smell of oil of bitter almonds. Its boiling point is 415°; the spec. grav. of the liquid is 1·209; of the vapour 4·376. It solidifies at 37½°, crystallizing in needles; soluble in ether and alcohol, insoluble in water. It is soluble in dilute acids, decomposed by SO_3, but not by alkalies. Its production is easily explained—

$$\begin{array}{lll} C_{12}H_6 + NO_5 = & C_{12}H_5 + HO \\ \text{Benzole} \quad \text{Nitric acid} & NO_4 \end{array}$$

where an atom of hyponitric acid is substituted for an atom of hydrogen. When nitrobenzide is heated with strong sulphuric acid, it is decomposed with the evolution of sulphurous acid. Potassium, when heated with nitrobenzide, occasions a detonation. Chlorine, bromine, and liquid caustic potash produce no change on nitrobenzide, but a solution of the hydrate of potash in alcohol decomposes it.

Dinitrobenzole.—$C_{12}2NO_4H_4$. Obtained by heating benzole with strong sulphuric acid and fuming nitric acid. It falls as a heavy oil, and solidifies on cooling into a mass of needles, which are purified by repeated solution in alcohol. This substance has for its basis benzole, in which 2 atoms of hydrogen are replaced by 2 of hyponitric acid. The following table shows the action of agents on benzole:—

Benzole,	$C_{12}H_6$
Chlorobenzole,	$C_{12}H_5Cl$
Bromobenzole,	$C_{12}H_5Br$
Azobenzole,	$\left\{C_{12}\begin{matrix}H_5\\N\end{matrix}\right.$
Nitrobenzole,	$\left\{C_{12}\begin{matrix}H_5\\NO_4\end{matrix}\right.$
Dinitrobenzole,	$\left\{C_{12}\begin{matrix}H_4\\2NO_4\end{matrix}\right.$
Sulphobenzole,	$\left\{C_{12}\begin{matrix}H_5\\SO_3\end{matrix}\right.$
Hyposulphobenzidic acid,	$\left\{C_{12}\begin{matrix}H_5\\S_2O_5\end{matrix}\right.$
Amidobenzole, or aniline,	$\left\{C_{12}\begin{matrix}H_5\\NH_2\end{matrix}\right.$
Nitramido-benzole, or nitraniline,	$\left\{C_{12}\begin{matrix}H_4\\NH_2\\NO_4\end{matrix}\right.$
Benzone, or Carbobenzole,	$\left\{C_{12}\begin{matrix}H_5\\CO\end{matrix}\right.$

Benzole may be viewed as the hydride of phenyle $C_{12}H_5H$, and is the type of a series of homologous compounds.

Benzolone. *Pyrobenzoline.* $C_{11}H_4O$. Crystalline body, obtained by heating hydrobenzamide with fused potash; probably a product of the decomposition of benzostilbine along with carbon and hydrogen.

Benzone. *Carbobenzole.* $C_{13}H_5O$. Pale yellow fluid, heavier than water, decomposed by sulphuric acid or chlorine, but not by nitric acid or caustic potash. It may be viewed as derived from benzole and carbonic acid by the separation of an atom of water, ($C_{12}H_6,CO_2$=$C_{12}H_5$,CO, HO) or as benzole in which 1 atom hydrogen is replaced by 1 atom carbonic oxide ($C_{12}H_5CO$). It is obtained by distilling benzoate of lime; the crude products heated gradually up as long as benzole passes over; the residue yields benzone from which a cold of —5° separates naphthaline in crystals.

Benzonitric Acid. $C_{14}H_4O_2,NO_5$. Plates or white masses resembling benzoic acid, obtained by acting on benzoic acid with strong hot nitric acid; it forms crystalline salts. The anhydrous acid as a white mass, is formed by heating oxichloride of phosphorus and dry benzonitrate of soda.

Benzonitrile. $C_{14}H_5N$. Colourless fluid heavier than water, with the smell of bitter almond oil; insoluble in water; soluble in alcohol and ether; boiling at 374°, obtained by distilling benzoate of ammonia, either alone or with anhydrous phosphoric acid. It is washed with water and dilute sulphuric acid, dried with chloride of calcium and redistilled.

Benzophenide. $C_{26}H_{10}O_4$. Pointed colourless prisms of 80°; fusing point 140°; distils without change; insoluble in water; soluble in alcohol and ether; decomposed by acids and alkalies; obtained by heating together chloride of benzoyle and phenylic acid.

Benzophenone. $C_{13}H_5O$. Large amber prisms; fusing point 115°; boiling point 600°; insoluble in water; soluble in alcohol; obtained along with benzone by distilling benzoate of lime; isomeric with benzone.

Benzostilbine. $C_{31}H_{11}O_2$. Small silvery crystals, insoluble in water; soluble in alcohol and sulphuric acid, with a red colour, and reprecipitated by water; formed by heating hydrobenzamide with fused potash till the mass becomes of the colour of gamboge; alcohol dissolves it up from benzolone, and it may be purified by destroying any oil present by a little chlorine.

Benzosulphuric Acid. $C_{14}H_4O_22SO_3$. Crystalline mass, deliquescent, obtained by the action of anhydrous sulphuric acid on benzoic acid.

Benzoyle. *Benzyle.* $C_{14}H_5O_2$ = Bz. The hypothetic radical of the benzoic acid series; that acid being the hydrate of the oxide of benzoyle.

Benzoyle Anilide. $C_{26}H_{11}N$ Crystals from equal volumes of aniline and oil of bitter almonds, insoluble in water, soluble in alcohol.

Benzoyle, Hydride of. *Hydret, Hydruret, Hydrobenzyle. Essential Oil of Bitter Almonds.* $C_{14}H_5O_2H$=BzH. A transparent colourless fluid, with a peculiar burning taste and smell. Spec. grav. 1·043; boiling point 356°; poisonous. By exposure to the air it absorbs 2 atoms oxygen and is converted into benzoic acid; burns with smoke; soluble in 30 parts water; mixes with alcohol and ether; passes through ignited tubes without decomposition. An alcoholic solution of caustic potash converts it into benzoate of potash and an oil; forms with chlorine and bromine chloride and bromide of benzoyle; mixed with prussic acid and caustic potash benzoine is formed, isomeric with itself but solid. The crude oil of commerce, which is prepared by distilling bitter almonds with water, contains cyanohydric acid, benzoic acid, and benzoine. Being used extensively in perfumery in France it is made in Paris on a large scale. No essential oil exists in the bitter almond, but it contains two substances, amygdaline and emulsine, which are prevented from acting on each other in the cells of the plant by the presence of a fat oil; as soon, however, as this fat is removed, the action commences, and the essential oil is the result. The almonds are bruised and freed from the fat oil by cold expression; the almonds further bruised, placed in the still with some cold water to form a paste, and allowed to rest for twenty-four hours and then distilled. The water which passes over contains much oil in solution, from which it is freed by redistillation with common salt. The crude oil is yellow, which deepens by keeping. To separate prussic acid it is distilled over red oxide of mercury with water after standing some days; it is finally distilled from fresh burned powdered lime, when it is obtained pure.

Benzoyle, Hydride of, Benzoate of. BzO, HO, 2 BzH. White crystalline powder, insoluble in water, soluble in alcohol and ether; obtained by saturating crude oil of bitter almonds with chlorine gas.

Benzoylureide. $C_{50}H_{28}N_8O_8$. White powder from 5 urea and 2 oil of bitter almonds.

Beraunite, or hydrous perphosphate of iron, a variety found at Beraun in Bohemia; colour, hyacinth-red; spec. grav. 2·878, H 2 to 2½.

Berberine. $C_{42}H_{18}NO_9$. Yellow silky needles or powder with a bitter taste; soluble in 500 parts cold water; in 250 cold alcohol; fusing point 266° with change; forms salts with alkalies and also with acids. It is used as a substitute for quinine in Bavaria; it colours cotton and silk without a mordant; it is obtained from the rasped root of the barberry (Berberis vulgaris) by extraction with water; and crystallizing the berberine from the extract by alcohol.

Berengelite. $C_{41}H_{16}O_8$. A yellowish resin, from Berengela in Peru, insoluble in water, soluble in alcohol and ether; unites with alkalies; employed for paying boats, &c.

Bergaptene. C_3HO. Colourless, tasteless crystals, fusing at 404°, subliming, neutral; a deposit in oil of bergamot; a product by pressure of the rind of the Citrus bergamia.

Bergmanite. A variety of scapolite.

Berries, *Persian* and *French.* The fruit of the Rhamnus infectorius, yielding a yellow colour.

Berthierine, or chamoisite, or carbonophosphate of iron.

Berthierite. *Haidingerite.* S 28·3, Sb 48·3, Fe 14·9, Zn ·3, SiO_3 3·2, FeS_2 3·2. Colour iron-black, in masses with the rudiments of crystals; fuses before the blowpipe; found near Chazelles in Auvergne, and at Braunsdorf; soluble in muriatic, and nitro-muriatic acids.

Beryl. See EMERALD.

Berzeline. Selenide of copper, in silver-white masses, consisting of Cu 64, Se 40, found in Smaland in Sweden.

Berzelite. A name given to petalite, dichloride of lead, and to an arseniate of lime and magnesia.

Betaorceine. $C_{18}H_{10}NO_8$. A blue substance obtained from archil.

Betaorcine. $C_{38}H_{24}O_{10}$. Square prisms, soluble in water, alcohol, and ether; sublimable; sweet taste; obtained by the destructive distillation of dry usnic acid.

Betaorsellic Acid. $C_{34}H_{17}O_{15}$. Crystalline body, nearly insoluble in cold water; soluble in hot alcohol; forms a red colour with chloride of lime; formed by heating archil lichen (Roccella tinctoria) from the Cape of Good Hope with milk of lime, and precipitating with muriatic acid.

Beta Orsellesic Acid. Not analyzed, but obtained by boiling beta orsellate of barytes with water.

Beta Resins. A mode of nomenclature by attaching the Greek letters when more than one resin is present in a mixture.

Betuline. $C_{40}H_{33}O_3$. A colourless resin, soluble in 120 parts cold, in 80 boiling alcohol; melts at 392°, and sublimes. Obtained by exhausting birch bark with hot water, and extracting from the aqueous residue with alcohol, and purifying with ether.

Beudantine. See NEPHELINE.

Beudantite. An arseniate of iron.

Beudantite. Black obtuse rhombohedrons, with resinous lustre, containing iron and lead, found at Hornhausen on the Rhine.

Bezoar Stones. Oval concretions, supposed to be voided in the East from animals of the deer tribe; they usually consist of *lithofellic* acid; in one I found a nucleus of a date stone.

Bezoaric Acid. Microscopic crystals, obtained from olive-green bezoars. Soluble in potash, and precipitated in stellated groups by chlorohydric acid. Soluble in sulphuric acid, and reprecipitated by water. It has not been analyzed. Also a synonyme of ellagic acid.

Bibromisatine. $C_{16}H_8Br_2NO_4$. Obtained by acting on isatine with bromine. When heated with potash it yields bibromisatic acid.

Bichlorisatine. C_{16},H_3Cl_2,NO_4. Yellow 4-sided prisms, obtained by passing chlorine through water, in which indigo or isatine is suspended; it forms with potash bichlorisatic acid.

Bichlorisatyde. $C_{16}H_5NCl_2O_3$.

Biethyl-Meconic Acid. $HO,2\,(C_4H_5O)\,C_{14}HO_{11}$. A compound of ethyle and meconic acid.

Bile. The bile, which is contained in the gall bladder, an appendage of the liver, possesses a weak alkaline reaction, and an oily consistency. Its colour is yellow, or greenish-yellow, which becomes darker by exposure to the air. When evaporated to dryness in the water bath, it dissolves readily in alcohol, leaving a quantity of mucus, which is insoluble in that fluid. It may be purified with animal charcoal, and by precipitating the colouring matter by means of barytes water. The cholesterin which it contains may be removed after it has been deprived of colour by the charcoal; if after concentrating the alcoholic solution a double volume of ether be added, the cholesterin dissolves in the ether, and the bile, which is insoluble in ether, separates in the form of a thick syrup. On removing the ethereal solution the cholesterin crystallizes out. The alcoholic solution now containing the bile, when evaporated to dryness, leaves a dry pulverulent residue, which is completely soluble in water and absolute alcohol. From its aqueous solution it is separated by caustic potash when added to saturation. Acetic and oxalic acids produce no change in the aqueous solution. By mineral acids, a milkiness is produced either immediately or by standing, and a syrupy fluid separates. A portion of the acid is found to be combined with soda; acetate of lead and nitrate of silver precipitate the aqueous solution of bile. A solution of pure bile is completely precipitated by the acetate of lead, but an excess of the salt redissolves a small portion of the bile. The portion in solution was considered a peculiar substance by Gmelin and termed biline, bile sugar, &c.

Test for Bile.—Add to the liquor containing bile its bulk of sulphuric acid. Pour in a few drops of a weak solution of sugar. A fine red colour is produced.

The view which has been supported of the composition of the bile by the most recent experiments is, that this fluid is composed of two different soda salts, formed of different acids, *cholic* and *choleic*, a sulphur acid. *Cholic* acid, when boiled with alkalies, is resolved into *cholalic acid* and *glycocoll*. *Choleic* acid, on the other hand, becomes *cholalic acid* and *taurine*. These acids are, therefore, conjugate acids.

Cholic Acid or Glycocholalic Acid.

Cholalic acid,........................$C_{48}H_{39}O_9$

Glycocoll, —1 atom HO, } ....$C_4\ H_4\ O_3\ N$

Cholic acid,........................ $C_{52}H_{43}O_{12}N$

Choleic Acid or Taurocholalic Acid.

Cholalic acid,....................$C_{48}H_{39}O_9$

Taurine, —HO, } ....$C_4\ H_6\ O_5\ NS_2$

Choleic acid,....... ....$C_{52}H_{45}O_{14}NS_2$

The human bile consists of—

Water,............................	85·96	86·84
Cholate of soda, Choleate of soda, }	8·18	11·09
Colouring matter,............... Oleine,............................		
Margarine,.......................	0·62	
Cholesterine,.....................	0·21	
Bile mucus,..............	2·82	
Inorganic matter, ...	14·04	13·16

Analysis.—A weighed portion of bile is exposed in a porcelain basin to the heat of an air bath, at the temperature of about 250°, until it ceases to lose weight; the loss is water. 2. The dry residue is pulverized and placed in a flask, and digested in repeated additions of boiling ether, until a drop of the fluid leaves no residue on a plate of glass. The ether, on evaporation, leaves the fatty matters (oleine and cholesterine). 3. The bile, from the action of ether, is boiled in alcohol and evaporated. The remainder is the biliary mucus; the alcoholic extract is the cholate and choleate of soda, with colouring matter. 4. To obtain the salts, a fresh portion is to be incinerated.

	Cholate of Soda.	Cholic Acid.	Choleic Acid.	Para-cholic	Cholalic Acid.	Choloidic Acid.	Dyslysin.
Carbon,	64 06	67·31	62·52	67·18	70·59	72·18	77·4
Hydrog.,	8·62	9·35	9·02	9·24	9·80	9·77	9·6
Nitrogen,	2·87	3·23	2·81	2·73			
Oxygen,	18·09	20·11	22·44	20·85	19·61	18·05	13·
Sulphur,			3·21				
Soda,.....	6·36						

Preparation of Crystallized Bile.—Ox bile is dried at 212°, dissolved in absolute alcohol, and treated with some ether; a syrupy dark mass falls, which is set aside. It consists of crystals and an amorphous mass. The clear liquor is drawn off and again treated with ether; pale yellow needles subside, which are transferred to a filter by means of alcohol, containing one-tenth of ether, with which they are washed. They are then dried in vacuo over SO_3, and consist of

C 60·50	S 2·50
H 8·63	N 2.8

Preparation of Cholic Acid.—The preceding crystals of bile, still moist, were dissolved in water and decomposed by sulphuric acid. In twelve hours colourless crystals formed, and oily drops. They were washed in water, when the oily drops disappeared. The turbid wash-water deposited a resinous precipitate after some days. The crystals were now snow-white. The precipitate which separated on the addition of SO_3 to the bile, purified by Theyer and Schlosser's plan, on the addition of ether was converted into a network of the same crystals. By dissolving in water and evaporating, the crystals again appear, and possess all the characters of Gmelin's *cholic acid* in delicate needles. 1000 parts of cold water dissolve 3·3, and boiling 8·3; reddens litmus in solution; readily soluble in alcohol, leaving a resin on evaporation; dissolves in acids, in NH_3, KO, and BaO water.

Another Method.—Precipitate fresh bile with acetate of lead. Wash well the yellowish precipitate; boil in alcohol of ·848 spec. grav. which partly takes it up. Pass SH through the solution; filter while hot. Add alcohol to the solution in air-tight vessels. In twelve hours a mixture of cholic and paracholic acids separate—208 grains were thus obtained from 10 bladders-full.

Paracholic Acid, isomeric with cholic acid, is obtained by boiling the latter with water, when 6-sided tables fall, insoluble in water. Both acids give a pink with SO_3 and sugar. When boiled with barytes, *cholalic acid*, or *cholic acid* of Demarcay, appears.

Cholalic Acid, when precipitated from its barytes salt by HCl, is resinous, but by dissolving in hot alcohol it crystallizes out on cooling in tetrahedrons; colourless; soluble in 750 boiling, and 4000 cold water.

Cholic acid, ...	$C_{52}H_{43}NO_{12}$
Cholalic acid, ...	$C_{48}H_{40}O_{10}$
Remainder, ...	$C_4H_3NO_2$
+ 2 atoms HO, ...	H_2O_2
Crystallized glycocoll, or glue sugar, ...	$C_4H_5NO_4$

After heating cholic acid with barytes, and separating the cholalic acid and barytes, glycocoll was obtained in crystals by Strecker.—With acids cholic acid is also changed: glycocoll and Demarcay's choloidic acid are formed. By continued action, choloidic acid disappears, and dyslysine of Berzelius appears.

Cholalic, ...	$C_{48}H_{40}O_{10}$
Choloidic, ...	$C_{48}H_{39}O_9$
Dyslysine, ...	$C_{48}H_{36}O_6$

Choleic Acid.—After the separation of cholic acid by acetate of lead, if we add to the liquor passing through the filter disacetate of lead, a precipitate falls consisting of basic cholate of lead, and choleate of lead. The latter contains N and S. Strecker has not been able to separate the cholate from the choleate; but the products of the decomposition of cholic acid by alkalies being known, he has examined the mixed products of these salts. After decomposing the salts of lead by solution of barytes, the soluble salts of barytes in water are boiled with excess of hydrate of barytes. In the products he has found *cholalic acid*, *glycocoll*, and *taurine*, from which he infers that, under the influence of alkalies, the cholic and choleic acids appear to undergo a metamorphosis. The first doubles into cholalic acid and glycocoll; the choleic into cholalic and taurine. He adopts for choleic acid the following formula:—

Choleic acid, ...	$C_{52}H_{45}NO_{14}S_2$
Cholalic acid, ...	$C_{48}H_{40}O_{10}$
Taurine, ...	$C_4H_7NO_6S_2$
	$C_{52}H_{47}NO_{16}S_2$
Subtract 2HO, ...	H_2O_2
Choleic acid, ...	$C_{52}H_{45}NO_{14}S_2$

The following table gives the relation of the carbon to the hydrogen in cholalic acid, with reference to some analogous acids:—

Cholalic acid	100 atoms carbon	to	83 ats. hydr.
Lichesteric	,,	—	86
Lithofellic	,,	—	90
Roccellic	,,	—	95
Oleic	,,	—	94
Fat acids	,,	—	100

Colouring matter of bile is separated from the urine in jaundice, by adding chloride of barium to the urine; a green precipitate falls, which is collected and washed. Digest this in alcohol, acidulated with muriatic acid, at a gentle heat; evaporate, wash the residue with water, and dissolve in a mixture of alcohol and ether. It is a dark green powder, nearly insoluble in water and ether, soluble in alcohol and alkaline solutions; the colour changing from green to brown. Muriatic acid produces the same change, which is that of oxidation.

		Change by Alkal.	Change by Acid.
C.......	68·182	62·086	61·837
H.......	7·437	6·567	6·464
N.......	7·074	7·101	9·080
O.......	17·261	24·246	22·619

Bile of the Inferior Animals. Bile of Fishes.—When dry fish bile is dissolved in absolute alcohol, filtered, and mixed with a small quantity of water, the colouring matter separates. The principal constituent of the bile is precipitated by the addition of a large quantity of ether, and increases by evaporation; and after exposure to the air it turns into crystals resembling wavellite. Colour slight, sweetish bitter taste, gives a deep purple with sugar and sulphuric acid. By barytes it is transformed into cholalic acid and taurine and glycocoll. It is therefore similar to the sulphur acid of ox bile (choleic acid).

Pig's Bile contains *hyocholic acid* ($C_{54}H_{43}NO_{10}$); differing from cholonic acid, a product of cholic acid by 2 atoms of water.

Dog's Bile consists principally of choleate of soda.

	Bile of Dog.	Serpent.	Hyocholic Acid.
Carbon,......	58·2	58·1	70·28
Hydrogen,...	8·2	8·5	9·33
Nitrogen,....		3·4	3·04
Sulphur,.....	5·9	6·2	
Soda,.........	5·8	11·5	
Oxygen,.....			17·35

The following table shows the relations of bile, its acids and derivatives:—

Ox Bile.

Cholic acid, ...	$C_{52}H_{43}NO_{12}$
Cholonic acid, ...	$C_{52}H_{42}NO_{11}$
Cholalic acid, ...	$C_{40}H_{40}O_{10}$
Choloidic acid, ...	$C_{48}H_{39}O_9$
Dyslysine, ...	$C_{48}H_{36}O_6$
Glycocoll, ...	$C_4H_5NO_4$
Choleic acid, ...	$C_{52}H_{45}NO_{14}S_2$
Taurine, ...	$C_4H_7NO_6S_2$

Pig's Bile.

Hyocholic acid, ...	$C_{54}H_{43}NO_{10}$
Hyocholalic acid, ...	$C_{50}H_{40}O_8$
Hyodyslysine, ...	$C_{50}H_{38}O_6$
Glycocoll, ...	$C_4H_5NO_4$
Hyocholeic acid, ...	$C_{54}H_{45}NO_{12}S_2$
Taurine, ...	$C_4H_7NO_6S_2$

Uses of the Bile.—When bile is injected into the rectum, it is all absorbed; but it does not make its way into the urine, while yellow prussiate of potash, or iodide of potassium, in the same state does so; and common salt in solution is also absorbed, and appears in the urine. The bile, therefore, seems to pass into the circulation, for the purpose, probably, of supporting respiration.

Use of Bile in the Dog.—Schellbach (Liebig's Ann. 79, 290) experimented on a dog weighing 12·3 lbs. avoird. (5·58 kilogram.) He found, on extirpating the spleen and introducing a tube into the ductus communis choledochus, so as to cause all bile formed to flow out externally, that the animal, when fed on flesh, did not fall off in weight. The amount of bile passed daily was (164·43 grammes) 2532·23 grains of fresh bile, or 107½ grains of dry bile; or for 2·2 lbs. of live dog, 460·4 grains of fresh bile are given daily. Bidder found only half this quantity, but the dog was unhealthy. On examining the fæces, and comparing it with a normal dog, it was found that not half of the fat of the food (lights and livers) was digested in the dog with the fistula, while nearly the whole disappeared in the normal dog. The digestion of albuminous matter was not impaired. The amount of fat in the system diminished, and none was found under the cutis after death. It is obvious, therefore, that the carbon in the case of this animal, usually supplied for respiration from the fat, must have been derived from muscular matter, and hence it consumed about 17 ounces of flesh, or nearly double the normal quantity. The largest amount of bile is secreted between thirteen and fourteen hours after eating. The fæces of a normal dog contained no bile. Pettenkofer likewise found in none of the constituents of normal fæces the constituents of bile, but Lehmann found in the alcoholic extract of fæces, by means of ether, small portions of cholalic acid, which gave a red colour with sulphuric acid and sugar.

Biliary Calculi. *Gall Stones.* These concretions in man are usually triangular bodies with a smooth surface, consisting of cholesterine, with a certain amount of colouring matter. They are soluble in boiling alcohol, brilliant plates depositing from the solution on cooling. From their appearance, they seem to have been detained in a pouch in the gall bladder, the bile supplying the matter of which they consist. Besides containing rhombic plates of cholesterine, when examined under the microscope, epithelial scales are observed, and mucus from the gall bladder. They are generally lighter than water, but their spec. grav. varies from ·803 to 1·06. Caustic potash dissolves up from them colouring matter, &c. and leaves white cholesterine. In oxen the gall stones consist principally of colouring matter used as pigment, and of margaric acid salts; in the dog, of chalk, some phosphate of lime, and colouring matter; in the pig, of cholesterine 6, resin 45, bile 4, animal matter, colouring matter, phosphate of lime, 45. The bezoars, consisting of lithofellic acid with a spec. grav. of 1·043, are supposed to be the gall stones of animals of the deer tribe; they are characterized by forming soaps with potash and soda, and being precipitated from the alcaline solution by acids.

Bilic Acid Impure cholic acid.

Biliphaine. $C_{32}H_{18}N_2O_9$. The brown colouring matter of the bile, probably identical with biliverdine.

Binary Theory. See BASE and SALTS.

Biogene. A name given to the yolk in the ovarian eggs of the lower animals.

Biotine. A variety of Anorthite.

Biphosphamide. PNO_2 or PNO_5. Grayish-white powder, obtained by heating phosphamide.

Birch Trees. Different species of Betula, supplying betulin, a resin or camphor, and essential oils; that of the B. alba has an odour resembling turpentine; spec. grav. ·847; of vapour 5·318; formula $C_{20}H_{16}$; that of the B. lenta is identical with the oil of Gaultheria procumbens; and salicylic acid is formed from it in crystals when it is acted on by oil of vitriol.

Bird Lime. A peculiar adhesive matter, due to the presence of viscin, extracted from the middle bark of the holly, mistletoe, and distaff thistle by boiling several hours in water; used for catching birds.

Birousa, or turquoise.

Bis, *Bi*, *Bin*, Twice. Latin affixes to denote the amount of oxygen, acid, &c. united to a base.

Bismuth. Bi 26·625; 213. *Character.*—Bismuth is found native in the form of a reddish-white metal, consisting of brilliant plates adhering to each other. *Process.*—The ore is separated from the rock by the hand, and thrown into red hot tubes *a b* placed in an inclined position, of which the lower extremity is closed by a clay plate with a small perforation, through which the metal flows (eliquates) into iron pans *c*, placed in a row at the mouths of the series of tube furnaces. After the iron pans have been nearly filled, the metal is removed by ladles into an iron mould, and formed into bars of from 25 to 50 lbs. The process for separating or eliquating 1 ton of ore lasts eight hours. In this state the metal contains small quantities of arsenic, sulphur, iron, &c. which do not interfere, however, with its applications. It may be purified by dissolving in nitric acid, evaporating excess of acid, adding water to throw down trisnitrate, taking up the arsenious acid by caustic potash, and reducing the oxide with black flux at a moderated temperature. It is used in the preparation of pewter, type metal. Fusing point 276° (Irvine), 496° (Crichton). When cooled slowly, it crystallizes in parallelopipeds, crossing each other at right angles. Its primary form is a rhombohedron. Spec. gr. 9·822 (Hatchett), 9·833 (Harepath), 9·654 (Karsten), 9·799 (Marchand). By exposure to the air it loses its brilliancy, in consequence of its union with oxygen. Obtained pure by heating the subnitrate with charcoal.

[*Mitscherlich.*]

Teroxide of Bismuth.—BiO_3, 29·625. Straw-yellow powder, without taste and smell, insoluble in water, soluble in NO_5. Melts by heat into an opaque glass. Soluble in caustic soda. The oxide may be obtained by burning the metal in the air, when a blue flame is produced, and the oxide sublimes—(flowers of bismuth)—or by igniting the trisnitrate of bismuth. At a red heat the colour of the oxide is brown, but on cooling pale yellow.

Trisnitrate of Bismuth. Flake white. Magistery of Bismuth.—3 BiO_3, NO_5. This is the only form of bismuth of interest. The process for preparing it forms a beautiful experiment. It consists in dissolving a portion of the metal in nitric acid in a flask, and then pouring the nitric acid solution into a pint or more of water in a tall jar; a fine white precipitate immediately falls, which is *flake white*, (3 $BiO_3 3NO_5$) being converted into trisnitrate (3 $BiO_3 NO_5$) and 2 atoms NO_5 are set free. It is much used in medicine in many dyspeptic cases, where the mucous membrane of the bowels seems to be affected, and particularly in the diarrhœa of children.

Quintoxide, Peroxide of Bismuth, Bismuthic Acid.—BiO_5. Formed by fusing hydrate of soda with oxide of bismuth, and is purified by boiling with caustic soda solution. A brown powder is obtained, which is washed with NO_5 and then with water. It resembles deutoxide of lead.

Fusible Metal, which dissolves in hot water, is formed by heating in a crucible 3 parts of tin, 8 bismuth, and 5 lead, the addition of a little mercury increases the fusibility. It is used to form casts for electrotype.

Bismuth, *Native.* Spec. grav. 9·737, H 2 to 2½. Colour silver-white, with a red tint; massive also in 8-hedrons and in acute rhomboids; texture in plates; surface rough, generally covered with oxide; lustre metallic, sectile, opaque, melts at the flame of a candle. B.B. is dissipated in yellow smoke, which falls on the charcoal; soluble in nitric acid, and precipitated white by water; found at Schneeberg in Saxony, and at Joachimsthal in Bohemia, also at Altenberg; in France, Sweden, Norway, Connecticut, and in Botallack and in St. Columb in Cornwall, and Carrock in Cumberland.

Bismuth Arsenical. As 96·785, Bi 3·001. Occurs at Palmbaum, near Marienberg.

Bismuth, Ferruginous Arsenide of. Bi 55·913, As 38·092, Fe 6·321. Specific gravity, 3·694, H 5·5. Dark brown externally, brownish-yellow internally, in plates; lustre resinous, brittle. B.B. decrepitates, and emits arsenical fumes, burning with a blue flame, being fused into a globule, or dissipated. Soluble in nitric and muriatic acids. Occurs at Schneeberg in Saxony.

Bismuthine. Bismuth glance, or sulphide of bismuth.

Bismuthite. Carbonate of bismuth.

Bismuth, Needle Ore of. Bi 43·2, Pb 24·32, Cu 12·1, Ni? 1·58, Te? 1·32, S 11·58, Au ·79. Spec. grav. 6·125, H 5·25. Steel-gray, with a copper-red tarnish, in 4 or 6-sided prisms, with the lateral faces longitudinally streaked. B.B. sulphur is driven off, the assay melting, and yielding metallic globules.

Bismuth, Oxide of. *Ochre.* Spec. grav. 4·361. Colour greenish-yellow, or yellowish-gray; massive and disseminated; fracture fine-grained, uneven; lustre adamantine; when fine-grained, glimmering; of the foliated, shining; of the earthy, dull. B.B. easily reduced on charcoal; consists of BiO_3 86·4, FeO 5·1, CO_2 4·1, HO 3·4. Found at Schneeberg and Joachimsthal.

Bismuth, Carbonate of. *Bismuthite.* A dirty-yellow earthy substance, found at St. Agnes, Cornwall—probably the same as the preceding.

Bismuth, Silicate of. *Bismuth Blende. Arsenical Bismuth.* BiO_3 69·38, SiO_3 22·23, PO_5 3·31, Fe_2O_3 2·4, Mn_2O_3 ·3, F and HO 1·01. Spec. grav. 5·912 to 6·006; H 5·5 to 6·. Clove-brown and reddish-brown rhomboidal 12-hedrons, and also in 4-hedrons, 6-hedrons, and pyramidal 12-hedrons; opaque to semitransparent, brittle. B.B. fuses to a dark yellow mass, and gives out an odourless sublimate; fuses easily on charcoal; fuses with carbonate of soda into a button, at first greenish-yellow, and then reddish-yellow, globules of metal appearing;

forms a yellowish-green glass with borax; with salt of phosphorus fuses, and leaves a skeleton of silica. Found at Schneeberg. 2 BiO_3 3 SiO_3.

Bismuth, Sulphide of. Bi 80·98, S 18·72, BiS_3, Sp 6·549, H 2·25. Lead-gray rhombic prisms of 91° and 89°; massive; and in minute needles or prisms; lustre metallic, opaque, rather sectile. B.B. volatilized, and covers the charcoal with a yellow ring. Very fusible; soluble in nitric acid, solution precipitated by water.

Bismuth, Telluride of. Bi 59·84, Te 35·24, S 4·92. Spec. grav. 7·5; H 2. Silver-white broad plates, laid on each other; lustre metallic, opaque; divisible into thin plates, and staining paper like sulphide of molybdenum; found in crystals of two different rhombohedrons, united together. B.B. melts, and gives out a yellow metallic vapour. Found at Deutsch Pilsen in Hungary.

Bismuth-Tri-Ethyle. $C_{12}H_{15}Bi = E_3$ Bi. Colourless fluid of spec. grav. 1·82; boiling point 122°. Smell disagreeable, taste burning; evaporates in the air in yellow fumes, and takes fire with a weak explosion, giving out a yellow smoke of bismuth; decomposed explosively by nitric acid, chlorine; insoluble in water, little soluble in ether; very soluble in absolute alcohol. With bromine and iodine, it forms in its alcoholic solution bromide and iodide; it unites with weak nitric acid. Bismuth-ethyle is prepared by the action of iodide of ethyle on bismuth-potassium. The latter is formed by heating slowly in a covered clay crucible 20 ounces of powdered bismuth with 16 ounces cream of tartar, and then raising the temperature for half-an-hour to an incipient white heat; it is then allowed to cool slowly; the bismuth-potassium is formed at the bottom of the crucible as a silver-white metallic mass. The iodide of ethyle is made by the action of iodine and phosphorus on 90 per cent. alcohol.

Bissolite. A variety of amphibole.

Bistre. A pigment prepared by boiling or grinding wood soot with urine or water into a smooth paste, and then diluting with pure water. The mixture is then allowed to stand until the larger particles of soot have subsided. The supernatant liquor is then poured off, and allowed to stand in another vessel for a few days, until a fine powder subsides, which is bistre. It is made into cakes by mixing it with a solution of gum in water.

Bisuccinamide. $C_8H_3O_5$, NH_4O. Colourless rhombic prisms, formed by dissolving succinamide in water and evaporating spontaneously.

Bitter Almonds. See BENZOYLE, HYDRIDE OF, and ALMONDS.

Bitter Principles consist of bodies extracted from plants by means of water, alcohol, or ether, and are the source of that peculiar taste possessed by the plants. They are generally neutral, uniting neither with acids nor bases. The nature of these substances is very imperfectly understood. Many of them are soluble in water, others only in alcohol and ether. The solutions in water and alcohol become darker coloured by exposure to the air and the absorption of oxygen. When evaporated to the consistence of syrup, they are termed *extracts*, and are divisible into watery or alcoholic, according to their relations to these substances. The alcoholic extract contains all the medicinal matters. By redissolving most extracts in water or alcohol, a dark matter remains, consisting of oxidized matter, albumen, mucus, and gallic acid. It is upon a knowledge of the behaviour of these substances with reagents that the mode of purifying bitter matters depends.

Absinthin.—From wormwood, by means of alcohol.

Aloin.—From aloe spicata.

Amanitin.—Obtained from poisonous mushrooms.

Antiarin.—Pearly plates; from the upas antiar, extracted by alcohol; poisonous.

Arthanitin.—White needles from the root of the cyclamen Europæum.

Asparagin.—($C_8N_2H_{10}O_8$) rhombic prisms from the althæa and asparagus.

Berberin, q. v.

Bryonin.—Yellowish-white mass, from the root of the bryonia alba.

Centaurin.—From centaurea benedicta.

Cetrarin.—From Iceland moss (Cetraria Islandica); by boiling the lichen with absolute alcohol, the cetrarin separates in grains by distillation. It is purified by washing with cold water, ether, and weak alcohol, and crystallizing out of absolute alcohol in fine white crystalline grains.

Colocynthin.—Yellow matter obtained by cold water and alcohol from colocynth, cucumis colocynthis, and is the aperient principle in the medicine of that name.

Cornin.—From cornus florida.

Coumarin.—A base yielding the odour of woodruff, hay, grass, and tonka-beans, derived from these plants.

Columbin.—Colourless rhombic prisms, bitter and neutral, extracted by alcohol from columbo root; scarcely soluble in cold water, spirit, and ether; sulphuric acid makes it red.

Cubebin.—White needles extracted from the marc of oil of cubebs (piper cubebas) by alcohol and precipitated by potash.

Cyclamin.—From Cyclamen Europæum.

Cytisin, Cathartin.—Both yellowish substances, apparently the same; the first extracted by alcohol from the seeds of the cytisus laburnum, the second from the cassia senna, in which plant it acts as the aperient.

Daphnin.—Colourless crystals from the daphne mezereum.

Elaterin.—An uncrystallized principle in the

momordica elaterium, much used as a powerful purgative.

Ergotin.—A brownish-red matter extracted from ergot of rye by ether and alcohol.

Fraxinin.—6-sided prisms from the bark of the ash or fraxinus excelsior, by precipitating the decoction by acetate of lead, the filtered liquor by SH, and evaporating.

Gentianin.—Golden bitter needles, subliming by heat, soluble in hot water, alcohol, ether, and acetic acid, obtained from the powdered root of the gentiana lutea, by treating with ether; distilling off the greater portion, and allowing the remainder to evaporate spontaneously; the residue is digested in alcohol of ·830; the gentianin crystallizes out.

Guaiacin.—Dark yellow mass obtained by alcohol from guaicum wood.

Hesperidin.—White needles from citrons.

Juglanin.—Dark brown bitter from the walnut, juglans regia.

Imperatorin.—Colourless rhombic prisms; by ether from the root of the imperatoria ostruthium.

Lactucin.—Yellowish or reddish matter from the common lettuce and other species of lactuca; obtained from the milky juice by alcohol.

Lapathin.—A hard bitter from rumex obtusifolius.

Liriodendrin.—Colourless scales, resembling boracic acid, from liriodendron tulipifera.

Lupinin, q. v.

Lupulin, bitter matter of hops.—Reddish-yellow powder obtained from hops by alcohol, mixing the evaporated extract with water, saturating the aqueous solution with lime, and treating the evaporated residue with ether and alcohol.

Meconin.—6-sided prisms obtained from an aqueous solution after precipitating morphin by ammonia. The meconin separates from the filtered liquor.

Menyanthin.—From the menyanthes trifoliata.

Melampyrin.—Crystals from melampyrum nemorosum.

Mudarin.—A pitchy matter obtained by water from the mudar, colotropis mudarii.

Ilicin.—Yellow crystals from holly.

Narcissin.—White substance obtained by alcohol from the narcissus pseudo-narcissus.

Olivin.—White needles by alcohol from the resin of the olive tree.

Peucedanin.—White prisms obtained by alcohol from the root of the peucidanum officinale.

Phillyrin.—Amorphous mass from the phillyrea media.

Phloridzine, q. v.

Picrotoxin. ($C_{12}H_7O_5$.)—Poisonous, white, 4-sided prisms; bitter, decomposing at high temperatures without melting; soluble in 150 cold, 25 boiling water, 3 boiling alcohol; soluble in ether; obtained from cocculus indicus by means of alcohol; the picrotoxin crystallizes out, mixed with oil, which is removed by pressing between paper, redissolving in alcohol, and filtering through animal charcoal.

Picrolichenin, or *Variolarin.*—Colourless octohedrons extracted by alcohol from the variolaria amara, a common lichen on trees.

Plumbagin. — Orange-yellow needles from plumbago Europæa.

Populin.—White hard needles from the populus tremula.

Porphyroxin.—Shining needles from opium by ether, water and ether. The last boiling with ether, gives codein, thebain, and porphyroxin. The first is precipitated by dissolving the three in HCl, and adding NH_3. Thebain and porphyroxin fall; spirit takes up the porphyroxin.

Quassin. (C 66·7, H 6·9, O 26·3.)—Bitter small white prisms, obtained from the wood of quassia amara, by boiling the wood in chips in water, filtering, evaporating to three-fourths of its bulk, throwing down pectin, &c. by slaked lime. After remaining one day in contact with the lime, the liquor is filtered and evaporated to dryness, and digested with alcohol of ·830 specific gravity. Impure crystals subside, which are further purified by absolute alcohol, ether, and evaporation. This is an important tonic in medicine.

Quercin.—Crystalline from Quercus robur (the oak) by alcohol.

Salicine, q. v.

Santonin.—White 6-sided prisms from chamomile and other species of artemisia.

Saponin.—White matter from the root of soapwort, saponaria officinalis, a common plant in gardens. By means of potash converted into saponic acid.

Scillitin.—Colourless matter derived from the sap of the scilla maritima, or squills, by alcohol.

Scoparin, q. v.

Senegin.—White substance from the polygala senega.

Smilacin.—Colourless tasteless needles, obtained by treating the root of the smilax sarsaparilla with alcohol and evaporation, and removing the colouring matter by animal charcoal.

Spartin, q. v.

Syringin.—An uncrystalline bitter from the syringa.

Tanghinin.—Crystals from Madagascar tanghin.

Xanthropicrin.—Greenish-yellow needles, obtained from xanthoxylum clava Herculis, by alcohol, water, and ether.

Bitter of Welter. Picric or Carbazotic acid.

Bitter Spar. A class of minerals, including the isomorphous varieties of carbonates of lime, magnesia, and iron, including magnesian limestone.

Bittern. The mother liquor which remains after the evaporated sea water has deposited its common salt. The deliquescent fluid, which does not crystallize, is employed in the manufacture of magnesia alba. The muriate of magnesia, which it contains, is decomposed by carbonate of soda, or by the purified ammoniacal liquor of gas works.

Bitumen. An expression used to denote a number of distinct bodies of organic origin, including amber, asphalt, berengelite, bog butter, boloretin, coal, dyssodyle, elastic bitumen, fichtelite, Highgate resin, Guayaquillite, hartite, Hatchettine, idrialine, ixolyte, Middletonite, naphtheine, ozokerite, petroleum, phylloretin, retinasphalt, Scheererite, tekoretin, xyloretin.

Bituminous Shale. The name of a slate containing bituminous combustible matter.

Bituminous Coals comprehend those coals which leave a comparatively small amount of coke when heated in close vessels. They are the best kind of coals for gas making.

Biuret. $C_4H_5N_3O_4 + 2HO$. White crystalline body, (Wiedemann Poggend. 74, 67), soluble in strong sulphuric, and not too concentrated nitric acid; not precipitated by metallic salts, nor by gallic and tannic acids; soluble in water and alcohol; forms a red solution with oxide of copper and potash, yielding a crystalline compound; by heat evolves ammonia, and then solidifies into cyanuric acid. It is formed from urea, by heating that substance for a long time at a temperature between 302° and 338°. The pasty mass is boiled with a small portion of water, and the filtered liquor precipitated with disacetate of lead; again filtered; the fluid separated from oxide of lead by sulphohydric acid, boiled and evaporated down to crystallize. Biuret is urea deprived of ammonia, or

$$\begin{array}{ll} 2\text{ Urea,} \ldots\ldots\ldots\ldots = & C_4H_8N_4O_4 \\ \text{Biuret,} \ldots\ldots\ldots\ldots\ldots & C_4H_5N_3O_4 \\ \hline & H_3N \end{array}$$

It may also be obtained by heating nitrate of urea to 305°; nitrate of ammonia, cyanuric acid, and biuret are formed.

Blacking. *Cirage*, Fr. Day and Martin's is made by incorporating bone black in powder with sperm oil; sugar, or molasses, and some vinegar are then added; strong sulphuric acid is gradually poured in. Sulphate of lime is thus formed from the bones, which produces a tenacious paste. After the action of the acid, the mixture is diluted with vinegar and bottled while warm.

Black Colours for Pottery. *Gray-black.* Gray flux 85; carbonate of cobalt 10; oxide of iron by ammonia 5. *Brown-black.* Gray flux 76; carbonate of cobalt 16; oxide of iron by ammonia 4; ditto. by heat 4. *Deep black.* Gray flux 78; carbonate of cobalt 11; oxide of iron by ammonia 11. *Iridium black.* Gray flux 75; sesquioxide of iridium 25.

Black Chalk. A bluish-black slate in Caernarvonshire and the island of Islay.

Black Dyes are chiefly formed by the action of tannic acid with a salt of iron.

Black Flux is prepared by igniting cream of tartar in a close crucible, previously lining it with paper to prevent the adhesion of the flux to the sides.

Black Jack. The miner's name for zinc blende or sulphide of zinc.

Black Lead. Plumbago or graphite.

Black Wad. A common miner's name for black oxide of manganese or an earthy variety of it.

Blanquette. The French term for kelp.

Blanquinine. An alkaloid said to have been detected in white cinchona bark.

Blanspath. A phosphate of alumina and magnesia from Styria.

Bleaching. *Blanchiment*, (Fr.) *Bleichen*, (Ger.) The process of bleaching, which consists in the removal of the natural colouring matter of cloth, either linen or cotton, used formerly to be conducted on the principle of oxidizing the colour by the agency of air and the light of the sun; but in consequence of the tedious nature of the operation, the process is generally performed by chloride of lime in the case of cotton, although linen is still exposed to the action of light. When cotton was thus bleached it was exposed to a fermenting action for about a week by steeping the cloth in water to remove the flour of the weaver's dressing, and after being washed in the dash wheel, it was spread or hung in the sun for a similar period and then boiled in caustic soda, the several processes being repeated until the cloth was sufficiently white. It was then soured or exposed to the action of sour milk or weak sulphuric acid, which had the effect of removing earths and alkalies from the cloth. In bleaching with chlorine, the preparatory stages vary with the description of goods. 1. They are often boiled with lime in the bucking boiler A. 2. They are transferred into similar boilers and boiled with carbonate of soda or weak soda ash. 3. They are washed with the dash wheel. 4. They are steeped in a weak solution of bleaching powder, or *chemic*, as it is technically called. 5. They are soured, and sometimes chemiced and soured again.

I. The first operation on the continent is to remove extraneous matter by steeping the cloth in water for twelve hours, at about 100°, when

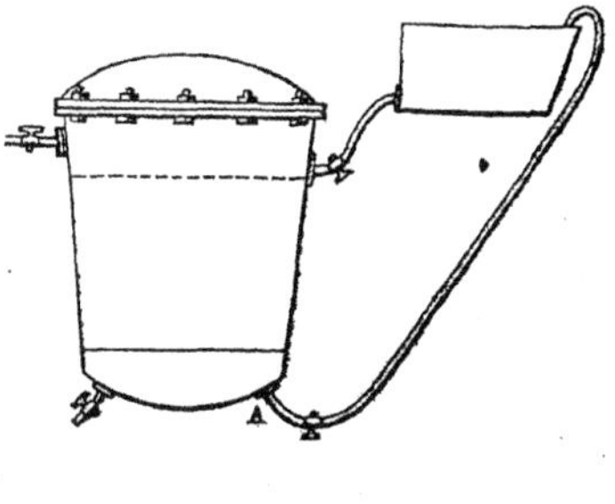

it is well rinsed out or washed in the dash wheel. In France this previous treatment with water is prolonged more than in this country. (1) The cloth there is first moistened with water, (2) then macerated for twenty-four hours at a temperature of about 80°, to remove the starch which may attach to it, by converting it into dextrin or sugar; it is then washed in the dash wheel. It is usual after the cotton is singed, steeped in cold water, and washed with the dash wheel, to put the goods in layers into a bucking boiler A, to cover each layer with a solution of lime and water of a creamy consistence, the quantity of lime being $\frac{1}{35}$ of the weight of the cloth. The boiling is thus continued for eight hours, when the cloth is again washed with the dash wheel in order to remove the lime. II. The cloth is next boiled in an alkaline lye. In France the strength of the lye is 44 lbs. carbonate of soda, 22 lbs. lime, 3,300 lbs. water, for 325 pieces of cloth, and the boiling is kept up for twenty-four hours. In this country the quantity of soda is often used to the amount of 10 lbs. to 100 lbs. cloth, with a sufficient quantity of water to allow the goods to boil freely, and the boiling is only continued for three hours. The goods are then allowed to drain, and hot water is poured over them to remove the alkali. III. The next operation consists in steeping the goods in a solution of bleaching powder. The bleaching liquor is placed in large troughs of stone or wood, termed bleaching vats, and must be carefully diluted to the proper strength, so as not to injure the goods. The best method of determining the strength of this liquid is that invented by Mr. Crum.

Crum's Chlorimeter.—A solution is formed of protochloride of iron by dissolving cast iron turnings in muriatic acid of half the usual strength, and the solution is saturated by keeping a piece of iron in it for some time at the heat of boiling water. One measure of this solution at 40° Twaddell (spec. grav. 1,200), is mixed with one of acetic acid, such as is sold in Glasgow. This forms the proof solution. If mixed with 6 or 8 parts of water it is quite colourless, but bleaching powder occasions with it the production of peracetate of iron, which has a fine red colour.

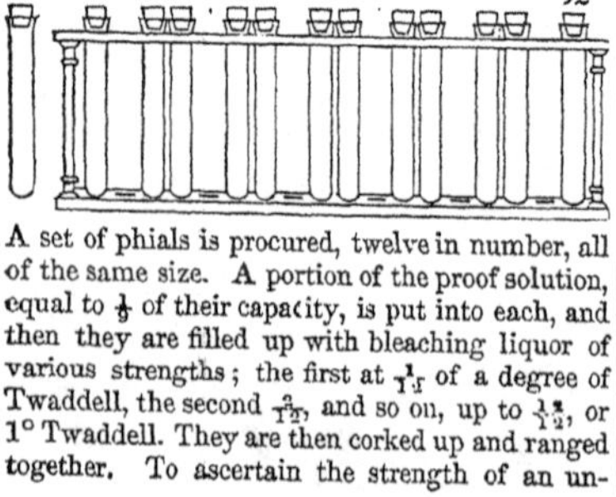

A set of phials is procured, twelve in number, all of the same size. A portion of the proof solution, equal to $\frac{1}{6}$ of their capacity, is put into each, and then they are filled up with bleaching liquor of various strengths; the first at $\frac{1}{12}$ of a degree of Twaddell, the second $\frac{2}{12}$, and so on, up to $\frac{12}{12}$, or 1° Twaddell. They are then corked up and ranged together. To ascertain the strength of an unknown and partially exhausted bleaching liquor, the proof solution is put into a similar phial up to $\frac{1}{6}$, and is filled up with the unknown liquor and placed beside that phial which it resembles. The number of that phial is its strength in 12ths of a degree of the hydrometer, and by inspecting a table we find at once how much of a solution of bleaching powder, which is always kept in stock at 60°, is necessary to raise the whole of the liquor in the steeping vessel to the desired strength. The theory of the operation is as follows:—

3 CaOCl	becoming	3 CaCl
6 FeCl	—	4 $FeCl 1\frac{1}{2} = 2\ Fe_2Cl_3$
	—	2 $FeO\ 1\frac{1}{2} = Fe_2O_3$

The peroxide of iron is taken up by the acetic acid, and produces the characteristic colour. IV. The goods after remaining some hours in the bleaching liquor are *soured*, that is, placed in weak sulphuric acid. The action of the acid is not sufficiently understood, but it has certainly the effect of removing a brown or yellow organic matter which resists the action of chlorine and alkalies. If necessary the processes of washing, steeping, and souring are again repeated. But different bleachers pursue different methods. In bleaching, besides the whitening effect which is produced, the fibres of the cotton or linen are shortened in breadth, so that the cloth does not measure the same breadth as before being subjected to the operation. This seems to be accounted for by the curving of the fibres as they cross each other. This shortening of the fibres is more particularly observable in some of the processes for finishing off cloth, in which the fibres are subjected to moisture and friction.

Measurement of Cloth before Bleaching and after Finishing.

	Wt. of Piece.	Breadth.	Length.
Grey state,......	5 lbs.	27 ins.	$29\frac{1}{4}$ yds.
After bleaching,	4 ,, $2\frac{1}{4}$ oz	$23\frac{1}{2}$	30 yds $23\frac{1}{2}$ in
After printing and finishing,		24	30 yds 4 in.

An interesting observation has been made by Mr. Mercer, that when steeped in caustic soda of 70 Twaddell (1·350), cotton cloth becomes of finer fibre, and produces finer colours in printing, but shrinks very much.

Bleaching by Sulphurous Acid.—In bleaching woollens chlorine is not sufficient; it is necessary to expose them to the action of sulphurous acid, in chambers built for the purpose. The goods are hung up, and the sulphurous acid evolved by placing a piece of hot iron in a pan of sulphur. The frames on which the goods are hung are generally of fir, supplied with hardwood pins, for catching the blankets or woollen goods. In the formation of this framework no nails are allowed to enter, as they are readily rusted by the damp and the goods damaged.

Thom's Sulphuring Apparatus.—A patent recently taken out by Mr. John Thom of Birkacre, Chorley, for a sulphuring apparatus, has greatly improved the process of bleaching by sulphurous acid, by reducing the period of contact of the sulphurous acid and goods to a few seconds. The sulphur apartment is supplied with the acid from an adjoining furnace, the goods being passed rapidly on rollers through the chamber, as represented in the drawing. The advantages of this process, according to the patentee, are, 1st, a considerable saving in fitting up sulphur rooms or chambers; 2d, economy of space and building; 3d, economy of sulphur, time, and labour; 4th, entire freedom from sulphur stains or drops; 5th, improvement in the quality of the cloth, and when printed the results are more certain, and the colours superior. He considers the most important precaution in using the patent successfully to be, to have the cloth in a proper state of humidity, neither too dry nor too damp; this condition being obtained "by using a certain weight upon the levers of the bowls through which the cloth passed as it finished being scoured." The piece is only in the chamber a few seconds; 50 to 70 lbs. crude sulphur are sufficient for 1000 pieces 36 yards long, and one lad manages two machines.

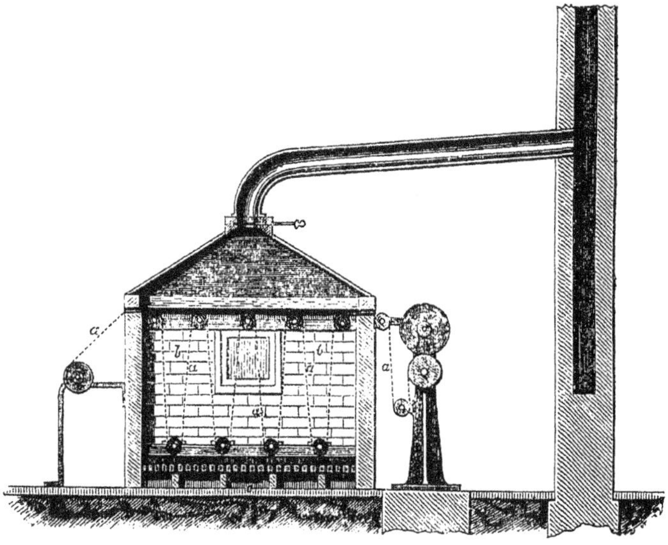

Theory of Bleaching. Action of Air.—When linen and cotton fabrics are exposed to moisture and the heat of the sun, a slow process of combustion takes place; the oxygen of the air uniting with the colouring matter converts it into carbonic acid, and the weight of the cloth diminishes. If the action be continued too long the linen and cotton lose their cohesion, and are converted into a pulp resembling that used for the manufacture of paper.

Action of Chlorine.—According to one view, chlorine, under the influence of light or oxidable bodies, including coloured substances, decomposes water, yielding chlorohydric acid, the oxygen forming chlorous acid, or binoxide of hydrogen. One or other of these compounds now gives oxygen to the colouring matter, and this goes on till all the chlorine is converted into chlorohydric acid. Another view is, that chlorine unites directly with the colouring matter, and forms a colourless chloride. But the most probable explanation is that the chlorine unites with the colouring matter by replacing hydrogen, while another atom of chlorine combines with the displaced hydrogen, and separates as chlorohydric acid.

Bleaching Powder. See Lime, Chloride of.

Blende. See Sulphide of Zinc.

Bleyniere. An impure arsenio-carbonate of lead.

Blistered Steel. Steel having blisters on its surface from the discharge of a gas from the interior of the bar, is so termed.

Block Tin, or metallic tin reduced from tinstone dug out of veins is less pure than *grain tin*, which is obtained from *stream tin*, or loose grains of tinstone in alluvial soil.

Bloedite. *Reussin* or *Reissite. Soda Sulphate of Magnesia.* From Ischel, in Austria, found along with *polyhalite*. It consists of sulphate of soda 33·34, sulphate of magnesia 36·66, water 29·, NaCl ·33, $FeOSO_3$ ·34, $MnOSO_3$ ·33. Efflorescences found at Seidlitz and Seidschutz in Bohemia, in flat 6-sided prisms and needles, contain $NaOSO_3$ 66·04, $MgOSO_3$ 31·35, MgCl 2·19, $CaOSO_3$ ·42.

Blood. A well-known fluid, amounting to from 22 to 26 lbs. in the human body, which, as seen by the naked eye, is red, and when viewed by the microscope, appears as a greenish-yellow liquid, in which red corpuscles or globules are floating. The fluid portion as it exists in the circulation is termed the *liquor sanguinis* or liquid part of the blood. When fresh drawn, the spec. grav. of the blood varies from 1051 to 1057; 1054 is about the mean of healthy human blood. The *size* of the human blood globules are differently stated by authors who have measured them. They were first noticed by Leuwenhock

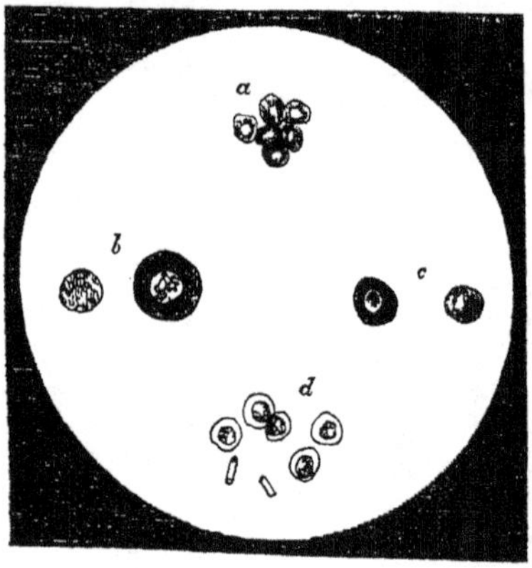

Human Blood Globules.
a, Carus. *b*, Dumas.
c, Bauer and Home. *d*, Wagner.

in 1674, who considered them 25,000 times smaller than a grain of sand. Sir E. Home found them $\frac{1}{1692}$th part of an English inch in diameter; Wollaston $\frac{1}{4992}$th part, and Young $\frac{1}{6062}$th part. But they differ much in different animals, those of man passing readily through a filter, and those of the frog being detained on it, and allowing a clear fluid to pass through the paper. The *form* of the globules is considered to be that of a flat disc in which a nucleus is seen. To view them under the microscope they must not be mixed with pure water, which partially dissolves them, but with saline or saccharine water. The *smell* of the blood is peculiar, and has been termed by physiologists *fragrant* and *alliaceous*, and is powerfully represented by Shakspeare as constituting the ever-present goad to the guilty conscience of Lady Macbeth, as she exclaims, "Here's the smell of the blood still; all the perfumes of Arabia will not sweeten this little hand." *The action on test paper* by the blood is alkaline—that is, it restores immediately the blue colour of reddened litmus. *Coagulation.* When the blood of the frog is drawn from the animal and thrown on a filter a clear liquid passes through, which is the liquor sanguinis, while the red globules remain on the filter. In a few minutes the clear fluid, however, gelatinizes. This is due to the separation of the *fibrine*, which in warm blood, as drawn from an animal, is in a state of solution. If we make the same experiment with the human blood, the red particles are so minute that they pass through the pores of the filter, and we cannot thus separate them from the fibrine. Human blood and all other kinds of red blood when freshly drawn therefore coagulate; that is, the fibrine separates as the blood cools and entangles the blood globules, which fall down together as a red *crassamentum* or *coagulum*, and an amber-coloured fluid separates from it, called the serum, in which the red cake is placed. It is important to distinguish between the liquor sanguinis and the serum, the former being the serum before the fibrine has separated by coagulation. Coagulation occurs in some animals much more rapidly than in others. With the sheep it is difficult to weigh out a portion of blood as it flows from a vein, since it coagulates in from ½ to 1½ minute; in the same time in the pig, rabbit, and fowl; in from 2 to 15 minutes in the horse; 2 to 10 in the ox; ½ to 3 in the dog; in mice and fish in a moment. Dr. Andrew Buchanan has contrived a very ingenious method of separating the blood globules in human blood from the fibrine. He throws a quantity of serum on a filter; allows fresh blood to flow into the serum; in general, the globules are detained on the filter; and the liquor sanguinis passing through coagulates by the separation of the fibrine which forms a clear jelly. Again, if we pour fresh blood in a tumbler into a quantity of serum, the red globules fall to the bottom of the vessel, and the fibrine separates in a coagulum above. *Composition of the clot.*—When the blood has been drawn from an animal it is usual to consider it as composed of two portions—a solid part, the clot or crassamentum floating in the serum or watery portion. The relative proportions of serum and clot I have found to be in healthy persons; in 1000 parts in man, serum 455·, clot 545; in woman, serum 424, clot 576. When 1000 parts of clot are heated a little above a steam heat until they cease to lose weight, it is found that 709 parts of water have evaporated, and 291 parts of solid matter, consisting of fibrine and blood globules, remain. From this experiment we learn that much of the serum is detained by the clot, and that the true composition of the clot, approximately, is serum and water of globules 793, globules, fibrin, and salts 207·. The globules consist of a modification of albumen, which has been termed *globuline*, of colouring matter or *haematine*, together with a portion of oxide of iron. The other constituents of the clot are *fibrine* and salts. *Composition of the serum.*—The serum is an alkaline fluid having a yellow or amber colour, and is quite clear when drawn from an animal which has not tasted food for some hours, and from whose food fat or oil has been excluded. Its specific gravity varies from 1027 to 1029. I have found it as a mean from many trials, about 1028. When heated (to 159°, Thomson,) as ascertained by Dr. Harvey, the serum coagulates, the *albumen* which it contains in solution becoming solid as with white of egg (Rouelle and Bouquet, 1776). When 1000 parts of serum are heated in a water bath, they lose in man 910 water and there remain 90 parts of solid residue. In the human female I have found 898 water, and 102 solid matter consisting of albumen and salts, with certain fatty substances, margarin, margaric acid, cholesterin, serolin, lecithine, cerebric acid, creatine, creatinine, &c. Besides these bodies, serum possesses a great absorptive power of gases, for while 1000 vols. of water only absorb $9\frac{1}{4}$ of oxygen, 1000 vols. blood absorb from 100 to 130 vols. oxygen. Magnus found 125 vols. of the blood of a horse to give out in vacuo, 5·4 carbonic acid, 1·9 oxygen, and 2·5 nitrogen. The following table presents a list of the various bodies which have been found in human blood:—

Clot.

Fibrinous Matter,....	Fibrine,....................	2·7
	Globules,..................	120·68
	Hæmatosine,..............	
	Globuline,.................	
	Caseine,....................	

Serum.

	Albumen,....................	69·41
	Water,........................	790·
Gases,.....	Carbonic acid,............	
	Nitrogen,....................	
	Oxygen,......................	
Fats,......	Fat with phosphorus,...	Fats, &c. = 10·96
	Cholesterine,.............	
	Seroline,....................	
	Oleic acid,................	
	Margaric acid,............	
	Cerebric acid,............	
Salts,......	Chloride of sodium,.......	Salts, = 6·45
	———— potassium,....	
	———— ammonium?..	
	Carbonates, soda,........	
	———— lime,........	
	———— magnesia,....	
	Phosphates of soda,......	
	———— lime,......	
	———— magnesia,.	
	———— iron,.......	
	Sulphate of potash,........	
	Lactate of soda,...........	
	Fatty acid salts,...........	
	Volatile fat salts,..........	
	Yellow colouring matter,....	
	Sugar,.........................	
	Urea,...........................	
	Hippuric acid?...............	
	Uric acid,.....................	
	Creatine,......................	
	Creatinine,...................	
	Licithine,.....................	
	Hæmatoidine,................	
	Albuminose?.................	

Salts of the Blood.

	Dog fed on Flesh.	Do. Bread and Potat.	Man 45 Years.	Woman 22 Years.	Ox.
Chlorine,....	30·25	30·94	37·50	33·76	32·60
Sodium,......	19·60	20·04	24·49	21·87	21·11
Soda,.........	5·78	2·02	2·03	6·27	14·40
Potash,......	15·16	19·16	12·70	11·24	8·76
Magnesia,...	0·67	4·38	0·99	1·26	0·59
Sulph. acid,.	1·71	1·08	1·70	1·64	1·16
Phosphor. acid,......	12·74	9·34	7·48	9·74	3·02
Ditto,.........	1·22	2·35	1·87	1·36	1·62
Lime,.........	0·10	0·70	1·68	1·85	0·70
Sesquiox. of iron,..	12·75	8·65	8·06	0·68	8·80
Carb. acid,..	0·53	0·37	1·43	0·95	6·49

Mean ash in blood, 6·45 per 1000.—The amount of carbonic acid increases with vegetable food, while the phosphoric acid diminishes. Phosphoric acid is largest in a dog fed with meat; the potash also diminishes with vegetable diet. The sesquioxide of iron in the blood from animal food is in great excess.

Ashes of Blood.

	Human.	Calves.	Sheep.
Phosphoric acid, ...	31·787	20·145	16·306
Alkalies and alk. earths,	58·993	66·578	60·576
Carbonic acid, ...	3·783	9·848	19·474

This is a proof that carbonate of soda can replace phosphate of soda in the blood, and both of these salts increase the power of the serum to take up carbonic acid.

White Blood.—That the serum of the blood frequently assumes a white appearance has been known for a long period, but it has been generally considered to indicate a state of disease. Hewson observed this phenomenon, and he rejected the view that it depended on the presence of chyle. He ascribed it to the presence of fat absorbed from the adipose tissue, which he supposed to be taken up more rapidly than the wants of the system require, and therefore to accumulate in the blood vessels. This opinion appears to be the common doctrine at the present day. John Hunter alone rejected it. He observed it most in the blood of breeding women. Dr. Trail observed it, and ascribed it to the same cause as Hewson; while Dr. Christison refers it to incipient disease of the kidney.—Dr. Andrew Buchanan first observed it to occur in healthy persons in 1844, in the following experiment:—

"A strong healthy young man, to whom a good dinner was an equivalent for the loss of a few ounces of blood, was easily prevailed upon to submit to the following regimen and treatment. He had no breakfast, and at four o'clock had for dinner one pound of beefsteak, half-a-pound of bread, sixteen liquid ounces of brown soup, and half-a-bottle of porter. Three ounces of blood were then taken from a vein in the arm at three different periods; the first time, half-an-hour after the meal; the second time, an hour and forty minutes after it; and the last time, next morning at eight o'clock, or sixteen hours after the meal, no food having been taken in the interval. The blood as it issued from the vein had the usual appearance, and the serum which separated from it was about the same in quantity each time. The first time the serum was whitish and turbid; the second time it was like whey; while the third time it was perfectly limpid. The crassamentum on the first occasion exhibited nothing peculiar, while on the last it was covered with a transparent fibrinous crust beautifully interspersed with white dots, which led the medical friend who assisted me in these investigations to compare it to a precious stone. As it might be supposed that this young man's blood was white before he took dinner, the two following trials were made to obviate that objection:—A vigorous man of about thirty-five years of age, after fasting nineteen hours, had for dinner twenty ounces of beefsteak, sixteen liquid ounces of brown soup, and eight ounces of bread. He was bled immediately before his meal, and three times after it, two ounces of blood being taken away each time. The serum obtained from the first bleeding before the meal was perfectly limpid; the serum from the second bleeding, three hours and fifteen minutes after the meal, was turbid; the serum from the third bleeding, eight hours and fifteen minutes after the meal, was still thicker; while that from the last bleeding, eighteen hours after the meal, was again quite limpid, although some supper had been eaten in the interval.—The young man first mentioned, after fasting eighteen hours, dined upon sixteen ounces of brown soup, four ounces of bread, eight ounces of potatoes, sixteen ounces of beefsteak, and sixteen ounces of London porter, and fasted eighteen hours after the meal. He had blood taken from his arm four times, to the extent of two ounces each time. The serum of the blood first taken, immediately before the meal, was of an amber-yellow, and quite transparent; the serum from the second bleeding, two hours and ten minutes after the meal, was turbid; the serum from the third bleeding, eight hours after the meal, was exactly of the colour of water gruel, and quite opaque; the serum of the blood last taken, eighteen hours after the meal, was still turbid, its limpidity not having been, as after his usual fare, restored by an eighteen hours' fast. In neither of the two last cases did the blood, as it issued from the arm, present white streaks or anything else unusual. The crassamentum of the blood drawn before the meal was in both cases of the usual red colour on the surface, as also that drawn first after the meal in the last case; but in all the other instances it exhibited the same *pellucid fibrinous crust* already described, although not dotted in the same remarkable way. We can scarcely avoid the conclusion that this pellucid crust is connected with finished digestion, when we reflect that out of nine bleedings practised within eighteen hours after a very full meal this crust was observed on every occasion, if we except those in which the blood was drawn within three hours and a quarter of the period of taking the meal.—Dr. R. D. Thomson was kind enough to examine a specimen of the white matter separated by the action of common salt and the filter, but too minute in quantity to admit of a satisfactory analysis. He found it quite insoluble in ether and alcohol, while it dissolved in caustic potash. On boiling it in a solution of sugar of lead, it gave traces of black sulphuret. He concluded, therefore, that it contained no fixed oil, and consisted most probably of an *albuminous compound*, like white of egg or muscle. A further opportunity was afforded of examining the chemical qualities of this serum in some specimens obtained for illustration. A man about thirty years of age, after fasting eighteen hours, dined upon twenty-four ounces of a pudding, consisting of two parts wheaten flour, and one part suet, seasoned with salt. Two ounces of blood taken before the meal yielded a perfectly limpid serum. Seven ounces were taken three hours after the meal, and the same quantity six hours after it. The serum from the former was like syrup, but a little white; that from the latter was milk-white. The white matter in the latter was separated by Dr. Thomson, by means of salt and the filter, and appeared similar to the substance before examined. It contained no fixed oil. The other specimen of serum threw up its cream spontaneously. It left upon the filter only a trace of white matter, but a notable proportion of a fixed oil, which was easily demonstrated by merely drying the filtering paper, and holding it between the eye and the light. It can scarcely be doubted that this oil was derived from the suet of the pudding, while the white albuminous substance most probably represented the gluten of the flour. Thus two of the three elements of which the food consisted were found in the blood, but the starch, the most abundant of all, was sought for in vain."—It was subsequently found, however, that all blood ferments with yeast. Hence it is possible that all starch is converted into sugar, and thus reaches the blood in solution. To ascertain if the same phenomena were constant also in the inferior animals, I caused a calf to be fed with milk and gruel at seven in the morning, and at ten it was killed. An oily substance separated from it, and a cream formed on its surface as in the preceding experiments. An ox also was fed on oats, and being killed in five hours afforded similar results. The quantity of white matter which was separated from the serum by means of salt was very considerable. It was afterwards found by Dr. Buchanan that arrowroot, jelly, and similar diet, unless when fat was present in the food, produced no whiteness in the serum.

Blood in Diseases.—This subject has attracted the attention of various physicians, and among others of Andral and Gavarret, who have made an interesting series of experiments upon the quantity of the materials of blood in different diseases. These they have determined by carefully separating and weighing them. Now, weighing is a most important consideration in medicine. It is common for surgeons to enter in their reports that the pulse of a patient is less frequent than yesterday, that the quantity of urine has diminished or increased. But this is a most indefinite mode of writing. If, on the contrary, the pulse is stated to be 90 one day and 80 the next, or that the quantity of urine was one day 2 lbs. and 3 or 1½ the next, a precise and definite nomenclature is employed; and if we are told at the same time the weight of the food and drink swallowed, most important data are afforded, from which interesting conclusions can be drawn. It is by such data that we are enabled to determine whether our patient is improving. For example, in diabetes the principal symptom is an unnatural discharge of saccharine urine. But if one is asked what he means by an unnatural discharge, many persons would say — a large quantity. This again is an indefinite expression. It is a comparative term, with an indefinite object of comparison. But if one is told that the average quantity of healthy urine daily is 40 fluid oz. and that in diabetes it may amount to, upon an average, 200 oz. this numerical statement is at once clear and obvious to the mind. The first mode of expression is equivalent to the rule of thumb, so prevalent among farmers and older manufacturers, while the latter method is expressed with mathematical precision. Weighing and measuring are most important objects in medicine, which remarkably distinguish the medical men of the continent from those of this country.—In considering the effect of diseases upon the blood, it is necessary to institute a standard of comparison—to determine the average amount of the constituents of the blood in health. Now, according to Lecanu, healthy blood contains—

Fibrine,	3
Globules,	127
Solid matter in serum,	80
Water,	790
	1000

Mode of Analysis by Dumas.—The blood is received into two vessels of equal capacity, each of which is capable of containing 5½ oz. Into one of the vessels the first fourth of the blood is drawn, and afterwards the last fourth; and this portion of blood is allowed to coagulate. Into the other vessel are drawn the second and third fourths. This portion is beat with a stick in order to obtain the fibrine, which is carefully washed. The two portions of blood thus separated ought to have the same composition. When the coagulation is completed the serum is to be carefully separated from the clot, and dried. There are thus obtained—(1) the fibrine; (2) the serum; (3) the clot. The dry fibrine is weighed. This gives the quantity of fibrine which the clot contained. The residue from the serum is weighed, and thus we obtain the quantity of water and solid material. The dried clot is then weighed. The quantity of water which it contained, represented by the loss sustained by the drying, enables us to calculate the quantity of solid materials of the serum which the dried clot contains. Subtracting from the weight of the dry clot the weight of the fibrine, plus the weight of the solid materials of the serum which the clot contained, and which must be calculated, there remains the weight of the globules which the clot contains. By this procedure we obtain (1) the weight of the fibrine; (2) the weight of the globules; (3) the weight of the solid materials of the serum; (4) the weight of the water.—The result of experiments upon different kinds of blood has shown the constituents to vary as follow:—

Fibrine, ...from	10½	to 1	per 1000
Globules, ...from	185	to 21	—
S. M. of serum, from	114	to 57	—
Water, ...from	915	to 725	—

The diseases which have been examined in regard to the composition of the blood appear divisible into four classes. I. Fibrine *increased.* This class comprehends articular rheumatism, pneumonia, capillary bronchitis, pleuritis, peritonitis, erysipelas, amygdalitis, cystitis, and inflammation of the lymphatic ganglia.—In acute rheumatism the quantity of fibrine varies from 10·2 to 4·1, the average quantity being about 7½. The intensity and extent of the pain in this disease appeared to exercise a greater influence in increasing the quantity of fibrine than the period or duration of the disease; and it is an important remark, that if a bleeding did not produce any effect upon the disease the fibrine did not diminish; but whenever the bleeding was beneficial the fibrine diminished. The following was the result of five bleedings:—

1 bleeding,	.........8·9	per 1000 fibrin.
2 —	.........9·8	—
3 —	.........8·5	—
4 —	.........6·4	—
5 —	.........2·8	—

The state of the globules in rheumatism appears

to be inversely that of the fibrine. In no case do these amount to their natural quantity, and often after the first bleeding their quantity has descended under the mean number.—In *pneumonia* the fibrine is always increased—the highest number is 10½, and the lowest 4. The quantity of fibrine increases with the intensity of the disease, and decreases with its amelioration.—The other diseases presented similar results.—In erysipelas the average of fibrine was 6. In one case, in which the disease was mild, the quantity was 3·6, and the globules were increased.—In *pulmonary* tubercles there is always a tendency to increase in the amount of fibrine and diminution of the globules. The increase of fibrine begins to show itself when the tubercles begin to soften. Even when the lung is filled with cavities, there is abnormal quantity of fibrine. It always happens that when the patient is reduced to a state of marasmus and exhaustion, the fibrine begins to decrease, the quantity then being as low as 2. These facts go to support the idea of the inflammatory nature of consumption. It also appears that every acute disease has a tendency to increase the quantity of fibrine in the blood.—II. The second class includes the *pyrexiæ*, which are characterized by the fibrine remaining normal in quantity or diminishing, while the same holds with regard to the globules; from which Andral concludes, that it is on the inflammatory action, and not on the fever which accompanies the diseases of the preceding class, that the increase of fibrine depends. The diseases of the present class include, continued and typhoid fevers, eruptive fevers (small-pox, measles, and scarlet fever), and intermittent fevers, and congestions.—III. Diseases in which the globules are diminished. Andral had proved, under the preceding classes, that, by loss of blood and deprivation of food, the globules of the blood diminish; he proved the same with regard to certain diseases which oppose the free and complete reparation of the blood, as cancer of the stomach and pulmonary tubercle. But there is still another state of body, termed *anæmia* (want of blood), into which some persons fall, who have either been weakened by accident or disease. This may happen after fevers or diabetes. But it is most particularly typified in delicate females in the form of chlorosis, where the waxy paleness of the whole body demonstrates the deficiency in the colouring matter of the blood. An interesting case is related by Andral of a young man who fell into a state of anæmia without any assignable cause. His blood when first examined contained 87·9 of globules, the fibrine being natural. Shortly afterwards he was again bled, the globules were 72·2. He was then treated with ferruginous medicines, and when the blood was again examined, the globules amounted to 86·9. The fibrine does not appear to diminish; from which Andral concludes that a chlorotic patient may contract an inflammatory disease, since the element of the blood, which is most modified in the inflammatory state, not only has not diminished, but predominates over the globules.—IV. The next class comprehends Bright's disease of the kidney, or albuminous nephritis, as it has been called. It is characterized by the presence of albumen in the urine; and Andral has found that, in proportion to the increase of albumen in the urine, that of the blood diminishes. But the same observation had previously been made by others. Indeed, it is almost self-evident that it should be so. This character of the blood may be detected by the diminished specific gravity of the serum. Healthy serum has a specific gravity of 1·027. If it falls much below this, we may conclude it to be abnormal.

Blood in Scurvy and Purpura. (R.D.T.)

	Scurvy.	Purpura.
Water, ...	807·82 }	992·074 (Water and Globules)
Globules, ...	185·74 }	
Albumen, ...		
Fibrin, ...		1·106
Salts, ...	6·44	6·82
	1000	1000

Total Salts in Blood. (R.D.T.)

Healthy Female, ...	5·84
Do. Male, ...	6·38
Typhus, ...	3·41
Cholera, lymphatic stage, ...	3·56

Blood in Cholera. (R.D.T.)—*Relations of Insoluble to Soluble Salts.*

	Ratio Insol.		Sol.
Healthy blood, Female, ...	1	to	4·4
Ditto, Male, ...	1	to	4·02
Cholera blood, Collapse stage, ...	1	to	3·81
	1	to	5·7
	1	to	2·59
Ditto, biliary, ...	1	to	4·3

Ratio Fibrine to Albumen.

		Globules and Water.
Cholera, ...	1 to 59·86	159
Health, ...	1 to 66·80	263

Blood in Diseases. (Le Canu.)

	Healthy.		Inflam.	Diabetes.	Typh.	Heart Dis.
Water, ...	790·		776·	816·1	805·	836·
Globules, ...	131.		144·	120·3	115·	80·
Albumen, ...	56·			55·5		
Fibrine, ...	2·7			2·4		
Oxide of iron,	1·7					
Carb. of soda,	1·00	7·1				
Chl. of sodium	4·1					
Do. potassium	2·1		80·	Salts 5·7	80	84·
Carb. of lime,	1·3	3·5				
Phos. lime, Phos. MgO.	·5					
Fats with P...	8·0					
Cruorine, ...	1·1					

	Chlorosis.	Cholera.	
Water,..............	862·	630	749·
Globules,...........	55· }	370	251
Albumen,..........	83· }		

Mean of 3 cases.

Bloodstone. A variety of quartz of a green colour traversed by blood-red veins, which owe their tint to sesquioxide of iron.

Blowpipe. As a knowledge of the results to be obtained by the use of the blowpipe can only be acquired by a good deal of practice, the student should provide himself with a blowpipe (which may be had very cheap when made of tin), and practise frequently, so as to be enabled to use the subsequent tables. The cheapest form

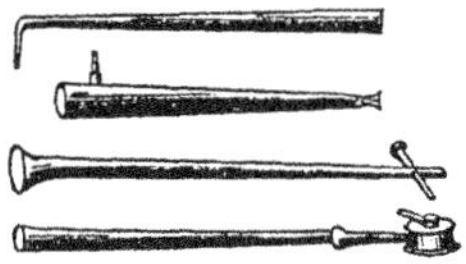

of blowpipe is No. 1, introduced by Dr. Black. No. 2, Wollaston's blowpipe, takes to pieces, and is, therefore, so far, convenient. In the absence of a gas or spirit-lamp, a common tallow candle will supply a flame capable of affording sufficient heat. The charcoal to be used should be well burned and free from fissures; a small cavity being made by means of the point of a knife, to retain the substance and fluxes.

Action of the Blowpipe upon Bodies.—By the influence of even comparatively low degrees of heat, and such as can be communicated by a gas or spirit-lamp, when the substance is placed on a spatula held over it, we are enabled to obtain much knowledge of the nature of many bodies. When greater degrees of heat are to be applied it is necessary to use the blowpipe, whether of glass or metal, bent at a right angle, and employed so to propel the flame of the lamp upon the body to be examined, as to enable the operator to observe distinctly the slightest alteration produced upon the body by the continued high temperature. When we propel a flame by means of a blowpipe, there are two portions of the flame which produce different effects, and therefore require particular attention. The flame of a common candle consists of an external red, and an interior blue or dark, cone. A flame is simply a hollow cone of light, consisting of gases, which merely burn at the surface, where they are in contact with the oxygen of the atmosphere. The point *a* is obviously in contact with the air, while the point *b* is farther removed from it. But when we blow through a flame the circumstances are altered: *b* in the first figure corresponds with *b* in the second, as in both unburned gases exist. In the second figure *a* is the oxidizing and *b* the reducing flame. To reduce an oxide to the state of metal, that is, to cause the oxygen of the oxide to combine with the gases and form flame, we place the body in the reducing flame. To oxi-

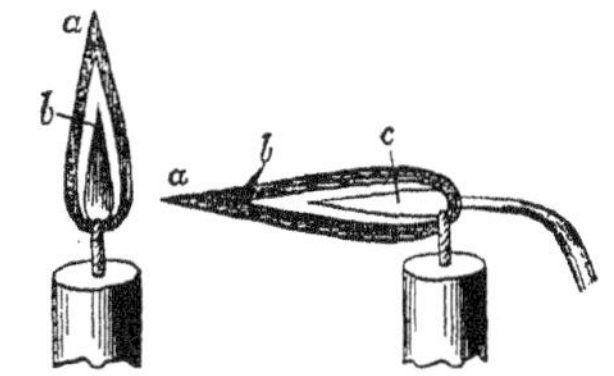

dize a metal, so as to cause the metal to combine with oxygen, we place the substance in the oxidizing flame. In exposing a body to the action of the blowpipe we first ascertain the effect (*per se*) by itself, simply by holding the body in the flame by means of a convenient instrument, either a thin platinum wire twisted round the substance or a pair of forceps tipped with platinum.

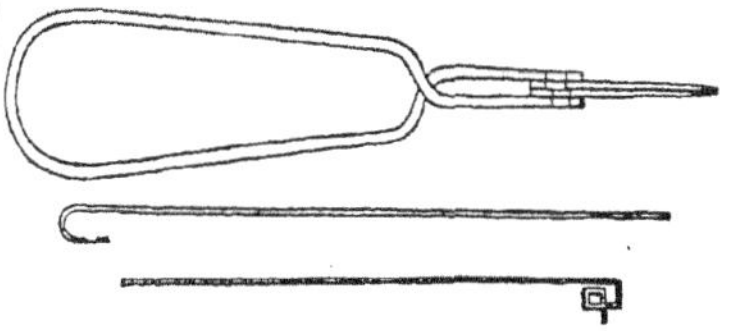

Action upon Bodies per se.—In making this experiment a long thin fragment of the body is fixed between the points of a pair of forceps, and held in the reducing flame. This method is particularly applicable to minerals, many of which have peculiarities which assist us in deciding on their species. A certain set of minerals, amounting to about thirty-six in number, are characterized by their frothing before the blowpipe, and hence they have been termed zeolites (from ζέω, I boil). They are well typified by Thomsonite. They contain from ten to twenty per cent. of water, and when heated they part with the water, and, boiling up during the process, either fuse into a white bead or fall to powder.—(For other characters of minerals, see Thomson's *Mineralogy*, vol. i. and this work.)

I. Colours produced by bodies *per se*.

Potash salts,................	Violet flame.
Lithia,........................	Purple red.
Strontian,....	Crimson.
Lime,.........................	Red.
Soda,.........................	Yellow.
Barytes,......................	Pale green.
Borates, with sulphuric acid,....................	Green.
Iodides, with salt of phosphorus and oxide of copper,................	Green.
Phosphates, with sulphuric acid,...........	Green.

Ammonia,	Dark green.
Tellurium oxide,	Green.
Copper,	Green, except Cl and Br (blue).
Zinc,	Whitish-green.
Bromides, with phosphorus and oxide of copper,	Blue, with green edges.
Arsenic,	Light blue.
Antimony,	Greenish-blue.
Selenium,	Blue.
Lead oxide,	Blue..

II. Odours produced.

Arsenic,	Garlic.
Selenium,	Horse-Radish.
Sulphur,	Peculiar.
Sulphurets,	Odour of SO_2.

III. *Blowpipe Action of Substances upon Charcoal.*—The *metals*, with the exception of platinum, palladium, iridium, rhodium, and osmium, are almost all capable of being melted before the blowpipe on charcoal.

Fusible Oxides.—The oxides are mostly infusible, with the following exceptions:—

Antimony, oxide.	Bismuth, oxide.
Vanadic acid.	Lead, ,,
Tellurous ,,	Copper, ,,
Molybdic ,,	

Volatile Metals.—The following metals sublime, leaving a powder on the charcoal:—

Arsenic.		Cadmium,	Yellow.
Antimony,	White Sublimate.	Bismuth,	Yellow.
Tellurium,	White Sublimate.	Lead,	Yellow.
Zinc,	White Sublimate.	Tin (trace),	White.

Mercury and osmium volatilize, without producing a sublimate.

Reducible Bodies.—Most metallic oxides are reduced to the metallic state on charcoal. The exceptions are, the alkalies, earths, cerium, uranium, manganese, iron, vanadium, chromium, tungsten, columbium, titanium.

IV. *Blowpipe Action with Carbonate of Soda.*—Bodies fusing into a bead on platinum wire in oxidating flame:—

Silica,	Colourless glass.
Antimony, oxide and acids,	White.
Tellurium, oxide and acid,	Colourless when hot, opaque when cold.
Chromium oxide,	Orange in oxidating, green in reducing.
Molybdic acid,	Milky.
Tungstic acid,	Yellowish.
Titanic acid,	Yellow glass when hot, white when cold.
Manganese oxides,	Green,
Cobalt oxide,	Red, and then gray.
Copper oxide,	Green when hot, colourless when cold.
Lead oxide,	Transparent when hot, yellow and opaque when cold.

Metals subliming on charcoal with carbonate of soda:—

Arsenic.	Tellurium.	Lead.
Antimony.	Zinc.	Bismuth.
Selenium.	Cadmium.	

Metals reduced on charcoal with carbonate of soda:—

Molybdenum.	Cobalt.	Silver.
Tungsten.	Copper.	Gold.
Iron.	Tin.	Platinum.
Nickel.		

Bodies neither Fused nor Reduced.—The alkaline earths, the earths, oxides of cerium, oxide of uranium, columbic acid.

V. *Blowpipe Action with Borax—Colour of Beads.*—Experiments with borax or carbonate of soda are generally conducted on platinum wire.

Oxidizing flame.		Reducing flame.
Colourless,	Lime,	Colourless.
	Barytes,	Colourless.
	Strontian,	Colourless.
	Magnesia,	Colourless.
	Alumina,	Colourless.
	Glucina,	Colourless.
	Yttria,	Colourless.
	Zirconia,	Colourless.
	Thorina,	Colourless.
	Silica,	Colourless.
	Columbic acid,	Colourless.
	Cadmium,	Colourless.
	Zinc,	Colourless.
	Tin,	Colourless.
	Tellurium,	Gray.
	Antimony,	Gray.
	Molybdic acid,	Reddish-brown.
	Tungstic acid,	Yellow.
	Titanic acid,	Yellow, hot; violet or dark blue, cold.
	Silver,	Gray.
	Bismuth,	Gray.
Green,	Chromium, when cold,	Green.
	Copper,	Brown.
	Cerium,	Colourless.
Red,	Iron,	Green.
	Nickel,	Gray.
Blue,	Cobalt,	Blue.
Violet,	Manganese,	Colourless.
Yellow,	Uranium,	Green.
	Vanadium,	Green.
	Lead,	Gray.

VI. *Blowpipe Action with Salt of Phosphorus—Colour of Beads.*—Microcosmic salt, or ammonia-phosphate of soda, when heated, parts with its ammonia, and is converted into biphosphate of soda.

Oxidizing flame.		Reducing flame.
Yellow,	Vanadium,	Green.
	Uranium, while hot, ... Uranium, green while cold, ...	Green.
	Bismuth,	Gray.
	Silver,	Gray.
Blue,	Cobalt,	Blue.
Violet,	Manganese,	Colourless.
Red,	Cerium,	Colourless.
	Iron,	Green.
	Nickel,	Red.
Green,	Chromium, when hot, yellow, ...	Green.
	Uranium, when hot, yellow, ...	Green.
	Molybdic acid, ...	Blue when hot, green cold.
	Copper,	Brown.
Colourless,	Tungstic acid, ...	Red, if iron present, blue without it.
	Antimonious acid, ...	Red with iron.
	Antimony oxide, ...	Gray.
	Titanic acid, ...	Red with iron, violet or blue without it.
	Tellurium,	Gray.
	Lead, yellow when hot, ...	Gray.

The alkaline earths, earths and other bodies not enumerated, give colourless or milky beads in both flames.

Blowpipe Oxyhydrogen.—An apparatus which acts on a similar principle to the preceding. Instead, however, of employing the force of the cheeks and common air, strong pressure is used, and a mixture of 1 vol. oxygen and 2 vols. hydrogen being propelled through the mouth of the blowpipe, the gases are ignited and produce intense ignition. This application of the two gases seems to have been invented about the same time by Dr. Thomas Thomson, (1800 or 1801,) and Dr. Hare, 1802; but as Dr. Thomson published no account of his experiments at the time, but merely showed them to his pupils, and as Dr. Hare gave a detail of his to the public, the latter has usually obtained the merit of the invention. The apparatus used by Dr. Thomson as subsequently improved is delineated in the woodcut under hydrogen.

Blowpipe Table.—In this form of blowpipe the air is propelled through the blowpipe by means of a pair of bellows pressed by the foot as in glass-blowing, that being the form of apparatus particularly valuable to the chemist for bending glass tubes, blowing bulbs, hermetically sealing tubes, &c.

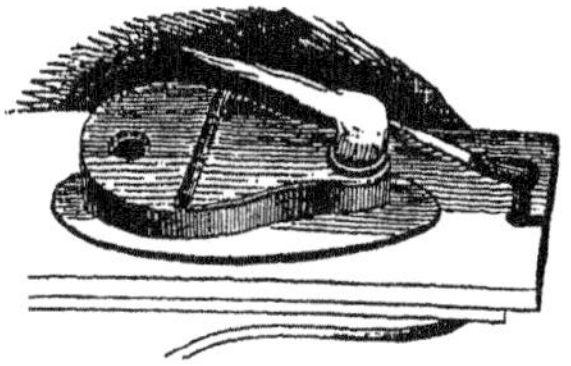

Blue, Azure. Consists of 67 gray flux (89 pebble flux, 75 minium, 25 sand,) and 11 fused borax. Another *azure blue* for skies is made by mixing on a glass slab 2 deep blue, 1 oxide of zinc, 4 lead glass (4 minium, 1 sand).

Blue, Berlin. See PRUSSIAN BLUE.

Blue, Brunswick, or celestial, is prepared as a pigment by adding to a solution of alum carbonate of soda to precipitate alumina, washing the alumina, and adding sulphate of iron, yellow prussiate of potash, sulphate of barytes, and a little bichromate of potash to supply oxygen. This colour is dried.

Blue, Cerulian. Gray flux 79, (see AZURE BLUE,) carbonate of cobalt 7, hydrous carbonate of zinc 14, for pottery.

Blue, Clear. 1 oxide of cobalt, 2 oxide of zinc, 6 glass of lead, (2 minium, 1 white sand,) 1½ lead glass, (2 minium, 1 sand, 1 calcined borax,) are mixed and fused.

Blue, Cobalt. See COBALT.

Blue, Damp. Is made by the same process as the cerulian, omitting the barytes and the drying.

Blue, Deep. For porcelain, 1 pure oxide of cobalt, 1 oxide of zinc, 1 lead glass (2 minium, 1 sand, 1 calcined borax), 4 glass of lead (2 minium, 1 white sand), are fused in a porcelain crucible at a red heat for three hours, washed, divided, and triturated on a glass slab.

Blue Colour of Flowers, or *Anthocyan.* Extracted by digesting blue flowers in alcohol of 85 per cent. away from light for some days; by water it is precipitated as an extract. The blue colour of the alcohol becomes red by acids, green by alkalies; hence some of the blue boragineæ by the secretion of an acid have frequently a red colour (echium, anchusa, pulmonaria).

Blue, Iron Earth. *Native Prussian Blue.* A fine blue powder occurring among the strata of bog iron ore, and in mosses in Europe and New Zealand. It is considered by many merely a form of vivianite. It contains PO_5 31·15, FeO 45·63, Al_2O_3 ·7, SiO_3 ·02, HO 22·5. B.B. melts into a black globule which is magnetic.

Blue, Indigo. For pottery, carbonate of cobalt 13, hydrous carbonate of zinc 26, gray flux 61. (See AZURE BLUE.) This is a fine colour, and may be modified into gray or black by proper mixtures.

Blue John. A miner's name for fluorspar.

Blue, King's. Carbonate of cobalt 29, sand 29, carbonate of potash 42. When fused it forms a beautiful blue, almost black. This glass is powdered and fused with half its weight of pebble flux (75 minium or litharge, 25 sand).

Blue, Leitner's, is formed by precipitating a solution of a salt of cobalt and alum, by an alkaline carbonate, and igniting. Pure clay when moistened with a salt of cobalt and ignited gives a fine blue.

Blue Litmus. See LITMUS.

Blue Malachite. Hydrous blue carbonate of copper.

Blue, Mineral. See PRUSSIAN BLUE.

Blue, Paris. See PRUSSIAN BLUE.

Blue, Prussian. *Berlin Blue. Sesquiferrocyanide of Iron.* $Fe_7Cy_9 = Cfy_3Fe_4$. (1704, by Diesbach and Dippel, by calcining blood and potash. Berlin Miscel. 1710, Stahl's Exper. p. 281.) A process was published for preparing it in 1720 by Dr. Woodward, obtained from Germany. (Phil. Trans. 1724.) *Preparation.*—1. When solutions of sulphate of iron and yellow prussiate of potash are mixed, a blue precipitate is produced which is Prussian blue. 2. Hochstätter recommends us to dissolve 6 parts of protosulphate of iron (green vitriol) and 6 parts of yellow prussiate separately in 15 parts of water, mixing the solutions; and adding 24 parts of concentrated hydrochloric acid, and 1 part of oil of vitriol, and agitating the mixture. After the expiry of some hours a clear solution of 1 part of bleaching powder dissolved in 80 parts of water is to be added gradually to the mixture, stopping always when chlorine is evolved. In a few hours the precipitate is to be washed and dried, either at the common temperature, or with a more elevated heat, or the blue precipitate may be treated with NO_5 until it becomes dark blue. 3. The blue which occurs in commerce under the title of *Paris blue* is obtained by precipitating a neutral salt of peroxide of iron, as of chloride of iron, by pure ferrocyanide of potassium. It contains a portion of the latter salt, which cannot be removed by washing. 4. Prussian blue is also formed by exposing green vitriol to the air, or oxidizing it by gently roasting it, dissolving it in water, and then precipitating by yellow prussiate; a portion only of the precipitate at first is blue. It is mixed with white cyanide of iron, but by exposure to the air it gradually becomes more blue, and finally assumes a fine deep tint. The process may be hastened by the addition of weak chlorine water or hydrochloric acid. 5. On the large scale, Prussian blue is often made by adding a solution of the impure cake of yellow prussiate to the dried protosulphate of iron; but as the cake contains carbonate of potash, it may be saturated by sulphuric acid, or, as is generally practised, by alum, in which case the alumina assists in giving more body to the Prussian blue. It is usual to add 2½ lbs. of alum for every pound of pearl-ash employed in the preparation of the yellow prussiate. On the continent, for the same purpose, moist carbonate of magnesia, carbonate of zinc, or sulphates of magnesia and zinc, have been substituted; the former are preferable to the latter, as in the latter case cyanide of zinc is formed, which renders the colour of the blue lighter; it is then termed *mineral blue*. Prussian blue is sometimes contaminated with starch, which injures the colour, but may be detected by forming an adhesive paste when mixed with water; and when boiled with water, the starch dissolves; on standing at rest the blue subsides and leaves a clear solution of starch which may be detected by iodine. Chalk is also sometimes added, which is easily detected by its effervescence with vinegar or weak acids. The lime may be separated by calcination and examination of the ash, or by boiling with dilute sulphuric acid, filtering and precipitating the lime with soda. Sulphate of lime is detected by calcination and digestion in water. *Characters.*—Prussian blue when dried spontaneously, possesses a fine soft dark blue colour, and when dried with a greater heat it assumes a coppery metallic lustre; when mixed with alumina it has a dull earthy fracture. It generally contains potassium, which has been considered by some essential to its composition; but it may be separated by careful washing. It is tasteless and odourless, and is not poisonous. It absorbs moisture from the air with great facility; it is converted by destructive distillation into hydrocyanic acid, carbonate of ammonia, and water, while carbide of iron remains in the retort. By the alkalies, Prussian blue is decomposed into peroxide of iron and alkaline ferrocyanides, a character which renders it less valuable than indigo as a dye, in consequence of its being affected by the alkali of soap. Sulphuric acid forms a paste with Prussian blue, from which the latter is again separated on the addition of water. Nitric acid dissolves the iron and destroys the blue. Cold hydrochloric acid does not change the colour. Sulphuretted hydrogen removes the colour, which it again acquires by exposure to the air. When heated to 400° in the open air, it catches fire and burns with difficulty like tinder, evolving a disagreeable odour and leaving a red ash. It is insoluble in water, alcohol, ether, oils, and dilute acids. Strong hydrocyanic acid takes up its iron and sets free hydroferrocyanic acid. Red oxide of mercury, when treated with Prussian blue and water, is converted into cyanide of mercury, which is soluble, and an insoluble mixture of oxide and cyanide of iron. Prussian blue decomposes by sunlight, and evolves cyanogen; but again recovers its colour in the dark (Chevreul), forming basic Prussian blue. The view generally taken of the formation of Prussian blue is, that 3 atoms of yellow prussiate and 4 of peroxide of iron, are converted into

Prussian blue, and 6 atoms of a potash salt. As Prussian blue, however, appears to contain the elements of water, it is probable that hydroferrocyanic acid exists in it—a view suggested by Dr. Thomas Thomson. Its formula will then be 4 FeCO + Cfy H_2. Prussian blue is the fine dark blue colour so frequent on muslin and woollen dresses, which is characterized by its being fugitive in presence of soap. It is commonly termed *blue*, and is used by washerwomen to conceal the yellow colour of linen, and hence it is a frequent application in the washing tub. It is also used for giving a blue tint to paper, and for communicating a green by means of chromate of lead to silk, termed *Raymond blue*. *Steam blue* is produced by mixing a solution of yellow prussiate and tartaric acid when cream of tartar precipitates. The liquid containing prussic acid, and the white Prussian blue, as it is termed, is then mixed with gum and printed on the cloth. It is now steamed, and the colour raised in a weak solution of bleaching powder, or of bichromate of potash. Oxalic acid is sometimes substituted for the tartaric acid. *Royal blue* is a steam colour, into which, in addition to the above ingredients, a small portion of cyanide of tin enters. *Mineral blue* is prepared with oxide of zinc or carbonate of magnesia, instead of alumina. It is usually sold in the form of cubic cakes. These are produced by placing the blue when it is moist on cloth filters, allowing it to dry, and then dividing it with a knife.

Blue, Saxon, consists of a solution of indigo in Nordhausen sulphuric acid, the process being discovered at Grossenhain, in Saxony, by Barth in 1810. The powdered indigo is dried at 135°; 1 part is then added by degrees to about 12 parts of acid, and the mixture heated up to 212°; and when the solution is complete, the vessel is covered up, and the acid poured into 20 times its bulk of water and filtered (T. Thomson). Another mode of preparing it is, to digest 1 ounce of indigo with 4 ounces sulphuric acid in a glass flask for an hour at 212°; then add 12 ounces of water, and when cold, filter (Ure). To dye with this colour, the cloth is steeped in alum, or in a hot solution of chloride of barium and bitartrate of potash mixed together, and then plunged into the blue liquor.

Blue, Old Sevres, for porcelain. Sand 39, calcined carbonate of soda 39, fused borax 3, calcined oxide of cobalt 19.

Blue Spar, or lazulite.

Blue Stone, or *Vitriol*. Sulphate of copper.

Blue, Thenard's. Is made by mixing 1 phosphate or arseniate of cobalt with 8 hydrous alumina and igniting, or by evaporating nitrate of cobalt and ammoniacal alum and igniting.

Blue Turquoise, for porcelain. 3 pure oxide of cobalt, 1 oxide of zinc, are dissolved in sulphuric acid, and mixed with a solution of 40 parts of ammoniacal alum; evaporated to dryness, the water expelled by heat, pulverized and ignited in a crucible for several hours at a red heat. It is composed of 4 atoms alumina, 3 oxide of cobalt, and 1 atom oxide of zinc. To give it a greenish shade, the alum solution is mixed with $\frac{1}{18}$ moist chromate of mercury; or as at Sevres, dissolve 2 carbonate of zinc, and 6 carbonate of cobalt in chlorohydric acid, and mix with 92 alum, dissolved in water; filter and add carbonate of soda till the solution is neutral. Wash, dry, pulverize and calcine at a cherry heat till the blue colour is developed; at too high a temperature it is destroyed and becomes violet. Mix it with $2\frac{1}{2}$ times its weight of flux (3 minium, 1 hydrous boracic acid, 1 sand), carefully triturated.

Blue Vitriol, or sulphate of copper.

Blue, Variegated, for porcelain. Fuse together 10 oxide of cobalt, 9 oxide of zinc, 25 glass of lead (2 minium, 1 sand), 5 lead glass (2 minium, 1 sand, 1 calcined borax). This is used to shade the other blues.

Blue, Azure. Mix on a glass slab 2 deep blue, 1 oxide of zinc, 4 lead glass (4 minium, 1 sand. Used for skies.

Boa Constrictor Fæces, or, more properly, urine, which are much used as a source of uric acid, consist essentially of urate of ammonia and bone earth.

Bodenite. Brown prisms, of 110 to 112°, accompanying felspar, near Marienberg (Kerndt). Sp. grav. 3·523. Silica 26·12, alumina 10·33, FeO 12·05; yttria, 17·43, CaO 6·32, MgO 2·33, MnO 1·61, KO 1·21, NaO ·84, protoxide of cerium 10·46, oxide of lanthanum, 7·56, HO 3·01.

Bog Iron Ore. A variety of brown hæmatite or sesquioxide of iron, containing phosphoric acid, so named from its habitat—apparently consisting of infusorial shields of *Gaillonella ferruginea*, invisible to the naked eye. See Animalcules.

Boheic Acid. $C_7H_5O_6$. Pale yellow powder in vacuo, fusing at 212° to a resinous body; deliquescent, soluble in water and alcohol; exists in tea leaves; obtained by precipitating a boiling decoction of tea leaves with acetate of lead and filtering; neutralizing the fluid with ammonia, decomposing the precipitate suspended in alcohol with sulphuretted hydrogen; when the filtered liquor is freed from sulphuretted hydrogen, and precipitated with an alcoholic solution of acetate of lead, boheate of lead results.

Bole. A term applied to a variety of coloured earths, used generally to colour other substances, by mixture. *Armenian bole* possesses a red colour, and is used principally for colouring tooth powders, but it has been employed for sophisticating the colour of anchovies. *Lemnos bole* has a pale red tinge, and consists of SiO_3 47·, Al_2O_3 19·, Fe_2O_3 5·4, $CaOCO_2$ 5·4, $MgOCO_2$ 6·2, HO 17·. *Blois bole* is a yellow calcareous earth. *Bohemian bole* is reddish-yellow. *French bole*, pale red. *Silesian bole*, pale yellow.

Boletic Acid. Colourless 4-sided prisms,

soluble in 180 water, 45 alcohol; sublimable; forms crystalline salts with bases; precipitates sesquioxide and not protoxide of iron; obtained from the *Boletus pseudo-igniarius*, by treating the evaporated expressed juice with alcohol; there remains a white residue, which is dissolved in water, precipitated with acetate of lead, and washed, the precipitate, suspended in water, and precipitated by sulphohydric acid; on filtering and evaporating, the acid crystallizes.

Bologna Stone, or *Phosphorus*. A peculiar radiated phosphorescent variety of sulphate of barytes, found in a clay slate at mount Paterno, near Bologna.

Bologna Vial is the name given to an unannealed tumbler, which breaks to pieces when an angular bit of stone is dropped into it.

Boloretine. C 75·5, H 11·7, O 12·8. Form. C_{40}, H_{32}, 3 HO. A fossil resin, found in the peat of Denmark on the remains of pines.

Boltonite. Spec. grav. 2·85, H 5·5. Yellowish or bluish-white grains, with traces of a doubly-oblique rhombic prism. Found at Bolton, Massachusets: consists of silica 56·64, Al_2O_3 6·07, MgO 36·52, FeO 2·46. Another analysis gave 46· silica, 5·66 Al_2O_3, MgO 38·15, FeO 8·63, CaO 1·52.

Bombic Acid. Obtained by distilling silk with sulphuric acid.

Bombite. Silica 50·, Al_2O_3 10·5, Fe_2O_3 25·, MgO 3·5, CaO 8·5, C 3·, S ·3.

Bonsdorfite. Greenish-brown, or dark olive-green 6-sided prisms; texture foliated, lustre like that of talc; translucent, in thin pieces, opaque in thick. H 3·5. Silica 45·05, Al_2O_3 30·05, MgO 9·, FeO 5·3, HO 10·6. Occurs at Biskopsokern, near Obo, in Finland.

Bone constitutes the skeleton, or framework of animals, to which the soft tissues are fixed. They are covered by a membrane, termed periosteum, which detaches itself when the bone is immersed for some time in water. Bones are divided into flat and cylindrical; the latter being distinguished by having a central perforation for the preservation of the marrow. When a bone is steeped in dilute muriatic acid for some days it becomes flexible, in consequence of the solution by the acid of the inorganic or earthy salts, the flexible cartilage alone remaining. The relative amount of organic and inorganic matter in bones varies, as indicated by the specific gravity. Human os humeri, 1·748. First phalanx of human great toe, ·9775. Os femoris of sheep, 2·0345. Tibia of sheep, 2·0329. Ileum of ox, 1·8353. Vertebræ of haddock, 1·635. The cartilage of a long bone consists of a congeries of long minute tubes. These tubes are formed of very fine circular plates, and the intervals between them are filled up by numerous circular plates, which encircle the tubes. The cartilages also contain oval-shaped particles, of minute size, the use of which is not well known. In infants, the bones are principally cartilage; but as age advances earthy salts are deposited, and ossification, as it is called, takes place. The mean amount of cartilage in the bones of a human fœtus is 42·8, and of the earthy matters 57·2; in the human adult, the mean organic matter is 41·, and the inorganic 59· (Rees). According to another determination, the organic matter in the bones of a fœtus is much less than this, or about 37 per cent. (Frerichs), and in the adult 34 per cent. The following table gives the composition of different bones (T. Thomson):—

	Human femur.	Sheep's Ileum.	Ox Ileum.	Haddock Vertebra	Saw-fish Snout
Cartilage,	39·12	43·30	48·50	39·49	46·31
Phosphate of lime, Fluoride of calcium,	43·67	50·58	45·20	56·08	42·55
Phosphate of magnesia	·49	·86	·24	·79	·10
Carbonate of lime,	14·	4·49	6·10	3·57	2·64
Alkaline salts,	2·06	·50	·31	1·58	·14
Water,					6·92

These analyses were made by digesting the bones dried at 212° in ether to take up fat, digesting in muriatic acid to remove earthy matters, and sometimes burning to extract completely. The effect of disease in bones is to remove inorganic matter, and to increase cartilage. In osteo malacia the constituents have been found in the dorsal vertebræ. Cartilage 79·75; phosphate of lime, 13·6; phosphate of magnesia, ·82; carbonate of lime, 1·13; alkaline and earthy soluble salts, 4·7 (Bostock). Compare with article Teeth.

Bone Ash is obtained by calcining bones in open vessels. It is white when properly burned, and is used for making cupels in the separation of silver from lead, the metallic lead when melted oxidizing, and partly sinking into the cupel, while the silver remains as a pure metal on the surface. It is also used for cleaning plate, under the name of powder of hartshorn, coloured with red vegetable matter, or Armenian bole, and also as a tooth powder. It consists essentially of phosphate of lime (3 CaO PO_5, or 8 CaO 3 PO_5). It is mixed with some phosphate of magnesia, carbonate of lime, and alkaline salts. It is used for the production of phosphorus.

Bone Black. Animal charcoal. See Charcoal.

Bone Earth. A synonyme of phosphate of lime, as found in bones.

Bone Gelatine, *Collin*, *Glutin*, *Gluc*. See Gelatin.

Boracic Acid. *Boronic Acid*, *Acid of Borax*, *Sedative Salt of Homberg*, *Narcotic Salt of Vitriol*, *Sassolin*. BO_3 4·375; 35·0; 4·35, 34·8 B = 31·1, O = 68·9. Specific gravity fused, 1·803 (Hassenfratz, Ann. Chim. 28, 11), 1·83 in vacuo at 39° (Royer and Dumas, Ann. Phil. 3, 392), discovered by Homberg in 1702 (Crell's Ann. 2·265). A colourless, translucent, hard, brittle glass, fusing at a red heat, and not volatile when free from foreign ingredients. When steam is passed over it, at a red heat, or when it is heated in contact with solutions of acids or alcohol, it volatizes. When a solution of it in water is boiled, it likewise is mechanically carried away. When fused in a crucible, it cracks on cooling, and evolves light along the course of the fissures, visible even in daylight (Dumas, Ann. Chim. 32, 335). Soluble in water and alcohol, it is destitute of smell, but has a slightly bitter though not sour taste. (E. Davy, Jameson's Journ. 6, 131.) It imparts to its solution in alcohol, and to its mixture with sulphur, a green flame; it is decomposed by sodium and potassium, and by the galvanic battery; not so by charcoal or phosphorus.

Source.—Boracic acid is brought from the interior of the earth by the *soffioni*, or jets of vapour which reach the surface through chalk and marl deposits, and are visible at a distance in the form of dense vapour in the Duchy of Tuscany, at Monte Rotondo, and other places. Boracic acid was first detected in the lagoon of this place by Peter Hoefer, apothecary to Leopold I. Grand Duke of Tuscany, who found 9 per cent. of acid in the water, a fact which was confirmed in 1779 by Mascagni (Journ. de Phys. 16, 364; 42, 37), who suggested the manufacture of boracic acid. Professor Gazzeri in 1807 found the boracic acid too small for this purpose (Dr. Bowring, Stat. of Tuscany, Phil. Mag. July, 1839, vol. 15, N. S. p. 21); but the first manufactory began in 1818, under the direction of M. Larderello (Count de Pomerance). According to this author, the boracic acid is not brought from the interior ready formed, as the vapours of the soffioni when condensed contain no trace of it; none is found in the fields surrounding the jets, nor in the earth at the mouths of the jets, nor in the earth at the bottoms of the lagoons which have been formed by the jets, and when examined after the jets cease to flow into them. Hence he supposes that at a great depth, there exists a deposit of boron and sulphur, which, when acted upon by water, evolves great heat. Vapour is thus formed, which forms a passage through the surrounding strata,—on one side sulphohydric acid, and on the other boron in the state of gas, for which the liquid becomes a creative agent of boracic acid; when quitting the interior of the earth, the gas passes into the water without any communication with the air. Payen supposes that in the interior, steam mixed with projected water, passing over boracic acid, would carry out the latter, which, by the reaction of its own organic matter on the sulphates, would produce sulphides, from which boracic acid would drive off sulphohydric acid. (Ann. Chim. 3, 1, 250.) Dumas conjectures that the water of the sea acting at a great depth on sulphide of boron would yield boracic acid, sulphohydric acid, and a high temperature which would carry off these products with the water; chlorohydric acid from the decomposition of earthy chlorides, and ammonia from organic matter. If limestone were near, the boracic acid would liberate carbonic acid; at a certain distance the sublimed boracic acid would form deposits; and according as the water of the lagoons descended to this point, or did not penetrate, the current would bring up fresh boracic acid or pass without volatilizing it. The air supplied by the sea to the fissure of the strata with sulphohydric acid would form sulphates of lime, alumina and iron from the rocks, and ammonia from the organic matter. (Ib. 250.) In order to collect the boracic acid, a wooden cylinder is placed over one of the soffioni so as to protect the workmen from the influence of the vapour, which has a temperature of 257° (Lardarello), 206° to 212° (Payen). A basin or lagoon of brick is then built round the jet, and lined with stones and lime. These lagoons A B C are of various sizes, from 100 to 500 and 1000 feet in circumference, and

from 7 to 20 feet deep. In the latter cases several jets terminate in one lagoon. The wooden chimney is then removed, and the basin filled with water from a neighbouring stream. This water never runs, according to Lardarello, into the soffioni in consequence of the tortuous nature of the passage and the pressure of the vapour. The water of the lagoon in immediate contact with the vapour jet gradually boils; but little, however, evaporates, the water in the lagoon generally remaining at the same level and exhaling a strong odour of sulphohydric acid. The water takes up boracic acid during twenty-four hours, and is then run off into another basin or lagoon of a similar kind. After passing through 6 or 8 of these basins, which are situated on the incline of a hill at different elevations, the water is charged to the extent of 1 to $1\frac{1}{2}$ per cent. It is then run into a depositing vessel, when a quantity of dark mud settles, containing little or no boracic acid. It is then placed in leaden pans 12 feet square and 14 inches deep, which are heated by the natural steam jets which are enclosed in pipes, and made to circulate under the pans. In seventy-two hours the liquors are transferred to crystallizing wooden tubs, and in threedays the acid is found attached to the sides A.

The mother liquors are run off into fresh concentrated fluid—the crystals placed in baskets to drain and spread out for a few hours on the floor of a drying chamber D, under which a jet of steam circulates. It is then packed in

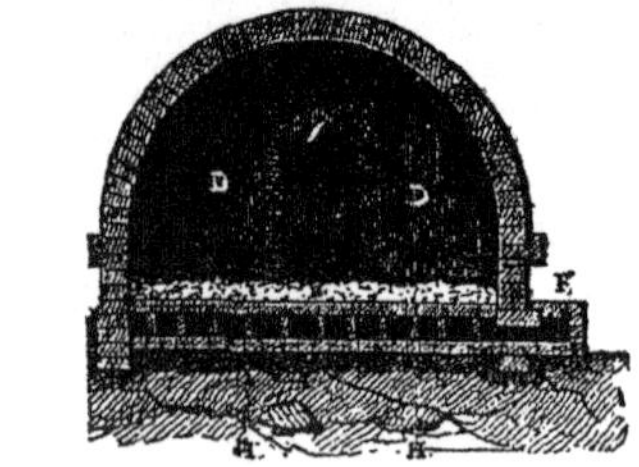

casks and sent to Leghorn. The product of the ten Tuscan works, in 1845, was one million kilogrammes, or $984\frac{1}{4}$ tons. The gases which issue from the soffioni are CO_2=57·3 N=34·81 O=6·57 SH=1·32. The condensed products are water, earthy matter, sulphates of lime, ammonia, alumina and iron, chlorohydric acid, organic matter with a sea smell, and little or no boracic acid. They deposit sulphur in all the narrow fissures and porous bodies which they traverse. (Payen. Ann. Chim. 3d, 1,247). Commercial boracic acid consists of boracic acid, 76·494; sulphuric acid united with boracic acid 1·322; the following sulphates:—of ammonia 8·508; of magnesia 2·632; of lime 1·018; of soda 0·917; of peroxide of iron ·365; of alumina ·320; of potash ·369; salammoniac ·298; silica 1·2; water 6·557. Organic matter containing nitrogen a trace.

Terhydrate of Boracic Acid. BO_3 3 HO 7·75, 62; 7,727,61·8; HO=43·69; spec. grav. 1·479 (Kirwan's Min. 2,4). Thin 6-sided scales of a silvery whiteness, having some resemblance to spermaceti, and the same greasy feel. It crystallizes in large crystals from a solution containing sulphuric acid. The crystals are then doubly-oblique prisms. P on M 80° 30′, P on T 84° 53′, M on T 118° 30′, P on K 75° 30′, M on K 120° 45′, T on K 120° 45′, (?). (W. H. Miller, Trans. Phil. Soc. Cambridge, 3, 365.)

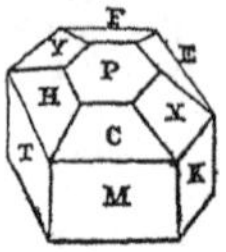

Taste bitterish, cooling, leaving an agreeable sweetness. When sulphuric acid is poured on it, a transient odour of musk is perceptible (Reuss de Sale. Sedat. 1778); reddens vegetable blues feebly, renders turmeric paper brown or reddens it (Faraday). Soluble in 25·66 parts water of 66°·2; in 14·88 at 77°; in 12·66 at 99°·0; in 10·16 at 122°; in 6·12 at 144°; in 4·73 at 167°; in 2·97 at 212°. A cold saturated solution has a density of 1·014. When distilled in a retort with water it crystallizes in the receiver. It is soluble in sulphuric acid, alcohol, and oils.

Preparation.—The terhydrate is usually prepared by dissolving 1 part of borax in 4 of boiling water, and 1 part of concentrated sulphuric acid is added. The acid deposits in crystalline scales as the liquid cools. To free it from adhering sulphuric acid it is ignited, redissolved in water, and crystallized. It may also be purified by solution in alcohol. It may also be prepared by subliming in a retort 16 borax, 2 water, and 5 sulphuric acid.

Borates.—Boracic acid is a feeble acid, and acts on very few of the metals; it however displaces carbonic acid; it combines with bases in a great variety of proportions, from $\frac{2}{3}$ to 6 atoms of acid to 1 of base, yet all of these combinations give a green colour to vegetable blues; they are soluble with difficulty in water except the alkaline salts; they fuse mostly into a colourless glass by heat, and dissolve metallic oxides before the blowpipe. Most acids separate boracic acid from the bases, and if alcohol be then added, the boracic acid dissolves and imparts to it a green flame when ignited. A borate also, when moistened with sulphuric acid, yields a green flame to the blowpipe flame.

Boracite. *Biborate of Magnesia.* 3 MgO 4, BO_3=MgO 33, BO_3 67. Small cubes, with their edges and alternate solid angles replaced by tangent planes, also in rhomboidal 12-hedrons; colour white, inclining to gray, yellow, or green; streak white, fracture conchoidal, uneven; lustre vitreous, with one axis of double refraction. H 4·25 to 7? spec. grav. 2·974. The solid angles become electric by heat. B.B. fuses and intumesces, and on cooling is covered with needles. With borax and salt of phosphorus fuses into a glass. Moistened with a drop of sulphuric acid, it gives a green flame. Found in gypsum near Luneburg, Brunswick; at Segenberg in Holstein; and at Taracapa, Peru.

Borax. *Biborate of Soda. Chrysocolla?* (Pliny). NaO $2BO_3+10HO$. 23·75. This salt is imported from Tibet, under the name of *tincal*, in the form of whitish-gray crystals. It was at an early period introduced by the Venetians from Asia, and purified from the soapy matter which it contains by mixing it with quicklime in cold water till the water became clear; it was then crystallized from hot water. Borax may also be obtained by dissolving 10 parts of boracic acid in hot water, and adding 12 parts of crystals of carbonate of soda and crystallizing. This is the method followed in France, but in this country much borax is still imported from India.

Characters.—Borax crystallizes in hexangular or octangular prisms, of which two sides are broader than the others. Spec. grav. 1·74. It converts vegetable blues to green; the taste is sweetish and alkaline. When heated it melts in its water of crystallization, and assumes a white opaque appearance, which is termed *calcined* or *burnt borax*. When strongly heated it fuses into a glass containing no water. The composition of borax is as follows:—

	Common Borax.	Octahedral Borax.
Soda,	17·8	69·36
Boracic acid,	35·6	
Water,	46·6	30·64
	$NaO2BO_3+10HO$	$NaO2BO_3+5HO$

Octahedral borax is obtained by dissolving so much of the salt in water at 212°, that the solution shall have a spec. grav. of 1·246. On cooling down to 174° the octahedrons begin to deposit, and this continues till the solution reaches the tempt. of 138°, when common borax is separated. (Buron and Payen.) The spec. grav. of the octahedrons is 1·815. Like common borax they become opaque in the air from loss of water. I have ascertained that native borax in Tibet is formed by boracic acid from hot springs saturating the natron of the soil.

Practical Application.—Borax is employed in medicine when mixed with honey, in ulcerations of the mucous membrane of the mouth, cancrum oris, &c. It is used also as a flux for metals, as in the assay of iron ores; and it enters as a constituent into the glazes employed for the finer kinds of porcelain.

Borech, *Trona*, or *Sesquicarbonate of Soda.*

Borneene, *Oil of Camphor, Fluid Borneo Camphor.* $C_{20}H_{16}$. Boiling point 320°. Lighter than water, without colour, smells of oil of turpentine, but the odour is more agreeable; obtained by fractional distillation from the native oil of Borneo camphor, which is procured by making incisions in the Dryobalanops camphora in Borneo and Sumatra, and receiving the oil which flows out; it is also formed artificially by fractional distillation from oil of valerian, and also by heating slightly Borneo camphor with dry phosphoric acid. It is used in perfumery.

Borneole. *Borneo Camphor.* $C_{20}H_{16}2HO$. White rhombic 6-sided prisms, heavier than water; fusing point 388°; sublimes at 414°; insoluble in water, soluble in alcohol; its odour differs from common camphor. Found in crevices of the wood of the D. camphora in Borneo, and is formed artificially by acting on borneene with caustic potash. It is much used in Japan, and seldom comes into the western market.

Bornine, *Bornite, Telluride of Bismuth.*

Borocalcite, *Hayesine.* CaO $2BO_3$ 6HO. Probably the same as *Boronatrocalcite.*

Borodiglione, *Cupferschaum, Subarseniate of Copper.*

Boron. Symbol, B. 1; 8 (Thomson), 1·35, 10·8 (Gmelin). The term borax first occurs in the works of Geber, written in the 10th century. It has long been used in Europe as a flux, being obtained from lakes in Tibet, and the borders of China. In 1702, Homberg, by distilling borax and green vitriol, procured boracic acid in small plates, which he termed *sedative salt*. In 1807, Davy succeeded when he exposed boracic acid to the action of the galvanic battery in obtaining the deposition of a black matter on the negative wire. In 1808, Gay Lussac and Thenard decomposed the acid by means of potassium into boron and oxygen; and Sir H. Davy, in 1809, obtained similar results. *Preparation.*—1. A small quantity of boron may easily be obtained by fusing some boracic acid in a platinum crucible to remove water, pulverizing it coarsely, and mixing it with twice its bulk of potassium, cut into small pieces, and free from potash, in a glass tube. The heat of a spirit or gas lamp is then to be applied for a few minutes; the mixture being then allowed to cool, is boiled with HCl to remove the potash; the boron being thrown on a filter, is well washed with water. The mixture is apt to detonate violently, when heated without the preceding precaution, from water being retained by the boracic acid; perhaps this could be removed by the addition of a drop of sulphuric acid and ignition (R. D. T.) 2. It may be prepared by passing fluoboric acid gas through a tube filled with crystallized boracic acid, and then through a tube filled with peroxide of lead, to free it from fluoride of silicon and sulphurous acid; and, lastly, over potassium heated. The mixture of boron and fluoride of potassium is then washed with water, solution of salammoniac and alcohol. 3. Fluo-borate of potash dried just under a red heat is heated with an equal quantity of potassium in an iron or glass tube. When the potassium melts, the mixture is stirred with a steel wire and ignited; boron is set free, and fluoride of potassium formed. The latter is to be washed out with a solution of salammoniac, often repeated; the boron is then heated with hydrogen gas to separate water and fluoboric acid; washed with water, and dried in a vacuum.

Characters.—Boron is a deep brown powder with a shade of green, destitute of taste and smell. When first prepared its spec. grav. is 1·183, but when strongly heated it sinks in sulphuric acid of spec. grav. 1·844. It is insoluble in water after strong heating (although somewhat soluble when freshly prepared, giving a greenish-yellow solution), also in alcohol, ether, and oils; it is a non-conductor of electricity. When heated to 600° it burns with great splendour, and becomes partially boracic acid. When heated with potash, borate of potash is formed.

Boronatrocalcite, *Tiza, Hayesine? Borate of Lime and Soda.* NaO $2BO_3$, 2CaO $3BO_3$ 10 HO. Spec. grav. 1·8, BO_3 49·5, CaO 15·7, NaO 8·8, HO 26·. Nodules of the size of a hazel nut, consisting of fine silky fibres, scarcely soluble in cold water; difficultly soluble in hot water; soluble in acids. B.B. fuses into a pearly bead; found at Iquique in Peru.

Borosilicate of Lime, or *Botryolite.*

Botany Bay Resin. $C_{40}H_{20}O_{12}$. A yellow resin, two-thirds of which are soluble in alcohol, obtained from the *Acarois resinifera*, a New Holland tree. When treated with nitric acid it yields picric acid.

Botryogene. *Native Red Iron Vitriol. Misy? Magnetic Oxide Sulphate.* SO_3 32·55, Fe_2O_3 23·86, FeO 10·71, HO 32·85=3FeO$2SO_3$ $3Fe_2O_3$ $2SO_3$ 36HO. Spec. grav. 2·039, H 2·25 to 2·5. Hyacinth-coloured oblique rhombic prisms of 119° 66′ and 60° 4′, collected in botryoidal masses (Βοτρυς, a bunch of grapes), found in the copper mine at Fahlun, in Sweden; it is only partially soluble in water.

Botryolite, *Biboro Silicate of Lime.* See Datholite.

Bottle Glass. See Glass.

Boulangerite. *Sulphide of Lead and Antimony.* 3PbS, SbS_3 Pb 53·9, Sb 25·5, S 18·5, Fe 1·2, Cu ·9. Spec. grav. 5·97. Colour bluish-gray; in black lead plumose masses. B.B. fuses, yielding fumes of sulphurous acid and oxide of antimony; on charcoal lead is reduced; soluble in chlorohydric acid; found at Molierés, France, and Nertschinsk.

Boulder. A mass of stone detached from its native rock and carried to a distance.

Bouquet of Wine. That peculiar flavour due to the presence probably of oenanthic ether.

Bourdaine. A name given by the French to commercial charcoal. When procured by distillation it contains 60 per cent. of carbonaceous matter, and 39·3 volatile matter; by combustion in the usual way, 74 per cent. carbonaceous matter.

Bournonite. *Disulphide of Copper, Sulphide of Lead,* and *Tersulphide of Antimony.* $3Cu_2S$, 6PbS, $3SbS_3$. Spec. grav. 5·79. Steel-gray right rectangular prisms; lustre metallic, opaque, brittle; fracture conchoidal. B.B. emits sulphurous fumes, leaving a crust of sulphide of lead; soluble in nitric acid. It is found at Neudorf and Clausthal in the Hartz, at Kapnik, in Transylvania, Servos, in Piedmont, in Cornwall, in Mexico, &c. and consists of sulphur 18·62, lead 41·37, antimony 24·71, copper 12·74.

Bouse. A technical term for lead ore in its mechanical preparation. See Lead.

Bovey Coal. *Wood Coal* or *Lignite*, as illustrated at Bovey, in Devonshire. Analogous to it is the Boghead coal at Bathgate, but in the coal series employed by Mr. Young for the distillation of his paraffine oil used in machinery. I find the latter under the microscope to consist of a continuous mass of organized matter.

Boyle's Fuming Liquor. Sulphohydride of ammonia, or hydrosulphuret of ammonia.

Brain. The blood with which the chyle is mixed in the right auricle and ventricle is the venous or dark blood which has been brought back to the heart from every part of the body. From the right ventricle it is sent by the pulmonary artery to the lungs, where it is exposed to the action of the air, assumes a red colour, giving out CO_2, and returns to the left side of the heart, thus completing the lesser circulation. It is now arterial blood, and is qualified to be sent through the arteries to the various parts of the body to supply them with the nourishment with which it is so amply endowed. The changes which the blood undergoes in the lungs are of vast importance in the human system. The blood proceeds from the left heart by the aorta, in the first instance, to the head and inferior parts of the body. The arteries are tubes of elastic membranes, which are readily dilated and contracted according to the amount of force applied to them from within. The blood sent to the head traverses the brain, and supplies it with nourishment. The *brain* is the organ of sensation and volition. The spinal marrow, which is a process of the brain, is the organ of the movements of ingestion and egestion while the ganglionic nerves constitute the nervous system of nutrition and secretion. With all of these the brain therefore is connected. One marked characteristic of the brain, as the seat of volition, is, that the brain sleeps; the spinal nerves and ganglionic nerves never sleep. If you watch a person who has fallen asleep in his chair it is impossible to fail being struck with this fact. The processes of secretion and nutrition are as active as ever—the sleeper nods and nods again—not because his brain is awake, and his will dictates the action, but because his brain is asleep, and his spinal nerves are brisk and active. It would be interesting to determine the chemical composition of the various systems of nerves. The brain, however, has alone been examined with any degree of care. The process of analysis is as follows:—The brain is cut into small fragments. It is several times heated with boiling alcohol. It is then allowed to remain in contact with this fluid for some days, in order to remove as much water from it as possible. If the alcohol has been properly applied, the albuminous part of the brain should be coagulated, have lost its elasticity, and be easily compressed. It is then to be submitted to pressure, rapidly triturated in a mortar, and treated with ether. The ether should be employed first cold and then hot. The ether on being distilled leaves a viscid matter, which may be termed the *ethereal product.* The alcoholic solution deposits a white substance on cooling, which contains phosphate, and was termed *cerebrote*, or cerebric acid, by Fremy and Vauquelin. *Cerebric acid* is extracted best from the ethereal product. This is treated with a large quantity of ether—a substance is left, which consists of cerebric acid and phosphates, &c. To purify it, the precipitate is to be heated with boiling absolute alcohol, slightly acidulated with sulphuric acid. Cerebric and oleophosphoric acids are dissolved, while sulphates of lime, and soda, and albumen, are suspended. Filter. On cooling the two acids separate. Cold ether takes up the oleophosphoric acid, and leaves *cerebric acid.* Cerebric acid is in white, small crystalline grains, swells like starch in boiling water; burns, giving out a characteristic odour. Cerebric acid consists of per cent.—

66C | 10·6H | 2·3N | 0·9P | 19·5O.

It combines with bases; with barytes it forms cerebrate of barytes, containing 7·8 per cent. of barytes. *Oleophosphoric acid* is separated from cerebric acid by its solubility in ether. It is in this state in union with soda. Decompose the salt with an acid, and dissolve the oleophosphoric acid in alcohol; evaporate; the acid remains; insoluble in water. It combines with potash, ammonia, and soda, forming soaps. When oleophosphoric acid is long boiled in alcohol or water, it is converted into oleine and phosphoric acid, so that it seems to be a compound of these bodies. It appears to be in consequence of the separation of these bodies that the softening of the brain is produced, which takes place soon after death, and also in diseases. *Oleine* has been found separately in the brain. *Cholesterine.* Gmelin first pointed out its presence in the brain. The ethereal product is boiled with alcohol and a little potash. A precipitate falls on cooling, from which the cholesterine may be taken up by cold ether. *Oleic* and *margaric* acids also exist in the brain. Besides, it contains water and albumen.

Constituents of Human Brain.

Water,	78·00
Albumen,	7·00
Cerebric acid, } Oleophosphoric acid, Cholesterine, Oleine, Margarine, Oleic acid, Margaric acid,	9·85
Soda, } Phosphate of lime, Phosphate of potash, Phosphate of magnesia? Common salt, Sulphate of potash, Free phosphoric acid,	5·15
	100·

I have found the different parts of the brain and nerves to present the following composition:—

	Entire Brain.	White.	Gray.	Spinal Marrow.
C.	56·48	55·47	56·31	55·42
H.		8·06	8·95	8·95
N. } O. } Salts. }		36·47	34·86	35·63

Ash of Human Brain.

Fresh Brain—Water,	78·48
Dry residue,	21·52
	100·

Of the dry residue ·027 was ash. It consists of—

Pyrophosphate of potash,	55·24
" soda,	22·93
" iron,	1·23
" lime,	1·62
" magnesia,	3·40
Chloride of sodium,	4·74
Sulphate of potash,	1·64
Free phosphoric acid,	9·15
Silica,	0·42
	100·

(Breed. Liebig's Ann. 80, 124). The free phosphoric acid enables us to detect traces of brain in cases of murder.

Brachytype. A variety of calcareous spar.

Brandisite, *Disterrite.* Leek-green, micaceous leaves. Spec. grav. 3·015. Constitution—SiO_3 20, Al_2O_3 43·22, Fe_2O_3 3·6, CaO 4·, MgO 25·01, KO ·57, HO 3·6 = 4, (MgO, Al_2O_3) (3RO) $2SiO_3$ 2HO.

Brachytypous Manganese, or Braunite, or Sesquioxide.

Bran. *Kleie* (German), *Son* (French). The exterior husk of grain is too often neglected as an element in bread, and given to the inferior animals. In 1843 I showed, in an analysis of Prussian brown bread, that when the bran was retained a larger quantity of nitrogen was present than when it was rejected, as in white bread. The amount of albuminous matter in bran I found to amount to from 12·01 to 16·031 per cent. At my recommendation bread made with the entire wheat was introduced into the prisons and charitable institutions of Glasgow. This view has been lately confirmed by Millon. Bran is used in calico printing for clearing goods after the application of the mordant. The mode of action is not well understood. Runge attributes its influence to the presence of an acid.

Brandy. *Branntwein (Burntwine)), Eau de Vie.* In this country this term is restricted to a reddish-coloured fluid obtained by the fermentation of grapes and distillation of the resulting fluid, to which a colouring and flavouring substance is added. On the continent it is the generic expression for various alcoholic fluids. *British brandy* is believed to be whisky flavoured with corn oil and one or more of its salts (acetate of amyle) and a vegetable colouring matter.

Braordite. Red silver ore.

Brass. *Messing* (Gr.), *Laiton* (Fr.) A familiar yellow alloy of copper and zinc; of spec. grav. 8·2 to 8·9, fusing point 1869°; was formerly made by igniting in a blast furnace or close crucible equal parts of calcined calamine or oxide of zinc, and half the weight of each of charcoal, which produced 7 or 8 parts of *arcot*. Another proportion is 30 copper, 20 calamine, 10 kiess, 16 charcoal. This was fused again with calamine and copper, and brass produced. It is in France now usually made by melting the two metals together in fire-clay pots or crucibles set in a circular domed furnace. On the sole of the furnace is placed a plate of cast iron pierced with eight apertures, which are fitted with as many cast iron tubes rising above the level of the plate. The plate is covered with a layer of fire clay. The eight crucibles being filled with the proper proportions of zinc and copper, the furnace is filled up with coals; in six or seven hours the temperature is at a white heat, and in ten hours the brass is formed. It is then poured for sheet brass of the proper thickness into granite moulds edged round with iron. The clay crucibles are made of good refractory fire clay about 16 inches deep, 9½ inches broad at the mouth, and 6 at the bottom, and from 1 to 1½ inches thick. The proportions used in Prussia are 41 old brass, 55 copper, and 24 zinc, a loss of 2 to 3 per cent. being sustained. Brass usually contains some lead and tin, probably derived from the solder of old brass. The usual composition of brass seems to be Cu_2Zn, or 8 copper, and 4·125 zinc, a dizincide of copper, but it varies considerably. The various alloys of copper and zinc are as follows:—

	Brass.	Pinchbeck or Tomba.	Princes' Metal.	Bath Metal.	Red Brass.	Dutch Gold.	Malleable Brass.
Copper,	66	85	72	55	2¾	11	60
Zinc,	34	15	28	45	1	2	40

Braunite. Sesquioxide of manganese.

Brazil-Wood. *Fernambuc* (Fr.) A red dyewood from Pernambuco, in Brazil, and the West Indies. That from the former place is said to be derived from Caesalpinia echinata. The other plants which supply it are C. *Crista* and C. *vesica*. It is much used in calico printing and for making red ink, and owes its value to the presence of Braziline.

Braziline. *Brazileine.* $C_{18}H_7O_7$? Orange-coloured needles; soluble in water, alcohol, and ether; its solution becomes yellow by acids, violet by alkalies; yields pink or purple precipitates with salts of tin and alum; is obtained by exhausting rasped Brazil-wood with water, evaporating to dryness; dissolving in water, agitating with litharge to remove some acid; evaporating to dryness, digesting in alcohol, filtering, and evaporating; diluting the residue with water, throwing down tannin by glue, filtering, evaporating, and taking up and crystallizing the Braziline by alcohol.

Brazing. *Brass Soldering. Hard Soldering. Soudure fort, Brazure* (Fr.), *Messing-löthung* (Gr.) The operation by which brass and other metals are united by means of a solder difficultly fusible, so that the blowpipe is required to effect the fusion and union of the two surfaces. The brass or hard solder consists of 18 brass, 3 zinc, 2 tin; or 12 brass, 4 zinc, 3 tin; or 16 zinc, 16 copper, 1 tin. The hardness or difficulty of fusion may be increased by diminishing the quantity of zinc. The surfaces to be united are filed quite bright, and strewed with dry borax powder, and in cases of great ignition the borax is mixed with powder of glass; the parts covered with charcoal, and the blowpipe applied or what is usual, the gas flame directed by the blowpipe upon the solder held over the surfaces.

Bread. *Pain* (Fr.), *Brot.* (Gr.) This term seems to have been originally applied to food in general, but is now restricted in a great measure to fermented loaves of wheat flour, although in common language we talk of unfermented or unleavened, and fermented or leavened, bread. Among primitive nations fermented bread is scarcely known. In the east, as in the Punjaub and Affghaunistaun, the flour is simply kneaded with water and made into thin cakes, similar to what are termed *scones* in Scotland. *Oat cakes* are made in a similar manner, and in the latter country, barleymeal and peasemeal *bannocks*, varying from half-an-inch to 1½ in thickness, are also composed of these meals kneaded with water. In Australia *dampers* are unfermented wheat flour bannocks. The *passover cakes* of the Israelites are thin unfermented cakes of wheat flour. The *corn bread* of America is made by kneading the Indian corn meal or maize with water. *Biscuits*, (*bis*, twice, *cuit*, baked,) (Fr.) *Zweibach*, (Gr.) are similarly prepared, but hardened to a greater extent by firing.

Bread formed by Leaven.—The earliest kind of fermented bread seems to have been produced by means of *leaven*, that is, a portion of sponge or previously fermented dough. On the continent, at the present day, it is usual to set a new sponge, that is, mix a new portion of dough made of flour and water with a portion of fermented dough called *leaven*, kept from the last baking. This is a necessary mode of baking when breweries from which yeast can be obtained are at a distance; but in this country the fermentation of bread is conducted entirely by means of yeast or barm. To prepare a sponge by means of leaven, the leaven, or portion of fermented dough from a previous baking, is placed in an oaken vessel. A sufficient amount of water is then added, and the leaven divided with the hands. The proper quantity of flour is next gradually mixed with it to form a dough or paste, the materials being carefully incorporated with the hands. If the mixing is not scrupulously attended to, knots of dry flour will be formed in the loaves. The sponge is now divided into a number of separate portions or loaves, and left for some time. They are then turned, and after being rolled in flour each is placed in a separate vessel of tin, where it is allowed to swell up. If the flour is of a good quality the sponge rises uniformly, but when it contains only a small portion of gluten, as when it is mixed with potato flour, the dough rises imperfectly and irregularly. The moist fermented loaves are now introduced into the oven upon a piece of wood, or wooden shovel, which is covered with a little bran or flour to prevent the adhesion of the loaves to the wood. They are exposed to the action of the oven, the heat being graduated in different parts of the oven according to the experience of the baker. The temperature of the baker's oven varies from 448° (Thomson), 482° to 572° (Claubry). The bread never, however, reaches but an insignificant part of this temp.; a thermometer when inserted in the interior of a loaf in the oven never rising so high as 212° (about 206°). When the bread has been exposed to the atmosphere for some time it loses its elasticity, and assumes the aspect of staleness, the cause of which is not evident, as it does not depend on its drying from loss of water. The objection to the use of leaven in bread making is the tendency of this ferment to become sour, which is frequently the case, especially with rye bread, as it is met with on the continent. The explanation of the various operations, however, in baking is precisely the same whether leaven or yeast be employed. The flour requires to be properly mixed with water in the preparation of the sponge in order to unite the water to the starch and to the gluten, and to dissolve the sugar and albumen. It is probable, too, that the water acting on the starch converts it into sugar, since rye meal, if mixed with water over night, yields to the latter a distinctly sweet taste. It has been frequently asserted that the amount of sugar decomposed by fermentation in baking does not exceed 1 per cent. This may or may not be correct; but at the same time it is an undoubted fact, as ascertained by the long experience of Mr. Dobson of the Borough, as communicated to me, that by the chemical process, as compared with the fermenting process, there is a difference of profit in favour of the bread, by the first proceeding, of 6½ per cent. That this was not due alone to the one species of bread containing an excess of water is easily proved by experiment. Attempts to collect the alcohol generated in the process of baking in a profitable manner, have proved fruitless, partly from the small amount of product, and partly, as in London, from a prejudice excited against bread from which alcohol has been obtained.

Baking with Yeast.—A sack of flour weighing 280 lbs. and measuring 5 bushels, is capable of forming 80 to 86 quartern loaves. The baker introduces the flour into the kneading trough, and then sifts it through a fine sieve; 4½ lbs. of salt are dissolved in a pailful of hot water in the *seasoning tub*, and sometimes a little is added, although this is not done by the best bakers. When this solution has fallen to 84°, three English pints of yeast are added, the whole well mixed, and strained through the seasoning sieve into a depression of the flour. The flour is then mixed up into a paste, and some dry flour being sprinkled over the surface, it is covered over with cloths, the operation being termed *setting quarter sponge*. After three hours an additional pailful of water is added and mixed, &c. This is called *setting half sponge;* it is left for five hours. Three pailfuls of warm water are now added; the mass is now kneaded for upwards of an hour. The dough, being then cut in pieces and sprinkled with flour, is left for four hours; it is then kneaded again for half-an-hour. The various kneadings are, however, modified by different bakers. It is then weighed out at the rate of 4 lbs. 15 ounces for every quartern loaf, and introduced into the oven. The *oven* is a square apartment from three to four feet high, with an arched circular roof and a brick or stone floor, and furnished with a door which may be shut close. It is often the custom to heat the oven with wood either of faggots or brushwood (Dr. T. Thomson). Loaves, according to act of parliament, must have a specific weight under a penalty. On this account it is necessary for the baker to calculate the amount of loss which the bread will undergo in firing. The following table gives the weights before and after firing:—

	Before Firing.		After Firing.			Avoirdupois. Loss.		
	lbs.	oz.	lbs.	oz.	drs.	lbs.	oz.	drs.
Peck loaf,	19	12	17	6		2	6	
Half-peck,	9	14	8	11		1	3	
Quartern,	4	15	4	5	8		9	8

A quartern loaf at the present time weighs about 4 lbs. Abroad or in our colonies it may be found impracticable to obtain yeast. In this case leaven may be formed by making a quantity of flour into a dough with water, and setting it in a warm situation for thirty-six hours, when it undergoes fermentation. A portion of this may be employed by mixture with flour to produce fermented bread. In India the juice of the toddy tree when allowed to stand ferments, and is used as a bread ferment. In the juice of a species of Eucalyptus, in Van Diemen's Land, I have detected alcohol, so that this or analogous fluids might be employed instead of yeast. I give these hints for the benefit of emigrants and others who may be thrown on their own resources.

Baking with Carbonate of Soda and Muriatic

Acid.—When none of the preceding ferments can be obtained, or even when they are within reach, by a little practice, an excellent bread may be made by mixing intimately with the flour, bicarbonate of soda, and adding the water with a quantity of muratic or chlorohydric acid. The effect of this operation is to disengage carbonic acid through every part of the sponge, and to form common salt to flavour the bread, without any interference with the constituents of the flour. The proportions for bread of this description are, 4 lbs. of flour, through which are mixed 320 grs. of bicarbonate of soda; dissolve then 300 grs. of common salt in 35 ounces of water, and add to it 6½ fluid ounces of muratic acid; mix the fluid with the flour rapidly, and heat in the oven. *Soda scones* are formed by mixing bicarbonate of soda with the flour, and then baking the mass up with butter milk; the lactic acid of the milk displacing the carbonic acid of the soda salt. *Gingerbread and light cakes* are often raised by carbonate of ammonia, which, being volatile, is dissipated like the carbonic acid in the common modes of baking. *Shortbread*, a sweet cake, is rendered brittle by means of butter.

Adulteration of Bread.—Bread in Belgium is adulterated with sulphate of copper. This substance is used for the purpose of improving the colour of bad flour. When present the dough requires less manual labour. The bread rises quicker, and the crust is finer. Two hundred loaves of 2¼ lbs. each contain about 1 ounce of sulphate. To detect it, we may take 3000 grs. of bread, and burn it in a platinum capsule; the residue is heated with half an ounce of nitric acid. After the dissipation of the acid, the residue is treated with water; the liquid filtered; an excess of ammonia is added, and some drops of a solution of carbonate of ammonia. It is allowed to cool, and again filtered, boiled down to one-fourth, acidulated with nitric acid, and being divided into two portions, one of them is tested with the yellow prussiate, and the other with sulphuretted hydrogen. *Alum* is mixed with flour to destroy the colour of pease and beans, which are sometimes mixed with it. It may be detected by burning the bread, and treating it as already described—testing for copper and precipitating the alumina by boiling with salammoniac—filtering and weighing it. The presence of potato starch may be detected under the microscope by the relative size of the starch granules. See STARCH.

Theory of Bread Making or Panification.—The first step of bread making consists in its fermentation, that is, the conversion of the sugar contained in the flour into carbonic acid and alcohol; the gaseous carbonic acid, as it is uniformly evolved throughout the mass of dough, forms vesicles or little bubbles of gas which expand by the heat of the oven, and along with the alcohol is dissipated, leaving a dry spongy mass constituting *a loaf.* The *raising* of the loaf depends upon the resistance to the escape of the gas offered by the gluten of the flour, which increases with the amount of that substance which is present.

Breane. $C_{30} H_{46}$ 3HO. A crystalline resin, extracted by alcohol from icica resin, a product of the Styrax officinale; it consists of C 84·06, H 11·73, O 4·21, exactly the composition of cholesterine.

Breccia. A conglomerate rock, composed of fragments of other rocks connected together.

Breidine. Rhombic prisms of 102° and 78°. Soluble in 260 water, easily in alcohol, less in ether; from arbol a brea.

Breine. Transparent rhombic prisms of 110° and 70°, obtained by crystallization from alcohol, from the resin of arbol a brea, a tree, *Canarium album*, growing in the Philippines. Insoluble in water; soluble in 70 parts of 85 per cent. alcohol; fusing point 369°.

Breislakite. Reddish-brown or chestnut-brown capillary crystals, resembling the form of augite. B.B. fuses into a slightly magnetic bead of a gold colour; from Capo di Bove.

Bremen Green. See VERDITER.

Breunerite. A synonyme of carbonate of magnesia.

Brevicite. White foliated or radiated mass in amygdaloid, from Brevig, in Norway; traversed by dark red streaks.

Brewing. This process of preparing fermented liquors does not seem to have been known at a very early period, so far as history informs us; but there is no doubt that it was discovered even by savage nations. The discovery of beer is usually attributed to the Egyptians, who made it from barley. It appears to have been long known to the ancient Germans, who believed that if they obtained the favour of their divinity Woden by their valour, they should be admitted after their death into his hall, and, reposing on couches, should satiate themselves with ale from the sculls of their enemies whom they had slain in battle. Dioscorides describes two kinds of beer—ζυθον and κουρμι, both made from barley. Pliny calls it *cerevisia* or *cervisia*. The process of brewing is usually conducted by converting barley into *malt*. See MALT. The malt is then infused in hot water; the infusion, when it is completed, being termed *wort*. This wort is fermented (see FERMENTATION), and is then called wash in distilling, or *beer*, *ale*, *porter*, when the process is carried no farther. The various stages of brewing are *mashing*, *boiling*, *cooling*, *fermenting*, *cleansing*, which are described under the various articles which are brewed.

Brewsterite. 6-sided prisms, terminated by two oblique and very low faces of 172°, 90°, and 93° 30′; the primary form being a right oblique prism, with an angle of 93° 40′. Sp. gr. 2·432, H 4·25. *Composition.* **Silica, 53·045; alumina, 16·540; barytes, 6·05; strontian, 9·; lime, ·8; water, 14·735. It consists of 1 atom silicate of**

barytes and strontian, 1 atom tersilicate of alumina, and 15 atoms water. B.B. loses water, becomes opaque, froths and swells up, and does

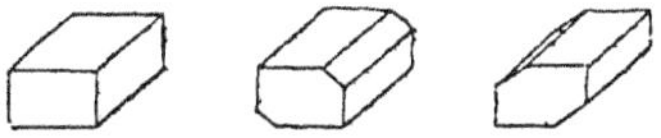

not easily fuse; leaves a silica skeleton when fused with microcosmic salt; it differs from heulandite in having lime replaced by other two earths. It occurs at Strontian, in Argyll, associated with galena, in grayish or yellowish-white crystals; near Freiburg; in Isere, and in the Pyrenees.

Brick. *Ziegelstein* (Gr.), *Brique* (Fr.) A well-known quadrangular form of clay used for building. The most important bricks are the *common* and *fire* bricks. The first are made of any adhesive clay, and are burned at a lower temperature than the second description. The clay employed in making fire bricks must contain but a small amount of oxide of iron and lime, otherwise, at the high temperatures to which it is exposed, it will be in danger of fusing, and forming a silicate of iron, with the silica which it contains. In selecting a clay for the manufacture of fire bricks, the less iron that is present the better. See CLAY. *Sun-dried* bricks are used in Germany, China, and in countries where the heat of the sun is sufficient to give the clay consistence. *Scouring* bricks are formed of washed clay, or of a mixture of fine sand and clay. *Floating bricks* were known to the ancients, and appear to be formed of the remarkable deposit now known as *mountain meal* and *kieselguhr*, consisting of the silicified remains of fossil infusoria, which are found in Italy, Germany, and even in the Hebrides.

Brightening. An operation in calico printing, performed by boiling printed goods in a solution of soda, &c. by which the colour is rendered more brilliant.

Brilliant. A term applied to the diamond when set.

Brimstone. The name of sulphur.

Brithyne Salt. Calcareo-sulphate of soda, or glauberite.

British Gum. A form of starch, soluble in cold water, used extensively in calico printing. It was formerly prepared by adding a small portion of nitric acid to common wheat or potato starch, but is now manufactured by passing pounded sago, as imported, through cylinders of iron, heated up to the proper temperature (600°?). The sago is introduced at one end of an open revolving cylinder, and by constant addition of the new sago it gradually traverses the length of the cylinder, and falls out prepared at the other extremity. It is the same as dextrin.

Britannia Metal. An alloy of tin, containing different quantities of other metals. One specimen consisted of tin 85·72; antimony 10·39; zinc 2·91; copper ·98. It is much used for making teapots, &c.

Brittle Silver Ore. See SILVER.

Brocatello. A variegated marble, composed of fragments of other marbles.

Brochantite. *Trisulphate of Copper. Konigine.* 3 CuO, SO_3 3 HO. Spec. grav. 3·78, H 3·5 to 4. Thin rectangular tables, bevelled on the edges, belonging to a right rhombic prism; colour emerald-green, transparent, lustre glossy, sometimes pearly; insoluble in water; soluble in acids. Found in England, Hungary, Siberia. *Composition.*—CuO 66·94, SO_3 17·43, HO 11·91, ZnO 3·14, PbO 1·. There are several minerals allied to this with the formula 4 CuO SO_3 3 HO, and 4 CuO SO_3 4 HO.

Bromachlonaphtone. $C_{40}H_9Cl_6Br$.

Bromachlonaphtose. $C_{40}H_8Cl_6Br_2$.

Bromachlonaphtune. $C_{40}H_7Cl_7Br_2$.

Bromal. C_4 Br_3 OHO. Spec. grav. 3·34. Boiling point above 212°; colourless oily fluid with a strong odour; affects the eyes; with caustic potash forms bromide of potassium, perbromide of formyle and formate of potash; formed by acting on alcohol with bromine as with chlorine in the preparation of chloral.

Bromaloine. C_{34} H_{15} Br_3 O_{14}, needles by acting on aloine with bromine.

Bromaniline. C_{12} H_6 Br N. 8-hedrons obtained by distilling bromisatine with potash. *Dibromaniline* C_{12} H_5 Br_2 N, 4-sided rhombic prisms; soluble in alcohol; formed by distilling bibromisatine with potash. The direct action of bromine on aniline yields *tribromaniline* C_{12} H_4 Br_3 N.

Bromanisal. C_{20} H_9 Br_3 O_2. Colourless, shining crystals, without smell, insoluble in water; little soluble in alcohol; very soluble in ether; formed by acting on the concrete essence of anise with bromine, and crystallizing from ether.

Bromanisic Acid. $C_{16}H_7BrO_3$. White brilliant needles, fusing at 399°. Scarcely soluble in water; soluble in alcohol and ether; forms crystalline salts with the alkalies; sublimes; formed by acting on anisic acid with bromine, and crystallizing out of alcohol.

Bromates. The salts of bromic acid. They deflagrate on charcoal by ignition, losing six atoms oxygen, and leaving oxides. When in solution muriatic acid renders the liquid red from the evolution of free bromine. Sulphurous, sulphohydric, and nitric acids have the same action.

Bromechlonaphtose. $C_{40}H_8Cl_4Br_4$.

Bromechlonaphtuse. $C_{40}H_6Cl_6Br_4$.

Bromeine. $C_{20}H_6Br_4$.

Bromenchlonaphtose. $C_{40}H_8Cl_5Br_3$.

Brome-thion-essile. $C_{26}H_7Br_2S$

Bromic Acid. BrO_5 118·4. Bromine 65, oxygen 35. Syrupy, colourless liquid, odourless, acid taste, reddening litmus and decolourizing it; volatile by heat in vacuo; precipitates nitrate of silver white, the precipitate being insoluble in dilute nitric acid, but soluble in ammonia; formed by acting on bromine with caustic potash; after boiling, crystals of bromate of potash fall, which is decomposed in its hot solution by fluohydric acid or bromate of barytes; may be decomposed by sulphuric acid.

Bromine. Br 10, 80, 78·4. *Syn. Muride. History.*—This liquid, termed bromine (βρωμος, fœtid,) from its disagreeable odour, was discovered by M. Balard in 1826, in the mother liquor of the salt marshes of Montpellier (Ann. Chim. 32, 337), and it is now known to exist in all sea waters and springs which have been connected with the ocean, in the form of bromide of magnesium, as in the salt springs of Germany (Liebig, Ann. de Chim. 33, 331), the water of the Atlantic (Graham), the Baltic (Wöhler), the Dead Sea (Gmelin), and marine plants. Spec. grav. of fluid 2·96 (Balard), 2·98 to 2·99 (Löwig), of vapour 5·54 (Mitscherlich), 5·3933 (Pierre). *Character.*—Bromine is a fluid of a brownish-red colour, a non-conductor of electricity. Its odour is very noxious, somewhat resembling euchlorine; but is stronger and more suffocating; taste sharp and strong; acts as a poison when taken internally; a single drop put into the bill of a bird causes instant death. It acts energetically on organic bodies, as cork, wood, &c. and on the skin, to which it communicates a yellow tinge. The colour disappears after some hours, unless the bromine remains long in contact with the skin, when it disorganizes it. It is fluid at 0°, but when cooled down to -4° it becomes solid and brittle (Serullas). It is very volatile: a drop put into a flask soon fills it with vapour, resembling fuming nitric acid. Boiling point 116°·6 (Balard,) 145°·4 (Pierre). The vapour is not decomposed by a red hot tube; when the vapour is breathed in mixture with much air, it appears to be harmless, but when inspired in a concentrated state, it excites oppressive cough, headache, &c. which are relieved by ammonia and alcohol. A taper plunged into the vapour is soon extinguished, but it burns for some moments with a flame green at the base, and red at the upper part, as with chlorine. When metals come in contact with the vapour, combustion results, as with iron, arsenic, antimony (Serullas, An. Chim. 38, 318). Phosphorus detonates with it. Bromine bleaches almost as well as chlorine. It dissolves in 33·3 water at 60°; the solution has a yellow colour, smells of bromine, and has a rough but hot acid taste. It loses bromine in the air. Bromine is more soluble in alcohol, and much more so in ether; olive oil acts on it slowly; it is scarcely soluble in sulphuric acid. It gives a fine orange colour with starch.

Hydrate of Bromine.—Br 10 HO 21·25, 170, 168·4, Br 46·56 HO 53·44. Hyacinth-coloured 8-hedrons, decomposing at 60° into bromine, and an aqueous solution of bromine (Löwig), obtained by cooling down a small portion of bromine with a large quantity of water to 32°, or by passing bromine vapour through a glass tube moistened with water at a temperature of 39°.

Process.—Bromine having a less affinity for bases than chlorine, the latter is used to displace bromine from its union with magnesium in sea water. 1. Balard's method of preparing bromine is to pass a current of chlorine through the fluid containing it, until it acquires a brown colour, without adding an excess, otherwise the quantity of bromine will be diminished. The chlorine unites with the magnesium, and sets the bromine at liberty. Ether is now added to the solution in a stoppered bottle, and the mixture agitated; the ether rises to the surface, and has acquired a hyacinth colour from the presence of bromine in solution. The ether is now poured off and agitated with caustic potash; the bromine unites with the base of the potash, and forms bromide of potassium ($K\,Br\,Mn\,O_2\,2SO_3 = KO\,SO_3\,MnO\,SO_3$ and Br), a salt crystallizing in cubes. This salt is pounded, mixed with binoxide of manganese, and introduced into a retort, the beak of which passes into a receiver filled with water. Oil of vitriol diluted with its own weight of water is poured on the mixture and heat applied. The bromine passes over in the form of red vapours, and condenses in the receiver. A portion of the bromine dissolves in the water, but the water is removed from it by distilling over chloride of calcium (An. Chim. 32, 342). To remove bromate of potash it is recommended to calcine the salt before mixing it with the black oxide of manganese (Löwig). 2. The plan of Desfosses (Jour. Pharm. 13, 252) is to boil the solution containing bromide of magnesium with milk of lime (Mg Br and CaO become Ca Br and MgO), in order to prevent the magnesian salt from being converted by the action of water into bromohydric acid, and into magnesia, which would precipitate. The bromide of calcium is then treated with binoxide of manganese and sulphuric acid. Protochloride of manganese (Mn Cl), water, and chlorine, result. The chlorine decomposes the bromide of calcium, forming chloride of calcium, and setting the bromine free. 3. The mother liquor may be at once heated with the binoxide of manganese and sulphuric acid, as in the case of iodine. This process may be applied to the residue of kelp salts, after the distillation of iodine, from which bromine may be procured (Barruel, Jour. Pharm. 23, 19).

Bromides.—These salts resemble the chlorides. They are easily distinguished in solution by the addition of chlorine water, which renders the solution yellow by the bromine being set free and dissolving in the water. When heated in a tube with bisulphate of potash or with binoxide of manganese and sulphuric acid, bromine vapour is evolved. If to a solution of a bromide, oil of vitriol and sulphate of indigo be added, the colour is destroyed, which does not occur with a chloride. The yellowish-white precipitate, with a bromide and nitrate of silver (AgBr), is insoluble in nitric acid but soluble in ammonia, which distinguishes it from iodide of silver. Bromide of barium differs from the chloride of barium by being soluble in alcohol, and hence may thus be separated.

Application.—When the daguerreotype was first invented, the plate of silvered copper was rendered sensible to the rays of light by the vapour of iodine alone; but as the action in this case was very slow, often half-an-hour, a combination of iodine and bromine was tried, and found to increase greatly the sensibility of the metallic plate (Foucaut), and to perfect the image instantaneously. In order to effect this object, the metallic plate previously acted on by iodine is exposed to the influence of the vapour of a very weak solution of bromine in water. Compound solutions of these two substances are also used, the preparation of which is generally kept secret. Bromine is used in medicine in the form of bromide of potassium, but as it has been given to the inferior animals in ounce doses without any effect, the therapeutic influence of this salt must be considered problematical. Bromine has been recommended as a disinfecting agent instead of chlorine, by allowing a few drops of it to evaporate in the air of the infected place (Löwig).

Estimation of Bromine in Solution.—Bromine is obtained from solutions along with the alkaline iodides by the process described under iodine. After the iodide of palladium has been collected on the filter and washed, the liquid which passes through the filter is to be precipitated by sulphuretted hydrogen to separate the excess of palladium. The filtered liquor is to be evaporated, the residue dissolved, and mixed with an ammoniacal solution of chloride of silver, prepared by adding 1 part of a saturated solution of chloride of silver in ammonia mixed with 1 part of ammonia and 1 part of water. Bromide of silver falls, which is collected on a weighed filter, dried and weighed. Every 23·5 grains contain 10 of bromine (Lassaigne).

Bromisatine. $C_{16}H_4BrNO_4$. Crystals by acting on isatine with bromine. By longer action bibromisatine is formed, and *bromisatic* acid by these with bases.

Bromisatinese. $C_{16}H_3NBr_2O_4$.

Bromitonic Acid. $C_6H_4Br_2O_4$. Long needles by the action of bromine on a solution of citraconate of potash with excess of alkali.

Bromlite. The baryto calcite of Bromley Hill. Crystallizes in doubly-oblique rhombic prisms. See BARYTO CALCITE.

Bromobenzine. $C_{12}H_6Br_6$.

Bromobenzinese. $C_{12}H_3Br_3$.

Bromobenzoic Acid. $C_{14}H_8BrO_4$. Colourless crystals, fusing at 212°, subliming at 482°; little soluble in water; very soluble in alcohol and ether; by acting on benzoate of silver with bromine and crystallizing from ether.

Bromobenzoyle. BzBr. Colourless plates by acting on oil of bitter almonds with bromine.

Bromobrucine. $C_{46}H_{25}N_2O_8$. Small slightly-coloured crystals, by acting with bromine on sulphate of brucine.

Bromocinchonine. $C_{38}H_{21}BrN_2O_2$. A base by the action of bromine on muriate or cinchonine.

Bromocinnamene. $C_{16}H_8Br_2$.

Bromocodeine. $C_{36}H_{20}BrNO_6$. White crystals, a base, by bromine on codeine.

Bromocomenic Acid. $C_{12}HBrO_8 2HO$. A bibasic acid, by acting on comenic acid with bromine.

Bromodinitronaphthaline. $C_{20}H_5Br(NO_4)2$. Crystals by nitric acid on dibromonaphthaline.

Bromodraconesic Acid. $C_{16}H_7BrO_6$.

Bromoform, or *Terbromide of Formyle.* C_2HBr_3. Clear fluid. Spec. grav. 2·9, of vapour 8·63, formed by a similar process to chloroform.

Bromohelicine. C_{26}, $H_{15}Cl$, O_{14}. A base formed by bromine on helicine.

Bromohydric, or *Hydrobromic Acid.* HBr. Colourless gas; taste acid; smell like muriatic acid. Spec. grav. 2·75; smokes in the air from its affinity for water; formed by making bromide of phosphorus by the union of these two substances, and adding water and gentle heat; or by the following process:—Put pieces of phos-

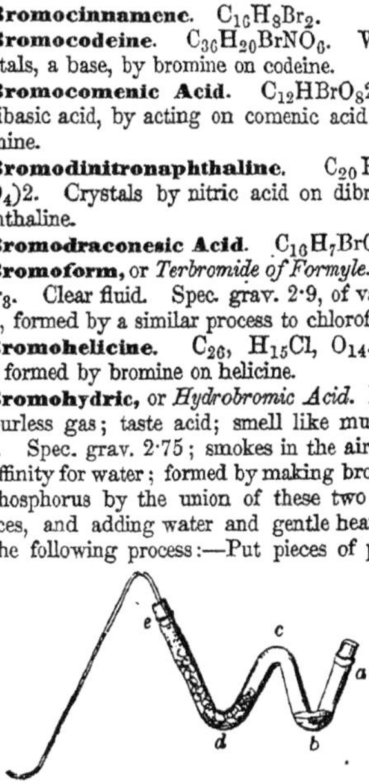

phorus in *d*, bromine into *b*, closing the tube with the cork *a*, and moistened glass into *d e;* heat *b* gently with a lamp, the bromine in vapour unites with the phosphorus at *d;* the bromide of phosphorus and water form phosphorus acid and bromohydric acid, which is collected over mercury.

Bromonaphthalase, or *Bronaphthase.* $C_{20}H_7Br$. Colourless oil by bromine, not in excess, on naphthaline.

Bromonaphthalase. $C_{40}H_{13}Br_3$.

Bromophenasic Acid. $C_{24}H_{10}Br_2O_2$. An oil by distilling bromosalicylic acid over sand and barytes.

Bromophenisic Acid. $C_{24}H_6Br_6O$. Colourless needles by bromine on hydrate of phenyle.

Bromopurreic Acid. $C_{40}H_{14}Br_2O_{21}$. Gold needles by bromine and purreic acid.

Bromorcine. $C_{14}H_5Br_3O_4$. Silky needles by bromine on solution of orcine.

Bromosalicylic Acid. $C_{14}H_5BrO_4$. Brilliant prisms by bromine on salicylic acid; by further action 2 and 3 atoms may replace the same number of atoms of hydrogen, and form two acids, the *bi* and *ter* bromosalicylic acids.

Bromosamide. $C_{14}H_5O_2N\frac{3}{2}Br$. Yellow crystalline plates by ammonia on bromide of salicyle.

Bromostrychnine. $C_{44}H_{23}BrN_2O_4$. A base by bromine on muriate of strychnine.

Bromoterebene. $C_{20}H_{12}Br_4$. Red viscid fluid. Spec. grav. 1·978 by bromine on terebene.

Bromotriconic Acid. $C_8H_5Br_2O_3$. Yellowish-white greasy scales, by the action of bromine on a neutral solution of citraconate of potash.

Bromous, *Hypobromous Acid.* BrO? When bromine is brought in contact with alkaline bases, bleaching compounds are formed analogous to the hypochlorites with the same odour.

Bromoxaform. $C_6\ H\ Br_5\ O_4$. Silky needles fusing at 167°; formed by bromine on citrate of potash.

Bronaphthene. $C_{40}H_{11}Br_5$.

Bronaphthese. $C_{20}\ H_6\ Br_2$. Crystalline by bromine in excess, on naphthaline; a bromide is formed by heating bromine with it.

Bronaphthise. $C_{20}\ H_5\ Br_3$. Crystalline by heat to the bromide of bronaphthise, a product of the action of bromine on naphthaline or on bronaphthase.

Bronaphthose. $C_{20}H_4Br_4$. Crystalline by distilling bromide of bronaphthese.

Brongniardite. Spec. grav. 5·95. PbS, AgS, Sb_2S_3. A black metallic mineral, with uneven fracture, from Mexico.

Brongniartine. $NaOSO_3+CaOSO_3$, or Glauberite.

Bronze. (E., Gr., Fr.) An alloy of copper, tin, zinc, and lead, used for statuary, casts, &c. The following is the composition of different kinds of bronze:—

	Antique Bronze.	Bronze Ware.	Bell Metal.	Gun Metal.	Gong.	Mirror Metal.
Copper,	85 to 97	82· 64·4	78 80	90· 91	80	66 66
Tin,....	15 to 3	3 ·25	22 10·1	10· 9	20	33·34
Zinc,....		18 32·4	5·6			
Lead,...		1·5 2·86	4·3			

The spec. grav. of bronze is 8·78 in gun metal, and 8·482 in statue of Louis XIV. Its colour s reddish-yellow, fine-grained in the fracture. I have found the strength of gun metal (which I found by analysis to consist of 91 copper 9 tin) to be to malleable iron as 1 to 2.

Bronzing consists in communicating the colour more rapidly, which bronze acquires in the course of time by the oxidation of the copper. This may be accomplished by placing the articles to be bronzed in a boiling very dilute solution of 2 acetate of copper and 1 salammoniac. A fine colour is also produced by the action of sulphuretted hydrogen which may be allowed to escape spontaneously in a room from a flat vessel containing an alkaline sulphuret in solution.

Bronzite, *Diallage.* Spec. grav. 3·154 to 3·3, H 4·5. Brown, dark green, or ash-gray, frequently resembling bronze plates, belonging to an oblique rhombic prism of 93° 30′. *Composition.*—Silica 56·81, magnesia 29·67, lime 2·19, FeO 8·46, MnO ·61, HO ·22. It seems a bisilicate of magnesia; found in Styria, Piedmont; Lizard, Cornwall; Girvan, Ayrshire. (?)

Brookite. *Jurinite, Prismatic titanium ore, Eumanite.* Spec. grav. 3.81, H 5·75. Primary form a right rhombic prism of 100° and 80°, the height being to the breadth as 11 to 30; the crystals are often 6-sided prisms with pyramidal summits; B.B. infusible on charcoal; fusible into a brownish-yellow glass with salt of phosphorus. It contains titanic acid 94·09, Fe_2O_3 4·5, Al_2O_3 trace. Loss by ignition 1·4. It is therefore slightly impure titanic acid; found in Dauphine, Savoy, Rutherford county, U.S.; Chesterfield, Massachusets; Snowdon, Wales.

Brown Coal. See COAL.

Brown Haematite. See HAEMATITE.

Brown Spar, a crystallized form of carbonate of iron, but perhaps more properly applicable to an isomorphous compound of carbonate of lime, iron, and magnesia.

Browns for Porcelain. *Light Brown.* 6 anhydrous copperas, 4 anhydrous sulphate of zinc, 13 nitre, are mixed and heated in a clay crucible till the nitre is decomposed. It is removed from the crucible by fracture and digested in water, which takes up the soluble parts. A yellowish-brown powder remains, composed of oxides of zinc and iron. This is made into a vitrifiable colour by triturating 2 parts of the brown powder, and 5 lead glass (12 minium, 3 sand, 1 calcined borax).

Another Light Brown.—2 anhydrous copperas, 2 anhydrous sulphate of zinc, are substituted for the above; or 1 copperas, 2 sulphate of zinc, yield another shade.

Brown Bistre.—1 calcined sulphate of manganese, 8 anhydrous sulphate of zinc, 12 anhydrous copperas, 26 nitre, are treated as the first light brown. Another shade is procured by substituting 4 copperas, and 4 sulphate of zinc; 12 nitre.

Brown Sepia.—1 calcined copperas, 1 calcined sulphate of manganese, 2 calcined sulphate of zinc, 5 nitre, are treated as for light brown, and the gray-brown residue mixed with $2\frac{1}{2}$ times its weight of the glass. Another shade is made by 1 copperas, 2 sulphate of manganese, 6 sulphate of zinc, 10 nitre.

Deep Brown.—1 dry sulphate of cobalt, 4 dry sulphate of zinc, 4 calcined copperas, 10 nitre, are mixed as for light brown, and afterwards with $2\frac{1}{2}$ times its weight of the lead glass.

Chrome Brown.—1 hydrous oxide of iron, 2 chromate of mercury are mixed on a glass slab, then ignited in a capsule placed in an open muffle to drive off the mercury; the residue fused with lead glass (5 minium, 2 sand, 1 borax), and triturated on glass.

Chesnut Brown is prepared by igniting red oxide of iron intensely till it becomes chesnut brown, and mixing 2 parts of this with 5 lead glass (12 minium, 3 sand, 1 calcined borax), and carefully triturating on glass.

Ochre Brown.—80 gray flux (pebble flux 89), fused borax 11, (for pebble flux see BLACK) 13 hydrous carbonate of zinc, 7 yellow oxide of iron; to be mixed without fusion. *Another.*—77 gray flux, hydrous carbonate of zinc 15, oxide of iron 8.

Deep Ochre Brown.—Gray flux 75, hydrous carbonate of zinc 12½, yellow oxide of iron 12½. Mix without fusing.

Brucine. $C_{46}H_{26}N_2O_8$. Colourless, transparent, oblique 4-sided prisms, becoming opaque by keeping, containing 16 per cent. of water, which they part with at 212°; soluble in 850 cold, 500 hot water; soluble in alcohol; not soluble in ether; its solutions are bitter and poisonous; tincture of iodine throws down an orange-brown precipitate; nitric acid turns it red, becoming yellow by heat, and violet by adding a tin salt; this test distinguishes it from strychnine and morphine. It is procured from the bark of the *Brucea antidysenterica*, by forming a decoction with boiling water, adding oxalic acid to the solutions, evaporating to an extract, digesting the extract in alcohol at 32°; the oxalate of brucine remaining is decomposed by boiling with magnesia, the brucine taken up and crystallized by alcohol. It is also obtained from the mother liquor of the strychnine solutions in the preparation of that alkaloid from nux vomica, by evaporating it to a syrup, adding dilute sulphuric acid slightly beyond saturation; crystals appear in a few days, which are pressed in a cloth, decolourized by animal charcoal, and the brucine precipitated by ammonia. It unites with acids and forms monobasic crystalline salts.

Brucite. A name given by American mineralogists to hydrate of magnesia and to chondrodite.

Brunolic Acid. An asphalt-looking matter obtained from coal oils in the preparation of carbolic acid. 12 coal oil being mixed with 2 lime and 50 water and allowed to stand; the liquor is saturated with muriatic acid, when carbolic, brunolic, and rosolic acids separate as an oil. When this oil is distilled with water, the first third which passes over is carbolic acid; the remainder is brunolic and rosolic acids. When saturated with lime and boiled with water, rosolate of lime dissolves, and brunolate of lime falls; this is repeated until all colour disappears; the brunolic acid is separated by muriatic acid, and is lastly dissolved in caustic soda, precipitated with muriatic acid, and dissolved in alcohol.

Brunone. A synonyme of sphene.

Brunswick Green. A name given to different compounds. The usual pigment consists of 3CuO, CuCl 4HO, obtained by exposing metallic copper to the air, and sprinkling it with a paste of sulphate of copper, common salt, and water; or with common salt and dilute sulphuric acid, or with a solution of salammoniac. A dicarbonate of copper ($2CuO\ CO_2\ HO$), also goes under this name, or *mountain green*, formed by decomposing sulphate or chloride of copper with an alkaline carbonate or carbonate of lime, and washing the precipitate with hot water. It is made to acquire various shades by mixing with sulphate of barytes, gypsum, &c.

Bryoidine. White silky fibres, fusing at 135°, from the watery solution of arbol a brea or resin of *Canarium album*.

Bryonine. Yellow-brown bitter extract, obtained by the process for *bitter extracts*, q.v. from the *Bryonia alba* and *dioica*, soluble in water, alcohol; insoluble in ether; sulphuric acid renders it blue and deep green.

Bucholzite. *Anhydrous Silicate of Alumina.* Spec. grav. 3·193, H 6·, silica 46·4, alumina 52·92. Colour grayish-white, with a slight tinge of yellow; structure fibrous in imperfect prisms; lustre silky, allied to *fibrolite* and *sillimanite*.

Bucholzite Hydrous. Spec. grav. 2·855. H 3. Silica 41·35, alumina 49·55, SO_3 2·01, CaO 1·105, HO 4·85. Colour light bluish-green, powder white, streak white, structure granular, formed of small scales; brittle. B.B. becomes snow-white and falls to powder; with carbonate of soda becomes an enamel; with borax fuses into a transparent bead; found in Sardinia.

Buchu. The leaves of several species of *Barosma* or *Diosma* from the Cape of Good Hope, much used in the form of tea in retention of urine, &c.

Bucklandite. *Scotine.* Spec. grav. 3·945 to 3·51. Soluble in nitric acid; 6-sided prisms, terminated by bihedral summits, derived from an oblique rhombic prism, with angles of 70° 40′, and 103° 56′; opaque, lustre vitreous. The crystal is the same as that of epidote. *Composition.*—Silica 36·97, alumina 21·84, Fe_2O_3 10·19, FeO 9·19, CaO 21·14, HO ·68, CO_2 .32. In some specimens have been found yttria, and ceria, and lantana, and this is explained by the fact, that orthite, allanite, and cerine are heteromeric with epidote.

Buckwheat. The seed of the *Polygonum fagopyrum*, contains resin ·36, gluten 10·47, albumen ·22, apotheme 2·53, saccharine 3·08, gum and mucilage 2·80, starch 52·29, fibre 26·93.

Buratite. *Cupro Carbonate of Zinc*, occurring at Campiglia in the Maremma Pisena. It consists of ZnO 32·02, CuO 29·46, CO_2 21·45, HO 8·45, CaO 8·62.

Bustamite. Spec. grav. 3·12 to 3·25 H 6·5. Pale gray, with a tint of green or red, radiated spherical masses from Puebla in Mexico. *Composition.*—Silica 48·9, protoxide of manganese 36·, lime 14·57, FeO ·81. *Formula*, 3 CaO 2 SiO_3, 2 (3 MnO $2SiO_3$).

Butter. E. and Gr. *Beurre*, Fr. The solid oil contained in the milk of the cow; but all solid oils are occasionally termed butters, whether of animal or vegetable origin. The butter of milk is diffused through the milk in the form of emulsion globules. These, by standing, from their less density, rise to the surface, mixed with milk and curd, in the form of cream. To separate the butter from the cream, see CREAM, the latter is agitated in a churn, when the butter separates from the globules, and swims in buttermilk. The temperature rises 4°, but no oxygen is necessary, as none is absorbed; the chemical change is confined to the separation into two parts of the lactin, or sugar of milk, in the form of lactic acid. The temperature for churning most advantageously has been found to be 55°. As soon as the churn works stiff the butter is known to be formed. The buttermilk is poured off from it, and the butter washed free from buttermilk by water and pressure. The best butter I have found composed of 12·79 water, 86·27 butter oil, and ·94 caseine or curd. It is the presence of the caseine which sets up a fermentation, and causes butter to become rancid. To *analyze* butter, it is heated at 212° till it ceases to lose weight; the butter is taken up by ether, and the caseine remains. To *preserve* butter, it may be heated in a pan to remove water, when it becomes a fine fluid oil; the oil, when strained through a muslin cloth, leaves caseine on the filter. If the oil thus deprived of water and curd be poured into close bottles, it solidifies, and may be kept for any length of time. In this way it is preserved in India, and called *ghee*, and likewise on the continent. To preserve it in its natural state, salt, nitre, and sugar are usually employed. Nitre is likewise used to deprive it of the peculiar taste which it possesses when cows are fed on turnips. The pure oil separated by ether or fusion consists of 68 margarine (margarate of oxide of glyceryle) 30 butyroleine (butyroleate of oxide of glyceryle), and 2 of caprine, caprylline, caproine, vaccine. The hardness of butter is produced by the predominance of margarine, which appears to be greatest in winter. Pure butter solidifies about 80°. Boiling alcohol of ·822 dissolves 3·46 per cent. of butter. The quantity of butter given by different kinds of food is various. The largest quantity given by a brown cow, on a grass diet in a fortnight, was 11 lbs. 4 oz.; by a white cow, on the same diet and the same time, 8 lbs. 2 oz. Five days on bean meal and hay, the brown cow produced 3 lbs. 11½ oz. butter; the white cow 3 lbs. 12 oz. butter. The following table gives the amount of butter in five days by different kinds of food, the mean of the two cows by my experiments:—

	lbs.	Nitrogen in food.
Grass,......,....................	3·5	2·32
Barley and hay,...............	3·43	3·89
Malt and hay,.................	3·20	3·34
Barley, molasses, and hay,..	3·44	3·82
Barley, linseed, and hay,....	3·48	4·14
Beans and hay,...............	3·72	5.27

This table shows that there is a connection between the nitrogenous matter in the food and the butter of the milk.

Butter is also the old name of certain chemical compounds, of the consistence of butter, as butters of antimony, tin, bismuth, zinc, &c.

Butter Bog. C 75·05, H12·56, O12·84, from the bogs in Ireland.

Butter Mountain. A name sometimes given to native ferrosulphate of alumina.

Butters. A name applied to solid oils, as *shea, palms, cocoa, nutmegs.*

Butylamine. *Petinine.* $NH_2C_8H_9$. An oil occurring in Dippel's animal oil along with ethylamine, methylamine, and propylamine.

Butyle. *Valyle.* C_8H_9. Colourless fluid with a pleasant odour, taste somewhat burning; boils at 226°, spec. grav. ·694, of vapour 4·053; obtained by decomposing valerianic acid by electricity; when oxidized it yields butyric acid; it is homologous with ethyle.

Butylene. *Carburet of Hydrogen, Tetarto-Carbohydrogen.* C_8H_8. An oil, spec. grav. ·627, of vapour 1·944, boils under 32°, occurs in the oil of oil gas, and in the decomposition of valerate of potash by galvanism.

Butyral. *Butyraldehyde, Hydrous oxide of Butyryle.* C_8H_7OHO. A fluid, boiling at 158° (Guckelberger), 203° (Chancel), spec. grav. ·8; it is homologous with aldehyde, and has similar properties; another compound isomeric with it is obtained by distilling butyrate of lime.

Butyramide. $C_8H_7O_2NH_2$. By acting on butyric ether with ammonia, in closed tubes.

Butyranilide. $C_{20}H_{13}NO_2$. Pearly plates, fusing at 194°, distilling without change, insoluble in water, soluble in alcohol and ether; by the action of anhydrous butyric acid on aniline.

Butyric Acid, *Anhydrous.* $C_8H_7O_3$, spec. grav. ·978, of vapour 5·38, B.P. 374°, strong odour of butyric ether, colourless liquid, very refractive, mobile, lighter than water; formed by allowing oxychloride of phosphorus to drop on dried butyrate of soda, distilling the product, or by distilling 5 chloride of benzoyle and 8 dry butyrate of soda.

Butyric Acid. $C_8H_7O_3HO$. Clear oily fluid, soluble in water, alcohol, ether, and oil of vitriol, possessing an acid, rancid odour; butyrate of lime distilled with sulphuric acid yields hydrous butyric acid; polymeric with aldehyde and isomeric with acetic ether; spec. grav. ·9739, of vapour 3·3, boils at 314°; still fluid at −4°; evaporates in air without residue; crystallizes in broad plates when exposed to the action of solid carbonic acid and ether; it unites with bases, and forms crystalline tasteless salts. *From butter* it is prepared by saponifying the butter with caustic potash; supersaturating with sulphuric acid; distilling half of the solution, pouring water on the remainder, and again distilling till the fluid passing is no longer acid. The distilled fluid is saturated immediately with barytes water, and kept in close vessels till the whole is collected at the end of the operation. The whole distilled fluid saturated with barytes is boiled down in the still to one-twentieth, and then distilled to dryness in a retort. This residue consists of 95 parts of butyrate and caproate of barytes, more soluble in water, and 5 parts of caprylate and caprate of barytes, with difficulty soluble in water. The whole residue is boiled with water, filtered, and crystallized. Caproate of barytes crystallizes out in silky needles like benzoate of lime, while butyrate of barytes remains in the mother liquor. Vaccate of barytes has in one instance been found alone instead of butyrate and caproate. The butyrate of barytes is decomposed by dilute sulphuric acid, and the butyric acid distilled over.

From Sugar.—It may also be prepared from sugar. Mix a solution of sugar with a small quantity of curd and a sufficient quantity of chalk to saturate all the acid which may be formed (100 sugar, 8 to 10 fresh curd, 50 chalk). This mixture, exposed to a temperature of 77° to 86°, soon undergoes great alterations. The fermentation, at first viscid, becomes next lactic, and then buttery. Much gas is evolved, consisting of hydrogen and carbonic acid; when the hydrogen is a third of the carbonic acid, the buttery fermentation is at its height. In the course of some weeks the action has ceased, and the liquid contains only butyrate of lime. The action is as follows, produced by the fermenting action of the caseine:—

$C_{12}H_{14}O_{14}$ = Sugar = $C_8H_8O_4$ Butyric acid.
C_4O_8 = 4 Carbonic acid H_4 = 4 hydrogen.
H_2O_2 = 2 water.

The lime may be separated from the butyric acid by adding weak muriatic acid or sulphuric acid in a retort and distilling. This production of butter from sugar must necessarily be an important point in the consideration of the question of the formation of fat in animals, as the transformation is effected without the agency of heat.

Butyric Ether. $C_4H_5O,C_8H_7O_3$ A fluid with a fine pine apple odour, used for flavouring rum and confections; by alcohol, sulphuric acid, and butyric acid.

Butyroacetic Acid. *Metacetonic acid, propionic acid.* $C_6H_5O_3HO$.

Butyroleic Acid. $C_{40}H_{30}O_4HO$. Colourless oil, specific grav. ·904; the oleic acid of butter; extracted by saponification from the butyroleine, which again is taken up by pressure from butter.

Butyrone. C_7H_7O or $C_{14}H_{14}O_2$; clear oil, spec. grav. ·83, of vapour 3·99, B.P. 144° C. (291¼°F.); taste burning; almost insoluble in water; very soluble in alcohol; obtained by distilling butyrate of lime or barytes.

Butyronitric Acid. $C_8, H_6 NO_4 O_3 HO$. Yellow oil; smell aromatic; taste sweet; burns with a red flame; insoluble in water; soluble in alcohol; forms bibasic salts, which when heated to 212°, decompose explosively; formed by heating gently in a retort equal parts of nitric acid and butyrone.

Butyronitryle. *Cyanide of propyle, Cyanide of metacetyle.* C_8H_7N or C_6H_7, C_2N. An oil, spec. grav. ·795, B.P. 245°. Smell similar to that of oil of bitter almonds; formed by distilling dry butyrate of ammonia with anhydrous phosphoric acid, or passing the vapour of butyramide through ignited lime.

Butyryle. C_8H_7. A hypothetic radical, homologous with formyle C_2H, acetyle C_4H_3, propionyle C_6H_5, which differ from each other by C_2H_2.

Buxine. A bitter alkaloid, in *Buxus sempervirens* or common box.

Byssolite *of Saussure*, probably asbestiform actinolite.

Bytownite. Spec. grav. 2·801 H6. Light greenish-blue, granular, amorphous mineral; translucent with splintery fracture and vitreous lustre, infusible before blowpipe, fusing with borax into a colourless glass. Silica 47·56, alumina 29·64, lime 9·06. Fe_2O_3 3·57, MgO ·4, NaO 9·6, HO 1·98; found at Bytown, Upper Canada.

C

Cabbage, or *Brassica Oleracea*, contains ·29 of albumen, 2·84 alcoholic and watery extract, ·63 green matter, ·05 resin, 2·89 gummy extract.

Cacao. The beans of the *Theobroma cacao*, containing 11·3 husk per cent. The kernel consists of oil 53·1, albumen 16·7, starch 10·91, gum 7·75, red-colouring matter 2·01, fibre ·9, water 5·28, ash 2·. From these beans *chocolate* is made, and likewise *coco*, q. v.

Cacao Butter. A yellowish fat, with the taste and smell of cacao beans; can be purified by solution in alcohol; is then colourless; it consists of C 76·6, H 11·9, O 11·5. Spec. grav. ·91; fusing point 85·1° (29·5°C.); it contains stearine, margarine, and some oleine; it is obtained by mixing the powdered kernel with 10 times its weight of boiling water, and expressing it between heated iron plates.

Cacholong. Spec. grav. 2·2. Opaque, milky, or yellowish form of quartz, externally dull, internally pearly; found on the Cach in Bucharia, Iceland, Nova Scotia, Greenland.

Cacodyle, *Kakodyle.* C_4H_6As or $AsMe_2$. Clear fluid, self-inflammable; crystallizing when cooled in large square prisms at 21°. B.P. 338° (170°C.). Spec. grav. of vapour 7·101; heavier than water; poisonous; it may be viewed as a coupled or conjugate compound of arsenic and methyle; formed by acting on chloride of cacodyle with zinc at 212°, or on sulphide or bromide of cacodyle with mercury between 392° and 572°.

Cacodyle of Butyric Acid. An oil, by

distilling butyrate of potash and arsenious acid.

Cacodyle Chloride. *Chlorarsine.* C_4H_6As Cl. Volatile fœtid fluid causing tears, formed by heating alkarsine with corrosive sublimate. Spec. grav. of vapour 4·56.

Cacodyle, Oxide of. *Cadet's Liquor, Alkarsine.* C_4H_6AsO or $AsMe_2O$. Limpid ether, a strong refractor of light; spec. grav. 1·462, of vapour 7·555. B.P. 302° (150°C). Smokes in the air; becomes solid at 13° in white scales; acts powerfully as an irritant and poison; obtained by distilling acetate of potash and arsenious acid together, and purifying by distillation in an atmosphere of carbonic acid $(2C_4H_3O_3)$, AsO_3= $4CO_2$, C_4H_6AsO; it is self-inflammable in air, and by water is converted into cacodylic acid; smells of garlic; insoluble in water; soluble in alcohol and ether; inflames in chlorine, bromine; unites with iodine; reduces oxide of mercury, cacodylic acid being formed.

Cacodylic Acid. *Alkargene.* $C_4H_6AsO_3$. F.P. 397°. Oblique 4-sided prisms, brittle and vitreous, forms amorphous salts; not poisonous; obtained by the action of water or air on the preceding.

Cacoplatyle Chloride. $C_4H_7AsPtO_2Cl$. Needles obtained by adding an alcoholic solution of bichloride of platinum to chloride of cacodyle in alcohol. Cacoplatyle has not been isolated.

Cacotheline. $C_{21}H_{11}N_2O_{10}$. Yellow plates, insoluble in water; soluble in alcohol; a base formed by the action of nitric acid on brucine.

Cacoxenite. *Diphosphate of Iron.* $2Fe_2O_3$, PO_5. Spec. grav. 2·336. Dull yellow small silky tufts, consisting of phosphoric acid 20·5, sesquioxide of iron 43·1, water 30·2, lime 1·1, MgO 0·9, silica 2·1.

Cactine. The red-colouring matter obtained from the *Cactus speciosus*, by expression or alcohol, has been so called.

Cadet's Fuming Liquor. Oxide of cacodyle or alkarsine.

Cadmia. The name given to the crust deposited on zinc ore furnaces; consisting of ZnO 86· to 92½, PbO 1 to 8·8, FeO 1½ to 3½.

Cadmium. (καδμεια, calamine.) Cd7, 6·9677; 55·7416. White metal. Crystals 8-hedrons. Spec. grav. 8·694 after, and 8·604 before hammering. F.P. 442° or 608°. Vapour point above 600°; fracture hackly, soft, and easily cut; sectile, ductile, and malleable; frequently occurs in zinc ores, and is volatilized in the furnace, collecting in the form of a crust on the roof of the arch. This is called cadmia; contains from 10 to 20 per cent. of cadmium. It is separated from the zinc by dissolving in sulphuric acid, and passing a current of sulphohydric acid through the solution; yellow sulphide of cadmium falls, and the zinc remains in solution, or the zinc may be dissolved by caustic soda, the oxide of cadmium falling; the oxide when heated with charcoal is reduced, and the cadmium sublimed.

Oxide of Cadmium.—CdO8·; 64; spec. grav. 8·183. Brownish-yellow powder, easily reduced, obtained by burning cadmium in the air, or by precipitating its salts by an alkaline carbonate, washing and igniting. The salts are colourless; reaction alkaline; disagreeable taste; act as emetics; a plate of zinc precipitates metallic cadmium.

Sulphide of Cadmium, Greenockite.—CdS. Cadmium 77·6. Sulphur 23·4. Sp. grav. 4·842. H 2·75. Yellow 6-sided pyramids, sometimes 6-sided prisms, belonging to the rhombohedral system. B.B. on charcoal; a yellowish-red ring is formed round the fragment. With soda the red ring is formed; with borax it forms a yellow glass; it is soluble in muriatic acid, with evolution of sulphohydric acid; crystallizes by evaporation, which distinguishes it from zinc; it occurs in an amygdaloid along with phrenite in the Greenock tunnel, and on the Kilpatrick hills. Artificially formed sulphide of cadmium is yellow with a tint of orange; crystallizing in a white heat in micaceous scales; it has been used in mixture with oil as a pigment.

Chloride.—Rectangular prisms falling to a white powder, after being heated in the air, instead of deliquescing as the chloride of zinc does; sublimes in micaceous scales.

Sulphate.—$CdOSO_34HO$. Rectangular efflorescing prisms; converted into a subsulphate by heat.

Caeruleum. A silicate of copper; a blue Roman pigment.

Caffeic Acid. ? White powder; soluble in water; insoluble in alcohol; reddening litmus. When heated it evolves an aromatic odour, resembling roasted coffee; obtained by precipitating the decoction of coffee with acetate of lead, filtering, and then with subacetate of lead. The last precipitate is thrown on a filter, mixed with water, and precipitated by sulphohydric acid; the filtered liquor evaporated to a syrup, and then mixed with its volume of alcohol; *caffeic acid* is thrown down, and caffeotannic acid dissolved.

Caffeine. *Theine, Guaranine.* $C_{16}H_{10}N_4O_4$. Fine white silky prisms or needles; soluble in water, alcohol, and ether; volatile and fusible; it acts as a weak base, but forms fine crystals with chlorohydric and sulphuric acids. It contains in its silky state 7·81 per cent. of water, which is removed at 248° (120°C.), while the caffeine loses its lustre, and is then easily pounded; fuses at 350½ (177°C.); sublimes at 723° (384°C.); soluble in 93 cold water, 158 absolute alcohol, 298 ether; anhydrous caffeine dissolves in 98 water, 97 alcohol, and 794 ether; boiled with barytes it yields ammonia, cyanic acid and carbonic acid, and by long boiling formic acid; it is found in coffee, Chinese, Paraguay, and Guarana teas (*Coffea Arabica, Thea bohea, Ilex Paraguayensis, Guarana officinalis* or *Paullinia sorbilis*), all from different plants, but yet employed as a drink by different nations. It is prepared by making a decoction with hot water of coffee or tea, precipitating with acetate of lead as long as a precipitate of caffeotannate of lead falls; this is filtered, the excess of lead precipitated by sulphohydric acid, and the filtered liquid evaporated; the caffeine crystallizes out, and may be further purified by solution and recrystallization in water or alcohol. From coffee from $\frac{1}{2}$ to 1 per cent. may be obtained, and from tea from $2\frac{1}{2}$ to $6\frac{1}{4}$ per cent. The anhydrous caffeine may also be sublimed from the evaporated decoction by the mode for benzoic acid, but the product is small. Caffeine by chlorine yields *amalic acid* $C_{12}N_2H_7O_8$, homologous with alloxantine, from which it differs by 2 (C_2H_2).

Caffeomurexide. *Murexoin.* $C_{36}H_{23}N_{10}O_{15}$. Cinnabar 4-sided prisms, with a golden tint: seems to be murexide, less 1 atom water; obtained by acting on *amalic acid*, the product of the action of chlorine on caffeine, with ammoniacal gas; the dry acid is exposed to this action, and the mass dissolved in hot spirit.

Caffeotannic Acid. $C_{14}H_8O_7$. Yellowish-white powder, with a sourish taste; soluble in water, not precipitated from a syrupy solution by alcohol; soluble in sulphuric acid, the solution becoming blood-red by heat; water, precipitating from it, flocks; soluble in caustic potash, with a reddish-yellow colour; soluble in ammonia, the solution by exposure becoming green; when burned it gives out the odour of roasted coffee; it does not precipitate gelatine, thus differing from tannic acid; it agrees with kinic acid in the fact that when mixed with 4 times its weight of black oxide of manganese, and 1 part of oil of vitriol, and twice its bulk of water, in a retort, crystals of kinone collect in the neck of the retort without the addition of much heat; the tannic acid of Paraguay tea, and common holly, likewise gives kinone; it is obtained by boiling pounded coffee beans in spirit of wine, precipitating the filtered solution with an alcoholic solution of acetate of lead, and washing the precipitate on the filter with spirit. The precipitate is then diffused through water, and decomposed by a stream of sulphohydric acid; the liquor is filtered from sulphide of lead, and evaporated to dryness in a water bath. Caffeotannic acid is analogous to boheic acid found in tea; the formula of the first being $C_{14}H_8O_7$, and of the second $C_7H_5O_6$; that of viridic acid is $C_{14}H_7O_8$, and that of catechin $C_{14}H_8O_8$. Caffeotannic acid, when added to nitrate of silver, yields a precipitate from which metallic silver separates, and by heating in a tube the silver is deposited on the sides of the tube. This appears to be the native acid united with caffeine. It agrees with gallic acid in becoming black in a solution of potash.

Cailcedrine. A bitter principle from the cailcedra bark (*Khaya Senegalensis.*)

Caincic Acid. $C_{16}H_{13}O_7$. White silky needles; soluble in water and spirit; forming no precipitates with salts of iron; dissolved in potash it becomes yellow; acetic acid precipitates from this solution a jelly (*chiococcic acid.*) Caincic acid is obtained from the alcoholic decoction of the bark of the root of the *Chiococca racemosa*, by precipitating with acetate of lead, which precipitates caffeotannate of lead; trisacetate of lead when added precipitates caincate of lead, which, when washed and decomposed by sulphohydric acid, yields caincic acid; this may be crystallized by evaporation and treatment with alcohol.

Cajeput Oil (*Cayoopooti*). Spec. grav. ·927, B.P 347°, $C_{10}H_9O$. A green oil, obtained by distilling the leaves and twigs of the *Melaleuca leucadendron*, a tree from the Moluccas; sulphuric acid makes it yellow; cold nitric acid and iodine have no action. It is sometimes adulterated with turpentine and lavender oil, rendered green by the resin of Achillea millefolium, but iodine then explodes with it.

Calaite, or *Turquoise.*

Calamine, or *Carbonate of zinc.*

Calamite. An asparagus rhombic tremolite, found at Normark, Sweden.

Calamus Oil. Spec. grav. ·979, B.P. 514°$\frac{1}{2}$, from the root of *Acorus calamus.*

Calcareo Sulphate of Strontian. Sp. gr. 3·81, H 2·75. Thin plates or radii, lustre silky, from Bristol. SO_3 45·93, CaO 7·16, SrO 46·88.

Calcareous Spar, or *Carbonate of lime.*

Calcedony, or **Chalcedony**, a stalactitic form of quartz from hot water.

Calchantum, or **Chalcantum**, probably a mixture of sulphates of copper and iron.

Calcination. The operation of exposing a substance to a red heat so as to drive off water or animal matter, or to oxidize a matter.

Calcite, or rhomboidal calcareous spar.

Calcium. Ca 2·5, 20. White metal like silver, from 4 to 8 times heavier than water, burning brilliantly in the air, and becoming quicklime, obtained by the same methods as barium and strontium.

Calcium, oxide of, and salts of. See LIME.

Chloride of Calcium.—7· 6·925, 55·4, Ca 35·72, Cl 64·28. Spec. grav. 2·040 (Karsten), 2·214 to 2·269 (Boullay). White opaque porous mass, or when fused somewhat crystalline, granular, non-electrical conductor (Davy). *Taste* pungent, bitter; very deliquescent. *Soluble* in 4 parts water at 60°, giving out heat. When exposed to the air it absorbs water, 100 parts in moist air, taking up 124 water in ninety-six days. (Brandes Schweig. J. 51, 433.) *Soluble* in alcohol. By ignition it becomes anhydrous, and loses some chlorine (Thomson), its solution being then alkaline (Liebig), the decomposition being increased by the presence of carbonic acid at a white heat (Petzholdt, Erdmann, J. 17, 464), it then shines in the dark, as observed by Homberg. (Homberg's phosphorus); decomposed by acetic acid (partially), and sulphuric acid in distillation. *Sources.*—It exists in sea water, and in many wells and springs. It was formerly called *fixed ammoniac*, from its being prepared by decomposing salammoniac by means of lime. It may be obtained for laboratory purposes by dissolving chalk, marble, or Irish limestone in chlorohydric acid, adding ammonia to precipitate iron and phosphate of lime, or the iron may be removed by a solution of bleaching-powder, so as to convert it into sesquioxide, and then adding an excess of chalk and boiling. The solution is then filtered, evaporated to dryness, and heated to redness in a clay crucible. When in a state of fusion it is poured out on a clean stone slab, or tile. *Use.*—It is employed to remove water from gases in organic analyses, and may then be dried at 450°; if heated to redness, it would contain free lime, which would absorb carbonic acid; if ammonia has been added, it must be ignited. Chloride of calcium is also formed by heating lime in chlorine and chlorohydric gases.

Sexhydrated Chloride.—Ca Cl 6 HO 13·75, 13·675, 109·4. Regular 6-sided prisms, often striated, fusing below 212°; in vacuo in twelve days, lose 4 atoms water, becoming opaque, and retaining 2 atoms, and their form with a lustre of talc (Graham). At 392° they also lose 4 atoms (Mitscherlich), leaving a white porous mass. The sexhydrate corresponds with the strongest chlorohydric acid Ca Cl 6 HO, and H Cl 6 HO (Kane). *Soluble* in ·5 water at 32°; in 0·25 at 60°·8; producing cold, and in any quantity of boiling water; easily soluble in alcohol. *Process.*—Obtained in cooling a hot concentrated solution of the chloride. When by evaporation the specific gravity of its solution is 1·45, it crystallizes by exposure to a temperature of 32°; if its specific gravity at 80° be 1·49, on cooling it becomes a hard, pearl-coloured mass (Walker Nicholson's J. 5, 226).

Oxychloride.—CaCl 3CaO 15HO. Prisms obtained by boiling a concentrated solution of chloride of calcium with an excess of slaked lime; the filtered liquor deposits the crystals on cooling.

Fluoride of Calcium. Derbyshire Spar. Fluor Spar.—CaF 4·75, 4·8375, 38·7; Ca 52·68, F 47·32. Spec. grav. 3·14 to 3·177. *Crystal.* Cubes and 8-hedrons, and modifications of the regular system. *Colour*, white, green, blue, &c. or as white flakes artificially prepared. By ignition it decrepitates, and phosphoresces strongly in the dark, even under water or in vacuo; it loses this phosphorescence after a continued heat, without loss of weight, which it only regains by decomposing by sulphuric acid, and reforming it (Scheele). By a heat of 51° Wedgewood, it fuses into a transparent glass (Saussure, Jour. Phys. 45, 16.) By a prolonged heat it decomposes, leaving pure lime (Smithson, An. Phil. 23, 101). It is partially decomposed by fusion with alkaline carbonates (Gmelin); vapour of sulphuric acid (HO SO_3), at a red heat decomposes it. Its solubility is promoted by carbonic acid. *Soluble* in 26545 water at 60°; a pint of water dissolves 0·15 grains (Dr. Wilson), more soluble in hot water. Cold sulphuric acid converts fluor spar into a syrupy mass, but no fluoric acid is evolved under 104°; if any silica is present, effervescence occurs. Boiling nitric and chlorohydric acids decompose fluor spar partially. When heated with oil of vitriol it gives out fluohydric acid, which decomposes glass—the test for fluor spar. *Sources.*—It occurs in veins of lead ore abundantly, it likewise exists in the teeth, bones, and also in some mineral waters and plants. It may be prepared by precipitating a lime salt by fluohydric acid in solution as jelly, or digesting carbonate of lime in solution of fluohydric acid.

Calcsinter, a stalactitic form of calcareous spar from the roofs of caves.

Calctuff, a calcareous deposit from springs, incrusting plants, &c.

Calculus. See BILIARY, GOUTY, SALIVARY, CONCRETIONS, and OSSIFICATIONS.

Caledonite, or Cupreoussulphate carbonate of lead.

Calendulinc. A mucilaginous matter, soluble in water and alcohol, extracted from the flowers and leaves of *Calendula officinalis.*

Caliche. Nitrate of soda from Peru.

Calico Printing. *Baumwollendruckerei, Farbenchemie*, Gr.; *Impression des tissues*, Fr. Consists in the art and process of applying colours by cylinders or blocks, as in letterpress and woodcut printing, to cloth of various kinds, as of cotton, silk, wool. The art seems to have been discovered in India, but it has also been long known in Egypt. Calico printing is skilfully carried on in Switzerland, France (Alsace), Germany, Belgium, and Great Britain.

Preliminary Processes.—The cloth after being woven is *singed*, that is, it is passed rapidly over a red hot iron plate to remove the protruding fibres of cotton which are conspicuous on its surface; it is then bleached, see BLEACHING; after which it is ready for *printing*. If we immerse a piece of cotton cloth in a solution of Brazil-wood, we shall find that we can remove the whole of the colouring matter by washing with water, but if we have previously placed the cloth in a solution of acetate of alumina before inserting it in the coloured liquid, the cloth remains permanently of a red colour. In this illustration there are two important substances; 1st, the *mordant* (from *mordeo*, I bite, or fix), which has an affinity for the cloth; and, 2d, the *colouring matter*, which has no affinity for the cotton cloth, but is attracted by the mordant. The mordant in this case is the alumina, and we can form this by precipitating a solution of alum by ammonia, throwing the precipitate on a filter, and washing it till the filtered liquor gives no precipitate with chloride of barium. If we now remove a portion of this moist hydrate of alumina by means of a spatula, and stir it in a solution of Brazil-wood for some time, and throw it on a filter, we shall find that the water passing through is colourless, thus affording evidence that the alumina (mordant) has united with the colouring matter, and removed it entirely from the solution.

Mordants.—1. *Alumina*, or *the alum mordant*, is made by dissolving alum in water, and adding acetate of lime or lead to the solution. See ACETATES. The liquid has a specific gravity of 1·08, and contains about as much alum decomposed as the liquid can hold in solution. In a dark Turkey red pattern the quantity of alumina attached to 1000 grs. or 2772 square inches of cloth, has been found 8·6 grs. $\frac{1}{370}$th of a grain to a square inch; and in a pale Turkey red, ·4 grs. or $\frac{1}{8000}$th to the square inch. This mordant is applied by immersion when the cloth is to be dyed red, but when the colour is to be applied to particular parts, the mordant is thickened with gum or starch, and applied to the cloth by means of the block or cylinder. By means of alum mordant, a *red* is produced with madder, Brazil, and peach-woods; a *pink* with cochineal; a *lilac* with madder; a *yellow* with quercitron and Persian berries. 2. *Oxide of tin* is applied in the form of *protochloride* or tin salts, and as *stannate of soda*. In the first case the salt is mixed with the colour, and both are applied by the block or cylinder to the cloth. Such colours are termed chemical, from the fugitive nature of the compounds formed in presence of light and soap. The stannate of soda is applied by immersing the cloth in a solution of that salt, and then passing it through water acidulated with sulphuric acid. Sulphate of soda is washed out, and peroxide of tin remains on the cloth. The colours required thickened with paste are then printed on the cloth, which is exposed to the action of steam. Colours thus fixed are called *steam* colours. 3. *Sesquioxide of iron*, acetate of iron, q. v. of the spec. grav. 1·05, is used for the immersion of the cloth for a *black* with madder; *purple* is formed by a weaker mordant; so likewise *red*. *Chocolates* are obtained by mixing the alum and iron mordants, and dyeing with madder. Indigo, oxide of manganese, catechu, do not require mordants, because they can be applied in solution to cloth, and afterwards become insoluble in water. 4. *Nitrate*, or *acetate of lead*, is employed in solution, thickened with gum, to those parts of the cloth intended to be yellow. The cloth is then passed through a solution of bichromate of potash, when yellow chromate of lead is formed.

Dischargers of Colours.—Those parts of the cloth which it is intended should not be coloured, are printed with an acid or body which dissolves the colour. These are called *dischargers*. 1. Citric acid is employed to dissolve alumina and sesquioxide of iron; it is mixed with gum for the cylinder, and with gum and pipeclay for the block. 2. *Tartaric acid*, when printed on cloth dyed Turkey red, and passed through a solution of bleaching powder, evolves chlorine, which bleaches the spots to which the acid is applied. 3. *Oxalic acid* discharges iron. 4. *Protochloride* of iron discharges manganese brown by depriving it of oxygen, forming protochloride of manganese, and leaves a buff of sesquioxide of iron. 5. *Protochloride of tin* discharges sesquioxide of manganese, and changes brown to a white; if mixed with Brazil-wood or cochineal, it discharges the manganese and leaves a *pink*; with logwood a *purple*, and with Prussian blue a *blue*. This salt also removes sesquioxide of iron and dichromate of lead.

Resist pastes are substances which, by parting with oxygen, restore the blue colour to dissolved indigo before the cloth immersed in the indigo vat reaches the air. 1. Blue or *vitriol paste* consists of a mixture of sulphate and acetate of copper, the solution being thickened with gum Senegal and pipeclay for the block, and with flour for the cylinder. When the cloth on which this paste has been printed is dipped into the indigo vat, the indigo is revived before it has time to reach the surface of the cloth. 2. *Mild paste* consists of sulphate of zinc, gum, and pipeclay. 3. *Red paste.* Alum mordant mixed with acetate of copper, gum, and pipeclay—a resist for pale blues. 4. *Neutral paste*, a compound of lime juice, sulphate of copper, gum, and pipeclay; a resist during a short dip in the vat. 5. *Chrome yellow resist paste*, a mixture of a salt of copper, to resist the blue vat.

Colours.—1. *Madder Red.* The alum mordant made into a paste with gum is printed on the cloth by the cylinder; the cloth is dried in a warm room, when the acetic acid is dissipated; it is passed through a hot mixture of cow dung and water; it is washed in cold water, and agitated again in the preceding mixture; it is then dyed in madder; it is boiled with bran or soap, exposed to light on the grass—cleared with bleaching powder solution. 2. *Madder purple* is formed in a similar manner, only substituting iron mordant. 3. *Cochineal pink* is formed with alum and cochineal. 4. *Logwood black* is formed with alum mordant and logwood. 5. *Prussian blue* may be formed by immersion in iron mordant, and then in yellow prussiate of potash solution and sulphuric acid. 6. *Buff from iron* is obtained by printing on a mixture of sulphate of iron and acetate of lead, exposing to the air, and immersing in lime water. 7. *Manganese bronze* is formed by printing on sulphate of manganese, when dry passing through caustic soda, and then immersing in bleaching powder solution (MnO becomes MnO 1½). 8. *China blue* may be applied by heating indigo with caustic soda and orpiment, when it becomes soluble, and may be printed on with starch gum; or it may be used by mixing indigo with orpiment in a solution of sulphate of iron and deoxidizing after being printed by alternate immersions in lime and copperas; or the indigo may be dissolved in a hot solution of soda and stannate of soda, and is then precipitated in the white state by muriatic acid; the precipitate being thickened and mixed with fresh chloride of tin is then printed on the cloth. When dry the cloth is immersed in carbonate of soda. 9. *Catechu brown* is formed by dissolving catechu in acetic acid, adding a solution of copper and salammoniac; the whole is printed on the cloth. 10. *Chrome orange* is formed by printing on the chromate of lead, and abstracting the half of the acid in hot lime water. 11. *White discharge* on *Madder red.* The cloth is first mordanted with alum, dyed in madder and discharged with citric acid. 12. *Madder and logwood* with alum mordant yields a brown. 13. *Cochineal pink* is dyed in cochineal after mordanting with alum. 14. *Black ground and white* produced by alum mordant and logwood, and white discharge by citric acid. 15. *Turkey red.* This mode of dyeing was introduced into France from the Levant, and into Glasgow by M. Papillon of Rouen, in 1785, and under Papillon's direction the firm of Dale and Macintosh was established at Dalmarnock, and continued till 1805. 1. The cloth is steeped in a weak alkaline lye *(rot steep)*, 4 or 5 lbs. caustic soda to 100 lbs. cloth, at 110° for twenty-four hours, in order to remove the weaver's dressing. 2. It is boiled in carbonate of soda, 7 to 10 lbs. carbonate of soda to 100 cloth. 3. 22 gallons of the following mixture are made to have a spec. grav. of 1020 to 1025:—1 gallon gallipoli oil, 1½ gallon sheep's dung, 4 gallons solution of carbonate of soda, spec. grav. 1060, 1 gallon solution of pearl-ash, spec. grav. 1040. This mixture is mixed in the *liquor tub*, and conveyed by pipes to the *padding machine*, where the cloth is immersed for fourteen days. The sheep's dung assists the bleaching process. 4. It is dried on the grass or in the stove. 5. It is immersed again in No. 3, and dried, and the process again repeated. 6. It is steeped in a solution of pearl-ash of 1007½ to 1010, at 120°. 7. It is immersed in the following mixture as in No. 3:—1 gallon gallipoli oil, 3 gallons soda lye, spec. grav. 1060, 1 gallon caustic potash lye, 1048 made up with water to 22 gallons; it is then dried on the grass, or in the stove. 8. The process 7 is repeated thrice. 9. It is steeped in caustic potash and soda, spec. grav. 1010 to 1012½ at 120°, drained and washed to remove superfluous oil. 10. *For galling.*—18 lbs. aleppo, or 33 lbs. sumach, or a mixture of 9 lbs. nutgalls, and 16½ sumach, are boiled for four or five hours in 25 gallons water down to 20 gallons, and the cloth immersed at from 80° to 100°. 11. *Mordanting or aluming.*—Precipitate from a solution of alum, spec. grav. 1040, the alumina by pearl-ash, soda, or chalk. The cloth is passed though this turbid liquor at 100° to 120°, and steeped for twelve hours. 12. The cloth is stove-dried, and then washed out of the alum liquor. 13. *Dyeing red.*—From 1 to 3 lbs. madder for every pound of cloth are placed in the boiler with the cloth in cold water. It is made to boil in an hour, and continued for two hours; the cloth being passed through the liquor by a winch; 1 gallon bullock's blood is added to every 25 lbs. cloth to produce a fine red. 14. *Clearing* removes the brown colour, and is performed by boiling the cloth for twelve or fourteen hours in 5 lbs. soda, 8 lbs. soap, and 16 to 18 lbs. residual liquor of No. 9, with sufficient water. 15. *Brightening.*—5 or 6 lbs. soap, and 16 to 18 oz. chloride of tin are boiled with water, and the cloth in a covered globular boiler at a pressure of 2 atmospheres, or at 250½°. 16. The cloth is spread out on the grass, and exposed to the sun for a few days. *Indigo blue.*—The blue vat is a solution of colourless indigo in lime water. To form it the indigo is ground with water, and mixed with sulphate of iron and an excess of lime. In a few hours the indigo is deoxidized and dissolved. The new products, sesquioxide of iron and sulphate of lime, subside, and a clear yellow solution of indigo remains. Cloth dipped in this vat attracts the indigo from the lime; on exposure to the air the indigo becomes blue. *Chemical colours* are such as are fugitive, and receive their permanence, such as it is, by being *raised* by *steam.* 1. *Chemical black.*—A salt of protoxide of iron is printed on the cloth with nutgalls or logwood and exposed to the air, when it becomes black. 2. *Chemical pink.*—A decoction of Brazil or peach-wood is thickened with gum Senegal and mixed with salt of tin, which forms a *lake* colour, and is printed on the block. 3. *Steam blue.*—To a solution of yellow prussiate of potash tartaric acid is added; *cream of tartar* precipitates, while hydrocyanic acid and white Prussian blue dissolve in water. This liquid is thickened with gum, and printed on the cloth. The print is now exposed to the action of steam in a closed box, and if the other colours allow, *raised* in a weak solution of bleaching powder or bichromate of potash or by exposure to the air. 4. *Steam green* is a combination of the preceding with a yellow colour, by a decoction of Persian berries, and fixed on the cloth by alum. The prussiate of potash is dissolved in water, and this liquor is mixed with a decoction of Persian berries and alum. On applying steam, sulphate of potash is formed, the alumina uniting with the yellow of the berries and fixing it. The heat also decomposes the white Prussian blue and deposits blue on the cloth. The blue and yellow thus produce a green. 5. *Annotto orange.*—Annotto is dissolved in potash or soda and printed on the cloth. 6. *Chrome yellow on Turkey red.*—Tartaric acid and nitrate of lead are dissolved together in water, which is thickened with gum and printed on those parts of the Turkey red cloth to be rendered yellow. By passing through bleaching solution the tartaric acid portions are deprived of colour, while the oxide of lead becomes fixed on the cloth, which is changed into yellow chromate of lead by passing through bichromate of potash. 8. *Chrome yellow on Indigo.*—A resist paste is made by mixing a solution of nitrate and acetate of lead and nitrate and acetate of copper with gum and pipe-clay. While the indigo is depositing itself on the cloth, the lime which holds it in solution is precipitating the oxide of lead on those parts of the cloth where the paste has been applied. After washing, the cloth is passed through a solution of bichromate of potash. To remove some copper deposited, the cloth is immersed in muriatic acid. 9. *Orange and yellow upon blue.*—In this case the cloth, prepared as in the last, is passed through a solution of dichromate of potash. When weak nitric acid thickened with gum is printed on portions of this orange, half of the oxide of lead is removed, and the orange is converted into yellow. 10. *Chrome yellow on bronze.*—Sulphate of lead mixed with chloride of tin is printed on bronzed cloth. Chloride of manganese is removed by washing, the oxide of tin and sulphate of lead remaining; the latter is decomposed by lime, the oxide of lead adhering to the cloth, and becoming yellow when passed through bichromate of potash. 11. *Red and chocolate resist on pale blue.*—If the aluminous mordant, with the addition of a little verdigris and a little soft soap, and thickened with gum and pipeclay, be printed on white cloth, the piece may be immersed in the blue vat without any of the indigo attaching itself to those parts so printed. When dunged and dyed in the usual way, a red is produced on the pale blue ground. The same mordant, with the addition of acetate of iron, gives a chocolate. A *neutral paste* of lime juice and sulphate of copper is employed in this case to defend the cloth against both the reds and chocolate and the blue. 12. *Pigment printing.*—The colours are the same as those used for painting or printing on paper, and are printed on dresses not intended to be washed.

Caliphite. A mineral composed of sesquioxide of iron 28·8, binoxide of manganese 28·13, water 19·01, silica 12·1, ZnO 6·3, CaO 2·55, TiO_2 1·2, Al_2O_3 ·6, MgO ·7.

Callus. The new bone or cement thrown out by the extremities of a broken bone to cause union; it is composed of animal matter 50, phosphate of lime 33, carbonate of lime 6, soluble salts 11.

Callutannic Acid. $C_{14}H_6O_8$. A yellow-colouring matter obtained by boiling Calluna vulgaris with spirit, precipitating it with lead salts, and decomposing.

Calluxanthine. $C_{14}H_5O_7$. Yellow amorphous body, obtained by acting on an aqueous solution of callutannic acid with mineral acids and heat.

Calomel. The old and still common name of subchloride or dichloride of mercury.

Caloric. A synonyme of heat.

Calorifiant Food. Food producing heat in the systems of animals.

Calorimeter. A measurer of heat. The oldest form of calorimeter is that of Lavoisier and Laplace, consisting of three cylinders within each other. The heated body being placed in the interior vessel and the spaces between the cylinders filled with ice; the amount of ice melted measured the heat evolved from the heated body.

Calorimotor. A galvanic apparatus named from its heating power, composed of several pairs of large copper and zinc plates.

Calotropis Mudarii, or *Gigantea.* Contains in its root bark 11½ per cent. of an emetic, *mudarine.*

Calp of Milton. A siliceous carbonate of lime.

Calstronbaryte. Right rhombic prisms and in plates. Spec. grav. 4·2, H 3·25, found in New York state; contains 65·5 $BaOSO_3$, 22 SrO,CO_2, 12 $CaOCO_2$.

Calumbine. $C_{22}H_{12}O_7$.? Colourless rhombic prisms, poisonous. Taste bitter, neutral; scarcely soluble in water, alcohol, ether; soluble in boiling alcohol of ·835, sulphuric acid, nitric, acetic acid, and alkalies; obtained from the root of Columbo (Cocculus palmatus) by boiling alcohol.

Calx. The old name for oxides of metals.

Calyptolite. H 6·5. Spec. grav. 4·34. A fluocolumbate from Connecticut in 4-sided prisms.

Cambium. A mucous viscid layer of minute transparent granules interposed in the spring beneath the bark and above the wood of trees.

Campeachy-Wood. *Logwood, Blauholz,* Gr., *Haematoxylon Campeachianum.*

Camphamic Acid. An amide acid from camphoric acid; little known.

Camphene. A general expression for essential oils composed on the type of oil of turpentine. C_5H_4, or a multiple of these atoms—$C_{10}H_8$ or $C_{20}H_{16}$. The camphenes and compounds are as follows:—

Oils.			Sp. Gr.	B.P.
Camphene,	turpentine,	$C_{20}H_{16}$	·860	150° C
,,	cubebs,	$C_{15}H_{12}$	·929	250° to 260°
,,	pepper,	$C_{10}H_{16}$	·864	167°
,,	lemon,	$C_{10}H_{16}$	·845	167° to 173°
,,	orange,	$C_{10}H_{16}$	·836	180°
,,	copaiva,	$C_{10}H_{16}$	·889	245°

Oxide of Camphene.—Common resin. $C_{20}H_{12}O_2$.

Oxide of Camphene.—Copaiva resin. $C_{10}H_6O$.

Hydrate of Camphene.—Stearoptene of turpentine. $C_{20}H_{12}4HO$.

Chlorohydride of Camphene.—$C_{20}H_{12}HCl$.

Camphereme. A term applied to a class of bodies produced by a simple chemical action from the camphenes, as by the action of acids. They consist of—

		Sp. grav.	Vapour.
Terebene,	$C_{20}H_{16}$	·864	4·763
Colophene,	$C_{20}H_{16}$	·939	9·526?
Chlorohydride of Terebene,	$C_{20}H_{16}$ HCl	·902	
Bichlorohydride of Terebene,	$C_{20}H_{16}$ 2HCl	1·017	
Dichloroterebene,	$C_{20}H_{14}$ Cl_2	1·137	

Camphileme. This class is formed from the campherenes by a simple reaction, or from the camphenes by a double action, and are isomeric both with the camphenes and campherenes. They are formed by passing the vapour of muriates of camphene or campherene through a tube filled with quicklime, heated by an oil bath up to near 400° F. till the oily product is free from acid, and rectifying over potassium. The type of them is *camphilene* or dadyle, the fluid produced by the decomposition of muriate of oil of turpentine (chlorohydride of camphene). Form. $C_{20}H_{16}$. Spec. grav. ·87, B.P. 135°. C.; colourless, aromatic like oil of turpentine. It does not possess its rotary power, and hence its molecular constitution must be altered.

Camphine. $C_{18}H_{16}$ by the action of iodine on camphor.

Camphine. A commercial name of purified oil of turpentine used for lamps.

Camphocreasote. By acting on camphor with iodine.

Camphogene. $C_{20}H_{14}$. Spec. gr. ·860, of vapour 4·78. Colourless; B.P. 175° C.; obtained by distilling Japan camphor with anhydrous phosphoric acid; it occurs naturally in the oil of cummin.

Camphogene Sulphuric Acid. *Sulphocampic Acid.* $C_{20}H_{13}S_2O_5$. Small crystals formed by distilling camphogene with fuming sulphuric acid.

Campholene. $C_{18}H_{16}$. A fluid. B.P. 135° C. Spec. grav. vapour 4·353; formed by distilling campholic acid with anhydrous phosphoric acid, by which it is deprived of 2 atoms water and 2 atoms carbonic oxide.

Campholic Acid. $C_{20}H_{17}O_3HO$. Spec. grav. vapour 5·938, B.P. 250° C. F.P. 80° C. formed by passing repeatedly the vapour of camphor through soda lime, at 300° to 400° C. (570°, 750° F.); the mixture heated with boiling water; on adding an acid it separates in white crystals.

Campholone. $C_{19}H_{17}O$. A fluid by distilling campholate of lime.

Camphor-Anile. $C_{32}H_{19}NO_4 = C_{12}H_7N, 2HO\ 2C_{10}H_7O_3$, F.P. 116° C. Slightly soluble in water; soluble in alcohol or ether; by acting with anhydrous camphoric acid on aniline with heat, and adding ammonia to the glacial product in which camphor-anile is insoluble.

Camphoranilic Acid. $C_{32}H_{21}NO_6$. Needles dissolved from the preceding glacial mass by ammonia.

Camphoric Acid. $C_{10}H_7O_3HO$. F.P. 338°, fine transparent plates or needles; taste acid, bitter; slightly soluble in cold, more soluble in boiling water; very soluble in alcohol and ether, sulphuric and nitric acids; formed by digesting nitric acid on camphor, and applying heat; when distilled, it yields water and anhydrous acid, with a spec. grav. of 1·194, F.P. 423° in rhombic prisms.

Camphor of Borneo. (*Solid*). $C_{10}H_9O$. 6-sided prism with rhombic base. F.P. 388°; B.P. 414°; taste between that of pepper and camphor; heavier than water—scarcely soluble in it; very soluble in alcohol and ether; heated with anhydrous phosphoric acid it yields a campherene; obtained from *Dryabalonops camphora.*

Camphor of Borneo. (*Fluid*). Resembles oil of turpentine. B.P. 320°, occurs in the same tree, and appears the same as the preceding campherene; it occurs in oil of valerian.

Camphor of Japan, or Common Camphor. $C_{10}H_8O$. Spec. grav. ·9857 to ·996, of vapour 5·317. F.P. 347°, B.P. 400°. Solid amorphous body in commerce, with a peculiar smell and burning taste; sublimes when exposed to light, and deposits on the sides of the bottle in 8-hedrons; soluble in 1000 water, nearly equal parts of alcohol, and precipitated by water; it evaporates more quickly in water than in air—moving about on its surface; soluble in sulphuric acid and nitric acid, without change—reprecipitated by water; is obtained by boiling the chipped wood of *Persea (Laurus) camphora* in a still.

Camphor, Oil of. $C_{20}H_{16}O$. Spec. grav. ·910, B.P. above 212°. Clear fluid occurring along with Japan camphor in the tree.

Camphorosine. $C_{22}H_{12}$. A substance formed by acting on camphor with iodine.

Camphoryle. $C_{10}H_7O$. A fluid, soluble in alcohol and ether; insoluble in water; a product of distillation of camphorate of lime at a high temperature.

Camphovinic Acid, or camphorate of oxide of ethyle.

Camphrone. $C_{30}H_{21}O$. B.P. 167°; colourless fluid, formed by passing vapour of camphor over red hot lime.

Camwood. A red wood from the coast of Africa, used instead of Brazil-wood in dyeing; from *Baphia nitida?*

Canaanite. Silica 53·36. Al_2O_3, 10·38, lime 25·80, FeO 4·5, MgO 1·62, CO_2 4; probably Scapolite.

Canada Balsam. A balsam from the *Abies balsamea,* much used for fixing microscopic objects on glass.

Cancer, Matter of. The cancerous part of the lip was found to consist of fibrinous matter 9·, water 85·, albumen 1·3, glutine ·83, extract 1·25, fat 2·7. The ash contained NaCl 10·, $NaOSO_3$ 2·, alkaline phosphate and carbonate 63·88, earthy phosphates 24·.

Cancrine. $C_{16}H_{13}O_4$? The colouring matter of crabs.

Cancrinite. Rose hexagonal crystals, spec. grav. 2·453, H. 5½ to 6·, soluble with effervescence in muriatic acid. B.B. fuses to a white vesicular glass; effervesces with microcosmic salt with a siliceous skeleton; silica 40·58, alumina 28·57, lime 20·2, soda 3·5, CO_2 6·4, Cl ·89. Ilmen mountains.

Candite. A variety of spinelle found at Candy, Ceylon.

Candles. Well-known instruments for affording light, formed of solid oils or fatty substances. There are various kinds. *Dip* candles are made by dipping the wicks into melted tallow. *Moulded* candles are made in a mould. *Stearic acid*, or *stearine*, and *wax* candles, are formed of these bodies. In China vegetable oil is used and obtained from the *Stillingia sebifera*. *Sperm* candles are made of Spermaceti. *Composition* candles, of various oils, of which cocoa nut oil is one frequently. *Paraffine*, which by Mr. Young's patent process can now be made in quantity, may also be used. It has been found that the relative consumption of various candles per hour is as follows, expressed in grains:—*Moulds* (10 to the pound) 132 grains per hour; *dips* (10 to the pound) 150 grains; *moulds* (8 to the pound) 132 grains; ditto (6 to the pound) 163 grains; ditto (4 to the pound) 186 grains. Argand oil flame 512 grains (Ure), but the light is much greater with an argand; 1 argand being equal to the following number of candles in each of the preceding classes: 5·7 candles; $5\frac{1}{4}$; $6\cdot\frac{6}{10}$; 5; $3\frac{1}{2}$.

Candy Sugar. Common sugar candy is obtained in rhomboidal crystals, by inserting strings in a strong solution of purified sugar. The term candy is likewise applied to preparations of molasses and syrup.

Cane Sugar. Produced from sugar cane. See SUGAR.

Cancelline. Needles, with a sweet taste, from canella alba.

Caniramine. (*Canis*, dog; and *ira*, anger), synonyme of Brucine.

Canna, *Tous le mois*. A starch from a species of Canna.

Cannabine. The bitter resin of *Cannabis sativa*.

Cannel Coal. *Candle coal*, *Parrot coal*, *Gas coal*. See COAL.

Cannelstein. SiO_3 38·8, CaO 31·25, Al_2O_3 21·2, FeO 6·5.

Cannon Metal. A species of brass or bronze, is composed of 91 copper, 9 tin.

Cantalite. Yellowish-green quartz.

Cantharidine. *Vesicolorine. Cantharides camphor.* $C_{10}H_6O_4$.? White crystalline plates like spermaceti; insoluble in water but soluble in union with the yellow matter of the Spanish flies; soluble in hot alcohol; soluble in ether and oils; obtained from *Lytta vesicatoria*, *gigas*, *violacea*, and other species of insects, by treating their watery extract with ether, and washing the mica-like scales with cold alcohol. From the blistering principle being soluble in ether, the improved blistering papers are prepared.

Canton's Phosphorus, or *Pyrophorus*, was prepared by calcining 3 parts oyster shells and 1 flowers of sulphur for an hour in a close vessel. When dropped in the air the sulphide of calcium takes fire.

Caoutchene. C.H. Needles mixed with a yellowish fluid obtained by distilling caoutchouc, and condensing the product by snow. Spec. grav. ·65. B.P. 58° ($14\frac{1}{2}$° C.); insoluble in water; soluble in absolute alcohol and ether.

Caoutchine. C_5H_8. Clear oil, smells of orange oil. Spec. grav. ·8433, of vapour 4·461, B.P. 340°; soluble in absolute alcohol, ether, fat oils, and bisulphide of carbon; obtained by fractional distillation of the oil of caoutchouc between 284°, (140°C.) and 536°, (280°C.); shaking the product with dilute sulphuric acid until the acid ceases to be coloured, washing with water and potash; frequently distilled; heated with muriatic acid; frequently distilled, and rectified over potassium.

Caoutchouc, *India rubber*, *Gum elastic*. C_8H_7. Spec. grav. ·9335. Colour pale yellow. F.P. 248°; soluble in ether, naphtha, and oil of turpentine, fixed and volatile oils, bisulphide of carbon; by long action in ammonia and alkalies; it is dissolved and decomposed by sulphuric acid with heat; decomposed by nitric acid. It is obtained in the form of a white juice from *Siphonia elastica*, *Cahouchu* in South America; *Curceola elastica* in Sumatra, Java; *Urostigma elastica Indica* and *religiosa* in India, &c. These trees, when punctured, emit a white juice which is caught in bottles. The fluid caoutchouc is similar to cream in appearance. Spec. grav. 1011·74. The juice consists of 31 caoutchouc, 56·37 water, 1·9 albumen, 7·13 peculiar azotized matter, soluble in water and alcohol. By heat and alcohol the albumen coagulates, while the caoutchouc is suspended as an emulsion. It is prepared in America for exportation by spreading thin layers on a mould and drying each deposit by a smoky fire. But if dried without smoke it is yellow instead of black or dark coloured. Caoutchouc is employed to remove stains from paper; to form elastic gas bottles, prepared from *sheet* caoutchouc by cutting from a block with a peculiar knife worked by machinery, which is kept moist with water. These sheets are most valuable for connecting pneumatic apparatus and forming rings for various purposes. They are *vulcanized* by exposing them to the action of sulphur either by immersion or in the form of vapour, sometimes previously dipping them in a salt of lead. *Vulcanized India rubber* is also made with a mixture of 25 caoutchouc, 5 sulphur, 7 white lead; the first being dissolved in some of the essential oils, and the whole ground together as pigments are. This preparation is used extensively for tubes, &c. as it is more completely elastic, stands a higher temperature without melting, and does not stiffen with cold. In *solution* in coal naphtha, produced by simple trituration, caoutchouc is applied to make cloth waterproof (Macintosh's patent).

Caoutchouc, Oil of. *Caoutchucine.* Colourless oil. Spec. grav. ·68. B.P. 93°, rising to 149°. Vapour very heavy; dissolves caoutchouc, resins, shellac, and amber. Obtained by distilling caoutchouc in iron retorts at 600°. The results of the decomposition are *caoutchene*, boiling at 58°; *heveene*, at 600°; *eupione*, at 124°; *Faradayine*, at 91° to 111°; *caoutchine*, at 336°.

Capacity. A term indicating relative amount,

as *capacity* for heat, or *specific heat.* Abroad, the expression *saturating capacity* is used in relation to acids and bases.

Caphopicrite. A name given to the bitter principle of rhubarb.

Capillary Attraction. The tendency which fluids have to be attracted by the sides of the vessels containing them, so that the surface of the fluid is concave and not horizonal.

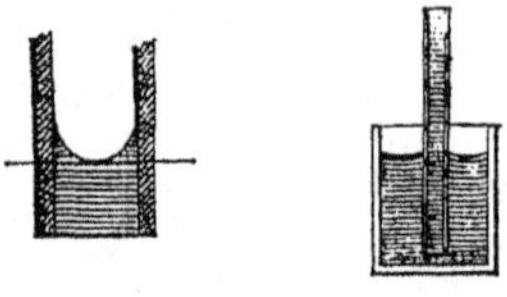

Capnite. Ferrocarbonate of zinc.

Capnomore. (καπνος, smoke; μοιρα, apart.) Colourless oil, with a smell of rum. Spec. gravity ·9775; B.P. 365°; fluid at -6°; insoluble in water, soda; soluble in alcohol, ether, and eupione; dissolves caoutchouc, fats, resins, colours; distinguished from creasote and picamar by its taste, insolubility in alkali, and dissolving caoutchouc; from eupione by spec. grav. and boiling point. Obtained from heavy oil of wood tar, along with creasote, picamar, and pittacal, by digesting with caustic potash of 1·2, heating for some time, decomposing by sulphuric acid, and distilling with potash. This process is repeated several times. The last portions distilled are capnomore.

Caporcianite. A reddish radiated zeolite, allied to chabasite, found at Caporciani, Tuscany, consisting of silica 52·8; alumina 21·7; soda ·2; potash 1·1; water 13·1; MgO ·4; Fe_2O_3 ·1; lime 11·3.

Capraldehyde, or *Capral.* $C_{12}H_{12}O_2$. A body accompanying caprone.

Capramide. $C_{20}H_{21},NO_2$. Brilliant scales by ammonia on caprate of ethyle. F.P. below 212°.

Capric Acid. $C_{20}H_{19}O_3HO$. Fine needles, smelling of sweat. Spec. gravity ·9103; F.P. 80°; it exists in butter, in oil of grain, and is formed by oxidating oil of rue, which is its aldehyde. It is prepared from butter by taking the less soluble caprate and caprylate of barytes, see BUTYRIC ACID, dissolving in boiling water, and filtering, when caprate of barytes falls on cooling in minute scales; when decomposed by sulphuric acid, the capric acid separates. Caprate of barytes in the air becomes carbonated, and gives the smell of Roquefort cheese.

Caprilyle. $C_{16}H_{15}$. A homologue of formyle.

Caproic Acid. $C_{12}H_{11}O_3HO$. Oily fluid, with a smell of sweat. Spec. grav. ·922. B.P. 390°; soluble in 96 parts water; miscible with alcohol, ether, and oils; soluble in sulphuric and nitric acids; produces a white spot on the tongue; obtained from the more soluble barytes salts of the volatile acids of butter; see BUTYRIC ACID; being less soluble than butyrate of barytes, along with which it occurs. It occurs in the oxidization of albumen, oleic acid, and œnanthole, and by treating capronitryle with an alcoholic solution of potash; caproates of ethyle, and methyle, have a pine-apple odour.

Caprone. $C_{11}H_{11}O$. A fluid obtained by distilling caproate of barytes.

Capronitryle. $C_{12}H_{11}N$. By heating potash-sulphate of amyle with cyanide of potassium.

Caprotylamine. $C_{12}H_{15}N\,? = C_{12}H_{13}$ NH_2? Probably in Dippel's animal oil.

Caprotyle. $C_{12}H_{13}$. Obtained by the galvanic decomposition of œnanthylic acid.

Capryle. $C_{20}H_{19}OHO$. The chief constituent of oil of rue.

Caprylic Acid. $C_{16}H_{15}O_3HO$. A fat crystallizing in needles below 50°, giving an odour to butter, obtained from the mother liquor of caprate of barytes, the caprylate of barytes being most soluble; it is decomposed by sulphuric acid.

Caprylone. $C_{15}H_{15}O$. Formed by distilling caprylate of barytes.

Capsicine. Crystalline, or a yellow or red-brown soft substance, like an oil or resin, evaporating when heated, and exciting cough and sneezing; slightly soluble in water and vinegar; very soluble in alcohol, ether, oil of turpentine; combines with barytes; obtained by heating the alcoholic extract of *Capsicum annuum* with ether.

Caput Mortuum Vitrioli. Colcothar. The red powder left by calcining sulphate of iron.

Caramel. $C_{12}H_9O_9$. Black powder, soluble in water, without imparting a taste; not fermentable; unites with lead; forms the basis of sugar; insoluble in alcohol; obtained by heating sugar to 410° to 430°; it is formed in the roasting of coffee and chicory, which contain sugar: it is used to colour wines, &c. yellow.

Carapine. A bitter white alkaloid (?) from the bark of *Carapa guianensis.*

Carat. A weight employed in estimating the value of the diamond, precious stones, and gold. 156 carats = 1 troy ounce, or 1 carat = 3·077 grains. Gold is estimated, as to purity, in carats; thus, in gold of 23 carats fine, it is meant that (24 carats indicating perfectly pure gold) in 24 carats of the sample 23 are pure gold.

Caraway, or *Carum Carui Oil,* is obtained by distilling the seeds of this plant. Spec. grav. ·938. B.P. 401°. See CARVENE, CARVACROL.

Carbamic Acid. *Bicarbamide.* C_2O_3 NH_2. An acid amidide.

Carbamate of Methyle. *Methylane.* C_2H_3 $O,C_2H_2O_3$. Fine crystals, by acting on pyroxylic spirit with chlorocarbonic acid, and then on this oily product with ammonia.

Carbamate of Amyle. *Amylo-urethane.*

$C_{10}H_{11}O,C_2NH_2O_3$. Crystals by the action of phosgene gas on amylic alcohol, and the addition of ammonia.

Carbamide. $CONH_2$. White volatile crystals by phosphene gas and ammonia.

Carbamide-Carbanilide. *Carbanilamide.* $C_{12}H_7N,HC_2NO_2$. Needles by the action of chloride of cyanogen in presence of water.

Carbamide—Nitrocarbanilide. $C_{14}H_7NO_4N_2O_2$. Yellow needles, by treating nitraniline, chloride of cyanogen, and basic dinitromelaniline, together.

Carbamide of Naphthalidine. $C_{21}H_8NO$. White body, distilling at 300°C.; obtained by distilling oxalate of naphthalidine.

Carbanilamide. See ANILAMINE UREA.

Carbanilic Acid. $C_{12}H_7N, C_2O_4$. Orange crystals by boiling carbanilamide with potash.

Carbanilide. $C_{12}H_6N$, CO. Satin needles, by phosgene gas on aniline.

Carbanilethane. *Carbanilate of ethyle.* $C_{18}H_{11}NO_4$. A liquid, by sulphide of ammonium on nitrobenzoate of ethyle.

Carbanimethylane. $C_{16}H_9NO_4$. Similar to the preceding, substituting the methyle salt.

Carbazotic Acid. See PICRIC ACID.

Carbo-methylo-vinide Bisulphide. *Sulphuretted Carbonate of Ethyle and Methyle.* $C_8H_8O_2S_4$. Yellow fluid, by distilling xanthate of potash with potash sulphate of ethyle and methyle.

Carbo-vino-methylide. *Carbonate of Ethyle and Methyle,* by distilling potash oxalate of ethyle and methyle with compounds containing potash sulphate of ethyle.

Carbohydrogens. A term applied to a series of compounds, in which the amount of carbon and hydrogen in each differs by an equal number of atoms, or by a multiple number, and on this account they are also said to be homologous. Their physical characters are likewise characterized by their amounts of carbon and hydrogen present, which affect particularly their boiling point. The same relation often exists in the boiling points of the acids derived from them.

	Boiling Point.
Pyroxylic spirit,.....$C_2H_4O_2$	138° ·2
Alcohol,..............$C_4H_6O_2$	172° ·4
Potato spirit oil,.......$C_{10}H_{12}O_2$	275°
Formic acid,..........210° ·2 }	34° ·2
Acetic acid,..........244°·4 }	
Valerianic acid,...... 347·	102·6

Every 2 atoms of CH raising the boiling point 34°·2.

Carbolic Acid. *Phenole, Hydrous Oxide of Phenyle.* $C_{12}H_5O$, HO. Colourless oil, or needles, sinking in water. Spec. grav. 1·062, 1·065; B.P. 368°; taste caustic and burning; acts on the skin; closely resembles creasote, but differs in being an acid, in being precipitated by acetate of lead, in not being altered by ammonia or the air, and in being changed by nitric acid into a red-brown matter; a splinter of fir dipped in carbolic acid and then in nitric acid becomes blue and brown; its salts are crystalline. Obtained from that portion of coal oils boiling or distilling between 300° and 400°; agitating this portion with twice its vol. of caustic potash, and decomposing by muriatic acid. Carbolic acid may be viewed as the alcohol of phenyle and chlorophenesic ($C_{12}H_3Cl_2O$, HO), and bromophenesic acids, as bodies in which chlorine, &c. replace hydrogen. Aniline ($C_{12}H_5NH_2$) is the amide of phenyle. This view connects the products of salicyle, indigo, &c.; it is also obtained by distilling *Castoreum canadense*; it is poisonous; it may be used like creasote in the treatment of toothache.

Carbon. C. ·75; 6·. Carbon, in the form of wood charcoal, was undoubtedly known in the earliest ages, from its importance as a fuel, and for the reduction of ores. The term coals in Isaiah (liv. 16), 712 years before Christ, obviously refers to charcoal; for although coal is said to have been recently discovered in Lebanon, it was quite unknown in the earlier ages of the world. The term Carbo, used by the Latin writers (Horace, Suetonius), refers to the same substance, since it was burned in the centre of the room, the Romans having no chimneys. Carbon in various forms occurs widely distributed in nature. It is presented in the purest form in the *diamond*, *coal gas carbon*, and more impure in *plumbago*, *anthracite*, *wood charcoal*, *animal charcoal*, *lamp black*, and *pitcoal*. In union with oxygen, in the state of carbonic acid, carbon constitutes an important ingredient in limestones; and it enters as a chief element into the formation of all vegetable and animal substances. According to Sorby (Brit. Assoc. Rep. 1850, 62), carbon crystallizes in four forms, or is tetramorphous. *Coke* has the regular form, but has a different spec. grav. 1·891 from diamond; *anthracite* (1·784), charcoal, and lamp black are square prisms, having nearly the same specific gravity; plumbago 6-sided prisms (2·3).

Carbonic Acid. *Gas Sylvestre, Spiritus Sylvestris, Fixed Air* (Black, 1754, Thesis), *Calcareous acid* (Keir), *Aerial acid* (Bergman, 1774, Opusc 1, 50), *Mephitic acid* (Bewdley), *Carbonic acid* (Lavoisier, 1781.)

	Vols.	Sp. Grav.	Atoms.	Wts.	Per Cent.
Carbon,	□ 1	·4234	1	·75	27·69
Oxygen,	□ 1	1·10563	2	2·	72·31
		1·52903		2·75	100·

Calculated sp. grav. =·55281×2·75=1·52023, weight of 100 cubic inches, that of air being 30·9451 grs. (Regnault) 47·3151 grs. Refracting power 1·526.

Source.—Carbonic acid exists in atmospheric air to the extent of $\frac{4}{10000}$ of its bulk; in calcareous districts it occasionally issues from the earth in considerable quantities, as in the Grotto del Cane at Naples, in the Upas Valley in Java, in the Pyrmont and Nassau mineral springs, &c.; and it exists in well waters.

Characters.—It possesses the mechanical properties of common air, and is destitute of colour; and when brought in contact with the nostrils a sharp pungent sensation is produced, similar to the effect produced by fermenting liquors. The *taste* is distinctly sour. It has no effect on aqueous infusion of red cabbage, but it imparts to infusion of litmus a purple-red colour, an effect, however, which speedily disappears by exposing the solution to the open air; the carbonic acid escapes, and the litmus resumes its original tint (Hey, 1772). From the high specific gravity of carbonic acid, it may be poured from a jar like water on a candle so as to extinguish it. Carbonic acid

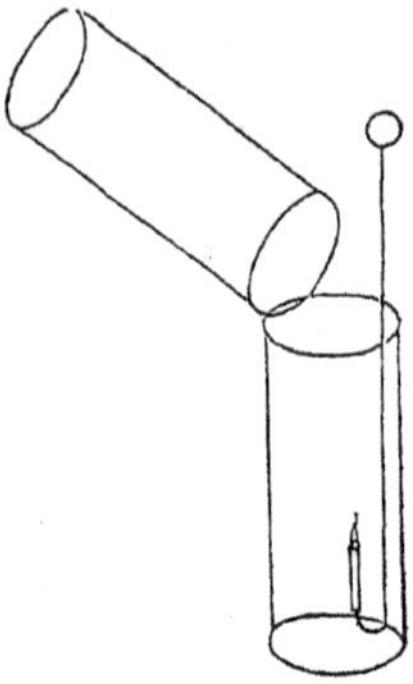

is incapable of supporting combustion, or of burning. Hence, when carbonic acid is poured on a taper, the light is extinguished, while with oxygen the flame is enlarged. It is also irrespirable, and speedily asphyxiates any animal which may be immersed in it; but when the animal is subjected to long respiration in an atmosphere containing much carbonic acid, poisonous effects are supposed to be produced. If, however, we introduce an animal into an atmosphere of this gas for a second or two until it appears to be just insensible, and then plunge it into cold water, it instantaneously recovers its sensibility, and if a bird it will fly from the hand. I was induced to try this experiment from a remark made by that quaint old Scottish traveller Lithgow, in 1609, in reference to the Grotto del Cane at Naples. This cave is situated in a limestone rock, from which carbonic acid is copiously evolved at the bottom of the cave. The gas, in consequence of its weight, remains principally at the lower part, so that a man entering, having his head above the region of the carbonic acid, is not affected, while a dog is speedily destroyed if not removed in time. Describing the occurrences at that cavern, Lithgow says, "The dog-keeper for an easy composition made trial of his dog, and having tied a string to his hinder leg, he cast the dog scarce half-way in the cave, where immediately his tongue hanging out he fell down dead, and forthwith his master repulling him back recast him in the lake, pouring in water in his ears; but he could never recover his life; whereupon the poor man cried out, 'Alas, I am undone! what shall I do, the dog that won my daily bread is dead.'" The oriental fable of the Upas Valley of Java is founded upon the fact of the evolution of large quantities of carbonic acid in a deep dell in that island. Travellers describe the neighbourhood of this spot as being strewed at intervals with the osseous remains of man and animals. The spirit of eastern romance had in former times ascribed the source of this mortality to the upas tree, which is still celebrated for the poisonous properties of its juices. But it is now ascertained that the lower strata of the atmosphere of the Upas Valley are contaminated with carbonic acid from a constant source, to such a degree as in some measure to afford countenance to the poetic tale. (Information communicated to me by C. Millet, Esq.) The destructive influence of carbonic acid upon animals may be taken advantage of by entomologists for killing butterflies and insects for preservation. It has been found that air containing from 5 to 6 per cent. of carbonic acid extinguishes a candle, and that life may continue in this atmosphere; but that a warm-blooded animal respires with great difficulty, and is seriously affected. In mines it is usual to test the salubriousness of the atmosphere by lowering a lighted candle or lamp. If the lamp is extinguished, the conclusion is drawn that it is dangerous for a human being to enter such a deteriorated atmosphere, while, if the taper continues to burn, it is inferred that the miner may proceed to his work in safety. It has been found that on some occasions a human being has been able to respire where a candle would not burn, but these were accidental cases, which are usually attended by fatal consequences. As plants derive their carbon from the carbonic acid of the air, its presence is essential to vegetation, which thus assists in purifying the air. According to Peclet (Ann. Chim. 5, 243), a ward containing 50 patients remaining shut for eight hours, should have a capacity of 50 cubic metres (1763·87 cubic feet) for each individual during the night. The actual quantity which we find allowed in hospitals is under this amount. Thus at the Salpetriere the ration of air granted to each patient is only $1\frac{1}{2}$ cubic metre (52·916 cubic feet) per hour. How much inferior it must be even to this in the houses of some of our ill-ventilated streets it is scarcely necessary to urge. Carbonic acid is copiously evolved in distilleries and breweries in fermentation, and hence the danger of the atmosphere over a tun.

Decomposition.—Carbonic acid is not decomposed by being passed through an ignited porcelain tube. It is partially decomposed by having electric sparks passed through it, into carbonic oxide and oxygen (Henry), but when the residual carbonic acid is removed by potash, the

CAR

carbonic oxide and oxygen unite by an electric spark. Equal volumes of carbonic acid and hydrogen, when passed through a red hot porcelain tube, are converted into carbonic oxide and water. When passed in a dry state over ignited potassium, it decomposes, charcoal being deposited (H. Davy). Phosphorus and boron have a similar action (S. Tennant, Gay Lussac.)

Preparation.—A convenient apparatus for the

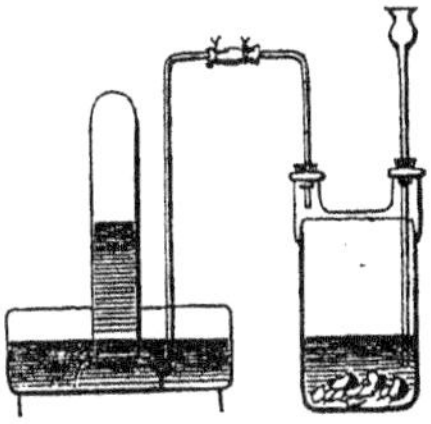

preparation of carbonic acid is a Wolfe's bottle and tube, by which the acid may be introduced from time to time; a number of pieces of marble are introduced into the bottle covered with water, and the acid is poured in by the funnel tube. See PREPARATION OF GASES.

Carbonic Acid in Water.—Carbonic acid is capable of being dissolved to a considerable extent in water—100 vols. absorbing 107 vols. of gas—or water may be said to absorb nearly its own volume of carbonic acid, and Dr. Henry (Phil. Trans. 1803) has shown as a general expression, that water takes up the same volume of compressed carbonic acid as of gas under ordinary pressure. Thus if water under common circumstances takes up an equal bulk of carbonic acid, under the pressure of two atmospheres, it will absorb twice its bulk; under 3 atmospheres 3 times its bulk, and so on: so that by increasing the pressure we may cause water to absorb any amount of carbonic acid. Water thus impregnated is freed from the carbonic acid which it may contain by being placed under the exhausted receiver of an air pump—by boiling, by freezing, or in a great measure by the solution of sugar and salts in the water. It is to the presence of carbonic acid in beer, soda water, champaigne, that the pleasant briskness and sparkling character of these beverages is due. Mineral waters, such as those of Selters and Seidlitz, owe their repute to the presence of a large quantity of carbonic acid in solution, which has been obviously condensed by pressure under the surface of the earth. Imitations of such natural products are, perhaps, of a more agreeable nature, and are familiar to us under the titles of soda water and lemonade. There are various methods of impregnating water with carbonic acid. The apparatus of Nooth is one of the simplest; but can only be employed on a limited scale. For preparing soda water in the large way, the soda water machine is used, of which there is considerable variety. Soda water was originally prepared with carbonate of soda and sulphuric acid, and hence originated the name. It does not now necessarily contain soda, although a small portion of the carbonate of soda is generally added. It is now usually manufactured from chalk or limestone and oil of vitriol, in consequence of the economy of the materials. Although chalk or marble is not soluble in water, yet it can be dissolved to a certain extent in water impregnated with carbonic acid. It is in this way that magnesia is rendered soluble, and sold in the form of liquid magnesia, as it is termed. The cause of this depends on the formation of a bicarbonate of the base, according to the following scheme:—

$$\left.\begin{array}{l} CaO\ CO_2 \text{ insoluble in } HO \\ \quad CO_2 \end{array}\right\} \text{Sol. in HO.}$$

When water containing carbonate of lime in solution is boiled, a deposit of chalk falls, which, with sulphate of lime deposited by the evaporation of the water, gives steam boiler deposits, which may be partially redissolved by muriate of ammonia.

Apparatus for the Preparation of Soda Water.—Fill B from A with water till nearly full,

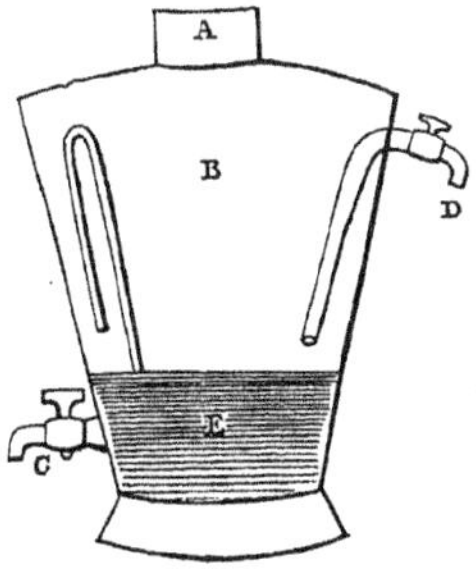

and screw on A again tight. Unscrew C and fill E nearly full with water; put in the bicarbonate of soda and tartaric acid, and screw on C tight, and the machine is then charged. If placed in a warm situation, the action will be instantaneous. A little wine or raspberry vinegar makes a delicious drink when mixed with the water put into the vessel. The proportions to be used of bicarbonate of soda and tartaric acid are the same as in an effervescing soda powder. See MINERAL WATERS.

Liquid Carbonic Acid.—This gas assumes the form of a liquid when subjected to sufficient pressure and cold (Faraday). Sulphuric acid and carbonate of ammonia were introduced into a strong glass tube, bent at an angle, the ingredients being separated by platinum foil. The tube was then hermetically sealed. The acid was made to act upon the carbonate of ammonia, and the gas evolved was condensed on the clean end of the tube, which was immersed in a freezing mixture.

Obtained in this way, carbonic acid is a colourless liquid, very fluid and light, with a refracting power inferior to water. At 32° its elasticity is equal to 36 atmospheres, specific gravity 0·838.

Solid Carbonic Acid.—Thilorier has succeeded (Ann. Chim. 60, 427) by imitating in the first instance the experiment of Faraday—using iron cylinders instead of glass—in condensing carbonic acid into a liquid, and then by an ingenious method of causing it to revolve when emitted by the stop-cock round a hollow hemisphere, it vapourizes and produces so much cold as to cause it to condense into a solid form like snow.

A more safe and convenient method is to prepare the fluid acid by means of a pump.

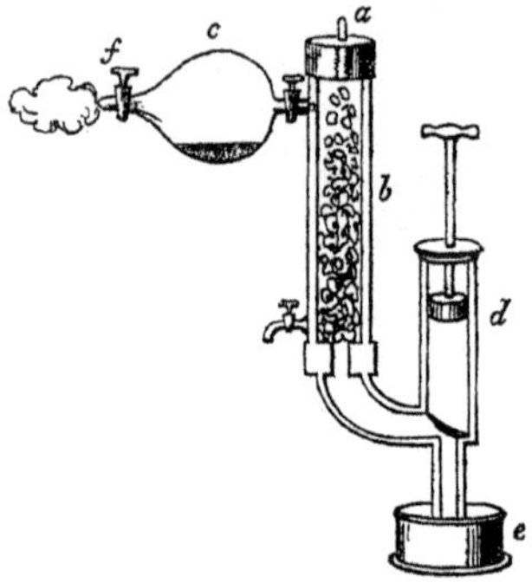

A strong cylinder of iron *b*, lined with lead, is filled with pieces of marble through an aperture *a* at the top. A force pump *d* draws dilute chlorohydric acid from *e* in contact with the marble. A stop-cock allows the evolved carbonic acid to pass into the strong iron receiver *c*, where it condenses into the fluid state. On opening the stop-cock, the fluid passes in the form of vapour. When received in Thilorier's receiver it is solid.

The temperature of this solid cannot be less than —148°. The effect of touching this cold mass with the tongue is remarkable. A sensation of burning is produced, and a blister results; when mixed with a small portion of ether and mercury, the metal may be frozen in considerable quantity. The elasticity of the vapour of carbonic acid as determined by Mr. Addams is as follows:—

Temperat.	Lbs. per square inch.	Atmospheres of 15 lbs. each.
0	280	18·1
10	300	20
30	398	26·5
32	413	27·6
50	520	34·7
100	935	62·3
150	1496	99·7

Liquid carbonic acid is capable of mixing in all proportions with oil of turpentine, bisulphide of carbon, alcohol, ether, naphtha; water and fatty oils are, however, incapable of dissolving it. When in union with alcohol, the compound may be frozen; it melts at —135, probably the lowest temperature ever measured by the thermometer. Spec. grav. at—20° C. is ·90 at 0° C. ·83 at + 30° C. ·60.

Analyses of Carbonates.—In a quantitative analysis of carbonates, it is usual to estimate the amount of carbonic acid by the loss sustained when a carbonate is acted on by a stronger acid, as sulphuric or chlorohydric acids; the former, from its fixity, is preferable in all cases when the resulting sulphate is soluble in water. Introduce into a light flask about 150 grs. of oil of vitriol, and dilute it with an equal bulk of water; weigh out in a small tube 50 grs. of the carbonate; suspend it by a thread in the flask, and close the mouth of the flask by a perforated cork and tube filled with chloride of calcium. Then weigh accurately the whole apparatus; allow the tube to descend into the acid, cork tightly, and gradually by agitation allow the acid to act upon the carbonate in the tube. On the cessation of effervescence, heat the flask gently; release the cork slightly, apply a suction tube to the top of the chloride of calcium tube, and suck out the carbonic acid; lastly, weigh the apparatus. The loss indicates the amount of carbonic acid in 50 grs. of the carbonate.

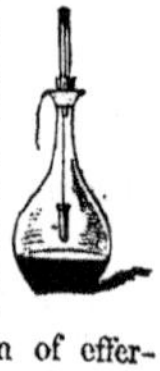

Carbonic Oxide.

	Vol.		Spec. Grav.	Atoms.	Wts.	
Carbon,	1	□	·4234	1	·75	43·38
Oxygen,	½	□	55281	1	1·	56·62
			·97621		1·75	100·

Calculated spec. grav. = 55281 × 1·75 = .9674; refractive power 1·157 (Dulong, Ann. Chim. 31, 166); weight of 100 cubic inches, that of air being 30·9451 grs. = 30·2089 grs. *Characters.*—Carbonic oxide is a colourless gas, and possesses the mechanical properties of common air; destitute of taste and smell. It possesses neither acid nor alkaline action on vegetable colours, and is a neutral body. It has not been condensed into a liquid; 100 vols. water absorb 2 of this gas (Davy, Thomson), 3·7 (Dalton), 6 (Saussure). Cold carbonic oxide when ignited gives a blue flame, but when previously heated a yellowish-red flame (Gmelin). In the formation of ball soda, the carbonic oxide evolved in the furnaces gives a yellow flame, due probably to the soda. Animals cannot exist in it, but when immersed in it are immediately asphyxiated. It is one of the most poisonous gases known (Leblanc) when existing in atmospheric air. All gases containing carbon have been found more or less to interfere with the natural physiological state when respired. When mixed with common air it does not explode like hydrogen, but burns brilliantly.

A mixture of 2 vols. of this gas with 1 of air explodes feebly by an electric spark, or when ignited by a red hot wire. When 1 vol. of carbonic oxide is fired with ½ vol. of oxygen, the mixture is converted into 1 vol. carbonic acid; equal vols. of carbonic acid and carbonic oxide contain the same amount of carbon, but in carbonic oxide the oxygen is only half of that in carbonic acid. A mixture of 1 vol. oxygen and 2 carbonic oxide, when exposed to the action of spongy platinum at the common temperature, is gradually converted into carbonic acid; while, if the heat be raised to 320°, the combination occurs immediately. When carbonic oxide is added in equal proportions to an explosive mixture of oxygen and hydrogen, it prevents the latter from exploding in presence of spongy platinum; but the gases still act on each other slowly; carbonic acid and water being formed. If the quantity of carbonic oxide is small, detonation occurs on the introduction of the platinum (Henry, Phil. Trans. 1824, 271). Equal vols. of carbonic oxide and hydrogen passed through a red hot tube, give origin to water and the deposition of charcoal. Potassium heated also decomposes it. *Preparation.*—1. If carbonic acid gas be transmitted over red hot charcoal in an iron or porcelain tube, the gas passing out at the opposite extremity of the tube is twice as bulky as before transmission, and instead of extinguishing a taper, it burns with a blue flame. In this process the second atom of oxygen in the carbonic acid unites with an atom of carbon from the charcoal, and there are 2 atoms of carbonic oxide produced. Half the oxygen may be removed by passing carbonic acid over ignited iron filings ($FeCO_2 = FeO,CO$).

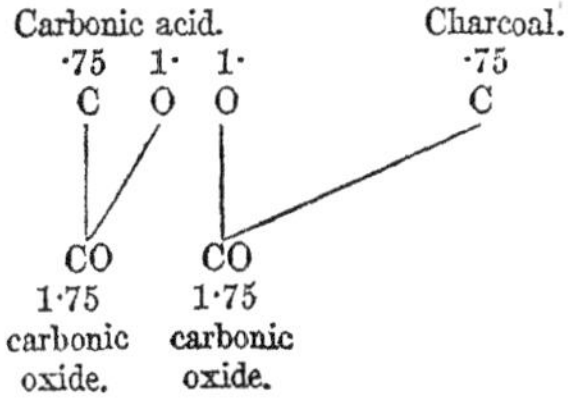

2. When oxide of zinc and charcoal are ignited in an iron bottle, or gun barrel, carbonic oxide is produced (Priestley, 1800). In the same way, oxides of iron, lead, or copper may be employed; but by this process carbonic acid is likewise evolved:—

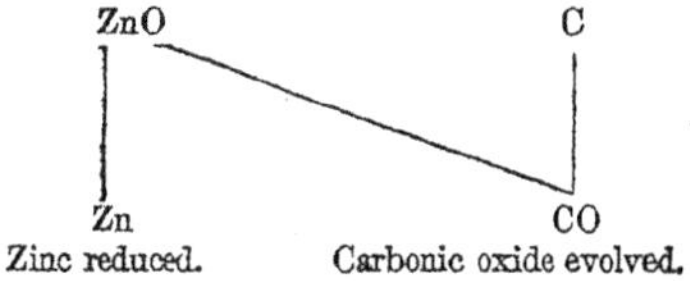

3. When 3 parts of carbonates of lime, barytes, or strontian are ignited in an iron bottle with 1 part of charcoal, carbonic oxide is evolved in abundance ($Ca\ OCO_2$, $C{=}Ca\ O$, $2\ CO$), and the caustic earths are left. Iron filings may be substituted for the charcoal ($CaOCO_2Fe = CaO$, CO,FeO). Any carbonic acid present may be removed by lime water, or milk of lime. 4. When oxalic acid and oil of vitriol are heated in a retort or flask, the acid removes water from the oxalic acid, and equal volumes of carbonic oxide and carbonic acid are evolved, from which the carbonic acid may be removed by an intermediate vessel containing milk of lime, or caustic soda, or by standing over water, if required; but without this precaution the blue colour of the oxide flame can be distinguished (Dobereiner, Ann. Chim. 1821, 19, 83).

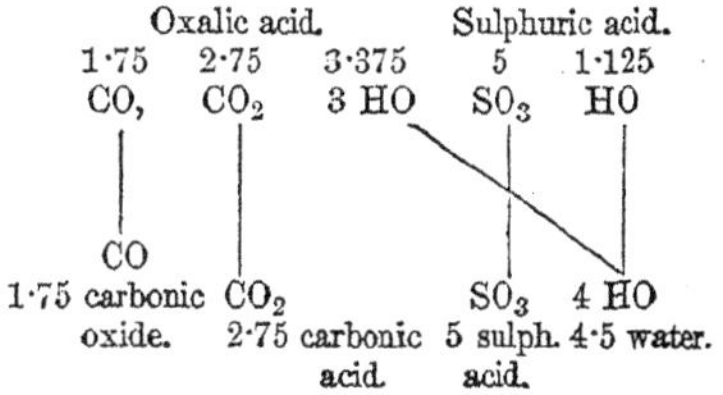

Binoxalate of potash, or oxalate of ammonia, may be substituted for the oxalic acid. The latter yields prussic acid also. (?) In these processes the carbonic acid may be absorbed in an intermediate vessel, containing milk of lime or

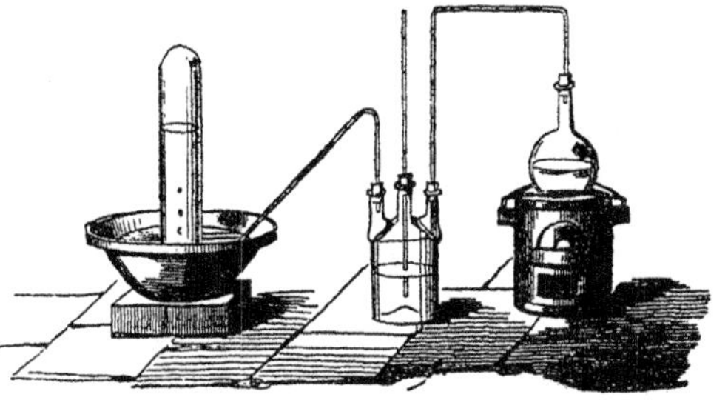

caustic soda (ib. 84). Dumas recommends the binoxalate (An. Chim. 33, 110, 1826). 5. Another method pointed out by Dr. T. Thomson (Ann. Phil. 12, 103, 1818, Inorgan. Chem. 1831, 2, 251) of procuring this gas is to pour sulphuric acid upon the ferrocyanide of potassium (yellow prussiate), and apply heat. Carbonic oxide passes over in abundance. *Practical Applications.*—Carbonic oxide is formed in many cases of intense ignition, in presence of carbon, at the expense of carbonic acid, as in lime kilns, where it may be distinguished by its delicate blue flame. It appears also when glowing coke or charcoal is burned in a furnace. The blue flame in these cases, which is popularly attributed to sulphur, is due to carbonic acid generated in the lower strata in contact with the air passing through the ignited coal. In smelting iron by the hot blast process, hot atmospheric air is brought in contact under great pressure with coal in a state of incandescence. The formation of carbonic acid is the first result. The gas passing upwards through the heated mass of carbonaceous matter, takes up an additional dose of carbon, and is converted into carbonic oxide.

Carbuncle. The ancient name of the precious garnet.

Carburets, Carbides. Compounds of carbon with the metals and simple bodies.

Carburetted Hydrogen. *Fire Damp* (1640), *Heavy Inflammable Air* (Priestley), *Marsh Gas*, *Light Carburetted Hydrogen*, *Bihydroguret of Carbon*, *Dicarbide of Hydrogen*, and *Proto-carbonated Hydrogen*.

	Vols.		Spec. Grav.		Ats.	Wts.	P.ct.	Wt. of 100 c. i.
Carbon,	1	□	·4234	·4166	1	·75	75	= 17·1747 grs.
Hydrog.	2	☐	·1384	·1384	2	·25	25	
			·5618	·5550		1·	100	

Experimental specific gravity ·5555 (Thomson); ·556 (Henry). Symbol, CH_2 = 1· or 8. Theoretical spec. grav. ·5528 × 1 = ·5528. From the want of knowledge of the exact specific gravity of the vapour of carbon there is a discrepancy between the per centage composition deduced from the specific gravity and from the atomic weight. *History.*—This gas is mentioned in the Phil. Trans. (1677) as having been observed in mines as early as 1640. Dr. Franklin, in a letter to Priestley, in 1774, mentions that in America it had been obtained by stirring stagnant pools with a stick, and that it had been fired. Dr. Priestley was previous to this unacquainted with it (On air, 1, 321). Dr. Dalton and Dr. Thomson first accurately examined it (Wernerian Trans. 1811, 1, 506). *Characters.*—A colourless gas, possessing the mechanical properties of common air; destitute of taste and smell; refracting power 1·504 (An. Chim. 31, 266). According to Perkins, it liquefies under a pressure of 1200 atmospheres (Phil. Trans. 1826, 544). It is a neutral body, since it does not change vegetable blues, and forms no compounds with acids or bases. It burns with a yellow flame; and exists in coal gas, although its illuminating power is contested; it requires a white heat to inflame it. This is illustrated by the well-known fact of steel mills being formerly employed in mines, which threw off red hot sparks sufficient to supply light, but not to explode the gas. It is upon analogous principles that the Davy lamp is constructed. Davy found that if he ignited this gas at one extremity of a very narrow metallic tube, and if at the other end there was an explosive mixture of carburetted hydrogen and air, the interior of the tube being quite free, the flame never could pass along the tube to the explosive mixture, but was so cooled that the temperature always subsided far below the degree required to ignite the explosive mixture. He found that wire gauze might be substituted for tubes; and hence that wire gauze is a network of capillary metallic tubes. If a lamp be constructed of wire gauze, the carburetted hydrogen may pass from the exterior air through the gauze into the lamp, and be ignited, but the flame will not pass from the interior of the lamp to the external air so as to ignite an atmosphere of carburetted hydrogen. The influence of wire gauze in checking the progress of flame may be illustrated by taking a piece of gauze, and holding it across a candle or lamp, when it will be found that the flame will not perforate the metallic meshes. If held over a gas lamp before ignition, so as to allow the gas to stream through the interstices, and if the gas be then lighted, the flame will be found never to descend below the gauze, unless the gas be forced

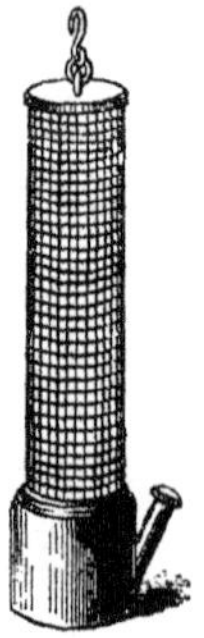

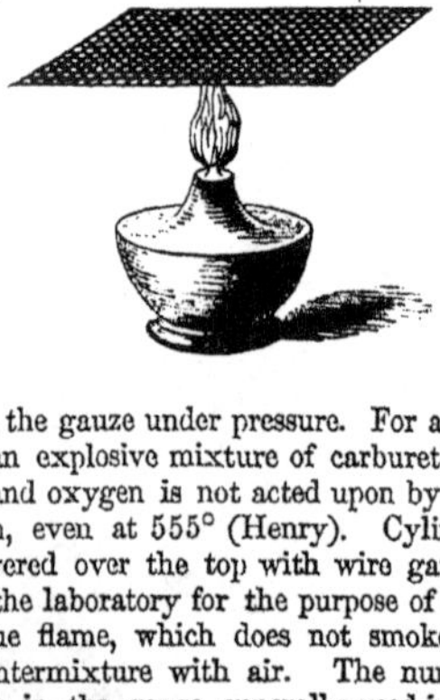

through the gauze under pressure. For a similar reason an explosive mixture of carburetted hydrogen and oxygen is not acted upon by spongy platinum, even at 555° (Henry). Cylinders of iron, covered over the top with wire gauze, are used in the laboratory for the purpose of obtaining a blue flame, which does not smoke, by a proper intermixture with air. The number of apertures in the gauze generally used in Davy lamps amounts to from 625 to 800 in the square inch. The principal accidents have occurred in the use of the Davy lamp not from any error in its principle, but from the carelessness of the miners. The security of the lamp is increased by immersing the gauze in caustic alkali. When an animal is immersed in this gas, insensibility is produced. "I once accidentally inhaled it; being anxious to empty a gas-holder filled with it as quickly as possible, I applied my mouth to the stop-cock to draw it out. After two or three inhalations, I fell insensible on the floor, but recovered again in about twenty minutes, without feeling any bad effects whatever, except a slight headache, which soon went off" (T. Thomson). Accidents of this kind frequently occur in gas works, but seldom prove fatal. Carburetted hydrogen is, therefore, an anæsthetic. When passed through chlorochromic acid, it carries the latter with it in vapour, and when fired, sesquioxide of chromium is deposited on a funnel held over the flame. *Preparation.*—1. Carburetted hydrogen may be procured by inverting a bottle or jar filled with water in a stagnant pool, supplied with an inverted funnel, and stirring under the mouth of the bottle the bottom of the pool with a stick. The gas bubbles up into the bottle. It

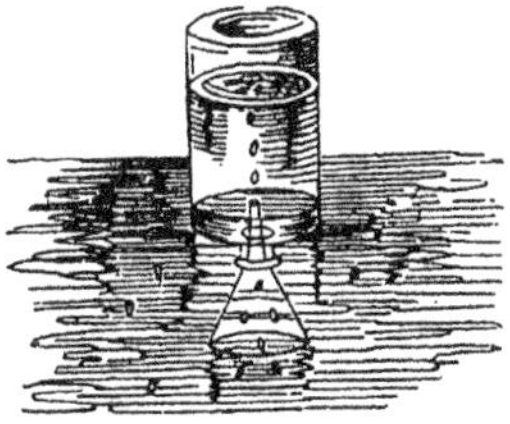

contains from this source some carbonic acid, which may be removed by caustic soda, or lime water, or prolonged washing with water. 2. It may be obtained by passing the vapour of alcohol through a red hot porcelain tube; carburetted hydrogen and carbonic acid pass over. 3. When coal is ignited in a close vessel the principal products are tar, water, &c.; and gases not absorbed

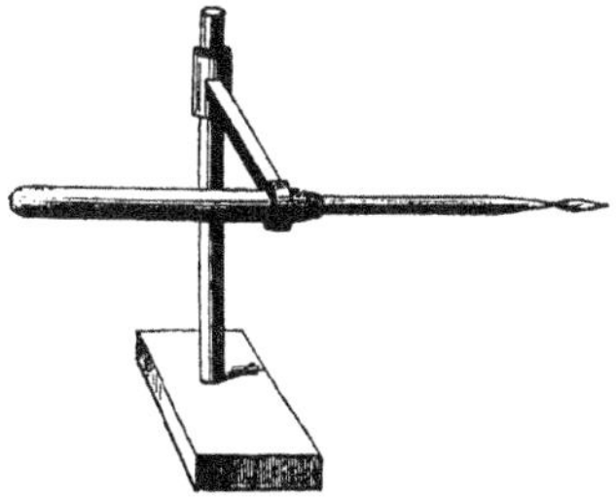

by water containing 70 per cent. of carburetted hydrogen. The experiment may be made as in the prefixed figure. A tube closed at one end is half filled with pounded coal, and united to a tube with a capillary opening. Heat being applied, the gas passes out and may be collected under water or ignited. It varies in quality according to the heat applied, and to the quality of the coal. When the gas is passed through a red hot tube, it is obtained more free from foreign mixture. 4. Acetate of potash and soda when heated with hydrate of potash or barytes yield carbonic acid which unites with the base, and carburetted hydrogen which may be collected pure (Perzoz). $NaO, C_4H_3O_3HO + NaO$ become $NaO\ CO_2NaO\ CO_2$ and $2\ CH_2$, or 4 parts acetate of soda, 40 solid potash, and 60 quicklime, heated in a glass retort, yield the same gas (Dumas). The soda is dissolved in a small quantity of water, and the lime is added in powder so as to form a paste. 5. This gas is found abundantly in deep coal mines, from whence it sometimes escapes by an aperture in the earth, as at one time at Bedlay, near Cumbernauld; near Lake Erie; at Bakou in Persia; perhaps in Lycia giving rise to the chimæra of the ancients. It frequently accompanies artesian springs, as at Dalmarnock. *Composition.*—When mixed with oxygen and fired by electricity, carburetted hydrogen explodes if the bulk of the oxygen prevails over that of the gas, and the detonation is prevented if the amount of oxygen surpasses that of the gas two and a-half times. When mingled with common air in the proportion of 1 gas to 7 or 8 common air, the mixture is capable of exploding most powerfully. With 3 or 4 times its bulk of air it does not detonate, and with 5 or 6 times its bulk it explodes in a slight degree. In any proportion between 7 and 14 times its bulk of common air it explodes. To burn it completely, 2 vols. of oxygen are required for every vol. of carburetted hydrogen, when it is resolved into its own volume of carbonic acid and into water. The action when fired with oxygen, will be as represented, ½ vol. of oxygen and 1 of hydrogen forming water—

Carbon.	Hydrogen.	Oxygen.	
C	HH	OO	By vols.
C	HH	OO OO	By atoms.
CO_2		2 (HO)	
Carbonic acid.		Water.	

When mixed with chlorine and water and exposed to the action of light, the carburetted hydrogen is gradually converted into carbonic acid, while the chlorine unites with the hydrogen and forms chlorohydric acid; 4 vols. chlorine are required to decompose 1 vol. carburetted hydrogen; 2 vols. hydrogen are supplied by the carburetted hydrogen, and 2 vols. by the water ($4\ Cl,\ CH_2\ 2\ HO = CO_2\ 4\ HCl$); an acid fluid is thus formed (Cruickshanks). With 3 vols.

chlorine a violent explosion occurs, chlorohydric acid being formed and carbon deposited. When the chlorine is allowed to act less powerfully by diluting it with its bulk of carbonic acid, and is passed through tubes into carburetted hydrogen, an oily liquid, bichloride of carbon (CCl_2 or C_2Cl_4), is formed, and by the continued action of chlorine the following series is obtained from pyroxylic spirit, which may be viewed as 2 atoms carburetted hydrogen (Dumas):—

Carburetted hydrogen, or methylene, C_2H_4
Chloride of methylene,.................. C_2H_3 Cl
Bichloretted methylene,.............. $C_2H_2Cl_2$
Terchloretted methylene or chloroform,.................................. $C_2H\ Cl_3$

Chloride of carbon,..................... C_2Cl_4

Applications.—As this gas is very prejudicial in coal mines, not only from the danger of explosions which its mixture with the atmosphere is apt to produce, but also from its unwholesome nature in respiration, it has been recommended that it should be absorbed by means of bleaching powder. According to Dumas, if 6 or 8 lbs. of chloride of lime be spread at night in an enclosed space containing the gas, in the morning the whole of the noxious gas will be found to have disappeared. But the true mode of removing all these dangerous agents to human life is by a proper system of ventilation in mines; because a Davy lamp, although it may prevent the poor miner from being instantly destroyed by an explosion, does not afford any precaution against a slow death from improper oxidation of the blood by the constant respiration of an atmosphere foreign to the animal system, instead of a pure mixture of oxygen and nitrogen in the natural proportions. For the purpose of drawing out the bad air mechanically, and thus properly ventilating the mine, various contrivances have been recommended. The method of Desaguiller is to extract the foul air

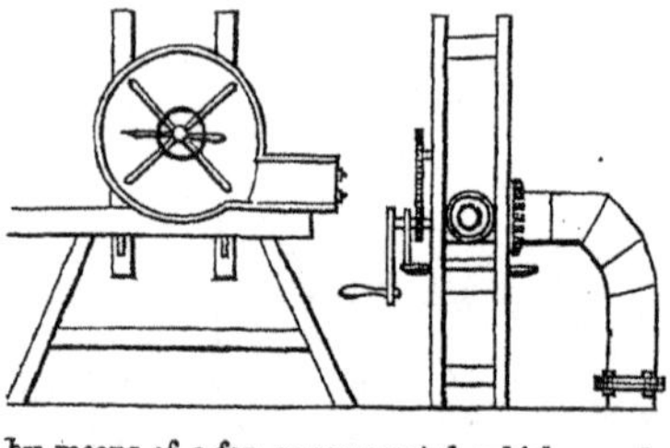

by means of a fan as represented, which may be driven either by the hand or by machinery.

Carbothialdine. $C_5H_5NS_2$. Brilliant crystalline base, by the action of bisulphide of carbon on an alcoholic solution of aldehyde ammonia.

Carbyle. CH. A base supposed to exist in the combination of sulphuric acid, when anhydrous sulphuric acid is allowed to be absorbed by absolute alcohol; sulphate of carbyle results = CH SO_3.

Carbylosulphuric Acid. *Aethionic Acid.* S_4, C_4 H_5, O_{13}. An acid liquor obtained by dissolving sulphate of carbyle in water. The aethionic acid of Magnus is a mixture of hyposulphate of ethyle and carbylosulphuric acid.

Cardamom Oil. $C_{10}H_8$. From the seeds of *Alpinia* (Ellettaria) *cardamomum; Amomum repens.* Spec. grav. ·943.

Cardole. $C_{42}H_{31}O_4$. Spec. grav. ·978. Yellow fluid from the liquor from which anacardic acid has been separated by protoxide of lead.

Caries. A disease of bones and teeth where bone earth is deficient. See TEETH.

Carinthine. A green hornblende from the Sau Alpe, Carinthia.

Carinthite. A synonyme of Molybdate of lead.

Carminic Acid. $C_{28}H_{14}O_{16}$. Purple-brown mass, the colouring matter of cochineal (Coccus cacti), soluble in alcohol and water; slightly soluble in ether; decomposes above 136°C.; obtained by precipitating an aqueous decoction of cochineal with acetate of lead, and decomposing the washed carminate by sulphohydric acid; evaporating to dryness and dissolving in absolute alcohol, digesting the alcoholic solution with carminate of lead, and lastly precipitating the last trace of nitrogenous matter with ether.

Carmine. The older name for carminic acid; commercially, the name is applied to a combination of cochineal with alumina, prepared by boiling 1 lb. cochineal for a few minutes in a gallon of water, adding $\frac{1}{16}$ of alum and boiling for a few minutes; the liquid is passed through fine sieves into porcelain vessels, where it is allowed to deposit. In a few days, a deposit of fine carmine takes place, and an additional quantity of inferior quality falls after some time longer.

Carmindine, Bibromide of. $C_{64}H_{15}N_7O_{10}Br_8$.

Carmufelic Acid. $C_{24}H_{20}O_{32}$. A crystalline body from cloves by nitric acid.

Carnelian. A white and red variety of quartz from Surat.

Carolinite. A synonyme of nepheline.

Carotine. Ruby crystals from the juice of carrot (*Daucus carota*) by ether.

Carpholite. See KARPHOLITE.

Carragheen. *Irish Moss.* The Chondrus crispus of the British and other shores. When the lichen is boiled in water, a solution is formed, which on cooling becomes gelatinous from the presence of a substance analogous to pectic acid. On examination, I have found the aqueous extract to be mixed with albuminous matter to the extent of 9·87 per cent. and to contain about 20 per cent. of ash, which accounts for its nutritive power. It is used for preparing blanc mange and puddings, and is probably much superior to arrowroots.

Carrot. The root of the *Daucus carota.* It contains 32 per cent. of sugar before and 11½ per cent after roasting. It is frequently used to adulterate coffee; but may be distinguished by its cells, which are small and very thin-sided.

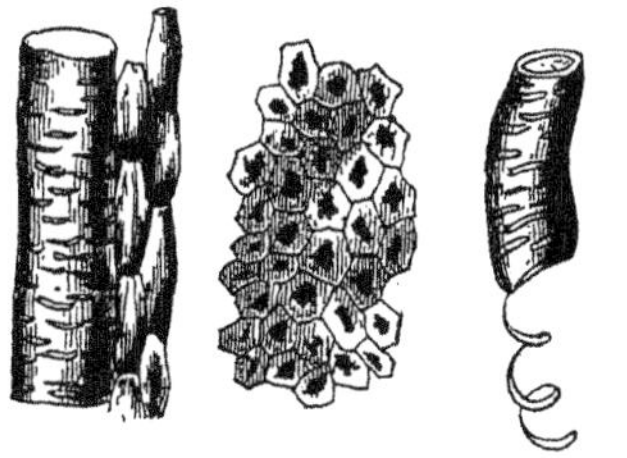

[*Hooker.*]

Dotted, barred, and spirally marked tubes occur abundantly among the cells, and are smaller than those of turnip and chicory. Boiled in weak caustic potash they give out the smell of carrots. Scarce any starch is present.

Carthamine. A dark red powder or needles, insoluble in water and dilute acids, slightly soluble in alcohol and ether; alkalies destroy its colour; obtained by exhausting safflower (*Carthamus tinctorius*) with cold water and acetic acid to remove the yellow-colouring matter. The flowers are then treated with water containing some carbonate of soda in solution, and the carthamine precipitated by an acid; citric acid is to be preferred.

Cartilage. See CHONDRINE.

Carvacrole. $C_{40}H_{28}O_3$. A colourless oil, heavier than water. B.P. 232°C.; soluble in spirit, ether, and potash; obtained in the preparation of carvene.

Carvene. $C_{10}H_8$. Colourless fluid, lighter than water. B.P. 343°. (173°C.) Spec grav. vapour 5·175; obtained by distilling oil of caraways (*Carum carui*), over hydrous phosphoric acid, until the distilled liquid loses the smell of caraways; carvene distils over, carvacrole remains.

Caryophyllic Acid. *Eugenic Acid.* $C_{24}H_{15}O_5$. Spec. grav. 1·079, B.P. 469°; colourless oil, tasting and smelling of cloves, forms a blue with iron from an alkaline solution; separated from oil of cloves by distilling with caustic potash; carryophyllate of potash remains. A light oil, of spec. grav. ·918, isomeric with camphene ($C_{10}N_8$), passes over.

Caryophylline. $C_{20}H_{16}O_2$. Silky crystals, insoluble in water; soluble in alcohol, ether, without taste and smell, volatile; obtained from cloves by boiling alcohol.

Cascarilla. The bark of the *Croton eleuteria* or cascarilla, yields a volatile oil, bitter resin, and *cascarilline*, a bitter matter. The oil is probably a camphene.

Case Hardening. The process of rendering the exterior of keys, &c. hard by heating them for an hour or two at a low red heat in a forge, wrapt in a paste of horns, hoofs, chopped hair, bone shaving, &c. and the whole enveloped in clay.

Caseic Acid. See APOSEPEDINE.

Caseine, *Curd, Caseum.* $S_2N_{36}C_{288}H_{228}O_{90}$ C=53·6, H=7·1, N=15·8, O 22·6 S=0·9. Coagulated caseine, when pure and dry, is hard, translucent, and yellowish; when pulverized, it is a yellowish-white powder; in water it softens and swells, but does not dissolve; soluble in strong oil of vitriol, and is reprecipitated by water; soluble in nitric and muriatic acids, the colour becoming blue with the latter above 60°. Solutions of caseine in acids, excepting in phosphoric and arsenious acids, are precipitated by yellow prussiate of potash; caseine is soluble in boiling alcohol, the excess precipitating as the solution cools; this property appears to depend on the presence of an alkali; all mineral acids, except trisphosphoric acid, coagulate caseine, uniting with it; the precipitate produced by acetic acid is soluble in excess of acid, but on the addition of a mineral acid again precipitates. In dilute solutions, mineral acids do not precipitate caseine; but lime water immediately produces a coagulum. The compounds of caseine with acids are soluble in weak acetates of soda and potash by heat. Caseine is very soluble in potash, soda, and ammonia; with earth and metallic oxides, as magnesia and oxide of zinc, it forms insoluble compounds. Caseine is precipitated from its solutions by all salts excepting those of potash, soda, and ammonia. Caseine decomposes in its aqueous solution into ammoniacal products, aposepedine, &c.; its compounds with acids are more permanent. The sulphate of caseine changes into aposepedine, sulphate of ammonia, &c. In oil of vitriol, caseine swells up and gelatinizes, and when washed with water it contains 8·4 per cent. sulphuric acid; nitric acid forms *xanthoproteic acid* (?); strong chlorohydric forms an indigo-blue solution, forming salammoniac and a brown matter; aqueous solutions of caseine are precipitated by tincture of nutgalls. Caseine exists in the milk of mammalia, and in many edible vegetables, as in potatoes, beans, pease, and leguminous plants generally, &c. and is a highly nutritive body. It is distinguished from albumen and fibrine by not coagulating, either spontaneously or by heat, and by forming a pellicle when its solution is evaporated. The products of the oxidization of caseine are, acetic aldehyde, metacetonic aldehyde, butyric aldehyde, acetic, butyric, benzoic, cyanohydric, metacetonic, caproic, formic, valerianic acids, oil of bitter almonds, and valeronitrile. It has been asserted that albumen and caseine are identical, but this is not strictly accurate, for it is only the fibrine of flesh and albumen of blood that are identical; the albumen of eggs differs from these, since it contains for the same amount of the other elements one-half more sulphur; it is certain that this sulphur must be separated when the albumen of eggs is converted into the albumen of blood. Caseine exhibits a similar, but reversed relation; for the same amount of sulphur it contains more carbon, hydrogen, and nitrogen than the albumen of blood; and it is absolutely certain, that a compound containing carbon, hydrogen, and nitrogen must be separated from this constituent of milk when it is converted into albumen of blood in the body of the young animal. From various considerations we may conclude, that the nutritive value of the caseine of milk is greater for the child and less for the adult than that of albumen; for it is certain that nature requires and applies to certain purposes in the body of the young animal the excess of the elements which caseine contains beyond those of albumen, and that these objects have no longer any importance on the adult animal (Liebig). *Process.*—Caseine is obtained from milk, by coagulating the milk by means of a few drops of acetic, or chlorohydric, or sulphuric acids, or alcohol, throwing the curd on a cloth filter, washing away the excess of acid by water, dissolving the curd in a weak solution of an alkaline carbonate, precipitating by an acid, washing the precipitate, and repeating these processes once or twice. As caseine is soluble in water, too much washing is not admissible. Vegetable caseine may be obtained from potatoes and pease, &c. by digestion in water and coagulation, &c. as described.

Cassava. The starch of the root of the *Jatropha manihot* is so named in the W. Indies.

Cassia Oil, or oils of cinnamon and senna.

Cassiterite. Native binoxide of tin, or tinstone.

Cassiteros. The Greek name for tin.

Cassius, Purple of. *Goldpurpur*, Gr. $AuOSn_2O_3$. (?) The composition of this purple pigment is somewhat doubtful; it appears to contain about 39·82 per cent. of metallic gold; but as it is formed by adding a solution of protochloride or bichloride of tin, in the proportion to form a sesquichloride of tin to a sesquichloride of gold, the above formula is perhaps accurate. For the methods of preparing gold purples, see PURPLE.

Castine. A bitter crystalline body, obtained by alcohol from the seeds of *Vitex agnus castus.*

Castor. An animal substance from the beaver (Castor fiber), contained in a bag between the genitals and anus of both sexes. It contains 43?per cent. of oil and resin.

Castor from Elba. A mineral containing 78 silica, 19 alumina, 2·76 lithia, and a trace of potash and soda.

Castoric Acid. The crystalline product of the action of nitric acid on castorine.

Castorine. Crystalline body from castor, by alcohol.

Castor Oil. A colourless slightly-tasted oil, of spec. grav. ·9575, at 77°, B.P. 509°; soluble in alcohol and ether in all proportions; its solubility in alcohol distinguishes it from all other fixed oils; it is obtained by expressing or boiling the seeds of the East and West Indian *Ricinus communis*, which contain about one-fourth their weight of oil. Water removes various impurities, and a fermenting principle. *Cold-drawn castor oil* is oil which has stood in contact with cold water. When distilled, it yields a *volatile oil*, like acroleine; by saponification, *ricinic*, *ricinoleic*, *ricinostearic* acids; by nitrous acid *palmine* and *palmic* acid; by nitric acid, *œnanthylic acid.* It is used extensively as a mild purgative.

Catalysis. (κατα, from; λυω, I loose;) a term sometimes used to denote a class of phenomena in which apparently the characters of a substance are changed by the contact of another; thus starch in contact with sulphuric acid is changed into sugar, a body of the same composition.

Catalysotype. Photographic paper prepared with water containing 2 drops of chlorohydric acid to 3 ounces of water. After drying with filtering paper, it is laid upon thin gum water in which $\frac{1}{80}$ of iodide of iron is dissolved, and 1 or 2 drops of tincture of iodine to each ounce. On drying it is laid on a solution of nitrate of silver 1 part to 4 parts water.

Catechine. *Catechuic acid*, *Tanningenic acid.* $C_{20}H_{10}O_9$. White crystals, soluble in hot water, alcohol, and ether, dilute mineral and acetic acids. F.P. 422° (217° C.); precipitates sesquichloride of iron green, lead salts white, sulphate of iron green; obtained by extracting Bengal catechu with cold water, which takes up mimotannic acid and leaves catechine and tannic acid. The latter are boiled in 8 times their weight of water. Catechine separates on cooling. By resolution, and treatment with animal charcoal, it is purified; by distillation it yields pyrocatechine. $C_6H_3O_2$, in scales fusing at 259°.

Catechu. *(Cate*, a tree; *chu*, juice.*)* *Cachou*, Fr. *Terra Japonica*, *Cutch.* An extract from different trees in India. The trees are felled; the interior coloured wood in chips is boiled with water in an earthen pot, and the decoction evaporated in the sun. Three kinds of catechu occur in commerce.

1. *Bombay catechu.*—Spec. grav. 1·39, taste astringent, from the *Areca catechu* or palm, by boiling the fruit; dark reddish-brown masses, with a shining fracture, almost all soluble in boiling water; contains 54·5 per cent. of tannic acid, which may be ascertained by dissolving a certain weight of glue in hot water, and adding the solution from an alkalimeter. The other constituents are:—Extract 34·, mucilage 6·5, inorganic matter 5·.

2. *Bengal Catechu.*—Sp. grav. 1·28. Chocolate-coloured pieces from *Acacia catechu*, leaving a large residue with cold water; by boiling almost entirely taken up—the solution containing less

tannic acid and more catechu than Bombay catechu. It contains tannin 48, extract 36, mucilage 8, resin 7.

3. *Gambir*, from the *Nauclea (Unearia) gambir*, in pieces of 1 to 1⅓ inch thick, of a brownish-yellow colour; fracture even, with shining points, slightly soluble in cold, almost completely soluble in boiling water. Catechu of all kinds is nearly all soluble in alcohol. It is used as an astringent in medicine, and for producing different shades of brown in calico printing.

Cathartine, *Bitter of Senna.* Neutral dark red, uncrystalline body; has a peculiar odour and bitter nauseous taste; soluble in water and alcohol; insoluble in ether; obtained from the decoction of senna leaves by the usual process for *bitter principles.*

Catlinite. Spec. grav. 2·54, H 1·5, SiO_3 48·2, Al_2O_3 28·2, MgO 6·, CaO 2·6, Fe_2O_3 5·, MnO ·6, HO 8·4. A red mineral from the Sioux country, North America, used for tobacco pipes by the Indians. Specimens sent to me proved to be claystone porphyry.

Cat's Eye. A gray, greenish, or reddish variety of opal or fibrous quartz, interspersed with thin filaments of asbestus, presenting, when cut, a peculiar opalescent appearance.

Cave. The space under the fire of a pottery furnace.

Cavear. The salted roe of *Accipenser huso*, or sturgeon.

Cavolinite. A synonyme of nepheline.

Cawk, (Chalk). Massive opaque sulphate of barytes, in lead mines.

Cecerite, or Cerite.

Cedrene. $C_{32}H_{24}$, B.P. 478° (248°C.), spec. grav. ·984, of vapour 7·9. Aromatic, peppery fluid, obtained by distilling cedrole with anhydrous phosphoric acid, and then over potassium.

Cedrine. A bitter, neutral, silky crystalline body, by ether and alcohol, from the fruit of the *Simaba cedron*, in New Granada.

Cedrirete. Red needles; insoluble in water, alcohol, and ether; forms a blue with pure sulphuric acid, purple with creasote; obtained by distilling beech tar, saturating the distilled oil with carbonate of soda and caustic soda; separating from insoluble part; neutralizing with acetic acid, and then distilling.

Cedrole. $C_{32}H_{24}2HO$, F.P. 165°, B.P. 539½°, specific grav. of vapour 8·4. Silky needles, the soluble part of oil of juniper (*Juniperus Virginiana*), by pressure and alcohol.

Celestine, or sulphate of strontian.

Cellular Tissue, in plants and animals, con-

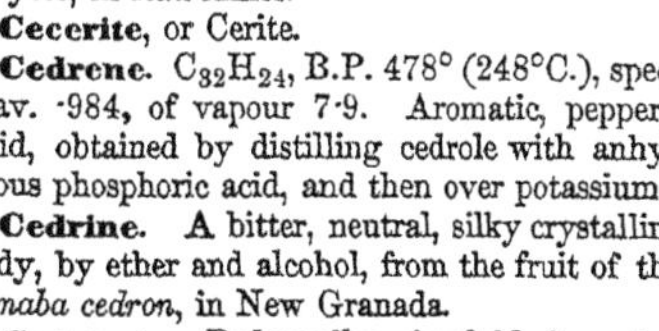

sists of spaces surrounded with walls; composed in animals of nitrogenous, and in plants of non-nitrogenous membrane (cellulose).

Cellulose, or *Celluline.* $C_{12}H_{10}O_{10}$. *Lignine* of Prout. C 44·8, H 6·2, O 50·. The idea now generally entertained by physiologists is that plants and animals originally exist in the form of a cell or vesicle, which is filled with fluid containing the elements for the production of other cells, and that these vesicles possess an absorptive power through their sides by which they can acquire new matter from without, and deposit it on the internal sides of the cells, or form new organs of a similar description. But the nature of the substance constituting the walls of the cells is quite different in the two kingdoms. Vegetable *cellulose* possesses the same composition as starch, or three elements, and is convertible into dextrine; it is not acted on by acetic acid and ammonia, and is very little altered by potash or soda. When distilled, it gives acid products. Animal cellulose, on the contrary, contains, besides carbon, hydrogen, and oxygen, a fourth element, nitrogen, and is often acted on by ammonia or acetic acid, and always by soda and potash; and when distilled it affords ammoniacal products. The following is the comparative amount of nitrogen in different animal envelopes:—

Skin,	20	per ct. Nitrogen.
Crustacea and insects (Chitine),	9	—
Envelope of Tunicata,	4·5	—
Vegetable cuticle,	2 to 2.5	—

Preparation of Cellulose. This substance is obtained in its purest form from cotton and linen. These substances are first to be washed with cold and then with hot water. They are then digested in a solution of caustic soda for two or three hours, at a boiling temperature; then placed in water, and a current of chlorine passed through it; after which the cellulose is digested in caustic soda. It is then washed with water, acetic acid, boiling water, ether, and alcohol, and lastly dried in vacuo at 212°. To extract cellulose from wood, it is necessary to pulverize it in the first place, and then to pound it into a paste with water. The paste is then heated in an oil bath with five times its weight of caustic soda, and the solution evaporated to dryness. It is then treated with chlorine, or a solution of chloride of lime, and washed with dilute chlorohydric acid. The chlorine acts by attacking the superficial layers of the substance, and thus destroys the adhesion of all foreign bodies. In a saturated solution of bleaching powder, paper pulp, when heated, gives out carbonic acid; the cellulose disagregates, burns, and disappears. It exists in certain animals, as *Phallusia* and *Cynthia papillata.* *Properties.*—Cellulose is white, solid, diaphanous, insoluble in cold water, alcohol, ether, fixed and volatile oils; its specific gravity is 1·525. It can sometimes be easily disagregated by boiling water, and the resulting substance possesses the characters and composition of *dextrine*. Strong *sulphuric* and *phosphoric* acids attack cellulose without heat, and render it soluble in water, and capable of being precipitated by alcohol. They convert it into starch, then into dextrine, and then into sugar. *Caustic soda* causes it to swell up, and acts upon it in a superficial manner. It is in this way that the effect is produced in bleaching. Sulphuric, phosphoric, chlorohydric, and nitric acids, when very dilute, do not dissolve it. Nitric acid, in a concentrated state, produces first *dextrine*, and then *xyloidine*, which may be precipitated by water. *Iodine* has no effect upon cellulose, except in some cryptogamous plants, where a violet tint is produced—a character by which cellulose in this class of plants seems to approximate to starch. The sporules of some mushrooms produce a violet with iodine, and hence they appear to contain cellulose.

Dextrine.—To obtain dextrine, we triturate in a mortar 100 parts of charpie with 140 parts sulphuric acid, added drop by drop, and stir until the mixture is formed into a gummy mass, then dilute with water, and saturate with chalk or barytes. The liquid is then filtered and concentrated; it is then precipitated and washed with alcohol. The product is dissolved in water and filtered, and the liquor contains dextrine possessing the same composition as cellulose, and being capable of being turned into grape sugar by sulphuric acid, and heat, and in every respect resembling a similar product from starch.

Xyloidine, or *Nitramidine.*—When cellulose is treated with strong nitric acid, it dissolves, combining with an atom of nitric acid. When water is added, this compound precipitates. If allowed to stand for some days, the precipitate disappears, and by evaporation, a new acid, deliquescent, white, and solid, remains. Xyloidine is very combustible; at the temperature of 356° it takes fire, and burns, leaving but a scanty residue. M. Pelouze has in consequence recommended it for the purpose of forming matches for the artillery. It can be easily made by plunging a piece of paper into nitric acid, of spec. grav. 1·5, and allowing it to remain for a few minutes. It may then be taken out and dipped in water, when it forms, after drying, a parchment impermeable to moisture, and of extreme combustibility. Its composition is—

$$\text{Xyloidine, } C_{12}H_9O_{10} \atop NO_4 \text{, or } C_{12}H_9O_9 \atop NO_5$$

an atom of the hydrogen of the cellulose being replaced by an atom of nitrous vapour. Nitric acid, when boiled with cellulose, produces oxalic acid. Soda or potash, when boiled with cellulose, displaces hydrogen, and oxalic acid remains in combination with the soda or potash. Chlorohydric acid, when boiled with cellulose, acquires a red colour, which alters to brown, and the liquid blackens, but does not dissolve. See GUN COTTON.

Cellulose Animal. The cellular tissue of animals resolved by boiling into gelatine.

Cellulostase. C 63·15, H 21·65, O 15·15, N ·05.

Cement. A term applied to various kinds of hydraulic mortars, &c.

Natural, or *Roman Cement*, consisting of limestone and siliceous clay. On calcination and mixture with water it hardens, becoming carbonate and silicate of lime.

Mastic, or *Artificial Stone*, consisting of sand, limestone, litharge, and linseed oil.

Asphalt Cement, formed by mixing coal asphalt with sand and chalk, or with powdered iron slag.

Cementation. The process by which carbon unites with the particles of iron to convert it into steel, by making its way from one atom to another, from the exterior to the interior of an iron bar, by a series of combinations or replacements.

Centaurine, *Cnicine*. A bitter substance, from *Carduus benedictus* root by ether.

Cephalote. An impure form of cerebric acid.

Ceraic Acid. $C_{20}H_{20}O_2$. A body said to exist in white wax, accompanying myricine.

Ceraine. The saponifiable portion of cerine.

Cerantic Acid. Granular crystals by alcohol, from the fuel in an antique lamp of the 4th century.

Cerasine. $C_{12}H_{11}O_{11}$. Spec. grav. 1·469 to 1·530. Solid, tasteless; insoluble in cold water and in alcohol; unfermentable. By boiling with water it is changed into arabine; it is the principal constituent of cherry tree gum, and of various species of *prunus*; it is obtained by digesting these gums in water; arabine dissolves and cerasine remains. When heated with nitric acid, it yields mucic and oxalic acids. This gum is of no value. See GUM.

Cerasite. (κέρας, horn). Chlor-oxide of lead.

Ceraunite, or Cerite.

Cerebric Acid. *Cerebrine*. C 67·04, H 10·85, N 2·24, P ·4, O 19·41. White crystalline grains obtained from the brain of animals. I have used two modes; drying the brain at 300°, boiling with alcohol, and then the extract with ether, which dissolves oleophosphoric acid and leaves cerebric acid, insoluble in cold ether, and crystallizing from boiling alcohol; or by saponifying the brain with dilute potash and gentle heat; albumen is dissolved and cerebrate of potash floats. The cerebrate is decanted and decomposed by tartaric acid, the cerebric acid dried, digested in cold ether, and crystallized repeatedly from boiling alcohol or ether. Cerebric acid swells up like starch when boiled in water; it yields by oxidation acids similar to the fats. It is found in the liver, blood, nerves, and brain.

Cerebroleine. *Cerebrol*, *Eleencephol*, *Brain Oil.* The oleine of the brain united to phosphoric acid, to form oleophosphoric acid.

Cerebrote. Impure cerebric acid.

Cererine, *Cererite*, or *Cerine*. A silicate of cerium.

Ceric Acid. C 64·9, H 8·7, O 26·4 A fatty acid. (?)

Cerine. Impure cerotic acid. The soluble part of bees' wax in alcohol, fusing at $161\frac{1}{2}°$, (72° C.) When saponified with potash it yields *cerotic* acid. Ceylon wax contains none; distilled it yields 2 fluid oils containing very little oxygen.

Cerine, sometimes confounded with ceraine, which is probably an impure cerine.

Cerine. A synonyme of *Allanite*.

Cerinic Acid. $C_{42}H_{34}O_{13}$. A fatty acid. (?)

Cerinine. $C_{20}H_{19}O_2$. (?) A waxy fat by alcohol and ether, from a brown coal at Merseburg.

Cerite, *Silicate of Cerium, Ochroite.* Spec. grav. 4·912, H 5·5. Clove-brown, cherry-red, or gray masses, granular; fracture splintery; streak white; lustre adamantine, translucent on the edges. B.B. infusible *per se;* forms with borax a yellow bead, which is paler on cooling. It consists of silica 16·, peroxide of cerium 26·55, oxide of lanthanum 33·38, water 9·1, CO_2 4·62, Al_2O_3 1·68, Fe_2O_3 3·53, CaO 3·56, MnO ·27. It occurs at Bastnäs, Redderhyttan, Westmanland, in a bed of gneiss.

Cerium, *Didymium, Lanthanum.* These substances have not been properly distinguished in consequence of the difficulty of separating them from each other. They occur in *Cerite* and *Allanite.* Ceria may be separated from lantana by igniting the mixture of these two bodies, and digesting the ignited mass in weak nitric acid, which has the property of dissolving the lantana, and leaving the ceria. The methods recommended in books for separating ceria, refer to all these three metals, which usually occur together, and have never been completely separated. See LANTHANUM, DIDYMIUM.

Cerolite. Soapstone from serpentine in Silesia.

Cerosic Acid. $C_{48}H_{48}O_3$. A wax acid. (?)

Cerosine. $C_{48}H_{50}O_2$. Pearly scales. F.P. $179\frac{1}{2}°$, the wax of sugar canes.

Cerotic Acid. $C_{54}H_{54}O_4$. F.P. 172·4° to $174\frac{1}{4}°$. Granular crystals, volatile without decomposing, except when in contact with the other constituents of wax; it is obtained by boiling alcohol on bees' wax to extract the cerine (impure cerotic acid). This is dissolved in much alcohol, and precipitated with a boiling solution in alcohol of acetate of lead and filtered while hot; the precipitate boiled with ether and alcohol to purify it, and then decomposed with acetic acid; the precipitate washed with boiling water, dissolved in hot alcohol and filtered while hot; it may be further purified by saponification, precipitation, and crystallization. Its salts are monobasic.

Cerotine. *Alcohol of Cerotic Acid and Cerotyle.* $C_{54}H_{56}O_2$. F.P. $174\frac{1}{2}°$. Crystalline; obtained by fusing Chinese wax with caustic potash; washing with water; precipitating by chloride of barium and treating the precipitate heated with alcohol or ether; cerotine is dissolved, and cerotate of barytes is left. Cerotine when heated with lime and potash, evolves hydrogen, and becomes cerotic acid.

Cerotene. $C_{54}H_{54}$. F.P. $135\frac{1}{2}°$. A crystalline body resembling paraffine; obtained by distilling Chinese wax; it is freed from cerotic acid by potash. Chlorine replaces 22 atoms. Cerotene is changed into a fluid when distilled in a sealed tube.

Cerotyle, Hydrous Oxide of. $C_{54}H_{55}O$. The basis united with cerotic acid in cerotine.

Ceroxyline. $C_{20}H_{16}$, O. F.P. 212°. The resin of palm wax.

Ceruline. See INDIGO.

Cerumen, *Ear Wax.* A yellow oil secreted by the glands of the auditory canal, concreting by exposure to the air; taste bitter; when heated it stains paper like a fixed oil, giving out an aromatic odour; forms an emulsion with water and putrifies. Cerumen appears to consist of stearine, oleine, otine, yellow matter soluble in water, uncoagulated albumen, coagulated albumen, lactates of lime, and potash or soda.

Ceruse, *Cerussa.* Carbonate of lead, white lead.

Cerussa Usta. Red lead.

Cetene. $C_{32}H_{32}$. An oil. B.P. 527°. Specific gravity, vapour 8·007; obtained by treating ethal with anhydrous phosphoric acid.

Cetine. *Cetylate of Oxide of Cetyle, Spermaceti. Wallrathfett,* Gr.; *Blanc de Baleine,* Fr. $C_{64}H_{64}O_4 = C_{32}H_{33}O, C_{32}H_{31}O_3$. White pearly scales. F.P. 120° (49° C.); volatilizes at 680° (360° C.); insoluble in water; 100 boiling alcohol of ·821 dissolve $2\frac{1}{2}$ cetine, mostly deposited on cooling; soluble in pyroxilic spirit, oil of turpentine, and fat oils; by distillation are obtained water, carbonic acid, oxide of carbon, olefiant gas, cetine, cetylic acid, and cetene; and in the retort charcoal; adipic acid is formed by oxidation by nitric acid; cetine is obtained by boiling spermaceti in alcohol of ·816 till the alcohol poured from the deposit yields no oil on evaporation.

Cetic Acid. $C_{32}H_{24}O_8$.

Cetraric Acid. $C_{34}H_{16}O_{15}$.

Cetrarine. White bitter powder; nearly insoluble in hot and cold water; 100 boiling alcohol dissolve $1\frac{3}{4}$; boiling ether ·93; does not fuse; chlorohydric acid changes it to a blue substance; by alkalies changed to ulmic acid; obtained by boiling Iceland moss (Cetraria Islandica) as long as it tastes bitter with alcohol, distilling off the alcohol, digesting the residue in ether, and crystallizing out of absolute alcohol.

Cetyle. $C_{32}H_{33}$. The hypothetic radical of cetine, &c.

Cetyle, Oxide of. $C_{32}H_{33}O$. Shining scales, by the action of sodium on ethal at 212°, and crystallization from alcohol.

Cetylic Acid. *Aethalic Acid.* $C_{32}H_{31}O_3$

HO. Colourless crystalline solid; fusing at 131°, volatile, insoluble in water; soluble in alcohol and ether; obtained by heating ethal or spermaceti by fusion with caustic potash, cetylic acid and oxide of cetyle being formed.

Cetylophenylamine. $C_{32}H_{33}, C_{12}H_5HN$. Reddish-white crystals, fusing at 107½° (42° C.); soluble in alcohol and ether; formed by the action of aniline on iodide of cetyle.

Cevadic Acid. Volatile white pearly needles, fusing at 68° in white hellebore seed (*Veratrum album*); extracted by hot ether and removing fatty matter by potash.

Cevadine, or Hordeine.

Ceylanite, or Spinelle.

Chabazite. (χαβαζιοι). *Cubic Zeolite, Cuboide.* Spec. grav. 2·088, 2·472, H 3·75. Obtuse rhomboids with angles of 94° 46′; fracture uneven, brittle; lustre vitreous; transparent to translucent. B.B. becomes a white spumous mass. It contains silica 49·20, alumina 17·91, lime 9·64, potash 1·92, water 20·4. Sometimes soda replaces lime; occurs in trap rocks in Scotland, Ireland, Ferroe, &c. *Form.* 3 (CaO NaO KO), 2 SiO_3, 3 (Al_2O_3 2 SiO_3), 18 HO.

Chaerophylline. A substance obtained by distilling with water the *Chaerophyllum bulbosum.*

Chalcantum. Probably a mixture of sulphates of copper and iron.

Chalcites. Probably sulphuret of copper.

Chalcolite, or Uranite.

Chalcopyrite, or Copper pyrites.

Chalcosine. Sulphide of copper.

Chalilite. (χαλις, flint). Spec. grav. 2·252, H 4·5. Reddish or brown resembling a flint, translucent on the edges. B.B. becomes white, and spreads out like a cauliflower; becomes a white bead with soda, with borax a colourless glass. Consists of silica 36·56, alumina 26·20, lime 10·28, Fe_2O_3 9·28, soda 2·72, water 16·66; from Sandybrae, county Antrim; allied to Thomsonite.

Chalk. *Kreie*, Ger.; *Craie*, Fr. *Carbonate of lime.* The name of the calcareous deposit distinguished by the presence of flints. It consists of microscopic calcareous shells of carbonate and trace of phosphate of lime. It constitutes the basis of whiting, Paris and Spanish white, crayons.

Black Chalk is a black carbonaceous soft slate, consisting of silica 57·5, Al_2O_3 12·98, alkali 4·02, charcoal 17·52, HO 6·3.

Red Chalk.—Bole or iron ochre.

Chalk stones in gout are *urate of soda.*

Chalybeate Waters. Waters containing iron. See WATER.

Chameleon Mineral, or manganate of potash, formed by melting 3 parts of nitre and 1 part of black oxide of manganese; when dissolved in water, it produces a variety of colours in the course of its decomposition.

Chamoisite. Spec. grav. 3· to 3·4. Magnetic dark greenish-gray masses; soluble with effervescence in acids; consisting of silica 12, FeO 50·5, alumina 6·6, water and bitumen 14·7, $CaOCO_2$ 14·4; found in limestone at Chamoisin in the Valais.

Chamomile Oil, from *Matricaria chamomilla,* by distillation with water; at first blue, then brown; with the smell of chamomile.

Champagne, an effervescing wine, deriving its name from the country of its production; it contains 4·9 per cent. of absolute alcohol.

Chantonnite. Spec. grav. 3·48, H 6½ to 7. Black veins in the meteoric stone of Chantonney.

Chapapote, or Asphalt.

Charcoal. *Vegetable* or *Wood Charcoal, Kohle,* Gr.; *Charbon,* Fr.; is of such great value on the continent, in consequence of the want of coal, that there are establishments for the purpose of managing forests. In this country much charcoal is derived from the iron stills in which wood is heated for the purpose of obtaining pyroligneous acid. The purest charcoal may be obtained by burning sugar, gum, or gelatine, as when prepared from wood it contains much ashes; but for the usual purposes to which it is applied, it is by preference made from oak or birch. The branches and por-

tions of trees are disposed in a conical form, and are covered with turf, with the exception of a few points at the bottom, by which the air is admitted. The pile is ignited from below, and as the combustion ascends, the lower holes are filled, and new ones opened higher up. When smoke ceases to be emitted from the conical piles, the process, it is concluded, has terminated, and all the apertures are closed, when the ignition of the charcoal ceases. Wood dried in the air consists nearly of carbon 38·5; combined water 35·5; free water 25; ash 1. The gases emitted in the carbonization of wood are carbonic acid 26·7, carbonic oxide 9·7, hydrogen 9·9, nitrogen 53·7. Charcoal possesses different properties, according to the circumstances under which it is formed. When it is burned at a temperature under redness, it is a non-conductor of electricity, a bad conductor of heat, and very easily burned. This form of carbon, therefore, does not answer for experiments with the galvanic battery. When charcoal is produced at a red heat, it conducts electricity and heat well, but is less easily burned; while, if it be heated to whiteness during its formation, it possesses a much superior conducting power to heat and electric fluid than in the previous circumstances, and is of very difficult combustion (Cheuvreusse, Ann. Chim. 29, 426). When heated away from air and moisture, charcoal is not altered, but becomes harder and more brilliant (Hoffmann's Observations, 298). Charcoal begins to burn at 460°; decomposes water at a red heat, hydrogen being set free, carbonic oxide and carbonic acid (CO and CO_2) being formed. Bunsen has found such gases to consist of hydrogen 56·03, CO 29·15, CO_2 14·65, CH_2 00·17. When charcoal has been exposed to a very high temperature, as in a blast furnace, it is such a good conductor of heat, that when one end is ignited it is impossible to hold the other in the hand. In masses it floats in water, but if reduced to powder it sinks. When free from air its spec. grav. is 3·5 (Leslie). The combustibility of charcoal depends in a great measure on the nature of the texture of the material from which it has been prepared. Thus the loose fabric of old linen affords a charcoal, so susceptible of ignition that it is the familiar companion in the tobacco pouch of the labourer. The same observation applies to the loose texture of charcoal made from the rapidly-growing trees, as poplars, willows, which absorbs more water, contains more pores, and is therefore of more easy combustion than charcoal of oak and birch. The property which charcoal possesses of absorbing water rapidly is one which interests the purchaser, and requires his attention to the nature of the wood from which it has been prepared. 100 parts of poplar charcoal have been found to absorb as much as 23 parts of moisture, while the same amount of guaiacum charcoal took up only 11 parts. Wood charcoal absorbs colours, dissolved in water, even in the cold state (Lowitz, 1792). Charcoal absorbs gases to a great extent. The following table expresses this absorption according to the experiments of Saussure on boxwood charcoal:—

1 vol. charcoal absorbs—

	Vols.
Ammoniacal gas,	90
Muriatic acid,	85
Sulphurous acid,	65
Sulphuretted hydrogen,	55
Protox of nitrogen,	40
Carbonic acid,	35
Olefiant gas,	35
Phosphuretted hydrogen,	10
Carbonic oxide,	9·42
Oxygen,	9·25
Nitrogen,	7·5
Hydrogen,	1·75

When charcoal, saturated with any gas, is put into another gas, it gives out a portion of its absorbed gas, and absorbs a portion of the new gas. These gases are again disengaged in a vacuum. It is by this property probably that the familiar recipe of burned bread as an antidote to the odour communicated to the breath by eating onions acts. (Lametherie, Jour. Phys. 30, 309; Marozzo, ib. 1783, 376; Rouppe and Van Noorden, Ann. Chim. 32, 3; Saussure, Annals Philos. 6, 241, 331.)

Animal Charcoal, Ivory Black, Bone Charcoal, Ebur ustum nigrum, Blood Charcoal, is prepared from bones, by placing them, deprived of their fat and grease as much as possible, in cast iron cylinders, one extremity of which leads by a tube into coolers, serving to condense the volatile and ammoniacal matters, while the other end opens by means of a disc capable of being opened and shut at pleasure, through which the crushed bones are introduced. Around these cylinders the fire of the furnace plays. The cylinders are heated red hot, and are kept in this condition for thirty-six hours, when they are emptied and recharged. For the usual purposes of the arts, animal charcoal does not require purification: but in the laboratory it is often necessary to remove from it the phosphate and carbonate of lime, which enter largely into its composition. This object is attained by digestion in dilute hydrochloric acid for some hours. The mixture should then be thrown on a filter and well washed with water, until the liquor passing through ceases to give a precipitate with nitrate of silver. Animal charcoal possesses the remarkable property of withdrawing from their solutions in water many substances, as lime from lime water, iodine from solution in iodide of potassium, soluble subsalts of lead, and oxides of the metals which are soluble in ammonia and caustic potash. It removes also nitrate of lead and most metallic subsalts. It also removes disulphate of quinine from its solutions, likewise acetate of morphine, strychnine, lupuline, sulphate of magnesia in part, chloride of barium, &c. It is also much used for removing colours from liquids, and thus purifying the substances which may be in solution. *Experiment.*—The effect of animal matter in decolourizing liquids can be illustrated conveniently by shaking a portion for a few minutes with a solution of Brazil-wood, and then filtering. A colourless fluid passes through the filter. The action of animal charcoal in depriving liquids of colouring matters with which they may be contaminated is increased by the application of heat, and its influence is likewise greatly heightened when the liquor to be decolourized is slightly acid. When an alkali is present, the solution, instead of losing any of its depth of colour, has its tint increased in consequence of the solution of a dark organic matter by the influence of the alkali. Animal charcoal is of great importance in the refining of sugar, in which process it is largely used to remove the colouring matters which usually render its solutions in the first periods of its purification exceedingly dull. It is also of great importance in the laboratory in the preparation of alkaloids and other delicately crystallizable bodies of organic chemistry. It is also used to purify solutions of tartaric acid.

Chay, or *Chaya Root*. The root of *Oldenlandia umbellata*, grown on the Coromandel coast, used for dyeing red by the natives, and considered to be finer than madder red.

Cheese. *Käse*, Ger.; *Fromage*, Fr. Cheese consists of the caseine of milk, accompanied by a certain amount of butter. According to the quantity of butter present, the richness of the cheese varies. Cheese may be made entirely from the cream, and then it contains all the butter of the milk; is soft, and known as *cream cheese*. When the cream has been removed, and the remaining milk employed to produce the cheese, it is termed *skim-milk cheese*. These are generally characterized by hardness and little flavour. When formed from the entire milk they are termed *sweet-milk cheeses;* this is the most common form of good cheese. When made from ewe's milk, they are termed *ewe cheeses*, and are distinguished by a peculiar sharp taste and high flavour, as in Roquefort cheese. The theory of cheese-making seems to be, to introduce into the milk a substance which causes the sugar to decompose into lactic acid; this acid immediately acts upon the caseine or curd, and causes it to curdle or coagulate, and form out of the milk a gelatinous opaque mass, from which the yellow whey separates by pressure. The usual coagulating medium is a solution formed by acting on the internal coat of the stomach of a pig or calf, with a solution of salt. The stomach is carefully washed, and then its internal surface is digested in a solution of common salt for many weeks. The solution is *rennet*. The milk to be converted into cheese is rendered lukewarm (loowarm, Scot.; lauwarm, Gr.) and mixed with the requisite quantity of rennet. In an hour or two the milk is found to be coagulated. The next object is to separate as much as possible of the whey. The curdled milk is first pressed on the surface with a cheese cloth. The whey passes through, and is removed. As it becomes drier, the curds are introduced into the cloth, which surrounds them in the form of a bag, and are carefully pressed. They are then rubbed with the hand so as to pulverize them, mixed with the requisite quantity of salt, placed in the cheese cloth, and deposited in a wooden mould of the required form of the cheese, which is supplied with draining holes at the bottom. It is then subjected to pressure in the cheese press, to remove the last traces of whey, and to consolidate the curds. The quality of the cheese depends on the process and the exposure to the air. The flavour is derived from the volatile acids of the butter. In commerce we have Stilton, Chedder, Cheshire, Gloucester, Dunlop, &c.; of continental, Parmesan, Roquefort; different kinds of Swiss cheese, as Gruyere, &c. Curd always contains a large quantity of salts. Gruyere has been found to leave an ash when burned formed of—alkalies 13·48; lime 39·22; magnesia 1·77; sesquioxide of iron ·35; phosphoric acid 45·; silica, ·18. What is called *Hand käse* in Germany, from its shape, is made from sour milk.

Chelidonic Acid. $C_{14}H_3O_{11}$. Colourless crystals from the aqueous extract of Chelidonium majus, by precipitating with a salt of lead, and decomposition.

Chelidonine. $C_{40}H_{20}N_3O_6$. Bitter alkaloid in *Chelidonium majus*.

Chelerythrine. Crystalline, obtained from the *Chelidonium majus*, after separating chelidonine by sulphuric acid.

Chelidoxanthin. Yellow needles, from Chelidonium majus.

Chelmsfordite. Said to be a variety of table spar.

Chemic. The commercial name of bleaching powder.

Chemistry. (*Chemie*, Gr.; *Chimie*, Fr.; *Kemi*, Schw.) The word χημεια first occurs in the dictionary of Suidas, a Greek writer of the 11th century, and is by him defined to be "the preparation of silver and gold." The term is by some supposed to be derived from an oriental word (see ALCHEMY), and by others from Chemi, Cham, or Chami, or χημια, chemia (Plutarche), the ancient name of Egypt. On the Rosetta stone is the word *Chmi*. In the present state of our knowledge, chemistry may be defined the study of the characters of the various elementary bodies of nature, their actions on each other, the products which they form, and the laws which regulate their actions. See Thomson's History of Chemistry.

Chenocholic Acid. A sulphur acid supposed to exist in the bile of the goose.

Chenocoprolite. An arseniate of silver and iron, from Clausthal, of a yellow or pale-green colour, resembling goose dung.

Chert. A variety of quartz, found in limestones, allied to hornstone.

Chiastolite. *Macle*, or *Wood Spar*. A variety of Andalusite, characterized by the remarkable appearance of its crystals in clay slate, as if the separate prisms, while imbedded in a soft medium, had not possessed sufficient freedom to allow them to unite perfectly.

Chica Red. The colouring matter of *Bignonia chica*, from Orinoco.

Chicory. See CICHORIUM.

Childrenite. Spec. grav. 3·247, H. 5·. Yellow rhombic 8-hedrons, with angles of 97° 50′, 120° 30′, 130° 20′. From the George and Charlotte mine at Tavistock. *Composition.*—Silica 3·82; phosphoric acid 28·24; alumina 18·06; protoxide of iron 29·58; protoxide of manganese 5·89; protoxide of copper ·65; water 16·35; 2 RO PO_5, $Al_2O_3PO_5$, 15 HO.

Chileite. An iron mineral.

Chiltonite. A variety of Prephite.

Chiococcine. A bitter alkaloid in *Chiococca racemosa*. See EMETINE.

Chiolite. *Fluoride of Aluminum and So-*

dium. From the Topaz mine of Miask, consisting of Al 18·16, Na 23·78, F 57·53.

Chinovic. See QUINOVIC ACID.

Chiretta. A yellow wood from Bengal.

Chitine. The principal part of the elytra of insects; a white body, insoluble in potash.

Chladnite. Spec. grav. 3·116. White oblique double prisms, consisting of SiO_3 65·69, MgO 27·11, CaO 1·39, CaO 1·02, HO ·68, in the meteorite of Bishopville, U.S.

Chlorabethamide. C 15·12, H 1·6, Cl 65·14, N 10·05.

Chlorabronaphthese. $C_{20}H_6Cl_6Br$.

Chloracetam. C 14·6, H 1·2, Cl 65·8, N 8·5.

Chloracetamide. *Chloroxethamide. Chlorocarbethamide.* $C_{10}Cl_7H_6N_3O_3$. By ammonia on chlor-oxalic ether.

Chloracetic Acid. $C_4Cl_3O_3HO$. Rhomboidal needles, deliquescent, irritating odour. F.P. 115°. Spec. grav. at 115°, 1·617. Corrodes the skin, and combines with bases; obtained by the action of chlorine on solid acetic acid in a jar, by sunlight. Formed on the type of acetic acid, chlorine replacing hydrogen. B.P. 387°.

Chloracetonitrile. C_4Cl_3N. Sp. grav. 1·444. B.P. 178°. A liquid by the action of anhydrous phosphoric acid on chloracetate of ammonia or chloracetamide.

Chloral. C_4Cl_3OHO. Spec. grav. 1·502; B.P. 202°, (94° C.) Oily fluid, with a peculiar smell, and acting on the eyes; mixes with ether and alcohol; dissolves sulphur, phosphorus, and iodine; obtained by passing dry chlorine through anhydrous alcohol for a long time, till the chlorine passes unchanged. Needles of hydrate of chloral deposit. These are treated with sulphuric acid and quicklime.

Chloralbine. Needles. $C_{12}H_6Cl_2$.

Chloraldehyde, (*Bichloride of Acetyle.*) $C_4H_4Cl_2$. Spec. grav. 1·174, of vapour, 3·478. B.P. 147°. Mobile fluid, by acting on chloride of ethyle with chlorine.

Chloraldehydene, (Chloride of Acetyle.) C_4H_3Cl. Spec. grav. of vapour 2·166; from the oil by the action of chlorine on olefiant gas.

Chloralide. C 18·61, H ·62, Cl 65·9, O 14·87.

Chloramilal. $C_{20}H_{17}O_4Cl_3$; B.P. 356°. Oily fluid by the action of chlorine on grain oil.

Chloranilam. *Chloranilamic Acid.* $C_{12}Cl_2H_3NO_6$. Black needles, obtained by mixing a solution of chloranilammon with chlorohydric acid.

Chloranilamide. $C_{12}N_2Cl_2H_4O_4$. Needles by heating an alcoholic solution of chloranil with ammonia.

Chloranilammon. $C_6H_3NClO_34HO$. Brown crystals by acting on chloranile with caustic ammonia.

Chloranile. $C_{12}Cl_4O_4$. Golden scales by acting on chlorisatine with chlorine.

Chloranilic Acid. $C_{12}Cl_2O_62HO$. Yellow crystals by dissolving chloranil in weak potash.

Chloraniline. $C_{12}H_6ClN$. A base by distilling chlorisatine with potash. Two other chloride bases are formed by the action of hlorine. Di and tri-chloraniline.

Chloranisal. $C_{20}H_6Cl_3O_2$. By acting on anisal with chlorine.

Chloranthracenese. $C_{30}H_{10}Cl_2$.

Chlorarsine, or chloride of cacodyle.

Chlorastrolite. A zeolite from Lake Superior. Spec. grav. 3·18, H 5·5. Bluish-green iridescent radiated fibres; silica 36·99, alumina 25·49, Fe_2O_3 6·48, CaO 19·9, NaO 3·7, KO ·4, HO 7·22.

Chlorazolitmine. $C_{18}H_{10}NClO_{10}$. Yellow insoluble substance by acting on azolitmine with chlorine.

Chlorebronaphthine. $C_{40}H_{11}Cl_4Br$.

Chlorebronaphthise. $C_{20}H_5Cl_2Br$. Solid by acting on chlornaphthise with bromine.

Chlorebronaphthone. $C_{40}H_9Cl_3Br_4$.

Chlorebronaphthose. $C_{20}H_4Cl_2Br_2$.

Chloribronaphthose. $C_{20}H_4Cl_3$, Br. All the remarkable isomeric forms in the action of chlorine on naphthaline are produced by the replacement of corresponding atoms of hydrogen.

Chloribronaphthuse. $C_{20}H_3Cl_3Br_2$.

Chloric Acid. ClO_5. Forms a colourless or yellowish solution in water. Obtained by decomposing chlorate of barytes by sulphuric acid. See manufacture of chlorate of potash under POTASH.

Chlorimetry. (Chlorine and μετρον). The process of estimating the amount of chlorine in bleaching powder. See BLEACHING for a convenient method in manufactures. One of the most accurate and rapid processes is that of Dalton, which depends on the conversion of protoxide of iron into sesquioxide of iron, by means of the chlorine of the bleaching powder, which decomposes water. For this purpose we take 4 atoms of good crystals of protosulphate of iron (FeO SO_37HO), or 69·5 grains previously well dried and pulverized, and dissolve them in 2 ounces of distilled water, with the addition of a few drops of sulphuric acid, placed in a stoppered bottle. We then weigh out 50 grains of the bleaching powder to be tested, and add it in small portions at a time to the copperas solution, shaking the mixture between each addition, and gently applying heat. As every atom of protoxide of iron requires half an atom of oxygen to convert it into the sesquioxide, this is equivalent to half an atom of chlorine in the bleaching powder, or 2·22 grains chlorine. When 4 atoms of the protoxide have been peroxidized, it is equal to 2 atoms of chlorine in the powder, or 8·88 grains of chlorine. The peroxidation of the solution is judged of by frequently taking out a drop with the end of a glass rod, placing it on a piece of white porcelain, or spotting it on blotting paper, and then adding a drop of a solution of red prussiate of potash. As soon as a blue precipitate of Prussian blue ceases to be formed, we infer that the whole of the protoxide, in the 69·5 grains of sulphate of iron, has been peroxidized. We then weigh the residue of the bleaching powder. Suppose it amounts to 15 grains; then this amount deducted from 50 grains, shows that 35 grains of the bleaching powder contained 8·88 grains of chlorine. We have then the proportion 35 : 100 :: 8·88 : 25·37 = the quantity of chlorine contained in 100 grains of the sample.

Chlorindatmine. $C_{12}H_4Cl_3N$, or trichloraniline.

Chlorindine. C_{16}, $H_5Cl\,NO_2$, 2 HO. A powder by the action of heat on chlorisatyde.

Chlorindoptene. $C_8H_2Cl_2O$. Crystalline; volatile; by chlorine on indigo.

Chlorindoptic Acid. See CHLOROPHENISIC.

Chlorine. Cl 4·5, 36, 4·437, 35·496. *Syn. Dephlogisticated marine acid* (Scheele, 1774), *Oxygenized muriatic acid* (Berthollet, 1785), *oxymuriatic acid* (Kirwan), *chlorine* (Davy, 1810). *Sources.*—Chlorine occurs abundantly in sea water, as chloride of sodium, or common salt, in rivers and wells, and in the soil; also in large deposits in the new red sandstone formation, as in Cheshire; Punjaub; Peru, Rio Negro, in South America; also as chloride of potassium in sea water. *Physical Characters.*—Chlorine is a greenish-yellow coloured gas, possessing a peculiar suffocating odour, resembling that of aqua regia (a mixture of chlorohydric and nitric acids). When it comes in contact with the lining membranes of the nostrils and windpipe, it induces great irritation, cough, watering of the eyes, and other symptoms analogous to a common cold, but accompanied with great prostration of strength if the inspiration has been longer than a second. Chlorine is about 322 times lighter than water. Its specific grav. air, being 1, is 2·5 (Thomson); 2·47 (Gay Lussac and Thenard, Rech. Phys.-Chim. 2, 125); 2·392 (Davy, Phil. Trans. 1810); 2·52 (ib. Elements, p. 236); 2·44 (Theoretic). The theoretic specific gravity may be deduced by multiplying by the atomic weight of the gas, half the spec. grav. of oxygen, $1{\cdot}10563 \div 2 = {\cdot}55281 \times 4{\cdot}437 = 2{\cdot}4528$. The weight of 100 cubic inches, that of air being 31 grs. and the spec. grav. of chlorine 2·44, is 75·64 grains; with a spec. grav. of 2·47, it is 76·57 grs.; with spec. grav. 2·5, it is 77·529 grains; with the weight of air at 30·9451 grs. (Regnault); and spec. grav. 2·44 it is 75·506. The refracting power of chlorine is to that of air as 2·628 to 1 (Dulong, Ann. Chim. 31, 166). Chlorine is an electro negative body, being deposited at the positive pole, when its compounds are decomposed except in the case of its union with oxygen, when the oxygen goes to the positive pole, and the chlorine to the negative.

Liquid Chlorine.—Greenish-yellow fluid, evaporating rapidly in air. Spec. grav. 1·33 (Faraday, Phil. Trans. 1823), non-conductor of electricity, procured by introducing crystals of hydrate of chlorine into one end of a bent tube, to be then

hermetically sealed at both ends, and applying hot water of 90° to 100°, so as to evolve the chlorine from the water. The farther extremity being immersed in ice, the gas by its own pressure soon liquefies; bleaches dry litmus paper (Kemp).

In the figure, *a* represents the end of the tube containing the hydrate of chlorine immersed in warm water, and *b*, the extremity with the condensed fluid chlorine, surrounded by a freezing mixture.

Aqueous Chlorine.—Chlorine is very soluble in water. 1 volume or measure of water dissolves 1·5 chlorine at 32°, 3·04 chlorine at 46°·4, its maximum absorption; at 57°·2, 2·50; at 122° it is 1·09 (Pelouze); at 60°, 2 (Dalton); at 212° almost nothing (Gay Lussac, Ann. Chim. 70, 407). A saturated solution at 42°·8 of chlorine in water; has a spec. grav. of 1·003 (Berzelius); the water has a greenish-yellow colour; the peculiar odour and the astringent taste of the gas. In collecting the gas over water, it is therefore desirable to raise the temperature of the water in the trough up to 80°, so as to diminish the absorption. When the chlorine solution is agitated with common air, almost the whole of the chlorine leaves the water in a few seconds, and hence no air should be left in a bottle of chlorine water. Light decomposes the solution, chlorohydric acid and some perchloric acid (HCl and ClO_7) being formed at the expense of the water. It should therefore be preserved in dark vessels, or bottles covered with black paper.

Decihydrate.— Cl, 10 HO (Davy), 15·687, 125·49, Cl 27·7, HO 72·3. Spec. grav. 1·2, rhombic 8-hedrons and needles or plates; non-conductor of electricity (Solly). Can be sublimed, loses chlorine at 90° (Regnault), and regains it if in a closed tube at a temperature of 70° (Faraday). When exposed to the light, the crystals attach themselves to the sides of the bottle next the light as is the case with camphor. When thrown into alcohol, the temperature rises 8 or 10 degrees, a violent action occurs, chloric ether and muriatic acid being formed, and a small portion of a compound of chlorine, carbon, and hydrogen. The hydrate exhibits the same actions as chlorine. It may be obtained by adding to a bottle full of chlorine one-fourth of water, stopping it up close and surrounding it with ice or snow; a temperature of 33° to 36° suffices. (Faraday, Roy. Inst. Journ. 15, 71)

Preparation of Chlorine.—1. The simplest method of preparing chlorine, and that generally employed, is to introduce into a retort 6 parts of chlorohydric acid of commerce, and then from 1 to 2 parts binoxide of manganese, according to its quality, so as to form a thin paste. The mixture is then shaken, the heat of a gas or spirit-lamp applied to the retort or flask, and the

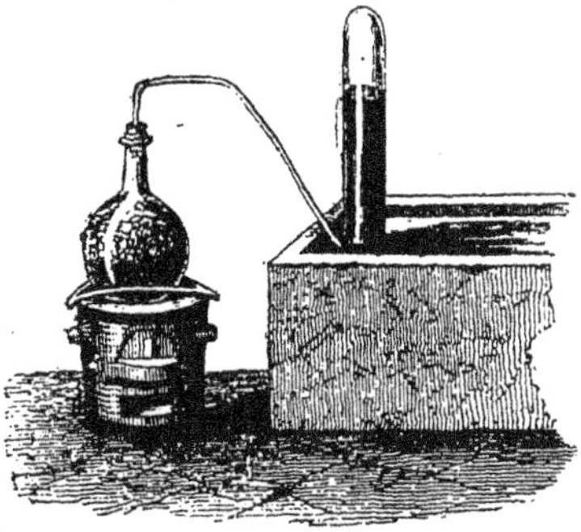

gas, which rapidly escapes, should be received over water heated to 80°, or containing salt in solution, in bottles supplied with greased stoppers, to prevent the pressure of the air by the absorption of the gas from retaining the stoppers. In consequence of the weight of the gas, it may be kept for some time in bottles. It may for the same reason be introduced into bottles by displacement of the common air. If the beak of the retort be introduced into a dry bottle, the chlorine will fall to the bottom, and press out the air above. The action is as follows:—

2. Chlorohydric acid. Binoxide of manganese.
4·437 4·437 ·125 ·125 3·447 1· 1·
Cl Cl H H Mn O O

Cl 2 HO MnCl
4·437 Chl. 2·25 Water. 7·884 Chl. of manganese.

In the manufacture of bleaching powder this process is usually adopted, the chlorine being evolved from chlorohydric acid and black oxide of manganese in a double stone box or still, which is previously boiled in tar, to render it less readily decomposed by the acid. Steam is admitted into the exterior casing, so as to assist the action. Another method is sometimes used where chlorohydric acid is not abundant—depending upon the evolution of chlorine from common salt by the assistance of the oxygen of the black oxide of manganese and sulphuric acid. If the manganese were quite pure, the proportions would be 7·309 common salt, 12·25 oil of vitriol, and 5·447 binoxide of manganese—the sulphuric acid being diluted with an equal bulk of water. The following scheme explains the action:—

Sulphuric acid. Binoxide of manganese. Sulphuric acid. Common salt.
5 3·447 1 1 5 2·872 4·437
SO_3 Mn O O SO_3 Na Cl

$MnOSO_3$ $NaOSO_3$ Cl
9·447 Sulph. of manganese. 8·872 Sulph. of soda. 4.437 Chlor.

Chemical Characters.—1. Chlorine cannot be inflamed, but if a taper be immersed in the gas it burns with a low red flame, giving out much

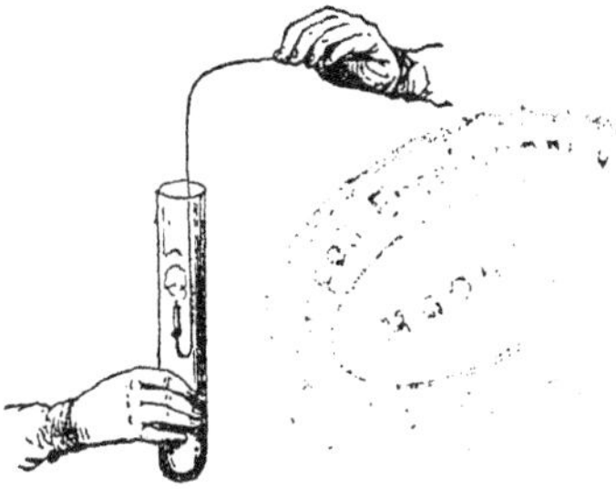

smoke, and is speedily extinguished. 2. Phosphorus when introduced by means of a dipper ladle into chlorine, spontaneously catches fire and burns with a yellowish-white flame. 3. Several of the metals, as antimony, arsenic, potassium, tin, iron, zinc, bismuth, and mercury, when finely divided, heated in a spoon, and then showered into a bottle full of the gas, or immersed in the gas, catch fire and produce *chlorides*. Hence chlorine may be considered a simple supporter of combustion. 4. A copper wire when heated and plunged into the gas burns with great rapidity. 5. Gold and silver leaf likewise burn, the gold with a green, the silver with a white flame. 6. When a feather moistened with oil of turpentine is introduced into chlorine, it bursts into flame with a copious evolution of lamp black. 7. When a piece of blue litmus paper moistened with water is introduced into chlorine, the paper becomes white without being previously reddened. Red, blue, and yellow flowers and green vegetables, are also bleached and cannot be recovered (Scheele, 1774). Dry chlorine has no action on dry litmus, but chlorine condensed into a liquid, bleaches colouring matters (Kemp); the presence of water enables the dry gas to act, possibly therefore by its being in a liquid form (Davy). 8. When equal measures of chlorine and hydrogen are mixed over water in a bottle, and fired by a taper, an explosion occurs, and chlorohydric acid is formed (HCl). 9. When an animal is plunged into chlorine it dies immediately.

Chlorides, Muriates, Hydrochlorates, Chlorohydrides.—Many of the chlorides are formed by the direct action of chlorine upon metals as stated above; many are produced by acting upon the oxides by means of chlorine, the oxygen being replaced by the chlorine; chlorides being formed and also chlorates and hypochlorites; the oxides which are not converted into chlorides without being ignited with charcoal, are alumina, glucina, thorium, and yttria. They are generally soluble in water, with the exception of chlorides of silver, dichloride of mercury, &c. The alkaline chlorides crystallize in cubes as well as the chloride of silver; chlorine at 212° replaces the oxygen in potash and soda. The actions of bromine, iodine, fluorine, and cyanogen, are analogous to those of chlorine.

Practical Application.—See BLEACHING.

Use in Medicine.—A solution of chlorine in water was recommended to remove the odours from the emunctories of Paris in 1785 by Hallé (*Rapport sur les fosses d'aisances*), and also by Fourcroy in 1791. In 1793 it was used in the army of the Rhine against hospital gangrene. It has since been used in a liquid form in scarlet fever; 2 drachms of a saturated solution being diluted with half-a-pound of water, have been administered during the day internally (Braithwaite), and also in putrid Spanish fever to the extent of an ounce per day. It is sometimes employed to spunge over the skin in fevers, and to immerse the hands in for the cure of itch (Thenard and Cluzel). It has been given internally with success in diarrhœa, chronic dysentery, and hydrophobia. To counteract infection it is used in the form of solution of bleaching powder, or chloride of soda, and sprinkled in the chamber. It is also used to remove the odour of sulphuretted hydrogen in sewers and cess pools by decomposition (SH and Cl becoming HCl and S). The vapour of ammonia may be breathed as an antidote where chlorine has been inspired, and alcohol has been employed for the same purpose. A French chemist, Pelletier, died from the irritation produced by the inspiration of this gas. According to experience in Glasgow, chlorine seems to possess no disinfecting power. In 1821 and 1822, the people employed at St. Rollox became infected with typhus, and suffered as severely from it as any others, although a temporary building was erected for some of them, and the use of chlorine on a very extensive scale tried. In 1827, Dr. John Couper thought that they perhaps suffered less than the inhabitants of other districts, but after studying the exhibition of chlorine in this instance as a matter of medical practice, he is most decidedly of opinion of its total inefficacy in preventing contagion. In small-pox, measles, &c. the people at St. Rollox take the infection as readily as any others (Life of C. Macintosh). Chlorine abundantly possesses the property of removing odours. Odours, however, are not characteristic of such poisons as we know to be capable of producing disease. On the contrary, when animal poisons decompose and give out smells, they are incapable of producing disease by inoculation.

Test and Analysis.—Chlorine is detected by its yielding with nitrate of silver a white curdy precipitate, insoluble in boiling nitric acid (AgCl), and thus distinguished from cyanide of silver; also a white curdy precipitate, which disappears when boiled with nitric acid. To determine the amount of chlorine in a solution, we render the solution acid by nitric acid, bring it nearly to the boiling point, and add the nitrate of silver, and stir the solution well—when dried, ignited, fused, and weighed, every 17·935 parts contain 4·437 chlorine, or very nearly one-fourth of its weight is chlorine. When the filter is burned with the last portions of precipitate, a portion of the silver is reduced. This is to be heated with a drop or two of nitric acid, and then chlorohydric acid, and again fused.

Chlorisamide. $C_{32}H_{12}N_4O_6Cl_2$.

Chlorisatic Acid. $C_{16}H_5Cl\,NO_5HO$. By potash on chlorisatine.

Chlorisatine. C_{16}, $H_4Cl\,NO_4$. Yellow prisms by chlorine on isatine.

Chlorisatyde. $C_{16}H_5Cl\,NO_4$. White powder by sulphide of ammonium on chlorisatine.

Chlorite. Spec. grav. 2·823, H 1·5. Dark green small scales or plates, cohering without cement, opaque; B.B. fuses into a black opaque glass; found in quartz veins in Bute and Arran, &c. Silica 27·62, FeO 27·54, alumina 23·71, magnesia 10·96, water 9·16; potash trace.

Chlorite Slate. A green soapy slate, forming a subordinate bed in mica slate, found abundantly on Loch Fine, at St. Catherine's. At Inverary, an ore of sulphuret of nickel occurs in it. It contains SiO_3 27·40, FeO 27·918, Al_2O_3 20·74, HO 8·8, potash 4·8, magnesia, lime, and loss, 7·322.

Chloritoide, or *Chlorite Spar*. Spec. grav. 3·55. A foliated greenish-black mineral from Ural, consisting of SiO_3 24·9, Al_2O_3 46·2, FeO 28·89. B.B. infusible.

Chloroacetamic Acid. *Chloracetam, Chlorocarbamic Acid.* $C_4NH\,Cl_4O_2$. Needles by chlorine on chloracetamide in sunlight.

Chloroacethyphide. C_4, PH_2Cl_3, O_2. White crystals by phosphuretted hydrogen on chloroaldehyde.

Chloroalbine. $C_{12}H_6Cl_2$. White needles, obtained along with terchloro-carbonic acid.

Chloroalide. $C_{10}H_2Cl_6O_6$. White crystals by oil of vitriol, and heat on hydrate of chloral.

Chloroazosuccinic Acid. C_8, NH_2H, Cl_4, O_4. 4-sided prisms by the action of ammoniacal gas on perchlorosuccinic ether.

Chlorobenzamide. From chlorobenzoate of ammonia.

Chlorobenzide and **Chlorobenzine.** See BENZOL.

Chlorobenzoenase. $C_{14}H_7Cl$.

Chlorobenzoenyle. C 27·6, H 0·7, Cl 71·7

Chlorobenzoyle, *Chloracetyle, Chlorocinnamyle.*

Chlorobutyral. C_8H_7Cl, O_2. Clear fluid by chlorine on butyral.

Chlorobutyrene. C_8H_7Cl. Colourless fluid by quintochloride of phosphorus on butyral.

Chlorocaffeine. $C_{16}H_9ClN_4O_4$. By acting on caffeine with chlorine.

Chlorocamphene. $C_{20}H_{12}Cl_4$. See CAMPHENE.

Chlorocarbethamic Acid. $C_{10}NH_2H_4Cl_7,O_{10}$. By ammonia on chlorocarbethamide.

Chlorocarbethamide. (*Chloracetamide?*) $C_{10}H_6Cl_7N_3O_4$. By ammonia on perchlorocarbonic ether.

Chlorocarbohyposulphuric Acid. *Terchlorobisulphite of Methyle.* $C_2HCl_3O_2$ 2 SO_2.

Chlorocarbonic Acid. CO Cl. *Phosgene gas.* By the action of carbonic oxide and chlorine by sunlight.

Chlorocarbonic Ether. CO,Cl,C_4H_5O,CO_2. By chlorocarbonic acid on alcohol.

Chlorocerotal. $C_{54}H_{41}Cl_{13}O_2$. The chloraldehyde of chlorocerotic acid.

Chlorocerotic Acid. $C_{54}H_{42}Cl_{12}O_4$. A resin by chlorine on cerotic acid.

Chlorocinchonine. $C_{38}H_{20}Cl_2N_2O_2$. By chlorine on cinchonine.

Chlorocinnamic Acid. $C_{18}H_6ClO_3HO$. Needles by potash on chlorostyracine.

Chlorocinnose. $C_{18}H_4Cl_4O_2$. By chlorine on oil of cinnamon.

Chlorocodeine. $C_{36}H_{20}ClN,O_6$. By acting on codeine with chlorine.

Chlorocomenic Acid. 2 HO, $C_{12}HCl,O_8$ 3Aq. By chlorine on comenic acid.

Chlorocuminole. $C_{20}H_{12}Cl_2$. By chlorine on cuminole.

Chlorocyanilide. $C_{30}H_{12}ClN_5$. Fine plates by aniline and chloride of cyanogen in alcohol.

Chlorocyanogen. CyCl. A gas by chlorine on cyanohydric acid.

Chlorodibromaniline. C_{12}, H_4ClBr_2, N. By chlorine and bromine on aniline.

Chlorodraconyle. $C_{32}H_{15}O_3Cl_6$. By chlorine on dragon's blood resin.

Chloroenanthic Acid. $C_{14}H_{11}Cl_2O_3$. By potash on chloroenanthic ether.

Chloroenanthic Ether. $C_{14}H_{11}Cl_2O_3$, C_4H_3O,Cl_2. From chlorine and oenanthic ether.

Chlorofilipelosic Acid. $C_{24}H_{13}O_9$.

Chlorofilixic Acid. $C_{26}H_{15}ClO_{10}$. Yellow fluid by chlorine on filixic acid.

Chloroform. *Terchloride of Formyle.* C_2HCl_3. Spec. grav. 1·480 to 1·5, of vapour 4·192, B.P. 141°·8. Peculiar agreeable odour, valuable as an anæsthetic in operations, and therefore largely prepared; but slightly soluble in water, to the bottom of which it sinks; mixes with alcohol and ether; insoluble in sulphuric acid, but gradually decomposed by it; not decomposed by solution of nitrate of silver; decomposed by boiling with potash and alcohol into formate of potash and chloride of potassium; passed over ignited lime there are formed carbon, chloride of calcium, and carbonate of lime; through an ignited copper or iron tube it yields a metallic chloride. *Process.*—Chloroform may be obtained by the action of chloride of lime (bleaching powder) on alcohol, pyroxylic spirit, acetone, acetate of lime or soda, oil of turpentine; or by distilling chloral or chloracetic acid with potash, lime, or barytes. I have found the most economical process to be to distil with a gentle heat in a half-gallon retort 8 ounces bleaching powder, 24 ounces water, 1½ ounce alcohol (·840). The mixture is apt to puff up, but it may be restrained by pouring in cold

water rapidly by a funnel tube, or by applying water exteriorly. Wash the product with its own bulk of water to remove chlorine. Add chloride of calcium to the oil, and allow them to stand for an hour. Pour off the oil, and distil with its own bulk of sulphuric acid. To detect the presence of alcohol, add a few small crystals of bichromate of potash, and then sulphuric acid; if alcohol be present, a green colour will appear in contact with the crystals. By the process above given, Mr. James King, in my laboratory, first obtained it of the spec. grav. 1·5 in 1848 (Glasg. Phil. Proc. 2,419).

Chlorogenic Acid. $C_{14}H_8O_7$. Obtained by precipitating the double salt of chlorogenite of potash and caffeine existing in coffee, by acetate of lead, and decomposing by sulphohydric acid.

Chlorogenine. A green powder from madder.

Chlorohelicine. $C_{26}H_{15}Cl\,O_{14}$. By chlorine on *helicine*.

Chlorohellenine. $C_{15}H_9Cl\,O_2$. By chlorine on *hellenine*.

Chlorohumic Acid. $C_{32}H_{26}Cl_2O_{17}$. By chlorine on humic acid.

Chlorohydranile. $C_{12}\,Cl_4\,O_2$, 2 HO. Pearly plates by sulphurous acid on chloranile.

Chlorohydric Acid. *Muriatic Acid.* See HYDROCHLORIC ACID.

Chlorohydrokinone. $C_{25}H_{10}Cl_2O_8$. By chlorine on kinone.

Chlorohyponitric Acid. NO_2Cl_2. Yellow fluid by chlorine on nitric acid.

Chloroiodoform. C_2H, ICl_2. Yellow oil by distilling iodoform with corrosive sublimate or quintochloride of phosphorus.

Chlorokinone. $C_{25}H_6Cl_2O_8$.

Chloromelal. $C_{60}H_{45\frac{1}{2}}Cl_{14\frac{1}{2}}O_2$. By chlorine on melissin.

Chloromelan, or Cronstedtite.

Chloromenthene. $C_{20}H_{18}$, HCl. By chlorine on menthene.

Chloronaphthalane. $C_{40}H_5Cl_{11}$.

Chloronaphthalase. *Hexachlonaphthine.* $C_{20}H_2Cl_6$. Prisms by chlorine on chlornaphthise.

Chloronaphthalese. Unknown.

Chloronaphthalise. *Octochlonaphthine.* C_{20} Cl_8. Crystalline solid.

Chloronaphthase. *Chlonaphthine.* $C_{20}H_7Cl$. An oil, by acting on subchloride of naphthaline with an alcoholic solution of potash.

Chloronaphthese. *Dichlonaphthine.* $C_{20}H_6$ Cl_2. Occurring in seven different isomeric forms.

Chloronaphthise. *Trichlonaphthine.* $C_{20}H_5$ Cl_3. In six isomeric forms.

Chloronaphthisic Acid. $C_{20}H_5ClO_6$. By acting with nitric acid on chloride or chloronaphthose.

Chloronaphthone. $C_{40}H_9Cl_7$.

Chloronaphthose. *Tetrachlonaphthine.* C_{20} H_4Cl_4. In four isomeric forms.

Chloronaphthuse. *Pentachlonaphthine* C_{20} H_3Cl_5. Unknown.

Chloroniceamide. $NH_2,C_{12}H_4Cl\,O_2$. By ammonia on chloroniceic ether.

Chloroniceic Acid. $C_{12}H_4ClO_3HO$. By chlorine on benzoic acid.

Chloronicene. $C_{10}H_5Cl$. By distilling chloroniceic acid with lime.

Chloronicine. $C_{10}H_6Cl\,N$. A product from chloronicene and nitric acid.

Chloronitrous Acid. NO_2Cl. By chlorine on nitric acid.

Chloropal. Spec. grav. 1·727 to 2·105, H 3 to 4. Greenish-yellow masses, infusible before the blowpipe, from Hungary and Ceylon. The latter contains silica 53, Fe_2O_3 26·04, MgO 1·4, Al_2O_3 1·8, HO 18.

Chlorophacite, probably chlorophæite from the Faroes.

Chlorophaeite. Spec. grav. 2·02. A green brittle silicate of iron, occurring in the Scuirmore, Rum, in amygdaloid. B.B. not altered.

Chlorophane. Green fluor spar, which when heated exhibits a green phosphorescence.

Chlorophenesic Acid. $C_{12}H_3Cl_2O$ HO. By chlorine on carbolic acid.

Chlorophenise. *Chlorobenzine.* $C_{12}H_6Cl_6$.

Chlorophenisic Acid. *Chlorindoptic Acid.* $C_{12}\,H_2\,Cl_3\,O$, HO. By further action on the chlorophenesic acid.

Chlorophenusic Acid. $C_{12}Cl_5OHO$. By further action on the preceding.

Chlorophenyle. $C_{12}H_5Cl$. By quintochloride of phosphorus on carbonic acid.

Chlorophtalisic. $C_{16}HCl_3O_6$. By nitric acid on chloronaphthalase.

Chlorophylle. $C_{18}H_9NO_8$? The green-colouring matter of plants.

Chlorophyllite. Spec. grav. 2·709, H 1·75. Greenish 6 and 12-sided prisms, pearly lustre, translucent, brittle, B.B., fuses on edges, with carbonate of soda a greenish enamel; silica 45·2, alumina or phosphate of alumina 27·6. MgO 9·6, FeO 8·24, MnO 4·08, HO 3·6. Found at Brevig, Norway, and Maine, U.S.

Chloropicrine. $C_2NCl_3O_4$. Clear oil by distilling picric acid with bleaching powder.

Chloroproteic Acid. A hypothetic body, supposed to consist of chlorine and proteine.

Chloropurreic Acid. *Chloreuxanthic Acid.* $C_{40}H_{14}Cl_2O_{21}$. Golden scales; insoluble in alcohol; by the action of chlorine on purreic acid.

Chlororceine. $C_{18}H_{10}N,O_8Cl_2$. From chlorine on orceine.

Chlorrubine. $C_{12}H_4O_3$. Red powder by heating rubichloric acid and HCl.

Chlorosalicine. $C_{26}H_{17}ClO_{14}$. By chlorine on salicine; a di and trichlorosalicine exist.

Chlorosalicylic Acid. *Chloride of salicyle.* $C_{14}H_5O_4Cl$. Rectangular plates, by chlorine on anhydrous hydride of salicyle.

Chlorosamide. $C_{14}\,H_5\,O_2\,N_{\frac{2}{3}}\,Cl$.

. **Chlorosalicylimide.** $C_{42}H_{15}N_2Cl_3O_6$. By ammonia on the preceding.

Chlorosaligenine. $C_{14}H_7ClO_4$. Produced by fermenting chlorosalicine, which consists of sugar and chlorosaligenine.

Chlorospinelle. Spec. grav. 3·592; H 8. Grass-green 8-hedrons in talc slate in the Ural. B.B. infusible; forms a green glass with borax. Silica 27·13, alumina 60·735, FeO_2O_3 11·731, CuO ·445, CaO ·135.

Chlorostilbase. $C_{28}H_{11}Cl$. By Cl on stilbene.

Chlorostrychnine. $C_{44}H_{23}ClN_2O_4$.

Chlorostyracine. $C_{60}H_{21}Cl_7O_6$.

Chlorosuccilamide. $C_4NH_2Cl_2O$? Silky needles by evaporating in vacuo a solution of chloroazosuccinic acid in aqueous ammonia.

Chlorosuccilic Acid. *Bichloroacetic Acid.* $C_4H_2Cl_2,O_4$. By boiling chlorosuccilamide with caustic potash.

Chlorosuccinic Acid. $C_6H_3Cl_3O_4$. Or metacetonic acid in which H_3 are replaced by Cl_3 ?

Chlorosulphokinone. *Brown.* $C_{25}H_8Cl O_8S_4O$? The *orange* has H_2 less.

Chlorosulphonaphthalic Acid. $HO, C_{20} H_6, SO_2, Cl, SO_3$.

Chloroterebenes. $C_{20}H_{14}Cl_2$ and Cl_4.

Chlorous Acid. ClO_3. A dark greenish-yellow gas, formed by mixing 1 tartaric acid, 2 chlorate of potash, in a retort, and pouring on the mixture 6 nitric acid and 8 water.

Chlorovalerisic Acid. $C_{10}H_7Cl_3O_4$. By chlorine on valerianic acid without light.

Chlorovalerosic Acid. $C_{10}H_5Cl_4O_3$. By Cl on valerianic acid in light.

Chloroxalic Ether. $C_{12}Cl_{10}O_8$. From chlorine and oxalic ether.

Chloroxamethane. $C_8H_2Cl_5NO_6$. By dry ammonia chloroxalic ether.

Chloroxenaphthalise, Oxide of. $C_{20}Cl_6 O_2O_2$. By nitric acid on chloronaphthalase.

Chloroxenaphthalisic Acid. $C_{20}HCl_5 O_6$. By potash on the preceding.

Chloroxenaphthose, Oxide of. $C_{20}H_4 Cl_2O_2,O_2$. By nitric acid on chloride of chloronaphthose.

Chloroxethide. C_4H_5O, $C_4 O_6$? Yellow oil by adding perchloroxalic ether to spirit, and decomposing by water.

Chloroxethose. C_4Cl_3O. An oil by perchloric ether on sulphide of potassium.

Chloryle. $C_6H_3Cl_3O_4$. A fluid by the action of chlorine on acetate of methyle.

Chocolate. A material for extracting a beverage from, prepared by pulverizing the kernel of the *Theobroma cacao*, mixing it with sugar, and compressing it into a cake.

Cholacrole. $C_8N_2H_5O_{13}$. Pungent fluid in the action of NO_5 on choloidic acid.

Cholalic, *Cholanic*, *Cholonic*, *Cholic*, *Choleic*, acids. See Bile.

Cholesteric Acid. $C_8H_4O_4HO$. Needles fusing at 136° by the action of nitric acid on cholesterine and choloidic acid.

Cholesteriline. $C_{32}H_{26}$, F.P. 464°. White needles by sulphuric acid on cholesterine. Other two hydrocarbons are formed at the same time; *b* crystalline $C_{32}H_{18}$, F.P. 491°; and *c* a resin $C_{27}H_{22}$ 260°.

Cholesterine. $C_{81}H_{69}O_3$, or $C_{36}H_{32}O$. Spec. grav. 1·03. F.P. 337°, 278°, C 85·09, H 11·85, O 3·06. White pearly scales, without taste and smell, unchanged by distillation free from air; soluble in 9 parts alcohol (·840), 5·54 (·816), in 12 cold, 22 boiling ether; soluble in pyroxylic spirit; with difficulty in oil of turpentine; 4 parts of dry soap, dissolved in water, take up one part cholesterine; strong sulphuric acid decomposes it into the three preceding hydrocarbons. It exists in the blood, bile, brain, yolk of egg, and may be taken up from biliary calculi by boiling alcohol, from which it is deposited on cooling. It has been suggested that cholesterine may be derived from cholalic acid.

Cholesterone (*a*). F.P. 154½. Rhombic prisms, containing C 87·7, H 12·1, by the action of phosphoric acid on cholesterine.

Cholesterone (*b*). F.P. 347°. Crystals with the same composition as the preceding.

Cholestrophane. *Nitrotheine.* $C_{10}2NH_2, H_2O_4,O_2$? White pearly plates and 8-hedrons. Soluble in 3 cold water; in ether and alcohol; obtained by the action of nitric acid or chlorine on caffeine; by potash, yields ethylamine.

Choloidanic Acid. $C_{16}H_{12}O_7$. Crystalline; a product of the oxidation of choloidic acid.

Choloidic Acid. $C_{48}H_{39}O_9$. Resin, insoluble in water, soluble in alcohol; obtained by boiling cholic acid or bile with chlorohydric acid; when longer boiled, it loses 3 atoms water, and leaves *dyslysine.* When heated with nitric acid it yields acetic, butyric, valerianic, caprylic, and capric acids, nitrocholic acid, and cholacrol, choloidanic acid, and cholesteric acid.

Chondrine. $C_{16}H_{13}N_2O_7$. Brownish elastic masses, insoluble in alcohol and ether; swells up in cold water, soluble in hot water, gelatinizing on cooling; from its solution it is precipitated by a drop of mineral acids, soluble in excess; by a stream of carbonic acid, by alcohol, diacetate of lead, chloride of tin, perchlorides of iron, platinum, and mercury, alum, sulphate of alumina. Caseine and chondrine are the only animal substances precipitated by acetic acid; it is obtained by boiling in water the permanent cartilages, as of the larynx, ribs, or joints, from twelve to eighteen hours. It is also yielded by skins, tendons, the cornea, &c.

Chondrodite. Spec. grav. 3·118 to 3·199; H 6 to 6·5. Yellow, brown, or greenish oblique prisms, from N. Jersey and Finland, in limestone. B.B. becomes black, white on charcoal, without fusing; dissolves in borax to a clear glass; gives out fluohydric acid with oil of vitriol. Silica,

36·; MgO 54·64, Fe_2O_3 3·97, F 3·75, HO 1·62, *Form.* MgF, 2 (3 MgO, SiO_3).

Chonicrite. A white massive mineral from Elba, consisting of SiO_3 35·69, Al_2O_3 17·12, MgO 22·5, CaO 12·6, FeO 1·46, HO 9·.

Chrismatine. A yellow resin on calc spar at Wettin, Halle.

Christianite. A name given to anorthite, and likewise to lime harmotome, or Phillipsite.

Chrome Iron Ore. Spec. grav. 4·321, H 5·5. Regular 8-hedrons or massive; colour

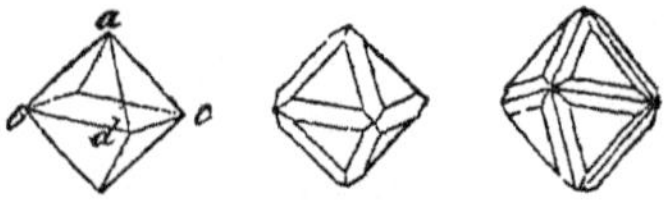

between iron-black or brownish-black, streak brown; lustre imperfect metallic; fracture uneven, imperfect, conchoidal; distinguished from magnetic iron ore by its not being attracted by the magnet, and by its resinous lustre. A specimen from Baltimore yielded Berthier protox. iron 35·, oxide of chromium 51·6, alumina 10·, silica 3·, but its composition seems to vary. By calculation, the composition would be, FeO 34·44, Cr_2O_3 57·4, Al_2O_3 8·15, if we represent the formulæ of magnetic iron, chrome iron ore, and Franklinite respectively as follows:—

2 FeO, 2 FeO, 3 Fe_2O_3, Fe_2O_3, Mag. iron ore.
2 ZnO, 2 FeO, 3 Fe_2O_3, Mn_2O_3, Franklinite.
2 FeO, 2 FeO, 3 Cr_2O_3, Al_2O_3, Chrome iron ore.

I have found the best method of decomposing this, one of the most refractory minerals, to be, to fuse it with bisulphate of soda. It is found in Shetland; near Baltimore; Roraas, Norway, &c.

Chrome Ochre. *Native Sesquioxide of Chrome*, *Wolchonskoite*, *Miloschine*, *Silicate of Chromium*, and *Iron*. Forms a rich green coating on certain rocks, but has often, I suspect, been confounded with native carbonate of nickel, which I first detected on the chrome iron ore of America. From Okhansk, it consists of SiO_3 27·2, Cr_2O_3 34, Fe_2O_3 7·2, MgO 7·2, HO 23·2. The iron is sometimes replaced partially by alumina.

Chromic Acid. CrO_3 6·5, 50·6. Pure chromic acid may be obtained by precipitating the solution of the fused mass of chrome iron ore and nitre, when saturated with nitric acid, with acetate of lead, or nitrate of barytes. The chromate of lead or barytes is then washed and ignited; 4 parts of chromate of lead, or $3\frac{1}{4}$ chromate of barytes, are then to be heated with 3 parts of pure fluor spar previously ignited, and 5 parts of sulphuric acid, as concentrated as possible in a lead retort with the aid of a spirit-lamp. Red vapours are evolved which, when condensed by water, are converted into fluohydric and chromic acids. By evaporation in a platinum vessel to dryness, pure chromic acid remains (Unverdorben). When cautiously evaporated, it is often obtained in 4-sided rich crimson-coloured prisms, with square or rectangular bases. If the vapours, consisting of fluochromic acid, be received in a large platinum crucible, the chromic acid deposits in large needles (Berzelius). It may be procured likewise by passing fluosilicic acid through a solution of bichromate of potash. The potash falls in union with the fluosilicic acid, the chromic acid remaining in the solution, and is procured by evaporating to dryness. Chromic acid is useful in the laboratory for testing, in consequence of the facility with which it parts with its oxygen. For this purpose it may be prepared nearly pure, by the following process. Form a hot saturated solution of bichromate of potash and allow it to stand at rest to cool, until its excess of salt is deposited. To 1 part of this cold saturated solution, add $1\frac{1}{2}$ part of oil of vitriol. The mixture on cooling gradually deposits crystals of chromic acid, from which the fluid is to be poured off, and the crystals dried on a porous tile. Chromic acid when evaporated to dryness is black; but when it cools it is deep red; when ignited, oxygen is evolved, and oxide of chrome left. Chromic acid decomposes alcohol, and gives origin to aldehyde, by substracting hydrogen from the spirit. I have found this to be an excellent test of alcohol in solutions. It affords such a delicate method of testing, that a drop or two of alcohol may be distinguished in an ounce of water. All that is necessary is to drop a few crystals of chromic acid into the liquor to be tested, when the odour of aldehyde will be perceptible, and oxide of chrome will be precipitated. A crystal of bichromate of potash, and a drop or two of sulphuric acid, answer the same purpose. According to Gay Lussac, chromic acid unites with sulphuric acid, and forms ruby-coloured prisms (Ann de Chim. xvi. 102); but this is now viewed as doubtful. Chromic acid is considered by Mitscherlich as isomorphous with sulphuric acid and selenic acid, and to consist of oxygen 46·03, and chromium 53·97. Many of the metallic chromates are insoluble in water; the alkaline salts are characterized by their brilliant colours, and are all soluble in water. The most important of these salts are the chromate and bichromate of potash.

Chromium. Cr 3·5, 28.26·7. *Hist.*—Chromium was discovered in 1797, by Vauquelin, in the chromate of lead (red lead ore), a beautiful red mineral from Beresow in Siberia, crystallizing in 4-sided prisms, and was so named from χρωμα, colour, from its power of communicating tints to other bodies. It was afterwards examined by Klaproth and others (Berzelius, Ann. of Phil. iii. 101). It occurs in nature in the form of chromate of lead, cupreo chromate of lead (2 (PbO CrO_3) + CuO), but most abundantly in the state of chrome iron ore, a compound of green oxide of chrome and of protoxide and peroxide of iron. The metal may be obtained from the green oxide by heating it in a forge, mixed with charcoal, at a white heat. It possesses a grayish-white colour, and a spec. grav. of 5·9; is brittle, easily pulverized, and not attracted by the magnet. It is a conductor of electricity. The acids can scarcely, if at all, be made to oxidize it, but it dissolves in fluohydric acid with the application of heat. When fused with the alkalies, or alkaline carbonates, or nitrates, it is converted into chromic acid. The metal may be procured with greater facility by passing a stream of ammoniacal gas over the chloride of chromium, salammoniac being formed (Liebig). The oxide may be obtained on the large scale by fusing chrome iron ore (a compound crystallizing in octahedrons near Baltimore, and consisting of 52·95 green oxide, 29·24 peroxide of iron and 12·22 alumina), with an equal weight of nitre, in a clay or iron crucible, boiling the fused mass with water, hydrochloric acid, and sugar, or oxalic acid, evaporating to dryness, dissolving in acid, neutralizing. If the liquor is now boiled, the iron falls, and the chrome and alumina may be separated by caustic soda, which dissolves the alumina, and leaves the chrome.

Oxide of Chromium, Sesquioxide of Chrome.—Cr_2O_3 10· is a green powder acquiring a brownish tint by ignition. It may be obtained by boiling a solution of bichromate of potash with a portion of oxalic acid or sugar and hydrochloric acid. The carbon of the sugar or oxalic acid unites with oxygen of the chromic acid, and is converted into carbonic acid, while the green oxide of chrome is dissolved by the mineral acid. The action is represented by the accompanying scheme:—

$$2\ (CrO_3)\text{ chromic acid, } 3\ C_2O_3\text{ oxalic acid,} \\ = 1\ Cr_2O_3,\ 6\ CO_2.$$

On the addition of caustic ammonia the hydrate of the green oxide precipitates. It may also be obtained from this salt by passing a current of sulphuretted hydrogen through its solution. It may likewise be formed by heating bichromate of potash with or without salammoniac, and washing the oxide free from potash. The oxide may also be procured by heating the chloride of chromium in the air, or by igniting the chromate of mercury, or by heating the chromate of potash with its own weight of muriate of ammonia in a close crucible (Wöhler). For commercial purposes the green oxide may be formed by fusing chrome iron ore with nitre, washing the fused mass, and boiling the liquor with sulphur: a sulphuret of potassium is formed which precipitates the green oxide (Frick). It may be obtained in rhombohedral crystals, by passing the vapour of chlorochromic acid through a red hot glass tube (Wöhler, Ann. de Chim. lvii). *Green oxide* when nearly heated to redness, glows or becomes ignited like tinder, a property possessed also by zirconia. After this it is no longer soluble in acids, but in order to render it soluble it must be fused with an alkali. It is isomorphous with sesquioxides of iron, manganese, and alumina. *Practical applications.*—The green oxide of chrome constitutes one of the best greens employed as a paint, especially for porcelain. It enters as an ingredient into the English pink used for porcelain, a colour formed of 100 parts peroxide of tin, 33 chalk, and 1 of oxide of chrome. The part which it plays in this mixture is not understood, since some other substances are said to answer equally well when substituted for it (Sims). It is supposed to act as the colouring substance of green serpentines and of the emerald (Vanquelin.)

Brown Oxide. Chromate of Chrome.—This oxide, which appears to be a mixture of oxide of chrome and chromic acid, may be obtained by passing a current of sulphurous acid through a solution of the chromate or bichromate of potash. It possesses a flea-brown colour, has no taste, and undergoes no alteration in the air. When washed with caustic ammonia it assumes the colour of the green oxide. It ordinarily consists of 3 $Cr_2O_3 + CrO_3$ (Thomson). It dissolves in acids, and forms the usual salts of oxide of chrome.

Protoxide of Chromium.—CrO 4·5. Obtained in combination with acids by Peligot, but has not been isolated, as it acts upon water, and is converted into the

Protosesquioxide.—Cr_3O_4, or $CrO\ Cr_2O_3$. Brown. Obtained by precipitating a salt of the protoxide with caustic alkali. When heated in the air it changes into sesquioxide.

Binoxide.—CrO_2 5·5, or chromate of sesquioxide of chromium $Cr_2O_3CrO_3 = 3\ (CrO_2)$, may be prepared by pouring, drop by drop, a solution of chromate of potash into a sesquioxide salt; another form is obtained by precipitating chrome alum by chromate of potash. Yellowish-green.

Chromochlorite. An American variety of serpentine.

Chromocyanogen. Cy_6Cr_2. Obtained by heating chromium with ferrocyanide of potassium in union with potassium.

Chromotype. This term is applied to the mode of preparing photographic paper by a solution of 1 part sulphate of copper, and 4 parts of a saturated solution of bichromate of potash.

Chromyle, or Xanthophylle.

Chryanisic Acid. $C_{14}H_4N_3O_{13}$. By nitric on anisic acid.

Chryiodamide. $C_{28}H_8N_3O_{13}$. Blue body along with oxide of chryiodammonium.

Chryiodammonium Oxide. $C_{28}H_{10}N_4O_{13}$. Blue by ammonia on chryiodine.

Chryiodine. $C_{56}H_8N_3O_{28}$. Violet body by sulphuric acid on chrysammic acid.

Chrysammic Acid. $C_{15}HNO_{13}, HO$. Greenish-yellow, shining scales; slightly soluble in boiling water; soluble in ether, alcohol, and strong acids; formed by the action of nitric acid on aloes.

Chrysammide. $C_{14}H_2NO_{11}NH_2$. By dry ammonia on chrysammic acid.

Chrysammin-amide. $C_{28}H_{10}N_7O_{21}$. By boiling ammonia with chrysammic acid.

Chrysanilic Acid. $C_{28}H_{11},N_2O_6$? By the action of potash on indigo.

Chrysatric Acid. $C_{24}H_6N_3O_{15}$. By alkalies on chrysammic or aloetic acids.

Chrysene. $C_{12}H_4$, F.P. 450°. Yellow powder; found in the last fifth during the distillation of wood tar; it is only slightly soluble in ether, by which it may be separated from pyrene.

Chrysindamide. $C_{28}H_9N_5O_{18}$. Blue by SH on chrysammic acid in ammonia.

Chrysindin-Ammonium Oxide. $C_{28}H_8N_5O_{13}$. By ammonia and SH on chrysammide.

Chrysite, or Chrysolite.

Chrysoberyl. *Cymophane.* Specific grav. 3·51 to 3·754 and 3·733, H 8·5. Right rectangular prisms; sometimes 8 or 12-sided, with 6 or 8-sided pyramidal terminations; sometimes it occurs as 3 prisms crossing, fig. 2; colour asparagus-green to greenish-white and gray; lustre vitreous, transparent to translucent. B.B. does not fuse *per se*, nor with soda; with borax into a glass. Alumina 76·75, glucina 17·79, FeO 4·49, HO ·48 (Brazil). Found in granite in Connecticut; but in Brazil in alluvium along with topaz.

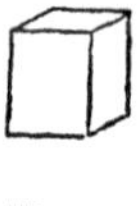

Fig. 1.

Fig. 2.

Chrysocolla. See SILICOCARBONATE of COPPER.

Chrysoidine. $C_{24}H_{22}O_8$. Yellow body from the berries of asparagus.

Chrysolepic Acid. $C_{12}H_2N_3O_{13}HO$. Golden scales; by nitric acid on aloes, believed to be the same as picric acid.

Chrysolite. *Peridote, Olivine, Chusite, Anhydrous Silicate of Magnesia.* Spec. grav. 3·338 to 3·41, H 6·75. Asparagus, grass-green or olive, right rectangular prisms from the Levant and as olivine in basalt. B.B. infusible; with borax a glass coloured by iron. Silica 41·54, magnesia 50·04, protox. iron 8·66, MnO ·25, alumina ·06.

Chrysolite of Saxony is Topaz.

Chrysopal, or Chrysoberyl.

Chrysophane, Clintonite, Seybertite, Holmesite.

Chrysophanic Acid. See PARIETIC ACID.

Chrysoprase. An apple-green variety of calcedony, owing its colour to *oxide of nickel.*

Chrysoretine. A brown resin, from senna leaves.

Chrysorhamnine. $C_{23}H_{11}O_{14}$. Yellow-colouring matter of Persian berries, *Rhamnus infectorius.*

Chrysotile, or Metaxite, a variety of serpentine. Specific grav. 2·223. From the Vosges. SiO_3 41·58, MgO 42·61, FeO 1·69, Al_2O_3 ·42, HO 13·7.

Chrysotype. Photographic paper, prepared with a solution of ammonia, citrate of iron, dried, and then, in the dark, with a solution of yellow prussiate of potash.

Chusite. A decomposed variety of Olivine, from Limbourg.

Chyle. The digested food in its passage from the stomach and intestines to the blood. For examination it has usually been taken from the thoracic duct. It is sometimes a colourless fluid, but generally a milky liquid, with a neutral reaction, and a specific gravity from 1021½ to 1022, and containing 4·8 to 7·8 of solid matter per cent. The following is the composition of chyle, according to various analyses:—1, Chyle of dog, from vegetable food (Prout); 2, chyle of dog from animal food (Prout); 3, chyle (Simon); 4, chyle of ass (Rees):—

	1.	2.	3.	4.
Water,............	93·6	89·2	94·0	95·5
Fibrine, ...	·6	·8	·03	·12
Albumen,.........	4·6	4·7	4·3	1·2
Albumen, with red matter,.....	·4	4·6	·04	
Sugar, ...	trace	—		
Oil, ...	trace	trace	0·12	trace
Salts, ...	·8	·7	1·01	·58
Extracts by alcohol and water,				1·5

The milky appearance of the chyle is said to be produced by the pancreatic juice on the fat.

Chyme. The pulpy condition of the digesting food in the stomach has been so named; it is obviously in a transition state from the insoluble to the soluble condition, and is therefore a heterogeneous mixture of various colours, according to the nature of the food.

Chynole. C_3 HO.

Chytophyllite. A slag consisting of bluish pearly scales, from iron works at Altenau. Spec. grav. 2·94, H 5·5. B.B. fuses into a greenish-black non-magnetic bead. SiO_3 54·89, Al_2O_3 5·07, FeO 20·79, CaO 20·34.

Cichorium Intybus. *Chicory. Wild Succory,* or *Wild endive. Chicorée.* This plant is extensively cultivated in Yorkshire and on the continent for its roots, which, when roasted and pulverized, are used in mixture with coffee. It must be considered as an adulteration, as it contains no principle analogous to that of tea or coffee. The permission by the legislature to use this article in competition with coffee, an article which pays duty, is to be regretted; but now that it has become a favourite with the public, the withdrawal of the admission would be unpopular in the highest degree. 100 parts of best chicory contain 10·11 water, 8·9 ash, and give up 72·3 to boiling water. 100 parts of second quality contain 10 water, 36·8 ash, and yield 48·5 to boiling water. 100 grammes chicory (1544¼ grs.) yield 35 grammes of extract to a litre (1·76 pint) of water, which contains ·574 gramme of nitrogen = 3·55 gramme albuminous matter. Yorkshire chicory contains in its ash sulphuric acid 10·29; chlorine 4·93; carbonic acid 1·78; phosphoric acid 10·66; silica 3·81; sand 9·32; potash 33·48; soda 8·12; lime 9·38; magnesia 5·27; sesquioxide of iron 3·81. Chicory contains a large quantity of sugar, from 23·76 to 35·23 per cent. in the dry root, and in roasted chicory are retained from 9·86 to 17·98; while coffee beans contain before roasting from 5·7 to 7·7 per cent. sugar (Graham, Campbell, Stenhouse). It is to the charring of the sugar, or its conversion into caramel, that chicory owes its bitter and colouring power. Chicory is distinguished from coffee—1. By its containing a much larger amount of sugar, 11·98 to 17·98 per cent. in roasted chicory, from 0 to 1 per cent. in roasted coffee. 2. By the large amount of silica and sand in its ash, 2 to 15 per cent.—in coffee less than ½ per cent.; also by less carbonic acid 2 to 3 per cent.—in coffee 15 to 18 per cent. and 3 by the red colour of the ash, from the large quantity of iron. When roasted chicory, in powder, is dropped into cold water, a pale amber-yellow colour gradually forms round each particle; but roasted coffee powder gives out no such colour. This arises from the presence of more soluble matter in chicory—100 grammes chicory give a fluid of specific gravity 2°·5, (Beaumé), and an intensity of colour, 130, while the respective numbers for coffee are 1°·50 B. and 100. When boiled in caustic potash chicory separates into two kinds of particles, pitch-brown and pale brown, while coffee has nearly a uniform colour, unless any fibres of skin or parchment be present. Chicory, likewise, contains lacticiferous vessels and coffee none. The following drawings are by Dr. J. D. Hooker:—

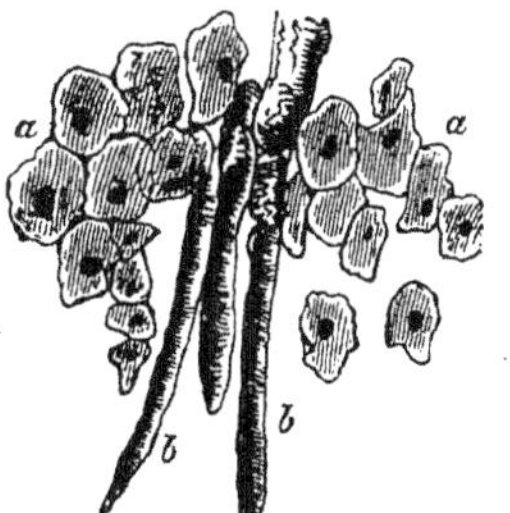

Lacticiferous Tissue.

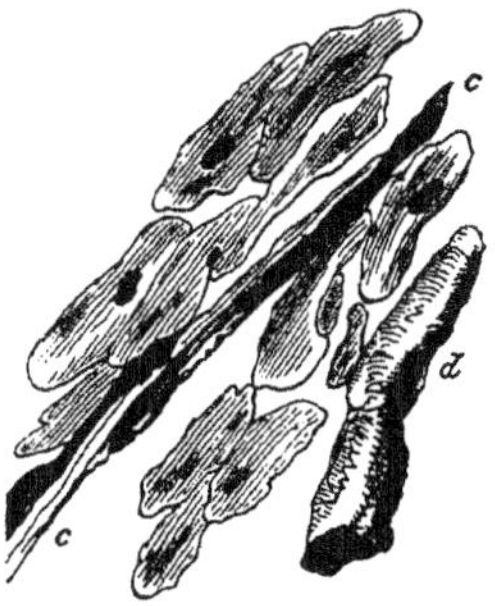

Vascular Tissue.

Cicutine. An unexamined principle from *Cicuta virosa.*

Cider. A fermented liquor from the juice of apples, extensively made in Gloucestershire and other parts of England.

Cimolite. A grayish or reddish clay from the island of Argentiera, (Cimola), consisting of SiO_3 63·53, Al_2O_3 22·70, HO 42·42 (from Alexandrowsk), or Al_2O_3 3 SiO_3 3 HO.

Cinchona. The barks of different trees appear to go under this designation, although no certainty seems to prevail on the subject. The barks known in commerce are the *pale*, *red*, and *yellow*. They contain certain bitter alkaloids, quinine, cinchonine, aricine, united or accompanied by certain acids, kinic and kinovic acids. They contain also tannic acid, colouring matters, and resins.

Cinchonicine. A new base isomeric with cinchonine, obtained by heating sulphate of cinchonine with a little water and sulphuric acid for three or four hours at 248°, 266°; sulphate of cinchonicine is thus obtained. If heated above this, a red resin is formed.

Cinchonidine. A new alkaloid existing in commercial quinine, which is not rendered green by chlorine and ammonia, and is isomeric with cinchonine.

Cinchonine. $C_{38}H_{22}N_2O_2 = \overset{+}{Ci}$. Colourless, transparent, 4-sided prisms; F.P. 165° C. (329° F.) sublimes like benzoic acid, soluble in 2500 boiling water; insoluble in ether, which dissolves quinine; 100 strong spirit dissolve 3 parts; forms a series of crystalline salts, often subsalts, insoluble in ether; it is obtained along with *quinine*, and is separated from that alkaloid by hot alcohol, which, on evaporation, allows the cinchonine to deposit in crystals, and retains quinine. With chlorine and bromine are formed, chloro and bromo cinchonine by the replacement of 1 atom of H by 1 atom Cl. Cinchonine and quinine both yield quinoline with hydrate of potash when heated. With bromine it forms *Dibromocinchonine.* $C_{38}H_{20}Br_2N_2O_2$.

Cinchotine, or *B. Quinine.* $C_{20}H_{12}NO_2$. Rhomboids in commercial cinchonine.

Cinchovatine. $C_{46}H_{27}N_2O_8$. A crystalline alkaloid from Cinchona ovata.

Cinine. A crystalline body in the seeds of Cina.

Cinnabar. Sulphide of mercury.

Cinnameine. $C_{36}H_{18}O_4$. An oil existing in balsam of Peru. When boiled with potash it yields peruvine and cinnamic acid, from which it has been compared to a fat, the peruvine being the glycerine. It has been found to contain cinnamic acid, $C_{18}H_7O_3$, and a new ether, $C_{18}H_7O$. At 26° it deposits in crystals of hydride of cinnamyle, $C_{18}H_8O_2$.

Cinnamene. $C_{18}H_9$. The hypothetic radical of cinnamyle.

Cinnamic Acid. $C_{18}H_7O_3HO$. F.P. 240°; B.P. 555°. Hard transparent prisms by exposing oil of cinnamon to the air; also by treating balsam of Peru with potash, when cinnamate of potash results.

Cinnamic Acid, Anhydrous. $C_{18}H_7O_3$. White needles. F.P. 260½°; insoluble in water and cold alcohol; rendered acid by boiling water; formed by acting on 6 parts dry cinnamate of soda with 1 oxychloride of phosphorus, or from chloride of cinnamyle and oxalate of potash.

Cinnamole. *Cinnamine.* $C_{16}H_8$. Spec. grav. ·88, of vapour 3·570°. B.P. 192° or 284°. Colourless fluid, insoluble in water, soluble in alcohol and ether, by distilling cinnamic acid with lime.

Cinnamon Oil. *Hydride of Cinnamyle, Purified Oil of Cinnamon.* $C_{18}H_9O_2$, or CiH. A fragrant oil obtained by purification of fresh oil of cinnamon, $C_{20}H_{11}O_2$. With 6 atoms oxygen from air, it becomes hydride of cinnamyle and 2 resins, and with 2 atoms more the hydride becomes cinnamic acid.

Cinnamon Stone. Cinnamon-coloured garnet.

Cinnamyle. $C_{18}H_8O_2$. The hypothetic radical of the cinnamon series.

Cinnanilide. $C_{30}H_{13}NO_2$. Needles from aniline and chloride of cinnamyle.

Cinn-anisidide Nitrated. $C_{32}H_{14}N_2O_8$.

Cipoline. A variety of carbonate of lime.

Cissampeline. See PELOSINE.

Citracamic Acid. *Citraconamic acid.* Formed by boiling citracimide with aqueous ammonia.

Citracanile. $C_{22}H_9NO_4$. From the action of anhydrous citracic acid on aniline.

Citracanilic Acid. $C_{22}H_{10}NO_5HO$. Crystals by boiling citracanile with dilute ammonia.

Citracartic Acid. $C_5H_2O_3$. See MESACIC ACID.

Citracic Acid, *Citraconic acid, Citribic acid.* $C_5H_2O_3HO$. Limpid oily liquid, distilling at 410°; volatile; by distilling itacic or itaconic acid in air; it attracts water, and forms a crystalline hydrate.

Citracimide. $C_{10}H_5NO_4$. By heating in a retort citracate of ammonia to near 356°.

Citrac-iodanile. $C_{22}H_8INO_4$. By boiling iodaniline with excess of citracic acid in water.

Citramide. *Citronamide.* By ammonia on citrate of ethyle.

Citric Acid. $C_{12}H_5O_{11}$ 3 HO. Rhombic prisms, with 8-hedral terminations. Soluble in water, soluble in alcohol; in the cold state lime water yields no precipitate to its solution, but when heated deposits citrate of lime. The usual crystals contain 5 atoms water, but part with 2 atoms at 212°. Fused with caustic potash, citric acid is split into 2 acetic acid, $C_8H_6O_6$, 2 oxalic acid, C_4O_6, and 2 water, H_2O_2. By sulphuric acid and heat it becomes 2 CO_2, 2 CO, 2 $C_4H_3O_3$ and 2 HO. It unites with 3 atoms of base, forming salts with lead, barytes, silver, &c. and an emetic with splendent crystals. When heated, citric acid fuses, and loses its crystalline water. At a higher temperature inflammable gas escapes, and aconitic acid is formed, $C_4H_2O_4$; while 4 carbonic oxide (C_4O_4) are evolved, with 1 acetone (C_3H_3O), and 1 carbonic acid (CO_2). It is obtained by neutralizing lime juice with chalk, and decomposing the citrate of lime formed by sulphuric acid. See ACONITIC ACID, ITACIC ACID.

Citricic Acid. *Itacic. Itaconic. Pyroaconitic.* $C_5H_2O_3HO$. Crystals condensing on heating aconitic acid from citric acid to from 356° to 392°; citracic acid occurs along with it.

Citridic Acid. See ACONITIC ACID.

Citrine, or Hyaline quartz.

Citronyle, C_5H_4, or *Citrene,* by acting on oil of lemons with muriatic acid gas.

Citroptene. C_2H_2O.

Civet. A substance analogous to musk, obtained from *Viverra zibetha* and *V. civetta.* It yields ammonia, oleine, stearine, mucus, resin, volatile oil, yellow-colouring matter, and bone earth, &c.

Clarification. A process of clearing a fluid from turbidity by the introduction of another substance into it, as when cold water is added to hot coffee, a sediment falls; solution of isinglass is used to fine or clarify malt liquors; turbid waters are clarified by a minute portion of alum, although none can be detected in the water.

Clausthalite, or Selenide of Lead.

Clay. *Argil; Thon,* Ger.; *Argile,* Fr. Earths which are adhesive and plastic from the presence of much alumina are termed clays. They are employed for the purposes of brick-making when of a coarser description, and for pottery when of a finer and whiter quality. The following analyses were made in my laboratory by Mr. Roderick A. Cooper:—

1. *China Clay.*—The finest clay (China clay) used for pottery in Great Britain is obtained from Cornwall, and is prepared artificially by running a stream of water over decomposed granite, which carries with it the finer particles of felspar, and is then received into catch pools, or ponds, where it is allowed to subside. The water is then run off, leaving a fine sediment, which is removed and

exposed to the atmosphere for four or five months, when it is ready for export.

2. *Sandy Clay* (stiff or ball clay) is found deposited in Dorset and Devonshires as an upper layer, and is used for making salt-glazed ware.

3. *Pipe Clay* is the second layer, which is used in making tobacco pipes.

4. *Blue Clay* is of a grayish colour, and is considered the best layer of clay in the whole series, owing to its turning perfectly white, and approaching in character nearest to China clay.

5. *Red* or *Brown Clay* is a surface clay, very abundant near Glasgow, containing much iron, used for common bricks.

6. *Yellow Clay* is obtained from various parts of the country. It may be imitated by mixing together sandy and red clay.

7. *Fire Clay* occurs both on the surface, and some fathoms under ground. It has generally a dark colour. It is employed for making fire bricks and firing furnaces, and is not easily fused from the absence of iron.

8. *Stourbridge Clay*, a very refractory clay, of which glass pots are made. The specimens were rendered anhydrous.

	Silica.	Alum.	Protox. iron.	Limo.	Mag.
1 Cornish China clay,	53·16	45·61	·31	·41	·51
2 Sandy clay,.........	70·29	27·47	1·33	·90	trace
3 Pipe clay,..........	61·39	36·61	1·54	·46	trace
4 Blue clay,..........	53·52	43·89	1·20	1·39	trace
5 Red clay,...........	52·04	36·19	8·17	1·56	2·04
6 Yellow clay,........	65·06	30·68	3·70	·56	trace
7 Fire clay,..........	69·33	23·62	5·56	1·49	trace
8 Stourbridge clay,....	72·	23·5	4·5		

When dried at 212°, the quantity of water varies in these clays from 13½ to 3 per cent.

Claying. An operation in sugar refining.

Clay Ironstone. Carbonate of iron, accompanied by clay and other impurities, found in the coal basin.

Clay Slate. An aluminous slaty rock, consisting of two species:—Primitive blue clay slate, or metamorphosed slate, a rock destitute of fossils, extensively distributed in Scotland. The Scottish slate is characterized by the presence of cubic crystals of bisulphide of iron. It consists, according to my analysis, of (spec. grav. 2·7); silica 58·47; water 3·12; sulphur ·66; protox. of iron 5·86; sesquioxide of iron with phosphoric acid 1·36; lime ·62; magnesia 1·2; potash 1·84; alumina 25·13. The second species in Silurian.

Cleansing, in brewing, consists in drawing off the fermenting worts from the gyle tuns into butts or casks, and continuing the fermentation until the yeast forms and separates from the beer. In Scotland cleansing is applied to the process of running off the beer from the gyle tun, and leaving the yeast behind.

Cleavage, in mineralogy, relates to the indications of the faces of the primary crystal which minerals exhibit when cleft by the stroke of a hammer or chisel. Calcareous spar, when struck, cleaves always with faces parallel to those of a rhomboid.

Cleavelandite. A laminated variety of albite, from Chesterfield, U.S.

Clingmanite. Spec. grav, 3·, H 3·. Occurring with corundum in N. Carolina. Silica 36·34; alumina 42·373; lime 10·14; MgO 4·46; HO 1·5; NaO 5.

Clinkstone, *Klingstein*, *Phonolite*, *Porphyry-slate.* So named from the metallic sound which it yields under the hammer. It is found on the Rhine, &c. Characterized by having a reddish tinge and felspar crystals interspersed. As it contains alkali when decomposed, it yields a fruitful soil. It is believed to consist of felspar and zeolite, being decomposed by HCl.

Clinochlor. An American variety of serpentine.

Clinometer. (κλινω, I lean; μετρον, a measure.) An instrument for measuring the dip of strata.

Clintonite, *Seybertite*, *Bronzite*, *Holmesite*, *Xanthophyllite?* Spec. grav. 3·, H 4·25. Copper-red or yellowish foliated mineral; primary form an oblique rhombic prism. B.B. infusible; a white bead with soda and borax; resembles mica, but is not so flexible; found near Amity, New York. Silica 17·; alumina 37·6; magnesia 24·3; lime 10·7; FeO 5; water 3·6; F trace.

Clitonite, or Prehnite.

Cloves. See CARYOPHYLLINE.

Cluthalite (Clutha, the river Clyde). Spec. grav. 2·166, H 3·5. Consists of a congeries of flesh-red imperfect crystals, apparently right rectangular prisms, in the amygdaloid of Kilpatrick, Dunbarton. Lustre vitreous, brittle, easily frangible. Silica 51·26; alumina 23·56; Fe_2O_3 7·30; NaO 5·13; MgO 1·23; HO 10·55.

Clyssus. The old name for the vapour of nitre with an inflammable substance.

Cnicine. $C_{28}H_{18}O_{10}$. White needles from *Cnicus benedictus*, &c.

Coagulum. A mass separated from a solution, as albumen by heat from the white of egg.

Coak or **Coke.** The black carbon which remains when coal is heated in close vessels. To determine the amount of coke afforded by a coal, a given weight of pulverized coal is heated in a close crucible. The residue is the coke and ash. Coke is extensively used where it is of importance to avoid smoke.

Coal. *Pit coal*, *Stone coal*, *Sea coal.* As this is one of the most important substances found in the earth, both in reference to arts, manufactures, and to the comfort of mankind, it is necessary that a correct idea should be entertained of its nature and properties. We have only to take a survey of a geological map of Great Britain to be convinced of the invaluable wealth possessed by those districts in which coal occurs. It is at once the source of our manufactures and commerce, by enabling steam power and machinery to be worked at the most economical rate. The varieties of coal which have been hitherto met with are anthracite, caking coal, cherry coal, splint coal, and cannel coal, which possess the following composition:—

	Welsh Culm or Anthracite.	Caking Coal.	Cherry Coal.	Splint Coal.	Cannel Coal.
Spec. grav.,.	1·348	1·280	1·268	1·307	
Carbon, ...	92·56	87·952	83·025	82·924	76·25
Hydrogen, ..	3·33	5·239	5·250	5·491	5·50
Nitrogen, ..	?	?	?	?	1·61
Oxygen, ...	2·53	3·806	8·566	8·847	13·83
Ashes, ...	1·58	1·398	1·549	1·128	2·81
	Regnault.	Richardson.			R. D. T.

Origin of Coal.—Coal is believed to be produced by the oxidation of vegetable matter, or by the influence of pressure and heat. If we subtract from woody fibre $C_{36}H_{22}O_{22}$, 3 atoms H, 3 atoms HO, and 9 atoms CO_2, we get $C_{24}H_{13}O$, the formula for cannel and splint coal. The gases in coal mines are 88 carburetted hydrogen, 1 nitrogen, and 11 oxygen. The evolution of these gases proves, according to the present views, the changes which are constantly proceeding in the coal. Now it so happens that the largest quantities of *fire damp* are met with in deep mines, as at Newcastle, at a great distance from the oxygen of the air or other obvious causes of decomposition, while in shallow mines, as at Glasgow and North Durham, the Davy lamp is scarcely required except in the oldest and deepest mines, a sufficient indication of the rare occurrence of fire damp. Carburetted hydrogen, when disengaged by processes of decay, appears to be produced in contact with oxygen as in stagnant pools, a locality very distinct from a deep coal mine. It is an interesting fact, that the gases met with in coal mines are in part identical with the products of the coal gas manufactory, that is, results of an elevated heat, not of a slow and gradual decomposition at a low temperature. The source of this heat may be traced to the volcanic agencies whose effects, in the form of whin dykes, constitute such a remarkable feature in the coal fields of this country. In the immediate vicinity of these dykes, the coal in the Glasgow basin resembles charcoal; at a greater distance from these volcanic streams it assumes the appearance of what is termed, in Scotland, *blind coal*, where the mineral approaches nearer to the character of true coal, and this improvement increases as we recede farther from the intercepting dike, until we arrive at the splint and cherry coal. These contain about 85 per cent. of carbon, or 8 or 10 per cent. less than anthracite, or blind coal; while they possess 8 or 10 per cent. more carbon than the gas or cannel coals, which, so far as observation at Glasgow goes, are situated at a distance from whin dikes. When cannel coal is heated to about 350°, a quantity of perfectly pure water distils over, amounting to 4 to 5 per cent. In specimens from Lesmahago, and when the heat is carried up to 450°, the quantity of water removed is very considerable, amounting to 10 per cent. and upwards. When the heat is elevated above 700°, tar, naphtha, ammonia, and other volatile products pass off, while nothing remains, if free access of air is kept up, save a quantity of ashes varying from 2 to 10 per cent. and upwards. Some of these products are of great value, and it is therefore of importance to be able to calculate after analyzing a specimen of coal, the product which ought to result from its destructive distillation. All of the products we cannot thus estimate; but there are some to which we can approximate. The ammonia may be adduced as an example. Now the quantity of nitrogen present in 100 parts of coal is about an atom, or 1·75. To convert this into ammonia, there are required ·375 of hydrogen, or 3 atoms. Therefore $1·75 + ·375 = 2·125$ = quantity of ammoniacal gas from 100 parts of coal. But a considerable amount of nitrogen distils over in union with oils. For manufacturing purposes coals are generally considered to consist of 2 parts—a volatile or bituminous portion, and a comparatively fixed part, usually known under the designation of coke. The mode of determining the amount of these various ingredients is simply to burn a weighed amount of coal, reduced to a coarse powder in a platinum crucible, which is closed by means of its lid. Lesmahago cannel coal, when examined in this way, was found to leave 57 per cent. of unvolatilizable matter. When this coke was burned with access of air, by removing the lid of the crucible—a tedious experiment requiring much caution in the ignition of the last portions of carbon—it was found that 2·2 per cent. remained of a white ash. The coke, therefore, afforded by Lesmahago coal amounted to 54·8 per cent. The ash, when fused with carbonate of soda, and examined in the usual way, was found to contain the following ingredients:—Silica 50·57, peroxide of iron 8·04, alumina 21·84, lime 16·10. Coal appears to contain a small quantity of phosphoric acid, although we might expect it to contain it in greater quantity, if coal is a substance of vegetable origin, as is the opinion universally entertained. Coke is extensively used in locomotive engines on railways, in consequence of its yielding no smoke, the volatile matter, or what forms the smoke of coal, being removed by ignition. It therefore bears in these respects a strong relation to charcoal. As the bituminous or volatile part of coal yields the gas used for purposes of lighting, it has been found that the heating power of the coal resides in the coke, and no heat is lost by first extracting the gas from coal by the usual methods of burning or rather distilling coal. Coal may be converted into coke in a manner somewhat similar to that adopted for the production of charcoal, by piling up the coal and covering it with turf or bricks. In this country, however, it is usual, especially on the railways, to burn the coal to be converted into coke in what are termed coking furnaces. These terminate with a horizontal flue, and the latter again, when the operation is conducted near towns, in a tall chimney. By this means the products of combustion are in a great measure burned, so that but an inconsiderable amount of smoke escapes by the chimney. The common process of procuring coal gas supplies a certain amount of coke. In this operation the coal is introduced into a cast iron or brick retort, and is heated by means of a furnace. The gas escapes, and is collected in appropriate apparatus, while a quantity of coke remains, which is drawn out to make way for a new charge. In a similar manner coke is sometimes prepared on a limited scale, especially on the continent.

Cobalt. Co 3·75, 30; 3·6875, 29·5. A steel-gray, hard, brittle metal, of spec. grav. 8·5, not magnetic (Faraday). Melting point about that of cast iron. It is obtained from arsenide of cobalt by the same process as that for nickel. After the separation of the arsenic, as there described, the insoluble oxide, consisting, it may be, of nickel, cobalt, and iron, is treated with an excess of oxalic acid, which dissolves the iron, and leaves the oxalates of cobalt and nickel undissolved. The iron is separated by filtration. The oxalate of cobalt is dissolved by the addition of ammonia, while the oxalate of nickel remains undissolved. Caustic potash precipitates the peroxide of cobalt, which may be reduced to the metallic state by a current of hydrogen.

Protoxide.—CoO 4·75. An olive-green powder, becoming brown in the air, obtained by heating the carbonate. Cobalt does not unite with oxygen at common temperatures, but at a red heat it oxidizes, and at a white heat burns with a red flame. It decomposes water at a red heat. When heated with borax before the blowpipe it gives a blue bead. It is precipitated from its solution blue by the alkalies; ammonia redissolves the precipitate; alkaline carbonates form a pink precipitate; SH does not precipitate it, but sulphohydride of ammonia forms a black sulphide.

Brown Oxide, Protosesquioxide.—CoO, Co_2O_3, 15·. Obtained by heating the protoxide.

Sesquioxide.—Co_2O_3. Black powder; obtained by heating the protoxide with bleaching powder or chlorine gas. The hydrate of the protoxide is suspended in water, and the gas passed through the mixture.

Silicate of cobalt, or zaffre, is used for making smalts. Commercial zaffre is simply a mixture of roasted cobalt and finely powdered quartz.

Cobalt occurs in commerce in the form of a blue powder, *Smalts*, made by fusing cobalt ore with sand and potash. The light *cobalt blue*, or *Thenard's blue*, is formed by precipitating a mixture of a solution of alum and cobalt salt by an alkali and ignition, or by calcining for half-an-hour in a covered crucible 1 part of moist phosphate of cobalt with 8 parts of gelatinous alumina, and reducing the calcined mass to fine powder. *Rinman's green* is produced by substituting zinc for alum in the preceding experiment. The finest smalts consist of—

Oxide of cobalt,	14·7
Oxide of iron,	4·2
Oxide of lead,	4·7
Alumina,	5·0
Potash,	14·0
Silica,	54·8

Chloride of Cobalt.—Co Cl, 8. *Sympathetic ink salt.* The solution is formed by acting on cobalt or the oxide by means of hydrochloric acid; when weak it is of a pink colour, and when concentrated blue. If we write upon paper with the pink solution the letters are invisible, but if heated so as to drive off the water the letters are blue. Hence it has been termed *secret* or *sympathetic ink.* Many other inks may be formed at one time invisible, and which may be rendered visible by chemical means. If we write, for example, on paper with a solution of chloride of iron, allow it to dry, and then brush the paper over with a solution of yellow prussiate of potash, we shall have a Prussian blue ink impression, which, however, is soluble in ammonia. A writing made with the colourless acetate of lead becomes black when brushed over with sulpho-hydride of ammonia.

Characters of Salts of Cobalt.—1. The greater number are soluble in water, the solutions having a reddish colour when the salts are neutral, and when concentrated they are often blue. 2. Ammonia gives a blue precipitate, soluble in excess, forming a reddish-brown solution. 3. Caustic potash and soda give a blue precipitate, a subsalt. 4. Sulphuretted hydrogen produces no precipitate. If acetate of soda be present, however, a precipitate falls. 5. Sulphohydride of ammonia; a black precipitate, insoluble in excess. 6. Carbonate of ammonia; a red precipitate, soluble in salammoniac. 7. Yellow prussiate of potash; a green precipitate, becoming gray. 8. Red prussiate of potash; a deep red. 9. With borax; before the blowpipe, a blue glass in both flames.

Separation of Nickel and Cobalt.—1. *Phillips's Method.* Add an excess of caustic ammonia to the two metals in solution, so as completely to dissolve them. Caustic potash or soda is then added in great excess, in a close flask, and heated, when the nickel is precipitated, and the cobalt remains in solution, and may be separated by evaporation to dryness to remove ammoniacal salts; dissolve the cobalt and precipitate by caustic soda. 2. *Rose's Method.*—Dissolve the oxides in muriatic acid, dilute with much water, and saturate with a stream of chlorine. The protochloride of cobalt becomes sesquichloride, while the protochloride of nickel does not change. Set aside for fifteen hours in a cold place, previously adding an excess of carbonate of barytes. The sesquioxide of cobalt is precipitated, and is mixed with the carbonate of barytes. Wash with water, and dissolve in muriatic acid; precipitate the barytes by sulphuric acid, and the cobalt by caustic potash. The solution containing the nickel is precipitated by sulphuric acid; sulphate of barytes falls; precipitate the nickel by potash. 3. *Liebig's Method.*—Boil the solution of the two metals in cyanide of potassium for a few minutes; allow the solution to cool; supersaturate with sulphuric acid; a green precipitate falls of potash cyanide of cobalt, while the nickel remains in solution. 4. *Laugier's Method.*—Precipitate the metals by carbonate of soda; digest the precipitate in oxalic acid, which dissolves oxide of iron, and forms insoluble oxalates of nickel and cobalt. Dissolve in ammonia, and expose to the air, the salt of nickel alone falls.

Cobalt from Manganese.—Precipitate the oxides by boiling with carbonate of soda. Wash the precipitate; treat it with cyanide of potassium. A double cyanide of cobalt dissolves, a portion of manganese is undissolved. Evaporate to dryness, and digest in water. Oxide of manganese remains, while the cyanide of cobalt is in solution.

Cobalt, Binarsenide of. *White Cobalt ore. Speiss cobalt* in part. Spec. grav. 6·466, H 5·5. Steel-gray cakes; fracture uneven; faces tarnished; lustre metallic, opaque, brittle. B.B. gives arsenical fumes, and melts into a metallic globule; gives a blue bead with borax and fluxes; gives a pink solution with nitric acid; arsenic 74·22, cobalt 20·31, iron 3·42, copper ·15, sulphur ·88; from Saxony.

Cobalt Bloom. *Diarseniate, Erythrine, Cobalt Crust, Prismatic Cobalt Mica.* Specific grav. 2·948, H 1·75. Red acicular crystals, the primary form being a right oblique prism, with an angle of 124°; lustre, pearly and adamantine; sectile, thin laminæ, flexible; arsenic acid 37·9, oxide of cobalt 39·2, water 22·9.

Cobalt Cyanic Acid. $H_3Cy_6Co_2$. Silky filamentous crystals; by decomposing cobalt cyanide of lead by sulphuric acid.

Cobalt Cyanide of Potassium. K, Cy_6 Co_2. Yellow rhombic prisms, isomorphous with those of the red prussiate of potash; precipitated of a fine rose colour by salts of protoxide of cobalt; formed by the action of cyanide of potassium on a protosalt of cobalt; hydrogen is evolved. It precipitates salts of nickel, the precipitate (Ni_3, Cy_6Co_2) being insoluble in dilute boiling acids, and hence it may be employed to separate cobalt and nickel from each other.

Cobalt Cyanogen. Cy_6Co_2. The radical of cobalt cyanide of potassium.

Cobalt Disulphate. *Red Vitriol* or *Tersulphate.* Rose-coloured crusts, and stalactites in oblique rhombic prisms, from the old mines of Breber in Hanau. Two formulæ have been obtained, $CoO3SO_3$ 6 HO, and 2 CoO, SO_3 9 HO.

Cobalt Ochre. *Earthy Cobalt. Peroxide of Cobalt.* Spec. grav. 2·2. Soft blue or brownish-black, botryoidal or stalactitic masses; fracture conchoidal, earthy, opaque, sectile; soils little, B.B. gives out an arsenical smell, and colours borax blue; is believed to consist of oxides of cobalt and manganese; found in Cheshire, Siberia, Hesse, Howth, Ireland, &c. CoO 32·, MnO_2 31·, O 6·78, HO 23·.

Cobalt Sesquiarsenide. *Radiated. White Cobalt. Smaltine.* Tin-white radiated ore; opaque, brittle, H 5·5. Arsenic 65·75; cobalt 28; iron and Mn 6·25.

Cobalt, Sulphide of. *Cobalt Kies,* in part *Coboldine, Cobalt pyrites.* Grayish-white kidney-formed masses; texture, compact; fracture uneven and granular; found at Bastnas. Cobalt 43·2; copper 14·4; iron 3·53; sulphur 38·5. It is a sesquisulphide of cobalt with copper pyrites.

Cobalt Sulphoarsenide. *Glance Cobalt, Bright White Cobalt Ore, Cobaltine, Gray Cobalt, Danaite, Shining Cobalt.* Spec. grav. 6·298, H 5·5. Silver reddish-white cubes; fracture uneven; lustre metallic, opaque, brittle. B.B. gives arsenical fumes, and fuses only after being roasted; from Tuneberg in Sweden, Modum, Norway. Cobalt 33·10, arsenic 43·46, iron 3·23, sulphur 20·08; probably CoS_2 $CoAs_2$.

Cobalt Terarsenide. *Bismuth Cobalt.* Sp. grav. 4·6, mixed with quartz, or without, 6· to 6·7; lead-gray radiated mass; lustre metallic; sometimes splendent; from Saxony; arsenic 77·96, cobalt 9·88, iron 4·76, bismuth 3·88, copper 1·30, nickel 1·10, sulphur 1·01, manganese, a trace.

Cobalus. A spirit supposed by the ancient miners in Germany to haunt mines; the source of the word cobalt.

Coca. For the following interesting account of this plant I am indebted to my friend, Mr. Whittall, R.N.:—"Coca is the dried leaves of

a shrub or tree (Erythroxylon coca), growing wild in Bolivia, or Upper Peru, and also in many provinces of Lower Peru, and which has been described to me as somewhat resembling the coffee plant. A few leaves are put in the mouth, with a small quantity of quicklime. It prevents hunger and sleep, and ultimately brings the user to as miserable an end as opium. A craving taste for it is acquired; its stringent qualities dry up the system; its victims eventually lose all energy; become paralyzed, and cannot exist without it. A great quantity of coca is annually cultivated in the Perus, entirely for home consumption, and principally for the miners—Indians with few exceptions—who are in the habit of working twenty and thirty hours at a stretch under its influence without sleep. The miners are victualled by their employers at a certain scale, and have their provisions issued to them once a-week—coca included. But coca they are not limited to, and it is given to them whenever they like to ask for it, if the weekly quantity issued is not sufficient. Some have been known to consume a pound a-week, which is considered an immense quantity. It is sold throughout Peru, in all general shops, at about 6 rials (3s.) the pound. It costs the proprietors of mines—who usually cultivate or purchase it in large quantities, about 4 rials (2s.) the pound. In many parts of Bolivia, the lines of communication are impracticable for a horse, and are travelled by Indians on foot. These tracks are marked at intervals by the used quids of coca as they have been cast out and renewed by the Indians, and form as it were distance marks on the journey. An Indian will exist on very little food with the use of coca, and will travel two and three days with scarcely anything else to sustain life. Food they seldom or never carry on their journies, which are invariably performed on foot at a sort of trot with the knees bent. A small bag of coca over their shoulder; the attendant bottle containing quicklime, with a piece of wood fixed in the cork, like the spoon in the stopper of a cayenne cruet; and a stick, which they carry in their hand on a balance,—form their travelling equipment. These, with food obtained on the road perhaps once a-day, and a pillow for their feet at night, form their chief wants. The coca is always taken from the mouth previous to going to sleep; indeed, it is asserted, that one cannot sleep while chewing it. La Paz in Bolivia, like New York in the United States, although not the capital, is the largest city,—and the distance from thence to Chuquisaca the capital—70 leagues—is travelled by the Indians on foot in three days. Soldiers marching in the interior and others, have owed their lives to this wild shrub and a blue clay, containing a large quantity of soda, found in many parts of Bolivia, which is used as a substitute for the quicklime, but in much larger quantities. One of my informants on this subject had travelled three years in the interior of Upper and Lower Peru, on mules accompanied by Indians on foot. He had travelled 300 leagues in a fortnight, during which period, the Indians consumed each about 2 lbs. of coca, and a ½ lb. of the blue clay. Mr. Lyons (a partner in the house of Lang & Co. Lima), was the first to give me any information on the subject, and which was confirmed by others with whom I afterwards conversed. Amongst them Mr. Whittingham, a surgeon practising in Callao (in May, 1850), who has lived three years in the interior as surgeon to a mining company, and who told me of an accident that occurred while he was there, by which two men were covered up in a mine by a fall of earth, and contrary to all hope and expectation were found alive, after eleven days' incessant work to dig them out. It appears they had sufficient room to live in, and there was water in the bottom of the mine. All they subsisted on was a small bag of coca, which they had with them,—and their raw hide sandals, commonly called Yankees, which they ate. Air of course they must have got a limited supply of, but it was the coca and water that saved their lives. They fainted on coming into the air, but only required cautious feeding at first, and they soon recovered. Mr. Whittingham has, when at the mines, been out of tea, and frequently used a decoction of coca leaves as a substitute; 4 or 5 leaves, he says, will make sufficient for half-a-dozen people; but he always put a little hot water on the leaves first to draw the acidity, and pouring that off, then poured some fresh water on them for the tea."

Coccolite. A greenish granular form of pyroxene.

Cocculus Indicus. The small berry of the *C. menispermum*, *C. levanticus*, and *C. piscatorius*, from Ceylon. It contains *picrotoxine*, *menispermine*, and *paramenispermine*. From its bitter taste it has been said to have been employed to adulterate porter; a practice, from the poisonous character of the substance, highly to be reprobated.

Cochineal. A scarlet-coloured insect, *Coccus cacti*, which feeds on the *Opuntia coccionellifer* or *cochinillifera* in Mexico. The colouring matter appears to be derived from the plant, as a scarlet juice exists abundantly in the flower. There are 70,000 of these insects in a pound weight. Water digested on cochineal becomes red, is changed to reddish-yellow by acids and tincture of galls, to violet by lime water, and to violet-blue by acetate of lead; alcohol likewise takes up the red colour. Cochineal consists of carmine and water 50, gelatine 10·5, mucus 14·, margarine? 10, skin 14·, salts 1·5. See CARMINE.

Cochlearia Armoracia. Horse-radish contains a volatile oil ($C_8H_5NS_2$) corresponding with that of mustard, and hence its use as a condiment. The oil is probably sulphocyanide of allyle, C_6H_5, C_2NS_2.

Cocic Acid. *Cocinic Acid*, *Cocostearic Acid?* $C_{26}H_{25}O_3HO$. F.P. 95°. Fine white body; distils without change, and forms salts. Obtained by saponifying cocoa nut oil.

Cocinone. $C_{42}H_{42}O_2$? Doubtful. B.P. above that of mercury. White scales formed by distilling cocinate of lime; soluble in ether and alcohol.

Cockle. A variety of tourmaline.

Coco, *Cacao, Cocao.* The fractured kernels of the *Theobroma cacao* are so termed, and are extensively used as a substitute for tea and coffee.

Cocoa Nut. The edible fruit of the Cocos nucifera. Contains oil 71·68; albumen 7·68; gum 5·58; mannite 1·59; yellow-colouring matter ·23; fibrine 14·95.

Cocoa Nut Milk. Consists of water 95, mannite 3·82, albumen ·75, gum ·25.

Cocoa Nut Oil, *Cocoa Butter.* An oil obtained from the nut of the *Cocos nucifera.* Soluble in ether. In India it is liquid, and is extensively used for burning in lamps. I have found it in India to become solid at 72½°. It is very difficult to saponify. It is used by the inhabitants of India for smearing their persons. It contains caproic, caurostearic, and caprylic acids.

Cocognic Acid. *Cocognidic.* 4-sided prisms from the seeds of *Daphne gnidium.*

Cocostearine. $C_{38}H_{36}O_4$?

Codeine. $C_{36}H_{21}NO_6$. F.P. 302°. Rhombic 8-hedrons, losing, when crystallized from water, 5·8 per cent. of water; alkaline reaction; bitter taste; 100 water dissolve 12·6 at 60°, and 59 at 212°; very soluble in alcohol and ether; with nitric acid, chlorine, bromine, cyanogen, it forms new bases, by replacement of hydrogen. When heated with bases it yields ammonia, methylamine, propylamine; codeine is obtained from the mother liquor after the separation of morphine by evaporation, when the double chlorohydride of morphine and codeine crystallizes; potash dissolves the morphine, and leaves the codeine as a viscid mass, which is purified by ether and water. Codeine and its salts seem narcotic, but produce an itching on the skin in a higher degree than morphine. By sulphuric acid it becomes amorphous.

Cod Liver Oil. Spec. grav. ·923. A clear oil when pure from the liver of the cod, *Morrhua vulgaris.* The two kinds of oil, white and brown, are procured, the first by spontaneous exudation, the second by expression. It is extensively used in diseases of the chest; its action, in some measure, depending probably on the iodine which it contains. It consists of fats 95·93, bile 1·31, iodine ·03, chlorine and bromine ·08, phosphoric acid ·05, sulphuric acid ·01, phosphorus ·007, lime ·08, magnesia ·004, soda ·002.

Coffee. The berries of the *Coffea Arabica,* a shrub of Arabia, but cultivated in India and the West Indies, Ceylon, &c. It consists of cellular tissue 34, absorbed water 12, fatty bodies 10 to 13, glucose, dextrin, and vegetable acids 15·5, legumine, caseine 10, chloroginate of potash and caffeine 3·5 to 5, nitrogenous substance 3·, free caffeine ·8, concrete and essential oil ·001, aromatic oil ·002, mineral constituents 6·69. Coffee, before roasting, contains from 5·7 to 7·7 per cent. of sugar; after roasting it is sometimes absent, and seldom above half a per cent. The composition of the ash is (Ceylon) potash 55·1, lime 4·1, magnesia 8·42, sesquioxide of iron ·45, sulphuric acid 3·62, chlorine 1·11, carbonic acid 17·47, phosphoric acid 10·36. The amount of silica present is usually under ·5 per cent. and very often absent. Rochleder has found in coffee the following substances:—Caffeic acid $C_{16}H_7O_6$, caffeine $C_{16}H_{10}N_4O_4$, legumine $C_{48}H_{36}N_6O_{14}$, glycerine, oleic acid, palmitic acid, cane sugar, woody fibre. The amount of nitrogen in coffee varies from 2½ to 3 per cent. and appears to be little affected by roasting. The coffee berry, as it occurs in commerce, consists of the skin and bean, the latter enclosed in the skin (see figs.) The richest inhabitants of Zemen do not use the beans of the coffee, but the inner shell or skin that contains them. In examining coffee by the microscope, so as to distinguish it from chicory, it is proper to use an achromatic microscope, having at least a half inch objective. The substance should be viewed in water, and previously boiled in weak caustic potash, which softens the tissue; salad oil may be used when the object is unusually opaque. In roasting coffee, the temperature should be about 482°, rapidly brought up; the effect is to set free some caffeine from the double compound; the sugar is converted partially into caramel, which gives a dark colour to the decoction, and increases the bitter taste. Roasting produces a loss of 18 per cent. when the colour is of a chesnut-brown, but if carried farther, chlorogenic acid and the soluble principles are carbonized, and a loss of 25 per cent. occurs. To distinguish coffee from chicory, mangel wurzel, turnips, carrots, beans, pease, lupines, acorns, malt, with which coffee is often adulterated, see these substances. Coffee is a powerful diuretic. I have found it produce 85 ounces of urine, the natural quantity being 56 ounces.'

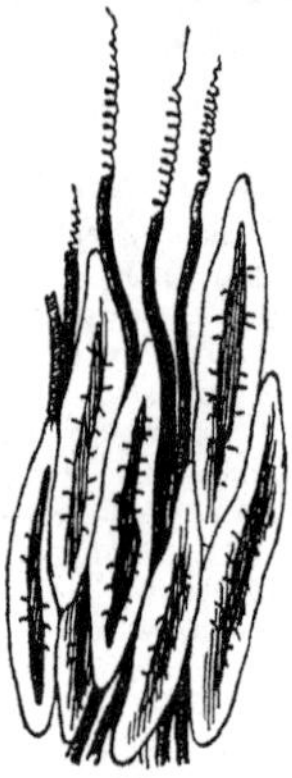

Skin.

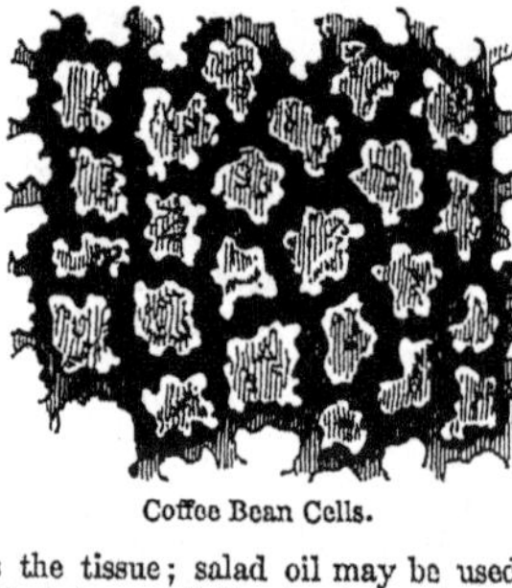

Coffee Bean Cells.

Cognac Brandy. A name derived from the town near Bordeaux where the brandy is distilled. Its genuine character is said to be recognized by a bluish-black colour produced on adding a drop of a solution of copperas from the presence of tannic acid; its appearance is imitated by adding to whisky some acetate of ethyle, and colouring it yellow with burned sugar.

Cohesion expresses the power by which the particles of the same kind of matter are held together. The cohesive action of lead may be shown by cutting two surfaces perfectly smooth, and causing them to cohere by pressure. The various qualities of bodies, as hardness, softness, toughness, depend upon the degrees of the cohesive power.

Cohobation. The frequent distillation of a fluid over a substance contained in a retort.

Coins, British. See MONEY.

Coke. See COAL.

Colchicine. Prisms, poisonous, bitter, soluble in water, alcohol, ether; coloured blue by nitric acid; from the seeds and bulb of *Cholchicum autumnale* by alcohol.

Colchicum Autumnale. *Autumnal crocus. Meadow saffron.* A well-known plant in gardens and meadows, particularly in Switzerland. The root consists of starch 1 oz. 4 drs. 57 grs.; crystallized sugar 9 grs.; uncrystallized sugar 3 drs. 28½ grs.; bitter extract 2 drs. 47 grs.; apotheme 40 grs.; soft resin 4½ grs.; extract by potash 39½ grs.; pectine 2 drs. 7 grs; fibre 2 drs. 4 grs.; water 12 oz. 6 drs. 48 grs. It is used in the form of wine of colchicum in gout and rheumatism with effect, especially when purging is induced.

Colcothar of Vitriol. *Caput mortuum. Rouge. Crocus.* The red oxide of iron which remains when copperas is calcined.

Cold. Absence of heat.

Colletiin. A crystalline bitter from *Colletia spinosa*, insoluble in water and ether.

Colline. *Colla, Common Gelatine, Glue.* A colourless, hard, cohesive substance, insipid, without smell; swells up in water, and changes to a tremulous jelly, but does not dissolve, becoming liquid at 93°; when mixed with water it decomposes gradually, and emits an odour. Insoluble in alcohol, ether, and oils; soluble in muriatic acid, acetic acid, alkalies; not precipitated by alum, persulphate of iron, acetate and diacetate of lead; precipitated by corrosive sublimate after some time, nitrate of mercury, protochloride of tin, sulphate of platinum, tannic acid (leather), consisting of tannic acid 46, colline 54 (Davy), 59¼ tannic acid, 100 colline (Schiebel), 10 tannic acid, 13 colline (Mulder). Colline is prepared by placing glue in cold water, and pouring off the water once in twenty-four hours, until all soluble matter is taken up. Heat the jelly to 123°, until it becomes fluid, and filter it; the jelly which passes through is colline. It differs from chondrine in not being precipitated from its solutions by acetic acid. It exists in, or is derived from, the skins and bones of animals.

Collodion, Collodium. A solution of gun cotton in ether, or in one part of alcohol and two of ether; it is used for wounds, for uniting tubes together, and for various purposes to which cements are applied.

Collyrite, *Scarbroite.* A white adhesive mineral from Schemnitz, consisting of SiO_3 15, Al_2O_3 45·5, HO 40·5. *Form.* 3 Al_2O_3 SiO_3, 5 HO.

Colocynth. The extract of the pulp and seeds of the *Cucumis colocynthis;* is used in medicine as an aperient. The peeled fruit consists of bitter matter 41·4, resin 4·3, pectine 18·6, nitrogenous matter 21·4, acetate of potash 5·7, deliquescent salts 7·1.

Colocynthine. A bitter yellow resinoid body, soluble in water, alcohol, and ether; obtained by alcohol from the pulp of the colocynth (Cucumis colocynthis), and then precipitation by cold water.

Cologne Water, *Eau de Cologne,* is made according to different receipts, of various qualities. The following are two:—

	1.	2.
Essence of Bergamot	13 drs.	12 drs.
— Portugal	11	12
— Lemon	1 oz.	1 oz.
— Neroli	10 drs.	12 drs.
— Rosemary	1 oz.	2 oz.
— Lavender	1 oz.	2 oz.
— Cedrat		1 oz.
— Benzoin		2 drs.
Rose Water	14 drs.	
Jessamine Water	13 drs.	
Orange Flower Water	15 drs.	

1. Durockereau's receipt:—The whole is mixed, and in twenty-four hours distilled from a water bath. The spec. grav. is 31° B. 2. Crozet's receipt:—The whole is mixed with 10 times its weight of spirit of wine, distilled and redistilled.

Cologne Yellow consists of 7 gypsum, 1 sulphate of lead, 2 chromate of lead.

Colophane, *Colophony, Colophone, Common rosin.* C 75·94, H 13·34, O 10·71. F.P. 275°. $C_{40}H_{26}O_4$. Obtained by distilling turpentine, when rosin remains, and oil of turpentine passes over. Cold alcohol takes up pinic acid and leaves sylvic acid. These acids unite with alkalies and form soaps; hence the use of rosin in yellow soap. It is used for the strings of violins, as a varnish, and for plasters.

Colophene. $C_{20}H_{16}$. Colourless. Specific gravity ·940; B.P. 595°. Obtained by mixing 20 oil of turpentine and 1 oil of vitriol for twenty-four hours; the fluid poured off, and distilled at 410° to 428°, terebene passes; above 430° colophene.

Colopholic Acid. Brown substance, obtained by distilling pinic acid.

Colophonite, *Yellow garnet, Topazolite, Allochroite, Succinite.* Specific gravity 3·871 to 3·965; H 6·75. Various shades of yellow rhomboidal 12-hedrons and other modifications. The fig. is formed of 12 rhombs, of which the angles are 109° 28′, and 70° 32′, similar to other garnets. Silica 35·64, protoxide of iron 30, MnO 3·02, MgO 2·35, lime 29·21. Found at Arendal, Norway; Ceylon; Piedmont, &c.

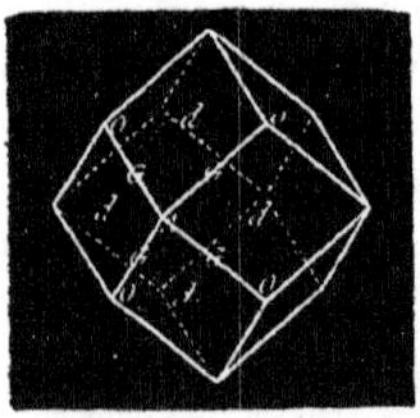

Colorine. The same as alizarine, apparently.

Colostrum. The name given to the milk of the cow, &c. immediately after calving, and having young. The following is the composition of different kinds:—

	Ass.	Dog.	Woman.	Cow.
Caseine, ...	14·50	14·6	3·43	15·19
Butter, ...	·85	13·3	2·53	2·60
Sugar, ...	2·86	3·	4·82	3·60
Salts, ...		1·48	2·30	·30
Water, ...	81·41	68·2	88·76	78·40

Colours used in painting and porcelain, &c. See Blue, Red, Purple, &c.

Colouring Matters are those substances generally of organic origin used in the processes of dyeing, &c.

Colrake. A shovel used to stir lead ore when washed.

Columbic Acid. CmO_3 or TaO_3. Spec. grav. 7·2. White powder, tasteless; insoluble in water, soluble in sulphuric acid, bisulphate of potash or soda, chlorohydric acid, binoxalate of potash; forms salts with bases. It occurs most pure in the tantalite or columbite of Finland. It is separated from pelopic and niobic acids by converting them into chlorides. The chloride of columbium requires a higher temperature to produce it than the others; it volatilizes at 291°; it dissolves partially in caustic potash; carbonate of potash has no action on it; chloride of pelopium is more soluble in caustic potash and is somewhat soluble in the carbonate of that alkali. Chloride of niobium dissolves completely in cold caustic potash, as well as in boiling carbonate, and at length in pure water.

Sulphocolumbic Acid.—TaS_3. Gray powder like black lead, formed by passing bisulphide of carbon over ignited columbic acid in a porcelain tube.

Columbine. $C_{42}H_{22}O_{14}$. Crystals by alcohol from columba root (Cocculus palmatus) accompanied by berberine. A yellow powder is extracted from the root by acids termed Columbic acid, $C_{42}H_{21}O_{11}$.

Columbite. *Bodenmais* or *Bohemian Columbite*, or *Tantalite.* Spec. grav. I have found from 6·18 to 6·19. Dr. T. Thomson found it 6·038; H 6 to 6·25. Colour black, even when in powder, becoming chocolate coloured by ignition. Structure foliated; crystals, rectangular 4-sided prisms, with angles of 90°. It was in consequence of the remarkable difference in the specific gravity of this and other minerals containing columbic acid, as pointed out by Dr. Thomson, (*R. D. Thomson's Records of General Science*, vol. iv.), that H. Rose was led to discover pelopium and niobium. The following table contains the specific gravity and constituents of different specimens of columbite from Bodenmais in Bohemia:—

Spec. grav.	6·038	6·39	6·021	5·976
Columbic acid, ...	79·65	81·07	80·64	79·732
Protox. of iron, ...	14·00	14·30	15·33	14·768
Do. manganese, .	7·55	3·85	4·65	4·772
Protoxide of tin, ..	0·50	·45		0·102
Protox. of copper,		·13	·105	1·512
Lime, ...		trace	0·21	
Oxide of nickel, ...		—	—	—
	(T. Thomson.)		(H. Rose.)	

Columbium, *Tantalum*, Cm or Ta 23·1, 184·9? Black powder, becoming possessed of a metallic lustre under a burnisher; takes fire in air under ignition, and burns, becoming columbic acid. Obtained by decomposing the double fluoride of columbium and potassium by potassium.

Columbium Binoxide. CmO_2 or TaO_2. Gray mass, scratches glass; insoluble in acids; burning by ignition in air, partly into columbic acid; prepared by heating columbic acid in a plumbago crucible in a blast furnace.

Colza Oil is obtained from the seed of a species of Brassica (Campestris) and is extensively raised in France. Spec. grav. ·9136.

Combining Weight. See Atomic Weight.

Combustible. The term, in chemistry, is applied to an element or compound which, in uniting with another element, gives out light and heat.

Combustion. This term is used to denote the phenomenon of the evolution of heat and light when a substance is burned. Common combustion takes place between the oxygen of the air and a combustible body, whether that body be an elementary or a compound body. 1. The original idea of the nature of combustion consisted in supposing that an element, *fire*, possessed the power of devouring certain other bodies, and of converting them into itself. 2. In 1665 (Micrographia, p. 103), Dr. Hooke proposed the view that a substance similar to that which exists in nitre, is fixed in common air, and that it has the property of dissolving all combustibles whenever their temperature is considerably raised. The solution occurs so rapidly, that heat and light, which he considered mere motions, are produced. This opinion was adopted about ten years afterwards by Mayow, without acknowledgment, and was supported with much ingenuity (De Sal Nitro, &c.) 3. The doctrine of Stahl, or the *phlogistic theory*, was originally suggested by Beccher (1690?). They considered all combustible bodies to be compounds of phlogiston, and another substance which gave the character to the combustible body. When phosphorus burns, phlogiston, one of the constituents of the phosphorus, escapes, exhibiting two of its properties, heat and light, while the other ingredient of the phosphorus, to which it owes its characters, remains in the form of white flocks, and is termed phosphoric acid. Phosphoric acid is, therefore, phosphorus deprived of phlogiston. As an additional evidence of the existence of phlogiston in all combustibles, they observed, that if phosphoric acid be heated with charcoal, lamp black, sugar, resin, &c. it is reconverted into phosphorus, by attracting and uniting again with phlogiston from these bodies. The reasoning was exceedingly ingenious, and probably could not have been refuted except by the introduction of weights and measures in the science. This was done by Boyle, who showed before 1691 (the Stahlian theory having been broached before this year), that when tin is melted on the fire, and converted by heat into a calx (oxide), it was considerably increased in weight. The oxidization could not, therefore, be due to the *loss* of phlogiston. 4. Priestley found that air, in which combustibles were burned, had undergone a remarkable change, partially noticed by Hailes in 1724; that combustibles would not burn in it; that as calcination goes on, "there is a diminution in the volume of the air, and that this diminution is nearly proportional to the augmentation in weight of the metal;" that animals could not exist in it; and the air diminished in bulk. He inferred that the air had united with phlogiston, and that air is necessary to combustion by attracting the phlogiston, for which it has a great affinity; but he did not account for the evolution of heat and light. Crawford made up this deficiency by applying Dr. Black's theory of latent heat. According to him, the phlogiston of the combustible combines during combustion with the air, and at the same time separates the heat and light with which that fluid had previously been united. The heat and light, therefore, previously exist in air. 5. In 1773, Lavoisier repeated the experiment of Boyle on tin, and on many other metals, with the same result; from which he concluded that metals, when calcined (that is, heated and converted into oxides), augment in weight by the *fixation* of air (Opuscules Physiques et Chymiques, p. 293; Rapport fait a l'Academiè, 7 Dec. 1773, published 1774). He thus advanced a step beyond Priestley. But in August, 1774, Priestley discovered oxygen gas, and soon afterwards showed Lavoisier his process for preparing it in his own laboratory in Paris. The knowledge of this discovery, and of that of nitrogen by Dr. Rutherford in 1772, with the paper of Bayen on the oxides of mercury in 1774, and others facts, enabled him to deduce the general law, that "in every case of combustion, oxygen combines with the burning body," and to establish the antiphlogistic theory. Combustion, according to him, consists of two stages; first, a decomposition; second, a combination. The oxygen of the atmosphere being in the state of gas, is united with heat and light. During combustion this gas is decomposed, its heat and light escape, while its base combines with the combustible, and forms the product. This product is incombustible, because its base being already saturated with oxygen, cannot combine with any further addition. This theory was not adopted generally in France till 1785, when the phlogistic theory may be said to have been completely overturned. The theory of Lavoisier only explains the conditions in which oxygen unites with combustibles, but it leaves the extrication of heat in numerous chemical combinations in obscurity. In the explosion of gunpowder, a mixture of solid bodies is expanded to at least 2,000 times its original volume. Hence, we should predicate the production of cold, rather than the extrication of heat upon the theory of latent heat of Dr. Black. Neither is the idea of the specific heat of the new compounds being less than that of the solid bodies applicable, as it has been shown to have little foundation in fact (Dulong and Petit, An. Chim. 10, 396; An. Phil. 14, 189); or at least, that in very energetic combinations, the heat evolved by the changes of constitution is but a small part of the total effect. If this explanation were correctly applied to the phenomena of combustion, it would follow that the latent heat of all oxides would be less than that of their constituents,—a position at direct variance with the results of Dulong and Petit. Nor does it appear that the heat depends either on the combustible body, or on the oxygen separately, but on both. According to Dalton's table, the quantity of heat evolved in combustion is closely connected with the amount of oxygen consumed; but, from the experiments of Despretz (An. Chim. 37, 180), it would appear that this affirmation holds only in a few instances; the same quantity of oxygen often giving double the quantity of heat with different bodies. The same observation applies to the disengagement of heat and light when sulphur unites rapidly with copper, or phosphorus with barytes, lime, or strontian, &c. The theory of Lavoisier being thus thoroughly defective, the only expression of the circumstances in the present state of our knowledge, appears to be, that the heat of combustion is a consequence of the rapid or sudden union of bodies with each other. Wilcke of Stockholm originally suggested that the heat of combustion is a compound formed by the two electricities,—an opinion supported by Grotthuss. Davy considered that all bodies having an affinity for each other, are in different electrical states—the one plus or positive, the other minus or negative (the combustible body positive, oxygen negative). The more intensely any body is positive and another negative, the greater will be the energy of their union, that is, combustion, and the greater the amount of heat extricated. Berzelius supported the view of Davy.

If the theory of the heat of combustion were properly elucidated, it is scarcely necessary to observe, that the results would be of incalculable benefit in manufactures, as upon this theory depends the economy of fuel. The following tables contain the results of experiments in this direction:—

Heat of Combustion.—15·438 grains (1 gramme) of the following substances, when burned in oxygen, give out the following units of heat, an unit being the quantity of heat which is necessary to raise the temperature of 15·438 grains of water 1°·8 F. (1° C.):—

	Favre and Silbermann. units.	Andrews. units.
Carbonic oxide,	2433	2431
Wood and sugar charcoal,	8086	7900
Gas coal,	8040	
Graphite (artificial),	7738	
Do. (natural),	7774	
Diamond,	7870	
Charcoal burned in NO,	10841	
Hydrogen,	34188	
Carburetted hydrogen,	13158	13108
Charcoal in do. do.,	5954	
Olefiant gas,	11900	11942
Pyroxylic spirit,	5304	
Alcohol,	7183	6850
Ether,	9027	
Spermaceti,	10342	
Wax,	10500	
Oil of turpentine,	10874	
Sulphur,	2208	2307

Slow, and Low Combustion.—In the manufacture of copperas or sulphate of iron from bisulphuret of iron, the latter undergoes a process of slow combustion by uniting with oxygen from the air and evolving heat (FeS_2 and 7 O become FeO SO_3 and SO_3). When a candle is blown out in the dark, a phosphorescence is observable for some seconds on the surface of the wick, especially if the wick is long, and when oils or wax are projected on a plate of metal heated below redness, a similar appearance is exhibited (C. J. B. Williams, An. Phil. 2d ser. 6, 44). The phosphorescence of phosphorus in the dark does not depend on oxygen, as it occurs likewise in hydrogen and carbonic acid (Berzelius, Marchand). Wood and vegetables decay by a slow process of combustion (carbonic acid and water being formed). In slow or rapid oxidation or combustion, no portion of the oxygen or combustible body is lost; the new product formed is exactly equal in weight to the amount of the combustible body and oxygen which have disappeared.

Combustion, Spontaneous, is said to occur when a body inflames when no one is present. Cases of this kind sometimes take place with lucifer matches, when the phosphorus which they contain unites suddenly with the oxygen of the air, either from its superabundance, or from a bundle of matches falling. The idea sometimes entertained of the possibility of the spontaneous combustion of the human body seems impossible. (Liebig's Letters, third edition, 280.)

Comenamic Acid. HO, $C_{12}H_4NO_7$, 4 HO. Colourless plates, obtained by boiling comenate of ammonia with excess of ammonia till the smell of ammonia ceases to be evolved, and adding chlorohydric acid.

Comenic Acid. $C_{12}H_2O_8$, 2 HO. Hard grains by boiling meconate of lime with dilute muriatic acid, and purifying by animal charcoal; it differs from meconic acid by containing 2 ats. carbonic acid less; it is changed by heat into carbonic acid, and *pyro* and *parameconic acids* (?); gives a red precipitate with sesquisalts of iron.

Commingtonite. Spec. grav. 3·201, H 2·75, grayish-white needles, infusible, from Commington, Massachusets. Silica 56·54; protoxide of iron 21·67; protoxide of manganese 7·80; soda 8·44; water 3·18.

Common Salt, or chloride of sodium.

Compound Radical. See RADICAL.

Comptonite. Spec. grav. 2·35 to 2·4, H 5·25. White right rectangular prisms from Vesuvius and Seeberg, Bohemia. It is a variety of Thomsonite.

Concentration. The process by which the particles of a body are brought into closer proximity. The concentration of a fluid is obtained by evaporation by heat, or in the vacuum of an air pump.

Concretions. See CALCULI and URINE.

Condensation. When the particles of a body are brought closer together by pressure, cold, or hammering, &c. the body is said to be condensed, as when steam is changed into water by cold air, or water, &c.

Condiments. Substances used along with articles of food to give them a flavour, as common salt, mustard, pepper, sugar, vinegar, &c.

Condurrite. Spec. grav. 5·204. Brownish-black mass, soft; B.B. gives out a white fume of AsO_3, leaving on charcoal a metal. It consists of copper 60·5, sulphur 3·06, arsenic 1·51, arsenious acid 25·94, water 8·99; from Condorrow, Cornwall; originally probably a sulphide of copper, and arsenide of copper..

Congelation. The solidification of water, &c. by freezing.

Conichrite. Silica 35·69, protoxide of manganese 22·5, alumina 17·12, lime 12·8, protoxide of iron 1·48, water 9·.

Conlic Acid. A suspected acid in *Conium maculatum.*

Conline. *Conia. Conicine.* $C_{16}H_{15}N$?

Spec. grav. ·878. B.P. variously stated 336° to 413°. Colourless oil, with disagreeable smell; vapour attacking the eyes; soluble in alcohol, ether, acetone, and oils; coagulates albumen; very poisonous. By oxidation converted into butyric acid, &c.; by nitric acid becomes blood-red. Extracted from Conium maculatum, or hemlock, in the same way as nicotine.

Conite. Spec. grav. 3·. Flesh-red dolomite, from Iceland and Hesse, from which it differs by scratching glass.

Contact. See CATALYSIS.

Convolvuline. Needle-formed alkaloid from the root of *Convolvulus scammonia.*

Copaiva, Balsam of. *Copaiba. Capivi.* Sp. grav. ·96 to ·95. A clear yellow fluid, obtained from species of Copaifera in South America; insoluble in water; soluble in alcohol, fat, and etherial oils; dissolves sulphur and phosphorus; combines with bases; consists of copaiva oil 41; yellow resin 51·38, brown resin 2·18, water 5·44. It is used extensively in discharges from the mucous membranes, and appears to act by its aperient influence. It is taken in the form of capsules, or enclosed in little bags of gelatine. Also dissolved in ether, &c.; used also for varnishes.

Copaiva Oil. $C_{10}H_8$. Spec. grav. ·878 to ·91, B.P. 473°. Clear oil, neutral, obtained by distilling the balsam with 6 to 8 parts water; soluble in 27 parts alcohol of ·850, and 2½ absolute alcohol; mixes in all proportions with ether.

Copaiva Resin, Brown. Insoluble in alcohol.

Copaiva Resin, Yellow. *Copaivic Acid.* $C_{40}H_{32}O_4$. Clear prisms, obtained by dissolving the balsam in ammonia, or the resin after the distillation of the oil in alcohol; soluble in alcohol, ether, fat, and ethereal oils.

Copaivene. $C_{20}H_{16}$. Not isolated, but obtained in union with chlorohydric acid.

Copaivilene. $C_{20}H_{16}$. Not yet obtained.

Copaivyle, Chlorohydride of. *Muriate of Copaivene.* B.P. 365°, $C_{10}H_9Cl$. White plates, like chlorate of potash; insoluble in water and cold alcohol; formed by the action of chlorohydric acid gas on copaiva oil.

Copal. Spec. grav. 1·045, 1·069 to 1·14. A clear yellow or white resin, without taste and smell; gives by distillation a fluid oil and water; soluble in nitric acid and alkalies. Copal when fused, and boiling alcohol, or oil of turpentine, poured on it, dissolves completely. When heated with alcohol of different strengths, it is resolved into four resins, two of which have been analyzed. *Alpha resin,* $C_{40}H_{31}O_5$. *Epselon resin,* $C_{40}H_{31}O_5$.

Copal Fossil. See HIGHGATE RESIN.

Copal Varnish. Take 4 parts of ground copal, boil it in a flask until water is no longer given out; then pour into it 1 part of linseed oil, and mix it with an equal weight of oil of turpentine (Mr. Lewis of Glasgow). This is used for clock faces.

Copiapsite, or native persulphate of iron. Fe_2O_3 33, SO_3 37, HO 30.

Copper. *Venus; Cuivre,* Fr.; *Kupfer,* Ger. Cu 4, 3·956; 32, 31·648. Spec. grav. 8·953, F.P. 1996°. Brownish-red metal; crystallizes in cubes and regular 8-hedrons; ductile, malleable. When strongly heated, communicates a green colour to flame; does not decompose water; insoluble in dilute sulphuric acid; very soluble in nitric acid. Copper raised in Cornwall and Devon in 1847, was 13,000 tons; in all England 22,500 tons; Wales 340 tons; and Ireland 14,000 tons. Copper was known from the earliest ages; it seems to have been employed by the North American Indians in the construction of arrow heads and cutting instruments, as appears from specimens preserved in my chemical museum, at St. Thomas's Hospital, which were dug up in the township of Perth, Upper Canada, in forming the foundation of a house. It was probably obtained by the Indians from the neighbourhood of Lake Superior, where it is found at the present day abundantly in a native state. The following are analyses of the copper composing the Indian implements, and of native copper from Eagle River, Lake Superior:—

	Copper from Lake Superior.	Copper composing Implements.
Copper, ...	99·20	99·71
Iron, ...	·112	·168
	99·312	99·878

The British ores of copper are brought from Cornwall, Devonshire, Anglesea, and Ireland; and the foreign ores from Australia and Chili. The ores of Cornwall when mined, are conveyed to the top of the mine, where they are broken into small pieces, and sorted or dressed, and are then carried to Swansea. But the foreign ores are imported to Swansea in masses, where they are either ground or broken by the hand into small pieces and put into lots; when this is done, a sample is assayed, and the ores are sold according to the per centage of copper they contain. A sale of ores thus assorted is held at Swansea every fortnight. In the smelting works there are a number of ore yards for the purpose of keeping the different ores separate; but, when the ores are of a low per centage, they are commonly mixed together in such a way that one ore will help to flux the other, care being taken that they do not contain more than 8 or 9 per cent. of copper. The number of processes through which the ores pass before the copper is ready for the market in the old way of smelting is seven.

I. *Process.—Calcination of the Ores.*—The effect of this process is to remove a portion of sulphur, and to substitute oxygen from the air in its stead. The ores are conveyed by the smelters in boxes (which are made to hold 1 cwt.) into a hopper or bin, which rests over a hole in the top of the furnace; the ore is let down into the furnace by drawing out a slider which is at the bottom of the hopper; it is then spread uniformly all over the bottom by means of an iron

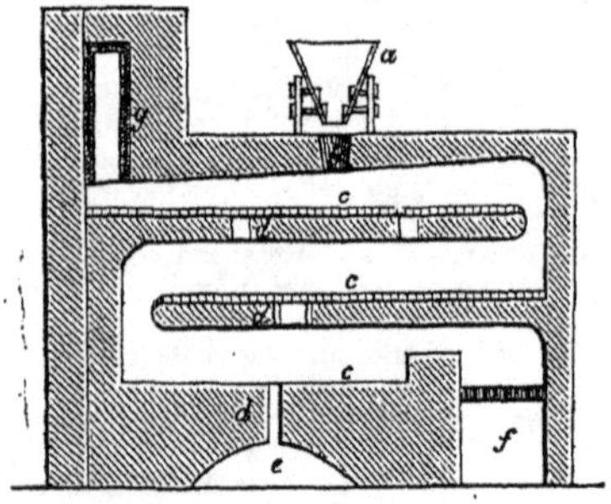

a Hopper or bin.
b Aperture for charging furnace.
c Bottoms of furnace, called beds.
d Apertures for passage of ore from upper to lower beds.
e Cubb for receiving calcined ore.
f Furnace and ash pit.
g Flue. *h* Bridge.

rake (*a rabble*). The doors of the furnace are now shut and the fire raised, care being taken not to apply too much heat, as the particles of ore are very apt to adhere (or clog) together. To prevent this adhesion of the particles, and to bring fresh portions in contact with the air, the doors are opened about once every hour, and the ore stirred with iron rakes. The sulphur in the ore in this state possesses a great affinity for oxygen, becomes sulphurous acid gas, which passes up the chimney; the copper and iron which were in combination with the sulphur, become oxidized; consequently the ore loses but little weight, for what is lost in sulphur is gained in oxygen. This calcination lasts from nine to twelve hours, when the ore is drawn out through a hole in the bottom or hearth of the furnace into a cub or pit under the furnace. This cub or pit is about 3½ feet below the level of the ground. Another charge is immediately dropped into the furnace and treated in the same way.

Composition of Ore before, and after Calcination.

This and the following analyses were made in my laboratory, by Mr. James Napier, jun.:—

	Before Calcining.	After Calcining.
Copper,	6·32	16·560
Sulphur,	14·20	20·688
Iron,	35·69	25·890
Silica,	37·70	33·700
Lime,	3·30	2·300
Water,	1·41	
	98·62	99·138

II. *Process.—Fusion of Calcined Ores.*—The men who have charge of this operation convey the calcined ore from the cubs of the calciners into the hoppers of the fusing furnaces, in wooden boxes which hold 1 cwt. When they

a Hopper. *c* Bridge. *f* Door. *g* Flue.

have as much in the hoppers as they think sufficient for a charge (which is generally about a ton or a ton and a-half), they let it down into the furnace, and spread it uniformly over all parts of the bottom; when this is completed, a quantity of foul slags obtained from the second fusion is thrown in, and also some *cobbing*, which consists of old furnace bottoms, pieces of brick, &c. This not only frees these scoria from any metal they may contain, but also increases the fusibility of the ores. The door of the furnace is now put up and luted tightly with clay, and the fire gradually raised till there is a white heat in the furnace. During the time which elapses between each period of firing, the workmen employ themselves in filling the hoppers with another charge of calcined ore. When they consider the charge in the furnace has had sufficient time to melt (which happens in about four or five hours), they take down the door and stir the liquid mass with a long rake to remove any pieces which may happen to adhere to the bottom of the furnace, and also to allow the metal to separate from the earthy matter. The slag or earthy matter being of a less specific gravity than the metal, swims on the top. After the stirring, the door is again put up for about twenty minutes to allow any metal which might have been mixed with the slag during the stirring to fall to the bottom. The slag now being skimmed off, another charge of calcined ore is let down into the furnace, and treated in the same manner. This is repeated till the metal at the bottom of the furnace is as high as the furnace will allow without flowing out by the door. At this period the tap hole is opened, and the metal allowed to run into an adjoining pit filled with water. On the metal's descent into the water it becomes granulated. These pits are about 8 feet deep, having a box at the bottom attached by ropes for elevating the granulated metal by a crane. By this fusion, a considerable quantity of earthy matter and iron is got rid of. When the ores are refractory, (which rarely happens), fluor spar is used as a flux. The composition of fused calcined ore I found as follows:—Copper, 33·600; sulphur, 20·115; iron, 33·470; silica, 9·000.

III. *Process.* — *Calcination of Granulated Coarse Metal.*—This operation is performed in the same kind of furnace as the first calcination. The chief object in calcining this metal is to oxidize the iron, an oxidation more easily completed in this state than in the first calcination, as the ores have been freed from a great quantity of their earthy matter by the first fusion. The doors of the calciners are opened every two hours, and the metal stirred to prevent the adhesion of the particles. This calcination lasts for twenty-four hours, when the metal is drawn out, and another charge let down into the furnace.

IV. *Process.—Fusion of Calcined Coarse Metal.* —The metal is conveyed by the smelters into the hopper and is dropped into the furnace, and spread uniformly. A quantity of slag obtained from a subsequent operation (which generally contains a large quantity of copper), is thrown in on the top of the metal. The door of the furnace now being raised and luted, the workmen proceed to bring up the heat. In the door of the furnace there is a small hole about half an inch in diameter, through which the men are enabled to see the state of its contents without displacing the door. This fusion lasts for about five hours, when the doors are taken down, and the liquid mass stirred. The door is again put up and allowed to stand for a short time, to give the fused metal time to fall to the bottom of the furnace, and to be separated thoroughly from the slag. The door now being removed, the slag is skimmed off with an iron rake (*rabble*). When as much as possible of the slag is removed from the surface of the metal, the tap hole is opened and the metal allowed to run into sand beds. The product of this operation is called *blue metal.* The analysis of blue metal gave me:—Copper, 77·408; sulphur, 18·041; iron, 4·200; insoluble matter, ·320.

V. *Process.* — *Roasting Blue Metal.* — The furnace in which this operation is carried on is termed a roaster. The difference between this furnace and a fusing one is trifling. The roaster is furnished with a side door, and some small holes in the bridge of the furnace to admit a current of air, which the fusing furnace has not.

The blue metal is introduced into the roaster by the side door in masses, the charge being about two tons and a-half. The doors are raised, and the fire increased gradually. The metal is kept in a sort of semi-fluid state, with a current of air passing over its surface. At intervals of several hours, the doors are taken down, and a quantity of floating slag skimmed off by the front door; this door being again put up when the slag is skimmed off. The metal is occasionally puddled with iron rakes at the side door to bring fresh portions in contact with the air. The metal is kept quite clear on the surface by the repeated skimming off of the slag as it forms. The object of this roasting is to oxidize the iron and other foreign matters. A quantity of sulphur is also driven away in the state of sulphurous acid which passes up the chimney. Shortly before the end of the operation, which lasts for about twenty-four hours, both doors are closed and luted tightly, and the heat raised so as to have the whole of the metal in a state of perfect fusion. It is now skimmed, and the metal tapped out into sand moulds. This is called *pimple* or *sponge regulus*, according to the state of advancement to the metallic condition. The composition of pimple or sponge regulus, was found to be—Copper, 75·402; sulphur, 21·140; iron, 2·128.

VI. *Process.—Roasting Pimple or Sponge Metal.* —This roasting is executed in the same way as in the preceding operation for blue metal; the process occupying about twenty-four hours. The product is coarse copper more or less purified according to the nature of the ores employed. The following is the composition of roasted *sponge metal*:—Copper, 97·11; sulphur, 1·58; iron, ·91; insoluble matter, ·40.

VII. *Process.—Refining or Toughening.*—The copper intended for refining, is introduced by the side door *a* of the refining furnace; the charge varies from 5 to 6 tons. For the first few hours, the fire is kept moderate to complete the roasting of the coarse copper, which might not have been effected in the roasting furnace. The copper is accordingly melted, and the front door taken down, and what slag has collected on its surface is skimmed off. This slag possesses a bright red appearance, and is very rich in copper. After about sixteen hours, if all things have gone on well, the metal will have a bright surface, and appears to evolve no gas. A ladleful is taken out and allowed to set. If it sets without spurting, it is ready to be refined. The men now insert into the fluid metal, poles of green wood and press them down. After a short time, the refiner takes out an assay in a small ladle and breaks it, on solidifying, in a vice. By inspecting the grain of the copper, he judges whether he may continue the poling, and, if so, what quantity of wood he will require. At the beginning of this operation the grain of the copper is coarse and sandy-looking. The surface of the copper is now covered with wood charcoal, and the poling continued. The escape of gases from the wood causes

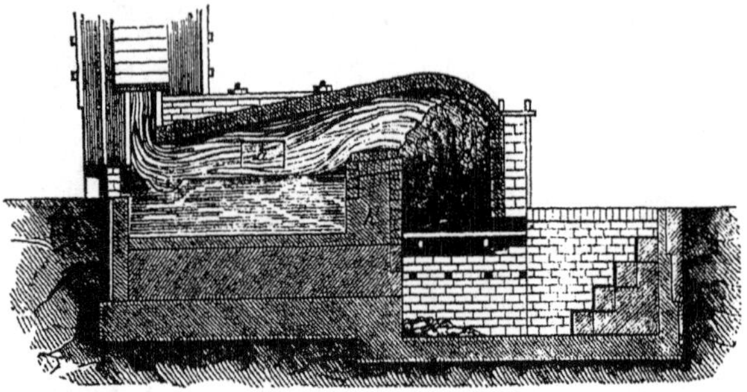

a violent boiling. The surface of the copper being kept covered with the charcoal, the refiner takes out assays in succession; the grain becomes finer and finer, and when it possesses a bright silky appearance when broken, the operation is considered to be finished. But to be more sure, a small portion is poured into a mould, and its malleability tried by hammering when red hot on an anvil. If it yields under the hammer without cracking on the edges, it is ready for taking out, or as the smelters say, it is at its proper *pitch*. The refiner now gives the workmen orders to lade (ladle) it out, which they do with iron ladles, previously washed inside with *tan ash*, and allowed to dry, which prevents the copper from adhering to the iron. The copper is poured into large iron moulds or pots, which are arranged in a circle before the front of the furnace. As the workmen are exposed to an intense heat during this operation, they require to be relieved.

There are generally six men at this operation, and three lading at a time. Three are necessary, because three ladlesful make a cake, and they are all poured in before the metal sets. When those three which are in the ring have laded once round the moulds, they are relieved by three others, and so on till the whole of the metal is out of the furnace. The size of the cakes is about 14 inches long by 8 to 10 broad and about 2 thick; the whole weighing about 90 lbs. This process of refining occupies a period of twenty-four hours. When the copper is obstinate and difficult to refine, a little lead is added. As this metal is very easily slagged, it assists the deoxidation of any oxide of copper which may be present. This is also a very delicate operation; it requires great skill and attention on the part of the refiner to observe the exact period when the metal is at its proper degree of purity. If the workmen cease to keep the surface covered with charcoal, the copper oxidizes, and the grain becomes coarse, and more poling would be required. To prevent this deterioration during the intervals of lading, they continue to throw in small pieces of wood to prevent the copper from becoming impure. It happens sometimes also that the refining is carried too far. When this is the case the copper is very brittle, depending it is supposed on its union with carbon. When this happens, the doors of the furnace are taken down, all the carbonaceous matter removed from the surface, and the copper wholly exposed to the air, which soon brings it back to its proper pitch. The composition of refined copper was found to be:—Copper, 99·965; iron, ·035. The carbonates and oxides are fused with process (No. 4). If the quantity of these ores used be great, the result will be regulus, but if the quantity which is used be minute, *blue metal* will be the product, only requiring one roasting of twenty-four hours, which brings it to the state of blistered copper, so called from its surface being covered with elevations in the form of blisters.

Napier's Patent Process for Smelting Copper.

An improvement has been made upon this mode of copper smelting by Mr. James Napier, who took out a patent for it a few years ago. The number of processes the ores have to undergo by his method of smelting before copper fit for the market is obtained, is five.

I. *Calcination of Ores.*—This calcination is performed precisely in the same way, and in the same time, as in the process first described.

II. *Fusion of Calcined Ores.*—The former part of this process is also performed in the same way as in the old process, but in the latter part there is a change. After the slag is skimmed off, a quantity of sulphate of soda and coal is added to the mineral which remains. After this mixture is stirred well and fused for about ten minutes, it is tapped out into sand moulds; when this is set or solidified (but not cold), it is thrown into large tanks filled with water, where it slakes and falls into a fine powder something like slaked lime. The sulphate of soda which was added to the metal when in the furnace, is reduced by the coal which was added at the same time into sulphuret of sodium, which combines with the sulphuret of iron and copper, and dissolves out what impurities the ores might contain, being carried away by the sodium into the water. The water in the tanks is now drawn off by taking out a plug, and the metal is got at the bottom of the tank. The following is its composition:—Copper, 34·56; sulphur, 24·05; iron, 39·90; silica, 1·50.

III. *Calcination of Washed Powder.*—This operation is carried on in the same way, and takes the same amount of time as the calcination of the granulated coarse metal previously described. But the metal in this case being in a powder, loses the most of its sulphur, and requires but one fusion to bring it to copper:—CuO 36·700, FeO 51·210, SiO_3 10·300, S 1·641.

IV. *Fusion of Calcined Powder.*—The calcined powder is mixed with a quantity of carbonaceous matter (coal), and introduced into the furnace through the hopper. When this is fused, the doors are taken down and the liquid mass stirred; it is then allowed to remain for a short time, when the slag is skimmed off, the tap hole is then opened, and the metal allowed to run into sand beds. The product is coarse copper.

V. *Refining or Toughening.*—This process is performed exactly as already described.

Copper, Dinoxide, *Suboxide, Red Oxide.* Cu_2O. Native in octahedrons. It may be prepared by boiling a solution of sulphate of copper with starch or sugar, and caustic potash or soda; the red oxide subsides; or it is formed by fusing chloride of copper with carbonate of soda, or by heating protoxide with copper filings. Specific gravity 6·. Absorbs oxygen at a red heat, and becomes black oxide. Dilute acids convert it into metallic copper, and a salt of the protoxide; dissolves in ammonia; colours glass red.

Copper and Alumina, Arseniate of. *Octahedral Arseniate, Liroconite, Linsenerz, Lenticular Copper Ore.* 12 CuO, 2 Al_2O_3, 3 AsO_5, 32 HO or CuO 37·4, Al_2O_3 10·09, AsO_5 22·4, PO_5 3·24, HO 25·44. Spec. grav. 2·926; H 2·25. Bluish-white, sky-blue, grass-green obtuse 8-hedrons with angles of 133° 30′, 60° 40′, 72° 22′. B.B. loses its transparency, and becomes a friable scoria with metallic globules. Occurs in Cornwall and Hungary.

Copper, Anhydrous Dicarbonate. *Mysorite.* CuO CO_2. Spec. grav. 2·62. H 4·25. Dark blackish-brown masses; fracture conchoidal; sectile, opaque; soluble in acids; CO_2 16·7, CuO 60·75, Fe_2O_3 19·5, SiO_3 2·, Mysore.

Copper Antimonial. See Gray Copper Ore.

Copper, Arseniide of. *White Copper Ore. Subsesquiarsenide, Arsenical Copper, Condurrite.* Cu_3As_2. Spec. grav. 4·5, H 3·25. Colour between tin-white and brass-yellow; lustre metallic, glistening; fine-grained; semi-hard; brittle; B. B. gives off arsenical fumes, melting into a grayish-black slag; contains 71·7 per cent. copper, 28·3 arsenic.

Copperas, *Couperose, Fr.* See Sulphate of Iron.

Copper, Bihydrous Pentarseniate of. *Erinite.* Spec. grav. 4·043, H 4·75. 5 CuO, AsO_5, 2 HO; oxide of copper 59·44, arsenic acid 33·78, water 5·01, alumina 1·77. Emerald-green mamillated crystalline groups, with arseniate of copper, from Limerick.

Copper, Bihydrous-tetraphosphate of. *Libethenite, Apherese.* 4CuO PO_5, 2HO or 2CuO 28·7, PO_5 63·9, HO 7·4. Spec. grav. 3·7, H 4. Olive or blackish-green right rhombic prisms, often so short as to have an 8-hedral form lustre resinous; fracture conchoidal, uneven, translucent on the edges, brittle; found at Libethen, Hungary. A variety with 3 HO is called *Tagilite,* from Tagilsk.

Copper Bismuth. 2 Cu_2S, BiS_3? Lead-gray, from Fürstenberg.

Copper, Chloride. See Atacamite.

Copper Chlorosulphate of. Blue 6-hedral prisms, Cornwall.

Copper Disulphide. *Glance Copper, Chalkosine, Vitreous Copper Ore.* Spec. grav. 5·702, H 2·75, Cu_2S; Cu 77·82, S 20·55, Fe 1·23, SiO_3 ·4 (Burra Burra, R.D.T.). Acute rhomboids, or regular 6-sided prisms, or massive, lustre metallic, opaque, sectile; B.B. melts and emits glowing drops with a noise; soluble in hot nitric acid, and in chlorohydric acid. *Locality.*—Redruth, Cornwall; Freyberg; Ayrshire; copper mines South Australia.

Copper, Gray Ore of. *Tetrahedral Copper pyrites. Trisulphoantimonite of Copper. Fahlerz, Panabase, Cuivregris.* 4 RS, SbS_3? Spec. grav. 4·4 to 5·2; H 2·75. Sulphur 25·77, antimony 23·94, arsenic 2·88, copper 37·98, iron ·86, zinc 7·29, silver ·62. Varies much in composition, owing to the replacement of isomorphous sulpho-bases, and sulphacids. Colour steel-gray or iron-black, streak unaltered, sometimes brown, massive, and in 4-hedrons. B.B. gives out vapours of arsenic or antimony; with soda gives, after a time, metallic copper; nitric acid digested on it becomes green, and dissolves it. *Locality.*—Alsace, Freyberg, Clausthal, Kapnick, Dillenburg, Fürstenberg.

Copper, Hydrotriscarbonate of. *Blue Copper Ore, Azurite, Azure Copper.* 3 CuO, CO_2HO, or 2 CuO CO_2, CuO HO. Specific gravity 3·831, H 4·25; blue rhombic oblique prisms, with angles of 91° 30′, and 98° 50′; CuO 69·08, CO_2 25·76, HO 5·46; texture foliated; fracture conchoidal; lustre vitreous, almost adamantine; transparent to translucent on the edges; brittle; becomes black when heated; B.B. melts on charcoal, and colours borax green;

soluble with effervescence in acids. *Locality.*—Occurs along with Malachite in Cornwall, South Australia, Chili, &c.

Copper, Hydrous Dicarbonate of. *Malachite, Mountain Green, Atlaserz.* Spec. grav. 4·008, H 4·. Green of various shades; massive or in oblique rhombic or right rhombic prisms, with angles of 112° 52′, and 107° 20′; lustre adamantine; B.B. becomes black; forms with borax a green glass; CuO 19·62, CO_2 70·25, HO 10·125. Form. 2 CuO, CO_2, HO.

Copper Indigo. *Blue Copper. Covellite.* CuS. Copper 64·77, sulphur 32·64, iron ·46, lead 1·04. Spec. grav. 3·8; H 2. From Sangerhausen in Thuringia and Vesuvius. Indigo-blue spheroidal masses. B.B. gives a blue flame.

Copper Manganese. Spec. grav. 3·15 to 3·25; H 1·5. A rare mineral in botryoidal masses, consisting of hydrate of oxide of manganese mixed with oxide of copper and gypsum. From Bohemia. MnO_2 74·1, CuO 4·8, HO 21·1, gypsum 1·05, SiO_3 ·3.

Copper with **Metals.** Copper unites with various metals, forming alloys. *Brass.*—Consists of 3 parts of copper and 1 of zinc, and sometimes of 4 or 5 parts of copper to 1 zinc. It is ductile, and is more fusible than copper. *Pinchbeck* or *Similor.*—Consists of equal parts of the two metals, and possesses a golden colour; hence the name. *Bronze* and the metal of cannon consist of 9 parts of tin, and 91 of copper; the strength of bronze of this composition is to malleable iron as 1 to 2. *Brunswick Green.*—Is an oxichloride, and may be formed by mixing copper turnings with muriatic acid, or with a solution of salammoniac exposed to the air; the green forms on the surface. *Bell Metal.*—Is also formed of copper and tin in different proportions; a common proportion is 3 copper to 1 tin. *Mirrors* of telescopes consist of 2 copper to 1 tin. *Tinned pans* are copper pans covered with a layer of tin in the inside to prevent the contents from acting on the copper. The copper pan is first rubbed over with salammoniac, and heated; pitch is thrown into it, and a piece of tin rubbed over the copper surface. *China silver* or *tutenag.*—Consists of copper, zinc, and nickel. The same substances enter into the composition of German silver, and forks of this description should not remain in contact with acid sauces, or pickles.

Copper Nickel. See NICKEL.

Copper, Novohydrous Pentarseniate of. *Copper Schaum, Bardiglione, Pharmacosiderite.* 5 CuO AsO_5, 9 HO or CuO 43·88, AsO_5 25·01, HO 17·46, CuO CO_2 13·65. Spec. grav. 3·098, H 1·25. Apple and verdigris-green masses and 8-hedrons; radiated or foliated; very sectile, thin laminæ, flexible; lustre pearly on one of the faces, vitreous on the others; translucent on the edges; accompanies copper ores, and especially blue copper ore, at Schwatz, Temeswar, and Matlock.

Copper, Protoxide of. CuO 5, 40; 4·957, 39·656; spec. grav. 5·868 to 6·414. Black powder formed by heating copper in the air, or by igniting the nitrate. When copper is exposed to moist air it becomes incrusted with a *green coating*, often called *verdigris*, which is a hydrate of the carbonate of the protoxide of copper. Copper salts are poisonous. Vegetable acids dissolve copper in the cold state and not in the hot state. Sauces containing vinegar, preserved fruits, or jellies, should not be allowed to remain in contact in the cold state with German silver spoons (which consist of copper, nickel, and zinc), nor copper stew-pans. Water containing common salt also dissolves copper. The protoxide colours glass green. I have for many years had large specimens of protoxide from Lake Superior. *Cupric Acid.*—Cu_2O_3. Dissolve 20 grains of black oxide in 70 grains nitric acid, specific gravity 1·35. Then 50 grains fresh hydrate of lime are mixed with 1 lb. solution of bleaching powder, spec. grav. 1·06, and added to the solution of copper. A purplish-black precipitate gradually subsides, which is washed with cold lime water. It is nearly black, but when placed in a thin layer on glass, it is rose-coloured. *Nitride.*—Cu_6N. A green powder, formed by passing dry ammonia over protoxide, turning at 570° into nitrogen and copper.

Salts of Copper.—These salts, from the facility with which copper oxidizes, are easily formed, and were known to the alchemists and earlier chemists by the name of *Salts of Venus.* 1. They are almost all soluble in water, or become so by the addition of an acid. The solution has a blue or green colour, or acquires it by exposure. They redden litmus. 2. *Caustic Ammonia* produces in small quantity a *greenish-blue* subsalt of copper, which is easily dissolved into a deep blue solution by excess, yielding a blue precipitate on standing with caustic soda of hydrous protoxide of copper. 3. *Caustic Soda* or *Potash* forms a blue precipitate, insoluble in excess, of hydrous protoxide, becoming black and anhydrous by boiling. When the alkali is insufficient, a greenish subsalt instead of black oxide is obtained. 4. *Sulphohydric Acid* in neutral acid solutions produces a *brown* when dilute; when concentrated, a black hydrous sulphide. 5. *Sulphohydride of Ammonia* produces a black or brown precipitate as in 4, insoluble in excess, and in ammonia. 6. *Alkaline Carbonates* yield a blue, the greenish precipitate with carbonate being soluble in excess, forming a blue solution. 7. *Ferrocyanide of Potassium* yields in neutral and acid solutions a reddish-brown precipitate, decomposed by ammonia, but insoluble in chlorohydric acid. 8. *Ferridcyanide of Potassium* yellow-green, insoluble in chlorohydric acid. 9. *A plate of iron* in slightly acid solutions is coloured yellow by the deposition of copper. 10. *Zinc* precipitates copper in a black spongy mass. 11. Before the blowpipe with borax, salts of copper are *green*, in the oxidizing flame (CuO), *red*, in the reducing flame (Cu_2O). *Chlorides.* — There are two chlorides, the dichloride, a brown powder, Cu_2Cl, formed by heating corrosive sublimate (HgCl) with copper; and the chloride, CuCl, green needles obtained by dissolving copper in hydrochloric acid; this salt, at a temperature above 400°, is converted into dichloride by losing half its chlorine. *Nitrate of Copper.*—$CuONO_5$, 3 and 6 HO. It is obtained by dissolving metallic copper in nitric acid. When ignited, black oxide of copper remains. When moistened and rolled up in tin foil, heat and light are given out, and peroxide of tin is formed. *Sulphate of Copper.*—CuO SO_3 5 HO. The blue doubly-oblique prisms, obtained by dissolving copper in sulphuric acid and crystallizing. It is used as an emetic in cases of poisoning, and as a caustic for sores under the name of *bluestone*. *Acetate of Copper* or *Verdigris.*—One of the cheapest salts of copper, is used for preparing Scheele's green by mixture with arsenious acid. *Estimation.*—Copper is usually separated as oxide by boiling with caustic soda or potash, and from iron by ammonia.

Copper Pyrites. *Yellow Copper Ore, Chalkopyrite.* FeS, CuS or Cu_2S, Fe_2S_3. Copper 33·64, iron 31·53, sulphur 34·65, earthy matter ·55. Spec. grav. 4·16 to 4·3; H 2·75. Colour brass-yellow, streak greenish-black, a little shining, tending to iridescence; massive, and in 4-hedrons, with the angles replaced by small triangular planes; the primary form is an 8-hedron, with a square base, the angles being 101° 52′ and 126° 30′. B.B. becomes black and red on cooling; melts into a globule, becoming magnetic. Soluble in dilute nitric acid, leaving sulphur. *Locality.*—This is the common ore of copper, and is found in Cornwall, Ireland, Cuba, Chili, Australia, &c.

Copper, Selenide of. *Berzeline.* CuSe. Silver white. Se 40, Cu 64. Silver crusts at Skrikerum, Sweden.

Copper, Semelhydrous Tetrarseniate of. *Olivenite, Prismatic Olivene Ore, Right Prismatic Arseniate.* 4 CuO,AsO_5HO = CuO 56·2 AsO_5 39·9, HO 3·9, PO_5 trace. Spec. grav. 4·166, H 3·. Olive-green, leek-green, to blackish-green; capillary flexible threads, and in right rhombic prisms of 111° 45′ or 110° 50′; most commonly in a rhombic prism of 84° and 96°; fracture conchoidal; lustre between vitreous and resinous; brittle. B.B. unchanged; on charcoal reduced; a white metallic globule being formed coated during cooling with red oxide. Soluble in nitric acid. From copper mines in Cornwall, as at Redruth, Tingtang, &c.

Acicular Olivene Ore, or *Oblique Prismatic Arseniate*, consisting of dark green oblique rhombic prisms, with angles of 56° and 124°. Spec. grav. 4·048; H 2·75. Possesses the same composition as prismatic olivene ore.

Amianthiform Arseniate, or *Acute 8-hedral Arseniate*, in olive-green threads; has also the same composition.

Copper, Septahydrous Tetrarseniate of. *Euchroite, Emerald Malachite.* Spec. grav. 3·389, H 3·75. 4 CuO, AsO_5 7 HO = CuO 47·85, AsO_5 33·02, HO 18·8. Apple-green 4-sided prisms, with angles of 117° 20′; cleavage indistinct; fracture conchoidal, uneven; lustre vitreous, transparent to translucent; refracts double pretty strongly, rather brittle; B.B. loses water, and becomes yellowish-green and friable; on charcoal instantly reduced with a kind of deflagration, leaving a globule of copper.

Copper, Sexhydrous Tetrarseniate of. *Copper Mica, Foliated Olivene Ore, Kupfer Glimmer, Hexahedral Arseniate, Rhomboidal Arseniate.* 4 CuO, AsO_5, 6 HO. CuO 58·, AsO_5 21, HO 21. Spec. grav. 2·5 to 2·6, H 2·. Emerald or grass-green 6-sided tabular crystals, the primary form being an acute rhomboid, with angles of 108° 40′, 110° 30′, 128° 18′, 124° 42′; cleavage perpendicular to the axis of the rhomboid; fracture conchoidal; lustre pearly, vitreous; B.B. decrepitates and becomes black and spongy, and then a black globule.

Copper, Silico Carbonate of. *Chrysocolla, Copper Green.* Spec. grav. 2· to 2·38, H 3·. CuO 49·63, SiO_3 28·37, CO_2 3·, HO 17·5. Probably an impure variety of the preceding; found in Cornwall, Cuba, Jersey, Tyrol, &c.

Copper Silver. See EUKAIRITE.

Copper, Sulphate of Alumina. *Lettsomite.* 6 CuO, Al_2O_3, 2 SO_3 12 HO? Blue fibrous globules, Moldawa.

Copper, Sulphate of. *Blue Couperose, Blue Vitriol, Bluestone, Cyanose.* CuO SO_3, 5HO. Spec. grav. 2·213, H 2·25. Blue doubly-oblique prisms, with angles of 127° 30′, 108°, and 123°; occurs as a result of the decomposition of copper pyrites at Rammelsberg, Neusohl, Anglesea, Wicklow, &c.

Copper Sulpho Arsenide. See TENNANTITE.

Copper, Terhydrous Pentaphosphate of. *Phosphoro-calcite, Ehlite, Dihydrite, Rhenite, Ypoleine, Hydrous subsesquiphosphate.* 5 CuO PO_5, 3 HO or CuO 67·25, PO_5 24·55, HO 8·2. Sometimes the water is a little deficient, varying from 2 to 3 atoms. Spec. grav. 4·2 to 4·3, H 5·. Emerald-green masses, or blackish-green oblique rhombic prisms, with angles of 97·30, and 37°·30; structure generally striated, from the crystals being aggregated; lustre adamantine or vitreous, translucent on the edges. *Locality.*—Rheinbreitenbach, near Bonn, and Tagilsk, Russia.

Copper, Terhydrous Subsesquisilicate of. *Dioptase, Achirite, Copper Emerald, Kiesel Malachite.* Spec. grav. 3·2 to 3·4, H 5; 3 CuO, 2 SiO_3, 3 HO = CuO 49·92, SiO_3 38·76, HO 11·32. Fine emerald-green rhombic 12-hedrons, from the Kerghese Steppes, on calc spar; insoluble in nitric acid (?), but soluble in muriatic acid, the silica gelatinizing; B.B. becomes black in the oxidizing, and red in the reducing flame; infusible; forms a green glass with borax.

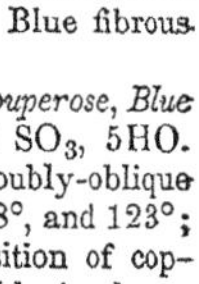

Copper, Tetrasulphate of. 4 CuO SO_3, $4\frac{2}{3}$ HO. A verdigris-green powder, found in Mexico.

Copper, Triple Sulphuret. See BOURNONITE.

Copper, Trisulphate of. See BROCHANTITE.

Copper Uranite. See URANITE.

Copper, Vanadiate of. *Volborthite.* Spec. grav. 3·55. Olive-green coloured 6-gonal crystals, sometimes in globular masses in Siberian ores of copper. CuO 44·15, CaO 12·68, VO_3 36·58, HO 4·62, MgO ·5, MnO ·4.

Copper, Variegated Ore of. *Buntkupfererz, Purple copper ore, Liver-coloured copper ore, Octahedral copper pyrites.* Spec. grav. 5·003; H 2·5 to 3. 2 Cu_2S, FeS or Cu_2S, FeS or $3Cu_2S$, Fe_2S_3=Cu 61·07, Fe 14, S 23·75, SiO_3 ·5. Colour intermediate between copper and pinchbeck-brown, with an iridescent tarnish, by which it is easily recognized; massive, and in cubes, with the solid angles replaced by triangular planes. Cleaves parallel to these faces; hence its primary form is the regular 8-hedron. Found in Cornwall, Ireland, Silesia, Norway, Sweden, Canada, Greenland.

Coprolites. The fossil excrements of extinct animals; found in the lias, chalk, and coal formations, as at Bristol, Lyme Regis, Auteuil,

Coprolites.

Connecticut, Burdiehouse, Fifeshire, &c. Consist of from 9·5 to 85 per cent. of phosphate of lime and magnesia; 10 to 61 of carbonate of lime; 3 to 13·5 of carbonate of magnesia; 7 to 25 organic matter; and sometimes small quantities of urate of ammonia, fluoride of calcium, gypsum, common salt, alumina, sesquioxide of iron, silica. When found in any quantity, they are valuable manures.

Copulated or **Conjugate Bodies.** Compound bodies, composed of two simpler compounds; thus salicine, the bitter matter of the willow, is formed of saligenine and sugar.

Coquimbite. $Fe_2 O_3 3SO_3 9HO$. Native sesquisulphate of iron, found in Chili; it is often a mixture.

Coracite. *Pitch Blende?* Spec. grav. 4·378, H 4·5. Black amorphous, found at the junction of trap and syenite, on the N. shore of Lake Superior; contains from 59·3 to 76·6 of protosesquioxide of uranium (U_3O_4), SiO_3 4·35, Al_2O_3 ·9, $Fe_2 O_3$ 2·24, Pb O 5·36, Ca O 14·44, CO_2 7·47, HO 4·64.

Coral. The name applied to various genera of sea animals, supposed to originate the chalk formation. They consist of carbonate and phosphorate of lime, magnesia, and animal matter. It

Astrea Viridis.

has been found that in the neighbourhood of coral islands the sea water is deficient in lime salts.

Cordierite. See IOLITE.

Coriander Oil. A colourless oil of spec. grav. ·759, from the seeds of Coriandrum sativum.

Coriarine. A crystalline substance from the leaves of *Coriaria myrtifolia.*

Cork. C 62·3, H 7·15, O 27·57, N 3·03. The modified liber or endophloeum of the Quercus suber or cork tree; but cork occurs in other plants, as in the potato, constituting the outermost layer of the stem, and forms a row of cells, readily distinguishable both by their structure and chemical reactions from the starch cells. The layer is easily separated from boiled potatoes. Strong sulphuric acid does not act for some time on cork. Nitric acid of 1·2 oxidizes it below 212°, and yields suberic acid, oxalic acid, &c.

Corkwax. $C_{25}H_{10}O_3$. *Cerine.* Needles obtained by alcohol and ether from cork, converted into an acid (ceric acid $C_{42}H_{17}O_{13}$) by nitric acid.

Cornaline. A variety of agate.

Corneine. *Hornstone, Compact Amphibole, Aluminous Hornblende.* Contains, from Nordmarck, silica 48·83, alumina 7·43, MgO 13·61, CaO 10·16, FeO 18·75, MnO 1·15, F ·41, HO ·5.

Corneous Lead, or Chlorocarbonate of lead.

Cornine. A crystalline resin from *Cornus florida.*

Corrosive Sublimate. See CHLORIDE OF MERCURY.

Corticine. A solid by alcohol from the bark of *Populus tremula.*

Corundellite. *Margarite.* Spec. grav. 3, H 3·5. Hexagonal whitish-yellow masses in plates, from Chester county, U.S.; SiO_3 35·71, Al_2O_3 53·13, CaO 7·27, KO 1·22, NaO ·41, HO and F 2·3.

Corundophyllite. Green plates from North Carolina, consisting of SiO_3 34·75, FeO 31·25, Al_2O_3 8·55, HO 5·47, NaO? 20.

Corundum. See SAPPHIRE.

Corydaline. $C_{34}H_{22}NO_{10\frac{1}{2}}$. Crystalline plates from *Corydalis tuberosa*, &c.

Cosmetics. (κοσμεω, I adorn). Preparations for the hair and skin.

Cotarnine. $C_{26}H_{13}NO_5$. Yellow radiated mass from narcotine by oxidation with binoxide of manganese and sulphuric acid.

Cotton. *Baumwolle*, Ger.; *Coton*, Fr. C

44·35, H 6·14, O49·51. The well-known wool or down surrounding the seeds of species of gossypium. It is nearly pure cellulose. Cotton

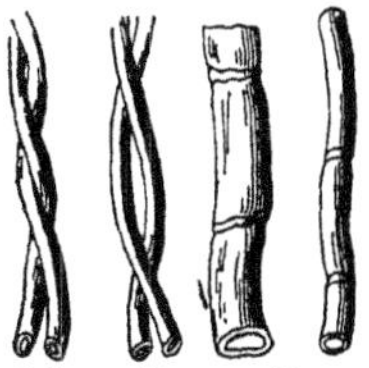

Cotton. (*Bauer.*) Linen.

consists of flattened, twisted tubes without joints, while linen consists of jointed cylindrical tubes. Hence, it is supposed that the flattened sharp edge of the cotton has an unfavourable irritating influence in dressing wounds, while the rounded form of the linen is destitute of any such effect.

Cotunnite. Chloride of lead in needles from the crater of Vesuvius.

Coulobrasine, or Selenide of Zinc.

Coumaric Acid. $C_{18}H_6O_5HO$. Plates by potash on coumarine.

Coumarine. $C_{18}H_6O_4$. Needles from Tonka bean, by alcohol, &c.

Couperose. (Ang. *Copperas*). The French name for sulphate of iron.

Couperose, White. Sulphate of Zinc.

Couzeranite. Spec. grav. 2·69, H 6·5. Black or indigo-blue lamellar, or in oblique rhombic prisms resembling pyroxene, from which it differs by its action before the blowpipe, and in its fracture, which is conchoidal; at Couzeran, Pyrenees. Silica 52·37, alumina 24·02, lime 11·85, MgO 1·4, KO 5·52, NaO 3·96. B.B. fuses to a white enamel.

Covellinite, or Nepheline.

Covellite, Covelline. Equisulphide of copper.

Cowdie. *Kaurie* or *Cowrie* resin. See Dammara Resin.

Crajuru. A red-colouring matter from Para, resembling chica red.

Crameric Acid. $C_{10}H_8O_5$. Crystals from Crameria triandra.

Craurite. Protophosphate of iron.

Crayons, Black. Are composed of fine charcoal from willows and other soft woods; white crayons from fine chalk; *red crayons* of sanguine 100, gum 4·41.

Cream. *Rahm*, Ger.; *Creme*, Fr. The oleaginous matter of milk which rises to the surface of that fluid after standing. It consists of 4·5 oil and 3·5 caseine, with 92 whey; it is therefore very liable to decomposition from the fermenting action of the caseine. The amount of cream which rises to the surface of milk, depends on the shallowness of the vessel in which the milk is deposited. The lactometers sometimes used, which are deep graduated cylinders, do not afford an indication of so much cream as would be obtained from a shallower vessel.

Cream of Tartar. Acid tartrate of potash.

Creasote, not *Creosote*, (κρέας, flesh; σώζω, I preserve). $C_{26}H_{16}O_4$? Spec. grav. 1·040. B. P. 398°·3; clear neutral oil, with a burning taste and peculiar odour, evaporating without residue, changing to a brown colour in light; leaves a white spot on the skin; it is soluble in alcohol, ether, and acetic acid, ammonia, potash; only slightly so in water; poisonous; it dissolves camphor, fats, essential oils, resins, and indigo, &c.; becomes purple by oil of vitriol; it forms weak combinations with bases; passed through an ignited tube it yields naphthaline, a gas and carbon; it approximates to hydrate of guiacyle in composition and character; it appears to be the active agent in the Aqua Binelli used as a styptic; it is a valuable remedy in toothache from its caustic influence on the pulp of the tooth; it preserves anatomical preparations, and likewise meat in summer; it is to its presence in pyroligneous acid that the preservative action of that substance is owing. It is derived from the wood tar obtained in the process for pyroligneous acid. When this acid in its crude state is saturated with a base, an oil floats on the surface. This oil is mixed with an alkaline carbonate and distilled with water, which frees it from acid. The oil is dissolved in caustic soda; the alkali saturated, the oil distilled; these processes being repeated until it is free from colour. It is often adulterated with carbolic acid. By the action of chlorohydric acid and chlorate of potash, yellow plates, *Hexachlorxylone*, $C_{26}H_6Cl_6O_6$, result; and by sulphurous acid this is converted into *hexachlorhydroxylone*. $C_{26}H_{10}Cl_6O_6$. *Creasotic acid.*—The product of the oxidation of creasote.

Creatine. See Kreatine.

Creatinine. See Kreatinine.

Crednerite, or *Psilomelanite*.

Crenic Acid. *Quellsaure*, Ger. $C_{24}H_{12}O_{16}$ or $C_7H_8NO_6$. A yellow mass, sour; reddens litmus; soluble in water and alcohol; with bases forms salts, soluble in water, insoluble in alcohol; obtained from the ochry sediment of ferruginous waters, by boiling it with caustic potash, saturating the filtered solution with acetic acid, and precipitating by acetate of copper, decomposing the crenate of copper by sulphohydric acid, and purifying the acid by alcohol.

Crichtonite. *Mohsite. Titaniate of iron.* Spec. grav. 4·; H 4·5. Protoxide of iron 46·53, sesquioxide of iron 1·2, titanic acid 52·27. Bluish-black acute rhomboids; allied to common titaniate of iron. B.B. infusible; with salt of phosphorus a red glass on cooling. *Locality.*—St. Christophe, Oisans.

Cripine. See Picryle.

Crispite. A synonyme of titanite.

Crocidolite. See Krokidolite.

Crocalite. A variety of mesotype.

Crocoic Acid. *Croconic Acid.* C_5O_4HO. Yellow crystals, forming yellow crystallized salts; it is obtained in the state of crocoate of potash, by heating rhodizoate of potash in water; fluosilicic acid takes the potash, and sets crocoic acid free.

Crocoise, or chromate of lead.

Crocus of Antimony, or oxysulphide of antimony.

Crocus Martial. *Martial Crocus,* or colcothar.

Crocus Sativus. See SAFFRON.

Croisette. See STAUROTIDE.

Cronstedtite. Spec. grav. 3·3, H 2·5. Black masses and 6-sided prisms; protoxide of iron 58·85, silica 22·45, water 10·7. MnO 2·88, MgO 5·08; B.B. swells up without fusing; with borax an opaque enamel; gelatinizes in muriatic acid; found at Przibram Bohemia, Cornwall, and Brazil.

Cross Stone. A variety of harmotome, in which the crystals cross each other lengthwise, and at right angles.

Crotonic Acid. *Jatrophic Acid.* A volatile acid, obtained by saponifying croton oil, and distilling the acid disengaged from the soap by a mineral acid.

Crotonine. Small crystals by alcohol, from the seeds and oil of the Croton tiglium.

Croton Oil. An amber-coloured oil, obtained by expression from the seeds of the Croton tiglium, used as a powerful purgative, and as a rubefacient externally; ether dissolves 60 per cent. of oil from the seeds; soluble also in alcohol.

Crucible. An important instrument in chemical operations, for fusing and heating substances.

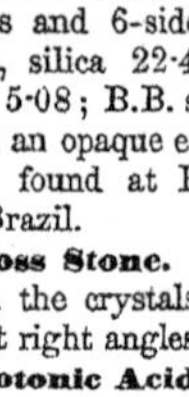

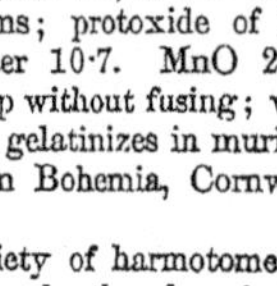

Crucibles are formed of refractory clay, porcelain, of platinum, silver, black lead, clay and charcoal.

Crucite. *Crucilite.* Spec. grav. 3·579 to 3·809, H 3·. Black 4-sided oblique prisms, crossing each other at angles of 60° and 120°. Sometimes forming a star with 6 rays; not magnetic; contains sesquioxide of iron 81·16, alumina 6·86, silica 6·, lime 4·, magnesia ·532; in clay slate, Clonmell, Waterford (R.D.T.)

Crusite, or Andalusite.

Cryolite. *Soda Fluoride of Aluminum.* (κρυος, cold, from its melting like ice when ignited.) Sp. grav. 2·95, H 2·25. White or brownish right rectangular prisms; translucent; lustre vitreous, brittle; affords fluorine 31·35, alumina 24·4, soda 44·25; found at Arksat in Greenland.

Cryometer. A thermometer filled with bisulphide of carbon, for very low temperatures, the scale ranging from 122° F. to about—250°.

Cryophorus. (κρυος, cold; φερω, I carry). A tube with a glass ball at each end to illustrate the formation of ice by the evaporation of water, one bulb containing water and the remainder of the tube vapour. When the empty bulb is immersed in a freezing mixture the vapour is condensed, and the cold produced by the evaporation from the surface of the water becomes so great as to freeze the water.

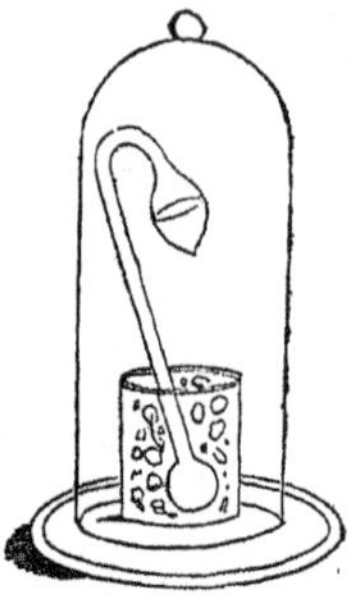

Cryptolite. *Phosphocerite.* Spec. grav. 4·6, 4·78, H 5. Red 8-hedrons from Arendal and Johannisberg in Sweden; composed of protoxide of cerium 73·7, protoxide of iron 1·51, phosphoric acid 27·37.

Crystal. (κρυσταλλος, ice). A regular formed figure with plane faces, increasing by an addition to these faces, with a definite chemical composition; in contradistinction to an amorphous body.

Crystalline. A white albuminous substance, of which the crystalline lens of the eye is formed; soluble in water after long digestion, and precipitated by alcohol; soluble in boiling alcohol, and separating on cooling; the solution in water begins to opalize at 163°, and coagulates at 200°, acetic acid coagulates it at 122°. It is composed of C 54·2, H 7·1, N 16·5, S 1·2, O 21·.

Crystalline. Synonyme of Aniline.

Crystallography. The established facts in natural science would lead us to infer that all bodies are capable of existing in three different states, depending upon the distance between the particles of matter. Thus, if we take some crystals of iodine and apply heat to them in a flask, we shall find that they first liquefy, and then gradually the area of the vessel becomes filled with a beautiful violet vapour. When the heat is removed, and the flask cools, the vapour condenses on the sides of the flask in the form of minute octahedrons. The effect of the heat has therefore simply been to cause the ultimate particles of the iodine to repel each other; first forming a fluid, and then a vapour, when these particles had obtained, under the circumstances, their greatest possible distance from each other. The removal of the heat was followed by the attraction of the particles towards each other, and the assumption by the iodine of its usual condition at the common temperature of the atmosphere. If, again, we take a crystal of common salt, a cube, and dissolve it in a cubic inch of water, we have simply removed its particles to a greater distance from one another; but we can again approximate them by the evaporation of the water, and thus recover the original cubic crystal. Now, it is one of the results of experience that, as a general law, all bodies which are capable of assuming a regular crystalline form, as in the case of common salt, are restricted to one crystalline shape, or to some form which is derived from or connected with it. If we attend to the mode in which a crystal is formed in a saturated solution, for example, we shall find that it grows or increases in size by additional layers to its exterior surface, a character which distinguishes inorganic from organized nature. To prove this, we may place a crystal of common potash alum in a solution of chrome alum, when we shall find that a dark layer has been deposited; and over this we can deposit a clear layer of ammonia, potash, or soda alum, by inserting the crystal in such solutions, the original form still remaining. By making a section of the crystal, these layers can be observed. It

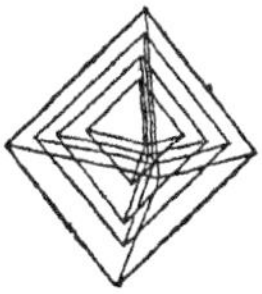

1.

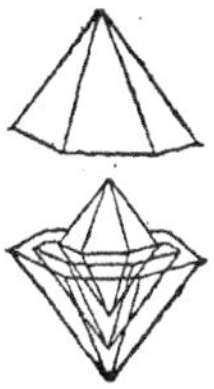

2.

then follows, according to the views of Hauy, that, in crystallization, every crystal possesses a nucleus. This nucleus is said to be the primitive form, because from it all secondary forms are derivable. Thus we find the same substance assuming a variety of shapes, as fluor spar, which occurs in octahedrons, cubes, and tetrahedrons; but it has been ascertained that the two latter can be derived from the first of these forms, and hence the octahedron is said to be the primitive form of fluor spar, in the language of Hauy. In order to understand the nature of the method by which the connection is traced, it is necessary to premise that a crystal is a solid body, bounded by plane surfaces which are termed the *faces* of the crystal. When 2 faces meet, they constitute an *edge*, B, which may be *blunt*, *sharp*, *equal*, or *unequal*. A cube or hexahedron has 8 edges. When 3 or more faces unite together in a point, they form a *solid angle*, A, which may have 3, 4, &c. *faces*, may be *regular*, *symmetrical*, *irregular*, *equal*, or *unequal*; thus a cube has 8 equal angles. *Cleavage*.—It is observed, that when we strike a crystal placed on a table with a hammer in a cautious manner the crystal splits into fragments, but these pieces are not fortuitous in their forms; thus a rhombic crystal of calcareous spar, when split by a hammer, is said to *cleave* into smaller rhombs, and not in any other manner.

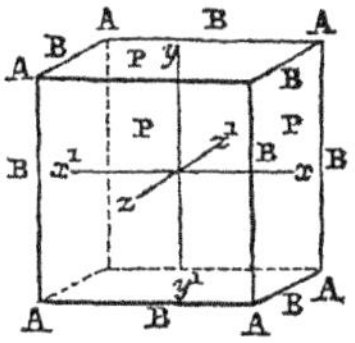

3.

Hence, as the surfaces of *cleavage* are in the same direction with those of the crystal, the rhomb is said to be the *primitive form* of calcareous spar. But frequently when a crystal is struck, the direction of the cleavage or the *natural joint* of the crystal is not parallel to its faces. Thus, if we take a cubic crystal of fluor spar, and cleave it with the aid of a hammer and knife, we find that it does not split parallel to the faces of a cube, but that the solid angles are removed, and an octahedron formed as in fig. 5, where the 8-hedron is seen in the centre of the cube. Then by continuing the cleavage we obtain from the 8-hedron the 4-hedron, as exhibited under ALUM; and, lastly, from the 4-hedron we recover the 8-hedral form, as in the 6th fig. The 8-hedron is therefore said to be the primary form of fluor spar, while the cube and 4-hedron are secondary forms. There is another form of crystallization which fluor spar frequently assumes, dependent on the replacement of the angles of the cube by *tangent planes*, and to this crystal, which has 12 faces, the term rhombic dodecahedron (fig. 11) is applied. Iron pyrites or bisulphide of iron is also, like fluor spar, usually found in the form of a cube; but when we attempt to cleave it, the crystal splits parallel to the faces of the cube; and hence its primary form is said to be that of a cube. This mineral also occurs in octahedrons, rhombic dodecahedrons, and in modifications of these forms. *Sulphide of zinc*, blende, or black jack, is met with of all the preceding forms. When we attempt to cleave the cubic form of blende, the direction of the cleavage is such as to truncate the edges of the cube, and to produce the rhombic dodecahedron (fig. 11), which is considered its primary form. Galena also affords a similar example. The system of primary forms started by Hauy, though ingenious, has been abandoned by mineralogists, for various reasons. One of these consists in the fact, that we are not enabled by cleavage, in many instances, to ascertain the primary form, but require to have recourse to other methods of investigation, which do not depend on experiment, but on calculation. The system of Weiss is founded on the relations of the axes of the crystals. The *axes* (x, y, z, fig. 3), of a crystal are imaginary lines which pass through the centre, and which symmetrically divide the faces. In the cube, the lines which unite the opposite faces are the axes. The system has thus been arranged, according to different ideas of nomenclature.

4.

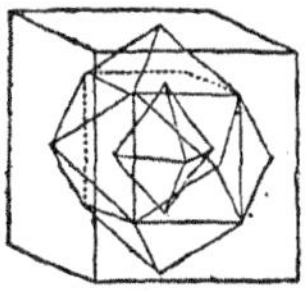

5.

Weiss.	Mohs.	English.
1. Regular system.	Tessular.	Octahedral.
2. 2 and 1 axed.	Pyramidal.	Pyramidal, or sq. prismatic.
3. 3 and 1 axed.	Rhombohedral.	Rhombohedral.
4. 1 and 1 axed.	Prismatic.	Right prismatic.
5. 2 and 1 membered.	Hemiprismatic.	Oblique prismatic.
6. 1 and 1 membered.	Tetartoprismatic.	Doubly-oblique do.

The faces of the crystals in regular octahedrons are *equilateral* triangles; in the second system *isosceles* triangles; and in the fourth, *scalene* triangles as in the figures. The bases of prisms are *square; rectangular*, the angles being right angles, and the opposite sides equal; *rhombic*, with all the sides equal and the angles oblique; *rhomboidal*, with the opposite sides equal and the angles oblique.

I. *Regular*, *Octahedral*, *Cubic*, or *Equiaxed System* (ὀκτω, eight; and ἑδρα, a seat), is characterized by 3 axes, which are equal, and at right angles to each other, and hence the examples of this class are very symmetrical. The peculiarity of this class is, that all the modi-

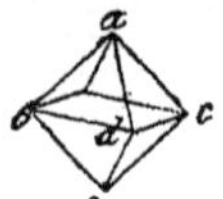

6.

7.

8.

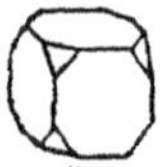
9.

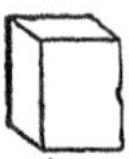
10.

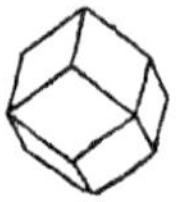
11.

fications are also distinguished by having equal axes. The homohedral or simple forms of this system are seven in number, depending on the only 7 possible intersections of the 3 axes by a plane; (fig. 6) 1. *Octahedron*, a figure contained under 8 equilateral triangles, consisting of 8 faces, 12 edges, and 6 angles, each angle having 4 faces, and the angle at the base of the pyramids 109° 28'. *Examples.*—Chrome iron ore, automalite, spinell, fluor spar, magnetic iron ore, alum, blende, salammoniac, diamond.

2. *Hexahedron*, or cube; a figure contained under 6 squares (fig. 10), 6 faces, 12 edges, and 8 angles; faces consist of 6 squares; edges and angles equal; the latter with 3 faces, formed by the plane cutting one axis, and being parallel to the other two. *Examples.*—Common salt, pyrites, galena, fluor spar, and various simple bodies, as copper, silver, gold, platinum.

3. *Rhombic Dodecahedron*, or *Monogrammic tetragonal dodecahedron*, termed generally *the* dodecahedron (fig. 11), when the plane cuts 2 axes at an equal distance from the centre, and is parallel to the third; it has 12 rhombic faces, 24 unequal edges, and 14 equal angles. *Examples.*—Garnet, sodalite.

4. *Icositetrahedron* (fig. 12), 24 faces, 48 edges, 26 angles; is produced by the plane cutting 2 axes at an equal distance, and the third at a less distance from the centre. The type of this form of crystal is *Leucite*, or white garnet, and hence it is termed the Leucite or leucitoid form, or leucitohedron. The garnet affords another example, and analcime a third. Various

12.

combinations of this, with simple and complicated forms, present themselves. Thus we have it combined with the dodecahedron and octahedron, in the magnetic iron ore from Piedmont.

5. *Triakisoctahedron* (fig. 13), (3 times 8 faces), 24 faces, 36 edges, and 14 angles, is produced by the plane cutting 2 axes at an equal distance, and the third at a greater distance from the centre. It seldom occurs alone, but generally in combination with other forms, as in the garnet from Piedmont, where it is united with the leucitoid and garnet forms.

13.

6. *Tetrakishexahedron* (fig. 14), (4 times 6 faces), 24 faces, 36 edges, 14 angles. When the plane cuts 2 axes at unequal distances from the centre, and is parallel to the third axis, as in fluor spar from Alston Moor in Cumberland.

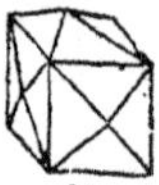
14.

7. *Hexakisoctahedron* (fig. 15), (6 times 8 faces), 48 faces, 72 edges, and 26 angles. When the plane cuts all the axes unequally, as in fluor spar from Cumberland and Derbyshire.

15.

To the regular, or octahedral system, belong more particularly what are termed hemihedral crystals (ἡμι, half; and ἑδρα, a seat), where the faces of one side of the crystal are more fully formed than those of the opposite side. Thus the tetrahedron, or hemi 8-hedral crystal, has 4 angles in all, 3 of which are equal, 4 faces and 6 edges, or it is a figure contained under 4 equilateral triangles. (See HEMIHEDRISM.) We sometimes observe this form in alum, frequently in blende, under various modifications; also in boracite, gray copper ore, specular iron ore. (For *hemihedral*, *hemitrope*, *holohedral*, and *transposed crystals*, see these words.)

II. *Pyramidal*, *square prismatic*, *quadratic*, or *2 and 1 axed or 4 membered system; 8-hedron*

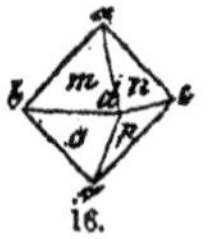

16.

17.

18.

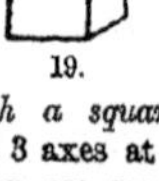
19.

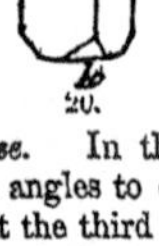
20.

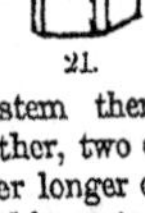
21.

with a square base. In this system there are 3 axes at right angles to each other, two of them are equal, but the third is either longer or shorter. The form of the type of this system may therefore be represented as two pyramids with square bases applied base to base (fig. 16). Each of the faces forms an isosceles, or scalene triangle; while, in the octahedral or regular system, each of the faces constitutes an equilateral triangle. When the angles at the base of this octahedron are replaced by planes, the crystal becomes a 4-sided square prism, or right square prism (figure 19), which may be terminated by 4-sided pyramids, as in the most common form of Harmotome (fig. 17). Frequently the edges of the prism are replaced by planes, and thus an 8-sided prism is produced (fig. 21). It is hence observable that the base of this system may alter in form without affecting the extremities, while in the regular system all sides must be equally altered. The value of *d* on *n* and on *o* will vary with different individuals of this class (fig. 16). Illustrations of this system are found in zircon, yellow prussiate of potash, peroxide of tin, edingtonite, arkansite, calomel.

III. *Rhombohedral,* 3 *and* 1 *axed system; Regular Hexagonal system.* Three of the axes

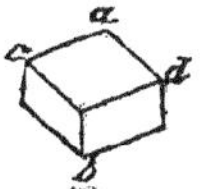

22.

23.

24.

25.

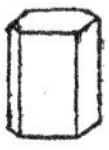

26.

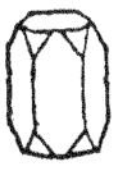

27.

in this system are equal, and intersect each other at angles of 60°. The fourth axis is unequal, and cuts the other axes at right angles. This system corresponds in many respects with the 2 and 1 axed system. Some of the common forms are the rhomboid of calc spar (fig. 22), and the hexagonal dodecahedron exhibited in the quartz double 6-sided pyramid, which has 12 faces, 18 edges, and 8 angles (fig. 23). The base often extends into a 6-sided prism, or 12-sided prism, containing, therefore, 2 sides more than the rhombohedron, which is the basis of the present system (figs. 24, 25). Two of the angles of the latter figure are truncated to form the hexagonal prism. Crystals of this system have 1 axis of double refraction, which is exhibited by turning a clear crystal of calc spar on its own axis over a cross written on paper. In all doubly refracting bodies there are one or more lines along which there is no double refraction, or where the ordinary and extraordinary rays coincide, as in the shorter diagonal of calc spar. Hence it is said to have *one axis of double refraction.* Crystals with this property belong to the second and third systems. Most crystals have 2 axes of double refraction. To this system belong carcareous spar, emerald, rock crystal, nitrate of soda, apatite, molybdena, arseniate of lead.

IV. *Prismatic, or* 1 *and* 1 *axed or right prismatic system; Right rhomboidal* or *Rectangular prismatic.* In this system there are 3 axes which intersect each other at right angles, and are all unequal in length. The primary form is a right rhombic prism (fig. 30). The forms of

28.

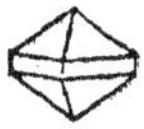

29.

30.

31.

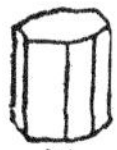

32.

33.

this system are principally three. 1. When their faces are inclined towards all the 3 axes, as in the rhombic octahedron, exhibited by sulphur (figs. 31, 33), which has 8 faces, 12 edges, and 6 angles, and by iodine. 2. When the faces are inclined towards 2 of the axes, and are parallel to the third, including vertical and horizontal 4-sided prisms, as in the topaz, ilvaite, arsenical pyrites, sulphate of barytes, q.v.; arragonite, andalusite. 3. When the faces are inclined towards 1 axis, and are parallel to the other 2 axes, as in desmin, topaz, sulphates of barytes and lead, chrysolite, tartar emetic.

V. *Oblique prismatic system,* 2 *and* 1 *membered system, oblique rhomboidal or rectangular prismatic system,* possesses 3 axes, which are all unequal, 2 of them being inclined at an acute angle towards each other, while the third forms

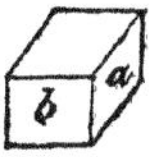

34.

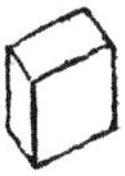

35.

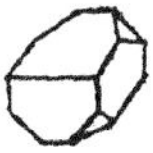

36.

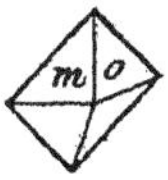

37.

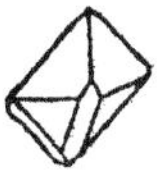

38.

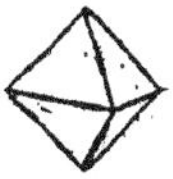

39.

a right angle with the other 2 axes. The oblique rhombic prism (fig. 35) is a quadrangular prism, whose bases are equal rhombs, and whose lateral faces are equal oblique angled parallelograms. The acuteness of the angle produced by the intersection of the first 2 axes distinguishes this system from the preceding. The principal axis is one of the oblique axes which is more developed than its neighbour. The primary form is (fig. 34) as in felspar. The various principal forms depend, 1st, on the inclination of the faces towards all the 3 faces, as in the oblique rhombic octahedron of gypsum (fig. 37), which has 8 faces, 12 edges, and 6 angles, the faces being in 4 pairs; and, 2d, on the inclination of the faces towards 2 of the axes, and their parallelism to the third, the feature of an oblique rhombic prism as in *mesotype.* Other illustrations are cream of tartar, titanite, augite, felspar, sulphate of soda, carbonate of soda, borax, sulphate of iron, oxalic acid, brewsterite, acmite, epidote, mesolite.

VI. *Doubly-oblique prismatic system, or* 1 *and* 1 *membered system of oblique parallelipipeds with oblique angles.*—This system possesses

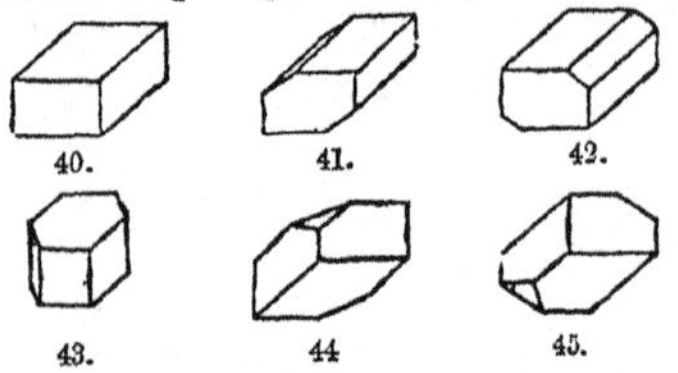

40. 41. 42. 43. 44 45.

3 axes, which are all unequal, cutting each other at right angles; and any of which may be viewed as the principal axis. The primary form is as fig. 40. The doubly-oblique prism is a quadrangular solid, whose bases are equal oblique angled parallelograms. The principal forms depend, 1st, on the inclination of the faces to all the 3 axes, as in the doubly-oblique octahedron; or 2d, on their inclination to 2 axes, and their parallelism to the third; or, 3d, on the inclination of the faces towards 1 axis, as in a combination of it occurring in axinite. Crystals belonging to this system are usually very complicated. Albite, sulphate of copper, cyanite, diaspore, quadroxalate of potash, paratartaric acid, axinite, labradorite, and sulphate of manganese afford examples of the doubly-oblique prism.

Measurement of Angles, Goniometry.—It is usual in crystallography to determine numerically the inclination of the planes or faces of a crystal. In a cube, for example, we can see with the eye alone that all the faces of the crystal are perpendicular to each other, and therefore form right angles with one another. The instrument for effecting the measurement of angles is termed the goniometer (γωνια, angle; μετρον, measure), or angle measure. The common instrument derives its principle from the fact, that if we suppose one diameter of a circle to revolve at a central point at right angles to a stationary line drawn through the centre of the circle, the revolving line, as it divides the circle in its revolutions always equally, will form with the stationary diameter equal opposite angles. This may be illustrated by the common goniometer; the arm *f* revolving on the centre

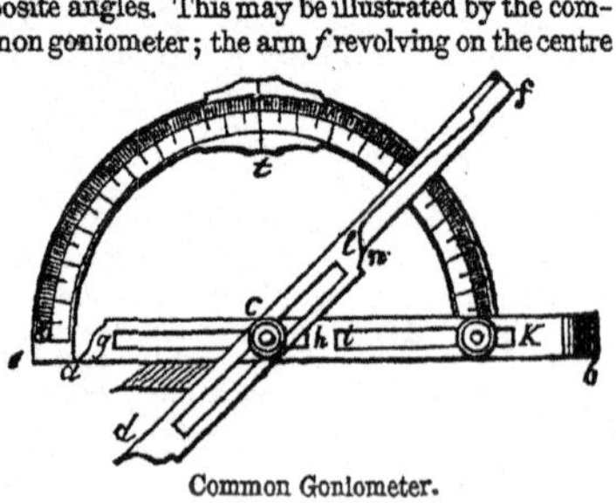

Common Goniometer.

c, the angles at *c* are always equal. If we take now a crystal of common salt, and place the fixed diameter *a b* of the goniometer on the upper plane, and then bring the moveable arm so as to be applied to the lateral plane, we shall find that while the arm will be vertical, the upper part of the arm will cut the semicircle at 90°—the value of a right angle. To facilitate descriptions of minerals, it is usual to place the crystals in such a position that the form can be at once appreciated, and this can be learned by a reference to figures and descriptions. The planes of a crystal are often distinguished by the letters P P″ M T, while the angles are designated by the vowels A E I O. When the planes are similar, the same letter is used accented frequently for precision. In the cube the planes are all marked P. For the sake of distinctness the upper plane is P, the front plane P′, and the plane to the right P″. When the solid angles of a cube are replaced by tangent planes, the latter being marked *a*, then the angle formed by *a* on P P′ or P″ is 125° 15′ 52″, expressed in degrees, minutes, and seconds. In the octahedron the angles formed by the planes P, P′ P″ are = 109° 28′ 16″. In the rhombic dodecahedron, P on P′ measures 90°, and P on P″ 120°. In the more regular forms the angles are similar, even in different individuals of the class, as, for example, in all cubes and octahedrons and other forms belonging to the regular system; but in the other systems the angles vary with different members of the class. For the purpose of obtaining approximations to the true value of the angles, the common goniometer will enable us to effect our object only when the planes of the crystals are perfectly formed and of sufficient extent. For perfect measurements it is necessary to have recourse to the reflective goniometer of Dr. Wollaston, a most ingenious and accurate instrument, and one of the greatest boons bestowed on chemical science by optics. It consists of a large

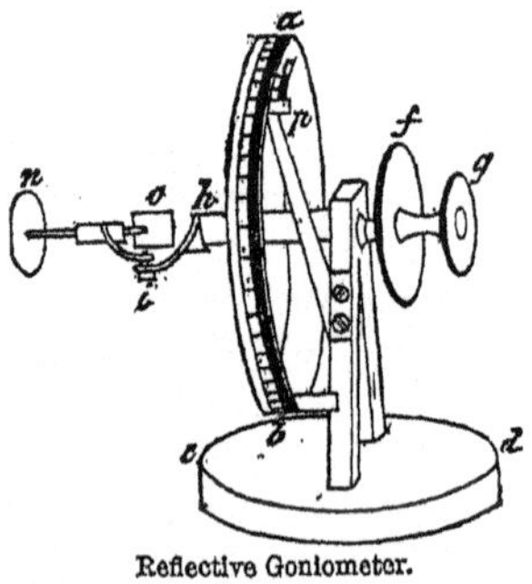

Reflective Goniometer.

graduated moveable circle, *a b*, and of two smaller ones, *f g*, and of a plate, *o*, for fixing the object to be examined. The instrument as sold is adapted to stand on the upper surface of the box which contains it. In this state it is placed on a table at from 6 to 12 feet from a *flat* window with the graduated circle perpendicular to the window, the plate being placed horizontally with its slit towards the observer. Place the crystal to be measured on the table, resting on one of the planes whose inclination is to be measured, with the plane which forms the angle with it farthest from you and parallel to the window. Fix a piece of bees' wax to copper plate, as in the figure, and bring the end of the wax in contact with the crystal. Press the crystal against the wax so as to make it adhere, and insert the copper plate in the cleft of the pin with that side uppermost which rested on the table. Bring the eye close to the crystal in an inclined manner, and adjust it by means of the plate so as to see the bars of the window reflected from the upper plane, then turn the exterior circle, *g*, until this reflected bar is brought down to cover the actual lower base of the window (not reflected). Then turn the crystal by moving the right exterior circle from you until the same bar of the window is reflected and brought to coincide with the base of the window as before. As soon as both planes are then adjusted accurately, the instrument is in a proper condition to proceed to measurement. Bring 180° on the graduated circle to *o* of the vernier at *p*, by turning the circular plate *f*. *Bring the reflected image of the bar from the first plane to coincide with the base of the window. Then turn the graduated circle from you by means of the middle circle until the reflection of the bar coincides also with the base of the window.* Observe the number opposite the *o* of the vernier. It is the value of the angle required.

Crystal, Rock. A variety of quartz in 6-sided prisms.

Cuban, or Copper Pyrites.

Cube Ore. *Pharmacosiderite*, or triarseniate of iron.

Cube Spar. See ANHYDRITE.

Cubebine. $C_{34}H_{17}O_{10}$, or $C_{17}H_9O_5$. Needles by alcohol from cubebs oil.

Cubebs Oil. $C_{15}H_{12}$. An aromatic colourless oil of spec. grav. ·929. B.P. 490°, obtained by distilling along with water the fruit of the Piper cubeba, or Java pepper.

Cubebs Stearoptene. $C_{15}H_{13}O$? Spec. grav. ·926; F.P. 154½°; B.P. 306°; rhombic crystals, tasting of cubebs.

Cubizite. See ANALCIME.

Cuboide. See CHABAZITE.

Cucumber. *Momordica elaterium.* The juice of this plant affords the purgative principle *Elaterine*, and the extract of elaterium.

Cudbear (*Persio*, Ger.); (named from Dr. Cuthbert Gordon), a red colour, obtained by steeping the *Lecanora tartarea* and *Parmelia omphalodes* in shallow vessels in ammoniacal liquor; dried and powdered. It was first prepared at Leith by Mr. Macintosh, in 1777, and afterwards at Glasgow, but it is now discontinued.

Cumanilide. $C_{20}H_{17}NO_2$. Needles by the action of aniline and chloride of cumyle.

Cumaric Acid. $C_{18}H_6O_6$.

Cumene. $C_{18}H_{12}$. A liquid, B.P. 291°, obtained by distilling cuminic acid with barytes.

Cumidine. $C_{18}H_{13}N$. Crystalline base, fusing and then resembling aniline; obtained by the action of nitric acid on cumole.

Cumin, Oil of. An oil derived from the seeds of the *Cuminum cyminum*, consisting of cuminole and cymole, and some cuminic acid.

Cuminamide. $C_{20}H_{13}NO_2$. White crystals by heating cuminate of ammonia.

Cuminic Acid. $C_{20}H_{11}O_3HO$. White plates or needles, F.P. 198°, B.P. 482°; sublimes; soluble in ether and alcohol; obtained by dropping oil of cumin on melted potash in a retort.

Cuminic Acid, Anhydrous. $C_{20}H_{11}O_3$. Colourless thick oil, with a feeble odour of fatty ethers; by the action of perchloride of phosphorus, or of oxychloride of phosphorus, on cuminate of soda.

Cuminole. $C_{20}H_{12}O_2$. Yellow fluid; spec. grav. ·969, B.P. 437°, one of the constituents of oil of cumin, being the least volatile oil; formed also by the reaction of phorone and anhydrous phosphoric acid.

Cummingtonite. Spec. grav. 3·2014, H 2·75. Grayish-white slightly diverging needles; lustre silky; found in mica slate at Cummington, Massachusets; B.B. infusible; with soda becomes a dark glass; silica 56·54, protoxide of iron 21·67, protoxide manganese 7·80, soda 8·44, water 3·18.

Cumole. $C_{18}H_{12}$. B.P. 298°·4. An oil by distilling cuminic acid and lime.

Cumonitrile. $C_{20}H_{11}N$. Colourless agreeably smelling oil, spec. grav. ·765; by boiling cuminate of ammonia.

Cumyle. *Cm.* $C_{20}H_{11}O_2$. The hypothetic radical of the cumin oil series.

Cupel. A shallow vessel made of bone earth, for the purpose of exposing a mixture of gold, silver, and lead, and other impurity, to a strong heat and oxidation, by which the lead is oxidized, and sinks into the cupel, and the silver is obtained alone in the metallic state. For common assays, a cupel is made by burning some bones, pulverizing finely the calcined phosphate of lime, and passing it through a sieve. It is then made into a paste with water, and introduced into a ring of iron, which is made 1½ inch high. The cupel mould is exhibited in the fig. The upper surface is then rendered concave, to admit the metals, and the whole

allowed to stand until it dries. On the large scale the cupel is made of much greater size.

Cupellation. The process performed by means of the preceding apparatus of purifying gold and silver from other metals.

Cuproangleaite. *Cupro plumbite.* See CUPREOUS SULPHATE OF LEAD.

Cupromanganesite. Spec. grav. 3·2, H 1·5. Black kidney-shaped masses with a resinous lustre. Seems a mixture of protoxide of copper, manganese, cobalt, magnesia, lime, barytes, and potash, found near Saalfeld, and in Missouri, U.S.

Cuprosulphide of Antimony. Cu_2S, SbS_3. Spec. grav. 4·47, H 3·. Black metallic masses, fracture conchoidal; B.B. fusible, evolving vapour of antimony, and becoming a globule, yielding a copper reaction.

Curarine. A bitter, alkaline, poisonous base, from Curara or Urara; an extract from a tree of the strychneae family, called Mavacure, in S. America.

Curcumine. C 69·5, H 7·46, O 23. Yellow powder, by alcohol and ether, from turmeric *Curcuma longa.*

Curd. The familiar name for caseine of milk.

Curry. A yellow powder composed of spices, and coloured yellow by turmeric; used as a seasoning to various kinds of meat.

Cusconine. From casco, identical with aricine.

Cusparine. *Angusturine, Galipeine.* Needles and irregular tetrahedrons by alcohol from the bark of *Bonplandia trifoliata*, and *Cusparia febrifuga.*

Cyamelide. *Insoluble Cyanuric acid.* C_2NHO_2. White porcelain mass, without taste and smell, insoluble in water, alcohol, ether, and dilute acids; converted by heat into liquid cyanic acid; by allowing cyanic acid to stand by itself it is soon converted, with great evolution of heat, into cyamelide.

Cyamelaric Acid. $C_{12}N_7HO_3$, or $C_{12}N_8HO_4$, an acid not isolated, obtained united to potash by heating mellonide of potassium with caustic potash.

Cyanamide. $C_2H_2N_2 = CyH_2N$. White crystalline body, obtained by passing chlorocyanogen into a solution of ammonia in anhydrous ether.

Cyanamylamide. $C_{12}H_{12}N_2 = Cy, C_{10}H_{12}N$.

Cyanethine. $C_{18}H_{15}N_3$. Pearly plates. F.P. 374°; B. P. 536°; obtained by acting on cyanide of ethyle with potassium.

Cyanethylamide. $C_6H_6N_2 = Cy, C_4H_6N$.

Cyanic Acid. C_2NO, HO. Colourless volatile fluid, with a smell like acetic acid, and an acid reaction; obtained by heating yellow prussiate of potash and black oxide of manganese, and decomposing the cyanate of potash thus formed by sulphuric acid; or by distilling cyanuric acid; monobasic.

Cyanides. Compounds of cyanogen and metals.

Cyanilic Acid. $C_6H_3N_3O_6$? 8-hedrons, or pearly scales by boiling mellone with nitric acid.

Cyaniline. $C_{14}H_7N_2$. Brilliant scales, but slightly soluble in any menstruum, forming salts with acids; obtained by acting with cyanogen on a solution of aniline in alcohol.

Cyanite. *Kyanite, Sappare, Disthene, Fibrolite, Sillimanite?* Spec. grav. 3·6, H 5·7, 3 Al_2O_3, 2 SiO_3. Doubly-oblique prisms, the terminations of which are nearly rhombs (6th Syst. See CRYSTALLOGRAPHY); the angles of the prism are 106° 15′, and 73° 45′; the colours are gray, white, blue, and brownish; lustre pearly; B.B. infusible; with borax becomes a colourless glass; silica 36·60, alumina 62·56, HO ·84, occurs in primitive rocks, in mica slate at St. Gothard, Tyrol, Chesterfield, United States.

Cyanmethylamide. $C_4H, N_2 = Cy, C_2H_4N$.

Cyanobenzile. Large crystals obtained by heating prussic acid and an alcoholic solution of benzile.

Cyanocumidine. $C_{20}H_{13}N_3$. Needles formed by treating an alcoholic solution of cumidine with cyanogen.

Cyanogen. (Gay Lussac, 1815. (Κυανος, blue; γεννω, I produce.) $C_2N = Cy$ 3·25; 26. Spec. grav. 1·8055 (Theory), 1·8064 (Gay Lussac, Ann. Chim. 95, 172), 1·80395 (Thomson); consisting of 2 vols. carbon, = ·8320, 1 vol. nitrogen, = ·9706.

		Spec. Grav.	Atoms.	Wt.		P. Ct.
Carbon,...2 vols.	□□	·8320	2	1·5	12·	46·16
Nitrogen, 1 vol.	□	·9720	1	1·75	14·	53·84
		1·8040		3·25	26·	100·

Theory and Sources.—When dry oxalate of ammonia (NH_2H, C_2O_2O) is distilled in a retort *oxamide* passes over, which is oxalate of ammonia, deprived of an atom of water (NH_2, C_2O_2 + HO). When oxamide is heated to a higher degree, it loses the elements of 2 more atoms of water, and becomes cyanogen gas (NC_2 + 2HO), which is said to be the nitrogen radical or *nitrile* of the original oxalate, for if we dissolve the gas in water, and allow it to stand, oxalate of ammonia is again re-produced (Pelouze and Richardson). Cyanogen is usually procured by heating bodies containing nitrogen, as woollen rags, with carbonate of potash, at a red heat; by passing ammonia, nitrogen, or atmospheric air over ignited carbonate of potash and charcoal (Desfosses, Journ. Pharm. 14, 280). This fact has been since confirmed by others, and has been applied successfully on the large scale in the manufacture of yellow prussiate of potash, but has been again abandoned from its expense. Considerable quantities of cyanide of potassium, derived probably from the nitrogen and potash of coal, have been obtained in the blast furnaces near Glasgow (Clark, Phil. Mag. 10, 729). It has been shown that cyanogen exists in the gases of the lower part of the furnaces. Mr. Rankine of Carluke has found considerable quantities of potash salts with cyanide to be sublimed at the top of the furnaces. Binoxide of nitrogen, nitrous acid, and nitric acid, produce, by acting on organic compounds, cyanohydric acid, cyanohydride of ammonia, or cyanide of potassium. 1 part nitre, and 2 cream of tartar, when ignited, yield much cyanide of potassium.

Characters.—A colourless gas, possessing the mechanical properties of common air; smell peculiar, very strong and disagreeable, somewhat similar to prussic acid; probably poisonous when inhaled in quantity. Inflammable, burning with a purplish-blue flame; not decomposed by a red heat. Water absorbs 4½ vols. of cyanogen, alcohol 25 vols. and acquires its smell and a peculiar taste. It reddens litmus solution; it is decomposed by electric sparks into charcoal and nitrogen. 2 vols. of oxygen convert 1 vol. of cyanogen into 2 vols. carbonic acid and 1 vol. nitrogen; spongy platinum causes cyanogen and oxygen to unite when heated. Dry chlorine in the light of the sun converts it into oils of chlorides of carbon and nitrogen (Serullas). Phosphorus, sulphur, and iodine may be volatilized in cyanogen without altering it; potassium, when heated in it, absorbs it with the evolution of a fine light. When passed over ignited iron, carbon is deposited, and nitrogen set free; when brought in contact with ignited carbonate of potash, cyanide of potassium and cyanate of potash are formed; caustic alkalies absorb it, and form metallic cyanide, cyanate, and azulmate; ammonia absorbs it, and gradually decomposes into cyanohydric acid, azulmic acid, oxalic acid, and urea (Wöhler). When dissolved in water, the solution gradually becomes brownish-black, and yields from $7NC_2 + 14HO$; cyanohydric acid $3C_2NH$; carbonic acid $4CO_2$; oxalate of ammonia $NH_3C_2O_3HO$; ammonia NH_3; urea $C_2N_2H_4O_2$ (Pelouze and Richardson). Alcohol and ether saturated with cyanogen likewise become dark coloured from the decomposition, much urea being precipitated.

Preparation.—1. Cyanide of mercury (HgCy) is heated to redness in a tube or retort, and the cyanogen gas evolved is collected over mercury; the mercury is volatilized, while a considerable portion of the cyanogen remains as a black solid in the retort in the form of paracyanogen (Johnston, Brewster's Journ. 2d, 1, 75), in which the carbon and nitrogen are present in the same ratio as in cyanogen; or, it may be, mellon (C_6N_4), uniformly mixed with charcoal. 2. 2 parts of dry yellow prussiate of potash, when mixed intimately with 3 parts of corrosive sublimate (Hg Cl), and ignited in a retort, yield cyanogen (2Cy), mercury (2Hg), cyanide of iron (FeCy), and chloride of potassium 2(KCl) (Kemp, Phil. Mag. 22, 179).

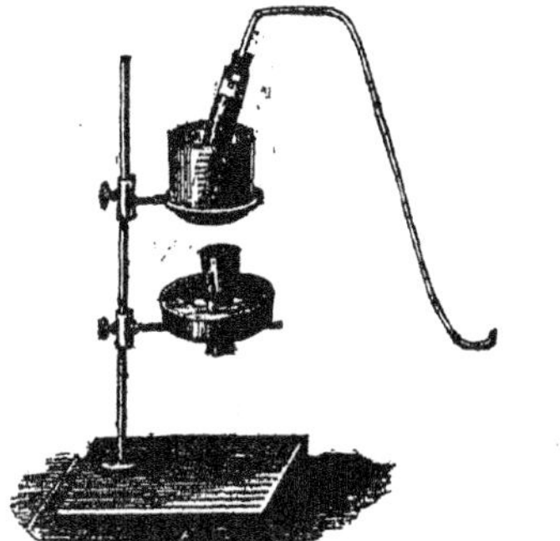

Analysis.—1 vol. cyanogen mixed with 2 vols. oxygen and fired by electricity, yields 2 vols. carbonic acid and 1 vol. nitrogen. The experiment is attended with a very violent detonation, and should be made on a small scale. It is observed that the cyanogen is more completely decomposed if a small portion of an explosive mixture of hydrogen and oxygen, prepared by decomposing water by the galvanic battery, is mixed with it. Cyanogen may also be analyzed by burning a cyanide with oxide of copper for the carbon, and with soda lime for the nitrogen.

Fluid Cyanogen.—Clear colourless fluid. Spec. grav. ·866, discovered by Davy and Faraday (Phil. Trans. 1823, 196). Refractive power 1·316 (Brewster). Non-conductor of electricity (Kemp); obtained by heating in one end of a strong closed tube cyanide of mercury, and immersing the other extremity in a freezing mixture (D. & F.), or simply by exposing cyanogen at the common pressure of the atmosphere in a tube to a cold of — 13° to — 22° (Bunsen, Pogg. Ann. 46, 101). It becomes solid some degrees under — 22°.

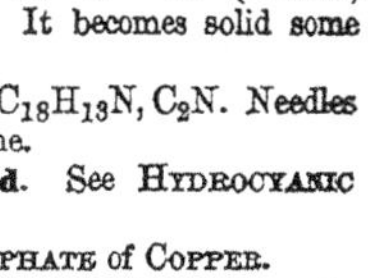

Cyanocumidine. $C_{18}H_{13}N, C_2N$. Needles by cyanogen on cumidine.

Cyanohydric Acid. See HYDROCYANIC or PRUSSIC ACID.

Cyanose. See SULPHATE of COPPER.

Cyanotoluidine. $C_{14}S_9N, C_2N$? By acting on toluidine with cyanogen.

Cyanotype. Photographic paper prepared by a solution of nitrate of mercury, and then with yellow prussiate of potash.

Cyanoxalic Acid, or *Uryle.* $C_8N_2O_4$. The hypothetic base of uric acid.

Cyanuramide, or *Cyanuramine, Melamine.* Formed by dissolving cyanamide in water and evaporating.

Cyanurenic Acid. *Kyanurenic Acid.* Colourless needles, sublimable in white satin crystals, soluble in alcohol, soluble in chlorohydric acid, potash, boiling dilute sulphuric and nitric acids; contains no nitrogen; soluble in strong cold sulphuric acids; on heating and adding water deposits a lemon precipitate. This acid is obtained by dissolving the coloured deposits of dog's urine in lime water, and adding chlorohydric acid when cyanurenic acid deposits in crystals.

Cyanuric Acid. Cy_3O_33HO. Anhydrous 8-hedrons, or hydrous prisms, possessing a weak acid taste; polymeric with cyanic and fulminic acids; obtained by distilling uric acid; by adding acetic acid to cyanate of potash.

Cyanurine. A dark blue substance, found in diseased urine, somewhat soluble in alcohol, from which a crystalline blue precipitate separates.

Cyclamine. See ARTHANITINE.

Cydonine. The peculiar gum of *Cydonia vulgaris*, or quince seeds; is not precipitated by alcohol, nutgalls, or oxalate of ammonia, by which it is distinguished from gum arabic.

Cymatine. Silica 57·98, lime 12·95, magnesia 22·38, protoxide of iron, 6·32, alumina ·58.

Cymene. $C_{20}H_{14}$. Spec. grav. ·861, ·857, of vapour 4·69; B.P. 347°, 341°. A camphogene, the more volatile oil of oil of cumin; unites with sulphuric acid, and forms cymene sulphuric acid; with nitric acid gives toluylic acid.

Cymidine. $C_{20}H_{15}N$. Fusible and volatile crystalline base from cymole.

Cyminic Acid?

Cymole. The less volatile oil of Cumin.

Cymolite. A variety of Halloysite.

Cymophane. See CHRYSOBERYL.

Cymyle. The hypothetic radical of the cymole series.

Cynapine. Rhombic prisms by alcohol from *Aethusa cynapium*, or fool's parsley.

Cynodine. *Asparagine?* Colourless prisms from a decoction of the root of Cynodon dactylon.

Cypomica. Arseniate of copper.

Cyprine. A blue variety of garnet from Norway.

Cystine. *Cystic oxide.* See URINARY CALCULI.

Cytisine. *Cathartine?* A bitter principle in the seeds of *Cytisus laburnum* by alcohol.

D

Dadyle. $C_{20}H_{16}$. A name given to the oil obtained by distilling muriate of oil of turpentine with caustic lime.

Daguerreotype. (From Daguerre, the inventor). The method of forming pictures on a silver or other surface by the action of light. A silver plate, or copper plate electroplated with silver, is well cleaned, and then exposed to the fumes of iodine in a box, and afterwards to those of bromine. The plate is now sensitive, that is, it is coated with a thin layer of iodide and bromide of silver, compounds which are easily decomposed by light when the plate is put into the camera obscura to take the likeness. The plate is then taken out of the camera and exposed to the fumes of mercury, which unites with those parts of the plate where the iodide and bromide have been decomposed. The iodide and bromide unacted on over the other portions of the plate are then removed by a solution of hyposulphite of soda, in which the bromide and iodide of silver are soluble. The mercurial surface is next prevented from deterioration by evaporation by being gilt. For this purpose, a solution of 1 part of sesquichloride of gold in 500 water, and 3 parts of hyposulphite of soda in 500 water is dropped on the moist plate, which is gently heated when the gold is deposited.

Dahline. See INULINE.

Damaluric Acid. $C_{14}H_{12}O_4$. Obtained from urine by treatment with caustic lime; it possesses the odour of butyric or valeric acid.

Damburite. See DANBURITE.

Dammara Resin. Spec. grav. 1·097 to 1·123. A light yellow resin resembling copal, said to be derived from the *Pinus dammara*, or *Dammara Alba* of *India*. It consists of a resin, soluble in alcohol 83·1 per cent. and insoluble 16·8 per cent. It forms a fine varnish with oil of turpentine. It is said to fuse at 163½°, and to consist of C 82·4, H 11·2, O 6·; it leaves ·215 ash. Ether extracts *Dammaryle*, $C_{45}H_{36}$; absolute alcohol, *dammarylic acid*, $C_{45}H_{36}O_3$; and weak alcohol, *hydrous dammarylic acid*, $C_{45}H_{36}O_3$, HO. These are all probably products of the operations.

Dammara Resin. *Cowdie*, *Kauri resin.* The yellow resinous product of the *Dammara Australis* of New Zealand. Spec. grav. I have found, 1·040 (light variety) to 1·062 (dark var.), ash ·424, containing sulphuric acid and lime. Fusing point 255°. A portion of it (57·33 per cent.) is dissolved in alcohol, with acid properties, *Dammaric acid*, $C_{40}H_{30}O_7$, HO. A portion remains undissolved, *Dammarane*, $C_{40}H_{31}O_6$. By distillation an amber-coloured oil, *Dammarol*, is obtained, $C_{40}H_{28}O_3$, or the resin deprived of 3 atoms of water. Distilled with lime it yields *Dammarone*, $C_{38}H_{30}O$, by the removal of 2 atoms carbonic acid, and 1 atom water. Cowdie resin is said to be extensively used in America as a varnish.

Damolic Acid. $C_{26}H_{24}O_4$. A volatile acid, obtained by the action of lime and distillation on urine.

Damourite. Spec. grav. 2·792, H 2·75. Yellow scales, the matrix of cyanite. Silica

45·22, alumina 37·87, oxide of iron, trace; potash 11·2, water 5·25.

Damp, Choke. The term applied by miners to carbonic acid, from its extinguishing flame and life.

Damp, Fire. The miners' name for carburetted hydrogen, or marsh gas, from its explosive nature.

Danaite. See GRAY COBALT ORE.

Danburite. Spec. grav. 2·83, H 7·5. Yellow crystalline body, containing 56 silica and 28·3 lime, probably allied to table spar or dysclasite.

Dandelion Root. The root of *Leontodon taraxacum*, used in extract as a diuretic. Its ash consists of silica 11·26, PO_5 11·21, CO_2 6·21, SO_3 2·37, Cl 3·84, Fe_2O_3 1·27, MgO 1·31, CaO 11·43, NaO 30·95, KO 17·95.

Daourite. A variety of Tourmaline.

Daphnine. Colourless bitter prisms by water and alcohol, from the bark of *Daphne mezereum* and *alpina*.

Dapicho. *Zapis.* An elastic bitumen, from the roots of *Siphonia elastica*, from South America.

Dates. The fruit of the date palm, containing much sugar, also mucilage, gum, albumen, and cellular matter; used extensively as food by the Arabs of Africa.

Datholite. *Borosilicate of Lime*, *Esmarkite*, *Humboldtite*. Spec. grav. 2·98 to 3·346, H 4·75; from Arendal in Norway, and New Jersey. White oblique rhombic prisms with angles of 103° 25′; the obtuse edges of this prism being replaced by tangent planes, lustre vitreous, inclining to pearly, translucent, streak white. B.B. becomes opaque at first, and crumbles between the fingers, and gradually fuses into a colourless glass, tinging the flame green; soluble in chlorohydric acid. Silica 37·36, boracic acid 21·26, lime 35·67, water 5·71. Form. 3 (CaO BO_3), 3 CaO 4 SiO_3 3HO.

Datisca Yellow. The colouring matter of Datisca cannabina.

Datiscine. See INULINE.

Daturine. Brilliant prisms. F.P. 212°; subliming; very poisonous; forms crystalline salts; dilates the pupil; obtained from the seeds of Datura *Stramonium*. See STRAMONINE.

Daucus Carota. *Carrot.* Contains cane sugar, uncrystallizable sugar and carotine, a glutinous body, fat and ethereal oil. See CARROT.

Davidsonite. A variety of beryl found in Aberdeenshire.

Davyne. A Vesuvian zeolite, probably nepheline.

Davy's Safety Lamp. See CARBURETTED HYDROGEN.

Davyte. Al_2O_3, 3 SO_3 11HO. Plumose or feather alum at Bogota, &c.

Decantation. The operation of pouring off a solution from a precipitate which has subsided.

Decay of Wood. Or EREMACAUSIS.

Dechenite. *Vanadiate of Lead.* PbO 57·57 Vanadic acid 49·27. Spec. grav. 5·81, H 3·75. Yellowish-red rhomboids from the Lauterthal.

Decoction. The solution obtained by boiling an organic substance with water.

Decoction Press is employed to remove the fluid from the organic matter in preparing the preceding decoction; see fig.

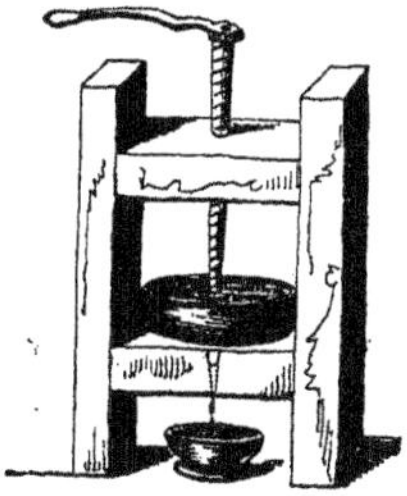

Decolorimeter. An optical instrument for determining the decolourizing influence of animal charcoal.

Decomposition. The separation of the elements of a compound. It may be *simple*, as when oxide of mercury is heated; it is resolved into oxygen and mercury (HgO becomes Hg and O); or *double*, as when sulphate of soda and chloride of barium undergo decomposition by being mixed and converted into sulphate of barytes and chloride of sodium (NaO SO_3, BaCl = NaCl, BaO SO_3).

Decrepitation. The crackling sound produced when common salt is thrown on the fire, proceeding from the sudden expansion and vaporization of the water between the plates of the crystals.

Deflagration, (*deflagro*, I burn). The peculiar rapid combustion of nitre and nitrate of ammonia when thrown on the fire.

Delessite. *Ferruginous chlorite*, from Oberstein and Planitz. It consists of SiO_3 29·45, Al_2O_3 18·25, Fe_2O_3 8·17, FeO 15·12, MgO 15·32, HO 12·57, CaO ·45.

Deliquescence, (*deliquesco*, I melt). The tendency to fluidity possessed by many bodies when exposed to a moist atmosphere, depending on their affinity for water, as chlorides of calcium and magnesium.

Delphine. $C_{27}H_{19}NO_2$? Yellowish-white powder, obtained by means of alcohol, from the gray and brown seeds of *Delphinium Staphisagria*, or *Stavesacre;* it is acrid and poisonous.

Delphinic Acid. *Phocenic acid.* The volatile acid of the *Viburnum opulus*, or guelder rose, and of whale oil, has been found to be valerianic acid.

Delphinite. A variety of Epidote.

Delvauxite. $2Fe_2O_3PO_5$ 24HO. Brown earthy phosphate of iron from Visé, Belgium.

Dendrites. (δένδρον, a tree). Tree-like crystals, as in the Arbor Dianae. The figures represent *serrated*, *plumose*, *triangular*, and *reticulated* dendrites.

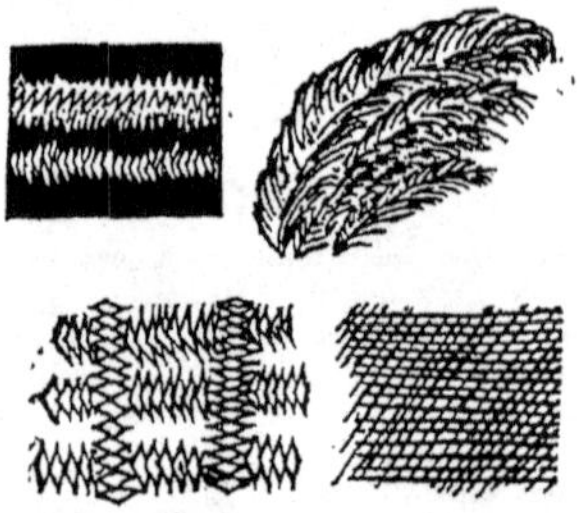

Density. See SPECIFIC GRAVITY.

Dentine. The ivory of the teeth. See TEETH.

Dephlegmation. An alchemistical term for the separation of volatile matter by distillation.

Dephlogisticate. To deprive a substance of phlogiston, a hypothetic body, afterwards hydrogen. Thus when oxygen was first obtained pure, it was supposed to be air deprived of phlogiston.

Derbyshire Spar. Fluor spar, or fluoride of calcium.

Dermatine. Spec. grav. 2·1. 3 MgO, Fe O, 2 SiO_3, 6 HO. Brown stalactitic masses from Waldheim, Saxony.

Desiccation. The process of removing water by drying.

Desmine. A variety of Stilbite.

Desrosne's Still. A still, the principle of which is, that in a mixture of vapour of alcohol and water, by partial cooling, the aqueous vapour will be condensed, and the alcoholic vapour rendered stronger. The same principle is carried out in Stein's apparatus. See STILL.

Detonation. The immediate combustion of a substance accompanied by sound and light is so termed, as the explosion of gunpowder, fulminate of silver, gun cotton, &c.

Deutoxide. A term originally applied by Dr. Thomas Thomson in 1804 to the second oxide of a metal. Afterwards, when he came to use symbols, he applied the Latin numerals to denote the exact number of atoms of oxygen or acid united to a base. See NOMENCLATURE.

Devonite. A synonyme of Wavellite.

Dew. The water deposited on the surface of plants, the earth, &c. during the night, especially in clear cold nights.

Dew-point. When the air is dry evaporation goes on rapidly and the wet bulb thermometer falls. When the air is saturated, the readings of the wet and dry thermometers will be the same. The important point is the dew-point, or that degree at which the atmosphere will part with its moisture, or will be cooled down to the point of saturation. The dew-point is, therefore, that degree of temperature at which saturation is attained and moisture deposited. A table of factors has been constructed by which the difference between the readings of the dry and wet bulb thermometer is to be multiplied in order to obtain the difference between the temperature of the air and that of the dew-point. This method affords a close approximation to the true number.

Mean temp. of dry bulb thermometer, November, 1851,............... =	38°·8
— wet bulb, —	36°·7
Difference,..................	2°·1
Factor,......... =	2°·5
	10 5
	42
Difference between air and dew-point,...	5·2 5

Then 38°·8—5°·25=33°·55=temperature of the dew-point. Or we find the dew-point and then find the elastic force of vapour corresponding to the dew-point (Greenwich Observ. 1842 and 1843). Then to obtain the elastic force from this, we have recourse to a table as at page 210 Thomson's Outlines of Heat, &c. where we find at 33°·5 the elastic force = 2°·10. The elastic force being obtained, all other problems can be solved.

Table of Factors.—(Glaisher.)

When the temperature of the air is below 24°.	The difference between the temp. of evaporation 8·5 and the temperature of the air multiplied by	Gives the difference between the temp. of the air and the temp. of the dew-point.
Between	24 and 25	7·3
—	25 " 26	6·4
—	26 " 27	6·1
—	27 " 28	5·9
—	28 " 29	5·7
—	29 " 30	5·
—	30 " 31	4·6
—	31 " 32	3·6
—	32 " 33	3·1
—	33 " 34	2·8
—	34 " 35	2·6
—	35 " 40	2·5
—	40 " 45	2·3
—	45 " 50	2·1
—	50 " 55	2·
—	55 " 60	1·8
—	60 " 65	1·8
—	65 " 70	1·7
Above...........	70	1·5

The interesting phenomenon of the formation of dew receives its explanation from the consideration of the subject of the radiation of heat. Dew is more abundant in spring and autumn than in summer, and especially during still and serene nights in autumn. Dew is always very copious on those clear and calm nights which are followed by misty or foggy mornings, the turbidness of the air in the morning showing that it must have contained during the preceding night a considerable quantity of moisture. Dew is unusually plentiful on a clear morning which has succeeded a cloudy night; for the air having in the course of the night lost little or no moisture, is in the morning more charged with watery vapour than it would be if the night had also been clear. Dew is also produced by preceding heat of the atmosphere, if the other circumstances are favourable. Such are some of the important facts observed by Dr. Wells in relation to the formation of dew. The theory of its production, then, is as follows:—All bodies have the power of radiating heat. During the day the heat lost by radiation is more than supplied by the sun, so that the temperature of bodies is increased instead of being diminished. But the contrary is the case during the night. The heat radiated by the bodies on the surface of the earth penetrates into the sky, and does not again return to them. The temperature of these bodies must, therefore, decrease, and must remain so during the night. If the sky be covered with clouds, the heat will be radiated back from them, so that the earth will not become colder. Hence we always find dew to be formed in clear nights, and not when the sky is overclouded. In accordance with this view, we find that the best radiators have always the largest amount of dew deposited on their surfaces. By this theory of dew we are enabled to explain many phenomena of common occurrence. 1. A pane of glass, covered by an inside shutter, is moister in the morning than one which is uncovered, because the shutter prevents the radiation of the heat from the room to the pane of glass, and thus the latter is colder from external radiation. 2. In a clear night we feel colder than on a cloudy night, in consequence of the greater radiation from our bodies. 3. Gardeners cover their plants with matting to prevent the radiation from their leaves. Dr. Wells found grass, on which a thin cambric handkerchief was placed, several degrees warmer than grass freely exposed to the sky. 4. The covering of snow upon the surface of the earth during winter is chiefly useful by preventing the radiation of the heat of the plants placed under it. 5. In tropical climates there is a prevalent idea that the exposure of animal substances to moonlight promotes their putrefaction. Now, as moonlight nights are clear, the consequence is, that dew, that is, water, the great auxiliary of putrefaction, is deposited upon the animal substances. 6. Ice is formed in Bengal by exposing water to the air, in shallow unglazed pans, placed upon dry straw or sugar canes, and sometimes sunk artificially below the surface of the earth. When the nights are clear and serene, ice is formed not by the evaporation of the water, but by the radiation of heat from its surface; the straw serving only to prevent heat from being conveyed into the water from the earth. Dr. Wells is inclined to attribute the whole of the cold to radiation, but that a certain amount of cold is produced by evaporation through the porous sides of the vessels, is obvious from the fact, that jugs of the same kind of ware are universally employed to cool the water contained in them for domestic purposes, under circumstances little calculated to ensure radiation, since they are close vessels often deposited in shaded positions. It is rather remarkable that such vessels are not more used in this country. Cool water, however, may be obtained during the night purely from the cold produced by radiation, by placing water bottles outside the window.—The practice of making ice in the Indian fashion appears to have been known to the Jews, since Josephus says (Wars of the Jews, iii. 10 and 7), "When the water of the lake of Gennesaret is left in the open air, it is as cold as that ice which the country people are accustomed to make by night in summer."

Deweylite. Spec. grav. 2·3, H 3·. White or greenish-white variety of serpentine, in veins, at Middlefield, Mass. U. S. containing silica 40·, magnesia, 40·, water 20·.

Dextrine. *Soluble Starch.* Soluble starch may be obtained by adding 2 parts of nitric acid to 300 of water, and mixing this liquid with 1000 parts of dry starch. The starch is then placed in a drying closet. A current of air is passed over the starch, which is then placed in a stove, the temperature being raised to about 176°. It is then passed through the common process for starch in a sieve, the temperature being retained at 212° to 230° (Payen). Dextrine is formed by heating starch up to 392°; also in a copper tube with water at 338°. *Applications.*—It is used under the name of gum substitute, soluble gum, &c. for thickening the colours in calico printing. It is also employed for making dextrine bandages for fractures. 100 parts of dextrine are mixed with 60 parts of camphorated spirit, the mixture being triturated until it acquires the consistence of honey. 40 parts of water are then added, and the bandage immersed in it (D'Arcet). For a fracture of the clavicle, 6,000 grains of dry dextrine are required; for that of the thigh 4,500; of the leg 3,000; of the forearm 2,220. Starch bandages are often employed in this country, being softened with warm water before being applied.

Dextroracemic Acid. Identical with tartaric acid.

Diabase. A granular variety of greenstone, with a predominance of hornblende.

Diabetic Sugar. A sugar, identical with glucose or grape sugar (which see), is found in the urine in diabetes. The urine in this disease greatly exceeds the normal quantity. If we take the usual daily evacuation of urine at 3·21 lbs. avoirdupois, that in diabetes rises to 10 lbs. and 20 lbs. and even 40 lbs. The presence of the sugar may be detected by boiling a portion of the urine in a tube with a solution of sulphate of copper and caustic soda; the brownish-red din-oxide of copper will be speedily deposited. To determine the amount of sugar contained in a given weight of urine, it is to be fermented at a temperature of 60° to 70°, by the addition of yeast, the alcohol distilled off, and estimated (see ALCOOMETRY); or the carbonic acid may be caught in a phial containing a mixture of solution of chloride of barium and caustic ammonia or barytes water, and the sugar estimated from the carbonate of barytes formed; every 13·50 grains of absolute alcohol are equivalent to 24·75 grains diabetic or grape sugar; while 49· grains carbonate of barytes are precipitated by the amount of carbonic acid evolved from the above weight of sugar when fermented. The sugar may be extracted from urine by evaporating to a syrup in a water bath, boiling the residue with alcohol (spec. grav. ·850), and evaporating in a water bath. The residue is washed with absolute alcohol as long as anything is taken up; it is then evaporated in a water bath; and lastly in the vacuum of an air pump over sulphuric acid. The cause of the formation of this sugar in the system is not known. It appears to depend on an abnormal change in the stomach, by which the sugar of the food is prevented from being converted into lactic acid and carbonic acid, as in the natural state. It probably depends on some diseased condition or functional derangement of the nervous system. Some sugar is found in the urine when particular spinal nerves are compressed or pricked; also in epileptics.

Diaclasite. Allied to hypersthene. SiO_3 53·74, MgO 25·09, CaO 4·73, FeO 11·51, MnO ·23, Al_2O_3 1·34.

Diadochite. Spec. grav. 2·035, H 3·. A variety of pitchy iron ore from Arnsbach, in Thuringia, consisting of perphosphate and sulphate of iron (3 Fe_2O_3 2 PO_5, 2 Fe_2O_3 2 SO_3 24 HO).

Diagonite. A synonyme of Brewsterite.

Diallage. A name given originally to an amphibolic mineral with a metallic aspect. Under this name are usually confounded *Bronzite*, *Schillerspar*, and *Hypersthene*. *Green Diallage.*—A mixture of thin plates of hornblende and augite. *Diallage metalloide* is Schillerspar. *Diallage rock*, a mixture of *diallage* and *Saussurite*.

Diallagite. A synonyme of carbonate of manganese; often mixed with lime and iron.

Dialogite. A synonyme of bisilicate of manganese.

Dialuric Acid. $C_8H_3N_2O_7HO$. Colourless prisms resembling alloxantine, obtained by adding an excess of sulphohydride of ammonium to a solution of alloxane, or alloxantine, and dissolving the dialurate of ammonia formed in hot chlorohydric acid. In solution this acid absorbs oxygen, and is converted into alloxantine.

Diamagnetism. A term employed to express the magnetic relations of certain bodies. All bodies are either magnetic or diamagnetic; for example, a magnetic bar oscillating between the poles of an electro needle takes an *axial* direction, that is, parallel to a plane following the direction of the axis of the needle; while a diamagnetic bar takes an *equatorial* position, or one which is perpendicular to the preceding plane. The magnetic metals in the order of their power are iron, nickel, cobalt, manganese, chromium, cerium, titanium, palladium, platinum, osmium. The diamagnetic metals are bismuth, antimony, zinc, tin, cadmium, sodium, mercury, lead, silver, copper, gold, arsenic, uranium, rhodium, iridium, tungsten.

Diamond. *Syn. Demant, adamas, adamantos. Physical Ch.*—Regular system, 1. octahedrons; 2. octahedrons with cubes; 3. cubes with dodecahedrons; 4. hexakisoctahedrons; 5. triakisocta-

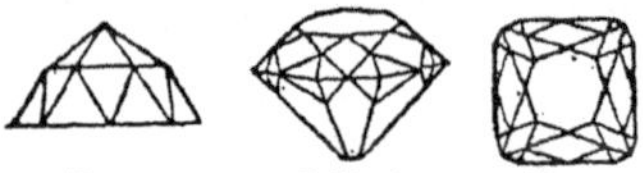

Roso. Brilliant. Freu.

hedrons. *Colour*, transparent, colourless, white, gray, blue, black, red, brown, yellow, green. *Hardness* 10·. Spec. grav. 3·48, 3·55. *Lustre*, splendent and peculiar. When rubbed becomes positively electrical. Heated in the sun, and brought into a dark place, it phosphoresces. *Blowpipe Ch.*—Unchanged, but when heated to redness in contact with air, it is converted into carbonic acid, leaving no residue. *Chemical Ch.*—Consists of pure carbon (C), and may be converted into carbonic (CO_2) by heating it in a glass tube with nitre or chlorate of potash, collecting the gas over mercury. *Uses.*—In 1456, Louis Berquen discovered that diamonds may be polished by rubbing them on their own powder (Hauy). When polished, they are sold in two forms—*rose diamonds* and *brilliants;* the former being conical and the latter flattened. In powder it is made into a paste, and being set on a handle is used for cutting glass by glaziers, and for sharpening cutting instruments. An unwrought diamond is worth £2 per carat (of 3·077 grains). The largest diamond known to exist belonged to the Great Mogul, and was found in the mine of Colone in 1550. It weighed originally 2,769 grains, and by cutting became 861 grains. It is now lost. The diamond purchased by Catherine of Russia formerly belonged to Nadir Schah, and weighed 593·86 grains, and cost £90,000, and an annuity of £6,000. The Pitt diamond, purchased by Mr. Pitt, governor of Bencoolen, was bought by the regent duke of Orleans for £130,000, and is now among the French crown jewels. It weighs 419¼ grains, and is worth half-a-million sterling. The Koh-i-noor is another large diamond, now in this country, formerly the property of Shah Soojah of Affghanistan. The mode of calculating the price of moderate-sized diamonds is to multiply the square of the weight in carats by 2. The Hunterian diamond weighs 3·518 carats (10·825 grains). Its value, if of the first water, that is, destitute of colour, is £24 15s. Large diamonds have the square multiplied by even more than 8. *Localities*.—In Europe the diamond is reported to be found in Ural. In Asia it has been obtained in alluvial soil in Golconda, Visapoor, Bengal, and Borneo. In Africa it has been found in the province of Constantine, in the sand of the river Gumil; and in America it has been obtained since 1740 in Serra Dofrio of Brazil, and at the present time abundantly in the Minas Geraes district, in strata derived from primary rocks (Gardner). *Source*.—In the Nalla Malla mountains of Madras the diamond occurs in an alluvial soil formed of a sandstone breccia, which consists of fragments of jasper, quartz, chalcedony, cemented together by a quartzose paste passing into a pudding-stone formed of rounded pebbles united together by clay and limestone (Voysey, Phil. Mag. 2d ser. i. 147). In *Brazil*, in the district near Rio, to reach the deposit in which the diamonds are situated, it is necessary to cut first through a reddish sandy soil twenty feet thick, then eight feet of tough yellow clay; then two and a-half feet of coarse reddish sand. Below this is the diamond bed, from one to four feet thick, consisting of pebbles of primitive rocks. This rests on hard clay, and below it are the slaty rocks. The formation in which the diamonds occur is washed in water by means of a large flat wooden plate. Specimens of diamonds have been found imbedded in the primitive rocks, and hence it is probable that they have been washed down into their present position along with the debris of these formations (Gardner's Brazil, 452). The sand in which the diamond is found in Bahia consists of quartz, a species of tourmaline (?), orthose, rutile, brookite, anatase, zircon, diaspore, hydrophosphate of alumina and lime, hydrophosphate of yttria, silicate of yttria, native gold, oxide of iron (Damour). *Chemical Characters*.—The diamond was long considered to be a perfectly fixed body till Newton conjectured, from its great power of refracting light, that it was combustible, an idea which was established in 1694 by the Florentine academicians, who succeeded in volatilizing several diamonds by means of a burning glass (Giornali di Litterati d'Italia, 8, 9), in presence of Cosmo, the third grand duke of Tuscany. In 1751, Francis I. emperor of Germany, witnessed the combustion of several diamonds (Das neueste aus der enmuthigen gelehrsamkeit, 1751, s. 540). These experiments were repeated by D'Arcet, Rouelle, Macquer, Cadet, and Lavoisier (Mem. Paris, 1766-70, 71, 72), who showed that air was necessary to its combustion. Lavoisier ascertained that carbonic acid is formed in the combustion (Macquer's Dict. 1, 337). When perfectly pure no residue is left on combustion, which can be effected as low as the melting point of silver (about 1900°).

Diamyle-amine. $2\,(C_{10}H_{11})\cdot NH$. B.P. 340°. Aromatic, burning, colourless oil, by the reaction of ammonia and iodide of amyle, or of amylamine and bromide of amyle; it is ammonia, in which 2 atoms hydrogen are replaced by two atoms amyle.

Diamyle-phenylamine. *Diamyle-aniline.* $N,2\,(C_{10}H_{11})\,C_{12}H_7$. B.P. 531°·5. A base, by the action of amylaniline and bromide of amyle, for some days at 212°.

Diaphanite. *Neolite.* Spec. grav. 2·77. Green silky fibres from Naskil. SiO_3 52·28, Al_2O_3 7·33, MgO 31·24, FeO 3·79, HO 4·0. A recent production.

Diaspore. *Hydrate of Alumina.* Al_2O_3 HO. Spec. grav. 3·43, H 6·25. Al_2O_3 76·06, HO 14·70, FeO 7·78. Greenish-gray plates, with a pearly lustre; also in doubly-oblique prisms; B.B. fuses with borax into a colourless glass. Oxide of iron is dissolved up from it by muriatic acid, but the alumina is unacted on. Found near Koroibrod, Orenbourg, and at Schemnitz.

Diastase. A white tasteless substance, obtained by moistening pounded malt with an equal weight of water, and pressing it through a bag. The turbid fluid is precipitated with alcohol, which throws down albumen, and is filtered. The diastase is now precipitated with a quantity of alcohol, redissolved in water, and precipitated with alcohol. It is probably a mixture of several bodies; but 1 part of it possesses the remarkable property of converting 2000 parts of starch into dextrine, and 1000 into sugar.

Diastatite. A synonyme of Amphibole.

Diaterebic Acid. $C_{14}H_{10}O_8$. By the action of nitric acid on oil of turpentine.

Dibenzoil-plenyl-amide (*Dibenz-anilide*). $2\,C_7H_5O,C_6H_5N$.

Dibromaniline. $C_{12}H_5Br_2N$.

Dibromobutylene. $C_8H_8Br_2$.

Dibromocinchonine. $C_{38}H_{20}Br_2N_2O_2$. Pearly needles, by bromine on cinchonine.

Dibromo-ethylamine. $C_4H_5Br_2N$.

Dibromoethylene. $C_4H_4Br_2$.

Dibromomelaniline. $C_{26}H_{11}Br_2N_3$.

Dibromonaphthaline. *Bromonaphthese.* $C_{20}H_6Br_2$.

Dibromopropylene. $C_6H_6Br_2$.

Dibromo-sulphonaphthalic Acid. $C_{20}H_5$, SO_2, Br_2, SO_3.

Dicetylophenylamine. $2C_{32}H_{33}$, $C_{12}H_5$.

Dichloraniline. $C_{12}H_5Cl_2N$.

Dichloro-ethylamine. $C_4H_5Cl_2N$. Yellow fluid.

Dichlorofilipelosic Acid. $C_{24}H_{12}Cl_2O_{10}$. By chlorine on filipelosic acid.

Dichlorohydrokinone. Colourless. $C_{12}H_4Cl_2O_4$. Violet species, $C_{12}H_3Cl_2O_4$ 2 HO, yellow species, $C_{12}H_3Cl_2O_4$.

Dichlorokinone. $C_{12}H_2Cl_2O_4$. Yellow prisms.

Dichloromelaniline. $C_{26}H_{11}Cl_2N_3$.

Dichlorophthalic Acid. $C_{16}H_4Cl_4$?

Dichloropropylene. $C_6H_6Cl_2$.

Dichroite. See IOLITE.

Dicyanocodeine. $C_{36}H_{21}NO_6,2Cy$. Shining 6-sided plates, by cyanogen on codeine in alcohol.

Dicyanomelaniline. $C_{26}H_{13}N_3,Cy_2$.

Didymium. Di 6·48. Discovered by Mosander in 1841 (Pogg. Ann. 56, 504; Phil. Mag. 28, 241). The metal has not been separated, but the oxide occurs along with oxide of cerium on cerite, Allanite, &c.

Oxide, Didymia.—DiO. Dark brown powder, irregularly tinted; dull white at a white heat; forms an amethyst bead with salt of phosphorus before the blowpipe, a whitish-gray mass with carbonate of soda; obtained by acidulating and evaporating the mother liquor from the sulphate of lanthanum, and thus obtaining sulphate of didymium in rose crystals. The rose crystals of sulphate of didymium in rhomboids, may be also picked out from the amethyst or violet prisms of impure sulphate of lanthanum when these salts exist together. The oxide is precipitated from the red sulphate by caustic soda, or potash, or oxalate of ammonia, washing and igniting. When precipitated it is violet, and absorbs carbonic acid.

Salts.—Oxide of didymium is a weaker base than oxide of lanthanum. The salts are pink or rose, and amethyst or violet. Ammonia precipitates a subsalt, which may be washed without passing through the filter. Potash and soda throw down the hydrous oxide. Sulphohydride of ammonia, unless in great excess, produces no precipitate.

Sulphate.—Rose-coloured rhombohedral crystals with secondary faces, when crystallized from a mixture of a lanthanum salt, but from its pure solutions in oblique rhombic prisms without secondary faces. Soluble in 5 water between 59° and 68°. A cold saturated solusion yields a deposit of crystals at 127½°, gradually increasing as the heat rises, and at 212° only 1 part of salt in 50½ water remains undecomposed at a low red heat; a white heat expels two-thirds of its acid. The *potash sulphate* of didymium is an amethyst salt, insoluble in sulphate of potash (Mosander).

Nitrate.—Red deliquescent salt, crystallizing with difficulty; loses acid by heat; forms in saturated aqueous solutions a red fluid; violet-blue by reflected light. Does not fall to powder, like nitrate of lanthanum. Formed by dissolving the oxide in nitric acid, and evaporating.

Diethyle-amine. $2(C_4H_5)HN$. Volatile oil, an amide base, where 2 atoms ethyle replace 2 atoms hydrogen in ammonia; from bromide of ethyle on ethyl-amine, and distilling the product with potash.

Diethyle-amylamine. $2(C_4H_5)C_{10}H_{11}N$. A nitrile base.

Diethyl-methylamine. E_2MeN.

Diethyle-phenylamine. *Diethylaniline.* $2(C_4H_5)C_{12}H_5,N$; by acting on ethyle-phenylamine by bromide of ethyle.

Diethyle-urea. $C_2H_2\,2(C_4H_5)N_2O_2$; by water on cyanate of ethyle.

Diffluan. $C_6N_2H_4O_5$. White powder by boiling alloxanic acid.

Diffusion of Gases. *Dispersion*, or *Transposition.* (The terms diffusion, transposition, first used by Priestley). A property of elastic fluids discovered by Priestley in a long series of experiments. He first found in the last century that if water be converted into vapour in an earthen retort, the vapour makes its escape through the pores of the retort, while the external air enters through the same pores, and may be collected by placing inverted jars over the beak of the retort plunged into a pneumatic trough. He filled a bladder with oxygen gas, and placed it for a month in a vessel filled with hydrogen. He found the gas in the jar and in the bladder to be mixtures of oxygen and hydrogen in the same proportion, "diffused through one another." On reversing the experiment he found the oxygen of inferior quality, and the bladder much shrunk. Both the air in the bladder and in the jar exploded with violence. Glass retorts allowed of similar transmission in time, and likewise broken jars mended with paint or cement. Dalton was of opinion, that gases dispersed into each other as into vacua, the particles of one gas offering no resistance to the passage of those of another; and corroborative of this, Mr. Faraday showed that gases, when pressed through capillary tubes, passed with a rapidity in connection with their levity (1818); it was shown by Graham that the ratio of dispersion depends on their specific gravity, the lightest gases travelling fastest, as was shown in Priestley's experiments, and the heaviest most slowly. The rate of diffusion Graham found to be inversely as the square root of the density of gases, or d being the density of the gas, then the formula $\sqrt{\frac{1}{d}}$ expresses the diffusion volume. This property of gases may be well illustrated by filling a large jar with hydrogen, placing it upright on a table covered with a glass plate. Another equal sized empty jar is then placed over it mouth to mouth; and on quickly removing the intervening plate of glass both jars will be found to explode when fired by a taper; thus proving that common air has entered into the lower jar, and hydrogen into the upper one. This discovery of Priestley has not always been properly attributed to him; but that he was aware of its vast importance is obvious from the following sentence, with which he commences his paper (American Trans. v. 14):—"One of the most extraordinary circumstances that ever occurred in the course of my experiments, is that of the vapour of water or of mercury changing places with any kind of air in vessels through which air could not be made to pass without great force, so that for most purposes they might be considered as air-tight. I had also observed that different kinds of air capable of forming a chemical union would do it through a bladder that was perfectly air-tight, that in this manner pure air was imbibed by the blood through the membrane of the lungs, while phlogiston was admitted into the air within it; and it is by means of this that respiration is carried on, that the atmosphere is retained in a state of purity sufficient to communicate to animals the proper mixture of gases required for healthy subsistence."

Diffusion of Fluids through Membranes.—Membranes, when applied to the extremities of tubes and exposed to the action of water, speedily putrefy and give way. They may, however, be rendered more permanent. The tube closed at one end by bladder is to be partially filled with alcohol and placed in water for some days, when it will be found to withstand the putrefying action for a considerable period. To *fix* the bladder on the tube it is moistened with cold water, and tied with a string to the open end of the tube. It is then allowed to remain twenty-four hours, when the bladder is thoroughly dry and glued to the tube. It is then to be tied to the tube with several turns of a linen thread. The tube may be about 6 inches long, and from ¾ to 1 inch in diameter, and it may be supplied with a border to retain the thread. The permeability of membranes to solutions depends on their being moistened, for when dry they are air-tight. Their affinity for water also is connected with their action. Membranes, tendons, muscles, cartilages, ligaments, &c. contain, in their natural state, more than half their weight of water. When dried the water evaporates, and they become hard horny bodies. But if they are immersed for twenty-four hours in water, they recover their former quantity of water. That the water is not chemically united, but merely diffused through the pores or tubes of the membranes by capillary attraction, is obvious from the fact, that mechanical pressure is sufficient to remove the water, as in a sponge, which is an analogous body. The tendency which fluids have to pass through membranes, depends in some measure on the nature of the animal substances, which absorb liquids with different degrees of power. 100 parts of ox bladder, in twenty-four hours, take up of water 268 vols.; saturated solution of salt 133 vols.; alcohol of spec. grav. ·8375 (84 per cent.) 38 vols.; oil of marrow 17 vols. 100 parts of pig's bladder take up in twenty-four hours 356 vols. of pure water, of brine 159 vols. oil of marrow 14 vols. Water passes most readily into membranes. Alcohol passes through membranes with great difficulty and slowness. The permeability of membranes may be shown by moistening one side of a membrane with yellow prussiate of potash and the other with chloride of iron: a blue spot of Prussian blue is formed in the centre. A tube containing urine, closed with a bladder and immersed in distilled water, in a few seconds gives out chlorine to the distilled water, and gradually the solution equalizes itself on each side. If bile, milk, urine, serum, syrup, solution of gum, be placed in the tube, they gradually diffuse into the distilled water, but in these cases more water passes into the tube than bile, &c. out of the tube, the denser fluid being always slowest in its progress. This gives rise to *endosmosis* and *exosmosis*. When the volume of a fluid increases by the lighter fluid passing into the denser, it is called *endosmosis*. When the volume diminishes, it is termed *exosmosis*. The diffusion of salts through water takes place as the formation of steel by cementation.

Diffusion of Liquids.—(1.) When solutions of common salt were placed in an open phial, with 1·25 inch aperture, and immersed in a large jar filled with water, the salt diffused into the pure water of the jar, in a period of eight days, as follows:—

Solutions NaCl Per Cent.	Salt Diffused.
1	1
2	1·99
3	3·01
4	4·00

The amount of salt which diffused amounted to ⅛ of the whole. (2.) The proportion of salt diffused increases with the temperature; an elevation of 80° F. doubles the amount of common salt diffused. (3.) Different salts have different diffusibilities; 20 parts of the following salts, in 100 of water, gave the following results:—common salt 58·68, sulphate of magnesia 27·42, sulphate of water 69·32, cane sugar 26·74, starch sugar 26·94, gum arabic 13·24, albumen 3·03. The small diffusion of albumen explains its deficiency in serous effusions, and in cholera fluids (R. D. T.) (4.) Urea is as highly diffusible as chloride of sodium. (5.) Isomorphous salts have an equal diffusibility, as chloride of potassium with salammoniac, nitrate of potash with nitrate of ammonia, and sulphate of magnesia with sulphate of zinc, not for chemical equivalents, but for equal weights. (6.) Acids differ greatly in diffusibility. Nitric acid is 4 times greater than phosphoric acid; nitric and chlorohydric acids equal; acetic and sulphuric acids equal. (7.) Soluble subsalts have a very low diffusibility. Sulphate of ammonia 8, sulphate of copper 4, ammonia sulphate of copper 1. (8.) When two different salts are dissolved in the diffusion cell, they diffuse into the water atmosphere according to the diffusibility of each, exactly as other gases diffuse in the air. An important consequence is, that in liquid diffusion we have a new method of separation or analysis for many soluble bodies. Thus, chlorides diffuse out from sulphates and carbonates, and salts of potash from salts of soda; and from sea-water the salts of soda diffuse out into pure water faster than the salts of magnesia. This circumstance is applied to explain the differences which have been found in the composition of the Dead Sea, the different salts diffusing into the sheet of fresh water with which the lake is frequently covered. (9.) Chemical decompositions may be produced by liquid diffusion. The constituents of alum separate in the diffusion cell, the sulphate of potash diffusing in largest quantity. (10.) One salt, such as nitrate of potash, will diffuse into another, as nitrate of ammonia, as rapidly as into pure water, the salts being naturally diffusible as gases are. (11.) The diffusibilities of the salts into water, like those of the gases into air, are connected by simple numbers. 4 per cent. solutions of the following salts diffused as follows: — carbonate of potash 10·25, sulphate of potash 10·57, sulphate of ammonia 10·51. The acetate of potash and yellow prussiate of potash belong to the same group. Nitrate of potash, chlorate of potash, nitrate of ammonia, chloride of potassium, and chloride of ammonium, form an equi-diffusible group. (12.) The *time* in which an equal amount of diffusion took place in these two groups, appeared to be as 1 for the second to 1·4142 for the first, or as 1 to the square of 2. Now, in gases, the squares of the times of equal diffusion are the densities of the gases. The relation between the sulphate of potash and nitrate of potash group, would therefore fall to be referred to the diffusion molecule and diffusion vapour of the first group, having a density represented by 2, while that of the second is represented by 1. These were named the solution densities of the salts in question. The corresponding salts of soda appear to fall into a nitrate and sulphate group, also related to each other, as the potash salts. The relation of the salts of potash to those of soda, in times of equal diffusibility, appeared to be as the square root of 2 to the square root of 3. If the times of diffusing are all squared, the following are the solution densities of salts:—$KOHO = 1$, $KONO_5 = 2$, $KOSO_3 = 4$, $NaONO_5 = 3$, $NaOSO_3 = 6$ (Graham).

Digenite. Spec. grav. 4·57, H 2·75. Lead-gray mass from Chili and Thuringia. Cu 70·20, Ag ·24. Loss (S?) 29·56; probably a disulphide.

Digester, Papin's. A strong metallic vessel, supplied with a screw for compressing steam, by which the temperature can be raised to upwards of 400°.

Digestion. The operation of exposing a body to the action of a fluid at a temperature under boiling.

Digestion, Animal. See STOMACH.

Digestive Salt, or chloride of potassium.

Digitalic Acid. Needles from Digitalis purpurea.

Digitalicrine. $C_{11}H_{10}O_3$. Yellow powder from digitaline by tannic acid.

Digitaline. $C_{10}H_9O_4$. White crystalline mass by alcohol from the aqueous extract of foxglove (*Digitalis purpurea*).

Digitasoline. $C_{19}H_{16}O_9$. Yellowish-white body in crude digitaline.

Digitoleic Acid?

Dihydrite. *Bihydrous pentaphosphate of copper.* $5\,CuO,PO_5, 2\,HO = CuO\ 68·, PO_5\ 24·$.

Di-iodo-ethylamine. $C_4H_5I_2N$. Blackish fluid by iodine and ethylamine.

Di-iodo-melaniline. $C_{26}, H_{11}I_2N_3$.

Di-iodo-methylamine. $C_2H_3I_2N$. Red powder by iodine on methylamine.

Dike. The appearance of a wall presented by trap rocks, as injected from below. The

figure represents a trap dike with the adjacent rocks worn away, as at Millport, Cumbrae.

Dilituric Acid. $C_8N_3H\ O_8, 2\ HO$. Obtained from the mother liquor of alituric acid.

Dillnite. *Collyrite.* Spec. grav. 2·574. White, adhering to the tongue, from near Schemnitz. SiO_3 23·53, Al_2O_3 56·4, MgO ·44, HO 21·13.

Dill Oil. Yellow oil. Spec. grav. ·881, from the seeds of *Anethum graveolens.*

Dimetaphosphoric Acid. $2\ HO, 2\ PO_5$, from metaphosphate of copper.

Dimethylamine. $2\ (C_2H_3)\ NH$, by bromide of methyle on methylamine.

Dimethyl-ethylamine. $2\ (C_2H_3)\ C_4H_5$, N, unknown.

Dimethyle-urea. C_2H_2, $2\ C_2H_3$, N_2O_2, by methylamine on cyanate of methyle.

Dimorphine. A species of rhombic sulphide of arsenic.

Dimorphism. (Δις, twice; μορφη, shape.) The laws of the axes of crystals would seem to render it impossible that a body can crystallize in more than one of the systems under which minerals have been classed. But it is remarkable that there are certain anomalies existing both among chemical bodies and minerals in this respect. Biphosphate of soda, for example, crystallizes sometimes in rectangular prisms and octahedrons, and sometimes in right rhombic prisms, although identical in composition. Calcareous spar also usually is found in rhomboids, but it occurs in the form of Arragonite in right rhombic prisms. The spec. grav. of the former is 2·721, and that of the latter 2·931. The hardness of the former is 3, that of Arragonite being 4. Idocrase has exactly the same composition as one of the varieties of garnet, but the first belongs to the 2d system, or is in right square prisms, while the garnet belongs to the 1st system, or is in dodecahedrons.

Dimorphous Bodies.

Diamond,	8-hedrons,	Graphite,	Rhombohedral.
Sulphur,	Prismatic,	By fusion,	Oblique prismatic.
Arsenious acid,	8-hedrons,	Glassy,	Prismatic.
Teroxide of antimony,	8-hedrons,	Native,	Prismatic.
Sulphate of potash,	Prismatic,	From kelp,	Rhombohedral.
Bisulphate of potash,	Prismatic,	By fusion,	Oblique prismatic.
Nitrate of potash,	Prismatic,	By evaporation,	Rhombohedral.
Calcareous spar,	Rhombohedral,	Arragonite,	Prismatic.
Carbonate of magnesia,	Prismatic,	Bitter spar,	Rhombohedral.
Cubic pyrites,	Cubes,	Cockscomb pyrites,	Prismatic.
Sulphate of nickel,	Prismatic,	Acid solution,	Pyramidal.
Carbonate of iron,	Prismatic,	Sparry mineral,	Rhombohedral.
Dinoxide of copper,	8-hedrons,	Fibrous red copper,	Rhombohedral.
Disulphide of copper,	Rhombohedral,	By fusion,	Regular.
Sulphide of silver,	8-hedrons,	Sulphocuprite of silver,	Rhombohedral.
Seleniate of zinc,	Prismatic,	Acid solution,	Square prisms.
Chromate of lead,	Oblique prismatic,	From Bannat,	Pyramid. or prismatic?
Carbonate of lead,	Prismatic,	Plumbocalcite,	Rhombohedral.
Iodide of mercury,	Square prisms (red),	By heat (yellow),	Prismatic.
Idocrase,	Right square prisms,	Garnet,	Rhombic 12-hedrons.
Baryto-calcite(Brooke's),	Oblique prismatic,	Bromlite,	Prismatic.

Dinitraniline. $C_{12}H_52NO_4N$.

Dinitranisic Acid. $C_{12}H_62NO_4O_6$, by sulphuric and nitric acids on anisic acid.

Dinitr-anisidine. $C_{14}H_72NO_4NO_2$, by NH_4S on trinitranisole.

Dinitranisole. $C_{14}H_62NO_4O_2$, by fuming nitric acid on anisole.

Dinitro-azobenzide. $C_{24}H_82NO_4N_2$, needles by nitric acid on azobenzide.

Dinitrobenzoic Acid. $C_{14}H_42NO_4O_4$, by boiling benzoic acid with sulphuric and nitric acids.

Dinitrobenzole. $C_{12}H_42NO_4$, by nitric acid on nitrobenzole.

Dinitrobenzophenide. $C_{26}H_82NO_4O_4$. Yellow plates by chloride of benzoyle on nitrophenisic acid.

Dinitrobenzophenone. $C_{26}H_82NO_4O_2$, by nitric acid on benzophenone.

Dinitrodiphenamic Acid. $C_{24}H_{12}$, $(NO_4,)_2N_2O_4$. Needles by NH_4S on nitrophenesic acid.

Dinitrogaultheric Acid. $C_{16}H_52NO_4O_5$, by sulphuro-nitric acid on salicylate of methyle.

Dinitromelaniline. $C_{26}H_{11}2NO_4N_3$, from aniline and chloride of cyanogen.

Dinitromesitilole. $C_{18}H_{10}2NO_4$, by sulphuro-nitric acid on mesitilole.

Dinitrosalicylic Acid. $C_{14}H_42NO_4O_6$. A product of dinitrogaultheric acid.

Dinitrosalithole, *Dinitrophenetole.* $C_{16}H_8 2NO_4O_2$. Needles by nitric acid on phenetole.

Dinitrosulphonaphthalic Acid. $C_{20}H_6 2NO_42SO_3$, by nitric acid on sulphonaphthalic acid.

Dinitrotoluole. $C_{14}H_62NO_4$.

Diopside. A synonyme of white augite.

Dioptase. See Hydrous Sesquisilicate of Copper.

Diorite. See Greenstone.

Dioxylite. See Sulphato Carbonate of Lead.

Diphanite. Silica 33·24, alumina 44·33, lime 13·11, FeO 3·04, MnO 1·13, water 5·18.

Diphene Compounds, are those derived by two conjugate atoms of benzole $C_{12}H_6$, or phenole $C_{12}H_6O_2$.

Diphenine. $C_{24}H_{12}N_4$ by NH_4S, on dinitro-azobenzide.

Diphosphamide. PNO_2, or PNO_5, by heating phosphamide.

Diplatinamine. $N_2H_4Pt_2$. The base existing in Gros' platinum salt.

Diplatosamine. N_2H_5Pt. The base existing in Reiset's platinum salt (*a*).

Diploite. See Latrobite.

Dippel's Animal Oil. *Rectified oil of Hartshorn.* An oil obtained in the distillation of bones for the preparation of bone black. In its crude state it is brown and viscid, but may be obtained colourless by redistillation. It contains *odorine* (picoline), *aniline, lutidine,* bases of the ethylamine series, and of the pyrol series, the latter characterized by giving a red resin when decomposed. According to Unverdorben, it contains *odorine, animine, olanine,* and *ammoline.*

Dipyre. See Scapolite.

Disacryle. $C_{10}H_7O_4$. A white powder from water and acroleine.

Discrase. See Antimonial Silver.

Disinfecting Liquor of Labarracque, or chloride of soda.

Disomose. Arsenio-sulphide of nickel.

Displacement. A sepcies of filtration in which an organic substance which acts as the filter has its soluble matter displaced by water or other fluid acting in the form of a column. Apparatus adapted for this purpose are termed displacement apparatus, and may be formed of glass or metal. For laboratory experiments they are most conveniently made of a long narrow glass cylinder, a foot or more in length, stoppered above, and with a narrow open outlet below, fitted by grinding into a glass receiver beneath. The substance to be exhausted is placed in the bottom of the cylinder *d*, water is poured on it, and the resulting infusion drops into the receiver below. In this way infusions are made on the great scale by a large tin apparatus, the glass cylinder being replaced by a tin tube many feet in length, which may be connected with a reservoir or headpressure of water.

Disterrite. Spec. grav. 3·042, H 5 to 6½. Hexagonal prisms. Silica 20·, alumina 43·22, Fe_2O_3 3·6, magnesia 25·01, lime 4·, potash ·57, water 3·60; from Fassa, Tyrol.

Disthene. See CYANITE.

Distillate. A fluid distilled, and found in the receiver of a distilling apparatus.

Distillation, (*destillare*, to drop), is a process which has been long practised. It consists in heating a fluid in a close vessel, and collecting and condensing the vapour which passes from it in a cool *receiver*. The apparatus required for distillation is a *retort* or *still*, and a *receiver* or *condenser*. The fluid to be distilled is placed in the retort or still which is closed. The retort is connected with the receiver. Heat being applied to the retort, enters the fluid and converts it into vapour. The vapour passes into the receiver, which is kept cool by a current of cold water running over it. The cold water deprives the vapour of the heat required to retain it in its elastic state; the vapour consequently condenses into a fluid state again. Distillation therefore supplies a method of removing volatile matters from more fixed materials when the volatile matter is not decomposed by the heat applied to it. A convenient method of distilling

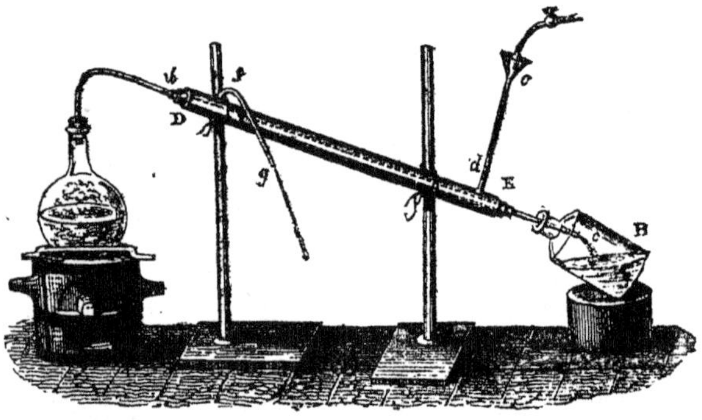

in a laboratory, is by means of Liebig's condensing tube. The condensing tube into which the beak of the retort enters, consists of an exterior tin tube, E D, and a narrower glass tube, *b*, which is enclosed in the tin tube, and communicates directly with the retort. There is therefore a considerable space between the glass and tin tubes for the reception of water. Cold water being admitted from a vessel standing on a stool by the funnel *c*, or from a reservoir, into the space mentioned, removes the heat from the vapour of the distilled fluid, becomes itself heated, and rising to the upper extremity of the tin tube *f*, runs off by the curved tube *g* into a receiving vessel. A constant flow of cold water is thus kept up from the water reservoir upwards from *d* to *f* along the tin tube. A convenient modification of this apparatus is the annexed. For pharmaceutical purposes, a convenient apparatus

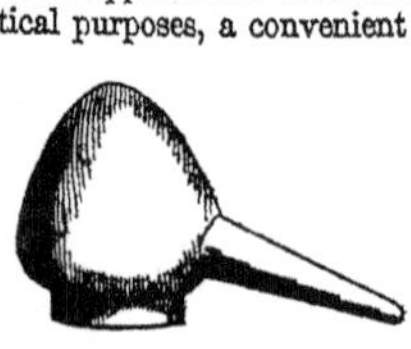

is to have a large globular vessel of green glass (bolthead), supplied with a *head* or *alembic* (fig.) For purposes of testing, a tube retort and receiver, *a b*, will be found convenient. A zig-zag tube, *m*, may also be employed for a similar purpose, cold being applied at the bend so as to produce

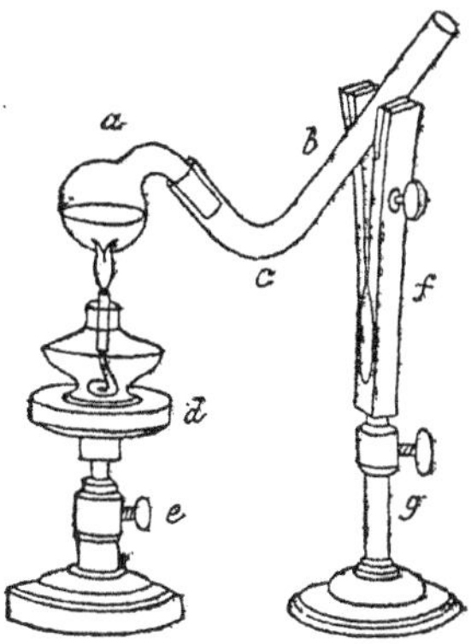

condensation of the fluid distilled from the closed end.

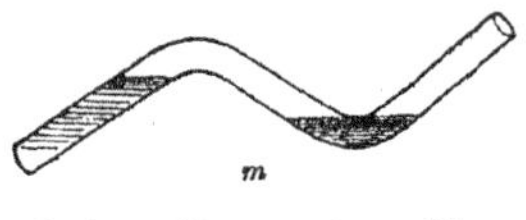

Distillation, Destructive. The result, it has been long known, of the application of heat to organic substances confined in close vessels, is merely a change of state, nothing being lost. Distillation was practised by the alchemists; but it is only in more recent times that the results of the distillation have been examined carefully with relation to the source of their production. So late as 1730 Hales distilled substances to ascertain if they contained air, no distinction being at that time recognized between one kind of gas and another. Soon afterwards it was observed by Beccaria, that flour contained two principles, one which yielded acid by distillation, and the other ammonia. These are now known to be derived, the acid from the starch, and the ammonia from the albuminous or nutritive matter. But besides these products, there are various gases and oils. Such results are well exhibited in the gas manufactory, and in the production of bone black. When substances consisting of carbon, hydrogen, and oxygen are distilled, the effect is to remove oxygen from the body, in the form of carbonic acid or water. After the removal of the oxygen, the carbon and hydrogen are removed in the form of the inflammable gases, olefiant gas and carburetted hydrogen. Sugar ($C_{12}H_9O_9$ 2 HO) for example, when distilled, yields first 2 atoms water, and is changed into soluble caramel ($C_{12}H_9O_9$). If the heat be continued, more water is given off with acetic acid ($C_4H_3O_3$), formic acid (C_2HO_3), carburetted hydrogen (CH_2), &c. and a vesicular charcoal remains behind. When many acids are distilled, they lose oxygen in the form of carbonic acid and water, and are converted into a new acid; 1 atom hydrate of salicylic acid by heat loses 2 atoms carbonic acid, and is changed into carbolic acid ($C_{14}H_6O_6 - 2\ CO_2 = C_{12}H_6O_2$); 1 atom meconic acid is deprived of 2 atoms water and 2 atoms carbonic acid, and becomes comenic acid ($C_{14}H_4O_{14} - 2$ HO, and $2\ CO_2 = C_{12}H_2O_8$). Nitrogenous bodies lose first ammonia, then cyanogen and cyanohydric or prussic acid. Urea when distilled loses ammonia, and is converted into cyanuric acid, which again is changed into cyanic acid. When organic acids are distilled in contact with strong bases, they lose the elements of carbonic acid, which unites with the base, and the acid less the carbonic acid passes over in the form of a new product, whose termination is in *one* (cetone). Acetate of lime becomes carbonate of lime and acetone ($C_4H_3O_3$, CaO = CaO CO_2, and C_3H_3O). The action in dry distillation has been compared to combustion from the tendency to the union of carbon and oxygen, and of hydrogen with oxygen; the products being carbonic acid and water.

Distillation of Spirits. See SPIRITS and STILL.

Distilled Waters are those waters used as cosmetics, &c. which have been formed by distilling water with plants containing essential oils.

Disuccinamide. $C_8H_3O_4NH_2$. By distilling bisuccinate of ammonia.

Dithionic Acid, or Hyposulphuric acid.

Dividivi. The fruit of the *Cæsalpinia coriaria* from Carthagena, containing tannic and gallic acids; used in tanning.

Doeglic Acid. $C_{38}H_{36}O_4$. Yellow fluid, the oleic acid of doegling oil.

Doeglic Oxide. $C_{24}H_{24}HO$. By distilling doegling oil and rectifying.

Doegling Train Oil. The oil of the bottlenose whale, *Balaena rostrata*, of spec. grav. ·868.

Dolerite. A trap rock consisting of labradorite and augite.

Dolomite. Magnesian limestone, a combination of 1 atom carbonate of lime, and 1 atom carbonate of magnesia, but often in other proportions; occurring crystallized, as rhomboidal bitter spar, and in large deposits as a distinct formation overlying the coal formation.

Domite. A variety of trachyte. Silica 51, Al_2O_3 24, MgO 7·8, CaO 2·, Fe_2O_3 8·8, MnO ·64, KO 4·6.

Dopplerite. A gelatinous substance from a peat bog in Styria.

Dracole. $C_{14}H_8O_2$. See ANISOLE.

Draconic Acid. See ANISIC ACID.

Draconine, Dracine. $C_{40}H_{21}O_8$. The pure red basic resin of dragon's blood (*Pterocarpus draco*), extracted by carbonizing the impurities of the crude resin by sulphuric acid. It is employed to colour varnishes; spec. grav. 1·196. When distilled it yields *dracyle*, $C_{14}H_8$, or *toluole*.

Dragonite. A variety of quartz.

Drawing Slate. See SLATE.

Dreelite. 3 ($BaO SO_3$,) $CaOSO_3$. Pearly rhomboidal crystals from Beaujeu, France; resembles chabasite, spec. grav. 3·3, H 3·25. B.B. fuses into a blebby glass coloured blue by nitre.

Drosometer. (δρόσος, dew.) An instrument for measuring the amount of dew.

Drummond Light. The intense light produced for signals by propelling a mixture of ignited hydrogen or coal gas and oxygen upon lime.

Drying. The process of removing moisture from bodies by means of the water or other baths, the air pump, &c. Gases are dried by passing them over asbestus moistened with sulphuric acid, by causing them to bubble through sulphuric acid, or by passing them through a tube filled with chloride of calcium.

Drying Oils. See OILS.

Dry Rot. The process of dry decay in timber, considered to be produced by the fermenting action of the albuminous matter still retained in the wood from want of proper seasoning. The antidote is to expose the wood to the air or running water, so that its juices may be washed out, or to immerse it in a metallic saline solution, or in creasote, derived from distilled resins which coagulate the albumen.

Ductility. The property possessed by metals of being drawn out into wires. This power is possessed by them in the following order:—gold, silver, platinum, iron, nickel, copper, zinc, tin, lead.

Dufrenite, *Alluaudite, Craurite.* Phosphoric acid 28·42, FeO 57·6, water 12·15. Olive or green colour. Spec. grav. 3·227; from Anglar, Limoges.

Dufresnoysite, *Arsenio Sulphide of Lead*, from Clausthal in the Hartz. 2 PbS, AsS_3.

Dukstein, or Trass. A recent volcanic mineral.

Dulcine, *Dulcose.* $C_{14}H_{14}O_{12}3HO$. A white sweet substance from an unknown source in Madagascar, crystallizing in oblique prisms. F.P. 359½° to 374°. At 527° it decomposes, giving out carbonic oxide, and perhaps acetic acid, and acetone. Almost insoluble in alcohol, which distinguishes it from mannite; it seems homologous with grape sugar, although it does not ferment; it has no action on polarized light; not precipitated from its solutions by acetate of lead. With sulphuric acid it forms a conjugate acid, the barytes salt of which is gummy.

Dumasine. $C_{10}H_8O$? Spec. grav. of vapour 5·204. Colourless empyreumatic fluid, obtained along with acetone in the distillation of acetate of lime.

Dutch Gold Leaf, or **Foil,** is composed of copper and zinc, or of bronze and copper leaf; it possesses a golden colour, and is often sold for gold leaf, from which, however, it may be distinguished by its solubility in either, muriatic or nitric acid, separately. Its frequent constitution is said to be 11 copper, 2 zinc, rolled from red bronze.

Dyeing. (*Teinture*, Fr.; *Färberei, Färbekunst*, Ger.) The term dyeing is usually restricted to that part of the art of communicating colours to cloth, which is accomplished by immersion in a fluid. It is the most ancient branch of the art, and is practised in the rudest states of society. Even at the present time it is not unfrequently a domestic art in the more sequestered parts of our island. The female Highlanders during the winter months often practise the art of dyeing their own dresses with vegetable materials immediately within their reach. Thus a *brown* colour is obtained by heating the yarn or cloth with a solution of copperas, which acts as the mordant with the colouring matters of sumac, logwood, and crottel (Parmelia omphalodes), a lichen which covers the surface of rocks and trees in a moist climate. *Black* is dyed with the bark of the alder (*Arn*, Scot.; *Fiarn*, Gael.) or Almus glutinosa, with copperas and a little sumac; *yellow* by means of the common heather (Calluna vulgaris) and alum; *red* with the roots of the galium verum and alum. It is obvious, therefore, that, as in calico printing, the application of colours to fabrics implies generally the use of a mordant to adhere to the cloth, and of a colouring matter which unites with the mordant, and thus forms a chemical combination. This treatment is particularly required in dyeing cotton; but between woollen and silk fibre and colouring matters an affinity exists which renders the operation of dyeing with these fabrics much simpler. The mordants used by dyers are usually termed *spirits*, and as they require different strengths for various colours, the peculiar mordant is designated by the colour for which it is used. *Brazil-wood*, or *red spirits*, is obtained by adding to 3 lbs. muriatic acid, and 1 nitric acid, metallic tin cautiously, to prevent too violent action, until the metal ceases to be dissolved. The proportions in *Barwood red spirits* are 6 lbs. muriatic acid and 1 nitric acid, the metal being added as before; while in *Plum spirits* the ingredients are 1 lb. nitric and 7 lbs. muriatic acid. To prepare the metal for these spirits, it is melted in an iron vessel, and while still fused it is run into cold water—a process technically called feathering. The details of the process of dyeing Turkey red has been given under CALICO PRINTING. It affords an illustration of one of the most difficult operations in the art. The best works on dyeing are by Bancroft, Berthollet, Chevreul, Persoz, Leuchs, &c.

Dyripe, from Mauleon, Pyrenees. SiO_3 60, Al_2O_3 24, CaO 10, HO 2·.

Dysclasite, *Okenite.* Spec. grav. 2·362, H 4·5. White fibrous crystals, difficult to fracture. B.B. fuses on the edges, and with soda into a semitransparent glass; gelatinizes with chlorohydric acid. Silica 57·69, lime 26·83, water 14·71, soda ·44, potash ·23, Fe_2O_3 ·32, MnO ·22. *Form.* 3 CaO, 4 SiO_3, 2 HO.

Dysintribite. A mineral from Diana, New York. Spec. grav. 2·76 to 2·81, H 3·75; dark green granular mass; fusing into a porcelain mass, SiO_3 47·68, Al_2O_3 41·5, FeO 5·48, HO 4·83.

Dyslitite, Schreibersite. Brownish-black powder remaining on dissolving many meteoric masses in acids.

Dysluite. Spec. grav. 4·551, H 4·5. Yellowish-brown 8-hedrons; texture foliated, opaque; lustre splendent and vitreous. B. B. becomes red, which it loses on cooling; does not fuse with fluxes, but the soda bead has a fine red colour, disappearing on cooling. It is very difficult of decomposition, bisulphate of potash or soda being the best agent; from Sterling, N. Jersey. Al_2O_3 30·49, ZnO 16·8, Fe_2O_3 41·93, MnO 7·6, SiO_3 2·96, HO ·4.

Dyslysine. $C_{60}H_{48}O_7$ or $C_{48}H_{36}O_6$. A resinous substance obtained by boiling choloidic acid with chlorohydric acid; so named from its difficult solubility.

Dysodile. Apparently used by many as a generic term to include a number of species. A variety occurring near Giessen consists of bitumen 49, carbon 55·5, sesquioxide of iron 11, silica soluble in potash 17·4, clay, &c. 1·. It has been described also as composed of siliceous infusoria.

Dyssnite. See SILICATE OF MANGANESE.

E

Eaglestone. A name given to a clay iron ore.

Earthenware. See POTTERY.

Earths. (*Erden*, Ger.; *Terres*, Fr.) A class of metallic oxides characterized by their insolubility in water. They are as follows:—*alumina, glucina, zirconia, thoria, didymia, lantana, ceria, yttria, terbia, erbia.* With the exception of alumina, which is one of the most abundant substances in nature, the earths are uncommon, and most of them have been very imperfectly examined in consequence of the close approximation of their characters. They are all precipitated white by ammonia from their solutions, and also by soda; but alumina and glucina are redissolved by soda or potash. They are all precipitated as oxides by sulphohydride of ammonium with the evolution of sulphohydric acid. They are not precipitated by sulphohydric acid.

Eau de Cologne. See COLOGNE WATER.

Eau de Javelle. A solution of chlorine in water.

Eau de Luce. A milky-coloured mixture of 1 part oil of amber, 24 alcohol, and 96 caustic ammonia.

Eblanine. (*Dublinine*) *Pyroxanthine.* $C_{21}H_9O_4$. Yellow needles, F.P. 291°, subliming at 273°; insoluble in water; soluble in ether, alcohol, and acetic acid; soluble in sulphuric and chlorohydric acids. It is obtained from crude pyroxylic spirit by distilling one-sixth of that fluid, saturating the distilled fluid with slaked lime, distilling off the spirit, treating the brown residue with dilute chlorohydric acid, and then with alcohol, which by distillation yields impure eblanine. This is purified by re-solution and evaporation.

Ebullition. The act of boiling produced by the rapid formation and escape of vapour from a fluid by the influence of heat. The temperature of the fluid as indicated by the thermometer is termed the boiling point.

Echynite of Miask. Columbic acid 33·39, titanic acid 11·94, zirconia 17·52, protoxide of iron 17·65, yttria 9·35, lanthana 4·76, protoxide of cerium 2·48, lime 2·40, water 1·56, traces of fluorine, tungstic acid, manganese, and magnesia.

Edelforsite. *Tersilicate of lime* from Edelforss. SiO_3 57·75, CaO 30·16, MgO 4·75, FeO 1·, MnO ·65, Al_2O_3 3·75.

Edelite, or Mesotype.

Edelithe, or Prehnite.

Edingtonite. *Antiedrite.* Spec. grav. 2·75, H 4·25. Primary form, right square prisms, angles of 129° 8′ and 92° 41′, with 8-hedrons with square base; along with Thomsonite in the Kilpatrick hills, Glasgow; lustre vitreous, streak white, brittle; B.B. fuses into a colourless mass. Silica 35·09, alumina 27·69, lime 12·68, water 13·32, alkali 10 to 12.

Edulcoration. (*Dulcis*, sweet). The process of washing a substance until the wash water no longer tastes saline but sweet.

Edwardsite. See MONAZITE.

Effervescence. (*Effervesco*, I boil.) The appearance of boiling produced by the rapid escape of a gas, as when an acid is poured on chalk or any carbonate.

Efflorescence. (*Effloresco*, I blossom, or shoot into flower.) The appearance often observable on brick walls, of a salt which renders the surface white; or when a salt, by losing water in the air, becomes powdery on the surface, as with crystals of carbonate of soda.

Egerane, Egrane. Streaked liver-brown idocrase from Eger, Bohemia.

Egg. (*Ei*, Ger.; *Œuf*, Fr.) The term egg is bestowed on the ovum of oviparous animals, but in common use it is applied to the egg of the hen, which consists of—1. the *shell*, a compound of animal matter 2, phosphate of lime and magnesia 1, carbonate of lime and a little carbonate of magnesia 97. 2. The *membrana putaminis*, a thin pellicle within the shell. 3. The *white*, or albumen of the egg, consisting of a glassy fluid deposited in membranous cells, coagulating at 160°. The white consists of water 80, albumen 15·5, mucus 4·5. The ash of the white consists in 1000 parts of SO_3 ·15, PO_5 ·46, Cl ·93, KO, NaO, and carbonates of do. 2·93, CaO, MgO, and carbonates of do. ·25. Total, 4·72. 3. The *yolk*, consists of water 53·78, albumen or vitelline 17·47, yellow oil 28·75. The ash consists of SO_3 ·06, PO_5 3·5, Cl ·28, KO and NaO and their carbonates ·27, CaO, MgO, and their carbonates ·61 (Prout). The oil contains phospho-oleic acid, the same as in the brain, and cerebric acid. The latest determination of the composition of the yolk is water 51·48, vitelline 15·76, margarine and oleine 21·30, cholesterine ·438, phosphoric body (containing 7·23 oleic and margaric acids, and 1·2 phosphoglyceric acid) 8·426, cerebric substance ·3, NH_4Cl ·03, NaCl, KCl, $KOSO_3$ ·28, $CaOPO_5$ and MgO 1·02, alcoholic extract ·4, colouring matter ·55. The spec. grav. of a new-laid egg is 1·080 to 1·090. An egg rapidly loses weight when kept, the aqueous portion being displaced by air. The loss is about ·744 grs. daily, the original weight of an egg being 907½ grs., in two years its weight was 363·2 grs. It is the presence of the air which promotes putrefaction. Hence the mode of preserving eggs, is to cover them with lard or butter. An egg also loses 20 to 30 per 1000 of its weight when boiled; some of its saline ingredients escape into the water. The relative weights of the 3 portions of the egg are shell and membrane 106·9, albumen 604·2, yolk 288·9. By keeping an egg for three weeks at a temperature of 100°, a chick is formed; in a week the average loss is 50 grs.; at the end of the second it has lost 130 grs.; the oil passes into the albumen, which diminishes; during the last week the yolk loses much of its phosphorus, which is found in the animal as phosphate of lime; at the period of incubation all the albumen has disappeared; the yolk is reduced in size, and is taken into the abdomen of the chick, the animal having attained a weight nearly equal to the original weight of the albumen.

Ehlite, or quintohydrous pentaphosphate of copper. CuO 65·99, PO_5 24·93, HO 9·06.

Ekebergite. *Sodaite.* Green, gray, or brown masses, transparent; lustre vitreous or resinous; with difficulty acted on by acids; B.B. yields water, and melts into a blebby glass; fuses with effervescence with borax. Silica 46, alumina 28·75, lime 13·50, soda 5·25, Fe_2O_3 ·75, water 2·25.

Elaene. $C_{18}H_{18}$. By distilling metoleic and hydroleic acids, and oleine together.

Elaeolite. *Fettstein, Lythrodes, Sodaite.* Blue, green, or brick-red, massive, with natural joints parallel to the planes of a right rhombic prism, with angles of 112° and 68°; fracture conchoidal; lustre resinous, translucent; H 6. Spec. grav. 2·54 to 2·62; B.B. fuses into a white enamel; SiO_3 44·19, Al_2O_3 34·42, NaO 16·87, KO 4·73, CaO ·52, MgO ·68, Fe_2O_3 ·65, HO ·6; gelatinizes in acids; from Laurvig, Stavern, and Frederiksvärn, Norway; and from the United States of America.

Elaidine, and **Elaidic Acid.** $C_{72}H_{66}O_5$. Tabular pearly crystals, by heating oleic acid with nitrous acid.

Elaine. See OLEINE.

Elaldehyde. Polymeric with aldehyde, from which it deposits by long keeping, and exposure to a temperature of 32°; elaldehyde then appears as ice-like needles; fusing at 35°.

Elaoptene. *Oleoptene.* The fluid portion of essential oils.

Elasmose. Bitelluride of Lead.

Elasticity, or *Spring*, denotes the tendency which exists in the particles of certain bodies to recover their natural position, when they have been mechanically or chemically displaced. When air has been compressed, and the particles of which it is composed are brought in closer juxtaposition, they will recover their former position on the removal of the pressure. When caoutchouc is compressed or extended, it returns to its original volume when the disturbing influences are at an end. Fluids possess the same power, but the particles of metals are less elastic. When hammered, metals retain the close proximity of their particles. Gases are, at equal temperatures and under equal atmospheric pressure, similarly elastic, except near their condensing points, when the reaction increases more than the pressure.

Elastic Bitumen. *Mineral Caoutchouc, Elaterite.* Blackish-brown, of various shades, internally shining and glistening, very soft, sectile, elastic, flexible. Spec. grav. ·905 to 1·233. It is said to consist of C 52· to 58, H 7·5 to 4·75, O 40· to 36·75; nitrogen, a trace; it is found in limestone, near Castletown, Derbyshire.

Elastic Gum. See CAOUTCHOUC.

Elaterine. *Elatine, Momordicine.* $C_{20}H_{14}O_5$. Silky rhombic prisms, with a bitter taste; fuses at 392° and volatilizes; soluble in sulphuric acid, with a blood-red colour; soluble in 5 parts alcohol, and 118 ether; obtained from the evaporated juice of the *Momordica elaterium* by strong alcohol (90 per cent.) evaporating, washing with ether, and again dissolving in boiling alcohol and evaporating.

Elathine. Brownish-yellow oil, lighter than water, partially decomposed by distillation, obtained by the action of ammonia and sulphur on acetone.

Elayle. A theoretic name given to olefiant gas.

Elecampane. The root of *Inula helenium*, containing 36·7 of inuline, a variety of starch.

Elecampane, Camphor. See HELLENINE.

Electricity. (ἤλεκτρον, amber). A term, originally used to express the fact that certain bodies, when rubbed, attract light bodies towards them; but now applied to an extensive branch of science, which not only includes such phenomena as are produced by friction, but also by chemical action. The consideration of this part of science belongs to natural philosophy, but in the present work the subjects of animal electricity and electro-metallurgy may be shortly noticed.

Galvanism. — *Animal Electricity.* — Sultzer, in his *Theorie Generale*, published in 1767, states that if a plate of lead be placed upon the tongue, and another of silver under the tongue, a styptic taste, similar to that of a chalybeate water, is immediately experienced. The same experiment may be performed with other metals. In 1786 Cotugno announced that a student of medicine at Bologna, while dissecting a living mouse, was surprised at receiving in his hand an electrical shock when he touched the phrenic or intercostal nerve with his scalpel. In 1790 Galvani, professor of anatomy at Bologna, had dissected a frog, and having placed it on a table where there was an electric machine, one of the persons who was assisting him, by accident touched the crural nerves of the frog with a scalpel near the conductor of the machine in action, when instantly violent contractions were produced. To make this experi-

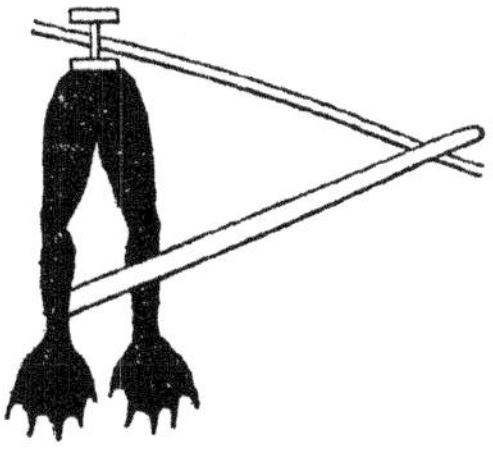

ment, it is necessary to lay bare the crural nerves, and to leave the legs attached to the spine by the nerves alone, and then a copper and zinc wire being either twisted or soldered together at one end, are to be applied, one under the nerve and the other to the leg. *Experiment.*—Place two glasses full of water together; into one put the thighs and legs of a frog, and lay the nerves over the edges of both glasses. Wrap a piece of tinfoil round the nerves. Place the fingers of one hand in the water of the glass that contains the legs, and holding a piece of silver in the other, touch the nerves; convulsions occur. Galvani considered the body of the animal to be in the state of a charged Leyden jar, of which the nerves and muscles were the internal and external coatings, and that on connecting these by the conducting wire the electricities were united, and renewed for an instant the phenomena of life. Volta, on the contrary, having repeated the experiment, attributed the phenomena not to an animal electricity, as Galvani had done, but to common electricity. He affirmed that when two metals are brought in contact, one acquires vitreous and the other resinous electricity; and it is the electricity thus induced that occasions the animal contractions. Fabroni, in 1792, advanced an opinion which is in all probability the correct one. According to him, the particles of every metal, when brought in contact with those of another metal, have a tendency to unite with each other, and they would do so if they were not restrained by the attraction of cohesion. Their mutual action increases their power of attracting oxygen from the atmosphere or other sources, and hence the effects produced upon the frog's limbs may be attributed to a chemical operation—the passage of oxygen from some combination to form a new one. In repeating the experiment of Sultzer, he observed that if he wiped his tongue as dry as possible, the effect was so much diminished as to be hardly distinguishable. He conceived, therefore, that the saliva must act a conspicuous part in the manifestation of the effect; and as he supposed that the taste was produced by the partial solution of the metallic bodies employed, he came to the obvious conclusion that the saliva acted the part of a menstruum in accomplishing the solution.

Electricity in the Frog.—Galvani first laid the basis of this study by his celebrated experiment. Volta repeated it, while Aldini, in his treatise on galvanism, records some experiments in which he obtained contractions from the prepared frog, by making a communication between the muscles and nerves of the animal through other animals and his own body. And it has since been ascertained that this statement is not only correct, but that every metallic or liquid conductor interposed between the muscles and nerves of a frog, and through which the circulation of the electric fluid can be established, determines this current, and consequently the contraction. Humboldt also obtained contractions by interposing between the muscles and nerves a portion of muscular substance, and according to him these contractions do not occur when the muscle touched is that of the thigh. Forty years elapsed without any addition to these facts, until Nobili showed that these contractions are produced by a current of electricity proceeding from the muscles to the nerves of the frog. In order to detect the electricity of the frog, Matteucci, to whom we owe many interesting facts in this important department of electricity, employs four porcelain capsules, which are filled with water slightly saline. Into the outer capsule, on each side, platinum plates are introduced, and connected with a galvanometer. The two extreme capsules are then united by moist cotton threads, and lastly the two middle capsules are connected by means of the frog, prepared or in a living state. To prepare it in the living state, it is necessary to raise the skin from the legs, to cut longitudinally that of the flanks, and to draw out with glass forceps the crural nerves. It is not necessary, however, in order to affect the galvanometer, that the two portions of the body of the animal plunged into the capsules should be only the nerves and muscles or tendons of the leg. Thus, without cutting the frog, as is done in preparing it, we may limit ourselves to raising the skin up entire, and plunging the legs into one capsule, the head and back into another. We have then as intense a current as that obtained with the nerves and muscles, always passing from the feet to the head.

Contractions in general only occur by the contact of the nerves and the muscles, although there are some animals possessed of such vivacity as to contract after the contact is destroyed. When a frog is placed in the capsules as described, Matteucci has found that the galvanometer at first is deflected 25° or 30°. The needle, however, soon begins to return, and becomes fixed at 3°, and in a quarter of an hour it remains at 2°. He has even kept frogs for six hours, and found them to produce an effect upon the galvanometer.

To account for these phenomena, two explanations have been suggested. One of these ascribes the contractions and currents to a thermo-electric source, or in other words, they are produced by the unequal temperature of the muscle and the nerve; an inequality occasioned, according to this hypothesis, by the difference of the evaporation in these two portions of the animal. But this view appears groundless when we consider that the currents are only produced when the galvanometer is supplied with a very long circuit of wire, and that they traverse liquids—conditions quite opposed to thermo-electricity. According to the second hypothesis, the currents are produced by the electro-chemical state of the body, some organs being alkaline, others acid. According to this view, as the current goes from the muscle to the nerve, the muscle ought to be alkaline and the nerve acid,—deductions which are not borne out by the chemical examination of these organs. Hence it is better to plead ignorance of the cause of these currents.

The conclusions to which Matteucci has come respecting amimal electricity are:—

1. That in the frog and in warm-blooded animals we find an electric current when the internal part of a muscular mass and its surface are placed in communication with a conducting arc.

2. That the nerve which belongs to a muscular mass, and the whole cerebral system, can perform the office of the internal part of the muscle in which the nerve is distributed.

3. That the current is directed in the animal from the interior of the muscle or of its nerve to its surface or to its tendon.

4. With the frog we obtain a current (the peculiar current) by placing in connection the muscles or the tendons of the leg with the muscles or the nerves of the thigh. This current is directed in the animal from the leg to the thigh or to the nerve.

Matteucci has added to these facts some very extraordinary ones, which seem to call in question the identity of the animal and electrical powers. He places in an isolated position a frog prepared in the usual way; then he prepares another frog, so as to have one leg with the nervous twig which goes from the spinal marrow to the muscles of the leg. It is necessary that this nerve should be perfectly cleared of the muscle. He then places the nervous twig over the thighs of the first frog, so that the filament of the leg to which the nervous twig belongs shall not touch the thighs. He waits till the nervous movements occasioned by the rapid preparation of the animal cease. Let the lumbar nerves be now touched with a pair of plates of copper and zinc, instantly the muscles of the thighs contract, *but at the same time the leg whose nerve is laid on the contracting muscles is observed to contract.* Instead of a voltaic pair, we may use a nail or any similar body with the same effect. The same experiment has been repeated by placing the prepared single leg on the muscles of a living rabbit's thigh with the same result. This phenomenon he terms *contraction by induction* or *induced contraction.* This is a remarkable experiment, and quite original, but the subsequent facts are not less so. When the muscles of the legs are covered with gold leaf, and the single leg placed as before, the voltaic pair produces the same contractions in the legs, but no effect is produced on the single leg. Hence it is obvious that metal is incapable of superseding the nerve in carrying the contracting power. But if we tear the gold leaf, and bring the animal surfaces in contact as before, the contractions recur. If in place of gold leaf we use unsized paper of very fine quality, the contractions take place. The phenomena never occur if the intervening body be an insulating plate, however thin it may be.

Electro-Magnetic Coil. Magnetism induced by a current of electricity, is termed electro-

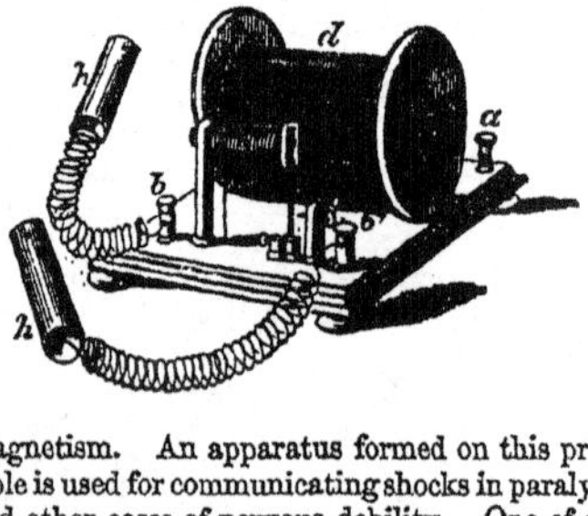

magnetism. An apparatus formed on this principle is used for communicating shocks in paralytic and other cases of nervous debility. One of the wires of a single cell battery is screwed in *a*, and the other at a corresponding coupling screw on the other side; the screw *g* must touch lightly on the spring *f*. By means of the handle *h*, the shock may be passed through the arms or through any part of the body by placing one handle at each extremity of the part to be acted on.

Electro-Metallurgy. The art of depositing metals by means of electric currents from solutions. The following account of a visit to an electro-plating establishment, is by my pupil Mr. Robert Kirkwood:—"Our conductor led us to the workshop, where the goods of German silver are fashioned. Some forms of vessels are made simply by hammering; the carved or ornamental work being done with small steel chisels of various forms. Other vessels are formed of several pieces, each of which is 'struck' by means of a die; they are then united, so as to form a vessel, by means of solder. One of the workmen was engaged carving or chasing on a kettle. The vessel was filled with a mixture of pitch and chalk, which, being poured in while hot, had solidified on cooling, thus giving solidity to the vessel, and enabling the workman to use his hammer and chisel freely, and without any risk of making indentations in the sides of the vessel. We also saw in progress some small round trays, the brims of which were being surrounded with raised ornamental work. This ornamental work is struck in lengths or pieces by means of a die, and the hollows on the reverse side are filled up with soft solder, thus giving solidity and strength, and serving also to attach the different parts to the brim. In another part of the workshop, a man and boy were busily engaged polishing various articles, previous to their being sent to the plating room for immersion in the solution of the metallic cyanides. The polishers are small discs 2, 3, or 4 inches in diameter, and from $\frac{1}{4}$ to an inch in thickness. Some are made of 'buffalo hide,' others are of wood, in which case a belt or band of leather surrounds the wooden block, that is to say, forms the circumference of the disc. They are placed on a spindle, or axis, and a rapid vertical rotatory motion is communicated, by means of a small engine outside the building. To render them sufficiently rough, they are besprinkled every few minutes with fine sand, which has been damped with oil; the sand is brought from Trent Valley, Staffordshire. Our conductor next led us to the room where the goods are plated; and here it may be as well to state briefly the conditions, and to explain as far as we are able to do so, the process, by which the electro-metallic precipitation is effected. Every body in nature is saturated with the electrical element, and whenever by any means the decomposition of a substance is effected, whenever, any change of state is induced, electricity becomes manifested. If, therefore, a piece of an oxidizable metal, as zinc, is placed in an acid solution, it becomes oxidized, and is gradually dissolved. During this process, a quantity of the electrical fluid is set free; and, if into the same acid solution we put another metal which is not so readily oxidizable, as copper, and by a wire connect the two metal plates, a constant circulation of electricity may be detected. This electricity, which was active in maintaining one body in a certain condition, is capable of reducing another compound to a similar state; or, in other words, the electricity evolved by the solution of an atom of zinc, is exactly proportioned to the task of precipitating from its solution one atom of silver, gold, copper, &c. It requires no difficult mechanical arrangement to accomplish this. In one

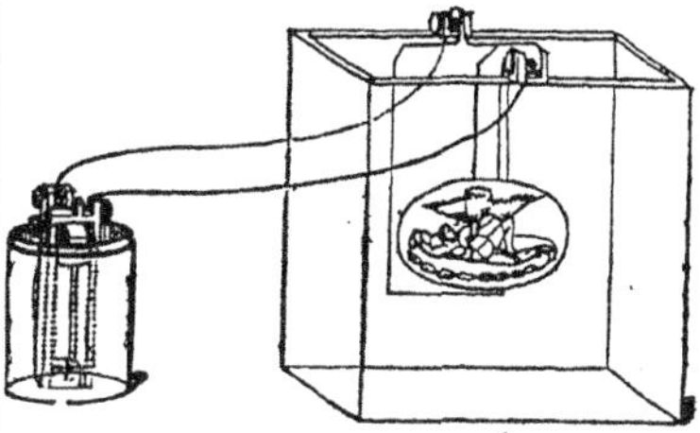

vessel filled with diluted sulphuric acid, a series of copper and zinc plates were placed, *a*, arranged in pairs. Two wires passed from the plates of this trough into a second trough, *b*, containing a solution of the metal to be precipitated (namely silver). One of the wires was connected with the articles to be plated, and to the other wire a piece of pure silver was attached, the use of which is to restore to the solution particles of metal as they are removed from it, and thus to maintain constantly an uniform strength. We were informed that the solution then in use had been made about a year and a-half previously, and that its strength was little if at all diminished. The electricity set at liberty at the zinc plate, is collected upon the copper plate, and conveyed by means of the wire to the goods in the decomposition trough; it passes then through the solution to the other wire, and returns by it, back to the zinc plate of the battery. Whilst performing this circuit, it effects the decomposition of one atom of the metallic solution (viz. AgCy), and the composition of another atom of the same salt (AgCy). Metallic silver combines with cyanogen, when subjected to the joint action of a solution of cyanide of potassium (KCy) and positive electricity; this is constantly going on in the trough, and it is thus that the strength of the solution is maintained. The zinc plates in the battery trough were coated over with mercury for the purpose of economizing the zinc, moderating the action of the acid, and obtaining a more steady or equable current of electricity. The fluid in the decomposing trough is a compound salt of cyanide of silver and cyanide of potassium, or a solution of cyanide of silver in cyanide of potassium. The cyanide of potassium is prepared by fusing 8 parts of dried ferrocyanide of potassium with 3 parts dried carbonate of potash in a covered crucible. The mixture becomes black, a clear fluid floats on the surface, and is to be poured off from the dark matter at the bottom, upon a clean slab. It is not a pure cyanide, both cyanate and carbonate of potash being present. The cyanide of silver is to be precipitated from nitrate of silver by cyanide of potassium, of which an excess is to be added in order to dissolve the precipitated cyanide of silver. A solution of a pure cyanide of silver might be used, but it is exceedingly difficult to decompose it, unless the battery be of very considerable power; the addition of KCy in slight excess greatly facilitates the operation, and renders a much weaker voltaic arrangement fully sufficient for the purpose. When the electrical force employed in the decomposition of the double salt is not too great, then only the AgCy undergoes decomposition, the KCy being merely set free; but if the electrical force is excessive, then the cyanide of potassium is also decomposed. A great excess of KCy should be avoided, for when this is the case, the solution is very unstable or is easily decomposed, and the silvering is not so well done. Again, if the electrical current is too powerful, the deposited metal is very brittle; and if it is deficient in power the deposit takes place slowly and imperfectly. The decomposition trough is made of wood, and is encased in lead; it is about 4 feet in length, 2 or 2½ feet in breadth, and 3 to 3½ feet in depth. In order to maintain an equable diffusion of the salts, so as to insure the uniform deposit of the metal, it is necessary that the articles to be plated be kept in constant motion. This is effected by means of a very simple mechanical contrivance. A brass frame of the same length and breadth as the trough is mounted upon wheels; it is connected, by means of a cord, to a spring kitchen-jack, and thus motion is communicated, and constantly kept up. Previous to introducing the goods into the trough, they are scrubbed with a brush and Isle of Wight sand; to remove grease, they are dipped in caustic potash, and then into dilute nitric acid. They are next rapidly washed in a solution of nitrate of mercury, just strong enough to give them a whitish appearance. Copper wires are attached, and the goods are immersed in nitric acid; the object being to have the surfaces as pure or clean as possible. After removing them from the acid, they are washed in water, and suspended by means of a copper wire from brass wires, which stretch across from the one side of the moveable frame to the other. After the goods have been immersed for a few minutes in the solution, they are taken out, brushed, and again introduced. After an immersion of for from four to six hours, a sufficiently thick coating of silver has been deposited; they are then taken out and brushed, the brush being moistened with stale beer. The brush employed is circular; motion is communicated by an apparatus resembling somewhat that of a turning-lathe. The process for gilding is exactly similar to that which has now been described, the double cyanide of gold being substituted for cyanide of silver. We were next conducted to the burnishing department; an operation performed by females. The tools employed are made of the finest steel; they are of various shapes, to suit the inequalities or irregularities on the surface of the vessel operated upon. Each operative is also provided with one or two burnishers made of blood-stone or heliotrope, with a leather strap to polish the burnishing instruments, and with a bowl or vessel containing a solution of soap. The object of burnishing is to obtain a smooth and brilliant surface, and this is done simply by friction."

Electrum. A term applied to native gold, as it occurs with silver.

Electrum. Amber.

Elements. The elements of nature, so called from their being incapable of further decomposition, have been enumerated under atomic weights. When arranged according to their electrical and isomorphous relations, they stand as follows:—1. *Sulphur*, selenium, tellurium, oxygen? 2. *Chlorine*, bromine, iodine, fluorine. 3. *Phosphorus*, arsenic, antimony, nitrogen? bismuth? 4. *Carbon*, boron, silicon. 5. *Potassium*, lithium? ammonium, sodium, silver. 6. *Barium*, strontium, lead. 7. *Alumina*, sesquioxides of iron, manganese and chromium, glucina, thoria, zirconia. 8. *Magnesium*, calcium, protoxides of manganese, zinc, cadmium, copper, iron, nickel, cobalt, cadmium, uranium, hydrogen. 9. *Titanium*, tin. 10. *Molybdenum*, tungsten, columbium, pelopium, niobium. 11. *Platinum*, palladium, iridium, osmium, ruthenium. 12. *Gold*. 13. Of uncertain position are vanadium, titanium, bismuth, mercury, yttrium, cerium, lanthanum, didymium.

Elemi Oil. C_5H_4. Spec. grav. ·849. Colourless ethereal. B.P. 345°, insoluble in water, soluble in alcohol and ether; forms a fluid and solid camphor with chlorohydric acid; obtained by distilling elemi resin with water.

Elemi Resin. *Elemine.* Spec. grav. 1·08. $C_{40}H_{34}O$. Yellow, translucent masses, with a fatty lustre, from Amyris elimifera and Ceylanica, the former from the West, the latter from the East Indies. It consists of two resins, one soluble in cold alcohol and crystallizable; the solution is not precipitated by metals, but by ammonia; changes into a jelly; it has the same composition as anime.

Eliquation. The separation of a more fluid from a less fluid metal when alloyed together, by heating them to the fusing point of one of the metals.

Ellagic Acid. *Bezoaric Acid.* $C_{14}H_2O_7$. Spec. grav. 1·667. Light yellow powder consisting of prisms, nearly insoluble in water and ether; slightly soluble in alcohol; soluble in sulphuric acid, from which it is separated by water; when the sulphuric acid solution is allowed to attract moisture from the air, the ellagic acid separates in long prisms; it forms a yellow solution with potash, which yields blue crystals of *glaucomelanic acid;* it forms black precipitates with salts of iron; ellagic acid is obtained from Oriental bezoars by digestion in potash, and precipitation by chlorohydric acid; it is formed by exposing infusion of nutgalls to the air, and remains in the insoluble matter from which the gallic acid is removed by boiling water.

Elutriation. (*Elutrio,* I clean). A mechanical process in the preparation of ores, by which their finer particles are suspended, and the heavier portions fall to the bottom. It is also extensively used in the pottery, and in preparing pigments.

Embolite. *Chlorobromide of Silver.* 2Ag Br, 3AgCl. Spec. grav. 5·80; 8-hedrons; cleavage cubic; externally olive-green, internally sulphur-yellow colour, from Copiapo, Chili; on calcareous spar.

Embrithite. See Boulangerite.

Emerald. *Beryl, Aquamarine, Agustite.* Spec. grav. 2·58 to 2·732, H 7.5 to 8. Emerald-green passing into blue, yellow, and white; regular 6-sided prisms. The bright greens are called *emerald,* the pale varieties *beryl.* B.B. the edges are rounded, and a vesicular slag is formed; it fuses with borax. It consists of SiO_3 66·858, Al_2O_3 18·406, GlO 12·536, Fe_2O_3 2·. *Form.* 3GlO $2SiO_3$, $Al_2O_3 2SiO_3$. The finest specimens come from Peru in Santa Fe, in primary rocks; it occurs in Upper Egypt, at Nertschinsk in Ural; in Cornwall, Caringorum, Wicklow, Brazil, Amity, New York.

Emerald Copper. See Dioptase.

Emery. Al_2O_3 86, SiO_3 3·, Fe_2O_3 4. Sp. grav. 4·. A gray variety of corundum from Naxos in the Archipelago, with a shining and adamantine lustre; opaque, very hard, and in powder is used to polish hard bodies.

Emerylite. *Corundellite, Clingmannite.* Sp. grav. 2·995, H 4·. White plates resembling mica, brittle or somewhat flexible, translucent; occurs along with corundum in Asia Minor, in Delaware, U. S. &c. Silica 30·18, Al_2O_3 51·4, CaO 10·87, MgO ·92, soda 2·23, water 4·52; 3 (CaO, MgO, NaO), SiO_3, 3 Al_2O_3 SiO_3, 3 HO. B.B. splits, gives out light and fuses on the edges.

Emetics. A series of compounds formed on the type of tartar emetic.—*Tartar emetic,* being KO SbO_3, $C_8H_4O_{10}$, *Iron emetic,* is KO Fe_2O_3, $C_8H_4O_{10}$, and *Lead emetic,* is PbO $SbO_3C_8H_4O_{10}$, &c.

Emetic Tartar. See Antimony.

Emetine. $C_{37}H_{27}NO_{10}$? White, bitterish powder, very poisonous; obtained by alcohol from the root of *Cephaelis ipecacuanha.*

Emmonsite. A variety of carbonate of strontian from the United States, contains 6· to 8·64 per cent. of carbonate of lime.

Empirical Formula. See Formulae.

Empyreuma. (ἐμπυρεύω, I kindle). The peculiar odour due to the partial decomposition of oils in the dry distillation of organic substances.

Emulsine. *Legumine.* C 48·79, H 7·73, N 18·83, O 24·65. (R. D. Thomson and Richardson). A fine white powder, giving out ammonia when boiled with alkalies; soluble in water, but insoluble in alcohol and ether; obtained by making an emulsion with water and sweet almonds, adding 4 times its volume of ether, allowing it to stand for three weeks, and drawing off the under clear fluid. On adding alcohol, the emulsine falls in white flocks, which are washed with alcohol, and dried in vacuo. It is coagulated by heat, and dissolved by acetic acid. It is the same substance termed legumine by Dumas, and found in many stone fruits. The substance examined by Robiquet, and termed synaptase, is quite different; the body analyzed by Ortliff is also different.

Emulsion. A name given to a thick gummy mixture, formed by trituration, of certain substances with water. One of the best examples of this kind is emulsion of almonds.

Enamel of Teeth. See Teeth.

Enamel, White, for pottery. Sand 53, calcined mass (consisting of 15 tin, 100 lead), 26, carbonate of potash 21, *Another.*—Fused borax 14, oxide of tin 14, arsenious acid 4, mixture (consisting of sand 45, minium 37, carbonate of potash 18), 68.

Enamelling. The process of covering the interior of iron vessels, &c. with an enamel of porcelain. The iron vessel is first heated gently with very dilute sulphuric acid, with which it remains in contact from half-a-day to twenty-four hours; it is then washed with water and rubbed with a brush. The first application of the enamel takes place; it consists of a frit of quartz and borax mixed with powdered felspar and clay free from iron, ground to a fine paste (100 lbs. flints, 50 calcined borax; 40 lbs. of this mixture are ground with 5 lbs. clay). The first applied coating is dried in a hot room, and then the second application is made. This consists of a mixture of 25 lbs. glass destitute of lead, 5 lbs. borax, and 4 lbs. crystals of carbonate of soda. To every 9 lbs. of this mixture are added 3½ oz. of carbonate of soda. It is mixed with hot water, dried, powdered, and the sifted powder dusted over the first coating, and dried at 212°. The pan is then exposed to a sufficient heat in a gloss kiln. These pans do not last long, in consequence often of the careless way in which they are used. They should always contain water, and should not be exposed to a higher heat than 212°.

Enargite. Spec. grav. 4·44, H 3·. Black rhombic prisms from Peru. S 32·22, As 17·59, Sb 1·61, Cu 47·20, Fe ·56, Zn 2·3, Ag ·02.

Enceladite. *Warwickite, Silicotitaniate of Iron and Magnesia.* Specific gravity 3·20, H 3½.

Dark brown rhombic prisms, with often a copper tinge, lustre metallic and vitreous. Titanic acid 28·2, silica 18·5, alumina 13·84, protoxide of iron 10·59, magnesia 22·2, lime 1·3. Occurs in limestone at Edenville, New York. By another description it possesses a specific gravity of 3·188, H 3 to 4. SiO_3 18·, Fe_2O_3 13·, TiO_3? 25·15, Al_2O_3 13·84, MgO 22·2, CaO 1·3, HO 7·35. Blue, or brownish-black rhombic prisms; lustre resinous or metallic; becomes red by heat. In magnesian limestone, Amity, New York.

Enchysiderite. A synonyme of pyroxene.

Endellionite. A synonyme of bournonite.

Endialite. From Kangerdluarsuk. Silica 53·33, zirconia 11·1, lime 9·79, soda 13·82, FeO 6·75, MnO 2·06, HCl 1·03, HO 1·8.

Endomorphism. A change of state in a plutonic rock. The endomorphic rocks are porphyry modified by slate and sandsone, claystone porphyry, and brown porphyry.

Endosmose. A term applied to express the diffusion of two fluids of unequal density into each other when placed on opposite sides of a

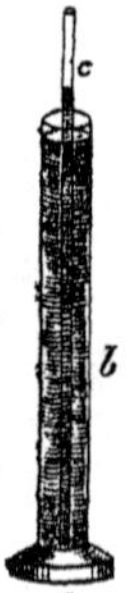

1

porous membrane; the term has, however, been restricted by some to the increase of volume produced by a fluid of less density passing from without inward to a heavier fluid. The phenomenon may be exhibited by filling a jar with water (fig. 1), and inserting in it a tube closed at the lower end with a piece of bladder, and filled to the same height with a saturated solution of common salt coloured by litmus; in the course of a few days, the fluid in the interior tube, *c*, will be found to have risen above that in the exterior jar, *b*, showing that more pure water has passed through the membrane than saline water. Another illustration is afforded by a bent tube (fig. 2), filled with a saline solution, or with pure water. One leg

3.

2.

of the tube may be longer than the other. When the extremities are covered with bladder, and the tube suspended, it will be found that the membranes become gradually concave, and that the curvature of the tube is occupied by a vacuum, or by air, proving that water has evaporated or diffused through the membranes into the air in the form of vapour. Hence there is constantly a vacuum forming in the tube. If the short end of this tube be inserted in coloured water, a current will pass from *a* to *b*, and the fluid in the tube will become entirely coloured. If a tube partially filled with salt water be closed with a membrane, and be brought in contact with one drop on the surface of water in another vessel, it will be found that the saline fluid augments in volume (fig. 3.) It is found that no fluids diffuse when the membranes are dry, and that in all cases of endosmose of this description the fluid must be in contact with the membrane. It is believed that in these phenomena we have an apt illustration of the probable nature of the ascent of sap in trees, and of the diffusion of fluids in the animal organism. For example, if we drink a glass of well water every ten minutes, until eight or ten glasses have been swallowed, if the water contained less salt than the blood, we shall find that the urine becomes nearly of the same density as the well water. If the same trial is made with water containing the same per centage of salt as exists in the blood, it will be found a very difficult matter to drink more than a glass or two of such water; while, if the salt greatly exceeds that of the blood, purging will be the result. See also DIFFUSION OF FLUIDS.

Energiatype. A term applied to a mode of producing photographic effects by the application of succinic acid and gum to paper, and afterwards washing with nitrate of silver.

Eocene. (ηως, aurora; καινος, recent). The lowest of the tertiary rocks containing a small number of shells now living—the dawn of the present state of the earth's surface.

Eoidine. $C_{24}H_{22}O_3$. The red-colouring matter of asparagus berries, similar to that of annatto.

Ephesite. Spec. grav. 3·2. White micaceous mineral. SiO_3 30·, Al_2O_3 56·45, CaO 2·11, FeO 1·, NaO 4·41, HO 3·, found near Ephesus.

Epichlorite. Spec. grav. 2·76, H 2·25. Dark leek-green masses with fatty lustre, fusing with difficulty; occurs in serpentine at Radanthal. SiO_3 40·88, Al_2O_3 10·96, Fe_2O_3 8·72, FeO 8·96, MgO 20·, CaO ·68, HO 10·18.

Epidermis, *Scarfskin*, *Cuticle*. (επι, upon; δερμα, true skin). C 50·03, H 6·8, N 17·22, O and S 24·93 (sole of the human foot, C 52·9, H 6·79, N and O 40·31, cuticle of Agave Americana). The animal epidermis consists of cells soluble in strong sulphuric acid and caustic soda. Nitrate of silver colours the cuticle black, partly by the formation of chloride of silver, and partly of sulphide of silver from the sulphur of the cuticle.

Epidote. (επιδιδωμι, I increase.) *Acanticonite*, *Achmatite*, *Arendalite*, *Bucklandite*, *Delphinite*, *Illuderite*, *Pistacite*, *Puschkinite*, *Scorza*, *Thal-*

lite, *Thulite*, *Zoisite?* Specific gravity 3·289 to 3·46, H 6·5. Massive and in right oblique prisms, the faces being elongated and streaked;

the edges so replaced with planes that the prism is 6, 8, or 10-sided; fracture uneven, translucent, lustre vitreous, inclining to pearly on perfect faces of cleavage, brittle; colour green, of various shades. BB. fuses with difficulty, and only on the thin edges, with borax into a clean globule. Silica 38·05, alumina 21·25, lime 20·20, protoxide of iron 17·3, protoxide of manganese 1·, water 2·45, 3 RO, SiO_3 2 $R_2O_3SiO_3$. Epidote was originally confounded with pyroxene; found at Arendal, Norway; Carinthia; United States.

Epiphyte. Plants which grow on the stems of others. My late pupil, Mr. John Thomson, M.A., found the composition of the Commelina Skinneri to be, water 79·64, organic matter 18·34, ash 2·02. The latter contained alkaline, sulphate, and chloride 42·72, silica, carbonate of lime, phosphate of lime and iron, &c. 59. In the Vanilla planifolia were found, water 89·06, organic matter 9·84, ash 1·1. The source of these salts is obviously the soil on which the plants grow, whether that be other trees or earth.

Epistilbite. A variety of stilbite. Specific gravity 2·2; white rhombic prisms. SiO_3 58·59, Al_2O_3 17·52, CaO 7·56, NaO 1·78, HO 14·48.

Epithelium. C 51·53, H 7·03, N 16·64, O 22·32, S 2·68. The scaly covering of the mucous membranes, of which a quantity exists in the saliva. The above is an analysis of the epithelium of the whale; almost insoluble in alcohol; slowly soluble in caustic potash, from which it is precipitated and redissolved by acetic acid; the acetic solution is precipitated by yellow prussiate of potash.

Epsom Salt. *Epsomite*, Sulphate of magnesia.

Equisalt, or *Semel Salt*, *Equioxide*. A salt or oxide containing an atom of each ingredient. Protoxide of iron FeO is a semel oxide, or equioxide.

Equisetic Acid, or aconitic acid.

Equivalents. A term introduced by Dr. Wollaston instead of atomic weight, which is of theoretic origin. But as the latter is in general use, it is a matter of little consequence which expression is employed.

Erbium, Er. The metallic base of the earth erbia which has not yet been thoroughly separated from yttria and terbia. It occurs along with these earths when they are dissolved from the minerals containing them in acids. When ammonia is added fractionally to nitrate of these earths, erbia first falls; then terbia; and lastly yttria, which is the strongest base. Erbia is dark yellow; the sulphate has a sweet taste.

Ercinite. A synonyme of harmotome.

Eremacausis. (ἠρέμα, gradually; καῦσις, combustion). The slow but incessant union in nature of the carbon and hydrogen of dead plants and animals with the oxygen of the air, to form carbonic acid and water, as exemplified in the rotting of trees, the decay of vegetables, and formation of mould. The presence of water and a suitable heat are required in the oxidizing process of decay, as in putrefaction and fermentation. Perfect dryness, or a temperature below freezing, puts an end to slow combustion. Wet hay or wood, if enclosed for some time, will be found to remove the oxygen from the air in which it is in contact, so that a taper placed in the atmosphere will be extinguished in it, just as if we had burned either of these substances in a close vessel, the resulting products being the same in both cases. The destruction of the colouring matter in the fibres of cotton in the process of bleaching is also an example of eremacausis.

Eremite. See MONAZITE.

Eretria. An unknown ancient substance.

Ergot. *Spurred Rye*, *Secale Cornutum*. A fungus growing on rye. It contains 34½ per cent. of a colourless oil, fungine 4 to 7¾ per cent. ergotine 1¼ per cent. It is extensively used in obstetrics, from the expulsive influence it is considered to possess on the uterus.

Ergotine. C 76·32, H 5·64, O 18·03. Bitter red powder, soluble in alcohol, sulphuric and acetic acids, and caustic potash; insoluble in water and ether, obtained by extracting oils from ergot by ether, taking up the ergotine by alcohol, and precipitating by water.

Ericinole. $C_{20}H_{15}O_2$. An oil from heaths by decoction; precipitating by lead, and decomposing by sulphuric acid.

Ericoline. $C_{16}H_{15}O_{21}$. A substance from heaths and ledum palustre.

Erinite of Thomson. (*Erin*, Ireland). Specific gravity 2·04, H 1·75. Yellowish-red, compact, fine-grained masses, in an amygdaloidal rock, four miles east of the Giant's Causeway. B.B. whitens, but does not fuse; with carbonate of soda into a blebby glass. Silica 47·03, Al_2O_3 18·46, CaO 1·, FeO 6·36, water 25·28, NaCl ·9.

Erinite of Turner. See PENTARSENIATE OF COPPER.

Erlanite. A variety of garnet.

Erucasic Acid. $C_{44}H_{42}O_4$. An acid in needles, from the oil of white and black mustard seeds.

Erucasic Acid. $C_{44}H_{41}O_3HO$. Needles, fusing at 93°, from white mustard oil.

Erucine. A whitish-yellow substance, insoluble in water and ammonia; soluble in alcohol, ether, carbide of sulphur; obtained from white mustard oil.

Erythrelesic Acid. *Erythretinic Acid.* Mica-like plates, obtained by boiling barytes with erythric acid.

Erythric Acid. *Orcino-lecanoric Acid. Erythrine.* $C_{34}H_{19}O_{15}$ 3 HO. Snow-white scales, silky and stelliform crystals, by boiling water and precipitating by alcohol, from Parmelia roccella and Roccella tinctoria, or by treatment with lime and acid. When boiled it is converted into carbonic acid and orcine; boiled with alcohol it yields orcine and lecanoric acid; hence it is probably a coupled acid; it unites with 4 atoms oxide of lead; when boiled with lime it yields picroerythrine and erythrelesic acid; when boiled with alcohol it gives lecanorate of ethyle; when heated it yields a sublimate of orcine; with ammonia it affords a red colour.

Erythrine, or triarseniate of cobalt.

Erythrite. A flesh-coloured mineral in the Kilpatrick hills, and in the amygdaloid at Bishopton. Spec. grav. 2·541; hardness about the same as that of felspar; texture compact; silica 67·9, alumina 18·, Fe_2O_3 2·7, lime 1·, MgO 3·25, KO 7·5, HO 1·. It seems to be allied to felspar.

Erythrobetic Acid. Crystals from the aqueous extract of dry beetroot, by acetate of lead.

Erythrogene. A name applied to two substances, one a crystalline matter from the bile in a peculiar case; the other the colouring matter of flowers rendered red by acids.

Erythroglucine. See PSEUDO-ORCINE.

Erythroleic Acid. $C_{26}H_{22}O_8$.

Erythroleine. $C_{26}H_{22}O_4$. Red oily fluid; but slightly soluble in water; somewhat soluble in alcohol and ether, with a red colour; soluble in ammonia, with a purple colour; obtained by heating the residue of litmus after extraction with water, with chlorohydric acid; the solution being filtered, a red mass remains on the filter, which is to be treated with alcohol and ether.

Erythrolitmine. $C_{26}H_{22}O_{12}$. Red powder, slightly soluble in water, very soluble in alcohol, with a red colour, yielding red grains from a hot saturated solution; forms a blue solution with potash; the alcoholic solution forms purple precipitates with metallic salts. Obtained in the preparation of erythroleine from the portion insoluble in ether.

Erythromannite. *Eryglucina.* $C_{22}H_{28}O_{22}$. Crystalline body by boiling picroerythrine with lime or barytes water. With nitric acid it forms an explosive compound, $C_{22}H_{18}O_{12}NO_5$.

Erythrophylle The red-colouring principle of flowers, obtained by extracting with alcohol; the solution precipitated by acetate of lead; the filtered liquor freed from lead by sulphohydric acid, evaporated, and a reddish-brown matter obtained.

Erythropicrine. *Picroerythrine, Amarythrine.* $C_{34}H_{24}O_{20}$. By boiling erythric acid with water.

Erythroprotide. $C_{13}H_8N_2O_5$. A doubtful body obtained by the action of alkalies on albuminous substances.

Erythroretine. Yellow resinous body by ether from rhubarb root.

Erythrose. A yellow body from rhubarb root by nitric acid.

Erythrosic Acid. Produced by alkalies and erythrose.

Erythrozyme. $C_{56}H_{34}N_2O_{40}$ 4 CaO. A reddish-brown matter by adding alcohol to an aqueous decoction of madder. When added to a solution of rubiane, the liquor becomes a jelly. The erythrozyme here acts as a ferment, and converts the rubiane into alizarine, verantine, rubiretine, rubiafine, rubiagine, rubiadipine, and sugar. When precipitated by alcohol it is united to lime; in water it putrifies and evolves gas.

Erythryline. $C_{22}H_{16}O_6$. White insoluble powder by alcohol from roccella tinctoria.

Erytraraine, or chloride of cacodyle.

Eschynite. See ÆSCHYNITE.

Esculine. *Esculic Acid.* $C_{52}H_{46}O_{24}$. Crystalline grains, insoluble in water, soluble in alcohol, by heating saponine from horse chestnuts with acids.

Esmarkite, or Datholite.

Esmarkite. Spec. grav. 2·709, H 3 to 5. Large prisms, covered with a glittering coat, from Brevig, Norway, in granite, accompanied by chlorite. Silica 46, alumina 32·1, magnesia 10·3, water 5·5, FeO 3·8, MnO ·4, PbO, CuO, CoO TiO_3 ·5

Essences. *Essential Oils.* The odours of plants usually depend on the presence of volatile oils peculiar to each plant. They are distinguished from fixed oils by the rapidity with which they evaporate when a drop is allowed to fall on blotting paper, and held before the fire. See VOLATILE OILS.

Essonite. *Cinnamon Stone, Canelstein.* Lime garnet. 3 CaO SiO_3,$Al_2O_3SiO_3$. Spec. grav. 3·631, H 6·5. Granular masses from Ceylon; colour between hyacinth-red and orange; lustre between resinous and vitreous; semitransparent to translucent; SiO_3 39·82, CaO 30·57, Al_2O_3 20·14, FeO 9·46. B.B. fuses into a greenish glass.

Ester. A name given by Gmelin to ox-acid ethers. *Vinester* is an ethyle ether, and *formester* a formyle or methyle ether.

Etching. The process of engraving on glass by fluohydric acid, or on copper by nitric acid.

Ethal. *Cetylic Alcohol. Hydrous Oxide of Cetyle.* $C_{32}H_{33}OHO$. Brilliant plates. F.P. 11.8°. Volatile; soluble in alcohol, insoluble in water; decomposed by nitric acid; unites with sulphuric acid; obtained by fusing in a silver basin spermaceti with hydrate of potash; the oil which floats is heated with potash, water, and acid; the mixed ethal and cetylic acid are digested in milk of lime, and the ethal separated by cold alcohol from the cetylate of lime.

Ethalic Acid. See CETYLIC ACID.

Ether. A term applied to a series of volatile compounds, consisting of oxide of ethyle as a base united to an acid, either inorganic or organic. They are usually prepared by boiling the acid with alcohol, with the addition of sulphuric acid in the organic ethers, when the acid replaces the water of the alcohol. Although they are termed ethers, they may be viewed as salts composed of an organic radical and an acid.

Acetic ether,	$C_4H_5O,\ C_4H_3O_3$
Chloracetic,	$C_4H_5O, C_4Cl_3O_3$
Perchloracetic,	$C_8Cl_8O_4$
Aconitic,	$C_3H_6O_4$
Adipic,	$C_{10}H_9O_4$
Allophanic,	$C_8N_2H_6O_6$
Arsenic,	C_4H_5O,HO,AsO_5
Anisic,	$C_{20}H_{12}O_6$
Azoleic,	$C_{18}H_{18}O_4$
Benzoic,	$C_{18}H_{10}O_4$
Benzoic, chloro-,	$C_{18}H_8Cl_3O_3$
Triboracic,	$3\ C_4H_5O,BO_3$,
Biboracic,	$C_4H_5O2\ BO_3$
Bromanisic,	$C_{20}H_{11}BrO_6$
Bromohydric,	C_4H_5Br
Bromic,	C_4H_4O,BrO_5
Butyric,	$C_{12}H_{12}O_4$
Butyroleic,	$C_{33}H_{30}O_5$
Camphoric,	$C_4H_5O,\ C_{10}H_7O_3$
Capric,	$C_{16}H_{16}O_4$
Carbonic,	$C_5H_5O_3$
Carbonic, bichloro-,	$C_5H_3Cl_2O_3$
Carbonic, quintochloro-,	$C_5Cl_5O_3$
Carbamic,	$C_6H_7NO_4$
Chlorocyanoformic,	$C_8H_6ClNO_4$?
Chlorohydric,	C_4H_5Cl
— semelchloro-,	$C_4H_4Cl_2$
— bichloro-,	$C_4H_3Cl_3$
— terchloro-,	$C_4H_2Cl_4$
— quaterchloro-,	C_4HCl_5
— quintochloro-,	C_4Cl_6
Chlorocamphoric,	$C_{14}H_{10}Cl_2O_4$
Chloro-formic,	$C_6H_5ClO_4$
Chloromesitylic,	C_6H_5Cl
Chloroniceic,	$C_{18}H_9ClO_4$
Chloronitric, azotized,	$C_{12}H_3ClNO_7, C_4H_5O$
Chloropyromucic,	$C_{14}H_8Cl_4O_6$
Chlorosuberic,	?
Chlorosulphuretted,	C_4H_3ClSO
Chloroxalic,	$C_{44}Cl_5O,C_2O_3$
Chloroxicarbonic,	$C_6H_5ClO_4$
Chrysanisic,	$C_{18}H_9N_3O_{14}$
Cinnamic,	$C_{22}H_{12}O_4$
Citracic,	$C_9H_7O_4$
Citric,	$C_8H_7O_5$
Cocostearic,	$C_{31}H_{34}1O$
Cuminic,	$C_{24}H_{16}O_4$
Cyanohydric,	C_6H_5N
Cyanic,	C_4H_5OCyO
Cyanuric,	$C_{18}H_{15}N_3O_6$
Elaidic,	$C_{39}H_{38}O_4$
Ether,	C_4H_5O

Fluohydric,	C_4H_5F
Formic,	$C_6H_6O_4$
— bichloro-,	$C_6\ H_4Cl_2O_4$
— sexchloro-,	$C_6Cl_6O_4$
Fumaric,	$C_8H_6O_4$
Hippuric,	$C_{22}H_{13}NO_6$
Hyponitrous,	$C_4H_5NO_4$
Indigotic,	$C_{18}H_9NO_{10}$
Iodohydric,	C_4H_5I
Iodomesitylic,	
Itacic,	$C_9H_7O_4$
Lactic,	$C_{20}H_{20}O_{12}$
Leconoric,	$C_{22}H_{13}O_9$
Margaric,	$C_{38}H_{39}\ O_4$
Mesitic,	C_6H_5O
Metacetic,	$C_{10}H_{10}O_4$
Methylic,	C_2H_3O
— semelchloro-,	C_2H_2OCl
— bichloro-,	C_2HOCl_2
— terchloro-,	C_2Cl_3O
Mucic,	$2\ C_4H_5O,C_{12}H_8O_{14}$
Myristic,	$C_{60}H_{60}O_8$
Nitroanisic,	$C_{20}H_{11}NO_{10}$
Nitrochloroniceic,	$C_{16}H_8Cl\ NO_8$
Nitrocinnamic,	$C_{22}H_{11}NO_8$
Nitric,	$C_4H_5ONO_5$
Nitrous,	$C_4H_5ONO_3$
Œnanthic,	$C_4H_5OC_7H_{13}O_2$
Œnanthylic,	$C_{18}H_{18}O_4$
Oleic,	$C_{48}H_{44}O_5$
Opianic,	$C_4H_5OC_{10}H_9O_9$
Oxamic,	$C_8H_7O_6N$
Oxalic,	$C_6H_5O_4$
Phosphoric,	$C_4H_5O_2HO\ PO_5$
Pyrocitric,	$C_9H_7O_4$
Pyromucic,	$C_{14}H_8O_6$
Pyrotartaric,	$C_{18}H_{16}O_8$
Racemic,	$C_6H_{11}O_{13}$
Salicylic,	$C_{18}H_{10}O_6$
— bibromo-,	$C_{18}H_8Br_2O_6$
— bichloro-,	$C_{18}H_8Cl_2O_6$
— binitro-,	$C_{18}H_8N_2O_{14}$
Silicic,	$C_4H_5OSiO_3$
Disilicic,	$2\ C_4H_5O,SiO_3$
Bisilicic,	$C_4H_5O,2SiO_3$
Stearic,	$C_{72}H_{72}O_7$
Stearophanic,	$C_{39}H_{39}O_4$
Suberic,	$C_{12}H_{11}O_4$
Succinic,	$C_8H_7O_4$
— perchloro-,	$C_{16}HCl_{13}O_8$
Sulphohydric,	C_4H_5S
— quaterchloro-,	C_4HCl_4S
— perchloro-,	C_2Cl_3S
Sulphocarbonic,	$C_4H_5O\ 2\ CS_2HO$
Sulphuretted,	$C_4H_3S_2O$
Sulphurous,	C_4H_5O,SO_2
Sulphuric,	$C_4H_5O\ 2\ SO_3HO$
— common,	C_4H_5O
Tartaric,	$C_4H_5O,C_8H_4O_{10}HO$
Tellurohydric,	C_4H_5Te,
Valerianic,	$C_{14}H_{14}O_4$
Veratric,	$C_{22}H_{14}O_8$

Ether, Sulphuric Ether. C_4H_5O, or C_4H_4HO. Ether is best obtained by what is termed a *continuous* process. Alcohol is used of specific gravity ·822 or 96 per cent. and is obtained by digesting proof spirits upon an equal weight of salt of tartar. Five parts by weight of this alcohol are mixed with 9 parts SO_3 in a copper or cast iron vessel, surrounded by cold water, and the mixture is then placed in a tubulated glass retort, which it fills one-half. The distillation is conducted by the heat of a sand pot. A glass tube is fixed by a cork in the tubulure of the retort, of which the extremity within the retort is drawn to a capillary point, and dips one inch in the liquor. It communicates with a vessel of alcohol by a stopcock, by which the supply of alcohol is regulated. Liebig says that when the process is properly conducted there is no product besides ether and water. The acid is not affected, and may be employed for making an indefinite quantity of ether. The effect of this operation has been to deprive alcohol of water.

Ether, ... C_4H_5O = EO
Alcohol, ... $C_4H_5O\ HO = EO,HO$

Process for Ether.—A mixture of 9 parts of oil of vitriol (SO_3HO), and 5 parts of alcohol of 85 per cent. heated to the boiling point, produce 1 atom of sulphate of ethyle, and 3 water. If we heat the mixture above 284°, the whole is decomposed into EO + Aq, which distils over, leaving SO_3 and water behind. It is remarkable that the escape of the ether is not facilitated by the addition of strong sulphuric acid, while more alcohol, when added, distils over without decomposition. The water in the mixture of ether is required, and seems to act as a stronger base, and thus displaces the oxide of ethyle in combination with the acid, setting it free as ether. When a great excess of acid is present, the decomposition of the $EOSO_3$ requires a temperature of 320°, which is supposed by Liebig to be due to the action of the SO_3 upon the elements of the EO itself. He explains the production of olefiant gas thus:—

1 atom $C_4H_5OSO_3 + HOSO_3$ become					
2 atoms sulph. acid,	S_2			O_4
½ atom olefiant gas,		C_2	H_2	
3 atoms water,			H_3	O_3
2 atoms carbon,		C_2		
	S_2	C_4	H_5	O_7

Mitscherlich has, however, suggested that ether and alcohol have different bases, for this reason, that, if we act upon alcohol with lime, oxide of potassium, or chloride of calcium, none of these bodies are able to remove the HO and convert the alcohol into ether; and further, that if ether were a base, the compound ethers ought principally to be formed by the solution of acids in ether, which never happens. If ether is added to SO_3, no sulphate of oxide of ethyle is formed till the temperature is raised to 280°, and maintained there for some time. Ether, again, is not converted into alcohol by remaining in contact with water. Mitscherlich is inclined to regard alcohol as a kind of compound ether, which contains water instead of an acid, and which is retained by a feeble affinity, but in which this cannot be removed by mechanical action, except in a certain condition. This condition is fulfilled for alcohol and compound ethers by different substances, which are more electro-negative, and which act by contact. Thus, fluoboric acid, some chlorides, as chloride of zinc, sulphuric and phosphoric acids, have this action. Ether is a limpid and colourless liquor. It evaporates rapidly, and produces cold. Boils at 96°. Used in cupping to produce a vacuum. Burns with a white flame. Ether may be separated from alcohol by washing it with water. Ether dissolves fats, and is much used for this purpose in chemical analyses. It acts as a stupefacient when breathed. Spec. grav. of ether ·7154, of vapour 2·586; non-conductor of electricity; it dissolves iodine, bromine, ammonia, potash, sulphur, phosphorus; water dissolves $\frac{1}{10}$ to $\frac{1}{14}$ its vol. of ether.

Etherification (Graham).—When 1 volume sulphuric acid and 4 to 8 volumes alcohol (83 per cent.) are heated in a sealed glass tube to 320° F. ether is formed by a polymerizing action, as it may be termed, of the acid. The hydrocarbon does not unite with the acid, but is merely increased in atomic weight and gaseous density, without any further derangement of composition. Sulphuric acid has a similar action on oil of turpentine, terebene and colophene being formed, one of which has a higher boiling and vapour point than oil of turpentine. Chloride of zinc has an analogous action upon alcohol (Mitscherlich). Bisulphate of soda may be substituted for sulphuric acid to produce ether. The formation of sulphovinic acid appears to impede etherization. According to another view, when sulphuric acid acts upon alcohol, half of the hydrogen of the water of the acid (H_2OSO_3) is replaced by carburetted hydrogen (C_2H_5), and is converted into sulpho-vinic acid; the water being viewed as formed of 2 atoms H, and 1 atom O corresponding with the volumes.

Sulphuric acid, $\begin{matrix}H\\H\end{matrix}SO_4 = \begin{matrix}H\\C_2H_5\end{matrix}SO_4 =$ Sulphovinic acid.

Then the sulphovinic acid, acting further on the alcohol, has its atom of hydrogen replaced by carburetted hydrogen, while sulphuric acid and water are formed.

Sulphovinic acid, $\begin{matrix}H\\C_2H_5\end{matrix}SO_4$ and $\begin{matrix}C_2H_5\\H\end{matrix}O$ = Alcohol, become—

$$\text{Ether, } \begin{matrix} C_2H_5 \\ C_2H_5 \end{matrix} O \text{ and } \begin{matrix} H \\ H \end{matrix} SO_4 = \text{Sulphuric acid.}$$

Water, alcohol, and ether are thus ternary homologous bodies, containing the same quantity of oxygen, and differing only in the form of the two molecules which are united to it.

Water.	Alcohol.	Ether.
$\begin{matrix} H \\ H \end{matrix} O$	$\begin{matrix} H \\ C_2H_5 \end{matrix} O$	$\begin{matrix} C_2H_5 \\ C_2H_5 \end{matrix} O$

(Williamson.)

To Purify Ether.—Agitate it with double its volume of water. The supernatant liquor is poured upon quicklime, where it should remain some days. The third of it distilled affords perfectly pure ether.

Etherine. *Camphor of Oil of Wine.* C_4H_4. Sp. grav. ·980; F.P. 230°, B.P. 500°. White transparent prisms; soluble in alcohol and ether; tasteless; separates from etherole by exposure to cold; isomeric with etherole and olefiant gas.

Etherole. *Ethene. Light Oil of Wine.* C_4H_4. Spec. grav. ·917; B.P. 536°. Solidifies at —31°. Colourless or yellowish oil, with a peculiar odour; soluble in alcohol and ether; obtained by distilling equal parts of sulphovinate of lime and caustic lime, and removing alcohol from the distilled fluid in vacuo, over oil of vitriol. This is heavy, or *sweet oil of wine* (sulphate of oxide of ethyle and of etherole). $C_4H_5OSO_3$, $C_4H_4SO_3$. A colourless oil, of spec. grav. 1·135; B.P. 536. When boiled with water, acids, and alkalies, it is decomposed into sulphovinic acid, etherole, and etherine. The two latter swim on the surface, and by exposing them to a cold mixture the etherine crystallizes, and the etherole remains fluid.

Ethionic Acid. C_4H_5O, $4SO_3$. By the action of anhydrous sulphuric acid on alcohol, or by the absorption of olefiant gas by sulphuric acid.

Ethiops Martialis. A mixture of protoxide and sesquioxide of iron.

Ethiops Mineral. *Ethiops Narcoticus.* A sulphide of mercury, by triturating sulphur and mercury.

Ethiops per se. Black oxide or dinoxide of mercury, formed by agitating mercury for some weeks or months.

Ethogene, or Nitride of Boron. BN? B = 43·76, N = 56·24? White light granular powder by igniting in a covered crucible 1 dry borax with 2 dry salammoniac, and boiling with water acidulated with HCl.

Ethylacemide. N,(C_4H_5,$C_4H_3O_2$,H.) An amide produced by the action of cyanic ether on protohydrous acetic acid. When cyanic ether and anhydrous acetic acid are enclosed in a sealed tube, and heated to 356°, there is formed *Ethyldiacetamide*, N($C_4H_5 2 C_4H_3O_2$), both being formed on the type of ammonia.

Ethylamine, *Ethylamide, Ethammine, Ethyliaque.* $C_4H_5ONH_2$. B.P. 68°. Spec grav. ·69, of vapour 1·5568. Volatile fluid, with the smell of ammonia; soluble in water; it occurs in coal tar, and in animal oil produced by the distillation of bones, and may be probably extricated in putrefaction; formed by heating cyanic ether, cyanuric ether, or ethlyo-urea with potash (C_4H_5O, C_2 NO, KO HO = $KO2CO_2$, C_4H_7N), or by acting on bromide of ethyle with ammonia. This substance is ammonia, in which the third atom of hydrogen is replaced by ethyle. It has been, doubtless, often obtained, but has been mistaken for ammonia from the similarity of their odours. It is an amide of ethyle. It acts on the chloride of platinum, and forms 2 platinum bases, H,E, Pt,N, and NH_2E,NH,E,Pt, in the second the atom is doubled.

Ethyl-amyl-phenylamine. *Ethyl-amyl-aniline.* E,Ay,$C_{12}H_5$,N. B.P. 503°·6. Colourless oil by the action of bromide of ethyle on amyl-aniline, or by bromide of amyle on ethyl-phenylamine.

Ethyl-aniline, or *Ethyl-phenylamine.* E, $C_{12}H_5$,HN. B.P. 399°. Spec. grav. ·954. Colourless liquid, refracting light; when dissolved in acids, turning fir wood yellow, with a smell of aniline; obtained by decomposing the hydrobromate of ethyl-aniline with potash; this salt being formed in 4-sided plates by heating aniline with bromide of ethyle.

Ethyle. (Ether and ὕλη, matter). C_4H_5 = E or Æ. The basis of the ethers. A colourless gas. Spec. grav. 2·00394; with 2½ atmospheric pressure it becomes a mobile fluid; formed by acting on iodide of ethyle with zinc. It is not certain that this is the base of the ethers, but it is at least isomeric with it.

Ethyle, Binoxalate of. *Oxalovinic Acid.* EOC_2O_3, HOC_2O_3. Obtained by adding a solution of potash in absolute alcohol, to a solution of oxalic ether in absolute alcohol; the quantity of potash being sufficient to decompose half of the oxalic acid, there is formed the potash oxalate of ethyle, $C_4H_5OC_2O_3$, KOC_2O_3. When fluosilicic acid is added to this salt the potash is removed, and oxalovinic acid is formed.

Ethyle, Bisulphate of Oxide of. *Sulphovinic Acid.* $C_4H_5OSO_3$, HO SO_3. Sp. grav. 1·319. A colourless oily fluid, with an acid taste, obtained by boiling alcohol and sulphuric acid together, or by passing the vapour of ether into oil of vitriol. When we boil alcohol and sulphuric acid together (EO HO, and $2SO_3$), we have formed a double compound in solution, of sulphate of ethyle and sulphate of water (EO SO, HO SO_3); and the proof of its formation is the fact, that when we saturate it with carbonate of lime or barytes, there is formed lime or barytes sulphate of ethyle ($EOSO_3$, $BaOSO_3$), where the basic water is replaced by barytes, and which is a salt crystallizing in shining rhombic prisms. The acid may be separated from the barytes, lime, or lead salts, by adding sulphuric acid to the first, oxalic acid to the second, and sulphohydric acid to the last, and evaporating the filtered liquor in vacuo over oil of vitriol.

Ethyle, Bromide of. E Br. Spec. grav. 2·164, of vapour 6·485. Colourless oil. B. P. 264°; by a similar process to that for the iodide.

Ethyle, Carbonate of. *Carbonic Ether.* C_4H_5O, CO_2. Spec. grav. ·975, of vapour 4·09; B. P. 260°. Clear fluid; burning with a blue flame; insoluble in water; soluble in alcohol and ether; obtained by acting on oxalic ether with sodium.

Ethyle-chloraniline. $E, C_{12}H_5Cl, N$. Yellow oil, by acting for some days on chloraniline with bromide of ethyle; a hydrobromate of this base is formed; by heating to 212° for two days, 2 atoms ethyle are united, and di-ethyl-chloraniline is formed, $2\ E,\ C_{12}H_4\ Cl\ N$.

Ethyle, Chloride of. E Cl. Colourless fluid. Spec. grav. ·874; B.P. 52°; obtained by saturating ether with chlorohydric acid gas, and distilling in a steam bath.

Ethyle, Cyanate of. *Cyanic Ether.* EO CyO. Volatile fluid. Spec. grav. of vapour 2·4, by distilling sulphovinate of potash, and cyanate of potash.

Ethyle, Cyanide of. E Cy. Volatile fluid, by distilling sulphovinate of potash, with yellow prussiate of potash.

Ethyle, Cyanurate of. *Cyanuric Ether.* $3\ EO, Cy_3O_3$. Crystals; F.P. 185°; B.P. 529°. Spec. grav. of vapour 7·4; soluble in alcohol and ether; formed by distilling sulphovinate of potash with tricyanurate of potash.

Ethyle, Hydrous Oxide of. See ALCOHOL.

Ethyle, Iodide of. EI. Clear neutral fluid. Sp. grav. 1·9206; of vapour 5·417. B.P. 161°; insoluble in water, not inflammable; formed by distilling alcohol, iodine, and phosphorus. The *biniodide* is in colourless needles.

Ethyle, Nitrate of Oxide of. *Nitric Ether.* $C_4H_5O\ NO_5$. Spec. grav. 1·112; B. P. 185°. An agreeable smelling fluid with a sweetish-bitter taste, burns explosively with a white flame; soluble in all proportions of alcohol; insoluble in water; dissolves iodine; decomposed by a solution of potash in alcohol into alcohol and nitre; obtained by distilling 2 volumes of alcohol with 1 vol. nitric acid (1·4), with 25 grains nitrate of urea by a gentle heat, removing the first fluid which comes over that contains alcohol; the product is washed with potash and then with water, and, lastly, digested with chloride of calcium and rectified.

Ethyle, Nitrite of Oxide of. *Nitrous Ether.* $C_4H_5O\ NO_3$. Sp. grav. ·947, of vapour 2·627; B.P. 16·4 C (61½°). Light yellow ethereal fluid, with an apple odour; becomes acid by keeping from the separation of nitric acid; burns with a white flame; it is decomposed by oil of vitriol, by sulphide of ammonium, into alcohol, ammonia, water, and sulphur. Nitrous ether is prepared by passing a current of nitrous acid vapour, by the action of nitric acid on starch (10 acid 1 starch), into dilute alcohol. The product is heated with water and chloride of calcium. Nitrous ether is soluble in 48 water, and in alcohol; *sweet spirits of nitre* is nitrous ether dissolved in alcohol; but as usually sold and prepared by the distillation of alcohol and nitric acid, it contains aldehyde and other impurities. It is used as an antispasmodic and diuretic.

Ethyle, Oxalate of. *Oxalic Ether.* C_4H_5O, C_2O_3. Spec. grav. 1·093; B.P. 364. Aromatic colourless fluid; remains permanent if pure, but if an excess of alcohol or oxalic acid be present, oxalic acid is deposited; it is formed by distilling 4 parts strong alcohol, 4 quadroxalate of potash, 5 sulphuric acid, washing the ether with water, and rectifying.

Ethyle, Oxalate of and Oxamide. $C_8H_7NO_6$. F.P. 212°; distils at 430°; soluble in water and alcohol; by ammonia on an alcoholic solution of oxalic ether.

Ethyle, Oxamate of. *Oxamic Ether.* $EO\ C_2O_3, AdC_2O_2$. Pearly plates; obtained by adding ammonia to oxalic ether; oxamate first falls, and the fluid yields oxamic ether.

Ethyle, Oxide of. See ETHER.

Ethyle, Silicates of. $3C_4H_5O\ SiO_3$, or C_4H_5O, SiO_3, or $2C_4H_5O\ SiO_3$. Spec. grav. ·932, of vapour 7·18; B. P. 324°½. Colourless neutral ethereal fluid, with a peppery taste; soluble in all proportions in alcohol and ether; obtained by pouring cautiously absolute alcohol into chloride of silicon. When the alcohol somewhat exceeds that of the chloride, no further action occurs. On distilling, chlorohydric ether first passes; then the greater part of the fluid distils over at 320° to 338°. This, when rectified, is silicate of ethyle. A *Bisilicate*, $C_4H_5O\ 2SiO_3$, or $C_4H_4O\ SiO_3$, a clear fluid, of spec. grav. 1·035; distils over after the separation of the preceding, between 338° and 572°.

Ethyle, Sulphate of Oxide of. $C_4H_5O\ SO_3$. A neutral oil, capable of being distilled, but decomposed by over heating; its taste resembles peppermint; it is formed by bringing anhydrous sulphuric acid in contact with ether.

Ethyle, Sulphide of. E S. Colourless fluid. B.P. 167°, with the smell of garlic; obtained by the reaction of chloride of ethyle, on sulphide of potassium.

Ethyle, Sulphocarbonate of Oxide of. *Oxide of Xantheline.* $HO\ EO_2CS_2$. Spec. grav. 1·0703; B.P. 415°; insol. in water; sol. in alcohol and ether. In a solution of potash in alcohol it changes into carbonate of potash and mercaptan; formed by heating *potash sulphocarbonate of ethyle* with iodine, filtering and distilling. The residue is shaken with water, the separating oil treated with chloride of calcium. This potash salt is produced by dissolving caustic potash in alcohol to saturation, and adding bisulphide of carbon, until the alkaline action disappears; the fluid congeals into a crystalline mass.

Ethyle, Sulphocarbonate of sulphide of. ES CS_2. Yellowish oil with the smell of garlic. B.P. 320°; burning with a blue flame; taste sweet and aromatic; formed by passing the vapour of chloride of ethyle into solution of sulphide of potassium in alcohol until all the chloride of potassium separates. The addition of water separates the oil, from which bisulphide of carbon is removed, by shaking it with a solution of sulphide of potassium in water.

Ethyle, Sulphohydride of sulphide of. *Mercaptan. Sulphide of ethyle and hydrogen.* E S, H S. Mobile colourless fluid; B.P. 97°. Spec. grav. ·842. Alliaceous smell. It is formed from alcohol by substituting sulphur for oxygen, by distilling sulphovinate of lime with a saturated solution of sulphohydride of potassium. The distilled fluid is digested with oxide of mercury and chloride of calcium, which remove sulphur and water. It acts on many oxides, the metal replacing the hydrogen of the sulphohydric acid, and forming compounds with the type E S, M S. According to Zeise, the basis of this compound is E S_2, or mercaptan (bisulphide of ethyle), which forms with metals *mercaptides.*

Ethyle, Triborate of. $3C_4H_5O$, BO_3. Sp. grav. ·871, of vapour 5·14; B.P. 246°. Smell agreeable, taste hot and bitter; burns with a green flame; soluble in water, ether, and alcohol; obtained by passing chlorine through a heated mixture of boracic acid and charcoal into a flask containing absolute alcohol, kept cold. The ether swims on the surface. It is distilled with some drops of alcohol; alcohol passes over first, and then the ether. A biborate is formed by heating and distilling absolute alcohol and boracic acid at 230°, treating the residue when cold with ether, and distilling the solution under 400°. There remains a yellowish, thick fluid, which gives out thick fumes at 390°, and decomposes at 570°.

Ethyle, Triphosphate of. *Phosphovinic Acid.* C_4H_5O2HO, PO_5. An acid fluid; soluble in water, ether, and alcohol; obtained by heating phosphoric acid and alcohol, saturating with carbonate of barytes and precipitating th barytes by sulphuric acid.

Ethyl-methyl-amine. E,M,H,N? By bromide of ethyle on methylamine.

Ethyl-methyl-aniline. E,Me,$C_{12}H_5N$. An oil by the action of iodide of methyle on ethylaniline.

Ethylocodeine. $C_{36}H_{20}C_4H_5NO_6$. A base supposed to exist in union with iodohydric acid, by the action of iodide of ethyle on codeine in a sealed tube.

Ethylo-comenic Acid. HO,E,$C_{12}H_2O_8$.

Ethylo-meconic Acid. 2HO,E,$C_{14}HO_{11}$.

Ethylomorphine. $C_{34}H_{18}C_4H_5NO_6$. A base supposed to exist in union with hydriodic acid, by the action of iodide of ethyle on morphine in a sealed tube.

Ethylo-narcotine. $C_{46}NH_{25}O_{14}$. Occurs in opium.

Ethylonitraniline. C_4H_5,$C_{12}H_5$,NO_4,N. Yellow needles by acting on nitraniline with bromide of ethyle, and treating the hydrobromate formed with potash.

Ethylo-propylamine. E,Pr,H,N.

Ethylo-stibyle. C_4H_5,Sb. A base obtained in the form of ethylostibylic acid, when stibethyle is slowly oxidized.

Ethyl-oxamide. $C_6H_6NO_2$, or oxamide in which 1 atom hydrogen is replaced by ethyle. Crystallizes; soluble; formed by acting on oxalic ether with ethylamine.

Ethyl-propyl-amyl-amine. E,Pr,Ayl,N.

Ethyl-urea. $C_6N_2H_8O_2$. Obtained by heating mixed solutions of cyanate of potash and sulphate of ethylamine.

Ethyl-urethane. $C_{10}H_{11}NO_4$. An amide by the action of cyanic ether on alcohol.

Euchlorine. A green gas, probably a mixture of quaterchlorous acid (ClO_4), and free chlorine or a binoxide, possessing bleaching powers; exploding when touched with a hot wire; formed by heating gently in a tube 1 chlorate of potash, 2 chlorohydric acid, and 2 water.

Euchroic Acid. (εὔχροος, beautiful). *Euchronic Acid.* $C_{12}NO_6$. 4-sided prisms, tasting like cream of tartar, obtained by heating mellate of ammonia in powder at 311°, till it becomes a yellow powder; water takes up euchroate of ammonia and leaves paramide; muriatic or nitric acid will remove the ammonia, and the acid falls as a white powder, which is to be purified by crystallization from boiling water.

Euchroite, or septahydrous diarseniate of copper.

Euchrone. A fine purple substance by zinc, or protoxide of iron and euchroic acid.

Euchysiderite. A synonyme of pyroxene, probably.

Euclase. Spec. grav. 2·907 to 3·098, H 7·5. Mountain-green, passing into blue and white, right oblique angled prisms of 130° 52′; fracture conchoidal, easily effected (hence the name *easily broken*); lustre vitreous, splendent, transparent. B.B. fuses into a white enamel, dissolves slowly in borax into a colourless glass, a white bead with soda, with salt of phosphorus effervesces, leaving a silica skeleton. Silica 43·22, alumina 30·56, glucina 21·78, Fe_2O_3 2·22, SnO_2 ·7; 2 Al_2O_3,SiO_3, 2 $G_2O_3SiO_3$. Found in Peru, and Villa Rica, Brazil.

Eucolite. Brown kidney-shaped mass, resembling Wöhlerite. SiO_3 47·85, Zr_2O_3 14·05, Fe_2O_3 8·24, CaO 12·06, CeO 2·98, NaO 12·81, MnO 1·94, HO ·94. Probably a Wöhlerite in which zirconia is replaced by Fe_2O_3; from near Brevig and Fredericksvarn, Norway.

Eudialite. Spec. grav. 2·9036, H 6. Red

rhomboidal 12-hedrons; structure foliated; fracture imperfect, conchoidal, or splintery; lustre vitreous. B.B. fuses into a slag; gelatinizes in acids. SiO_3 53·325, Zr_2O_3 11·102, CaO 9·785, NaO 13·822, FeO 6·754, MnO 2·062, HCl 1·034, HO 1·801; from Kangerluarzue, in Greenland.

Eudiometer. (εὐδιος, good; μετρον, measurer of goodness of air). This instrument is an accurately graduated tube, first used by Priestley and Fontana. Priestley introduced a definite quantity of air, and mixed with it a certain amount of binoxide of nitrogen which united with the oxygen, but this method was found to be incorrect, in consequence of the formation of different nitrogen compounds. Perhaps the most convenient eudiometer is that usually called *Volta's*, for firing oxyen and hydrogen, and consists of a straight tube accurately graduated, closed at one end and open at the other—the mouth being ground smooth as recommended by Bunsen. It is from $23\frac{1}{2}$ to $27\frac{1}{2}$ inches in length, of equal calibre throughout; its interior diameter being about $\frac{3}{4}$ inch, and the thickness of the glass $\frac{3}{80}$ of an inch (·0590 inch). At the closed end of the tube two fine platinum wires are made to perforate the glass, one on each side, and are made to ascend, lying flat on the sides of the tube till they approach within $\frac{1}{10}$ (·118) of an inch of each other, at the summit. To fix the wires with accuracy, the tube is heated at *s* with a pointed flame from the blowpipe, and the softened glass is drawn out to *s b* by means of an ignited platinum wire; the portion *a a* is then cut off with a triangular file. The apertures are then heated by the blowpipe till they are nearly

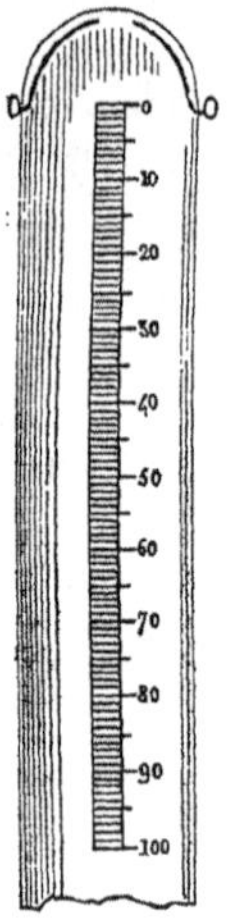

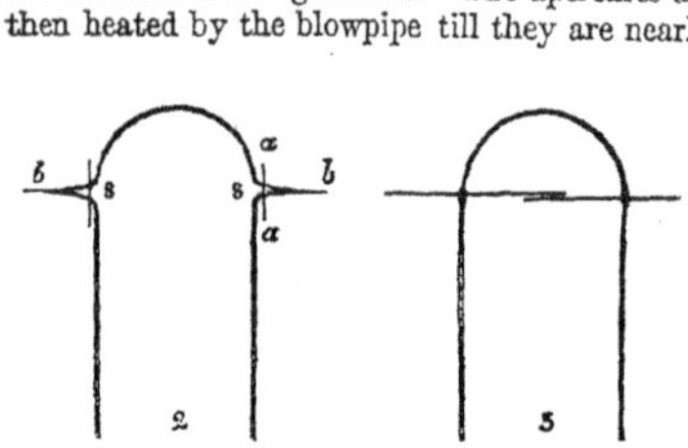

of the diameter of the wire to be introduced. Two platinum wires of the thickness of a horse's hair are then adjusted, and are melted into the glass by the blowpipe. When the tube has cooled, the extremities of the wire are pressed up to the top of the tube by means of a cylindrical piece of wood rounded at the end, and which fills up the interior of the tube. For the graduation of the tube, Bunsen employs a peculiar instrument, which also answers for thermometers. See GRADUATOR. Another form of eudiometer is that of Ure, *b*. It is not so readily graduated with accuracy, but is convenient of application. The figure exhibits the principle on which the gaseous mixture is exploded. The Leyden jar is charged; a chain conducts from one wire of the eudiometer to the outside coating of the jar, while the other wire is brought in contact with the knob of the jar. In reading off the amount of gas contained in the eudiometer, the tube is to be depressed in the water or mercurial trough, until the exterior fluid is on a level with that in the interior of the eudiometer, and the upper curvature of the fluid is that which is usually considered the point to be noted (Cavendish recommended the lower circle). The trough in which the experiment is made should therefore be sufficiently deep to admit of the proper depression of the tube.

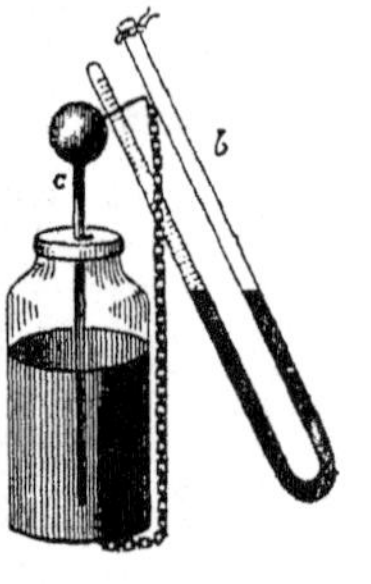

Eudiometry. The mode of analyzing atmospheric air, or generally of determining the nature of mixed gases. See ATMOSPHERIC AIR.

Eudnophite. A form of analcime belonging to the rhombic system, found on the island of Lamö, near Brevig

Eugenesite. An alloy of gold, palladium, silver, and selenium from Tilkerode.

Eugenic Acid. *Caryophyllic acid, Heavy oil of Cloves.* $C_{20}H_{12}O_5$. Oily fluid, specific grav. 1·079; B.P. 469°; is separated from the light oil of cloves (both coming over when cloves are distilled with water); by adding potash and distilling, the light oil passes over, and the heavy oil united to potash remains; by then adding sulphuric or phosphoric acid, and distilling, the heavy oil passes over.

Eugenine. $C_{20}H_{12}O_4$. Yellow pearly scales from the distilled water of cloves.

Eukairite. *Argento Selenide of Copper.* AgSe, Cu_2Se. Lead-gray and silver-white masses, lustre metallic, opaque, scratched by the knife, impressed by the hammer. B.B. fuses, giving out the odour of horse-radish, and leaving a globule of lead; the globule melted with borax gives the colour of copper; dissolved in nitric acid, white selenide of silver falls. Se 26, Cu 23·05, Ag 38·93, earthy matter 8·9; from Skrikerum, in Smoland.

Eukolite. *Brown Wöhlerite.* See EUCOLITE.

Eulebrite. *Riolite.* Selenide of zinc.

Eulysite. A granular mixture of augite,

garnet, and a mineral related to olivine, forming strata in gneiss, in Sweden.

Eulytine, or Silicate of Bismuth.

Eumanite. A synonyme of Brookite.

Eupatorine. A bitter white powder extracted by water, acidulated with sulphuric acid from *Eupatorium cannabinum*, precipitating by slaked lime, and treating by alcohol.

Euphorbic Acid? In the juice of *Euphorbia cyparissias*, but doubtful; by means of acetate of lead and sulphohydric acid.

Euphorbium Resin. The dried milky juice of several species of euphorbia in Arabia; in yellowish-brown masses, internally white, mixed with foreign matters, taste burning, smell when melted by heat agreeable. When dissolved in alcohol and precipitated by acetate of lead, *alpha resin* falls with the oxide of lead, and *beta* resin (euphorbine) remains in solution. The *alpha* resin is separated from the lead by sulphohydric acid and dissolved in alcohol; it is dark brown, taste bitter; the *beta* resin is separated from its lead base by sulphuric acid, and the resin purified by alcohol; it is colourless and bitter; *gamma* resin is obtained by heating the insoluble part of euphorbium in water; first with cold and then with boiling alcohol; it separates from the last in white masses; its formula is $C_{40}H_{33}O_3$. Euphorbium is used in medicine as a stimulating resin.

Euphotide. *Gabbro.* A rock having some species of felspar for its basis; at Mount Genevre it consists of saussurite, and diallage (smaragdite), the whole mass being SiO_3 45, Al_2O_3, and Fe_2O_3 26·83, CaO 8·5, MgO, NaO, and KO 13·9, HO and CO_2 5·78.

Euphyllite. Spec. grav. 2·693, H 3·5 to 4·5. Pearly brittle scales like mica at Unionville, Pennsylvania. SiO_3 43·69, Al_2O_3 44·69, CaO 3·98, MgO ·75, NaO ·98, KO ·82, HO 5·6.

Eupione. (εὖ, good; and πιων, oil). C_5H_6. Spec. grav. ·655 to ·740; B.P. 117°. Colourless, agreeably smelling, tasteless volatile liquid, insoluble in water, very soluble in alcohol; obtained from wood tar oils, by distilling and separating the first and the last distillates, and rectifying separately over oil of vitriol. These, although having different boiling points, seem to be homologous.

Eupyrchroite. A massive radiated variety of apatite or phosphate of lime, occurring in America.

Eurite. *Whitestone.* A granular mixture of felspar and quartz.

Euxanthic Acid. See Purreic Acid.

Euxanthone. See Purrenone.

Euxenite. *Polycrase.* Spec. grav. 4·6 to 4·76, H 6·5. Brownish-black rhombic prisms, lustre metallic, in thin splinters. Columbic niobic, and pelopic acids 57·6, yttria 25·09, protoxide of uranium 6·34, protoxide of cerium 3·14, lime 2·47, MgO ·29, HO 3·97; from Jölster, Norway. B.B. infusible, becomes a yellow enamel with borax, a yellow thin colourless pearl with salt of phosphorus.

Euzeolite. A synonyme of Heulandite.

Evaporation. (*Abdünstung*, Ger.) The process of removing the volatile portions of a solution either by heat, by spontaneous action, or by removal of atmospheric pressure. When the evaporation is performed for the purpose of obtaining organic extracts, the process is termed *inspissation*, or when the solution is to be rendered more condensed, the process is called *concentration*. Evaporation of chemical solutions is usually performed in the open air, in large or small flat vessels termed evaporating basins. The best are made of Berlin or Dresden porcelain. When the residue from the evaporation is to be weighed, the evaporating basin should be small, or the contents of a large one as they diminish, may be emptied into a smaller one weighing a few hundred grains. Heat may be applied to the evaporating vessel by any of the baths described under that title. When it is desired to evaporate for the purposes of crystallization in a dry atmosphere, the solution may be concentrated in the open air by heat, and may then be introduced under a glass receiver in convenient vessels, standing over sulphuric acid, which absorbs the aqueous vapour as fast as it rises.

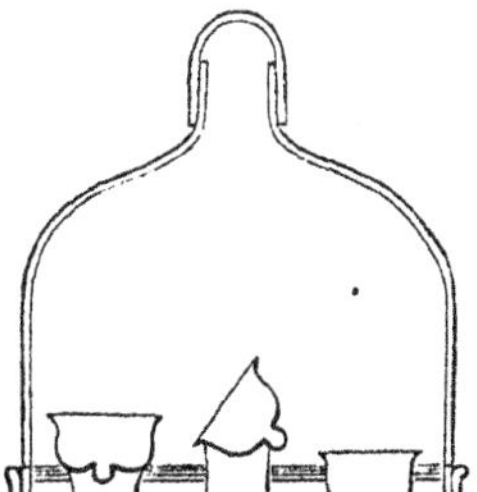

Drying and evaporation over sulphuric acid.

When it is desirable to evaporate a solution in a vacuum, the object is effected by placing a flat vessel containing sulphuric acid on the plate of the air pump. Within it the vessel containing the solution on a support, and over all a receiver, ground, and smeared on its ground surface with lard. The exhaustion when made is to be kept up by frequently working the pump from time to time. Evaporation in vacuo is much employed in the laboratory, and in the purification of sugar upon the great scale in sugar works. *Evaporat-*

ing vessels are made of porcelain, the best being Berlin porcelain, of glass when they are only used for evaporation in vacuo, and of platinum when the substance contained in the solution does not affect that metal. The latter are very convenient, especially when evaporation and calcination can be performed in the same vessel.

Evernesic, or Everninic Acid. $HO,C_{18}H_9O_7$. Silky needles by boiling evernic acid with potash or barytes, adding chlorohydric acid. It appears to be formed from evernic acid, along with orcine and carbonic acid ($C_{34}H_{16}O_{14}+2HO$ = orcine $C_{14}H_8O_4$ and $C_{18}H_{10}O_8+CO_2$).

Evernic Acid. $HO,C_{34}H_{15}O_{13}$. Crystals from the lichen evernia prunastri, by milk of lime and chlorohydric acid; is not coloured red by chloride of lime; yields orcine when heated, and affords a red colour slowly with ammonia when exposed to the air.

Exanthalose. Synonyme of sulphate of soda.

Excrements. *Solid Excreta, Fæces.* These consist of all the matters ejected at the lower extremity of the intestinal canal of animals, and are composed essentially of matters destitute of the capability of solution in the stomach and intestines, and consequently of nutritive power. Their nature depends entirely on the constitution of the food. In a cow the excrement is formed essentially of straw or grass, with the green-colouring matter and wax so characteristic of green food. In the human subject the peculiar colour is derived from the decomposed colouring matter of bile, and the odour from altered caseine. They usually contain little matter which can be of service as manure, the most valuable manure being the urine. Hence the richness of guano, which is the urine of birds, discharged in that class of animals into a common passage with excrement. The following table contains a view of the composition of some fæces:—

	Human.		Cow.	Cow.		Boa Constrictor.
Water,	73·3	Water,	70·	85·900	Mucus and colouring matter,	2·94
Organic remains,	7·	Fibre,	24·		Uric acid,	90·16
Bile ?	·9	Green wax,	0·76		Ammonia,	1·7
Albumen,	·9	Sugar,	3·	12·352 (Fibre to Resin and salts)	Potash,	3·45
Extract,	2·7	Albumen,	2·		Sulphate of potash,	·95
Biliary matter, &c.	14·	Resin and salts,	17·		Chlor. sodium,	trace.
Salts,	1·2			1·748	Phos. of lime, carb. of lime, and magnesia,	·80
(Berzelius.)		(Morrin.)		(Liebig.)	(Prout.)	

With a grass and farinaceous diet mixed, I have usually found the excrement of cows to yield 88·33 to 86 per cent. of water, 1·37 to 1·35 per cent. ash, and 12·½ to 9·92 organic matter. The ash of human fæces has been found—potash 6·1, soda 5·07, lime 26·46, magnesia 10·54, Fe_2O_3 2·5, phosphoric acid 36·03, sulphuric acid 3·13, CO_2 5·07, NaCl 4·33. The fæces contained 6·7 cent. of ash (Porter).

Exitele. A synonyme of white oxide or teroxide of antimony.

Exomorphism. The case in which a stratified rock has undergone alteration by the influence of plutonic rocks. Examples of this kind are—slate, modified sandstone, green porphyry, black passing into brown porphyry.

Exostoses. Excrescences of bony matter from other bones are so termed. I have found a bony tumour from Guy's Hospital, given me by the late Alex. Nasmyth, Esq. to consist of—cartilage 21·12, water 8·72, phosphate of lime and magnesia 65·802, carbonate of lime 4·848, and to have a spec. grav. of 1·910.

Explosive Cotton. See GUN COTTON.

Explosive Mannite. See ERYTHRO-MANNITE.

Explosive Sugar. A solid, brittle substance, becoming soft and pliable when heated; semifluid at a boiling temperature, and when more ghly heated explodes without residue; it is tasteless, inodorous, and colourless, and acts to solvents as a resin; formed as a glutinous insoluble mass by the action of a mixture of sulphuric and nitric acid on cane sugar, and washing with water and drying.

Extract. The constituents of a plant which are soluble in some fluid which is capable of extracting them. When a plant is digested in cold water, the aqueous infusion poured off and evaporated; the residue is termed the *cold watery extract;* if the water were used hot and boiled with the plant, the remainder would be the hot aqueous extract. By means of alcohol and ether, are obtained the *alcoholic* and *ethereal* extracts. The best mode of making extracts is by means of *displacement* apparatus (which see), the expression of the fluid thoroughly from the plant by means of a press, and evaporating the fluid in a vacuum, to prevent the oxidation of essential oils, &c.

Extractive Matter. A name given to a mixture of bodies which precipitate during the evaporation of watery extracts, and which become by oxidation insoluble in water. It is analogous to the humine, &c. formed by the decay of vegetable matter in soils.

Extract of Meat. When butcher meat is chopped, and exhausted with water by pressure, the water dissolves 16 to 17 per cent. of the weight of the dry flesh, and all the savoury constituents of the juice; the fibrine which remains being destitute of flavour. When the aqueous infusion is heated to 133°, flocks of albumen separate; at 158° the red-colouring matter coagulates; the liquor is pale yellow and reddens litmus. When the albumen separates, amounting in old animals to 1 to 2 per cent., and in young to 14 per cent., the liquid has all the savoury characters of soup, and when evaporated, leaves a brown mass amounting to 12 or 13 per cent. of the weight of the original dry flesh. When this extract is dissolved in about 32 parts of hot water with some salt, the water acquires the taste, and all the peculiar properties of an excellent soup. It is in this extract that the peculiar flavour of each kind of meat resides, as it is impossible to distinguish the insoluble fibrine of one animal from that of another (Liebig).

Eye. The eye consists of two chambers, one behind and the other in front of the crystalline lens. The posterior chamber is filled with the vitreous humour—see HUMOURS; and the anterior with the aqueous humour. The lens consists of globuline or crystalline, a modification of albumen. See GLOBULINE.

F

Fæces. See EXCREMENT.

Fæcula. *Fecula.* The meal of different plants.

Fahlunite. *Triclasite, Pyrargillite.* Spec. grav. 2·6 to 2·7, H 3. Reddish-brown, green, or black oblique rhombic prisms of 6 sides; angles 109° 28′ and 70° 32′; occurs in chlorite at Eric Matts mine, Fahlun. B.B. turns gray, and fuses on the thin edges; with borax fuses into a glass coloured by iron. Silica 43·51, Al_2O_3 25·81, MgO 6·35, FeO 6·35, KO and NaO 5·39, HO 11·66 (black var.), 3 (MgO, NaO, KO FeO), 2 SiO_3, 3 ($Al_2O_3SiO_3$), 6 HO.

Fahlunite, Hard. See IOLITE.

Faience. The French term for the finer kinds of pottery.

Faradayine. A volatile oil from caoutchouc, which passes over at the temperatures 91° to 111°.

Farina. A term similar to faccula.

Farina, Fossil. *Agaric Mineral, Mountain Milk.* Probably silicified animalcules. From China it consists of SiO_3 50·6, Al_2O_3 26·5, MgO 9·1, CaO ·4, FeO ·2, HO and organic matter 13·2.

Fassaite. Grass-green pyroxene, from Fassa.

Fats. (*Fette*, Ger.) The term applied to solid oily bodies, found in the vegetable and animal worlds. The best known of the fats are as follows:—

Vegetable Butters.	Fusing Point.
Cacao butter,	85°
Carapa oil,	50·
Chinese vegetable tallow,	80
Cocculus oil,	?
Cocoa nut oil,	72·5
Galam butter,	73
Laurel oil,	?
Minia batta,	95·
Nutmeg butter,	?
Palm oil,	80·5
Shea butter,	110·
Valteria butter,	95·

Animal Fats.	Fusing Point.
Badger fat,	86°
Beef tallow,	98½
Calf,	136°·8
Camel,	131
Cochineal fat,	104
Cow's butter,	79·7
Duck fat,	77
Dog,	79¾
Fox,	129
Hare,	117½
Hog's lard,	80·5
Horse grease,	140
Human fat,	77
Pheasant,	109
Turkey,	113
Stearine (human),	120
,, (sheep),	109
,, (oxen),	111
,, (hog),	109
,, (duck),	109
Cetine,	120
Cholesterine,	278
Cantharides fat,	93¼
Margarine (butter),	105
Palmitine,	115

Vegetable fats are principally met with in the seeds of plants contained in cells, in the form of small globules. In order to express them it is necessary to apply heat and pressure. Animal fats are likewise similarly deposited in cells, and are equally difficult of expression without heat. The fats in different plants and animals are somewhat different. When pressed between blotting paper, a fluid oleine is absorbed, and a solid stearine remains; but the stearine possesses a different acid in different plants. The stearine, or a portion of it in palm oil, is, for example, *palmitine.* When saponified by means of potash, a soap is formed containing palmitic acid, and oxide of glyceryle is set free; the stearine of nutmeg butter is myristine, and the acid contained in it myristic acid. The amount of these peculiar acids, which have been obtained from many vegetable butters, is small compared with the quantity of solid oils usually operated on. From my own experience, I am much disposed to think that these oils require much more minute investigation, and that the acids may then be probably diminished in number. By the present modes of investigation, the composition and fusing point of a fat acid depend very much on the number of times that it has been crystallized out of alcohol. If the acid is examined at an earlier stage, it will almost certainly turn out of different composition from any previously examined, while the fusing points of the same fatty body are found to vary very considerably under different circumstances. Mr. Duffy has shown that mutton stearine has three different fusing points, and different specific gravities.

Origin of Fat in Animals.—According to some views, the fat of animals is supposed to be derived from fat existing in their food, while, according to others, it is formed from the starch and sugar of the food. In my experiments made in 1845 for the government upon a great scale, and extending over a period of several months, I found that the largest amount of butter in cows was not produced by food containing oil, but by such kinds of fodder as contain no oil at all. Grass, for example, when used as food, produces much butter, although it cannot be said to contain a trace of oil. The only substance allied to oil which it contains is chlorophylle, a green wax, which, however, is believed to contain less oxygen than an oil, and hence could not be converted into an oil in the animal system, which is a deoxidating apparatus. The production of butter from sugar by the action of caseine or curd is, on the contrary, a process with which chemists are familiar, and is therefore more readily admissible into physiological theories, than the idea of the formation of butter from wax, since we are unacquainted with any analogous example. That there is a decided connection between the nitrogenous constituents of food, and the butter of cows, is obvious from the following table:—

Food.	Butter in Five Days.	Nitrogen in Food in Five Days.
Grass,	3·5 lbs.	2·32 lbs.
Barley and hay,	3·43	3·89
Malt and hay,	3·20	3·34
Barley, molasses, and hay,	3·44	3·82
Barley, linseed, and hay,	3·48	4·14
Beans and hay,	3·72	5·27

(*Experimental Researches on Food,* 174.)

Fatty Acids are those acids which exist in nature, in union with oxide of glyceryle or lipyle, and which combine with alkalies in saponification. Those belonging to the solid fats are usually crystalline in pearly scales, and afford salts of fine lustre. The acids, on the other hand, of the fluid oils are liquid at the usual temperature. The fatty acids are homologous, that is to say they differ from each other by an equal number of atoms of carbon and hydrogen.

Acids.		Fusing Point.
Formic,	$C_2H_2O_4$	
Acetic,	$C_4H_4O_4$	
Propionic,	$C_6H_6O_4$	
Butyric,	$C_8H_8O_4$	
Valerianic,	$C_{10}H_{10}O_4$	
Caproic,	$C_{12}H_{12}O_4$	
Oenanthylic,	$C_{14}H_{14}O_4$	
Caprylic,	$C_{16}H_{16}O_4$	
Azoleic or Pelargic,	$C_{18}H_{18}O_4$	
Capric,	$C_{20}H_{20}O_4$	86°.
Margaritic,	$C_{22}H_{22}O_4$	
Laurostearic,,	$C_{24}H_{24}O_4$	109·4
Cocic?	$C_{27}H_{27}O_4$	69·8
Myristic,	$C_{28}H_{28}O_4$	120
Benic,	$C_{30}H_{30}O_4$	109½
Ethalic and Palmitic,	$C_{32}H_{32}O_4$	131 & 140
Margaric	$C_{34}H_{34}O_4$	136½
Stearophanic,	$C_{35}H_{35}O_4$	154½
Bassic,	$C_{36}H_{36}O_4$	160
Stearic,	$C_{38}H_{38}O_4$	158
Behenic,	$C_{42}H_{42}O_4$	172½
Cerotic,	$C_{54}H_{54}O_4$	171
Melissic,	$C_{60}H_{60}O_4$	192
Ceric,	$C_{68}H_{68}O_4$	149

Fat Lute. A composition of pipeclay and linseed oil.

Faujasite. Spec. grav. 1·923. Brown 8-hedrons, with a square base, with angles of 74° 30′, 111° 30′, 105° 30′. B.B. swells up and fuses into a white enamel, with soda a transparent glass. Silica 49·36, alumina 16·77, lime 5·, soda 4·34, water 22·49, 3 Al_2O_3, 2 SiO_3, $(CaNaO)_4SiO_3$. 8 HO. In amygdaloid at Kaiserstuhl.

Fault. The interruption which frequently

occurs in the continuity of a stratum in the same plane, is so called by miners. It generally proceeds from the sinking, or elevation of one portion of the stratum.

Fayalite. A volcanic-looking slag from Fayal. It consists of an insoluble and soluble portion in muriatic acid. The insoluble part consists of SiO_3 58·11, Al_2O_3 12·53, FeO 18·55, MnO 6·67, CuO 2·28. The soluble part consists of SiO_3 24·93, Al_2O_3 1·84, FeO 65·84, MnO 2·94, CuO ·6, sulphate of iron 2·77.

Feather Ore of Sulphide of lead and antimony, 2 (PbS) SbS_3, fusing before the blowpipe and giving white fumes; soluble in muriatic acid.

Feathers. The light portion of the wings and plumage of birds; appear to be prolongations of the cuticle, analogous to nails and horns. Their composition is similar to that of horn. The light portion contains much silica in its ash, the quill being destitute of this substance. The formula of feathers is $C_{48}H_{39}N_7O_{16}$; but they also contain sulphur.

Fellic or **Fellinic Acid.** A resinous acid from the decomposition of bile by chlorohydric acid. When the resin produced by this action is treated with cold alcohol, fellic and cholinic acids are dissolved, and dyslysine remains. The addition of water to the alcoholic solution precipitates cholinic acid, while fellic acid remains in solution; it is a very doubtful substance.

Felspar, Common. *Amausite, amazonstone, adularia, felstein, felsite, fusible hornstone, ice spar, lemanite, lodulite, murchisonite, moonstone, Napoleonite, necronite, orthose, orthoclase, fusible petrosilex, pegmatolite, potash felspar, sunadin.* Spec. grav. 2·394 to 2·581, H 6·. White, gray,

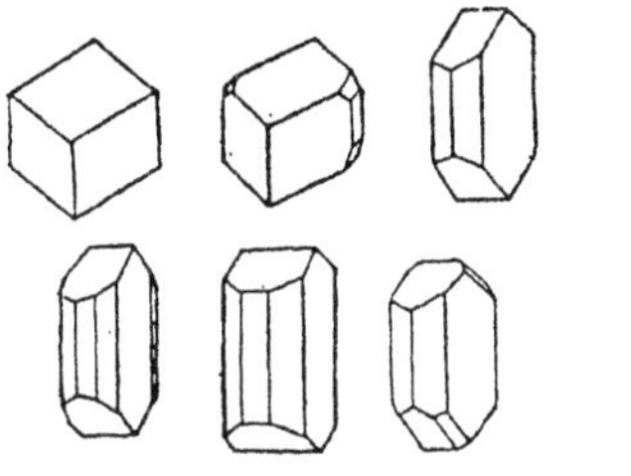

flesh-red, or green doubly-oblique prisms, with angles of 90°, 67° 15′, and 120° 35′, or massive. Structure foliated, fracture conchoidal, lustre vitreous, brittle, transparent to translucent on the edges. B.B. fuses with difficulty, and only on the edges, with borax into a transparent glass, and also with salt of phosphorus, leaving a skeleton of silica; with soda, forms a glass full of vesicles. Silica 64·2, alumina 18·4, potash 16·95, soda a trace. *Form.* $KO\,SiO_3, Al_2O_3\,3\,SiO_3$ or KO3SiO, 3(AlO 3 SiO). Felspar is an essential constituent of granite. Fine-grained granite from Goatfell in Arran, I have found to have a spec. grav. of 2·546, and to consist of silica 74·30, alumina and oxide of iron 15·24, lime 3·68, magnesia and alkalies 6·78. Porcelain clay and *Halloylite* are products of the decomposition of felspar.

Felspar Glassy. *Ryacolite.* Specific grav. 2·576 to 2·582, H 6. Gray or grayish-white, of the same form as common felspar. The crystals are usually much cracked, small, and imbedded in the lava of Vesuvius, Laiecher, &c. in the pitchstone porphyry of Arran and Rum; it is distinguished by its strong, splendent, and vitreous lustre: its optical properties are distinct from those of common felspar. It consists of SiO_3 66, Al_2O_3 19·13, KO 7·45, NaO 3·85.

Felspar, Apyrous. See ANDALUSITE.

Felspar, Aventurine, or *Sunstone.* A variety with a peculiar lustre from the presence of imbedded mica spangles.

Felspar Opaline, or Labradorite.

Fennel, Oil of. The stearoptene of this oil has the same composition as that of anise, and as cuminole.

Fergusonite. Spec. grav. 5·8, H 5·75. Brownish-black scales, or 8-hedrons with a square base, the terminal edges of the pyramids being 100° 28′, and their lateral edges 128° 27′. B.B. infusible; dissolves with difficulty in borax; the glass when hot being yellow, but a white portion remains undissolved; with salt of phosphorus slowly dissolves, the undissolved part remaining white; found at Kikertaursak, near Cape Farewell, Greenland. Columbic acid 47·75, yttria 41·91, protoxide of cerium 4·68, zirconia 3·02, oxide of tin 1·, oxide of uranium ·95, sesquioxide of iron ·34.

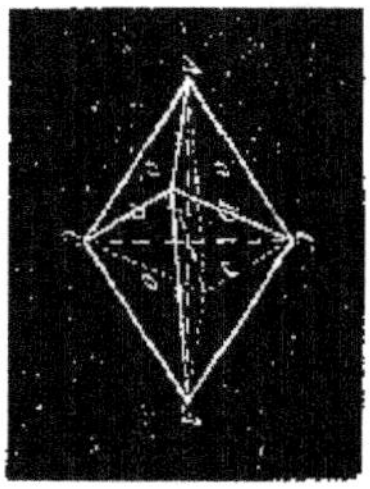

Ferment. *Yeast, Barm, Leaven.* (*Hefes Gährungsmittel,* Ger.; *Levain, Levure,* Fr.) See YEAST and FERMENTATION.

Fermentation. The conversion of a compound body into simpler compounds by means of a second substance which is not changed.

Fermentation Vinous.—By this term is implied the conversion of sugar into alcohol and carbonic acid ($C_{12}H_{12}O_{12}$, become 2 $C_4H_6O_2$, and 4 CO_2). Leeuwenhoek had noticed the formation of cells (bläschen) in fermented fluids without attributing anything to their presence (Carus Müller's Archiv. 1839, 3, 245). In 1827, Desmazieres (Ann. Scien. Natur. x. 42-67) describes globules accurately in yeast, as *Mycoderma cerevisiae,* or vini, glutinis, farinulae malti, juniperi malti, not as yeast nor as a fungus, but as *animalcula monadina.* In April, 1835, Cagniard Latour, in examining fresh yeast in water, observed that it is almost entirely composed of small diaphanous globules, some round, others slightly oval, which he considered to be of an animal nature. In August of the same year, in a report upon his paper, by Turpin, Thenard, and Becquerel, it appears that he was not committed to the opinion of their animal nature, but conceived that they might be vegetable; and accounts for the increase of the yeast in the fermented wort by the multiplication of the globules developed end to end—the number being augmented sevenfold after the yeast is added to the wort, exhibiting a growth by buds. The evolution of the carbonic acid during the fermentation, he attributes to the vegetation of the fermentation fungus (Geiger's Pharm. Botanique, 37). According to Quevenne (Journ. Pharm. 24, 265), when yeast is washed several times by subsidence, the residue consists entirely of globules, and is the ferment, the watery part having but a weak action. He found that oil of turpentine, prussic acid, corrosive sublimate, acetate of copper, &c. prevent fermentation, by poisoning the plants. Morphine and strychnine had no such action, which may be viewed perhaps as evidence of their vegetable and not animal nature. Schwann (Pogg. Ann. 39, 487, and 41, 184) also observed these fermentation globules. Turpin termed them *Torula cerevisiæ*. Meyen viewed these globules as distinct in different kinds of fermentation, and has described *saccharomyces*, vini, cerevisiæ, pomorum (Iahresbericht, 1837, 99) fungi as occurring in wine, malt, and cider. Kützing (Erdmann's Journ. xi. 390, and Physologia Generalis, 4to, 1843, 148) in 1834, states that he first observed the yeast globules, but published no account till 1837. He noticed, as was well known previously, that a solution of pure sugar, destitute of a nitrogenous substance, never ferments, but that every vegetable juice or sap containing sugar soon begins to undergo this process when withdrawn from the living vegetable. At the same time yeast globules can be detected. These he describes as a species of *cryptococcus*, a genus characterized as *mucous, hyaline or clear globules, aggregated into an undeterminate mucous layer*. *Cryptococcus Fermentum* (Kützing), *Torula cerevisiæ* (Turpin), *Mycoderma cerevisiæ* (Desmazieres), submersed, consisting of solid elliptic globules, with one or two points in the centre. This he considers an alga of the lowest kind, but a fungus of high development. If yeast is placed with the fermenting fluid in a flat vessel in the air, yeast globules collect on the surface. These lengthen, reach each other, and increase without limit (fig. 2). These globules,

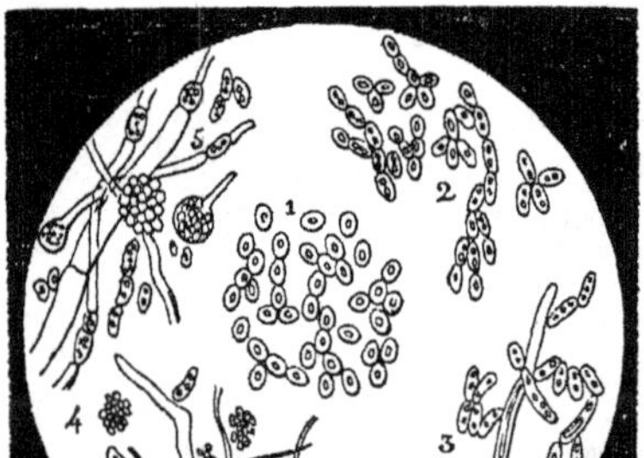

1st, 2d, 3d, 4th, and 5th stages.

which in some measure resemble starch granules, possess a somewhat opaque round nucleus (fig. 1). After some days a pellicle forms on the surface, composed of yeast globules, and several globules adhere to each other (fig. 2). They gradually unite, and are converted into oblong bodies (fig. 3), forming the third stage of their progress, similar to Diatoma and Exilaria. The whole mass forms into a mould, and a *sporotrichum* appears (fig. 4); on a moister part of the vessel a *mucor* forms (fig. 5).

Chemical Character of Yeast.—When yeast is washed successively with water, hot and cold alcohol, and then by ether, a white substance remains, which is incapable of producing fermentation, although the globules appear to have undergone little change. When the residue is heated with dilute caustic potash, a *nitrogenous* substance dissolves, and a matter insoluble remains, destitute of nitrogen. The nitrogenous matter contained in the interior of the globules is precipitated from its alkaline solution by acids. Dried at 212°, it has an amber colour, and leaves no residue when burned.

It consists of	While the insoluble matter is
C 55·53,	45·45
H 7·50,	6·87
N 14·01,	
O 22·96,	47·68
S —	

Or the non-nitrogenous body has exactly the composition of cellulose obtained from lichens. Schmidt obtained the mother of vinegar by adding yeast to some beer, and allowing it to stand for three weeks; alcohol, acetic, and lactic acids were formed. The mother of vinegar was at the surface, while the yeast layer was at the bottom. *No trace of difference* of form could be distinguished between the one set of globules and the other. In the yeast were found crystals of 8-hedrons, with a square base, and also crystals of phosphate of lime, which appeared to be secretions of the cells, just as we find similar salts secreted by the mucous membrane of the kidneys. I have found yeast used in Mr. Neil Thomson's bakery, near Glasgow, to be composed entirely of globules under the microscope, mixed with a few starch grains floating in water, and to have the following composition:—

	1 Lb.	Per 1000.
Water,	5474 grs.	950·348
Organic matter,	262	45·486
Alkaline phosphate,	8·30	1·440
Phosphate of lime and magnesia,	14·55	2·527
Siliceous matter	1·15	·199
	5760	1000·

The process of fermentation has been compared by M. Turpin to that of vegetation. He states that all yeast derives its origin from the separation of minute particles from organic tissues, which rise to the surface of the fluids in which they are produced and germinate, this vegetation being produced by the heat and the oxygen of the air. M. Turpin considers the addition of yeast to a fluid equivalent to sowing a number of seeds in a productive soil. He calls the yeast plant *Torula cerevisiæ*, the CO_2 and alcohol being the excretion of the plant. The idea of the presence of a plant in fermentation has been supported by the observations of Mitscherlich. This chemist considers that there are three kinds of sugar,—cane sugar, grape sugar, and caramel. The first is obtained from the sugar cane, the second from grapes and starch, and the third by pounding cane sugar with water, and heating it in a chloride of zinc bath to the temperature of 320° F. This sugar is identical with the black matter obtained by fusing sugar at the same temperature, but it is less coloured. The same substance may also be obtained by heating a solution of sugar to an elevated temperature. When yeast is added to the caramel it ferments. The sugar found in fruit possesses the composition $C_{12}H_{12}O_{12}$, and that of grapes $C_{12}H_{11\frac{1}{2}}O_{14}$. Mitscherlich believes the three species of sugar capable of fermentation. Indeed this has been proved with respect to the sugar of fruit and grapes; but it was supposed that cane sugar, before fermenting, was converted into grape sugar. It is easily proved that the sugar of grapes and fruit ferments, by examining them by polarized light. The rotatory power of grape sugar is to the right, and fruit sugar to the left; and during their fermentation the action upon light continues the same, only diminishing with the decomposition. In plants cane sugar is converted by acids into the sugar of fruits, and then by crystallization into grape sugar; while starch, by means of acids and diastase, becomes first *dextrine*, and then grape sugar. According to Mitscherlich, fermentation is produced by a vegetable, putrefaction by an animal being. He has only been able to detect, after examining a great many putrefying substances, one species of infusory animal, which consists of globules, sometimes solitary, sometimes united together, to the amount of twenty or more, forming then a mass, the diameter of each individual being ·000393 English inch, or ·001 millimetre. Their movement is undulating. According to all the observations which he has made, the other animalcules which have been observed in putrefying matter have been carried thither either by the air or by insects, or by some other means. It is necessary for the existence and development of vibrios that a certain quantity of oxygen should be present. It results, from Mitscherlich's experiments, that the phenomena of putrid fermentation depend upon the presence of a certain quantity of atmospheric air. The maceration of vegetable substances in winter, in a warm apartment, is speedily attacked by these vibrios. In this action nitrogen is developed. These vibrios exist abundantly in the intestinal canal, through the whole of its extent, as well as in the esophagus and stomach. They can be easily observed, by examining under the microscope the matter attached to the teeth. They are also observable on the skin; but Mitscherlich has not yet discovered them in the blood, milk, urine, bile, nor other fluids of the human body. If we add to the liquids in which these animals are formed a small quantity of sugar, these vibrios are produced in much greater abundance still, and at the same time there is formed a vegetable being, that is to say, yeast. If we now add a greater quantity of sugar, the formation of these animals is arrested, or suddenly suspended, and that of the yeast augments. Mitscherlich has never seen yeast form in a liquid which did not contain sugar. One may satisfy himself that the cryptogamic plant belongs to the yeast, by the fact that, when added to sugar, under the microscope it immediately excites fermentation. In a clear liquid, where yeast is forming, we observe first a turbidness, and under the microscope globules or spheres of various diameters are observable. From day to day these globules increase in size, and many new ones make their appearance. In these liquids—for example, the juice of the grape—we observe only solitary globules, generally of an oval form, and sometimes a second forms upon them of inferior size. Yeast acts in a different way when it has been produced for some time by means of another yeast. It propagates itself for a series of years, always with a constant character. In brewing we distinguish two kinds of yeast—the *frothy* and *precipitated* yeast; the former rising to the top, and the latter falling to the bottom of the vessel in which they are placed. The frothy yeast increases at the temperature which should not exceed 44°½, nor fall below 32°. It is the fermenting principle of the Bavarian beer. The finest frothy yeast is that of white beer, which is propagated at a temperature of 77°. The *precipitated* yeast consists of isolated globules, of the most variable dimensions. The small globules are always dispersed through the liquid, and are produced at the same place with the large ones. In the frothy yeast, small solitary globules are never observed, but only large ones, at the extremities of which small ones are developed, assuming a branched appearance. These globules are propagated by buds; the *precipitated* yeast, on the contrary, by small isolated globules swimming in the liquid. Mitscherlich has been enabled to observe their structure at different periods. In old yeast there can be distinguished an envelope and an interior granular substance, which appears more distinct when an aqueous solution of iodine is poured over it. By means of a compressor, invented by him, the granular body can be expelled from the envelope under the microscope, and he considers it probable that in precipitated yeast the globules become flattened in order to drive out this granular matter, and that each grain thus expelled gives origin to a globule. Yeast is thus multiplied by spores. Substances, which act as poisons upon cryptogamic vegetables, exercise the same influence upon yeast, such as corrosive sublimate, and analogous bodies. Substances, on the contrary, which react energetically on the animal organism, as tartar emetic, in whose solutions we observe species of mucedones, do not affect the progress of vegetation. There are many cryptogamic plants which have been considered as diseases in plants, which act in a similar manner to yeast. It is thus that the boletus acts on the fibre of wood; and, on the same principle, it is possible that we may be able to explain the decomposition which the roots of plants produce in the soil, and hence in this view fermentation acquires a double interest.

Experiments on Fermentation with the Vinegar Plant. (R. D. T.)—The vinegar plant, or mother of vinegar, belongs, according to Kützing, to the genus *Ulvina*, characterized by consisting of a compact lubricous layer of very minute granules (fig. 1st, 2d, and 3d stages), *Ulvina aceti*, at first mem-

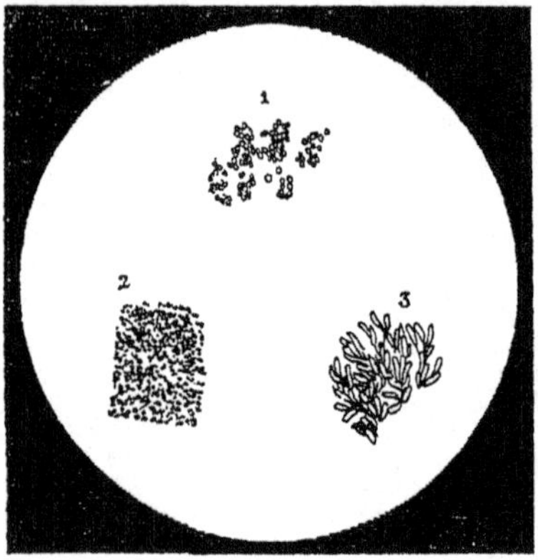

branaceous, then forming a compact stratum, vertically divided into dichotomous branches densely aggregated. He describes it as occurring always in the vinegar fermentation, upon the *surface* of the vinegar pot. The formation of the vinegar plant commences with that of the vinegar. It begins as a thin pellicle on the surface of the fluid, with little consistence. Under the microscope it consists of small globules, which are six times smaller than those of yeast. This pellicle becomes thicker, more compact, and coherent, and in fourteen days it begins to grow on the exterior border. It then presents the aspect of a *chætophora*—a gelatinous, fucoid appearance. To endeavour to throw some light upon the mode of action of the vinegar plant, I inserted a portion of a plant, well washed with distilled water, in a solution of sugar, and exposed the whole to the influence of the air. The liquid, when first formed, had no action on litmus paper, but in a few days it was characterized by a distinct acid reaction. After some weeks I took a portion of the fluid, saturated it with carbonate of soda, and distilled it in a glass retort. A liquid passed over which possessed the odour of alcohol, and which gave aldehyde and green oxide of chromium when treated with bichromate of potash and sulphuric acid, according to a mode of testing which I described some years ago (Proc. Phil. Soc. Glas. ii. 94). After two-thirds of the fluid in the retort was distilled, the receiver was changed, and sulphuric acid was poured into the retort, heat being applied cautiously. An acid liquid passed over into the receiver, which possessed the odour of vinegar, and rendered yellow a colourless solution of sesquichloride of iron. It was therefore acetic acid. Another experiment was made to determine the nature of the products of the vinegar plant, in the absence of oxygen. An ounce of sugar was dissolved in twenty ounces of water, a vinegar plant was introduced into the solution, the bottle was stopped close with a ground stopper, carefully waxed, and inverted in a glass full of distilled water. After some weeks, only a small portion of fluid was found in the bottle, smelling strongly of alcohol, and yielding aldehyde and green oxide of chromium to bichromate of potash and sulphuric acid. The stopper was still fixed in the bottle, but the wax had given way in one place by the pressure of the gas, so as to allow of the expulsion of the fluid into the exterior vessel filled with water. The greater portion, two-thirds of the bottle, was filled with a gas, which upon examination was found to precipitate lime water abundantly, and to be absorbed by caustic soda. It was, therefore, carbonic acid.

Formation of Alcohol.—From these experiments it seems undoubted that the vinegar plant possesses the power of breaking up sugar in its solutions into alcohol and carbonic acid, and as the plant during the process appears to be increasing in bulk, it seems scarcely legitimate to ascribe the decomposition of the sugar to any process of decay in the plant itself. I am not prepared from my own observation to state that, in absence of air, the vinegar plant when immersed in a solution of sugar, does increase in bulk, or, in other words, grow, although I have no reason to doubt the fact. But from the observations of Schlossberger and Schmidt, there can be little hesitation in concluding that the vinegar plant is merely a modification of the yeast globules, and therefore that it is capable of augmenting in size under the same conditions as the latter form of vegetation. When a plant of this description is found to vegetate in a position of immersion in a liquid, it must possess a power of extracting nourishment from the fluid atmosphere with which it is surrounded, just as sea and water plants effect their object. It is, however, impossible that in a solution of pure sugar the vinegar plant can increase from the influence of vinegar and any albuminous substance, as has been asserted (Mulder) to be the mode of its propagation, since the presence of the vinegar plant precedes the formation of the vinegar in the trials detailed. If the numerous experiments of various chemists are to be depended on, it is certain that the cellular structure of the vinegar plant, consisting of cellulose chiefly, must derive its carbon from the carbonic acid of the sugar, in absence of common air, or possibly from the atmosphere, in its presence, which may also supply it with nitrogen for its albuminous constituent; or the nitrogenous principle, like the salts, may be capable of a greater diffusion, without any considerable increase in its original amount.

Formation of Vinegar.—The circumstances most favourable to the production of vinegar from sugar are, when a vinegar plant is introduced into an open shallow vessel, containing a solution of sugar or treacle. The plant exposes its upper layer near the surface of the solution, and augments by the deposition of a new layer or stratum above the old plant, to which it is attached, but both can readily be separated simply by lifting up the superincumbent layer. It is thus worthy of notice, that the new growth takes place between the old plant and the atmosphere, that is, in closer contact with the air. My observations tend to show that when the vinegar plant falls to the bottom of a deep vessel filled with a saccharine fluid, the progress of the acetification is much more slow than when the plant is in contact with the air. The action of a cellular plant is, under these circumstances, analogous to that of a porous body which is capable of condensing oxygen from the atmosphere, to a condition approximating to fluidity upon the area of its wall cells. The action of spongy platinum and platinum black in the absorption and condensation of oxygen, are sufficiently well known; and through this power, of oxidizing and acidifying, hydrous oxide of methyle into formic acid, and alcohol into vinegar. The absorption and retention of air in the cells of the vinegar plant may assist in explaining the fact, which is particularly noticed by Kützing and Schmidt, that it is distinguished by its floating at the surface of the fermenting fluid, while the yeast globules remain at the bottom of the liquid. The influence of cellular plants in decomposing the higher oxides, in consequence of their absorptive action on oxygen, is well exemplified in the case of yeast globules, which, when brought in contact with binoxide of hydrogen, speedily cause the removal of the second atom of oxygen. This action seems quite parallel to that of porous paper on certain coloured solutions, as the red solution of permanganate of potash, and the amethyst solution of ferrate of potash, the former of which is slowly, the latter with great rapidity, deprived of tint when passed through common filtering paper. The yeast globules and vinegar plant, when introduced into a solution of permanganate of potash, remove its fine colour rapidly, but the action is not so instantaneous as when these globules are placed in the ferrate of potash. A similar action I find to occur when spongy platinum is placed in a solution of permanganate of potash, or of ferrate of potash. The change is, however, somewhat more slow than that of the influence exerted by the yeast plant. It is in this way also that the cellular matter, in the form of chips of birch, acts in the formation of vinegar from alcohol in the quick vinegar process.

Analysis of the Vinegar Plant.

107·05 grs. gave 101·2 water.
6·85 solid residue.

The constitution of 100 parts was found,—

Water,	94·53
Cellulose,	5·134
Alkaline salts,	.336
	100·

The salts, when dissolved in water, indicated the presence of chloride, sulphate, and a trace of phosphate. The plant, when digested in weak caustic soda, gave a turbidity when the alkaline fluid was saturated with acetic acid, indicating the presence of some albuminous matter.

Fermentation, Acetous, consists in the conversion of alcohol into acetic acid by means of the oxygen of the air, and the presence of a ferment (C_4H_5OHO and O_2, become $C_4H_3O_3$ HO and 2 HO).

Fermentation, Benzoic.—When emulsine from sweet almonds is mixed with amygdaline from bitter almonds, no change is produced on the emulsine, but the amygdaline is converted into prussic acid, oil of bitter almonds, sugar, formic acid, and water. See Benzoyle Series.

Fermentation, Butyric.—Sugar in solution when mixed with cheese (the ferment), is converted into lactic, and then into butyric acid. ($C_{12}H_{12}O_{12} = C_8H_7O_3$, 4 CO, H_4, HO).

Fermentation, Lactic, consists in the conversion of sugar into lactic acid; 1800 sugar in 9500 of water with some curd of cheese, and milk of lime are converted in four or five weeks into lactate of lime. The atom of sugar is divided into two parts, lactic acid having the composition $C_6H_5O_5$.

Fermentation, Mucous or *Viscous*, is induced in saccharine juices, as of carrots, beetroot, onions, &c. lactic acid ($C_6H_7O_6$), and mannite ($C_6H_5O_5$), being produced from sugar.

Fermentation, Madder.—By the action of erythrozyme on rubian; 6 new bodies are formed, viz. alizarine, verantine, rubiretine, rubiafine, rubiagine, rubiadipine.

Fermentation, Sinapic.—In the seed of black mustard no oil seems to exist, but when it is moistened, the odour of mustard begins to be evolved.

Fermentation, Salicylic.—When yeast is mixed with salicine, the latter body is resolved into saligenine and grape sugar ($C_{26}H_{18}O_{14}$ 2 HO $= C_{14}H_8O_4$, and $C_{12}H_{12}O_{12}$).

Fermentation, Urinous.—In this phenomenon, 1 atom fused urea by taking up 2 atoms water, is converted into 2 atoms volatile carbonate of ammonia ($C_2O_2N_2H_4$, and 2 HO = 2 NH_3 CO_2).

Fermentation, Oils from.—Certain oils appear to be produced by fermentation. Oil of spirit and potatoes is probably one of these. When the leaves of Centaureum minus are macerated with water and distilled, an oil passes over. Similar oils are formed from oak leaves (spec. grav. ·795), Echium vulgare, Urtica urens, Chaerophyllum silvestre, apples, millfoil, &c.

Fern. The roots of Aspidium filix mas and Asplenium filix femina have been examined by Bock. The first contains fibre 45·, ash 2·1, starchy fibre 1·5, tannic and gallic acids 10·, pectine 2·1, sugar 11·, starch 10·, albumen 3·5, gum 3·3, mucus ·4, resin 4·, stearine 1·, fat oil 6·, volatile oil ·04. The second contains fibre 60·24, ash 4·94, starch fibre 1·5, tannic and gallic acids 11·0, pectine 4·, starch 7·5, albumen 5, gum 2·, mucus 1·3, resin ·4, fat oil 1·2, volatile oil ·02. A variety of acids have been obtained, all, however, amorphous, as filimelisic, filipelosic, chlorofilipelosic, dichlorofilipelosic, sulphofilimelisic, filixolic, filosmylic, tannaspidic, pteritannic, &c.

Ferric Acid. See IRON.

Ferricalcite, or Cerite.

Ferridcyanogen. 6Cy+2Fe=Cfdy. When a solution of the common yellow prussiate of potash (ferrocyanide of potassium) is treated with a current of chlorine, a new radical is formed containing double the amount of cyanogen and iron existing in ferrocyanogen, and instead of uniting with 2 it combines with 3 atoms of base.

Hydroferridcyanic Acid.—3H+Cfdy. Obtained by diffusing ferridcyanide of lead in water, and passing a stream of SH through it, or by adding SO_3 to the salt. On filtering, the solution deposits without heat by spontaneous evaporation crystals of this acid. The compounds of this radical are as follows:—

Ferridcyanogen, C_6Fe_2 = Cfdy
Hydroferridcyanic acid, Cfdy+3H
Ferridcyanide of potassium, ... } Cfdy+3 K

Ferridcyanide of Potassium, discovered by Gme-1, and usually termed red prussiate of potash, prepared by acting with chlorine on the yellow ussiate of potash, until it yields no longer a ecipitate with persalts of iron. It is usually prered in stone or lead chambers resembling those bleaching powder, in which the yellow prussiate spread out in powder. The chamber being filled ith chlorine, 1 atom of potassium is speedily reoved. If the salt is now dissolved in water, and aporated in earthen or glass vessels, crystals of d prussiate appear. It is of importance that no etal should be present during its concentration, herwise yellow prussiate again is formed. If e represent the composition of the yellow prussiate according to the following formula, the action is easily understood:—

$Cy_6Fe_2K_3K$ yellow prussiate.
$Cy_6Fe_2K_3$ red prussiate,

the fourth atom of potassium being removed by Cl, or by the following scheme:—

2 Ferrocyanogen	4 Potassium	1 Chl.
Cy_3Fe, Cy_3Fe,	K_3 K	Cl
Cy_6Fe_2 + K_3		K Cl
Ferridcyanide of potas.		Chloride of potas.

The manufacture of the red prussiate may easily be studied in a large flask, to represent the chamber; the rapid action is striking.

(1). Yellow prussiate is converted into red by placing a piece of lead in the solution, (2), or by passing electricity through it. If mixed with a few drops of protoxide salt of iron, the Prussian blue may be detected speedily on the conductor.

Characters.—Transparent red rhombic prisms, anhydrous, soluble in 3¾ parts of cold water, more soluble in hot water. When held in the flame of a lamp, the salt burns, giving off sparks. When ignited in a close crucible, cyanogen and nitrogen are given out, and a black matter remains, consisting of carbide of iron and ferrocyanide of potassium. This salt has been extensively used for producing the dark blue on woollen dresses, and it is a valuable test in the laboratory for the protoxide of iron, with which it forms a fine brilliant blue.

Ferrocyanogen, or *Iron Cyanogen.* Cfy (Gay Lussac), C_6N_3Fe, or $FeCy_3$ — 13·25, 106·. This substance has not been obtained in a separate state, but it appears to exist in the ferrocyanides. The compound which it forms with hydrogen, is known under the appellation of—

Hydroferrocyanic Acid. — $FeCy_3$ + 2 H, or 2 H, Cfy. 1. Mr. Porrett discovered this acid in 1814 (Phil. Trans. p. 580). By dissolving 58 grains of tartaric acid in alcohol, and 50 grains of ferrocyanide of potassium (yellow prussiate) in as little warm water as possible, and mixing the liquids, cream of tartar is precipitated, and hydroferrocyanic acid remains in solution, which, by spontaneous crystallization, separates in small yellow cubes. 2. Berzelius recommends decomposing the ferrocyanide of lead or copper by SH, filtering the liquor and obtaining crystals of the acid by evaporation. Dr. Posselt (Ann. der Pharm. xlii. 164) has pointed out a method of obtaining the acid at once from its aqueous solution, simply by agitating with ether in a flask a concentrated solution procured by the process of Berzelius. When sufficiently concentrated, the whole mass becomes thick, and after some time the acid becomes suspended in the ether, and swims on the surface of the fluid. The mixture is thrown upon a filter, washed once with a mixture of ether and alcohol, pressed between blotting paper, and lastly dried in vacuo. 3. Dr. Posselt has, however, proposed a still simpler method of procuring this acid. A concentrated solution of yellow prussiate of potash is prepared in boiling water, and being covered, is allowed to cool. If into the ferrocyanide solution containing chlorohydric acid, we pass the vapour of ether, bluish-white crystalline scales of the acid fall, apparently depending on the solution of the vapour of ether by the water, which thus loses its power of dissolving hydroferrocyanic acid. It is then mixed with an excess of HCl free from air, and the whole shaken with ether; the acid separates as previously described. It may then be purified by dissolving it in alcohol, separating by ether, and drying it as before. *Characters.*—The acid thus prepared is a white powder with a faint yellow or bluish shade. When quite dry it may be preserved for some time in the air without altering. When moist it is more rapidly decomposed; the acid gradually becomes blue, and is at length completely changed into Prussian blue. It can be kept for some time at a temperature of 212°, without decomposition when covered from the air; but even in this state it gradually alters into Prussian blue. When strongly heated, hydrocyanic acid escapes, and compound cyanide of iron remains. When CO_2 is passed over it when heated at 212°, hydrocyanic acid escapes, and white cyanide of iron remains. Hydroferrocyanic acid is more soluble in alcohol than in water. The composition of the acid is C 33·33, H 1·84, iron 25·06, N 38·89.

Hydroferrocyanic Acid appears to afford the type according to which all the compounds of ferrocyanogen are formed. It contains 2 H, and these 2 atoms are represented by the same number of atoms of other bases, according to the following table:—

Ferrocyanogen, ...	Cfy	$=FeCy_3$
Hydroferrocyanic acid, ...	Cfy	H_2
Ferrocyanide of potassium, ...	Cfy	K_2
Ferrocyanide of potassium and iron,	2Cfy	$\begin{cases} Fe_3 \\ K \end{cases}$
Prussian blue, ...	3 Cfy	Fe_4
Basic Prussian blue, ...	3 Cfy	$\begin{cases} Fe_4 \\ Fe_2O_3 \end{cases}$
Soluble Prussian blue, ...	2 Cfy	$\begin{cases} Fe_2 \\ K \end{cases}$
Ferrocyanide of zinc and potassium	2 Cfy	$\begin{cases} 3\ Zn \\ K \end{cases}$

Ferrocyanides.—The ferrocyanides are all decomposed by a red heat, generally giving rise to a cyanide of the base, with a carbide of iron and evolving nitrogen. The soluble ferrocyanides are decomposed by being boiled with red oxide of mercury into cyanide of mercury, free alkali and cyanide of iron with oxide of iron. They generally contain crystalline water with NH_3. The ferrocyanides of zinc, copper, and mercury form double compounds. They are decomposed by NO_5, cyanogen and ferrocyanides being formed.

Ferrocyanide of Potassium, phlogisticated alkali, yellow prussiate of potash, ferroprussiate of potash.—Formula, K_2 Cfy+3 HO, or 2 K $(C_6 N_3Fe)+3HO$. At. Wt. = 26·625. This salt was discovered by Macquer in 1750?, and was for a long time used merely as a test for iron. It has now been introduced into the arts for many years. It was first begun to be manufactured at Campsie, near Glasgow, by Charles Macintosh, Esq. in 1807-8. It was usually sent out to the printers in the fluid state, but in 1809 it was crystallized (R. Clarke); it is still made there in large quantities. *Manufacture.* — This salt is prepared in commerce by heating together azotized animal or even vegetable substances, with 2 or 3 times their weight of carbonate of potash. For this purpose, blood, horns, hoofs of cattle, hair, hides, woollen rags, fish, guano, rape cake, cabbages, &c. have been employed. Animal substances are preferred when a sufficient supply can be obtained, but in consequence of the great extent to which the use of this salt is carried, the matters mentioned, and many others containing azote, have been used, when the supply of substances of animal origin was insufficient to meet the demand. The animal matter to be employed in the manufacture of this salt, is first sorted or sifted, according to its nature, to remove dust, and other unnecessary matters adhering to it. It is then dried in a hot stove to remove water from it. The temperature should, however, not be so high as to drive off ammonia, although, as commonly managed, the odour of ammonia is generally perceptible to a great extent in the drying room. Some manufacturers, however, calcine the animal substances in cast iron cylinders before bringing them in contact with potash, and when such a process is followed, the method of drying the matters is entirely omitted. The product, however, appears to be smaller after calcination than when the materials are used in their natural state after careful drying. The next stage of the manufacture consists in introducing the animal and alkaline ingredients into a cast iron pot or furnace, to which the heat is applied. These pots are of various forms. Sometimes they resemble an oval retort, which is built into brickwork, and has no communication with the air, except by a narrow mouth which opens horizontally, and admits of being closed. This form is generally made use of when the animal matter is previously carbonized. Another shape of pot is simply that of a small hemispherical boiler of cast iron, covered by a cast iron lid, which is perforated with an aperture for admitting an iron stirrer for the purpose of keeping the matters contained in it in a state of intimate mixture. The pearl-ash is first introduced into the pot in equal proportions, when the animal matter has previously only been dried; but when it has been calcined, 10 parts of pearl-ash are used for every 8 parts of charry matter. The animal substances are thrown upon the pearl-ash in the pot; and when the proper quantities have been added the cover is placed on the pot, and the materials kept in intimate union by the stirrer, which revolves by means of machinery connected with a steam engine. A copious evolution of combustible gases takes place, which escape under the edges of the cover, and burn with a brilliant illumination. The heat is continued until the melted mass exhibits the charcoal diffused equally throughout. To restrain the temperature of the pot, and the consequent decomposition of the contents, Mr. Young recommends cooling it by passing cold water through a tube which perforates the boiler, and which acts as the stirrer by rotating perpendicular to the bottom. After several hours, and when no more gases pass off, and the action is considered to be at an end, the liquid mass is removed from the pots by means of iron ladles, and is allowed to cool. The mass has then a dark appearance, and is called *black cake*. I have found its composition to be—

Soluble Salts.	
Cyanide of potassium,..................	10·95
Chloride of potassium,..................	1·58
Hydrate of potash,..................	10·93
Carbonate of potash,..................	29·30
Sulphate of potash,..................	6·62
Sulphate of lime,..................	11·42
Insoluble Matter.	
Carbonaceous matter,..................	18·
Iron,..................	
Sulphide of iron,..................	8·56
Carbonate of lime,..................	·72
Silica and titanic acid,..................	3·30
	101·38

The cake is then introduced into cold water deposited in iron pans, where it is allowed to remain till it dissolves, heat being applied by means of steam. The dissolved portions are then drawn off, and evaporated at a temperature below 200°. The salt thus obtained is impure; but it is purified by again dissolving it in water, evaporating the solution, and, when sufficiently concentrated, introducing it into wooden casks, where it is allowed to crystallize spontaneously on strings suspended in the fluid, when it appears of a colour resembling sugar-candy. In consequence of the complicated nature of animal matter, this process does not admit of a very easy explanation; at the same time, the experience of the manufacturer, and the theories of chemistry, have elucidated the subject in a considerable degree. The general theory consists in the production of cyanogen at the expense of the carbon and nitrogen of the animal matter, the consequent production of cyanide of potassium, by the action of the cyanogen upon the potash, while a portion of iron is removed from the pots either in the state of sulphide or oxide, or in the metallic state. To facilitate the production of the last substance, it is usual on the continent to add to the mixture a quantity of iron filings; but this is not necessary, and is frequently not practised in this country. It appears that when the fused cake, which has cooled, after being taken from the pots, is treated with alcohol or cold water, the solution contains nothing but potash and cyanide of potassium. Hence it may be inferred that the impure cake consists of cyanide of potassium, iron in some form of combination, and charcoal. When the cake is dissolved in water, it becomes then a solution of cyanide of potassium, capable of producing the reactions previously described, and of forming a double compound with iron. According to Liebig iron dissolves in cyanide of potassium with the evolution of hydrogen.

Iron. 3 Cyanide of potassium. Water.

Fe 2 K Cy, Cy K OH

Cy_3FeK_2 KO H

Potash. Hyd. evolved.

Sulphide of iron dissolves in K Cy with the production of sulphide of potassium, while the protoxide undergoes a similar solution with the formation of caustic potash. It is sometimes recommended that pearl-ash should be free from sulphate of potash; but Liebig has endeavoured to show that this salt acts most beneficially in the course of the process. He considers that the sulphate is deoxidized by the charcoal into bisulphide of potassium, and carbonate of potash. The bisulphide of potassium takes up an atom of iron from the pot, and forms a fusible double sulphide, which is diffused through the melted mass. Liebig ascribes the sources of deficiency of product in the manufacture of the yellow prussiate, either to the absence of a sufficiency of iron in the melted contents of the pots, which causes the cyanogen to be converted into formate of ammonia by the caustic potash when the cake is dissolved, or by the production of sulphocyanide of potassium by the union of the sulphur of bisulphide of potassium with the cyanide of potassium. The important considerations in this manufacture, therefore, are, to have a sufficient quantity of iron present, and to exclude the air effectually. *Characters.*—Ferrocyanide of potassium crystallizes in fine lemon-yellow 4-sided plates

with bevelled edges, or deeply truncated octahedrons with square bases; spec. grav. of the crystals 1·832. Taste saline and cooling; structure foliated; splitting into thin plates, which are flexible. It is soluble in 4 parts of cold, and 2 of hot water; insoluble in alcohol; at 212° it loses 12·82 per cent. of water. When ignited in close vessels, it is converted into cyanide of potassium and carbide of iron, but the former by access of the oxygen becomes cyanate of potash (KO CyO).

When it is acted on by sulphuric acid, hydrocyanic acid is evolved, with carbonic oxide and some nitrogen and ammonia (Thos. Thomson). Ferrocyanide of potassium may be used as an oxidizing agent, by immersing cloth dyed with indigo in yellow prussiate solution, and then in caustic potash or soda.

The following table shows the action of yellow prussiate on various salt solutions:—

Alkalies, ...	0
Alkaline earths, including magnesia, ...	0
EARTHS—	
Alumina, ...	0
Glucina, ...	0
Yttria, ...	White.
Thorina, ...	White.
Cerium oxides, ...	White.
Zirconia, ...	White.

METALS—(*white.*)

Antimony.	Manganese, then red.
Lead.	Zinc oxide.
Tin.	Cadmium, yellowish.
Bismuth.	Nickel, greenish.
Mercury.	Silver.
Chromium, .. 0	Rhodium, ... 0
Osmium, ... 0	Iridium, ... 0
Platinum, .. 0	Tellurium, ... 0

Uranium, ...	Reddish-brown.
Molybdenum, ...	Reddish-brown.
Vanadium, ...	Yellow, then green.
Copper, ...	Brown.
Iron protoxide, ...	Light blue.
Iron sesquioxide, ...	Prussian blue.
Cobalt, ...	Green.
Palladium, ...	Greenish on standing.
Gold, ...	Emerald-green.

Ferrocyanide of lead,.....Pb_2 Cfy, white powder.
Ferrocyanide of zinc, ...Zn_2 Cfy, white.
Ferrocyanide of copper,..Cu_2 Cfy, reddish-brown.
Ferrocyanide of mercury, Hg_2 Cfy, white.

Ferrocyanide of potassium with ferrocyanide of zinc, 2 Cfy KZn_3. This is the salt obtained by adding yellow prussiate to a salt of zinc, and has been used as a medicine.

Ferruginous. Containing iron.

Fibrine. When prepared from fish and dried, fibrine is white; from blood and flesh it has a dirty yellow colour; in water it imbibes that fluid and swells up. It is destitute of taste and smell, the flavour of meat being due to matters soluble in water; it dissolves in caustic alkalies, and is precipitated by acids and alcohol from the alkaline solution. Acetic acid forms with fibrine a jelly, which dissolves on the addition of hot water, from which yellow prussiate of potash throws down a white precipitate. As long ago as 1835, it was observed by Dr. Rainy of Glasgow that fibrine from different sources possesses somewhat different chemical characters. *Fibrine from blood* he found to be soluble in twenty-four hours in a dilute solution of common salt without heat; the solution coagulates at 130°; is not precipitated by corrosive sublimate, while white of egg solution, mixed with common salt solution, does not coagulate. Salammoniac, chloride of calcium, chloride of barium, nitrate of potash, sulphate of soda, potash tartrate of soda, have the same effect. *Fibrine from muscle* is not completely dissolved in such solutions (R. D. Thomson's Records of Science, vol. ii. p. 316). Liebig has recently directed his attention to the subject, and has obtained the following results:—*Fibrine of blood*, if laid in water, and a drop of chlorohydric added to every fluid ounce, swells up to a stiff jelly without dissolving. When the quantity of water is not too large, it is absorbed by the swollen fibrine as by a sponge, and if strong HCl be added to the mass, the fibrine shrinks to its original bulk. If again laid in water, it again swells up and shrinks to its original bulk. It contains 2 per cent. of ash (Fe_2O_3, $3CaO\ PO_5$). *Fibrine of muscle*, under the same circumstances, dissolves into a clear liquid, only rendered turbid by drops of fat suspended in it (Liebig). When fibrine of the blood is exposed in close vessels and in water to a cold temperature, it putrefies and dissolves in three weeks. The solution possesses now the character of that of albumen, coagulating by heat, and the coagulum has exactly the composition of albumen. The composition of the fibrine of the blood approaches more nearly to that of albumen, than to muscular fibrine.

	Fibrine of Muscle.	Fibrine of Blood.	Fibrine of Oysters.
C, ...	54·46	53·90	53·67
H, ...	7·28	6·99	7·00
N, ...	15·84	15·58	16·89
S, ...	1·21	1·50	
O, ...	19·81	21·83	22·44
Ash, ...	1·40	·28	(R. D. T.)

Muscle fibrine, ...$S_2N_{27}C_{216}H_{169}O_{68}$.
Blood fibrine, ...$S_2N_{40}C_{298}H_{228}O_{92}$.

So early as near the beginning of the present century, Dr. Thomas Thomson viewed fibrine in solution in caustic potash, as analogous to albumen. From fish, fibrine is prepared by washing it carefully with cold water to remove albumen, heating it with very dilute muriatic acid, to remove as much as possible of insoluble salts. It is then boiled repeatedly in alcohol and ether, digested for a considerable time in ether, to separate the last traces of fat and oil (R. D. T. Phil. Mag. 3d, 28, 368). It is obtained from fresh-drawn blood by agitating it before it has coagulated with a glass or metallic rod, throwing it on a cloth and washing it carefully with cold water until it is quite deprived of colour. For analysis it is usually dissolved in chlorohydric acid and precipitated by ammonia. When fibrine is oxidated by means of bichromate of potash and sulphuric acid, it yields much butyric acid, butyric aldehyde, acetic acid, aldehyde, oil of bitter almonds and metacetonic aldehyde, benzoic acid and prussic acid. These products differ, therefore, from those of albumen in its giving little acetic acid, while albumen affords a greater quantity. Fibrine combines both with acids and bases. Its atomic weight seems to be about 64·, that weight uniting with 5 of oxide of copper to form fibrinate of copper.

Fibroferrite. *Native Sesquisulphate of Iron, Fibrosiderite*, from Chili.

Fibroine. *Fibrine of Silk.* C 47·99, H 6·57, N 17·35, O 28·09. Purified silk forms this substance. Raw silk boiled successively in water, alcohol, ether, and acetic acid, till all soluble matter is taken up, leaves *fibroine;* in yellow silk it amounts to 53·37 per cent.; in white raw silk to 54·04. It is softer and much more brittle than natural silk, as it breaks very easily; heavier than water; soluble in strong sulphuric acid, from which it is precipitated by nutgalls; soluble in chlorohydric and nitric acids; soluble in phosphoric and pyrophosphoric acids, assisted by heat; soluble in strong caustic alkalies; heated with dry caustic potash it is converted into oxalic acid. After digestion in acetic acid, Mulder found it to leave ·3 ash. Without this digestion I have found it to leave ·688 per cent.

Fibrolite. Spec. grav. 3·214. Harder than quartz. White or gray right prisms with rhombic bases, having angles of 100° and 80°; internal lustre vitreous; texture fibrous; becomes negatively electric by friction. Silica 38, alumina 58·25. Allied to cyanite (Dr. Thos. Thomson). Accompanies corundum in the Carnatic. $4Al_2O_3 3SiO_3$.

Fichtelite. *Mountain Butter.* C_4H_3. Pearly needles. F.P. 115°. Slightly soluble in alcohol; easily soluble in ether; in brown coal at Usnach, Fichtel Mountains.

Figs are composed of sugar 62·5, fibrous matter 15, water 16, gum and a phosphate 5·2, extractive and CaCl ·4.

Figure Stone. See Agalmatolite.

Filimelisic Acid. $C_{24}H_{13}O_8$. Ochrey powder, by heating filixic acid with dilute caustic ammonia or potash, out of the air; when heated with dilute potash in air it forms

Filipelosic Acid. $C_{24}H_{13}O_9$. A bright insoluble powder; chlorine forms with it *chlorofilipelosic acid*, $C_{24}H_{12}ClO_9$, and *dichlorofilipelosic acid*, $C_{24}H_{12}Cl_2O_{10}$, a similar powder. When filixic acid is dissolved in fuming sulphuric acid, and the solution dropped in some hours into a solution of sulphate of soda, a golden powder falls of *sulphofilimelisic acid*, $C_{24}H_{13}O_8SO_3$; soluble in alcohol and ether.

Filixic Acid. $C_{26}H_{15}O_9$. Yellow crystalline powder by alcohol and ether from fern roots. When acted on by chlorine it yields *chlorofilixic acid*, $C_{26}H_{15}ClO_{10}$ in brown drops; by an excess of chlorine it gives a yellow powder, $C_{26}H_{13}Cl_3O_{10}$, *terchlorofilixic acid.*

Filixolic Acid. $C_{42}H_{41}O_5$. Red oil by alcohol, ether, and water, from fern root.

Filosmylic Acid. $C_3H_2O_3$. Obtained as a volatile acid of the brown heavy oil from fern root, by saponification.

Filtration. The method by which substances mechanically diffused through fluids are separated. The simplest kind of filter is a paper or calico filter placed in a funnel or filler. Paper filters are formed as in the figures. A sheet of filtering foolscap paper is folded into three parts lengthwise, and then twice on itself, *a*, *b*. The extremities being rounded off, six filters are formed, *c*. This may be more neatly performed by means of a filter cutter made of tin or pasteboard, cut into the shape of half a filter. The filter is then folded into four, *d*, and placed in the funnel. Besides impurities, precipitates are collected in this way, and in such cases it is requisite to pour the fluid without spilling a drop. This may be effected in various ways. The filter being moistened, a little lard may be rubbed on the outside of the lip of the vessel from which the liquid is to be poured, which prevents the fluid from running over,—or, as is more convenient, pouring by means of a glass rod.

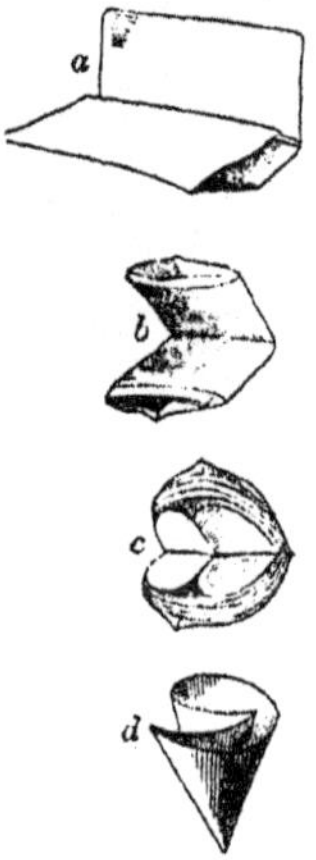

My pupil, Mr. Eustace C. Summers, has suggested the following ingenious method of continuous filtration. It consists of a reservoir, fig. *a*,

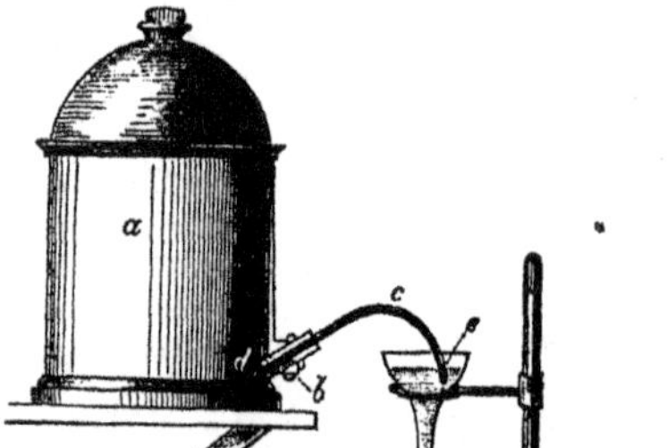

having the mouth placed at its base. This, when the vessel is filled with water, is closed by a cork, pierced by a straight tube, *b*, passing through it obliquely downwards. It is now obvious that the water will rise in the tube *b* so long as air can gain access to the interior of the reservoir; which, however, is prevented when the water in the tube is sufficiently high to cover the whole of its lower extremity. A siphon is then introduced through this tube, so as to project into the reservoir slightly beyond the lower extremity of the tube, care being taken to leave sufficient space to allow of the passage of air between the siphon and the surrounding tube. The funnel is placed beneath the long leg of the siphon, so that the edge of the filter is a little above the level of the water in the straight tube. In this condition the apparatus works. As long as the level of water in the filter is lower than the level of that in the tube *b*, the column of water *ce* will be heavier than the column *cd*, and consequently the siphon will keep up a supply of water, bubbles of air obtaining access to the interior of the reservoir, as the level of the water in the tube *b* is reduced by the action of the siphon. But no sooner does the water in the filter attain the same level with that in the tube, than the column of water *cd* is equal in pressure to the column *ce*, and consequently the action of the siphon is stopped, the water in the tube *b* remaining at just sufficient height to exclude the air from the interior of the reservoir. This simple apparatus may be affixed to *any* convenient vessel having an aperture at the bottom; all other openings being of course closed, so as to prevent the ingress of air except by the tube *b*. A common salt glaze earthenware washing-bottle, or a Griffin's gas-holder, will be found to answer the purpose perfectly.

Continuous Filtration.—On a larger scale, filtration of potable water may be managed by means of a cask, which is to have a false bottom fixed near its lower extremity. This consists of a bottom pierced with holes. The portion of the cask above this is divided by a vertical partition. The space is filled with alternate layers of coarse and fine sand, leaving about a foot on each side at top free; on one side for pouring in the water, and on the other side for its escape after it has been filtered. To remove the colour from mossy water, it may be found advantageous to place a layer of charcoal on the false bottom. Such a contrivance I have found to answer well for the removal of colours from Scottish Highland waters. It seems unnecessary to describe the various drip-stones by means of porous stones, sponges, &c. which may be seen abundantly at the present time exhibited in the streets; but the figure of the Glasgow Gorbals Water Works filter is so simple, that it may be strongly recommended for filtration of water in the supply of towns, especially upon the gravitation principle. The water meanders from the conduit on the top of a compartment of gravel, through which it sinks and passes into the well adjoining by an aperture at the bottom, to the top of which it rises, and then runs into the coarse sand, then in a similar manner into fine sand, and, lastly, is discharged into the storing tank.

Fiorite, or *Pearl Sinter.* Pearly globular and botryoidal masses of silica in cav tseiiof volcanic tuff.

Fire. The term applied to the flame or red appearance presented during combustion.

Fire-Damp. Carburetted hydrogen.

Fireworks, or Pyrotechnics.

Fish Scales. The scales of fishes have been examined both in the recent and fossil state, as follows. Their composition is:—

	Recent Fish.	Senegal Crocodile.	Megalichthys (Fossil.)
Organic matter, and bituminous matter,	55·00	1·15	·12
Phosphate of lime, ...	37·80	70·92	50·94
Phosphate of magnesia,	·90	1·20	trace
Carbonate of lime, ...	3·06	10·27	11·91
Carbonate of soda, ...	·90	—	·47
Phosphate of soda, ...	—	3·25	—
Liquid fat, ...	·40	—	—
Siliceous matter, ...	—	13·21	33·10

Fischerite. Spec. grav. 2·46. Colourless or green 6-sided prisms at Nischne Tagilsk, composed of Al_2O_3 38·47, PO_5 29·03, Fe_2O_3, and Mn_2O_3 1·2, CuO ·8, 3 CaO, PO_5 3·. Seems a variety of Wavellite.

Fixed Air, or Carbonic acid.

Fixed Bodies. Substances which are not volatilized, nor decomposed by very high temperatures.

Fixed or **Ammonium Bases.** Under this may be comprised a number of bases discovered by Hofmann, of which ammonium (NH_4) is the type, and in which the hydrogen is replaced by various organic radicals. They stand a considerable temperature without decomposing, and correspond in many of their characters with the fixed alkalies; precipitating metallic oxides, and having a caustic taste. They are tetra-methyl ammonium, tetra-ethyl ammonium, methyl-tri-ethyl ammonium, &c. See Bases.

Fixed Oils. See Oils.

Flake White, or Carbonate of lead.

Flame. The light evolved by rapid combustion. See BLOWPIPE, COMBUSTION.

Flask. Matrass. A globular glass vessel for the digestion of chemical solutions.

Flavequisetine. A yellow matter from Equisetum fluviatile.

Flavindine. $C_{16}H_6NO_3$? C 72·5, H 3·8. Yellowish needles from the mother liquor of hydrindine by an acid. See INDIGO.

Flavine. A name given to a vegetable extract from America. It is a light brown powder, and is used to replace quercitron bark. It is rich in colouring matter. It seems to contain much tannic acid, as it gives black precipitates with iron salts.

Flesh and **Juice.** The flesh or muscular part of animals consists of solid fibrine, which constitutes the basis of the soft parts of the body, and of a juice (blood and lymphatic fluids, &c.) diffused through it. When the juice has been squeezed out by pressure, and washed with water, the tasteless insoluble fibrine remains, amounting to 70 per cent. of dry muscle. In dead animals the juice possesses an acid reaction, but I am not aware of any facts which prove that in the flesh of living animals any free acid exists. If this suspicion be correct, then it may turn out that lactic acid, the partial cause of the acidity in dead muscle, does not enter as a constituent into any part of the body, but is merely present in the stomach during digestion. If we dry a piece of muscle without washing it, the ultimate constitution is found to be the same as that of blood, $C_{48}H_{33}N_6O_5$. Its proximate analysis (that is, the substances which have been detected as entering into its composition) shows it to contain, water 79·, fibrine and cellular matter 17·, albumen 3·, lactic acid ?, phosphoric acid, fat, creatine, inosic acid, phosphate of soda, phosphates of lime and magnesia, chlorides of potassium and sodium 1·. Fresh meat when calcined leaves 3½ per cent. (of the weight of the dried flesh) of salts; meat exhausted by boiling, leaves scarcely 1 per cent. Hence we see that in making soup, unless the meat be used at the same time, only a portion of the substances necessary to form blood is taken, the object of the use of all food being, in the first place, to produce blood, from which the solid parts of the body originate. The same result occurs with salted meat. When salt is sprinkled on flesh a brine is formed, that is, the water of the flesh, along with a great part of the salts contained in it, passes into the containing vessel. The flesh is thus rendered by so much dissimilar to blood. If the meat in this condition is eaten, the blood formed from such nutriment must be destitute of the salts removed in the brine. Hence it is not surprising that diseases, such as scurvy, depending on the incapacity of the blood to form the solids of the body from deficiency of materials, should result from the use of such food. Lime juice and succulent vegetables are valuable antidotes to such diseases, because I find them to contain the very salts which have been removed from the meat, viz. alkaline phosphates, sulphates, chlorides, and phosphate of lime. The ashes of flesh consist of phosphoric acid 36·6, potash 40·2, earths and oxide of iron 5·69, sulphuric acid 2·95, chloride of potassium 14·81. The salts removed in the brine might be replaced, by adding to the salted meat the extract of meat which is preserved extensively for use, and contains its soluble materials.

Flint. A siliceous nodule, found in the chalk formation. Flints are of various forms, and probably have been produced by the deposition around sponges of silica from a state of solution in water,—good examples of this kind of formation I have in my possession. They usually have a darkish colour, which is probably dependent on the presence of organic matter; this they lose by ignition, and then become perfectly white. Flints are used sometimes for guns, and extensively in the manufacture of pottery.

Flint Glass. See GLASS.

Flinty Slate. See BASANITE.

Floatstone. *Nectic Quartz.* A light vesicular form of quartz.

Flos æris, or red oxide of copper.

Flosferri. A coral-like form of Arragonite, coating cavities.

Flour. The pulverized grain chiefly of wheat. See WHEAT and BREAD.

Flowers. An alchemistical term applied to certain sublimates, as flowers of sulphur, antimony, and benzoic acid.

Fluellite. *Fluoride of Aluminum,* AlF ?, from Stenna-gwyn, Cornwall. H 3. White rhombic 8-hedrons, rare.

Fluid, Fluidity. The mobile condition of the particles of matter as exhibited in water, and distinguished from its firm state in ice and other solids.

Fluoboric Acid. *Terfluoride of Boron.* BF_3. Spec. grav. 2312·4. Colourless incombustible gas, with acid reaction and pungent smell; attracts moisture from the air like chlorohydric acid gas; water absorbs 700 times its volume; obtained by igniting boracic acid and fluor spar, or by heating boracic acid, fluor spar, and sulphuric acid.

Fluocerine. Ce_2O_3 84·2, FH 10·85, HO 4·95. Yellow rhombic 12-hedrons, Finbo.

Fluocerite. Ce_2O_3 82·64, YO 1·12, FH 16·24. Red 6-sided prisms. B.B. infusible; dissolves in borax and salt of phosphorus; blood-red in oxidizing flame; not fusible in soda; Finbo and Broddbo.

Fluohydric Acid. *Fluoric Acid, Hydrofluoric Acid.* HF. Spec. grav. 1·0609. B.P. 59°. Colourless fluid, with a strong odour; when falling on the skin producing corrosion; acid reaction; obtained by heating fluor spar and

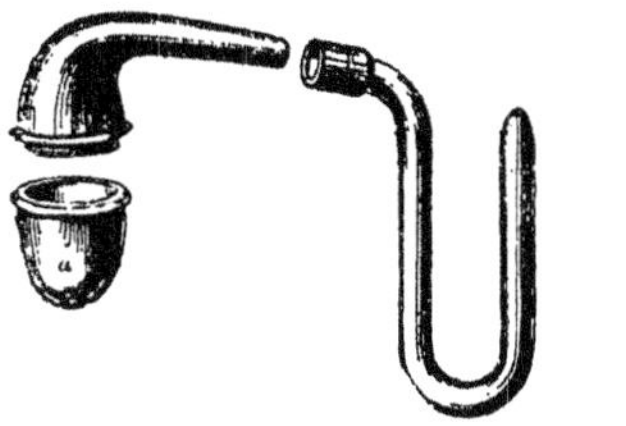

sulphuric acid in a lead or platinum still, and condensing in a similar receiver. It is used in etching on glass. The materials are placed in *a*; the two portions of the retort cemented with fat lute, the neck being screwed tight.

Fluorine. *Sources.*—Exists in fluor spar, in the teeth and bones of animals, in Carlsbad water (Berzelius), in sea water (Wilson), in many rocks and minerals, as phosphate of lime or apatite, in some species of mica, karpholite, hornblende, fluate of cerium, fluoride of cerium, topaz, cryolite, yttrocerite, wagnerite, wavellite or phosphate of alumina, lepidolite, apophyllite, felspar. *Characters.*—Fluorine is, according to Louyet, a colourless gas with a smell; not bleaching vegetable colours; decomposing water at the ordinary temperatures, and under the influence of light; attacking glass very feebly or probably not at all; acting upon almost all metals, with the exception of gold and platinum, at least not when it is in the nascent state. Its affinities are closer to those of oxygen and sulphur, than to chlorine, bromine, or iodine. Chloride, &c. of silver are stable and insoluble; fluoride of silver is deliquescent, decomposed by heat; contains combined water which is only removed by heat. Chloride, &c. of calcium are deliquescent; fluoride of calcium is insoluble; on adding a solution of fluoride of silver to chloride of platinum or of gold, chloride of silver, and oxide of platinum or of gold, fall; the water giving hydrogen to the fluorine, and its oxygen to the platinum or gold. The mixture is not changed when digested with concentrated fluohydric acid; but of the precipitates of chloride of silver, and peroxide of gold, or by the hydrogen acids of chlorine, bromine, or iodine, a chloride, bromide, or iodide of gold is formed, and the chloride of silver is bleached. *Process.*—Louyet in his experiments used vessels of fluor spar, placed at his disposal by Messrs. G. & T. Knox. He repeated many of the experiments of these chemists and confirmed their results. The gas was liberated by acting upon dry fluorides, as fluoride of silver by means of chlorine gas (Knox, Phil. Mag. 9, 107; 16, 192). *Fluorides.*—These are distinguished from chlorides, bromides, and iodides by their forming with silver a very soluble salt in water. A fluoride in powder when mixed in a platinum crucible with sulphuric acid, corrodes glass placed over the mouth of the crucible. A method of quantitative estimation is that pointed out by Wöhler, of determining the amount of loss when a fluoride is heated with sulphuric acid in a flask supplied with a chloride of calcium tube. *Test for Fluorine.*—The simplest test for fluorine depends on its property of corroding glass. The substance supposed to contain the fluorine or fluoride, is placed in powder in a platinum crucible, sulphuric acid is poured on it, and a plate of glass is laid across the mouth of the crucible. Gentle heat being applied, fluohydric acid comes in contact with the glass and forms fluosilicic acid, which corrodes it (FH and SiO = HO and FH). The quantity of fluorine in an analysis is inferred by the loss, there being no direct method of estimation.

Fluor Spar. *Fluoride of Calcium. Derbyshire Spar.* CaF. Spec. grav. 3·16, H 4. Regular cubes and 8-hedrons, of various colours, and colourless, red, violet, green, blue, &c. found in connection with lead ores. See FLUORIDE OF CALCIUM.

Fluosilicic Acid. See SILICON, FLUORIDE OF.

Flux. (*Fluo*, I flow). A substance employed to cause the fusion and reduction of a metallic ore. *Black flux*, formed by the calcination of cream of tartar, and consisting of carbonate of potash and charcoal, is a valuable flux for the reduction of various metals, the carbon removing the oxygen. In the reduction of iron, limestone is the flux employed; it acts by its lime uniting with the silica of the impure ironstone, and forming a slag. Fluor spar is often a useful flux, and so likewise is protoxide of lead, from the facility with which they unite with silica and other acids, forming compounds, which are fluid at a red heat.

Fluxes, for pottery, used at Sevres. *Pebble flux.*—Minium or litharge 75, sand 25. *Gray flux.*—Pebble flux 88·88, fused borax 11·11, or minium or litharge 66·66, sand 22·22, fused borax 11·11. *Carmine flux.*—Fused borax 55·55, sand 33·33, minium or litharge 10·12. *Purple flux.*—Minium or litharge 37·5, sand 12·5, crystallized boracic acid 50. *Violet flux.*—Minium or litharge 67·56, sand 5, crystallized boracic acid 27·5. *Green flux.*—Minium or litharge 73, sand 9, crystallized boracic acid 18. *Metallic flux.*—Subnitrate of bismuth 11, fused borax 1. According to circumstances, $\frac{1}{12}$ or $\frac{1}{15}$ of this flux is mixed with 1 of gold.

Foliated Tellurium. *Red Telluride of Lead.*

Food of Animals. The food of domestic animals consists of matter prepared for them by plants, and having the same composition as the solid parts of animals. An animal may be defined to be, a being which appropriates food similar to the matter of which its own body is composed, while a plant produces solids from the air and surrounding gaseous matter.

Food of Plants. The carbon of plants is derived from the carbonic acid of the atmosphere, absorbed by the leaves and roots of plants, the hydrogen and oxygen from water and the atmosphere, the nitrogen from ammonia in the air, and the inorganic matter from the soluble salts of the soil taken up by the roots.

Forge Furnace, or blacksmith furnace, consisting of an open furnace, urged by a powerful bellows, which often answers well for fluxing.

Formal. See METHYLAL.

Formanilide. $C_{14}H_7N\ O_2$. Rectangular prisms, resembling urea. F.P. 114¾°. Obtained by heating oxalate of aniline at 338°.

Formaniline. See METHYLE-ANILINE.

Formates. Formic acid has a great affinity for bases, with which it forms crystallizable salts; *formate* of ethyle, or formic ether, obtained by distilling an alkaline formate with alcohol and sulphuric acid, is a volatile fluid with an aromatic odour.

Formation. An epithet applied in geology to a group of rocks believed to have been deposited at the same epoch.

Formester. A German name for *ethers* from pyroxilic spirit.

Formic Acid. (*Ameisensaure*, Ger.) $C_2H\ O_3$ HO. Spec. grav. 1·2353; B.P. 209¼°. Crystals below 32°, brilliant plates. Clear colourless fluid, giving off a slight fume, and a peculiar odour; its vapour burns with a blue flame; it corrodes the skin; mixes with water; heated with alcohol, formic ether is formed; heated with sulphuric acid it yields water and carbonic oxide (C_2HO_3, HO and $HOSO_3 = 3$ HO, SO_3 and 2 CO). When heated with red oxide of mercury, or oxide of silver, the metal is set free, and carbonic acid escapes; with 2 atoms water it forms the dihydrate. B.P. 221°. Spec. grav. 1·11. Formic acid was obtained originally by Fisher of Leeds, in 1670; and in 1671 by Dr. Hulse, by distilling red ants. It may be prepared by boiling 1 starch with 4 sulphuric acid and 4 water, allowing the liquor to cool, then adding gradually 4 black oxide of manganese and distilling; formic acid comes over by the following action:—⅛ of an atom of sugar $C_2H_2O_2$, produced by the action of sulphuric acid on starch, and 2 MnO_2 with 2 SO_3 become $C_2H\ O_3$ and 2 MnO SO_3 and HO. To purify it, carbonate of lead is added to saturation; formate of lead is produced which resembles acetate of lead in being in needles, but is much more insoluble in water. When this is distilled with sulphuric acid, or the lead is precipitated by sulphohydric acid, and the fluid distilled, the acid is obtained strong and pure. It may also be prepared by oxidating pyroxilic spirit by spongy platinum. When prepared by distillation from starch sugar as above, the copper still should be very capacious, as abundance of carbonic acid is evolved at first, which will carry over a quantity of acid into the receiver.

Formobenzoilic Acid. (*Mandelsäure*, Ger.) $HO,C_{16}H_7O_5$, or $C_2HO_3,C_{14}H_6O_2$,HO. Large tabular rhombohedrons, obtained by evaporating to dryness some water and oil of bitter almonds with a little muriatic acid; ether takes up the formobenzoilic acid. The cyanohydric acid, contained in the oil, in presence of a mineral acid, is converted into formic acid and ammonia; the formic acid unites with the oil of bitter almonds, and forms the acid. It is an example of a *coupled* or *conjugate* acid.

Formomethylal. See METHYLAL.

Formonitrile. NH_4O,C_2HO_3 less 4 HO $= C_2NH$, or cyanohydric acid.

Formula consists of a number of symbols, or initial letters of simple bodies, united together so as to express the composition of a substance containing more than one element. Symbols of various kinds, chiefly hieroglyphical, were used at an early period by the alchemists, and were subsequently variously modified. But the present system of symbols was introduced by Dr. Thomas Thomson, in 1796-99, before the promulgation of the atomic theory, and was first applied to the composition of minerals. "All those minerals composed of the same ingredients, he arranged under the *same* genus. According to this plan, there must be as many genera as there are varieties of combinations of earths, &c. existing in nature. The varieties in the *proportion* of the ingredients constitute species." He denoted each genus by a symbol. This symbol was composed of the first letter of every substance which entered in any considerable quantity into the composition of the minerals arranged under the genus denoted by it. The symbols which he used were as follows:—A, alumina; M, magnesia; S, quartz or silica; I, iron; Z, zirconia; G, glucina; C, chromium; L, lime; W, water; P, potash; B, barytes. In the year 1807, in giving an account of Dalton's theory, he used the hieroglyphics introduced by that chemist, and in the same year he employs regular formulæ (Phil. Trans.) in his paper on oxalic acid. For example, he says, "If an atom of oxygen be *w*, an atom of carbon *c*, and an atom of hydrogen *h*, an integral particle of oxalic acid may be represented by $4\ w + 3\ c + 2\ h$," and so on throughout the paper. In 1814, Berzelius published his new system of mineralogy in Swedish, a copy of which he sent to Dr. Thomson, asking him as a favour to endeavour to have it translated into English. The Swedish work and the letter are still extant. Dr. Thomson prevailed on his friend John Black, Esq. to execute the translation. In this work, Berzelius says, in reference to formulæ for minerals, I "shall for this purpose follow the rules given by Thomson in his System of Chemistry. He arranges the initial letters of the name of the earths in such order, that he begins with that of which each fossil contains the most, and so on in succession till that of which it contains the least. I cannot make use of the same letters as Thomson, because they are merely related to the English name, and as these formulæ ought everywhere equally to be understood, I consider they ought to be founded on the Latin nomenclature." In the same year, in his paper on atomic proportions, he modified some of these symbols; and in his letter, which I possess, dated 12th May, 1814, he observes:—"Vous trouverez quelques changement faites dans les signes chimiques employés dans l'essai sur la mineralogie J'ai cru devoir preferer l'employer des exposants algebraiques au lieu de mettre les numeros audessus les lettres pour ne pas oter une ligne." Formulæ, at the present time, are usually constructed on the plan adopted in this work, and the symbols of which they are composed are the same as those given under *Atomic Weights.* A formula represents the relations in which the atoms of the elements of a compound exist in respect to each other, and is obtained in the following manner:—The first object is to determine the composition of a compound in 100 parts. If we divide by the atomic weight of each substance, the amount of each, contained in 100 parts, we shall effect this object, and then all that remains is to reduce these numbers to their lowest terms. In the following table the 1st column gives the symbols; the 2d column represents the per centage composition of sulphate of oxide of iron by analysis; 3d, atomic weights of each substance; 4th, by dividing 3d by 4th; 5th, obtained by dividing the two last terms by the first or lowest:—

1.	2. Per Cent. Analysis.	3. Atomic Weights.	4. Atoms.	5. Reduced to Lowest Terms.	6. Calculated	7. Composition.
FeO 1 at protoxide of iron,...	25·7	4·5	5·71	1·	4·5	25·89
SO_3 1 ,, sulphuric acid,......	28·9	5·	5·78	1·012	5·	28.77
7HO 7 ,, water,................	45·4	1·125	40·35	7·06	7·875	45·34
	100·				17·375	

The formula, therefore, is nearly $FeOSO_3$ 7HO. There is a slight excess in the case of the acid and water, but trifling. It is obvious that from the formula may also be calculated the per centage composition, by looking for the atomic weight in the table, and constructing column 7. Instead of using the initial letter for oxygen, sometimes dots are used; thus, the above formula would be $\dot{Fe}\,\dddot{S}\,7\,\dot{H}$; but this plan is seldom adopted in England, France, or Germany. A theoretic method sometimes employed is to calculate the oxygen in the acid and base, of which the former should be a multiple of that in the base. In the above example the oxygen in the acid is three times that of the base. But this method is never used in this country, as it is quite inapplicable in organic chemistry, and in many basic salts. Sulphur compounds are sometimes represented by a comma; thus, bisulphide of iron would be $\overset{,,}{Fe}$. In organic chemistry a horizonal or minus line over a body indicates that it is an acid, as $\overline{A}$ acetic acid. A plus over a body shows that it is a base, as $\overset{+}{M}$ = Morphine.

Formylamine. *Formyliaque.* NH_2C_2H. B.P. 289°. Colourless fluid, with a caustic taste, and slightly ammoniacal odour; saturates acids; by the action of bromated Dutch liquor on ammonia.

Formyle. C_2H = Fo. The hypothetic base of formic acid, which is the teroxide; it is methyle deprived of 2 atoms hydrogen, by means of 2 oxygen converting the hydrogen into water. An important compound is terchloride of formyle, or *chloroform.*

Formyline. See METHYLAMINE.

Fornacite. A crystallized slag from a lime furnace at Tanndorf. SiO_3 46, CuO 22·5, MgO 7·5, FeO and MnO 8, Al_2O_3 14·. A somewhat analogous black slag, gelatinizing with muriatic acid, was examined by my late pupil, John Brown, Esq. from St. Rollox. SiO_3 36·47, CaO 28·89, FeO 12·68, Al_2O_3 18·88.

Forsterite. Right rhombic prisms, in brilliant crystals, with green pyroxene, at Vesuvius; contains silica and magnesia.

Fosresenic Acid. Yellow powder, formed by acting on Highgate resin, with nitric acid.

Fossils. A term applied to organic remains existing in strata in the earth.

Fossil Resin. *Fossil Copal, Copalite.* See HIGHGATE RESIN.

Fowlerite. Apparently a silicate of manganese.

Fracture. The irregular surface formed when a mineral is broken. The species of fracture are conchoidal, even, uneven, splintery.

Francolite. A variety of phosphate of lime, or apatite, from near Tavistock, in irregular aggregated or mammillated crystals.

Frankincense. See OLIBANUM.

Franklinite. Spec. grav. 5·069, to 4·87, H 6 to 6·5. Iron-gray masses and 8-hedrons, somewhat magnetic; fracture conchoidal; lustre metallic; opaque, crystals semitransparent, with a blood-red tint. Found at Franklin, N. Jersey. Fe_2O_3 66·, Mn_2O_3 14·96, ZnO 17·425, SiO_3 ·204, HO ·56. It is isomorphous with magnetic iron ore. For its formula, see CHROME IRON ORE. B.B. infusible. Gives oxide of zinc on charcoal at a high temperature.

Fraxinine. 6-sided prisms in Fraxinus excelsior or ash.

Freezing Mixtures. Mixtures or solutions of salts to produce cold, are employed for cooling wine, and for experimental purposes. All bodies which produce cold by mixture act chemically on each other. They are either both solid, or, at least, one of them is solid, and they begin to become fluid as soon as they are mixed; the degree of cold is proportioned to the rapidity of the liquefaction. Very considerable degrees of cold, however, are produced by anhydrous salts, as nitre or nitrate of potash, the cause of which is not so apparent. In those cases where solid water is liquefied, the cold is produced by the latent heat required to convert the solid into the fluid. Intense cold is produced by sprinkling muriatic acid over pulverized sulphate of soda (from 50° to 0°). Carbonate of soda, in crystals, is a convenient salt for the production of cold; 9 phosphate of soda and 4 dilute nitric acid reduce the temperature from + 50° to — 12°. 5 sulphate of soda and 4 dilute sulphuric acid lower the temperature to 3°. 9 phosphate of soda, 6 nitrate of ammonia, 4 dilute nitric acid, produce a cold of — 21°. 2 snow 1 common salt produce a cold of — 5°. 5 snow, 2 common salt, 1 salammoniac, lower the temperature to — 12°. 3 snow 2 dilute sulphuric acid from 32° to — 23°. 8 snow 5 muriatic acid from 32° to — 27°. 4 snow, 5 chloride of calcium from 32° to — 40°. 3 snow 4 potash to — 57°.

Freezing Point. The point as indicated on the thermometer, at which water freezes. The freezing point on Fahrenheit's thermometer is 32°; Centigrade, 0·; Reaumur, 0·.

French Berries. See BERRIES.

Friesland Green, or *Brunswick Green,* or ammonia chloride of copper.

Frit. An imperfectly fused mass of silica and bases in the preparatory stage of glass-making.

Frost, Hoar. Frozen dew which has been deposited on the cold surface of vegetables, and bodies exposed to radiation on a clear night.

Frugardite. A variety of idocrase, from Frugard, Finland.

Fuchsite. Biaxial, oblique, or potash mica. SiO_3 48·, Al_2O_3 34·5, KO 10·75, Fe_2O_3 1·8, MgO ·72.

Fuchs' Soluble Glass is made by heating 15 pure sand, 10 carbonate of potash, and 1 charcoal. This, when dissolved in 4 or 5 parts of boiling water, and evaporated to the specific gravity 1·24, allows the water to evaporate from it when it is spread on a surface, and leaves a glassy glaze, which has been used to protect wood from fire.

Fuci. A term often applied to the class of sea weeds. In chemistry the term includes particularly those sea weeds which are burned for the value of their ash. Of all vegetable productions sea weeds contain the largest amount of inorganic matter. The *Padina pavonia* contains as much when dried as 34¾ per cent. ash; the Sargassum vulgare, or Gulfweed, 22·58 per cent.; the Chondrus crispus 20·61.

	1.	2.	3.
Sulphuric acid,......	2·86	5·05	1·45
Chlorine,............	·23	4·77	2·41
Phosphate of lime,..	1·05	1·42	1·47
Lime,................	2·35	1·87	0·95
Magnesia,............	1·19		0·78
Potash,..............	0·98	4·24	2·68
Soda,................	1·05	5·48	1·66
Silica, &c...........	1·20	·11	·08

1. Fucus vesiculosus, per cent. 2. Laminaria digitata. 3. L. latifolia. The amount of iodide of sodium contained per cent. in the dried plant is as follows:—Fucus digitatus 3·34, F. vesiculosus ·32, F. nodosus ·34, F. serratus 1·18. The respective quantities of ash in these species being 20·4 per cent. 16·39, 16·19, 15·63.

Fucusamide. $C_{15}H_6NO_3$. Needles, by adding NH_3 to fucusole.

Fucusine. $C_{30}H_{12}N_2O_6$. Small plates, isomeric with furfurine; soluble in 2400 water (twice as insoluble as furfurine), less soluble in spirit than furfurine; forms crystalline salts with acids. Obtained by boiling fucusamide with potash, forming nitrate of fucusine with the product, crystallizing it out, and decomposing by NH_3.

Fucusole. $C_{15}H_6O_6$. Spec. grav. 1·15; B.P. 341°. Colourless oil, becoming green with HCl, yellow with NO_5, greenish-brown with SO_3. Obtained by distilling sea weeds, ferns, and Iceland lichen with water and sulphuric acid, washing the product with water, and rectifying over chloride of calcium. When treated with SH in a solution of alcohol it yields *Thiofucusole,* which is converted by dry distillation into *Pyrofucusole.*

Fuel. (*Combustible,* Fr.; *Brennstoff,* Ger.) On the continent of Europe, in addition to coal, which is sparingly used, turf and wood charcoal are the usual kinds of fuel. In this country, turf or peat is only used in the boggy parts of Ireland and the mountainous regions of England and Scotland, while coal is the almost universal article of consumption. The value of a coal is indicated by the amount of steam which its heat can raise in a given time.

Loss of Gas in Coking Coal.—It is an opinion entertained by some, that in using coal for raising steam the bituminous matter is dissipated, and that it is from this matter that the gas is obtained, so that in coking coal the gas is actually thrown away. Attention was directed to this point in the recent experiments upon coal for the steam navy. If we multiply the coke of coal by its heating power 13268, and divide by 965·7, the latent heat of steam, we obtain the number of lbs. of water which the coke could evaporate without the aid of the bituminous matter. On comparing the results of this calculation with the actual work done by different coals, we find that it is actually greater, with certain exceptions.

$$\left(\frac{C \times 13268}{965 \cdot 7}\right) + \left(\frac{H - h \times 6247}{965 \cdot 7}\right) = x$$

C = carbon, H = hydrogen in an unit of fuel, and h the hydrogen corresponding to the hydrogen contained in the coal. These multiplied by their heating powers (Dulong), and divided by the latent heat of steam, indicate the number of lbs. of water evaporated.

Coal.	No. of lbs. of Water converted into Steam by 1 lb. of Coal. Practical.	No. of lbs. of HO convertible into Steam by the Coke left in the Coal. Calculated.
Anthracite (Welsh), ...	9·46	12·554
Pontypool, ...	7·47	8·144
Grangemouth, ...	7·40	7·292
Slievardagh (Irish), ...	9·85	10·895
Wylam's patent fuel, ...	8·92	8·378

Fuliginous. (*Fuligo*, soot). A combustible body is said to be fuliginous when it gives out much smoke or soot when burning.

Fuller's Earth. *Argiles smectiques*, *Terres a foulon.* A clay so named from the property which it possesses, owing to its porosity, of removing stains of grease from cloth. It occurs in the green sand and oolitic formations. Colour dull greenish-gray; texture earthy; fracture uneven, opaque, rather tough, sectile; feel soapy, lustre dull; hardness 1. Spec. grav. 2·4448; adheres very slightly to the tongue; falls to powder in water. SiO_3 44, HO 24·95, Al_2O_3 23·06, CaO 4·08, MgO 2·, FeO 2. B.B. fuses into a gray enamel; with soda into a bead; transparent and colourless when hot, opaque when cold; does not unite with borax; forms a frit with microcosmic salt.

Fulminating Cotton. See GUN COTTON.

Fulminating Powders. The most important of these are commonly termed detonating powders, and are salts of fulminic acid, viz. fulminate of mercury, and fulminate of silver.

Fulminic Acid. (*Knallsäure*, Ger.) $C_4N_2O_2 = Cy_2O_2$ 8·5, 68·. Bibasic acid, known only in the state of union with bases; discovered by Howard. The fulminate of mercury is obtained by dissolving 1 part of mercury in 12 nitric acid (Sp. grav. 1·36), and adding 11 alcohol (·8475) with gentle heat. The action becoming violent, the heat should be withdrawn; a white precipitate falls, which, when washed and dried, is *fulminate of mercury* ($2\,HgO,Cy_2O_2$). In this action aldehyde and nitrous ether are evolved, and the fulminate precipitates according to the reaction. $C_4H_6O\ 2NO_3 = C_4N_2O_2,\ 5\ HO$. It is used for filling percussion caps, but must be warily used from its liability to explosion. 10 parts are mixed on a marble slab, with 30 parts water, by means of a wooden pestle. To this paste are added 6 parts of saltpetre. The caps are then filled to the proper extent with the mixture. The salt may be obtained in silky needles by dissolving in hot water. *Fulminate of silver.*—$2\ AgO,Cy_2O_2$. Crystalline powder obtained by a similar process, using silver instead of mercury, and 10 parts nitric acid with 20 alcohol. According to E. Davy this acid contains hydrogen, which would account for its explosive character. Fulminate of silver may also be formed by passing the vapour of nitrous acid into a solution of nitrate of silver in alcohol.

Fumaramide. $C_8\ 2\,NH_2H_2O_2,\ O_2$. White insoluble powder, by adding ammonia to fumaric ether.

Fumaric Acid. *Paramaleic Acid*, *Lichenic Acid.* HO,C_4HO_3. Rhombic mica-like scales, obtained by keeping malic, or maleic acid at the fusing point for a long time; soluble in 200 water. It is obtained also from *Fumaria officinalis*, and *Cetraria Islandica.*

Fumaric Ether. $C_4H_5OC_4HO_4$. Oily fluid by passing chlorohydric acid gas into malic acid in absolute alcohol, and rectifying.

Fumaroles. *Soffioni.* Terms applied to volumes of steam, which issue in large quantities through openings in the earth, and carry with them, mechanically, boracic acid; occurring in Tuscany, and other parts of Italy and Sicily.

Fume. A term applied to the sulphate of lead, &c. carried by the draught into the flues of lead furnaces.

Fumigation. The process of exposing infected places to the action of chlorine, bleaching powder, sulphurous acid, &c. to destroy the supposed cause of infection.

Fuming Liquors. A term applied to such fluids as the chloride of soda, or fuming liquor of Libavius, from the vapours which it evolves in the air.

Fungic Acid. A colourless syrup with acid taste, mixing with water and alcohol in all proportions; nearly insoluble in ether, from fungi; it is fumaric acid (Bolley).

Fungine. The basis of the fungi, probably a mixture of cellulose, with a nitrogenous body.

Funkite. Green coccolite or pyroxene.

Furfurine. $C_{30}H_{12}N_2O_6$. F.P. under 212°. White silky needles resembling caffeine, without smell, and scarcely any taste; nearly insoluble in cold, soluble in 135 hot water; very soluble in alcohol and ether, with an alkaline reaction; forms monobasic crystalline salts. Obtained by boiling *furfurolamide* for ten to fifteen minutes, with a large quantity of dilute caustic potash. A heavy yellow oil falls, which becomes crystalline on cooling. It is purified by dissolving it in a boiling dilute solution of oxalic acid; on cooling, binoxalate of furfurine separates in crystals. These are dissolved in hot water, and purified with animal charcoal. Ammonia separates the base by uniting with the acid.

Furfurolamide. *Furfamide.* $C_{30}H_{12}N_2O_6$. Isomeric with furfurine. Needles, or a white soluble powder; decomposed by boiling in water and alcohol into ammonia and furfurole.

Furfurole. *Oil of Ants.* $C_{10}H_4O_4$. Spec. grav. 1·1648; B.P. 323¾°. Clear volatile oil; soluble in water, alcohol, and ether. First obtained by Dobereiner by distilling sugar, black oxide of manganese, and sulphuric acid. Emmett showed that it could be procured by distilling meal, bran, sawdust, &c. with sulphuric acid, without any oxide. The best proportions are, 6 parts bran, 5 parts sulphuric acid, diluted with water, distilled in a capacious retort. Bran yields 2¾ per cent. of this oil. Compare *fucusole.*

Furnace. (*Fourneau*, Fr.; *Ofen*, Ger.) An apparatus in chemistry for the application of high temperatures. The most important are the *Reverberatory furnace, Flamm ofen,* where the heat is applied to the substance by a flame playing on its surface (see SODA MANUFACTURE, and COPPER FURNACE); *Wind furnace, Windofen,* where, by means of a tall chimney and strong draught, air is drawn through or over a fire with great force.

Fusaine. A preparation of charcoal used in Paris, composed of charcoal 82·8, calcined cinders 1·6, volatile matters 15·6.

Fuscine. A brown base, becoming red in air. Obtained in rectifying bone oil, by Unverdorben.

Fuscite. A variety of Scapolite.

Fusel Oil. *Potato and grain oil.* In the distillation of potato and grain spirit, a quantity of solid oil is obtained, and likewise a fluid portion. Dumas described the fluid portion from potato spirit as an amylic alcohol in 1836, and in the same year I described a similar body from grain spirit, which I obtained from the late Mr. Coffey, patentee of the well-known still (Brit. Annual, 1837). The solid oil has been shown to contain margaric acid, oenanthic acid, capric acid, and bassic acid (?) The presence of the amyle oil in spirit renders it noxious; it should therefore be carefully removed by water.

Fusibility. The capacity of fusing or melting.

Fusible Metal may be formed of 31 lead, 19 tin, 50 bismuth, to melt at 203°, or of 34 lead, 19 tin, 47 bismuth.

Fusing or **Melting Point** of oils and oily acids is determined by introducing them into small tubes, with a fine thermometer inserted, plunging into water, which is gradually heated, and noting the point at which the thermometer stands when the oil fuses.

Fusion. The conversion of a solid into a liquid state, depending upon the absorption of latent heat by the body; in certain substances the fusing point seems to vary after one fusion, as in the crystalline organic compounds. An analogous result occurs with stearine and other fats.

Fusteine. Orange-coloured varnish, formed from fustine by absorption of oxygen.

Fustet. A name for sumac.

Fustic. *Yellow-wood.* Two varieties of sumac are known in commerce. *Old fustic*, from *Morus tinctoria; new fustic*, from *Rhus cotinus*, from Venice. Fustic is a yellow-coloured wood employed in dyeing; it gives a yellow colour to water, with a slightly acid reaction; this colour becomes purplish or reddish by alkalies. Alum and chloride of tin give orange precipitates. Some greens are dyed with the assistance of fustic; the yarn being first dyed blue by indigo, and then passed through pyrolignate of alumina, and lastly into a hot solution of fustic, which gives it a fine green shade.

Fustine? Colourless crystals from fustic, by making a decoction in water, precipitating tannic acid by gelatine, filtering, evaporating to near dryness, extracting by ether, adding water, precipitating by oxide of lead; the yellow precipitate decomposed by sulphuric acid. Fustine deposits from the liquor. It resembles *rhamnine.* Morine is another colouring principle of fustic.

G

Gabbro. *Ratchel, Diorite, Euphotide.* A rock consisting of a mixture of Labradorite and Diallage, with compact albite.

Gabronite. A synonyme of Wernerite or Scapolite.

Gadolinite. H 6·5. Spec. grav. 4·1493, and 4·028 to 4·238. Greenish-black, velvet-black, massive, or in oblique rhombic prisms, with angles of 115° and 98°; fracture flat, conchoidal; lustre vitreous, inclining to resinous, slightly translucent on the edges, almost opaque. BB. intumesces, forming cauliflower excrescences, becoming white with loss of water; fuses with borax into a dark glass coloured by iron, becoming bottle-green in the reducing flame; with soda becomes a slag. It sometimes is mixed with platinum. A specimen from a Swedish cabinet yielded silica 24·33; glucina 11·60; yttria 45·; cerium oxide 4·33; protox. iron, 13·59; water ·9. (T. Thomson, A. Steel, 1835.)

Gaduine. $C_{35}H_{22}O_8$. Dark brown powder, soluble in strong sulphuric acid and potash, but not in nitric and muriatic acids; soluble in alcohol and ether; insoluble in water, without taste and smell; obtained by saponifying cod liver oil, decomposing the soap with acetate of lead, extracting the lead soap with ether; the margarate of lead remains, while the oleic acid and gaduine dissolve. The solution is decomposed by dilute sulphuric acid, and the gaduine purified.

Gahnite. *Automolite* (αυτομολος, deserter.) *Zinc Spinel; 8-hedral Corundum.* Specific grav. 4·261; H 7·25.

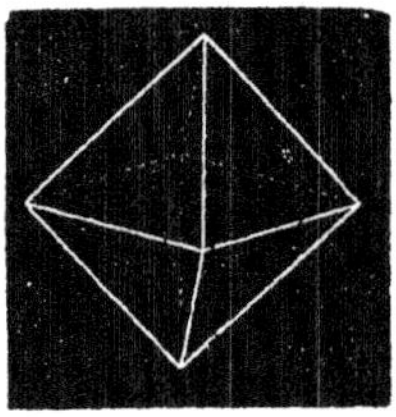

Dark green regular 8-hedrons, similar to magnetic iron ore; cross fracture conchoidal; opaque, or only translucent on the edges; occurring near Fahlun, imbedded in talc slate. B.B. infusible; fuses with borax into a clear glass; greenish when hot, colourless when cold; with salt of phosphorus fuses into a clear glass; does not fuse with soda; silica 3·84; alumina 55·14; zinc oxide 30·02; magnesia 5·25; protox. iron, 5·85.

Galactine. $C_{33}H_{29}O_2$. Spec. grav. ·969; F.P. 117° to 137°; B.P. 640°. Snow-white flakes when moist, and yellow when dry; resembling wax; tasteless, plastic between the teeth, and adheres to the teeth like a resin; insoluble in fixed alkalies; soluble in nitric acid, and reprecipitated by water. Obtained by alcohol from the extract of the milk of the cow tree (*Galactodendron utile*) of S. America. (T. Thomson, 1829.)

Galactite. A mineral occurring at Kilpatrick, a mesotype or Thomsonite.

Galam Butter. A reddish-white fat from Africa, said to be from the *Bassia butyracea.* F.P. 72°; insoluble in alcohol and water.

Galanga. The root of the *Alpinia Galanga.* It affords a crystalline body to alcohol and ether. It yields also a yellowish volatile oil, $C_{10}H_9O$, hardening and thickening in air.

Galapectite. A variety of *Halloysite.*

Galbanum Resin. $C_{40}H_{27}O_7$. Yellowish-brown resin. The Levant resin is in tears and masses; the Persian in masses with a strong smell; soluble partly in alcohol, with an acid and rather a bitter taste; insoluble in oil of turpentine. By distilling with water, a colourless oil comes over. Specific grav. ·912; soluble in alcohol and ether. By dry distillation it yields a blue oil.

Gale, Sweet Oil of. Specific grav. ·876. Dark yellow oil, obtained by distilling the leaves of *Myrica gale* with water, consists of ¼ Oleoptene and ¾ Stearoptene; very soluble in ether.

Galena, or *Sulphide of Lead.* See LEAD.

Galipot. A mixture of resin from *Pinus abies*, and *sylvestris*, or sometimes common turpentine deprived of some of its oil.

Gall. See BILE.

Gall Nuts. See NUTGALLS.

Gall Stones. See CONCRETIONS and CHOLESTERINE

Gallic Acid. $C_7HO_3 2HO$. Silky needles, of rhombic prisms, without smell; slight acid astringent taste; precipitates neither glue nor organic bases; but it forms a black precipitate with iron; its solution is not changed in air; in contact with bases and air, it becomes black; and like humus (*tannomelanic acid*, $C_{14}H_4O_7$), it is precipitated from its solutions by salts of lead, alumina, magnesia, &c.; soluble in 100 cold, 3 hot water; soluble in alcohol, less soluble in ether. When mixed with 5 parts of strong sulphuric acid it becomes red; and on the addition of water, *rufogallic acid*, $C_7H_2O_4$, or *Parellagic acid*, falls, partly flocky and partly in granular crystals of a brownish colour. Gallic acid by dry distillation at 410°, gives off crystals of pyrogallic acid, C_2HO; by a strong heat, a black matter, *metagallic acid*, $C_{12}H_3O_3HO$, is formed. Gallic acid exists ready formed in plants, and may be obtained from nutgalls, divi divi, &c. by boiling with water, precipitating the tannic acid from the solution by gelatine, evaporating the decanted liquor to extract, treating the extract with boiling alcohol, evaporating the extract and crystallizing from ether, decolourizing by animal charcoal. It deposits likewise from a decoction of nutgalls by standing. It is used in photography, and in testing to detect iron.

Gallinace. A synonyme of Obsidian.

Gallipoli Oil. Olive oil purified at Gallipoli by cold water.

Gallizinite. A synonyme of sulphate of zinc and anatase.

Gallon, Imperial, has a capacity of 277·274 cubic inches. A gallon of distilled water weighs 10 lbs. or 70,000 grains. 1 French litre is equivalent to ·22009687 gallon, or 1 gallon = 4·543458 litres.

Galmei. Calamine, or carbonate of zinc.

Galvanism. See ELECTRICITY.

Gambir, or *Catechu.*

Gamboge. (*Gomme gutte*, Fr.; *Gummigutt, Gutti*, Ger.) Yellow resin, used as a paint and purgative, from the *Gambogia gutta*, and other species from India. It occurs in four states in commerce, which are as follows, according to Christison:—

	Pipe G.	Cake G.	Common G.	Ceylon G.
Resin,...	74·2	64·3	51·4 to 65·	68·8
Gum,...	24·	20·7	17·2	20·7
Starch,......		6·2	7·8	
Water,....	4·5	4	7·2	4·6
Insol. matter,		4·4	7·8	6·8

The resin is soluble in ether, without smell or taste; soluble in alcohol, with acid reaction; soluble in caustic potash, forming a dark solution, a compound separating by evaporation; soluble in ammonia with a red colour, and precipitating by an acid. The precipitated resin when dried consists of C 72·05, H 7·23, O 20·72.

Gangue or **Matrix.** That portion of a rock in which an ore is deposited.

Ganomatite. See CHENOCOPROLITE.

Garancine. A reddish-brown powder, prepared from madder root. The madder is ground and digested in diluted sulphuric acid, washed with water and dried. The action of the acid seems to be to remove lime with which the colouring matter is united. Spent madder which has been used in dyeing is found to yield an additional amount of colour when converted into garancine. 1 part of garancine is estimated usually to be equal to 3 to 3½ madder; but it sometimes varies to 2, 2½, 5, and 7. The value of garancine is determined by its dyeing powers, either upon a small or large scale.

Garlic, Oil of. *Sulphide of Allyle.* C_6H_5S. B.P. near 300°. Yellow oil, decomposing at 302°, evolving suffocating vapours, and leaving a brownish mass; insoluble in water; very soluble in alcohol and ether; not altered by dilute acids nor by alkalies; nitric acid forms with it sulphuric and oxalic acids; dissolves in strong sulphuric acid with a purple colour, reprecipitated by water; it forms double compounds with solutions of platinum, mercury, silver, palladium, &c.; when distilled over potassium, oil of garlic is perfectly colourless, and there remains in the retort sulphide of potassium. It is obtained by distilling the roots of onions (*Allium Sativum*) with water, and rectifying the oil.

Garnet. *Precious Garnet, Almandine, Aplome, Greenlandite, Pyrenaite, Carbuncle of the Ancients.* Specific grav. 3·157 to 4·208; H 6·5 to 7·5. Red rhomboidal 12-hedrons; the edges are often replaced by tangent planes which ultimately produce the

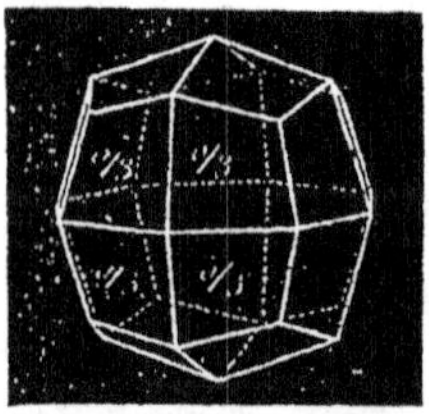

leucite crystal of 24 trapezoidal faces; structure foliated, and the cleavage faces parallel to those of the primary form; fracture uneven or conchoidal; lustre vitreous; transparent to translucent. B.B. fused easily into a dark-coloured globule. This mineral is a mixture of various garnets, or silicates of various bases, isomorphous with each other. These will be now described. Garnets are found in mica slate and primary rock.

Garnet, Black. *Melanite, Iron magnesia Garnet.* Spec. grav. 3·157 to 3·73; H 6·75. Velvet-black 24-hedrons; lustre inclined to resinous; fracture flat and imperfect conchoidal; opaque. Found at Arendal, Norway. SiO_3 42·45, Al_2O_3 22·475, FeO 9·292, MnO 6·273, MgO 13·43, CaO 6·525.

Garnet, Bohemian. *Pyrope.* Spec. grav. 3·78; H. 7·5. Blood-red rounded or angular grains of right prisms with a square base (?); fracture conchoidal; streak white. B.B. fuses into an opaque black bead with difficulty. Found in the mountains on the south of Bohemia, at Elie in Fife in trap tuff. SiO_3 42·, Al_2O_3 20, MgO 20·199, FeO 9·096, Fe_2O_{32} 1·507, MnO ·32, CrO_3 3·013, CaO 1·993 (Kobell); SiO_3 43·7, Al_2O_3 22·4, MgO 5·6, FeO 11·48, MnO 3·68, CrO_3 7·68, CaO 6·72. (From Meronitz, by Wachtmeister).

Garnet, Brown Manganese. Spec. grav. 3·829; H. 6·75. Tombac brown garnet crystals; brittle and easily frangible. SiO_3 33·716, CaO 25·884, Al_2O_3 7·972, FeO 15·84, MnO 16·704, HO ·08 (T. Thomson); SiO_3 39·, Al_2O_3 14·3, FeO 15·44, MnO 27·9 (D'Ohsson).

Garnet, Green. *Grossularite, Lime Garnet.* 3 $CaOSiO_3$, $Al_2O_3SiO_3$. Spec. grav. 3·372 to 3·64; H. 6·75. Asparagus-green rhombic 12-hedrons, or generally 24-hedrons, similar to leucite crystals; lustre resinous, dull to shining; fracture between conchoidal and uneven; translucent. B.B. thin splinters are rounded on the edges, but not completely fused. SiO_3 40·55, Al_2O_3 20·1, FeO 5·, MnO ·48, CaO 34·86. It has the same composition as idocrase, but a different hardness and crystalline form. Found in Siberia, near the Wilni river.

Gas. Chemical writers before Van Helmont (1577-1644), speak of an elastic fluid disengaged from bodies by fermentation, combustion, and effervescence, under the name of *spiritus sylvestris;* but they do not seem to have distinguished it from respirable air. Van Helmont gave it the name of *Gas*—(*Geest*, Dutch; *Gaist*, *Gast*, Scottish; *Geist*, Ger.)—which he defines a *non-coagulable spirit, as from fermenting wine*, likewise from hydromel, verjuice, bread, salammoniac, in combinations, from vegetables, and from gunpowder. He attributes the deadly influence of the Dogs' Grotto near Naples, and the accidents from suffocation in mines and in cellars, to this *gas.* He drew the conclusion that *gas* is a different body from the air we breathe. Boyle afterwards (1627-1691) used the term *artificial air*, instead of the word *gas;* and he showed that artificial air is not always the same when obtained from various sources. Hales followed in the same track, and showed that an almost infinite number of bodies yields air by heat. But it was not till 1758, when Dr Black characterized *fixed air* (carbonic acid), in its true light, that genuine pneumatic chemistry, so fertilely cultivated afterwards by Cavendish, Priestley, &c. can be said to have been indubitably based. The term is applied at the present time to those elastic fluids which are permanent at the usual pressure and temperature of the atmosphere.

Coal Gas.—In fixing on the position for the manufacture of gas, it is obvious that coal gas being lighter than air, much inconvenience would be experienced if the gas required to descend. It is necessary, therefore, that a gas work should be situated in the lowest part of the city which it is intended to supply. The essential apparatus required for the preparation of gas, are (1st), iron or fire-clay retorts, *a*, (see fig.) into which the coal is introduced and surrounded by fire at *c*, to disengage the gas; (2d), pipes to carry off the gas, in order that it may be purified; (3d), condensers, or vessels for removing the tar, and volatile matter by cooling, from the gas; (4), lime purifiers to condense carbonic acid and sulphuretted hydrogen, which injure the lighting properties of the gas; and (5th), gasometers for retaining the gas after purification.

1. *Retorts.*—The retorts are made of iron or of fire-brick. Their shape is cylindrical or flattened, from 6 feet 6 inches to 9 feet long, and from 12 inches wide by 12 high, to 20 inches wide by 14 high. They are placed in a horizontal direction, sometimes 1 and sometimes 3, 5 or 11 together, in the same furnace for the sake of economy. Between the furnace and the retort is placed an arch of brickwork to save the iron, and the flue turns back upon the retort, so as to economize the heat. The arch and flue are seen in the diagram, the arch being darkened over and below the retort. Retorts are sometimes 12 to 18 feet long; they are then open at both extremities for charging, and have a furnace at each end. The coal used in Scotland, is the Parrot or cannel coal. It affords a much larger supply of gas than the cherry or splint, or caking coals. A ton of Lesmahago coal produces in Glasgow 10,880 cubic feet of gas; while the Newcastle Wallsend coal yields only 7,400 cubic feet; in London, 8,500 to 9,500. The cause of the inferiority appears to be, that the quantity of carbon in the cannel, is more nearly in the relation to the hydrogen, of equal atoms, than we find existing in the domestic coal. The weight of the 10,880 cubic feet of gas, supplied by the Lesmahago coal, amounts to about one-fifth of the whole of the coal employed; so that a ton of coal will yield 4 cwt. of gas. The remainder of the coal is spent in forming $\frac{3}{5}$ coke, $\frac{2}{5}$ tar, water, &c. That coal contains more water than is usually supposed, is obvious from the results which I have obtained with some kinds of cannel coal. When distilled at 350°, it lost 4·097 per cent. of pure water, which I collected and examined. In another instance, when the coal was heated up to between 400° and 500°, the loss in a week exceeded 14 per cent. When heated a little above 600°, paraffine passes over without any tar or gas (Young). The quantity of fuel required to heat the retorts, amounts to from 30 to 40 per cent. of the coal required for producing the gas; but this is usually derived from the coke extracted from the retort. From the retort, the gas is conducted by the ascension (*h*) and dip (*a*) pipes, to the *hydraulic main*, E, a cast iron pipe possessing a diameter of 12 or 15 inches. The hydraulic main is supported by pillars placed in front of the retorts, and serves the double object of condensing tarry matter, and also, as it is half filled with fluid, and the dip pipes, *a a*, penetrate under the surface of the water, to prevent any gas from returning into the retorts when the workmen are charging them. At the upper part of the dip-pipe is a plug for allowing tarry matter to be extracted, which is apt to accumulate and detain the gas.

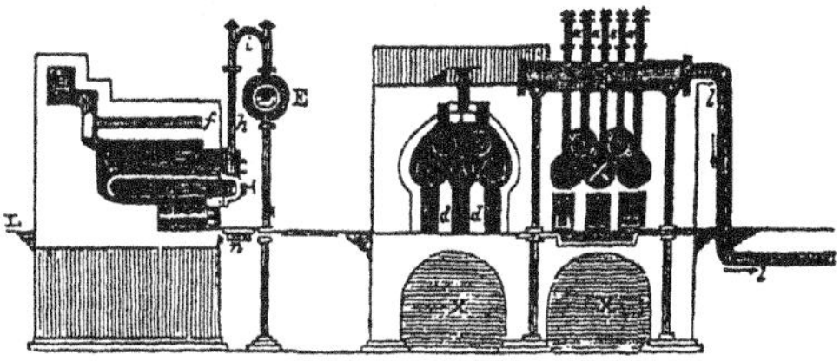

The hydraulic main leads by *l* into *s*, a vessel for receiving the tar and ammoniacal liquor.

2. *Condensers.*—The gas then passes into lofty syphon-like tubes, not exhibited in figure 2 at *a*, which it traverses, the tubes being usually moistened by water to remove the heat, and condense tar and ammoniacal water which run into the tar well. In some works, the gas now passes up a vertical vessel filled with coke, which is moistened with a stream of cold water to condense ammonia, &c. In other works, the gas is washed by being passed over the surface of water. To overcome the resistance to the progress of the gas produced by these processes, in large gas works it is usual to have a machine termed an exhauster, by which the gas is pumped from the hydraulic main as rapidly as it is generated. It is believed that the carbonaceous deposits in the retorts are thereby diminished, and the leakage of the gas to the extent of 12 per cent. prevented.

3. *Purification.*—The coal gas, as it comes from the condensers, contains several impurities, viz. sulphuretted hydrogen, carbonic acid, ammonia, &c.; and in order to remove these, so far as practicable, the refrigerated gas is passed into a wet lime purifier, D, which is a cylindrical iron vessel with an air-tight cover. The vertical axis of the agitator, *t*, works in a stuffing box in the cover, with two arms at the bottom for the purpose of keeping the lime suspended in the water. The gas enters by a tube, and on passing the milk of lime it is freed from its impurities. The milk of lime is made by mixing 1 part of lime with 25 of water, and is introduced by the tube V, and is removed by *n*. Sometimes the gas is purified by what is termed the dry lime process.

The lime is slaked and moistened, till, on being squeezed in the hand, the whole adheres into a solid mass, when it is in a condition for the proper purification of the gas. The purifiers are either square or oblong vessels, with generally three or four tiers of trays. The lime is laid on the trays to the depth of about 12 inches, and after being made level, it is watered again with a watering-pot, so as to make it moist, to prevent the gas from passing up the sides of the

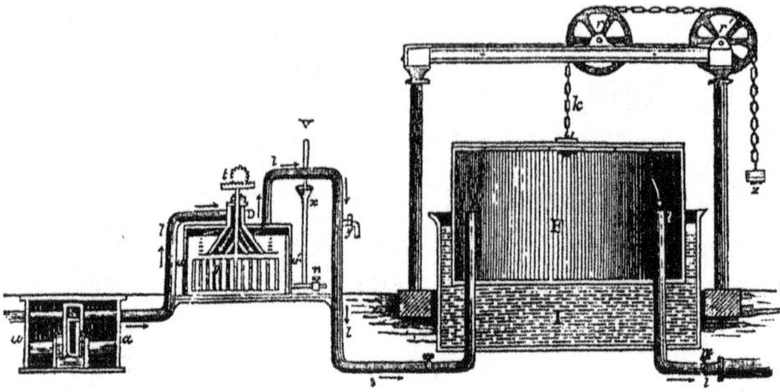

vessel without being acted on. The gas enters at the bottom, and after passing through the lime escapes at the top. The impurities in gas are removed in a variety of ways. Mr. Croll introduced the use of muriate of manganese and muriate of iron in 1840. The processes about to be detailed are those most commonly in use. Mr. Croll, by his patent dated in 1840, removes the ammonia by means of a dilute solution of sulphuric acid. The gas, after leaving the condensers, and before entering the lime purifiers, passes through a series of two vessels, made either of wood or iron and lined with lead, and provided with proper inlet and outlet pipes. The vessels are filled to about $\frac{1}{4}$ with the dilute acid, made by mixing 7 lbs. of sulphuric acid with 100 gallons of water. The acid, on becoming neutralized with the ammonia from the gas, is supplied with a continuous stream of acid from a reservoir placed on the top of the saturator, till the solution attains a specific gravity of 1·170, or as near the crystallizing point as is practicable. The stream of acid is then discontinued, and the gas passed through, till all the free acid is neutralized, when it is withdrawn and evaporated in iron pots, and yields good commercial sulphate of ammonia. The gas is afterwards passed through lime to free it from its carbonic acid, and sulphuretted hydrogen. Mr. Croll also removes the ammonia from the gas by means of muriate of manganese dissolved in water, in vessels similar to those above described. The gas on passing through the solution forms sulphide of manganese, which precipitates, and the muriate of ammonia remains in solution. The liquid is drawn off, evaporated and crystallized, and forms good commercial muriate of ammonia. Muriate of ammonia obtained in this way is very well adapted for the sublimation of salammoniac. In this case the greater part of the sulphuretted hydrogen is removed, and when the ammonia is sufficient to saturate it the whole is removed. Impure gas can also be freed from its ammonia, and the equivalent of carbonic acid by means of a saturated solution of muriate of lime absorbed into sawdust (Laming). The mixture is put into a dry lime purifier, and the gas passed through it, which gradually converts the muriate of lime into muriate of ammonia and carbonate of lime. There are generally two vessels employed for this purpose, that while the one is in action, the other may be emptied and refilled. The saturated material on being taken out is washed, and the solution of muriate of ammonia evaporated and crystallized. It is afterwards put in baskets, drained, and forms a good commercial article. When the above process is adopted, the principal remaining impurity in the gas, viz. sulphuretted hydrogen, is removed by means of sesquioxide of iron. The sesquioxide of iron is mixed with sawdust for the purpose of giving greater porosity to the mass, and the mixture is placed in dry lime purifiers to the depth of about 12 inches. The gas is allowed to ascend through it, which, in the first instance, reduces the sesquioxide to the state of protoxide with the formation of water, and one atom of sulphur is precipitated; as the action is continued the protoxide is reacted on, forming sulphide of iron and water, so that the products in the end are water, sulphur, and sulphide of iron. The material on being saturated, is taken out, spread on a floor, exposed to the air, and becomes reconverted into sesquioxide of iron, while the sulphur that was in combination with it is precipitated. The mixture is again put into the purifiers, and purifies the gas as before, and this process is repeated for twenty or thirty times, till the sulphur accumulates so as to prevent the action of the gas on the sesquioxide of iron (Laming).

4. From the purifier the gas passes to the meter, where it is measured, and thence to the gas-holder, F, which is a cylindrical iron vessel, supplied with a counterbalance, *k z*, and floating in a tank, I, of water. In large works these holders are constructed on the sliding principle of the telescope.

Ground Plan of a Gas Work, by Mr. H. T. Lowe.

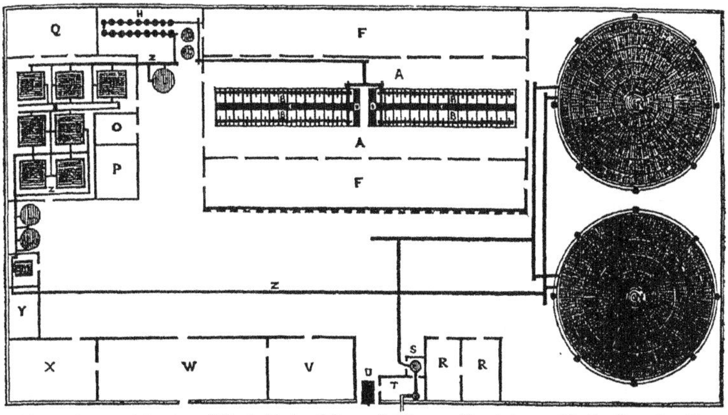

A Retort house 220 feet long;* *B* retort beds; *C* flues; *D* chimney; *F* coal stores; *G* hydraulic main; *H* condensers; *I* tar and liquor tanks; *J* ammonia purifiers; *K* purifiers for patent material; *L* scrubbers; *M* meter and house; *N* gas-holders; *O P* boilers and engine-house; *Q* revivifying shed; *R* smith's and carpenter's shops; *S* governor; *T* gate and valve-room; *U* weigh bridge; *V* house; *W* offices; *X* store-house; *Y* laboratory; *Z* mains.

Preparation of Gases in general.—When gases are prepared in large quantities, it is usual to employ a gas-holder for their reception, and a retort for the generation of the gases.

The *gas-holder* has various forms. One of the most economical kinds is Griffin's, consisting of glazed earthenware of the annexed shape. It has

two apertures at the top, one of which, *b*, is connected with a tube which passes down to the bottom of the vessel, and is surmounted by a funnel for the introduction of water; the other aperture, *d*, is at the summit, and is intended for the escape of the gas. There is a third opening at the bottom, *e*, by means of which the gas-holder is filled with gas. In order to fill it with water, which is the first step in preparing it for the reception of a gas, the two apertures at the top are opened, that at the bottom, *e*, is closed. Water is then poured in by the funnel at the top, until the vessel is completely full, the air escaping by the middle aperture. When filled, the two openings are closed, one by a cork, the other by a stop-cock; the lower aperture is now opened, and the beak of the retort containing the substance for generating the gas is inserted in it. The water cannot run out of the vessel, because the openings at the top are closed, and the atmospheric pressure is therefore withheld from the surface of the water; neither can the air enter from below, because the lower aperture is placed obliquely like a bird's fountain glass, where the water exposed to the air is pressed upon by the atmosphere, upon the principle exhibited in Cooper's tube for collecting gases over mercury. When we fill this

tube with water or mercury, retaining the stopper in it, the tube may be held vertically without any escape of fluid; and the beak of a retort or tube connected with a vessel for generating gas, being inserted in the lower opening, the gas enters and displaces the fluid. The inferior aperture of the gas-holder is constructed in a similar manner. The beak of the retort being introduced into the gas-holder, as described, as soon as the gas is disengaged, water begins to flow from the vessel in proportion to the amount of gas which enters; and as soon as all the water has been displaced by gas, we close the aperture, and set aside the gas-holder for use. In order to remove gas from the vessel as it is required, a bent tube is fixed to one of the apertures at the top, connected with a pneumatic trough, which is simply in this case a cylindrical vessel of earthenware, filled two-thirds with water, and having an inverted cylinder standing in it with an opening on one side for the extremity of the tube to enter, and another on its upper surface to admit of the escape of the gas from the end of the tube, and convey it into an inverted glass jar filled with water. The glass jar is thus placed by filling it in its natural position with water, covering its mouth closely with a plate of glass, and while holding it firmly, inverting it, and placing it on the shelf. The apparatus being thus arranged, the stop-cock at the top of the gas-holder is opened, the funnel is adjusted to the other aperture and opened. Water being poured in by the funnel, the gas escapes by the bent tube and passes into the inverted glass jar in the pneumatic trough. Thus a succession of jars may be filled until the whole of the gas has been expended. A very convenient gas-holder is one made of copper, or of japanned tin or zinc, usually called Pepys' gas-holder. This apparatus consists of a cylinder (A) for holding the gas, and of an upper vessel (C) for holding the water which is intended to displace the gas. They are connected by two tubes, one (b) dipping to the bottom of the cylinder, and the other (a) entering at the top. In order to fill the gas-holder with water, the stop-cocks on the two tubes (b, a) are opened, and water is poured into the upper vessel, until no more air escapes by the tube which enters at the top of the cylinder. There is also a stop-cock (c) at the side, which may be opened to facilitate the escape of the air. When full, the stop-cocks are closed. The gas-holder is filled with gas in the same way as the earthenware gas-holder. To procure the gas for use after it has been collected, an inverted jar, filled with water, is placed in the upper vessel (C) over the opening of the short tube (a), and the two stop-cocks being opened, the gas rises and displaces the water in the inverted jar. Connected with the cylinder there is a glass tube or gauge, which exhibits the amount of gas in the vessel in proportion to the rise or fall of the water contained in it. Gases, however, may be collected in smaller quantities, and then a gas-holder is unnecessary. The substances to supply the gas are placed in a flask (see Chlorine), communicating with an inverted jar, placed in a pneumatic trough. Heat being generally required to facilitate the evolution of the gas, is applied cautiously by means of a spirit-lamp, or a gas flame, covered with a wire gauze cylinder; or, when the amount of gas required is minute, a wide test-tube of Bohemian glass (a, b, c), supplied with a perforated cork and exit tube, may be used. English glass does not answer for such experiments, in consequence of its liability to fusion from its containing lead.

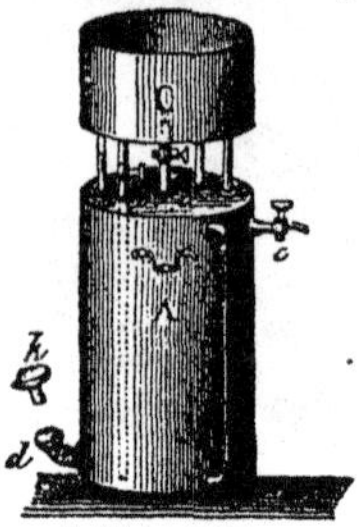

C Water-holder.
A Gas-holder.
b Water tube.
a Gas tube.
h Gauge.

Gastric Juice. The fluid found in the stomach. The following are analyses of it:—

	Dog. Without Saliva.	Dog. With Saliva.	Sheep.
Water,	973·062	971·171	986·147
Dry residue,	26·938	28·829	13·853
Ferment,	17·127	17·336	4·055
Chlorohyd. acid,	3·050	2·337	1·234
KCl,	1·125	1·073	1·518
NaCl,	2·507	3·147	4·369
CaCl,	0·624	1·661	0·114
NH_4Cl,	0·468	0·537	0·473
Phosph. of lime,	1·729	2·294	1·182
Phosph. MgO,	0·226	0·323	0·577
Phosph. of iron,	0·082	0·121	0·331
	26·938	28·829	13·853

When both par vagums were cut it was found that digestion went on, but more slowly. Thus albumen dissolved, but more slowly than when the nerves were not cut. The gastric juice had the same composition as in the healthy state, but more salammoniac, 4·517 per 1000 in one, and 3·745 in another case of a dog.

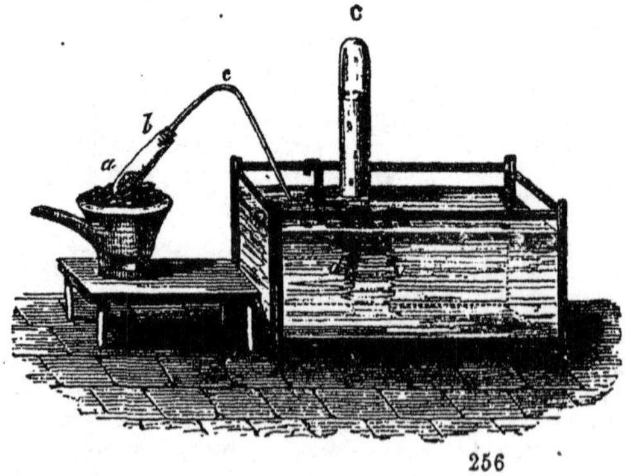

Gaultheric Acid. *Salicylate of Methyle. Gaultheria Procumbens Oil. Winter-green Oil. Birch Bark Oil.* $C_{16}H_8O_6$, or C_2H_3O $C_{14}H_5O_5$. Spec. grav. 1·180, of vapour 5·42. B.P. 428°. Colourless oil, becoming reddish on exposure to air; with difficulty soluble in water; mixes, in all proportions, with alcohol and ether; taste and smell agreeably aromatic. The aqueous solution of the oil produces a violet colour with salts of sesquioxide of iron; distilled with potash it yields pyroxylic spirit and salicylate of potash. It is therefore a salt of an organic radical, with acid properties, as it unites with bases and forms crystalline salts. It is obtained by means of alcohol from the *Gaultheria procumbens* of New Jersey, also by distilling with water the bark of the *Betula lenta*. Bromine displaces hydrogen, and forms a new compound.

Gaultherilene. $C_{10}H_8$. Spec. grav. of vapour 4·699. B.P. 320°. Colourless oil, with an agreeable odour; converted by nitric acid into a resin. Obtained by distilling commercial oil of gaultheria with caustic potash; water, pyroxylic spirit, and gaultherilene pass over. The product is purified by washing with water, caustic potash, and treatment with chloride of calcium.

Gaultherinic Acid. A crystalline compound obtained by boiling gaultherine with barytes water, precipitating the excess of barytes by carbonic acid, and decomposing by sulphuric acid, boiling the syrupy mass with carbonate of lead, and decomposing by sulphohydric acid.

Gay Lussite. *Natro Calcite.* Spec. grav. 1·928 to 1·99, H 2·5. Yellowish-white oblique prisms with angles of 68° 50′, 96° 30′; the obtuse lateral edges being in general replaced by tangent planes. When heated, the crystals de-

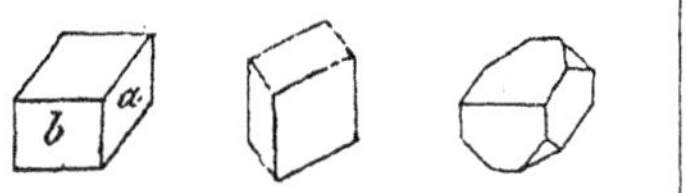

crepitate and become opaque. B.B. melts rapidly into an opaque globule, which becomes infusible, with an alkaline taste; soluble with effervescence in acids. It consists of $NaOCO_2$ 33·96, $CaOCO_2$ 31·39, HO 32·2, Al_2O_3 1·, CO_2 1·45. *Form.* $NaOCO_2$, $CaOCO_2$ 5HO. Found in the Natron Lake in Columbia.

Gedrite, from Gedré, Pyrenees. SiO_3 38·81, Al_2O_3 9·31, FeO 45·83, MgO 4·13, CaO ·67, HO 2·3. Hornblende, resembling anthophyllite.

Gehlenite. *Stylobite.* Spec. grav. 2·9166 to 3·029, H 6·5. Gray rectangular prisms with square bases (2d system ?); fracture uneven, passing into splintery, lustre resinous, inclining to vitreous, usually opaque, brittle. B.B. fuses with difficulty in thin splinters; slowly soluble in borax; gelatinizes with chlorohydric acid. SiO_3 29·386, Al_2O_3 24·924, CaO 36·34, FeO 5·455, HO 3·92. *Form.* $3(3RO_2SiO_3)\ 3R_2O_3\ SiO_3$. Found on Mount Monzoni, Fassa, Tyrol.

Geic Acid. A product of the decomposition of vegetable matter in the soil, a species of *Humus* or mould. Its formula has been given as $C_{40}H_{12}O_{14}2NH_3HO$. But it is probably not constant.

Gelacine. A variety of cell substance coloured green by muriatic acid. Found in Cuactis, Scytonema, and Lyngbya.

Gelatine, or **Glue.** (*Leim*, Ger.; *Colle*, Fr.) Gelatine includes two varieties, termed *Colla* and *Chondrin*. See these articles, and likewise Glue, Isinglass.

Geline. A name given to the matter constituting the cells of fuci, gelatinizing by boiling.

Gems. (*Edelsteine*, Ger.) *Precious stones.* These terms are applied to certain minerals which have become valuable from their colour, hardness, and lustre. The principal of these are, diamond, ruby, sapphire, emerald, topaz, hyacinth or jacinth, garnet, amethyst, aquamarine, opal, turquoise, blood-stone, carnelian.

Gems, Artificial. These are prepared by means of various kinds of coloured glass, which are made to resemble the precious gems. Glasses which serve for the bases of these gems are made as follows:—1st, 100 silica or sand, 136 red lead, 25 carbonate of potash, 9 borax, $\frac{1}{8}$ white arsenic. 2d, 75 silica, 100 red lead, 10 carbonate of potash. 3d, 100 silica, 156 red lead, 55 caustic potash, 7 borax, $\frac{1}{3}$ white arsenic. 4th, 112 litharge, 16 silica, 36 boracic acid. When a gem is to be formed, 100 parts of one of these kinds of glass is pulverized and mixed with the proper proportion of colouring mixture, some of which are as follows:—*Reddish-yellow topaz*, 4 parts glass of antimony, $\frac{1}{10}$ gold purple. *Yellow topaz*, 1 red oxide of iron. *Garnet*, 50 glass of antimony, $\frac{2}{5}$ gold purple, $\frac{2}{5}$ black oxide of manganese. *Turquoise*, 100 parts of glass whitened with oxide of tin, 3 oxide of copper, $\frac{1}{3}$ zaffre, $\frac{1}{10}$ black oxide of manganese. *Emerald*, $\frac{7}{10}$ glass of antimony, $\frac{1}{25}$ oxide of cobalt. *Sapphire*, $1\frac{1}{2}$ oxide of cobalt. *Amethyst*, $\frac{4}{5}$ black oxide of manganese, $\frac{1}{2}$ oxide of cobalt, $\frac{1}{30}$ gold purple. The glass with which these colours are mixed must be first made by exposure in refractory clay crucibles to the heat of a pottery or wind furnace for twenty-four hours, and after the mixture is made, the crucible must be again subjected to a similar temperature for twenty-four to thirty hours.

Gentianine. The bitter principle of *Gentiana lutea*, which has not been obtained in a pure state.

Gentinine. *Gentisine. Gentisic acid.* A new body in yellowish needles with the formula $C_{14}H_5O_5$; has no bitter taste. It is obtained from the aqueous extract by alcohol. It unites with soda and barytes; with nitric acid *nitro gentianine* is formed, $C_{14}H_5\ NO_{10}$, a green powder.

Geocronite. *Kilbrickenite.* 5 PbS, Sb S_3. Spec. grav. 6·5, H 2 to 3. Lead-gray metallic masses, and granular or earthy; fracture uneven; Pb 68·87, Fe ·38, Sb 14·39, S 16·36. Kilbricken, Clare; Sala, Sweden. Sometimes the antimony is replaced partly by arsenic.

Geode. A round hollow mass of rock occurring in mountains, with the interior cavity lined with crystals.

Geognosy. (γεα, earth, and γινωσκω, I know.) The description of the earth as distinguished from speculations.

Geology. (για, earth, and λογος, a discourse.) A term restricted by some to the theoretical view of the earth, but generally used to imply the science of the structure of the earth. The rocks of which the earth's crust is composed may be divided into igneous and sedimentary, or into non-fossiliferous and fossiliferous, or those in which the remains of organic beings cannot be detected, and those in which they are abundantly present. Whether the earth be a solid, or consist of a crust, has been disputed by physicists. The depth to which it has been penetrated amounts only to 21,358 feet or 3·6 miles. Chimborazo is rather more than as much above the level of the sea, so that 45,000 feet or $\frac{1}{435}$ part of the earth's semidiameter is all that we know of its crust; an amount of knowledge insufficient to solve the preceding problem. The earth is one of numerous planets revolving round the sun. Its mean semidiameter is 8,286 miles or 8,000 statute miles; its mean circumference being 24,858 miles. It performs its daily revolution in 365 days, 48 minutes, and 49·7 seconds. It is an oblate spheroid, being compressed at the poles. Its equitorial radius is 3963·025 miles; its polar radius is 3949·8 miles. The polar radius is therefore 13·225 miles shorter than the equatorial. The mean density of the earth is, according to Cavendish, corrected by Bailey, 5·448. The rocks of which the earth is composed, as far as they have been examined, beginning at the lowest and ascending, are comprehended in the following table:—

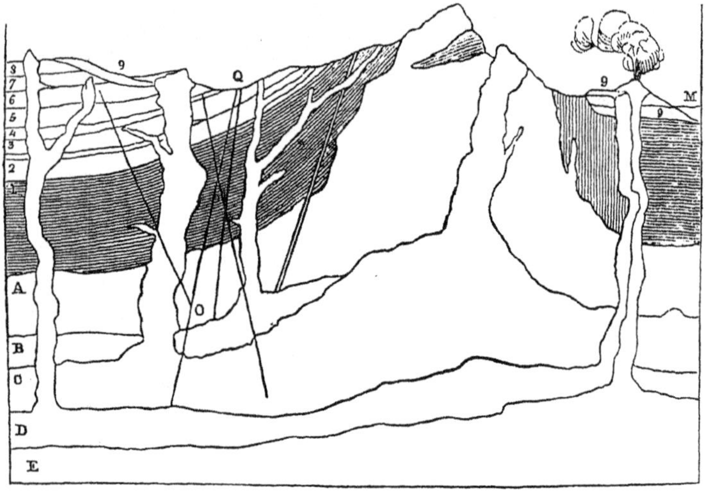

Illustration of the Arrangement of the various Groups of Rocks.

Stratified Rocks.
- Quaternary Rocks, ... 9. Diluvial and Alluvial } Deposits.
- Tertiary, ... 8. Molasse.
- Secondary, ... 7. Chalk. 6. Jura. 5. Trias. 4. Zechstein. 3. Coal.
- Transition Rocks, ... 2. Graywacke.
- Primitive Rocks, ... 1. Slate.

Plutonic Rocks.
- A. Granite.
- B. Greenstone.
- C. Porphyry.
- D. Basalt.
- E. Volcanic Rocks.

M. The Sea.
O. Mineral Veins.

Interior mass of the earth.

Arrangement of Rocks.

Non-Fossiliferous Rocks. Crystalline, or Plutonic Rocks.

Granite, ...(Felspar, quartz, mica).

1. *Talc Granite*, or protogine (felspar, quartz, talc).
2. *Graphic Granite, Pegmatite.* Crystallized mixture of felspar and quartz.
3. *Granular Granite. Eurite, Granulite, White Stone.*
4. *Schorly Granite.* Granite containing schorl.
5. *Porphyritic Granite*, with large felspar crystals.
6. *Porcelain Clay*, from graphic granite.
7. *Greissen*, or *hyalomite*, granite without felspar.

SYENITE,.......(Felspar, Quartz, Hornblende.)
Syenitic Granite; with mica.
Slaty Syenite.
Metamorphic Rocks.

GNEISS,.........Ingredients of granite in layers, giving the rock a ribbond aspect.
Talc Gneiss, Stratified Protogine (felspar, quartz, and talc in layers.
Syenitic Gneiss, the ingredients of syenite arranged in layers).

MICA SLATE,..(Mica, quartz).
Chlorite Slate (green chlorite, with quartz and felspar.

HORNBLENDE SLATE, ... (Hornblende, with quartz and felspar).

CLAY SLATE,.. (Silica & alumina).
Quartz Rock.
Talc Slate.
Potstone (talc with chlorite and mica).

Elastic Sand Slate.

PRIMARY METAMORPHIC, or HYPOGENE LIMESTONE.

1. *Newer Igneous Rocks.* (*Traps.*)

GREENSTONE, Hornblende and albite (diabase, whinstone).
(Diorite), aphanase, trap.
Hornblende-Albite-Porphyry (Dioritic porphyry).
Labrador Augite Rock (Labrador-greenstone dolerite).
Hypersthene-Labrador Rock.

Amphibolite,...or *Hornblende Slate;* when compact, ophanite.

DIALLAGE, ...Diallage and Saussurite, or compact felspar (or jade, and felspar), gabbro, euphotide.

Serpentine,......Silica and magnesia (ophiolites).
Ophite (serpentine and limestone).
Schillerspar Rock, Schillerspar, and Labradorite.

AMYGDALOID,.An earthy rock, containing almond-shaped cavities.

BASALT,........A black siliceous rock, containing olivine and augite, or peridote, labradorite and magnetic iron.

CLINKSTONE,..*Porphyry Slate. Phonolite. Petrosilex* (felspar and mesotype).

LHERZOLITE,..Consisting of bright green pyroxene.

WACKE,........A greenish-gray clay

TRAP TUFF, ...A coarse sandstone, containing fragments of igneous rocks.
Laterite. (Coarse sandstone, peroxidized by heat like brick.)

PORPHYRY, ...A rock having crystals distributed in a base.
Augite Porphyry. (Labradorite and augite).
Felspar Porphyry. A felspar basis, with crystals of felspar interspersed.
Pitchstone Porphyry. Pitchstone, with crystals of felspar.
Hornstone Porphyry. A siliceous rock resembling horn.
Claystone Porphyry. A basis of clay, with felspar crystals.
Felspar, or Petrosilex, when the crystals disappear.
Pitchstone, in the same circumstances.

2. *Recent Volcanic Rocks.* (*Lavas*).

TRACHYTE, ...A felspar basis, with glassy felspar.
Trachytic Porphyry, without scoria, with quartz in grains or crystals.
Domite. Fine-grained trachyte, friable and rough to the touch.
Pearlstone, with the lustre of pearl.
Obsidian. Deep greenish-black; vitreous like bottle glass.
Millstone Trachyte. Hard and cellular.
Trachytic Conglomerate. Pumice, &c., agglutinated.
Leucite. Rock of leucite, augite, and magnetic iron ore, in grains.

TUFF, ...Agglutinated mass of volcanic scoria.
Stone Tuff.
Granular Tuff.
Earthy Tuff.
Fossiliferous Rocks.

1. OLDER PALÆOZOIC PERIOD. *Lower Silurian.* Caradoc sandstone, Landeilo flags, shales, graywacke and slate. *Upper Silurian.* Ludlow and Wenlock, Cumbrian and Cambrian, graywacke and slaty sandstones of Germany, &c.

2. MIDDLE PALÆOZOIC. Devonian system, or old red sandstone; slates and limestone of Devon, conglomerate, &c.

3. NEWER PALÆOZOIC PERIOD. Carboniferous system, magnesian limestone. Permian system, lower new red sandstone, Zechstein.

4. OLDER SECONDARY PERIOD. Upper new red sandstone, salt marls, red sandstone, and conglomerate.

5. MIDDLE SECONDARY PERIOD. *Lias group.* Lower lias slate and limestone. Upper lias shale. *Oolite system, lower oolite*, cornbrash, forest marble, Bradford clay, Stonesfield slate. *Middle Oolite*, coral crag, Oxford clay, Kelloway clay. *Upper Oolite*, Portland stone and sand, Kimmeridge clay. *Wealden formation*, Weald clay, Hastings sand, Purbeck beds.

6. Newer Secondary Period. *Cretaceous* or *chalk system*. *Lower chalk* without flints, green sand, chalk marl, gault. *Upper chalk* with flints.

7. Tertiary Rocks. *Older tertiary*, *Eocene*, Bagshot sand, London clay, characterized by containing a small per centage of living species among its fossil shells. *Middle tertiary*, *Miocene*, coralline crag, red crag, with a minority of its fossil shells referable to living species. *Newer tertiary*, *Pliocene*, Norwich crag, Clyde till, with the largest part of fossil shells of recent species. *Superficial deposits* (Pleistocene) or diluvium, alluvium, &c.

German Silver. An alloy of about 60 copper, 25 zinc, 15 nickel, with a spec. grav. of 8·5.

Germination. Consists of the development of the seed embryo, a fleshy body occupying the interior of the seed, and constituting the rudiment of a future plant. When the embryo is placed in circumstances favourable for germination, the caulicle will extend so as to separate the cotyledons from the radicle by an interval, the extent of which varies in different plants; the radicle lengthens downwards forming the root, while the cotyledons either rise above the earth and unfold, or remain under ground, lose their starch, and shrivel up, and the plumule elongates upwards and gives rise to stem and leaves.

Gersdorffite. *Nickelglance*, or arseniosulphide of nickel.

Geyserite. The siliceous incrustation deposited by hot springs.

Gibbsite. *Terhydrate of Alumina.* Al_2O_3 3HO. Spec. grav. 2·385, H 3· to 3·75. Grayish, greenish, or reddish-white hexagonal crystals or stalactitic, lustre pearly or faint, translucent; giving, when breathed on, a peculiar odour. Al_2O_3 64·8, HO 34·7. Richmond, Massachussets, and Ural. B.B. infusible, phosphoresces; gives a blue with solution of cobalt. It is said sometimes to contain phosphoric acid.

Gieseckite. Spec. grav. 2·832, H 3·5. Externally brownish, internally greenish 6-sided prisms, probably pseudomorphous of nepheline; fracture splintery, uneven, lustre resinous, almost dull, opaque or feebly translucent on the edges. SiO_3 46·079, Al_2O_3 33·828, MgO 1·203, FeO 3·358, MnO 1·155, KO 6·200, vol. matter 4·886. From Akulliarasiarsuk, Greenland.

Gigantolite. Spec. grav. 2·8, H 2·5. A variety of Pinite, according to some. SiO_3 46·25, Al_2O_3 25·1, Fe_2O_3 15·6, MgO 3·8, MnO ·89, KO 2·7, NaO 1·2, HO 2·. Named from the size of its crystals; occurs at Tammela in Finnland.

Gilbertite. Spec. grav. 2·648, H 2·75. Whitish-yellow micaceous plates lying irregularly on each other, lustre silky, translucent, sectile, easily cut with a knife. SiO_3 45·155, Al_2O_3 40·11, CaO 4·17, MgO 1·9, FeO 2·43, HO 4·25; Stonagwyn, St. Austle, Cornwall.

Gillingite. See Hisingerite.

Ginger. The root of the *Amomum Zingiber*, contains a volatile oil lighter than water, an aromatic resin, &c.

Gingseng. The root of the *Panax quinquefolium*, much used medicinally in China.

Ginkoic Acid. An organic acid found in the fruit of *Gingko biloba*, but little known.

Giobertite, or Carbonate of Magnesia.

Gismondine. *Abrazite*, *Zeagonite*, *Phillipsite* of Levy, *Harmotome* of Levy. Spec. grav. 2·265, H 4·5. SiO_3 35·88, Al_2O_3 27·23, CaO 13·12, KO 2·85, HO 21·1. *Form.* $2Al_2O_3$ SiO_3, 3(CaOKO) SiO_3 9HO. Found on Vesuvius in dew-looking drops.

Githagine. See Saponine.

Glairine. *Glairidine.* See Baregine.

Glance. A German term applied to certain sulphides of metals.

Glaserite or sulphate of potash.

Glass. (*Glas*, Ger.; *Verre*, Fr.) The term *glass* has by some been supposed to be derived from the Latin *glacies*, ice; and by others from *glastum*, *blue* or *woad* (Isatis tinctoria), from glass usually having a blue shade. Glass beads are frequently found on the Egyptian mummies. It is mentioned as ὕαλον in the "Clouds" of Aristophanes, a play acted 423 years before the Christian era. A large plate of glass was discovered in Herculaneum, and is supposed to have been used as a looking-glass. Pliny ascribes the discovery of glass to an accident which occurred to some Phœnicians who, when encamped on the river Belus, found that the ashes of their wood fire had vitrified, by combining with the sand of the river. In his time it is obvious that the art of making stained glass was understood. The theory of glass-making depends on the union of an acid, silica, with a base, and the formation of a translucent fused mass which, under certain conditions, is perfectly transparent and colourless. The glass varies in quality according to the nature of the base, and consists of four kinds as manufactured in this country. 1. Bottle glass. 2. Crown or window glass. 3. Flint or common table glass. 4. Plate glass. Excess of soda, potash, or lead, renders the glass more soluble in water or in acid, and if the excess of oxide of lead be great it renders the glass easily fusible. It is said that the lustre and refractive power of glass, increase with the atomic weights of the bases contained in it. The silicates of some of the metallic and earthy oxides have a tendency to crystallize, and they may be deprived of this property by means of the silicates of soda or of potash. The following table exhibits the composition of several varieties of glass:—

	Silica.	Potash.	Lime.	Ox. Lead.	Alumina.	Water.
Bohemian,	69	11	9	0	10	0
Crown,	63	12	12	0	3	0
Window,	69	11 NaO	18	0	7	0
Bottle,	54	5	29	6 FeO	0	0
Flint,	45	12	0	43	0	0
Crystal,	61	6	0	33	0	0
Strass,	38	8	0	53	1	0
Soluble,	62	26	0	0	0	12

Some kinds of glass when slowly cooled in the pots, separate into two portions, the one crystalline, the other vitreous. According to the following analysis by Dumas, the crystalline portion parts with some of its alkali to that which remains vitreous.

	Crystallized Portion.	Vitreous Portion.
NaO,	14·9	19·8
CaO,	12·	12·
Al_2O_3,	4·9	3·5
SiO_3,	62·2 = 100	64·7 = 100

Some kinds of glass when exposed for several hours to an intense heat, assume a condition resembling porcelain. This happens most readily in glass in which the quantity of alumina is great. Glass so changed has been named Reaumur's porcelain, while the change itself has been named devitrification. The specific gravity of glass varies between 2·5 and 3·6.

1. *Bottle Glass*, or *Green Glass*, is formed in this country generally by using the insoluble portions of kelp, which consist principally of carbonate of lime; 3 parts of this matter, soaper's waste as it is termed, are mixed with 9 of river sand, and heated to redness in a reverberatory furnace for a day. The mixture is then deposited in clay pots which are fixed in the fusing furnace, and ignited for half-a-day, or eighteen hours. By this heat, it is converted into a fused mass of green glass, which is then blown into bottles by means of a long iron tube, or in moulds according to fancy or desire; of this kind of glass wine bottles and carboys are made. It contains no lead, and therefore stands a very high degree of heat. A frequent mixture for bottle glass in this country, is 80 parts of spent soaper's ash, 48 kelp, 20 of potash, 50 clay, 50 broken glass. Champagne bottles are formed of 100 felspar, 10 lime, 7½ common salt, 62½ iron slag. Common French bottle glass consists of 20 parts barilla, 85 spent ash, 20 kelp, 50 clay, 5 broken glass. Glass requires to be annealed, otherwise it is apt to fly to pieces on the sudden application of heat. Annealing consists in exposing the glass to a long continued regulated heat in an oven or heated chamber.

The following description of a visit to Mr. Macadam's Bottle Glass Work, in Glasgow, is by my pupil, Mr. Andrew Buchanan:—"The manufacture of bottle glass differs chiefly from the flint glass manufacture in the crude materials which are employed, and the less careful precautions which have to be taken against whatever can interfere with the transparency or fineness of the products. The following are the chief particulars which deserve notice respecting it:—1. To supply the silica common river sand is used, which often contains a large admixture of iron and other impurities; but here, as colour is of little moment, it is judged sufficiently pure. 2. Quicklime, prepared from magnesian limestone, is the principal base employed; but as this is not specially purified, it contains much iron and magnesia besides. In addition to these, kelp waste, or kelp after being exhausted of soluble alkaline salts, but which still contains some alkaline matter, is introduced among the materials. Soda ball is sometimes here used with advantage, but is rather expensive for ordinary bottles. These ingredients are mixed in definite proportions, and are then ready for the furnace. 3. That the materials may fuse with greater readiness they are calcined in a furnace apart before being introduced into the crucibles, to remove water, carbonic acid, and organic impurities. There is a large conical chimney in which these crucibles, four in number, are inserted at regular intervals, being heated by a central fire; and the furnaces alluded to are likewise worked by the help of the same fire, with which they are connected by apertures entering the large chimney from behind. After being thoroughly dried in these, the materials are removed by a long iron shovel to the crucibles. 4. The crucibles themselves are not covered, but are simply large round pots of earthenware. Common fire-clay enters most largely into their composition; though it has been found expedient to mix this with a quantity of plastic or pipeclay. The clays, being intimately mixed together, are placed in lumps upon the floor and trodden down with the naked foot to expel all the bubbles of air, as well as to detect any hard substance, such as gravel, the presence of a very small particle of which would cause the crucibles to crack. For further security it is next worked with the hands, and rolled into oblong cakes about half a foot in length. One of the workmen taking these now builds up the crucibles with his hands; and these, upon being completed, are transferred to an apartment above the glass furnace house, and preserved at a temperature of about 90° Fahr. But before use they must be heated in a separate furnace by themselves, from which they are conveyed with great care, and at a high temperature, to their places. As the crucibles require to be renewed frequently, there are new ones constantly making, and a store kept in readiness; they are, moreover, very apt to give way, (which they invariably do when the glass is in a state of fusion and the heat consequently greatest,) but sometimes admit of being patched up so as to serve out their time. The constant wear of these crucibles makes them the most expensive articles in the glass work. 5. The raw materials being shovelled into the crucibles are now permitted to fuse; the heat being first raised very high, till the glass acquires a watery consistence and the air bells rise, and then gradually reduced that it may become tougher. Upon being ready for working, the heat is maintained as steadily as possible. 6. Bottles are made in the following manner:—A hollow rod or *punty* is introduced into the melted mass, and the glass wound up like wax upon its extremity. When a sufficient quantity is upon it, the rod is removed by the workman, and carried to a horizontal bar of iron, or any other convenient erection, across which the middle of the punty is laid, and rolled along it from end to end. Some superfluous glass generally drops off here in long threads upon the ground; and the rod during the operation is a little inclined that the red hot mass may sink towards or rather a little below the ends of it. After this the workman rolls the glass backwards and forwards on a smooth stone, occasionally thrusting it beyond the farther edge, and pulling the rod gently towards him. He is thus able to elongate the mass, or diminish its circumference at pleasure; and the operation is continued until a bottle's length of glass remains below the extremity of the punty, and this of such circumference that, when inserted in a mould the size of an ordinary bottle and inflated, the sides shall have a proper thickness. Moulds are made of iron, in two pieces, like a bottle cleft down the middle; and these pieces are so constructed as to open or separate when a spring is touched by the foot. The glass at the one extremity of the punty is dipped into the mould and the spring let go, which had been previously touched to let it in; while at the same time the workman blows through the other end and so forms a bottle. But inasmuch as the bottoms of these moulds are plane, and it is customary to make the base of the bottle convex, it is afterwards set in a small mould, likewise worked by the foot, where the requisite impression is made. The neck of the bottle is next cut to a right length, and the rim properly rounded with forceps. Finally, it is placed in a separate furnace to be annealed. These furnaces may be heated at first to within a little of the melting point of glass; and after being filled, when at this temperature, with bottles, the fire is allowed to subside of its own accord. 7. Carboys for acids, as well as bottles, are sometimes made at this kind of manufactory. The process of moulding them is very similar; but, as it would be tiresome and injurious to inflate such large vessels with the lungs, a mouthful of water is taken by the workman and injected into the interior, while he at the same time places his finger on the extremity where he introduced it."

2. *Crown, Sheet*, or *Window Glass.*—The mixture for this glass in this country is often one part of fine Lynn or Isle of Wight sand and two parts of kelp—or soda is generally substituted. Perfectly white glass is also made from 50 lbs. sand, 26¼ potash, 7¼ chalk, black oxide of manganese 4 oz., broken glass 62½ parts. These substances are intimately mixed and placed in a reverberatory furnace. The mixture is well stirred, and in four hours, the heat being gradually increased, it is fused into what is termed a frit. The frit is then introduced into clay pots placed in the melting furnace. In about ten hours the frit has melted down, and a fresh charge is placed over the fused mass, and the heat continued. In the course of a few more hours broken glass is added to the pot, and in forty hours the glass is ready for working. English window glass consists of silica 69·, soda 11·1, lime 12·5, alumina 7·4.

German Sheet Glass is made of the same materials, the name depending on the mode of blowing; a cylinder is formed, which is split and flattened out in a furnace. *Crown* glass is made by the centrifugal process. Sheet glass has been much more extensively employed since the removal of the duty. The same composition is used for water and gas pipes, which are connected by tin coupling screws. There is an inferior kind of window glass, termed *broad* or *spread* glass, made by fluxing soaper's waste, kelp, and sand, but it is less frequently sued than formerly.

3. *Flint*, or *Common Table Glass*, is so termed, because it was originally made with flint. It is often made by mixing 1 part of pearl-ash, 2 parts of red lead, and 3 of sand. Sometimes a portion of saltpetre is added, together with manganese and arsenious acid, for the purpose of oxidizing carbonaceous matter which may be present in the metal. The pearl-ash is previously purified by dissolving it in water, allowing it to stand at rest, drawing off the liquor and evaporating. The other materials are then mixed with it, and the whole introduced into the crucibles or clay pots. After the first portions are melted, additional quantities are added, so as to keep the pots full. The heat is continued from thirty to forty-eight hours, broken glass being added as the matter in the pot subsides. The mixture is known to be completely formed when the silica has disappeared and no air bubbles are distinguishable. This is determined by taking out portions of the *metal*, as it is termed, at various intervals, and blowing it at the end of a tube to ascertain whether it has fused into a perfect glass. The composition of flint glass is, silica 44·30, potash 11·75, lead oxide, 43·05. Tumblers, wine glasses, decanters, and all our crystal glasses in this country, are made of flint glass. It is very fusible in consequence of the lead which it contains, and can be more readily freed from air bubbles and other imperfections than the harder kinds of glass. The following account of a visit to Mr. Cochrane's Flint Glass Work, in Glasgow, is by my pupil and assistant, Mr. Robert Kirkwood:—"Only one kind of glass is made at these works, viz. flint glass. The silica is obtained from fine sand brought from Alum Bay, Isle of Wight. In order to free it from the salts, chalk, and carbonaceous matter present in it, it is first washed repeatedly in large wooden troughs, then dried, and calcined in a reverberatory furnace. When properly calcined it is a very pure and almost white sand. When the sand has been thus properly prepared, it is mixed either with soda ash, or with pearl-ash, or with both, according to the nature of the glass desired, and also with some oxide of lead; definite pro-

portions of all the ingredients being employed. In order that the glass may be of fine colour and transparent, it is necessary that some oxidizing agent should be added; binoxide of manganese is usually employed for this purpose. The manganese serves to oxidize the carbon contained in soda, or pearl-ash, and which would otherwise have rendered the glass brown. It also serves to peroxidize the iron contained in the sand. The iron exists in the sand as a protoxide, in which state it would give the glass a dark green colour, but by means of the binoxide of manganese it is converted into sesquioxide of iron, in which state it communicates a faint yellow tint. It is necessary, however, to exercise caution in adding the manganese; for if too much be employed, so that the carbon and protoxide of iron cannot reduce it all to the state of protoxide, in which condition it dissolves without imparting any colour to the glass, a portion of it is taken up, and stains the glass of a purplish or reddish colour. These ingredients may be called the raw materials; it is a common practice to mix along with these one-third or one-fourth part their weight of broken or refuse glass. The building we next entered was the glass-house; it is built somewhat in the form of a cone. The glass pots arranged round the furnace occupied the centre; on one side there was a low brick building for annealing the glass; on the other side a furnace and a single pot for making vessels of large size. The furnace in the centre had an arched or dome-shaped roof; and the pots, of which there were eight or nine, were arranged in a circular form round the fire, which occupied the centre.

"*The crucibles* or *pots* are built of Stourbridge clay, and require to be formed with great care. The clay is made into rolls, which are applied in successive layers, and firmly kneaded together. The crucibles, after being finished, are allowed to remain for many months, for the purpose of being dried, in a warm apartment. They are then fired in a peculiar furnace, which is situated as near as convenient to the glass-house furnace, to which when they are transferred at a red heat, all communication with the external air must be carefully avoided, since the effect of cold air would be to fracture them. When fixed in their places, glass dust is thrown in, and this, fusing over the soft material, glazes the internal surface. Some of these pots crack almost immediately, while others last for more than a year. When they do give way, it is generally when cooling; and the experience of the workmen who go round for this purpose at stated intervals and examine them minutely, detects almost immediately any injury they may have sustained. In each pot there was a semicircular opening, called the mouth, or working hole, by which the materials are introduced, and the soft or ductile glass is taken out. The mixed ingredients formerly mentioned are heated so as to cause the expulsion of the carbonic acid and water, and thus diminish the subsequent swelling up or frothing in the pot. This is called fritting, and the product is named a fret or frit. In this state it is shovelled into the pot and fused, more being added when the preceding portion has subsided by melting, and this is repeated until the pot is filled. The impurities and 'sandiver' rise to the surface, and are removed by skimming. About forty-eight hours are required for the complete vitrification of the mass. It has then become glass, and is in a state fit for working or forming into vessels. This is seemingly an easy matter, but great skill and dexterity are necessary in order to obtain proficiency in the art, and it would be vain to attempt fully to describe all the manipulations that are required to form a vessel. It may be as well to notice here the glass-blower's instruments. These are few and exceedingly simple. First there is an iron tube of about from 5 to 5½ feet in length. This is called the blowiron, or blowing-tube. There is also a solid rod of iron of about the same size or length; it is called the punto or punty rod. Again, there are one or two pairs of forceps of different sizes; a ruler or scale to indicate the proper sizes; and lastly, a pair of scissors. Having noticed the instruments employed by the workman, I will endeavour to describe the mode of making a tumbler, this being one of the vessels made during our visit. The operatives wrought in parties of three; one dipped the end of the blowing-tube, previously heated at an opening in the furnace, into the melted mass or metal; and having gathered upon the end of it sufficient metal to make the vessel intended, he took it out, and by blowing through the tube and swinging it once or twice in the air, he caused it to expand, and to assume a form resembling somewhat that of a pear. He then rolled it on the marver, a smooth iron plate or table, occasionally blowing into it, until he thought it had become of sufficient size; he then sat down upon his chair or stool, and laying the rod over his thigh, he communicated a rotatory motion to it with his hand. A little boy then approached, and held the flattened part of a pair of forceps to the bottom of the vessel, while the workman continued the rolling motion. When the bottom was properly formed and flattened, the workman, still continuing the rotatory motion, began with a pair of pincers to smooth and taper slightly the sides of the vessel. It had now assumed something of the form of a bottle, when the workman, having smoothed the sides of the vessel, went with it to a small wooden table where he touched its neck with a piece of cold iron, by this means breaking off the neck and thus detaching his tube. Another man having a punty rod with some hot metal on the end of it, applied it to the bottom of the vessel and thus lifted it, and placed it in the working hole of the pot. When sufficiently heated, he lifted it out, and by swinging it once or twice in the air caused it to open out.

He next cut it round with a pair of scissors, and again introduced it into the furnace, and when sufficiently heated brought it out, swinging it as before. He then communicated a rotatory motion to it by rolling the punty rod along his thigh, and with a piece of wood opened out and smoothed the inside of the tumbler. This done, he detached it from the punty rod by a blow (the mark caused by the separation being named the punty mark). The little boy then appeared, and with a pair of wooden forceps lifted the tumbler and carried it away to the annealing furnace or oven. I observed that in making wine glasses the foot of the glass was blown as a separate bulb, and then attached to the body of the glass and rapidly opened out or flattened by means of a pair of forceps. Glass in this condition is very brittle; change of temperature being often sufficient to cause it to crack and fly in pieces. To remedy this defect, the vessels are made to undergo what is called the annealing process, which consists in keeping them for several hours, or it may be days, at a temperature progressively lowered from the highest degree which the glass can bear without softening, to the temperature of the atmosphere. In some cases it is necessary, when the vessels are intended for particular purposes, to anneal them by boiling them in oil; they are thus enabled to bear a higher range of temperature. The peculiar properties of unannealed glass are well illustrated by those glass tears known as Prince Rupert's drops. These are formed by letting drops of melted glass fall into cold water; they have a rounded head which tapers to a slender tail. If struck on the thick end they will sustain no injury, but if a small point of the tail be broken off, the whole mass will fly asunder with violence. The oven into which the glass vessels were put had some resemblance to a flue; the floor was elevated above the level of the floor of the glass-house. At one end of the oven there was a fire, from which the hot air and smoke were wafted or sucked along through the oven, and escaped through a chimney near the opposite extremity. There was also an opening in the side of the oven, and through it the vessels to be annealed were introduced, and set upon large iron trays. The articles were placed on the tray next the fire, and as soon as that tray was filled it was removed farther from the fire, and another tray put in its place. These trays are hooked to each other in a series, and are gradually drawn from the fire towards the cold end by means of a windlass and chain. It is neither easy to understand this process of annealing, nor to explain it, unless we consider heat as a material substance. When the trays are taken out of the furnace they are all coated with soot. Our conductor next led us to the grinding shop. Here the grinding lathes were arranged in two rows along the whole length of the building. A shaft extended along the ceiling, and, from the pulleys of the shaft, driving belts descended and passed round the driving pulleys near the end of each lathe. The turning lathe was of a very simple construction, consisting of a spindle or axis, and a grinding stone, the motion of which was vertical. A vessel containing water was suspended above the grinding stone, upon which it constantly trickled down. Below this again there was a trough which served the double purpose of receiving the water and containing the sand, or other polishing agent required by the grinder. A handful of wet sand was thrown upon the stone, and the vessel to be ground was held to it; when required of a fine polish it was taken to a finer stone, and, in some instances, these grinding stones or discs seemed to be metallic. In some cases, instead of using a grinding stone, the article to be ground was attached to the spindle of the lathe, and motion communicated by throwing on the gearing. The workman having wet the glass, placed a handful of sand on it, and holding a brush made of iron wire to it, speedily rendered its surface equal. We also saw the mode of stoppering bottles; it is very simple. A piece of wood of a conical shape is attached to a spindle, which is put in motion by means of a driving belt or gearing as in the other lathes. Upon the conical piece of wood a quantity of paste made of emery, &c. is put; it is then introduced into the neck of the bottle to be ground, which is held firmly in the hand of the operator. The stopper is next fastened to the spindle and set in motion. The operator holds in his hand a semicylindrical piece of tin, coated also with emery paste, and applies it to the revolving stopper. In a short time he lays the tin aside, substituting for it the bottle which he intends to stopper. The stopper is then placed in the neck of the bottle, the apparatus set in motion, and thus they mutually grind each other. Passing from this part of the premises, we next entered the show-rooms. Here we saw productions of numerous and varied shapes and colour, many of them possessing great beauty of form, as well as richness of colouring. We saw some vessels which were opaque, and of a milk-white colour. Bone earth and arsenious acid are employed to form this very pretty glass. The glass produced is named enamel if it be free from any tinge of colour, but if it possesses a slightly reddish tint it is named opal. There were also some beautifully coloured transparent specimens. To produce a blood-red colour dinoxide of copper is used; for a violet-red, binoxide of manganese; for a ruby-red, purple of Cassius and oxide of tin; for a blue, oxide of cobalt; for a green, sesquioxide of chrome."

Plate Glass is that which is employed for mirrors and large panes of glass. The proportions used in the preparation of this glass are, 14 parts of Lynn or Alum Bay sand, $4\frac{3}{4}$ parts of anhydrous carbonate of soda, and two parts of quicklime. A little nitre is also useful to remove carbonaceous matter. The mixture is introduced into the crucibles, and in ten hours melts; the heat is still continued till the air bubbles are dissipated. The crucibles are kept filled by the addition of broken plate glass, or cullet, as it is technically called. The glass is then ready for being worked and rolled out on a table to the requisite size of plate, an operation requiring great dexterity. Plate glass consists of silica 75·9, soda 17·5, lime 3·8, alumina 2·8. The principal plate glass-houses in Great Britain at present are at South Shields, Sunderland; St. Helens, Lancashire; Smethwick, Birmingham; Blackwall, London and Ravenhead, near Prescot. From the difficulty of forming such large masses of glass, accidents being frequent, a great capital is required to establish such glass-houses. French mirror glass of St. Gobin is made from 1 lime, 7 carbonate of soda, 20 pure sand, 20 broken glass, and some manganese or nitre.

Technical Terms used in Glass-making.

Supplied by my friend and pupil Mr. Roderick A. Couper:—

Annealing, or tempering the glass, is the act of passing the goods through the leer slowly, so that they gradually cool as they are drawn from the one end to the other.

Batch is the several ingredients of glass mixed well together, and in the state ready to be put into the pot for fusion.

Blown Goods are hand-made goods, or such as are blown into by the workmen.

Blow-irons are hollow iron rods about 5½ feet long, on the end of which the workmen gather the metal out of the pots.

Cased Goods are formed by putting coloured metal on the top of flint glass, or any other colour.

Cave, The, is a passage of 4 or 5 feet wide to give draught, and also to clean the grate-room or work the furnace; in length it is generally carried 10 feet outside from the glass-house on both sides.

Charging the Pots is filling them with the batch and cullet.

Collars are of the same shape as stoppers, but with the centre cut out; they are used also like stoppers for the mouth of the pots.

Cullet is broken glass.

Cut Goods. Glass cut on the surface to improve it, either in utility or as an ornament.

Cutter's Frames are two uprights, well stayed, between which a spindle revolves with an iron, stone, or wooden wheel, as circumstances may require; a trough lying directly under the spindle betwixt the uprights.

Dragading or *Dragaded Metal* is the batch fused into a pot and then ladled out into water, for the purpose of being used over again, making in this way the finest metal or glass.

Flint Glass is so called from flint being used in its manufacture in its earliest age, and still keeps the name, although Lynn and Isle of Wight sand have been substituted for flint.

Furnaces vary in size, containing from 6 to 12 pots, and are built within the cone, the material being either fire-clay blocks or freestone got in Newcastle-on-Tyne. A furnace consists of its pillars and crown; an 8-pot furnace may have an inside diameter of from 11 to 12 feet, and height in centre of 4½ feet, the crown resting on 8 pillars, and between these 8 pillars are 8 spaces like 8 low-arched doorways, in which the pots are placed; at the back of each of these pillars are flues for the conveyance of smoke, and to cause a draught.

Glass Cone is a chimney outside the furnace, height from 50 to 100 feet, for carrying away the smoke and keeping up a draught.

Glass-House is where the glass is manufactured.

Glassmaker's Chair is a long stool with two arms attached to it, for the purpose of rotating the blow-iron on it.

Harbours are large chests with covers 6 or 7 feet long to hold the batch and cullet in, previous to being put into the pots.

Hoppers are conical vessels hung from the ceiling over the frame, and filled with sand and water to supply the cutter's wants.

Leer, The, is a long arched building from 30 to 40 feet long, 6 feet broad, and 28 inches high, with an opening at the end next the glass-house for putting in the goods, and also the pans upon which the goods are placed. Close to this opening is the fireplace or fireplaces, while the chimney is at the other extremity, causing the heat thus to travel along the whole length of the long arched building.

Marvers are iron plates planed on their surface for the glass-makers to roll their glass on.

Metal is glass in a state of fusion.

Mixing House is where the materials are kept and mixed up as batch to make glass.

Obscuring is putting the glass on a revolving spindle, then holding a brush of iron wire to it, and feeding it with loamy clay, which breaks the surface of the glass, making it opaque or to appearance frosted.

Picking House is where the cullet or broken glass is picked, washed, and dried.

Polishing is performed by rubbing the smoothed glass over a wooden revolving wheel, which is supplied with a mixture of ground pumice stone and rottenstone moistened together; this takes away all opaqueness, making the glass quite translucent.

Pots are made generally of Stourbridge fire-clay, closed at the top, with an opening called the mouth of the pot about 25 inches from the bottom, by which the pot is both filled and emptied; around the mouth is a projecting portion of clay, the top part of which is called the *hood*, and the under part the *pat*; the mouth of the pot is 10 inches wide by 9 inches high, the pot itself being about 3 feet wide by 3¼ feet high.

Pot Arch is a small kiln for firing one or more pots previous to their being set in the furnace.

Pot Carriage is like an iron three-pronged fork, but of large dimensions, placed on two wheels.

Pot Setting is the act of first taking the broken pot out of the furnace, with the aid of the pot carriage, then by the same means taking a pot out of the pot arch which has been for six or seven days gradually brought up to a red heat, and placing it in the arch of the furnace from which the broken pot was taken out, and building up with fire-brick as quickly as possible to protect it from the cold air. This is a remarkably warm operation, and requires to be done very expeditiously.

Pressed Goods are made by pressing glass into a mould by means of a machine.

Proofs are little cylindrical pieces of metal blown to judge of the colour of the glass.

Punties are solid iron rods about 5½ feet long, used to stick to the foot or bottom of any article so as to be able to turn it to finish the mouth or top of it, such as a wine glass, tumbler, &c.

Puntying is the cutting away the marks of the punty from the bottoms of glasses, &c.

Putty is a powder made from tin and lead melted together in an open shallow pot; the oxides formed on the surface are skimmed off, being the putty.

Puttying is holding the polished glass to the wood wheel, supplying it with moistened putty, which gives the cutting a brilliancy or lustre.

Roughing. A roughed glass is produced by holding the glass over an iron mill or wheel that is kept revolving, over the edge of which runs a stream of sharp sand, which breaks the surface of the glass, and roughens it to the necessary style; the cut parts become quite opaque.

Sand Arch is a reverberatory furnace to calcine the sand in, previous to its being used in the mixture for glass.

Shoes are little clay vessels of a semicircular form about a foot long and 4 inches wide, placed at the side of the pots in the furnace to put the blow-irons in, to keep the ends of them hot.

Siege, The, is the floor on which the furnace stands; it has an opening in the centre called the eye or grate-room, which communicates with the cave underneath; the coals lie in the grate-room and upon the siege.

Skumming or *Skimming.* When the metal is a little rough or strong, the top is skimmed by gathering it on the end of a blow-iron.

Smoothing is effected by holding the rough glass to the edge of a revolving stone, on which trickles a stream of water, the glass being still opaque.

Stoppers are flat pieces of fire-clay to fit the mouth of the pot.

Strockles are shovels with turned up edges, adapted to the filling of pots from the harbours.

Teaze-hole is the opening in the furnace through which the coals are put in to keep the heat up.

Teazer is the fireman who attends the furnace and leer.

Topping off is when the pots have been twice charged, or received two fillings, and been allowed to settle down. A sample proof is taken out, so as to judge of the colour, and to alter it if required. It is then filled up the third time, which last filling generally brings the metal, when fused, to a level with the pat of the mouth of the pot.

Glauberite, *Brogniartine.* $CaOSO_3NaOSO_3$ = gypsum 49, glauber salt 51. Spec. grav. 2·807, 2·73, H 2·5 to 3. Yellowish or grayish-white flat rhomboidal prisms, of the oblique prismatic system, with angles of 83° 20′; 104° 15′; 112° 20′; 131° 35′; 116° 20′; lustre vitreous, brittle; taste saline, astringent and weak; fuses when heated with decrepitation into a white enamel, which becomes white and opaque in water. From Ocagna, New Castile; Aussee, Austria; Tarapaca, Peru.

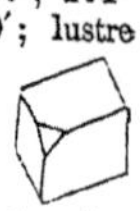

Glauber's Salt, or sulphate of soda.

Glaucene. C_4N_3H. A product from sulphocyanide of ammonium.

Glaucine. A crystalline alkaloid from the leaves of *Glaucium luteum.*

Glaucodote. $(CoFe)S_2(CoFe)As$. Spec. grav. 6, H 5·. *Cobalt mispickel,* one-third of the cobalt being replaced by iron.

Glaucolite. Spec. grav. 2·72 to 2·9, H 5. Lavender-blue or green rhombic prisms of 143° 30′; fracture splintery; lustre vitreous. SiO_3 54·58, Al_2O_3 29·77, CaO 11·08, KO 4·57. From Menge, near Lake Baikal, Siberia.

Glaucomelanic Acid. Obtained by the action of oxygen on bezoaric acid dissolved in caustic alkali; it forms bluish-black salts.

Glauconite. (γλαυκος, sea-green.) *Green earth, chlorite baldagée, Talc chlorite zographique.* Spec. grav. 2·598 — 2·632. Celadine green globular masses; structure earthy; fracture small-grained, uneven; dull, streak feebly glistening; feels greasy; adheres slightly to the tongue; not acted on by chlorohydric acid, but by aqua regia. It is a common constituent of amygdaloid, and the green sand formation; and is employed as the mountain green in water painting. SiO_3 57·8, FeO 7·5, Al_2O_3 6·5, MgO 19·5, KO 4·, HO 4·7.

Glaucophane. Spec. grav. 3·108, H 5·5. SiO_3 56·49, Al_2O_3 12·23, FeO 10·91, MnO ·5, MgO 7·97, CaO 2·25, NaO 9·28. Light blue 6-sided prism, translucent, opaque, slightly attracted by the magnet. B.B. fuses easily into an obscure olive-green glass. From the island of Syra. 3(FeO, MnO, MgO, CaO, NaO) 2 SiO_3, 2(Al_2O_3, 2 SiO_3.)

Glaucopicrine. White bitter plates from the roots of *Glaucium luteum.*

Glaucosiderite. See VIVIANITE.

Glazes for pottery. See POTTERY.

Glazing gunpowder. Giving the grains a smoothness by friction.

Gliadine. A transparent and reddish matter, separated by cold alcohol from gluten; soluble in tartaric acid; reprecipitated by tannic acid.

Globules of Blood. See BLOOD.

Globuline. *Crystallin.* C 54·50, H 6·90, N 16·50, S 1·23, O 20·87 (Mulder). *Characters.*—Globuline, a modification of albumen, forms the principal constituent of the blood globules, and of the crystalline lens. When dried at 120° it is a yellowish mass, which becomes a white powder when pounded. Without taste and smell; soluble in water, from which it is precipitated insoluble in water; soluble partly in boiling alcohol, separating in flocks on cooling. Aqueous solution coagulated by ether; and by heat at 200°. When a drop of acetic acid is added to a solution of globuline it becomes turbid, and if the acid be accurately saturated with ammonia, a flocky precipitate falls. Dilute acetic acid renders solution of globuline opaline; when heated to 120°, a milky coagulum separates; the solution rendered turbid by a small quantity of acetic acid becomes clearer by an additional quantity, but is always milky; about 212° a coagulum separates. By excess of boiling acetic acid, globuline loses its capacity of coagulation; coagulated by creasote; more easily decomposed than other albuminous bodies. Even by boiling with water ammonia is disengaged; coagulated by mineral acids. *Preparation.*—*Soluble* globuline is obtained from the crystalline lens of the eye by coagulating it by acetic acid under 120°, evaporating to dryness, and exhausting with ether and dilute alcohol. *Coagulated* globuline is obtained by precipitating its solution by boiling, and digesting in alcohol and ether (Mulder), or by precipitating by chlorohydric acid, washing with the same acid, dissolving in water, precipitating by carbonate of ammonia, and washing with water, alcohol, and ether (Lehmann). It is distinguished from other albuminous bodies by its tendency to precipitate, when its solution in water, rendered acid or alkaline, is neutralized. The globuline of the blood is precisely similar, but it has never been obtained quite pure.

Glottalite. (*Glotta*, the Clyde). Sp. grav. 2·181, H 3·5. Regular 8-hedrons or 4-sided pyramids, the faces of which appear to be equilateral triangles; lustre vitreous, strongly translucent, brittle. B.B. swells up and melts into a white enamel; fuses into an opaque white bead with soda, into a translucent glass with borax. SiO_3 37·014, CaO 23·927, Al_2O_3 16·308, Fe_2O_3 ·5, HO 21·25. *Form.* 3 CaO, 2 SiO_3, $Al_2O_3SiO_3$ 9 HO. In the hills above Port-Glasgow.

Glucic Acid. $C_{12}H_8O_8 3HO$. Obtained by the action of alkalies and lime on sugar by boiling; soluble in water; not crystalline; forms neutral soluble salts with barytes, lime and lead.

Glucina. *Oxide of Glucinum. Oxide of Beryllium.* GO 3·25, 26; 3·222, 25·77; 1·5808, 12·6464; O = 63·25; G = 36·72 (Awjedew, L'Institut. 461). Spec. grav. 2·976 (Ekeberg, Ann. Chim. 43, 277). Soft light white powder, without taste or smell, neutral, adheres to the tongue; insoluble in water, but forms a paste with water, having some ductility; does not combine with oxygen and combustibles; it unites with sulphohydric acid (Fourcroy, 2, 159); fusible before the oxyhydrogen blowpipe (Davy); soluble in the hydrous state in caustic potash and soda; insoluble in caustic ammonia, soluble in carbonate of ammonia, which distinguishes it from alumina, and is five times more soluble in that carbonate than yttria; forms sweet-tasted salts with the acids. *Process.*—The beryl, when reduced to a fine powder, is fused in a platinum crucible, with 3 times its weight of carbonate of soda or potash; the fused mass digested in dilute chlorohydric acid, and the silica separated in the usual manner; the filtered liquor from the silica is precipitated by caustic ammonia. The precipitate, consisting of oxide of iron, alumina, and glucina, is digested in a strong solution of carbonate of ammonia, which takes up the glucina. The whole is then filtered, when the liquor is boiled, glucina separates; or the solution may be saturated with chlorohydric acid, and precipitated with ammonia. The alumina may also be separated from the glucina by dissolving the mixed earths in a solution of sulphurous acid, and boiling the clear liquid till the excess of acid is removed, when the sulphite of alumina precipitates, and the sulphite of glucina remains in solution (Berthier, Ann. Chim. 50, 371).

Hydrate of Glucina.—GO HO, 2·7125, 21·7, or 3 GO 4 HO at 212°. Gelatinous mass when freshly precipitated; when dry, a white powder, resembling alumina.

Glucinum. From γλυκυς, sweet. *Beryllium. History.*—Vauquelin, in 1798, analyzed the *Beryl* at the request of Hauy, to ascertain whether it contained the same ingredients as the emerald, a suspicion entertained by that mineralogist from crystallographic considerations. The result of the analysis confirmed that idea. A new earth was found in it, which Vauquelin called *Glucina*, from the sweet taste of its salts (Ann. Chim. 26, 155). Klaproth confirmed this result (Beitrage, 3, 215). It occurs also in *Euclase*, *Chrysoberyl*, *Helvine*, *Leucophane*, *Phenakite*, *Gadolinite*, and has been examined by Berzelius (Schweigg. Journ. 15, 296; Wöhler, Pogg. Ann. 13, 577; Gmelin, ib. 50, 175; Schaffgotsch, ib. 183; Steel and Thomson, Rec. of Gen. Science, 1; Awjedew, Pogg. Ann. 56, 101). *Characters.*—G 2·25, 18 (Thomson); 2·22, 17·7 (Berzelius); ·5808, 4·6464 (Awjedew). Dark gray powder, black mass or gray scales, with the metallic lustre when burnished, difficult of fusion; does not absorb oxygen from the air at the usual temperature, but burns with splendour.

When heated in air or oxygen gas leaves glucina; not altered by water, even when boiling; soluble in strong hot sulphuric acid, with evolution of sulphurous acid; readily soluble in dilute sulphurous acid, and in chlorohydric acid, with evolution of hydrogen; also in nitric acid, binoxide of nitrogen being liberated; soluble in caustic potash, but not in ammonia. *Process.*—1. Obtained by passing the vapour of potassium over glucina, at a white heat, in a platinum tube; a mixture appears of potash, glucina, and glucinum, which, when heated with air, is converted into glucina (Sir H. Davy). 2. By means of potassium or sodium from the chloride of glucinum, as with aluminum (Wöhler, in 1828; Pogg. Ann. 13, 577; Bussy, Journ. Chim. Medic. 4, 455).

Chloride. — GCl 5·0125, 40·1, G = 11·74, Cl = 88·26. Fine white silky needles or flakes, very volatile, deliquescent, fusing at a gentle heat into a brown fluid. Obtained by passing dry chlorine gas, in a glass or porcelain tube, over a mixture, at a red heat, of glucina and charcoal (H. Rose, Pogg. Ann. 9, 39); the chloride sublimes into the colder parts of the tube. It is also formed by heating glucinum in chlorine. From this salt glucinum is obtained. *Hydrous Chloride.*—GCl, 4 HO, in crystals, is formed by dissolving glucina in dilute chlorohydric acid, and evaporating; by ignition glucina is left.

Glucose. See GRAPE SUGAR.

Glue. *Colla. Glutine.* (*Leim*, Ger.; *Colle*, Fr.) $C_{82}H_{67}N_{13}O_{32}$? C 49·3, H 6·9, N 18·3, O and S 25·2. Glue or colla is obtained from hides, which are boiled in a copper or iron boiler with pure water, being previously, however, digested in lime water for three weeks to dissolve blood and soft parts The impurities are skimmed off as they rise, and when the whole is dissolved, a little alum, or finely-pounded lime, is thrown in. It is then strained through baskets, and allowed to settle. The clear liquor is boiled again, and poured into wooden frames, where it concretes into a jelly. It is then cut by a spade into square cakes, which are again cut by a wire, as with soap, into thin slices. These are put into a kind of coarse network, and dried in the air. For convenience it is usual to have a reservoir with water at a higher level, communicating with the boiler by a tap, while a stop-cock at the bottom of the boiler allows the solution of glue to be drawn off into a boiler, at a still lower level, for the purpose of concentration. But it is obvious that, instead of the direct application of fire to the boilers, a preferable plan is that often adopted of heating the materials by passing steam through the water covering the hides, and thus avoiding the risk of singing the animal matter by over heating. The quantity of glue obtained from hare and rabbit skins amounts to 54 per cent., parchment scraps 62, tanners' clippings and refuse 38 to 42, the epidermis of hides, used for manufacturing belts, 30, pieces of glove skins 45. Glue is also made from pelts from furriers, the hoofs and ears of horses, oxen, calves, sheep. It may also be procured from bones, muscles, tendons, ligaments, membranes, &c. the best, however, being obtained from skins. It may be formed from bones by digesting them for a week in chlorohydric acid, diluted with 4 parts of water. The bone earth is dissolved, and may be used as a manure, or for the manufacture of phosphorus. The insoluble cartilage is now boiled in water, by passing steam through it, and when dissolved is cooled, &c. as already described. The best glue is hard and brittle, with a dark brown colour, and an equal degree of transparency and chonchoidal fracture, without black spots. In cold water it swells up, and becomes gelatinous, but is not soluble except when weak. Its solution precipitates tannic acid, and forms leather. The mode of purifying glue, so as to make it pure colline, is given under that substance. It seems to unite with chlorine when that gas is passed through its solution. When boiled for six days with water, it no longer gelatinizes. When digested with sulphuric acid, or boiled with caustic potash, it yields leucine and glycycoll. When 1 part glue is oxidized with 2 bichromate of potash, 3 oil of vitriol, and 30 water, there are formed benzoic, formic, acetic, butyric, and valerianic acids, prussic acid, oil of bitter almonds, and valeronitrile. Oxidized with 3 binoxide of manganese, 4½ oil of vitriol, and 30 water, it gives aldehyde, butyric, benzoic, formic, acetic, metacetonic, valerianic, and capric (?) acids. Glue is extensively used for its adhesive quality. It is not adapted as nutritious food, but may answer for the purposes of respiration.

Glue, Marine. A solution of caoutchouc in naphtha, or tar oil, with the addition of shellac.

Gluten. (*Kleber*, Ger.) The term applied to the adhesive matter which remains when the flour of wheat is washed on a cloth. It consists of fibrine, caseine, glutine and oil, which may be separated by boiling with alcohol. In cooling, caseine separates in flocks, while glutine and oil are obtained by evaporation. The matter insoluble in alcohol is vegetable fibrine, identical in composition with animal fibrine or muscle. It is to the presence of gluten that the superior quality of wheat bread is due, from its power of detaining carbonic acid when fermented, and forming a spongy, porous, light article of food. When flour is injured with moisture, the gluten becomes soluble in water. Many salts appear to render gluten again insoluble, by forming with it a chemical combination; and the bakers of Belgium discovered how to bake with damaged flour by adding sulphate of copper (a poison) to the dough. Alum has a similar effect (Liebig). It is probable, however, that alum is principally used for the purpose of bleaching the flour.

Glutine. C 53·27, H 7·17, N 15·94, O 23·62. Obtained by evaporating the alcoholic decoction from gluten. It has the same composition as albuminous bodies.

Glyceric Acid. A synonyme of metacetic acid.

Glycerides. Fats yielding glycerine by saponification.

Glycerine. *Sweet principle of oils. Lipyle oxide and water.* $C_6H_8O_6 = C_6H_7O_5HO$. *Hydrate of oxide of Glyceryle.* Spec. grav. 1·28. A colourless, transparent, sweet, odourless, thick syrup, capable of being heated to 302° without decomposition, and without losing its combined water; at a higher temperature part of the glycerine distils over, but a portion is decomposed into acroleine; soluble in water and alcohol in all proportions; insoluble in ether; is capable of being kept without change, but its colour becomes dark, which may be removed by digestion with oxide of lead. It unites with bases as sugar does; it dissolves oxide of copper with a green colour; with lime, barytes, and strontian, compounds are formed, which are not decomposed by carbonic acid. By yeast it becomes acid by the formation of formic and metacetic acids; chlorine and bromine unite with it; iodine does not act on it. It unites with phosphoric acid and forms phosphoglyceric acid, which exists ready formed in the yolk of egg. It unites with benzoic, sebacic, camphoric, and chlorohydric acids, forming *benzoycine*, an oil, *sebine*, crystalline, *camphorine*, thick fluid like turpentine, and *chlorohydrine*, a neutral oil. B.P. 448½°. With acetic acid it forms *acetine*, an ethereal fluid ($C_{10}H_8O_8$), and *acetidine.* Glycerine is obtained in the purest state by heating 2 litharge, 3 fat or oil, and 1 water with steam; lead plaster is formed, and the glycerine separated; water is added, which dissolves the glycerine. Sulphohydric acid is passed through the solution to remove any lead which may be present. It is then filtered and evaporated in the water bath. Glycerine, as a commercial article, is a product in the manufacture of stearic acid candles. The theory of its separation from fats is explained by some by supposing that, at the moment of its separation, 2 atoms of oxide of lipyle (C_3H_2O) unite with 4 atoms water to form glycerine $= C_6 H_7 O_5 HO$.

Glyceryle. C_6H_7. The hypothetic radical of glycerine.

Glycocoll. *Glycycoll, Glycocine, Glue sugar.* $C_4H_4NO_3HO$. Colourless, sweet, rhomboidal prisms, fusing at 352°; soluble in 4·3 cold water; neutral reaction; insoluble in ether and absolute alcohol; heated with caustic potash it yields ammonia. It unites with acids, and forms crystallized salts. It has been supposed to be a compound of urea and ½ atom glucose = 2 glycocoll, or 1 atom cane sugar and 4 ammonia = 4 glycocoll. The presence of sugar in it is inferred from its action on polarized light. It is obtained by pouring on 1 part glue, 2 parts oil of vitriol, and in twenty-four hours adding 8 parts water, boiling for five hours, neutralizing with chalk, filtering and evaporating, treating the residue with alcohol and evaporating. Glycocoll and leucine crystallize. By solution and treatment with animal charcoal it may be obtained colourless (Braconnot, 1819). It may be also formed by boiling glue with caustic potash as long as ammonia is evolved, neutralizing by sulphuric acid, and treating with alcohol (Mulder). Another method is to treat 3 to 4 ounces hippuric acid with 12 to 16 ounces strong chlorohydric acid with heat till it is dissolved; in half-an-hour water is added, when benzoic acid separates in oily drops, hippuric acid, consisting of benzoic acid and glycocoll; while the glycocoll unites with the chlorohydric acid. The liquid is filtered and boiled to dissipate benzoic and chlorohydric acids; ammonia being added, a white precipitate of glycocoll in small crystals falls, which are filtered and washed with absolute alcohol till the filtered liquor gives no precipitate with nitrate of silver. It is isomeric with aconitate of ammonia. It appears to be the amide of glycolic acid.

Glycolic Acid. $C_4H_3O_5HO$. A syrup, like lactic acid, decomposed by heat; it is a homologue of lactic acid ($C_6H_6O_6$) differing by C_2H_2 in the latter; it is also a homologue of leucic acid ($C_{12}H_{12}O_6$). It is obtained by boiling benzoglycolic acid with acids, when it is resolved into benzoic acid and glycolic acid.

Glycyrrhizine. *Glycione, Liquorice Sugar.* $C_{16}H_{12}O_6$ or $C_{36}H_{24}O_{14}$. Brown translucent mass, yielding a brownish-yellow powder; taste sweet and bitter; soluble in boiling water and alcohol, with acid reaction; rendered more soluble by alkalies; insoluble in ether; not fermentable. It is obtained by precipitating a hot decoction of *Glycyrrhiza glabra* and *echinata* with basic acetate of lead, washing the precipitate and decomposing with sulphohydric acid, filtering, evaporating, and purifying by absolute alcohol (Vogel). It may also be separated from the cold aqueous infusion by adding dilute sulphuric acid as long as a precipitate falls. The pitchy mass is kneaded with acid, and then with pure water freed from acid. It is then evaporated and dissolved in absolute alcohol. Nitric acid produces no nitropicric acid, but a yellow powder, insoluble in water, soluble in alcohol, which is bitter and unites with bases.

Gmelinite. See HYDROLITE.

Gneiss. A metamorphic rock. See GEOLOGY.

Goethite. *Lepidocrokite.* Brownish-red rhomboidal or rectangular prisms of sesquioxide of iron from Hollerterzug, Westerwald.

Gokumite. A variety of Idocrase from Gökum.

Gold. Au (Aurum). (*Sol*, Lat.; *Or*, Fr.) 12·25, 98· or 24·5875, 196·7. Spec. grav. 19·2 (fused), 19·3 to 19·4 (hammered); F.P. 1298° (Dijon, A.), 2590° (Davies), 2517° (G. Morveau), 2589°·8 (Daniell), 2192° (Pouillet). Rather softer than silver. It is exceedingly ductile, and may be beaten out so thin that 1 grain will cover $56\frac{3}{4}$ square inches, the leaves being $\frac{1}{280000}$ of an inch in thickness, and a grain may be drawn out into a wire 500 feet long; an ounce of gold or silver wire will extend 1,300 miles. *Crystal.*—Native gold occurs in the form of the regular system in cubes. Its *colour* is orange-red or reddish-yellow; it has no taste nor smell; its lustre is considerable, yielding only to that of platinum, steel, silver, and mercury; its tenacity is less than iron, copper, platinum and silver; a gold wire of ·078 inch diameter supports a weight of 150·07 without breaking (Sickingen). When melted it assumes a bright bluish-green colour; it expands in the act of fusion and contracts in solidifying, hence it is less proper for casting in moulds than other metals. It requires a very high temperature to volatilize it (Claveus, Kunkel, Parker). It has been seen, however, to volatilize (Homberg, Macquer, Lavoisier). After fusion it crystallizes. It is not altered by exposure to the air. *Sources, &c.*—Gold is found in a native state united with silver and copper. It sometimes occurs on the surface of rocks in a dendritic form, as if deposited by electrical action; more generally it is obtained in the form of gold dust among the sand of rivers, and latterly it has been procured in large quantities in the alluvial quartz derived from primary rocks, principally granite, which has been broken up and disintegrated. It is found metallic and in sulphides. The density affords some evidence of the value of the native gold. I found the density of a mass of Californian gold, weighing 2572·13 grains, shown to me by Dr. Stewart of Glasgow, to be 15·808. According to a mean of analyses, Californian gold contains about $88\frac{3}{4}$ per cent. of gold. I found the density of gold brought from the rivers of Khoten and Cashgar, and given me by Dr. Thos. Thomson, Bengal army, and now in my museum, to be 17·224. It consisted of 89·3 gold, 7·3 silver, 3·4 copper. African gold with a density of 14·457, consisted of gold 78·, copper 11·8, silver 9·48 (Dr. T. Thomson). Native gold occurs in *Europe* in Transylvania, Hungary, the Rhine, Spain, Portugal, Wicklow, Cumberland, Leadhills, Glenturrit, Glencoich. *Asia.* Eastern side of Ural to Ekatherinenberg in Tomsk and Yeneseisk, Cashgar, Australia, &c. *Africa.* Between Darfur and Abyssinia, from the Niger, Gambia, and Senegal. *America.* Brazil, in Minas Geraes, Mexico, Virginia, South Carolina, California. *Alloys.*—Gold in an alloyed state is said to be of the value of so many carats. Pure gold is 24 carats, which is the unit in this mode of estimation. Each carat is divided into twelfths. Standard gold is 22 carat gold, or in every 24 parts 22 are gold and 2 alloy, or in every pound Troy there are 22 carats or $\frac{11}{12}$ fine gold and 2 carats or $\frac{1}{12}$ alloy. A sovereign weighs 123·274 grains, and contains 10·273 grains of alloy. See MONEY. *Gold amalgam* is a combination of gold and mercury for gilding metals. *Green gold*, used in jewellery, is an alloy of 30 silver and 70 gold. *Jewellers' standard gold* is 18 carat gold, or contains 75 per cent. gold, that is, three-fourths of its weight. When less rich in gold the Goldsmiths' Hall stamp is not attached to it. Gold containing much copper is liable to tarnish, which, however, may be removed by immersion in ammonia and friction.

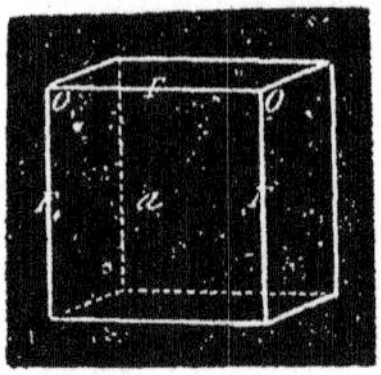

Preparation.—Native gold being found mixed with many foreign substances, it is necessary to remove much extraneous matter from it by washing. When the gangue is only sand, the gold particles being heavier fall to the bottom while the sands are mixed with the water, and by frequent washing the sand is run all off. For this purpose inverted conical pans of iron or zinc are used, or inclined tables traversed by grooves, or the auriferous sands are placed in iron boxes perforated with holes, upon which water is run. The fine particles pass through the apertures and fall upon inclined tables, where the gold, by means of a broom, is detained while the sand is washed away. It is probable that some of the adaptations of the mechanical preparation of lead ores might be rendered available in gold sand washing, such as the use of the revolving endless cloths instead of the brooms in the stage now detailed. Much of the gold washing is performed in California and Australia by means of cradles, so called from the rocking motion which is given to them, their bottoms being rounded like their prototypes. A grating at the upper end of the cradle receives the gold and matrix. Over them are poured water, which carries the particles of gold and sand through the grating, the larger portions being detained, and deposits the gold immediately underneath; but the sand is washed off, and passes with the water out by an aperture at the lower end of the cradle, whose bottom is inclined for this purpose. When pure particles of gold are mixed with oxides of iron and comminuted particles of earth, it is usual at some works to amalgamate the gold with mercury in revolving barrels, and to distil off the mercury and condense it in water. When the gold is mixed with arsenical pyrites, as at Reichenstein, another process is adopted. These ores contain 200 grains of gold in the ton. The ore is roasted in a reverberatory furnace, surmounted by a large condensing chamber, in which the arsenious acid is condensed, leaving in the furnace oxide of iron, some arsenic, and the whole of the gold. A current of chlorine is now passed over this residue in an appropriate apparatus, by which the gold and iron are dissolved. Water and chlorohydric acid are added, and the gold separated by sulphohydric acid gas. The sulphide of gold is then roasted in a porcelain crucible, and reduced in the usual way.

Parting. — All native gold contains silver, and usually copper. To remove these the operation of parting is required. This process may be performed with nitric acid, which dissolves the silver and copper, and leaves the gold. But as nitric acid is expensive on the large scale, sulphuric acid is now usually substituted. It has been long used in India to separate the gold and silver of the Sycee silver of China; the occurrence of gold in the Chinese metal having been first discovered by Dr. Andrew Steel about 1826. In parting, the alloy must not contain more than 20 per cent. of gold nor 10 per cent. of copper. The best proportions for facility in parting are 72·5 per cent. silver, 20 gold, 7·5 copper. If more gold is present, it may escape being acted on; and if more copper exists in the alloy, much anhydrous sulphate of copper surrounds the metal, and shields it from the action of the acid. The alloy being fused and granulated in water, is placed in a porcelain or cast iron boiler, and covered with 2 to 3 times its weight of sulphuric acid of spec. grav. 1·848. The boiler is covered with a head, which conducts into a draught, to remove the sulphurous acid formed. Heat is applied by a direct fire for from four to eight or twelve hours, till the solution is completed. The acid liquor is cooled, and drawn off into lead vessels, and the gold boiled with some fresh acid. But considerable variation occurs in this stage of the process; sometimes weaker fresh acid being added before the first acid solution is poured off. The gold is well washed, removed, and ignited in a black lead crucible. From the solution the silver is reduced by the insertion of copper bars in it.

Assay of Gold.—The ore, especially when much gangue is present, is mixed with a flux consisting of borax, alkaline carbonate, common salt, charcoal, and litharge, and fused in a clay crucible. The reduced metals are then cupelled as with lead, when a button of gold and silver alloy alone remains, the litharge and copper having been oxidized. The alloy, which must contain 3 parts silver to 1 gold, is then boiled with nitric acid, diluted with thrice its weight of water. The residual gold is then ignited and weighed. Gold is reduced from its solutions by protosulphate of iron, with the addition of a few drops of chlorohydric acid. Another method is to mix solutions of protochloride of tin and sesquichloride of iron until a pale green is formed, and then add the mixture to a solution of chloride of gold freed from nitric acid.

Dinoxide, suboxide.—Au_2O 25·5, 204. Green powder obtained by precipitating the dichloride (Au_2Cl) by caustic soda.

Sesquioxide, Auric acid.—Au_2O_3 27·5, 220. Brownish powder, easily decomposed by light; does not combine with acids, but forms feeble compounds with bases. It is formed by adding a slight excess of caustic magnesia to a solution of gold in nitro-chlorohydric acid; the oxide precipitates in union with magnesia; wash the precipitate; digest in nitric acid, which removes the magnesia, and leaves the oxide in the state of a hydrate, which parts with its water at 212°. When heated with chlorohydric acid, sesquichloride is formed; an oxide is formed by passing a spark of electricity through gold leaf placed between two plates of glass; the gold burns, and leaves the trace of the spark; gold burns also when exposed to the action of the battery: the *oxide* forms fulminating gold when digested with ammonia. Gold has less affinity for oxygen than any other metal. Hence its value in retaining its colour without tarnishing.

Purple of Cassius. Gold purple.—$AuOSn_2O_3$? A fine purple powder, a compound of oxide of gold and sesquioxide of tin, formed by adding to a solution of chloride of gold a solution of protochloride of tin. It is used as a pigment for glass and porcelain, and may be prepared by dissolving one part of grain tin in 9 of nitric acid, diluted with 2 of water; 3 parts of common salt are added, as the addition of salts promotes the separation, and then 80 parts of water. The chloride of gold is then mixed with it in drops. See PURPLES.

Sesquichloride.—Au_2Cl_3 37·8, 302·4. Pale yellow needles, formed by igniting gold leaf in chlorine gas, or by dissolving gold in aqua regia (a mixture of 1 NO_5 and 3 H Cl), evaporating away the excess of acid, and dissolving in water; the crystals are deliquescent, and form a yellow solution. Gold is precipitated in the metallic state from this solution by phosphorus, sulphate of iron, iron, zinc, copper, &c. oxalic acid, and the vegetable acids, and salts of the alkalies. Pure gold is often obtained by adding protosulphate of iron to this salt, and washing with dilute chlorohydric acid and water. This salt is used in the electrotype for gilding, as a caustic in surgery, as an application to cancerous surfaces, and also as an internal remedy in syphilis. *Fulminating gold.*—2 $NH_3Au_2O_3$, or NH_3Au_2N, 3 HO. Yellow-brown powder, formed by precipitating sesquichloride by NH_3. *Dichloride.*—Au_2Cl. Yellow mass, insoluble in water, obtained by heating the *sesquichloride* up to the melting point of tin. Converted by water into gold and sesquichloride.

Sesquisulphide.—Au_2S_3 30·5, 244·. Formed by precipitating the sesquichloride by SH. It is a yellow powder, much employed in painting on pottery. It is made for commercial purposes, by fusing in a crucible 3 parts of potash, 9 sulphur, and 1 gold; pouring out the melted mass and digesting in water with access of air. Acetic acid is then poured on to divide it finely.

Diniodide.—Au_2I. Yellow powder by adding iodide of potassium to sesquichloride of gold in solution. *Sesquiiodide.*—Au_2I_3. Dark green, formed by reversing the preceding process, and adding sesquichloride of gold to iodide of potassium. *Sesquicyanide.*—Au_2Cy_36HO. Fine tables, soluble in water, alcohol, and ether, obtained by decomposing the potassium cyanide of gold ($KCy\ Au_2Cy_3, 1\frac{1}{2}\ HO$) by nitrate of silver. The latter salt is formed by adding 46 cyanide of potassium to 35 gold converted into neutral sesquichloride. By evaporation fine colourless crystals are obtained, which become milky in the air from loss of water.

Goldbeater's Skin. This substance is the peritonæal or serous membrane separated from the intestinal tube of the ox, and other animals, rendered thinner by hammering, and prepared to resist putrefaction.

Goniometer (angle measurer). For a description of the *common* and *reflective* goniometers, see CRYSTALLOGRAPHY. The goniometer of Babinet is sometimes used in France. To measure an angle with it, stick the crystal, a cube, for example, as shown in the figure with wax on the centre piece, turn the moveable telescope *b* till its wires become oblique to those of the fixed telescope *a*, for example, at 45°; place the cross piece *c* at 180°, and the moveable telescope on the opposite side of the circle; move the support of the crystal so as to allow the crystal to reflect the wires into the moveable telescope, and to throw the point of crossing of the wires in this manner upon the vertical wire of the fixed telescope; turn then the cross piece *c* until the other face of the crystal produces the same coincidence of the point of crossing with the vertical wire of *a*; the angle is then read off.

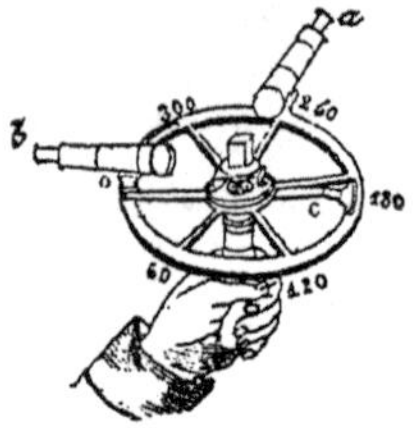

Goulard's Lotion and **Extract.** See LEAD.

Gouty Concretions, or **Calculi, Chalk Stones.** *Urate of Soda* (Wollaston, 1797). These are found in the joints of persons of a gouty disposition. They are insoluble in cold, but somewhat soluble in boiling water; soluble in caustic potash with heat. The uric acid may be separated by nitric or sulphuric acid. The deposition of these calculi augurs want of oxidation in the system, carbonic oxide, the basis of uric acid, instead of the higher oxidation carbonic acid, being formed.

Grain Oil, *Potato Oil, Fusel Oil.* The oil which comes over in the distillation of raw grain consists of a solid portion (margarine, caprine, &c.), and a liquid part, hydrate of oxide of amyle. These I first pointed out in 1836 (Thomson's Brit. Annual, 1837).

Graduator, Bunsen's, for dividing thermometer tubes, &c. into degrees. It consists of a mahogany frame, $5\frac{1}{2}$ feet long, and 7 inches wide, $\frac{3}{4}$ inch thick. Throughout its centre is a groove an inch wide, half-an-inch deep, arched at the bottom for the reception of tubes. At one part, five inches from the end, is placed a brass plate, $1\frac{1}{2}$ foot long, and 2 inches wide, in such a position that when screwed down its edge comes one-half over the groove. It is furnished with four screw-nuts, passing through a cut portion of the plate, $\frac{1}{4}$ inch long, so as to allow a certain advancement or withdrawal of the plate at pleasure. C and D are two similar plates, placed at the other end of the wooden board C, having the same amount of motion as B, and being precisely similar. D is a brass plate of the same dimensions as B and C, but the screws go through holes of the same size as themselves, into the wood. It is cut, at intervals of 5 millimetres, into notches, every alternate one being $\frac{1}{20}$ and $\frac{1}{10}$ inch deep. E is a wooden rod, 3 feet long, 1 inch broad, and $\frac{1}{2}$ inch thick. It has two steel points, placed by screws at half-an-inch from either end. One of these, F, is in the form of a knife, the other, G, of a bradawl. The tube to be graduated is covered with a layer of melted wax and turpentine, and is placed in the groove between C and D, which are then screwed down in their places. A standard tube, previously mathematically divided, is now placed in the groove under B, which is then screwed upon it. The rod E is now used, the point G being put in one of the graduated marks on the standard tube; the knife-formed steel F is now upon the waxed tube, and is made to make a mark upon it, the length of which is regulated by the distance between the edges of C and D. The pointed steel is now removed back one division on the standard tube, and the corresponding mark made on the waxed one, and so on. The waxed tube is now exposed to fluohydric acid, and etched in a lead box.

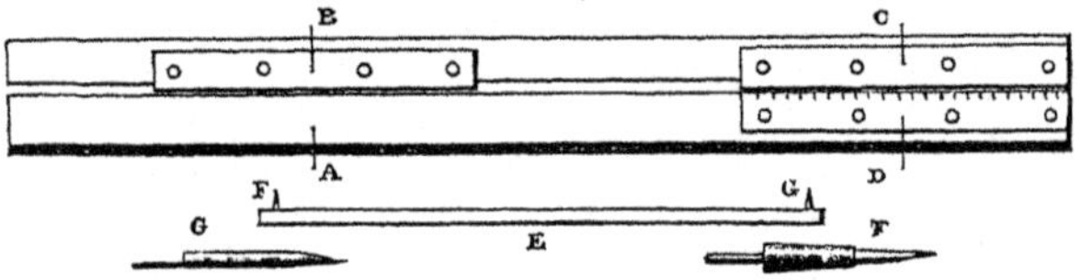

Grammatite. *Granatite.* A synonyme of tremolite or hornblende.

Grammite. A synonyme of table spar.

Granite (*granum*, a grain). A plutonic or

igneous rock (see GEOLOGY), consisting of quartz, red or white felspar, and mica in black plates. The composition of small-grained granite, from the milldam at Goatfell, Arran, I have found as follows:—Spec. grav. 2·546, silica 74·3, sesquioxide of iron 6·3, alumina 8·94, lime 3·68, magnesia, water and alkalies 6·78. The large-grained granite of Goatfell has a spec. grav. of 2·576 to 2·573, and consists of silica 73·48, sesquioxide of iron 9·92, alumina 6·68, lime 2·11, magnesia, water and alkalies 7·81. Granite mountains have usually peaked summits, as in the figure.

Granular. Consisting of grains. This structure of rocks or minerals may be produced either by the particles having been agglutinated together, or by the action of heat under pressure, as with granular limestone and greenstones.

Granulation. The conversion of metals into grains, by projecting them in a fused state into water.

Granulite. *Eurite*, *Leptynite*. A granular mixture of quartz and felspar.

Grape Sugar. See SUGAR.

Graphic Gold. See TELLURIUM.

Graphic Granite. Composed of a base of felspar, with quartz distributed through it, presenting the appearance of letters or writing.

Graphite. *Plumbago*, *Plumbagine*, *Black Lead*. *Physical Characters*. — 3 and 1 axed system. Seldom in 6-sided prisms (Greenland and Philadelphia). Commonly its texture is foliated and amorphous, the laminæ being flexible, in flat 6-sided tables. *Hardness*. 1. Spec. grav. 1·9 to 2·25, 2·32. *Lustre* metallic. *Streak* splendent and metallic, sectile; feel greasy; writes readily on paper. *Fracture* uneven. *Blowpipe Character*. — Unchanged, infusible. *Chem. Character*.—When graphite is boiled with aqua regia a portion of iron is dissolved. When ignited in a crucible with 10 times its weight of nitre it deflagrates. Conducts electricity. Spec. heat ·20187 (Regnault). *Source*.—It occurs in pieces about the size of the fist, in a porphyry traversing graywacke slate, at Seathwaite, in Borrowdale, Cumberland. This mine was discovered in the reign of Queen Elizabeth, when gold and silver were worked in the vicinity by a company of Germans. The top of the hill, called the Moor, in which the mine is situated, is 1,500 feet above the sea. A new level has been lately driven through a porphyry branch dyke, lower down the hill, for the purpose of reaching the old workings, several fathoms below. It is now (1849) 460 yards long. Indications of plumbago have been frequently obtained, but no collection of masses or *sops* has been met with. No new graphite has been come upon since 1841. What is now used is taken from old stock on hand, which was sold last year to the extent of £1,500. The mode in which the sale of plumbago was conducted was peculiar. Once a-year the mine was opened, and a sufficient quantity of graphite extracted to supply the purchasers during the year. The mine was then closed, and the product carried to London, where it was exposed to sale at the black lead market, which was held on the first Monday of each month at a public-house in Essex-Street, Strand. The buyers, who amounted to seven or eight, examined every piece with a sharp instrument to ascertain its hardness, those which were too soft being rejected. The individual who had the first choice paid 45s. per lb. the others 30s. But as there used to be no addition made to the stock during the year, the same portions were examined over and over again until the lot was exhausted. Graphite has been observed to float on the refined iron in the iron manufacture (Schafhaeutl). The following table gives the composition of different graphites:—

	Borrowdale.	Pennsylvania.	Artificial.	Lehigh. Anthracite.
Carbon,.........	95·8	95·4	85·45	90·10
Water,.........	3·0	0·6		6·60
Silica,.........		2·6	4·03	1·20
Alumina,.......				1·10
Oxide of Manganese,..... Oxide of Iron,	1·2	1·4	9·50	0·20

Black lead is likewise found in Ceylon, apparently in granite, and also in the United States and in Greenland, from which some specimens have recently been imported. Much of the lead now used is, however, manufactured of a mixture of plumbago powder and tersulphide of antimony. These are ground together by means of large stones revolving in tubs containing water. This mixture, when in sufficiently fine powder, is introduced into a large clay crucible, and is exposed to the intense heat of a reverberatory flame. The crucible is taken out, cooled, and the material removed to a strong iron box, where it is subjected to great pressure. It is then sawn into slices for pencils. *Uses*.—There are three kinds of pencils occurring in commerce, common cedar pencils, ever-pointed pencils, and plummets. The two first often consist of pure plumbago, the last of one-third tersulphide of antimony and two-thirds plumbago. In the manufacture of these pencils, the first process consists in sawing out the cedar into long planks, and then into tops and bottoms; the grooves are sawn out by means of a fly-wheel; the lead is then scraped on a stone, having been previously cut into thin slices to suit the groove, and heated in contact with the fumes of sulphur. It is next introduced into the groove, and the side scratched with a sharp point, so as to cause it to break off above the groove and fill it. The fourth process consists in glueing the tops and bottoms together, and turning the cedar cases in a gauge. They are then French polished and stamped by a machine. The ever-pointed pencils are first cut into thin slabs, and next into square pieces, by means of a steel gauge. They are then drawn through three small holes, armed with rubies, in order to render them fine. The friction destroys the rubies in three or four days. Genuine plumbago pencils must cost two shillings per half-dozen, and a genuine cedar pencil at least sixpence. On the continent pencils are often made of a mixture of plumbago, fuller's earth, and vermicelli.

Grass Oil. This amber-coloured oil from Bengal, I first described in 1836, and found it to begin to give off bubbles of vapour at 120°, the boiling continues with the thermometer rising to 370°. Its smell resembles oil of cajeput. It is the product of a species of andropogon, probably *A. ivaracusa.* It appears to be a mixture of oils analogous in composition to oil of turpentine (Brit. Annual, 1837). Sulphuric acid forms with it a fine crimson colour. It is applied to various economical purposes in India.

Gratiolacrine. $C_{23} H_{20} O_5$. Reddish-brown resin from *Gratiola officinalis.*

Gratioline. $C_{21}H_{18}O_7$. White bitter crystalline powder from the same plant.

Gratiosoline. $C_{18}H_{16}O_{10}$. A soluble bitter principle from the same.

Gray Antimony. See ANTIMONY.

Gray Copper. See COPPER.

Graywacke. *Traumate, Psammite.* (*Grau*, gray; *wacké*, clay). A name given by the Freyberg miners to a sandstone composed of quartz, felspar, lydian stone and clay slate, and filled with metallic veins. It constitutes the lowest of the fossiliferous rocks, and is included in the Silurian system. See GEOLOGY. This rock usually forms fine rounded pastoral hills, as in Scotland and the north of England.

Grease. *Fat.* (*Graisse*, Fr.; *Fett*, Ger). Soft melted tallow, employed for the axles of wheels. The yellow grease for railway carriages in England consists of tallow 253, palm oil 88, soda 25, water 634; in Belgium, tallow 83, palm oil 207, soda 14, water 696.

Green Colours. Many of the following green colours are diluted by such substances as gypsum, sulphate of barytes, and lead, and then go by different names.

Green, Bluish, for pottery. Oxide of chrome 50, carbonate of cobalt 25, hydrous carbonate of zinc 25. Calcine for a quarter of an hour, and take 25 parts and mix it with 75 *green* flux, which is formed of 73 minium or litharge, 9 sand, and 18 crystals of boracic acid fused.

Another Bluish Green.—10 chromate of mercury, 1 pure oxide of cobalt, are triturated on glass, and then heated in a porcelain tube to expel mercury. The powder is next poured into a porcelain crucible, on which the cover is luted; it is intensely heated in a porcelain kiln. The crucible is broken, and the mass digested in water, to remove chromate of potash. The vitrifiable colour is then made by triturating 1 of this residue, ½ oxide of zinc, 5 lead glass (2 minium, 1 sand, 1 calcined borax). By mixing this bluish-green with citron-yellow, all intermediate shades may be procured.

Green, Bremer. *Verditer.* A fine pigment for oil or water colours. 225 lbs. common salt, and 222 lbs. sulphate of copper, are mixed in the dry state, and made into a paste with water on a stone. Sulphate of soda and chloride of copper result. 225 lbs. old ship's copper, in square inch pieces, freed from impurity by washing with dilute sulphuric acid, and afterwards with water, are placed with the salts in oaken chests in a cellar. By the absorption of oxygen oxychloride of copper is formed (3 CuO, CuCl 4 HO), which is a green insoluble sub-salt. It is allowed to remain for three months, and is stirred weekly. It is then washed free from salts, the wash-water evaporated to a paste, and the insoluble matter added to the green salt. The green salt is heated with dilute chlorohydric acid for thirty-six hours, and then to the acid liquor is added caustic potash to decompose the neutral salt partially formed by the acid. After thirty-six hours, it is washed free from potash, thrown on filtering cloths, exposed to the air for some weeks, pressed in cloth, and dried in the air, or by gentle heat under 88°.

Green, Brunswick, occurs of various shades. A common mode of manufacturing it (3 CuO, CuCl 4 HO) is by mixing metallic copper with a paste of sulphate of copper, common salt, and water, mixed either with common salt and dilute sulphuric acid, or sprinkled with a solution of sal-ammoniac, and exposed to the air. A common form of it is dicarbonate of copper (2 CuO CO_2 HO), and called also *mineral* and *mountain* green. It is prepared by precipitating sulphate or chloride of copper with carbonate of lime, washing the precipitate with hot water, and then drying it. It may be diluted to different shades with barytes, gypsum, &c.

Green, Deep, for pottery. Oxide of chrome 75, carbonate of cobalt 25, intimately mixed on glass with water, calcined, and 25 parts mixed with *green* flux, as above (Sevres). Chromate of mercury alone is treated in the same manner as for its mixture with oxide of cobalt, and 1 fine green oxide of cobalt thus obtained is mixed with 3 parts lead glass (2 minium, 1 sand, 1 borax).

Green Earth, *Green Sand.* See GLAUCONITE.

Green Hydrokinone. See HYDROKINONE.

Green Iron Stone. See PHOSPHATE OF IRON.

Greenlandite. An iron garnet.

Green Malachite. Carbonate of copper. See COPPER.

Greenockite. *Sulphide of Cadmium.* CdS. Spec. grav. 4·82 to 4·534, H 2·75, Cd 77·6, S 22·4. Yellow hexagonal prisms or combinations of truncated 6-sided prisms, with several 6-sided pyramids. The only specimens of this mineral, which is very rare, have been found in the amygdaloid of the Bishopton tunnel, Renfrew, and I have small specimens in my museum on prehnite from the Kilpatrick hills.

Greenovite. A variety of sphene or silico-titaniate of lime, from Piedmont.

Green, Scheele's. See ARSENITE OF COPPER.

Green, Schweinfurth. *Vienna Green.* See ARSENIO-ACETATE OF COPPER.

Greenstone. A mixture of green hornblende and albite, commonly termed *whin* and *trap.*

Green, Ultramarine. A green which is formed by a similar process to that for ultramarine blue, and which is now printed on cloth by means of a paste.

Green, Variegated. 8 chromate of mercury, 1 oxide of cobalt, mixed and placed in a flat capsule, and heated in a porcelain kiln. This is mixed with twice its weight of lead glass (2 minium, 1 sand, 1 calcined borax).

Green Vitriol, or sulphated protoxide of iron.

Gregorite. A synonyme of titaniate of iron.

Grenatite. A synonyme of staurolite.

Grengesite. A variety of green-earth, from Dalarne, Sweden.

Groppite. Spec. grav. 2·73, H 2·5. Crystalline; colour rose-red; fracture splintery, splinters transparent. B.B. shows only incipient fusion; swells up with borax and dissolves; forms a glass with soda. SiO_3 45·008, Al_2O_3 22·548, Fe_2O_3 3·063, CaO 4·548, MgO 12·283, KO 5·237, NaO ·215, HO 7·11; from Gropptrop, Sweden, in limestone.

Groroilite. See PYROLUSITE, WAD, or ANHYDROUS BINOXIDE OF MANGANESE.

Gros's Salts. See PLATINUM.

Grossularite. See GREEN GARNET.

Grossuline. A synonyme of Pectine.

Guacine. A resinous bitter principle by means of alcohol, from the leaves of *Guaco,* a species of *Eupatorium,*

Guaiacene. $C_{10}H_8O_2$. Spec. grav. ·874, of vapour 2·92. B.P. 244½°, colourless oil. Obtained by distilling gum guaiacum.

Guaiacic Acid. $C_{12}H_7O_5HO$. Brilliant needles (benzoic acid?), obtained by distilling the tincture of guaiacum to one-fourth, neutralizing with barytes water, evaporating to one-half, adding sulphuric acid, evaporating to a syrup, and extracting the acid by ether.

Guaiacine. Dark yellow bitter resin obtained from the root and wood of *Guaiacum officinale* by alcohol; soluble in alcohol and hot water; less soluble in cold water; insoluble in ether; unites with sulphuric acid, not with alkalies.

Guaiacum Resin. $C_{40}H_{23}O_{10}$? Spec. grav. 1·22. Yellowish or reddish-brown amorphous pieces, partly translucent, but usually covered with a greenish-gray powder, from the *Guaiacum officinale,* a West Indian tree; sweetish bitter taste; soluble in alcohol and ammonia; less soluble in ether and oil of turpentine. It is easily detected in solution by the fine green or blue colours produced when a solution of bleaching powder is added to its alcoholic solution. Commercial resin consists of 80 per cent. resin, 16·5 bark, 1½ gum, 2 per cent. extract. The resin appears to consist of 2 or 3 different resins. It is used in medicine in the form of tincture, and in the solid state.

Guaiacum Soap. A pharmaceutical preparation, made by heating powdered guaiacum resin in caustic potash and evaporating; soluble in water and alcohol.

Guaiacyle, Hydride of. *Pyroguaiacic Acid.* $C_{14}H_8O_4$. Spec. grav. 1·119, of vapour 4·9; B.P. 410°. Colourless peppery oil, becoming red and then black in the air; very soluble in alcohol, ether, and caustic potash. It unites with alkalies; it reduces salts of silver and gold, and peroxide salts of iron, and copper, to the lowest state of oxidation; it resembles creasote, from which it differs by 2 atoms more oxygen.

Guanic Acid. $C_{10}H_3N_4\ O_7\ 2\ HO$. Colourless rhombic prisms, gritty between the teeth, without taste and smell; obtained by gently heating for twenty-four hours, 5 chlorate of potash, 3 guanine, 25 water, and 30 muriatic acid.

Guanine. $C_{10}H_5N_5O_2$. White insoluble powder, becoming wax-like by friction; not decomposed at 482° when distilled with water; soluble in acids, forming crystalline salts; soluble in alkalies, and uniting with them; by the action of chlorohydric acid, and chlorate of potash, it is converted into oxalic acid and ammonia; obtained from guano by digesting it with milk of lime till the solution is no longer brown, but greenish-yellow; it is then filtered and neutralized with chlorohydric acid; equal quantities of uric acid and guanine subside in some hours; by boiling with chlorohydric acid, guanine dissolves, and uric acid remains; on cooling chlorohydride of guanine falls, from which the guanine is precipitated by ammonia.

Guanite. Spec. grav. 1·65, H 2; (MgO 2 NH_3, PO_5, 5 HO). Insoluble right rhomboidal prisms, found in guano.

Guano. *Huano.* The urine of sea-fowls deposited on the coasts of Peru, Africa, &c. of great value as a manure, in consequence of its containing the salts and nitrogen necessary for the food of plants. It is usually mixed with a certain amount of granite powder from the soil on which it is deposited. Sometimes it contains bones of birds, feathers, &c.; and often salts, such as ammonia phosphate of magnesia, bicarbonate of ammonia. It has been used for ages as a manure by the Peruvians, and was first brought into prominent notice in Europe by Humboldt and Boussingault; but more recently by Liebig, who has done great service to our agriculturists in this and other respects. When genuine it should have a light brown colour, with a smell of ammonia; should, when burned, leave a fine white ash of bone earth, &c. This ash, when heated with dilute muriatic acid, should not effervesce, indicating the absence of chalk or carbonate of lime, and should dissolve with the exception of less than 2 per cent. of *white* sand and microscopic shells. When treated by water alone, the best Peruvian guano I have found to yield fine crystals of oxalate of ammonia. As the other constituents are pretty equally diffused in different shipments of guano, one of the most important tests of its value depends on the amount of uric acid and ammonia which is present. But in all analyses the amount of the different constituents should be determined. From a long experience in the examination of guano, imported at Glasgow and elsewhere, I have constructed the following tables. Although I have frequently had submitted to my examination adulterated samples, I am happy to say, for the credit of importers, that I have scarcely ever found sophistication to be attempted in their hands, and I have therefore strongly to recommend farmers to unite and purchase their stocks at the ship's side, the price to be determined by chemical analysis:—

	A	B	C	D	E
1,......	11·90	9·40	25·02	37·72	50·30
2,......	5·20	9·43	15·07	4·65	1·00
3,......	53·90	51·58	28·98	30·13	13·60
4,......	27·80	27·98	29·65	26·00	12·10
5,......	1·20	1·59	1·28	1·5	15·00
6,......	18·38	17·28	6·85	5·31	3·47

1, water; 2, soluble salts, phosphate and sulphate of potash, chlorides of potassium and sodium; 3, organic matter, containing uric acid, ammoniacal salts, &c.; 4, phosphate of lime and magnesia; 5, siliceous matter from the soil; 6, ammonia. A and B, fine specimens of guano from Chincha, Peru, imported by John Thomson, Union Street, Glasgow. C, Inferior Peruvian guano. D, African guano. E, Guano from Ailsa Craig, coast of Ayrshire; along with the phosphate of lime is included 1·5 carbonate of lime. A specimen of adulterated guano I found to consist of water 26·60, volatile matter containing 4½ ammonia 27·4, phosphate of lime 12·09, carbonate of lime 10·58, sulphate of lime and peroxide of iron 1·84, sulphate of soda and some potash 2·3, chloride of sodium 5·06, river sand 12·56, cinders 1·57. Independently of its river sand and other substances, the presence of so much soda salts is a suspicious circumstance, as potash is the principal alkali in guano soluble salts.

Guaranine. A synonyme of Theine and Caffeine found in Brazilian *Guarana (Paullinia sorbilis.)*

Gums are exudations from plants which soften in water, have the same composition as starch and sugar, are insoluble in alcohol, but differ from starch in not being coloured by iodine, and from sugar by not being sweet, and not fermenting with yeast. They are converted by nitric acid into mucic acid. The principal species of gums are—1. Cerasine, insoluble in cold water, found in the plum tree and bird cherry, first distinguished by this name by Dr. Thos. Thomson. 2. *Arabine*, soluble in cold water, in gum Arabic. 3. *Bassorine*, *gum tragacanth*, forming a mucilage in water without dissolving. The following bodies appear also to be connected with this class. *Calenduline* from marigold, *saponine* from saponaria.

Gum Lac. See LAC RESIN.

Gum Resins are more opaque than resins, and consist of a mixture of gums, resins, oils, &c. in various proportions. They do not melt like resins when heated; have a strong odour; are partially soluble in water and in dilute alcohol and caustic alkalies. They have been divided into—1. *Fetid*, including ammoniac, galbanum, asafœtida, opoponax, sagapenum. 2. *Stimulating*, comprising olibanum, myrrh, euphorbium, and bdellium. 3. *Cathartic*, aloes, gamboge, scammony. 4. *Sedative*, opium, lactucarium, upas.

Gun Cotton. *Pyroxyline. Ternitramidine.* $C_{12}H_7N_3O_{22} = C_{12}H_7O_7 3NO_5$. The explosive compound known under this name resembles common cotton wool. The best specimen which I have seen was prepared by Mr. Crum, who found it to have the preceding composition (Proc. Phil. Soc., Glasg., ii. 165), or cellulose, in which 3 atoms water are replaced by 3 nitric acid. The cotton used was fine Sea Island; it was thoroughly carded and bleached by boiling in caustic soda and steeping in solution of bleaching powder, then in caustic soda again, and afterwards in weak nitric acid. It was well washed and beaten in a bag after each operation. When burnt 10000 parts left 9 of ash. The cotton dried and carded after bleaching was exposed in parcels of 10 grains each for several hours to the heat of a steam bath. Each parcel was immersed while hot in 1 measure of a mixture of 1 measure sulphuric acid of 1·840, 3 measures pale lemon-coloured nitric acid of 1·517. After one hour it was washed in successive portions of water till no trace of acid remained, and dried in the open air. By this process 100 of dry cotton produced 177·9 gun cotton. It often happens that a portion of cotton remains unacted on. Mr. Crum overcame this difficulty by immersion for an hour in nitric acid alone. The varying analyses which have been given by others obviously depend on this circumstance. The tendency of this substance to explode by friction has prevented its introduction into the arts. Its solution in ether, known as collodion, is valuable in photography and surgery.

Gunjun. An Indian balsam from *Dipterocarpus levis.* Spec. grav. ·934; B.P. 240°; soluble in alcohol and ether.

Gunpowder was known to the Chinese at a very early period. A Franciscan, Berthold Schwartz, is reported to have known it at an early period. Roger Bacon, who died in 1292, was acquainted with it. Brother Ferrarius, a Spaniard, in the 13th century, gives a receipt for gunpowder, or flying fire; 1 sulphur, 2 charcoal, 6 nitre (Sir F. Palgrave). It seems to have been used at the siege of Niebla in 1257; it is mentioned by an Arabic author in 1294. At Amberg there is a piece of ordnance with the date 1303. Guns, or "crackies of war," appear to have been employed at the battle of Werewater in 1327, at Algesiras in 1343, and at Cressy in 1346. *Characters.*—Gunpowder should be slate-coloured, or dark-bluish gray, so hard as not to fall into powder by the pressure of the finger. When powder has not been glazed, the slightest touch detaches dust from its surface; and, when reduced to fine powder, gunpowder by agitation in carriage would separate into its constituents according to their gravity. Granulated powder burns instantaneously, while powder in mass gives a long fire. When powder evolves its gas too rapidly, as in detonating powders, the greater part of the force is spent on the sides of the gun. Good powder inflames at a temperature of about 572°. It ought to explode on paper without leaving any appreciable residue, and without inflaming the paper. 1 measure of powder has been found to give 450 measures of gas, but as a temperature of 2200° (4352° Brianchon), is produced, the expansion of these 450 measures has been calculated to amount to 2000 measures. When heated above the melting point of sulphur over a lamp, the sulphur may be distilled off in absence of air without explosion. Gunpowder, as used in different countries, has the following composition (Pelouze):—

	Nitre.	Charcoal.	Sulphur.
Spain,................	76·47	10·78	12·75
Austria, ...	76·	11·5	12·5
Switzerland,.........	76·	14·	10·
England,.............	75·	15·	10·
France,........... America,......... Prussia,..........	75·	12·5	12·5
Sweden,.............	75·	9·	16·
China, ...	75·	14·4	9·6
Russia,...............	73·78	13·59	12·63
Holland,............	70·	16·	14·
French blasting powder, ...	65·	15·	20·

Manufacture. Mixture.—The nitre must be as free as possible from chlorides, which is effected by repeated crystallization and careful washing, until the washings give little or no cloud with nitrate of silver. The charcoal is made of the younger branches of bird cherry, elder, poplar, maple, and walnut, previously deprived of bark. The charring is performed in an iron retort or cylinder resembling a gas retort, at a high temperature. The sulphur on the continent is usually employed in the condition of roll sulphur. The materials are separately introduced into oak mortars, which are set in a series, and pulverized by means of pestles or stampers, which are worked by machinery like the flint mill at a pottery, or stampers at a bleachfield. The mortars are 14 inches in diameter. The materials are moistened with water into a paste. In France the pulverization is made in revolving wooden or leathern drums containing metallic balls of different sizes. A third, and more common method, is to pound the materials separately by means of millstones formed of cast iron, sometimes enclosed in a ring of bronze. *Pressing.*—The powder, properly ground and thoroughly mixed, is conveyed to the press. The powder is arranged in layers separated by metallic plates, and is subjected to intense pressure in a Bramah press, or it is placed on cloths and pressed between rollers on endless cloths. The nature of the metals employed as separating media is important; Muntz's metal has been found most durable in standing the pressure. The condensation by this process is so great as to reduce the cake formed to one-fourth of its original bulk. *Graining* or *Corning.*—The broken cake is made to pass between a series of rollers; the first armed with long teeth, the second series having finer teeth, and the last being smooth. The powder thus formed is sifted in brass wire sieves, having apertures suited to the kind of powder required; or the broken cake is first placed in parchment or leathern sieves and pressed upon by a disc of hard wood, till the grains pass through the apertures; the coarse grains and dust are then sifted in brass wire sieves moved by machinery. *Dusting.*—From the corning house the powder is carried to the *dusting* house, where the dust is removed from the grains by sifting. *Glazing* or *polishing.*—The corned powder has a dull leaden aspect. To give it a shining appearance, it requires to be subjected to a polishing process. This is effected in a somewhat similar manner to that employed for removing the asperities from steel pens. The corned powder is placed in the glazing casks or barrels with projections in their interior, which are deposited horizontally, and made to revolve on their axes. The barrels are three-fourths filled. The glazing process, which is attended with an increase of temperature, is prolonged according to the degree of glazing required, the rapidity of rotation being increased as the process proceeds. In France the glazing process lasts 36 hours, in Austria 8 to 10 hours. When the blasting powder is to be kept for a long time, or to be exported, it is now customary to glaze it with black lead to enable it to withstand hygrometric water. For this purpose a quantity of finely pounded plumbago, prepared in London, from Trieste, is introduced into the glazing barrels along with the corned powder. *Drying.*—The powder is dried at a temperature of 110° to 120° in an apartment heated by means of steam, which passes in a wide pipe or chamber at the bottom of the apartment, a current of dry air passing over the powder. The powder is placed in wooden trays with canvas bottoms, which are deposited on shelves. In France the powder is spread on woollen cloth in a box heated by steam, through which a current of dry air is drawn by means of a ventilator. The *force* of powder may be *measured* by the distance to which a given weight will propel a ball, as a 60 lb. ball in a mortar by 2 ounces of powder. *Analysis.* The moisture is determined by drying 100 grains at 212° till it ceases to lose weight. The nitre is dissolved out by means of water, and by washing through a weighed filter until the washings leave no residue on a plate of glass when evaporated, drying at 212° and weighing. The sulphur is estimated by deflagrating 1 part with 1 nitre, 1 carbonate of potash, and 4 common salt, to prevent spirting (Gay Lussac); the sulphur being converted into sulphuric acid, is precipitated by chloride of barium; every 14½ sulphate of barytes are equal to 2 sulphur; or the sulphur and charcoal may be boiled with strong nitric acid. The charcoal is estimated by the loss. The sulphur and charcoal may also be ascertained by placing them in a glass ball, to which a second bulb is attached. Heat being applied to the mixture, and hydrogen passed over it, the sulphur sublimes into the empty bulb, while the charcoal remains in the first ball, and both may then be weighed.

Gunpowder, White. 1 sugar, 1 yellow prussiate of potash, 2 chlorate of potash mixed with water and triturated in a mortar for fifteen minutes, and granulated by passing through a sieve (Augendre). I have found 6·1 grains of this powder, at five feet distance, carry a bullet through 100 pages of paper, while black powder in the same quantity did not rupture one page; even 36·6 grains of black powder only made an indentation.

Gurhoffite, Gurhofian, or Calcareo-carbonate of magnesia.

Gurolite. H 3·5. Small white globular masses of pearly-like plates along with laumonite, and apophyllite, in amygdaloid at Storr in Skye. SiO_3 50·7, Al_2O_3 1·48, CaO 33·24, MgO ·18, HO 14·18, or 2 ($CaOSiO_3$), 3 HO; decomposed by chlorohydric acid, loses water at 212°, and becomes $CaOSiO_3$ HO. It is often mixed with other zeolites, or with carbonate of lime (Anderson).

Gutta Percha. *Niato*, or *Pertsha.* C 87·8, H 12·2. The milky juice of the *Isonandra gutta* (Sapotaceæ), a tree growing in Malacca and Borneo, each tree yielding from 20 to 30 lbs. It has been long known in these countries, and I recollect in 1831 of seeing articles made of it at Singapore. Spec. grav. ·979. Softens at 122°; at 158°—176°, it is very workable, and can be made into moulds; insoluble in water, alcohol, dilute acids, and alkalies. Ether and ethereal oils swell it up as with caoutchouc; it dissolves in oil of turpentine, oil of tar, and oil of caoutchouc. It can be vulcanized by means of sulphur in the same manner as caoutchouc. Its various uses are so well known that it is unnecessary to mention them. Three million pounds weight were imported in 1848.

Guyaquillite. $C_{40}H_{16}O_6$. Specific grav. 1·092; F.P. 157°—212°. Pale yellow fossil resin; soluble in alcohol with a bitter taste; soluble in sulphuric acid and reprecipitated by water; soluble in caustic potash; precipitated by nitrate of silver; found near Guayaquille in America.

Gyle Tun, or fermenting tun. See ALE.

Gymnite. 2 MgO SiO_3 3 HO. A variety of serpentine found in the Bare hills, Baltimore. Colour pale yellow with resinous lustre; fusing B.B. into a white opaque mass with soda, a colourless glass with borax. SiO_3 40·16, Al_2O_3 1·16, MgO 36, CaO ·8, HO 21·16 (Thomas Thomson).

Gypsum, or sulphate of lime.

Gyrophoric Acid. $C_{36}H_{18}O_{15}$? Colourless crystals from *Gyrophora pustulata* by milk of lime, precipitation by chlorohydric acid, and crystallizing from alcohol; probably lecanoric acid.

H

Hæmaphæine. The yellow-colouring matter of blood serum.

Hæmateine. $C_{40}H_{15}O_{16}$. Reddish-brown when moist, when dry dark green; soluble in hot water, alcohol, potash, and acids; unites with bases; with ammonia forms hæmateine-ammonia, violet 4-sided prisms, containing 2 atoms ammonia. Hæmateine is obtained by dissolving hæmatoxyline in ammonia and allowing it to stand, and precipitating by acetic acid.

Hæmatinone. Spec. grav. 3·5. A red glass used by the ancients; its colouring matter appears to be red oxide of copper.

Hæmatite. See SESQUIOXIDE OF IRON.

Hæmatoidine. *Xanthose.* Ruby oblique rhombic prisms; soluble in caustic potash with a bright red colour; changed to green, blue, rose and yellow by sulphuric acid. It is always found in the sanguineous extravasations, by the bursting of the Graafian vesicles in menstruation, frequently in old extravasations in the brain, hæmorrhage of the spleen, and purulent abscess of the extremities.

Hæmatosine, *Hæmatine.* $C_{44}H_{22}N_3O_6Fe$? The colouring matter of blood. Dark brown mass, without taste and smell; insoluble in water, alcohol, ether, fat, and volatile oils; soluble in weak alcohol acidulated with sulphuric or chlorohydric acid; insoluble in strong acids, which, however, take up some iron; very soluble in dilute caustic alkalies; precipitated from its solution in ammonia by salts of silver, lead, and copper; decomposed by nitric acid and heat. When allowed to remain in contact with sulphuric acid for some time, and then diluted with water, hydrogen is evolved and copperas dissolved, and by repeating this process the whole iron may be removed. When treated with chlorine, a compound of 1 hæmatine and 6 chlorine is formed. Chlorohydric acid gas forms a violet mass, soluble in water and alcohol. Hæmatosine is obtained by treating blood with 8 times its volume of a solution of an alkaline sulphate or of common salt, filtering and washing with the same solution. The residue deprived of serum is dissolved in water and coagulated by heat. It is then boiled in spirit containing sulphuric acid, till a colourless fluid passes through the filter. The filtered fluid is saturated with ammonia and evaporated to dryness, some globuline and sulphate of ammonia which deposit, being separated; the residue is digested with water, alcohol, and ether, and may be again dissolved in ammoniacal spirit for greater purity.

Hæmatoxyline, *Hæmatine, Hematine.* $C_{40}H_{17}O_{15}$. Brilliant yellow needles or oblique rhombs, yielding a white or yellowish powder with a sweetish taste; slightly soluble in cold, very soluble in boiling water; soluble in alcohol and ether; not volatile but decomposed by heat; soluble in sulphuric and chlorohydric acids without decomposition; nitric acid forms oxalic acid; precipitated white by isinglass; it has a tendency to unite with bases, but then becomes blue by access of oxygen. Pure hæmatoxyline is obtained by ether from the watery extract of logwood (*Hæmatoxylon campechianum*). The commercial extract may be employed for this purpose, 2 lbs. of the extract yielding 3 to 4 ounces of crystals. I have obtained several ounces of it, in beautiful crystals, during frosty weather, from the sides of the evaporators of logwood liquors.

Haidingerite. Ferro-sulphide of antimony. See BERTHIERITE.

Hair. C 49·78, H 6·36, N 17·14, O 21·72, S 5·. The hair consists of tubes formed of an internal and external layer which probably differ in composition. The exterior contains fat which is considerable in wool, and has been termed *stearerine* and *olaerine* (ἔρια, wool). Hair contains much sulphur, brown 4·98 to 5·44, black 4·85 to 5·22, red 5·02, gray 4·63. The ash consists chiefly of oxide of iron and earthy salts, brown 1·1 per cent. to ·32, black 1·02, red 1·3 to ·5, gray 1 to ·75. When boiled with water for a long time, hair is converted into a gelatinous mass, and a similar mass is formed in a Papin's digester with the evolution of much sulphohydric acid. From the quantity of sulphur it is easy to dye hair with nitrate of silver.

Hair Pyrites. See SULPHIDE OF NICKEL.

Hair Salt. Sulphate of magnesia in capillary crystals, also Ferro-sulphate of alumina (FeO 13·56, Al_2O_3 7·127, SO_3 35·6, HO 43·713), found in the alum slate at Hurlet and Campsie alum works.

Hallite. See ALUMINITE.

Halloylite. *Halloysite.* Spec. grav. 1·8 to 2·1. SiO_3 39·5, Al_2O_3 34·, HO 26. White compact mineral; lustre waxy; fracture conchoidal; absorbs water when immersed in it, and gives out air; the alumina is dissolved by sulphuric acid and the silica left; found at Angleure by d'Halloy. Attempts have been made by some writers to classify the silicates of alumina obviously without proper attention to the subject; minerals with a spec. grav. of 1·8 being placed along with others of 2·55. Such arrangements are mischievous to true mineralogy, which must be based on chemistry.

Haloide Bodies. (ἅλς, salt; εἶδος, form). A term for chlorine, iodine, and fluorine, which produce salts with metals.

Haloide Salts. (ἅλς, salt; εἶδος, form). Salts formed by a haloide body and a metal. They are distinguished from *amphide* salts by their two constituents being generally elementary bodies.

Halotrichite. A synonyme of feather alum.

Hamathionic Acid. See EUXANTHIC ACID.

Hampshirite. A synonyme of steatite.

Hardness of minerals is determined by an ascending scale of 10 degrees, each degree being represented by a harder mineral; 1 talc, 2 rock salt, gypsum, 3 calcareous spar, 4 fluor spar, 5 phosphate of lime, 6 felspar, 7 rock crystal, 8 topaz, 9 corundum, 10 diamond. When a mineral scratches talc, but not rock salt or gypsum, its hardness is said to be 1·25 to 1·5 or 1·75, according to the impression which it makes on the talc and gypsum.

Harmaline. $C_{27}H_{14}N_2O_2$. Brownish transparent bitter crystals or rhombic prisms. Soluble in alcohol; slightly soluble in water and ether; fusible, partly volatile; forms yellow salts with acids. The harmala red of commerce is the powdered seeds of *Peganum harmala*, which contain phosphate of harmaline. By the action of prussic acid a new base, *hydrocyanharmaline*, is formed. By oxidization *harmine* and *chrysoharmine* are formed.

Harmine. *Leucoharmine.* $C_{27}H_{12}N_2O_2$. Colourless rhombic prisms, less soluble in cold alcohol than harmaline; soluble in ether; forms salts with acids. Obtained with harmaline from the seeds of *Peganum harmala*, which grows in the Steppes of Russia.

Harmotome (ἁρμος, a joint; τεμνω, I cut). *Andreolite, Andreasbergolite, Cross stone, Ercinite Barytes harmotome.* 3 BaO2SiO$_3$, 4 (Al$_2$O$_3$ 2 SiO$_3$) 18 HO. Spec. grav. 2·390 to 2·45, H 4·25. White, occasionally gray, yellow, red, and brown, right rectangular prisms. The face of the prism is often replaced by four faces on the solid angles of the primary form; fracture imperfect, conchoidal, lustre vitreous; sometimes transparent, but commonly only translucent, brittle. SiO$_3$ 48·753, Al$_2$O$_3$ 15·1, BaO 14·275, CaO 3·18, KO 2·55, HO 14·. Occurs in amygdaloidal rocks at Kilpatrick, Strontian, Hartz. B.B. fuses with difficulty without boiling. For Lime Harmotome, see PHILLIPSITE.

Harringtonite. Spec. grav. 2·217, H 5·25. Snow-white crystalline or earthy masses, with almond lustre, opaque, tough. A mean of five analyses gave me for its composition SiO$_3$ 40·85, Al$_2$O$_3$ 29·41, CaO 11·06, NaO 2·80, KO trace (volatile matter) 15·1, FeO ·63. From the north of Ireland. It is often mixed with carbonate of lime.

Hartine. C$_{20}$H$_{17}$O$_2$. Spec. grav. 1·115 at 32°; F.P. 410°. White insoluble resin in the brown coal of Hart, Austria; without taste and smell; slightly soluble in alcohol and ether; soluble in naphtha, from which it separates in long needles; by distillation it is resolved into acid water, carbonic oxide, olefiant gas, tar, and a white crystalline body.

Hartite. C$_6$H$_5$. Spec. grav. 1·046; F.P. 165°. White resin with a fatty lustre; soluble in alcohol and ether; distils without decomposition; separates from its ethereal solution in silky crystals; decomposed by sulphuric acid; from the coal of Oberhart.

Hatchettine. *Mineral tallow, Mountain tallow.* CH. Spec. grav. ·8035 to ·983; F.P. 118°; B.P. 290°. Yellow translucent pearly plates; soft like wax; without odour till heated, when it smells like fat; distils without change; less soluble in alcohol than in ether; becomes dark in the air; decomposed by sulphuric acid, not apparently by nitric acid. Found on the coast of Finland, at Strasburg, Inverary, Oban, Merthyr Tydvill. A greenish-yellow substance found in the trap rocks near Glasgow, and used by the country people as a balsam for wounds, seems to belong to this class. The hatchettine of Conybeare fuses at 170°.

Hauerite. *Bisulphide of Manganese.* MnS$_2$. Spec. grav. 3·463, H 4·. Reddish-brown; 8-hedrons or cubes; lustre metallic, adamantine, streak brownish-red. B.B. gives off sulphur and becomes MnS; from Kalinka, Hungary.

Hausmannite. *Red Oxide of Manganese, Black Manganese, Pyramidal Manganese.* MnO 1⅓ or MnO Mn$_2$O$_3$. Spec. grav. 4·722, H 4·75. Brownish-black masses, and 8-hedrons with square bases, with angles of 105° 45′, 117° 30′, the summits of the pyramids being often replaced

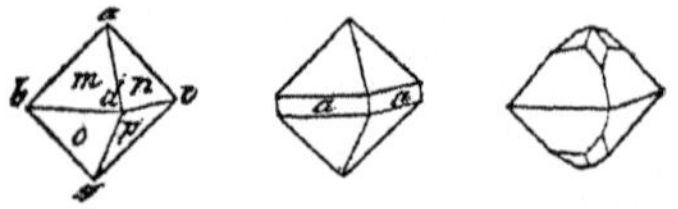

by low 4-sided pyramids; lustre imperfect metallic, opaque; streak reddish. In porphyry at Ihlefeld, Hartz, Ilmenau, Framont, and Pennsylvania.

Hauyne. *Berzeline, Marialite, Ittnerite.* Spec. grav. 2·45, H 5·75. Blue or green crystalline grains, or 12-hedrons; lustre glassy, translucent; fracture conchoidal, irregular. B.B. fuses into a white opaque glass with decrepitation; gelatinizes with muriatic acid, SiO$_3$ 35·48, SO$_3$ 12·39, Al$_2$O$_3$ 18·87, CaO 12·, KO 15·45, Fe$_2$O$_3$ 1·16. Found in Vesuvian lavas, in basalt at Andernach, and at Albano and Marino. Allied to sodalite and spinellane.

Haydenite. Spec. grav. 2·136 to 2·265, H 3·. SiO$_3$ 49·5, Al$_2$O$_3$ and Fe$_2$O$_3$ 23·5, HO 21·, CaO ·27, MgO trace, KO ·25; oblique rhombic prisms, found near Baltimore; probably Chabasite.

Hayesite. *Hayesine.* A synonyme of borate of lime and soda.

Haytorite. Pseudomorphous quartz with the crystal of Datholite, Devon.

Heat. *Caloric.* See *Cyclopædia of Natural Philosophy.*

Heavy Spar. See SULPHATE OF BARYTES.

Heavy Wine Oil. *Sulphate of Ethyle and Etherole.* C$_8$H$_9$O 2 SO$_3$. Spec. grav. 1·133; B.P 540°. Colourless neutral oily fluid, with an ethereal odour and cooling taste; mixes with alcohol and ether, distils unchanged when dry; decomposed by water and alkalies, forming bisulphate of ethyle and etherole; it gives mercaptan with sulphide of potassium. It is obtained whenever ether and anhydrous sulphuric acid come in contact; it is readily formed by distilling an intimate mixture of equal parts of quicklime and dry sulphovinate of lime or potash, at a heat not above 540°; or by distilling basic sulphovinate of lead. It is freed from alcohol by evaporation in air, or in vacuo.

Hebetine. A synonyme of silicate of zinc.

Hedenbergite. *Bihydrous Bisesquisilicate of Iron.* Spec. grav. 3·154, H 3·5. 7 FeO 7 SiO$_3$ 2 HO = SiO$_3$ 40·62, FeO 35·25, HO 16·05, MnO ·75, CaO 3·37, Al$_2$O$_3$ ·37, CO$_2$ 1·56. Greenish-black masses, with rhomboidal cleavage resembling calcareous spar; fibrous, opaque, streak olive. B.B. loses its lustre, becomes black, and is attracted by the magnet, fusing into a black glass; with borax an iron glass; with soda a grayish-green bead. Allied to *Black Pyroxene.* Near Tunaberg, Sweden.

Hederic Acid. C 66·46, H 9·45, O 24·09. Fine needles, insoluble in water and ether, without smell; tastes of ivy. Obtained from the seeds of *Hedera helix*, or ivy, by boiling alcohol.

Hederine? Bitter substance from ivy seeds, by acid water; neutralizing with slaked lime, and treating the precipitate with alcohol.

Hedyphane. Spec. grav. 5·46 to 5·493, H 4·75. White amorphous masses or 6-sided pyramids, translucent, lustre adamantine. B.B. fuses into an opaque globule, tinging the flame greenish-blue; on charcoal, lead is reduced and arsenious acid evolved in fumes; with salts of phosphorus it froths, and muriatic acid is given out. PbO 52·95, CaO 14·034, Cl 2·029, AsO_5 22·78, PO_5 8·207. Found at Longbanshyttan, Sweden. It is an arseniophosphate of lead.

Helenine. $C_{42}H_{28}O_6$; F.P. 161½°; B.P. 530°. White 4-sided prisms, lighter than water; insoluble in water; soluble in alcohol, ether, and ethereal oils; soluble in caustic alkalies, and reprecipitated by muriatic acid without change; soluble in sulphuric acid, with which it forms a coupled acid, sulphohelenic acid remaining insoluble when water is added. With nitric acid it forms *nitrohelenine*. Helenine is obtained by hot alcohol from the root of the *Inula helenium*.

Helicine. $C_{26}H_{16}O_{14}$, 1½HO. White needles, without smell, taste bitterish; little soluble in cold, very soluble in hot water; soluble in spirit; insoluble in ether; at 212° loses 4⅓ per cent. water; at 347° fuses into an oil, which, by continued heat, yields the odour of salicylous acid; it is split by acids, and fermentation, into salicylous acid and sugar ($C_{12}H_{10}O_{10}$, $C_{14}H_6O_4 = C_{26}H_{16}O_{14}$). It is obtained by acting on 1 part of salicine with 10 nitric acid, spec. grav. 1·16, for twenty-four hours, and crystallizing out. With chlorine and bromine it forms *chlorohelicine* and *bromohelicine*. Along with *helicine* is usually formed a nitrogenous body, *anilotic* acid, which strikes a red with iron salts.

Helicoidine. $C_{52}H_{34}O_{28}$ 3 HO = 2 sugar, 1 salicylous acid, 1 saligenine. Crystals obtained by treating salicine with nitric acid of spec. grav. 1·09. They are purified by washing with cold water, and crystallizing from hot water.

Heliotrope, or *Blood Stone*. Green quartz, with blood-red spots and veins.

Hellebore, White. *Veratrum album*. The root contains *Veratrine*, *Barytine?* and *Jervine*.

Helleborine. Resinous crystals from *Helleborus hyemalis* by alcohol.

Helmintholite. A variegated marble at Hall, Tyrol.

Helvine. Spec. grav. 3·166, H 6·5. SiO_3 33·258, GO and Al_2O_3 12·029, FeO 5·564, MnO 31·817, MnS 14·, Vol. matter 1·155. Yellow wax or sisken-green tetrahedrons or acute rhomboids; lustre vitreous or resinous; translucent on edges; fracture uneven. B.B. melts on charcoal into a globule, with borax into a transparent glass; decomposed by chlorohydric acid with evolution of sulphohydric acid gas. Its composition has been given as MnO, MnS, 3(2MnO2GO 2FeO) SiO_3.

Hemathionic Acid. C_{14} H_7 O_{12} SO_3. Gummy mass, along with purrenone by the action of sulphuric acid on purreic acid.

Hemidesmus Indicus. *Ununtamul* of Bengal. The root is used in Madras as a substitute for sarsaparilla.

Hemihedrism is exemplified when half the similar alternate angles or edges of a crystal are altered independently of the other half, or when the similar angles or edges are altered but by half the regular number of planes.

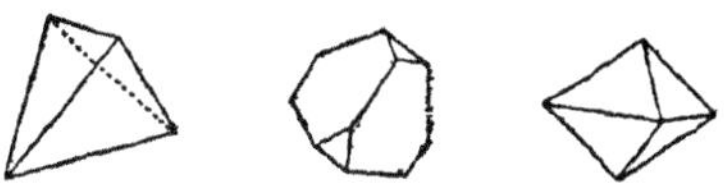

Hemimorphism. A case in which bodies possessing the same or analogous chemical composition, crystallize with several angles almost identical, and with others varying considerably.

Hemipic Acid. *Hemipinic acid*. $C_{10}H_4O_5$ HO. F.P. 356°. 4-sided prisms; sublimes; little soluble in water, more soluble in alcohol and ether. When its hot aqueous solution is mixed with sulphuric acid, and brown oxide of lead, it is resolved into carbonic acid and water; obtained by mixing a boiling aqueous solution of opianic acid with brown oxide of lead, and dilute sulphuric acid dropped in, till carbonic acid begins to be evolved; the dissolved lead is precipitated by sulphuric acid, and the hemipic acid crystallized out.

Hemitropism, *Hemitrope Crystals*, *Macle*, *Twin crystallization*, is the appearance in a crystal as if it had been slit, and then one portion turned half round on the other upon an imaginary axis in the figures. The 8-hedron

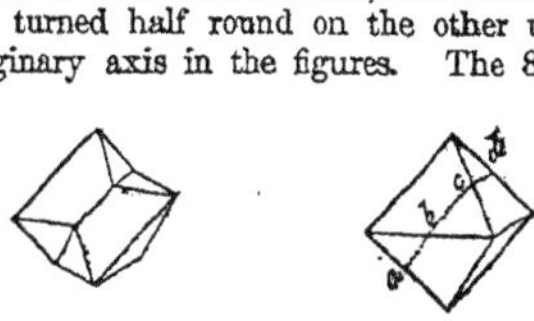

has been cut at *a*, *b*, *c*, *d*, and one half being turned round on the other, presents the aspect of the other figure. When the evolution has been only one-sixth, it is called *transposition*, which occurs in calc spar, gypsum, &c.

Hemlock. *Conium Maculatum*. See CONIINE.

Hemp, Common. *Cannabis Sativa*, contains a bitter resin, soluble in alcohol and ether (*Cannabine*).

Hemp, Indian. *Cannabis Indica. Gunjah. Bang.* An alcoholic extract is formed from the tops of the plant, and is used as a narcotic.

Hepatite. An impure variety of sulphate of barytes, with a fœtid smell on friction.

Herbeckite. A synonyme of Hisingerite.

Hercinite. A variety of spinell. Al_2O_3 61·17, FeO 35·67, MgO 2·92; Bohemia.

Herderite. *Allogonite.* Spec. grav. 2·985, H 5· Prismatic system; colour yellow and greenish-white; streak white, translucent; lustre vitreous, brittle; fracture conchoidal. Supposed to contain phosphate of alumina with fluorine. One specimen only found at Ehrenfriedersdorf, Saxony, in fluor spar.

Herrerite? Spec. grav. 4·3, H 4·5. Green rhombohedrons? or kidney-shaped masses; cleavage in three directions; streak yellowish-gray, translucent. CO_2 31·86, TeO (ZnO?) 55·58, NiO 12·32. Albarradon, Mexico.

Herschellite. Sp. grav. 2·06. SiO_3 47·39, Al_2O_3 20·9, NaO 8·33, KO 4·39, CaO ·38, HO 17·84. *Form.* 3 NaO,3 Al_2O_3 8 SiO_3 15 HO. 12-hedrons, the summits of which are deeply replaced, and in streaked regular 6-sided prisms; colour white, translucent, or opaque; cleavage parallel to the base of the prism. In amygdaloid from Poonah; at Aci Reale, Catania. Allied to chabasite.

Hesperidine. White bitter silky plates; soluble in water and boiling alcohol; insoluble in ether; converted into a wax by boiling with water; obtained by water and alcohol from the white spongy part of unripe bitter oranges.

Hessite. A synonyme of Telluride of silver.

Hessonite. A variety of garnet.

Heteroclite. Silicate of manganese in oblique rhombic prisms.

Heteromerism. A property supposed to exist by which all bodies possessing a similar crystalline form, although dissimilar in the nature, number, and grouping of their atoms, are capable of crystallizing together.

Heteromerite. A synonyme of Idocrase.

Heteromorphism. The property of two or more compounds with an equal number and grouping of atoms, and having a different crystalline form. Thus, copperas is clinorhombic, and white vitriol rhombic.

Heteromorphite. See FEATHER ORE OF LEAD.

Heterosite. *Hetpoizite. Heteposite.* Spec. grav. 3·524. Greenish-gray or violet rhomboidal prisms, usually in foliated masses; softer than quartz; soluble in acids. B.B. fuses into a brown enamel, with semi-metallic lustre. FeO 35, PO_5 4·777, MnO 16·18, HO 4·4, SiO_3 2·2. Hureaux, Haute Vienne. 2 (5 $FeO_2 PO_5$) 5 $MnO_2 PO_5$ ·5 HO.

Heulandite. *Foliated Zeolite, Stilbite of Breithaupt* and *Rose.* Spec. grav. 2·195, H 3·5. Snow-white or flesh-red right oblique prisms with angles of 130° 30′, 112° 15′, sometimes massive and in round balls; lustre vitreous and pearly, translucent, brittle, streak white. B.B. yields an opaque vesicular bead; decomposes but does not gelatinize with chlorohydric acid. SiO_3 59·145, Al_2O_3 17·92, CaO 7·652, HO 15·4. *Form.* 3CaOSiO_3, 4$Al_2O_3$3SiO_3 18HO. Kilpatrick; Bombay; &c.

Heveene. A volatile oil boiling at 600°, obtained by fractional distillation of caoutchouc oil.

Hexachlorhydroxylone. $C_{26}H_{10}Cl_6O_6$. Fine long, blonde, vitreous needles resembling benzoic acid; soluble in alcohol and ether; little soluble in water; decomposes by boiling with water, to which it imparts a violet colour; soluble in acetic acid with heat; obtained by passing sulphurous acid through water in which *hexachloroxylone* is suspended, and crystallizing the crystals obtained out of alcohol and ether. When heated with nitrate of silver and evaporated, *Violet hexachlorhydroxylone* ($C_{26}H_8Cl_6O_6$4HO) in crystals is obtained, and when treated for a long time with hypochlorite of soda, plates are formed of *yellow hexachlorhydroxylone.* $C_{26}H_8 Cl_6O_6$.

Hexachloroxylone. $C_{26}H_6Cl_6O_6$. Golden scales, or rhombic tables, beginning to sublime at 248°, but subliming best at 365°; at a higher temperature decomposing and giving copper needles; insoluble in water; soluble in 171 parts of hot alcohol; very soluble in ether; decomposed by sulphuric acid; soluble in nitric acid without change; decomposed with solution by caustic potash and ammonia; formed by the action of chlorohydric acid and chlorate of potash, on creasote ($C_{26}H_{16}O_4$?).

Hidantoic Acid. $C_8H_9N_4O_9$. White mass or syrup by allowing allantoine to remain some days in contact with strong caustic potash.

Highgate Resin. (T. Thomson, 1813.) *Fossil Copal.* $C_{40}H_{32}O$? Colour muddy yellowish light brown, semitransparent; lustre resinous, and appears as if it had been under water. Softer than copal; smell when heated resinous and aromatic; fuses into a liquid like water; partially soluble in ether and alcohol; nitric acid converts it into a reddish substance; insoluble in caustic potash; from the London clay of the Highgate tunnel.

Hippuric Acid. (Liebig, 1829). *Urobenzoic Acid, Urinic Acid, Glycobenzoic Acid.* HO,$C_{18}H_8NO_5$ = *Benzoic Acid* and *Glycocoll.* $C_{14}H_5O_3 + C_4H_4NO_3$, or *Benzoic Acid* and *fumaramide.* Colourless rhombic prisms; soluble in 600 water at 32°; very soluble in hot water and alcohol, much less so in ether; reaction acid; fuses at a gentle heat; distinguished from benzoic acid by its crystalline form; less solubility in water; reaction with alkalies and by distillation; by chlorohydric acid and boiling it is resolved into benzoic acid and chlorohydride of glycocoll; with nitric acid it yields nitrohippuric acid where 1 hydrogen is replaced by NO_4; boiled with alkalies it separates into glycocoll and benzoic acid;

oxidized with brown oxide of lead it yields benzamide; by binox. manganese and sulphuric acid it gives sulphate of ammonia and benzoic acid. It is obtained from the urine of the horse or cow by evaporating gently to the consistence of a syrup and acidulating with chlorohydric acid, when, on standing, brown crystals fall. These are boiled with excess of milk of lime to decolourize them, and sufficient water to dissolve the hippurate of lime. The addition of chlorohydric acid causes the deposition of hippuric acid; benzoic acid may be separated from it by ether. From human urine it may be obtained by evaporating to a syrup in the water bath, adding chlorohydric acid and a volume of ether, which takes up the hippuric acid, and in an hour $\frac{1}{20}$ of alcohol is mixed with it. Some urea is dissolved by the alcohol, but the addition of water removes the urea and alcohol, and the hippuric acid remains dissolved in the ether, from which by evaporation the acid crystallizes. Mr. Alex. Ure showed that benzoic acid, when swallowed, is converted into hippuric acid in the system.

Hircic Acid. A volatile acid, probably Valerianic acid, from the fat of the goat.

Hisingerite, or THRAULITE.

Hoganite. A synonyme of *Natrolite.*

Holmesite. *Clintonite. Rhombohedral Pearly Mica, Chrysophane, Seybertite? Xanthophyllite?* Spec. grav. 3·08, H 4·25. Yellow or reddish-brown hexagonal oblique rhombic prisms in granular limestone, Orange County, New York. SiO_3 19·35, Al_2O_3 46·80, Fe_2O_3 4·8, MgO 9·05, CaO 11·45, MnO 1·35, HO 4·55, Fl ·9. B.B. infusible; with borax and soda fuses into a white pearl (T. Thomson).

Holohedral Crystals, (ὅλος, all; ἕδρα, seat), are such as have all the similar parts modified in

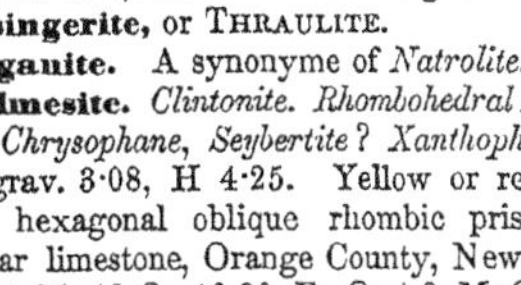

the same way, as in the tetrahexahedron or modified cube.

Homberg's Phosphorus. An ignited mixture of chloride of calcium and lime, which, in a closed vessel exposed to the sun, gives out light in the dark.

Homberg's Pyrophorus. An ignited mixture of 3 alum and 1 meal kept in a close vessel. Sulphide of potassium is formed, which takes fire when shaken out in the air.

Homohedrism. A crystal with similar faces, as in the cube and 8-hedron.

Homolactic Acid. $HOC_4H_3O_5$. Spec. grav. 1·197. Syrupy fluid, homologous with lactic acid; attracts moisture; completely soluble in alcohol and ether; coagulates milk; dissolves iron and zinc with evolution of hydrogen; in a close tube heated to 392° white vapours are sublimed; the silver salt is in plates. Obtained from the mother liquor of fulminate of mercury by saturating with carbonate of lime and distilling the filtered liquor in a water bath. The residue consists of nitrate, formate, and homolactate of lime. The nitrate of lime is removed by alcohol; the lime precipitated by oxalic acid; the formic and acetic acids being displaced, are distilled off; the liquor saturated with lime and homolactate of lime, purified by animal charcoal and crystallization, and decomposed by oxalic acid.

Homomorphism. (Similar shape.) It has been observed that sulphates of potash, barytes, strontian, &c. approach each other in specific vol., relation of axes and angles; to these the above term has been applied.

	Spec. Vol.	Angles.	
$KOSO_3$	32·98	120° 29′	106° 46′
$BaOSO_3$	26·35	116° 22′	105° 6′
$SrOSO_3$	23·48	117° 10′	103° 58′

Honeystone. See MELLITE.

Honey Sugar, *Glucose, Grape Sugar,* is the sugar extracted by bees from the aromatic part of plants, and is by them converted into wax for the purpose of nutrition. It is obtained pure by digesting in alcohol to remove fluid sugar, and washing with alcohol; dissolving in hot water, heating with animal charcoal, and crystallizing out of alcohol.

Hopeite. Spec. grav. 2·76, H 2·5. White 4-sided prisms terminated by a truncated 6-sided low pyramid, the primary form appearing to be a right oblique prism with angles of 81° 34′; transparent with two axes of double refraction. B.B. becomes milk-white and fuses into a colourless globule, giving the flame a greenish tint; forms a colourless glass with borax and salt of phosphorus; with soda a slag, oxide of zinc being condensed round it; soluble without effervescence in chlorohydric acid; probably a hydrous phosphate of zinc (Brewster).

Hop Resin. Reddish-yellow resin; soluble in alcohol and ether; obtained from the hop tincture by the addition of water.

Hops. *Volatile Oil.* Spec. grav. ·91. Yellow; soluble in water, alcohol, and ether, becoming a resin by standing; obtained by distilling hops with water. Hops, the fruit of *Humulus lupulus,* contain 2 per cent. of oil, 10 per cent. of *lupuline,* and 55 of resin.

Hordeine. The starch of barley, *Hordeum distichon.*

Horn is analogous to epidermis and epithelium, and consists of C 51·99, H 6·72, N 17·28, O 20·68, S 3·33.

Hornblende. See AMPHIBOLE.

Horn Mercury. *Native Calomel, Corneous Mercury.*

Horn Silver, or native chloride of silver.

Hornstone. A form of quartz resembling horn.

Horse-radish. The root of the *Cochlearia armoracia;* owes its taste to the oil of mustard.

Hudsonite. A black variety of Augite.

Humboldilite. Spec. grav. 2·9—3·104, H 5·. Brown-yellowish or greenish-yellow right square prisms of 6, 8, or 16 sides, or in masses; lustre vitreous, translucent, and in thin layers transparent. B.B. melts with effervescence with difficulty into a blackish glass. I found the composition of this mineral several years ago SiO_3 43·16, Al_2O_3 4·76, CaO 30·56, Fe_2O_3 12·72, MgO 5·88, KO 1·2. From the lava of Vesuvius.

Humboldtine. *Oxalate of Iron, Eisen Resin.* Spec. grav. 2·13, H 2·. Yellow masses, consisting of a congeries of crystals or fibres; insoluble in alcohol and water; easily soluble in acids; decomposed by ammonia yielding oxalate of ammonia, FeO 41·13, C_2O_3 42·40, HO 16·47. *Form.* FeO, C_2O_3 $1\frac{1}{2}$, HO. B.B. yields an organic smell, and is decomposed, becoming red. From Kolosoruk, Bohemia; Gross Almerode, Hesse.

Humboldtite. A synonyme of borosilicate of lime, or datholite.

Humic Acid. $C_{40}H_{12}O_{12}$? Brownish-black; obtained by acting on sugar with sulphuric acid; it exists also in the mould of soil, and is produced by the decomposition of vegetable matter; but no definite composition can be attached to it.

Humine. *Humus.* $C_{40}H_{15}O_{15}$? Similar to the preceding.

Humite. A synonyme of *Chondrodite.*

Humopic Acid. *Humopinic Acid.* $C_{40}H_{20}O_{14}$? Dark brown substance by heating narcotine to 428°; insoluble in water and dilute acids; soluble in alcohol with a reddish tinge.

Humours of the Eye. These consist of the anterior humour or aqueous, the crystalline lens, and the posterior or vitreous humour. The principal constituents are water and globuline.

Hureaulite. Spec. grav. 2·27, H 5·. Reddish-yellow right oblique prisms with angles of 117° 30′ and 62° 30′; lustre vitreous, transparent, PO_5 38·, FeO 11·52, MnO 33·305, HO 18·. B.B. fuses into a bluish bead with the metallic lustre. Hureaux Limoges, Haute Vienne, forming a small vein in granite.

Hurine. A soluble crystalline body from the milky juice of *Hura crepitans.*

Huronite. Spec. grav. 2·8625, H 3·25. Light yellow foliated or granular masses, translucent on the edges, in cavities in boulder hornblende from Lake Huron. B.B. infusible; fuses with borax into a greenish glass. SiO_3 45·2, Al_2O_3 35·9, FeO 4·99, CaO 15·.

Hyacinth. See ZIRCON.

Hyalite, *Muller's Glass.* Colourless or white botryoidal or stalactitic quartz.

Hyalomicte, *granitic* or *slaty.* A variety of granite associated with tin ores, and characterized by the absence of felspar.

Hyalosiderite. Spec. grav. 2·875, H 5·5. Yellow or reddish-brown 8-hedrons, with a rectangular base, the summits being commonly replaced by planes parallel to the base, with angles of 99° 22′, 77° 50′; internal lustre glassy, of surfaces metallic, translucent on the edges. B.B. becomes black, and fuses into a black bead, attracted by magnet; with borax becomes a clear bead. SiO_3 31·634, FeO 29·711, MgO 32·403, Al_2O_3 2·211, MnO ·48, KO 2·788. In amygdaloid near Sasbach, Brisgau. Allied to chrysolite.

Hydracids. Acids containing hydrogen as an essential constituent, along with another element or compound having the characters of such an element. See ACIDS.

Hydramides. A class of amides typified by hydrobenzamide.

Hydrargillite. A mineral which appears to be a hydrate of alumina, with some phosphoric acid; allied to gibbsite. Found in talc slate near Slatoust, Ural.

Hydrargyromethylnine. *Methose of Mercury.* C_2H_3Hg. A base supposed to exist in union with iodine, when iodide of methyle and mercury are exposed to sunlight.

Hydrarsine. A mixture of cacodylic acid and cacodylate of oxide of cacodyle.

Hydrate (ὕδωρ, water). A compound of a substance with water.

Hydride. *Hydruret. Hydret.* A compound of hydrogen with a metal or organic radical.

Hydrindine. $C_{32}H_{13}N_2O_5$. Yellow crystalline body, resolved by heat into 2 indine $2(C_{16}H_6N,O_2)$ and water. Obtained by heating indine, sulphasatyde, or isatyde, with potash.

Hydriodic Acid. See IODOHYDRIC ACID.

Hydroaloetic Acid. $C_{14}H_4N_2O_{11}$. Dark violet powder formed by the action of water and protochloride of tin on aloetic or chrysamic acid.

Hydrobenzamide. $C_{42}H_{18}N_2$. F.P. 230°. 8-hedrons or rhombic prisms, without taste, smell, or colour; insoluble in water, soluble in alcohol, slightly soluble in ether; converted by HCl into hydride of benzoyle and salammoniac. Obtained by acting on 3 hydride of benzoyle or oil of bitter almonds with 2 ammonia, in a stoppered bottle, at 105° to 120°.

Hydrobenzoinamide. White powder, isomeric with the preceding by heating benzoine and ammonia.

Hydroboracite. *Hydrous Calcareo-biborate of Magnesia.* Spec. grav. 1·9, H 2·, 3 CaO 4 BO_3 3 MgO 4 BO_3 18 HO. White fibres and plates resembling gypsum, spotted with red sesquisilicate of iron. CaO 13·519, MgO 10·57, BO_3 49·571, HO 26·33. Caucasus.

Hydrobromic Acid. See BROMINE.

Hydrocarbon. Compounds of equal proportions of these elements.

Hydrochlorates, or *Chlorohydrides.* Compounds of hydrochloric or chlorohydric acid with oxides.

Hydrochloric Acid. *Chlorohydric Acid. Muriatic Acid. Spirit of Salt. Marine Acid.*

	Vols.	Sp. Grav.	Atom.	Atomic Weight.	Per Cent.
Chlorine,.....	½	1·22	1	4·437	97·26
Hydrogen,...	½	0·0346	1	·125	2·74
		1·2546		4·562	

Physical Characters.—A colourless gas, elastic, like common air, with a peculiar smell and very sour taste. Water absorbs it with great avidity, so that it can only be preserved over mercury. It smokes in the air by uniting with water. No combustible body will burn in it, and it destroys life when an attempt is made to respire it; indeed it cannot be introduced into the lungs, the glottis being spasmodically closed whenever it comes in contact with this gas. When compressed by 40 atmospheres at 50°, it becomes a colourless fluid of specific gravity 1·27 (Faraday, Phil. Trans. 1826, 544). It even liquefies under the pressure of a single atmosphere when placed in a vacuum with solid carbonic acid and ether.

Preparation.—1. Chlorohydric acid gas is obtained by mixing equal volumes of hydrogen and chlorine in a strong bottle. On bringing a taper to the mouth of the bottle an explosion occurs, the gases unite, and chlorohydric acid is formed equal in volume to the two gases. It was first observed by Gay Lussac and Thenard, and at the same time by Dalton, that this union is effected by the agency of the light of the sun in a strong tube (Rech. Phys.-Chimiques, 2,129), but not in a dark place. Although the experiment was first made by the French chemists, the true explanation was first given by Sir H. Davy. When the same gaseous mixture is fired by electricity, union occurs. Davy ascertained that a mixture of chlorine and hydrogen burns at a much lower temperature, and produces a greater degree of heat during the combustion, than oxygen and hydrogen gases. These last gases ceased to explode when rarefied 18 times. But the mixture of chlorine and hydrogen still burns when rarefied 24 times. 2. When 7·308 of common salt (NaCl are heated in a retort with 6·125 sulphuric acid) chlorohydric gas comes over, and may be collected over mercury ($NaCl,HOSO_3 = HCl,NaO SO_3$). Sulphate of soda or glauber salt remains in the retort, and chlorohydric acid gas is evolved.

Chemical Characters.—When oxygen is mixed with chlorohydric acid, and electric sparks are passed through the mixture, water is formed and chlorine set free (HCl and O becoming HO and Cl). When electric sparks are passed through the pure gas, a part is decomposed into hydrogen and chlorine. Spongy platinum, when heated with this gas mixed with oxygen, forms water and chlorine. Dry chlorohydric and sulphurous acid, decompose each other, and form chlorine, sulphur, and water. Some metals decompose chlorohydric acid gas, either at common or elevated temperatures, as potassium, zinc, tin, mercury, and iron, setting hydrogen free, and being converted into chlorides. Some metallic oxides, as oxides of silver and lead, form chlorides and water. Caustic barytes and strontian, when heated in chlorohydric acid, become red hot; and lime evolves much heat. It has no action on non-metallic bodies. Silver decomposes it at a high temperature.

Analysis.—If we take a tube bent into the

form of a retort, fill it with mercury, replace the mercury by dry chlorohydric acid gas, and let up a globule of potassium on the end of a platinum wire into the gas, and apply heat to the potassium, the gas is decomposed, the chlorine unites with the potassium, and the hydrogen is liberated. When the gas has cooled it will be found to have decreased by one-half its bulk, proving that every volume of chlorohydric acid contained half a volume of each element.

Practical Applications.—Guyton de Morveau, in 1773, recommended muriatic acid gas as a disinfecting agent, and employed it at Dijon, and in Paris. He considered 10 ounces of common salt, and 8 of sulphuric acid sufficient to disinfect a ward of twenty beds. This gas has also been used in the treatment of malignant cancer of the face by Zugenbühler.

Chlorohydric Acid in Water.—When chlorohydric acid gas is allowed to escape into the air, it becomes visible in the form of a cloud; the gas combines with the watery vapour which always exists in the atmosphere, and the resulting compound, possessing a feebler tension than pure aqueous vapour, is consequently deposited in the form of a cloud. When a piece of charcoal soaked in water is introduced into a tube containing this gas inverted over mercury, the gas is absorbed, and the tube speedily filled with mercury; if air be mixed with the gas, the absorption is very much retarded. Water at 32° dissolves 500 times its volume of chlorohydric acid gas; the solution possesses a spec. grav. of 1·2109, and contains 6 atoms water. When this is exposed to the air, it smokes, loses half its acid, and then contains 12 atoms of water (HCl 12 HO), having a spec. grav. of 1·128 at 60°, and boiling at 220°; the solubility diminishes with the temperature of the water. 100 cubic inches of water at 69°, are capable of absorbing 41,782 cubic inches of gas; sthe liquid acquires the specific gravity 1·1958, and expands to 134·38 cubic inches. It contains 40·39 per cent. of acid, and 59·61 of water (Thomson). This condensation is accompanied with a powerful evolution of heat, which must be absorbed by ice-cold water, or cold water constantly flowing over the condenser. *In the Laboratory.*—Chlorohydric acid may be formed by mixing in a retort or flask 7½ parts common salt, 9·2 sulphuric acid, and 2 parts of water; the beak of the retort or flask passes into a Wolfe's bottle containing a small portion of water in the bottom for the purpose of washing the gas; a second Wolfe's bottle is connected to the first by a tube which dips into water occupying the lower half of the bottle; other bottles are connected in a similar manner. Each bottle is supplied with a safety tube which passes from the air just under the surface of the water by a central aperture in each bottle. The object of this tube is to prevent the water in one bottle from rushing into that which precedes it, when,

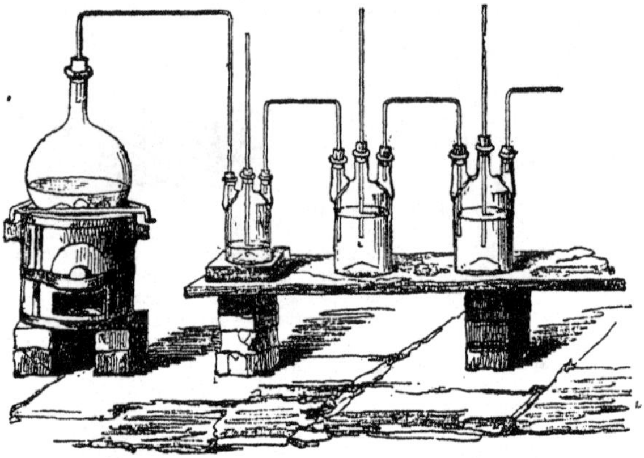

by the rapidity of absorption, a vacuum has been formed in any part of the connected apparatus. Instead of this form, a Welter's tube may be introduced between the first and second bottles.

Commercial Acid.—1. A great source of the muriatic acid of commerce is the preparation of the sulphate of soda for the ultimate production of carbonate of soda, in the soda manufacture. Common salt and sulphuric acid are introduced into a pan and strongly heated. An action identical with that previously described occurs; sulphate of soda is formed and chlorohydric acid set free, which passes up a tower filled with coke, through which a current of water trickles from a reservoir placed at the summit of the tower. The gas meeting the current of water is condensed, and carried into appropriate receivers. As thus prepared it is very impure, being mixed with sulphuric acid, &c. which may be got rid of by distilling the acid, previously adding some chloride of barium to retain the sulphuric acid. 2. In a purer state chlorohydric acid is manufactured on the large scale by introducing the common salt into a cast iron cylinder, *c*, which is closed at one extremity with a circular mouthpiece, *a*, *b*, of siliceous sandstone luted on with lime. At the other extremity is attached a tube, *d*, for the evolution of the gas. The sulphuric acid is run in by an aperture, *k*, in the cylinder, and heat applied below; the gas passes into stoneware condensers or carboys, A, which are half filled with water. The cylinders

are often placed in pairs, so as to be heated by one furnace. The cylinders are about $5\frac{1}{2}$ feet long, $1\frac{1}{2}$ to 2 feet in diameter, and $\frac{1}{4}$ inch thick, and are capable of containing 179 lbs. and upwards of common salt. Upon this are run in about $148\frac{1}{4}$ lbs. of oil of vitriol of spec. grav. 1·80. Mr. W. Blythe recommends the following proportions:—common salt 600 lbs., oil of vitriol (1·750) 700 lbs.

Characters.—Liquid chlorohydric acid, or the solution of the gas in water, should be colourless, if pure. The yellow colour of the commercial acid is frequently due to the presence of chloride of iron derived from the still. When that is the cause the acid gives a blue with yellow prussiate of potash. A colour may be communicated by organic matter, as cork, straw, &c. As the colour disappears on placing a piece of metallic tin in the acid, it may be due to bromine (Thomson). Chlorohydric acid boils at a low temperature. Acid of spec. grav. 1·203 boils at 107°. Hence it may be easily boiled over when mixed with sulphuric or nitric acids. The strongest acid of 1·21 consists of HCl 6HO, and by exposure to the air becomes HCl 12HO losing acid, and then has a spec. grav. of 1·128 (Bineau), boiling at 222°. A solution of chlorohydric acid, when distilled, gives off much acid at first, but its boiling point speedily becomes constant at 230°, with a spec. grav. of 1·094, the formula being HCl 16HO. There are thus 3 hydrates of chlorohydric acid, containing respectively 6, 12, and 16 atoms water.

The following table by Dr. T. Thomson gives the strength of hydrochloric acid with different proportions of water:—

Atoms of Water to 1 Acid.	Dry Acid Per Cent.	Spec. Grav.
6	40·66	1·203
7	37·00	1·179
8	33·95	1·162
9	31·35	1·149
10	29·13	1·139
11	27·21	1·1285
12	25·52	1·1197
13	24·03	1·1127
14	22·70	1·1060
15	21·512	1·1008
16	20·442	1·0960
17	19·474	1·0902
18	18·590	1·0860
19	17·790	1·0820
20	17·051	1·0780

Hydrocyanic Acid, *Cyanohydric acid, Prussic or Prussian acid, Berlin Blue acid.* H Cy 3·375.

Preparation.—1. The method of Gay Lussac for the preparation of this acid, is to place in a small tubulated retort or flask a mixture of cyanide of mercury and hydrochloric acid, and to fit to the beak of the retort a glass tube 2 feet in length, and $\frac{6}{20}$th of an inch wide; the first third of the tube is filled with fragments of marble, to saturate any HCl which might pass over, and the remainder of the tube occupied with chloride of calcium, to abstract the water from the hydrocyanic acid. To the extremity of the tube, a small receiver surrounded with a freezing mixture is adapted; a moderate heat is then applied to the retort; the acid passes over and condenses at first on the marble; but it can be easily removed by a gentle heat, and caused to pass over the chloride of calcium into the receiver.—2. Trautwein recommends for obtaining the strongest acid, to distil with a gentle heat in a retort, 15 parts of finely powdered ferrocyanide of potassium (yellow prussiate of potash), with 9 parts of oil of vitriol ($HOSO_3$), and 9 parts of water. The distilled liquor is to be collected in a strong receiver (placed in a freezing mixture), which contains 5 parts of chloride of calcium in large pieces; the distillation may be stopped when the CaCl is covered with the liquid. The latter is then to be poured into a well-stoppered bottle.

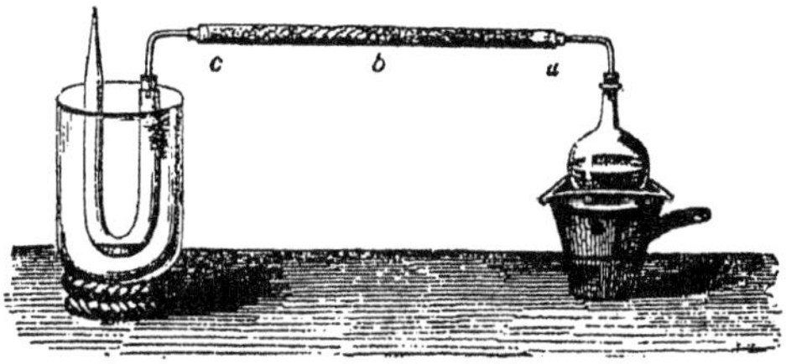

a Flask with HgCy and HCl.
b Portion of tube filled with marble.
c Portion of tube filled with CaCl, conducting into the curved tube surrounded by freezing mixture in a beaker.

Characters.—Hydrocyanic acid prepared by the preceding methods is a colourless liquid, having a strong smell, similar to that of peach blossoms. The taste is sharp, and at first cooling; but it gradually produces a sensation of burning in the mouth, and acts as a powerful poison. Its specific gravity at $44\frac{1}{2}$° is ·7058, and at $64\frac{1}{2}$°, ·6969. It boils at the temperature of 79°·7 and congeals at 5°, assuming then the appearance of prisms like nitrate of ammonia. The spec. grav. of its vapour, according to Gay Lussac, is ·9476. This corresponds nearly with the following composition:—

	Vols.	Atoms W.	S. G.	Per Cent.
Carbon	1	1·50	·4160	44·44
Nitrogen	$\frac{1}{2}$	1·75	·4861	51·85
Hydrogen	$\frac{1}{2}$	·125	·0346	3·71
		3·375	·9367	100·

Hydrocyanic acid scarcely reddens litmus paper. When fired it burns with great facility, with a white flame. If 1 vol. of the vapour of the acid be mixed with $1\frac{1}{4}$ oxygen, it detonates when electric sparks are passed through it, and produces 1 vol. carbonic acid, and $\frac{1}{2}$ vol. nitrogen; the remaining $\frac{1}{4}$ vol. oxygen combining with $\frac{1}{2}$ vol. hydrogen. Anhydrous hydrocyanic acid decomposes rapidly even when kept in a dark place, sometimes in a few hours, at other times it may be preserved for several weeks without change.

Hydrous Hydrocyanic Acid, Medicinal Prussic Acid.—1. *Scheele's method*, improved by Geiger, consists in distilling 4 parts crystallized ferrocyanide of potassium with 18 parts of water, and 2 parts of oil of vitriol ($HOSO_3$, specific gravity 1·842). In the receiver 20 parts of water are placed, and the distillation is carried on until the quantity of liquid in the receiver amounts to 38 parts. The distillation should be conducted by the aid of a chloride of calcium bath.—2. *Clark's method* consists in dissolving in a stoppered bottle $8\frac{1}{4}$ drachms of pure cyanide of potassium in 100 drachms of water, adding to the solution $18\frac{3}{4}$ drachms of crystallized tartaric acid previously dissolved in 20 drachms of water; double decomposition ensues; bitartrate of potash falls, and 120 drachms of water remain, holding in solution 2·81 per cent. of hydrocyanic acid. The action in this process is exhibited in the following scheme:—

$$\begin{array}{l} HO,\ KCy,\ C_8H_4O_{10} = \\ HCy,\ KO\quad C_8H_4O_{10} \end{array}$$

This solution contains in the ounce from $2\frac{1}{2}$ to 3 grains of bitartrate of potash (cream of tartar). The dose for medicinal purposes is 8 to 10 drops. The addition of a little alcohol tends to preserve it. —2. The London pharmacopœia formerly directed extemporaneous acid to be prepared from the cyanide of silver, by decomposing it by means of hydrochloric acid. A fluid ounce of distilled water was placed in a stoppered phial, and to it were added $48\frac{1}{2}$ grains cyanide of silver, with $39\frac{1}{2}$ grs. hydrochloric acid.

Tests.—Whenever hydrocyanic acid is prepared by distillation, in consequence of the volatility of the acid it always varies in its strength. It is therefore necessary to determine the strength of the acid after it has been obtained by distillation. A simple method is to dissolve 6·3 grains of nitrate of silver in an ounce of distilled water, to which is added some ammonia, and to add hydrocyanic acid from a graduated measure in drachms, drop by drop, agitating between each addition until the silver is completely precipitated. Note the quantity of acid used; it contained exactly a grain of dry hydrocyanic acid. By weighing the acid before and after the experiment, we can determine with accuracy the per centage by weight of prussic acid present, and then by simple proportion, the quantity of water to be added to bring it to any strength may be calculated. But for medicinal purposes, the acid prepared by Dr. Clark's method will be found the best in practice, provided the cyanide of potassium be pure, and it can be now obtained genuine prepared by various makers. The cyanide when dissolved in water should not effervesce when mixed with acetic acid. Perfectly pure hydrocyanic acid should leave no residue when evaporated to dryness. Clark's acid contains, unless alcohol be added, about 3 grains in the ounce of cream of tartar. It should not become black, nor be precipitated by sulphohydric acid. It should be colourless like water. When quite pure it deposits a dark sediment in consequence of decomposition; but if a minute portion of sulphuric or chlorohydric acid be present, the decomposition is prevented, or a similar object may be attained by adding a small portion of alcohol. The most permanent acid which I have met with, is that prepared by Clark's process, with the addition of a little alcohol. I have preserved it for many years without any decomposition whatever. The presence of sulphuric acid may be detected by chloride of barium, and that of chlorohydric acid by precipitating by nitrate of silver, and boiling the precipitate with nitric acid; if the precipitate does not disappear, the conclusion may be drawn that chlorohydric acid is present. These acids may be removed by rectifying the hydrocyanic acid containing them, over carbonate of lime or chalk.

Antidotes.—Hydrocyanic acid is one of the most powerful poisons with which we are acquainted; a single drop of the concentrated acid when applied to the eye of an animal, producing very rapid death preceded by violent convulsions. When a weak dose of this acid has been administered the symptoms are merely giddiness, slow pulse, and other signs of weakness, or anæsthesia. When the dose is larger, however, these appearances are aggravated and convulsions occur, in which the head is drawn backwards towards the heels, a scream or peculiar cry is generally uttered, and death occurs with great rapidity. I have never found birds when the acid of $2\frac{1}{2}$ per cent. has been introduced into the mouth to the extent of a few drops, begin to be affected until after the expiration of fifteen seconds.

Dashing cold water on the patient has been found a successful restorative, and I have observed that bleeding in the lower animals gave immediate relief, when used within a few seconds of the assumption of the poison.

Mode of detecting Hydrocyanic acid when taken into the system.—On opening the stomach of a person who has swallowed any considerable quantity of prussic acid, the peculiar odour is generally distinctly perceptible. I have even been able to distinguish it in the blood vessels of the extremities, and above all in the brain; but if the body has undergone any amount of decomposition, the odour of the prussic acid may be obscured by sulphuretted hydrogen, and the smells evolved by the mass of decomposition. Generally, however, it appears that when the body is recent the odour of the acid in the situations enumerated is recognizable, and besides it is often perceptible at the mouth, where commonly a quantity of frothy mucus is protruded, engendered during the violent struggling of the victim. To obtain the acid, a portion of cold distilled water is to be added to the contents of the stomach, and the whole carefully filtered, as much as possible under cover, to prevent the volatilization of the acid. The filtered liquor is to be distilled in a water bath. If hydrocyanic acid is present it may be recognized—(1), by its smell; (2), by the liquid giving a precipitate with nitrate of silver (cyanide of silver), soluble only in nitric acid when boiled, with the evolution of prussic acid which may be distilled over; (3), by its forming a white precipitate with a salt of copper (cyanide of copper); (4), add to the suspected fluid a few drops of a solution of decomposing copperas (which contains both peroxide and protoxide of iron); neutralize by a few drops of ammonia or caustic soda, and then add a few drops of hydrochloric acid so as to render it acid, Prussian blue falls; (5), add to the suspected liquor in a test-tube, half a drachm of sulphohydride of ammonia, and a few grains of flowers of sulphur. Boil till the smell of sulphuretted hydrogen disappears. Filter the solution, and add a drop of it to a solution of sesquichloride of iron: a blood-red sulphocyanide of iron forms (Liebig); (6), by its destroying a bird in a state of convulsions when a few drops are introduced into its mouth.

Hydrocyano-harmaline. $C_{27}H_{14}N_2O_2$ HCy. Rhombic plates, decomposed at 356° into harmaline and prussic acid; by HCy on harmaline.

Hydroferridcyanic Acid. See FERRIDCYANOGEN.

Hydroferrocyanic Acid. See FERROCYANOGEN.

Hydrofluoric Acid. See FLUOHYDRIC ACID.

Hydrogen. *Source.* Hydrogen constitutes one-ninth part by weight of water. It exists likewise abundantly in vegetable and animal substances.

Physical Characters.—Hydrogen possesses the mechanical properties of common air; it has no taste, colour, nor smell. It is the lightest body in nature. Its specific gravity has been found ·0732 (Biot and Arago, Mem. de l'Institut. 1806, 320), ·0694 (Prout, Ann. Phil. 6,322), ·0693 (Thomson, An. Phil. 16,168), ·0688 (Berzelius and Dulong, An. Chim. 15,386), ·0693 (Dumas), ·06926 (Regnault). By theory $\cdot 5528 \times \cdot 125 = \cdot 06911$. It is nearly 16 times lighter than oxygen, and 14½ times lighter than air. Its lightness may be exhibited, by filling a bladder with hydrogen, and affixing a tobacco pipe to it: plunge the bowl of the pipe into a lather of soap, and press out the hydrogen so as to form soap bubbles filled with hydrogen. These are so light that they rise in the air.

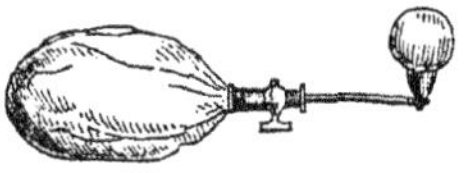

Dr. Black seems to have first suggested the employment of hydrogen to fill balloons (Thomson's Hist. Chemistry, vol. ii.), about 1767-68. In 1782, Mr. Cavallo read a paper on ærostation to the Royal Society; but the first voyage made in a hydrogen balloon was that of Messrs. Roberts and Charles of Paris, 27 August, 1783. For illustrating this experiment small balloons are made of goldbeaters' skin. The balloon is filled from a jar with an opening at the top, supplied with a stop-cock; the balloon is tied on to this, a portion of cotton being placed in the tube to prevent water from being projected into the balloon. These balloons should not be moistened, otherwise they readily tear. On the 8th October, 1776, Macquer, and Sigaud de Lafond, collected the product of the combustion of inflammable air in common air, upon a porcelain saucer, and observed that no soot was deposited, but a clear liquor like water. 100 cubic inches weigh, according to Regnault's specific gravity (100 cubic inches of air weighing 30·945 grains), 2·13324 grains, at 30 inches pressure, and 60° temperature. Its refractory power is ·47, that of air being 1· (Dulong, An. Chim. 31,1176); and hence, considering its lightness, it refracts light more powerfully than any other gas. 100 vols. of boiled water absorb 1·53 vols. of hydrogen gas (Dr. Henry, Phil. Trans. 1803, 274), and alcohol 5·1? (Saussure, An. Phil. 6, 340). A pressure of 1000 atmospheres has no effect in condensing hydrogen gas. The intensity of sound diminishes with the rarefaction of the medium in which it is produced; but in hydrogen it is much greater than could have been anticipated. Even hydrogen, when mixed with common air, blunts the sound. Sound moves at least thrice as fast in hydrogen as in common air.

Preparation.—The source of hydrogen, by whatever process it is prepared, is water, which, according to the synthetic experiment of Warltire as explained by Watt (1781), and demonstrated by Cavendish (1781), consists of oxygen and hydrogen. 1. *By electricity from water.*—

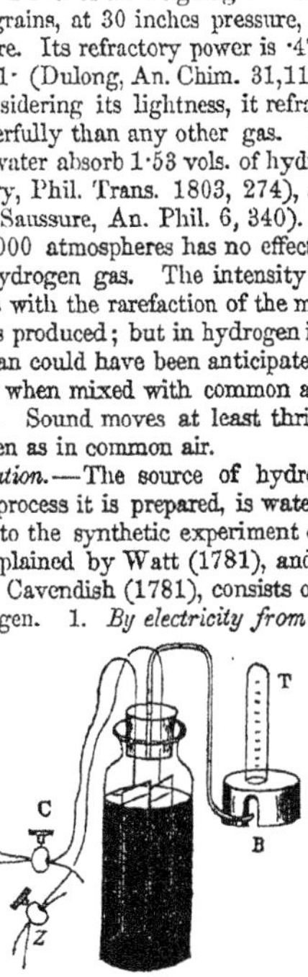

Hydrogen is obtained in the experiment, by which we prove by analysis, that water is composed of oxygen and hydrogen. One or two tubes filled with water, in which some common salt is dissolved to render its conducting power greater, are inverted over two platinum wires connected with the two wires of a battery. The tube in contact with the negative pole of the battery, is speedily filled with hydrogen, while the tube at the positive pole is only half filled with oxygen. 2. *By potassium from water.*—If we rub up mercury and potassium in a mortar into an amalgam, and introduce the mixture into an inverted tube filled with water, hydrogen is evolved which has no smell (HO and K = H and KO) (Berzelius). 3. *By passing steam over red hot iron.*—An economical method of preparing a large amount of hydrogen, is to convert water in a retort or flask, into steam, by means of a lamp or fire placed beneath; the steam passes through an iron tube, *a*, *b*, partially filled with iron filings kept at a red heat in a furnace. The retort or flask is fitted into the iron tube by means of a cork, and

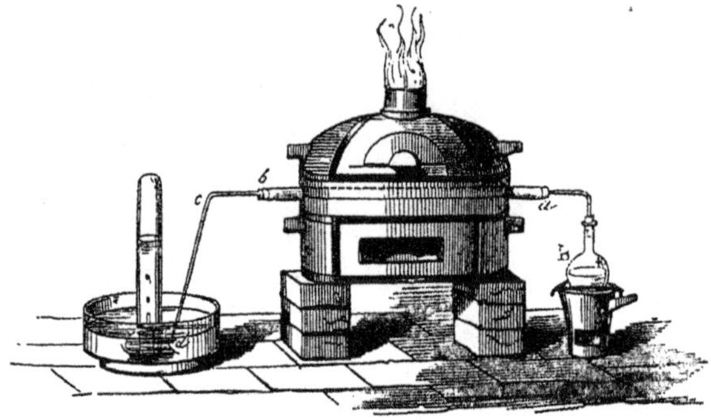

cemented by pipeclay. An educt bent tube, *c*, *d*, conveys the gas at the other end into an inverted vessel filled with water in a pneumatic trough. The steam, on coming in contact with the ignited iron filings, yields up its oxygen to the iron, while the hydrogen is set free, and displaces the water in the inverted jar, 4 atoms of water and 3 of iron giving out 4 atoms of hydrogen, and yielding 1 atom of magnetic oxide of iron (4 HO and 3 Fe = 4 H and Fe_3O_4 or FeO, Fe_2O_3). 4. *From zinc and sulphuric acid.*—The most convenient method of preparing hydrogen gas is to pour upon zinc filings, in a retort or Wolf's bottle, sulphuric acid, and 8 times its bulk of water. Any of the glass apparatus, described under GAS PREPARATION, may be employed. The hydrogen is evolved in large quantities, while the oxygen unites with the zinc, forming oxide of zinc; and this oxide again combines with the sulphuric acid, producing sulphate of zinc.

Water.		Sulphuric acid.	Zinc.
·125	1·	5	4·125
H	O	SO_3	Zn
H		$ZnOSO_3$	
·125 Hydrogen.		10·125 Sulphate of Zinc.	

For the zinc, iron filings may be substituted, but in this case, the hydrogen is very impure.

Chemical Characters.—1. Hydrogen extinguishes a candle when immersed in it, but the gas takes fire at the mouth of the bottle in which it is contained, burning at the expense of the oxygen of the air, and forming water. It is therefore a combustible body, and not a supporter of combustion. This may be shown comparatively with oxygen by inverting a bottle of oxygen over a candle; the candle increases in brilliancy of illumination; and then a bottle of hydrogen in the same way; the candle is extinguished, but the hydrogen takes fire, and burns at the mouth of the bottle.

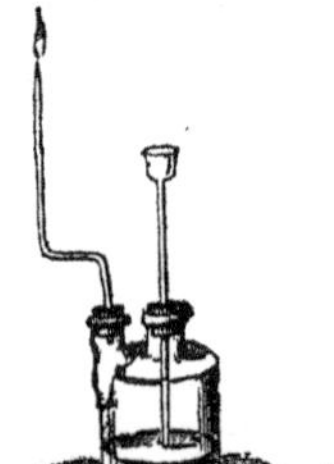

2. Its lightness, and combustibility, may be exhibited by placing the mouth of an inverted jar filled with common air, over the mouth of a jar filled with hydrogen. In a few seconds the gas in each will be found to explode when a taper, or red hot iron, is brought in contact with it. This experiment also affords an illustration of the diffusion or transposition of gases, since the light hydrogen mixes instantaneously with the heavier common air placed above it. The rapidity of the dispersion is proportionate to the lightness of the gases (Priestley). 3. When a small flask is fitted with a perforated cork and capillary

tube, and hydrogen is generated in the flask from water, sulphuric acid, and zinc; after waiting till the common air has been replaced by the hydrogen, ignite the hydrogen arising from the capillary jet; it will then be found that if a long tube be inserted over the lamp musical notes will be heard, varying in tone according to the length of the tube (Dr. Higgins, 1777, Nicholson's Journ. 1, 130). The cause of these sounds was first explained by Mr. Faraday, who showed that small successive explosions take place in the tube by the rapid union of mixed portions of air and hydrogen. These explosions occur so rapidly that they produce a continuous sound. 4. The influence of hydrogen in producing heat is shown by a simple experiment. Take a phial or flask supplied with a rectangular tube terminating in a capillary opening, insert the other end in a perforated cork, and disengaging hydrogen in the flask, secure the cork and tube in its mouth, then cause the hydrogen arising from the capillary extremity to play on some previously heated spongy platinum placed on a stand (Döbereiner); the platinum speedily becomes red hot, and will ignite paper immersed in ether. A simple apparatus frequently used

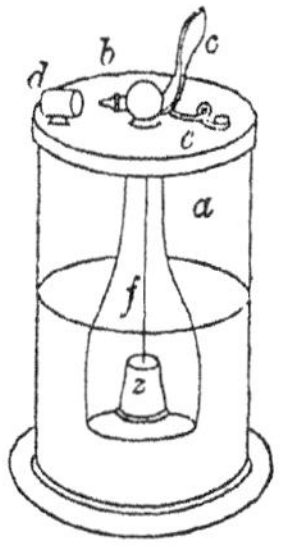

consists of a gas lamp chimney, *f*, inverted, and suspended from a metallic lid in a jar, *a*, surrounding a piece of zinc, *z*, also suspended. Into the jar is introduced diluted sulphuric acid, which acting on the zinc, evolves hydrogen. This gas fills the lamp chimney, and on touching the valve, *c*, inserted in the lid, the hydrogen escapes by a jet, *b*, and is projected on spongy platinum placed opposite to it in a holder, *d*.

Applications. — Hydrogen was used at the commencement of ærostation to fill balloons, but its employment has been superseded by that of coal gas. Notwithstanding the greater density of coal gas, its greater cheapness renders it more available. Hydrogen is principally applied in the laboratory to deoxidize or remove oxygen from metallic oxides.

Analysis.—The quantity of hydrogen in a mixture is determined by introducing into a tube over mercury the mixed gas, and then an excess of oxygen, noting the amount of the gases, then transferring it to a *eudiometer*, q.v., firing it by elec-

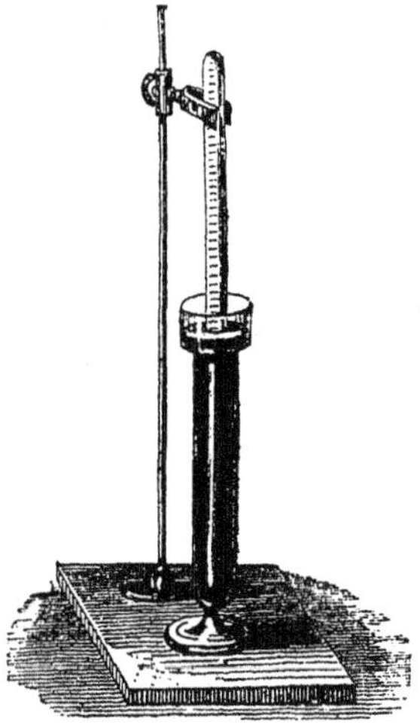

tricity, and returning it to the tube, and noting the amount of diminution; two-thirds of the gas which has disappeared was hydrogen, and the remaining one-third oxygen.

Hydroleic Acid. $C_{36}H_{34}O_4HO$. Yellowish oil, by the action of oil of vitriol on oleic oil; it is soluble in cold alcohol.

Hydrolite. See SARCOLITE.

Hydromargaric Acid. $C_{34}H_{35}O_5HO + C_{34}H_{33}O_3HO$? Milk-white and solid; fusing at 140°; by the action of sulphuric acid on margarine.

Hydromargaritic Acid. $C_{34}H_{34}O_4HO$? Colourless prisms, insoluble in water; soluble in ether and hot alcohol; found along with the preceding, and hydroleic acid, &c. by the action of oil of vitriol.

Hydromel. *Mead.* A fermented mixture of water and honey.

Hydromellonic Acid. C_6N_4H. Snow-white powder, decomposed only at a strong red heat; formed by mixing a hot solution of mellonide of potassium, with strong chlorohydric acid and cooling.

Hydrometer. (ὕδωρ, water; μέτρον, a measure.) (*Aræometer*, Ger.; *Areometre*, Fr.) An instrument employed to determine the specific gravity of fluids. It depends on the principle that a body floating in a fluid will be depressed in proportion to the lightness or diminution in the density of the liquid, and upon the law that a floating body displaces an amount of fluid equal to its own weight. They are usually made of glass or of brass. Those in common use are *Twaddell's*, which is valuable for taking the specific gravity of sulphuric acid, alkaline and kelp salt solutions, &c. It is in general use in all the manufactories of this country. It is said to have been the contrivance of the late John Wilson, Esq. of Thornley, near Glasgow. Others ascribe its invention to the late Charles Macintosh, Esq. The common name by which it is designated, is that of the glass-blower of Glasgow by whom it was originally prepared. In using this or any other hydrometer, the instrument is immersed in the fluid, and the point of the scale at which the fluid cuts it, is read off. The temperature of the fluid is usually taken at 60°. They are commonly

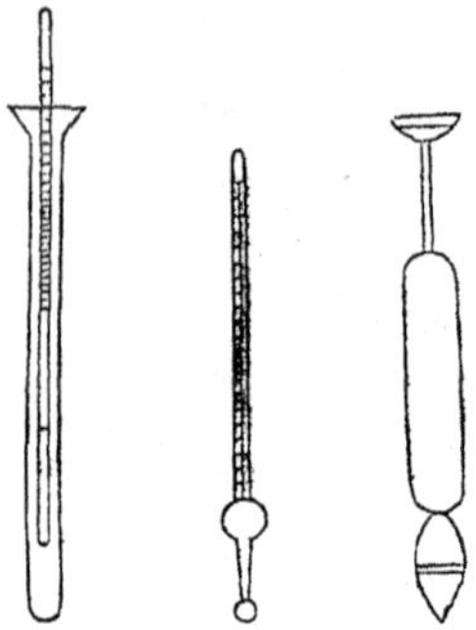

sold in boxes with 6 spindles, ranging from 0° to 170°. The scale is formed on the proportion that every degree is equal to 5 degrees of specific gravity, and therefore to convert Twaddell's degrees into the comparative standard with water, it is only necessary to multiply by 5 and add 1000. Thus 5 T $= 5 \times 5 = 25 + 1000 = 1025$, water being 1000, and 50 degrees T $= 50 \times 5 = 250 + 1000 = 1250$. Hydrometers are sometimes made for particular fluids, so as to enable the operator to read off at once the true specific gravity. Thus with urine the *Urinometer* is employed, with a scale ranging from 0° to 100°, somewhat beyond the limit required. With such a scale, all that is necessary is to add 1000 to the degree indicated by the point at which the fluid cuts the immersed scale. If the point be 30°, the specific gravity of the urine is 1030. *Sikes's hydrometer*, which is used in the English excise, is formed of brass, and is so arranged as to allow a weight to slip into the bottom of its spindle, according to the density of the fluid. Allan's saccharometer, invented by Dr. Thos. Thomson, is used by the Scottish excise, and allows the additional weights required, to be adjusted at the top of the spindle out of the water. See SACCHAROMETER. *Beaumé's hydrometers* are arranged for determining the density of fluids lighter and heavier than water, but as they have no reference to the true specific gravity as reckoned from water, they have never been in use in this country.

Hydronitro-prussic Acid. $Fe_5Cy_{12}3NO, 5H,6HO$. Dark red deliquescent crystals, soluble in water, alcohol, and ether; decomposed by boiling in water. Obtained by decomposing nitro-prusside of barium, or silver, by HCl or SO_3, or nitro-prusside of potassium, by an alcoholic solution of tartaric acid.

Hydrophane. A variety of opal becoming transparent in water.

Hydrophite. Spec. grav. 2·65, H 3·5. Green fibres. SiO_3 36·193, FeO 22·729, MnO 1·66, MgO 21·082, Al_2O_3 2·895, H 16·080, VO_3 ·115. A variety of serpentine, Taberg, Sweden.

Hydrorhodeorhetine. $C_{42}H_{30}O_{21}$. Brown resinous mass; obtained by dissolving rhodeoretine in caustic ammonia, precipitating by acetate of lead, and decomposing the precipitate with sulphohydric acid gas; it is soluble in water, alcohol, and acetic acid, not in ether.

Hydrosalts. Salts containing hydrogen as a constituent.

Hydrosulphocyanogen. See SULPHOCYANOGEN.

Hydrosulphuric Acid. See SULPHOHYDRIC ACID.

Hydrotalcite. H 2·. Pearly mineral in leaves with a greasy feel, Al_2O_3 12·, Fe_2O_3 6·9, MgO 36·3, CO_2 10·54, HO 32·06. Snarum, Norway.

Hydrothionic Acid. A synonyme of sulphuretted hydrogen.

Hydrous. Bodies containing combined water are said to be hydrous.

Hydurilic Acid. $C_{12}H_5N_3O_{11}$. White powder in needles; insoluble in cold water and alcohol; soluble in hot water; obtained by evaporating gently the mother liquor from alloxane.

Hygroscopic. The tendency to absorb water from the atmosphere.

Hyocholalic Acid. $C_{50}H_{40}O_8$. Soluble in alcohol and ether; insoluble in water; obtained by boiling hyocholic acid with alkalies.

Hyocholeic Acid. $C_{54}H_{45}NO_{12}S_2$? An acid occurring along with the following, but not yet sufficiently examined.

Hyocholic Acid. $C_{54}H_{43}NO_{10}$. Consists of hyocholalic acid and glycycoll. F.P. above 248°; white or yellow resin, becoming soft and silky in boiling water; soluble in acids, and reprecipitated by water; soluble in alkalies, slightly soluble in ether, soluble in alcohol; obtained from pig's bile, by evaporating to dryness in a water bath, dissolving in absolute alcohol, decolourizing by charcoal; precipitating by ether, the hyocholate of soda falling. The hyocholate is dissolved in water and precipitated by acetic acid. The precipitate is washed in water, dissolved in alcohol, and precipitated by water and dried.

Hyoscyamine. Silky needles, alkaline, narcotic; dilates the pupil; not decomposed by the air; sublimes partially; obtained from the seeds of *Hyoscyamus niger*, by hot alcohol with 2 per cent. of sulphuric acid; filtering, neutralizing with lime, adding a slight excess of sulphuric acid, and distilling off the alcohol; diluting the residue with water, and adding carbonate of potash. It is then washed several times with ether, which takes up the hyoscyamine.

Hypargite. A synonyme of red silver ore.

Hyper. (ὑπερ, above.)

Hyperchloric Acid. *Perchloric Acid.* ClO_7. White crystalline solid, in 4-sided prisms, very deliquescent; by decomposing perchlorate of potash with sulphuric acid.

Hyperite. A syenite in which the felspar is replaced by labradorite, and the hornblende by hypersthene.

Hypermanganic Acid. *Permanganic acid.* Mn_2O_7. A blood-red solution, obtained by decomposing manganate of barytes by sulphuric, or carbonic acid.

Hyperoxymuriatic Acid. A synonyme of chloric acid.

Hypersthene. (ὑπερ, above; σθενος, power). A mineral, so named from its hardness and power of withstanding the decomposing action in the air, over amphibole, as is remarkably exemplified in the beautiful Cuchullin (Coolen) range in Skye, where it was first discovered and named by Dr. Maculloch; spec. grav. 3·338—3·355, H 4·75. Rhombic prisms with angles of 93° 30′ and 86° 30′. Colour grayish or greenish-black, sometimes nearly copper-red; lustre metallic and vitreous, SiO_3 51·91, MgO 18·640, FeO 20·350, MnO 3·88, CaO 2·410, HO ·30. It occurs mixed with felspar, in Skye in hypersthene rock.

Hypo. (ὑπο). The Greek term for *under*.

Hypoazotic Acid. A synonyme of *Hyponitric acid* (NO_4).

Hypomargarylic Acid. A synonyme of Stearic acid.

Hyposclerite. A variety of Augite.

Hypostilbite. A variety of Stilbite from Faroe.

Hyposulphethylic Acid. C_4H_5,S_2O_5HO. Obtained by heating sulphocyanide of ethyle with nitric acid.

Hyposulphobenzidic Acid. $C_{12}H_5S_2O_5,O$. Found in the fluid, from which sulphobenzide separates, when benzole has been acted on by anhydrous sulphuric acid.

Hyposulphobenzoic Acid. $C_{14}H_4O_3,S_2O_5,2HO$. Formed by acting on benzoic acid with anhydrous sulphuric acid.

Hyposulphoglutic Acid. A hard vitreous mass, formed by acting on naphthaline with an excess of anhydrous sulphuric acid.

Hyposulphoindigotic. $C_{16}H_4NO_2,S_2O_5$. Contained in the filtered liquor in preparing sulphopurpuric acid, which is separated by dissolving its ammonia salt, in alcohol.

Hyposulphomethylic Acid. $C_2H_3S_2O_5HO$. By acting on sulphocyanide of methyle with nitric acid. The salts are monobasic; chlorine forms 3 new acids, mono-, bi-, and terchlorohyposulphomethylic acid ($C_2H_4Cl_3S_2O_5$).

Hyposulphonaphthalic Acid. $C_2H_8S_2O_5HO$? Crystalline mass, by the action of hot oil of vitriol on naphthaline along with the following acid; its salt of lead is soluble in alcohol, while that of the lead salt of the following acid is insoluble.

Hyposulphonaphthic Acid. $C_{11}H_4S_2O_5HO$. Obtained along with the preceding by the action of sulphuric acid on naphthaline.

Hyposulphosuccinic Acid?

Hyposulphuric Acid. *Dithionic acid.* SO_2SO_3. Obtained in combination with oxide of manganese, by passing sulphurous acid through water containing binoxide of manganese suspended in it. From this salt the various other salts may be formed. It is decomposed by sulphuric acid, into sulphurous and sulphuric acids.

Hyposulphurous Acid. *Dithionous acid.* SO_2S. Found only in combination with bases; obtained by passing sulphurous acid through carbonate of soda, and adding sulphur and evaporating; decomposed into SO_2 and S, by an acid.

Hypoxanthine. $C_5H_2N_2O$. White crystalline powder with 1 atom less oxygen than in xanthic oxide; soluble in 1090 cold, 180 boiling water; obtained from the spleen of the ox, by boiling with water, precipitating with barytes water, evaporating the filtered liquid; the excess of barytes separates as carbonate, along with 2 organic bodies which dissolve in boiling dilute caustic potash, and can be precipitated by chlorohydric and carbonic acids; one of these is uric acid, and the other hypoxanthine. The uric acid is separated by solution in caustic potash and adding salammoniac, when urate of ammonia falls. The filtered solution deposits hypoxanthine as yellow crystals. These are dissolved in ammonia and evaporated; the resulting foliated crystals are dissolved in dilute caustic potash, and carbonic acid passed through the solution. A white crystalline powder falls, which is washed free from potash with cold water.

Hyssopine. An alkaloid said to exist in the root of the Hyssopus officinalis.

Hystatite. Titaniferous iron from Arendal, Norway.

I

Iberite. Spec. grav. 2·89, H 2·5. Greenish 6-sided prisms; lustre glassy and pearly, opaque; SiO_3 40·9, Al_2O_3 30·74, FeO 15·47, MnO 1·33, CaO ·397, MgO ·81, KO 4·58, NaO ·046, HO 5·57. B.B. fuses into a dark bead; Montalvan, Toledo, Spain.

Ice. Specific gravity ·920 (Thomson), ·918 (Brunner). Water congeals at the temperature of 32°. Dr. Nettis, at Middleburg, in 1740, observed and figured ninety-one forms of snow and ice, each particle one-twentieth of an inch in size, the angles being 60°. Ice crystals belong

to the rhombohedral system, the usual form being 6-sided prisms, terminated by 6-sided pyramids. As the density of ice is less than that of water, it is obvious that water expands in freezing. Hence the cause of the bursting of bottles and water pipes by their contents freezing.

Iceland Moss. *Cetraria Islandica.* A well-known lichen used in medicine. It yields when boiled a starch or gum, known as *Lichenine*, q. v.

Iceland Spar. Transparent calcareous spar, the finest specimens being obtained from Iceland.

Icespar. Spec. grav. 2·4365, H 3·. Grayish or yellowish-white thin longish 6-sided tables or right oblique 4-sided prisms; lustre shining, vitreous, strongly translucent, crystals transparent. B.B. on charcoal becomes glassy, semitransparent and white, and fuses on the edges with difficulty into a glass, SiO_3 63·56, Al_2O_3 24·06, KO 10·03, CaO ·94, Fe_2O_3 ·92, HO ·37; allied to felspar; in Vesuvian lavas.

Ichthyocholine. A crystalline body from the bile of different species of Cyprinus.

Ichthyocolla. Isinglass.

Ichthyophthalmite. See Apophyllite.

Icicane. $C_{160}H_{137}$.

Icica Resin. *Breane.* $C_{80}H_{67}O_3$. F.P. 314½°. White crystalline needles corresponding in composition with cholesterine; slightly soluble in alcohol; insoluble in water and alkalies; soluble in sulphuric acid with a red colour, and reprecipitated by water; from an African species of Icica.

Idocrase. *Cyprine*, *Egrane*, *Frugordite*, *Loboite*, *Vesuvian*, *Wiluite*. Spec. grav. 3·349 to 3·399, H ·6. Brown, passing into leek, pistachio, olive, and oil-green; right square prisms; the lateral edges of the prism are frequently replaced by tangent planes, which sometimes increase so much as to nearly obliterate the primary planes of the prism; the new faces making angles of 135° with the primary faces. When the new faces obliterate the base of the prism, they form a 4-sided terminal pyramid; lustre vitreous and resinous, fracture uneven, semitransparent to

translucent on the edges. B.B. fuses into a yellow glass. SiO_3 37·658, Al_2O_3 17·695, FeO 6·489, CaO 31·896, MgO 4·537, MnO ·499. *Form.*—3 RO SiO_3, $R_2O_3SiO_3$? It has the same composition as garnet, but a different crystalline form. It derives its name from ιδω, I see; and κρασις, a mixture, from its resemblance to other minerals.

Idrialine. *Elastic Bitumen.* $C_{42}H_{14}O$. F.P. 482° to 572°. Colourless pearly masses of minute rhombic plates; insoluble in water; with difficulty soluble in ether, alcohol, and acetic acid; very soluble in acetone, oil of turpentine, creasote, olive oil; decomposed at 572° partially; forms a compound with chlorine, which dissolves in sulphuric acid with a purple colour; nitric acid replaces 3 atoms hydrogen and forms a red acid compound; it unites also with sulphuric acid; obtained from Idrialite by dissolving the latter in oil of turpentine, from which it crystallizes, or by sublimation in a stream of carbonic acid.

Idrialite. Grayish-black resinous body, found in the slate of Idria, consisting of idrialine, gypsum, iron pyrites, silica, and alumina.

Idryle. $C_{42}H_{14}$. F.P. 186¾°. Colourless rhombic plates, with a greenish lustre; soluble in hot alcohol, ether, and oil of turpentine, with a blue tint; little soluble in water; soluble in sulphuric acid, with a golden colour, and unites with it, forming a coupled acid. Obtained from a fatty or resinous matter called *stupp*, sublimed in the reduction of mercury at Idria, by boiling with alcohol of ·8475, and dissolving the residue left by its distillation in hot acetic acid, and crystallizing out of alcohol.

Igasuric Acid, or lactic acid, found in the seeds of *Strychnos ignatii.*

Iglesiasite. A synonyme of *Carbonate of Lead*, from Sardinia.

Igloite. A synonyme of *Arragonite.*

Ilicine. Brown translucent crystals, bitter; soluble in water and alcohol, not in ether. Obtained from the leaves of the *Ilex aquifolium* by boiling with water, precipitating by acetate of lead, and filtering; precipitating the filtrate by sulphohydric acid gas, evaporating, and crystallizing out of alcohol.

Illipe, Oil of. A species of oil from the African coast, apparently distinct from Galam and Shea butters, and palm oil.

Illuderite. A synonyme of *Epidote.*

Ilmenite. Spec. grav. 4·766 to 4·808, H 5·75. Brownish-black crystals, of the same form as anhydrous sesquioxide of iron; lustre metallic, shining, translucent, brittle, slightly magnetic. TiO_3 46·67, Fe_2O_3 11·71, FeO 35·37, MnO 2·39, SiO_3 2·8, MgO ·6, CaO ·25, CrO_3 ·38. Ilmen Sea, near Minsk.

Ilmenium. A synonyme of niobium and pelopium.

Ilvaite (from Elba), *Yenite*, *Lievrite*. Spec. grav. 3·994 to 4·2, H 4·75 to 5. Black, with a

shade of brown or green; right rhombic prisms,
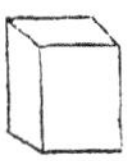
with angles of 112° 37′ or 112°, or 111° 30′. The usual form is a long 4-sided prism, terminated by a low 4-sided truncated pyramid; the inclination of the side of the pyramid, and corresponding face of prism, is 128° 50′, lustre imperfect metallic, opaque; streak black, fracture uneven. B.B. fuses into a black globule, and with fluxes into a black glass. SiO_3 29·28, FeO 52·542, CaO 13·779, MnO 1·587, Al_2O_3 ·614, HO 1·268. Island of Elba in limestone, Silesia, Fassum Norway, Siberia, N. America, Greenland.

Imabenzyle. $C_{28}H_{11}NO_2$. By ammonia on benzile.

Imabibromisatine. $C_{32}H_7Br_4N_3O_6$. Orange crystalline powder, by acting on bibromisatine in alcohol with dry ammonia.

Imachlorisatine. $C_{32}H_9Cl_2N_3O_6$. Brownish-yellow crystals, by boiling chlorisatine with alcohol and ammonia.

Imasatic Acid. *Isamic Acid. Isatinamic Acid. Rubindic Acid.* HO, $C_{32}H_{12}N_3O_7$. Red rhombic or 6-sided plates insoluble in water and alcohol, soluble in ether and alkalies, by adding sulphate of ammonia to isatate of potash in alcohol, and decomposing the imasate of ammonia with muriatic acid.

Imasatine. $C_{32}H_{11}N_3O_6$. Grayish-yellow plates, insoluble in water and ether, soluble in sulphuric acid, caustic potash. Obtained by adding ammonia to a solution of isatine in alcohol in which some isatine is suspended.

Imechlorisatine. $C_{16}H_5Cl,NO_2$. Yellow insoluble prisms, or 6-sided plates, formed by passing dry ammonia gas into a solution of chlorisatine in alcohol.

Imesatine. $C_{16}H_6N_2O_2$, or isatine with O_2 replaced by NH. Dark yellow rectangular prisms, soluble in boiling alcohol, little soluble in ether; by dry ammonia on isatine in alcohol, with suspended isatine.

Imide. A substance formed on the type of ammonia (NH_3), in which 2 atoms of hydrogen have been replaced, and only NH remains of the original model. Thus *Diethylamine* is considered to have been originally ammonia (NHH_2); the two last atoms being replaced by ethyle. Its formula then is NH, C_4H_5, C_4H_5, and it is said to be an imide base.

Imperatorine. $C_{24}H_{12}O_5$? Spec. grav. 1·192; F.P. 167°. Crystalline resin from *Imperatoria ostruthium*, by distilling most of the ethereal solution of the root, and then evaporating spontaneously. The crystals are separated from the fat oil by pressure in paper, and treatment with caustic potash, and crystallizing from alcohol of ·8475. It is insoluble in water, but soluble in ether, alcohol, ethereal and fat oils; little soluble in ammonia; soluble in caustic potash, and reprecipitated by acids without change; soluble in sulphuric acid without decomposition; soluble in nitric acid with a yellow colour, and precipitated orange by dilution.

Imponderables. A term applied to heat, light, electricity, and magnetism, which have not been ascertained to possess weight.

Incandescence. The visible glowing appearance which metals present when intensely heated.

Incineration. The removal of organic matter by heat, so as to leave nothing but ash.

Incombustible Cloth. Cloth made of asbestus, or cotton or linen impregnated with a solution of certain salts, as of alum, or covered with a coating of Fuchs' soluble glass.

Indelibrome. $C_{32}H_8N_3O_8Br_4$.

Indianite. See ANORTHITE.

Indic Acid. *Indinic Acid.* Indine, when moistened with alcohol and digested in strong caustic potash, unites with the alkali and forms black crystals, which contain 32 atoms carbon to 1 potash, but which are decomposed by water and acids.

Indicolite. A synonyme of *Tourmaline.*

Indigo. A blue-colouring matter in commerce, obtained from the leaves of the *Indigofera tinctoria, anil,* and *argentea,* in India. It is found in smaller quantities in many other plants, as *Mercurialis perennis, Isatis tinctoria.* It was known to the Greeks and Romans as a paint. But its value was not estimated till the middle of the 16th century, in Europe, when it was first imported from India into the west, by the Dutch. It was, however, prohibited in England in the time of Queen Elizabeth, and also in Saxony. In the edict it is spoken of as a corrosive substance, and called "food for the devil." An acre of ground produces about 10 lbs. of indigo. In Bengal the plant is cut before it flowers. When cut it is tied up in sheaves about 5¼ feet in circumference, and carried to the factory, where a large vat is filled with them. They are strongly pressed down with bamboos and a cross beam, covered with water, and allowed to steep for twelve hours. They must not, however, be steeped too long, otherwise the indigo is injured. The liquor becomes yellow, and is drawn off by means of a plug at the bottom of the vat into a similar vessel, where it is agitated by means of bamboos for two or three hours. Grains form and separate. When they have ceased to deposit, a few pailsful of cold water or of lime water are added, and the whole gently stirred with a circular motion. The liquor is drawn off, the indigo at the bottom washed with water and placed in a copper boiler, where it is allowed to ferment; it is then placed on a bamboo frame, covered with a cloth in the form of a filter, and the liquor allowed to drain off. It is then pressed by means of screws, and cut into cakes. In this state it is impure. I have found Bengal indigo to yield of volatile matter 96·4, soluble salts, principally sulphate of lime and common salt ·05, silica ·61, phosphate of lime 1·73, oxide of manganese and magnesia ·33, carbonate of lime ·86. In Spanish indigo I have found as much as 18·6 to 19·5 per cent. of ash. Indigo from the bazaars of Japan, yielded of water 4·8 per cent. and 51·8 ash. Of the organic matter, a large portion is destitute of colouring power. The constituents are, 1, a substance analogous to gluten, which may be obtained by digesting indigo in dilute sulphuric acid; 2, a brown matter which may be dissolved in strong caustic potash; and, 3, a red resin, soluble in alcohol. *Testing.*—The best test for indigo is its dyeing power. Bolley's method of testing is:—Dissolve 1 gramme indigo (15·438 grs.) in 10 grms. fuming SO_3, diluted to 200 cub. centimetres. Dilute 50 cc. with 200 cc. water; heat in a porcelain basin to 122°; mix with 50 cc. of a solution of sulphate of soda, and drop in a solution of chlorate of potash (4 grms. to 400 cc. water), until the solution is decolourized.

Indigo Blue. *Indigotine.* $C_{16}H_5NO_2$. Sp. grav. 1·35. Purple and coppery plates, and 6-sided prisms, with a rhombic base of angles 32° and 148°; sublimes at 554°; the vapour is violet-coloured; indigo volatilizes when cautiously heated without any residue; it is without taste and smell, and is neutral; insoluble in water, ether, dilute acids, and alkalies; soluble in sulphuric acid, forming a compound; nitric acid forms indigotic and carbazotic acids. It is obtained in the crystalline state by cautiously subliming commercial indigo between two platinum crucible lids, and in the granular state by the *blue vat*, as described under white indigo; but in the latter state it is not pure, and for accurate purposes requires to be sublimed.

Indigogene. *White Indigo.* $C_{16}H_5NO_2H$. Grayish silky white powder, without taste and smell, neutral; insoluble in water; soluble in alcohol and ether, with a yellow colour, indigo blue depositing gradually. Soluble in alkalies, with a yellow colour. Indigo white probably exists ready formed in plants, and is converted into indigo blue by uniting with oxygen. It is obtained from commercial indigo, by mixing 1 part of powdered indigo, 3 slaked lime, 150 water, and 2 of protosulphate of iron (the *Indigo*, or *Blue Vat*), and placing it in a stoppered bottle to which no air has access. A yellow solution is formed which contains all the indigo. If this liquor be drawn off by means of a syphon filled with hydrogen, and mixed with dilute chlorohydric acid, to which sulphate of ammonia has been added, a white flocky precipitate of indigogene falls, which must be kept away from the air, and washed with boiling water containing sulphate of ammonia. The above formula represents white indigo as blue indigo with the addition of 1 atom hydrogen; the formula is similar to that of cyanide of benzoyle, or of benzile. Dumas views the relation of white to blue indigo to be similar to that of hydride of benzoyle to benzole, or of alloxantine to alloxan.

Action of Sulphuric Acid on Indigo. Sulphoindigotic Acid. Saxon Blue. Cerulin. Sulphindelic Acid. — $C_{16}H_4NO2SO_3HO$. When 1 part of blue indigo is left in contact with 8 parts of oil of vitriol for some days at a common temperature, or at 130°, a blue solution is produced of sulpho-indigotic acid, while a purple substance remains—*Sulphopurpuric Acid.*—$C_{32}H_{10}N_2O_42SO_3$. The first acid is formed by the indigo losing an atom of water, while in the second acid the atom is doubled.

Oxidation and Chlorination of Indigo.—When indigo is dissolved in diluted NO_5, colourless needles of *indigotic* or *anilic acid* separate on concentration. They consist of $C_{14}H_4NO_9$. When the preceding acid is acted on further by NO_5, yellow prisms of *picric*, or *carbazotic acid*, or *bitter of Welter*, $C_{12}H_2N_3O_{13}$, are formed. When purified indigo blue is heated with a mixture of equal parts of sulphuric acid, and bichromate of potash, previously dissolved in 25 parts of water, the mixture gently heated until there is a slight disengagement of carbonic acid, the liquid assumes a deep brown-yellow colour. Crystals make their appearance on cooling, which are reddish-yellow rhombic prisms of *Isatine.*—It is soluble in hot alcohol and water; the solution colours the skin. It consists of $C_{16}H_5NO_4$, or simply indigo, with 2 atoms oxygen derived from the chromic acid. When isatine is treated with caustic potash, a compound of isatic acid ($C_{16}H_6NO_5$) with potash is formed. When isatine is dissolved in sulphohydride of ammonia a white powder, *isatyde*, $C_{16}H_5N_2O$, separates on cooling. Sulphohydric acid passed through a solution of isatine in alcohol yields *sulphesatyde*, $C_{16}H_6NO_2S_2$, in grayish-yellow crystals. By adding to the latter in solution in alcohol, caustic ammonia cautiously, *sulphasatyde* is procured, $C_{16}H_6NO_3S$. By the action of potash on sulphesatyde, sulphasatyde, or isatyde, there is formed *indine*, $C_{16}H_6NO_2$, red crystals. By heating it with potash, *hydrindine*, $C_{32}H_{13}N_2O_5$, a yellow crystalline body results. By nitric acid, and chlorine, *nitrindine*, $C_{16}H_4N_2O_7$, and *chlorindine*, $C_{16}H_5Cl\,NO_2\,2HO$, are produced. If a current of chlorine be passed through a solution of isatine, an atom of hydrogen is replaced by Cl, and a new compound, *chlorisatine* ($C_{16}NH_4ClO_4$), is produced, a substance crystallizing in orange-yellow 4-sided prisms, dissolving in 50 parts of boiling water and alcohol; it is converted into chlorisatic acid by caustic potash ($C_{16}H_9Cl,NO_5$). By the action of dry ammonia on chlorisatine are formed *imechlorisatine*, $C_{16}H_4Cl,N,2NH,O_2$, and *imachlorisatine*, $C_{16}H_4Cl,N,NH,O_3$, yellow crystals. By the further action of chlorine, *bichlorisatine* ($C_{16}NH_3Cl_2O_4$) is produced in red prisms, more soluble in alcohol and water than the preceding body. It forms an acid with potash. When chlorisatine is dissolved in hot sulphohydride of ammonia, *chlorisatyde*, a yellowish-white powder, separates on cooling ($C_{16}H_5N,Cl$

O_4). When chlorine is passed through an alcoholic solution of chlorisatine, pearly scales of *chloranile* ($C_6Cl_2O_2$) make their appearance, which are converted by potash into *chloranilic acid.* When chloranile is dissolved in hot NH_3 the solution becomes blood-red, and crystals of *ammonia chloranile* ($C_{12}H_6Cl_2N_2O_6$), in the form of chestnut needles, appear. After the separation of this body, if chlorohydric acid be added, *chloranilam* in needles separates. When chlorine is passed through water containing indigo blue in suspension, the whole becomes a yellow mass. When this is distilled, white needles, consisting of a mixture of *chlorindatmine*, or terchloraniline, and *chlorindoptic acid*, condense in the neck of the retort, while chlorisatine and bichlorisatine remain in the retort. Bromine has a similar action to chlorine in all these cases of replacement, forming *bromisatine*, $C_{16}H_4BrNO_4$, and *dibromisatine*, $C_{16}H_3Br_2NO_4$. The latter, by dry ammonia, gives *imabromisatine*, $C_{16}H_3BrN, NHO_3$, an orange crystalline powder. The formation of the preceding bodies from isatine, affords an example of the replacement of hydrogen, as in the following table:—

Isatine,	$C_{16}\,H_5\quad NO_4$
Chlorisatine,	$C_{16},H_4Cl,\ NO_4$
Bichlorisatine,	C_{16},H_4Cl_4,NO_4
Bromisatine,	$C_{16}\,H_4Br\quad NO_4$
Bibromisatine,	$C_{16}\,H_4Br_2\,NO_4$

Action of Alkalies on Indigo.—When indigo blue is boiled with caustic potash (spec. grav. 1·45) it dissolves into a yellowish fluid, and by evaporation crystals of *chrysanilic acid* ($C_{28}H_{11}N_2O_6$) and potash are obtained. When caustic potash (spec. grav. 1·35) is boiled with indigo blue, and finely powdered black oxide of manganese is added in small portions at a time, *anthranilate* ($C_{14}H_6NO_3$) of potash is formed, which may be separated by crystallization. If indigo be decomposed by strong caustic potash, or soda, and then distilled, *aniline*, a colourless oil, appears ($C_{12}H_7N$), with a spec. grav. 1·020, the boiling point at 356°, soluble in water and alcohol. It is obtained likewise among the oils of coal tar. When dry ammonia is passed through boiling alcohol containing isatine in solution, *imesatine* ($C_{16}H_5NO_2NH$) in dark yellow crystals forms, and when aqueous ammonia is added to a solution of isatine in alcohol with isatine suspended, *imasatine*, $C_{16}H_5N,NH,O_3$. *Amasatine* appears along with the preceding, a yellow substance, $C_{16}H_5N,NH_2O_3$.

Indigotic Acid. A synonyme of nitrosalicylic acid.

Indine. See INDIGO.

Infusion (*in* and *fundere*, to pour). An infusion is made by pouring cold, or hot water on vegetable or other substances, so as to extract soluble matters.

Ink (*encre*, Fr.; *tinte*, Ger.) Common ink is a finely diffused precipitate through water, of tannate and gallate of iron, of a deep blue shade. When it is deposited on paper, the water evaporates, and leaves the precipitate. To remove it, all that is necessary is to apply to it a weak acid, such as a solution of oxalic acid, in which the tannate of iron is soluble. Ink stains on tables, and cloths are thus removed. Hence ink or writing, exposed to the action of acid fumes, must necessarily be injured. To form a permanent ink, such as printers' ink, the black ingredient must be lamp black, which is not influenced by acids or other chemical agents.

Printers' Ink.—3 gallons of linseed, or nut oil are boiled down to the state of a varnish, and while hot 12 lbs. of rosin and 3 lbs. of brown soap are to be incorporated with it; 10 lbs. of lamp black, and 5 oz. of indigo, and 5 of the variety of Prussian blue called Paris blue, are then to be added.

Writing Ink.—A good black writing ink is made by boiling bruised nutgalls 2¼ imperial lbs. in 28 lbs. of water, filtering and dissolving in it, in an open vessel, 1 lb. of green vitriol (sulphate of iron) and 1 lb. of gum arabic. When it has acquired a dark blue colour it may be carefully bottled up. Dr. Traill has recommended a permanent ink to be made by dissolving fresh gluten in hot pyroligneous acid; a soapy liquor is thus formed, which is to be diluted until it acquires the strength of ordinary vinegar; a pound of this liquor is then to be mixed with 120 grains lamp black and 16 grains indigo. This ink is not decolourized by chlorine.

Stephen's Blue Ink is said to be prepared by triturating Paris Prussian blue, with one-sixth of its weight of oxalic acid in water. It is then diluted to the proper shade. Blue ink is also made with sulphate of indigo.

Red Ink is prepared by digesting 50 parts of Brazil-wood in 200 of vinegar for three days. It is then boiled for an hour, filtered, and to the liquid are added 6 parts of each, gum arabic, sugar, and alum. Another *red ink* is made by infusing 50 parts of Brazil-wood in 125 of alcohol for a day, filtering and evaporating the liquid till it becomes 50 parts; 32 parts of alum and 16 of gum arabic are then added to it.

Green Ink is made by dissolving in 7 ounces of water 75 grains of acetate of copper and 1⅔ ounce bitartrate of potash (cream of tartar). The liquid is boiled down to one-half its bulk, and filtered.

For writing on Zinc.—Braconnot recommends an ink composed of 5 parts of water, 2 parts each of salammoniac and acetate of copper, and ½ part lamp black.—To prevent ink from moulding, cloves, creasote, corrosive sublimate, &c. are sometimes added in small quantity.

Inolite. A variety of calcareous spar.

Inosic Acid. $C_{10}H_6N_2O_{10}HO$. Amorphous white mass, with a strong acid reaction and fleshy flavour; soluble in water; insoluble in alcohol and ether. It is obtained by cutting up the flesh of the fowl and turkey, mixing it with water in a bag, and expressing in a press. The expressed juice is boiled to remove albumen; barytes water is added as long as a precipitate of phosphate of barytes falls, the filtered liquor is evaporated in the water bath, and allowed to stand, that the creatine may crystallize out; lastly, alcohol being added, a milky fluid is thus obtained, which deposits crystals of inosate of barytes in silvery scales, from which the barytes may be separated by sulphuric acid.

Inosite. C 40·, H 6·66, O 53·34. *Form.* $C_{12}H_{12}O_{12}$. *Hydrous.* $C_{12}H_{12}O_{12}4HO$. F.P. 410°. Clear rectangular prisms resembling gypsum, becoming opaque in dry air, and losing 17 per cent. water at 212°; taste sweet; very soluble in water; with difficulty soluble in strong spirit; insoluble in absolute alcohol; not decomposed by dilute chlorohydric and sulphuric acids; does not reduce oxide of copper; does not ferment with yeast; when mixed in solution with cheese and chalk, it yields lactic and butyric acids, and therefore agrees with sugar of milk. It is obtained from the mother liquor of creatine, by precipitating the barytes by sulphuric acid; filtering, distilling the filtrate to separate volatile acids; agitating with ether and a little caustic soda to neutralize fluid acids; add strong spirit till it becomes turbid; sulphate of potash and inosite crystallize out together.

Intestinal Calculi. These concretions are very common in horses, and seem to be formed round a nucleus, which may be a nail, or any metallic or hard body swallowed by the animal Large numbers of these may be easily seen where old horses are killed. They consist of vegetable matter, nearly one-half their weight, but the proportion varies, and of phosphates of lime and magnesia, and carbonates of lime, magnesia, potash, and siliceous matter. Other intestinal concretions are composed, 1, of phosphate of lime; 2, phosphate of magnesia; 3, ammonia phosphate of magnesia; 4, biliary matter; 5, resinous; 6, fungous; 7, hairy; 8, ligniform; 9, lithofellic acid, or bezoar stones.

Inuline. *Alantine, Dahline, Datiscine, Helenine.* $C_{14}H_9O_9$? $C_{24}H_{21}O_{21}$ (Parnell). Specific gravity 1·356. White powder or crystalline grains, of nearly the same composition as starch, but differs from it in becoming only brown, not blue, in presence of iodine; tasteless; insoluble in cold, soluble in hot water; it is converted by sulphuric acid into grape sugar; obtained by boiling the roots of *Inula helenium, Dahlia variabilis, Colchicum autumnale, Leontodon taraxacum, Cichorium intybus, Helianthus tuberosus*, &c.; the inuline separates from the water on cooling.

Iodaniline. $C_{12}N,IH_6$. White crystalline powder, prisms, and 8-hedrons.

Iodides. The combinations of iodine with bases bear this name. They resemble chlorides and bromides in many respects. The alkaline iodides and bromides are anhydrous, and crystallize in cubes. The bromide of silver is similar to the iodide of silver, but it is soluble in ammonia. Iodides are precipitated red by corrosive sublimate, and yellow by salts of lead. They are decomposed by chlorine which sets the iodine free; the addition of starch then gives a blue colour, but an excess of chlorine converts the iodine into iodohydric acid, which does not precipitate starch.

Uses.—Iodine is used in medicine when dissolved in alcohol. It is also used extensively for the same purpose in the form of iodide of potassium, and it is said is now employed on the continent in the arts. When an overdose of it has been taken, the best antidote is iron filings.

Iodic Acid (IO_5 20·75).—If we boil 80 parts of iodine, 75 chlorate of potash, 1 nitric acid, and 400 water in a flask, for some time, and then add nitrate of barytes, a precipitate of iodate of barytes falls. The iodate of barytes is washed two or three times with water by decantation. It is then boiled for half-an-hour with 40 parts of SO_3 diluted with 150 parts of water. It is filtered and evaporated. On cooling the iodic acid is obtained in the form of a white crystalline powder, or in 6-sided crystalline plates, which appear to be segments of an octahedron; without smell; sour astringent taste; heavier than sulphuric acid, very soluble in water; with a solution of morphia, gives a reddish-brown colour, from which starch precipitates iodine. The iodates closely resemble the chlorates.

Test.—Oxygen is evolved when the acid is heated, and iodine dropped. Sulphurous acid evolves iodine, which may be detected by starch. There are various other oxygen acids of iodine of no peculiar interest, except from their analogy to the chlorine compounds, as hypoiodic acid (IO_4), periodic acid (IO_7), the former obtained by heating iodic acid with SO_3 in a crucible, and the latter by heating iodate of soda with hypochlorite of soda.

Iodine. I 15·75, 15·887, 126·, 127·. (ἰώδης, violet). *Source.*—Iodine, besides being derived from the ashes of sea-weeds, is found also in nature united with metals, as iodides of mercury, silver, and in the white lead ore of Catorce, and in the zinc ore of Silesia.

Characters. — Specific gravity 4·948 (Gay Lussac); of its vapour 8·678 to 8·716 (Gay Lussac and Dumas, An. Chim. 91, 17; 33, 346); (theory ·5528 × 15·75 = 8·706); specific heat 0·05412; 100 cubic inches weigh $270\frac{1}{4}$ grains. Its colour is grayish-black, when pure in the form of scales, or when crystallized in acute rhombic 8-hedrons, the axes of which are to each other as 2, 3, and 4 (Wollaston, An. Phil. 5, 237); somewhat similar to the prismatic form of sulphur. It sometimes occurs as double 6-sided pyramids (Plisson and Soubeiran). When volatilized it deposits usually in plates, but when dis-

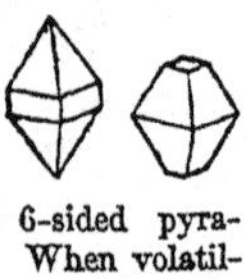

solved in a solution of iodohydric acid in water, or in ether, it may be obtained in crystals. It is very soft and friable. It melts at 224½, and boiling, volatilizes between 347° and 356° with a fine violet colour (Gay Lussac, An. Chim. 91, 7); but like essential oils when mixed with water and the liquid boiled, it may be distilled over with the water; non-conductor of electricity. *Odour* similar to that of chlorine, though weaker. *Taste* acrid and hot. It acts as a poison (Orfila, Toxicol. 1, 290). Like chlorine it destroys vegetable colours such as sulphate of indigo, but with less energy. It stains the skin of a deep yellow colour which speedily disappears. Paper is permanently corroded by iodine. *Soluble* in 7,000 parts of water, forming a yellow solution which has the smell of iodine; its solubility in water is promoted by the presence of iodides; in 9 parts of alcohol, spec. grav. ·842. When kept, the alcohol is reacted on by the iodine, and acquires a peculiar smell, while iodohydric acid is formed.

Preparation of Iodine from Kelp or ashes of sea-weeds.—See KELP.—1. The kelp broken into pieces is introduced into cast iron vats or keaves, and digested in water for twelve hours; the solution is then run off and fresh water is added. Earthy salts remain undissolved. 2. It is then placed in a boiler and concentrated to 60° Twaddell (1·300). Sulphate and muriate of soda fall, and may be removed occasionally. This concentrated fluid when run off into a cooler, deposits muriate of potash on its sides during five days. 3. The fluid is drawn off and boiled down to 68° Twaddell (1·340) when it deposits sulphate, muriate, and carbonate of soda. When run into a cooler it again deposits muriate of potash. 4. It is again placed in a boiler and brought to 74 Twaddell (1·370), when it deposits sulphate, muriate, and carbonate of soda, with, on cooling, muriate of potash. The latter is used in alum-making, and the carbonate of soda is sold to soap-makers. Sometimes the liquor is brought up to 74° Twaddell (1·370) at once. The concentrated liquor is then saturated with sulphuric acid of 1·700 spec. grav. (140° Twaddell), and allowed to stand from one to two days, till the sulphur from the oxysulphurous acids collects on the top. The iodine retort is filled to three-fourths. The fluid is brought up to near the boiling point but is not boiled, otherwise the condensers would be filled with steam (Meldrum). The usual iodine still is an iron vessel, either lined with lead or without this precaution, covered by a lead or iron top or by a dome, which terminates in a tube conducting into receivers in which the iodine condenses. This is shown in the figure. Sometimes there are three or more sets of receivers communicating with the still, by as many tubes passing through the lid or into the dome. The liquor, by the action of the sulphuric acid, loses sulphohydric, chlorohydric, carbonic, and sulphurous acids, while sulphur, from the decomposed hyposulphites and sulphides, collects on the top (Macintosh); but no iodine escapes if the acid has not been added in too great quantity, as the iodine being united with the metal, sodium,

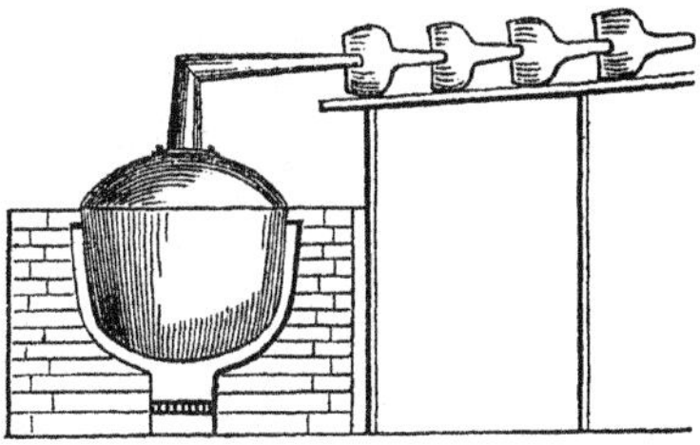

cannot be evolved without being displaced by oxygen. This is effected by the addition of binoxide of manganese, which is mixed with the liquor in the still (Wollaston, Ann. Phil. 3, 314, 1814), as explained by the following scheme:—
$NaI\,MnO_2\,2\,SO_3 = NaOSO_3MnOSO_3I$, and when an excess of acid is added, and heat applied previous to the addition of manganese, iodine is given off, in consequence of the excess of acid yielding oxygen from the water which it contains, to unite with the sodium. When heat is applied to the still, the iodine passes over, but is often in union with chlorine, if the common salt of the solution has not previously been decomposed. This, together with inattention to the proper regulation of a moderate temperature, frequently causes a great diminution in the product.

Test.—The most delicate test for iodine, is starch. A solution of starch is made in boiling water, and added to the iodine solution. If the iodine is in a free state, and the fluids are cold, a blue colour is immediately produced; but if the iodine is in union with a base, no change occurs. It is necessary to add nitrous acid (Millon), or commercial nitric acid (Dr. A. Buchanan), which contains nitrous acid, when a colour appears; or if sulphuric acid be added, a colour may not appear, as substances evolved by the sulphuric acid (as SO_2 and SH) may convert the iodine into iodohydric acid (HI), which does not precipitate starch. In this case it is advisable, after adding the acid, to hold over the mixed solution the mouth of an open bottle of chlorine gas, or chlorine water, when the blue colour appears at the surface proceeding downwards (Balard, An. Chim. 28, 178). If too much chlorine be added, it may convert the iodine into iodohydric acid, which does not affect starch. The blue precipitate is an iodide of starch. It may be formed by triturating starch with an excess of iodine, dissolving the mixture in potash, and then precipitating by acetic or tartaric acid. It has been shown that starch acquires a perceptibly blue tinge when the iodine does not exceeed $\frac{1}{450000}$ of the liquid. It is soluble in water of 150°. On cooling, a portion of the iodine again falls; but if the heating and cooling be frequently repeated, the cold solution at last remains colourless. Nitrate of silver gives with iodine a yellowish curdy precipitate insoluble in nitric acid, and in caustic ammonia. When mixed with a chloride, and precipitated by nitrate of silver, the iodide remains insoluble, on adding ammonia, and may be thrown on a filter. The chloride of silver passes through dissolved in ammonia; but reappears on adding an excess of nitric acid.

Estimation.—The best method of determining the amount of iodine in kelp, and in mixed solutions, is to evaporate to dryness; exhaust the dry salts with alcohol so as to take up all the alkaline iodide. The alcoholic fluid is distilled down to an ounce and mixed with a few grains of carbonate of soda; evaporated to dryness in a platinum crucible; the residue gently ignited to destroy organic matter, dissolved in water and filtered. The solution is carefully neutralized with chlorohydric acid, and precipitated by chloride of palladium, or potassium chloride of palladium. A blackish precipitate of iodide of palladium falls, which should be allowed to stand for twelve hours, when the supernatant liquor will be found quite clear. The precipitate is then thrown on a weighed filter, washed; dried at 212° and weighed; 22·413 grains of this salt (6·663 Pd + 15·75 I), contain 15·75 iodine, or 1804· contain 127·1 iodine.

Iodocodeine. *Teriodide of Codeine.* $C_{36}H_{21}NO_6,I_3$. Doubly-oblique prisms.

Iodocyanogen. White needles by iodine on cyanides.

Iodoform. *Teriodide of Formyle.* C_2HI_3. Yellow crystalline volatile solid body, obtained by the reaction of alcohol, potash. and iodine.

Iodohydric Acid. *Hydriodic Acid.* HI. 15·8875, 128. Spec. grav. 4·409 to 4·433. This, the heaviest known gas, is colourless, and smokes in the air; is rapidly absorbed by water and decomposed by mercury; prepared by heating in a tube 1 phosphorus, 16 iodine; covered with

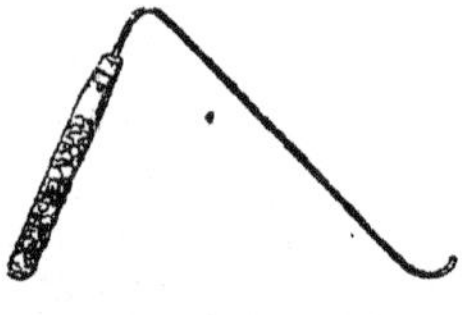

moistened powder of glass. A solution may be prepared by mixing 40 iodine, 40 finely divided lead, with 24 water, heating the mixture and agitating until the fluid has lost its smell, passing sulphohydric acid through the solution until the iodide of lead is decomposed (PbI and SH = PbS and HI), when iodohydric acid remains in solution. It speedily becomes brown.

Iodolite. A name given to an American meteorite.

Iodosalicylic Acid. $C_{14}H_5O_4I$. By the action of iodine on hydride of salicyle.

Iodostannates. Compounds of iodide of tin with other iodides.

Iolite. (ιον, a violet; λιθος, a stone.) *Cordierite*, *Dichroite*, *Peliome*, *Steinheilite*, *Sapphire d'eau.* Spec. grav. 2·5969 to 2·6643, H 7·. Various shades of blue, inclining to black, massive or in regular 6-sided prisms, usually with the lateral edges replaced by tangent planes; the terminal edges are frequently replaced by planes making angles of 137° 46′, with the corresponding lateral faces of the prism; fracture conchoidal; lustre vitreous translucent, sometimes transparent in the direction of the axis; the colour is blue; perpendicular to the axis it has a yellowish-gray colour. B.B. the edges fuse with difficulty into a glass of the same lustre as the mineral. Found at Cape de Gata; Bavaria; Greenland; Norway. Silica 49·62, Al_2O_3 28·72, MgO 8·64, CaO ·228, FeO 11·58, MnO 1·508. *Form.* $3(MgO\ FeO)\ 2SiO_3$, $3(Al_2O_3Fe_2O_3)\ SiO_3$.

Iolite Hydrous. H 3·75. Greenish-brown or olive-green 6-sided prisms, the lateral edges replaced so as to form nearly a cylinder; structure straight, foliated, the plates at right angles to the axis of the prism; cross fracture conchoidal; lustre of the fracture talc-like, of the surface of the plates waxy. B.B. infusible. Found in granite at Abo, Finland. Silica 45·05, Al_2O_3 30·05, MgO 9·, FeO 5·3, HO 10·6; allied to Pinite.

Ipecacuanha. The root of the *Cephaëlis ipecacuanha*, containing from 1 to 16 per cent. of emetine, the emetic principle for which this remedy is so extensively used in medicine.

Ipomic Acid. See HYDRORHODEORETINE.

Iridescence. A display of colours similar to the rainbow.

Iridiocyanogen. Cy_3Ir. Contained in the iridiocyanide of potassium, but not isolated; a salt in colourless crystals by the action of cyanide of potassium on protochloride of iridium.

Iridium. *Tennant*, 1803 (from *iris*, the rainbow) Ir 12·375; 99·. *Characters.*—White metal; the least fusible of all metals. Spec. grav. 16·26 to 23; resists the action of all acids; very hard. *Process.*—It exists in the black matter which remains on treating the ore of platinum, in which it exists to the extent of from 1 to 5 per cent. with nitro-hydrochloric acid. This matter is a compound of osmium and iridium. They are separated by fusing the black substance with potash in a silver crucible; dissolving the potash out with water, when the osmium is taken up of an orange colour. The residue is to be alternately treated with potash and HCl till it dissolves. On the evaporation of the red chloride of iridium, the salt crystallizes in octahedrons. When a plate of iron or zinc is placed in the solution of this salt, the metal falls. *Protoxide* (IrO).—Greenish-gray when moist, black when dry; obtained by precipitating IrCl by potash. *Sesquioxide* (Ir_2O_3).—Dark brown as hydrate, bluish-black when dry, formed by igniting the potash chloride of iridium with an alkaline carbonate. *Bitoxide* (IrO_2). — Imperfectly examined, but probably an acid. *Teroxide* (IrO_3?) brownish-yellow hydrate, formed by precipitating the double potash salt with an alkaline carbonate. The *blue* oxide is supposed to be composed of IrO and Ir_2O_3. *Protochloride* (IrCl).—Formed by passing Cl over ignited iridium: forms double salts with the alkaline chlorides. *Sesquichloride* (Ir_2Cl_3),—obtained as a black mass by igniting iridium with nitre and digesting in HCl. *Bichloride* ($IrCl_2$),—formed by dissolving the last chloride in nitric and hydrochloric acids; by evaporation it is reconverted into the sesquioxide. *Terchloride* ($IrCl_3$)—has only been obtained in union with chloride of potassium in black octahedrons. Iodine unites with iridium, forming an iodide (IrI) a black powder; it forms four combinations with sulphur. Similar bases with ammonia seem to exist, as with platinum.

Separation of Osmium and Iridium.—The ore is dissolved in aqua regia, and both metals are precipitated from the solution as double salts, by means of salammoniac. They are diffused in water, and through the mixture a current of sulphurous acid is passed. The chloride of iridium is decomposed into sesquichloride, which dissolves in the solution, while the ammonium chloride of osmium is not decomposed, and remains undissolved in the solution when it contains much salammoniac. The solution deposits by evaporation crystals of ammonium sesquichloride of iridium, which, when carefully heated, yield metallic iridium in place of crystals. The double salt of osmium is heated in a current of hydrogen, and changed into metallic osmium. To obtain the metals in the first instance in solution, it is necessary to fuse them with three times their weight of nitre in a porcelain crucible, and ignite for an hour at a red heat. The mass is poured on a metallic plate, and it is necessary to avoid the action of the fumes of osmic acid on the skin; a small amount of osmic acid is dissipated. The mass is heated in a retort with nitric acid, and the osmic acid condensed in a strong solution of potash. The residue of this distillation being treated with water which takes up the nitre, contains still oxide of iridium, and a certain quantity of oxide of osmium. The two oxides are then dissolved in aqua regia, and treated as above.

Iridosmine. *Native Iridium.* Spec. grav. 19·5 to 21·118, H 6 to 7. Tin-white irregular flattened grains or in 6-sided prisms, terminated by truncated 6-sided pyramids. There are 3 varieties: 1, flat plates, Ir Os, spec. grav. 19·25; 2d, in plates, spec. grav. 18·645 to 19·25, $IrOs_3$; 3d, very rare 6-sided plates, spec. grav. 21·118, Ir Os_4. This ore is found in South America and Siberia.

Irite. Spec. grav. 6·506. Black magnetic grains or scales found in Ural with native platinum, &c.; containing peroxide of,iridium 62·86, protoxide of osmium 10·3, FeO 12·5, sesquioxide of chromium 13·7.

Iron. *Sources.*—Upwards of forty species of iron ores have been described. Iron occurs also in all rocks, into the constitution of which it usually enters as a base to silica. The most valuable ores of iron are the oxides and carbonate.

Metallic Iron.—Fe 3·5; 28·. Sp. grav. 7·645 (Shaw); 7·788 (Brisson); 7·7 (Hatchett); 7·817 (Swedenburg); (7·6 Muschenbrœck, softened by heat); 7·763 (if hammered hot); 7·875 (Wasserberg, hammered cold); 7·843 (T. Thomson); 7·79 (Karsten, soft bar iron). An iron wire of 0·078 inch diameter supports a weight of 449·34 lbs. (Rennie, Phil. Trans. 1818, p. 126. See also Ann. Phil. 7, 320. Ann de Chim. 33,225). Grayish-white metal. Its cleavage is sometimes parallel to the faces of a cube, especially in iron plates heated to whiteness under the hearth of a blast furnace; and it has been observed in 8-hedrons in the hollows of a large cast iron roller (Wöhler, Pogg. Ann. 26,182; Breithaupt; Erdmann's Journ. 4,425). *Taste* styptic. When rubbed, iron emits a *smell*. *Hardness* greater than most metals; attracted by the magnet, under an orange heat; malleable at all temperatures, but increasing with the heat. Ductility greater than silver, gold, or copper, as it may be drawn out into a wire at least as fine as a human hair. In the state of wire its strength is 1½ times greater than that of hammered iron.

Reduction.—Iron exists rarely in nature in a free state, although it is thus found occasionally in meteorolites or metallic masses which fall from the atmosphere. A common form in which it presents itself is that of *Hæmatite*, or hydrous sesquioxide of iron (Fe_2O_32HO). Magnetic iron ore ($FeOFe_2O_3$), and carbonate of the protoxide of iron ($FeOCO_2$). The theory of the production of iron consists in removing the oxygen from the oxides by charcoal or coal, and the extraction from the carbonate of carbonic acid in the first instance, and then of the oxygen. The process of iron-making from the clay ironstone, as it is termed, or carbonate of iron, may be divided into 4 stages. 1. *Decarbonating, calcining,* or *removing the carbonic acid.* 2. *Deoxidizing* or *smelting.* 3. *Decarbonizing* or *refining*, or *removing carbon.* 4. Further *decarbonizing* or *puddling.*

1. *Decarbonating or Calcining.*—This process consists in mixing a certain amount of coal with the ironstone in heaps, and allowing it to burn slowly; the proportion of coal to ore varies from 1 to 5 to 1 to 8. But in certain species of ironstone, termed *black band*, the amount of coal diffused through the ore is so considerable, that little or no coal is required to be intermixed with it; but it burns spontaneously after being once ignited. In the course of a few weeks the ore has lost its earthy appearance, and has acquired a metallic aspect, in consequence of the removal of the carbonic acid. More recently the ore has been calcined by a kiln by the aid of the waste gases of the iron furnace. A red oxide of iron, techni-

cally called *mine*, remains, which is then ready for the second stage of the process. When coal is used, it is spread on the ground in a layer of 8 or 9 inches thick, over a surface of 10 feet by 8. Over this is laid a stratum of ore, 5 or 6 feet thick, mingled with coke dust and cinders; and over the whole is placed coal. It is then set on fire. The loss of weight varies from one-fourth to one-third, according to the amount of carbonic acid in the ore. Every 7·25 grs. FeO CO_2 become 4·833 magnetic oxide, or 100 parts become 66·.

2. *Deoxidating or Smelting.*—This process is conducted in the *blast furnace*, a tall building, from 50 to 60 feet high, containing within it a

a Blowing machine with its valves opening and shutting alternately above and below.
b Throat, or tunnel hole.
c Nose-pipe of tweers.
d Arches.
e Boshes.

conical space, into which the roasted ore is introduced, mixed with coal and limestone. The coal and ore are mixed in equal proportions, while the quantity of limestone amounts to about a fifth of the coal. These being thrown in at the top of the furnace, *b*, make their way to the bottom, *c*, where they are exposed to the action of a powerful blast of hot air from a blowing apparatus connected with a steam engine. The air thus driven into the furnace is never at a lower temperature than 612°, the melting point of lead, which is frequently used as a test of its being at a proper heat. An intense temperature is produced in the furnace, the oxide is reduced to the metallic state, while the impurities of the ore, consisting of silica and alumina, unite with the lime and form a slag, which floats on the surface of the melted iron, in the hearth or lowest part of the furnace. After the furnace has been kept in a charged state at an intense heat for twenty-four hours, it is *tapped*, and the melted iron allowed to flow off into small canals, made in sand at the bottom of the furnace. These, from their being formed into main channels, and smaller furrows at right angles to them, have been termed by the workmen the *sow* and *pigs*. Hence the iron cooled in these channels is called *pig iron*. There is sometimes found in cavities in the cast iron a substance in scales of a glistening lustre; and sometimes floating on the surface of the melted iron as it flows into the sand-channels, a substance termed *keesh* by the workmen. It is a carbide of manganese, according to Dr. Wollaston. Spec. grav. 2·982 (T. Thomson). When these ingots or pigs are examined, they are sorted according to the quantity of charcoal which they contain, mechanically mixed, as detected by inspecting their cross fracture. The following is the constitution of cold and hot blast iron, No. 1, and of slags:—

	Cold.	Hot.
Spec. grav.,	6·703	7·06
Iron,	91·15	95·58
Manganese,	2·03	0·87
Carbon,	3·85	2·09
Silicon,	1·17	1·08
Aluminum,	1·65	0·42

By Dr. T. Thomson, and Mr. John Tennent.

Slags.

	R.D.T.		Berthier.
Spec. grav.,	2·848		
Silica,	43·75	38·12	40·4
Alumina,	12·50	16·97	11·2
Lime,	31·22	32·77	38·4
Magnesia,	7·15	6·22	5·2
Protoxide of manganese. Protoxide of iron,	4·88	5·92	3·8

Black cast iron, or No. 1, is the most valuable product of the blast furnace. It contains the smallest amount of carbon visible to the eye, diffused through the iron. Spec. grav. 6·83—6·90. Texture granular; used in castings. Its slag is uniform in colour and appearance, glassy and feebly translucent. No. 2.—*Gray or mottled cast iron.*—Spec. grav. 7·07. Harder than No. 1; it contains carbon, chiefly in the centre, in the form of black dull particles, easily distinguishable from the white, metallic, glistening aspect of the metal itself; artillery is usually made of it. The slag is opaque, heavy, of a yellowish-green colour, exhibiting bands of blueish enamel. No. 3.—*White cast iron*, resembling silver. Spec. grav. 7·68. Particles of charcoal may be observed interspersed over the whole fractured surface. It is therefore the least valuable species of cast iron. It is very hard and brittle, and is only useful for bullets, weights, and bombs. It consists of iron 95, carbon 5 (or Fe_4C). The slag is black, vitreous, and blebby, and gives out the smell of sulphuretted hydrogen. It sometimes happens that in smelting, white cast iron free from interspersed carbon, is at once procured. If this result could be obtained invariably, a great improvement in iron-making would be accomplished, as it would save the next process. The impurities in the iron ore unite with the lime and form slag, which can be tapped over the metal. When thus drawn off, the quality of the slag as described, enables the iron smelter to predict the nature of the iron in the furnace.

Theory of Iron Smelting.—A blast furnace consists of two columns, one of air, ascending from below upwards, and the other of solids, thrown in from above, and descending. At the top, the amount of watery vapour is 9 to 14 per cent. But the vapour in the gases rapidly diminishes. At the top the gases are nitrogen 57·79, carbonic oxide 23·51, carbonic acid 12·88, hydrogen 5·82. At between 8 and 17 feet from the top, the carbonic acid and hydrogen diminish, and the water completely disappears. The oxygen of the air is first changed into carbonic acid, and then by the coal into carbonic oxide. The hydrogen is produced by the vapour of the water, which is decomposed by carbon into hydrogen and carbonic oxide. As the column of gas ascends, the carbonic oxide diminishes, and the carbonic acid increases,—a change produced by the action of carbonic oxide upon sesquioxide of iron, which yields carbonic acid and metallic iron. It seems remarkable that hydrogen should exist in contact with sesquioxide of iron at a red heat without deoxidizing it; but it has been shown that an electric spark passed through carbonic oxide, oxygen, and hydrogen, produces carbonic acid, while hydrogen remains. Between the neck of the furnace and the boshes the minerals lose their moisture, and the oxide of iron $\frac{28}{33}$ of its oxygen to the deoxidizing power of carbonic oxide, while the $\frac{5}{33}$ remaining are removed between the boshes and tweers by the carbon of the coal. At a foot above the tweers the temperature diminishes, and to this the line of fusion is limited, and here the change of carbonic acid into carbonic oxide is complete.

Composition of the Gases in Iron Furnaces.

Depth from the top in Feet.	5	8	11	14	17	23	24	34
Nitrogen,	55·35	54·77	52·57	50·95	55·49	58·28	56·75	58·08
Carbonic acid,	7·77	9·42	9·41	9·10	12·43	8·19	10·08	——
Carbonic oxide,	25·97	20·24	23·16	19·32	18·77	29·97	25·19	37·43
Carb. hydrogen,	3·75	8·23	4·57	6·64	4·31	1·64	2·33	——
Hydrogen,	6·73	6·49	9·33	12·42	7·62	4·92	5·65	3·18
Olefiant gas,	0·43	0·85	0·95	1·57	1·38	——	——	——
Cyanogen,	——	——	——	——	——	trace	trace	1·34

(Bunsen and Playfair, Rep. Brit. Ass. 1845.)

The occurrence of the cyanogen is interesting, although its formation had been long ago pointed out by Dr. Clark. When an iron tube was introduced through a hole 2 feet 9 inches above the hearth, a gas passed out burning with a yellow flame, similar to that occurring during the preparation of potassium. The tube was soon obstructed by cyanide of potassium. The cyanogen is produced by the nitrogen of the air or coal, while the potassium is derived from the potash of the coal (·07 per cent.) and from the iron ore (·75 per cent.), amounting to 270 lbs. in twenty-four hours. The cyanide of potassium is volatilized, and is decomposed by the air into carbonate of potash and cyanide, which is gradually decomposed as it ascends. That the cyanogen is not altogether decomposed is, however, proved by the results obtained at Coltness Ironworks, where, in saving the waste gases for the purpose of calcining the ore, a considerable quantity of alkali, containing cyanide of potassium and carbonate of potash, &c. is sublimed in the common furnace at the summit. The amount of alkaline salts sublimed may be about 2½ tons per furnace during the year, according to the estimate of Mr. Hunter. For this observation I am indebted to Mr. Rankin of Carluke.

3. *Decarbonizing or Refining.*—When iron is to be made into castings, it is used in the state of No. 1. But when it is to be converted into bar and malleable iron it undergoes the present process, which is intended to remove the carbon which is still artificially mixed with it. 6 pigs are introduced into the refining furnace, mixed with coke. The refining furnace is a crucible or hearth,

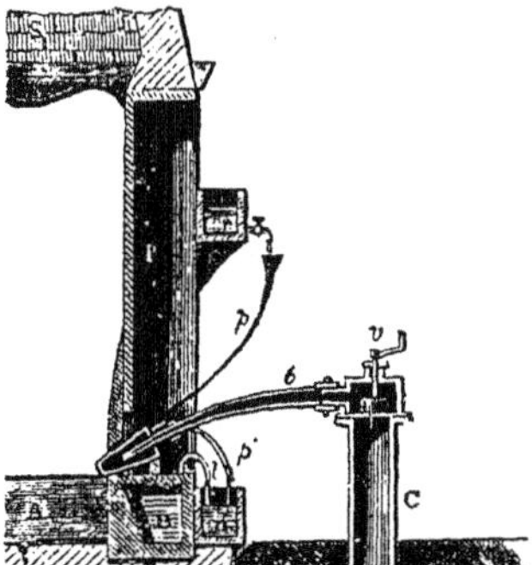

sunk 2½ feet deep in the ground, 3 feet square, and composed of 4 plates of cast iron. There are generally two tweers, one at each side, occupied each by one or two nose-pipes. In the *figure* one side of the refining furnace only is seen. These are kept cool by a current of water which passes from a cistern above into one below in a continuous stream; the nose-pipes are thus preserved from being fused. Air is propelled by these pipes into the heated mass; the intense heat is continued for two or three hours, during which carbonic oxide and carbonic acid are evolved. The metal is then tapped, and allowed to run into a cavity 10 feet long, 3 feet wide, and 2½ inches deep. The iron fills this space and acquires its form. Water is then run upon it. When cool it is *white cast iron*, and very hard; fracture fibrous or radiated, and it is frequently filled with spherical cavities like those which occur in amygdaloidal rocks. The metal loses in this process from 12 to 17 per cent. The slag from this process consists of 35 silica, 61 FeO, and Al_2O_3, with from 1½ to 4 per cent. of phosphoric acid. It is dark coloured with a bluish shade, often fibrous and crystallized.

4. *Further Decarbonizing or Puddling.*—Refined iron still contains a quantity of carbon united with it, which is removed to a great extent by the puddling process, contrived by Mr.

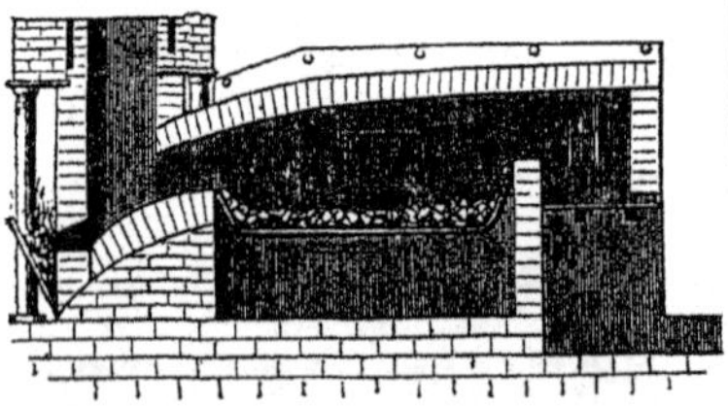

Cort, in 1785. It consists in placing the refined iron, broken in pieces, in a reverberatory furnace, where it is heated by a strong flame playing on its surface. In the course of half-an-hour it is brought up to a white heat, when it begins to soften and to melt. It is then continually stirred or puddled, so as to keep it in a fine state of division, by means of a long iron instrument, which is introduced through a door in the furnace. Much carbonic oxide is evolved, and burns with a blue flame. After the gas has ceased to burn, the metal is still puddled till it is reduced to the state of sand. As soon as it begins to *work heavy* it is collected into balls, and the heat raised so as to *weld* the particles together. It is removed in the form of large balls from the furnace by means of a pair of tongs after it has remained 2½ hours in the furnace. It is then carried to be *welded*, being first hammered and then drawn by rollers into *mill bar iron*. In Wales 10 tons of coals are used in puddling 9 tons of refined or *fine metal;* and by this process a loss is sustained of 10 per cent. Various methods have been recommended of facilitating the removal of the carbon in the puddling furnace, such as the addition of black oxide of manganese, common salt, and potter's clay (Schafhaeutl). The slag obtained from the puddling furnace is black with a shade of blue, sometimes crystalline, and corresponding in composition with pyroxene (Mitscherlich), that is, nearly a sesquisilicate of iron.

Steel is made from bar iron, by placing it, surrounded by finely pounded charcoal, in a furnace covered up with sand, and heated intensely for a week or ten days. The carbon unites with the iron, and forms a carbide of iron, or steel. When the steel bars are broken and frequently *welded* or hammered together, they are termed German or *shear* steel, because they were originally thus prepared for making *shears* to dress woollen cloth. *Cast steel* is obtained by breaking up the bars, fusing them in a crucible, and pouring the melted steel into moulds. *Steel* is whiter than iron; texture granular, fracture whitish-gray; more brittle than iron. Its hardness is increased by heating and plunging into water. A drop of nitric acid leaves on steel a black spot (carbon); on iron a whitish-green stain (oxide). *Case hardening* consists in converting the surface of iron into steel by heating it with charcoal for a short time. Keys and gun-locks are thus treated. *Tempered* steel is required for knives and other cutlery instruments. This is effected by heating the articles in oil raised to the proper temperature. Penknives and razors require a heat of 460°, and have a straw-yellow colour; watch-springs are made deep blue at 580°. These colours are obviously produced by a partial union with oxygen of the air, for when heated in hydrogen no change of colour occurs.

Protoxide of Iron, FeO 4·5, 36, may be obtained as a hydrate; that is, in union with water, by precipitating a solution of well crystallized *green vitriol* or *copperas* ($FeO\ SO_3 7\ HO$), by caustic ammonia, or potash, or soda. It is first thrown down white and then becomes green; by exposure to the air it absorbs oxygen, and is converted into the sesquioxide or brown oxide ($Fe_2\ O_3$). The dry protoxide has not been isolated.

Sesquioxide, peroxide, red oxide, deutoxide, rust, $FeO_{1\frac{1}{2}}$ or $Fe_2\ O_3$, 5, 10; 80, may be formed from sulphate of iron or green vitriol, by dissolving the salt in water, adding nitric acid and heat until the black colour, which at first shows itself, is changed into a bright yellow. On the addition of caustic ammonia to the sulphate of the peroxide of iron, the peroxide falls in the form of a brownish-red flocky precipitate. It may also be prepared by dissolving iron in nitric acid, and precipitating by alkalies or alkaline carbonates. When dried it is a fine brownish-red powder, insoluble in water, tasteless, soluble in hydrochloric acid. When freshly precipitated it constitutes the antidote to arsenious acid, and when administered is to be diffused through water by means of a knife or spoon, and drank freely. It is used also as a tonic in nervous diseases. Peroxide of iron occurs in nature abundantly, in the form of hæmatite, an iron ore ($Fe_2 O_3$ 2 HO). *Rust* is formed when iron is exposed to the atmosphere. It consists of the hydrous red oxide, mixed with some carbonate of the protoxide of iron.

Protosesquioxide, Octahedral iron ore, Magnetic oxide, Loadstone, Black oxide, Ferroso-ferric oxide. $FeO_{1\frac{1}{3}}$ 4·833 or FeO Fe_2O_3 14·5, 116, occurs abundantly in nature, as magnetic iron ore, in octahedral crystals. It occurs in Sweden, and forms the iron from which the best steel is made. It enters into the composition of the roasted ironstone, or carbonate of iron, which is magnetic. It is prepared artificially by burning iron wire in oxygen gas, or by pouring a mixed solution of proto and sesquisulphate of iron into caustic ammonia.

Ferric acid, Fe O_3 6·5,52·, is obtained by passing a current of chlorine gas through a strong solution of potash, in which the sesquioxide of iron is suspended. When the mixture is filtered it possesses a fine amethyst colour—ferrate of potash. *Protochloride of iron,* Fe Cl 8;64, a gray, variegated, metallic solid, in plates, melting at a red heat; soluble in water; obtained by dissolving iron in hydrochloric acid, and igniting the residue in a tube away from the air. *Sesquichloride,* Fe_2 Cl_3 20·5, 164, produced when iron wire is burned in chlorine gas. A solution of this salt in alcohol is used in medicine as a tonic, termed Tinctura ferri sesquichloridi. *Iodide of iron.* (See Iodide of potassium.) *Sulphide of iron,* Fe S 5·5, is much used in the preparation of sulphohydric acid. It is made by heating an iron bar to redness, and bringing it in contact with a roll of sulphur, over water; sulphide of iron is formed and drops into the water. It is essential that the iron be intensely hot. It may also be formed, but of inferior purity, by heating sulphur and iron filings in a clay crucible.

Sulphate of iron, Green vitriol, Copperas, Fe O SO_3 7 HO, 17·375; 139, may be formed by dissolving iron in sulphuric acid, with the addition of water. In commerce it is prepared on a large scale in the alum manufacture, the sulphate of iron being crystallized out from the sulphate of alumina. Copperas crystallizes in green rhomboidal prisms. When heated it loses water and acid, and leaves a red powder (*colcothar of vitriol*), which is an impure peroxide. *Persulphate of iron,* Fe_2 O_3 3 SO_3 25·, formed by oxidating the protosulphate by means of nitric acid, as already described. Green vitriol is used as a tonic in nervous diseases, and where the blood is supposed to be deficient in iron, as in chlorosis and anæmia. *Potash tartrate of iron,* obtained by boiling the sesquioxide with cream of tartar (bitartrate of potash), and evaporating. Used as a tonic in medicine.

Gun barrels are browned by sponging with the following mixture:—3 oz. alcohol, 3 oz. tincture of iron, 1 oz. corrosive sublimate, 3 oz. of nitric ether, 2 oz. sulphate of copper, 1½ oz. nitric acid; and after drying, rubbing with a scratch card or brush.

Estimation of Iron.—Iron is always estimated in the form of sesquioxide; every 5 parts being equivalent to 3½ of metallic iron. It is usually precipitated from its solution by caustic ammonia.

Analysis of Iron Ores.—The *hydrates of iron* are analyzed by heating to redness; the loss of weight indicates the water with which the oxide was united. A fresh portion of ore finely pulverized is dissolved in aqua regia; the sesquioxide of iron precipitated by ammonia; the alumina separated from the iron by caustic soda; the lime by oxalate of ammonia, and the magnesia by phosphate of soda. Magnetic oxide requires to be fused with carbonate of soda. *Carbonate of iron* is analyzed by dissolving in aqua regia, and proceeding as with the oxide; or it may be heated in a platinum crucible, with bisulphate of potash, when the object is merely to determine the amount of iron. *Cast iron and steel* are analyzed by pulverizing them in a steel mortar. The carbon is then mixed with twenty times its weight of chromate of lead, and an equal weight of chlorate of potash. The mixture is introduced into an organic analysis tube, and to it are adapted a chloride of calcium tube, and a potash bulbed tube. The carbon is converted into carbonic acid, and is absorbed by the potash. The *silicon* is obtained by dissolving the iron in muriatic acid, evaporating the solution to dryness, and filtering the silica which remains. The *sulphur* is obtained by dissolving the iron in aqua regia, and precipitating the sulphuric acid by chloride of barium. The *phosphorus* is obtained by dissolving in aqua regia, precipitating the iron by sulphohydride of ammonia, and the lime by oxalate of ammonia; filtering, boiling off the alkaline sulphide, and precipitating the phosphoric acid by sulphate of magnesia and ammonia. The solution contains salammoniac, and one-eighth of caustic ammonia. The manganese and iron, being precipitated as sulphides, are to be separated by dissolving in aqua regia, neutralizing, precipitating the iron by benzoate of ammonia; the manganese from the filtered liquor by carbonate of soda.

Separation of Iron from Alumina.—Dissolve the oxides of iron and alumina in chlorohydric and nitric acids, free the solution by evaporation from excess of acid, and then add an excess of caustic potash or soda, and boil for twenty minutes. The alumina is dissolved, and the sesquioxide of iron remains. Fresenius recommends that the solution of the alumina and iron should be boiled, and then with sulphite of soda, to reduce the salt of iron to the state of protoxide. It is again boiled and neutralized with carbonate of soda. It is then boiled with caustic soda until the precipitate becomes black and granular. The precipitate is filtered and washed with hot water. The filtered liquid is then acidulated with chlorohydric acid. Some chlorate of potash is added to destroy organic matter from the paper, and then precipitated with ammonia; on standing for some hours the alumina is filtered.

Iron from Manganese.—Dissolve the two metals or oxides of the metals in chlorohydric acid, with the addition of some nitric acid, add caustic ammonia until the greater portion of the acid is neutralized, and then add an excess of chalk or carbonate of barytes, and boil the mixture. The whole of the sesquioxide of iron is precipitated with the excess of chalk or barytes, while the manganese remains in solution. The precipitate is thrown on a filter, and the oxide of iron determined by a solution in acid, and precipitation by caustic ammonia. The manganese is then precipitated by sulphohydride of ammonia, the precipitate washed, dissolved in chlorohydric acid, and precipitated by carbonate of soda; evaporating the whole to dryness, digesting in water, filtering, igniting, and weighing. The protosesquioxide, or brown oxide, remains, containing $1\frac{1}{3}$ atom oxygen. The usual method of separating iron and manganese, is to saturate the solution as accurately as possible with ammonia, and then to add benzoate or succinate of ammonia. The sesquioxide of iron falls, in union with the organic acid, as a buff-coloured precipitate, which leaves sesquioxide on ignition. The manganese may then be precipitated by evaporation with carbonate of soda.

Iron Ores. *Native iron. Bolide.* Spec. grav. 5·95 to 6·72, H 4·5. Bluish-white masses, attracted by the magnet. It has been doubted if native iron exists. A specimen examined by Klaproth contained Fe 92·5 Pb 6· Cu 1·5 (Great Kamsdorf); and others have been found in veins at Canaan, Connecticut. Demarcay states that he found metallic iron in cerite from Bastnas; and Andrews in Irish basalt.

Iron, Meteoric. Spec. grav. 7·3. Silver-white masses, not so liable to rust as common iron, with a granular structure. A specimen from Blanskö contained Fe 93·816, Ni 5·053, Co ·347, Sn and Cu ·46, S ·324. This variety of iron ore is supposed to be derived from igneous meteors which are occasionally seen to fall through the atmosphere. Their source is unknown. They are found on the surface of the earth in Siberia, South and North America, Cape of Good Hope, &c.

Iron Specular Ore. *Anhydrous sesquioxide*, or *peroxide of iron, Oligiste iron, Iron froth, Elba iron ore, Iron mica, Red hæmatite, Red clay, Iron stone, Red silicious iron stone, Red ochre, Reddle, Red chalk.*—Fe_2O_3 = Fe 70 O, 30. Specific gravity 5·251 to 4·232, H 5·5 to 6·5. Red crystalline masses, or steel-gray and iron-black acute rhomboids, with angles of 86° 10′ and 93° 10′, or 8-hedrons, triangular 12-hedrons, with their apices truncated very deep, or near the summit; lustre metallic, opaque, brittle, sometimes acting slightly on the magnet; in thin plates, blood-red. When in micaceous scales in volcanic countries, it is termed *iron mica.* Spec. grav. 4·491 to 5·059, in thin 6-sided plates. *Iron froth* consists of friable, soiling scales, with a cherry or brownish-red colour. *Red hæmatite.* Spec. grav. 6·305, H 7·; in masses, stalactites or fibrous kidney-shaped balls, opaque, with a semi-metallic lustre, is found in Wales, &c. *Red clay iron stones* are found in an impure state, mixed with clay, &c. in Great Britain. The finest crystals are found in the island of Elba.

Iron, Crucite, or **Crucilite.** See CRUCITE.

Iron, Sesquihydrous Sesquioxide, or *Peroxide of iron, Bog iron ore, Bonerz, Brown fibrous hæmatite, Brown ochre, Brown clay iron stone, Gœthite, Lepidocrokite, Ligniform hæmatite, Limonite, Pisiform iron ore, Pyrosiderite, Rubinglimmer, Stilpnosiderite, Umber, Yellow clay iron stone.* Fe_2O_3 83·38, HO 15·01, SiO_3 1·61. *Form.* Fe_2O_3 $1\frac{1}{2}$ HO. Spec. grav. 3·922 to 4·04, H 4·5 to 5. Various shades of brown masses, stalactites, or botryoidal pieces, found in veins and beds in primary and secondary rocks in England, Carinthia, Bohemia, Bonn, Brazil, North America; does not act on the magnet. B.B. becomes black and magnetic; soluble in nitrochlorohydric acid.

Iron, Protohydrous Sesquioxide. *Gœthite* in part, *Pitchy iron ore* in part. Fe_2O_3 91·7, HO 8·5=Fe_2O_3 HO. Spec. grav. 4·375, H 4·5 to 5· (T. Thomson). Reddish-brown needles, or right rhombic prisms, with angles of 130° 40′ and 49° 20′; lustre imperfect, metallic, and silky, brittle, opaque, streak brown. In nodules in trap near Greenock; in crystals at St. Just, Cornwall, and St. Vincent; rocks, Bristol.

Iron, Manganesian Ore. Sp. grav. 5·079, H 7·. Black splendent masses, or 8-hedrons; streak red; texture foliated; surface smooth, brittle, opaque; acts feebly on the magnet. Fe_2O_3 75·5, Mn_2O_3 22·65, TiO_2 1·15. Stirling, North Jersey; allied to Franklinite.

Iron, Franklinite. See FRANKLINITE.

Iron, Magnetic Pyrites. *Sulphide of iron, Pyrrhotin.* Spec. grav. 4·631, H 5·5. Fe 59·85, S 40·15. *Form.* 5 FeS, FeS_2. Bronze-yellow or copper masses, or irregular 6-sided prisms, variously modified; affording by cleavage the regular 6-sided prism; lustre metallic, opaque, brittle; acts slightly on the needle; streak grayish-black; occurs, along with other minerals, in several rocks, Cornwall, Wales, Galloway, Bodenmais, Bavaria, Geyer, Saxony. It readily decomposes in air, and cannot long be preserved in cabinets.

Iron, Bisulphide of. *Bisulphuret of iron, Iron pyrites, Cubic pyrites, Mundic, Sulphur.* Spec. grav. 4·83 to 5·031, H 6·5. Sulphur 54·26. Iron 45·74. *Form.* FeS_2. Bronze-yellow cubes and regular 8-hedrons, and in derived figures, as the pentagonal 12-hedron and icosahedron; cleaving parallel to the faces of a cube; fracture conchoidal, opaque; lustre metallic; streak brownish-black; strikes fire with steel. Hence its name pyrites (fire stone). This mineral is extensively distributed in nature, being found in old crystalline rocks, in clay slate, and abundantly in the coal formations, from which it is obtained for the manufacture of sulphuric acid. B.B. gives off sulphurous acid, and leaves an

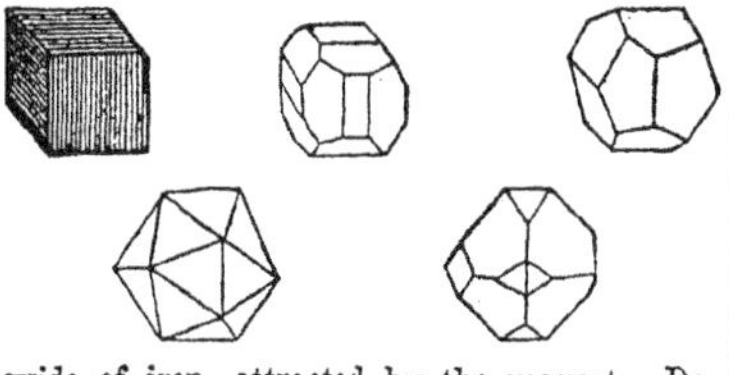

oxide of iron, attracted by the magnet. Decomposed by nitromuriatic acid. Wicklow pyrites, used in manufactures, I have found to consist of S 47·41, Fe 41·78 Cu 1 93, As 2·11, SiO_3 3·93; insoluble matter 1·43; Zn 2.

Iron, White Bisulphide of. *Radiated pyrites*, (Zn variable), *Cockscomb pyrites*, *Kammkies*, *Leberkies*, *Zelkies*, *Spear pyrites*, *Hepatic pyrites*, *Rhombischer eisenkies.* Spec. grav. 4·678 to 4·847, H 6·, S 53·35, Fe 45·07, Mn ·7, SiO_3 ·8. *Form.* FeS_2. Possesses the same composition as the preceding, but its crystalline form is different. It is therefore an isomeric form of iron pyrites. Pale bronze-yellow, sometimes inclining to green or gray, right rhombic prisms, with the angle 106. The most common form is a flat crystal, consisting of similar portions of five crystals united together, having, at first sight, the aspect

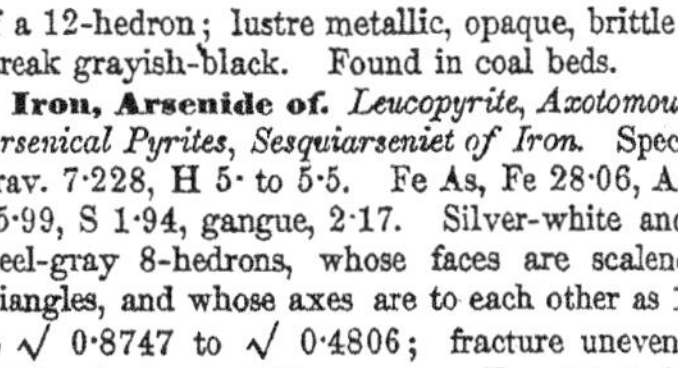

of a 12-hedron; lustre metallic, opaque, brittle; streak grayish-black. Found in coal beds.

Iron, Arsenide of. *Leucopyrite*, *Axotomous Arsenical Pyrites*, *Sesquiarseniet of Iron.* Spec. grav. 7·228, H 5· to 5·5. Fe As, Fe 28·06, As 65·99, S 1·94, gangue, 2·17. Silver-white and steel-gray 8-hedrons, whose faces are scalene triangles, and whose axes are to each other as 1 to $\surd$ 0·8747 to $\surd$ 0·4806; fracture uneven, brittle; lustre metallic, opaque. Found in beds with carbonate of iron, at Löling, Carinthia, and imbedded in serpentine, at Reichenstein, Silesia.

Iron, Arsenio Sulphide, or *Arsenical Pyrites*, *Mispickel*, *Marcasite*, *Sulphoarsenite of Iron.* Spec. grav. 6·127, H 4·75 to 6·. As 45·74, Fe 33·98, S 19·6. *Form.* FeAs, FeS_2. Silver-white or steel-gray right rhombic prisms, with angles of 111° 12′, the obtuse angles of the prism being often replaced by triangular planes. These are sometimes so large as to efface the bases of the prism, and present the aspect of an elongated 8-hedron; lustre metallic, streak grayish-black; fracture uneven, opaque, brittle. B.B. yields arsenical vapour, and leaves a magnetic residue. Found in Saxony and Bohemia, Hartz, Tunaberg, Sweden, Cornwall, &c.; often accompanies metallic ores.

Iron, Carbonate of. *Brown Spar*, *Argillaceous Iron Ore*, *Clay Iron Stone*, *Fer Spathique* (*Spathic Iron*), *Sparry Iron Ore*, *Spathose Iron*, *Sphærosiderite*, *Strahlstein.* Spec. grav. 3·7317 to 3·829, H 3 to 4·5. FeO 54·57, MnO 1·155, CaO 3·176, HO 2·63, CO_2 35·9. *Form.* FeO CO_2. Yellowish-gray or yellowish-white and red, &c. obtuse rhomboids, resembling calcareous spar, with angles of 107° and 73°; also in double 3-sided prisms, and regular 6-sided prisms; fracture imperfect and conchoidal; lustre vitreous or pearly, streak white, translucent, to opaque, brittle. B. B.

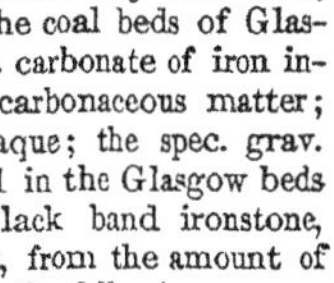

gives off carbonic acid, and leaves a magnetic oxide, which affects, and is affected by the magnet. The valuable ore termed *clay iron stone*, found in connection with the coal beds of Glasgow and Staffordshire, is a carbonate of iron intermixed with clay and carbonaceous matter; the fracture is earthy, opaque; the spec. grav. varies from 2·936 to 3·471 in the Glasgow beds (Dr. H. Colquhoun). Black band ironstone, so named from its colour, from the amount of coal which it contains, has the following composition:—FeO 53·03, CO_2 35·17, CaO 3·33, MgO 1·77, SiO_3 1·4, Al_2O_3 ·63, Fe_2O_3 23, Coal 3·03. In many specimens of common clay ironstone, sulphur is present in the form of pyrites, and phosphorus as phosphate, and likewise arsenic, all of which are prejudicial to the iron. From a very extensive series of analyses of these iron ores made for various ironmasters, the largest amount of protoxide of iron which I have met with was about 54 per cent., the specimen being from the neighbourhood of Airdrie.

Junkerite is a carbonate of iron, with a different crystalline form. See—

Iron, Sulphated Sesquioxide of. *Form.* Fe_2O_3 $2\frac{1}{2}SO_3$ 9HO? Yellow grains covered with thin 6-sided plates, translucent, pearly, mixed with siliceous matter. Near Copiapo, Chili.

Iron, Tersulphated or **Bisulphated Sesquioxide of.** *Form.* Fe_2O_3 $3SO_3$ 9HO. White granular masses, or regular 6-sided prisms, terminated by a truncated 6-sided pyramid. Completely soluble in water. Found at Copiapo, Chili, in a bed on granite from decomposed pyrites.

Iron, Sexhydrous Trisphosphate of. *Blue Iron Earth*, *Native Prussian Blue.* *Form.* 3FeO PO_5HO. Grayish-white, becoming, by exposure, smalt-blue. Soils slightly, feels harsh. B.B. becomes reddish-brown, and fuses into a black brilliant globule, attracted by the magnet. It is found along with bog iron ore, and in mosses. A specimen from New Zealand, yielded my pupil, Mr. R. T. Pattison, 28·4 HO, 2·8 organic matter, 5·2 silica, and 62·8 phosphate of iron.

Iron, Phosphates of. These are—*Blue iron earth, Delvauxite, Dufreynite, Vivianite, Pseudo-triplite, Mullicite, Cacoxenite, Childrenite, Triphylline, Karphosiderite, Alluaudite, Hetopizite.*

Iron, Carbono-phosphate of. Sp. grav. 3·71. Greenish-blue grains, like chamoisite, magnetic. Soluble, with effervescence, in chlorohydric acid, leaving some silica. Fe_2O_3 41·12, FeO 29·98, CO_2 11·873, PO_5 3·38, CaO 2·14, HO 2·9, SiO_3 6·99, MgO ·775. In Jura limestone, at Vignes.

Iron, Tetrarseniated Sesquioxide of. *White Iron Sinter, Diarseniate, Pitticite.* AsO_5 30·25, Fe_2O_3 40·45, HO, with trace of SO_3, 28·5. *Form.* $4Fe_2O_3AsO_5$ 13HO. Yellow-gray reniform pieces, soft, friable, adhering to the tongue. Found near Freiberg.

Iron, Triarseniated Sesquioxide of. *Cube Ore of Iron, Pharmacosiderite.* Sp. grav. 3·, H 2·5, Fe_2O_3 39·2, AsO_5 37·82, PO_5 2·53, CuO ·65, HO 18·61, insoluble matter 1·76. *Form.* $3Fe_2$ O_3AsO_5 7 HO. Olive-green, of various shades, to brown and grass-green cubes,

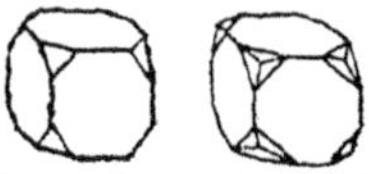

the angles sometimes replaced by tangent planes; cleavage difficult and imperfect, fracture conchoidal, uneven, lustre adamantine. B.B. becomes red, on charcoal gives out fumes of arsenic, leaves a magnetic scoria. Found in Cornwall, France, and Saxony.

Iron, Arseniate of. *Scorodite, Neoctese.* $Fe_2O_3AsO_5$ 5HO. Spec. grav. 3·1 to 3·3, H 3·75, Fe_2O_3 34·30, AsO_5 49·60, HO 16·90. Pale or greenish porous masses, or 4-sided prisms, with square bases? powder white, streak white; with caustic potash it becomes rust coloured. B.B. becomes yellow without change of form; on charcoal, fumes with the smell of arsenic; insoluble in nitric, soluble in chlorohydric acid. Found at Villa Rica, Brazil, and at Loaysa, Popayan. *For other Arseniates,* see ARSENIOSIDERITE, SYMPLESITE.

Iron, Silicates of. The silicates of iron are—*Sideroschisolite, Chamoisite, Anhydrous silicate of iron, Cronstedtite, Hedenbergite, Chloropal, Wehrlite, Anthosiderite, Polyhydrite, Thuringite, Hisingerite, or Thraulite, Krokidolite, Chlorophaeite, Achmite, Arfvedsonite, Knebelite, Polylite, Nontronite, Commingtonite, Pyrosmalite.*

Iron, Hydrous Bisilicate. *Sideroschisolite.* Spec. grav. 3·, H 2·5. FeO 75·5, SiO_3 16·3, Al_2O_3 4·1, HO 7·3. *Form.* 6FeO SiO_3 3 HO? Velvet-black microscopic 4-hedrons. Found at Conghonas do Campo, Brazil. Soluble in chlorohydric acid.

Iron, Anhydrous Silicate of. SiO_3 29·6, FeO 68·73, MnO 1·78. *Form.* 3FeO $5SiO_3$ or FeO, $2\frac{1}{2}$ SiO_3. Spec. grav. 3·8846, H 4. Dark brown foliated obscure 4-sided prisms, attracted by the magnet. Soluble in chlorohydric acid, leaving silica; B. B. infusible, in the reducing flame becomes metallic looking, and like magnetic iron ore. Slavcarrach, Morne (Doran).

Iron, Titaniate of. See ISERINE.

Iron, Magnesia Carbonate of. 2(FeO CO_2) $MgOCO_2$. Light yellow rhomboids, from Grande Fosse, Vizille.

Iron, Phosphate of, and Manganese. Spec. grav. 3·562, H 5·25. FeO 31·9, MnO 32·6, PO_5 32·8. *Form.* 2 MnO 2 FeO PO_5. Blackish-brown masses, or with cubic cleavage. Limoges.

Iron, Alumina Sulphate of. *Hair Salt.* SO_3 33·92, FeO 11·23, Al_2O_3 11·48, MgO ·27, HO and KO 43·1 (R.D.T.) *Form.* Al_2O_3 3 SO_3, FeO SO_3 24 HO? Capillary or hair crystals, with a sweet and astringent taste, soluble in water; often mixed with portions of sulphates of iron and magnesia. Found on the surface of alum shale in coal beds.

Iron, Titaniferous Ore of. Spec. grav. 4·488 to 4·787, H 5 to 5·5. Iron-gray masses and rhomboids, texture foliated, opaque, brittle. B.B. infusible, but the edges are rounded. TiO_2 41·, Fe_2O_3 56·2, SiO_3 2·5. Arendal, Brazil, Bohemia, North America.

Isamic Acid. See IMASATIC ACID.

Isatane. $C_{16}H_6NO_3$. Insoluble white powder formed by the action of bisulphite of ammonia on sulphesatyde; when heated it yields isatine and indine.

Isatic Acid. *Isatinic Acid.* $C_{16}H_6NO_5$ HO. White powder, obtained by dissolving isatine in potash.

Isatilime. $C_{48}H_{16}N_4O_{10}$. Amorphous flocks by evaporating the liquid from which isatimide deposits.

Isatimide. $C_{48}H_{17}N_5O_8$. Yellow 4-sided plates, insoluble in water, alcohol, and ether; obtained by passing dry ammonia over isatine moistened with absolute alcohol.

Isatine. $C_{16}H_5NO_4$. Beautiful, splendent reddish-brown hexagonal prisms, with a rhombic base, with an orange-red powder. It may be obtained in fine crystals by dissolving it in caustic potash, with some alcohol added, and precipitating it while hot by chlorohydric acid, and evaporating; volatile; slightly soluble in cold water, more soluble in hot water, soluble in boiling alcohol and ether, which have no action on litmus. Isatine unites with bases, forming peculiar combinations. It is formed by adding common nitric acid to powdered indigo, and applying heat until the blue colour disappears and a yellow solution is formed; or by heating indigo with sulphuric acid and bichromate of potash. It differs from indigo in containing two additional atoms of oxygen, derived from the oxidizing agent.

Isato-sulphurous Acid. $C_{16}H_5NO_4$ 2 SO_2HO. By acting on isatine, or isatates with sulphuric acid.

Isatyde. $C_{16}H_6NO_4$. Gray crystalline

powder, by acting on a solution of isatine in alcohol, with bisulphohydride of ammonia.

Ischelite, or Polyhalite.

Iserine. *Titaniate of Iron. Gregorite. Menachanite.* TiO_2 57·187, FeO 39·78. Spec. grav. 4·427, 4·491, and 4·65. It varies considerably in constitution. Grayish-black angular grains, imperfectly lamellar, lustre glistening, between adamantine and metallic, opaque, yields to the knife, not attracted by the magnet, but is often mixed with magnetic iron sand. B.B. fuses into a blackish-glass, slightly attracted by the magnet; with borax forms a green bead, becoming brown on cooling. Found at Menaccan, Cornwall; river Don, Aberdeen; Seacombe Ferry; Liverpool; the Iser, in the Riesengebirge.

Isethionic Acid. $C_4H_5O\ 2SO_3$. By the action of anhydrous sulphuric acid on alcohol or olefiant gas; isomeric with sulphovinic acid.

Isinglass. A form of nearly pure gelatine, from the sounds of certain species of sturgeon, found in the Danube and Russian rivers. 1 part dissolved in 100 hot water becomes a jelly on cooling; but 1 part in 150 water does not solidify, though the solution is somewhat gelatinous. It is soluble in rectified spirits, by which it is distinguished from colline.

Isodimorphism. A term applied to designate such bodies as are both dimorphous, but isomorphous in both modifications (SbO_3 and AsO_3). It is also used to indicate the property of a substance to crystallize in two forms, which closely resemble each other, but belong to different systems.

Isolusine. A bitter resin from the root of Polygala senega, amara, &c.

Isomerism. (ἴσος, same; μέρος, part). A term first applied by mistake by Berzelius to explain the circumstance, discovered by Clark, that phosphate of soda gives a yellow precipitate with salts before, and a white precipitate after ignition; the cause of which has been demonstrated by Graham. But the expression is now used to classify such bodies as have the same chemical composition in 100 parts, but different characters. Thus, gum and sugar are isomeric, but have totally different characters. Gum is not sweet, and does not crystallize and ferment, two characters possessed by sugar. Many oils have the same composition, as oil of turpentine, $C_{10}H_8$, but have an agreeable odour and different properties. There are different varieties of isomerism, as when the formulæ of one body is a multiple by 1, 2, or 3, &c. of the formula of another, that body is said to be *polymeric.* Thus, cyanic acid, which is CyO, cyanuric acid, Cy_2O_2, and fulminic acid, Cy_3O_3, are polymeric bodies. *Metameric bodies* have the same per centage composition, the same empirical formula, but a different rational formula, the constituents being arranged in a different manner. Cyanate of ammonia $NH_3\ HO$, $CyO = NH_3O\ HO\ C_2\ NO$, has the same empirical formula as urea, but the elements are arranged differently ($C_2H_4N_2O_2$).

Isomeromorphism is exemplified when in two compounds the same atoms are arranged in the same proportions in the same form, but certain atoms of the one compound being found, when compared with the other compound, to have changed places, as in bichlorohydrate of bibromated cinchonine.

Isomethionic Acid. $HO,C_2H_3O_2,S_2O_5$, formed by passing the vapour of solid sulphuric acid into pyroxylic spirit; forms rhomboidal crystals with barytes.

Isomorphism. (ἴσον, the same; μορφή, shape). Indicates those cases in which different elements, or compound bodies, assume the same crystalline form, with the same number of atoms in the compound, one element being replaced by another. Thus arsenic acid (AsO_5), and phosphoric acid (PO_5), are isomorphous. They have each 6 atoms of matter present; they unite with soda in the same proportion, and form compounds with the same number of atoms, $2\ NaO\ HO\ PO_5\ 24\ HO$, phosphate of soda; $2\ NaO\ HO\ AsO_5\ 24\ HO$, arseniate of soda, and the same crystalline form, or oblique rhombic prisms. And so decided is the tendency of these two acids to assume the same forms, that they may both crystallize together in various quantities, in the same salt, without change of shape; hence the difficulty of obtaining an arseniate free from a phosphate, if both acids have access to the same solution. We have thus afforded to us an explanation of the presence of many impurities in what appear to be physically pure salts. Alum, a familiar salt of great external purity, is seldom free from protoxide of iron, in consequence of its isomorphous tendencies with potash. The following minerals have the same number of atoms entering into their composition, with a different atomic volume, and their angles nearly the same, but not identical:—

		Vertical.	Horizontal.	Atom. Vol.
Carbonate of Barytes,	$BaOCO_2$	118° 30′	106° 50′	13
" Strontian,	$SrOCO_2$	117° 16′	108° 12′	11⅓
" Lead,	$PbOCO_2$	117° 14′	108° 13′	11⅓
Arragonite,	$CaOCO_2$	116° 16′	108° 27′	10

In the subsequent table of minerals belonging to the rhombohedral or hexagonal system, the number of atoms is the same, while the value of the angles slightly differs:—

			Atom. Vol.
Carbonate of Lime, ...	$CaOCO_2$	105° 5′	10½
Magnesia, carbonate of Lime, ...	$CaOCO_2MgOCO_2$	106° 15′	9
Carbonate of Manganese, ...	$MnOCO_2$	106° 51′	9
,, Iron, ...	$FeOCO_2$	107° 0′	8½
,, Magnesia, ...	$MgOCO_2$	107° 25′	8
,, Zinc, ...	$ZnOCO_2$	107° 40′	8

Complete identity in the measurement of the angles appears to pertain only to those isomorphous bodies which belong to the regular system. The following classes include a number of the elements:—*Regular system.*—Carbon, phosphorus, titanium, potassium, sodium, iron, copper, silver, cadmium, zinc, lead, tin, gold, platinum, palladium, iridium, manganese, calcium, lithium, ammonium, nickel, cobalt, osmium. *Rhombohedral system.*—Carbon, arsenic, antimony, tellurium, osmium, iridium, palladium, bismuth. *Right rhombic*, or *Prismatic system.*—Sulphur, selenium, chlorine, iodine, bromine, fluorine, arsenic? oxygen. Kopp and Schrœder have endeavoured to work out the relation of the atomic volume, (the diameter of the atom of Dalton), or relative size of the different atoms of bodies to their isomorphism, and have succeeded in pointing out some remarkable connections. The differences in certain particulars have given rise to the formation of a number of sub-species of isomorphism, of which the following is an example. There are three peculiarities which are usually compared:—

	Crystalline Form.	Atomic Formula.	Atom. Vol.
		Isomorphism.	
Isotomism, ...	Equal,	Equal,	Equal.
Homo-omorphism,	Equal,	Equal,	Unequal.
	Equal,	Unequal,	Equal.
	Equal,	Unequal,	Unequal.
		Anisomorphism.	
Heteromorphism, ..	Unequal,	Equal,	Equal.
	Unequal,	Equal,	Unequal.
	Unequal,	Unequal,	Equal.
Anisotomism, ...	Unequal,	Unequal,	Unequal.

Isomorphism Polymeric. Simple isomorphism indicates the case in which 1 atom is replaced by 1 other atom. Polymeric isomorphism is exemplified when 1 atom is replaced by a number of other atoms, and is illustrated in mineralogy. It is obvious that deductions of this kind must be received with caution. It is but justice to Scheerer, however, to observe that his views on this subject are very plausible. 1. He has endeavoured to show that 1 atom of the magnesian group of metals, including MgO, MnO, FeO, CaO, NiO, &c. may be replaced by 3 atoms water. Thus *Cordierite* and *Aspasiolite* have many of the same physical characters, but are different in other respects. Their formulæ are as follows:—

Cordierite, 3RO (= FeO, MnO, CaO) $2SiO_3$ + $3Al_2O_3SiO_3$.
Aspasiolite, 3RO (= 3HO) $2SiO_3$ + $3Al_2O_3SiO_3$.

In cordierite the protoxides of the radical (RO) are mixtures of FeO, MnO, CaO isomorphous protoxides, while in aspasiolite these are replaced by water. 2. He has given an example where 1 atom oxide of copper is replaced by 2 atoms water, as in olivinite and libethenite. 3. 2 atoms silica may replace 3 atoms alumina, as in some varieties of hornblende. In Babingtonite, which seems to be a species of hornblende, I found 6·48 per cent. of alumina, and 47·46 silica; and Arppe found only ·3 alumina, with 54·4 silica. Calculating the alumina as replacing the silica, the formula, 3RO SiO_3, 3RO $2SiO_3$, is the same in both analyses.

Isopyre. (ἴσος, equal; πῦρ, fire.) Spec. grav. 2·912, H 6, 6·5, SiO_3 47·09, Al_2O_3 13·91, CaO 15·43, Fe_2O_3 20·07, CuO 1·94. Colour velvet-black with red spots like heliotrope, fracture conchoidal, lustre vitreous, streak light greenish-gray; resembles obsidian and opal; found embedded in granite in Cornwall in small amorphous pieces. B.B. fuses into a magnetic bead; not dissolved by acids, but decomposed by fusion with alkalies.

Isotartaric Acid. A name applied by Laurent and Gerhardt apparently to Fremy's tartralic acid. They state that when tartaric acid is perfectly fused, *metatartaric acid* is formed, and by longer heat *isotartaric acid* is produced, the neutral salts of which are isomeric with the acid tartrates. The isotartrate of lime is $CaOC_8H_5O_{11}$, dried at 320°.

Isotartridic Acid. $C_8H_4O_{10}$. Probably the *tartrelic acid* of Fremy, by subjecting tartaric acid to a higher temperature than for the preceding acid. The composition of its salts is $MOC_8H_3O_9$.

Isoterebenthene. $C_{20}H_{16}$. Spec. grav. ·8432; B.P. 350°. Colourless fluid; strongly refracting light; rotating light to the left; obtained by heating English essence of turpentine to 570° for two hours, and distilling the product.

Isotungstic Acid. RO, Tn_2O_6. Formed by boiling metatungstate of ammonia with excess of ammonia and separating by an acid; slightly soluble in water.

Itacanilic Acid. $C_{22}H_{11}NO_6$; F.P. 372¼°. Colourless needles by acting on itacic acid with aniline.

Itacanilide. $C_{17}H_8NO_2$; F.P. 365°. Crystalline mass by heating aniline and hydrate of itacic acid in a retort to above 360°; it remains in the retort, and may be crystallized out of alcohol.

Itacic Acid. *Itaconic*, *Pyroaconitic*, *Pyro-*

citric, Citricic Acid. HO $C_5H_2O_3$; F.P. 320°. Colourless 8-hedrons with a rhombic base, soluble in hot water and alcohol; becomes citracic acid when heated to its subliming point; obtained by heating aconitic acid or citric acid to the melting and decomposing point (356 to 392°), in a retort. An oily liquid distils over, sinking to the bottom of water, and becoming crystalline on standing; it is monobasic.

Itacolumite. A peculiar species of elastic mica slate from Itacolumi in Brazil.

Ittnerite, or Hauyne.

Ivory, Animal. See TEETH.

Ivory Black, or animal charcoal

Ivory, Vegetable. The fruit of the *Phytelephas macrocarpa*, or *Taqua nut.* A South American palm resembling ivory in appearance, consisting of cellulose 81·34, HO 9·37, gum 6·73, caseine 3·8, oil ·73, albumen ·42, ash ·61.

Ixolyte, or mountain tallow, resembling hartite; softens at 169°, and is tenacious at 212°.

J

Jade, Common. A green variety of nephrite or serpentine.

Jade, Oriental, or compact tremolite.

Jade, Tenacious. *Jade*, or *Saussurite.*

Jalapine. *Rhodeoretine.* $C_{42}H_{35}O_{20}$. A resin extracted from the root of Jalap by alcohol.

Jamaicine. Yellow bitter square prisms, from Geoffroeya inermis.

Jamesonite. Spec. grav. 5·564 to 5·8, H 2 to 2·5. Steel-gray 4-sided right rhombic prisms of 101° 20′; lustre metallic, opaque, sectile.

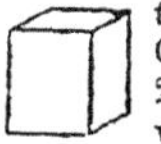

It consists of S 22·34, Pb 40, Cu 14, Fe 2·64, Sb 34·26 = 3PbS $2Sb_4S_3$. Soluble in chlorohydric acid, with deposition of chloride of lead; fuses B.B. on charcoal, giving out antimonial vapour.

James's Powder. A mixture of antimonious acid and bone earth, sometimes used in medicine, and acting on the infinitesimal dose system, as it has no physiological influence. It consists of antimonious acid 38, bone earth 62.

Japan. A preparation for varnishing metallic and other vessels. The common brown japan is composed of 1 gal. linseed oil boiled, 1 lb. umber, and afterwards adding about ½ of a gallon of oil of turpentine.

Japan Wax. $C_{53}H_{55}O_{10}$? Spec. grav. ·97; F.P. 104°. Whitish-yellow; soluble in hot alcohol and ether, and deposited on cooling; saponifies with soda, and yields an acid apparently peculiar (T. Thomson).

Japan Wood, or *Red Wood.*

Japonic Acid. $C_{12}H_4O_4$. Flocky precipitate of a blackish-brown colour, a black mass on drying; when moist reddens litmus, and is slightly soluble in water; insoluble when dry; insoluble in alcohol; forms dark solutions with alkalies, and not precipitated by acetic acid; obtained by dissolving catechuic acid in caustic potash, and exposing it to the air for some days, and then adding chlorohydric acid.

Jargon. See ZIRCON.

Jasmine, Oil of. Yellowish oil from the flowers of *Jasminum officinale;* deposits at 42° a stearoptene or camphor in plates, which is lighter than water, and fuses at 257°; easily dissolved by ether, alcohol, and oils.

Jasper. *Sinople, Ribbon Jasper, Egyptian Jasper.* Spec. grav. 2·6. A variety of quartz, with a deep brown, yellow, or red colour; lustre resinous. The brown-coloured jasper is termed *Egyptian*, from its abundance in that country. When stripes of green, yellow, and red occur on the same mineral, it is called *striped jasper. Jasper agate* is reddish-white. Jasper seems to be coloured by iron.

Jatrophic Acid, or *Crotonic Acid.*

Jeffersonite. Spec. grav. 3·51 to 3·55, H 4·5. Dark olive-green rhomboidal prisms, lustre resinous, semi-metallic; streak gray, powder light green. B.B. fuses into a dark globule. SiO_3 56, CaO 15·1, MnO 13·5, Fe_2O_3 10, ZnO 1·, Al_2O_3 2, HO 1. Found at Franklin, New Jersey; allied to pyroxene.

Jelly, Animal. See GELATINE, GLUE, and ISINGLASS.

Jelly, Vegetable. The familiar name for pectic acid, as exhibited in the preserves made from gooseberries, currants, apples, oranges, grapes, &c.

Jervine. $C_{60}H_{45}N_2O_5$. White crystals, soluble in alcohol, insoluble in water, fusing into an oil when heated along with water; not decomposed at 374°, but beginning to decompose at 392°. The salts are soluble in acetic acid, and again precipitated by alkalies; obtained from the root of the *Veratrum album*, by extracting with water containing chlorohydric acid, and precipitating the acid solution with pure carbonate of soda. The precipitate, consisting of jervine and veratrine, is washed with water, dissolved in alcohol, and crystallized out. It is then combined with sulphuric acid. The sulphate of jervine separates from the solution, while the salt of veratrine remains dissolved. The jervine is then separated by carbonate of soda.

Jet. A variety of bituminous coal, admitting of a fine polish.

Jewellers' Gold. An alloy of gold with copper and silver, in various proportions. Trinket gold consists of 75 gold, 25 copper.

Johannite, or Subsulphate of Uranium.

Johnite, or Turquoise.

Julep. (*Julap*, Arab.) A solution of sugar in an aromatic water.

Junkerite, or *Prismatic Carbonate of Iron.* Spec. grav. 3·815, H 3·5. Yellowish-gray right oblique rhombic prisms, differing from common carbonate, which occurs in obtuse rhomboids. FeO 53·6, CO_2 33·5, SiO_3 8·1, MgO 3·7. B.B. fuses into a yellowish-green glass. Found in the pit Koenig, at Poullaouen mine.

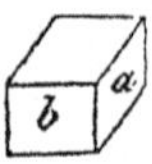

Juniper, Oil of. A volatile oil from *Juniperus communis.* It consists of two oils mixed together, formed on the turpentine type ($C_{10}H_8$); one oil, with spec. grav. ·847, boils at 318°; the other has a specific gravity of ·868, and boils at 536°. They are separated by distillation.

Jurinite. A synonyme of Brookite.

K

Kakodyle. See CACODYLE.

Kakoxene. See CACOXENITE.

Kalaite. See TURQUOISE.

Kalamite. A species of amphibole.

Kaliphite. Spec. grav. 2·8, H 1·. Fibrous brown mineral, apparently a mixture containing iron and manganese, zinc, silica, &c.

Kalium. The Latin name for potassium.

Kammererite. Spec. grav. 2·76, H 1·75. Violet-red 6-sided prisms; lustre pearly; translucent, sectile, and flexible. B.B. infusible, allied to serpentine. SiO_3 37·, MgO 31·5, CaO 1·5, FeO 1·5, Al_2O_3 14·2, HO 13·, Cr_2O_3 1·.

Kaneelstein. A synonyme of garnet.

Kaolin. The Chinese term for porcelain clay.

Karabe of Sodom. Asphalt from the Dead Sea.

Karpholite. (Strawstone). Spec. grav. 2·935, 2·923, 2·9365, H 2·5. Straw-yellow tufted minute diverging crystals; lustre silky, glistening, opaque. B.B. becomes white, and fuses imperfectly; with soda into a green enamel; not acted on by chlorohydric acid. SiO_3 36·154, Al_2O_3 28·669, MnO 19·16, FeO 2·29, CaO ·271, F 1·47, HO 10·78. Found in granite at Schlaggenwalde, Bohemia.

Karphosiderite. Sp. grav. 2·5, H 4. Yellow resinous mineral, found in mica slate on the coast of Labrador. Probably a subsesquiphosphate of iron.

Karstenite. See Anhydrite.

Keaves. (Cuves). Vessels employed to contain salts while exhausting them of their soluble parts by water.

Keilhauite. *Yttrotitanite.* Sp. grav. 3·69, H 6·5. Brownish-black mass, with a resinous and vitreous lustre. B.B. fuses into a black scoria swelling up, with soda into a bead coloured by manganese. SiO_3 30·, TiO_2 29·01, Fe_2O_3 6·35, Al_2O_3 6·09, Mn_2O_3 ·67, Ce_2O_3 ·32, CaO 18·92. In felspar, near Arendal, Norway.

Kelp. Kelp is the commercial name of the ash obtained by the incineration of sea weeds. It is valuable from the potash salts and iodine which it contains. The largest amount of iodine is obtained from drift weed kelp. The drift weed is so named from its being drifted ashore, from deep water in which it grows, and is thus easily secured by the natives of Ireland, and the Hebrides. The cut weed is obtained at low water, and is cut by means of sickles. The plants are spread out thinly in the rays of the sun for two days, and are then brought to the kiln, which consists of a parallelogram of about 15 feet long and 2 feet wide, surrounded with a rough wall of stones about 9 inches high. The burning lasts for twelve hours. The kiln is kindled with a layer of dry heath or straw, and overlaid with a stratum of dry weed,—a succession of layers of weed is thus burned. When the operation is completed, kelp presents the appearance of a dark-coloured fused mass of salts; the dark colour being due to portions of charcoal interspersed through the mass. The greater portion of kelp is made from the *Fucus nodosus, F. serratus, F. vesiculosus, Laminaria digitata, L. saccharina, Halidrys siliquosa, Alaria esculenta, Rhodomenia palmata.* The best analysis of kelp which has been made so far as I know, is a very careful one executed under my eye by my pupil Mr. George William Brown. The amount of iodine present by three trials was respectively ·283, ·306, ·287 = mean ·292 per cent. (Proc. Phil. Soc. Glasgow, vol. iii.)

Table of per centage Composition of Orkney Kelp.

Insoluble Salts.		
Carbonate of lime,..................	2·591	
Phosphate of lime,..................	10·556	
Oxysulphuret of calcium, 3 CaS, CaO,..............................	1·093	
Silicate of lime,..........................	3·824	
Carbonate of magnesia,............	6·554	
Sand,..	1·575	
Alumina,......................................	·142	
Carbon,.......................................	·920	
Hydrogen,....................................	·144	
Oxygen,......................................	1·152	
Nitrogen,.....................................	·658	29·209
Soluble Salts.		
Sulphate of potash,..................	4·527	
Sulphate of soda,......................	3·600	
Sulphate of lime,........................	·279	
Sulphate of magnesia,...............	·924	
Sulphite of soda,.........................	·784	
Hyposulphite of soda,..................	·220	
Sulphide of sodium,....................	1·651	
Phosphate of soda,......................	·540	
Carbonate of soda,......................	5·306	

Chloride of potassium,	26·491	
Chloride of sodium,	19·334	
Chloride of calcium,	·229	
Iodide of magnesium,	·316	
Bromide of magnesium,	trace.	
Water,	6·800	71·000
		100·209

Keramohalite, or Sulphate of Alumina.

Kerargyre, or Chloride of Silver.

Kerasite. *Kerasine.* Chlorocarbonate of lead.

Keratine. The basis of horns, epidermis, and epithelium, &c. It appears to contain more oxygen than albumen or fibrine.

Keratophyllite. A variety of Amphibole.

Kermes. *Grana Kermes.* An insect from Quercus coccifera, containing a colouring matter similar to cochineal.

Kermes Mineral. S 30·93, Sb 63·28, K 5·79. A brick-red powder, formed by just fusing 1 carbonate of potash and 2⅔ tersulphide of antimony, boiling with water and filtering, when the kermes separates on cooling; formerly used in medicine.

Kerolite. *Cerolite.* Spec. grav. 2 to 2·2, H 2·25. White or green mineral, transparent, translucent, feel greasy, vitreous or resinous, not adherent to the tongue. B.B. blackens but does not fuse. SiO_3 46·96, MgO 31·26, HO 21·22. Found in Silesia and Zöblitz along with serpentine.

Ketones, or *Cetones.* A term sometimes applied to a class of bodies of which acetone may be taken as the type. They are formed by distilling organic acids with a base, such as lime, which removes the elements of carbonic acid from them. Staedeler considers them to be all ether combinations. If we view acetic acid as a compound of methyle, like formic acid, in which 1 atom hydrogen is replaced by methyle, C_2 (C_2 H_3, H) O_4, then acetate of potash will be C_2 (C_2H_3K) O_4. When heated, carbonic acid is evolved, and potassium methyle formed, a compound similar to zinc methyle (2 CO_2, KC_2H_3), which, with a second atom of acetate, is transformed into potash and acetone, K, C_2H_3 and C_2 (K, C_2H_2) O_4 = 2 KO and C_2 2(C_2H_3) O_2 = acetone. In the same way all the cetones may be viewed as ether combinations. Propione is C_2 2 $C_4H_5O_2$; butyrone C_2 2 $C_6H_7O_2$; valerone C_2 2 C_8H_9, O_2, and even butyral is probably C_2, 2 C_2H_3,O_2. When acetone is boiled with dilute alkalies, *acetonic acid*, $C_8H_8O_6$, is formed, consisting of acid prisms; monobasic. By a mixture of acetone, sulphide of carbon and ammonia, *carbothiacetonine*, $C_{20}H_{18}N_2S_4$, in yellow prisms is formed, and by the action of sulphohydric acid and ammonia on acetone, *thiacetonine*, $C_{18}H_{19}$ NS_4, identical with akcethine, is produced.

Kieselghur. The siliceous remains of animals deposited in the form of Tripoli or polishing slate, at Bilin, composed of silica 72·, alumina 2·5, Fe_2O_3 2·5, HO 21· (Klaproth). Similar remains I have examined from the island of Raasay, from near Aberdeen, and from Bengal, containing some organic matter.

Killbrickenite. *Geocronite.* Spec. grav. 6·5, H 2·5. 5 PbS, $SbAsS_3$. Metallic lead-gray grains, or earthy masses and crystalline; found at Kilbricken, Clare, and at Sala, Sweden.

Killinite. Specific gravity 2·711, H 3·5. Brownish-yellow, sometimes tinged with green masses, or 4-sided prisms, similar to spodumene; lustre waxy, opaque, streak yellowish-white. B.B. fuses into a white opaque bead, with soda into a glass. SiO_3 49·08, Al_2O_3 30·6, KO 6·72, FeO 2·27, CaO 68, MgO 1·08, HO 10·. Found in granite at Killiney, Dublin.

Kinic Acid. *Quinic Acid.* $C_{14}H_{11}O_{11}$ HO? Colourless crystals, resembling tartaric acid. Spec. grav. 1·637; soluble in 2½ water; acid taste; converts starch into sugar by boiling. It occurs in Peruvian bark in the form of kinate of quinine, and appears during the process for separating that alkaloid, as kinate of lime, in white rhomboidal plates, with angles of 78° and 112°, from which the lime is separated by oxalic acid. It unites with bases, and forms apparently bibasic salts; their formulæ being $C_7H_4O_4$ 2 MO (?) When distilled with dilute sulphuric acid and binoxide of manganese, it yields *kinone* in yellow plates; by adding to kinone sulphuric acid, we obtain *white* and *green hydrokinone.* It is thus easy to detect the presence of kinic acid in a vegetable mixture.

Kino. The commercial name of reddish-brown dry masses, resembling catechu, occurring in four forms—African kino, from *Pterocarpus erinaceus;* Asiatic kino, from *P. marsupium;* Australian kino, from *Eucalyptus resinifera;* American kino, from *Coccoloba uvifera.* It is used in medicine in the form of tincture in diarrhœa, &c. The African is most abundant, and contains 75 per cent. tannic acid, 24 red gum, and 1 lignine. It precipitates salts of sesquioxide of iron, green, and is converted by nitric acid into oxalic acid. Soluble in cold, more soluble in hot water.

Kinone. $C_{25}H_8O_8$. Golden-yellow prisms, fusible and volatile, soluble in water, with a pungent odour. Obtained by distilling kinic acid, or a kinate, with dilute sulphuric acid and binoxide of manganese; the presence of kinic acid is thus easily detected.

Kinovic Acid. *Chinovic Acid.* $C_{38}H_{30}$ O_{10}? White light matter, little soluble in water; easily soluble in alcohol and ether, from which it is precipitated by water. The alkaline kinovates are soluble; it bears a resemblance to stearic acid. Obtained by heating the alcoholic extract of the China nova bark with alcohol and magnesia; colouring matter, &c. precipitate, and kinovate of magnesia dissolves, from which the kinovic acid is precipitated by an acid in white flocks.

Kirwanite. Specific gravity 2·941, H 2·. Colour dark olive-green, texture fibrous, fibres diverging from a centre and forming brushes; opaque. B.B. becomes black, and partially fuses, with fluxes into a dark brown glass. SiO_3 40·5 to 38·9, FeO 23·91, Al_2O_3 11·41, CaO 19·78, HO 4·35. *Form.* 3(2CaO 2 FeO) SiO_3, $Al_2O_3SiO_3$ 2 HO. In basalt, north-east coast of Ireland (R.D.T.)

Klaprothine. A synonyme of Lazulite.

Klinoclase. A synonyme of aphanese, or variety of olivenite.

Knebelite. Specific gravity 3·714. Gray, spotted, dirty-white, red, brown, and green, massive; external surface uneven and full of holes; lustre glistening, opaque, hard, brittle. B.B. infusible; with borax forms a dark olive-green bead. SiO_3 32·5, FeO 32·, MnO 35 (Döbereiner). Locality unknown. 3 FeO SiO_3, 3 MnO SiO_3.

Kobellite. Spec. grav. 6·29 to 6·32. Shining leaden metallic masses, along with cobalt ore, at Hvena, Sweden. S 17·86, Sb 9·24, Bi 27·05, Pb 40·12, Fe 2·96, Cu 8, rock 1·45 = 3 FeS, 2 SbS_3, 3 PbS, Bi_2S_3.

Koccinic Acid. Yellow crystalline grains by nitric acid on purreic acid.

Koenigite. See BROCHANTITE.

Kollyrite. See COLLYRITE.

Konichalcite. Spec. grav. 4·123, H 4·5. Green kidney-shaped masses. AsO_5 30·68, 4 (CuOCaO) (PO_5AsO_5) PO_5 8·81, VO_3 1·78, CuO 31·76, CaO 21·36, HO 5·61. Cordova, Spain; allied to olivenite.

Konite. An ash-gray dolomite ($CaOCO_2$) 3($MgOCO_2$), from Freiberg, Hesse.

Konlite. C_2H. Spec. grav. ·88; F.P. 237°; 193½°. Soft white crystalline plates and grains, slightly soluble in cold alcohol; more soluble in boiling alcohol and ether. In brown coal. Uznach, Bavaria

Koreite. A synonyme of Agalmatolite.

Koscine. Colourless needles by alcohol from the flowers of *Brayera anthelminthica* (Flores Koso).

Koupholite. A synonyme of Prehnite.

Koumiss. An alcoholic fluid obtained by the fermentation of mare's milk by the Russian Tartars.

Krablite. Spec. grav. 2·389. A variety of felspar from Iceland, containing SiO_3 74·83, Al_2O_3 13·49, Fe_2O_3 4·4, CaO 1·98, MgO ·17, NaO 5·56.

Krisuvigite. See BROCHANTITE.

Kupaphrite, or *Cupaphrite.* See NOVOHYDROUS PENTARSENIATE OF COPPER.

Kuteera, Gum. Occurs in commerce in the form of loose wrinkled drops or pieces without smell or taste, and mostly transparent. In water it slowly forms a pulp or jelly like gum tragacanth, but if pounded well in a mortar, and then boiled in water for fifteen minutes with constant agitation, it is said to be completely dissolved. In India it enters into the composition of some varnishes; in this country it is used by calico printers and veterinary surgeons. It is a product of the *Sterculio urens*, in Hindostan. It seems a variety of *bassorine.*

Kyanite. See CYANITE.

Kymatine. A variety of amphibole.

Kypholite. A variety of serpentine.

Kyrosite. ·32 FeS_2, Cu S or S 52·63, Fe 45·63, Cu 1·69, As ·93. A variety of iron pyrites.

L

Labarracque's Disinfecting Liquor, or hypochlorite of soda, prepared by passing chlorine gas through caustic soda, or boiling carbonate of soda with bleaching powder.

Labdanum, or *Ladanum resin.* $C_{40}H_{33}O_7$? Dark, almost black masses, very soft, with a fra grant odour and bitterish taste; partially soluble in water; obtained by scraping the surface of the leaves of *Cystus Creticus;* in Syria and the Greek islands.

Laboratory (pronounced *labratory*), (*Laboratoire*, Fr.); (*Laborare*, to labour.) The experimental room of the chemist.

Labradorite. *Labradro felspar.* Sp. grav. 2·69 to 2·75, H 6·. Doubly-oblique prisms, resembling albite, of dark ash and smoke-gray; in certain positions reflecting a great variety of colours, as blue, green, yellow, red, and brown, in iridescent forms. Several cargoes of this ore were brought from the neighbourhood of Hudson's bay, by Sir Martin Frobisher, under the impression of its being gold ore; lustre of the sides of the prism vitreous, of the base pearly, translucent in thin pieces. B.B. behaves like felspar. SiO_3 53·48, Al_2O_3 26·46, CaO 9·49, NaO 4·10, KO ·22, FeO 2·69, MgO 1·74, MnO ·89, HO ·42. *Form.* CaO SiO_3,Al_2 O_3SiO_3, the lime being replaced by magnesia and soda. It is found in Labradore; in green stone as at Campsie; Gleniffer, Paisley; Finland; Ingermannia.

Lac. *Gum lac*, *Lac dye*, *Stick lac*, *Shell lac*, *Seed lac.* Stick lac is the commercial name of a resin which flows in the form of a milky juice by the puncture of an insect (*Coccus ficus*), from the *Ficus Indica*, *F. religiosa*, and *Rhamnus jujuba.* Stick lac is the substance in its natural state with a deep red colour, encrusting small twigs. *Seed lac* is brown, and is prepared by breaking off the resin from the twigs, and boiling in water, by which the red colour is removed. *Shell lac* is brown, and is obtained by melting it, and reducing it to the state of a thin crust. The following is the composition of lac in these different states (Hatchett):—

	Stick Lac.	Seed Lac.	Shell Lac.
Resin,....................	68	88·5	90·9
Colouring matter,......	10	2·5	0·5
Wax,....................	6	4·5	4·
Gluten,..................	5·5	2 .	2·8
Foreign bodies,.........	6·5		
Loss,.....................	4·0	2·5	1·8
	100·0	100·	100·

Water dissolves ½ to 15 per cent. colouring matter. Alcohol takes up most of the resin; ether is less powerful; sulphuric acid dissolves and chars lac; nitric acid dissolves and oxidizes it; soluble in an aqueous solution of borax (20 grs. borax, 100 lac, 4 oz. water); this solution, mixed with lamp black, forms an Indian ink, and may be used as a varnish; soluble in potash and soda. The resin of lac consists of five species of resins. It is the principal constituent of sealing wax, and varnishes. The following are the varieties in the characters of the resins:—*a*, A resin soluble in alcohol (·879) and ether, amounting to ½ per cent.; it is brown, soluble in potash, which becomes violet. *b*, Soluble in alcohol of ·860, and of greater strength; insoluble in ether; from alcohol precipitated by water as a jelly; decomposes carbonate of magnesia when in solution in alcohol, and dissolves the magnesia. *c*, Reddish-yellow; soluble in cold and hot ether, alcohol, and oil of turpentine; obtained by boiling the alcoholic solution of lac with carbonate of magnesia, when it precipitates with the magnesia, from which it is separated by acid. *d*, Soft and flexible; fusing at 212°; very soluble in alcohol and ether, potash and ammonia. *e*, May be kneaded in hot water, decomposed by fusion, and changes into a true resin; while cold insoluble in alcohol, ether, and volatile oils; soluble in potash and ammonia.

Lac Dye. The form in which lac is used for dyeing red colours, especially with wool. To prepare the dye tub, 4 parts lac are digested with strong sulphuric acid for twenty-four hours in summer, two days in winter, and then diluted with 32 parts boiling water. The clear liquor is poured into a lead pan, and mixed with the washings from the residue; it is then saturated with lime to the extent of ⅘ the sulphuric acid employed. The liquor poured off is used for dyeing.

Lac Sulphuris. A form of sulphur obtained by boiling with caustic soda or lime, and decomposing the sulphide thus formed with chlorohydric acid. As found in the shops it often consists of sulphate of lime 50·73, sulphur 35·85, water 13·42 (Schweitzer, Murdoch), being then precipitated by sulphuric acid.

Laccine. Brown, brittle, translucent resin, insoluble in alcohol, not altered at 212°; insoluble in oils; soluble in hot potash; obtained by exhausting lac with alcohol; the residue is laccine; it may then be dissolved in alcohol mixed with some chlorohydric acid; water is added to this solution, the alcohol distilled off, and the laccine remains.

Lackmus. German name for Litmus.

Lacquer. A solution of lac in alcohol, and coloured with saffron, or some vegetable colour.

Lacsatine. A crystalline body by alcohol, from the leaves of *Lactucca sativa.*

Lactamic Acid. $HO\,C_{12}H_{12}NO_9$. Obtained in union with ammonia (lactamide); by acids, with dry ammonia, on anhydrous lactic acid.

Lactamide. $C_6H_7N\,O_4 = C_6H_5O_4, NH_2$. Colourless rectangular prisms, isomeric with urethane, alanine, and sarcosine; soluble in water; decomposed by weak acids, alkalies, and heat into lactic acid and ammonia. Obtained by treating lactide with dry ammonia gas or by boiling lactamate of ammonia with water, and crystallizing from alcohol.

Lactarine. A term which has been applied by my friend and pupil, Mr. Robert Thomson Pattison, to a preparation of caseine from milk, discovered and patented by him, and now in extensive use among calico printers. I am indebted to him for the following account of it:—Lactarine is pure caseine, made by adding an acid to milk, and precipitating the curd, which is purified, dried, and ground into a powder. A solution of this lactarine, in an alkali mixed with any pigment colour, and printed upon cloth and then exposed to the action of steam, is fixed so as to resist the action of water. The theory is, that caseine combines with oxygen at a certain temperature, forming a compound insoluble in water. The first idea of this invention was from the fact of observing the film formed upon the surface of boiling milk. Lactarine is more completely fixed so as to resist the action of soap (as in washing), by the addition of an alkaline earth to the lactarine, with which it forms an insoluble compound in alkalies. This principle has been applied by winching calico in a solution of lime water for some time, and afterwards, when dry, printing on a solution of lactarine in ammonia mixed with the pigment wished to be fixed. The ammonia is driven off during steaming, and the lime takes its place, forming a compound insoluble in boiling soap and water. Lactarine is chiefly used to fix ultramarine and other beautiful pigments where they are desired in combination with other colours; the ultramarine pigment gives a colour especially brilliant to all combinations with other colours, such as cannot be obtained by the use of any other blue. This blue is used by almost the whole of the calico printing trade, in preference to the Prussian blue. The following are the best receipts for using lactarine.:—

For Block Printing.—Mix up 1¼ lb. lactarine in 4 pints water, and then add half a pint ammonia; this will soon dissolve the lactarine. Now add the pigment mixed in water, and make up to one gallon.

For Machine Printing.—Mix up $1\frac{1}{2}$ lb. lactarine in 4 pints water, then add 1 pint caustic soda at 50 Twaddell (1·25); when dissolved add the pigment mixed in water; then make up to one gallon. Steam as usual.

Lactate of Zinc. ZnO $C_6H_5O_5$ 3HO. 4-sided truncated prisms; soluble in 58 cold, 6 boiling water; almost insoluble in hot and cold alcohol; loses all its water at 212°; it is not altered at 464°.

Paralactate of Zinc.—ZnO $C_6H_5O_5$ 2 HO. Thin plates; it loses its crystalline water (12·9 per cent.) at 212°, and begins to decompose at 302°; soluble in 2·88 boiling, 5·7 cold water, in 2·23 cold alcohol. This is the form of salt obtained from animal fluids. Hence the animal lactic acid has been termed *paralactic acid.*

Lactic Acid. *Nanceic Acid. Zumic Acid.* $C_6H_5O_5HO$. Specific gravity 1·215. A clear syrupy fluid, becoming anhydrous at 266° to 284°. At 464° it yields carbonic acid, carbonic oxide, lactide and lactone, aldehyde, and citracic acid. It is still fluid at —11°. Lactic acid in its pure concentrated state, is a colourless fluid, of the consistence of syrup, without smell, of a strong acid taste; it absorbs water from the air; mixes with water and alcohol in all proportions; very soluble in ether, which removes it from its aqueous solutions; when boiled with water a portion is carried over with the steam; it displaces acetic acid in acetates, and chlorine from chloride of calcium, but not from common salt; it dissolves precipitated triphosphate of lime and coagulated albumen. The following test has been given for lactic acid (Pelouze):—Lactate of copper is not completely precipitated by an excess of milk of lime; hence, if lactic acid is converted into lactate of copper, and an excess of milk of lime added, on filtering the liquor a copper-coloured solution should pass through. Many substances interfere with this action (Strecker). It is obtained by dissolving 6 lbs. cane sugar, $\frac{1}{2}$ oz. tartaric acid, in 26 lbs. boiling water, which are allowed to stand for two days, when the cane passes into starch sugar. 4 oz. of decayed cheese are then added, diffused in 8 lbs. of acid skimmed milk. 3 lbs. of powdered chalk are mixed with it, and the whole allowed to stand at a temperature of 86° to 95°. In eight or ten days the whole becomes a thick paste of lactate of lime, which is to be dissolved in 20 lbs. boiling water, with $\frac{1}{2}$ oz. quicklime, filtered and evaporated to a syrup. In four days the lactate of lime becomes granular crystals. It is pressed three or four times with one-tenth of its weight of cold water; dissolved in twice its weight of boiling water; and for each pound of lactate of lime are added $3\frac{1}{2}$ ounces of sulphuric acid, diluted previously with its own weight of water. The whole is filtered; the filtered solution containing the free lactic acid is boiled for a quarter of an hour with $1\frac{3}{8}$ ounce carbonate of zinc for every ounce of sulphuric acid. Lactate of zinc deposits from the filtered liquor in crystalline crusts, which is to be washed with cold water, and then dissolved in $7\frac{1}{2}$ times its weight of boiling water. The zinc is precipitated by sulphohydric acid gas; the liquor filtered, and evaporated in a water bath. The acid may be further purified by solution in ether. 2. To procure lactic acid from a fluid containing it:—The fluid is evaporated almost to dryness in the water bath, the residue taken up by alcohol (·820); the filtered fluid decomposed by an alcoholic solution of tartaric acid, as long as a precipitate falls. The alcoholic solution is digested with carbonate of lead, until some oxide of lead has dissolved; the liquid is filtered, the alcohol evaporated, and the residue saturated with hydrate of barytes. The solution of lactate of barytes is decomposed by sulphate of zinc as long as a precipitate falls, and evaporated, crystallized, and decomposed by sulphohydric acid gas 3. Another method is to evaporate the lactic or lactate fluid in the water bath; dissolve the residue in alcohol, and evaporate; decompose the residue with dilute sulphuric acid; add 3 to 4 times its volume of alcohol; add to the solution ether as long as a turbidness occurs; filter, distil off the alcohol and ether, and evaporate in the water bath to the consistence of a syrup. To this syrup add $\frac{1}{2}$ volume alcohol and 5 volumes ether; pour off the solution, and evaporate. It may be further purified by uniting it with oxide of zinc, as above. It occurs in sour milk, in decomposing animal and vegetable fluids, in the yolk of egg, in the human stomach, in the oxidation of alanine by nitric acid, in the decomposition of caseine and gluten, in sauerkraut.

Lactic Fermentation. The term applied to the formation of lactic acid from sugar by the agency of decomposing caseine or cheese. 1 atom lactine or sugar of milk is in this case split into 2 atoms lactic acid or $C_{12}H_{12}O_{12} = 2C_6H_5O_5$ 2HO. It is believed that no other substance is produced.

Lactide. $C_6H_4O_4$; F.P. $224\frac{1}{2}$°; B.P 482°. White rhombic tables, without taste and smell; sublimes by heat above 248°; obtained by heating lactic acid at 280°, as long as water passes over, and then to 484° to 500°, and evaporating the distilled fluid in the water bath; lactide separates on cooling in crystals. It is lactic acid deprived of water.

Lactidic Acid. *Anhydrous Lactic Acid.* $C_{12}H_5O_5$, Yellow mass, with bitter taste, fusing below 212°; converted in moist air and alkalies into lactic acid; dissolves by boiling with water; soluble in alcohol and ether; obtained by distilling lactic acid at 266°, or rapidly at 356° to 392°.

Lactine. (*Lactose*, Fr.; *Milk Sugar; Milch Zucker*, Ger.) $C_{12}H_{12}O_{12}$. Spec. grav. 1·534. Colourless 4-sided prisms terminated by 4-sided pyramids; soluble in 5 to 6 parts cold, $2\frac{1}{2}$ hot water; insoluble in cold alcohol and ether. At

212° it is not altered; at 266° loses 12 per cent. or 2 atoms water, when it is colourless; at 302° it becomes a brown extract. An aqueous solution turns the plane of polarization to the right, 201·9 parts in this respect being equal to 164·7 cane sugar; it unites with bases, as potash, soda, lead, &c.; it absorbs $12\frac{1}{2}$ per cent. ammonia; soluble in milk of lime; the addition of alcohol renders the solution turbid, and precipitates in strong solutions a white mass, which, when washed with alcohol and dried over sulphuric acid, contains $11\frac{1}{4}$ to $15\frac{5}{8}$ lime per cent. It is obtained from whey by evaporating to the consistence of syrup, and allowing it to stand, when the sugar separates in crusts, and may be purified from water by animal charcoal and recrystallization; when fermented it is resolved into carbonic acid and alcohol; converted by dilute acids into grape sugar, and by nitric acid into mucic acid.

Lactometer. An instrument for measuring the value of milk.

Lactone. $C_{10}H_8O_4 = 2(C_6H_5O_5$, 2 CO_2 2HO. B.P. $197\frac{1}{2}$. Colourless oily fluid, becoming dark in the air, with burning taste and aromatic odour, lighter than water, and very soluble in it; burns with a blue flame; obtained by distilling lactic acid at 248°, after being heated to 482°. On mixing the distillate with water, a light oil separates, which is to be rectified over chloride of calcium.

Lactucarium. The dried milky juice of different species of Lactuca, as *L. sativa*, *elongata*, *virosa*. It is obtained by piercing the stem of the plant about the time it flowers, and collecting the juice. When dry it is a dark substance softening in hot water, narcotic, with a bitter taste; only partially soluble in water, spirit, and ether. It consists of 42·6 lactucone, 3·98 wax, 2 lignine, 6·98 albumen, 27·68 lactucine, 14·96 watery extract, 1·75 lactucine.

Lactucic Acid. Perhaps a mixture of lactucine and ether, obtained as a bitter amorphous matter by precipitating the juice of Lactuca virosa with acetate of lead, and decomposing the precipitate with sulphohydric acid gas.

Lactucine. Pearly scales like boracic acid from water, or needles from ether; it is without smell, with a bitter taste. Obtained by the process for lactucic acid.

Lactucone. *Lactucerine.* $C_{40}H_{32}O_3$. F.P. 300° to 400°. Colourless 4-sided prisms in the form of warts, without taste, smell, and physiological action; soluble in alcohol, ether, and fat oils, scarcely so in water; sublimes in a current of carbonic acid; potash and chlorine have no action on it; it differs from betuline by an atom of hydrogen. Obtained by exhausting lactucarium with hot alcohol, and treating the resulting deposit with alcohol and animal charcoal.

Lagonite. Supposed borate of iron.

Lakes. A name applied to vegetable colouring matters united with metallic or earthy oxides, as alumina, tin, or zinc oxides, and used as pigments. They are prepared by adding to a solution of alum or tin salts a solution of colouring matter, and then enough of caustic soda to precipitate the lake. *Orange lake* consists of alumina and turmeric; *red lake* of cochineal and alumina; *madder lake* of madder and alumina; *blue lakes* of ultramarine, indigo, cobalt blue, &c. and alumina.

Lamas Poison, from Peruvian trees, probably *Tabernœmontana*, *Capsicum*, &c.

Lamp. The lamps used in chemical manipulation are—1 and 2, spirit-lamp; 3, argand lamp, with a double current of air.

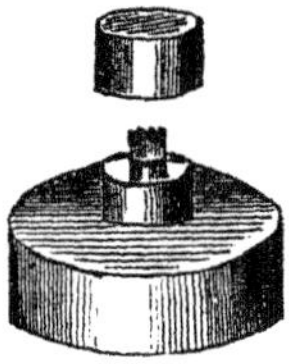

1.

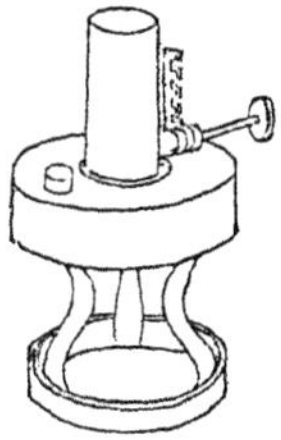

3.

2.

Lampadite. A variety of wad. MnO_2? 82·, CuO 13·5, SiO_3 2·.

Lamp Black. *Smoke Black*, *Peach Black*, *German* and *Spanish Black*, *China Ink*. Specific gravity 1·782. All of these varieties of carbon or smoke are procured by burning woods or vegetable matters away from the access of air, so that the carbon, instead of being converted into carbonic acid, as in ordinary cases of combustion, is deposited as a soft, smoky residue. In Germany, the pitch, tar, resin, oil, or wood employed for the production of lamp black is burned and received in a close vessel, which is lined with leather, sheep's skin, or soft material, upon which the carbon falls gently without losing its lustre, and to facilitate its deposition; or in this country it is deposited in rooms or flues. Lamp black is by no means a pure form of carbon, as it contains only 79·1 per cent. of that body, the remaining ingredients being—resinous and bituminous matters = 7 per cent. ulmine, sulphates of ammonia, potash, and lime, water, and phosphate of lime forming the remainder. Lamp black constitutes the basis of lithographic ink, and when used for this purpose it is recommended to be ignited in close vessels, in order to remove the resins, bitumen, and water, which are injurious to the ink. The *adulteration* of lamp black by bone black may be detected by burning it in a platinum crucible; lamp black leaves but little residue, while ivory black leaves phosphate of lime. When oil of turpentine is distilled over from the impure turpentine, a quantity of resin remains in the retort, which may be advantageously used for the production of lamp black. The purest form in which, in the opinion of Davy, this substance can be obtained, is by passing oils, or spirits of wine through tubes in a state of ignition. Carbon is then deposited in the shape of an impalpable powder destitute of taste and smell, a conductor of electricity, and more than twice as heavy as water. Lamp black is used as a water-colour pigment in the form of *bistre*. See BISTRE. Common lamp black is extensively used as an oil paint of a very permanent nature. It is the colouring matter of common printer's ink, which is prepared by mixing lamp black with concentrated nut or linseed oil.

Lampic Acid. An acid produced in the combustion of the lamp without flame, probably aldehydic acid.

Lana Philosophica. The volatile flocks of oxide of zinc, prepared by burning zinc in air.

Lanarkite. A synonyme of sulphato-carbonate of lead.

Lancasterite. $MgO\,CO_2, MgO, HO$. Spec gravity 2·34. White crystals like stilbite. North America.

Langstaffite. A synonyme of chondrodite.

Lantanuric Acid. $HO\,C_6H_4N_2O_6$? White amorphous mass, very soluble in water; insoluble in alcohol; obtained by dissolving allantoine in cold potash, adding acetic acid to produce a slight acid reaction, evaporating the fluid to a syrup in the water bath; then adding absolute alcohol as long as a precipitate falls; acetate of potash may be separated from the lantanurate by washing with alcohol. The lantanurate of potash is dissolved in water and precipitated with acetate of lead; the precipitate decomposed by sulphohydric acid gas.

Lanthanite. *Carbonate of Cerium, Carbonate of Lanthanum.* H 2·75, LaO 75·7, CO_2 10·8, HO 13·5. Grayish or yellowish-white 4-sided plates, or earthy. Lustre dull or pearly. B.B. infusible; found coating cerite at Bastnäs, Sweden.

Lanthanum. La 5·5375, 44·3; 4·52, 36·16? Dark gray metallic powder, the particles feeling soft and cohering; prepared by dissolving oxide of lanthanum in chlorohydric acid, evaporating to dryness, heating the residue in a current of chlorohydric gas, to make it anhydrous. The chloride then formed is decomposed by sodium; the common salt dissolved out by alcohol (·833), and the lanthanum freed from oxide by washing, and dried between blotting paper in vacuo over sulphuric acid.

Oxide of Lanthanum, Lanthana.—White powder, and brownish when it contains oxide of didymium; not decomposed by sodium or potassium; prepared by taking common hydrous oxide of cerium (consisting of oxides of cerium, lanthanum, and didymium), dissolving in nitric acid, evaporating to dryness, and removing the acid by ignition. The residue is pulverized and digested with weak nitric acid (1 acid and 75 water) assisted by some heat, when nitrates of lanthanum and didymium are formed; but as nitrate of lanthanum is more readily formed than nitrate of didymium, the first digestion consists principally of a lanthanum salt. When cold dilute nitric acid is used, oxide of lanthanum alone dissolves with traces of didymium. But to separate lanthanum and didymium in the most complete manner yet known, the nitrates of these substances as above obtained, are precipitated by caustic potash, and the washed precipitate dissolved in sulphuric acid; the sulphates are then evaporated to dryness in a platinum capsule, and cautiously heated to remove the excess of acid, and obtain anhydrous sulphates; the residue is digested in about 5 times its weight of water of the temperature of 35½° to 37½°, or the salt may be pulverized and added in small portions to the water, the vessel containing which is to be surrounded with ice-cold water to prevent the increase of temperature of the solution, which would then crystallize. The solution being completed, the liquid is heated in a water bath to 104°, when the sulphate of lanthanum crystallizes, leaving the didymium salt in the mother liquor. When perfectly pure, the salt is white; if it contains didymium it has an amethyst tint. To purify it thoroughly, it is ignited gently, rendered anhydrous, dissolved in cold water as before, and crystallized, the process being repeated until the salt is white.

Salts.—Oxide of lanthanum is easily soluble in acids even after powerful ignition; the salts formed are destitute of colour when pure; their taste is styptic. They are precipitated as gelatinous hydrate, and must be dissolved by caustic ammonia, potash, or soda, and by sulphohydride of ammonia. The oxide is soluble when boiled with salammoniac, a chloride being formed and ammonia evolved. Sulphate of potash yields no precipitate, as a salt of cerium does; the precipitate from ammonia is mixed with a subsalt which is soluble in water. When left on a filter, the subsalt becomes a carbonate, and when washed, the carbonate remains and a neutral salt dissolves; carbonate and bicarbonate of ammonia and potash, phosphate of soda, oxalate of ammonia, yield white precipitates, insoluble in excess of reagent.

Sulphate.—White right rhombic prisms (Bolley, Liebig, Ann. 83, 126), amethyst coloured (Rammelsberg Pogg. Ann. 55, 65), when containing didymium; 6-sided prisms, with 6-gonal summits containing 3 atoms water. Soluble in 6 water at 36⅔°, in 42½ water at 73½°, in 115 at 212°. When anhydrous, it contains 56·64 oxide per cent. At a red heat it loses half its acid.

Potash sulphate of lanthanum.—Reddish-yellow precipitate, consisting of KO = 27·48, LaO = 25·69, SO_3 = 46·68.

Carbonate of lanthanum. See PARISITE.

Lapathine. A bitter substance from Lapathum acutum.

Lapis Causticus, or solid caustic potash.

Lapis Infernalis. Nitrate of silver.

Lapis Lazuli. A synonyme of Hauyne.

Lard. The fat of the pig.

Lardite. A synonyme of Agalmatolite.

Laricine. $C_{14}H_{24}O_4$. White powder from Agaricus albus by alcohol.

Lasionite. A synonyme of Wavellite.

Latent Heat. The insensible heat which enters into a body before it can change from the solid to the fluid state, or from the condition of fluidity to that of vapour. When we heat ice it becomes fluid, but when it is just melted its temperature is still 32°, although heat has been pouring into it. The latent heat of water is 140° (Black), 142°·3 (Regnault). To determine it, Black mixed equal weights of water at 172° and ice at 32°. He found that the ice instantly melted, and the resulting temperature was 32°. It is therefore obvious that 140° of latent heat were required to convert the ice into water, without adding to the heat appreciable by the thermometer. The latent heat of steam at 212° is 950° (Watt), 998°·6 (Regnault); at 248°, 973°·6; at 302°, 941°·5. According to Person, the latent heat of vaporization is the same in substances which boil at the same temperature. The following table gives the latent heats, in column 1 referred to the bodies; column 2 referred to water:—

	1.	2.
Ice,	140·°	140·°
Sulphur,	143·68	27·14
Spermaceti,	145·	
Lead,	162	{ 5·6 10·93
Bees' wax,	175	
Zinc,	493	48·3
Tin,	500	{ 33· 24·85
Bismuth,	550	23·25

Latent Heat of Fluids.

		Boiling Points.	Latent Heat Of Atomic Weight.	Latent Heat Of Unit of Weight.
Sulphuric ether,	C_4H_5O	96·0	421°·3	91°·1
Pyroxylic spirit,	CH_2O	151·7	527·7	263·8
Acetic ether,	$C_4H_4O_2$	165·2	582·	105·8
Alcohol,	C_2H_3O	173·	598·8	208·3
Butyrate of Methylene,	$C_5H_5O_2$	199·4	556·5	87·3
Water,	HO	212·	603·	536·
Formic acid,	$C_2H_2O_4$	212	694·	120·7
Valerianic ether,	$C_{10}H_{11}O$	235·4	685·	69·4
Acetic acid,	$C_2H_2O_2$	248·	382·	101·9
Valerianic alcohol,	C_5H_6O	269·6	606	121·4
Oil of turpentine,	$C_{20}H_{16}$	312·8	584	68·7
Terebene,	$C_{20}H_{16}$	312·8	571	67·2
Oil of lemons,	$C_{20}H_{16}$	329·	595	70·
Butyric acid,	$C_4H_4O_2$	327·2	632	114·9
Valerianic acid,	$C_5H_5O_2$	347·	660	103·5
Carbohydrogen,	$C_{12}H_{12}$	388·4	629	59·9
Ditto,	$C_{15}H_{15}$	491·	783	59·7
Ethalic alcohol,	$C_{16}H_{17}O$	680·?	884	58·4

Latrobite. Spec. grav. 2·8. Pink doubly-oblique prisms, with angles of 98° 30′, 91°, 93° 30′. SiO_3 44·653, Al_2O_3 38·814, CaO 8·291, MnO 3·16, KO 6·575. From Amitoke island, Labrador.

Laumontite. *Laumonite, Lomonite, Efflorescing Zeolite.* Spec. grav. 2·3; H 3·5. White, with a shade of red, yellow or green oblique rhombic prisms, with angles of 113° 30′ and 86° 15′; lustre vitreous, translucent; falls into powder in the air. B.B. fuses into a white spumous mass; gelatinizes in acids, and becomes negatively electric by friction, if insulated. SiO_3 51·98, Al_2O_3 21·12, CaO 11·71, HO 15·05. *Form.* 3 CaO, SiO_3, 3 (Al_2O_3 2SiO_3) 12 HO. Huel Goet, Brittany; Kilpatrick hills; Greenland, and North America; generally in trap rocks.

Laurel Oil. $C_{20}H_{16}O$. Spec. grav. ·914. White, viscid oil, consisting of two isomeric oils from laurel berries.

Laurel Water. A preparation obtained by distilling the leaves of the *Cerasus laurocerasus* with water.

Laurelic Acid. An uncrystalline substance from the pericarp of *Laurus nobilis.*

Laurine. A bitter crystalline body from the kernel of *Laurus nobilis,* by boiling alcohol.

Laurostearic Acid. HO $C_{24}H_{23}O_3$; F.P. 108¼°. Spec. grav. ·883. Crystals obtained by saponifying laurostearine; readily soluble in alcohol and ether.

Laurostearine. $C_{27}H_{25}O_4 = C_3H_2O, C_{24}H_{23}O_3$; F.P. 112°. White crystalline grains, by alcohol from laurel berries; soluble in boiling alcohol and ether; also in cocoa nut oil.

Laurostearone. $C_{46}H_{46}O_2$. White scales by distilling laurostearate of lime.

Lava. A name applied to the products of active volcanoes, and which seem to be mixtures of various minerals, especially pyroxene. A homogeneous lava, from Krafla, Bunsen found to consist of SiO_3 76·38, Al_2O_3 11·53, FeO 3·59, CaO 1·76, MgO ·4, KO 1·88, NaO 4·46. Lava from Stromboli was found by Abich to consist of SiO_3 50·25, Al_2O_3 13·09, FeO 10·55, MnO ·38, CaO 11·16, MgO 9·43, KO and NaO 4·92.

Lavender Oil. $C_{15}H_{14}O_2$; B.P. 397°·4. Spec. grav. ·892 to ·917. Yellow oil, becoming thick in the air; mixes with alcohol in all proportions; obtained by distilling lavender leaves (*Lavendula spica, augustifolia, and latifolia*) with water.

Lavendulane. Spec. grav. 3·014; H 2·75. Lavender-coloured masses, with a blue streak; contains arsenic, cobalt, nickel, copper, and water; from Annaberg, Saxony.

Lazulite. *Azure Stone, Prismatoidal Azure Spar, Hydrous Diphosphate of Alumina and Magnesia.* Spec. grav. 3·057; H 5 to 5·5. Pale blue, of various shades, in granular masses, or pieces the size of a hazel nut, or in regular rhombic prisms, with angles of 121° 30′; streak white; surface smooth, all the faces alike; lustre vitreous, translucent on the edges, sometimes opaque. B.B. intumesces, and becomes glassy, but does not melt; with borax a colourless bead. PO_5 41·81, Al_2O_3 35·73, MgO 9·34, SiO_3 2·10, FeO 2·64, HO 6·06. Werfen, Salzburg; Voran, Austria, &c. 2 (3 MgO FeO + P O_5) 4 Al_2O_3 3PO_5 6HO.

Lead. *Metallic Lead.* Pb (plumbum) 13·, 12·945, 103·56. Lead is a bluish-white metal, in masses or regular 6-hedrons. It has no taste, but a peculiar smell by friction. It leaves a mark on paper, and is a poison when taken internally. Spec. grav. 11·35 (Fahrenheit), 11·357 (Crichton), 11·352 (Herepath and Brusor), 11·407 (T. Thomson), 11·445 (Berzelius). Melting point 613° (Gay Lussac), 606° (Crichton, jun.) The specific gravity is not increased by hammering. Very malleable, and may be thus reduced into thin plates. It cuts readily with the knife. Lead is principally obtained from Galena. The following valuable account of lead manufacture has been drawn up by my pupil, Mr. James Taylor. The experiments were made by him under my eye in my laboratory:—

Mechanical Preparation of the Ore.—When the ore is dug out of the mine it is generally associated and blended with extraneous substances, which are separated from it by a mechanical process of cleaning or dressing before being fit for smelting. In this condition the ore is called by the miners *bouse.* With a view, therefore, to free the ore as much as possible from the matrix, it is washed until it has attained the degree of purity necessary for the subsequent metallurgic operations; previous to which it is pulverized, by crushing and stamp mills, moved either by water or steam power. The crushing mill consists essentially of a series of iron rollers, between which the ore is made to pass. The hopper which supplies the ore to the rollers, is placed at an elevation, and under it is a trough called *a shoe*, which is constantly moved by the machinery, so as to cause the ore to fall gradually upon the rollers. To prevent the rollers from becoming hot by their action, a stream of water is made to flow into the shoe, which is dispersed over them, keeping them always cold. The stamp mill consists of a wooden framework, in which the stampers are moved up and down. The bouse ore to be stamped is placed on an incline behind the stampers, and a stream of water from an elevation made to fall upon it, as the ore gradually descends under the stampers, to be pulverized.

Grating.—The grate consists of bars of iron 1 inch square, and from 24 to 32 inches long, placed horizontally and parallel to each other, 1 inch apart. The bouse ore is placed upon the grating, and a stream of water, conveyed by a wooden channel, is allowed to flow over it; the water and ore that pass through are led by an inclined plane into a hemispherical basin underneath, where they are collected. The pieces of ore that remain on the grating, mixed with the stones and other matters, are separated and broken into pieces suitable for the smelting-house, and conveyed to the *bing-stead*, being a place near to the works, where all the dressed ore is deposited. If any ore still remains mixed with the gangue on the grating, it is carried to the crushing mill and reduced to a coarse powder. After being crushed it is again placed upon the grating, and subjected to the same treatment as before. If any ore still adhere to the stony matter which remains on the surface of the grate, it is subjected to a second process of crushing and washing, and if not it is cast away. The ore that passes through the grate, and collects in the basin under it, receives the name of *dashed* or *crushed* ore.

Bouse Best.—In order to purify the crushed ore further, it is washed upon the *brake sieve*, which is similar to the grating, but the apertures are smaller, and is suspended at the end of a wooden lever. On commencing the process of washing, the brake sieve is placed in a *cistern* filled with water, and a quantity of crushed ore put upon it. The end of the lever, in connection with the sieve, is agitated in such a manner as to cause the finer portions of ore to pass through the openings into the cistern below. The contents of the sieve are thereby separated into layers. The upper stratum or portion that rises to the surface, called *cuttings*, is removed by a scraper made of sheet iron, called by the washers a *limp*, and carried to the cutting washers. A second stratum, or centre portion, called by the miners *chats*, is taken out, and conveyed to the crushing mill, ground, and adapted for the subsequent treatment. A third stratum, or portion, next the meshes of the sieve, is pure ore, which is also taken out by the limp and transported to the bing-stead. This last is technically called by the miners *bouse best*. The following table gives the analyses by Mr. Taylor:—

	1. Bouse Best.	2. Tails Best.	3. Bouse Common.
Gangue,...............	6·90	8·00	8·80
Sulphuret of lead, ...	89·00	85·80	83·80
Tersul. of antimony,.	0·60	0·80	0·90
Copper, ...	trace	trace	trace
Carbonate of iron,....	1·20	3·48	3·00
Alumina, ...	0·25	0·80	0·90
Carbonate of lime, ...	1·90	1·00	2·40
	99·85	99·88	99·80

Tails Best.—The *chats*, after being ground, are again transferred to the brake sieve, and treated in the same manner as above described. The gangue which collects on the surface of the sieve, called gritty chats, is taken out by the limp, placed under the stampers, and pulverized. The pulverized gangue, now called *bouse smiddum*, is carried to the cutting washers, and further purified. The under or third stratum of pure ore lying on the sieve, and called by the miners *tails best*, is likewise removed and deposited in the bing-stead.

Bouse Common.—The ore that passes through the brake sieve, and known as bouse smiddum, is placed in the *running buddle*, which is a large box into which a stream of water flows at what is called the eye of the buddle. The smiddum to be buddled is placed on the bottom, and while the water is flowing in, it is agitated with a colrake (shovel), and by that means resolved into two qualities, the purest, which remains at the high end, and the sludge or slimy portion, called *smiddum tails*, which flows to the lower end; they are both taken out, the former, which is the purest, is deposited by the side of the buddle, and the latter is carried to the trunk buddle. The purest portion is again put into the buddle, and washed properly before being conveyed to the bing-stead. After being washed and sorted in the manner before described, it receives the name of *bouse common*.

Tails Common.—The smiddum tails are next put into the trunk buddle, which is composed of two parts—a box, into which a current of water flows, and a large tank with a smooth bottom. The ore to be trunked is placed in the box, and agitated with an iron shovel turned up at the sides, and the coarse particles are removed from time to time, while the finer particles pass along with the water, and are deposited on the level area. The portion at the upper end of the buddle, called sludge, is taken out and deposited at the side for further purification, and the other portion at the low end, containing a very small quantity of ore, is taken to the cutting washers. The ore taken out from time to time by the shovel is placed upon what the washers call a *bedding*, consisting of a layer (about 2 inches thick) of ore on the brake sieve. The sieve is moved in the water by means of the lever, and the pulverized ore, owing to its great specific gravity, and the minuteness of its particles, passes through the bedding, while the stones and gangue are taken off by the limp, and transported to the cutting washers. This operation is called "*letting in*," from the ore being let into the cistern under the sieve. The ore, after being let in, is taken out and put into the running buddle, and washed until it is clean, when it is removed to the bing-stead. It is now known by the name of Tails Common, and on analysis was found to have the following composition:—

	1.	2.
Gangue, ...	10·00	8·40
Sulphuret of lead, ...	81·80	77·80
Tersulphuret of antimony,....	0·80	0·87
Copper,.......	trace.	trace.
Carbonate of iron, ...	2·90	4·06
Alumina,.........................	0·80	1·20
Carbonate of lime, ...	3·60	7·60
	99·90	99·93

1, *Tails common.* 2, *Cuttings common.*

Cuttings Common.—The sludge taken from the high end of the trunk buddle is washed over again in the same manner. During the time it is being agitated, the washer moves his shovel uniformly, from the low to the high end of the buddle, and causes the separation of the materials into two parts. This operation is repeated until the slimy matters are in a suitable condition to be *dollied*. The dolly-tub is a cylindrical vessel, with an upright shaft which supports a perforated board, called a dolly, which, on being quickly turned by a winch handle, communicates a circular motion to the contents of the tub. The ore to be washed is put in gradually by a shovel, and, as soon as it is thoroughly mixed with the water, the dolly is quickly withdrawn, and the sides of the tub are smartly struck with a mallet for a considerable time, in order to hasten the subsidence of the heavy particles of ore; while the slimy or light portion remains suspended in the water, but on the fluid coming to rest it subsides also. The water is then decanted off, and the slime, having little or no ore in it, is taken out as refuse. The pure ore at the bottom of the tub, called *cuttings common*, is next removed and placed in the bing-stead. The cuttings, called *sieve toppings*, are placed on the brake sieve and treated in the same manner as the bouse ore. Upon the removal of the stony matter from the surface of the sieve, the pure ore is taken out and deposited in the bing-stead. The smiddum which passes through the sieve is taken out of the tub, trunk *buddled and let in*, as before noticed respecting the bouse smiddum. The slimy matters that collect at the lower end of the apparatus, when buddling the bouse smiddum, are carried to the trunk buddle and trunked. It is afterwards let in, and lastly treated in the running buddle until the ore is pure, when it is carried to the place of deposit. The fine slimy matters produced by the last operation, are taken to the *stirring buddle*, which is a box in which the sludge is thoroughly mixed with water, till it assumes the consistency of slime, and where much of the impure matter is washed out. They are next removed to the *nicking buddle*, which has a board called the *nicking board*, placed at a slight incline. Upon this the slime is spread and water is made to trickle in small streams down its surface, which, as it descends, comes in contact with the small ridges of slime made by the washer's shovel, and this process is continued until all the ore is ultimately washed down into the trunk, where it becomes deposited in regular layers on the floor. This process is repeated as often as is necessary to fit the ore for dollying, as before described. Cuttings common are composed as above.

Smelting of Lead Ore. Roasting.—It is only within the last half-century that the ore to be smelted in the Scotch furnace was subjected to the preparatory step of roasting, in order to effect its partial desulphuration and oxidation. The form of furnace generally employed varies considerably in its dimensions, but it always consists of a long flat hearth, with an arched roof, and heated by a fire placed at one end. There are two doors on either side for the removal and working of the ore. The charge of the furnace varies from 9 to 12 cwt. of galena, and requires from two and a-half to three hours for roasting. The ore is introduced into the furnace without the addition of any flux—spread even over the surface of the sole—and heated cautiously, taking care that the temperature is always kept under the fusing point of galena. If at any time during the operation, the ore presents the appearance of being fused, a fresh surface is exposed to the action of the flame. By this operation a large portion of the sulphur, &c. is driven off, and the ore so agglutinated as to be enabled to resist the action of the blast furnace, without being carried into the flues. On analyzing two specimens of the roasted ore, the following results were obtained:—

	First Specimen.		Second Specn.	
	I.	II.	I	II.
Insoluble matter,.........	10 80	9 60	8·80	6·40
Oxide of lead,............	32·11	32·40	33·20	32·60
Sulphuret of lead,.........	41 90	41 48	44·10	44·40
Sulphate of lead, ...	8·60	8·80	7·10	7·20
Tersulphuret of antimony,	0 3	0·40	0·5	0·60
Copper,.................	trace	trace	trace	trace
Peroxide of iron,.........	3 80	4 40	3·20	3·60
Alumina,................	0·60	0·80	0·60	0 90
Carbonate of lime,........	1·80	2·00	2·40	4·20
	99·91	99·88	99·90	99·90

Scotch Furnace or Ore Hearth.—This furnace is used in Durham, Cumberland, Northumberland, and Lanarkshire. It is of a rectangular form, 24 inches long, 12 inches in breadth, and from 22 to 25 inches in depth. The bottom and

sides are lined with cast iron consisting of one casting; surrounded with a ledge 2½ inches in thickness, and 4½ in breadth, except on the side facing the work-stone. In front of the hearth there is another cast iron plate called the *work-stone*, *a*, which is 2 feet 10 inches in breadth, and 1 foot 6 inches in length; which is surrounded with a ledge, *b*, 1 inch square, except at the side opposite the sole of the furnace. It is placed at an incline of 6 inches, its upper end is elevated 4 or 5 inches above the hearth bottom, and the intermediate space filled with bone ashes, previously powdered and moistened with water. At the back edge of the hearth is placed a prism of cast iron called a *back-stone*, on which rests the nozzle of the tuyeres, over which is placed another cast iron prism called the *pipe-stone* of equal length, and 8 inches in height. The tuyere is introduced into the cavity, and made to project 2 inches into the hearth. On the pipe-stone is placed another back-stone of the same dimensions as the first, which completes the back part of the hearth. Along the two sides are placed two prismatic castings, called *bearers*, 26 inches long and 5 inches square, which project slightly over the upper margin of the work-stone. At an elevation of 5 inches above these bearers, and at a distance of 12 inches from the back of the hearth, is placed another bar of cast iron, called the *fore-stone*, which rests on fire-bricks, and has the same dimensions as the back-stone on which the tuyere of the blowing apparatus rests. It is closed at either end with a cube of cast iron measuring 10 inches square, called a *key-stone.* The intervening space between the key-stone and the back part of the hearth is filled up with two other castings of similar dimensions. The *lead-pot, E*, which is of cast iron, is placed in the front of the work-stone. It is enclosed in a cast iron jacket, *G*, built in masonry, into which the melted lead from the hearth is conducted by means of a gutter in the work-stone. In order to prevent the escape of fumes in the smelting house, which would be injurious to those employed, a hood of arched masonry covers the whole work. The draught is regulated by means of an iron plate, and the blast communicating with the tuyere is also regulated by a valve placed in a pipe approached by an arched doorway, which always remains open. This arrangement was first employed by Dr. Richardson of Newcastle.

Smelting or Reduction, or Deoxidation and Desulphuration.—The hearth is prepared for working by filling it with peats or billets of wood loosely thrown in, but in the front next the work-stone, they are built up in the form of a regular wall. The bellows are put in action, and an ignited peat placed before the nozzle, when combustion throughout the whole mass quickly ensues. Over it is placed a quantity of coal for the purpose of giving greater solidity to the mass, and also to increase the temperature. A quantity of *browse*, which is a species of slag, that remains on the hearth in a semi-reduced state, from a previous operation, is now thrown on the surface of the burning mass, and after fusion the contents of the hearth basin are taken out on the work-stone, by means of a rake, and the metallic lead allowed to run into the lead-pot, while the refuse or gray slag known by its shining appearance remains, and is removed and deposited at the right side of the furnace in water tanks to prevent the escape of noxious vapour. In order to prevent the browse from entering the tuyere, half a peat is placed in front, which from its porosity, and the facility with which it ignites, communicates the blast equally through the whole mass. The browse after having been freed from the *gray slag* is again placed on the hearth, with a quantity of finely pulverized coal if it requires further reduction, or with lime if the slag is not sufficiently disengaged, which is ascertained by the mass exhibiting a soft appearance, and tendency to fuse. The lime from its affinity for the siliceous matter, forms a silicate of lime, which greatly facilitates the extraction of the lead, by causing the scoriæ to pass into *lumps or balls*, called *gray slag*, containing from $\frac{1}{10}$ to $\frac{1}{60}$ of the entire lead in the ore. The gray slag is afterwards fused on the slag hearth, and as much of the lead extracted as possible. After all the browse has been introduced on the hearth, a quantity of roasted ore is spread upon its surface, and the blast worked for about twenty minutes, when the contents of the hearth are again taken out on the work-stone, and the metallic lead allowed to run into the lead-pot. The gray slag is removed as before noticed; and another portion of peat placed before the tuyere. The browse, together with the requisite quantities of coal and lime, are again thrown on the hearth; and also the requisite quantity of roasted ore. These operations are repeated for a period of twelve or fifteen hours, and form what is technically called a smelting shift. The produce varies from 1 to 2½ tons, according to the quality of the ore. The lead obtained by this process is invariably purer than that from the reverberatory furnace, from the heat not being sufficiently great to cause the reduction of some of the foreign metals. On analyzing two specimens of metallic lead, from different qualities of ore, the following results were obtained:—

Bouse Best.—Lead, 99·16; iron, 0·56; antimony, 0·28; copper, trace. Total, 100·00.

Bouse Common.—Lead, 98·88; iron, ·84; antimony, ·28; copper, trace. Total, 100·00.

Gray Slag.—Combustible matter, 33·60; siliceous matter, 14·00; oxide of lead, 40·36; teroxide of antimony, ·98; oxide of copper, 2·40; peroxide of iron, 3·60; alumina, 0·40; carbonate of lime, 4·40. Total, 99·74.

Reverberatory or Smelting Furnace.—This furnace is now in extensive use both in England and Scotland. It is usually upwards of 8 feet in length, and 6 feet in breadth, and about 2 feet in height from the hearth to the roof. The fireplace is separated from the furnace by a fire-bridge

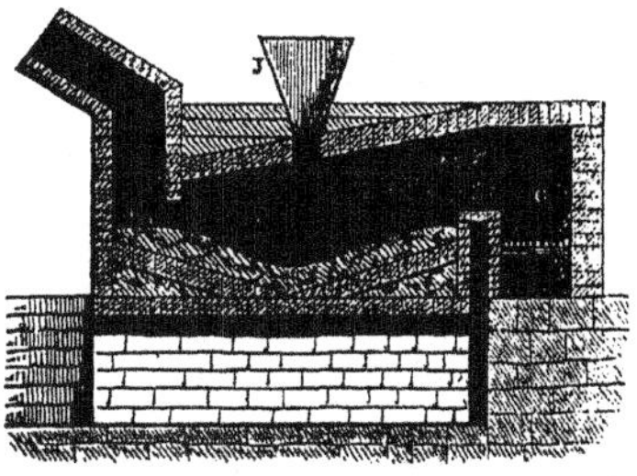

about 1 foot 4 inches in height. The arch forming the roof spans the distance between the fire-bridge and the opposite end of the furnace, where it reaches within 6 inches of the floor. The flue is divided into two openings, which, together, are the whole width of the furnace. The openings are formed by a triangular block of fire-stone. The flue, after traversing a channel covered by flat stones, loosely jointed with fire-clay, for the deposition of the lead fumes, and their ready removal, is conducted into a lofty chimney, by which the smoke and other products escape. The sole is formed of fused slag, obtained in previous operations, which, being semi-fused, is worked into the proper form, with a depression in the centre for the accumulation of the fused metal.

At the bottom of this hollow is situated the tap-hole for allowing the reduced lead to flow out. The furnace is divided into what is called the "labourers' side" and the "working side." At the labourers' side are the door, *G*, for supplying the fire with coals, and also three small openings, *h*, about 10 inches by 8 inches, covered with iron plates, which may be removed at pleasure, to admit of a free current of air when required. At the opposite, or working side, are three similar openings, stoppered in the same manner for cooling the furnace and working the charges. The sole of the furnace, when in a semi-fluid state, is raised to nearly the level of the small doors on the labourers' side, where it slopes down towards the working side to about 18 inches below the middle door, where the tap-hole is situated by which the lead flows into a level cast iron pan, placed in a niche a little lower, at the side of the furnace. In the centre of the arch is an aperture called the crown hole, in which is placed a hopper made of sheet iron. Into this the charge of ore is always put before it passes into the furnace. The tube leading down to the hearth is provided with a damper, or sliding valve, which can be opened and shut at pleasure by means of an iron lever. When the charge is drawn off, and the slag removed, the damper is pulled out, and the ore in the hopper is allowed to fall into the furnace. The weight of the charge varies according to the quality of the ore. In the north of England it is about 12 cwt., and in Wales from 20 to 24 cwt. The charge, on being let down through the hopper, is spread over the bottom of the furnace by means of an iron rake. The doors are all closed and also the dampers communicating with the flues. At certain intervals, the ore is turned over with an iron paddle, and fresh portions exposed to the action of the flame. During the first two hours after the introduction of the charge, very little fuel is added, the heat being kept up by the declining fire of a previous operation, and the roasting carried on principally by radiation. At the end of the above period, the doors are opened, and the roasted ore turned over with an iron paddle, and the slags rich in lead, obtained from a previous operation, introduced, and the doors again shut. In the course of half-an-hour, the tap-hole is opened, and the metallic lead run off. The charge is again turned over, and the doors closed, when a quantity of fuel is added to the fire, and the furnace brought quickly up to a dull red heat. About three quarters of an hour afterwards, another portion of fresh slag is thrown in, and the whole turned over, and a little caustic lime added through the central opening, when the doors are again shut. After a certain interval the charge is further mixed, and the slags placed directly under the fire-bridge, and the doors now left open, in order to facilitate the separation of the metallic lead from the slag. The slags are frequently pushed back underneath the fire-bridge, and small quantities of lime occasionally added, for the purpose of facilitating the liberation of the last portion of lead, and giving greater tenacity to the slag. The fire is now renewed, and the doors shut and allowed to remain for three quarters of an hour, when they are re-opened, and the slag pushed under the fire-bridge, and the whole made level, and a small portion of lime thrown on the surface, for the purpose of covering the lead, and preventing further oxidation, and, at the same time, to dry up the slag. In the process of reduction, the first effect produced upon galena, by roasting, is the oxidation and conversion of a small portion into sulphate of lead. Another portion is, at the same time, converted into oxide of lead; the sulphur passing off in the form of sulphurous acid, while the last portion of ore remains unaltered, or very little changed, in its characters. In the subsequent operations, these three salts of lead react on each other, on increasing the temperature of the furnace. The sulphuret acts on the oxide, forming sulphurous acid, which is evolved, and the metallic lead is set free, while another portion of the sulphuret acts in like manner on the sulphate, forming sulphurous acid, which is given off, and the lead liberated. The lime which is added performs a double function, viz: first, it sets free the last traces of lead; and, secondly, it acts mechanically in thickening the scoriæ. The fume from the furnace-draughts has the following composition:—

Fume from Reverberatory Furnace Draughts.—Siliceous matter, 10·00; oxide of lead, 47·12; teroxide of antimony, ·68; oxide of iron, 3·20; alumina, 2·40; lime, 11·84; sulphuric acid, 22·80; carbonic acid, 1·20. Total, 99·24.

Slag Hearth.—The slag hearth is a species of blast furnace, having the form of a rectangular prism 26 inches in length, 22 inches in breadth, and 33 inches in height. The bottom is composed of cast iron plates 2 inches in thickness, placed at a slight incline towards the front of the furnace. On either side of the bottom plate are placed cast iron bearers similar to those employed in the ore hearth, and on these is placed the *fore-hearth*, which consists of two strong cast iron plates, about 12 inches in breadth and 22 inches in length, leaving a space of upwards of 5 inches between the front stone and the bottom of the furnace, and by that means an additional height of 2½ inches is obtained by placing a row of fire-brick between them. In front of the fire-hearth is a large iron cistern, into which water is constantly flowing, so that the slag which escapes through the opening at the breast passes over the surface of a pot of peculiar construction into the cistern of cold water, in which it is shattered to pieces, and rendered fit for the operation of washing, to which it is afterwards subjected. On commencing to work the slag hearth, it is filled up with small cinders beaten closely together till within 4 or 5 inches of the orifice of the tuyere.

The pot for the reception of the lead is also filled with cinders, and intended, in both cases, to act the part of a filter in separating the metallic lead from the less fusible portions. Above the cinders are piled masses of peat, similar to those employed in the Scotch furnace, and above that coke. After the whole is fully ignited, a stratum of gray slag is thrown over, and afterwards added alternately. The heat is kept up by means of the blast, and the metallic lead percolates through, while the slag, from its viscidity, is retained. When the slag has become perfectly liquid, a hole, about 1 inch in diameter, is made in the centre of the cinders, through which the fused silicates flow into the cistern of water. The lead, in consequence of the elevated temperature at which it is obtained, is always inferior to that produced directly from the ore.

Separation of Antimony and other impurities.—The lead obtained by either of the above processes frequently contains silver in sufficient quantity to admit of its being profitably extracted. It is, besides, generally associated with antimony, tin, copper, and various other impurities, which have to be removed before the silver can be separated with advantage. The lead,

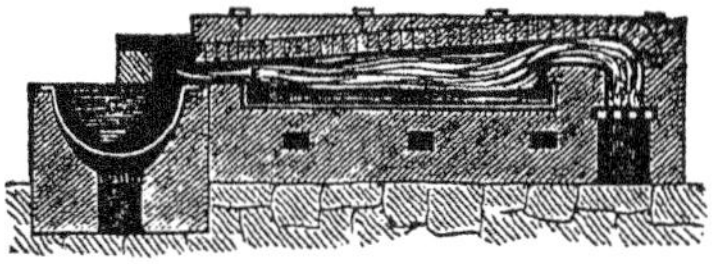

however, from some foreign ores, is much more impure than from British ores, particularly that from Spain. The process of separating these impurities consists in fusing the metal containing them in a reverberatory furnace, having for its bottom a large cast iron pan, *G*, 10 feet in length, 5 feet in width, and 9 inches in depth, and elevated 1 inch at one of the ends. The pan is placed in a bed of sand on arches of brick 2 feet 6 inches above the floor, in order that it may be all equally supported. The sides on which the arch of the furnace rests, are not in contact with the pan, so as to admit of free expansion. The arch at the farther extremity is about 6 inches above the level of the solum, while at the other end it is 1 foot 6 inches. The fireplace is 5 feet in length, and 1 foot 6 inches in width, divided from the lead-pan by a fire-bridge 2 feet 3 inches in width, which rises about 8 inches beyond the margin of the pan. At the other extremity are the flues, *E*, which lead into the main chimney. The pan is provided with an iron lip, for running the contents off into a *tripod*, called a receiving pot. There is also, at the side, a large iron pot, set in brick-work, for fusing the pigs of impure lead, from which it is laded out into the calcining furnace. The sole plate is heated to incipient redness before the fused metal is introduced. The charge varies from 8 to 10 tons. The flame is now made to play over the surface of the lead, to oxidize the antimony, tin, copper, and any other impurity which may be present, together with a small portion of lead. The surface becomes covered with a thick granular scum, which is removed by an iron rake, and a clean metallic surface kept constantly exposed to the action of the flame. The length of time required for the purification of the lead, depends wholly upon the nature and amount of impurities. In some varieties of hard lead, twelve hours are found sufficiently long, while in other varieties the operation has to be continued for three or four weeks consecutively. After the lead has been freed from all extraneous matters, and in a fit state for tapping, which is ascertained by placing a portion, by means of a ladle, in a small mould, and observing, as it cools, if it assumes, on the surface, a peculiar flaky crystalline appearance, it is run off by a moveable iron spout, into a pot supported on a tripod stand, from which it is subsequently laded out into moulds.

Concentration of Silver in Lead.—In the year 1829, H. L. Pattinson, Esq. of Newcastle-on-Tyne, noticed the circumstance, that lead containing silver, on being melted, and allowed to cool slowly, with constant agitation, deposits at a temperature a little under the fusing point, small crystals of metallic lead, which fall to the bottom, and, on being drawn out, are found to contain much less silver than the original, and, by that means, the mother liquor becomes proportionally richer in silver. The discovery and application of this fact constitutes the groundwork of a patent, generally known under the name of "Pattinson's Process," which is extremely simple in its nature. A series of 10 cast iron pots are placed in a row, with a fireplace underneath each. At the extreme right of the series is the *market*-pot, from which the poor lead—that is, after it has been desilverized—is laded out into moulds. It is a little elevated above the adjoining pots, which are called *working-pots*, from the lead being crystallized in them. Each pot is capable of holding about 5 tons of molten lead. On commencing the operation, the lead is introduced into a pot in the middle of the series and melted; and, after being fused, it is carefully skimmed with a ladle perforated with small holes, and the fire withdrawn from under the pot. The lead is now allowed to cool slowly, and during this time it is constantly stirred with a long iron paddle, and in a short time crystals of lead begin to fall to the bottom, which are removed by means of a large perforated ladle, and deposited in the adjacent pot to the right in the series as fast as they accumulate. This operation is continued until about 4 tons of crystals are removed from the original lead. The remaining one ton is now greatly richer in silver. The enriched lead is laded into the next pot to the left in the series, and the same process repeated on a fresh quantity of lead in the same pot, and the crystals removed to the pot to the right, while the enriched alloy is passed to the left. The pots in either direction, after being filled with lead of their own quality—that is, as regards silver—are liquefied and crystallized; the poor lead still passing to the right, and the rich to the left, and so on, until the poor lead reaches the market-pot, and the rich lead the other extremity. It follows, that the pots to the right must become gradually deprived of silver, and those to the left must become proportionally richer. In the one instance, the lead to the right ultimately contains a mere trace of silver, while that to the left becomes very rich, and yields a considerable quantity on cupellation. Between every two pots there is a small pot called *temper-pots*, containing fused lead at a considerably elevated temperature, for the purpose of introducing the ladle employed for taking out the crystals when it becomes *chilled*—that is, the lead adheres to it so firmly as to stop up the perforated holes. They are generally 2 feet in diameter, and therefore just admit the ladle in an upright position. The following analyses are annexed of the skimmings and the lead obtained from them by reduction, and also of market lead, market crystals, and rich lead from the crystallizing process:—

Skimmings.—Lead, 98·10; antimony, 0·28; copper, trace; iron, 1·4; lime, ·22. Total, 100·00.

Skimming Reduced.—Lead, 99·72; antimony, trace; copper, trace; iron, ·28. Total, 100·00.

Market Crystals.—Lead, 99·9; iron, ·1; antimony, trace; copper, trace. Total, 100·0.

Market Lead.—Lead, 99·9; iron, ·1; antimony, trace; copper, trace. Total, 100·0.

Rich Lead from Crystallizing Process.—Lead, 99·8; iron, ·2; antimony, trace; silver, (?). Total, 100·0.

Extraction or Refining of Silver.—This operation is performed in a cupel, which forms the bottom of a reverberatory furnace, called a *refinery*. Instead of the litharge produced being absorbed, as formerly, by materials composing the cupel, it is run off in the fluid state. The fireplace is about 2 feet square, separated from the body of the furnace by a fire-bridge, 18 inches in breadth, and the flame and heated air made to pass directly over the surface of the cupel into the main chimney. The cupel, or test, is of an oval form, surrounded by an iron ring about 4 inches in depth, its greatest diameter being 4 feet, and its lesser 2½ feet. The frame is provided with cross bars for supporting the cupel. The bone-ash—the substance employed for forming the test—is finely ground to powder, and moistened with water which contains a small quantity of carbonate of potash, for agglutinating and giving consistency to the ash when heated. After being thoroughly mixed, it is put into the test frame, and beaten with an iron rammer. The centre portion is now scooped out by means of a small trowel, until the sides are about 2 inches thick at the top, and 3 inches at the bottom, while the sole itself is reduced to the thickness of about 1 inch. At the fore part of the test, called the *breast*, the width of the margin is about 5 inches, and, in its centre, there is a cavity which communicates with an aperture for leading off the fluid litharge into a pot underneath. The test is now placed in the refining furnace, and wedged at a proper height against an iron ring firmly built in the wall. On first lighting the furnace the heat is applied with great care, to allow the test to get thoroughly dried, as, by too rapid heat, it would be liable to go in pieces. The temperature is gradually increased until it reaches a cherry-red heat, when the test is filled with rich lead previously fused in a pot, at the side, arranged for that purpose. The fused lead, on being introduced into the cupel, becomes covered on the surface with a grayish dross, which disappears on increasing the heat to incipient whiteness, being changed to ordinary oxide of lead or litharge. The blast is next put in motion, and the litharge at the back part of the cupel forced to the breast, and, through the aperture, into an iron pot placed for receiving it. The blast not only serves to remove the litharge from the surface of the lead, but also supplies oxygen for its formation—care being taken as to the proper degree of heat. As the pure lead diminishes by the continual oxidation and removal of the litharge, another portion from the pot is added, until the lead in the cupel is increased to its original volume, and the operation is continued till five or six tons have been introduced into the test. The last charge, which contains the silver from the previous charges, is wrought off until the whole of the silver is left with only three or four cwt. of lead. A hole is then made in the centre of the bone-ash forming the bottom of the cupel, called the *tap-hole*, and the whole is run into a pot underneath. The tap-hole is again closed up by a pellet of moistened bone-ash, and another charge introduced, and the same process repeated, until a quantity of rich lead has been obtained, capable of yielding, on assaying, several thousand ounces of silver. In some instances when the lead is sufficiently rich, the silver is extracted directly from it, but when it is poor it is always concentrated as above. The test employed for assaying the rich lead is similar to the other, but only farther hollowed out in the centre, so as to give greater thickness to the resulting plate of silver, and admit of the removal of the slags from the edge after the operation has been completed. The silver becomes, suddenly, highly lustrous, as the last traces of lead disappear, of which the *brightening*, as it is called, is a test. The silver afterwards becomes white. The litharge invariably carries with it a notable quantity of silver, and, to obtain it, the oxide of lead, together with the cupel bottoms, are reduced, and the resulting lead melted and treated in the cupel in the manner before noticed. On analysis, the various products were found to have the following composition :—

Unfallen Litharge, common.—Oxide of lead, 99·7; peroxide of iron, ·3; antimony, trace; copper, trace. Total, 100·0.

Fallen Litharge, common.—Oxide of lead, 99·6; peroxide of iron, ·4; antimony, trace; copper, trace. Total, 100·0.

Fallen Litharge, best.—Oxide of lead, 99·7; peroxide of iron, ·3; antimony, trace; copper, trace. Total, 100·0.

Test Bottom.—Siliceous matter, 0·40; oxide of lead, 59·60; teroxide of antimony, 1·04; copper, trace; iron, trace; lime, 20·00; magnesia, 0·40; phosphoric acid, 18·52. Total, 99·96.

Fume from Refining Furnace.—Siliceous matter, 1·60; carbonate of lead, 97,00; teroxide of antimony, 0·16; peroxide of iron, 1·10; alumina, 0·10; lime, trace. Total, 99·96.

Reduction of Litharge.—The litharge from the preceding operation, also the pot dross, and other compounds of metallic oxides which occur during the separation of the antimony, are all reduced to the metallic state in a reverberatory apparatus similar to the smelting furnace, but only smaller in its dimensions. The sole, however, instead of having a depression under the middle door, slopes gradually from the fire-bridge to the other extremity, where it is hollowed out, in the centre of which hollow is a tap-hole for conducting the lead from the furnace, as it becomes reduced, into small iron pots placed for its reception, from which the lead is laded into proper moulds. In order to prevent the fused litharge from acting on the sole of the furnace, it is covered over with bituminous coal to about 2 inches in depth, which very soon ignites, and is burnt to a spongy mass. The litharge is mixed with a quantity of small coal, and is then placed on the red hot embers immediately under the fire-bridge. The gases play on the mixture and cause the reduction of the lead, which flows gradually through the porous cinders into the hollow near the flues, and thence, by the tap-hole, into an iron pot. During the operation the mass is frequently stirred by an iron rake, for the twofold purpose of exposing fresh portions of it to the action of the flame, and for facilitating the descent of the metallic lead, which, if allowed to remain long exposed to the action of the flame, would experience loss from sublimation. As the metallic lead escapes, fresh portions of the mixture of litharge and small coal are added, until, at the end of the shift, which lasts generally about twelve hours, the cinder floor, together with other matters, are broken up, and the temperature of the furnace is considerably elevated, with a view to remove the last traces of lead. A furnace with a sole 6 feet in length and 5 feet in breadth will, from ordinary litharge, produce about 3½ tons of metallic lead in twenty-four hours. On analyses of the various products, the undernoted results were obtained :—

Litharge Cinder or Slag.—Siliceous matter, 6·4; carbonate of lead, 75·52; oxide of lead, 12·60; teroxide of antimony, 0·68; peroxide of iron, ·80; alumina, 1·60; carbonate of lime, 2·40. Total, 100·00.

Lead from unfallen Litharge.—Lead, 99·8; iron, ·2; antimony, trace. Total, 100·0.

Black Slag.—Siliceous matter, 40·80; oxide of lead, 7·08; teroxide of antimony, 0·68; peroxide of iron, 21·96; alumina, 3·20; lime, 24·64; magnesia, 1·2; phosphoric acid, trace. Total, 99·56.

Fume from Slag Hearth.—Sulphate of lead, 21·60; carbonate of lead, 70·64; oxide of lead, 3·76; teroxide of antimony, 1·72; peroxide of iron, ·50; alumina, ·3; carbonate of lime, 1·2. Total, 99·72.

Lead from Slag Hearth.—Lead, 99·43; antimony, 0·29; iron, 0·28.—Total, 100·00.

Suboxide of Lead, Dinoxide.—Pb_2O 21·627·. Black powder obtained by heating oxalate of lead in a retort, and by exposing lead to the air.

Lead Oxides and Salts :—

Protoxide of Lead. Yellow Oxide, Massicot, Litharge, Plumbous Acid. PbO 14·112. This oxide is formed in cupellation by melting lead, and blowing air upon it in its fused state (Massicot). When lead is acted on by water, a white border of oxide and carbonate forms at the surface of the water; and if the solution be examined, it gives a black precipitate with SH, from which it has been concluded that the oxide is soluble in water; but if the water be previously filtered through a double filter, most of the oxide, it will be found, is detained on the filter, for if the interior of the filter is exposed to the action of SH it becomes black, while the exterior of the filter is unchanged (R. D. T.); and in water which has passed through pipes the oxide can be detained by a filter of sand (Dr. Clark). Pure oxide of lead may be procured by calcining the nitrate or carbonate of lead. When dissolved in caustic soda, and left in an open vessel, it crystallizes in 12-hedrons.

Characters.—Yellow scales or flakes, or powder. When melted or slowly cooled, it crystallizes in octahedrons with a rhombic base. In the melted state it is a fine yellow glass. Specific gravity 9·20 (Karsten), 9·277 (Herapath). At a white heat it volatilizes. It is tasteless and insoluble in water. When heated on charcoal before the blowpipe, the oxygen unites with the carbon to form CO_2, while the metallic lead remains in the form of globules on the charcoal; 2PbO and C, becoming 2Pb and CO_2. Oxide of lead forms salts with acids which are generally colourless, partly soluble and partly insoluble in water. The soluble salts are poisonous when taken in large doses, from their coagulating the tissues and fluids with which they come in contact. Their action is neutralized by means of SO_3, or an alkaline sulphate, a consequence of the formation of insoluble sulphate of lead.

Sesquioxide of Lead.—(Winkelblech), Pb_2O_3,

29·232; obtained by adding caustic soda or potash to acetate of lead, till the precipitate which at first appears is again dissolved, and then mixing this alkaline solution with a cold solution of hypochlorite of soda as long as a precipitate falls. When washed and dried, this oxide is a reddish-yellow powder, and does not combine with acids. It is soluble in HCl; reduced by formic and oxalic acids to protoxide.

Binoxide, Peroxide, Deutoxide, Brown or *puce oxide, Plumbic acid.*—PbO_2, 15·. When nitric acid, specific gravity 1·26, is poured upon 20 parts of *minium* ($PbOPb_2O_3$) 18½ parts dissolve, while 1½ part remains in the state of *brown* oxide. It may also be obtained by passing chlorine into water through which red lead is diffused, until the oxide is dissolved, and then precipitating by caustic soda. It is also formed by precipitating a boiling solution of acetate of lead with hypochlorite of soda, or by passing a current of chlorine through acetate of lead, or by the action of bleaching powder on the same salt (W. Crum). The brown oxide is a tasteless insoluble powder of a flea-brown colour. Spec. grav. 8·902 (Herapath); not acted on by sulphuric or nitric acids. By heat it becomes protoxide. It is used as an oxidizing agent sometimes in organic chemistry. It is decomposed by the light of the sun into oxygen and minium.

Plumbate of Lime.—If we dissolve 1 lb. of nitrate of lead in water, and add it with 3 atoms of lime to 16 lbs. of a solution of bleaching powder, specific gravity 1·08, and heat the mixture to 160° F., frequently stirring for five hours; pour off the clear liquor, and add 16 lbs. more of the solution, and continue the heat three hours longer, a compound of lime and binoxide of lead is obtained, which may be termed *plumbate of lime.* If we digest it in nitric acid, the oxide is left of a jet-black colour (Crum), but the composition is not easily made out.

Protosesquioxide, Red Oxide, Minium, Red Lead.—PbO P_2O_3 43, 344, is formed when oxide of lead is placed in a reverberatory furnace for forty-eight hours, and constantly stirred. It is a tasteless, insoluble scarlet powder of specific gravity 9·09 (Herapath). When digested in acetic acid, protoxide of lead dissolves, and a brown powder remains (binoxide?) Hence minium may be 2(PbO) PbO_2. By heat it is decomposed into oxygen and protoxide. It is sometimes adulterated with colcothar, the residue on heating copperas, from which it may be distinguished by its becoming yellow (PbO) when heated. It does not unite with acids. With sulphurous acid (SO_2) it is converted into sulphate of lead ($PbOSO_3$).

Chloride of Lead.—PbCl 17·5, 140. Specific gravity 6·; occurs in nature as horn lead. It is easily prepared by dissolving galena or sulphuret of lead in hydrochloric acid; it crystallizes in 4-sided prisms, more frequently in white scales and needles; tasteless, dissolving in 22 parts of boiling, in 135 to 150 cold water. Oxide of lead unites with PbCl in several proportions; mixtures of 2 to 3 atoms PbO to 1 atom PbCl, are white or pale yellow; 1 PbCl with 7 PbO, when melted, forms Turner's yellow. It may be also obtained by heating 10 parts minium with 1 salammoniac. Oxychloride as a pigment is patented by Mr. Pattinson.

Iodide of Lead.—PbI, 28·75, 230·. Yellow powder or scales formed by precipitating a salt of lead by iodide of potassium, dissolving in boiling water or in hot dilute acetic acid, and allowing the solution to cool.

Sulphide of Lead, Sulphuret, Galena.—PbS. Lead ore. This ore occurs in the primary rocks of Cornwall and Argyllshire, in the graywacke rocks of Wanlockhead, Leadhills, and Carsphairn, and in the north of England in the mountain limestone upon which the coal formation reposes. Its *colour* is pure lead-gray; it is apt to tarnish. It sometimes occurs in a massive or fibrous form, but it is generally crystallized; the primary form is a cube; but it occurs in octahedrons, and in the shape of the leucite crystal, and sometimes as an icositetrahedron. Hardness 3·; spec. grav. 7·5. When heated before the blowpipe it gives off sulphur; and if on charcoal, a metallic bead of lead is left. When sulphide of lead is heated in contact with atmospheric air, the sulphur is oxidized and sulphate of lead is produced (PbS + 4 O become PbO SO_3). In the reduction of lead, therefore, this is the first product obtained. If an atom of sulphuret of lead and one of sulphate of lead be exposed to a powerful heat, they are resolved into metallic lead and sulphurous acid, PbS and $PbOSO_3$ become 2Pb and $2SO_2$. Galena is decomposed by iron when they are heated together, PbS and Fe becoming FeS and Pb.

Carbonate of Lead, White lead, Cerussa, Magistery of lead.—$PbOCO_2$, 16·75, 134. Known to the ancients. As it occurs in commerce, it contains always an excess of oxide, its composition being CO_2 14·00, and PbO 86·, instead of CO_2 16·43, Pb 83·57. It occurs in nature white, with intermediate shades down to a black. Crystalline form, right rhombic prism; specific gravity 7·235. Diamond lustre, fracture resinous. It may be formed artificially by precipitating acetate of lead by an alkaline carbonate; but in this state is not fitted for a pigment. It is usually prepared for commercial purposes by the Dutch method, which consists in spreading out thin sheets of lead in an apartment, the floor of which is covered with tan or horse-dung; the lead is laid over earthen pots containing weak acetic acid; the temperature of the dung or tan rises to 150°; the lead becomes corroded, being brought into a fit state by the acetic acid for combining with CO_2 from the dung. White lead is also made by bringing carbonic acid in contact with litharge, mixed with a small portion of acetate of lead in water.

Acetate of Lead. See ACETATES.

Sulphate of Lead.—$PbOSO_3$, 19·, 152. Specific gravity 6·3. White rhombic prisms when native, or heavy white powder when prepared by precipitating acetate of lead by sulphuric acid, or an alkaline sulphate; soluble in 22,816 water, in 480 sulphuric acid, in 172 nitric acid.

Nitrate of Lead.—$PbONO_5$ 20·75, 166·; colourless 8-hedrons. May be obtained by dissolving litharge or white lead in nitric acid. When the nitric acid is properly neutralized, the process of boiling down may be conducted in a copper vessel.

Chromate, Chrome Yellow.—$PbOCrO_3$ 20·5, 164; specific gravity 6·. Red oblique rhombic prisms; yellow powder formed by precipitating 100 bichromate of potash with 176 acetate of lead, and washing by subsidence. It is used in organic analysis; as a pigment; and in calico printing.

Dichromate, Chrome Orange.—An orange powder obtained by acting on chrome yellow with caustic soda. When chrome yellow is printed on cloth, if the cloth be passed through lime water, the yellow becomes orange, 2 (PbO CrO_3), and CaO become 2 $PbOCrO_3$, and CaO CrO_3.

Assay of Lead—Dry Assay.—It is impossible to arrive at completely accurate results in the assay of lead in the state of sulphides, more particularly when mixed with other sulphides. An approximation, however, may be attained by mixing 100 parts of the pounded ore (containing, for example, galena, antimony, and selenium), with 30 parts of fine iron filings and placing them in a Hessian crucible filled up to three-fourths. The mixture is covered with a layer of common salt or carbonate of soda, or black flux. A cover is then fitted on without heating, and the crucible introduced into a moderate heat, which is to be urged towards the close. When the crucible is fractured on cooling, we find a button of metal, which, on being struck with a hammer, divides into 2 parts; the inferior layer consisting of pure lead, and the upper containing magnetic oxide of iron. Sometimes particles of lead are interspersed through the fused mass, which are to be removed by pounding and passing through a fine sieve. Another method is to fuse 100 parts of galena with 150 carbonate of soda, 30 iron filings, and 5 of charcoal, or black flux may be substituted for the soda and charcoal. To reduce the oxides they are to be mixed with 2 parts of black flux; and if they contain lime, with 2 parts of borax or carbonate of soda, increasing the borax in proportion to the amount of lime. To reduce the sulphate, the powder is mixed with 4 parts of black flux; and iron filings may be further added to remove the sulphate more readily. Other fluxes are:—2 galena, 2 black flux, 1 native iron pyrites; or, 10 galena, 20 black flux, 3 oxide of zinc. For frequent assays, it is best to use a crucible made of hammered iron (cast iron is soon destroyed), of a cylindrical form, about 4 to 6 inches high, and covered with an iron lid.

Wet Assay.—To obtain the lead from an impure sulphuret, 100 grains of the ore are to be boiled in pure hydrochloric acid, until the fumes cease to strike a dark colour upon a piece of filtering paper dipped in acetate of lead. The acid should be free from sulphuric acid, otherwise an insoluble sulphate will be formed; and it should contain no nitric acid to prevent the conversion of the sulphur into sulphuric acid. The liquid is then to be mixed with water, and thrown on a weighed filter; the insoluble matter is washed as long as the liquor passing through the filter affords a precipitate with nitrate of silver. The insoluble matter is dried along with the filter in a water bath at 212°, till it ceases to lose weight. The chloride of lead being very insoluble in cold water, will deposit as the filtered liquid cools. This is dissolved by heating the fluid, and a current of sulphuretted hydrogen is passed through it until the lead is precipitated. The sulphuret of lead is thrown on a filter and washed with distilled water. It is then digested off the filter in dilute nitric acid, precipitated by sulphuric acid, or an alkaline sulphate from a concentrated solution, then filtered, the sulphate washed, dried, and weighed on the filter; 19 parts contain 13 of lead. In using nitric acid there is always a tendency to the formation of sulphate of lead, which, however, may be taken up by caustic potash or soda.

Separation of Lead.—To determine the amount of lead in pure galena, when it is mixed only with silver, we digest the ore in strong nitric acid; sulphate of lead is formed, which is thrown on a filter, washed, and weighed. The silver remains in solution, and may be precipitated by hydrochloric acid, and weighed as chloride of silver; 18 grains contain 13½ of silver. When the quantity of silver is very minute, 2 ounces of the lead must be placed on a cupel in a muffle, and heated with access of air till the lead is oxidized, and the silver remains in a state of purity. A few hundredths of a grain in 2 ounces may thus be appreciated.

Lead Minerals, or **Natural Salts**:—

Lead, Selenide. Specific gravity 7·187, H 3·. Pb Se; Se = 27·59, Pb = 1·81. Lead-gray granular masses, resembling sulphide of lead. When heated in a glass tube open at both ends crystals of selenic acid are formed, which deliquesce; B.B. burns in charcoal with a blue flame. Zorge, Hartz.

Lead, Bitelluride. *Foliated ore of Tellurium, Tellurium glance.* Spec. grav. 7·085, H 1 to 1·5. Lead-gray tables, or right square prisms; foliated, opaque, flexible when in thin plates. B.B. fuses into a malleable button; with borax gives a bead of gold, containing some silver; soluble in nitric acid. Te 32·08, Pb 54·74, Au 8·72, Cu 1·22, S 3·03. Nagyag and Offenbanya, Transylvania.

Lead, Chloride. PbCl. Spec. grav. 1·897?

White needles. Scratched by the nail. Soluble in cold water. Crater of Mount Vesuvius.

Lead, Dichloride. *Mendipite.* Pb_2Cl, or Pb ClPbO, or Pb 83·2, Cl 13·77, CO_2 1·03, SiO_3 1·46, HO ·54. Spec. grav. 7·, H 2·75. White 4-sided prisms on earthy black ore of manganese in the Mendip Hills.

Lead, Chloroxide. *Matlockite.* Pb 83·667, Cl 8·45, CO_2 1·03, SiO_3 1·46, HO ·54. Straw-yellow, cleaving with a foliated texture; the cleavage planes meeting at an angle of 102° to 103°. B.B. easily fuses, and is reduced on charcoal to a metallic globule. Soluble partially in water; soluble in nitric acid, with little or no effervescence. From Mendip (Berzelius); pure specimens from Cromford, near Matlock, in fine splendent right square prisms, consist of Pb Cl 55·177, PbO 44·3, HO ·072, with a spec. grav. of 7·21 (Greg and Smith).

Lead, Oxychloride, is prepared as a pigment by Mr. Pattinson, by forming chloride of lead from galena and muriatic acid, and mixing its solution in water, with sufficient lime water to remove half the chlorine.

Lead, Chlorocarbonate. *Murio-Carbonate, Corneous Lead.* Spec. grav. 6·056, H 2·75. Pb Cl PbO CO_2=Pb 39·17, Cl 13·56, PbO 43·32, CO_2 8·51. White rectangular square prisms, with pale tints of gray, yellow, and green; streak white; partially soluble in water; soluble, with effervescence, in nitric acid. Cromford, near Matlock; Alston moor; Badenweiler, Germany; Southampton, Massachusets.

Lead, Tungstate of. Spec. grav. 8·, H 3·. PbO TnO_2. Yellowish-gray 8-hedrons, with a square base, resembling molybdate. B.B. fuses, yielding a crystalline globule.

Lead, Molybdate. See MOLYBDATE OF LEAD.

Lead, Trismolybdate. 3 PbO, MoO_3. Spec. grav. 6·. Greenish-yellow concretions; from near Pamplona, South America.

Lead, Cupreous Sulphate of. $PbOSO_3$, CuOHO. Spec. grav. 5·3, H 3·. Deep blue right oblique prisms Leadhills.

Lead, Sulphato-carbonate of. *Lanarkite.* PbO SO_3, PbO CO_2. Spec. grav. 6·8 to 7·, H 2·25. Greenish or yellowish-white oblique angled 4-sided prisms, with curved faces, terminated by two planes; lustre adamantine, inclining to resinous; translucent, sectile. Partially soluble in nitric acid, with effervescence, leaving sulphate of lead. Leadhills.

Lead, Sulphato-tricarbonate of. *Leadhillite.* $PbOSO_3$, $3PbOCO_2$. Spec. grav. 6·3 to 6·5, H 2·5. Pale greenish rhomboids, or yellowish or colourless 6-sided prisms. More readily soluble in nitric acid than the preceding. Leadhills.

Lead, Phosphate of. *Polychrome, Pyromorphite, Brown lead ore, Green lead.*—Spec. grav. 6·5781 to 6·915, H 2·75. PbO 69·639, PO_5 17·6, Pb 7·668, Cl 2·656, FeO 2·, HO ·8. *Form.* Pb Cl; 6 to 8 3PbO, PO_5. Green, yellow, or brown masses and 6-sided prisms, frequently terminating in 6-sided pyramids, the edges of the prisms frequently being replaced by tangent planes; lustre resinous, brittle; fracture imperfect, conchoidal; semitransparent to translucent. B.B. melts on charcoal; soluble in warm nitric acid. Leadhills; Huel Goet, Brittany; Saxony.

Lead, Vanadiate of. Spec. grav. 6·6623 to 7·23, H 2·75, Pb 7·063, Cl 2·446, PbO 66·326, VO_3 23·436, Fe_2O_3 and SiO_3 ·163 (R.D.T.), 3 (3 PbO VO_3), Pb Cl. Light brownish-yellow in spheres or 6-sided prisms; opaque or slightly translucent on the edges; brittle; fracture even, or flat conchoidal; lustre resinous. B.B. fuses on charcoal into a black scoria; reduced with carbonate of soda; with borax, a blue or green bead. Soluble partially in sulphohydride of ammonia, becoming blue or green on boiling and the addition of muriatic acid. Wanlockhead; Wicklow; Mexico.

Lead, Arseniate of. *Mimetene, Nussierite, Campylite, Hedyphane.* Sp. gr. 6·41 to 6·9 and 7·25, H 2·75 to 3·5. PbO 75·59, AsO_5 21·2, PO_5 1·32, HCl 1·89 (Wöhler), 3(3 PbO, AsO_5 PO_5), Pb Cl. Pale or orange-yellow 6-sided prisms; texture foliated; lustre resinous, sectile. B.B. gives out fumes of arsenic on charcoal. Soluble in nitric acid; the acid is taken up by sulphohydride of ammonia, sulphuret being precipitated on the addition of muriatic acid to the solution. Cornwall; Cumberland.

Lead, Cupreo Chromate of. *Vauquelinite.* Spec. grav. 5·5 to 5·78, H 2·75. PbO 60·87, CuO 10·80, CrO_3 28·33. *Form.* 2 (PbO CrO_3) CuO. Dark or olive-green masses, or oblique rhombic prisms, with nearly the same angles as chabazite; fracture partly compact, partly earthy; streak siskin-green. B.B. melts into a dark gray bead with some metallic globules; on platinum foil passes into a green and then yellow bead, which forms a yellow solution with water. Accompanies Siberian chromate.

Lead, Plumbaluminite. *Binaluminate, Sexaluminate, Plombgomme.* Spec. grav. 4·88 to 6·35, H 4·25. PbO 40·14, Al_2O_3 37·, HO 18·8, CaO, MnO, FeO 1·8, SiO_3·6. *Form.* PbO, 2 Al_2O_3 6 HO. Yellowish and reddish-brown reniform masses, translucent, resembling Muller's glass. B.B. decrepitates and becomes white and opaque, with borax a transparent colourless glass; on charcoal becomes an enamel. Sometimes contains phosphates of alumina and lead. Huel Goet, Brittany.

Lead, Cupreous Sulphato Carbonate of. *Caledonite.* Spec. grav 5· to 6·4, H 2·75. PbO SO_3 55·8, PbO CO_2 32·8, CuO CO_2 11·4. Deep verdigris-green radiating crystals of right rhombic prisms of 95° and 85°; streak greenish-white; lustre resinous, translucent, rather brittle. Leadhills.

Lead, Nagyag Tellurium Ore of. Spec. grav. 6·84. Au 6·7, Te 13·, Pb 63·1, Sb 4·5,

Cu 1·, S 11·7. Iron-black curved plates interlaced together; disseminated in silicate of manganese; lustre splendent. From Nagyag.

Lead, Antimoniate of. Spec. grav. 3·93 to 4·76, H 4·. Gray, brown, or yellow reniform masses, opaque. SbO_5 31·72, PbO 61·83, HO 6·46. Nertschinsk.

Lead, Seleniate of. H 3 to 4. Sulphur-yellow masses in spheres and grape-like clusters. Frederichsglück.

Leather. (*Leder*, Ger.; *Cuir*, Fr.) A mode of preparing skins so as to fit them for domestic and other purposes. Skins or hides are prepared in various ways:—1. *Tanned leather* is formed by digestion in a solution of tannic acid, in the form of oak bark. 2. *Tawed* or *white leather*, from the skins of sheep, goats, and calves, used for kid gloves, &c. is prepared by means of alum and common salt. 3. *Chamois* or *wash leather*, formed with oil or fat. 4. *Parchment*, obtained by immersing skins in lime water. Tanned leather is obtained by steeping skins in water, to deprive them of fat, blood, &c.; they are then digested in lime water or caustic soda, to soften and deprive them of their hair. They are afterwards steeped in a solution of oak bark for several weeks, to unite the tannic acid with the gelatine of the skins. Russian leather is acted on with birch, tar, and alum. Morocco leather is dyed with madder or other colouring matter, mixed with tin salts, and alum.

Leaven. Fermenting dough.

Lecanoric Acid. *Lecanorine*, *Variolarine*. $C_{18}H_8O_8$. White silky needles, scarcely soluble in cold water; soluble in 2500 hot water, in 150 cold spirit, 15 boiling spirit, in 80 ether; soluble in caustic and carbonated alkalies. The solutions, when boiled, yield orcine and carbonic acid. Obtained by treating powdered Lecanora, Parmelia, or Variolaria by displacement with ether, evaporating the solution; treating the residue with ether and boiling water, and then dissolving and crystallizing with hot alcohol; or by treating the lichens with spirit containing ammonia, precipitating by acetic acid, and crystallizing from alcohol. Erythric acid appears to be orcino lecanoric acid, a coupled acid, consisting of orcine and lecanoric acid.

Lecithine. A phosphoric fat from human venous blood, not crystallized; similar to that of the egg, brain, &c.

Ledererite. SiO_3 49·47, Al_2O_3 21·48, CaO 11·48, NaO 3·94, HO 8·58, PO_5 3·48, Fe_2O_3 ·14. A variety of chabazite from Nova Scotia, with half the quantity of water.

Lederite. A variety of sphene.

Leditannic Acid. 3 HO C_{14}, H_6, O_6. An amorphous substance from Ledum palustre, by alcohol and lead salts.

Ledixanthine. $C_{14}H_6O_6$. A red or yellow body by the action of mineral acids on solutions of leditannic acid.

Ledum Camphor. From *Ledum palustre.*

Leelite. Spec. grav. 2·606, H 6·25. Deep flesh-red masses; structure, compact; fracture splintery and conchoidal; translucent like horn. SiO_3 81·91, Al_2O_3 6·55, FeO 6·42, KO 8·88. Grythittan, Nerike; allied to felspar.

Legallic Acid. A form of gallic acid by frequent sublimation; does not precipitate iron, black.

Legumine. C 50·68, H 6·74, N 18·76, S ·482, Ash ·6. The principal nitrogenous constituent of peas and beans, extracted by cold water. Seems identical with caseine.

Lehuntite. Spec. grav. 1·953, H 3·75. Flesh-red or white mass, like lump sugar, but often consisting of a congeries of scales or crystals under the microscope. B.B. fuses into a white enamel. SiO_3 47·33, Al_2O_3 24, NaO 13·2, CaO 1·524, HO 13·6. Allied to natrolite. Glenarm, Antrim.

Leiocome. A synonyme of British gum or soluble starch.

Lemanite. A synonyme of Saussurite.

Lemons, Essential Salt of. *Salt of Sorrel*, or binoxalate of potash.

Lemons, Oil of. $C_{10}H_8$. Sp. grav. ·847, of vapour 4·835. B.P. 343½°; deposits crystals at — 4°; obtained by distilling the skin of Citrus limonum; by HCl gas citrene is formed, a camphor united to HCl.

Lenzinite. Spec. grav. 1·8 to 2·1. SiO_3 37·5, Al_2O_3 37·5, HO 25·. Soft and soapy feel; adhering to the tongue; polishes with the finger; sectile, fracture conchoidal, dull; in amorphous masses, covered with oxide of iron. Kall, Eifeld, St. Sever.

Leonhardite. SiO_3 54·9, Al_2O_3 22·5, CaO 9·1, HO 13·5. *Form.* 3 ($CaOSiO_3$), 4 (Al_2O_3 2 SiO_3), 15 HO. Similar in appearance and form to Laumontite.

Lepidine. Yellow bitter mass or powder from the seeds of *Lepidium Iberis*.

Lepidocrocite. *Protohydrous Sesquioxide of iron.* Fe_2O_3 HO. A synonyme of Göthite.

Lepidolite. *Lithia Mica*, *Peach blossom mica.* Spec. grav. 2·8 to 3·, H 3·. SiO_3 47·99, Al_2O_3 23·75, MnO 7·06, MgO ·42, KO 14·63, LO 3·16, HFl 2·71, HO 1·5. Peachy scales, or oblique rhomboidal 8-gonal prisms. See MICA. Found in granite, &c.

Lepidomelane. Sp. grav. 3·, H 3. Black or leek-green hexagonal plates or scales; lustre adamantine; streak green. SiO_3 37·4, Al_2O_3 11·6, Fe_2O_3 27·66, FeO 12·43, MgO,CaO ·26, KO 9·2, HO ·6. B.B. becomes brown, and fuses into a black enamel, with borax a green glass. Persberg, Wermland.

Lepolite. Spec. grav. 2·76, H 6·. Colourless, gray, or green prisms. SiO_3 42·8, Al_2O_3 35·12, Fe_2O_3 1·5, CaO 14·14, MgO 2·27, NaO 1·5, HO 1·56. Finland.

Leptynite. *Whitestone.* Granite, sienite, or gneiss, deprived of their quartz, mica, and amphibole, with felspar alone remaining.

Lettsomite. *Velvet Copper Ore.* Blue globules in fibres. SO_2 15·38, Al_2O_3 11·06, Fe_2O_3 1·18, CuO 48·16, HO 23·06. Moldawa Bannat.

Leuchtenbergite. Spec. grav. 2·71, H 2·5. Yellowish or white imperfect rhombohedrons; feel greasy. SiO_3 32·35, Al_2O_3 18·, MgO 32·29, FeO 4·37, HO 12·5.. B.B. fuses into a green thin colourless bead with salt of phosphorus, colourless glass with borax. Slatonsk. Allied to chlorite.

Leucic Acid? $C_{12}H_{12}O_6$. Oily solution by the action of nitrous acid on an aqueous solution of leucine.

Leucine. (λευκος, white). *Aposepedine, Oxide of caseine.* $C_{12}H_{13}NO_4$. Small pearly plates, resembling margaric acid. Soluble in 27·7 water, 658 alcohol; insoluble in ether; soluble in acids and alkalies; neutral reaction; sublimes by heat in a current of air without decomposition; obtained by dissolving 1 part glue in 2 strong sulphuric acid; allowing them to stand for a day, diluting with 8 water; boiling for some hours, and neutralizing with chalk; evaporating the filtered liquor to a syrup, boiling with alcohol and crystallizing. By a similar process it is procured from albumen, fibrine, &c. or by boiling these substances with strong caustic alkali for several hours; neutralizing with sulphuric acid; evaporating and crystallizing from alcohol.

Leucite. *White garnet, Amphigene, Vesuviane, Leucolite.* Spec. grav. 2·49, H 7. Reddish, yellowish, generally white tetrakishexahedrons, or icositetrahedrons, or modifications of the 8-hedron, having 24 faces, with angles of 131° 48′ 16″, and 146° 26′ 33″. To this form in other minerals, the term *leucite crystal* is given; fracture conchoidal; lustre vitreous. B.B. infusible; fuses with borax or carbonate of lime into a clear globule. SiO_3 56·1, Al_2O_3 23·1, KO 21·15, Fe_2O_3 ·95. In basalt and lava in Italy and Bohemia; between Rome and Frascati.

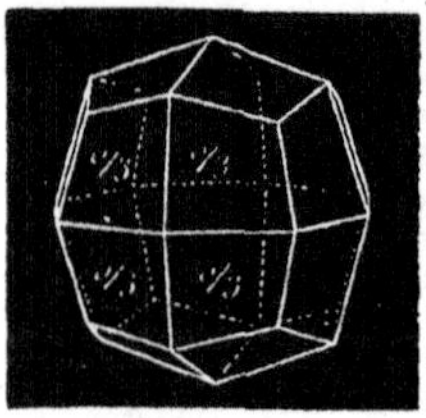

Leucoharmine. A synonyme of Harmine.

Leucoline. *Leucol. Quinoline.* $C_{18}H_7N$. Spec. grav. 1·081. Refracting power 1·64. B.P. 462¼°. Colourless fluid, with a disagreeable taste like bitter almond oil; becomes brown in air; sparingly soluble in water; soluble in alcohol, pyroxylic spirit, acetone, fat, and ethereal oils; it dissolves sulphur, phosphorus, camphor, and rosin; soluble in acids, with which it unites, and forms crystalline salts; obtained from coal gas heavy oils which pass over last in the distillation of coal oils. These oils are shaken with strong chlorohydric acid, and distilled with milk of lime in a copper still. Ammonia, odorine, aniline, and leucoline, pass over. The distilled fluid is treated with chlorohydric acid, and evaporated; caustic potash being added, the oils swim on the surface. On distillation ammonia and odorine pass first, then aniline, and lastly leucoline. Aniline can be separated by distillation, as it boils at 359½°, and gives a blue with bleaching powder. It is also obtained by heating quinine, cinchonine, strychnine, and thialdine, with potash, by the separation of carbonic acid and hydrogen.

Leucopetrine. $C_{50}H_{42}O_3$. Resinous needles by alcohol from an ethereal extract from brown coal.

Leucophane. Spec. grav. 2·974, H 3·5. Pale green to wine-yellow, crystalline, or massive and columnar; cleavage imperfect in 3 directions; lustre vitreous; powder white and phosphorescent; when heated electric. B.B. fuses into a violet glass. SiO_3 47·82, CaO 25·, GlO 11·51, MnO 1·01, NaO 7·59, KO ·26, F 6·17. In syenite, Langesundfiord, Norway.

Leucophorum. A pigment of the ancients, consisting of 6 red chalk, 10 siris(?), and 2 melinum (probably chalk).

Leucopyrite. A synonyme of arsenide of iron.

Leucoturic Acid. $C_6H_3N_2O_6$. White crystalline acid, obtained by boiling a solution of alloxanic acid to dryness, heating pretty strongly in a capsule, and adding water; difluan dissolves, and leucoturic acid remains.

Levigation. The mode of separating the finer parts of powders by suspension in water.

Levyine. Spec. grav. 2·161, H 4·. SiO_3 44·48, Al_2O_3, 23·77, CaO 10·71, KO 1·61, NaO 1·38, HO 17·41. Has been confounded with chabazite; it intumesces like scolezite before the blowpipe. It may be considered a natrolite with a double quantity of water. *Form.* (RO SiO_3, $Al_2O_3SiO_3$) 4 HO. In rhombohedrons similar to those of chabazite. Scotland, Ireland, Iceland.

Leyden Jar. A glass jar or bottle, the lower half of which is coated on the out and inside with tin foil. Into the mouth is passed a rod, terminating in a knob at top, while the lower end splits into three legs, or terminates in a chain passing to the bottom of the jar. When the jar is held in the hand, and the knob brought near the prime conductor of a machine, a spark passes to the knob, and a shock is experienced in the arm holding the jar, which is explained as follows:—When we pass a spark of positive electricity from the prime conductor into the interior of a Leyden jar, the person holding the jar receives a shock in the arm. This is a spark of positive electricity passing from the exterior of the jar to the earth, repelled by every spark of

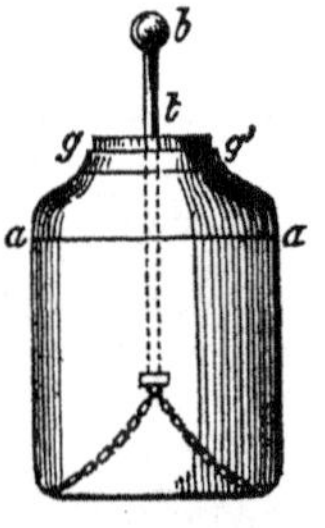

the same electricity which enters the interior of the jar. For every spark that enters the jar, therefore, a corresponding spark of the same electricity passes from the exterior coating until the positive electricity of the exterior of the jar is removed, and an equivalent amount of the same electricity has accumulated in the interior. When we connect the exterior and interior of the jar by a discharger, the accumulated positive electricity of the interior flies to the exterior, and supplies the place of that which passed to the earth, and thus the equilibrium is restored or everything returns to its natural condition. The principles involved in the preceding statement may be simply illustrated:—If we insulate a Leyden jar on a glass stool or glass tumbler inverted, connect it with the prime conductor and work the machine, we shall find that we cannot charge it, because the electricity of the same name with that of the machine cannot escape to the earth, glass being a non-conductor of electricity. The electricity, therefore, cannot accumulate in the interior of the jar, which therefore merely acts as a continuation of the prime conductor.

Lherzolite. Green pyroxene from Lake Lherz, Pyrenees.

Libavius's Fuming Liquor. Bichloride of tin.

Libethenite. Hydrous tetraphosphate of copper.

Lichens are familiar plants growing on rocks and trees. They consist principally of cellulose 17 per cent.; starch or lichenine 70; cetraric acid, lichesteric acid, fat, thallochlore, lichulmic acid, substance soluble in alcohol; unmixed nitrogenous matter; often sugar which I have crystallized, 3 per cent. In 1843, I showed that all lichens contain inorganic matter as an essential part of their composition, and that when burned they leave more ash than any other plants, except sea weeds; although previously it was the general expressed opinion of botanists that lichens derived their nourishment solely from the air (Proc. Phil. Soc. Glasgow, vol. i.) When lichens are boiled with spirit containing some carbonate of potash, lichesteric, and cetraric acids, thallochlore, lichulmic acid, fat, &c. are dissolved, while lichenine or lichen starch, with cellular matter, remains. Various crystalline bodies and powders have been obtained from lichens, but no evidence has been supplied that these are anything else than products of the processes employed; every new experimenter having rendered the subject more unintelligible by the multitude of such decompositions. Lichens are employed for the production of litmus and cudbear, by digesting them in ammoniacal liquor.

Lichenine. *Lichen Starch.* $C_{12}H_{10}O_{10}$. Colourless transparent mass, swelling up in water, and coloured blue by iodine; forming a clear solution in boiling water, but only swelling up and yielding a greenish colour with iodine; the starch separated by alcohol yields a blue with iodine; insoluble in alcohol and ether; 1 starch converts 23 hot water into a jelly; by long boiling it ceases to gelatinize, being probably converted into dextrine or sugar; nitric acid changes it into oxalic and saccharic acids, not into mucic acid. It does not exist in the state of granules, but is spread out between the fine cells. It is prepared by digesting for twenty-four hours Iceland moss (Cetraria Islandica) in cold water, containing $\frac{1}{300}$ of its weight of carbonate of potash; filtering the brown bitter solution off, and treating the residue with cold water till its bitter taste and alkaline action are removed. It is then boiled with 9 times its weight of water, the solution evaporated to one-third, the liquor strained through calico while boiling hot, and pressed; on cooling the liquid gelatinizes. The starch gradually contracts on cooling, and separates from the water.

Lichesteric Acid. HO $C_{28}H_{23}O_5$; F.P. 248°. White pearly mass, or 4-sided plates, without smell, but with a peculiar taste; insoluble in water, soluble in ether, alcohol, and fat oils; unites with bases, forming unibasic salts; obtained by boiling for half-an-hour, in a retort, lichens with strong spirit and some carbonate of potash; filtering while boiling through a cloth, acidulating with chlorohydric acid, and then adding 4 or 5 times its volume of water. The precipitate which falls is filtered and washed, and boiled with 9 times its volume of alcohol (·930). The crystalline precipitate which separates on cooling consists principally of lichesteric acid, in 4-sided tables; the cetraric acid mixed with it is in needles, which can be distinguished under the microscope, while a fatty body is in granules. To purify it completely, it is repeatedly crystallized from petroleum or mineral naphtha, which dissolves the lichesteric acid, but not cetraric acid or fat.

Lichulmic Acid. $C_{30}H_{13}O_{13}$. Brown substance, formed by the action of oxygen on cetraric acid in presence of an alkali.

Liebigite. H 2·25, A_2O_3 38·, CaO 8·, CO_2 10·2, HO 45·2. Apple-green warty transparent masses; loses water by heat, and becomes gray and black and orange on cooling. On pitch blende from Adrianople.

Liebnerite. Spec. grav. 2·814, H 3·5. Greenish-gray 6-sided prisms. B.B. not fusible. SiO_3 44·66, Al_2O_3 36·51, FeO 1·75, MgO 1·4, KO 9·9, NaO ·9, HO and CO_2 4·49. Mount Viesena, on felspar porphyry.

Lienine. C 53·71, H 8·95, N 4·82, O 32·52. Crystalline body from the fluid of the spleen.

Lievrite. See ILVAITE.

Ligament. The bands or cords which retain the joints in their situations. They are partially soluble in water, affording gelatine; insoluble in alcohol; soluble in sulphuric, nitric, and chlorohydric acids; the solutions are not precipitated by potash or yellow prussiate of potash, but are by nutgalls. Ligaments are soluble in potash, the solution yielding a slight precipitate with acetic acid. They contain much water, which can be partially removed by pressure and by heat, and when again immersed in that fluid they absorb it like a sponge.

Lignine. *Woody matter*, *Incrusting matter.* $C_{34}H_{24}O_{20}$ (Payen). The matter deposited on the interior of the cells of plants, consisting of a mixture of bodies which have different relations to solvents. *Lignose* is insoluble in water, alcohol, ether, and ammonia; soluble in potash and soda. *Lignone*, insoluble in water, alcohol, and ether; soluble in potash, soda, and ammonia. *Lignine*, insoluble in water and ether; soluble in potash, soda, ammonia, and alcohol. *Lignireose*, insoluble in water; soluble in potash, soda, ammonia, alcohol, and ether. According to some chemists lignine has the same composition as cellulose; but it is better to confine this name to the incrusting matter of the cells.

Lignite. Wood coal.

Ligurite. From Stura, in the Appenines, in oblique rhombic prisms, containing SiO_3 57·45, CaO 25·3, Al_2O_3 7·36, MgO 2·56, FeO 3·. Supposed by some to be sphene.

Ligustrine. A bitter extract by alcohol from the bark of Ligustrum vulgare, or privet.

Lilaicne. Bitter needles from the Syringa vulgaris.

Lilalite. A synonyme of lepidolite.

Limacine. A white substance obtained by decoction from garden slugs (Limax agrestis). Soluble in water, alcohol, and alkalies.

Limbelite. A decomposing chrysolite or olivine from Limbourg.

Lime. *Quicklime*, *Oxide of Calcium*, *Calcia.* (*Chaux*, Fr.; *Kalkerde*, Ger.) CaO 3·5, 28. Spec. grav. 2·3 (Kirwan), 3·08 (Royer and Dumas, An. Phil. 3,392), 3·1605 (Karsten), 3·2 (Richter). White mass, moderately hard, but easily pulverized. *Taste* hot and burning. It destroys the texture of those organic bodies to which it is applied; tinges vegetable blues green and then yellow. Lime is so refractory that it cannot be melted in a furnace; but it fuses before the oxy-hydrogen blowpipe into a brilliant limpid glass with a red light. Lime is not volatile; it is decomposed at a red heat by potassium, and at a white heat by chlorine. *Soluble* in 656 water at 32° (Phillips); in 778 at 60°, 758 (Thomson); at 130°; 972 at 212°; 1270 (Dalton); 1280 (Phillips); 1532 (Winsten). Hence lime is 1·688 times more soluble in cold than in hot water, and when a cold saturated solution of lime is boiled, it deposits lime.

Process.—1. Pure lime is obtained by heating calcium in air or oxygen. 2. By exposing limestone (CaO CO_2), to a strong heat, carbonic acid is evolved, and *quicklime*, or lime, or oxide of calcium (CaO) remains. It was long known that limestone lost something by exposure to heat. Some thought the loss due to water (Van Helmont, Ludovicus, Macquer), and that it regained it by exposure to the air. Stahl attributed the new properties of quicklime to the more minute division of its particles by fire. Boyle ascribed them to *fixation* of fire in the lime, a view adopted by Newton, Hales, and Meyer. In 1756, Dr. Black, then of Glasgow, showed that the quantity of water lost is not sufficient to account for the difference in weight of limestone before and after calcination; but remembering that Dr. Hales had proved *air* to be disengaged from limestone by the action of acids, he conjectured that this might be what is lost in calcination. By the application of weights and measures, he found that the loss was exactly due to the water and air removed. By exposing it again to the air, he found that it regained its weight by absorbing air. He termed this air, *fixed air*, which was afterwards examined by Priestley and others, and called carbonic acid. On the great scale, limestone is burned for mortar and manures in lime-kilns, funnel-shaped chambers, in which the limestone is mixed with coal and ignited; the carbonic acid is converted first into carbonic oxide, which burns and changes into carbonic acid (CO_2, C $=$ 2 CO, then 2 CO and $O_2 = 2\,CO_2$). When it contains too much clay (silica and alumina), the lime unites with the silica and becomes *dead-burnt*, that is, it will not slake. Oyster shells are burned into quicklime, which is much used in India for building and whitewashing, under the name of *chunam*.

Hydrate. Slaked lime.—CaOHO 4·625, 37·, CaO 75·67, HO 24·33. Fine white powder, or transparent 6-sided prisms or tables, cleaving in planes parallel to the faces of a rhomboid. It loses water at a low red heat, but does not fuse. Quicklime does not absorb carbonic acid; but as soon as it is in a state of hydrate, the carbonic acid of the air partly replaces the water (Scheele), which in this case seems to occupy the function of an acid. Soluble in 384 water at 60°; 729 at 130°; 952 at 212° (Dalton). *Process.*—When we pour water upon a piece of quicklime, we observe that the water is absorbed, and air filling the pores which have been formed by the expulsion of the carbonic acid, makes its escape with an audible sound. The absorption of the water, however, at last ceases; the lime feels hot to the hand; vapour begins to rise; the lime swells up, crackles, emits much heat, and gradually falls into powder. This process is termed *slaking* lime. When the lime is considerable in quantity, the heat is so great as to inflame combustibles, and even vessels loaded with it; in a dark place, light is perceptible.

Salts of lime.—1. A considerable number of the salts of lime are insoluble in water. Some of those which are soluble, cannot easily be crystallized. 2. Insoluble salts of lime when boiled with carbonate of soda or potash, leave an insoluble carbonate of lime. 3. Caustic ammonia does not precipitate lime, while soda and potash precipitate it. 4. Sulphate of lime does not precipitate lime, but oxalate of ammonia precipitates it in neutral and alkaline solutions, the precipitate being insoluble in water and acetic acid.

Sulphate, anhydrous sulphate, anhydrite. CaO SO_3 8·5, 68·. Specific gravity at 39°·2, in vacuo (natural) 2·96, (Royer and Dumas); (artificial) 2·9271 (Karsten); hardness equal to calcareous spar. *Crystal* right rectangular prism, the lateral edges often replaced by

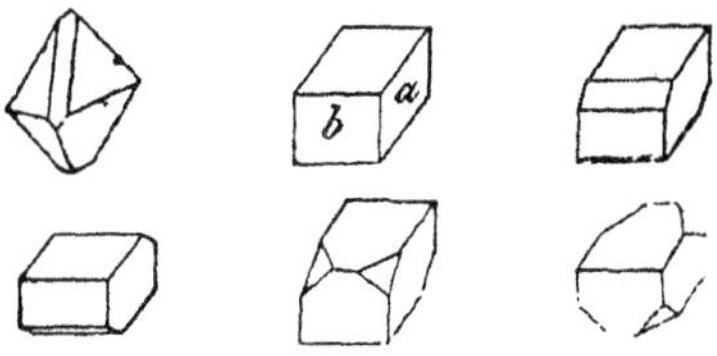

tangent planes, or amorphous; phosphoresces when heated; fuses at a red heat, and becomes a white crystalline mass; converted by ignition with charcoal into sulphide. Quicklime absorbs the vapour of sulphuric acid at a red heat (Rose). Oil of vitriol in large quantity poured on quicklime, renders it red hot (Kastner); and when diluted or concentrated acid is dropped on quicklime, light is emitted in the dark. *Insoluble* in water; vapour of solid sulphuric acid, and oil of vitriol, do not decompose carbonate of lime; the acid must be diluted (Marx, N. Br. Arch. 15, 103). *Sources.*—It may be obtained by heating hydrous sulphate of lime to the temperature of 400° (Millon). It occurs also native, as first pointed out by Hauy; a specimen was described by Fleurieu in 1798 (Journ. Physique), and analyzed by Vauquelin, who first ascertained its composition; afterwards described by Bournon (Jour. Min. An. x. 2, 355); Chenevix (ib. 418); and Klaproth (Gehlen Jour. 2, 355).

Semihydrated sulphate.—CaO SO_3, ½HO 8·0625 72·5. Specific gravity 2·757; needles with a dark coating, obtained from the water of a boiler worked under two pressures (Johnston, Phil. Mag. 13, 325).

Bihydrated sulphate.—CaO SO_3 2HO 10·75. 86·. Specific gravity 2·322, at 32° in vacuo (Royer and Dumas). *Crystal* native right oblique prism, the faces inclined at angles of 113° 8′, and 66° 52′. The most common shape is a thin rhomboidal prism with bevelled edges; lustre silky, very soft; artificially, it is in fine needles, or granular, scratched with the nail. At 212° in vacuo, over oil of vitriol, loses 1 atom of water or 10·46 per cent.; becomes anhydrous at 270° (Graham); 259° (Mitscherlich); near 400° (Millon). When dried at the temperature of 270°, and mixed with water, union takes place with the evolution of heat; the salt *sets*, and is thus used to take plaster casts and impressions. But if the heat be greater, the water does not readily unite. Hence gypsum *over-burnt* does not again set, and is of no value. *Soluble* in 460 water at 60° (Bucholz, Gehlen, J. 5, 165), 380 (Giese), and 450 boiling water (B). 388 (G.). *Taste* slightly nauseous, scarcely perceptible, except in considerable quantity (Macquer); solubility increased by salammoniac (Vogel); common salt (Trommsdorf); sulphate of soda (O. Henry); and succinate of ammonia (Wittstein); soluble in sulphuric acid. Melts before the blowpipe, giving an opaque glass at a temperature of 51° Wedgewood (Saussure, Jour. Phys. 45, 16). The solution saturated under pressure with hydrogen, contains in six months much sulphide (Paul, Phil. Mag. 15, 63).

Hypochlorite, Chloride of lime, Bleaching powder, Tennant's bleaching powder.—CaO ClO, Ca Cl 2 HO. 18·25, 18·1, 144·8. White powder with the peculiar odour of hypochlorous acid; partially soluble, leaving hydrate of lime; attracts water and carbonic acid slowly in the air, and evolves chlorine; heat converts it into chloride of calcium and chlorate of lime, oxygen being evolved and sometimes chlorine, and the bleaching power lost (18 (Ca Cl + CaO ClO) = 12 O + 17 Ca Cl + CaO ClO_5) (Morin); becomes moist after long exposure in the air, giving out oxygen slowly, losing its bleaching power, and being converted into chloride of calcium. The solution of bleaching powder is transparent and colourless, with the characteristic odour. When boiled, it evolves oxygen, and the same action occurs under the influence of metals and light, one-third being changed into chloride of calcium and chlorate of lime, and two-thirds into chloride of calcium and oxygen gas. By exposure to the direct solar rays, the hypochlorite changes into chlorite of lime.

Preparation.—On the 23d January, 1797, Charles Tennant of Darnley, near Glasgow, obtained a patent for preparing bleaching liquor, and on 30th April, 1799, for forming solid bleaching powder. This important article has been ever since manufactured on an increasingly great scale, by Messrs. Charles Tennant & Co. of St. Rollox. 1. It was originally made in the fluid form, and slaked lime was mixed with water and placed in a receiver. 2. The method of preparing dry bleaching powder at the present time, is to slake quicklime with water, since lime and carbonate of lime do not absorb chlorine. The slaked lime is spread out on the floor, and sometimes on shelves, in large stone chambers, to the depth of 3 or 4 inches. The doors of this apartment, formed of flags, are closed up air-tight by a cement of pitch, rosin, and gypsum. A lead tube conducts from the bleaching powder chambers to the stills in which the chlorine is prepared. These are formed of sandstone, and consist of a double box. Into the interior is introduced the finely ground black oxide of manganese, and upon this is run from reservoirs, chlorohydric acid, which is derived in soda works from the gas proceeding from the decomposition of common salt by sulphuric acid. The exterior box, of comparatively much smaller dimensions, is filled with steam, in order to assist the action of the acid upon the binoxide of manganese. The chlorine thus liberated, passes by the lead tube into the chamber by an aperture in the top or sides. If the chlorine be admitted too rapidly, the temperature rises, and may reach 246°, which will have the effect of forming chloride of calcium and chlorate of potash. The chlorine therefore ought to be passed slowly. An atom of lime slaked with half an atom water, absorbs only $\frac{1}{4}$ atom chlorine, with 1 or 2 atoms water, $\frac{1}{2}$ atom chlorine; but when fully slaked with an excess, and the redundancy removed by heat, 1 atom lime absorbs nearly 1 atom chlorine (53 per cent.). The general action is expressed by the formula 2 CaO 2 Cl = Ca Cl CaO ClO. It has been found that when the slaked lime is mixed with $\frac{1}{20}$ of common salt, or sulphate of soda, the absorption is promoted. The lime remains exposed to the action of the chlorine for forty-eight to ninety-six hours; the doors are then opened, the excess of chlorine allowed to escape, and the bleaching powder packed in casks. For mode of testing, see CHLORINE.

Mercer's Method of Testing Bleaching Powder, with Yellow Prussiate of Potash, depends on the conversion of the yellow into the red prussiate, as indicated by pouring a solution of a given strength into a solution of bleaching powder, and dipping a piece of pale sesquioxide of iron-buff calico into the mixture. When the cloth ceases to become blue, it is a proof that the yellow has become red prussiate, which does not form Prussian blue with sesquioxide of iron. A convenient apparatus for this test is sold by J. Ronchetti, Manchester, where it is commonly employed.

Carbonate, limestone, marble, calcareous spar, chalk, marl, oolite, travertine, tufa.—CaO CO_2. 6·25, 50; spec. grav. of Iceland spar 2·717 in vacuo (Le Royer and Dumas); 2·721 (Mohs); 2·75 (Neumann); of acute calc spar 2·6987; less acute 2·700; most obtuse 2·7064 (Karsten). *Dimorphous.* 1. Calc spar crystallizes in obtuse rhombohedrons, with angles of 105°

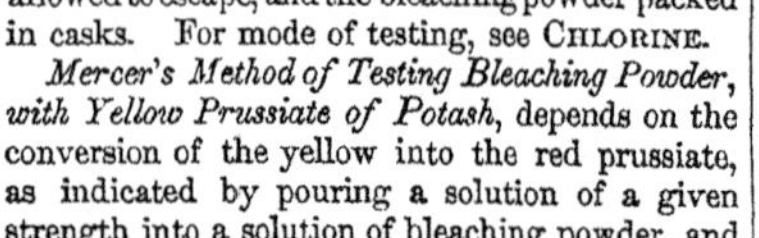

5' (Wollaston); of this form 680 modifications have been described. 2. Arragonite right rhomboidal prism, usually as 6-sided prisms, hardness 3·75; specific grav. 2·931 (Haidinger); 2·920 (Bournon); 2·9267 (Biot); 2·885 to 2·9304 (Stromeyer); 2·7647 to 2·9467 (Beaudant). Carbonate of lime in all its forms is

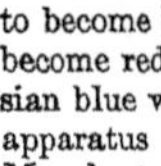

tasteless; it is soluble in 10,601 water at 60°, and 8834 at 212°, and in 65,246 water, having alkaline salts in solution (Fresenius). When newly precipitated, it dissolves easily in a strong cold solution of salammoniac, but with difficulty after standing a day; it is also soluble in solutions of carbonate, sulphate, nitrate and succinate of ammonia. When the precipitate is allowed to stand, it is much less soluble (Vogel, Wittstein); at a red heat it loses all its carbonic acid, and if steam be present, less heat is required. When heated in a glass tube with half its weight of sodium, carbon is deposited. It dissolves readily in water through which carbonic acid is passed, or under pressure (CaO CO_2 and CO_2, become CaO 2 CO_2), and on boiling the water carbonic acid is evolved, and carbonate of lime is reprecipitated.

Trisphosphate, Common Phosphate, Apatite, Asparagus stone, Bone earth, Moroxite, Phosphorite.—3 CaO PO_5 19·5, 155·4, or 8 CaO 3 PO_5 55, 440 (Berzelius). White powder as precipitated from guano and bones, or in low 6-sided

prisms with a white, blue, or green colour; in a native state with a spec. grav. of 3·0989 to 3·2 (Hauy). *Soluble* to a small extent when newly precipitated, in pure water, the solution becoming cloudy on the addition of nitrate of lead (Phillips); in water containing starch or glue; in solution of common salt (Thomson); in ammoniacal salts (Wöhler); chlorohydric and nitric acids dissolve it by removing 2 atoms of lime and converting it into protophosphate. From these solutions, acetate of lead precipitates phosphate of lead; 100,000 parts of water saturated with carbonic acid, at 50° dissolve 75 parts of pure phosphate (Lassaigne, Comptes Rendus, 23, 101); from this property we may explain the mode in which bone earth is absorbed into plants and animals, and how bones decay when exposed to the air. Dilute sulphuric acid removes the lime, and leaves a soluble compound with a small portion of lime; 1 atom trisphosphate, and 2 atoms sulphuric acid, give sulphate of lime and protophosphate of lime; some sulphate remains in solution, which may be precipitated by the addition of alcohol; 10 parts of bone earth, 5 silica or sand, and 5 clay, ignited in a charcoal crucible, yield 3 phosphorus when distilled with water, silicate of lime and carbonic oxide being formed (Berthier).

Natural Sources.—1. Bone earth occurs in animals and plants, constituting the basis of the bones and teeth, and of those portions of plants used as food. It is prepared pure from bones and guano, by calcining them in a clay crucible, dissolving the white residue in dilute chlorohydric acid, precipitating the solution by caustic ammonia in the form of a jelly, and filtering rapidly without access of air to prevent carbonate of lime from being formed. It occurs in the form of apatite or phosphorite mixed with chloride and fluoride of calcium in Norway, Spain, and Tyrol. Its hardness is 5·; it fuses before the blowpipe into a white enamel, and into a glass with borax and biphosphate of soda.—*Artificially*, bone earth is obtained by precipitating an ammoniacal solution of chloride of calcium with phosphate of ammonia (Fuchs). It then contains 54 (Fuchs) per cent. of lime; or 52·95 (R.D.T.); 53·34 (Berzelius); approximating to the formula 3 CaO PO_5. Formerly, it was found to contain 51·68 lime (Berzelius); and its formula was 8 CaO 3 PO_5.

Lime Silicate. See TABLE SPAR.

Lime, soda silicate of. See WOLLASTONITE.

Lime, silicate of, Tersilicate of lime.—Specific grav. 2·2055; H. 3·5. Snow-white, resembling tremolite; texture fine radiated, phosphoresces when rubbed; decomposed by acids; easily frangible. B.B. acts like table spar, fusing into a colourless glass with difficulty. SiO_3 55·2, CaO 34·284, Al_2O_3 4·16, FeO 2·896, HO 3·4. *Formula* CaO SiO_3 (T. Thomson); a similar mineral from Girvan in Ayrshire, I have found to have a specific grav. of 2·845, and to contain SiO_3 55·5, CaO 33·81, with some soda, magnesia, iron, and alumina. For other silicates, see DYSCLASITE, WOLLASTONITE, or SODA TABLE SPAR, TABLE SPAR, APOPHYLLITE, GUROLITE.

Lime, arseniate of.—Specific gravity 2·848; H 2 to 2·5. White, 8-hedrons, with an oblique base. Locality unknown. AsO_5 67·781, CaO 17·9, HO 14·319, or CaO AsO_5 85·681, HO 14·319. See also PHARMACOLITE.

Lime, Tungstate of, Scheelate of Lime, Tungsten.—Specific grav. 5·959 to 6·076, H 4·5 to 5· White or yellow, gray or brown, 8-hedrons, with a square base more or less modified, found on quartz; with angles of 100° 40′, 128° 46′; structure imperfectly foliated; lustre glassy. B.B. crackles, but does not fuse; with borax becomes a clear glass; CaO 19·06, TnO_3 78, SiO_3 2·8. *Form.* CaO TnO_3. Zinnwald, Bohemia; Saxony; Pengelly, Cornwall; Coquimbo.

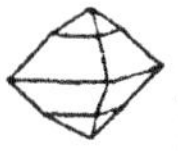

Limonine. $C_{42}H_{25}O_{13}$. F.P. 471°. Rhombic crystals, slightly soluble in water, ether, and ammonia; soluble in mineral acids, alcohol, and acetic acid and potash. By alcohol from orange and lemon seeds.

Limonite. Bog iron ore.

Linarite. Cuprosulphate of lead.

Lincolnite. A synonyme of Heulandite.

Lindsayite. Probably a pseudomorph of Anorthite, consisting of SiO_3 42·22, Al_2O_3 27·55, Fe_2O_3 6·98, FeO 2·, MgO 8·85, KO 3·, NaO 2·53, HO 7·.

Linea Vulgaris. The flowers of this plant are used as a yellow dye, due to the presence of *Aethokirrine*, a yellow amorphous substance, soluble in ether and alcohol.

Linine. White powder from the root of Linum catharticum, by alcohol.

Linoleic Acid. $C_{46}H_{38}O_5$ HO? The oleic acid from linseed oil by saponification.

Linseed Oil. Yellow oil; specific gravity ·9347; soluble in ether and alcohol; from *Linum usitatissimum.*

Lipic Acid. $C_5H_3O_3$. F.P. 284° to 293°. Plates or needles, soluble in alcohol and ether; less so in water; isomeric with pyrotartaric acid; forms crystalline salts with alkalies; lipate of ammonia yields precipitates with salts of lime, strontian, barytes, iron, copper, and silver; obtained by boiling oleic, stearic, or margaric acid, for twelve hours in a retort with nitric acid, heating the acid liquor four or five times with fresh nitric acid, and evaporating to one-fourth of its volume. On standing, suberic and azoleic acids crystallize out. The mother liquor by evaporation, gives more suberic acid and pimelic acid. The supernatant fluid contains adipic and lipic acids, which separate by evaporation at a gentle heat. Warm ether dissolves the two acids, and boiling alcohol dissolves the ethereal extract, and by spontaneous evaporation allows the adipic acid to separate in warty, the lipic acid in foliated crystals.

Lipyle. C_3H_2. The hypothetic radical of oxide of Lipyle.

Lipyle, Oxide of. C_3H_2O. A hypothetic body supposed to form the basis of oils. When saponified, 2 atoms of it unite with 4 atoms water to form glycerine ($C_6H_7O_5$ HO).

Liquefaction. The conversion of a solid, vapour, or gas, into a liquid.

Liquor of Cadet. Protoxide of Kakodyle.

Liquor, Dutch. ($C_4H_4Cl_2$). Bichloride of olefiant gas. An oil by the action of chlorine on olefiant gas.

Liquor, Fuming, of Libavius. Bichloride of tin.

Liquorice. See GLYCYRRHIZINE.

Liriodendrine. F.P. 181½°; rectangular tables or needles with a bitter taste; soluble in alcohol and ether; from the root bark of Liriodendron tulipifera.

Liroconite. A synonyme of octahedral arseniate of copper.

Lithamaric Acid. An acid resembling chinovasic acid, and said to exist in bezoar stones.

Litharge. Protoxide of lead.

Lithia. *Oxide of Lithium.* LO. Lithia exists in petalite, likewise in spodumene, lepidolite an amethystine variety of mica, tourmaline. It has been likewise detected in the mineral waters of Carlsbad, Franzbad, and Marienbad (Berzelius, Pogg. Ann. 4, 248); Pyrmont (Brandes Schweig. 45, 369); Kissingen, Hall, Bilin, Kreuznach, Lavey, Slintsch.

Hydrate of Lithia.—LO HO? White, with a crystalline fracture; not volatile; changes vegetable blues to green, with a caustic taste; not deliquescent, but absorbs carbonic acid in the air; soluble in water giving out heat; equally soluble in hot and cold water, but less soluble than potash or soda; scarcely soluble in alcohol of specific grav. ·85; corrodes platinum. When alcohol in which lithia is diffused is ignited, it burns with a *carmine* flame.

Process.—1. The finely pulverized mineral is mixed with a quantity of fluor spar equal to 2½ times the amount of silica contained in the mineral. The mixture is placed in a silver crucible, and made into a paste with sulphuric acid. On the application of heat, silicofluoric acid passes away carrying off all the silicon; as soon as fumes cease to come off, and the mass is tolerably dry, the heat is gradually raised to ignition to drive off excess of sulphuric acid. The mass is digested in water, which takes up sulphate of lithia and sulphate of lime. The latter may be removed by boiling with oxalate of ammonia, filtering, and evaporating. The lithia may be thrown down from the sulphate by carbonate of ammonia, and the carbonate of lithia is then deprived of carbonic acid by means of caustic lime, in the same way as with the other caustic alkalies. 2. The finely powdered mineral may be fused with 4 times its weight of carbonate or nitrate of barytes, the fused mass dissolved in muriatic acid (HCl), evaporated to dryness, moistened with muriatic acid, water added, and the silica separated by filtering; the alumina and earths separated by boiling with carbonate and oxalate of ammonia, filtration, and the filtered liquid ignited. Instead of barytes, caustic lime may be employed.

Lithia Salts.—1. The salts of lithia are all soluble in water, but they are less soluble than potash or soda salts. 2. Carbonate of potash precipitates concentrated solutions of lithia salts, which disappear on the addition of water, or by boiling. 3. Bichloride of platinum, caustic potash, prussiate of potash, infusion of gallnuts, do not precipitate salts of lithia. 4. Lithia is *characterized* by its conduct with phosphate of soda. When we mix this salt with a salt of lithia and evaporate, the solution becomes turbid. On the mass becoming dry by evaporation, if we add water, an insoluble white powder remains, which is a phosphate of lithia and soda (NaO, LO HO PO_5). 5. An alcoholic solution of a salt of lithia burns with a carmine flame. 6. When a mineral or salt in powder containing lithia, is mixed with fluor spar and bisulphate of potash, and heated before the blowpipe, the flame of the lamp is tinged of a purple red.

Lithic Acid. Uric acid.

Lithiospore. Sulphate of barytes.

Lithium. L. ·8125; 6·5. A metal resembling potassium obtained by the galvanic battery from lithia (Davy).

Chloride.—LCl 5·25, 5·225, 41·8. Cubes with a taste resembling common salt; fuses by ignition, and volatilizes by a white heat; less volatile than common salt, more so than chloride of potassium (Gmelin, Rose); obtained by dissolving the carbonate or hydrate of lithia in muriatic acid, and crystallizing in the sun or from an alcoholic solution in vacuo. When it deliquesces in the air, it gradually forms rectangular prisms with 4-sided summits, which are bi- or quater hydrated crystals (LCl 2 HO or 4 HO. Rammelsberg, Poggend. Ann. 66, 79; Hermann, ib. 15, 480).

Iodide.—Deliquescent needles, with 6 atoms water, often yellow from free iodine.

Fluoride.—Opaque crystalline grains, scarcely soluble in water. *Bifluoride* or *Fluohydride.*—Small acid crystals, becoming fluoride on heating.

Sulphide.—Soluble in water and alcohol; obtained by heating sulphate of lithia with charcoal. *Bisulphohydride.*—By passing a current of sulphohydric acid through the sulphide, and evaporating, long prisms result.

Sulphate.—$LOSO_3HO$ 7·925, 63·4. Shining rhombic prisms, with dihedral summits (Hermann); long efflorescing tables (C. Gmelin); equally soluble in hot and cold water; soluble in alcohol; fuses with difficulty by ignition. When gypsum is present, it fuses below a red heat.

Lithofellic Acid. *Bezoaric Acid.* F.P. of crystallized 401°, of amorphous acid 225°. HO $C_{40}H_{35}O_7$. Colourless 6-sided prisms; insoluble in water; soluble in 29 cold, 6½ boiling alcohol; in 444 cold, 47 boiling ether; soluble in strong acetic acid; obtained by digesting oriental bezoars in soda, precipitating by an acid, and crystallizing the precipitate out of alcohol. By dry distillation *pyrolithofellic acid*, $C_{40}H_{33}O_5$, is obtained, and by nitric acid *nitrolithofellic acid*, $C_{40}H_{28}N_2O_{22}$, results.

Lithomarge. Spec. grav. 2·457, H 2·5. Yellowish earthy substance. SiO_3 55·5, Al_2O_3 24·3, Fe_2O_3 2·5, CaO 1·55, MgO ·95, KO 9·25, HO 5·8.

Lithrodes. A synonyme of Nepheline.

Litmus. (*Turnsole*, Fr.; *Lackmus*, Ger.) Blue cubical cakes imported from Holland, &c. It is prepared from the Leconora tartarea, and perhaps other lichens. They are pulverized and mixed with putrid urine, when the blue colour is developed. It is used in the form of infusion and tincture to detect acids, and when reddened to test for alkalies, and also in the form of litmus test paper, formed by immersing paper in the infusion, and drying it.

Litre or **Liter.** A French measure of capacity; 4·54345794 litres are equal to 1 imperial gallon of 277·274 cubic inches; 1 litre is therefore equivalent to 61·02709 cubic inches; 1 pint = ·567932 litres; 1 quart = 1·135864 litre; 1 bushel = 36·347664 litres. The litre is = 1·760773 pint = ·2200967 gallon; a decalitre

$= 2{\cdot}200967$ gallons; a hectolitre $= 22{\cdot}00967$ gallons.

Liver. A term applied to certain roasted metallic sulphides from their liver colour.

Lixivium and **Lixiviation.** Terms applied to the water and process used in separating soluble from insoluble salts by digestion.

Lizaric Acid. $C_{30}\,H_{10}\,O_9$. Crimson needles from Zealand madder, subliming; soluble in sulphuric acid with a blood-red colour.

Loam. A term usually applied to soil of a dark colour, due to vegetable matter.

Lobelic Acid. From the decoction of the leaves of Lobelia inflata.

Lobeline. Yellow bitter matter from the root of Lobelia inflata by alcohol.

Loboite. A synonyme of idocrase.

Logwood. The wood of the Hæmatoxylon Campeachianum, growing on the American continent. It sometimes contains fine crystals imbedded in it, of the colouring matter *hæmatoxyline*, q.v. Logwood is used by the calico printer for black and brown colours, with an alum mordant. The colouring matter is usually sold in the form of logwood liquor, a decoction in hot water strongly concentrated.

Lonchidite. Spec. grav. 4·925, H 6·5, 25 FeS_2, FeAs or S 49·61, As 4·4, Fe 44·23, Co ·35, Cu ·75, Pb ·2· A variety of radiated pyrites: Freiberg, Schneeberg, Cornwall.

Lophine. $C_{46}H_{16}N_2$. Silky crystals, without colour, taste, or smell; insoluble in water, nearly so in alcohol and ether; soluble in petroleum, turpentine, and in alcoholic solutions of potash; forms salts with acids; obtained by distilling hydrobenzamide; ammonia is evolved and an oil. The residue solidifies, and consists of two bodies, one of which, lophine, is soluble in ether. By nitric acid it is converted into trinitrolophyle, $C_{46}\,H_{13}\,3NO_4N_2$, in orange crystals.

Lotalite. A variety of amphibole.

Loxoclase. Spec. grav. 2·61, H 6·25. A variety of felspar or albite, from Hammond, New York, containing SiO_3 63·5, Al_2O_3 20·29, Fe_2O_3 ·67, CaO 3·22, NaO 8·76, KO 3·03.

Loweite. *Bloedite?* Spec. grav. 2·376, H 2·75. 2 MgO, SO_3, 2 NaO, SO_3, 5 HO? Yellowish-white tetragonal pyramids; taste feebly saline and astringent; from Ischl. Contains NaO 18·97, MgO 12·78, SO_3 52·35, HO 14·45, Fe_2O_3, and Mn_2O_3 ·66.

Luchonine. A substance found in the sulphur wells of Luchon, identical with Baregine (?); greenish; coagulates at 178°, and by alcohol.

Lucifer Matches are made by dipping wooden matches in a mixture of 2 chlorate of potash, 4 phosphorus, 7 gum arabic, and 2 glue or isinglass, coloured with indigo or minium.

Lunar Caustic. Nitrate of silver.

Lupine Seeds. The seeds of various species of Lupinus. In the figs. are represented the cells with starch, which resemble those of peas and beans. The skin completely resembles that of the bean. The Levant lupine has been found

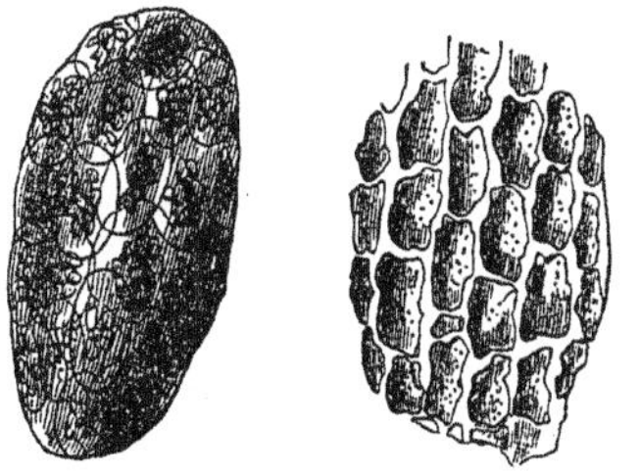

Cells with starch. Skin cells.
[*Hooker.*]

to afford an ash consisting of SiO_3 ·87, PO_5 25·53, CO_2 ·56, SO_3 6·8, Cl 2·11, MgO 6·18, CaO 7·75, Na 17·75, KO 33·54.

Lupinine. A bitter substance from different species of lupine.

Lupuline. *Lupulite.* Yellow bitter principle of hops (Humulus lupulus); soluble in 20 boiling water; insoluble in ether; soluble in alcohol. It is a neutral body; obtained by forming a tincture by means of alcohol with hops, distilling off the alcohol, adding water, which precipitates resin; precipitating from the liquor the tannic acid by lime, evaporating, removing resin by ether, and taking up the lupuline by alcohol.

Lustre. (*Eclat*, Fr.) A term referring to the amount of light reflected by crystals. The species are metallic, diamond or adamantine, resinous, vitreous or glassy, pearly, or that of mother-of-pearl, fatty, silky. The intensity of lustre is said to be, splendent, shining, glistening, glimmering, dull.

Lute. (*Kitte*, Ger.) Cement for tube joints, usually made with fats. Fat lute is formed with pipeclay and linseed oil, and stands a high heat. A good lute for steam boilers and stills consists of equal parts of chalk and flour made into a paste with water.

Luteogallic Acid. The yellow colouring matter of nutgalls, insoluble in water, alcohol, and ether.

Luteoline. Yellowish needles, sublimable, slightly soluble in water; soluble in alcohol, ether, and alkalies, barytes and lime waters, with a yellow colour; soluble in acid without decomposition, as it is reprecipitated by water; acts as a dye with alum; obtained from the Reseda luteola or woad, by boiling for a quarter of an hour with water; luteoline separates in yellow flocks.

Lycopodium. A yellowish powder found in the fruit vessels of the club moss (Lycopodium clavatum), used in fireworks, and sometimes in pharmacy for covering pills.

Lydian Stone. *Basanite*, *Touchstone.* A

dark-coloured quartz found in Lydia, Saxony, and Bohemia, used to determine the value of gold alloys used in jewellery, by judging of the impression made by touching the metal.

Lymph. The fluid contained in the lymphatic or absorbent vessels of animals. These vessels absorb the lymph from the serous cavities, and carry it to the thoracic duct, where it is mixed with the chyle. When the lymphatics cease to act, the lymph accumulates in the cavities and causes dropsy. Lymph from a wound in the foot consists of spec. grav. 1037, water 969·26, fibrine 5·2, albumen 4·34, fat 2·64, Na Cl, KCl, $KOCO_2$, $CaOSO_3$, 3CaO, PO_5, FeO= 15·44 (Marchand and Colberg). From a young ass, water 965·36, albumen 12·, fibrine 1·2, extract (alcoholic) 2·4, aqueous extract 13·19, fat trace, salts 5·85 (Rees). The fluids left in the serous cavities in dropsy appear all to contain fibrine, for when mixed with serum of blood, or with a small morsel of moist fibrine, or many solids, they gelatinize (Dr. A. Buchanan), and I have collected large quantities of fibrine in this way. The fluids from the intestines in cholera I have found to be of an analogous character, and to be distinct from serum, as in the following table:—

	Salts.	Organ. Matter.	Water.
1. Serum, ...	11·	89	900
2. Cholera F, ...	6·75	6·75	986·5
3. Hydrocele F,......	7·82	——	——
4. Hydrocephalus F,	8·66	1·52	989·82

M

Mace. The exterior covering of the nutmeg. (Myristica officinalis).

Macerate. To steep in water, alcohol, &c.

Macle. A synonyme of Andalusite. The term macle of Brittany, has been given to a crystal deposited in mica slate in Bretagne. The black matter of these slates is placed regularly in the centre of the crystals and in the angles. See fig.

Maclurite. Chondrodite.

Madder. (*Krapp*, Ger.; *Garance*, Fr.) The root of the Rubia tinctorum, principally grown for the use of dyers at Avignon in Alsace near Mastricht in Zealand, Holland, and in the Levant at Smyrna and Cyprus. In Europe the root is used in the third year, in the Levant in the fifth and sixth years. The amount of ash in Alsace madder is 8·14 to 8·42 per cent., in Levant madder 9·8, dried at 212°. The composition of the Alsace ash is KO 26·64, NaO 11·67, CaO 29·25, MgO 3·68, Fe_2O_3 3·36, PO_5 4·62 NaCl 13·25, SO_3 2·14, SiO_3 5·36. The nature of the origin of the colouring matter of madder has given rise to much investigation, but hitherto without any very definite result. According to Robiquet, the colouring principle is alizarine, which exists in the root. Runge considers that there are several colouring matters, madder red, purple, orange, and brown. Debus ascribes the properties of madder to the presence of lizaric and oxylizaric acids. Higgin found in madder, alizarine and purpurine distinguished by its solubility in solution of alum. According to Schunck, when 1 lb. of madder is exhausted with 16 quarts of boiling water, and sulphuric acid added to the filtered liquid, a brown precipitate falls, which consists of six substances. It is washed with cold water, and treated with boiling alcohol, which dissolves *Alizarine*, *Rubiacine*, *Rubiane*, and two resins. Freshly precipitated alumina is added until the liquor loses its colour; the coloured alumina is filtered and washed with alcohol. It is boiled three times in a solution of carbonate of potash, which leaves a compound of alizarine and alumina, and dissolves the other bodies. Boiling in chlorohydric acid takes up the alumina and leaves the alizarine, $C_{14}H_5O_4$, as a red crystalline powder. It is obtained in reddish-yellow prisms by solution in alcohol; it sublimes at 420°, and is soluble in ether, and slightly so in boiling water; by nitric acid it is converted into *Alizaric* or *phthalic acid*, $C_{14}H_5O_7$. The brown precipitate from the decoction of madder, when treated with perchloride or pernitrate of iron, chlorohydric or sulphuric acids, yields *Rubiacine*, $C_{31}H_9O_{10}$, in yellow plates and needles. This substance, when oxidated, affords *Rubiacic acid* $C_{31}H_9O_{17}$, a lemon-yellow powder. By washing the brown precipitate of madder decoction with cold water, *Rubiane*, the bitter principle of madder, is obtained. Madder also contains sugar, pectic acid, and xanthine, a brownish syrup procured by evaporating the decoction. From further experiments Schunck states that *rubiane*, by the action of a ferment existing in madder, *erythrozyme* q. v. is decomposed into *Alizarine*, *Verantine*, *Rubiretine*, *Rubiafine*, *Rubiagine*, and *Rubiadipine* q.v. Emulsine is the only substance which acts likewise as a ferment in decomposing rubiane. He considers alizarine the only colouring matter in madder, all the other colouring substances described containing alizarine in a state of greater or less purity.

Madic Acid. $HOC_{32}H_{30}O_3$; F.P. 130°. Fatty needles from the oil of Madia Sativa by saponification.

Magistery of Bismuth. Trisnitrate of Bismuth.

Magistral. Roasted sulphur and copper pyrites in the amalgamation of silver.

Magma. A pasty mass.

Magnesia. *Hydrocarbonate.* See HYDROMAGNESITE.

Magnesia, Calcareo Carbonate of. *Anthraconite*, *Brown Spar*, in part, *Bitter Spar*, *Dolo-*

mite, Conite, Gurhofian, Magnesian Limestone, Miemite, Muricalcite, Pearl Spar, Picrite, Tharandite.—$CaOCO_2MgOCO_2$. Spec. grav. 2·815 to

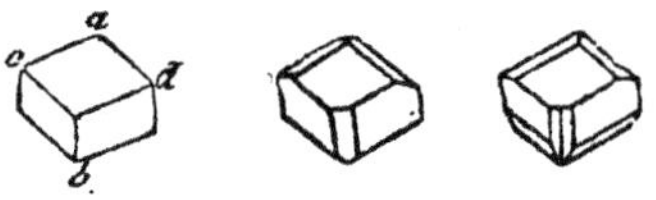

2·884, H 3·5 to 4·. White inclining to red or green, also brown, gray, black masses, and rhombohedrons, with angles of 106° 15′; streak grayish-white; fracture conchoidal; structure foliated; lustre vitreous, inclining to pearly; B.B. acts like calcareous spar. The two carbonates are often present in various proportions, affording an example of the isomorphism of lime and magnesia, but the angle slightly various from calc spar. It occurs amorphous in the magnesian limestone strata of England.

Magnesia, Carbonate of. *Magnesite, Baudisserite.*—Spec. grav. 2·808 to 2·95, H 4·5. $MgO CO_2$. White or yellowish masses, and rhombohedrons or needles, sometimes radiated. Occurring in hills in Salem, Madras, and in North America. Slowly soluble in acids; B.B. shrinks; with borax fuses into a glass; becomes red with nitrate of cobalt.

Magnesia, Ferrocarbonate of. *Brown Spar, Misitine Spar.* Spec. grav. 3· to 3·112, H 4·25. White, gray, yellow, or brown rhomboids, with angles of 107° 22′. From Salzburg, Tyrol, and Switzerland. Consists of various proportions of carbonate of magnesia and carbonate of iron, and is a good example of isomorphism, although the angles are somewhat different from calc spar.

Magnesia, Hydrous Calcareo Carbonate of. *Dolomite Sinter.* $4(CaO MgO)2CO_2$ 4HO. Stalactitic yellowish-white globular masses from Vesuvius.

Magnesia, Potash Silicate of. Spec. grav. 2·87. Yellowish-white crystals like tremolite; easily reduced to powder. SiO_3 58·16, MgO 26·48, KO 6·32, Fe_2O_3 7·6, Al_2O_3 ·4, CaO ·64, HO ·4. From St. Yrieix.

Magnesite. Spec. grav. 2·096. Grayish-white unctuous mass in the marly limestone of Paris. SiO_3 50·7, MgO 23·65, Al_2O_3 3·55, FeO 1·7, HO 20·6 (John Tennant).

Magnesium. Mg 1.5, 12·11, 12·36, 8-hedrons. Specific gravity 1·87 (Pelouze), 2·24 (Playfair and Joule). White solid metal resembling silver, sinks rapidly in water; is permanent in dry air, but in moist air becomes covered with a crust of oxide; it may be washed with cold water without being acted on; boiling water is slowly decomposed by it, giving out a few bubbles of hydrogen less rapidly than the other metals; due to the insolubility of the magnesia, for when the water is acidulated with sulphuric acid, the decomposition is accelerated; and some salts have a similar effect. It is very ductile, and capable of being beat out with a hammer; it is hard, but yields to the file. When heated in the air it burns; it is volatile at a white heat.

Process.—Davy, in 1808, succeeded in separating magnesium by means of the galvanic battery. In 1828, Bussy obtained it by igniting chloride of magnesium with potassium or sodium. The sodium may be placed in a porcelain crucible, and pieces of the chloride over it; heat being applied, the vapour of the sodium passes through the salt, and decomposes it (Mg Cl and Na becoming Mg and Na Cl). The metal is washed first with water, and then with alcohol. The sodium may be placed in the closed end of a narrow tube, and the tube itself filled with pieces of chloride. The salt being ignited, the sodium is likewise heated, and its vapour made to pass through the salt.

Magnesia, Oxide of Magnesium, Calcined Magnesia.—MgO 2·5, 20=Mg 60, O 40. Spec. grav. 2·3 (Kirwan), 3·2 (Karsten), 3·07 (Richter), 3·61 (Rose). A soft white powder, without taste and smell; becomes somewhat agglutinated at a high temperature (D'Arcet); fuses in a stream of oxygen into a white enamel, and by electricity (Davy); more *soluble* in hot water than in cold water (Butini); *soluble*, according to different authorities, in 5760 water at 60°; in 36,000 water at 212°, in 5800 water at 59°, 7900, 16,000, in 53 to 56,000 parts of hot and cold water; changes reddened litmus paper to blue. With nitrate of cobalt, when heated before the blowpipe, magnesia acquires a red tint; it is fixed at a red heat, but at very high temperatures volatile (Liebig). With fuming sulphuric acid it incandesces and unites.

Hydrate, Brucite.—MgO HO 3·625,29. Hardness 1. Spec. grav. 2·35 (Haidinger), 2·336 (Brewster). *Crystal,* occurs native in Unst, Scotland, in low 6-sided prisms, needle-form crystals, rhombohedrons; colour white; texture foliated; lustre pearly. Before the blowpipe it loses water, but does not fuse. Artificially prepared, by precipitating a salt of magnesia by caustic potash or soda, it is a fine soft white powder.

Source of Magnesia.—As usually prepared it is obtained from the bittern or mother liquor of sea water, which remains after the greater portion of common salt crystallizes out. This liquor is precipitated by an alkaline carbonate, generally carbonate of ammonia, obtained in working up the products of the distillation of coal and bones. Hence magnesia is frequently made by bone black makers. The product first obtained is magnesia alba, of the shops, $3(MgOCO_2)$ MgOHO,3HO, which, by calcination at a strong heat, loses its water and carbonic acid, and becomes caustic magnesia. It is likewise obtained from magnesian limestone, which is a double carbonate of lime and magnesia. The mineral is decomposed by sulphuric acid; the insoluble sulphate of lime separated from the soluble sulphate of magnesia, and the magnesia precipitated by an alkaline carbonate. By calcination the carbonic acid is removed. It may also be prepared from native carbonate, which occurs in the Salem district, Madras, or from serpentine.

Salts of Magnesia.—1. The greater number of the salts of magnesia are soluble in water, and crystallizable. 2. Ammonia precipitates magnesia, but not when ammoniacal salts are present in the solution, in consequence of the formation of an ammoniacal salt of magnesia. Caustic soda or potash precipitates it completely, except in presence of some organic substances. 3. If salammoniac be added to a solution of a salt of magnesia, then caustic ammonia to the extent of one-eighth of the volume of the solution, and while the solution is clear, if phosphate of ammonia or soda be dropped in a crystalline precipitate of ammonia phosphate of magnesia falls (Wollaston). 4. A salt of magnesia mixed with nitrate of cobalt, when fused before the blowpipe, gives a flesh-coloured bead. When the acid is volatile or organic, it is necessary to mix with it borax or biphosphate of soda.

Sexhydrated Chloride.—Mg Cl 6 HO, 12·75, 12·68, 101·4. Large 4-sided prisms (Thomson), needles and prisms (Gmelin). *Taste* bitter, hot, and biting; deliquescent; *soluble* in ·6 part cold and ·273 hot water, in 5 alcohol (spec. grav. ·900), 2 alcohol (spec. grav. ·817). By heat it liquefies and changes to a white mass; at 222°·8 acid begins to be evolved; fuses at 246° (Brandes), is converted by ignition into magnesia, and a small portion of chloride of magnesium (Davy); the presence of common salt prevents this decomposition, but if chlorate of potash be present it is decomposed, and hence this method has been proposed for estimating magnesia.

Process.—It exists in sea water, and in several mineral waters, from which it may be extracted. It may be formed by dissolving native carbonate or magnesia alba in muriatic acid, and evaporating. For the preparation of magnesium, it may be formed by charring a mixture of equal parts of starch and magnesia, and passing chlorine over the ignited mass in a tube. It is best prepared by dissolving the carbonate in chlorohydric acid, and adding salammoniac, until, when evaporated to the consistence of syrup, it yields no precipitate with ammonia; it is then evaporated to dryness, and ignited till the ammonia be dispelled.

Use.—It is employed to prepare magnesium, and in sea water it affords much of the magnesia alba of commerce.

Septahydrated Sulphate. Commercial Epsom Salt.—MgO SO_3, 7HO 15·375, 123, MgO 16·25, SO_3 32·52, HO 51·21. Right prisms, the bases of which to the eye appear squares, but when the crystals are very large the angles obviously deviate from right angles slightly, being 90° 30′ and 89° 30′, M on M′ 90° 30′, M on *h* 134° 45′, M on *e* 129°, a on *a* 120°, the crystals refract double (Brooke, Ann. Phil. 2d Ser. 6, 40). Permanent in air; effloresces in warm dry air; *soluble* in ·9 water at 55°, in 1· at 60°, in ·66 at 212° (the volume of the water being increased $\frac{4}{10}$ by adding the salt), (Ann. Phil. 2d Ser. 12, 403), in ·799 water at 65¾°, having a specific gravity of 1·2932; a saturated solution at 46°·4 has a density of 1·267 (Anthon). *Taste* intensely bitter. By heat it melts in its crystalline water; at 300° it loses 6 atoms of water, and does not part with the remaining one below 390°.

Process.—Grew described sulphate of magnesia in 1675, and in 1723 Mr. Brown published an account of the process for extracting it from springs and purifying it. It was extracted from the springs of Epsom and Seydschütz. 2. At La Guarda near Genoa it was made from serpentine (silicate of magnesia) rocks, containing sulphur and magnesia. These rocks are burned with a limited supply of air, as in alum-making. The calcined ore is digested in water, the metallic substances precipitated by lime, and the salt crystallized. To remove iron and copper, it should be heated nearly to the decomposing point of the sulphate, and the oxides of these metals are thus rendered insoluble. It is then dissolved in water and crystallized. 3. It may also be formed from the magnesian limestone which is largely developed in England, commencing at Sunderland and extending southwards. It is digested in dilute sulphuric acid, the solution of sulphate of magnesia removed, evaporated to dryness, calcined to remove iron, and then dissolved and crystallized. It is sometimes contaminated with iron, which may be detected by bisulphohydride of ammonia, or by yellow or red prussiate of potash.

Use.—Epsom salt is used extensively as a purgative, and sometimes also for agricultural purposes.

Magnet. *Loadstone.* A form of iron or its oxides which has the property of attracting light pieces of iron towards it. The term magnet is derived from μαγνης, the name of a shepherd who first discovered the loadstone on Mount Ida. Pliny mentions it under the name *magnes*, and gives an account of its property of attracting iron. Its polarity was unknown to the ancients. It occurs most frequently in primitive mountains, especially in gneiss, mica slate, and limestone, under the name of magnetic iron ore. It is found in Europe, Asia, America, and Australia. Mr. Crichton of Glasgow has described a very powerful magnet which was sent from Virginia in 1776 by Dr. Franklin to Professor Anderson, and is now in possession of Mr. Crichton. It weighs 2½ grains, and is capable of sustaining a load of 783 grains, or 313 times its own weight. It was known at an early period in China (British Annual). A magnet may be formed artificially by drawing a magnet or loadstone along a bar of steel in the same direction several times, or by placing a bar of steel parallel to the axis of the earth, or parallel to the dip of the needle; and by giving it some smart blows while thus placed—it acquires magnetic properties. Iron railings are magnetic. If the steel bar thus magnetized be suspended in the middle on a pivot, one extremity points to the north and the other to the south. Scoresby recommends the following method of forming magnets without the aid of other magnets:—Two bars of steel are taken 30 inches in length with two other plate bars of steel 8 inches long and ½ inch broad and a long bar of iron, all of them destitute of magnetic power. The large bar of iron is first struck in a vertical position and then placed, without changing its direction, upon the steel bars, which have also been struck. They are then struck upon each other. Each of the small bars suspended vertically to the summit of one of the large steel bars is successively struck, and in a few minutes they acquire a considerable magnetic power. Two more of the small bars united by two small parallelopepids are rubbed with the four bars, and are then replaced by two others, and these again by the two last. Each pair of bars being thus treated for a certain time is found magnetized.

Attraction and Repulsion of Poles.—When a magnet is placed in iron filings, the latter are attracted to the extremity of the bar and not to the middle; and as each particle of iron becomes a magnet, and attaches itself to its neighbour, on a careful scrutiny we find that the particles are arranged in magnetic curves.

Use of the Magnet in the Manufacture of Needles.—In forming the points of needles, small fragments are broken off, and are apt to be inhaled by the operatives. By placing magnets at the mouth and nose, the dangerous nature of this trade has been ameliorated.

According to Mr. Barlow, the intensity of the action of iron on a magnetic needle is proportional to its surface and not to its solid contents, so that a hollow sphere of iron affects a magnet as much as a solid mass of the same volume. This is an important fact in reference to the action of masses of iron on board ship upon the needle. Barlow also found that *white* hot iron had no influence on the needle, but as the iron cools it begins to act, and possesses the greatest action at a blood-red heat. He also observed that if a magnetic needle be placed at a certain distance from a mass of iron, the needle is deflected, when the iron is put in rapid motion. Faraday has shown that while magnetic bodies, when freely suspended, point in the magnetic meridian, all the other bodies arrange themselves at right angles to the meridian, and are therefore termed *diamagnetic.* He has also shown that all bodies are more or less affected by magnetism.

Magnetism. It appears from the varied experiments of electricians, that every current of electricity is capable of communicating magnetic properties to metals, and that all metals are susceptible of magnetism while under the influence of currents, but iron is the only metal which appears capable of becoming a permanent magnet; nickel also seems magnetic.

Magnus' Salt. $PtClNH_3$. A green crystalline compound; insoluble in hot water; soluble in a boiling saturated solution of sulphate or nitrate of ammonia, and depositing on cooling in yellow crystals; obtained by digesting protochloride of platinum in ammonia.

Maize. *Indian Corn.* The grain or seed of the Zea Mays, termed in America *corn;* its flour being corn meal. I have found it to contain 10·93 albuminous matter.

Malachite. See GREEN CARBONATE OF COPPER.

Malacolite, Green. Greenish-gray pyroxene.

Malacolite, White. A synonyme of alalite, or white augite.

Malacone. *Hydro Silicate of Zirconia.* 2 $Zr_2O_3SiO_3HO$. Sp. grav. 3·9 to 4·047, H 6·5. Brown, internally bluish-white square 8-hedrons, with angles of 124° 49′, and 124° 57′; lustre vitreous; somewhat resinous; powder reddish-brown. B.B. infusible; acted on by boiling sulphuric acid. SiO_3 31·31, Zr_2O_3 63·6, Fe_2O_3 ·41, YO ·34, CaO ·39, MgO ·11, HO 3·03. Hitteroe, Norway, and Haute Vienne.

Malamide. $C_8N_2H_{10}O_8$. See ASPARAGINE.

Malamidic Acid. See ASPARTIC ACID.

Maleic Acid. 2 HO $C_8H_2O_6$. F.P. 266°. Clear colourless plates, or sharp rhombic prisms, with 8-hedral terminations; soluble in its own bulk of water; by a continuous heat in a long-necked vessel at 275°, it is converted into fumaric acid, which is isomeric with maleic acid; heated in a retort at 320° water is given out; if the residue be heated to 349° anhydrous maleic acid, fumaranhydride, or maleide, pass over, and some fumaric acid remains. Anhydrous maleic acid fuses at 134½°; maleic acid is obtained by heating malic acid quickly up to 392° in a retort, when it loses 2 atoms water, and maleic acid distils over; if the temperature be only 350° fumaric or paramaleic acid is also formed, which has the same composition. It is a dibasic acid.

Malic Acid. *Sorbic Acid.* $C_8H_4O_8$ 2 HO, F.P. 181½°. Granular crystalline mass; deliquescent, the fluid decomposing; at 248° it loses no water; at 302° it begins to change; at 349° maleic and paramaleic acids are formed in equal quantities; malic acid reduces salts of gold, and is converted by nitric acid into oxalic acid. By sulphuric acid it is decomposed into acetic acid and carbonic oxide; it does not precipitate lime water; it yields succinic, acetic, and carbonic acids by the action of yeast on malate of lime; it is bibasic; it is obtained by expressing the juice of the berries of the Sorbus aucuparia, mountain ash or rowan tree, nearly saturating with milk of lime; by the application of heat, malate of lime forms (2 CaO,$C_8H_4O_8$), which is removed by a colander; more milk of lime is added, when more malate falls. The malate is washed with cold water, and dissolved in 1 nitric acid and 10 water, assisted by heat. Acid malate of lime (CaOHO,$C_8H_4O_8$ 6 HO) crystallizes out, which may be purified by solution in boiling water and crystallization. The acid may be separated by sulphuric acid, or the bimalate is precipitated by acetate of lead, and the lead salt decomposed by sulphohydric acid. Malic acid is also contained in apples, rhubarb, and other acid fruits.

Malleability. The capacity which metals possess of being extended under the hammer.

Malt. (*Malz*, Gr.) The term applied to grain which has been allowed to germinate, and the vegetation suddenly checked. The appearance presented by the various parts of malt under the microscope is that exhibited in the fig. The process of malting consists essentially in converting

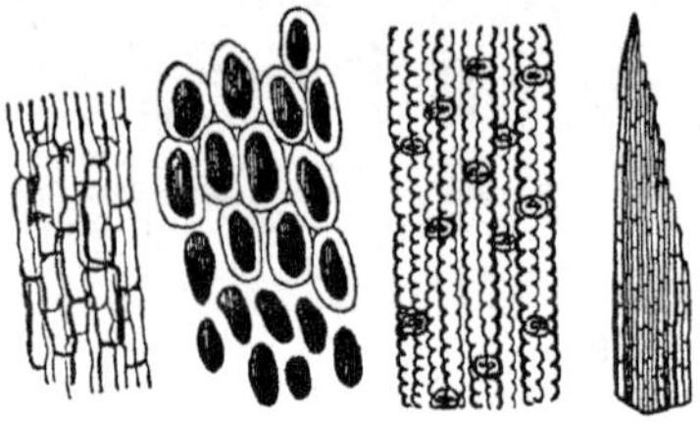

Cellular Membrane. Skin. Husk.

[*Hooker*]

the starch of grain partially into sugar, and therefore any species of the cerealia will answer for the purpose; but barley (Hordeum vulgare) is usually preferred, characterized by having two rows of seeds on the spike. *Big* (Hordeum hexastichon) with six rows, and *bere* (Hordeum tetrastichon) with four rows of seeds, have been employed in Scotland, and likewise other species of grain. 1. The first process in malting is *steeping*. The barley is put into the steep, a square cistern lined with stone, and is covered with water, where it remains forty to eighty or even one hundred hours; 100 lbs. of barley weigh when taken out of the steep 147 lbs. and its bulk increases from 100 to 124 measures. The grain loses organic matter, soluble and insoluble salts, which remain in the steep water, which is thus found to be a good manure. 2. It is then placed on moveable wooden boards termed the *couch* for twenty-six hours. 3. Afterwards it is spread on the malt floor to the depth of three or four inches. Here it begins to germinate, which is marked by the development of the *acrospire*, or stem and root. It is turned three or four times a-day for a period of ten days or a fortnight. It increases in temperature 10°, and *sweats*, moisture being abundantly exhaled. 4. Germination having thus been encouraged is suddenly stopped, by transferring the malt to the kiln, which consists of an apartment lined with plates full of minute holes, or with wire or haircloth. The malt is spread on this surface to the depth of from three to six inches, and a fire kindled below. The temperature is at first about 100°, but gradually it rises to 140°. The darker kinds of malt are obtained by raising the temperature to nearly 200°. After being forty to eighty hours in the kiln, while still warm, it is trodden by the workmen in order to break off the *commings*. It is then passed through the fanners, or the rootlets are separated by the *harp*. Barley diminishes in weight when thus malted and dried. In four returns from maltsters, I found 100 barley became 80·4 malt, showing a loss of 20 per cent. or one-fifth. The relative composition of barley and malt I have found to be:—

	Barley.	Malt.
Carbon, ...	41·64	33·95
Hydrogen, ...	6·02	5·31
Nitrogen, ...	1·81	0·88
Oxygen, ...	37·66	34·46
Ash, ...	3·41	1·34
Water, ...	9·46	4·06
	100·	80·

If 100 measures of barley weigh 100, then the same bulk of malt weighs 76·65. The loss sustained by barley in malting may perhaps be stated as follows:—Per cent. water 7·, saline matter ·48, organic matter 12·52. Total = 20. That albumen is removed in the steep water is obvious from the quantity of nitrogen obtained by analysis from the residue. The steep residue lost 40 per cent. of its weight of organic matter by ignition (R. D. Thomson on the Food of Animals, p. 122). The result of these facts is, that barley is deteriorated by malting as an article of food for animals.

Malthacite. Spec. grav. 2·, H of wax. White or yellowish plates, and in masses in basalt and greenstone in Bohemia. B.B. infusible. SiO_3 50·2, Al_2O_3 10·7, Fe_2O_3 3·1, CaO·2, HO 35·8. Allied to Halloylite.

Mancinite. A feathery silicate of zinc, from Mancino, Leghorn.

Manganese. Mn 3·5, 3·459, 28, 27·674. Specific gravity 6·85 (Bergman, Opusc.), 7·0 (Hjelm), 8·013 (John, Gehlen Jour. 3, 460), 7·95 (Berthier). Colour gray, somewhat whiter than cast iron; texture finely granular, or prismatic; hard, and so brittle, that it may be reduced to powder in a mortar; not attracted by the magnet, and not magnetic; fuses at the highest temperature of an assay furnace; not volatile; it gradually absorbs oxygen from the atmosphere, and at the ordinary temperature decomposes water, when reduced by charcoal in the blast furnace, but when reduced in contact with silica, so as to contain 6 or 7 per cent. of silicon, it does not oxidize even at a red heat, and is not soluble in nitro-chlorohydric acid (Sefström). This has been considered an allotropic state of manganese.

Preparation.—The pure oxide is to be obtained by dissolving crystallized binoxide of manganese in pure chlorohydric acid, neutralizing the solution, and boiling to precipitate oxide of iron. The oxide of manganese is precipitated by an alkaline carbonate, filtered, well washed, and ignited. The residual oxide is then mixed with one-tenth of charcoal, and heated in a crucible in an assay furnace for some hours.

Protoxide, Manganoxydal.—MnO 4·5, 4·459; 36, 35·672. Light green powder, with a shade of yellow, gradually becoming brown by exposure to air; fuses in the blast furnace into a fine green mass, but when precipitated from its solution by caustic soda, it is a white flocky precipitate, containing 24 per cent. of water (H. Davy), becoming brown by exposure to air, forming red or protosesquioxide. It unites with acids, forming protosalts, which have usually a flesh colour, due to hypermanganic acid (Pearsall), or to the presence of a higher oxide, and an astringent taste. Pearsall observes that red solutions of manganese decolourize indigo, and lose their colour. A solution of bleaching powder slowly precipitates from the protoxide solution, the hydrous binoxide. Caustic ammonia precipitates the protoxide only partially; alkaline carbonates precipitate protoxide in the state of carbonate, which retains its white colour, and is slightly soluble in salammoniac. Carbonate of lime, when boiled with protoxide salts, does not cause a precipitate according to Fuchs, but a precipitate occurs according to Demarçay with earthy carbonates, likewise with carbonate of magnesia and calcined magnesia.

Process.—1. Protoxide may be obtained by igniting pure carbonate of manganese in a close crucible (Scheele), or oxalate of manganese (Lassaigne); 2. by heating carbonate or binoxide of manganese in a tube, and passing dry hydrogen over it; 3. by fusing equal parts of chloride of manganese and carbonate of soda with a small portion of salammoniac, and exhausting with water when cold.

Sesquioxide, Deutoxide, Manganoxyd, Braunite.—Mn_2O_3 or $MnO_{1\frac{1}{2}}$ 10·, 9·918 or 5·, 4·959; 40·, 39·672. Spec. grav. 4·82, 8-hedrons with a square base. Mn = 70, O = 30. Colour black with a shade of brown, tasteless and insoluble in water; soluble in sulphuric and chlorohydric acids, forming red solutions; cold chlorohydric acid does not decompose it, but when hot, chlorine is evolved and protoxide remains in solution; nitric acid, when strong and boiling, dissolves protoxide and leaves binoxide, which remains undissolved; sulphurous acid forms sulphate and hyposulphate of the protoxide. The vegetable and mineral acids, along with sugar, reduce it to the state of protoxide, which dissolves, carbonic acid being evolved. At a white heat it loses 33·1 per cent. of oxygen, and changes into protosesquioxide or red oxide. It is isomorphous with alumina, sesquioxide of iron, and sesquioxide of chromium. It exists as *braunite* in nature, and is prepared artificially by exposing the nitrate of protoxide or the binoxide to a low red heat.

Protosesquioxide, Red oxide, Manganoso-Manganic oxide, Hausmannite, Deutoxide.—$MnO_{1\frac{1}{3}}$ 4·833, 38·666, or MnO Mn_2O_3 14·5, 116, or 14·377, 115·016. Spec. grav. 4·722, 8-hedrons with a square base. Colour brownish-black, streak dark reddish or chestnut-brown. It occurs also massive, granular; artificially prepared it is reddish-brown. All the oxides of manganese, when ignited, are converted into this oxide. When mixed with charcoal and ignited in an assay furnace it is reduced. It varies in colour according to its mode of preparation. Prepared by means of chlorine it has a violet shade. Nitric acid, when strong and boiling, dissolves up from it protoxide and leaves binoxide, from which its composition might be represented as $2MnO + MnO_2$. With chlorohydric acid and heat it yields protochloride and chlorine.

Binoxide, Peroxide, Pyrolusite, Black oxide, Gray oxide, Wad, Braunstein, Anhydrous binoxide.—MnO_2 5·5, 44; 5·459; 43·672; Mn = 63·65, O = 36·35. Spec. grav. 4·819; 4·7583 (Berthier); 4·97; 4·94; hardness 2· to 2·8. Rhombic prisms like those of manganite, usually in radiated needles or indefinite prisms. Colour, iron-black, powder, gray. Occurs on the Rhine, in Devonshire, Norway, Cape of Good Hope, New Zealand, United States, &c., and is a most valuable ore for the evolution of chlorine and supply of oxygen. When heated gently it begins to part with oxygen, at an incipient red heat it loses half an atom of oxygen, and is converted into sesquioxide (MnO_2 5·5 becoming $MnO_{1\frac{1}{2}}$ 5·). As the temperature rises it loses 12 per cent. and becomes protosesquioxide or red oxide ($MnO_{1\frac{1}{2}}$ 5· becomes $MnO_{1\frac{1}{3}}$ 4·833.) When heated with charcoal or in a charcoal crucible, binoxide loses 18 per cent. of oxygen and leaves 82 per cent. of protoxide (Berthier). Cold nitric acid does not act on it; aided by heat it forms protonitrate and disengages oxygen. Sulphuric acid acts in a similar manner. Cold sulphuric acid dissolves the hydrate and forms a red solution (Berthier). Hot sulphuric acid yields oxygen and protosulphate of manganese. Cold sulphurous acid dissolves it readily, forming sulphate and hyposulphate of manganese. Chlorohydric acid dissolves it, disengaging much chlorine, and forming protochloride of manganese. When treated with oxalic acid and sulphuric acid, carbonic acid is evolved and protosulphate dissolves ($MnO_2, C_2O_3, SO_3 = MnOSO_3, C_2O_4$ or $2CO_2$). Native black oxide contains generally impurities, as sesquioxide of iron, chloride of calcium, sulphate of lime, carbonate of lime. Oxygen, produced from such specimens will contain carbonic acid, which must be removed by an intermediate vessel containing milk of lime. When chlorides are present and sulphuric acid is added, chlorine is evolved ($CaCl, MnO_2, 2SO_3 = CaO\,SO_3, MnO\,SO_3, Cl$). When fused with caustic soda, or potash, in a crucible, manganate of potash and sesquioxide of manganese are formed. When ignited in an atmosphere of hydrogen, binoxide loses 1 atom oxygen, and is converted into protoxide. The amount of oxygen thus given out measures the value of the commercial binoxide and the amount of chlorine which it is capable of supplying by decomposing chlorohydric acid. The finest Giessen manganese yields 17·79 oxygen = 78·93 chlorine. Kilpatrick, 17·19 oxygen = 76·27 chlorine. Thuringia, 14·44 O = 64·07 Cl. New Zealand, 14·02 O = 62·20 Cl. Table Mountain, 11·39 O = 50·53 Cl (G. W. Brown, Proc. Phil. Soc. Glasgow, 3, 188). *Process.*—It may be obtained artificially by boiling red oxide or sesquioxide in strong nitric acid, by slightly igniting protonitrate of manganese, and boiling the residue in powder with nitric acid to dissolve sesquioxide. By igniting protocarbonate in the air to 500°, and digesting in cold dilute chlorohydric acid to remove undecomposed carbonate; binoxide remains.

Testing value of Black Oxide.—Into a light glass flask, weighing a few hundred grains, are to be introduced about 600 grains of water, 150 of sulphuric acid, and 75 grains of oxalic acid; 50 grains of the sample of black oxide of manganese, finely pounded, are weighed out in a small tube, weighing about 30 or 40 grains, which is to be suspended in the flask by means of a thread, retained in its place by a cork, through which passes a small tube filled with chloride of calcium, the tube being closed at each end by a piece of cotton. (See apparatus under CARBONIC ACID.) The whole is then weighed accurately; the tube containing the oxide is then lowered into the mixture in the flask. An effervescence takes place by the extrication of carbonic acid (CO_2 CO, MnO_2, SO_3 becoming 2 CO_2 Mn O SO_3). Every atom of black oxide (MnO_2), or 5·5 grains, will give rise to the extrication of 2 atoms of carbonic acid (2 CO_2) or 5·5 grains. When the action has ceased, which is denoted by the brown colour of the residue in the flask, the apparatus is again weighed. The loss of weight indicates the amount of CO_2 extricated. But as the weight of 2 atoms of carbonic acid (5·5) is equal to that of 1 atom of black oxide (5·5), it is obvious that the loss of weight is exactly equal to the amount of black oxide present in the sample (Dr. T. Thomson).

Weight of black oxide,.........	50 grs.
Weight of flask and materials,	1025
	1075
Weight after effervescence,.....	1041
Loss of carbonic acid,...........	34 grs. = 68 per cent. MnO_2.

The following method of testing manganese by means of copperas and bichromate of potash has been used by Mr. William Blythe since February, 1837, in his manufactory:—1. Weigh out 600 grains of pure and dry crystals of copperas; put the salt into a flask with about 5 or 6 oz. of water, and about 2 oz. of muriatic acid. Let the flask be carefully heated till the copperas is dissolved, the flask being repeatedly shaken, to prevent the copperas "settling" to the bottom of the flask by the heat. 2. Weigh 100 grains of the sample of manganese crushed to powder, and put it into the flask. Let the mixture in the flask be boiled for five minutes, or till there is no black residue having the appearance of manganese. 3. The solution is now poured into a tall glass, and diluted with four times its volume of water. 4. A solution of bichromate of potash of a known strength is now poured from a graduated tube (a common alkalimeter is very convenient) till the solution does not change a solution of red prussiate of potash to blue. This is ascertained by putting a drop of a weak solution of red prussiate upon a bit of cotton cloth, and then a drop of solution from the end of the glass-stirrer. The quantity of bichromate required corresponds to the quantity of copperas not peroxidized by the action of the manganese. *Example.* — Copperas, 600 grains; manganese, 100 do.; bichromate, 32; As 1 atom of bichromate 19 : 6 atoms copperas 104·25 :: 32 : 175; 600 — 175 = 425. 2 atoms copperas 34·75 : 5·5 :: 425 : 67 = per cent. of pure ore.

Varvicite.—2MnO_2, Mn_2O_3, HO 22·125, 177·, or MnO, 3 MnO_2 HO. Spec. grav. 4·283 to 4·531. Consists of thin plates and fibres, without any regular crystalline shape; hardness 2·5. It contains 81·12 protoxide; oxygen 13·48; water 5·4.

Manganic Acid. Manganesic Acid.—MnO_3 6·5, 52; 6·459, 51·672 (Phillips's Phil. Mag. 2d, 5, 209). Scheele first discovered that when binoxide of manganese is strongly heated with potash or nitre, and the mixture dissolved in water, a fine red solution is obtained, which, from its tendency to decompose and change colour, he called *chameleon mineral.* This acid has not been isolated, but it forms salts with potash, soda, barytes, and strontian, which have a bluish-green colour. The manganate of potash is isomorphous with the corresponding salts of sulphuric, selenic, and chromic acids. Caustic soda, fused with binoxide of manganese, yields manganate of soda, which, however, is too soluble to be freed from the carbonated and caustic soda by crystallization. Nitrate of barytes, fused with binoxide of manganese, yields manganate of barytes, but it may be best formed by adding to a solution of permanganate of barytes a solution of barytes, and allowing the liquid to stand a long time in a jar which it about half fills; green crystals of manganate of barytes separate, which, like sulphate of barytes, are insoluble in water. The manganates of potash and soda form green solutions which are not decomposed if an excess of alkali be present, but if it be deficient, the solution reddens, and the salt is converted into permanganate and carbonate, while binoxide of manganese falls.

Permanganic Acid. Hypermanganic Acid.—Mn_2O_7 14·, 112; 13·918, 111·344. This acid has not yet been isolated, although, according to Frommherz, it is a carmine fluid, and according to Unverdorben a red gas. The permanganate of potash may be obtained as described under POTASH. When a manganate of potash is treated with a solution of caustic potash, and evaporated in vacuo, fine crystals of manganate are reproduced; but if crystals of manganate are dissolved in water, the salt decomposes, a brown crystalline precipitate falls, which appears to be a combination of binoxide and potash; by washing, the potash is removed, and hydrous binoxide remains. The solution is deep red coloured, and when evaporated till crystals appear on the surface, the solution poured off from any precipitate which may have appeared into a warmed dish, then deep red crystals are obtained when it cools. The same happens when a solution of manganate of potash is exposed to the influence of air and carbonic acid. Hence we obtain sometimes a mixture of these red crystals with manganate, if the latter solution is exposed to carbonic acid during evaporation, $3(KOMnO_3)$, $2\ CO_2 = KOMn_2O_7 + MnO_2 + 2\ KOCO_2$).

Protochloride.—MnCl 8·, 64; 7·896, 62·168, Mn = 44·25 Cl = 55·75. Light pink or flesh-coloured broad thin plates, melting at a red heat without alteration, in close vessels and fixed; but in the open air it decomposes, muriatic acid being given out, and red oxide remaining; chlorine is not evolved; converted into protosulphate by sulphuric acid. Chlorine and heat yield binoxide; bleaching powder solution changes the solution to red and violet; then carbonate of potash or soda renders it green, and lets fall carbonate of lime (Pearsall). Hydrogen does not decompose it at a red heat. The brown aspect which it sometimes possesses, may be due to iron or sesquioxide by the action of air. *Process.*—Obtained, 1st, by igniting manganese in chlorine gas; 2d, by passing chlorohydric acid gas over pure carbonate or protoxide of manganese, first when cold, and then at a low red heat; 3d, by dissolving binoxide of manganese in chlorohydric acid, evaporating and exposing the white salt to ignition in a glass tube with a capillary extremity; 4th, by passing chlorine over a strongly heated mixture of charcoal, and protoxide, needles of protochloride are formed (Rose, Pogg. Ann. 27, 574).

Quaterhydrous protochloride.—MnCl, 4 HO 12·5, 100; 12·396, 99·168 (Brandes); Mn = 28·17, Cl 35·61, HO = 36·22; spec. grav. 1·56 (John). Fine flesh-coloured short 6-sided prisms, two of the opposite sides of which are much larger than the other four.

Salts of Protoxide of Manganese.—Most of the combinations of protoxide of manganese with acids are soluble in water, and form colourless or flesh-coloured solutions; the latter depending probably on the presence of permanganic acid (Pearsall). Taste bitterish and saline, but not so disagreeable as that of most metallic salts. When sufficiently concentrated, many of the salts of manganese crystallize. The crystals generally have a flesh colour. 1. Ammonia forms a white gelatinous precipitate with protoxide salts, which speedily becomes brown by oxidation, in presence of air; partially soluble in neutral, but wholly soluble in acid solutions. 2. Caustic soda, potash, barytes, lime, strontian, form a similar precipitate, not soluble in excess. 3. Alkaline carbonates precipitate protoxide of manganese, in the form of carbonate, which retains its colour in air. The oxide is only completely thrown down in these cases by boiling. 4. Sulphohydric acid produces no precipitate, except a slight white turbidness of sulphohydride, which is dissolved by acids. 5. Sulphohydride of ammonia precipitates sulphohydride of manganese, of a flesh colour, insoluble in excess, but soluble in acetic acid. 6. *Yellow prussiate* (ferrocyanide) of potash forms a white precipitate. *Red prussiate* gives a brownish-gray precipitate.

Quaterhydrous Sulphate.—MnO SO_3 4 HO 14, 112; specific gravity 2·092 (Kopp), 1·834 (John). Contains 32·14 per cent. of water. Colourless or flesh-coloured right rhombic or 6-sided prisms; the colour being due according to Brandes and Frommherz (Pogg. Ann. 31, 677) to the presence of a higher oxide. The red salt is obtained by igniting binoxide with fuming sulphuric acid, and the colourless salt by using common oil of vitriol, a statement not confirmed by Gmelin. The red colour may be removed by igniting and dissolving in water, or by boiling it when pulverized with alcohol or ether, and then dissolving in water. Soluble in 1·78 water at 43¼°, in 2 parts water at 59°, in 1 at 122°; insoluble in alcohol. The colour is not removed by boiling with sugar, sulphurous acid, sulphohydric acid. It is obtained by evaporating the solution of protosulphate between the temperatures of 68° to 73° (John), or 68° to 86° (Mitscherlich, Regnault). In vacuo over oil of vitriol it loses 1 atom of water, and becomes terhydrate, and likewise when boiled in powder with absolute alcohol; cold alcohol and boiling ether having no action (Brandes). This salt is isomorphous with sulphate of iron.

Quintohydrated Protosulphate.—MnO SO_3 5 HO 15·125, 121; spec. grav. 2·877 (T. Thomson). MnO = 47·63, SO_3 = 52·37. Doubly-oblique prisms, with angles nearly of 148° and 32° (T.T.) resembling sulphate of copper and axinite.

Manganese, Hydrous Binoxide of. MnO_2HO. Brownish-black dull masses; chocolate powder; loses 24 per cent. by heat, becoming reddish; slowly soluble in sulphuric and sulphurous acids. Groroi, Cautern, Vecdessos, mixed usually with other oxides.

Manganese, Hydrous Sesquibinoxide of. Spec. grav. 3·312, H 4·25. Brownish-black pieces, of the size of a pea; lustre imperfect metallic, opaque; silica 22·9. Mn_2O_3 23·48, MnO_2 17·22, Fe_2O_3 28·64, HO 8·05; Cork.

Manganese, Sulphide of. *Manganese Blende, Black Earth.* 16 MnS, MnS_2. Spec. grav. 3·95 to 4·014, H 3·5 to 4. Iron-black; streak dark green; said to occur in cubes. B.B. fuses with difficulty. Nagyag, Transylvania, and Cornwall.

Manganese Diarsenide. Mn_2As. Specific gravity 5·55. Grayish-white foliated mass; fracture in one direction uneven, fine, granular, and shining; in the opposite direction dull and warty, and in that direction easily frangible. B.B. burns with a blue flame, and falls to powder. Quite soluble in aqua regia. Saxony.

Manganese Carbonate. *Dialogite, Rhodocrosite, Allagite, Photizite, Rhodonite.* MnO CO_2 7·25, 58, MnO = 62·07, CO_2 37·93. Specific gravity 3·55 to 3·59. Found native in the Saxon mines, in rose-red to brown obtuse rhomboids, with angles of 107° 20′. Changes to gray, brown, or black, before the blowpipe, infusible; with borax forms a blue bead; soluble in nitric and chlorohydric acids; ignited in a close crucible it leaves green protoxide. By long exposure to moist air it becomes brown (Berthier).

Semihydrous Carbonate. — MnO $CO_2\frac{1}{2}$HO, MnO=57·6, CO_2=35·2, HO 7·2. White tasteless insoluble powder, forming colourless solutions with acids. Obtained by precipitating protosulphate.

Manganese Subsesquisilicate. 3MnO, 2 SiO_3 = SiO_3 48·, MnO 48·98, CaO 3·12, MgO ·22 Spec grav. 3·538 to 3·63, H 6·5. Rose-red masses, or oblique rhombic prisms, resembling augite, with angles of 87° 5′; fracture flat, conchoidal; lustre between pearly and resinous; translucent on the edges, brittle; B.B. becomes dark brown. The silicates of manganese appear often to contain quartz interspersed, as the massive varieties from America I have found to contain, in the same mass, different proportions of silica.

Manganese, Carbosilicate of. *Horn Manganese.* Spec. grav. 3·89 to 3·1. MnO 54·86, SiO_3 34, CO_2 8·, HO 2·, FeO ·5. Chestnut-brown masses; B.B. phosphoresces.

Mangano-Calcite. Rhombic prisms, from Schemnitz. MnO CO_2 67·48, FeO CO_2 3·22, CaO CO_2 18·81, MgO CO_2 9·97.

Manganese, Ferrophosphate of. *Triplite, Zwieselite.* 4 (MnO FeO), PO_5. Spec. grav. 3·65, H 5·25. Brown imperfect crystals, at Limoges; streak yellowish-gray. See also HUREAULITE.

Mangano-Cyanogen. Cy_6Mn_2. A probable organic radical.

Mangel Wurzel. *Mangold Root. Beta vulgaris,* var. *campestris.* The cells of this root are large and transparent, and thin-sided. The roots contain pitted and dotted tubes, accompanying which are long prismatical cells. I find the

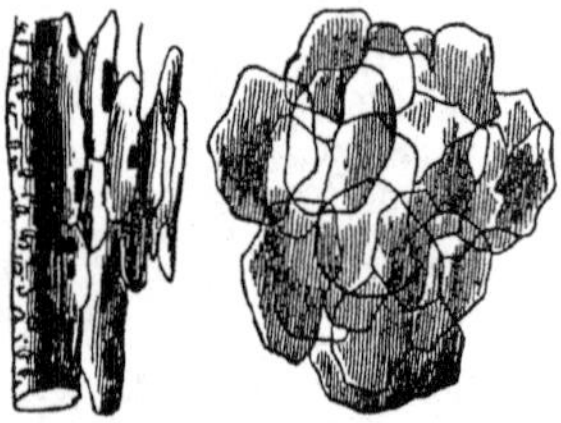

Prismatic Cells. Cells.
[*Hooker.*]

roots in the red variety to contain 91·57 per cent. of water, and 8·43 solid matter; and the white variety 87·63 water, and 12·37 solid matter. It has been found to contain nearly 25 per cent. of sugar.

Manna. *Mannite.* $C_6H_7O_6$, or $C_{12}H_{14}O_{12}$. Colourless 4-sided prisms, with a sweet taste; soluble in 5 cold, very soluble in hot water; little soluble in cold, very soluble in hot alcohol; insoluble in ether; unfermentable; does not reduce oxide of copper; soluble in strong sulphuric acid, forming manna sulphuric acid ($C_{12}H_{11}O_9$ 6 SO_3?) without charring. Mannite is extracted from commercial manna usually employed as an aperient, which consists of mannite 60, mucilaginous sugar 5·5, gummy extract ·8, gum 1·5, albumen ·2, water and loss 32·. Manna is extracted by incision from the *Fraxinus ornus*, which grows in Sicily and Calabria. It occurs in commerce in the form of *flake manna, manna in sorts* and *in tears.* It occurs also on the surface of certain sea weeds, as Laminaria saccharina, Halidrys siliquosa, as ascertained, in 1815, by Gaultier de Claubry. It also exists in Agaricus cantharellus, Phallus impudicus, Peziza nigra, Hydnum repandum, &c. It is also formed from grape sugar by fermentation at the temperature of 86°, lactic acid, gum, and mannite being formed. Mannite is obtained from manna by boiling alcohol; the crystals on separating are washed with cold alcohol, and again dissolved in boiling spirit. When sugar occurs along with it, yeast should be added previously, so as to destroy that body. It unites with metallic oxides, the compound with lead being 2 $PbOC_6H_5O_4$, and forms conjugate acids with sulphuric and formic acid (Manna formic acid, $C_6H_7O_6$,C_2HO_3). When manna is dissolved in sulphuric acid, and added to olive oil, it yields no colour; while sugar gives a crimson.

Manna Croup. This is the name under which a preparation of wheat is imported from Russia. It is the same as Semolina.

Mannitrin. *Nitromannite.* C_6,$H_4$3NO_4,O_6. Colourless silky needles; insoluble in water; slightly soluble in cold alcohol; soluble in hot alcohol and ether; carefully heated it fuses with slight evolution of red fumes; explodes when struck with a hammer, but not when ground in a mortar; formed by digesting 1 part of mannite in powder with nitric acid (spec. grav. 1·5) until it is dissolved; sulphuric acid is then added until the proportions are $4\frac{1}{2}$ nitric and $10\frac{1}{2}$ sulphuric acid; it is decomposed by metallic iron in its alcoholic solution; by nitric acid converted into oxalic and saccharic acids.

Manures. Substances applied to land for the purpose of serving as nourishment to crops. As plants remove from the soil a certain amount of inorganic salts, the study of the ash of plants affords a key to the proper manure for each plant. The best manure is obviously the decayed matter which has been grown on the soil, while all artificial manures should possess the same constituents. One of the most important manures is guano, which is valuable from its variety of constituents. The presence of organic matter, especially of ammonia, in manures, has a powerful influence in stimulating the growth of plants, although it is not decided that it is indispensable.

Marble. White variegated limestones employed for ornamental purposes.

Marcasite. A synonyme of iron and arsenical pyrites.

Marceline. A syn. of silicate of manganese.

Marekanite. A synonyme of obsidian.

Margaramide. *Elamine.* $C_{34}H_{33}O_2NH_2$. F.P. 140°. White silky mass, soluble in alcohol; obtained by digesting olive or almond oil in an alcoholic solution of ammonia for some days at a gentle heat, and then heating to 90° to remove ammonia.

Margarane. $C_{68}H_{65}O_5$. F.P. $144\frac{1}{2}$°. Soft waxy mass, forming a brown fluid when heated; obtained by heating anhydrous phosphoric and margaric acids together, boiling with water and alcohol to remove margaric acid. Nitric acid converts it into $C_{68}H_{64}O_7$.

Margaric Acid. *Margarylic Acid, Metastearic Acid.* $HOC_{32}H_{33}O_3$, or $HO(C_{32}H_{33})C_2O_3$. F.P. 140°, 126° (Duffy). White pearly scales or fine needles, without taste and smell; insoluble in water; very soluble in ether and absolute alcohol; soluble in hot weaker spirit, separating on cooling; solution with an acid reaction; distils over at 572° without decomposition; obtained by the saponification of human or goose fat; the soap thus formed is dissolved in hot water, precipitated with acetate of lead, the dried salt treated with ether, and the residue decomposed with chlorohydric acid in alcohol; chloride of lead falls, and margaric acid dissolves. It may also be obtained from butter, and by boiling stearic acid for ten minutes with an equal weight of nitric acid (spec. grav. 1·28), and crystallizing out of alcohol; by nitric acid it is converted into pimelic, suberic, succinic, adipic, lipic, and azoleic acids; by brown oxide of lead it is oxidized into hypermargaric acid ($HOC_{34}H_{33}O_4$?); fusing at $109\frac{1}{2}$°; by anhydrous phosphoric acid it becomes margarane ($C_{68}H_{65}O_5$?). Margaric acid is probably monobasic.

Margarine. *Margarate of Oxide of Glyceryle* or *Lipyle.* $C_2H_3OC_{34}H_{33}O_3$? F.P. $118\frac{1}{2}$°; 104° to 132° (Duffy); solidifies at 106°. White grains or porcelain-like mass; slightly soluble in boiling alcohol; easily soluble in ether; by heat it is converted into acroleine, margarone, and free margaric acid; obtained by frequently crystallizing human fat out of alcohol, or by melting butter, washing it with water and crystallizing, repeatedly out of ether and alcohol. By the action of sulphuric acid on margarine two acids are formed, glycero-sulphuric acid, and margaro-sulphuric (sulphomargaric) acid, the last of which has not been obtained free from oleosulphuric (sulpholeic) acid; by the action of water on these acids, metamargaric or paramargaric acid is formed.

Margarite. (μαργαρον, a pearl.) *Diphanite?* Spec. grav. 3·032, H 4·. A variety of mica, consisting of grayish, yellowish, or reddish-white micaceous plates, consisting of SiO_3 33·45, Al_2O_3 58·0, Fe_2O_3 0·42, CaO 7·5, MnO 0·03, MgO 0·05. In chlorite at Sterzing in the Tyrol. Soluble in acids.

Margaritic Acid. *Stearoricinic, Ricinostearic.* $C_{35}H_{31}O_6$. F.P. 266°. Pearly scales, by the saponification of the stearine of castor oil.

Margarodite. *Slaty Talc from Zillerthal.* Spec. grav. 2·872, H 2·75. Silvery-white micaceous scales, biaxial. SiO_3 47·05, Al_2O_3 34·90, Fe_2O_3 1·5, MgO 1·95, NaO 4·07, KO 7·96, HO 1·45.

Margarone. *Acetone of Margaric Acid.* $C_{33}H_{33}O$. F.P. $170\frac{1}{2}$°. White pearly scales, decomposing by distillation; soluble in 50 parts alcohol; 100 absolute alcohol dissolve 15 parts; ether dissolves $\frac{1}{4}$ of its weight; acetic ether and oil of turpentine have a similar action; decomposed by chlorine, and converted into a colourless fluid. Obtained by distilling margarate of lime or margaric acid with $\frac{1}{4}$ of quicklime, the distilled matter boiled with caustic potash, to remove free acid, the residue dissolved in ether and crystallizing repeatedly from the same solvent, which is preferable to alcohol.

Margaryle. $C_{34}H_{33}$ or $(C_{32}H_{33})C_2$. The hypothetic radical of margaric and stearic acids. If this substance be viewed as the base of the series, the following table may be constructed:—

Margaryle	= M
Margaryle oxide	= MO
Stearic acid	= $MO_{2\frac{1}{2}}$
Margaric acid	= MO_3
Hypermargaric acid	= MO_4

Margaryle, Oxide of. *Stearone, Margarone, Stearene.* $C_{34}H_{33}O$ (Bussy), $C_{33}H_{33}O$ (Redtenbacher), $C_{28}H_{28}O$ (Rowney). F.P. 168°·8. C 82·2, H 13·8, O 4·. White pearly plates, soluble in alcohol and ether. Obtained by distilling stearic acid with $\frac{1}{4}$ or $\frac{1}{3}$ its weight of quicklime and crystallizing out of ether. Considerable doubt still exists respecting the nature of this body and others derived from the fats, from the remarkable changes which it seems to undergo in the process of preparation as regards its fusing point, and probably its specific gravity.

Margarylene. *Hydride of Margaryle.* $C_{32}H_{33}H$. A fluid obtained in the last stage of the distillation for margarone, procured by being retained in solution by ether while the margarone deposits.

Marianite. A synonyme of nitrate of soda.

Marine Acid. A synonyme of muriatic or chlorohydric acid.

Marine Metal. Spec. grav. 11·1. An alloy prepared for ships, consisting of 94·4 lead, 4·3 antimony, and 1·3 mercury.

Marl. A mixture of carbonate of lime and earthy matter, frequently found in nature, probably connected with organic deposits, and valuable as a manure.

Marmatite. Black zinc blende, from Marmato, near Popayan, consisting of 77·5 per cent. ZnS and 22·5 FeS = 3 ZnS, FeS.

Marmolite. A variety of serpentine, with an additional amount of water.

Marrow. The fatty matter which fills the hollow portions of the long bones of animals, consisting, in the ox, of pure marrow 96; skins and blood-vessels 1·; albumen, gelatine, extractive, peculiar matter, water 3·. Water takes up from it a substance with the smell of roast meat. When melted in water, and passed through a cloth, it is white, with a shade of blue; taste insipid and sweetish, melts at 113°; when cooled slowly it crystallizes in sphericles, like olive oil; by distillation it yields a yellowish oil, with carbonic acid, water, and an inflammable gas, and afterwards a white solid oil, with a disagreeable smell, reddening vegetable blues, and when boiled yielding some sebacic acid. The white oil amounts to about one-third of the marrow. Marrow is soluble in strong nitric acid; it saponifies with alkalies.

Martial Ethiops. A synonyme of black oxide of iron.

Martinsite. Common salt, containing 9 per cent. Epsom salt (Karsten).

Martite. A synonyme of sesquioxide of iron, which is found in 8-hedrons in mica, and clay slates in Brazil, Tyrol, and Saxony.

Marum Camphor. Colourless plates, the stearoptene of the oil from *Teucrium marum.*

Mascagnine. Native *sulphate of ammonia.* $NH_3SO_3\ 2\ HO$. Found in the fissures of lava at Etna, Vesuvius, and Lipari.

Mash Tun, or *Wort Vessel.* See BREWING.

Masonite. Spec. grav. 3·45, H 5·75. SiO_3 28·27, Al_2O_3 32·16, FeO 33·72, MgO ·13, HO 5·. A variety of chlorite, found in Rhode Island.

Masopine. $C_{12}H_9O$. F.P. 311°, after once fusion 158°. Snow-white powder or needles, feeling glutinous between the fingers; yields, by distillation, a brown oily mass, with an acid reaction. On fusing masopine an agreeable odour is evolved; it is insoluble in water; very soluble in ether. Obtained from the dry sap of the *Dschilte*, a Mexican tree, by crystallizing out of alcohol.

Massicot. Yellow protoxide of lead.

Masticine. *Beta resin of Mastich.* $C_{40}H_{31}O_2$. That portion of mastich resin insoluble in alcohol.

Mastic Resin. Spec. grav. 1·074. Yellow transparent brittle grains, from the *Pistacia lentiscus*, a tree growing in the island of Chios; softens in the mouth, and has been used for stopping decayed teeth; partially soluble in alcohol; (resin α.) $C_{40}H_{32}O_4$; the insoluble portion is masticine, $C_{40}H_{31}O_2$.

Materia Perlata. The alchemical name of hydrous antimonic acid.

Matico, Maticine. The leaves of a plant from Peru, which have been recommended in medicine as an astringent. When boiled in water a camphor oil separates, and a bitter substance (maticine) remains in solution, which is soluble in alcohol, but not in ether.

Matrass. The old name for a globular flask.

Maynas Resin. Spec. grav. 1·12; F.P. 221°. $C_{14}H_9O_4$. Yellow acid crystals by solution in alcohol from the resin of the *Calophyllum calaba*, an American tree; insoluble in water; very soluble in alcohol, ether, oils, cold sulphuric and acetic acids; strong nitric acid forms a yellow acid; acid of 1·31 forms with it butyric acid?

Measures of Capacity. In the imperial weights and measures authorized in Great Britain, the following are the principal data:—

Cubic inch of distilled water at 62° in vacuo,				= 252·72 grains.	
Cubic inch	ditto,	ditto,	in air,	= 252·458	„
Cubic foot	ditto,	ditto,	in vacuo,	= 62·3862 lbs. avoird.	
Cubic foot	ditto,	ditto,	in air,	= 62·3206	„
1·73298 cubic inch water,				= 1 ounce avoirdupois.	
277·296	ditto,	at 62°,		= 10 lbs.	„
277·274	ditto,	at 60°,		= 10 lbs.	„
Standard quart,				= 2·5 lbs.	
Ditto, pint	= 20 fluid ounces,			= 1·25 lbs.	
Ditto, bushel	= 8 gallons,			= 80 lbs.	
Ditto, sack	= 3 bushels,			= 240 lbs.	
Ditto, chaldron	= 12 sacks,			= 2880 lbs.	

As the capacity of the measures depends on the weight of water at a certain temperature, the specific gravity of that fluid at different temperatures is an important consideration. The difference of temperature between 62° and 39°, where water attains its greatest density, will vary the bulk of a gallon of water rather less than the third of a cubic inch; and assuming, from the mean of numerous estimates, the expansion of brass ·00001044 for each degree of Fahrenheit, the difference of temperature from 62° to 39° will vary the contents of a brass gallon measure just one-fifth of a cubic inch. The specific gravity of clear water from the Thames exceeds that of distilled water in the proportion of about one-sixth of a cubic inch of a gallon, or in that of 1·0006 to 1°.

Table of Specific Gravity of Water at different Temperatures.

70°...·99913	56°...1·00050	44°...1·00107
68°...·99936	54°...1·00064	42°...1·00111
66°...·99958	52°...1·00076	40°...1·00113
64°...·99980	50°...1·00087	38°...1·00113
62°...1·	48°...1·00095	
58°...1·00035	46°...1·00102	

Measures of Capacity.

Cubic Inches.	Gills.	Pints.	Quarts.	Gallons.	Pecks.	Bushels.	Quarters
8·6648076103	1	0·25	0·125	0·03125	0·015625	0·00390625	0·00048828125
34·6592304412	4	1	0·5	0·125	0·0625	0·015625	0·001953125
69·3184609825	8	2	1	0·25	0·125	0·03125	0·00390625
277·2738435700	32	8	4	1	0·5	0·125	0·015625
554·5476871400	64	16	8	2	1	0·25	0·03125
2218·1907485601	256	64	32	8	4	1	0·125
17751·2598848179	2048	512	256	64	32	8	1

Measures of Capacity.

ENGLISH.	FRENCH.
Pint ($\frac{1}{8}$ of a gallon), ...	0·567932 litres.
Quart ($\frac{1}{4}$ of a gallon), ...	1·135864 litres.
Imperial gallon, ...	4·54345794 litres.
Peck (2 gallons), ...	9·0869159 litres.
Bushel (8 gallons), ...	36·347664 litres.
Sack (3 bushels), ...	1·09043 hectolitres.
Quarter (8 bushels), ...	2·907813 hectolitres.
Chaldron (12 sacks), ...	13·08516 hectolitres.

FRENCH.	ENGLISH.
Litre, ...	1·760773 pints. 0·2200967 gallons.
Décalitre, ...	2·200967 gallons.
Hectolitre, ...	22·009667 gallons.

For French measures, see METRE.

Mecca Balsam. *Opobalsam, Balm of Gilead, True balsam.* Spec. grav. ·95. A balsam without much colour when fresh, from the *Balsamodendron Gileadense*, a tree growing in Arabia. By keeping, it becomes whitish and turbid, and in the air dries completely up. It consists of volatile oil, hard resin, soft resin, a bitter colouring matter.

Mechloric, or Mechloic Acid. $C_{14} H_7 O_{10}$? F.P. 320°. White pearly plates or 4-sided prisms, soluble in water, alcohol, and ether. Obtained by acting on fused meconine with chlorine—heating the chlorine compound to 377° till it becomes oily, dissolving this in caustic potash, saturating with nitric acid, when mechloic acid crystallizes out.

Meconamidic Acid. $C_{84} H_{39} N_7 O_{78}$. White substance by ammonia on ethylomeconic acid.

Meconic Acid. $C_{14} HO_{11}$ 3HO, 6HO crystallized, and $C_{14}HO_{11}3HO$ at 212°. Slightly acid tasted white mica-like scales, needles, or rhombic prisms; at 212° loses 21·3 per cent. (6 atoms) water; very little soluble in cold water; the dried acid is soluble in 4 parts boiling water; soluble in alcohol and ether; changed when boiled in water into comenic acid; meconic acid is readily detected in most minute quantity by salts of sesquioxide of iron, which yield a blood-red colour with it. It is prepared from the meconate of lime, the refuse of the preparation of muriate of morphine, by dissolving in 20 parts almost boiling water, to which 3 chlorohydric acid are added. The mixture is agitated and heated until all is dissolved; on cooling, bimeconate of lime crystallizes, which is thrown on a cloth and pressed. The salt is dissolved in the same quantity of water and acid as before. By repeating the process frequently, the acid is obtained pure. The colour is got rid of by dissolving it in hot water, and adding caustic potash nearly to saturation (if in excess, the meconic acid is converted into oxalic and carbonic acids). The meconate of potash deposits on cooling, while the colouring matter remains in the mother liquor. It is dissolved in 16 hot water, and 2½ muriatic acid added; bimeconate of potash crystallizes out; a second similar solution yields meconic acid. Meconic acid is a bibasic acid. The most important salts are the meconates of lime and iron. The lime salt is formed by adding an alkaline meconate to chloride of calcium, and is a crystalline powder. The bimeconate of lime is in pearly scales. Sesquimeconate of iron contains 28 per cent. of sesquioxide of iron; it is soluble in water, and therefore does not fall when meconic acid is added to a salt of iron, but forms a red solution. The presence of opium in any solution may always be recognized by this colour. This colour is not destroyed by weak acids, even on boiling, nor by chloride of gold, which is the case with sulphocyanide of iron.

Meconine. $C_{20}H_{10}O_8$. F.P. 194°; not volatilized at 518°. Fine colourless 6-sided prisms; without odour; with a sharpish taste; soluble in 266 cold, 18½ boiling water, fusing into oily drops; soluble in alcohol, ether, ethereal oils, acetic, muriatic acids, caustic alkalies, except ammonia; precipitated from its solutions by carbonate of ammonia and by trisacetate of lead; by nitric acid it is converted into nitromeconic acid ($C_{20}H_9NO_4O_8$). Chlorine changes it into mechloic acid, and sulphuric acid apparently decomposes it. Meconine is obtained by exhausting opium by cold water, evaporating till the solution has the spec. grav. 1030° (6° Twad.), and adding dilute ammonia as long as a precipitate falls; the supernatant fluid is decanted in forty-eight hours, and the precipitate well washed with water. Additional meconine may be obtained by evaporating the mother liquors and wash waters. It is further purified by solution and crystallization out of alcohol, and decolourization by animal charcoal. Ether dissolves meconine but not morphine, by which the two alkaloids may be separated.

Meconium. A dark pitchy matter found in the intestinal canal of the fœtus. It consists of cholesterine 16, extractive and bile resin 10·4, caseine 39, picromel 6, green biliary matter 4·, epithelial scales and probably albumen 26 (Simon), or water 72·7, mucus and epithelium cells 23·6, cholesterine and margarine 7, biliary colouring matter and oleine 3· (J. Davy).

Mecono-ethylo-meconic Acid. *Ethylo-bimeconic Acid.* $3HO_2(C_{14}HO_{11})2HO,C_4H_5O$.

Medal Alloy. Gold medals contain often 91·6 gold, 8·4 copper. Silver medals and plate contain 95 silver and 5 copper.

Medjidite. H 2·5. Amber transparent masses, somewhat crystalline; fracture with glassy lustre; found on pitchblende near Adrianople. $U_2O_3SiO_3,CaOSiO_3,15HO$.

Medulline. Cellulose, constituting the principal constituent of the pith of plants.

Meerschaum. *Magnesite, Hydrous Tersilicate of Magnesia.* Spec. grav. 2·127, H 2·. Colour snow-white, or yellowish; fracture fine, earthy, or conchoidal; surface smooth, fine, granular, dull, opaque. SiO_3 42·, MgO 30·5, CaO 2·3, Al_2O_3 2·, HO 23·. Found in Natolia in carbonaceous or magnesian beds; of the consistence of wax when first dug.

Meine. An oil by alcohol and ether from the root of *Athamanta meum.*

Meionite. A synonyme of scapolite.

Melaine. A black matter resembling the black pigment of the eye, and obtained by deposition from the ink of the Sepia, or cuttle fish; it is insoluble in most menstrua, and is decomposed by nitric acid.

Melam. $C_{12}H_9N_{11} = 2\,C_6N_4,3\,NH_3 = 2$ atoms mellone and 3 ammonia. Heavy white powder; insoluble in water, alcohol, and ether; soluble in potash, nitric, and sulphuric acids; obtained by moderately heating sulphocyanide of ammonium, dissolving the gray residue in boiling potash, and filtering; melam deposits on cooling.

Melamine. $C_6H_6N_6$. 8-hedrons with a rhombic base; easily soluble in hot, less soluble in cold water; insoluble in alcohol and ether; sublimes with heat, a small portion being decomposed into mellone and ammonia, ammeline and ammelide; fused with potash, cyanate of potash is formed, and ammonia is evolved; combines with acids, and forms acid salts; obtained by boiling melam in 20 parts water, containing 1 part hydrate of potash in solution, and evaporating.

Melampyrine. Colourless and tasteless rhombic prisms; soluble in water, a little soluble in spirit; almost insoluble in absolute alcohol and ether; contains no nitrogen; obtained by forming a decoction of the Melampyrum nemorosum, and evaporating, when crystals of this substance are formed. Additional quantities may be procured by precipitating by acetate of lead, filtering, precipitating the excess of lead by sulphohydric acid, decolourizing by animal charcoal, filtering, and evaporating.

Melanchlor. A synonyme of Dufrenite.

Melangallic Acid. *Metagallic Acid.* A product of the action of heat on gallic acid.

Melanic Acid. A name given by Prout to a black substance found in the urine of a child; soluble in caustic alkalies, and reprecipitated by acids; insoluble in water and alcohol; soluble in nitric and sulphuric acids, and reprecipitated by water.

Melanic Acid. $C_{10}H_4O_5$. A black matter formed by the action of the air on the potash salt of hydrosalicylic acid. Water takes up acetate of potash, and leaves melanic acid.

Melaniline. $C_{20}H_{13}N_3 = C_{12}H_7N,C_{12}H_6$ Cy,N. A crystalline base, obtained as chlorohydride, by acting on aniline with gaseous chloride of cyanogen; scarcely soluble in water; soluble in alcohol and ether. It has been viewed as aniline united to another atom of aniline, in which an atom of hydrogen is replaced by cyanogen. With chlorine, bromine, iodine, and nitric acid, &c. it forms the new bases dichloromelaniline, dibromomelaniline, di-iodomelaniline, dinitromelaniline, and dicyanomelaniline.

Melanisic Acid. A black matter formed by boiling nitraniside with strong caustic potash.

Melanite. *Black Garnet.* Specific gravity 3·157 to 3·73, H 6·75. Velvet-black to grayish-black rhomboidal 12-hedrons, the edges being replaced by tangent planes, making a 24-sided figure; lustre resinous; fracture flat and imperfect, conchoidal; opaque. SiO_3 42·45, Al_2O_3 22·475, FeO 9·292, MnO 6·273, MgO 13·43, CaO 6·525. B.B. the edges are rounded, but not fused. Frescati, near

Rome, Monté Somma, near Naples, Bohemia, United States.

Melanochroite. *Subsesquichromate of Lead.* $1\frac{1}{2}$ PbO, CrO_3. Spec. grav. 5·75, CrO_3 23·31, PbO 76·69. Cochineal and hyacinth masses, becoming yellow in air, or rhombic prisms, with two of the opposite faces larger than the others; lustre resinous, glimmering; translucent on the edges, almost opaque; very soft, slightly brittle; easily pulverized. B.B. on charcoal cracks, but does not fly off; fuses into a dark mass, crystalline on cooling; in the reducing flame lead is sublimed, and oxide of chromium and grains of lead are left on the charcoal; gives green beads with fluxes. Beresofsk, Ural.

Melanolite. A green variety of groppite, occurring in America.

Melanoximide. *Oxamelanile.* $C_{30}H_{11}N_3O_4$, or *binoxalate* of *melaniline*, deprived of 4 atoms water. Yellowish crystalline crusts, obtained by dissolving dicyanomelaniline in common chlorohydric acid; the liquid soon becomes turbid from the separation of the imide.

Melantannic Acid. See Tannic Acid.

Melanterite. A synonyme of protosulphate of iron.

Melaphyre. A porphyry occurring in the Alps at Fassathal and Klausen, in the Tyrol, containing labradorite, augite, mica, hornblende, and pyrites, with calc spar, in cavities. SiO_3 51·13, Al_2O_3, Fe_2O_3 29·73, CaO 4·73, MgO, and alkalies, and loss 10·73 CO_2 and HO 3·68.

Melassic Acid. *Molassic Acid.* $C_{24}H_{12}O_{16}$? or $C_{24}H_{10}O_{10}$? Brown matter obtained by melting hydrate of barytes in its crystalline water, with grape sugar, and neutralizing with an acid. Its composition appears nearly the same as japonic acid.

Melathine. A black substance giving out ammonia when treated with ammonia, the residue containing sulphide of potassium; obtained by distilling the product of the action of ammonia and sulphur on acetone.

Melene. $C_{60}H_{60}$. F.P. 143·°6. White crystalline body, resembling paraffine, which however fuses between 110° and 126°. It is obtained by saponifying bees' wax, removing the soap, pressing the residue in paper, crystallizing out of ether, expressing, distilling over potash, expressing, and crystallizing out of ether.

Meliceris. An encysted tumour, consisting of coagulated albumen 52·5, cholesterine 3·12, oleine and oleate of soda 28·54, stearine 1·96, noncoagulated albumen 9·17, lime 1·88, magnesia ·92.

Melinophane. A variety of Leucophane.

Melinose. A synonyme of molybdate of lead.

Melinum. White powder from Melos and Samos, used by the ancients, probably chalk or limestone. Also a synonyme of cadmium.

Melissic Acid. HO $C_{60}H_{59}O_3$. F.P. 191°·3. Crystalline body, soluble in alcohol; obtained by heating melissine with a mixture of hydrate of potash and lime, and decomposing the salt in dilute acid.

Melissine. *Melissic Alcohol.* $C_{60}H_{61}OHO$. F.P. 185°. Silky crystals obtained by saponifying the myricine of bees' wax with caustic potash, dissolving the soap in water, precipitating with chloride of barium, exhausting the barytes salt with ether, from which melissine deposits; this is further purified by repeated solution in ether, or by decomposing the fresh soap with acid, and crystallizing the separated fat out of alcohol and naphtha. Melissine is partially decomposed by distillation, by chlorine, a resin, *chloromelal*, is formed, $C_{60}H_{44\frac{1}{2}}Cl_{14\frac{1}{2}}OHO$. Sulphuric acid combines, forming probably sulphate of melissyle.

Mellane. A synonyme of mellone.

Mellite. *Mellilite, Mellate of Alumina, Honeystone.* Spec. grav. 1·55 to 1·597, and 1·66, H 2·75. Honey-yellow, reddish or brownish 8-hedrons, with a square base or rhomboidal 12-hedrons, with angles of *m* on *n* 118° ·3, *m* on *o*

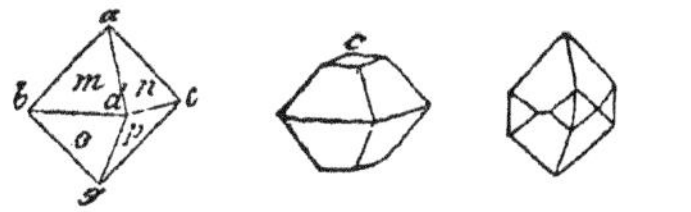

93°; lustre resinous; streak white. B.B. whitens, but does not flame; soluble in nitric acid; decomposed when boiled in water, the alumina being separated nearly pure, and the acid still retaining a little alumina being dissolved in the water; soluble in caustic alkalies. It consists of mellitic acid 41·4, Al_2O_3 14·5, HO 44·1. *Form.* Al_2O_3, 3 C_4O_3, 15 HO. In brown coal at Artern, Thuringia. I have also specimens from Ireland; Bilin, Bohemia; Walchow, Moravia.

Mellitic Acid. *Honeystone Acid.* HOC_4O_3 7·125, 57·. Fine silky needles, very soluble in water and alcohol; the crystals contain 1 atom basic water, not removed at 392°; not altered by the air; very acid taste; burns with an aromatic smell, and smoky flame; when distilled sublimes as *pyromellitic acid;* unacted on by cold, soluble in boiling sulphuric acid without change; insoluble in boiling nitric acid; by boiling with alcohol converted into *mellit-ethylic acid.* The salts of mellitic acid are monobasic; *mellate of ammonia*, $HONH_3C_4O_33HO$, formed by boiling mellite with carbonate, crystallizes. When heated to 302° the salt is converted into acid *euchroate of ammonia* and *paramide*, ammonia and water being evolved; *bimellate of ammonia*, with 6 water, is obtained by heating paramide with water at 392°; *termellate* with 6 water, by heating euchroic acid to 392° in contact with water.

Mellone. *Mellane.* C_6N_4. A grayish or yellowish powder, insoluble in water and alcohol; dissolved and decomposed by hot caustic potash; decomposed by a strong heat into 3 vols. cyanogen and 1 nitrogen; by chlorine it yields a white volatile matter acting on the eyes; by nitric acid it affords cyanilic acid, isomeric with cyanuric acid; caustic potash converts it into cyamelurate of potash; it unites with bases, and forms mellonides. It is prepared by heating to low redness sulphocyanide of ammonium ($C_8H_{16}N_8S_8$, or 4 sulphocyanide of ammonia become 2 bisulphide of carbon C_2S_4, 4 ammonia $4NH_3$, 4 sulphohydric acid 4 SH, and 1 mellone C_6N_4). It is also formed by passing chlorine over a mixture of 1 sulphocyanide of potassium, and 2 common salt, at a red heat; chlorides of sulphur and cyanogen are evolved, while mellone remains with chlorides of potassium and sodium ($4C_2NS_2 = C_6N_4$, 2 CS_2,4S). It is also procured by heating melamine, ammeline, and ammelide, with the following reaction:—Melamine $C_6H_6N_6 = C_6N_4,2NH_3$. Ammeline $C_6H_5N_5O_2 = C_6N_4NH_32HO$. Ammelide $C_{12}H_9N_9O_6 = 2C_6N_4NH_36HO$.

Mellonide of Potassium. KC_6N_4 5 HO. White efflorescent silky needles, with a bitter taste and neutral reaction; insoluble in alcohol; soluble in water; the crystals contain 25·4 per cent. water, of which $\frac{4}{5}$ are removed at 248°, the remainder at 302°; heated above its fusing point in close vessels it evolves nitrogen and cyanogen, and leaves cyanide of potassium; fused in the air it is converted into cyanate of potash; by chlorohydric acid it yields *hydromellonic acid*, C_6N_4 H, a snow-white powder; when mellonide of potassium is boiled with an excess of potash, ammonia is evolved, and, on the addition of acetic acid, a white crystalline body falls, unexamined; strong chlorohydric acid converts it into cyanuric acid and salammoniac. Mellonide of potassium is prepared by heating gently in a crucible 2 parts of dry yellow prussiate of potash, and 1 part of sulphur, till all the cyanide of potassium is converted into sulphocyanide; the heat is then increased until all the sulphide of carbon is removed, and cyanogen begins to burn with a purple flame. Liebig, but not Gmelin, recommends 5 per cent. of carbonate of potash to be added towards the end of the fusion.

Melopsite. Greenish-white lithomarge, Neudeck, Bohemia.

Melting Points. The temperatures at which solid substances assume the fluid form.

Membranes. These parts of animals are usually divided into *Serous*, or thin transparent tissues, which form close sacs, and are kept moist by an albuminous fluid secreted or diffused from the blood. When boiled they yield colline, or gelatine. The *mucous* membranes line open cavities, as the mouth and intestinal canal, and are moistened by mucus, which is a fluid diffused from the blood, and mixed with epithelial scales, &c.

Menachanite. *Menachane.* The original name of titaniate of iron given by Mr. Gregor when he discovered titanium, from the locality in Cornwall.

Mendipite. *Chloroxide of Lead. Oxychloride of Lead.* PbCl2PbO. Rhombic prisms from the Mendip Hills, Somersetshire.

Meneghinite. 4PbS, SbS_3. A variety of Jamesonite.

Mengite. A synonyme of Ilmenite, and also of Monagite.

Menilite. A brown opaque variety of opal in kidney-shaped masses from Menilmontant.

Menispermic Acid? A crystalline body from Menispermum cocculus, from the alcoholic mother liquor after the separation of picrotoxine.

Menispermine. $C_{18}H_{12}NO_2$. F.P. 248°. White 4-sided prisms terminated by 4-sided pyramids, resembling cyanide of mercury; insoluble in water; soluble in alcohol and ether; nitric acid forms with it oxalic acid, and a yellow substance; sulphuric acid forms with it a compound which fuses at 329°, and at a higher temperature evolves sulphohydric acid. Menispermine is obtained by exhausting cocculus indicus, or the fruit of the menispermum cocculus with boiling alcohol of ·833; the alcohol is distilled off, and the residue boiled with distilled water, and filtered while hot. From the filtered liquor another alkaloid also separates (picrotoxine), when a few drops of acid are added to it before cooling. The portion which is not taken up by hot water is extracted with dilute acid, and the extract precipitated by ammonia. The brown precipitate is treated with water containing some acetic acid; a blackish-brown matter remains undissolved. The solution is again precipitated with alkali, and a granular precipitate obtained. By agitation with cold alcohol it is freed from a yellow basic substance; it is then extracted with ether; by evaporation the menispermine remains. The portion insoluble in ether is paramenispermine, which is isomeric with menispermine, crystallizing in rhombic 4-sided prisms, fusing at 482°.

Menstruum. (*Mensis*, a month.) The fluid used for the solution of a substance, so named from the length of time during which it was formerly allowed to remain in contact with the solid.

Menthene. $C_{20}H_{18}$. B.P. 325½°. Spec. grav. ·851 Clear colourless fluid with an agreeable taste and smell; insoluble in water; soluble in alcohol, ether, oil of turpentine and pyroxylic spirit; coloured yellow by cold, red by hot chlorohydric acid, violet by bromine, red by iodine; chlorine converts it into $C_{20}H_{15}Cl_5$. Menthene is obtained by distilling the stearoptene of oil of peppermint ($C_{20}H_{20}O_2$) with anhydrous phosphoric acid; the result of the action is the removal of 2 atoms water.

Menyanthine. Yellow bitter granular substance, very soluble in water and alcohol; almost insoluble in absolute alcohol and ether. Obtained from the root of the *Menyanthes trifoliata* by exhausting with rectified spirit, and distilling off the spirit; a resinous matter separates; the fluid is filtered, fermented with yeast, and then digested with hydrous oxide of lead, filtered, the excess of lead removed by sulphohydric gas, and evaporated in a gentle heat, the dark extract taken up with alcohol, decolourized by animal charcoal, and evaporated in vacuo.

Mephitic Air. The name given to nitrogen by its discoverer, Rutherford, in 1772.

Mercaptan. A term now applied sometimes to a series of compounds with the type C*n* H*n* S, SH. The first of these discovered was the sulphohydride of the sulphide of *Allyle*, or alcohol in which the oxygen is replaced by sulphur, and the water by sulphohydric gas. C_4H_5S,SH. Spec. grav. 0·842; B.P. 97°. Colourless mobile fluid with a strong and disagreeable smell of garlic. It acts on red oxide of mercury, forming mercaptide of mercury, and an oxide of lead forming lemon-yellow crystals (ES, PbS). Mercaptan is formed by distilling equal measures of fluid sulphovinate of lime (spec. grav. 1·28) and of caustic potash of the same density, previously saturated with sulphohydric acid, or by passing chlorohydric ether through bisulphide of potassium. When the product is distilled over oxide of mercury and chloride of calcium it is pure.

Mercaptan-amyle. $C_{10}H_{12}S_2$. Spec. grav. ·835; B.P. 242½°. Colourless, oily, with the odour of garlic; by sulphide of ammonium on potato and grain oil.

Mercurius Cinereus Blackii. A mixture of dicarbonate of mercury with mercurius solubilis Hahnemanni.

Mercurius Cinereus Moscati. Black oxide or dinoxide of mercury.

Mercurius Cinereus Saunderi. Gray or black precipitate of mercury by ammonia on calomel (Hg_2Cl,Hg_2NH_2).

Mercurius Dulcis. A synonyme of *calomel*, or dichloride of mercury.

Mercurius Phosphoratus Fuchsii. A white crystalline precipitate, generally a mixture of phosphate of dinoxide and protophosphate of mercury, by precipitating nitrate of mercury with phosphate of soda.

Mercurius Precipitatus Albus. HgCl $HgNH_2$. White precipitate of mercury by precipitating corrosive sublimate by ammonia.

Mercurius Precipitatus Ruber. HgO. Protoxide of mercury.

Mercurius Solubilis Hahnemanni. Hg_2ONO_5,Hg_2NH_2. Black powder, by adding ammonia to nitrated dinoxide of mercury.

Mercurius Violaceus. Obtained by subliming a fused mixture of 4 sulphur, 6 mercury, 4 salammoniac.

Mercury. (*Hydrargyrum*), *Quicksilver*, *Argentvive*. Hg 12·5,100. Known from the earliest ages in consequence of being easily reduced. Is a white silvery fluid, destitute of taste and smell. Spec. grav. 13·568 at 60° (Cavendish and Crichton); 13·596 (Regnault); when solid at—38·66, 14·465 (Biddle); when frozen it is malleable, and can be hammered without breaking. B.P. 660° (Crichton). When heated it readily volatilizes at this temperature, and can thus be freed from many impurities. Spec. grav. of its vapour 6·976 (Dumas). Mercury readily unites with many metals and forms amalgams, and from this property it is capable of permeating masses of tin, zinc, cadmium, lead, silver, gold, copper, bronze; but not iron, platinum, palladium.

Preparation.—Mercury occurs in large quantities in nature in union with sulphur, under the name of cinnabar, in Carinthia, Saxony, in gneiss; in Transylvania in graywacke; the largest deposits are, however, found in Idria, and Almaden in Spain. Its colour is red. Crystallizes in acute rhomboids and 6-sided prisms. Sp. grav. 8·098. When pure it consists of 84·5 mercury, and 14·75 sulphur (HgS). Mercury also occurs in nature with silver under the form of *native* amalgam, which contains 72·5 per cent. mercury, and 27·5 silver. The theory of the process for the separation of mercury from cinnabar, consists in combining the

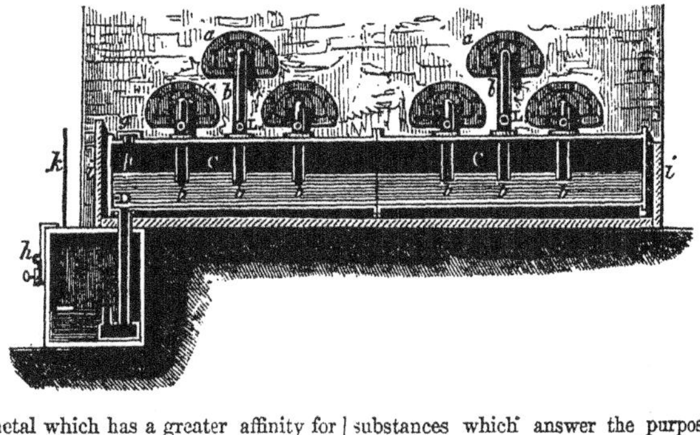

ore with a metal which has a greater affinity for the sulphur than the mercury has, so that by the application of heat the mercury passes over. The substances which answer the purpose best are lime and iron. In the Palatinate the ground ore and lime are mixed and introduced into a clay retort, a series of which is placed in a furnace. The mercury is distilled into a receiver of the same nature filled with water; the whole apparatus being similar to that used in the preparation of Nordhausen sulphuric acid. The other methods at present employed are equally rude, but might be easily improved. Retorts, such as are used in the preparation of nitric acid, might be employed with great advantage, or cylinders of iron, similar to those for preparing muriatic acid. The apparatus exhibited in the figure has been employed several years. It consists of *a*, a number of retorts resembling gas retorts, which are filled to two-thirds with a mixture of 4 parts cinnabar and 1 quicklime; *b*, pipes for the escape of the mercury dipping into water in C, an iron inclined condenser; L, screw plugs for the introduction of a wire to clear the pipes. The condenser is placed in *i*, a wooden trough, through which a current of water flows. The mercury passes into E by the pipe D, and its amount is indicated by the float *k*. To purify the quicksilver it may again be distilled with iron filings in iron retorts. When red oxide is distilled, the metal passes over with a small portion of oxide, which may be removed by agitation with dilute nitric acid.

Dinoxide of Mercury, Suboxide, Black Oxide, Ethiops per se, Mercury of Moscati.—Hg_2O 26, 208. First obtained by Homberg, in 1699, by attaching a bottle of mercury to a mill wheel; it was first described by Boerhaave. Mercury is not oxidized by water, but on exposure to the air it tarnishes, a black powder being formed, Hg_2O. It may be prepared most conveniently by precipitating calomel (Hg_2Cl), the subchloride of mercury, by an alkaline solution, as potash, soda, or lime water. In the formation of the black wash of surgeons, the suboxide is precipitated by lime water by the following action:—$Hg_2Cl+CaO$ become Hg_2O, CaCl. Caustic soda precipitates it very readily. The black precipitate is washed with water, and constitutes a pure suboxide. It is grayish-white, tasteless, and insoluble in water, and when obtained by the preceding process often contains metallic quicksilver in globules, which may be easily separated. When boiled with water it is converted into the red oxide (HgO). It dissolves in acids without effervescence, and is completely volatile. Spec. grav. 10·69. It forms salts with acids, which are generally colourless or white, and difficultly soluble in water. By alkalies the grayish-black oxide is precipitated.

Oxide of Mercury, Protoxide, Red Precipitate, Deutoxide of Mercury, Peroxide, Precipitate per se.—HgO 13·5, 108. When formed by heat it crystallizes in regular octahedrons. Colour deep red; may be heated nearly to 800° without decomposition, but above a red heat is converted into mercury and oxygen (Priestley, 1774). It may be formed by heating mercury to 600° in the open air. The form of the oxide, formerly termed precipitate *per se*, was obtained by placing some mercury in a flask, the neck of which is drawn out to a capillary point, and subjecting the flask to a boiling temperature. The mercury gradually becomes red on the surface, and after some weeks the whole metal is converted into deep red crystals, octahedrons. By dissolving mercury in nitric acid, and heating the nitrate cautiously, it becomes *red precipitate*. The taste of this oxide is acrid and disagreeable. When pure it should give off no fumes if heated in a tube. It is decomposed by heat and light. With sulphur, when heated, it deflagrates; heated with zinc or tin it sets them on fire; slightly soluble in water, but probably only when nitrate is present. It is used in medicine as an application to sores, as an escharotic, to clean ulcers, by giving them a new surface, either by being sprinkled over the parts, or in the form of ointment. The salts of this oxide are colourless or yellow; some of them soluble, others insoluble in water, and are poisonous. The solutions of the salts are precipitated white by caustic ammonia, a hydrate of an amide and an oxide of mercury being formed ($HgNH_2+2$ HgO 3 HO). When corrosive sublimate is precipitated by lime water, a yellow powder falls, the mixture being termed in pharmacy yellow wash, and used as an application to syphilitic ulcers. The action in this case is produced by HgCl and CaO becoming HgO and CaCl. The yellow oxide is isomeric or allotropic with the red oxide. The *ammoniacal oxide of Mercury*, 3 $HgOHgNH_2$ 3 HO, may be formed by placing oxide of mercury in a flask, and filling it with ammoniacal solution, and stopping the flask. For this purpose either the yellow or red isomeric forms of oxide may be employed, but the yellow answers best. The hydrate, in a vacuum, loses its water, and also at the temperature of 266°. The hydrate is insoluble in water and alcohol; cold caustic potash scarcely acts on it; when boiling, ammonia is evolved. It absorbs carbonic acid as readily as lime and barytes; its carbonate is not decomposed at 212°, but loses its water at 284°, and becomes the anhydrous carbonate. When sulphate of mercury is thrown into ammonia it dissolves in large quantity. When the solution is diluted a white powder falls, which is *ammoniacal turbith*, or ammonia sulphate of mercury, 3 HgO $HgNH_3SO_3$, but it is not constant in composition. The following combinations have been obtained with the ammoniacal oxide acting as a base:—

Hydrated base,.........	3HgO,	$HgNH_2$	3HO
Intermediate hydrate,	„	„	HO
Anhydrous base,........	„	„	
Sulphate,................	„	„	SO_3
Hydrous carbonate,....	„	„	CO_2HO
Carbte. dried at 275°,	„	„	CO_2
Oxalate,.................	„	„	C_2O_3
Nitrate,.................	„	„	NO_5HO
Bromate,................	„	„	BrO_5

Dichloride of Mercury, Subchloride of Mercury, Protochloride of Mercury, Calomel, Draco mitigatus, Sweet Mercury. — Hg_2Cl 29·437, 235·5. Known to the alchemists; the term *calomel* was first used by De Mayenne about 1616. Usually a dull white mass. When slowly sublimed it crystallizes in 4-sided prisms, terminated by pyramids, the primary form being a square prism (Brooke). Spec. grav. 7·175 to 6·992. 1 part is soluble in 12,000 parts of water. Tasteless, not poisonous, but acts as a purgative; it is diffused by being converted into the chloride, which is a soluble salt. When calomel is mixed over night with a solution of an ammoniacal chloride, we find mercury in solution in the morning, indicated by the liquor giving a yellow precipitate with caustic soda or potash, and a white with ammonia, showing that the calomel has been converted into HgCl.

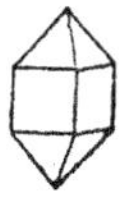

Preparation.—Calomel may be obtained either by dry distillation or by precipitation. 1. When 4 parts of corrosive sublimate, and 3 of mercury, are intimately mixed in a mortar, and sublimed in a flask, calomel is produced. 12·5 parts of mercury, when converted into sulphate of mercury, by heating with SO_3, the dried salt mixed with 2·5 parts mercury and 8½ parts of common salt dried, yield, by sublimation, calomel. If the sublimed calomel is allowed to come in contact with steam, it falls as a fine powder (Howard and Jewel). This process consists of three stages:—1, the formation of sulphate of mercury by boiling mercury with sulphuric acid; 2, mixing each atom of sulphate with an atom of mercury; 3, subliming the whole with an atom of common salt. The process is expressed by the following formulæ:—

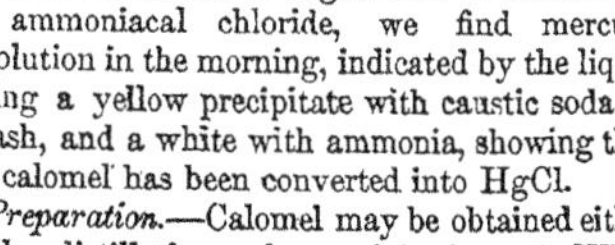

The result is calomel sublimed, and sulphate of soda left as a residue. 2. It may also be obtained by dissolving the acid nitrate of mercury, in 16 parts of water, and adding a dilute solution of common salt, mixed with some HCl to remove nitrate, as long as any precipitate falls. This is to be thoroughly washed with cold water.

Gray Precipitate. — $Hg_2ClHg_2NH_2$. Is obtained as a dark gray powder, when calomel is digested in caustic ammonia, the action being as follows:—

$$2\,Hg_2Cl,\ 2\,NH_3 = (Hg_2ClHg_2NH_2)\,NH_3HCl$$

Chloride of Mercury, Corrosive Sublimate, Muriate of Mercury, Bichloride.—HgCl 16·937; 135·496. It is mentioned by Geber, in the seventh century, and was early known to the Chinese. Fine white semitranslucent mass of prisms, soluble in water, crystallizing in cubes or rhomboids on evaporation, the primary form being a right rhombic prism, the angles $\acute{u}$ on t being 138°·8, and $\acute{u}$ on u 93°·44 (Brooke). Sp. grav. 5·139 to 5·42. Taste acrid and caustic, with a styptic impression; virulent poison, of a corrosive nature, destroying the integrity of the structures with which it comes in contact. It dissolves in 18 parts of cold and 2 or 3 parts of boiling water. It is very soluble in hydrochloric acid, and also in ether and alcohol. The solution reddens litmus paper.

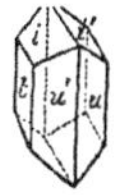

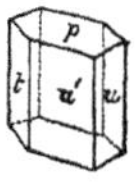

Preparation.—1. Corrosive sublimate may easily be obtained by dissolving red precipitate in hydrochloric acid. 2. Another process is to form the sulphate by heating SO_3 and Hg, and then subliming 5 parts of the dry mass, 5 parts of dry common salt, and 1 part of black oxide of manganese in a retort. In this process $HgO\ SO_3 + NaCl$ become HgCl and $NaOSO_3$. The manganese is added to convert any sulphate of suboxide of mercury into the sulphate of the oxide, as frequently happens in the dry sulphate, as made in the manner recommended. 3. Another method for the preparation of corrosive sublimate is to add to a solution of nitrate of mercury hydrochloric acid, as long as a precipitate falls; an equal amount of acid is then added, and the whole is boiled. The precipitate formed is again dissolved, and allowed to crystallize out.

Tests.—When pure it is white, dissolving in water, alcohol, and in from 4 to 6 parts of ether; when heated it sublimes entirely. When dissolved in HCl and a solution of protochloride of tin in the same acid is added, a precipitate falls, white at first, then becoming black, of metallic mercury. Alkalies, with the exception of ammonia, throw down a yellow precipitate (HgO HO); sulphohydride of ammonia a black precipitate; iodide of potassium a scarlet precipitate. A plate of copper becomes brown in a solution of corrosive sublimate. Gold becomes white.

Poisoning.—When taken in over-doses it produces violent irritation, pain in the pit of the stomach and bowels, and diarrhœa, with vomiting. When the dose is larger, these symptoms show themselves as soon as the salt is swallowed. The taste of the substance distinguishes it from arsenic, a fact of importance, since the other symptoms are closely analogous.

Antidotes.—The best antidotes for poisoning by mercury are albumen or white of egg dissolved in water. The curd of milk likewise unites with corrosive sublimate and neutralizes it. Iron filings may also be given with advantage.

White Precipitate, Chloramide of Mercury.—$Hg\,Cl\,Hg\,NH_2$, or N,HgH_3Cl. A white powder prepared by adding caustic ammonia to a solution of corrosive sublimate, or by mixing 1 part of salammoniac and 2 parts of corrosive sublimate in solution in water, and precipitating with carbonate of soda or potash. The salammoniac promotes the solution of corrosive sublimate in water, and hence they are often prescribed in connection. The nature of the action which occurs in the formation of white precipitate appears to be as follows:— $2 HgCl 2 NH_3 = HgCl HgNH_2, NH_3 HCl$. It is a white poisonous powder, with a metallic taste. It is very insoluble in water and alcohol. It may be distinguished from calomel, which it resembles in appearance, by its becoming a canary-yellow powder when boiled with water or alkalies, and salammoniac being dissolved in the water. $2HO,2(HgClHgNH_2)=(HgClHgNH_2 2HgO) NH_3 HCl$. It may also be distinguished from calomel by the circumstance that when boiled or heated with caustic soda, ammonia is evolved, and a yellow powder formed, which, under the same circumstances with calomel, is black.

Diniodide, Subiodide.—Hg_2I 40·75, 326·. Yellowish-green powder, formed by dropping an iodide into a solution of the subnitrate of mercury, or by rubbing in a mortar 25 parts of mercury with 15¾ parts of iodine, with some alcohol.

Iodide.—(HgI 28·25, 226·). Scarlet, crystalline powder, formed by adding iodide of potassium to corrosive sublimate, or by rubbing in a mortar 12½ parts Hg with 15·75 I, with the aid of alcohol, so as to form a red powder, which is to be boiled with a strong solution of common salt until it dissolves. On cooling it deposits in fine carmine scales. Slightly soluble in water, very soluble in iodide of potassium. Specific gravity of its vapour 15·68. Becomes a yellow fluid at 400°. Both of these iodides are used in medicine, internally and in the form of ointment.

Cyanide of Mercury.—(Hg Cy 15·75, 126.) Square prisms prepared by dissolving 2 parts yellow prussiate of potash in 15 boiling water, adding 3 parts dry sulphate of mercury, boiling for fifteen minutes, filtering while hot, and crystallizing out the cyanide. Additional crystals may be obtained from the mother liquor. It is used for preparing cyanogen, and is poisonous.

Black Disulphide of Mercury.—(Hg_2S.) Black powder, obtained by precipitating the protonitrate of mercury by sulphohydric acid.

Sulphide, or *Red Sulphide.*—HgS. *Cinnabar, Vermilion.* Occurs in nature, but it may be prepared artificially by fusing 1 part of sulphur and 6 of mercury in a covered crucible, occasionally removing the lid and stirring the mixture. The residue, when sublimed, possesses a red colour, and constitutes the vermilion of commerce, although before sublimation it is black. We can prepare it at once of a red colour by moistening white precipitate with sulphohydride of ammonia; the black sulphide is produced; but it passes after a little time into the red colour.

Sulphate of Mercury ($HgO SO_3$) is formed in the process for calomel. When boiled with water, it becomes yellow, being converted into *Turbith mineral* ($3HgO SO_3$). *Subnitrate of mercury.*—($Hg_2O NO_5 \cdot 2HO$.) Colourless prisms formed by dissolving mercury in cold dilute nitric acid. *Dinitrate of mercury* ($2HgO NO_5 2HO$) is obtained by dissolving mercury in hot nitric acid.

Dry Assay.—To determine the amount of mercury when in the state of metal or oxide, all that is necessary is to distil in a tube retort 100 grs. or 150 grs. of the specimen to be examined into

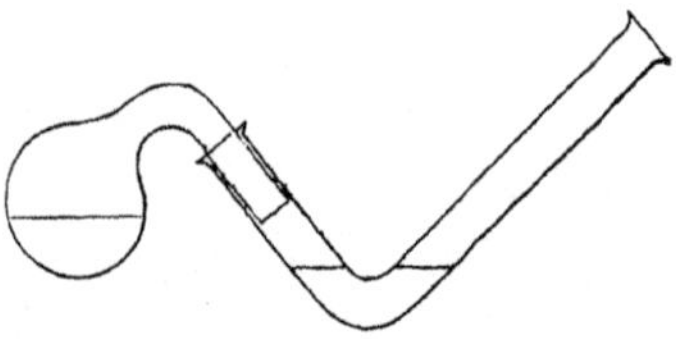

a receiver accurately fitted, and containing water. The tube should be made of hard Bohemian or English green glass. The retort is heated to redness in a charcoal fire, and the upper part kept hot to prevent the condensation of the mercury. The metal collects in the neck of the retort. If the water has been previously expelled by drying the mixture carefully, the amount of mercury may be estimated by weighing the retort before and after the experiment. But as the mercury collects in the neck of the retort, it is necessary to cut it off, to weigh it, and then to remove the mercury by means of a feather, and to weigh it again. The mercury is made to run into the water of the receiver. The water is poured off, the mercury dried and weighed. When the ore is in the state of sulphide or selenide, it is necessary to mix it with iron filings, to detach the sulphur. 100 grains of the cinnabar are intimately mixed with 50 grains of iron filings introduced into the tube retort, the surface covered with a layer of iron filings, and treated as before described. Instead of iron filings, 66 per cent. of black flux may be used, or 33 per cent. of caustic lime mixed with 10 parts of powdered charcoal.

Wet Assay.—To determine the amount of mercury in an ore, the specimen is digested with pure strong hydrochloric acid. It must not be heated too strongly, otherwise the mercury will be volatilized. The liquid is then decanted from the residue, and the latter washed. A solution of protochloride of tin is then added, which has been rendered transparent by hydrochloric acid, and the mixture is boiled quickly in a flask, corked, and set aside. The mercury falls as a black powder, the fluid is poured off, the mercury placed in a porcelain or platinum crucible, well washed, dried with blotting paper and with gentle heat, and weighed. Nitric acid should not be present in the preceding analysis, but should be expelled by boiling with hydrochloric acid. To analyze calomel, it is necessary to dissolve it by the aid of pure nitric acid, to precipitate the chlorine by nitrate of silver, and the mercury by protochloride of tin. Corrosive sublimate (HgCl) may be analyzed in the same manner, and as it is soluble in water, the addition of any acid is scarcely necessary. White precipitate (Hg_2 Cl NH_2) should be dissolved in pure nitric acid, the chlorine precipitated from one portion, the mercury from a second, and likewise the ammonia, if required, by bichloride of platinum.

Mercury. *Mercurius.* A name given by the alchemists to all volatile bodies. The metal was termed *common mercury;* alcohol, *vegetable mercury.*

Mercury, Ores of. *Native Mercury.* Spec. grav. 13·568. Tin-white liquid, opaque; lustre metallic. B.B. entirely volatile; soluble in nitric acid. Idria, in Carniola; Almaden, Spain, &c.

Native Amalgam.—Sp. grav. 13·755, H 3·25. Hg 72·5, Ag 27½· = 3 Hg, Ag. Silver-white rhombic 12-hedrons and 8-hedrons; the edges being frequently replaced by planes; lustre metallic; streak white, opaque, brittle; gives a grating sound when cut; one amalgam is solid, another fluid. B.B. mercury is volatilized, and pure silver remains. Palatinate; Hungary; Sala, Sweden; France; Spain.

Sulphide, Sulphuret, Liver Ore, Cinnabar.—Hg S. Spec. grav. 8·098, H 1 to 4·25. Hg 83·8, S 14·25. Cochineal-red, sometimes inclining to gray; streak scarlet; in acute rhomboids, with angles of 72°, or 6-sided prisms, but generally massive. There is a dark impure variety termed liver ore; fracture conchoidal, semitransparent,

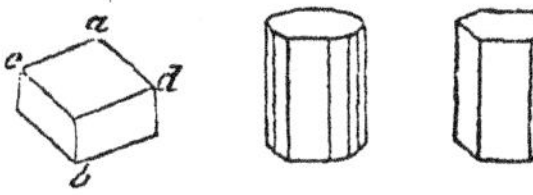

translucent on the edges, or opaque; sectile. B.B. volatile; soluble in nitric acid. In gneiss at Reichenau; in bituminous shale at Idria and in Spain.

Dichloride of Mercury, Native Calomel, Corneous Mercury.—Spec. grav. 6·482, H 1·5. Hg 84·74, Cl 15·26, Hg_2 Cl. Ash-gray or yellowish crusts and 4-sided square prisms, terminated by 4-sided pyramids; the lateral edges of the prisms are often replaced by tangent planes; lustre adamantine; translucent on the edges; sectile. Moschellandsberg, Deuxponts; Idria; Almaden.

Iodide.—HgI. Reddish-brown particles on selenide of mercury. Casas Viegas, Mexico.

Selenide.—H 2·5, HgSe 23·1, HgS 75·1. Lead-gray masses, giving out the odour of selenium B.B. San Onofre, Mexico.

Antimonite.—Red powder from Chili. SbO_4 12·5, SiO_3 26·5, HgO 14, Fe_2O_3 22·3.

Meroxydic Bodies. Those organic substances the inorganic constituents of which are present, partly in an oxidized, partly in an unoxidized condition; such are peas, rapeseed, and wheat.

Mesaconanile. A hypothetic body not yet obtained.

Mesaconic Acid. *Mesacic acid.* 2 HOC_{10} H_4O_6. F.P. 392°. Fine white shining needles; sublimable above 400°; soluble in 37 cold, in less than 1 hot water; ·306 dissolves in 1 cold alcohol (90 per cent.) ·957 when boiling; the solutions decompose carbonic acid salts; it forms bibasic salts; prepared by heating near to boiling from a quarter to half-an-hour, a dilute solution of citracic (citraconic) acid, with about a sixth part of strong nitric acid; gas is evolved, and on cooling the acid separates in crystals; from the mother liquor oxalic acid crystallizes; it is isomeric with citracic acid; in the above action *nitracic acid* $C_{10}H_3NO_4O_6$ 2 HO? is also formed.

Mesidine. $C_{18}H_{11}NH_2$? A hypothetic body, probably derivable from Nitromesitylole.

Mesite. A fluid from the products of the distillation of wood, supposed by some to be a mixture of acetone and acetate of methyle.

Mesitene. A fluid obtained by distilling xylite (a mixture of pyroxylic spirit, acetate of methyle, and methol), with an equal portion of sulphuric acid. It is probably a mixture of methol with acetate of methyle.

Mesitic Alcohol. A synonyme of Acetone.

Mesitic Aldehyde. $C_6H_4O_2$? A reddish fluid obtained by gently heating acetone with half its volume of strong nitric acid. By this process binitromesitylole alone has subsequently been obtained.

Mesitic Chloral. *Bichloracetone.* C_6H_4 Cl_2O_2. Specific grav. 1·33; B.P. 160°. A heavy oil smelling of chloroform and Dutch liquor; blistering the skin. With potash it yields pteleylic acid.

Mesitine Spar. A lenticular form of brown spar, or ferrocarbonate of magnesia.

Mesityle, Chloride of. C_6H_5Cl. By the action of chlorohydric acid on acetone.

Mesitylene. *Oenyle.* C_6H_5. The hypothetic radical of Acetone.

Mesitylole. *Mesitylene, Oenole.* $C_{18}H_{12}$. B.P. 325½° or 507⅗°. Specific grav. of vapour 4·31; colourless fluid, insoluble in water; burning with a smoky flame; isomeric with cumole; obtained by distilling a mixture of 2 vols. acetone, and 1 vol. oil of vitriol; two layers are obtained, the under aqueous, the upper oily, which is distilled; the portions boiling between 248° and 300°, being reserved and rectified. With chlorine are formed *trichloromesitylole* $C_{18}H_9Cl_3$ (chloride of pteleyle of Kane); with bromine, *tribromomesitylole* $C_{18}H_9Br_3$; with nitric acid, *dinitromesitylole* $C_{18}H_{10}$ 2 NO_4; and *trinitromesitylole* $C_{18}H_9$ 3 NO_4. Various acids act on acetone and form salts of mesityle.

Mesitylosulphuric Acid. $HOC_{18}H_{11}S_2$ O_5? Formed by the action of fuming sulphuric acid on mesitylole.

Mesole. *Flabelliform Kouphone Spar.* Spec. grav. 2·370, 2·382. White uniform and globular masses, radiated, with various shades of colour, grayish-white, yellowish-gray; occurring in basalt and amygdaloid, along with apophyllite, in Scania, Sweden; Faroe and Greenland. Its composition is SiO_3 42·17, Al_2O_3 27·, CaO 9·, NaO 10·19, HO 11·79 (Sweden, Berzelius). *Formula* NaO 2 SiO_3, 2 $CaOSiO_3$, 9 $Al_2O_3SiO_3$, 8 HO. It differs from mesolite, which is a tersilicate.

Mesoline. A granular zeolite from Faroe, without any determinate marks of crystallization; fracture grained and exhibits irregular facets; filling the cavities of lava in the form of the head of a pin. SiO_3 47·5, Al_2O_3 21·4, CaO 7·9, NaO 4·8, HO 16·19 (Berzelius). *Formula* $NaOSiO_3$, 2 $CaOSiO_3$, 9 $Al_2O_3SiO_3$, 14 HO.

Mesolite. Specific gravity 2·125 to 2·333. White 4-sided prisms terminated by 4-sided pyramids. Needlestone from Faroe, consisting of SiO_3 46·8, Al_2O_3 26·5, CaO 9·87, NaO 5·4, HO 12·3 (Berzelius).

Mesotype. A synonyme of natrolite and Thomsonite. It is much to be desired that writers on mineralogy who are not chemists, should preserve the names originally given by experimenters. The preceding minerals on this account have in some works been lost sight of, and the labours of a valuable chemist sunk.

Mesoxalic Acid. C_3O_4. Soluble, very acid crystals, unchanged by air and boiling; gives white precipitates with lime and barytes; monobasic and bibasic salts; obtained by boiling alloxanic acid salts; by this action hydrous alloxanic acid $C_8H_4N_2O_{10}$ is split into two bodies, mesoxalic acid C_6O_8, and urea $C_2H_4N_2O_2$.

Meta. (μετα, along with.) Originally applied to metaphosphoric acid, to indicate that it was not simply phosphoric acid, but the acid together with water. It is now used apparently at random, to denote a distinction between two approximate acids.

Metacetamide. A synonyme of Propionamide.

Metacetamide. $C_6H_5NO_2$? Obtained by acting on propionate of ethyle with liquid ammonia; *acetamide* $C_4H_5NO_2$ is obtained by the action of ammonia on acetic ether, and is a fine white solid. F.P. 172½°; B.P. 430°.

Metacetamine. A synonyme of Propylamine.

Metacetic Acid. *Metacetonic acid.* A synonyme of Propionic acid.

Metacetone. *Propione.* $C_{12}H_{10}O_2$ or C_6H_5O. B.P. 183°. A colourless oil with an agreeable odour, insoluble in water; soluble in alcohol and ether, obtained by distilling 1 part sugar, 3 caustic lime (10 to 15 lbs. at a time), at 285°, and rectifying the product; it also occurs in the dry distillation of lactate of lime, sugar, starch, gum, mannite, with caustic lime; it has the same relation to metacetic or propionic acid that aldehyde has to acetic acid; it is changed into propionic acid by distillation with bichromate of potash and sulphuric acid; it is isomeric with oxide of mesityle.

Metacetonitrile. A synonyme of Propionitrile.

Metacetyl-urea. *Ethyle-urea.* C_4H_4, $C_2H_4N_2O_2$. Fine prismatic crystals, fusing easily, soluble in water and alcohol; obtained by the action of a solution of ammonia on cyanate of ethyle; heat is evolved, and by evaporation crystals separate.

Metachlorite. A variety of chlorite from the Harz, containing SiO_3 23·77, Al_2O_3 16·43, FeO 40·36, MgO 3·1, CaO ·74, KO 1·37, NaO ·08, HO 13·75.

Metacinnameine. $C_{18}H_{16}O_2$. A crystalline body often accompanying cinnameine, from which it may be separated by treatment with weak alcohol, and cooling below 32°.

Metafurfurole. An aromatic oil accompanying crude furfurole, possessing a higher boiling point, becoming by distillation a brown resin, and yielding a purple colour; with chlorohydric, nitric, or sulphuric acids, it may be separated from furfurole by rectification; it forms no amide with ammonia, but becomes a resin.

Metagallic Acid. *Gallohumic acid, Melangallic acid.* $HOC_{12}H_3O_3$? Black lustrous amorphous mass, tasteless, odourless; obtained by heating gallic acid to 482°, or pyrogallic acid above its vapour point. It may be viewed as pyrogallic acid deprived of 2 atoms water; insoluble in water, alcohol, and ether; easily soluble in alkalies, aud reprecipitated in black flocks by acids; it displaces carbonic acid from the alkaline carbonates but not from carbonate of barytes; heated above 500° it is charred; it is monobasic.

Metalbumine. A form of albumen contained in the fluid of ascites; it is coagulated by boiling, and by acetic acid with heat; but is not precipitated by yellow prussiate of potash and acetic acid, and the coagulum by alcohol dissolves in water.

Metaldehyde. Long needles or 4-sided prisms; volatilizing at 230°, and condensing in snowy flocks; insoluble in water, soluble in alcohol; possesses the same composition as aldehyde; obtained by preserving aldehyde in close vessels at the usual temperature.

Metalepsy. *Theory of substitutions* (μεταληψις, *commutation*, exchange). In all cases of chemical reaction where certain elements of compound bodies change, it was formerly ascribed to what was called affinity. Thus, sulphate of soda in solution, when added to a solution of nitrate of barytes, produced sulphate of barytes and nitrate of soda. The cause of the change was ascribed to the greater affinity existing between the sulphuric acid and barytes, than between the sulphuric acid and soda. All that we in reality know, however, of the cause of the reaction may be as well expressed by saying that the barytes replaces or is substituted for the soda in presence of sulphuric acid. But the best examples of substitution are afforded by organic chemistry. 1. Carbon has never been replaced by any other element; and hence its presence has been considered as the essential evidence of organic constitution. 2. Hydrogen is often replaced by chlorine, iodine, and bromine, as in acetic acid, which when acted on by chlorine becomes chloracetic acid ($C_4H_4O_5$ and 6Cl become $C_4HCl_3O_4$ 3 HCl). For examples of substitution by iodine and bromine, it is only necessary to refer to the compounds of these elements. The reverse action may also occur, and hydrogen may be made to replace chlorine, iodine, and bromine; thus chloro-acetic acid is changed back into acetic acid by the influence of the amalgam of potassium. Amidogene also replaces these elements, as when chlorobenzoyle with ammonia is converted into benzamide and chlorohydric acid ($C_{14}H_5ClO_2$, and $NH_3 = C_{14}H_5$, NH_2O_2 and HCl). Sulphur also replaces chlorine as in Dutch liquor ($C_4H_4Cl_2$ and 2KS = $C_4H_4S_2$ and 2 KCl), where bisulphide of olefiant gas is formed. 3. Oxygen likewise replaces hydrogen when alcohol is converted into acetic acid ($C_4H_6O_2$ and O_4 becomes $C_4H_4O_4$ and 2 HO), although a different view has been taken of this action (C_2H_6O and $O_2 = C_2H_4O_2$ and H_2O), when we view water as composed of H_2O. 4. Hydrogen is replaced by metals, as when oil of vitriol is poured on zinc ($HOSO_3$ and Zn become $ZnOSO_3$ and H); or when dry ammoniacal gas is passed over heated potassium (NH_3 and K become NH_2K and H), hydrogen is evolved by the substitution of potassium and potassamide is formed. 5. Hydrogen is replaced by hyponitric acid, as when nitric acids act on cellulose or cotton, nitramidine is formed ($C_{12}H_{10}O_{10}$ and NO_5 become $C_{12}H_9NO_4O_{10}$ and HO); also when nitric and sulphuric acids act on cotton, *ternitramidine* or gun cotton is formed ($C_{12}H_{10}O_{10}$ and 3 $NO_5 = C_{12}H_7$ 3 NO_4O_{10} and 3 HO). No case has yet occurred where more than 3 atoms of hyponitric acid have replaced hydrogen; the compounds have then the oxygen present in such proportion as to burn the carbon. 6. Hydrogen is replaced by sulphurous acid, as in benzole, when acted on by sulphuric acid ($C_{12}H_6$ and $SO_3 = C_{12}H_5SO_2$ and HO). 7. Hyponitric acid is replaced by amidogene, as in the change of nitrobenzole into aniline ($C_{12}H_5$, NO_4, and 6HS = $C_{12}H_7N$ and 4 HO and S_6). 8. Hyponitric acid is replaced by nitrogen, as in nitrobenzole, which, when heated with a solution of potash in alcohol, is changed into azobenzole ($C_{12}H_5NO_4$ become $C_{12}H_5N$ and O_4 which oxidizes the alcohol). 9. Oxygen is replaced by amidogene, as when salicylate of ammonia is heated ($C_{14}H_6O_6$ and $NH_3 = C_{14}H_5$, NH_2, O_4 and 2 HO), it becomes salicylamide and 2 water. 10. Oxygen seems likewise to be replaced by nitrogen, but the examples of this kind are doubtful; so also are those of the substitution of imidogene (NH or $\frac{1}{2}$N $\frac{1}{2}$H) for oxygen. 11. Oxygen is replaced by sulphur, as when sulphovinate of barytes and sulphide of hydrogen and barium form mercaptan ($C_4H_6O_2$,SO_3BaOSO_3, and BaSHS = 2 $BaOSO_3$, HO, $C_4H_6S_2$.) 12. Oxygen is replaced by tellurium when telluride of ethyle is formed from telluride of sodium and sulphovinate of barytes. 13. Sulphur is replaced by oxygen as when oxide of silver decomposes oil of garlic, or sulphide of allyle, and forms oxide of allyle and sulphide of silver (C_6H_5S and AgO = C_6 H_5O and AgS.)

Metalloids (*Resembling metals*), or non-metallic bodies, are twelve in number, viz. oxygen, chlorine, bromine, iodine, fluorine, hydrogen, nitrogen, carbon, boron, silicon, phosphorus, sulphur; selenium seems also to belong to this class, although from its termination (in um), it is usually classed as a metal. The metalloids differ from the metals in the characters which are given under those bodies.

Metallurgy. (μεταλλον, metal; and ἐργειν, to work.) The art of separating metals from their ores, and preparing them for useful purposes.

Metals. A name applied to certain elementary bodies. 1. They generally possess the peculiar *lustre* termed metallic, except when in the state of black or gray powders, which assume a lustre when subjected to friction. 2. Their colour is usually gray, of different depths, to white and yellow. 3. They have seldom *taste* and *smell*; tin, copper, iron and lead, however, give out an odour when they are rubbed. 4. *Heat.* When heated in atmospheric acid, most of the metals have a tendency to combine with oxygen and to be converted into oxides. 5. They are insoluble in water; but certain of them (potassium, sodium, lithium), decompose cold water; others do so when they are heated. 6. *Crystallization.* They are all susceptible of crystallizing, but they have not all yet been placed in favourable circumstances for accomplishing this object. Their most general crystalline form is the regular system, either in the cube or octahedron. Antimony and bismuth, however, crystallize in the 3d system as rhombohedrons. 7. *Spec. grav.* They are all heavier than water, with the exception of potassium and sodium. See ATOMIC WEIGHTS. 8. The *hardness* of metals varies. Lead and tin are soft; but iridium and rhodium are supposed to be the hardest bodies in nature, especially rhodium, which is said to exceed that of the diamond. 9. *Ductility* is the property which metals possess of being drawn out by pressure into wires. The metals are ductile in the following order:—Gold, silver, platinum, iron, nickel, copper, zinc, tin, lead. 10. *Malleability* is the character which metals possess of becoming denser by the stroke of the hammer, and of being reducible to leaves. They are malleable in the following order:—Gold, silver, copper, tin, platinum, lead, zinc, iron, nickel. 11. *Tenacity* is indicated by the weight required to break a wire or metallic rod. The following weights are required to break different wires of 0·07875 English inch diameter:—

Iron,	551·37 lbs.
Copper,	302· ,,
Platinum,	275·65 ,,
Silver,	187·46 ,,
Gold,	149·97 ,,
Zinc,	110·25 ,,
Nickel,	105·86 ,,
Tin,	35·28 ,,
Lead,	26·46 ,,

12. *Chemical action.* Dry oxygen unites with some metals when the latter have been previously heated. Iron and copper are illustrations; but when water is likewise present, silent union takes place with the formation of an oxide. Thus iron when exposed to moist air rusts. *Chlorine* unites with many metals, as antimony and arsenic, even in the cold state.

The metals amount to forty-nine in number, exclusive of *Norium* and *Aridium*, which has been announced as existing in Swedish chrome iron ore; but the existence of both these metals seems doubtful.

Metamargaric Acid. *Paramargaric Acid.* $C_{68}H_{68}O_8HO$. Crystalline acid, insoluble in water; very soluble in alcohol and ether, becoming solid at 122°; distils mostly without change. When sulphuric acid is allowed to act on margarine, sulpholeic and sulphomargaric acids are formed, and when these are allowed to stand they soon decompose, yielding a fluid and a solid acid; the fluid acid is separated by pressure at 40°, the solid or metamargaric acid by crystallization from alcohol. It forms basic and acid salts.

Metameconic Acid. A synonyme of Comenic acid.

Metamerism expresses the fact that when two organic compounds unite, a new compound is often the result, having the same number of atoms of elements present as in another substance formed from two other organic compounds. The empirical formulæ are the same, although the rational formulæ differ. Thus cyanate of ammonia HO, NH_3, C_2N,O, contains the same number of atoms as urea $C_2H_4N_2O_2$. The empirical formula $C_{14}H_7NO_4$ is the same for (1) anthranilic acid, (2) benzamic or carbanilidic acid, (3) salicylamide, (4) nitrotoluole, while the rational formulae are (1) $C_{14}H_7NO_4$; (2) $C_{14}H_5$, NH_2O_4; (3) $C_{14}H_5$, NH_2O_2, O_2; (4) $C_{14}H_7NO_4$?

Metamilene. A form of Amilene.

Metamucic Acid. *Modified Mucic Acid.* $C_{12}H_8O_{14}$ 3 HO? Square plates, more soluble in water than mucic acid; soluble in alcohol, forms a different set of salts from mucic acid, soluble in 17·2 boiling water, in 73 cold water; when its boiling, saturated solution is allowed to cool, it deposits common mucic acid; its salts are more solid than the mucates; obtained by boiling down a solution of mucic acid or evaporating at a high temperature.

Metanaphthaline. *Retisterene.* C 93·7, H6, O 6·5. F.P. 152½°; B.P. 617°. White crystalline tasteless substance, insoluble in water, soluble in hot alcohol and cold ether, also in naphtha and oil of turpentine; it is isomeric with naphthaline and is neutral; chlorine and nitric acid convert in into resins; it is the last product in the distillation of common rosin; it is decolourized by pressure in paper, by the action upon it of sulphuric acid and by crystallization out of alcohol.

Metantimonic Acid. See ANTIMONIC ACID.

Metaœnanthole. $C_{14}H_{14}O_2$. Crystalline solid at 50°, an isomeric form of œnanthole, by the action of cold nitric acid on that substance.

Meta-Oleic Acid. M,Ol 2 HO. Yellowish fluid, mixing in all proportions with ether, insoluble in water and almost insoluble in alcohol, by which last character it is distinguished from oleic and hydroleic acid; decomposed by distillation into a fluid carbohydrogen and carbonic acid; it forms acid salts; it occurs along with metamargaric acid, from which it is separated in the action of sulphuric acid on margarine by alcohol, which takes up metamargaric acid.

Metapectic Acid. $C_8H_5O_7$ 2 HO. Uncrystalline body, soluble in water, forming soluble salts, and precipitated only by basic acetate of lead; when boiled yielding acetic acid and a black matter resembling ulmic acid; decomposes boiling solutions of the potash tartrate of copper, as sugar does, but it does not ferment nor act on polarized light; is formed by allowing a solution of pectine to stand for some days, by the action of strong acids on pectine or of an excess of alkali on pectine, pectosic and pectic acids.

Metapectine. $C_{64}H_{46}O_{62}$. By boiling parapectine with dilute acids, this substance is formed, which is uncrystalline, soluble in water, reddens litmus; insoluble in alcohol, and forms with bases salts of pectic acid; it yields with acids compounds soluble in water which are precipitated by alcohol.

Metaphosphoric Acid. *Glacial Phosphoric Acid.* $HOPO_5$ or HPO_6. Fusible, transparent deliquescent glass, obtained by evaporating and igniting trisphosphoric or common phosphoric acid, or by igniting phosphate of ammonia; it unites with only 1 atom of base; it coagulates albumen and precipitates salts of barytes and silver as monobasic phosphates, in bulky gelatinous precipitates; the basic atoms of water cannot be expelled by heat; but the whole acid sublimes when strongly heated in close vessels and rapidly in the open air below ignition. There are four modifications of metaphosphoric acid:—1. When syrupy phosphoric acid prepared from phosphorus by nitric acid is fused with 2 parts of nitrate of soda, a soda salt $NaOPO_5$ is formed as a white powder, insoluble in water. 2. Another modification is obtained by heating pyrophosphate of soda to low redness; it fuses and becomes on cooling a transparent glass very soluble in water and deliquescent. 3. A third form is obtained by heating ammonia phosphate of soda (microcosmic salt) to a gradually increasing temperature, when water and ammonia are evolved and a white saline mass remains which is soluble in water, and is acid; when further heated, it loses more water and its acid character. 4. Rose considers it probable that the acid obtained by combustion of phosphorus characterized by producing a large precipitate in a solution of chloride of barium is a fourth modification. All these forms precipitate white of egg.

Metastannic Acid Hydrous. $5(2HOSnO_2)$, or $10HOSn_5O_{10} = SnO_2$, 80·64, HO 19·36. White, fusible, semitransparent, with a vitreous fracture, or as a white powder; obtained by heating tin with diluted nitric acid, washing the precipitate, and drying in a dry current of air at the common temperature; it is quite insoluble in caustic ammonia; in vacuo it loses 5 atoms of water, and becomes $5HOSn_5O_{10}$, containing 11·3 per cent. of water; at 266° it loses another atom of water, having 8·8 per cent. of water; at 320° it loses another atom. If any of the preceding hydrates be dissolved in an alkali, and precipitated by an acid, metastannic acid results, but it is soluble in caustic ammonia. The most constant hydrate is $5HOSn_5O_{10}$, which, when acted on by soda, becomes $NaOSn_5O_{10}4HO$.

Metastearic Acid. A synonyme of Margaric acid.

Metastyrole. $C_{14}H_7$. A solid vitreous form of liquid styrole, obtained by exposing styrole in a closed tube to a temperature of 400°, slowly at 212°, and by sunlight; with nitric acid it forms nitrometastyrole, C_{14},H_6NO_4.

Metasulphammonic Acid. $S_6NH_3O_{16}3HO = 4SO_32SO_2NH_33HO$. An acid formed in union with potash, along with bisulphate of potash, by allowing sulphammonate of potash to stand. The latter salt is formed by the action of an excess of sulphurous acid on nitrite of potash. When metasulphammonate of potash is boiled in water, it is decomposed into bisulphate of potash and sulphamidate of potash ($S_4NH_3O_{10}2KO = 2SO_3,2SO_2,NH_3$ 2KO).

Metasulphazic Acid. $7SO_22NO_36HO$. An acid formed by the action of sulphazite and sulphazate of potash on each other.

Metasulphazilic Acid. $S_6NH_3O_{20}3HO = 4SO_22SO_3NO_33HO$. An acid formed by the action of the brown oxide of lead, or of oxide of silver on sulphazotate of potash.

Metasulphazotic Acid. $S_8N_2H_6O_{28}6HO + 2HO =$ Sulphazous acid $S_3NH_3O_{12}$, and sulphazotic acid $S_5NH_3O_{16}$. Obtained by passing sulphurous acid through nitrate of potash, when plates are often deposited, consisting of metasulphazotate of potash. This salt is very soluble in water; reaction strongly alkaline; it is only stable in an alkaline fluid, and by pure water is decomposed into sulphazite, and basic sulphazotate of potash; acids evolve from it binoxide of nitrogen.

Metatartaric Acid. $2HOC_8H_4O_{10}$. Crystallized tartaric acid when exposed to the heat of an oil bath just until perfectly fused, is converted without loss of weight into metatartaric acid, resembling a transparent deliquescent gum, forming with potash and ammonia salts of different forms of crystallization, and of more solubility than the tartrates; it does not precipitate lime salts by itself, and only slowly when neutralized by ammonia, producing a precipitate which is soluble in much water, and is of a crystalline form different from that of tartrate of lime; metatartrate of ammonia, $NH_4O,HO,C_8H_4O_{10}$, in needles, separates when ammonia is added to fused tartaric acid; it is converted by boiling into tartrate.

Metatitanic Acid. Ti_3O_6. Small shining plates, separating when anhydrous bichloride of titanium is saturated with carbonate of barytes, adding water and boiling rapidly; it is obtained amorphous by heating common titanic acid, or it is precipitated by boiling from its solution in sulphuric acid, the only acid that dissolves it. The acid precipitated by boiling, when dried at 284°, has 12 per cent. water, and consists of Ti_3O_62HO. The metatitaniates do not crystallize, and are insoluble.

Meta-toluidine. *Metatoluole.* $C_{30}H_{17}N_3$. Crystalline plates; a resinous base; obtained by passing gaseous chloride of cyanogen over toluole ($2C_{14}H_9N,C_2NCl = C_{30}H_{17}N_3,HCl$), little soluble in cold, more soluble in hot water; it is purified by crystallization out of alcohol.

Metatungstic Acid. HO,W_3O_9. The ammoniacal salt of this acid is obtained by boiling for some hours bitungstate (paratungstate) of ammonia for some hours, and evaporating the solution to the consistence of syrup; on cooling it deposits very soluble 8-hedrons; its solutions are not precipitated by strong chlorohydric acid.

Metaxite. Sp. grav. 2·4212, H 3·5. Greenish-white silky oblique prisms, half-an-inch long, with angles of about 99° and 81°; rather sectile. B.B. fuses without colouring the flame; yields water by heat; with soda fuses into a white bead; gives a lilac with nitrate of cobalt, SiO_3 45·04, MgO 34·, Fe_2O_3 5·28, HO 15·4 (Dr. T. Thomson, jun.) It is allied to serpentine.

Met-elaic Acid. A synonyme of Meta-oleic acid.

Met-ellagic Acid. *Rufiogallic Acid.* See Gallic Acid.

Meteoric Dust. A sandy matter which has been observed to fall on vessels in the Atlantic. SiO_3 37·18, Al_2O_3 16·74, Fe_2O_3 7·65, Mn_2O_3 3·44, $CaOCO_2$ 9·59, MgO·1·8, KO 2·97, NaO 1·9, CuO ·25, HO, and organic matter 18·53 (Gibbs). It often contains infusoria.

Meteorites. See Aerolites.

Meter. An instrument for measuring the amount of gas consumed. There are two kinds—*water* and *dry* meters.

Meter. *Metre.* The unity of the French measures, amounting to the 40 millionth part of the length of the earth's meridian, or the 10 millionth part of the distance from the equator to the pole. The toise of Peru, which is 6 Paris feet in length, or 443,296 lines on a platinum rod at 61°·25 F. is the length of the metre, and forms the basis of the French measures. The following tables give the corresponding English and French measures:—

1 yard = 144 lines.
1760 yards = 1609·315 metres.

English.	French.
·03937 inches	Millimetre.
·393708 „	Centimetre.
3·937079 „	Decimetre.
39·37079 „ } 3·2808992 feet } 1·093633 yards }	Metre.
1093·63 yards	Kilometre.
6·2138 miles	Myriametre.

French metres and English feet.

Metre.		English Feet.
1	=	3·280899
·3248394		1·065765
·3047945		1·
·3138535		1·029722
·2918592		0·9575606
·2920947		0·9583333
·2831901		0·9291180
·2853624		0·9262453
·2876991		0·9439117
·2864903		0·9399459
·3000000		0·9842697
·3161109		1·037128
·2846000		0·9337440
·2500000		0·8202247

French cubic metres and English cubic feet.

Cubic Metre.		English Cubic Feet.
1·	=	35·31658
·03427727		1·210556
·02831531		1·
·03091584		1·091842
·02486109		0·8780087
·02492133		0·8801360
·02271088		0·8020706
·02323755		0·8206707
·02381307		0·8409962
·02351418		0·8304405
0·02700000		0·9535476
0·03158774		1·115571
0·02305179		0·8141105
0·01562500		0·5518216

1 cubic metre = 1 stere for measuring wood.
·001 „ = 1 litre = 1 cub. decimetre.
1 litre = 1000 cub. centimetres = 50·412416 Par. cub. inch.
100 litres = 1 hectolitre.

Methamine. A synonyme of Methyl-amine.

Methionic Acid. Formed, when acting on solid sulphuric acid, with the vapour of ether, so as to form sulphate of ethyle; on boiling with water isæthionic and methionic acids are formed; when the fluid is neutralized with carbonate of barytes and evaporated, crystals of methionate of barytes separate, while isæthionate of barytes remains in solution. Methionate of barytes consists of 2 BaO $C_2H_6O_2$ 4 SO_3. The acid may be separated from it by sulphuric acid, and is obtained as an acid solution not decomposed by boiling.

Methole. $C_{12}H_9$? Colourless aromatic oil, unchangeable in air, distilling at 347°; taste burning; smell resembling that of oil of turpentine; by sulphuric acid it is acted on, forming, on adding water, in some days, 3 layers; the lower aqueous layer consists of metholo-sulphuric acid, the two upper, undecomposed methole and methole resin ($C_{24}H_{16}O$).

Metholo-Sulphuric Acid. $C_{12}H_9SO_3HO$ SO_3? Obtained on one occasion in a mixture of methol and sulphuric acid, in the form of small crystals. When saturated with lime, a salt results with the formula $CaOSO_3$, $C_{12}H_9SO_3$.

Methostann-ethyle. $Sn_2,3C_4H_5$. A radical formed in union with sulphuric acid in prisms by action on Stannethyle.

Methylal. *Formal, Formomethylal.* $C_6H_8O_4$; specific gravity ·8551, of vapour 2·625; B.P. 107½°. Colourless aromatic fluid, soluble in 3 parts water, from which it is separated by potash and partly by chloride of calcium; miscible in all proportions in alcohol and ether; formed by distilling equal parts of pyroxylic spirit and binoxide of manganese with 1½ part of sulphuric acid, diluted with an equal weight of water; the product consists of methylal, formate of methyle, aldehyde, formic acid, and undecomposed pyroxylic spirit. The first portion of the distilled fluid alone is rectified over chloride of calcium, till it boils at 107½° constantly.

Methylamine. *Methylamide, Methamine, Methyliaque.* C_2H_3,H_2N. Spec. grav. 1·08, 1·0731. Colourless gas, with smell of ammonia, mixed with that of putrid fish; reaction alkaline; burns with a yellowish flame; yields a white vapour when mixed with chlorohydric gas or carbonic acid; it liquefies below 32°, and is then heavier than water; it is the most soluble of all gases; 1 vol. water dissolves 1150 vols. gas at 54°·5, 960 vols. at 82°; the solution having all the characters of the gas; the solution precipitates salts of chromium, uranium, magnesia, iron, manganese, copper, bismuth, zinc, and tin, as ammonia does; it redissolves zinc and copper, giving a fine azure blue with the latter. It precipitates without redissolving cadmium, nickel, and cobalt; acetate of lead is scarcely rendered turbid, but nitrate of lead is precipitated; mercurial salts are precipitated. A black substance falls with silver salts, which is analogous to fulminating silver, but is not explosive; chloride of gold gives a yellowish precipitate soluble in excess; bichloride of platinum gives an orange precipitate, crystallizing in plates in a concentrated solution; when passed through an ignited porcelain tube it is converted into cyanide of ammonium, prussic acid, carburetted hydrogen, and hydrogen; by potassium it is converted into cyanide of potassium and hydrogen. It forms generally equiatomic compounds with acids. The chlorohydride (C_2H_5N,HCl) one of the most important, is a colourless salt in plates, fusing at 212°, and volatilizing in white vapours, and is obtained by saturating the gas with chlorohydric acid, or by boiling cyanate or cyanurate of methyle with potash, saturating with chlorohydric acid, and evaporating. *Process.* —Methylamine is obtained by decomposing cyanate or cyanurate of methyle with alkalies, ($C_2H_3O,C_2NO,2HOKO = 2KOCO_2,C_2H_5N$). The cyanate of methyle is formed by passing cyanic acid into pyroxylic spirit. Methylamine is likewise obtained by the action of ammonia on iodide of ethyle; in the action of chlorine on caffeine; in boiling caffeine with potash, by heating morphine at 392° with excess of potash, and mixed with propylamine by heating codeine to 248° to 347°, and probably in the destructive distillation of gas asphalt, as I have observed a large amount of ammoniacal smelling gas disengaged in this process with an alkaline reaction. It is distinguished from ammonia by its ready inflammability, burning with a yellowish flame into carbonic acid, water, and nitrogen, mixed sometimes with cyanogen and prussic acid. *Dimethylamine* and *trimethylamine* are formed in small quantity in the action of iodide of methyle on ammonia.

Methyle. C_2H_3. Spec. grav. 1·0365 = 1 vol. carbon, ·8292, and 3 vols. hydrogen = ·2073. Colourless and odourless gas, almost insoluble in water, and little soluble in alcohol; not condensed at 3°·2; burns with a bluish flame; not acted upon by sulphuric or fuming nitric acids, or by caustic potash; chlorine, mixed with it in equal volumes in daylight, unites without change of bulk; the product consists of equal bulks of chlorohydric gas, and of a gas (C_4H_5Cl) isomeric with chloride of ethyle, or of a mixture of unacted on methyle and of a gas C_2H_2Cl. There is some doubt as to the formula for methyle; some considering it to be C_4H_6, and it has been contested whether it is the basis of the methyle compounds. It is isolated by exposing iodide of methyle to the action of zinc in a hermetically sealed tube, at 302°, iodide of zinc being formed together with methyle and zinc methyle; it is also formed by passing a galvanic current from a Bunsen's battery of 4 plates through a strong solution of acetate of potash, when hydrogen separates at the negative and methyle with carbonic acid at the positive pole, and also oxygen as the action advances. It forms compounds with various bodies; by displacing hydrogen in ammonia we have methylamine, $C_2H_3NH_2$. By displacing hydrogen from phosphuretted hydrogen, there are formed *phospho-trimethyle* $3(C_2H_3)$ P and *phospho-dimethye* $2C_2)H_3)$ P obtained by the action of chloride of methyle on phosphide of calcium. *Zinc methyle,* C_2H_3Zn, a colourless fluid, is formed in the action described.

Methyle-amyle-phenylamine. $C_2H_3,C_{10}H_{11},C_{12}H_6N$. An oily alkaloid, nearly insoluble in water, with a high boiling point, by heating methyl-ethyle-amyle-phenylammonium.

Methyle-aniline. *Methyle phenylamine.* $C_2H_3,C_{12}H_5,H,N$. B.P. 377⅓°. Transparent oil, becoming blue with chloride of lime; formed by mixing aniline and bromide of ethyle.

Methyle-butylamine. C_2H_3,C_8H_9,H,N.

Methyle-butyle-amylamine. $C_2H_3,C_8H_9,C_{10}H_{11},N$.

Dimethyle-Propylamine. $2C_2H_3,C_6H_7N$.

Methyle, Cyanide of. See ACETONITRILE.

Methyle-diethyle-amyle-ammonium. $C_2H_3,2(C_4H_5),C_{10}H_{11}N$. An alkaline fluid, formed by treating diethyle-amylamine with iodide of methyle; a compound of the above base, with iodine results, from which the iodine is removed by oxide of silver. Formed on the model of ammonium.

Methyle-ethyle-amylamine. $C_2H_3,C_4H_5,C_{10}H_{11}N$. B.P. 275°. Clear oil, with an agreeable smell and taste; little soluble in water, to which it gives an alkaline character; by heating the methyle-diethyle-amyle-ammonium.

Methyle-ethyle-amyle-phenylammonium. $C_2H_3,C_4H_5,C_{10}H_{11},C_{12}H_5N,HO$. The hydrate is an oily base, obtained by heating ethyle-amyle-aniline, and iodide of methyle, in a hermetically sealed tube for three or four days; two layers are formed, the lower containing the iodide of this base; when this is treated with oxide of silver, ethyle-amylamine remains, mixed with the iodide of silver while the ammonium base remains in solution. It is formed on the type of ammonium (NH_4), each atom of hydrogen being replaced by an alkaloid.

Methyle-ethyle-aniline. *Methyle-ethyle-phenylamine.* $C_2H_3,C_4H_5,C_{12}H_5N$. Volatile oil by the reaction of iodide of ethyle, and ethyle-aniline. Formed on the type of ammonia, all the hydrogen being replaced by alkaloids.

Methyle-ethyle-Campheric Acid. $HO,C_2H_3O,C_{20}H_{14}O_6$. F.P. 154⅓°. Long needles or plates decomposing by heat into campheric acid, carbon, and a glutinous fluid; obtained by substituting pyroxylic spirit for spirit in the preparation of ethercampheric acid.

Methyle-ethyle-urea. $C_2,H_2,C_2H_3,C_4H_5,N_2O_2 = C_8H_{10}N_2O_2$, by the action of cyanate of ethyle on methylamine.

Methyle, Hydrous Oxide of. *Methylic Alcohol, Pyroxylic Spirit, Bihydrate of Methylene, Wood Spirit, Wood Naphtha.* Said to have been observed in 1812 by Philip Taylor. Dr. Thomas Thomson used it in lamps in 1816, and found it at Glasgow in 1817. In confirmation of this, Mr. John Turnbull informs me that it was manufactured as an article of commerce in that year by Turnbull & Ramsay of Glasgow. It was first described in print by P. Taylor in 1822. C_2H_3OHO, or C_2H_2 2 HO. Spec. grav. ·798 (Dumas), at 68°; ·807 at 48° (Deville), ·8207 (Pierre), ·81796 (Kopp); of the gas 1·12. B.P. 150° (Thomson), 151·3 (Pierre), 149°·9 (Kopp), 150·4 (Andrews), 151°·7 (Person). Tension of vapour 3·27 inches mercury. Colourless fluid, with a peculiar odour; soluble in all proportions in water, alcohol, ether, and fat oils; burns with a pale blue flame; it dissolves many substances like alcohol, as caustic alkalies, salts, alkaloids, resins, &c.; it dissolves a small portion of phosphorus; it dissolves chloride of calcium, forming a crystalline compound (CaCl, 2 $C_2H_4O_2$), sesquichloride of iron, bichloride of tin. The following table gives the specific gravity and composition of pyroxylic spirit of different strengths (Ure):—

Spec. Grav.	Per Centage.
·8136,	100·
·8418,	90·90
·8674,	82·
·8822,	77
·8984,	70·42
·9116,	65·
·9218,	60·24
·9296,	56·18
·9414,	50
·9484,	46
·9564,	40
·9620,	35·71

When dropped on spongy platinum, pyroxylic spirit is converted into formic acid by oxidation ($C_2H_4O_2$ and $O_4 = C_2HO_3$ and 3HO); converted by distillation with sulphuric acid and binoxide of manganese into formic acid and aldehyde(?) with sulphuric acid it forms hydrosulphate of methyle ($HOSO_3,C_2H_3OSO_3$), which, by heating, yields gaseous oxide of methyle (C_2H_3O), which burns with a blue flame; nitric acid converts it into formic acid and nitrate of methyle; by chlorine an oily fluid is formed (C_6H_3 Cl_3O_2 ?), by chlorides of tin, iron, and antimony it is converted into chloride of methyle; by phosgene gas, oxichlorocarbonate of methyle ($C_2H_3OC_2$ ClO_3) is formed, a colourless fluid, yielding with ammonia needles of carbonate of methyle (urethylane) C_2H_3O, CO_2, NH_2, CO; with chloride of lime it forms chloroform; with solid potash and lime, gently heated, it gives hydrogen and formate of potash. Dr. T. Thomson first showed, in 1826, that the basis of pyroxylic spirit was a carbo-hydrogen (Edin. Roy. Soc. Trans.)

Methyle, Iodide of. C_2H_3,I. Spec. grav. 2·199, of vapour 4·883; B.P. 110¾°. A colourless fluid, neutral, not miscible with water; burns with difficulty, giving out iodine fumes; soluble in pyroxylic spirit, alcohol, and ether; by chlorine converted into chloride of methyle.

Methyle-narcotine. $C_{44}H_{23}NO_{14}$. From opium.

Methylene. C_4H_4, or C_2H_2. (μέθυ, wine; and ὕλη, wood.) The hypothetic base of the methyle series (Dumas).

Methyle-piperidine. $C_2H_3,C_{10}H_{10}N$. Sp. grav. of vapour 3·544. B.P. 244½°. Colourless oil with ammoniacal odour; soluble in water; by acting on piperidine with iodide of methyle in closed tubes, and heating the crystalline body formed with caustic potash. *Ethyle* and *amyle-piperidine* may be similarly formed.

Methyle-Quinoline, or **Quinine?** C_2H_3 $OHO,C_{18}H_8N = C_{20}H_{12}NO_2$. Quinoline C_{18} H_8N is obtained by the action of fused potash on quinine, and if we subtract this formula from that of quinine, or $C_{20}H_{12}NO_2$, we obtain C_2H_3O, HO, or pyroxylic spirit.

Methyle-triethyle-ammonium. C_2H_3, $3C_4H_5$,N. The hydrate is a strong alkaline base, soluble in water; crystalline in vacuo; by heating triethylamine with iodide of methyle, and removing the iodine from the resulting compound by oxide of silver. *Tetramethyle-ammonium.* $4C_2H_3N$. The hydrate is a crystalline body, attracting carbonic acid rapidly from the air; obtained by heating iodide of methyle, with strong solution of ammonia, in a close tube; after the reaction has continued hydriodides of methylamine, dimethylamine, trimethylamine, and iodide of tetramethylammonium, are formed; the latter remains after evaporation, and is obtained as a hydrate by treatment with oxide of silver.

Methyle-Urea. *Dimethyle Urea.* $C_2H_2,2C_2$ H_3,N_2O_2. F.P. 206½°. White crystalline mass; volatilizing by heat; soluble in water and alcohol, and forms a nitrate ($C_6H_8N_2O_2HONO_5$).

Methyle Urea. *Cyanate of Methylamine.* C_2, $H_3,C_2H_3N_2O_2$. 4-sided prisms, soluble in water; by evaporating a solution of sulphate of methylamine and cyanate of potash.

Methylocitric Acid. *Citrotrimethylic acid.* $3C_2H_3O,C_{12}H_5O_4$. Prisms obtained by passing dry chlorohydric gas into a saturated solution of citric acid in pyroxylic spirit, and heating to 192°, pressing the crystals in paper, and drying in vacuo.

Methylodithionic Acid. See HYPOSULPHOMETHYLIC ACID.

Methylol. A synonyme of carburetted hydrogen, or Marsh gas.

Methyloxamic Acid. $C_6H_4NO_5$. A crystalline sublimate, by heating binoxalate of methylamine.

Methyloxamide. C_2H_4N,C_2O_2. Long needles, by the action of methylamine on oxalic ether, or by subliming oxalate of methylamine in a retort.

Metylene Stannethyle. $Sn_22C_4H_5$. See STANNETHYLE.

Miargyrite. (μείων, less; ἄργυρος, silver.) *Bisulphoantimoniate of silver.* *Hemiprismatic ruby* *blende.* Spec. grav. 5·234, H 2 to 2·5. Iron-black flat 4-sided prisms, one of the angles at the base being replaced by a 3-angular face; streak dark cherry-red; lustre between metallic and adamantine; opaque except in thin splinters, when it is blood-red. B.B. behaves nearly as dark red silver ore. Sulphur 21·95, antimony 39·14, silver 36·4, copper 1·06, iron ·62. *Form.* AgS,SbS_3.

Mica. (*Micare*, to shine). (*Glimmer*, Ger.) *Common Mica, Binaxial Mica, Muscovy Glass, Glacies mariae, Chlorite Earth.* Sp. grav, 2·824, 2·949, 3·08, H 2 to 2·5. Various shades of gray passing into green, brown, or black, scales, or oblique rhombic prisms, with angles of 98° 40′, 81° 20′, and 60°. The acute edges of the prism

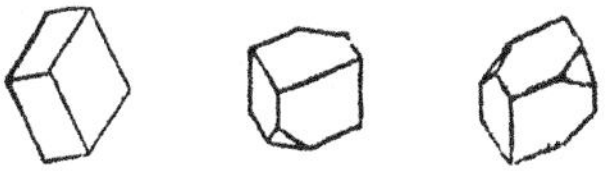

are often replaced by tangent planes, which convert the crystal into a 6-sided prism; it occurs also in right rectangular prisms; lustre pearly; structure foliated. SiO_3 47·19, Al_2O_3 33·8, Fe_2O_3 4·47, MnO 2·58, CaO ·13, F ·29, KO 8·35, HO 4·07. *Form.* $KOSiO_3,4,Al_2O_3SiO_3$? B.B. fuses into a yellow blebby glass; fuses with salt of phosphorus and with borax.

Micaceous Iron. *Iron Mica.* A variety of sesquioxide of iron, or specular iron ore, found in connection with igneous rocks.

Mica Lithia. See Lepidolite.

Mica Magnesian. *Uniaxial Mica, Hexagonal Mica, Rubellane, Black Mica of Siberia.* Spec. grav. 2·8 to 3·, H 2·5. Black, brown, or dark green rhomboidal plates, with a pearly lustre; thin foliæ, flexible; transparent, opaque; B.B. fuses into a black glass. SiO_3 40·, Al_2O_7 12·67, Fe_2O_3 19·03, MgO 15·7, KO 5·6, F 2·1, Mn_2O_3 ·63, TiO_3 and Fe_2O_3 1·63. *Form.* 3 RO, SiO_3, $Al_2O_3Fe_2O_3SiO_3$.

Micaphyllite. A synonyme of Andalusite.

Micarelle. A synonyme of Pinite.

Mica Slate. Spec. grav. 2·74. A metamorphic slate occurring abundantly in Scotland and other primitive countries. It consists essentially of quartz and mica, but has intermixed with it garnets. From Ben Lawers it consists of SiO_3 66·63, Al_2O_3 17·12, FeO 13·34, CaO 1·16, MgO ·95, KO ·29, HO 1·, PO_5 trace (R. D.T.)

Michaelite. *Randanite, Gelatinous Silica.* HO, 4 SiO_3 and $HO2SiO_3$. Occurs as a fine earth, consisting of infusoria at Ceyssat, Algiers, and probably at Farnham, in chalk. The silica is soluble in potash; it consists of HO, CO_2 10·, sand ·8, gelatinous silica 87·2, oxides, iron and alumina, 2·, lime and magnesia traces.

Microcosmic Salt. *Salt of Phosphorus, Native Salt of Urine, Ammonia-Hydro-trisphosphate of Soda.* NaO, $NH_4OHOPO_5 + 8$ HO. Colourless fine rhombic efflorescent tables, soluble in 1 part hot, 6 cold water; loses ammonia when ignited, and leaves acid phosphate of soda in the form of a clear bead, and hence its use in blowpipe experiments. It is formed by neutralizing a portion of phosphoric acid with ammonia, and another equal portion with soda, and evaporating the united solutions; or dissolve 6 parts crystallized phosphate of soda in 2 parts water, add 1 part salammoniac, and crystallize. It is deposited also from ammoniacal urine.

Microlite. Apparently the same as pyrochlore.

Middletonite. Spec. grav. 1·6. C_{20},H_{11} O. A fossil resin of a reddish-brown colour, soluble in sulphuric and boiling nitric acids, with decomposition; colours ether, alcohol, and turpentine yellow. Found on the surface of Yorkshire and Staffordshire coal.

Miemite. A variety of brown spar from Miemo, Tuscany.

Miesite. A synonyme of phosphate of lead.

Mildew. (*Mehlthau*, Ger.) A white dewy aspect which the leaves of plants and grapes assume by the deposition of a fungus.

Milk. The white secretion from the mammary gland of the mammalia, for the support of their young. It consists of minute white globules, suspended in a clear fluid. These globules are formed of butter, surrounded by a vesicle, and when treated with dilute acetic acid, allow the butter to separate, as may be seen under the microscope. After parturition the milk contains, in addition to the milk globules, irregular drops of fat, termed *colostrum* (*beist*, Scot.) Milk has an alkaline reaction. The specific gravity of cow's milk, from experiments repeated twice daily on two different cows, for several months, I have found to be 1030·. The specific gravity of human milk is 1020·3, of the mare 1034·6, ass 1035·5, goat 1034·1, sheep 1040·9. When milk is allowed to stand, the cream rises in the form of a white oily layer to the top, increasing in quantity in proportion to the shallowness of the vessel; the milk soon becomes acid, from the conversion of the milk sugar into lactic acid, which is hastened in warm electric weather, and speedily it becomes curdled from the coagulation of the caseine by the action of the acid. The same effect is produced by rennet, see Cheese, Acids, Alcohol. It then separates into *curds* and *whey*, the former constituting, when pressed free from the whey, *cheese.* The following is the composition of different kinds of milk:—

	Cow.	Woman.	Ass.	Goat.
Water, ...	871·9	884	916·5	856
Sugar, ...	43·5	47	60·8	50
Butter, ...	37·	25	11·1	42
Caseine, ...	41·6	37	18·2	45
Soluble salts, ...	1·5	2	{ —	{ —
Insoluble salts, ...	4·4	—	{ 3·4	{ 6·8

Cow, R.D.T.; *woman*, Simon; *ass* and *goat*, Henry and Chevallier.

The salts of cow's milk are—soda ·042, chloride of sodium ·024, chloride of potassium ·144, phosphate of lime ·231, phosphate of magnesia ·042, sesquiphosphate of iron ·007 (Haidlin); or per 1000, KCl 14·18, NaCl 4·74, KO 23·46, NaO 6·96 united to PO_5, SO_3, and CO_2. For the influence of food on milk, see my *Experimental Researches on the Food of Animals.* The amount of milk given daily by a cow may amount in summer to 20 imperial pints, or 25½ lbs. weight; but it often much exceeds this, and at other seasons falls below it. Milk is now sold in white cakes as portable milk; it is understood to be prepared by adding 4 oz. of sugar to a gallon of milk, and evaporating in vacuo, or at a temperature of 120°; and pressing it into cakes. *Analysis of milk.*—Milk may be analyzed by evaporating 1000 grains to dryness at 212°; the loss, when it ceases to lose weight, is water; the residue is digested in ether, which takes up the butter; on igniting the butter, any soluble salts dissolved by the ether will be obtained; the soluble salts and sugar are removed from the residue by water and by ignition, the caseine disappears, and the insoluble salts remain.

Milk of Cow Tree. A white fluid derived from the Galactodendron utile of Venezuela, South America. Spec. grav. 1012·42. It consists of water 573, albumen 40, wax ($C_{35}H_{33}O_3$) 58, resin (*galactine* $C_{38}H_{29}O_2$) 314, gum and sugar 47, salts 4·.

Milk Sugar. See LACTINE.

Millerite. A synonyme of nickel pyrites.

Millet. The seed of the Milium Sorghum, used in Africa as food, and as a source of spirits.

Miloschine. Green oxide of chromium (Cr_2O_3).

Mimetene. *Mimetesite.* A synonyme of arseniate of lead.

Mimotannic Acid. See CATECHUIC ACID.

Mindererus, Spirit of. The solution of acetate of ammonia in water.

Mineral Alkali. Old name of soda.

Mineral Caoutchouc. *Elastic Bitumen.* $C_{35}H_3O_2$ and C_{41}, H_2O_2? Spec. grav. 1·233 to ·9053. Black, glistening, soft, sectile, elastic, flexible substance, catches fire readily, and burns with a yellow flame, and bituminous odour. Found in the lead mine of Odin, Mamtor, Derbyshire; Montrelais, France; Neufchatel; Zante.

Mineral Charcoal. A form of charcoal associated with coal. I have fine specimens from the coal of Glasgow, from the neighbourhood of trap dykes and blind coal.

Mineral Green. *Mountain Green.* Hydrous dicarbonate of copper. See GREEN, BRUNSWICK.

Mineral Oil. *Rock Oil.* See NAPHTHA MINERAL.

Mineralogy. (Mineral and λογος, a discourse). That branch of chemistry which treats of the form and composition of native minerals and salts. It must not be confounded with mere crystallography, which is equally applicable to chemistry. The total number of species of minerals known at the present time is upwards of 600; of these 45 belong to the acid radicals, 557 to the basic radicals, and 5 are neutral. The different species are described in this work under their appropriate names.

Mineral Pitch. See ASPHALT.

Mineral Waters. See WATERS.

Minium. *Red Lead.* See LEAD.

Mirabilite. A synonyme of glauber salt.

Mirror Metal. Ancient mirrors were composed of Cu 62, Sn 32, Pb 6. Modern speculum metal for telescopes is formed of 2 copper, 1 tin.

Misenite, or native bisulphate of potash, from Cape Misen, Rome.

Mispickel. *Arsenical Pyrites, Marcasite, Danaite.* Spec. grav. 6·127, H 5·75. FeS_2, Fe As. Right rhombic silver-white prisms, yielding by heat fumes of tersulphide of arsenic, and smell of garlic.

Misy. $2Fe_2O_3 5SO_3 6HO$. Rhombic pearly yellow plates, accompanying sulphates of zinc, magnesia, and potash, near Goslar.

Mizzonite. A variety of Meionite or Scapolite.

Mocha Stone. *Moss Agate.* A species of quartz, or chalcedony, with the appearance of mosses imbedded in it.

Mohringa Oil. See OIL OF BEN, BENIC, and MORINGIC ACIDS.

Mohsite. *Mohsile.* A synonyme of Ilmenite.

Molasses. A dark syrupy fluid, not readily crystallizable, which constitutes the mother liquors of sugar in the West Indies. It is now largely imported into this country, and from it is extracted the crushed sugar of commerce; the uncrystalline mother liquor forming treacle.

Molecule. (*Molis*, a hill). An assemblage of atoms. It has been suggested by Ampere and Prout that the atoms of simple gases are not all isolated and repulsive of each other, but that a certain number of the atoms may be associated into a molecule, as *a b c d*, and that this many-sided solid may then repel a similar formed molecule. This view might enable us to explain the fact that many bodies of the same per centage composition have totally different properties.

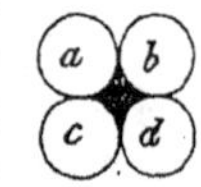

Molybdate of Lead. *Yellow Lead Ore.* PbO MoO_3. Spec. grav. 5·706 to 6·760, H 2·75. Wax-yellow, siskin-green, olive-green, orange-yellow, yellowish-gray, grayish-white 8-hedrons

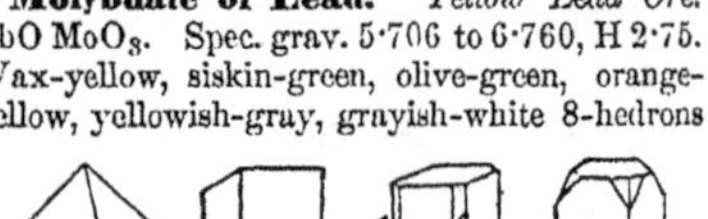

with a square base, with angles of 99° 16′ and 131° 15′, and in crystals approaching to cubes, in tables, and in 4, 6, and 8-sided prisms terminated by 4-sided pyramids, structure lamellar, fracture conchoidal. B.B. assumes a darker colour, leaves some metallic lead on charcoal; with borax fuses into a transparent glass, dark and opaque on cooling; with salt of phosphorus into a green glass or black in greater quantity. I have found in it PbO 59·60, MoO_3 40·4; Carinthia; Mendip.

Molybdena. *Molybdenite, Bisulphide of Molybdenum.* MoS_2. Spec. grav. 4·569 to 4·7385, H 1·25. Lead-gray 6-sided prisms, or foliated metallic-looking mineral; the plates are easily separated, and are flexible, but not elastic; splendent; streak unchanged; opaque. B.B. is not reduced, but on charcoal it emits sulphurous fumes, which distinguishes it from plumbago, to which it bears a close resemblance—even staining paper when rubbed on it. Occurs in granite or gneiss at Altenberg, Saxony; Zinnwald, Bohemia; Cornwall; United States; Norway.

Molybdenum is obtained by reducing molybdic acid by means of charcoal at a very high temperature. It appears in the form of globules or a porous mass resembling spongy platinum. The source of the metal in general is molybdena, which is first roasted and passed through a seive. The powder is dissolved in ammonia, the solution filtered and evaporated to dryness; the residue moderately heated with some nitric acid, becomes a white powder, the oxide of molybdenum (MoO). The oxide is then made into a paste with some oil or charcoal, and exposed to a powerful heat. It may be procured by passing a current of dry hydrogen over molybdic acid, heated to redness in a porcelain tube. It is a grayish powder, and has not been procured in a button. Its colour when in cohering masses is, however, silver-white, sp. grav. 8·615. By roasting it becomes brown (the oxide), then blue (binoxide), and lastly white (molybdic acid); nitric acid converts it into the acid. Sulphuric acid in contact with it is converted into sulphurous acid and oxide.

Protoxide of Molybdenum. Brown oxide.—MoO 6·75, 54. Is formed when the metal is heated in the air. It may be prepared in a pure state by dissolving a molybdic acid salt in a small portion of water, adding hydrochloric acid till the precipitate which at first appears is redissolved, and then digesting the solution with distilled zinc; the liquid becomes gradually blue, reddish, brown, and finally black. Ammonia is then added, when the hydrate precipitates in the form of black flocks. This is washed, away from the air, in water containing NH_3, and then dried in vacuo. It is a black powder, insoluble in acids when it has been ignited, but soluble in the hydrous state. When heated in air it takes fire, and is converted into molybdic acid. The hydrate also dissolves in carbonate of ammonia, but the other alkaline carbonates have no action on it.

Binoxide, Deutoxide.—MoO_2 7·75, 62. When molybdic acid is allowed to digest with chlorohydric acid and copper the liquid assumes a red tint. On adding ammonia the binoxide precipitates. It may be prepared also by mixing 2 parts of molybdate of soda with 1 of salammoniac in a red hot crucible, which is to be covered. The soda is converted into common salt, and the molybdic acid into the deutoxide. The salt may be taken up by water. When dried it has the colour of the peroxide of iron, and by exposure to the air becomes green; it is slightly soluble in water, giving it a yellow tint, and hence in washing, it is necessary to use water holding salammoniac in solution, and it should be dried in vacuo. It is insoluble in caustic alkalies; the alkaline carbonates and bicarbonates readily dissolve it; it reddens litmus paper, but does not possess acid properties. What was formerly termed molybdous acid is a bimolybdate of the deutoxide, spec. grav. 3·46. When ignited in the air it volatilizes and may be sublimed in yellow scales; sparingly soluble in water, 570 parts of that fluid dissolving 1 part of acid, and reddening litmus paper; dissolves before ignition in acids, and forms salts; it dissolves in cream of tartar. The alkaline molybdates are colourless and soluble in water; the other salts are insoluble in this fluid, but are readily soluble in strong acids, and the alkaline molybdates are in general precipitated by strong acids; the precipitate is soluble in excess of acid, with the exception of nitric acid. *Molybdate of Ammonia* is very soluble in water. Molybdates of potash and soda are prepared by dissolving the acid in solutions of carbonate of potash or soda, the latter resembles tungstate of soda; chromate and molybdate of lead are isomorphous; hence there is a connecting link between tungsten, chromium, and molybdenum.

Molybdenum, Chlorides of.—Chloride of molybdenum (Mo Cl) is obtained by dissolving the hydrous protoxide in hydrochloric acid; a deep blue substance remains on evaporation. The bichloride (Mo Cl_2) is produced in black crystals resembling iodine when dry chlorine is passed over the heated metal; it deliquesces in the air, becoming blue, red, and yellow. It unites with salammoniac. *Chloromolybdic acid* (MoO_3 + Cl or MoO_2Cl), a substance corresponding with chlorochromic acid is formed when chlorine is passed over binoxide of molybdenum, at a red heat; it condenses in yellowish-white scales. *Sulphides.*—The bisulphide has been already described as molybdena. The *tersuphide* (MoS_3) corresponding with the acid, is procured by saturating an alkaline molybdate with sulphuretted hydrogen gas; on the addition of an acid the potash sulphate of molybdenum is decomposed, and the tersulphide falls, in the form of a blackish-brown compound. The *quatersulphide* (MoS_4) is procured by passing SH through the bimolybdate of potash. Salts of molybdenum are distinguished by their affording before the blowpipe with salt of phosphorus a green bead, in the reducing flame with borax a reddish-brown bead.

Bimolybdate of Soda.—NaO 2 MoO_3 7 HO. Colourless crystals, obtained by fusing in an iron crucible 5 parts of molybdate of lead with 2 of carbonate of soda, and treating the fused mass with water. The lead remains partly reduced and partly oxidized. The solution is evaporated to separate the carbonate of soda by crystallization. The liquor is saturated with acetic acid, when acetate of soda is formed, and molybdate separates in crystals.

Trismolybdate. See LEAD.

Molybdic Acid. MoO_3 8·75, 70·. Spec. grav. 3·5. White porous mass or powder, fusing into a yellowish mass; separating into silky crystalline scales in water. It forms by heat and access of air into fine white plates and needles; acid reaction; sour taste; with difficulty soluble in water; it unites with strong acids, and forms a base to them; these salts are yellow coloured, with a metallic taste. The purity of molybdic acid is ascertained by its completely volatilizing. The molybdates are white or yellow, partly soluble, partly insoluble, in water. It is obtained by roasting native molybdena (MOS_2) until the sulphurous acid ceases to be evolved, the colour changing from black to yellowish-white; the residue is dissolved in ammonia and crystallized; the resulting molybdate of ammonia by heat leaves the acid; it may be also obtained by the action of nitric acid on the bisulphide, or on the binoxide. It occurs native (*molybdic ochre*) with bisulphide.

Momia. *Mummy.* A specimen of mummy has been found to consist of water 10·4; azulmic acid 35·17; azulmate of ammonia, carbonate of potash, chlorides of potassium and sodium, sulphate of potash, phosphate of soda 2·72; ammoniacal soap, resin, 12·71; phosphate of lime, carbonates of lime and lead, silica and sand 39·.

Momosite. A synonyme of Dolomite.

Monazite. *Mengite, Edwardsite, Eremite.* Spec. grav. 4·922 to 5·019, H 4·5. Clove-brown, reddish-brown, dirty brownish-red oblique rhombic prisms with angles of 92° 30′ and 100° 25′ 13″. B.B. loses colour, becoming pearl-gray, and fuses on the edges with great difficulty to a glass; with borax it turns white and forms a yellow-green glass when hot, colourless when cold; tinges the flame green with sulphuric acid. Oxide of cerium (thorina?) 26; oxide of lanthanum (thorina?) 23·4, thorina 17·95, SnO 2·1, MnO 1·86, CaO 1·68, PO_5 28·5, TiO_2 trace, KO trace. Slataoust, Ural.

Monazitoide. Spec. grav. 5·281, H 4. Prisms of 92°. Associated with monazite near Lake Ilmen. CeO 49·35, LO 21·3, CaO 1·5, PO_5 14·94, columbic acid? 6·27, HO 1·36, MgO and FeO traces.

Monesine. A bitter gummy substance, by alcohol, from the bark Monesia, occurring in commerce, from South America; its source is not known.

Monometaphosphoric Acid. A name given to the common form of metaphosphoric acid, RO PO_5, to distinguish it from a series of other compounds which have been termed dimeta, trimeta, tetrameta, and hexameta phosphoric acids, where the atoms of acid and base are multiples of the atoms in the monometaphosphate by 2, 3, 4, and 6.

Money. Stamped or cast pieces of metals used as a circulating medium. I am indebted for the following tables of British money to Mr. Mushet of the Mint:—

English Copper Coins.

Denomination of Coin.	Number of Pieces per Pound Avoirdupois.	Value of One Pound.	Number of Pieces in a Ton.
Pence	24	2*s.*	53,760
Half-pence	48		107,520
Farthings	96		215,040
½ Farthings...	192		430,080
⅓ Farthings...	288		645,120

Value of 1 Ton of Coined Copper, £224.

English Silver Coins.

Number of Pieces in the Pound Troy.	Standard Weight of each Piece. dwt.	gr.		Fine Silver in each Piece. dwt.	gr.		Alloy in each Piece. dwt.	gr.	
792......Pence	...	7	·2727	...	6	7272	...	...	·5455
396......2 Pences......	...	14	·5454	...	13	4545	...	1	·0909
264......3 Pences......	...	21	·8181	...	20	1818	...	1	·6363
198......4 Pences......	1	5	·0909	1	2	9090	...	2	·1819
132......6 Pences......	1	19	·6363	1	16	3636	...	3	·2727
66......Shillings......	3	15	·2727	3	8	7272	...	6	·5455
26 & 1 Shilling } ½ Crowns...	9	2	·1818	8	9	8181	...	16	·3637
13 & 1 Shilling } Crowns.....	18	4	·3636	16	19	6363	1	8	·7273
33 Florins....................	7	6	·5454	6	17	·4544		13	·0910

BRITISH STANDARD OF SILVER.

Oz.	dwt.		
11	2	Fine Silver	per Pound Troy.
	18	Alloy	

N.B.—The assays of silver are reported in pennyweights—the lowest report being ½ pennyweight.

English Gold Coins.

Number of Pieces in the Troy Pound.		Standard Weight of each Piece. Oz.	dwt.	gr.		Fine Gold in each Piece. Oz.	dwt.	gr.		Alloy in each Piece. Oz.	dwt.	gr.	
$93\frac{9}{20}$	½ Sovereigns..	...	2	13	·637	...	2	8	·500	...	...	5	·137
$46\frac{29}{40}$	Sovereigns....	...	5	3	·274	...	4	17	·001	...	...	10	·272
$23\ \frac{9}{60}$	£2 pieces.....	...	10	6	·548	...	9	10	·003	...	...	20	·545
$9\frac{69}{200}$	£5 „	1	5	16	·372	1	3	13	·008	...	2	3	·364
178	Guineas.......	...	1	8	3595	...	1	5	6629				
133½	⅓ „	...	1	19	1460	...	1	15	5505				
89	½ „	...	2	16	7190	...	2	11	3258				
44½	Guineas.......	...	5	9	4382	...	4	22	6516				
22¼	2 Guineas....	...	10	18	8764	...	9	21	3034				
8½	5 Guineas....	1	6	23	1910	1	4	17	2584				

GOLD STANDARD.

22 Carats, or $\frac{11}{12}$ fine gold } per Pound Troy.
2 „ „ $\frac{1}{12}$ alloy }

N.B.—The assays of gold are reported in carats and carat grains.—The lowest being ⅛ of a carat grain=7½ troy grains.

Monophane. A synonyme of Epistilbite.

Monoxy-mercur-azotine. HgO (Hg_3N 2HO). A base by the action of aqueous ammonia on oxide of mercury. A tri-base also exists with 3 instead of 1 atom of mercury.

Monradite. Spec. grav. 3·2673, H 5·75. Yellowish massive or foliated; cleavage inclined to 130°, lustre vitreous. B.B. gives out water; with borax the presence of iron is indicated; with phosphate a silica skeleton; with soda a grayish pearl. Bergen, Norway. SiO_3 56·17, MgO 31·63, FeO 8·56, HO 4·04=4 (MgO FeO), 3 SiO_3 3 HO.

Monrolite. Spec. grav. 3·045 to 3·096, H to 7·25. Greenish, gray, and whitish mineral resembling sillimanite, SiO_3 40·92, Al_2O_3 56·618, MgO, 285, HO 3·091.

Monticellite. Yellow crystals of 132° 34′, gelatinizing with acid; in granular limestone at Vesuvius, supposed to be olivine or chrysolite.

Moonstone. Adularia, which when polished reflects a variegated surface.

Mordant. (*Mordere*, to bite). The chemical base employed to fix a colour on cloth in calico printing.

Moreine. A volatile oxide of morine in yellow plates, obtained by boiling morine in water.

Morindine. $C_{28}H_{15}O_{15}$. Yellow silky crystals; slightly soluble in cold, more soluble in hot alcohol; insoluble in ether; slightly soluble in cold, more soluble in hot water; soluble in alkalies with a red colour; in sulphuric acid with a purple colour; precipitated by lead, barytes, strontian, lime, and iron salts; from the bark of the root of Morinda citrifolia by alcohol; it is nearly allied to ruberythric acid in madder.

Morindone. $C_{28}H_{10}O_{10}$. Red needles, insoluble in water; soluble in alcohol and ether, alkalies and oil of vitriol; precipitated by alum cobalt blue; it is obtained by sublimation by heat from morindine, and seems identical with alizarine.

Morine. White crystalline powder, yellow by ammonia; soluble in 4000 water. One of the colouring principles of the Morus tinctoria or fustic; soluble in acids with a yellow colour; sublimes; obtained from the deposit of the aqueous decoction by solution in alcohol, and the addition of oxalic acid to decompose the compound of lime and morine.

Moringic Acid. $C_{30}H_{28}O_4$. Spec. grav. ·908; F.P. 392°. Colourless oily acid obtained by saponification from the oil of Mohringa or ben; solidifies at 32°; it is homologous with oleic acid; it contains 2 atoms hydrogen less than benic acid; soluble in alcohol.

Morintannic Acid. $C_{18}H_8O_{10}$. The yellow crystalline colouring matter of the Morus tinctoria; obtained by treating the central part of the wood with boiling water with a little chlorohydric acid; in the course of time the acid separates as a yellow crystalline powder; soluble in 6·4 water, in 2·14 boiling water; very soluble in alcohol, ether, and sulphuric acid; when heated up to 518° it loses carbonic acid, and a clear fluid smelling of phenole or carbolic acid, passes over, speedily crystallizing into pyromorintannic acid ($C_{12}H_6O_4$), which seems to be identical with pyrocatechine; it has been termed oxyphenic acid.

Morione. Black resinous quartz.

Moroxite. *Moropite*, *Apatite*, or *Phosphate of Lime.*

Morphine. *Morphia.* $C_{34}H_{19}NO_6$. Colourless pearly 4-sided rectangular prisms, the primary form being a right rhombic prism; soluble in 500 boiling water, in 40 cold and 30 boiling alcohol, in fat and essential oils, and in caustic alkalies; insoluble in ether; taste bitter; it differs from codeine by C_2H_2 less; it forms crystalline salts with acids, which constitute narcotic poisons; nitric acid turns morphine and its salts red, a characteristic property; iodic acid renders it brown by the separation of iodine; sesquichloride of iron renders it blue, which soon disappears; by sulphuric acid it is converted into *sulpho-morphide?* when heated it loses 2 atoms water, and at 570° is decomposed; heated with potash it yields methylamine. *Process* I. *Mohr's.*—1. Macerate the raw opium, cut into pieces, with 3 times its weight of water, press the solution through a cloth, and repeat the maceration and filtration three or four times; meconate, with some sulphate of morphine which exists in opium, is thus obtained in solution, with colouring matter. 2. The solution, after concentration, is mixed with one-fifth of caustic lime, previously made into a cream with water, and boiled for a few minutes. The colouring matter is thus precipitated, meconate of lime formed, while morphine remains in solution. 3. The mixture is then filtered and washed with boiling water, and the liquor evaporated till it reaches double the weight of the opium employed, filtered rapidly through paper, boiled, and an ounce of salammoniac added for each pound of opium. The muriate of morphine crystallizes out; it may be further purified by boiling with milk of lime and reprecipitating by salammoniac. To obtain the morphine from the muriate, a solution of the salt is exactly saturated with ammonia, when the alkaloid falls. II. *Robertson and Gregory's Method.*—1. Dissolve out the meconate of morphine from opium by frequent maceration in water. 2. After concentrating the infusion, add a slight excess of chloride of calcium; meconate of lime is thus formed and precipitated, while muriate of morphine remains in solution, contaminated with much colouring matter. 3. The muriate is crystallized out by evaporation, and filtered in a bag, subjected to pressure; the dark mother liquor thus squeezed out contains narcotine, while the brown cake consists of impure muriate of morphine. 4. The *cake* is dissolved in hot water, and passed through a filter of animal charcoal, and frequently crystallized. The morphine may be separated from the salt by ammonia, and thus isolated from codeine, which is usually contained in the salt. The *muriate*, M,HCl 6 HO, in silky needles, is used in medicine extensively as a narcotic, in doses of from one-eighth to one-fourth grain; it is soluble in 18 cold water, in its own weight of boiling water; soluble in alcohol; 1 lb. opium yields $1\frac{1}{2}$ ounce muriate. That it contains codeine is obvious, from the fact that I have sometimes found it in commerce to contain less nitrogen than it ought to possess, as follows:—

	Atoms.	At Weight.	Theory.	Expt.
C, ...	34	25·5	54·32	55·36
H,....	19	2·375	5·05	5·00
N, ...	1	1·75	3·73	3·19
O,............	6	6·	12·78	13·53
HCl, ...	1	4·569	9·74	9·92
HO,...........	6	6·75	14·38	13·
		46·944	100·00	100.

In six determinations the following numbers have been obtained for the nitrogen:—2·74, 2·901, 2·969, 2·81, 3·19, 2·881. This is perhaps sufficiently explained by the fact, that the formula of morphine has hitherto been estimated too high. It loses almost exactly 13 per cent. water at 212°.

Sulphate of Morphine.—$\overset{+}{M}SO_3 6HO$. Feathery soluble crystals; *acetate*, difficultly crystallizable in prisms; altered by keeping; taste bitter; very soluble in water; formed by dissolving morphine in acetic acid; dose $\frac{1}{8}$ to $\frac{1}{4}$ grain.

Morpholite. A name given to the marls of Sweden, containing nearly half their weight of carbonate of lime.

Morsine. Animal matter from the nests of the swallow, consisting of C 54·81, H 7·02, N 11·64, O 26·53.

Mortars. Common mortar may be distinguished from hydraulic mortar or that kind of mortar which sets under water by the fact that when treated with chlorohydric acid it yields no gelatinous silica, which is always present in hydraulic mortars. The solidity of common mortar depends on the union of carbonic acid slowly with the lime. When it unites too rapidly the building is not so strong. Sometimes the mortar in the interior of a wall does not solidify, but remains soft. Common mortar in contact with water falls to pieces. Hydraulic mortars, which contain 8 to 12 per cent. of clay, yield a mortar which hardens in fifteen or twenty days under water. When it contains 15 to 18 per cent. it hardens in eight days; with 25 per cent. it hardens on the third or fourth day; when 30 to 40 per cent. of clay are present, the mortar is Roman cement, and hardens in a quarter of an hour; first manufactured in London in 1796 from the *Septaria* in the Isle of Thanet. When these clayey limestones are calcined, the theory is that the silica unites with the alumina, lime, and magnesia forming a silicate, which when mixed with water takes up water of crystallization, like plaster of Paris, and becomes as hard as rock. The temperature during calcination must not be raised too high, otherwise a *fritt*, as in glass-making, would be produced, which is incapable of uniting with water.

Artificial Hydraulic Mortars are made in Paris by treating in water 1 part of Passy clay and 4 parts of chalk by means of millstones. The deposit is formed into small bricks which are dried in the air, and then moderately calcined. This mortar when slaked increases in volume only two-thirds, while common burned lime triples its volume. Lava and Puzzolano are also mixed with lime to form hydraulic mortar. A hydraulic mortar is made by mixing liquor of flints or silicate of potash with lime. Hamelin's mastic consists of 62 Portland stone, 35 sand, and 3 litharge.

Morvenite. A species of harmotome from Strontian, consisting of SiO_3 64·75, Al_2O_3 13·425,

CaO 4·16, FeO 2·59, HO 14·47. Damour has analyzed what he calls morvenite, but upon what authority he does not say. SiO_3 47·59, Al_2O_3 16·71, BaO 20·45, FeO ·56, HO 14·16.

Mosaic Gold. The old name for bisulphide of tin.

Mosandrite. Spec. grav. 2·96, H 4·. Reddish flat prisms, and in masses fusing to a greenish-brown bead; it contains silica, titanic acid, oxides of cerium and lanthanum, manganese, lime, magnesia, potash, and water. Brevig, Norway.

Moss, Iceland. The Cetraria Islandica.

Moss, Irish. The Chondrus crispus, a sea weed. See CARRAGHEEN.

Moss, Reindeer, or **Lichen.** The Cynomyce (Cladonia) rangiferina.

Mother Liquor. The manufacturing term for the fluid from which crystals have deposited.

Mother-of-Pearl. The iridescent shell of certain mollusca, containing 66 $CaOCO_2$, and 34 animal membrane.

Mother of Vinegar. A species of plant, formed in impure acetic acid. See FERMENTATION.

Mould. *Humus* and *Ulmic Acid.* In perfectly dry air wood keeps perfectly entire. When wood is exposed to moist air it gradually decomposes by an action which Liebig has termed eremacausis. It is a slow process of combustion in which the combustible parts of a plant unite with the oxygen of the atmosphere. The lignine in contact with air or oxygen gas converts the latter into an equal bulk of carbonic acid, and its decay ceases with the disappearance of the oxygen. If the carbonic acid is removed and oxygen replaced, its decay recommences. Woody fibre in the state of decay is termed *humus.* The property which lignine has of converting oxygen into carbonic acid diminishes as its decay advances, and at last it becomes brown, when it is termed mould. Its decay may be prevented not only by oxygen, but by metallic salts, as in Kyan's patent, HgCl, Burnett's patent, ZnCl. We can accurately observe the nature of the decay by comparing the formulae:—

Oak wood,............. $C_{36}H_{22}O_{22}$
Humus,................. $C_{34}H_{18}O_{18}$
2 4 4 $= 2CO_2 + H_4$

The humus has lost 2 CO_2 and 4 H, oxidized by the air to form water. If humus be examined at different stages of decay, the elements will be found diminishing in number, and it has been found that in vegetable moulds, these different states are acids containing various quantities of ammonia in combination. Wood coal is a stage of decay under certain circumstances, consisting of $C_{33}H_{21}O_{16}$. It seems to have lost 1 atom hydrogen, and 3 carbonic acid. The decay of plants is a subject of extreme interest in agriculture; and who is there who is not interested in this subject, indebted as we are to it for our comforts and daily bread? We observe decay well exemplified in old oak and elm trees, which are hollow and rotten in the heart. The rottenness is merely the common term for the absence of oxygen and hydrogen, and the preponderance of carbon. Next in point of decay we may range the organic constituents of soils which have their origin in the manure thrown on the soil, and in the decayed crops. We can render these apparent by simple experiment. *Experiment.*—Digest a soil in caustic soda, filter and precipitate by acid. Turf is another example of decayed vegetable matter. Mulder has given the following formulæ:—

Acid from the soil of a garden,........ $C_{40}H_{12}O_{14}$
Humic acid from willow,............... $C_{40}H_{12}O_{12}$
Ulmic ,, from fresh turf,.............. $C_{40}H_{14}O_{12}$

These acids were obtained by treating the substances with water and alcohol, and then with carbonate of soda, and precipitating by means of an acid. It is an interesting fact that in the neighbourhood of the brown coal deposits, as in Germany, mineral waters containing abundance of carbonic acid are particularly met with. I refer to the districts north of Francfort, immediately east of the Rhine. The carbonic acid of these wells may be derived from the gas formed by the union of the oxygen of the air with the carbon of the wood. It is obvious, however, that the carbonic acid of volcanic springs must be derived from a different source. Similar products in composition are obtained by the action of sulphuric acid upon sugar, but they are not identical, although very closely approximating. The products of the decomposition of vegetable matter are described under HUMUS, HUMINE, HUMIC ACID, GEIC ACID, ULMINE, ULMIC ACID, CRENIC and APOCRENIC ACID.

Mountain Green, or *Malachite.* Green carbonate of copper.

Mountain Leather. Spec. grav. 1·334. Light buff spicular crystals, obvious to the naked eye, but most distinct under the microscope (R. D.T.); felted so as to resemble leather; absorbent of water; tough, soft, opaque. B.B. curls up and fuses into an opaque bead; with soda into a transparent yellow bead; with borax into a colourless glass. SiO_3 51·65, Al_2O_3 9·505, CaO 10·, MgO 2·065, FeO and some MnO 5·8, HO and organic matter? 21·7. Strontian, Argyllshire.

Mountain Soap. A synonyme of bole or soapstone.

Mountain Tallow, or Hatchettine.

Mountain Wood. A mineral from Sterzing, resembling asbestus in appearance, but differing in composition, containing water; it is decomposed by acids, leaving silica of the form of the mineral. $Fe_2O_3 3SiO_3$, 2 MgO SiO_3, MgO 5 HO.

Mucamide. $C_{12},NH_2,H_8O_{10},O_2$. Sp. grav. 1·589. Rhombic 8-hedrons; little soluble in water; insoluble in alcohol and ether; by adding ammonia to mucic ether.

Mucic Acid. *Saccholactic Acid. Saclactic Acid.* $C_{12}H_8O_{14}2HO$. Spec. grav. 1·0015 (Morveau). White crystalline powder, with a slightly acid taste; soluble in 60 boiling water (Scheele), one part in 80 (Morveau), one part in 6; obtained by treating sugar of milk, gum, or mannite, with moderately strong nitric acid, the strongest or weakest acid forming oxalic acid; 100 parts of milk sugar produce 36 (Hagen), 60 to 65 (Guckelberger). The mucates are bibasic. Mucate of potash ($2KO,C_{12}H_8O_{14}HO$) is in white crystalline grains. When dissolved in water, and heated for a long time, mucic acid becomes *modified* and more soluble in water and alcohol, from which it separates in square plates; when its solution is saturated at 212°, it deposits as common mucic acid; its combinations are more soluble than the mucates; when exposed to dry distillation, pyromucic acid is formed in white scales, by the removal of 2 CO_2 and 6 HO.

Mucilage. $C_{12}H_8O_{10}$? A gummy matter obtained from several plants; that from linseed is obtained by digesting the seed in hot water, and concentrating the infusion in a steam bath; it forms brown brittle crusts, with a peculiar odour; it reddens vegetable blues and thickens water; insoluble in alcohol; one portion being soluble in water resembles arabine, and forms mucic acid with nitric acid; the other is insoluble, and gives no mucic acid when oxidized; under this term have been classed the products from Irish moss, lintseed, quinceseed, althaea root, symphitum, and salep.

Mucine. Transparent grains obtained by boiling water on the gluten of wheat, and filtering while hot. Mucine deposits on cooling; it burns like animal matter, and is more soluble in water than gluten (1 mucine in 25 hot water); insoluble in ether; precipitated by infusion of nutgalls; converts starch into dextrine and sugar.

Mucus. The slimy fluid matter secreted by the mucous membranes of the body, as in the mouth, nose, trachea, stomach and intestines, gall bladder, urinary organs. All of these secretions contain an animal body, which does not dissolve in water, but which absorbs that fluid, swelling up, and becoming soft and viscid, and which appears to possess nearly the ultimate composition of an albuminous body; it is usually alkaline, from the presence of soda; acids, by neutralizing the alkali, cause it to coagulate, and hence, in all probability, produce false membranes. There seems strong evidence of the existence of an acid and alkaline secretion from the mucous membrane, which, in the healthy state, is characterized by a predominance of alkali, and in the inflamed condition by a preponderance of acid. Of the latter fact I have had abundant evidence, and have long ago suggested the use of alkalies as a counteracting agent, especially in the seminal and urinary organs; I believe the experience of others has confirmed the justice of my recommendation. The mucus of the bile consists of C 52·84, H 7·09, N 15·4, O and S 24·67.

Mudarine. A brown extract obtained by digesting the bark of Calotropis (procera, Hamiltonii, and gigantea), Indian trees, in alcohol; insoluble in ether, essential, and fat oils. Its aqueous solution gelatinizes when heated to 95°, and becomes fluid on cooling. It coagulates at high temperatures, a pitchy matter separating. It is used as a diaphoretic in India.

Mudesic Acid. $C_{12}H_5O_{10}$. A brownish-yellow soluble and deliquescent acid, giving yellow precipitates with metallic salts; obtained by oxidizing mudesous acid with nitric acid.

Mudesous Acid. $C_{12}H_5O_8$. Dark brown soluble organic acid, united to alumina in the granite caves of Cornwall; in the mineral pigotite a brown incrustation, insoluble in water and alcohol; the acid on separation unites with bases, and gives with metallic salts brown precipitates.

Muffle. A hollow vessel of fire-clay, with a flat bottom, arched roof, and apertures at the side for the admission of air, closed behind, and open in

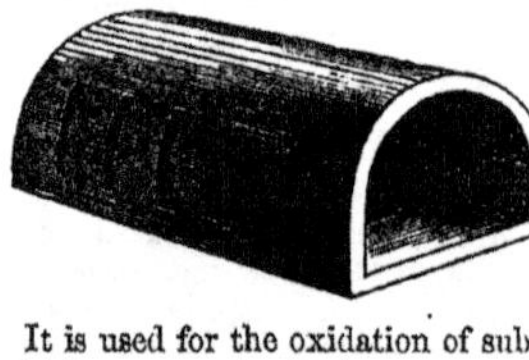

front. It is used for the oxidation of substances, over which a current of air passes when the muffle is placed in a furnace.

Mulberry Calculus of Oxalate of Lime. See URINE.

Muller's Glass, or **Hyalite.** Stalactitic opal.

Mullerine. *Mullerite.* White or yellow ore of tellurium.

Mullicite. Spec. grav. 1·787, H 1·. Bluish-black radiating needles, from the centre of a cylinder; opaque, sectile. PO_5 24, FeO 42·65, HO 25, sand 7·9, 3 FeO, PO_5 8 HO. Allied to vivianite.

Mundic. A synonyme of bisulphide of iron, or iron pyrites.

Munjeet. The root of the Rubia munjista (cordifolia), used in India for similar purposes to madder. It occurs in commerce from 1 to 3 feet long. By the natives of India it seems to be preferred to madder for many purposes.

Muntz's Metal. When first issued, its composition was, copper 60, zinc 40; afterwards it was 50 copper, $40\frac{3}{4}$ zinc, $3\frac{3}{4}$ lead.

Muracite. A synonyme of anhydrous sulphate of lime, or anhydrite, when it is crystallized in broad plates.

Murchisonite. Specific gravity 2·509, H 6·. Yellowish 4-sided prisms, with angles of 90°,

106° 50′, 90°; one of its cleavage faces has a pearly lustre, with a spotted golden-yellow reflection; texture foliated; in thin laminæ, transparent. SiO_3 68·6, Al_2O_3 16·6, KO 14·8. In new red sandstone at Dawlish, near Exeter. Allied to felspar.

Murexane. *Purpuric Acid of Prout*, $C_6N_2H_4O_5$, is prepared by dissolving murexide in potash, applying heat till the blue colour disappears, and saturating with sulphuric acid. It consists of silky scales, insoluble in water and dilute acids; soluble in cold alkaline solutions without neutralizing them; soluble in strong sulphuric acid, and precipitated without change by water. A solution of murexane in ammonia, when exposed to the air, becomes purple, and deposits murexide in crystals; with excess of ammonia the solution becomes colourless, and contains oxalurate of ammonia. 2 atoms murexane 1 NH_3 and O_3 contain the elements of 1 murexide and 3 water.

Murexide. *Purpurate of Ammonia.* $C_{12}N_5H_6O_8$. When uric acid is dissolved in dilute nitric acid, and evaporated until it becomes fleshred, it is allowed to cool to 160°; it is then treated with dilute ammonia, slightly in excess, diluted with half its volume of boiling water, and allowed to cool. It is proper, in this process, occasionally to saturate a small portion of the uric acid solution with ammonia; if it becomes turbid, and deposits a red powder, it is proper to add a little nitric acid to the uric acid solution; if it gives a yellow slimy precipitate, it will be necessary to pass SH through the solution before murexide will separate. It may be procured by dissolving 7 grains of hydrous alloxane, 4 of alloxantine, in 240 grains of water, and adding 80 grains of a cold saturated solution of carbonate of ammonia. Crystals of murexide precipitate (Gregory). If it does not succeed at once, add a little water, and try it again, and so on till a solution of the carbonate is obtained, which gives a good result. In this case 1 atom alloxane, 2 alloxantine, and 4 ammonia, contain the elements of 2 murexide and 14 of water. A solution of uric acid in nitric acid contains alloxantine, urea, and nitrate of ammonia. When evaporated to a red solution, a portion of the alloxantine becomes, by the action of the nitric acid, alloxane. The crystals of murexide are short 4-sided prisms, with a fine green colour by reflected light, resembling cantharides; by transmitted light the crystals are red. Insoluble in ether and alcohol; soluble in boiling water, with difficulty in cold water; decomposed by the acids and alkalies.

Murexoine. A synonyme of C olestrophane.

Muriatic Acid. See CHLOROHYDRIC and HYDROCHLORIC ACID.

Muricalcite. A synonyme of dolomite, or calcareo-carbonate of magnesia.

Muromontite. Spec. grav. 4·265, H 7·. Black or greenish grains, found in felspar at Boden, Marienberg, mixed with Bodenite; streak grayish-white; lustre glassy. SiO_3 31·089, yttria 37·143, FeO 11·231, MnO ·905, LaO 3·536, CeO 5·544, Al_2O_3 2·235, glucina 5·516, MgO ·424, CaO ·707, NaO ·651, KO ·17, HO and loss ·849. B.B. glows like gadolinite; slightly fuses on edges; shows soda in flame. Allied to Bodenite.

Muscle. *Flesh.* (*Fleisch*, Ger.; *Chair*, Fr.) The name of the fibrinous parts of animals, constituting their motory apparatus, from which it may be inferred that the force capable of being exercised by an animal is proportional to his muscular mass. Muscle or flesh has been found to consist of—

	Human.	Ox.
Fibrine, vessels, nerves,	15·54	15·43
Albumen and hæmatosine,	1·93	1·99
Cellular tissue,	2·07	—
Extractive matter,	3·71	4·98
Fat,	2·30	
Ash,	—	·88
Water,	74·45	76·72

Besides these substances, we find in minute quantity present, lactic and phosphoric acids, creatine, inosic acid, inosite. The ash amounts to about ·88 of the fresh muscle, or 3½ per cent. of the dry muscle. The ash consists of PO_5 36·27, KO 38·87, Cl 8·63, K 9·4, SO_3 3·59, RO 4·23, phosphate of lime 3·06, phosphate of magnesia 5·76, phosphate of iron ·57. Muscular matter, or fibrine, appears to characterize even the lower classes of fish, as I have found it in oysters. The following is the result of my analyses of different fish, when they were washed with water and treated with alcohol and ether:—

	Oysters.	Herrings.	Cod.
C,	53·98	53·77	53·67
H,	—	7·44	7·00
N,	—	16·23	16·89
O,	—	22·56	22·44
S,	—		

Muscoide. A synonyme of moss-green phosphate of lead.

Muscovy Glass. A synonyme of mica.

Mushroom. (*Champignons*, Fr.) Well-known ephemeral cryptogamous plants. The Agaricus arvensis contains 90·62 water, 9·4 solid matter, including 1·08 salts, 0·77 nitrogen. The dry plant therefore possesses 8·3 per cent. of ash and 11·6 per cent. of nitrogen. Various other species have been examined with analogous results. The cellular portion has sometimes been termed fungine; it appears to be a modification of cellulose or lignine.

Mushroom Sugar. According to analyses it appears that mushrooms contain both fermentable sugar and mannite.

Musite. A synonyme of *Parisite* or carbonate of lanthanum.

Musk. A substance formed in the follicle of the prepuce of the musk deer (Moschus moschiferus). It is imported from China, Tibet, Russia, and Bocharia. It contains 45 per cent. of water, 36·5 per cent. of an organic substance united with ammonia, also fats, cholesterine, and a peculiar odoriferous substance which is supposed not to be very volatile, but to be carried in small portions along with ammonia evolved from the musk.

Musk, Artificial. $C_{15}H_8NO_7$? A resin obtained by the action of 3 fuming nitric acid on 1 part of oil of amber; soluble in alcohol, ether, and essential oils.

Mussel. A molluscous animal consisting chiefly of fatty matter and fibrine.

Mussite. A grayish-green variety of Pyroxene from Mussa in Piedmont.

Must. The unfermented juice of the grape.

Mustard Seed. The fruit of the Sinapis nigra contains myronic acid, united with potash, myrosine, fatty matters, gum, colouring matter, sugar, sinapine. The essential oil of mustard ($C_8H_5NS_2$) does not exist in the seeds, but is formed by distillation with water by the fermenting action of myrosine, which is liberated from its union with potash and lime. The same oil is obtained from the root of the horse-radish.

Mustard, Black Oil of. *Sulphocyanide of Allyle.* C_6H_5, C_2NS_2. Spec. grav. 1·01, of vapour 3·54; B.P. 298°. Clear oil, insoluble in water; soluble in alcohol and ether; obtained by digesting the seeds of mustard in cold water and then distilling and rectifying. By the action of barytes solution, or hydrous protoxide of lead, sinapoline is obtained; by ammonia thiosinnamine is formed, which, when acted on by oxide of lead, is converted into sinnamine.

Mustard, White Oil of. Contains sulphosinapasine, erucine, and a fixed oil.

Mycomelic Acid. $C_8N_4H_5O_5$. When a boiling solution of alloxane is mixed with an excess of ammonia and neutralized with dilute sulphuric acid, the boiling being continued for a few minutes, mycomelic acid falls as a yellow gelatinous precipitate, which, on drying, becomes a -ellow absorbent powder; it is soluble in hot 'ater and gives an acid reaction; it decomposes ıe alkaline carbonates, and dissolves readily in ıustic alkalies. With oxide of silver it forms a ellow compound insoluble in water. There appear to be produced by the action of ammonia ı this reaction, 1 atom mycomelic acid and 5 rater.

Myricine. F.P. 149°. That portion of ees' wax which is insoluble in cold alcohol was ırmerly so called. When saponified it yields almitic acid and hydrous oxide of melissyle, the cid having been in union with oxide of melissyle ı the myricine.

Myriospermic Acid. An acid by saponication from myriospermine, allied to cinnamic cid.

Myriospermine. A fluid oil soluble in boiling alcohol, existing in the Balsam of Peru, affording by saponification with potash myriospermic acid.

Myristic Acid. *Sericic Acid.* $C_{28}H_{27}O_3$ HO. F.P. 119°. White pearly plates, soluble in alcohol and ether; obtained by the saponification of myristine, the solid part of the butter of nutmegs.

Myristine. $C_{118}H_{113}O_{15} = 4C_{28}H_{27}O_3, C_6H_5O_3$. F.P. 88°. A white vegetable fat, obtained from butter of nutmegs (Myristica officinalis) by heating it with cold alcohol, and crystallizing the residual myristine from ether; decomposed by saponification into myristic acid and glycerine.

Myristone. $C_{50}H_{50}O_2$. Pearly scales by distilling myristate of lime.

Myrobalanus Citrina. This tree in Bengal supplies a fruit containing tannin, which comes to this country under the names of yellow, black, brown, ash, &c. myrobalans.

Myronic Acid. An acid syrup having a bitter taste, consisting of carbon, hydrogen, nitrogen, sulphur, and oxygen, and forming crystallizable salts; it is soluble in alcohol, slightly so in ether; obtained by expressing the fat oil from flour of mustard seed, treating the residue with alcohol, which removes foreign matters; the residue is then treated with water, which takes up myronate of potash; this is evaporated, and the crystals washed with alcohol, and decomposed by tartaric acid; the acid is taken up by alcohol.

Myrosine. An albuminous substance analogous to emulsine and synaptase, existing in black and white mustard, and many other seeds, as Raphanus sativus, Brassica napus, Erysimum alliaria, Cheiranthus cheiri, Draba verna, Cardamine pratensis, &c.; obtained by digesting white mustard with cold water, evaporating under 100° to a syrup, precipitating by alcohol, dissolving the precipitate in water, and evaporating to dryness. A solution of myrosine, when mixed with myronic acid salts, evolves the odour of oil of mustard in five minutes.

Myroxiline. A thick oil, in Peruvian balsam, insoluble in alcohol, changed by alkalies into cinnamic acid and a resin.

Myroxocarpine. $C_{48}H_{35}O_6$. Rhombic prisms by alcohol, from the balsam of a species of myrospermum; insoluble in water; soluble in ether.

Myrrh. Spec. grav. 1·36. The gum resin of the Balsamodendron myrrha from Arabia, in the form of tears or irregular pieces; transparent, colour reddish-brown, brittle, taste bitter and aromatic; does not fuse by heat; forms with water a yellow opaque solution soluble in alkalies; distilled with water it yields an oil. Occurs as, 1st, myrrh electa (selected), 2d, myrrh in sorts of second quality, and, 3d, as indica or nova in fragments. It contains oil, gum (half its weight), resin, and mucilage. It is used in medicine as an antispasmodic and astringent.

Myrrh Resin. $C_{21}H_{16}O_3$? F.P. 200°. Obtained from myrrh by alcohol; yellowish-brown, hard and brittle, without taste and smell; converted by heat into an acid resin.

Myrrh, Oil of. $C_{22}H_{17}O_2$? Colourless or yellowish by keeping; smells of myrrh; thickens in the air into a varnish; soluble in alcohol ether, and the fixed oils; forms a red liquor with the mineral acids by which myrrh is distinguished from bdellium.

Myrtle Wax. Spec. grav. 1·015; F.P. 109°. Pale green wax; alcohol dissolves $\frac{1}{20}$; but ether dissolves one-fourth, depositing it on cooling in white plates like spermaceti, the ether retaining the green colour; obtained from the berry of the Myrica cerifera, a shrub of North America.

Mysorine. A synonyme of native anhydrous carbonate of copper.

N

Nacrite. *Anhydrous Bisilicate of Alumina, Talcite, Earthy Talc.* Sp. grav. 2·788 to 2·793, H 2·75. Silvery-white, or light greenish-white, soft, silky, flexible, non-elastic scales, or long 4-sided prisms, with angles of about 89° to 91°, intermixed with scales; lustre splendent, translucent. SiO_3 64·44, Al_2O_3 28·844, FeO 4·428, HO 1·. In mica slate at Brunswick, in Maine; Wicklow, Ireland.

Nanceic Acid. *Zumic Acid.* A synonyme of lactic acid.

Naphtha. *Petroleum, Rock Oil, Stone Oil, Mineral Naphtha.* Persian naphtha is nearly colourless. Spec. grav. ·753; B.P. 320°, rising to 352°; smell bituminous. Rangoon naphtha is of the consistence of lard at the usual temperature; spec. grav. ·880; but is liquid at 80° or 90°; by distillation it yields one-sixth naphtha and paraffine; rectified naphtha has a specific gravity of ·765, and boils at 200°? When treated with sulphuric acid and strong caustic potash, its specific gravity was ·744, and its boiling point 180°. Its specific gravity has also been found ·775, and its formula C_6H_5. Derbyshire naphtha, or petroleum, according to Mr. James Young, is of a dark brown colour as it comes out of the rock (a coal measure sandstone); its specific gravity is ·900. When cooled to the freezing point of water it becomes very thick from the paraffine crystallizing. On distilling there first came over a small quantity of very light naphtha, spec. grav. ·715; next a light oil of spec. grav. ·778; this was used for burning in lamps, and contained no paraffine; next came the oil containing the paraffine, the spec. grav. of which was ·860; this was used for lubricating machinery. The first two products were colourless, the last was reddish-yellow. German naphtha has been found to consist of at least three oils, one distilling at 203°, a second at 233°, and a third at 600°. Its spec. grav. is ·753; it evaporates in the air; insoluble in water; miscible in all proportions with absolute alcohol and ether, fat and volatile oils; decomposed by chlorine, forming an oil. It is used to preserve sodium and potassium, to prevent their oxidation. Naphtha is believed to be produced by the action of heat and pressure in the earth on organic matter. Coal naphtha consists principally of benzole.

Naphthalase. $C_{20}H_7O$. Yellow substance by heating nitronaphthase with 10 moistened lime; with sulphuric acid it produces a fine blue; insoluble in water and alcohol.

Naphthalic Acid. *Phthalic Acid.* $C_{16}H_4O_6 2HO$ F.P. 221°. Silky plates but slightly soluble in water; unites with bases, and is converted by ammonia into *pthalmide,* $C_{16}H_6NO_5$; it is formed by acting on chloride of naphthaline with nitric acid and heat; when heated with lime it becomes benzole and carbonic acid.

Naphthalidine. $C_{20}H_9N$. *Napthalidam* = $C_{20}H_7, + NH_2$. F.P. 86°; B.P. 582°. A strong base in white needles, forming crystalline salts. Obtained by acting on nitronapthalase with sulphohydric acid in presence or absence of ammonia. When heated it remains fluid at 32°, and absorbing oxygen becomes violet.

Naphthalidine-Carbamide. $C_{28}H_8NO$. Light white body, in small needles, distilling at 572°; insoluble in water, slightly soluble in boiling alcohol; formed by heating dry oxalate of naphthalidine when it distils over.

Naphthalimide. $C_{20}H_5O_6N = C_{20}H_4O_6$, NH = an imide. Colourless fibrous mass, subliming into crystalline plates; nearly insoluble in cold water; somewhat soluble in boiling water, crystallizing on cooling in needles; soluble in sulphuric acid, leaving, on adding water, naphthalic acid; potash disengages ammonia. Obtained by heating naphthalate of ammonia, when napthalimide sublimes.

Naphthaline. $C_{10}H_4$ or $C_{20}H_8$. Spec. grav. 1·048, of vapour 4·528; F.P. 174; B.P. 410° to 413°·6 and 428°. Crystalline plates or scales with a peculiar odour; burns with a smoky flame; reaction neutral; insoluble in cold water; slightly soluble in hot water; soluble in 4 boiling alcohol; very soluble in ether, fat and volatile oils; separates from its solution in oil of turpentine in prisms with pyramidal terminations; it distils over with water, and evaporates in the open air like camphor. It is obtained in distilling the last portion of coal tar and crystallizing out of alcohol, or by acting on the coal tar oils with chlorine, which destroys most of the oils, leaving solid naphthaline. When subchloride of naphthaline is acted on by potash in alcohol, it is converted into chlonaphthase or *chlornaphthaline,* C_{20}

H_7Cl, where 1 atom chlorine replaces 1 atom hydrogen. Chlonaphthese, or *diclornaphthaline*, or *dichlornaphthine*, $C_{20}H_6Cl_2$, is formed by the further action of chlorine. In the second compound it is remarkable that it occurs in seven different isomeric forms, but that no two of the forms are alike in physical characters, proving that different atoms of hydrogen are replaced by chlorine in each. It was in acting on naphthaline by various substitutional bodies that Laurent laid the basis of the theory of substitutions which has been so fertile in the development of organic chemistry. Other compounds with chlorine are *trichloronaphthine*, $C_{20}H_5Cl_3$; *tetrachloronaphthine*, $C_{20}H_4Cl_4$; *penta* and *octochloronaphthine*, $C_{20}Cl_8$. With bromine 1, 2 and 3, hydrogen are replaced by bromine, and sometimes a new body is formed by replacement of hydrogen by bromine and chlorine at the same time, as *chlorebronaphthise* or *dichlobronaphthine*, $C_{20}H_5Cl_2Br$; *chlorebronaphthese* or *dichlodibronaphthine*, $C_{20}H_4Cl_2Br_2$; *bromechlonaphthose* or *trichloro-dibronaphthine*, $C_{20}H_3Cl_3Br_2$. Sulphuric acid forms with naphthaline hyposulphonaphthalic, hyposulphonaphthic, hyposulphoglutinic acids, sulphonaphthaline and sulphonaphthalide. With nitric acid are formed nitronaphthase, nitronaphthese, naphthionic acid, thionaphthamic acid, nitronaphtheise, nitronaphthise, nitronaphthale, nitronaphthesic, nitronaphthaleisic, and nitronaphthisic acids; oxide of chloroxenaphthose, chloronaphthisic, oxide of chloroxenaphthalise, chloroxenaphthalesic acid, phthalic or naphthalic acid, nitrophthalic acid, chlorophthalesic acid. In distilling benzoate of lime naphthaline is obtained, and two isomeric solid compounds, one fusing at 198° and the other at 149°.

Naphthameine. Purple powder similar to orceine; insoluble in water and alkalies; little soluble in alcohol; very soluble in ether; forms a blue solution with sulphuric acid. Obtained by precipitating muriate of naphthaline with chloride of iron.

Naphtheine. A kind of asphalt, found on limestones on the Maine and Loire.

Naphthene. $C_{16}H_{16}$. B.P. 233°. That portion of mineral naphtha which comes over second in distillation, depending on its boiling point.

Naphthidine. *Seminaphthalidam.* $C_{10}H_5N$. Yellow needles by SH on nitronaphthalese.

Naphthionic Acid. $HOC_{20}H_8NS_2O_5,HO$. White bulky, silky powder or needles, soluble in 2000 cold water; scarcely soluble in alcohol; soluble in boiling water, separating in needles; soluble in oil of vitriol; forms crystalline salts with a reddish tinge; obtained from the mother liquor of thionaphthamic acid as naphthionate of ammonia, which is decomposed by chlorohydric acid.

Naphthole. $C_{24}H_{22}$? B.P. 374°? 600°? The last oil in naphtha which comes over in distillation.

Naphtylamine. See NAPHTHALIDINE.

Naphtyle. $C_{20}H_8$. The hypothetic radical of naphthaline, which would be a hydride.

Naples Yellow. A pigment prepared by calcining together 1 lb. antimony, 1½ lb. lead, and 1 oz. alum and salt.

Napoleonite. A variety of felspar.

Narceine. $C_{46}H_{29}NO_{18}$. F.P. 197°·6. Silky needles and 4-sided prisms, with a bitter taste, and without smell; decomposes about 230°; decomposed by strong acids. Obtained from the mother liquor of morphine, in separating that base from opium; it is rendered blue by chlorohydric acid, but not by chloride of iron; unites with acids, and forms salts.

Narcitine. A white body, soluble in water, alcohol, and acids, obtained from different species of Narcissus.

Narcogenine. $C_{36}H_{19}NO_{10}$. A body known only in combination; obtained by acting on narcotine by the bichloride of platinum, with which it unites, and forms orange needles; 2 atoms narcotine give, with 5 oxygen, 2 atoms narcogenine, 1 opianic acid and 3 water; resolved, when separated, into narcotine and cotarnine. It is very probably a double salt of chloride of platinum and narcotine with chloroplatinate of cotarnine.

Narcotic Acid. Isomeric with narcotine, or differing from it only by 1 or 2 atoms water; it is known only in combination, and is formed by heating narcotine with hydrate of potash.

Narcotine. $C_{40}H_{20}NO_{12}$, or $C_{46}H_{25}NO_{14}$. F.P. 338. White pearly needles, or powder; a weak base; insoluble in water and alkalies; soluble in alcohol, ether, and acids; it has no bitter taste, and strikes no blue with chloride of iron; nitric acid does not turn it red, by which it is distinguished from morphine; obtained by digesting the aqueous extract of opium in ether, which takes up narcotine; digesting in muriatic acid and animal charcoal, and precipitating by ammonia, or from the mother liquor of muriate of morphine by ammonia, or by digesting the insoluble part of opium in acetic acid, and precipitating by ammonia; it is purified by crystallization from alcohol. It is the most abundant base in opium next to morphine; it forms crystalline salts with bases. It is poisonous. When oxidized it affords opianic acid, opiammon, xanthopenic acid, opiano-sulphurous acid, sulphopianic acid, hemipic acid, cotarnine, humopic acid, apophyllic acid, opianyle, teropiammon.

Natrocalcite. *Calcite.* $CaOCO_2$ 96·4, CaO SO_3 1·9, FeO and MnO 1·3. From Sangerhausen.

Natrolite. *Crocalite, Edelite, Feather Zeolite, Hoganite, Soda Mesotype.* Spec. grav. 2·139 to 2·2303, H 4·5. White, with a shade of brown, right rhombic prisms, of 91° 10′, terminated by a 4-sided pyramid, or frequently by 8 faces; lustre vitreous, transparent to translucent, brittle; electric by heat; B.B. gives a phosphorescent light, and fuses into a white enamel; gelatinizes in acids even after ignition. SiO_3 47·56, Al_2O_3 26·42, FeO ·58, NaO 14·932, CaO 1·4, HO 10·44, NaO SiO_3, Al_2O_3 SiO_3 2 HO. The minerals allied to it are Lehuntite, Mesolite, Scolezite, Antrimolite, and Harringtonite.

Natron. *Nitrum of the ancients, Trona, Native Sesquicarbonate of Soda, Urao.* The natron of Tibet, which is a white crystalline powder, and no doubt produced as likewise in S. America, by the reaction of common salt on carbonate of lime, I have found to consist of insoluble micaceous matter 13·12, chlorine 1·26, carbonic acid 36·07, soda 30·57, water 18·98.

Necronite. A variety of felspar, which exhales a disagreeable odour by friction.

Necrosis. *Ulceration of Bone.* The radius has been found to consist in this disease of cartilage 19·58, phosphate of lime 72·63, phosphate of magnesia 1·93, carbonate of lime, 4·03, NaCl ·61.

Needle Ore of Bismuth. See BISMUTH.

Needle Spar. A synonyme of Arragonite.

Needlestone. The old name of Thomsonite and scolezite.

Nemalite. *Brucite, Hydrate of Magnesia.* MgOHO; spec. grav. 2·35, H ·1 to 2. White or green plates, or low 6-sided prisms or needles, translucent on the edges, sectile. B.B. gives out water; infusible except by oxy-hydrogen blowpipe. MgO 67·98, HO 30·96, FeO 1·57. In serpentine, at Hoboken, New Jersey; Swinaness, Unst.

Neocatechu consists of tannin 32·2, gallic acid 35·, colouring matter 18·8, fibre, &c. 12·.

Neolite. Spec. grav. 2·77, H 1 to 2. Stelliform green silky needles, from a mine at Näskel. SiO_3 47·35, Al_2O_3 10·27, MgO 24·73, FeO 7·92, MnO 2·64, HO 6·28.

Neoplase. A synonyme of red sulphate of iron, and also of arsenide of nickel.

Nepheline. *Beaudantite, Davyne, Covellinite, Carolenite, Fatstone, Pinguite, Rhomboidal Felspar, Sommite.* Spec. grav. 3·27, H 2·5. Grayish or greenish-white grains, or regular 6-sided prisms; fracture conchoidal, lustre vitreous; transparent to translucent; brittle. B.B. the edges are rounded off, giving a colourless and vesicular glass; becomes nebulous in nitric acid (νεφέλη, a cloud). SiO_3 44·11, Al_2O_3 33·73, NaO 20·46, HO ·62. Found in the cavities of granular limestone at Monte Somma, Vesuvius, and in the lava of Capo di Bove, near Rome.

Nephrite. (νεφρός, the kidney). *Nephritic Jade, Beilstein, Ceraunite, Pierre de hache.* Sp. grav. 2·9 to 3·1, H 6 to 7. Leek-green, greenish or whitish stone, worked into ornaments in China, New Zealand(?) and North America. SiO_3 50·5, MgO 31·, Al_2O_3 10·, FeO 5·5, Cr_2O_3? ·05, HO 2·75. Allied to serpentine. The nephrite of Iona has a specific gravity of 2·595, H 3·5, SiO_3 44·85, MgO 36·05, FeO 3·6, Al_2O_3 1·3, HO 13·55.

Neroli Oil. *Orange Flower Oil.* Colourless, becoming red in presence of air, obtained by distilling orange flowers. It seems to consist of two oils, which have not been carefully examined, and a stearoptene which fuses at 122°.

Nerves. The nerves are analogous to the brain in function and in composition. As in the brain the amount of water varies from 70 to near 80 per cent. The cerebric matter is about 4·4 per cent., the albumen 22·07, gelatinous matter 2·75, salts ·4. But there is difficulty in separating the true nervous matter from the adjacent membranes.

Neurolite. (νεῦρον, a tendon). Sp. grav. 2·476, H 4·25. Greenish-yellow imperfectly foliated thin fibres, brittle; fracture uneven, opaque, or only translucent on the edges. B.B. yields water, becomes white, infusible; with soda fuses into a transparent slightly yellow glass; with borax leaves a snow-white opaque matter in the centre of the colourless globule. SiO_3 73·, Al_2O_3 17·35, CaO 3·25, MgO 1·5, Fe_2O_3 ·4, HO 4·3. Stamstead, Lower Canada.

Neutralization. The condition of an acid and a base by proper mixture, in which they exhibit neither acid nor alkaline properties.

Neutral Salts. This term was originally applied to salts in which there is no acid nor alkaline reaction to test paper, but the term proto or equisalts would be preferable, since several of the so-called neutral salts have an acid or alkaline reaction, as sulphate of zinc which is acid, and carbonate of soda which is alkaline.

Newkirkite. Spec. grav. 3·824, H 3·25. Brilliant black metallic-like needles in right rectangular prisms with square bases, occurring on red hæmatite at Newkirchen, in Alsace. MnO_2 56·3, Fe_2O_3 40·35, HO 6·7.

Nicene, Chloro-. $C_{10}H_5Cl$. Sp. gr. 1·141, of vapour 7·52. B.P. 559°. Amber liquid by distilling chloroniceic acid with lime.

Nicene, Nitrochloro. $C_{10}H_4ClNO_4$. Amber silky needles by the action of nitric acid on chlornicene.

Nicine. A hypothetic alkaloid, to be derived from niceic acid.

Nicine, Chloro-. $C_{10}H_6ClN$. An amorphous alkaloid isomorphous with nicotine, obtained by passing sulphohydric acid through nitronicene dissolved in alcohol saturated with ammonia, and uniting the product with chlorohydric acid and crystallizing.

Nickel. Ni 3·75, 30; 3·6875, 29·5. *Sources.*—Nickel occurs in the form of sulphide of nickel, but it is generally associated with arsenic, as in the different species of arsenical nickel, antimonide of nickel, sulphoarsenide of nickel, diarseniate of nickel, and as carbonate of nickel on the surface of chrome iron ore (R.D.T.)

Metallic Nickel. Spec. grav. 8·279, 7·807 (Brisson), when melted, but when hammered 8·666 (Richter). Tourte found the spec. grav. of Richter's nickel 8·402, and when hammered 8·932. Lampadius found this nickel to contain cobalt and arsenic 8·38 fused, 8·82 hammered (Tupputi), 8·51 (Phillips), 8·637 (Brunner), 9· (Vauquelin, Hauy), 8·477 (Baumgartner), 8·237 (Pelouze), 7·832 finely divided (Playfair and Joule). *Colour.*—Pure white, resembling silver, inclining to steel-gray, and, like silver, it leaves a white trace when rubbed upon the polished surface of a hard stone (Fourcroy). It is rather softer than iron; wears a file very rapidly. It is malleable both cold and hot, and may without difficulty be hammered out into plates not exceeding $\frac{1}{100}$ of an inch in thickness (Richter), and drawn out into wires $\frac{1}{58}$ of an inch in diameter. It is magnetic, and capable of being converted into a magnet (Bergman, Klaproth, Tassaert, Hauy, Sage), and when delicately suspended points to the north; its magnetic energy is to that of iron as 35 to 55 (Lampadius); according to Wollaston, as 2 or 3 to 8 or 9; ignition destroys its magnetic power (Tourte). It has not been crystallized; it is not altered by exposure to the air, nor by being kept under water (Richter); at a red heat it oxidizes; it does not decompose water at the usual temperature; it decomposes water at a red heat, but more slowly than iron; it decomposes water feebly, assisted by heat and sulphuric, chlorohydric, and phosphoric acids, hydrogen being evolved; it is readily soluble in nitric acid, with extrication of binoxide of nitrogen, and in aqua regia; it is oxidized by saltpetre and chlorate of potash; it forms alloys with many metals. The preparations of nickel appear to be irritant and poisonous (Tupputi, C. G. Gmelin.

Preparation.—For the preparation of the pure metal, arsenical nickel may be employed, but an impure metallic substance, called *Speiss* in commerce, is usually employed for this purpose. *Speiss* is a deposit which occurs in the formation of smalts, or silicate of cobalt. Cobalt and arsenical nickel are fused with carbonate of potash and pounded quartz, so as to form a blue glass; the speiss collects below this glass as an alloy—the nickel not being so readily oxidized and united with silica as the cobalt. It contains about 50 per cent. of nickel, 30 of arsenic, and portions of sulphur, cobalt, iron.

Extraction of Nickel. Wöhler's plan.—Heat in a crucible, at a low red heat, 1 part of arsenide of cobalt, 3 carbonate of potash, and 3 sulphur. Treat the fused mass with water; sulphate of potash, sulphide of potassium, and sulphoarsenide of potassium dissolve, while sulphide of nickel remains. Treat this with nitric acid or sulphuric acid and calcine; or if copper is present, precipitate first with sulphohydric acid. The speiss which collects at the bottom of the smalts crucibles, consisting of an arsenic sulphide of nickel, may be readily purified in this way; and also Kupfernickel, the arsenide of nickel. *Thomson's plan.*—Digest pulverized speiss, an impure arsenide of nickel, in dilute sulphuric acid, mixed with some nitric acid—concentrate and crystallize. Dissolve the crystals in water, and pass sulphohydric acid through the solution to separate copper. Crystallize again, and precipitate the oxide of nickel by an alkaline carbonate. When cobalt is present, it must be separated by the process given under COBALT. The metal is obtained from the oxide by mixing it with 3 per cent. of resin, and making it into a paste with oil, and then exposing it to a strong heat in a forge in a black lead crucible (Tupputi, Ann. de Chim. 78, 133).

Protoxide of Nickel.—NiO 4·75, 38; 4·6875—37·5. Spec. grav. 5·597 (Playfair and Joule) in powder; 6·605 in crystals (Genth). *Colour.*—Blackish-gray (Tupputi), green or olive-green powder (Berthier), greenish-yellow (Erdmann), ash-gray (T. Thomson). It is not magnetic. By ignition converted into the black sesquioxide. Sulphohydric acid at a red heat converts it into sulphide. Ammonia dissolves it with difficulty, forming a peculiar *violet-blue* solution, and only in small quantities. It is very easily reduced by hydrogen, carbon, sulphur, phosphorus, and arsenic. Before the blowpipe on charcoal with carbonate of soda it is reduced; with borax and a small quantity of oxide an olive-yellow glass is formed, and with an excess, a red glass. Salt of phosphorus gives a similar glass. *Preparation.*—The protoxide is formed by calcining the nitrate of nickel, latterly in a closed vessel, at a white heat (Berthier), or by dissolving nickel in nitric acid and precipitating by potash, or an alkaline carbonate, washing, and calcining. When prepared by the last process it is apt to contain sesquioxide, which may be removed by passing hydrogen over it at a temperature of 212° (Erdmann).

Sesquioxide.—N_2O_3. Black powder by treating the hydrous protoxide with a chlorite.

Chloride.—NiCl. Yellow crystalline scales like Mosaic gold, formed by passing chlorine at a red heat over nickel, or by heating oxide of nickel with salammoniac.

Sulphide, Protosulphide, Sulphuret, Haarkies, Capillary Pyrites, Hair Pyrites.—NiS 5·75, 46·; 5·6875, 45·5. Sp. grav. 5·76 (Berthier), 5·278 (Miller), 5· (Breithaupt), 5·65 (Rammelsberg). Steel-gray or brass-yellow capillary regular 6-sided prisms, derived from an acute rhombohedron. First examined by Klaproth, who considered it pure nickel. Melts before the blowpipe, giving out sulphurous acid, leaving magnetic nickel (Berzelius); insoluble in chlorohydric and sulphuric acids; soluble in nitric acid with heat; easily soluble in aqua regia; loses little sulphur when heated in steam (Regnault), not decomposed by cold chlorine, or hydrogen at a red heat (Rose). Occurs native in Cornwall and Germany. It may be formed artificially by heating the oxide of nickel to redness in a glass tube, and passing sulphohydric acid over it. It is then dark gray, and not magnetic (Arfwedson). By igniting the sulphate with carbon, or by heating nickel, the oxide or arsenide with an alkaline sulphide (Berthier).

Ferrosulphide of Nickel.—2FeS,NiS. Spec. grav. 4·6. Bronze masses or 8-hedral? not magnetic, usually mixed with magnetic and copper pyrites. Fe 41·95, Ni 22·1, S 35·95. Norway, Germany, and Inverary, Scotland.

Arsenide, White Nickel Pyrites.—NiAs 13·125, 105; 13·065, 104·53 (Hoffmann. Pogg. Ann. 25,491). Spec. grav. 6·735 (Rammelsberg). Colour tin-white, massive. Before the blowpipe it acts like copper nickel (Berzelius). It occurs in 6-hedrons with the angles and edges truncated at Riegelsdorf in Hesse, united with cobalt and iron, with the formula 6 NiAs, CoAs, Fe As, with a tin-white colour. Before the blowpipe it yields arsenious acid, and fuses into a brittle button. With borax this button gives a blue glass, indicating cobalt; if after this the button be fused with biphosphate of soda, we have the reaction of nickel, viz. a clove-brown transparent glass in the outer, and a brown opaque glass in the inner flame (Booth, Silliman's Journ. 29, 241, and Pogg. Ann. 32, 395), not magnetic.

Diarsenide, Kupfernickel, Copper Nickel, Nickeline.—Ni_2As 16·875, 135; 16·75, 134; spec. grav. 7·655, H 5·5. Usually in copper-red masses, but also in double 6-sided pyramids and right prisms. Fracture conchoidal, uneven. Before the blowpipe it emits arsenical fumes, and melts into a bead, which darkens by exposure to the air. When roasted in the air, a yellow-green remains, which gives, with carbonate of soda and borax on charcoal, a magnetic button of nickel. Soluble in nitric acid and in aqua regia. Ni 44·2, As 54·72, Fe ·34, S ·4, Pb ·32. Occurs in Saxony, Cornwall, Lanarkshire?

Triarsenide, artificial Cobalt Speiss.—Ni_3As or Ni = 52·7, As = 44·05, MnFe,CuS = 3·25. Occurs in pinchbeck, yellow 8-hedrons, with a square base, the summit of the pyramid generally being truncated so deeply as to give the crystals the appearance of a table with bevelled edges, like yellow prussiate of potash. Brittle, not magnetic. Occurs in smelting houses, formed by the long-continued fusion of copper nickel.

Diantimonide, Antimonial Nickel.—Ni_2Sb 23·5, 188·. Spec. grav. 7·541. Ni 27, Sb 59·7, Fe ·84, PbS 12·36. Copper-red short 6-sided prisms, sometimes terminated with 6-sided pyramids or 6-sided plates, yielding a reddish-brown powder, and an antimonial sublimate before the blowpipe on charcoal; soluble in aqua regia, less easily in nitric acid; not magnetic.

Nickel Antimony Sulphide.—Spec. grav. 6·097, H 5·5. White cubes, or 8-hedrons, with the metallic lustre. S 15·76, Sb 55·11, Ni 27·7.

Bisulphoarsenide. Nickel Glance.—Ni_2AsS_2 or NiS_2 + NiAs 20·75, 166; Ni = 30·8, S = 19·29, As = 48·06, Fe 2·99, SiO_3 1·. Spec. grav. 6·097 to 6·9. Tin-white massive, yielding a grayish-black powder. When heated, sulphide of arsenic sublimes, while a mass like copper nickel remains; it is found at Loos in Helsingland, &c.

Characters of Salts of Nickel.—The soluble salts of protoxide of nickel have an emerald-green colour, while that of the insoluble salts is usually light green, and in some cases leek-green. 1. Ammonia precipitates au apple-green hydrate of protoxide and redissolves it in excess, forming a violet-blue solution. 2. Caustic potash and soda precipitate nickel salts green, becoming dark by boiling. 3. Sulphuretted hydrogen yields no precipitate with acid salts of nickel, but completely precipitates the acetate, or other salts of nickel, when an alkaline acetate is present. 4. Sulphohydride of ammonia gives a black precipitate, slightly soluble in excess. 5. Carbonate of ammonia gives a green precipitate, soluble in excess; carbonate of soda, a similar precipitate, insoluble in excess. 6. Yellow prussiate of potash, a greenish-white. 7. Red prussiate, greenish-yellow. 8. With borax, salts of nickel give a red glass in the oxidizing flame—a gray glass in the reducing flame.

Sulphate of Nickel.—NiO SO_3 7HO; NiO = 26·30, SO_3 28·16, HO 45·54 (Phillips); 17·625, 141; 17·5625, 140·5. Spec. grav. 2·037 (Kopp). *Crystallizes* in two forms—1st, Green rectangular prisms with square bases, from acid solutions; and 2d, as right rhombic prisms resembling the crystals of sulphate of zinc; the inclination of the lateral faces being to each other as 91° 10′; separates from neutral solutions. In the air, the rhombic crystals lose 1 atom water, and become white; not so the square prisms, which contain 30 per cent. of acid, or 2 per cent. more than the rhombic crystals. When heated, the crystals swell up, but do not melt, and assume a yellow colour when the water is expelled. At 218° they lose 6 atoms of water, retaining the seventh till the temperature reaches 535°, when the salt becomes anhydrous; by ignition the acid is removed. Soluble in 3 cold water; insoluble in alcohol and ether; precipitated by the strongest acetic acid from its solution in water (Persoz). When the rhombic crystals are exposed in a close glass vessel to sunshine, or when placed on paper and exposed to the sun, they change their crystalline form and become 8-hedrons. Obtained by dissolving oxide of nickel in oil of vitriol and crystallizing.

Carbonate, Native.—Thin green crystalline layers in prisms on the surface of chrome iron ore from Texas, Pennsylvania, and New Jersey (R.D.T. 1847), generally contaminated with magnesia and lime. Silliman, jun., has stated its formula to be $NiOCO_2$ 2(NiO 3HO), and its spec. grav. 2·57 to 2·693.

Nickel-Bismuth, Sulphide of.—Ni 40·65, Bi 14·11, S 38·46, Fe 3·18, Co ·28, Cu 1·68, Pb 1·58. Gray 8-hedrons and cubes (Grünau).

For *estimation of nickel* see COBALT.

Nicotianine. The nitrogenous essential

principle supplying nicotine, by distillation with potash, has been so named. C 71·52, H 8·23, N 7·12, O 13·13.

Nicotine. $C_{20}H_{14}N_2$ or $C_{10}H_7N$. Spec. grav. 1·027, of vapour 5·618; B.P. 585°. Colourless, limpid oil; powerful base and poison; alkaline reaction; mixes with water; soluble in alcohol, ether, and oils; forms with acids difficultly crystallizing salts; 100 nicotine dissolve 10·58 sulphur at 212°; chlorine makes it blood-red, with evolution of chlorohydric acid, and depositing needles at 46½°; nicotine rapidly absorbs water from air. Different tobaccos contain different quantities of nicotine: Virginia, 6·87 per cent.; Kentucky, 6·09; Maryland, 2·29; Havannah, under 2·; Lot (France), 7·96; Alsace, 3·21. It is best prepared by concentrating the aqueous decoction of tobacco to the consistence of syrup, and mixing while hot with 2 vols. alcohol; a thin layer collecting on the top is separated by decantation; the alcoholic fluid is distilled, and the residue treated again with alcohol. This extract contains the nicotine; it is separated from the spirit, agitated while hot with potash, and after cooling with ether, which takes up the nicotine, with some yellow-colouring matter; the nicotine is precipitated from the ether solution by powdered oxalic acid as oxalate in a syrup, and washed with ether, agitated first with potash, then with ether, and lastly the ether is distilled off in the water bath, and the residue exposed to a current of dry hydrogen at 284° for several days, and then distilled at 356°; at the boiling point it decomposes; dry snuff affords 2 per cent. nicotine. A method of determining the amount of nicotine in tobacco is simple (Schloesing); 10 grammes (154·38 grs.), the moisture of which is previously ascertained, are exhausted with ammoniacal ether in a displacement apparatus (see DISPLACEMENT), having a tubulated balloon of the capacity of one-sixth litre; the ethereal extract is boiled to remove ammonia, and the remaining ether evaporated in the air; the residue is neutralized with sulphuric acid of known strength. If a be the sulphuric acid used and b the nicotine ($C_{40}H_{14}N_2$), we have the proportion $5 : 20{\cdot}25 : a :: b$.

Nigrine. A black variety of titanic acid, or rutile.

Nihil Album. An old term for flowers, or oxide of zinc.

Ninaphthase. $C_{20}H_7\ N\ O_4$.

Ninaphthese. $C_{20}H_6\ N_2O_8$.

Ninaphthine. $C_{40}H_{11}N_5O_{20}$.

Ninaphthise. $C_{20}H_5\ N_3O_{12}$.

Niobic Acid. NiO_3? Spec. grav., *amorphous*, 5·2545 to 5·262; *crystalline* 4·664 to 4·7633. Yellow powder when hot, colourless or white when cold; after heating it is highly shining, while columbic acid is a dull powder; converted by ammonia gas into a black powder, by sulphohydric acid into a black sulphide; little soluble in chlorohydric acid; slightly soluble in oxalic and fluohydric acids; soluble in fused bisulphate of potash or ammonia; expels carbonic acid from carbonate of soda when fused with it; alkaline niobates are soluble; precipitated by acids at usual temperatures, while columbates require heat; a blue colour is formed in niobate of soda with HCl and zinc. Fused with alkaline carbonates forms acid salts; soluble in water and in potash or carbonate of potash; obtained from the *chloride*, which is formed by heating the Bavarian columbite (tantalite), mixed with charcoal in a current of chlorine; two chlorides are thus formed, *yellow* chloride of pelopium, which is very fusible and volatile, and *white* chloride of niobium, which is infusible and less volatile; and these chlorides, when placed in water, yield *pelopic* and *niobic* acids; the acid obtained from the infusible white chloride is again mixed with charcoal and heated in chlorine, and the process repeated. B.B. with borax niobic acid yields a colourless bead, becoming opaque on flaming; grayish-blue in the inner flame; with salt of phosphorus a colourless bead in oxidizing flame, a violet and blue in inner flame.

Niobite. A synonyme of Torrelite, or Niobic-columbite.

Niobium. Black powder converted into niobic acid when heated in air; attacked by a mixture of fluohydric and nitric acids, not by aqua regia; obtained by passing dry ammonia over chloride of niobium.

Nitracrole. Colourless oil, decomposing at 212°; by the action of nitric acid on choloidanic acid and oenanthole.

Nitramidine. *Xyloidine.* $C_{12}H_9O_9NO_5$, or $C_{12}H_9NO_4,O_{10}$. A substance produced by the action of nitric acid on paper. Gun cotton is *ternitramidine*, $C_{12},H_7 3NO_4O_{10}$, or $C_{12}H_7O_7$ $3NO_5$.

Nitraniline. $C_{12}H_6NO_4N$. F.P. 230°; B.P. 545°. Yellow needles heavier than water, subliming at 212° in plates; reaction neutral; colours pine wood and the skin yellow; smell when heated like aniline; a weak alkaloid, converted by bromine into nitrodibromaniline; by nitric acid by boiling into picric acid; by chlorocyanogen into dinitromelaniline; little soluble in cold, more soluble in boiling water; quite soluble in alcohol and ether, with a red colour; obtained by dissolving dinitrobenzide in alcohol, saturating with ammonia gas and then with sulphohydric acid; chlorohydric acid is added and the liquor concentrated and filtered; potash precipitates the base which is crystallized out of alcohol and ether; the dinitrobenzide is prepared by adding benzine drop by drop to equal volumes of nitric and sulphuric acids, and crystallizing from alcohol.

Nitraniline-Urea. $C_{14}N_3H_7O_6$. Yellow needles by acting on an ether solution of nitraniline with chloride of cyanogen.

Nitranisic Acid. $C_{16}H_7NO_4O_6$. Crystalline; soluble in hot water and in alcohol; by the continued action of nitric acid of 1·33 on anise camphor. Another acid, binitranisic acid, $C_{16}H_6 2NO_4O_6$, exists.

Nitraniside. $C_{20}H_{10}N_2O_{10}$. By the action of nitric acid on anise camphor.

Nitranisidine. $C_{14}H_8NO_4NO_2$. Red needles, insoluble in cold, soluble in hot water and ether; forms crystalline salts; by the action of sulphide of ammonium on dinitranisole.

Nitranisole. $C_{14}H_6,2NO_4,O_2$. Yellow liquid by nitric acid on anisole; by fuming acid a *di-* and *tri-*ole are formed.

Nitranthracenese, or nitrate of anthracenase, &c. See ANTHRACENE.

Nitrates. Combinations of nitric acid with oxygen bases. They are formed on the type of semelhydrate of nitric acid ($HONO_5$), having the formula $BONO_5$, and without water of crystallization. With magnesia and copper the salts are formed on the type of the quaterhydrate ($4HONO_5$), prismatic nitrate of copper being $CuO\ NO_5\ 3\ HO$. The nitrates are all soluble in water, and can be decomposed by heat; some of them yielding oxygen and nitrates; others oxygen and hyponitric acid, or protohydrate of nitric acid. When heated with charcoal the nitrates deflagrate. When mixed with metals and sulphuric acid, they yield binoxide of nitrogen, which changes in contact with the oxygen of the air into orange fumes of quaternitrous acid (NO_4). With muriatic they form aqua regia, which dissolves gold leaf. When nitrates are mixed with sulphate of indigo and sulphuric acid, the blue colour is removed. In the neutral nitrates the oxygen of the acid is 5 times that of the base, but there are numerous basic salts in which the oxygen of the base is 2, 3, and 6 times greater than in the neutral nitrates. There is no acid nitrate. To distinguish the presence of a minute quantity of a nitrate, the protosulphate of iron is dissolved in water containing sulphuric acid. This is added to the nitrate, and a plate of iron introduced into the mixture. If a nitrate is present, the liquid is coloured pink or brown (Pelouze).

Nitrazobenzide. $C_{24}N_3H_9O_4$. Yellow needles by nitric acid on azobenzide; in the same action a dinitrazobenzide occurs.

Nitrazoxybenzide. $C_{24}N_3H_9O_6$. Yellow needles along with the preceding.

Nitre. For potash and soda nitres, see POTASH and SODA SALTS.

Nitrethionessile. $C_{25}H_7N_2O_8S$.

Nitric Acid. *Dissolutive water of Geber, Acid Spirit of Nitre, Aquafortis, Azotic acid.*

Source.—Nitric acid occurs extensively in nature, but always in union with bases in the form of nitric acid salts. Friestley and Cavendish showed that the oxygen and nitrogen of the atmosphere may be converted into nitric acid by connecting a eudiometer on one side with the prime conductor of a machine and the other by a chain with the earth, and placing in the tube litmus solution or potash. On turning the machine nitric acid is formed. Nitrates were detected in well water in 1754 by Margraf.

Preparation.—Nitric acid is usually prepared from nitrates of potash and soda. 1. *Protohydrate of Nitric acid.*—$HO\ NO_5$ 7·875, 63· = 86· acid, 14· water; spec. grav. 1·522; boiling point 186°·75. In preparing this strongest form of the acid if we use only one atom of sulphuric acid the mixture in the retort will yield, after heating, bisulphate of potash, protohydrate of nitric acid, and the half of the nitre not decomposed. When it is distilled the half of the nitric acid passes over in the form of the protohydrate at a temperature of 269°·5, after which bisulphate of potash and nitre react upon each other. The residue in the retort becomes solid and cannot be distilled as nitric acid, because no water is present and nitric acid cannot exist without water at common temperatures. If we were to continue the heat to the solid mass binoxide of nitrogen and oxygen in union with it as hyponitric acid (NO_4) would alone pass over. The proper proportions and nature of the reaction are exhibited in the following scheme:—

2 Sulphuric acid. Nitrate of potash.
5 1·125 5 1·125 5·90 6·75
SO_3 HO SO_3 HO KO NO_5

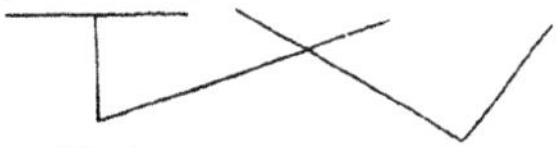

$KO\ SO_3$
$HO\ SO_3$
17·025 Bisulphate of potash.

$HO\ NO_5$
7·875 Protohydrate of nitric aid.

The proper amounts of the ingredients are $12\frac{3}{4}$ parts of oil of vitriol and 12·65 of nitrate of potash, and the product amounts to $7\frac{7}{8}$ of nitric acid (or 100 parts nitre and 98·6 acid). Convenient quantities in a laboratory will be found to be $1\frac{1}{2}$ lb. oil of vitriol, $1\frac{1}{2}$ lb. pure saltpetre. Süersen and Bucholz (1819 Taschenb. 201) first pointed out the importance of using 2 atoms of sulphuric acid. When nitrate of soda is employed the proportion of the ingredients is $12\frac{3}{4}$ oil of vitriol and 10·622 (3·872 NaO·6·75 NO_5) nitrate of soda. When 2 atoms of nitrate of potash are distilled with 2 atoms of sulphuric acid only 1 is decomposed ($2\ KONO_5\ 2\ HOSO_3 = KOSO_3, HOSO_3 + HONO_5 + KONO_5$). A high temperature is required to decompose nitrate of potash by means of bisulphate of potash; so high, indeed, that the resulting nitric acid is decomposed into hyponitric acid and oxygen ($KONO_5, HOSO_3KOSO_3 = 2\ KOSO_3, NO_3, O, HO$). The retort in laboratory experiments should be of glass, and the receiver a Wolfe's bottle with a safety tube, or a common receiver surrounded with a cloth upon which cold water is allowed to drop continuously from a vessel supplied with a stop-cock. The distillation should be performed by means of a sand heat, and the temperature need not exceed 260° or 300°. *Characters.*—When

nitre completely destitute of water is employed in this operation the nitric acid produced contains only 14 per cent. of water, and possesses a spec. grav. of 1·522. Fuming nitric acid ($HONO_5$) may also be prepared by distilling the nitric acid of commerce (spec. grav. 1·33) with about 3 times its volume of oil of vitirol ($HOSO_3$). Only half of the acid employed is distilled over. It is distilled a second time with its own volume of sulphuric acid, and a third time with some nitrate of barytes to retain sulphuric acid which may have passed over. If 5 parts of oil of vitriol be used in the first operation the nitric acid decomposes into nitrous acid and oxygen. Convenient quantities are 8 oz. nitric acid of commerce and 24 oz. oil of vitriol. It is then to be redistilled with its own volume of oil of vitriol; it becomes solid at — 40°; it is a colourless liquid, but when exposed to the light it becomes yellow, giving out oxygen, while the nitrous acid unites with another part of the nitric acid. When the colour is yellow from the presence of nitrous acid, the acid may be rendered colourless by gentle heat, but the density is reduced to 1·521, or the nitrous acid may be removed by distilling the coloured acid with nitrite of lead in fine powder. The portions containing nitrous acid pass over first, when the receiver is charged and the pure nitric acid collected in a second receiver. The last portions may also contain nitrous acid. It fumes when exposed to the air, has a peculiar smell, an intensely acid taste, and reddens vegetable blues. When placed on the skin it deprives it of all organization, and leaves a yellow stain, which only disappears when the cuticle desquamates. From this property it has been employed to afford a yellow border to the blue or green woollen cloths used in this country as table covers. There is no doubt that in this action there is a chemical union between the gelatinous part of the tissue and a portion of the nitric acid. It is on the same principle that nitric acid acts when it is used in surgery to destroy morbid growths and to clean the surface of phagedenic ulcers. 2. *Quaterhydrate of Nitric Acid.*—$4HONO_5$ 11·25, 90·. Spec. grav. 1·422; boiling point 253½°; nitric acid 60, water 40. If we add to the fuming nitric acid ($HONO_5$) or protohydrate a small quantity of water and distil, introducing a thermometer into the retort, the fluid which comes over at first contains more acid than the liquid in the retort. The thermometer goes on ascending till it reaches 253°·4, when it remains stationary, and the acid in the retort contains 40 per cent. of water. We may obtain the same quaterhydrate in the following manner:—If we add 20 per cent. of water to the protohydrate and distil, the thermometer will at first stand at 212°, but gradually it will rise to 253°·4, and remain stationary, showing that the acid is a permanent compound. The first portions distilled are nearly pure water, but it gradually becomes stronger, and when the boiling point in the retort is 253°·4, the acid which passes over is of the same strength.

Other Hydrates.—Another hydrate containing 2 atoms of water ($2HONO_5$) obtained to the extent of 2 ounces by distilling 2¼ lbs. of very concentrated acid, has been described, with spec. grav. of 1·484 and 25 per cent. of water, but its existence is doubtful. A third hydrate procured by distilling the last third of the nitric acid of commerce, has been stated to contain 4½ atoms of water or 42·77 per cent. of water with a spec. grav. of 1·405.

Pharmaceutical Acid.—The London pharmacopoeia (1836) recommended 2 lbs. nitre and 2 lbs. oil of vitriol to be distilled together. The acid passing over has a density of 1·503 to 1·504. The dilute acid of London consists of 1 ounce strong acid and 9 ounces water, spec grav. 1·080. 100 grains saturate 31 grains of crystals of carbonate of soda; each fluid drachm contains nearly 9 grains of concentrated acid and saturates 19½ grains of crystals of carbonate of soda (Phillips).

Commercial Acid.—On the larger scale nitric acid is prepared in iron cylinders identical with those described as being used in the preparation of chlorohydric acid. The charge for a cylinder of the dimensions described is 190 lbs. potash nitre, upon which are run on by a funnel 184 lbs. of sulphuric acid ($HOSO_3$). The mouth-piece is luted on in the usual manner. The heat is so applied as to render the inferior portions of the cylinder red hot, and when red vapours appear the fire is to be urged. These shortly disappear and the process terminates. Soda nitre has long replaced potash nitre in this country. My friend Mr. William Blythe recommends the proportions 600 lbs. nitrate of soda and 500 lbs. oil of vitriol (spec. grav. 1·750).

Anhydrous Nitric Acid.—(NO_5). 6-sided right prisms with a rhombic base, fuses at 85°, and boils at 113°; obtained by passing dry chlorine over fused nitrate of silver in close vessels.

Tests.—1. When chlorohydric acid is mixed with nitric acid, gold leaf is readily dissolved in the cold state. Neither of the acids singly has any effect on gold. The mixture of 1 nitric acid and 3 to 4 chlorohydric acid constitutes *Aqua regia* or *Nitro Muriatic Acid.* With reference to the action by which it dissolves metals, Davy considered the efficient cause was chlorine, which is liberated as can be detected by the odour. According to Millon aqua regia if not concentrated or heated does not attack arsenic and antimony, even a current of chlorine has no action; nitrous acid must be present. Chlorohydric acid by acting on nitric acid produces nitrous acid. Platinum is not dissolved. According to E. Davy, chloronitrous acid (NO_2Cl_2) is formed and dissolves the gold. According to Gay Lussac, gold is dissolved by the free chlorine liberated in the action of nitric on chlorohydric acid (NO_2O_3 and 3 HCl become NO_2 3 HO and 3 Cl). Two gases occur in this action, NO_2Cl_2 and NO_3Cl. 2. When a dilute solution of indigo in sulphuric acid is boiled with free nitric acid, the blue colour disappears. But as similar results occur with chloric, bromic, and iodic acids, it is necessary to secure the absence of these substances before drawing any conclusion. 3. An amalgam of 1 atom of zinc and 5 of mercury digested with protochloride of iron produces on the addition of a nitric acid salt a black spot (Runge). 4. If the liquid supposed to contain the nitric acid be mixed with a few drops of protosulphate of iron solution and then an equal volume of pure sulphuric acid be poured in cautiously so as to form a heavy layer at the bottom of the test-glass, when nitric acid is present, even in very minute quantity, a pink or blackish ring will be formed at the point of junction of the sulphuric acid with the aqueous solution. 5. Oil of vitriol with a few drops of a nitrate solution when mixed with a minute quantity of powder of brucine becomes blood-red. Narcotine likewise yields first a yellow and then a red colour. 6. When the nitrate is in a solid state it is to be introduced into a test-tube and heated with sulphuric acid, fumes of nitric acid are soon perceptible to the smell. 7. If a nitrate be mixed with twice its weight of caustic potash or soda and a small quantity of charcoal, and be heated to redness in a test-tube, carbonate of the alkali is formed, and ammonia is disengaged, which can be detected by the smell and by its changing moistened turmeric paper to a brown colour. 8. When nitric acid is heated with copper turnings binoxide of nitrogen is evolved.

Table of the Strength of Nitric Acid.

Sp. Grav.	Acid p. c.	Sp. Grav.	Acid p. c.
1·522	86·0	1·215	29·5
1·498	84·2	1·152	21·5
1·478	72·9	1·122	17·5
1·434	62·9	1·105	15·14
1·422	60·0	1·0708	10·36
1·376	51·9	1·0485	7·17
1·353	48·7	1·0375	5·58
1·300	40·6	1·0212	3·18
1·283	38·5	1·0159	2·39
1·252	34·2	1·0053	0·79
1·234	31·8		

Analysis.—The amount of free nitric acid in a solution may be determined by weighing out 100 grains of the acid fluid, introducing it into a large stoppered flask of hard glass. 1000 grains of dry oxide of lead are then placed in the flask, and the lead dissolved in the nitric acid. The whole is evaporated carefully to dryness. The last traces of water may be vaporized and sucked out by means of a dry tube. The nitric acid unites with a portion of oxide of lead. Hence the increase in the weight of the original lead oxide gives the amount of nitric acid in the 100 parts of fluid.

Nitricinnolic Acid. $C_{14}N_5NO_8$. By boiling NO_5 on cinnamic acid.

Nitric Oxide. Binoxide or deutoxide of nitrogen or azote.

Nitrification. The process in nature of the formation of nitric acid. At the present time it is supposed that ammonia is in the first instance evolved from animal matter which seems requisite for the production of nitre, and by the oxidation of ammonia nitric acid results. The formation of nitric acid from ammonia is abundantly illustrated by the amount of this acid present in city wells, which appears to be derived from the oxidized ammonia of urea filtering with the urine into the wells through the porous earths and rocks.

Nitriles. This term is applied to those nitrogen compounds which may be represented as ammoniacal salts, deprived of all their hydrogen and oxygen, which can be removed in the proportion to form water, and is the next step after an amide. An ammonia salt, as oxalate of ammonia, becomes when heated, by the removal of HO, oxamide ($NH_2H,C_2O_2O = HO,NH_2,C_2O_2$), by further heat oxamide loses 2 HO and becomes cyanogen ($NH_2C_2O_2 = 2$ HO, NC_2). Cyanogen is the *nitrile* radical of oxalate of ammonia. The more compound nitriles have been viewed as cyanogen compounds in which the cyanogen as chlorine in chloride of ethyle cannot be detected by usual reagents, and thus *acetonitrile*, $C_4H_3N =$ acetate of ammonia — 4 HO will be C_2H_3, C_2N, or cyanide of methyle. In confirmation of this view it has been noticed that the boiling points of valeronitrile ($C_{10}H_9N$ 257° $= C_8H_9$, C_2N, cyanide of butyryle), and cyanide of amyle, ($C_{12}H_{11}N$ 294°·8) differ by 37°·8, which corresponds with a difference of C_2H_2 in composition. Some of the other nitriles are chloroacetonitrile, benzonitrile, butyronitrile, cumonitrile, &c. The term *nitrile* bases is applied to such organic alkaloids artificially prepared as are formed on the type of ammonia, when all the hydrogen has been replaced by other bases, and nothing remains of the original ammonia except the nitrogen. Thus NH_3, typified in ammonia-trimethyle-amine, is NMe_3.

Nitrilophile. $C_{46}H_{14}N_5O_{12}$ 4 HO.

Nitrindine. $C_{16}H_4N_2O_7 = C_{16}H_4NO_4O$, NO_2. Violet powder by nitric acid on indine and hydrindine.

Nitripicryle. $C_{42}H_{11}N_4O_{16}$.

Nitrites. The nitrites of potash and soda may be obtained by heating the nitrates of potash and soda till their solution precipitates nitrate of silver, brownish. The difficultly soluble salt of nitrite of silver is then dissolved in water and filtered; it crystallizes out on cooling. By mixing the nitrite of silver with the metallic chlorides the nitrites of these bases may be obtained. Nitrite of potash may be formed in a hard glass retort by heating the nitrate of potash above 600°, stopping the action when nitrogen begins to be given off; the nitrite may be dissolved out from the nitrate by alcohol. Nitrite of potash is deliquescent; nitrites of barytes and strontian may be obtained in crystals by forming first the nitrite of soda by heat from the nitrate, decomposing it by nitrate of silver and treating the crystallized nitrite of silver by the chlorides of barium and strontium.

Nitroangelic Acid. *Nitrovalerianic?* $C_{10}H_7NO_8$, or $C_{10}H_9NO_8$. Rhombic plates, subliming at 212°, soluble in water; by the action of nitric acid on valerianic acid

Nitrobenzamide. $C_{14}H_6NO_4, NO_2$. Yellow needles by long fusion of nitrobenzoate of ammonia.

Nitrobenzanisidide, or ine. $C_{28}H_{12}NO_4, NO_4$. Yellow insoluble needles by chloride of benzoyle on nitranisidine; it is analogous to benzamide or benzanilide.

Nitrobenzide. *Nitrobenzole.* $C_{12}H_5NO_4$. See Benzole.

Nitrobenzoenase. *Protonitrated Benzoene*, or *Toluidine.* See Nitrotoluide.

Nitrobenzoenese. *Dinitrotoluole.* $C_{14}H_6N_2O_8$.

Nitrobenzoic Acid. See Benzonitric Acid. *Nitrobenzoic acid, benzoated anhydrous.* By chloride of benzoyle on nitrobenzoate of soda.

Nitrobenzoile. *Nitrobenzoyle.* $C_{14}H_5N$. White crystals; the residue insoluble in alcohol in the preparation of azobenzoyle.

Nitrobenzole. See Benzole.

Nitrobenzoyle Hydride. $C_{14}H_5,NO_4,O_2$. White needles by nitric acid on oil of bitter almonds.

Nitrobichlorophenic Acid. *Nitrobichlorocarbolic Acid.* $C_{12}H_2Cl_2NO_5HO$. Yellow oblique rhombic prisms, slightly soluble in water; more soluble in boiling alcohol and ether; by acting on coal naphtha, which boils between 356° and 392°, with chlorine, and then with nitric acid.

Nitrobromophenisic Acid. *Binitrobromocarbolic Acid.* $C_{12}H_2Br,N_2O_9$. By the action of bromine on nitrophenisic or nitropicric acid.

Nitrobutyric Acid. See Butyronitric Acid.

Nitrocaprylonic Acid. *Nitroenanthylic Acid?* Yellow oil by nitric acid of 1·4 on caprylone.

Nitrocarbolic Acid. *Nitrophenole.* C_{12},H_5NO_4,O_2. Fine crystals by nitric acid on carbolic acid. See Nitrophenesic Acid. *Ternitrocarbolic Acid*, or *Nitrophenisic Acid.* See Picric Acid.

Nitrocholic Acid. $C_2HN_4O_9$. An oily acid by the action of nitric acid on choloidic acid.

Nitrocinnamic. C_{18},H_6NO_4,O_3HO. By cold nitric acid on cinnamic acid; its salts detonate.

Nitrocinnanisidide. $C_{32}H_{14}NO_4, NO_4$. Yellowish needles by nitric acid on chloride of cinnamyle.

Nitrococcussic. $C_{16}H_3N_3O_{16}$ 2HO. Yellow rhombic prisms by nitric acid of 1·4 on carminic acid; it is bibasic.

Nitrocodeine. $C_{36}H_{20}NO_4NO_6$. Crystalline powder by heating codeine, under 212°, with nitric acid of 1·060.

Nitrocoumarine. C_{18},H_5NO_4,O_4. F.P. 338°. Fine needles, little soluble in water, alcohol, and ether, by cold nitric acid on coumarine.

Nitrocumidinamide. $C_{14}H_5O_2,C_{18}H_{11}N_2O_4$. By chloride of benzoyle on bromide of nitrocumidine.

Nitrocumidine. $C_{18}H_{12}NO_4,N$. F.P. below 212°. Radiated yellow scales after fusion; insoluble in water; soluble in alcohol and ether; from binitrocumole.

Nitrocuminic Acid. $C_{20},H_{11}NO_4,O_4$. Yellowish needles, insoluble in water, soluble in alcohol and ether, by fuming nitric acid on cuminic acid.

Nitrocumole. $C_{18}H_{11}NO_4$. By nitric acid on cumole.

Nitrodibrom-aniline. $C_{12}H_4Br_2NO_4,N$. By the action of bromine on nitraniline.

Nitro-erythro-mannite. $C_{10}H_65NO_4,O_{11}$. F.P. 142°. Plates by cold fuming nitric acid on erythro-mannite; burns when heated, and detonates when struck, corresponding therefore with nitro-mannite.

Nitro-euxanthic Acid. See Nitropurreic Acid.

Nitrogen. N 1·75, 14. Mephitic air (Dr. Rutherford, 1772), Azote (Lavoisier, 1787), Alkaligene, Nitrigene (Lavoisier, 1789), Nitrogen (Chaptal, 1790?).

Physical Characters.—A colourless gas, without taste and smell. Spec. grav. ·9722 (Thomson), ·972 (Dumas), ·9713 (Regnault). Theoretically $1{\cdot}75 \times {\cdot}5528 = {\cdot}9674$; 100 cubic inches weigh 30·076 grs. (Thomson), 30·056 grs. (Regnault); 854 times lighter than water. Refractory power 1·02, that of air being 1; 100 cubic inches of water previously boiled absorb 1·47 cubic inch or 2½ of nitrogen. It has not been condensed by a pressure of 50 atmospheres. It is not combustible or respirable.

Preparation.—As common air consists of oxygen and nitrogen, if we cause a combustible body to burn in a limited portion of air over water, the oxygen will be removed, and the nitrogen

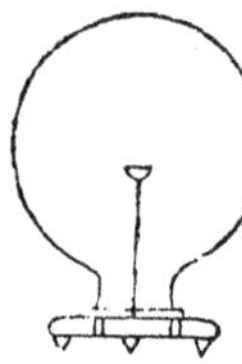

will remain. 1. A bit of phosphorus may be placed in a small cup of copper supported on a stand, in a water trough; when touched with a hot wire it will ignite. A jar of common air is then inverted over it, the mouth of which just dips under water. The oxygen unites with the phosphorus, and white fumes of phosphoric acid are produced which gradually dissolve in the water; the water of the trough rises in the jar, and occupies about one-fifth of the bulk of the jar. The residual gas amounting to $\frac{4}{5}$ is nitrogen gas. 2. A cork is floated in water; a piece of copper foil is placed on its upper surface to receive a piece of ignited phosphorus, and rapidly covered by a jar or receiver as in the previous experiment. 3. The oxygen may be removed by slow combustion by inserting a piece of dry phosphorus at the end of a platinum wire in common air over mercury in a graduated tube in a warm room, the oxygen disappearing, and the mercury rising in the tube. 4. Chlorine gas may be passed through caustic ammonia (NH_3). The chlorine is evolved from a flask containing black oxide of manganese and chlorohydric acid, and passes into a second flask by means of a bent tube partially filled with dilute caustic ammonia. The chlorine unites with the hydrogen of the ammonia forming chlorohydric acid, and the nitrogen is liberated. $4NH_3 + Cl_3 = N + 3(NH_3HCl)$. When the ammonia is strong slight explosions occur, which are avoided by dilution.

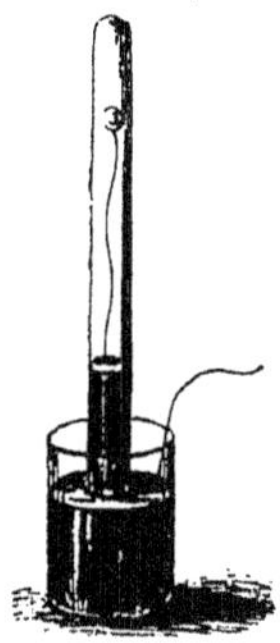

Protoxide of Nitrogen or *Azote* (Thomson), *Dephlogisticated Nitrous Gas* (Priestley, 1776), *Nitrous Oxide* (Davy, 1800), *Oxidized Azote*, *Laughing Gas*.

	Vols.	Spec. Grav.	Ats.	Wts.		P. C.
Nitrogen	□ 1	·9713	1	1·75	14·	63·63
Oxygen	□ $\frac{1}{2}$	·5528	1	1·	8	36·37
		1·5241		2·75	22·	100·

Theoretical spec. grav. $·5528 \times 2·75 = 1·5202$; refractory power 1·710; Ann. Chim. 31·166. Experimental spec. grav. 1·5269 (Thomson).

Characters.—A colourless and invisible gas, possessing the mechanical properties of common air. It has a peculiar and agreeable smell, and a sweetish taste. 1. It is not a combustible gas, but is eminently a supporter of combustion. A glowing taper without flame, or piece of red charcoal, bursts into flame when introduced into the gas. 2. Phosphorus, when touched with a hot wire, or previously ignited, burns with great brilliancy in this gas. 3. Sulphur, and 4. iron wire, burn in protoxide of nitrogen as in oxygen gas. 5. Charcoal likewise burns at the expense of the oxygen. 6. A solution of chloride of strontium in alcohol, when ignited and immersed in the gas by means of a spoon, burns with a red flame. 7. A solution of muriate of copper in alcohol gives a blue flame. In all these cases the substances require to be heated to a higher temperature before immersion than in the case of oxygen. When this gas is exposed to a strong heat, it is resolved into nitrogen and quaternitrous acid ($4NO = NO_4$ and N_3). When heated with spongy platinum, it changes into nitric acid and nitrogen ($5NO$ becoming NO_5 and N_4). Protoxide of nitrogen is not changed by admixture with air; spongy platinum, when inserted in a mixture of equal volumes of this gas and hydrogen, becomes red hot, and forms water and nitrogen. 100 vols. of water absorb 78 of gas.

Fluid State.—When compressed by a pressure of 50 atmospheres at 45°, it becomes a colourless fluid (Faraday), or by a pressure of 50 atmospheres at 59° (Natterer), conducted by means of an iron pump. The liquid is sweet, amounting to $\frac{1}{400}$ of its original bulk, destroys the skin, and may be kept some hours under the usual pressure of the atmosphere; when placed on the hand it burns it. Metals, when dipped into it, produce a hissing sound; potassium is not altered by it; it mixes with ether and alcohol; sulphuric and nitric acid and water freeze in contact with it. *Solid.*—By exposing the liquid to a temperature of—150°, it becomes a clear crystalline solid, blistering the hand.

Effects on Animals.—Like oxygen, the protoxide of nitrogen may be respired, but the effects of the two gases are very distinct upon the animal system, although the result varies in individuals of the same species. A rabbit, placed in the gas, exhibits no symptoms of uneasiness for some time, but after the lapse of a short period it becomes excited, pants, falls prostrate, and dies. But a somewhat similar effect is produced even by oxygen. The effects on man are quite different. Mr. Southey says—When he took the bag from his mouth he immediately laughed; the laugh was involuntary, but highly pleasurable, accompanied by a thrill, and a tingling in the toes and fingers, a sensation perfectly new and delightful.

Preparation.—1. The protoxide of nitrogen may most conveniently be prepared from nitrate of ammonia, a salt which is formed by saturating the carbonate of ammonia of the shops with nitric acid. On evaporation the nitrate crystallizes in long prisms. The nitrate deprived of free acid by the addition of an excess of carbonate of ammonia, is to be introduced into a retort, and heated up to the temperature of about 400°. It speedily melts and evolves a gas in great abundance, which is the protoxide of nitrogen. If the salt is decomposed with too great rapidity, nitrogen, or binoxide of nitrogen, will be mixed with the protoxide, and an explosion may even occur. The action is as follows:—$NO,O,O_3NH_3HO = 2NO,4HO$; where an atom of nitric acid, and an atom of ammonia, with 1 of water, are resolved into 2 protoxide of nitrogen, and 4 of water, or 10 grs. of the salt yield $5\frac{1}{2}$ of the gas. *Analysis.*—If we fill a graduated tube retort with 1 vol. of protoxide, introduce into it potassium, and apply heat to the

metal, the potassium unites with the oxygen of the protoxide, but the gas does not diminish in volume, and pure nitrogen remains.

Binoxide or *Deutoxide of Nitrogen*, or *Azote* (Thomson). *Nitrous Air* or *Gas* (Priestley). *Nitric Oxide.*—

	Vols.	Spec. Grav.	Ats.	Wts.		Per Cent.
Nitrogen,	$\frac{1}{2}$ □	·4856	1	1·75	14	46·67
Oxygen,	$\frac{1}{2}$ □	·5528	2	2·	16	53·33
		1·0384		3·75	30	100·

Theoretical spec. grav. = (·5528 × 3·75) ÷ 2 = 1·0365. Refractory power 1·03. *Characters.*—It is colourless, and similar to atmospheric air in its physical characters. It is impossible to ascertain whether it has any taste or smell, as it combines with an additional dose of oxygen whenever it is exposed to the air, and forms red fumes, which are very corrosive. It has no action on litmus, and seems to be a neutral body. This gas is not combustible, but it supports the combustion of some bodies. Phosphorus, when introduced into this gas in a state of active inflammation, burns with almost as much vividness as in oxygen (Davy, Research. p. 135). A taper and burning sulphur are extinguished in it. If some drops of sulphide of carbon are poured into a jar filled with the binoxide, and a taper applied, combustion is produced, accompanied with a blue flame, and the deposition of sulphur. Homberg's pyrophorus, when introduced into it, takes fire spontaneously. When mixed with hydrogen it burns with a green-coloured flame.

Preparation.—1. The method of obtaining this gas depends on the mode in which nitric acid decomposes when brought in contact with metals, and on the stronger affinity of 2 atoms of oxygen than of the remaining 3 for nitrogen. Nitric acid can only exist at common temperatures in union with water as a base. If we heat nitric acid so as to drive off the water, the acid decomposes and yields orange-coloured fumes of hyponitric acid (NO_4). The usual method of preparing binoxide of nitrogen is to bring cold nitric acid diluted to the specific gravity 1·131, or 1·25, in contact with mercury, silver, or copper. The latter is commonly employed, but the purest gas is obtained from mercury. It is necessary to use dilute acid, since nitrogen is evolved when the acid is concentrated, together with protoxide of nitrogen and nitrous fumes. The decomposition may be effected in a retort, or in a flask with a bent tube, or in a Wolfe's bottle, care being taken not to collect the gas until the red fumes which first appear cease to be formed. The action is as follows:—4 atoms of nitric acid, and 3 of metallic copper are resolved into one of binoxide and 3 of nitrate of copper—

Nitric acid.	3 Copper.	3 Nitric acid.
1·75, 2 3	12·	20·25
NO_2O_3	3 Cu	$3\ NO_5$
NO_2	$3\ (CuONO_5)=$	

2·75 Binoxide of nitrogen. 35·25 Nitrate of copper. *Analysis.*—This gas may be analyzed in the same way as the protoxide over mercury.

Nitrous Acid, Ternitrous Acid, Hyponitrous Acid, Azotous Acid.—NO_3. (Discovered by Gay Lussac).

	Vols.	Spec. Grav.	Ats.			
Nitrogen,	1	·9713	1	1·75	14·	36·84
Oxygen,	$1\frac{1}{2}$	1·6584	3	3	24·	63·16
		2·6297		4·75	38·	100·

Characters.—A blue fluid, boiling nearly at 32°; it decomposes readily when water is brought in contact with it into nitric acid and binoxide of nitrogen ($3NO_3$ become $NO_5,2NO_2$). *Preparation.*—1. When $\frac{1}{2}$ vol. oxygen is mixed in a wide tube with 2 vols. binoxide of nitrogen, a compound is formed which is immediately absorbed by water, and therefore consists of NO_3, because, as it contains 1 vol. nitrogen and $1\frac{1}{2}$ oxygen, and as $\frac{1}{2}$ vol. oxygen corresponds with 1 atom, it is obvious that 1 vol. nitrogen, and $1\frac{1}{2}$ vol. oxygen, must be equivalent to 1 atom nitrogen, and 3 atoms oxygen. 2. It may be prepared also by dissolving 1 part of starch in 8 parts of nitric acid, of specific gravity 1·25, and allowing the disengaged gases to pass through a chloride of calcium tube 2 feet long, and then into a dry glass tube, surrounded with a freezing mixture (see apparatus, HYDROCYANIC ACID); a liquid condenses which at a low temperature is colourless, but at the usual temperature of the atmosphere is greenish.

Hyponitric Acid, Quaternitrous Acid, Hypoazotic, Nitrous Acid, Quadroxide, Peroxide of Nitrogen.

	Vols.	Atoms.		Wts.	
Nitrogen, ½ □	·4856	1	1·75	14	30·43
Oxygen, 1 □	1·1056	4	4	32	69·57
	1·5912		5·75	46	100·

Theoretical specific gravity 5·75 × ·5528 ÷ 2 = 1·5893. Experimental spec. gravity = 1·62. Weight of 100 cubic inches = 49·2335 grains.

Physical Characters.—In a gaseous state this acid occurs in the form of orange-coloured fumes, usually called nitrous fumes. In the fluid form it is orange-yellow at 60°, fawn coloured at 32°, almost colourless at 14°, and quite destitute of colour at —4°. Taste very caustic; smell very strong; specific gravity 1·451 to 1·42; it reddens litmus, possibly only from its decomposition by water; stains the skin orange, and destroys its texture. It boils at 82°·5, becoming an orange-coloured gas. Spec. grav. of gas 1·70 to 1·62 by experiment.

Preparation.—Hyponitric acid is most conveniently prepared from nitrate of lead ($PbONO_5 = NO_4$, O,PbO), which is dried at a temperature of 300°. It is then pulverized, and introduced into a clay or bottle glass retort, to which is luted a U tube which terminates in a capillary opening to present some resistance to the escape of the

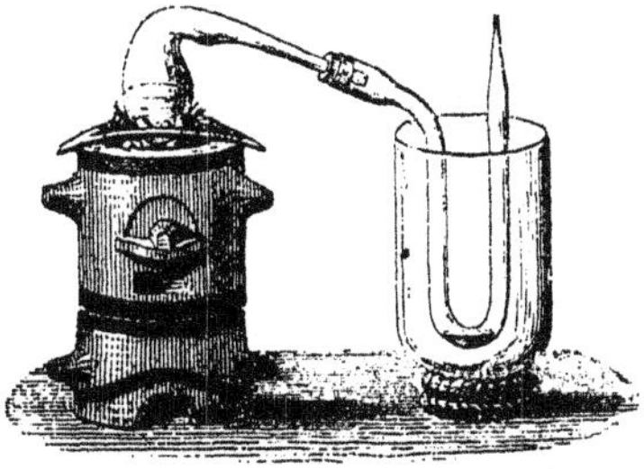

gas. The tube is surrounded with a freezing mixture.

Nitrogentianine. C_{14}, H_4NO_4, O_5HO. Green powder by nitric acid of 1·43 on gentianine; it is coloured red by ammonia.

Nitroharmalidine. *Nitroharmaline, Chrysoharmine.* C_{27}, $H_{13}NO_4$, N_2O_2. Yellow prisms by nitric and sulphuric acids on harmaline in alcohol, and decomposing the bisulphate by an alkali.

Nitrohellenine. C_{15}, H_9NO_4, O_2. By nitric acid on hellenine.

Nitrohematic Acid. *Deoxidized picric Acid.* Brown crystallized grains by acting on picric acid with hydrate of barytes, and then with SH.

Nitrohippuric Acid. C_{18}, H_8NO_4, NO_6. F.P. 306°. Fine needles, soluble in 271 water; by heat yields nitrobenzoic acid as a sublimate; converted by chlorohydric acid into nitrobenzoic acid and glycocoll; by the action of nitric and sulphuric acids on hippuric acid.

Nitrohumic Acid. By NO_5 on humic acid.

Nitrohydrilnric. *Nitrohydurilic Acid.* C_8, H_2NO_4, N_2O_{10}.

Nitrohydrochloric Acid. See AQUA REGIA and NITRIC ACID.

Nitroleucic Acid. $C_{12}NH_{13}$,HO,NO_5. Needles by nitric acid on leucine.

Nitroline. C 57·2, H6·32, N12·2, O 24·28? The residue after treating rotten wood with alkaline carbonates,

Nitro-Mannite. See MANNITRIN.

Nitromeconic Acid. *Nitromeconine.* C_{20}, H_9NO_4,O_8. By nitric acid on meconine.

Nitromesidine. C_{18},$H_{12}N_2O_4$. F.P. below 212°. Golden volatile needles, isomeric with dinitrocumole by acting with sulphohydric acid on dinitromesitilole in alcohol; another base occurs by the action of SH on trimesitilole.

Nitromesityleue. $C_6H_3NO_4$.

Nitrometacetic. *Nitrometacetonic, Nitropropylic.* HO, C_6, H_4 NO_4 O_3. Aromatic inflammable oil, by nitric acid on butyrone.

Nitrometastyrole. $C_{14}H_9NO_4$. By NO_5 on metastyrole.

Nitronaphthalase. *Nitronaphthaline.* C_{20}, H_7NO_4. F.P. 110°. Sulphur prisms, sublimable; by SH and NH_3 converted into *naphthalidine*, and by lime into *naphthalase*; by nitrous acid on fused naphthaline. The method of distinguishing the addition of the displacing acid by the vowels, should be supplanted by a more precise nomenclature. This has accordingly been presented in the subtitle.

Nitronaphthale. C_{38} H_5 N_3 O_{11}. F.P. 419°. Colourless crystals, by nitric acid on the mother liquors of the preceding.

Nitronaphthaleise. C_{40} $H_{11}N_5O_{20}$ (= 5 NO_4) or $C_{20}H_{6\frac{1}{2}}$ 2½ NO_4. Very fusible. Yellow feathery needles along with the succeeding on adding in small quantities naphthaline to much hot nitric acid.

Nitronaphthalese. *Dinitronaphthaline.* C_{20}, H_6 2 NO_4. F.P. 365°. Crystalline colourless powder, insoluble in water, little soluble in alcohol; by the action of nitric acid on the preceding.

Nitronaphthaleisic Acid. C 51·5, H 2·6, N 14·4, O 31·5 = $C_{24}H_8N_3O_6$. From nitronaphthaleise on nitronaphthalesic acid.

Nitronaphthalesic Acid. $C_{32}H_9N_3O_8$. Blackish-brown, amorphous powder; by boiling nitronaphthalese with potash in alcohol and precipitating by nitric acid.

Nitronaphthalic Acid. $C_{16}H_3NO_{10}$. Yellowish rhombic tables or 6-sided plates, from the mother liquor of nitronaphthalese. It is naphthalic or phthalic acid with NO_4 in place of H.

Nitronaphthalidam. *Nitronaphthalese.* A carmine body by SH on dinitronaphthaline.

Nitronaphthalimide. C_{16}, H_4NO_4, NO_4. By heat on acid naphthalate of ammonia.

Nitronaphthalise. *Trinitronaphthaline.* C_{20}, H_5 3 NO_4. F.P. above 212° or 410°. Yellow rhombic prisms or plates; insoluble in water;

little soluble in alcohol and ether; by boiling for many days naphthaline with nitric acid.

Nitronaphthalisic Acid. $C_{12}H_3NO_5$. From nitronaphthalise.

Nitronicene. $C_4H_5NO_4$.

Nitropeucedanin-amide. $C_{24}H_{10}$ 2 NO_4, NH_2? Insoluble, rhombic prisms; soluble in ether; by heating nitropeucedanine in ammonia gas at 212°.

Nitropeucedanine. $C_{24}H_{11}NO_4$, O_6? Colourless crystalline plates, by nitric acid on peucedanine; nearly insoluble in water; soluble in alcohol and ether.

Nitrophenamiglidine. A crystalline base homologous with nitranisidine by acting on phenamylole with fuming nitric acid and treating the resulting oil with sulphide of ammonium in alcohol.

Nitrophenesic Acid. *Binitrocarbolic acid.* C_{12}, H_3 2 NO_4OHO. F.P. 237°. Brownish-yellow doubly-oblique prisms by the mild action of nitric acid on carbolic acid, or by boiling dinitranisole with potash in alcohol.

Nitrophenetidine. C_{16},$H_{10}NO_4$,NO_2. A base by passing SH through dinitrophenetole in alcohol.

Nitrophenetole. C_{16},H_9NO_4,O_2. An oil along with *dinitrophenetole* in needles distilling over; by nitric acid on phenetole.

Nitrophenisic Acid. *Chrysolepic acid, Ternitrocarbolic acid, Yellow bitter of Welter, Nitrophenic acid.* See Picric Acid.

Nitrophthalic Acid. $C_{16}H_5NO_{10}2HO$.

Nitropicric Acid. A synonyme of Picric acid.

Nitropopulic Acid. HO,$C_{14}H_3N_2O_{13}$ 2 HO. Yellow needles; soluble in water, alcohol, and sulphuric acid; by boiling with nitric acid converted into picric acid; with HCl and chlorate of potash changed into chloranile; it is distinguished from oxypicric acid (styphnic acid) by the last yielding with bleaching powder chloropicrine, and by the nitropopulate of potash being with difficulty soluble in cold water, and easily soluble in alkaline solutions, while picrate, chrysammate, and styphnate are less soluble in alkaline solutions than in water; by heat it sublimes and its salts explode; obtained by treating the watery extract of the bark of Populus balsamifera with nitric acid and heat for twenty-four hours, evaporating to dryness, dissolving in hot water, saturating with potash; a yellow precipitate falls of picrate and nitropopulate of potash, from which the nitropopulate is separated by cold water (Stenhouse).

Nitroprusside of Sodium. C_5Cy_{12} 3 NO, 5 Na, 10 HO. Ruby rhombic prisms with angles of 105° 17′, 136° 32′, and 124° 52′; soluble in 2½ cold water; obtained by pouring 5 atoms nitric acid (1·33) diluted with an equal volume of water on 2 atoms yellow prussiate of potash; a coffee-coloured fluid forms with the evolution of binoxide of nitrogen, cyanogen, prussic acid, nitrogen, carbonic acid, and cyanic acid (?) The solution contains red prussiate, nitroprusside of potassium, and nitre, and sometimes oxamide; the mother liquor is poured off from the nitre and neutralized in the cold with carbonate of soda, heated to boiling, filtered and crystallized; the nitre which separates being removed. *Nitroprusside of Potassium.*—Fe_5Cy_{12} 3 NO, 5 K, 3 HO. A salt similar to the preceding but deliquescent; hence the sodium salt is best for preservation; it is obtained by a similar process, neutralizing with carbonate of potash instead of soda. The nitroprussides are valuable reagents for the detection of the most minute quantities of sulphides, with which they form a magnificent red-purple or blue colour of a very ephemeral nature, but more permanent in alcohol. The composition of the precipitate was found to be $Fe_5Cy_{12}N_3O_3$, 5 Na, 3 NaS, 6 HO (?) (Playfair).

Nitroprussine. $Fe_2Cy_5NO_2$, 2 M or $Fe_5Cy_{12}N_3O_3$, 5 M or Fe_2Cy_5NO 2 M. The base of the nitroprussides.

Nitropurreic Acid. *Nitro-euxanthic acid.* C_{40}, $H_{15}NO_4$,O_{21}. By the action of nitric acid on purreic acid.

Nitroquercite. White explosive resin by nitric and sulphuric acids on quercite.

Nitrosaccharic Acid. *Nitrate of Glycocoll.* $C_4H_6N_2O_{10}$ = GlHO, NO_5HO. By the action of nitric acid on glycocoll; insoluble in water, soluble in hot alcohol; distinguished from nitromannite by not crystallizing.

Nitrosalicylamide. $C_{14}H_6N_2O_8$. Yellow crystals by allowing nitrosalicylic ether to stand in close vessels.

Nitrosalicylic Acid. *Nitrosalicylide, Indigotic Acid, Anilic Acid.* HO, $C_{14}H_4NO_9$. Yellow prisms by boiling nitrosalicylamide with potash, by nitric acid on hydride of salicyle or salicylic acid; forms explosive compounds with bases.

Nitrostyrole. $C_{16}H_7NO_4$. By nitric acid on styrole.

Nitrosulphalic Acid. S_2O_6,NO_4,H. Supposed to exist in the crystals of the sulphuric acid chambers.

Nitrosulphobenzidic Acid. C_{12}, H_5NO_4 2 SO_3. By acting on sulphobenzidic acid with nitric acid.

Nitrosulphonaphthalic Acid. $C_{20}H_6NO_4$ S_2O_5. By sulphuric acid on nitronaphthaline and by nitric acid on sulphonaphthalic acid.

Nitro-Sulpho-naphthalidamic Acid. C_{20}, H_6NO_4, NH_2 2 SO_3? By SH on dinitrosulphonaphthalic acid.

Nitro-Sulphuric Acid. *Nitro-Sulphurous Acid.* NO_2SO_2. An acid uniting with bases and forming nitrosulphites; by the action of 2 volumes sulphurous acid and 4 binoxide of nitrogen on a base; in water it changes into protoxide of nitrogen and sulphuric acid.

Nitrotartaric Acid. Silky crystals by the action of nitric and sulphuric acids on tartaric acid.

Nitrotheine. A synonyme of Cholestrophane.

Nitrotoluide. *Nitrobenzoene, Nitrotoluole.* $C_{14}H_7NO_4$. A fluid by acting on benzoene or toluole ($C_{14}H_8$) with nitric acid; along with it occurs *dinitrotoluole.* By ammonia and SH it becomes *toluidine.*

Nitrotyrosine. $C_{18}H_{10}, NO_4, NO_6$. Yellow stellated needles by diffusing tyrosine through water and adding nitric acid drop by drop, stopping when the liquid is yellow; in some hours yellow nitrate of nitrotyrosine falls; it precipitates with nitrate of silver; and the silver salt being decomposed by SH yields nitrotyrosine in crystals.

Nitrous Acid. See under NITROGEN.

Nitrous Ether. *Hyponitrous Ether.* EO, NO_3 ($C_4H_5ONO_3$). B.P. 62°; spec. grav. ·947. Light yellow fluid with an agreeable odour; obtained by passing nitrous acid (NO_3) by the action of nitric acid on starch into dilute alcohol and condensing in a Liebig's tube; it is deprived of water by chloride of calcium. Nitrous ether of the shops, or sweet spirits of nitre made by boiling alcohol and nitric acid, contains some nitrous ether mixed with alcohol, aldehyde, &c.

Nitrous Oxide. A synonyme of Protoxide or Semeloxide of Nitrogen.

Nitrovalerianic Acid. $C_{10}H_9NO_8$ or *Nitroangelic Acid* $C_{10}H_7NO_8$. Rhombic tables subliming at 212°; very soluble in hot water. By boiling valerianic acid or caprone with nitric acid, &c.

Nitroxylole. $C_{16}H_9NO_4$. By nitric acid on xylole ($C_{16}H_{10}$).

Nomenclature, or system of naming chemical compounds. In alchemistical times names were given to bodies according frequently to their obvious characters or imagined functions. Thus mercury or quicksilver derived its name, which it still retains, from its liquid form and volatile character, and a similar plan was followed afterwards by scientific chemists. This was probably the best mode of proceeding in reference to simple or elementary bodies, but it was not applicable to compound substances. It was in the year 1787 that Lavoisier, as the head of a commission of the French Academy, laid the basis of the scientific system of nomenclature, which has been generally adopted, and in 1804 Dr. Thomas Thomson extended this system by endeavouring to represent the constitution of compounds in the most concise language by naming the formulæ which he had introduced in 1798. *Nomenclature of simple bodies and oxides.*—The termination of metals is in *um*, as chromium, uranium, &c., especially when their Latin designations are employed. Names ending in *on* denote elements which are not metallic, as carbon, silicon, boron. In naming the oxides it is desirable to express the number of atoms of oxygen present by means of the Latin language when these atoms are equal or greater in number than the metal with which the oxygen is united, and to apply Greek numerals when the metallic atoms are in excess over those of oxygen. This can be well illustrated in the case of the oxides of lead. Pb or plumbum represents an atom of lead, and O an atom of oxygen.

Ratio of Lead to Oxygen. Atoms.		
2 to 1	Dinoxide of lead, ...	Pb_2O
1 to 1	Proto, equi, or semeloxide,	Pb O
1 to $1\frac{1}{2}$	Sesquioxide, ...	Pb $O1\frac{1}{2}$
	or	Pb_2O_3
1 to 2	Binoxide, ...	Pb O_2
3 to 4	Red lead, ...	Pb_3O_4
or 1 to $1\frac{1}{3}$ or	....	Pb $O1\frac{1}{3}$
1 to 1 + 2 to 3	Protosesquioxide,	PbO, Pb_2O_3

It is very much to be desired that the use of Latin and Greek numerals should not be employed as if they were members of the same language, as is often done at present. Sometimes the expression *per*, thorough or highest degree, is employed. This term was introduced in 1804, when the number of oxides was exceedingly limited, but it is scarcely admissible at the present time, when the great object should be to have a concise expression for a formula indicative of the composition of the body. Any other method in the present state of the science will lead to confusion, and have a retrograde tendency.

Nomenclature of the Acids.—The highest degree of oxidation of an acid in the early state of the science, when such acids were limited in number, had the termination *ic* appended to it, while the lowest oxide terminated in *ous;* but as the number of acids in any series increased, other affixes were required, and although in some instances they answered their purposes sufficiently well, in others the confusion is considerable. The following illustration is from the oxygen salts of sulphur:—The original known acids were sulphurous acid (SO_2), sulphuric acid (SO_3); but afterwards other compounds were discovered, hyposulphurous acid, S_2O_2 or SO_2S, and subsequently hyposulphuric acid, SO_2SO_3, or S_2O_5. To the first the Greek prefix hypo (ὑπο,) *under*, was applied, to denote its containing less oxygen than sulphurous acid, and the next new compound was termed hyposulphuric acid, as it contained less oxygen than sulphuric acid. The table will stand thus:—

Ratio of Sulphur to Oxygen.		
2 to 2	Hyposulphurous acid,............	SO_2S
1 to 2	Sulphurous acid, ...	SO_2
1 to $2\frac{1}{2}$	Hyposulphuric acid, ...	SO_2SO_3
1 to 3	Sulphuric acid,.....................	SO_2O

Should any acid occur with a higher degree of oxidation than any in the table it would be termed hyper (ὑπερ), above. It is obvious, however, that this is an attempt to accommodate new discoveries to a nomenclature formed when the present extent of the science could not have been contemplated. The scientific designations for the preceding acids would be semelsulphurous, bisulphurous, bisesquisulphurous, tersulphuric.

Nomenclature of Oxides and Salts.—When simple bodies unite with others, it is usual to terminate the most electronegative element in *ide;* thus oxide, chloride, bromide, iodide, fluoride, hydride, nitride, carbide, boride, silicide, phosphide, sulphide, selenide, arsenide, antimonide, telluride, chromide, vanadide, uranide, molybdide, tungstide, columbide, titanide, are compounds of oxygen, chlorine, &c. with more electro-positive elements, as potassium, sodium, &c. The terms employed to indicate the different degrees of oxidation have already been given. When acids unite with oxides, the resulting salt termination is *ite*, when that of the acid is *ous*, and *ate* when that of the acid is *ic*,—thus the preceding acids form hyposulph*ites*, sulph*ites*, hyposulph*ates* and sulph*ates*. To indicate the relative number of atoms of acid present in a salt the Latin numerals—1, semel; 2, bis; 3, ter; 4, quater; 5, quinque; 6, sex; 7, septem; 8, octo; 9, novem; 10, decem, are employed; and to express the number of atoms of base the Greek numerals are used for the basic atoms—1, apax, protos, or monos; 2, dis; 3, tris; 4, tetrakis; 5, pente; 6, hex; 7, hepta; 8, octo; 9, ennea; 10, deka; 11, endeka. Examples of this will be found among the acetates.

Nomenclature of Organic Bodies.—In such a novel branch of chemistry as that which treats of organic bodies, much confusion must be expected to prevail in reference to the names which they at first bear, from the impossibility of appropriating to every substance on its discovery its proper position in a scientific arrangement. But even at the present time there are certain terminations which indicate the characters of certain classes of compounds. The termination in *ine* it would be well to restrict to alkaloids, which correspond with the metallic oxides. Sometimes the termination *ia* is used in this country, as morphine or morphia, brucine or brucia, &c. Neutral bitter principles should perhaps end in *ite*, as *gentianite*, &c. The terminations in *yle*, as in organic radicals; in *ile*, as nitrile; in *amide* and *imide*, as in the classes of amides and imides; in *one*, as in the series of *cetones* or *ketones*, designate compounds of peculiar character, which are described under these words, and the individual species of their classes.

Nomenclature of Minerals.—The names of minerals terminate in *lite* or *ite*, (λιθος, a stone); and it is desirable that this method should be adhered to.

Nontronite. A synonyme of chloropal.

Norium. A name proposed for a metal contained in Zirconia, from Zircons, which is thought to be a mixture of earths. The oxide of norium appears to be a sesquioxide; its oxalate and chloride are less soluble in acids than the oxalates of the others; its acid sulphite is more easily crystallized than that of the others.

Noscan. *Nosine, Hauyne, Spinellane, Sodalite.* Spec. grav. 2·25 to 3, H 5·5. SiO_3 35·99, SO_3 9·17, Al_2O_3 32·57, CaO 1·12, NaO 17·84, Cl ·65, HO 1·85. Gray, blue or brown, or black, like Hauyne; fuses B.B. on the edges and gelatinizes with acids; it seems to be sodalite.

Novaculite. A clay slate used as a hone-stone.

Nucleus. The central point of a concretion; likewise applied to the base or radical of organic compounds in the nucleus or kernel theory.

Nussierite. Arsenio-phosphate of lead; Nussiere, Rhone, France.

Nutgalls. *Gallnuts.* Spherical concretions produced by an insect, a species of cynips, on the bark of the Quercus infectoria and cerris. Aleppo galls contain about 75 to 80 per cent. of tannic, gallic, ellagic, and luteogallic acids, starch $1\frac{1}{2}$, woody starchy matter 11. The tannic acid is extracted by displacement. Infusion of galls in water is employed to detect iron, and extensively in dyeing.

Nutmegs, Butter of. A solid oil extracted by expression from the nutmeg (Myristica aromatica), in the form of cakes, consisting of tallow like oil (Myristine) 7, yellow oil $8\frac{1}{4}$, volatile oil $\frac{1}{8} = 16$. The yellow and volatile oils are soluble in cold alcohol and ether, not the tallow; boiling ether or alcohol dissolves the whole, and deposits the tallow on cooling. Nutmegs yield by distillation with water a volatile yellowish oil. Sp. grav. ·920 to ·948; soluble in alcohol and ether; deposits a solid stearoptene (myristicine of John).

Nutrition. See ALIMENTARY BODIES and ANIMAL CHEMISTRY.

Nuttalite. Spec. grav. 2·748, H 7·. White 8-sided prisms, cleaving in the direction of a right square prism; lustre vitreous. SiO_3 37·8, Al_2O_3 25·10, CaO 18·33, Fe_2O_3 7·89, KO 7·3, HO 1·5. Bolton, Massachusets; allied to scapolite.

Nux Vomica. *St. Ignatius's Beans.* The seeds of the Strychnos nux vomica, and ignatia. It is used for poisoning rats, and contains strychnine, brucine, strychnic or igasuric acid, tartaric acid, gum, starch, resin, vegetable fibre.

O

Oats. (*Avoine*, Fr.; *Hafer*, Ger.) *Avena Sativa.* The seed of the oat is extensively used as food in consequence of the large amount of nutritive matter which it contains, and from the comparative hardiness of the plant in a northern climate. Oatmeal contains as much as 15·61 or even more per cent. of albuminous matter. This circumstance is in some measure due, when compared with wheat flour, to the retention of the greater portion of the bran by the mode of grinding, while in flour, as commonly prepared, the bran is in a great measure excluded. Scotch oats, dried at 212°, contain about 2 per cent. of ash, French oats nearly 4 per cent. The ash consists of silica 54·25, SO_3 2·15, PO_5 1·94, Cl 1·5, Fe_2O_3 1·41, CaO 7·29, MgO 4·58, KO 12·18, NaO 14·69. The straw contains 5 per cent. of ash.

Obsidian. Spec. grav. 2·363 to 2·372, H 6·5. Velvet-black amorphous mass, resembling bottle glass; fracture conchoidal; lustre vitreous, splendent, translucent on the edges; easily frangible; streak gray. B.B. fuses into a vesicular glass. The mean of two specimens from Iceland and Mexico gave me silica 83·38, FeO 5·46, Al_2O_3 3·79, NaO and KO 4·53, CaO 2 42, HO ·15. My pupil, Mr. J. Murdoch, found a specimen from New Zealand to consist of SiO_3 73·2, Al_2O_3 6·86, Fe_2O_3 6·54, CaO and MgO 3·83, NaO and KO 7·57. It was named after Obsidius, a Roman, who first brought it from Ethiopia. The Mexicans made knives, sword blades, and razors of it. In the Canaries lance points were formed of it, and in Quito looking-glasses. It is a product of volcanic action; it gelatinizes with acids.

Ochran. A species of halloylite, consisting of SiO_3 31·3, Al_2O_3 43, Fe_2O_3 1·2, HO 21·. Orawitza.

Ochre. A term applied to various states of sesquioxide of iron, and oxides of other metals. *Red ochre* is a form of specular iron ore, or anhydrous sesquioxide of iron. *Brown ochre* is a variety of brown hæmatite or sesquihydrous sesquioxide of iron. The native oxides of cerium, molybdenum, lead, tungsten, chromium and uranium, are known as ochres. My pupil, Mr. Aitkin, found the ochre of New Zealand, used as a paint by the natives, to consist of Fe_2O_3 64·36, SiO_3 13·92, HO 20·2, organic matter 4·72, with traces of alumina and lime; it gelatinizes with acids.

Ochroite. A synonyme of cerite.

Octahedrite. A synonyme of anatase.

Odmyle. $C_8H_8S_2$, or $C_{16}H_{18}S_4$? B.P. 160°. A clear fluid by distilling linseed oil and sulphur; unites with corrosive sublimate and bichloride of platinum.

Odorine. *Picoline.* $C_{12}NH_2H_3,H_2$? Spec. grav. ·955; B.P. 272°. Thin fluid oil, peculiar rancid smell, bitter taste, alkaline reaction; forms a cloud with chlorohydric acid, not poisonous; isomeric with aniline; obtained by fractional distillation from bone oil (Dippel's oil), or by distilling about 311° piperine with 4 parts soda lime; it is also contained in coal oils; it forms crystalline salts.

Odour. The well-known impression produced on the olfactory nerves by peculiar volatile matter.

Oedelite. A synonyme of mesolite.

Œnanthic Acid. (οἶνος, wine; ἄνθος, flower). $C_{14}H_{13}O_2HO$. A solid resembling butter, found in union with ether in wines, to which it supplies the bouquet. It is obtained by distilling the ether with potash, and then distilling the potash salt with sulphuric acid; it is soluble in alcohol and ether.

Œnanthic Ether. $C_4H_5O,C_{14}H_{13}O_2$. Sp. grav. ·862; B.P. 440°. Colourless fluid, with the smell of wine casks; obtained by distilling the oils of wine lees with carbonate of soda to remove acid, and then with chloride of calcium.

Œnanthole. *Hydrous Oxide of Œnanthyle, Œnanthal.* $C_{14}H_{14}O_2$. Spec. grav. ·8271; B.P. 314°. Colourless fluid, refracting light strongly; strong penetrating aromatic odour; taste sweet, acrid; soluble in all proportions in alcohol and ether; little soluble in water; it is isomeric with butyrone; it absorbs oxygen, and becomes œnanthylic acid. Meta-œnanthal is solid. Œnanthole is obtained by distilling castor oil; the first part of the distillate consists of elaiodic and ricinic acids, and then œnanthole; it is first distilled with water, and then rectified over barytes. It is the aldehyde of œnanthylic acid.

Œnanthylic Acid. *Azoleic Acid?* $C_{14}H_{13}O_3HO$. B.P. 298½°. Volatile oily acid, colourless, aromatic, inflammable; obtained by oxidizing castor oil, wax, stearic acid, oleic acid, albumen, &c. by nitric acid, and by exposing œnanthole to the air or hot nitric acid.

Œnole. A synonyme of mesitylene.

Œnothionic Acid. A synonyme of sulphovinic acid.

Œnylamine. C_4H_7,NH_2. A fluid base by the distillation of narcotine with potash.

Œnyle. A synonyme of mesityle.

Œrstedite. Spec. grav. 3·269, H 5·5. Reddish-brown square prisms, resembling the crystal of Zircon; lustre splendent, adamantine; opaque. SiO_3 19·716, CaO 2·612, MgO 2·055, FeO 1·141; titanic acid and zirconia 68·96, HO 5·33. B.B. infusible; with borax a colourless glass; not fused with soda. Arendal, Norway.

Œtlite. A synonyme of brown hæmatite.

Oil of Vitriol. A synonyme of commercial sulphuric acid, consisting of 1 atom acid and 1 atom of water ($HOSO_3$). Spec. grav. 1·848.

Oils. The term was originally applied to those bodies which left a greasy stain on paper when they sink into it. They may be subdivided into—1, solid oils, butters, or fats; 2, into fluid fixed oils, which are not dissipated from paper by heat; and 3, volatile oils or essences, which are volatilized at a comparatively low temperature.

Extraction of Oils.—The whole of the processes put in practice in the extraction of oils may be included under the pounding of the grain or substance from which the oil is to be extracted, and the extraction or expression of the oil. For the purpose of bruising or grinding the seeds, various methods are adopted in different countries. In the neighbourhood of Lille, in the north of France, there are not less than 400 wind-mills employed in extracting oil, by means of a sort of pestles, which are formed of oak, armed with a head of iron. Instead of this method, in some places a species of coffee-mills has been introduced for the purpose of crushing the seed. But the best method for grinding the seed is the common mill. After this operation the bruised grain is heated in vessels of copper or cast iron by means of steam. In a few minutes sufficient heat has been applied; the object of this operation being to liquefy the oil and render it more easy of extraction. The oil is next extracted by pressure. Of course the hydraulic pressure is that to be recommended; but on the continent this is seldom if ever employed, wind-mills being generally the moving force. The best presses employed are horizontal, and hollow in the middle, so as to admit of the introduction of steam into their interior, which keeps up the heat. After being pressed once they are again ground, heated, and pressed as before. The oil obtained from the second pressure is more impure than the first extraction. Oils thus extracted always contain in suspension a quantity of mucilage, colouring matter, and resins. By allowing it to remain at rest, a portion of this impurity separates. But chemical means are required to purify completely. $\frac{1}{200}$th part of sulphuric acid is added, which gives the oils a greenish colour. In twenty-four hours a quantity of water, at the tempt. 167°, equal to $\frac{2}{3}$ of the oil is added. It becomes milky. In two or three weeks the oil is clear, and is then drawn off.

Nature of Oils.—If we take mutton fat and expose it to pressure between leaves of blotting paper, an oily stain is left on the paper. If the paper be now immersed in boiling water, the stain disappears, and a fluid oil is observed to float on the surface of the water. This substance is termed *oleine*, or *oleate*, of the oxide of glyceryle (GlyO Ol). There remains unacted on by the paper a white solid substance termed *stearine* or *stearate* of oxide of glyceryle (GlyO St_2). The cause, then, of the solid nature of mutton fat is the predominance of the solid stearine in it, while it contains only a small portion of the liquid oily body, oleine. Hence the difference between a liquid and a solid oil or butter depends upon the presence in the liquid oil of a large quantity of the liquid oleine, and a minute quantity of stearine, while in a solid oil the stearine is most abundant. To prove that a liquid oil contains a solid substance in solution, we have only to expose it to the action of snow or ice, when the stearine is deposited, and the oleine floats above. Thus it is that in winter olive and castor oils become thick and deposit a white solid matter. As the oleine and stearine consist each of an acid united to a base the oxide of glyceryle, they may be considered as oily salts, and on this view the process by which they are converted into *soaps* becomes easy of appreciation The change produced on mutton fat depends on replacing the oxide of glyceryle by an alkaline base potash, and thus converting the solid oily stearate of oxide of glyceryle insoluble in water into a stearate of oxide of potassium, a *soap* which is quite soluble in water, GlyO St_2 and KO becoming KO St_2, and GlyO set free. *Saponification*, or the formation of soap, is effected by boiling oils with a solution of caustic potash or soda. The experiment may be made on a small scale in an iron or silver basin. The boiling is kept up, using an excess of alkali until the whole of the stearic or other oily acid has been separated from the oxide of glyceryle, and has united with the potash. The matter contained in the basin should therefore dissolve completely in hot water if the soap is thoroughly formed. From this solution the soap may be precipitated by adding a strong solution of common salt to remove from it oxide of glyceryle, and may be again dissolved in water, from which solution the fatty acid may be separated by the addition of HCl, or tartaric acid, which will unite with the potash, and set the acid free. Soaps thus made with soda, being firm, are called *hard* soaps; those made with potash being soft and deliquescent, are termed *soft* soaps. Instead of the alkalies the fat acids may be united with metallic oxides, and *plasters* or metallic soaps formed. Litharge or *diachylon* plaster, which forms the basis of medical plasters, is produced by boiling 1 part of litharge (PbO), with 2 of olive oil.

Oils, Drying. Fixed oils are termed drying oils when they retain their transparency on solidifying either by being dried up by the action of the oxygen of the air, or by the application of heat into a hard transparent varnish. The drying oils are those which are employed for printer's ink. The principal drying oils are castor oil (Ricinus communis), spec. grav. ·9611 or ·9575, solid at 0°; cress oil (Lepidium sativum); croton oil (Croton tiglium); grape seed oil (Vitis vinifera); henbane oil (Hyoscyamus niger); hemp oil (Cannabis sativa), spec. grav. ·9276, solid at 17°; linseed oil (Linum usitatissimum), spec. grav. ·9347, solid at 17°; nightshade oil (Atropa Belladonna); poppy oil (Papaver Somniferum), spec. grav. ·9249, solid at 0°; sunflower oil (Helianthus annuus); tobacco oil (Nicotiana tabacum); walnut oil (Juglans regia), spec. grav. ·9283, or ·923, ·947, solid at —17°.

Oils, Fat, are oils which become solid by long exposure to the air and absorption of oxygen; and then present the appearance of tallow. The following are the principal:—Almonds (Amygdalus communis), spec. grav. ·918, ·932; alyssum or camiline (Myagrum or Alyssum sativum); beech (Fagus sylvatica), spec. grav. ·9225; ben (Hyperanthera moringa), spec. grav. ·917; bays (Laurus nobilis); cardamoms; colza (Brassica campestris), spec. grav. ·9136; cotton seed (Gossypium Barbadense); chestnut (horse) (Aesculus hippocastanum), spec. gra. ·927; cherry (Prunus cerasus), spec. grav. ·9239; dolphin (Delphinus globiceps), spec. grav. ·918; hazelnut (Corylus avellana), spec. grav. 926; mustard (sinapis (1), alba (2), nigra), (1) spec. grav. ·9142, (2) ·917; neat's foot oil, spec. grav. ·8795; olive oil (Olea europaea), at 77°, ·9176, solid at 36°; plum (Prunus domestica), spec. grav. ·9127; porpoise (Delphinus phocena), spec. grav. ·937; peach; rapeseed, spec. grav. ·9109; rape (Brassica rapa ·9167 and napus ·9128), solid at 28°; tea (Camellia oleifera and probably Thea bohea), spec. grav. ·927; train or whale (Balena misticetus) spec. grav. ·927; B.P. 620°.

Oils, Solid. These oils are solid at the usual temperature of the atmosphere from the preponderance of the solid stearine. See FATS.

Oils, Sulphur Volatile. Black mustard, $C_8H_5NS_2$. Horse-radish, $C_8H_5NS_2$. Cochlearia, $C_8H_5NS_2$? Garlic, C_6H_5S.

Oils, Volatile. *Essences, Essential oils.* These oils are found in various parts of plants, and are contained in vessels or pouches. They are extracted by distilling the plants or flowers with water or suspending the plant in a bag or net to prevent it from coming in contact with the sides of the retort. They consist of a solid portion stearoptene or camphor and of a liquid oleoptene; the former may be separated by the application of cold and pressure. *Hydrocarbon oils.*—Certain of the essences possess the same per centage composition, and are said to be formed on the type of oil of turpentine. They contain carbon and hydrogen alone. Their formula is $C_{20}H_{16}$ or C_5H_4 or $C_{10}H_8$ or $C_{15}H_{12}$.

Hydrocarbon Oils.	S.G.	B.P.
Amber,	·879	297°
Athamanta,	·843	325
Andropogon,		297
Cedar,	·980	507
Cedrat,		
Copaiva,	·960	473
Cubebs,	·929	490
Elemi,	·852	330
Juniper,	·839	311
Lemons,	·850	343
Neroli,	·930	
Oranges,	·835	356
Pepper,	·864	333
Pine,	·856	333
Portugal (orange peel),	·835	356
Sabine,	·915	315
Storax,		
Turpentine,	·864	312

Oxygen Volatile Oils.—The following have been analyzed:—

Asarabacca,	C_8H_4O
Anise,	$C_{20}H_{12}O_2$
Basil,	$C_{20}H_{22}O_6$
Bergamot,	$C_{30}H_{26}O_2$
Camphor,	$C_{20}H_{16}O$
Cassia,	$C_{20}H_{11}O_2$
Cedar (solid),	$C_{30}H_{26}O_2$
Cinnamon,	$C_{20}H_{11}O_2$
Cloves,	$C_{24}H_{15}O_5$
Cajeput,	$C_{10}H_9O$
Cumin,	$C_{20}H_{12}O_2$
Fennel,	$C_{20}H_{12}O_2$
Galangal,	$C_{10}H_9O$
Juniper,	$C_{32}H_{26}O$
Lavender,	$C_{15}H_{14}O_2$
Marjoram,	$C_{50}H_{40}O$
Meadow Sweet,	$C_{10}H_9O_2$
Parsley,	$C_{12}H_8O_3$
Peppermint,	$C_{24}H_{20}O_2$
Pennyroyal,	$C_{10}H_8O$
Sassafras,	$C_{10}H_5O_2$
Sweet Bay,	$C_{20}H_{16}O$
Spearmint,	$C_{35}H_{92}O$
Rue,	$C_{28}H_{28}O_3$
Rosemary,	$C_{60}H_{38}O_2$
Tarragon,	$C_{24}H_{16}O_2$
Valerian,	$C_{12}H_{10}O_2$
Wormwood,	$C_{20}H_{16}O_2$?

Ointment. An oleaginous compound usually having as its basis lard or suet, sometimes flavoured by an essential oil. A cerate is a compound of wax and oil or lard and wax.

Oisanite. Anatase from Oisans.

Okenite. A synonyme of Dysclasite.

Olanine. A base found by Unverdorben in bone oil.

Oleene. $C_{12}H_{12}$. Spec. grav., vapour 2·875; B.P. 131°. Colourless ethereal fluid, lighter than water, very inflammable; very soluble in alcohol and ether; poisonous to man; obtained along with elaene by distilling metoleic and hydroleic acids.

Olefiant Gas. *Bicarbonated Hydrogen, Hydroguret of Carbon, Heavy Carburetted Hydrogen, deuto* or *tetartrocarbohydrogen, Hydride of Acetyle, Elayle.*

	Vols.	Spec. Grav.		Atoms.	Wts.	Per Cent.
Carbon,	2	·8468	·8333	4	3·	85·71
Hydrogen,	2	1384	1384	4	·5	14·29
		·9852	·9717		3·5	100·

Probable composition, C_4H_8H.

Calculated spec. grav. 3·5 ÷ 2 = 1·75 × ·5528 = ·9674; refractory power 2·302.

Weight of 100 cub. in. = 30·487 grains or 30·0692 grains. 100 water dissolve 15⅓ olefiant gas; oil of vitriol 83 times its bulk.

Olefiant gas (*oleum*, oil; and *fieri*, to be made;) has no colour, but it possesses an ethereal odour, and burns with a fine white flame, which affords much light. A red heat in a porcelain tube decomposes the gas into charcoal, which is deposited (Hecht and Vauquelin), and into carburetted hydrogen or hydrogen. By passing electric sparks through it, carbon is deposited and hydrogen set free. When sulphur is heated in olefiant gas, sulphohydric acid gas is formed and carbon deposited. When chlorine gas is mixed with an equal volume of olefiant gas, an oil is formed (the oil of the Dutch chemists) which is chlorohydride of the chloride of acetyl ($C_4H_3Cl + HCl$). If this oil be left in contact with potash dissolved in alcohol, for a few days, we have a gas formed, which is olefiant gas, in which the fourth atom of hydrogen is replaced by chlorine (C_4H_3Cl). This oil may be prepared by passing chlorine and olefiant gas from separate retorts into a common vessel, when it collects at the bottom; or a more convenient method is to pass the gas into chloride of antimony until the latter is saturated, and then to distil over from it the Dutch liquor. When 2 volumes chlorine and 1 olefiant gas are rapidly mixed in a glass vessel under water and a taper is immersed in the mixed gases, combustion takes place with a red light, and carbon in the form of lamp black flakes is copiously deposited (C_4H_4 4 Cl become C_4, 4 HCl), and chlorohydric acid is formed.

Preparation.—This gas may be obtained by the dry distillation of many organic substances, such as fats or in the ignition of coal in close vessels. It is usually prepared by distilling 1 part of alcohol with from 4 to 7 parts of oil of vitriol in a retort or gas bottle with the assistance of heat. In order to procure the gas free from carbonic and sulphurous acids, it may be passed through milk of lime. The fluid in the retort becomes first brown, and then black (melanic acid, Erdmann), and the gas is latterly contaminated with sulphurous and carbonic acids.

Fluid Olefiant Gas.—To obtain it in a liquid state this gas is exposed to a cold mixture of solid carbonic acid and ether in glass tubes under the exhausted receiver of an air pump. To these tubes are attached two force pumps sufficient to apply a pressure of 40 atmospheres. When this is obtained, it is a clear colourless transparent fluid; but it does not become solid. It dissolves resinous bodies, bituminous, and oleaginous substances. Aimé has observed, that when introduced in a vessel sunk in the sea to sufficient depth, olefiant gas suffered diminution of volume by a pressure of 124 atmospheres, and has calculated its density in a liquid state at ·44.

Composition and Analysis.—If we mix 4 volumes oxygen, 1 olefiant gas, and fire the mixture by an electric spark, 3 volumes remain. Caustic potash will absorb 2 of carbonic acid, while 1 volume oxygen remains, which may be removed by phosphorus. 1 volume olefiant gas therefore yields 2 volumes carbonic acid or 2 of vapour of carbon, while 1 volume oxygen has been consumed by 2 of hydrogen to form water. In every volume of this gas, there therefore exist 2 volumes of carbon vapour and 2 volumes hydrogen.

Oleic Acid. $C_{36}H_{33}O_3HO$ (Gottlieb), $C_{26}H_{23}O_4$ (Heintz). F.P. 57? solid at 40°. Colourless neutral oil; insoluble in water; soluble in alcohol and ether; oxidizes in air; obtained by saponifying mutton fat with potash, decomposing the soap, by dilute sulphuric acid; the fat acids dissolved in hot alcohol, the solution on cooling is expressed, and the operation repeated frequently; the acid dissolved in alcohol, and precipitated by barytes and purified. Oleic acid when distilled gives sebacic, capric, caprillic, valerianic, butyric, and carbonic acids. By nitric acid it gives elaidic acid; oxidized with nitric acid it yields azelaic, azoleic, adipic, lipic, and pimelic acids; also, acetic, metacetic, butyric, caprillic, caproic, capric, œnanthylic, pelargonic, and valerianic acids. Sulphuric acid produces sulpholeic, metoleic acid; with alkalies it unites and forms soaps or oleates.

Oleine. *Oleate of Oxide of Glyceryle* or *Lipyle.* C 79·5, H 11·9, O 8·6? Spec. grav. ·91. Colourless fluid, without taste and smell; insoluble in water; soluble in ether, alcohol, essential and fat oils; needles at 19°; obtained by boiling tallow in alcohol; on cooling most of the solid fats (stearine) separate, while the oleine remains in solution; it is also obtained from olive oil by exposing it to intense cold, expressing the oleine, and purifying it by solution in alcohol. It has not been obtained in a state of purity. It constitutes the more fluid portion of oils.

Oleophosphoric Acid. Viscid yellow oily solid, insoluble in water, swelling up somewhat in boiling water; insoluble in cold; soluble in boiling alcohol and in cold ether; decomposed by boiling in water into oleine and phosphoric acid; it is purified from oleine by absolute alcohol, which does not dissolve this acid; obtained from the brain by treating it with hot ether; on cooling cerebrate and oleophosphate of soda separate. These are dissolved in hot alcohol acidulated with sulphuric acid; on cooling the two acids separate; oleophosphoric acid dissolves in cold ether, while cerebric acid remains. It contains about 2 per cent. of phosphoric acid. The softening and decomposition of the brain may be owing to the separation of the oleine from the phosphoric acid.

Oleoptene. *Elaoptene.* The fluid portion of essential oils corresponding with the oleine of fat oils as contrasted with the solid stearoptene of essential oils, and the stearine of fat oils.

Olibanum. *Frankincense.* Dull resin, *a*, $C_{40}H_{32}O_6$. Clear resin, *b*, $C_{40}H_{32}O_4$. Whitish-yellow resin, usually covered externally with a whitish powder; taste acrid and aromatic; giving an agreeable odour when burned. It consists of volatile oil 8·, resin 56, gum 30, insoluble substance like gum 5·2. The oil ($C_{35}H_{28}O$) has a smell resembling oil of lemons; the resin consists of two, one, *a*, of which exists in the rounded, opaque, dull, olibanum of commerce; and the other, *b*, clearer yellow, generally less brittle, and in long tears. Olibanum is a stimulating gum resin, used in unnatural secretions of the mucous membranes, and as a perfume. It comes from various trees; Amyris gileadensis, Boswellia serrata.

Olidic Acid. An acid obtained by the action of fused potash on oleic acid, identical with palmitic acid.

Oligiste Iron Ore. A variety of specular iron ore.

Oligoclase. *Soda Spodumene.* Spec. grav. 2·64 to 2·66, H 6. SiO_3 63·55, Al_2O_3 23·32, MgO ·6, FeO ·55, CaO 3·07, KO 1·1, NaO 8·03. Yellowish and greenish-white oblique rhomboidal prisms; lustre vitreous, pearly, and greasy; transparent to translucent; fracture conchoidal, uneven. $CaONaO,SiO_3$; $Al_2O_3SiO_3$. Danvickszoll, near Stockholm, in granite; Arendal, Norway.

Oligon Spar. *Ferro Carbonate of Manganese.* 3, $FeOCO_2$,2,$MnOCO_2$. An isomorphous variety of spathic iron from Ehrenfriedersdorf.

Olivene. The olive variety of chrysolite common in basalts and lavas.

Olivenite. See Copper Arseniates.

Olive Oil. *Florence, Gallipoli, Lucca, Sweet, Salad Oil.* Spec. grav. ·9109 at 77°; solid at 36°. Pale yellow oil, soluble in 1½ time its weight of ether at 59°; obtained by expression from the pericarp of the fruit of the olive or Olea Europæa. Virgin oil is obtained by gentle pressure when cold; by stronger pressure and hot water an oil of second quality is obtained; while a third oil of inferior quality is procured by boiling with water, and used only for making soap. It is purified at Gallipoli by standing in contact with cold water. The peculiarity of olive oil is, that it becomes solid (elaidine) in contact with nitrous acid, which distinguishes therefore its adulteration with any other oil. A simple test is to dissolve 6 parts mercury in 7½ nitric acid, and mix 2 parts of this with 96 oil, and agitate; in 7 hours it becomes a thick magma, and in twenty-four hours hard. The presence of poppy and other oils prevents solidification. It is an article of food, and used in liniments and calico printing.

Oliviline. $C_{12}H_9O_4$. F.P. 158°, 248°. White needles without smell; taste sweetish bitter; obtained by alcohol from the resin of the olive tree (Olea Europæa), insoluble in ether.

Olivine. *Olivite.* Colourless crystals by alcohol from the leaves of the olive tree.

Omichmyle. The hypothetic base of a brown organic matter, with a strong smell of castor in concentrated urine.

Omphasite. Leek-green pyroxene.

Onchosine. An apple-green massive mineral, allied to agalmatolite; lustre greasy. Spec. grav. 2·8, H 2·. SiO_3 52·52, Al_2O_3 30·88, MgO 3·82, FeO ·8, KO 6·38, HO 4·6. B.B. fuses to a blebby glass; soluble in SO_3; not in HCl.

Onegite. A variety of pitchy iron ore from Lake Onega, Siberia.

Onion. The bulbous root of the Allium cepa. The juice undergoes, when kept at 60° to 70°, fermentation, while mannite and acetic acid (lactic?) is formed, which do not exist naturally; it contains gum, gluten, sugar, an oil with sulphur.

Onyx. A variety of agate, consisting of alternate horizontal bands of brown and white.

Oolite. (ὠον, small egg or roe.) The name of a limestone, consisting of spherical granules, extensively distributed in the oolite formation.

Oosite. A decomposed pinite, from Oos, Baden.

Opal. *Alumocalcite, Cacholong, Gyrasol, Geyserite, Hyalite, Hydrophane, Menilite, Michaelite, Opal jasper, Semiopal, Sinter, Wood opal.* Spec. grav. 2·015, 2·21, H 6·75. A variety of siliceous minerals possessed of some transparency, but rendered in some measure opaque by a different colour floating, as it were, in the stone. *Common opal* is milk-white, and also yellow, brown, red, and green; *Precious opal* exhibits a beautiful play of colours—of blue, green, yellow, and red, when held in the light; *Fire opal* exhibits a red play; *Wood opal* is fibrous; *Semiopal* has a flat chonchoidal fracture and resinous lustre; *Mother-of-pearl opal* has a pearly lustre. Common opal consists of silica 20·, water 1·125; precious opal of 93·9, SiO_3 and HO 6·1. Opal is generally soluble in caustic potash.

Ophite. A variety of serpentine, which appears to have been produced by the passage of a greenstone into a limestone; it seems also to be sometimes converted into gypsum by the action of sulphohydric gas, since gypsum usually accompanies ophite.

Ophitone?

Opiammon. $C_{40}H_{19}NO_{16}$. Pale yellow powder by heating opianate of ammonia till the ammonia escapes. It loses NH_3 and 4HO. Insoluble in cold water; soluble at 302°. *Teropiammon*, $C_{60}H_{29}NO_{26}$. Needles by nitric acid on narcotine.

Opianic Acid. $HOC_{20}H_9O_9$. F.P. 284°. Thin colourless prisms soluble in water; displaces carbonic acid; above its fusing point it has the same composition, but is insoluble in water, alcohol, and dilute alkalies; by the action of sulphuric acid and binoxide of manganese on narcotine.

Opianine. $C_{66}H_{36}N_2O_{21}$. Colourless rhombic needles; insoluble in water; soluble in much boiling alcohol, with an alkaline reaction; precipitated from its solutions by ammonia; soluble in NO_5, not in strong SO_3; narcotic; from Egyptian opium by precipitating by ammonia the aqueous solution and crystallizing from alcohol.

Opiano Sulphurous Acid. $C_{20}H_9O_{11}S_2$ HO. Crystalline bitter mass by the action of sulphurous acid on opianic acid; forms crystalline salts.

Opianyle. $C_{20}H_{10}O_8$. Needles, soluble in

alcohol and ether along with chlorine, by the further action of nitric acid on narcotine.

Opium. The inspissated juice of the Papaver somniferum or poppy. The following account of opium is by my friend and former pupil Dr. Eatwell of the Bengal army:—The variety album is alone cultivated in Benares and Behar. The plant reaches a height of 4 feet, and requires 3½ months to attain maturity, the seed being sown broadcast in November. When ripe, the capsules are scarified with a sharp iron instrument from top to bottom through the pericarp, and on the following morning the juice is scraped off with a kind of trowel; the opium is now dried in the air for several weeks until it contains only 30 per cent. of water when dried at 200°; it is then trod and formed into cakes. The juice of opium consists of morphine 5·5, narcotine 16·3, substances soluble in alcohol 260·5, insoluble matter or dry marc 112·6, water driven off at 200° 605. When expressed at 200° till it acquired a standard consistence, opium was found to give morphine 24·9, narcotine 30·9, soluble in alcohol 546·7, insoluble in alcohol 215·, water 182·5. The average morphine per cent. in Benares commercial opium is 3·21, and of narcotine 4·06; while the portion soluble in cold water is 47·99, and the residue per cent. after a temperature of 200° 75·5. Turkey opium contains from 6 to 9 per cent. morphine, Egyptian 3 to 4 per cent. The alkaloids in opium are, codeine, morphine, narceine, narcotine, opianine, papaverine, porphyroxine, pseudomorphine, thebaine, and besides meconic acid, brown extract, resin, fat, volatile oil, bassorine, caoutchouc gum, ligneous matter and salts. The presence of opium in solution can always be detected by the red colour given by a sesqui salt of iron with the meconic acid.

Opobalsam. *Meccabalm, Balm of Gilead.* Spec. grav. ·950, ·876 at 71½°. Yellow fluid of the consistence of honey, with an odour like oil of lemons or rosemary, becoming a resin in the air; it consists of volatile oil 30, hard resin 64, soft resin 4·, bitter colouring matter ·4. The volatile oil is obtained by distilling with water; odour agreeable; the hard resin has a spec. grav. 1·333, fusing at 196°; soluble in alcohol; the soft resin fuses at 233½, and is insoluble in alcohol (*burserine*). Opobalsam is obtained from the *Amyris Gileadensis* and *Opobalsamum.*

Opoponax. Spec. grav. 1·622. A gum resin in reddish-yellow lumps, white within, with a peculiar odour and bitter, acrid taste, forming a milky solution with water, one-half dissolving; by distillation a brown oil is obtained with acetic acid; it consists of resin 42, gum 33·4, wood 9·8, starch 4·2, malic acid 2·8, extractive 1·6, caoutchouc trace, wax ·3, volatile oil and loss 5·9. From Pastinaca opoponax, Levant, by wounding the roots of the plant.

Opsimose. Altered Fowlerite or silicate of manganese.

Orange Flowers, Oil of, or **Neroli Oil** of the South of Europe. Spec. grav. ·835, of vapour 4·65; B.P. 356°. C=87·22, H 11·66, or $C_{20}H_{16}$. Very fluid yellowish or colourless oil, reddening on exposure, lighter than water, with the odour of orange flowers; obtained by distilling orange flowers (Citrus aurantium) with water; it causes polarized light to deviate to the right. With chlorohydric gas it forms a solid ($C_{20}H_{16}$ 2HCl) and liquid camphor. When the oil is mixed with alcohol of ·842 *Aurade* falls in white scales, which is soluble in ether and alcohol, not in water. The oil called petit grain comes from the berries and leaves of Citrus bigarea.

Orangite. Spec. grav. 5·35, H 4·5. Orange coloured, translucent, foliated masses from near Brevig, Norway. SiO_3 17·69, Thorina 71·24, $CaOCO_2$ 4·04, Fe_2O_3 ·31, Mn_2O_3 ·21, KO NaO ·3, HO 6·9. The thorina at one time was supposed to be a new earth, and called donaria.

Orceic Acid. A synonyme of Orceine.

Orceine Alpha-. $C_{18}H_{10}NO_5$. Red powder, little soluble in water; colours it red, but is completely separated by a neutral salt; very soluble in alcohol, giving it a scarlet colour, and precipitated by water; scarcely soluble in ether; very soluble in potash and ammonia, with a fine purple colour; obtained by rendering orchil (*Parmelia roccella*), slightly acid by chlorohydric acid; evaporating to dryness; exhausting the residue by boiling alcohol; distilling the alcoholic fluids to dryness; washing the crimson residue with cold water to remove salammoniac, and exhausting by ether.

Orceine Beta-. $C_{18}H_{10}NO_8$. Mixed with the preceding, the proportions varying with the age of the lichen; it unites with 3 atoms oxide of copper and lead.

Orchil. *Archil, Orseille, Purple of Amorgos, Amorgis.* A dyestuff prepared from the lichen Parmelia roccella from the Canaries, &c. The plant is pulverized, sifted, and moistened with stale urine, the mixture stirred daily, and soda added till it becomes clove-brown; it is then covered in a wooden cask with urine, lime water, and a solution of gypsum. Orceine is extracted from orchil as described; the ether solutions are then evaporated, an oil remains, *erythroleic* acid. The residue of the lichen resisting alcohol is boiled in water, which takes up yellow-colouring matter, and there remains azoerythrine, which is purified by dissolving the residue in potash, saturating with an acid, and evaporating to dryness, and washing away the potash salt.

Orcine. $C_{14}H_8O_4$2HO. Colourless flat 4-sided prisms, with dihedral summits; taste sweet, disagreeable; soluble in water and alcohol; becomes blood-red by nitric acid, and deep violet by ammonia and air; reddish-brown by potash and soda; obtained by boiling lecanorine with barytes water; carbonate of barytes is deposited; the excess of barytes precipitated by carbonic acid, the fluid filtered and evaporated, when orcine crystallizes.

Orcino-lecanoric Acid, or Erythric acid.

Oreoselone. $C_{14}H_5O_3$. F.P. 374°. Crystals isomeric with dry benzoic acid; tasteless and odourless; soluble in alcohol and ether; obtained by uniting athamantine with dry chlorohydric acid, and boiling with water; crystals of hydrate of oreoselone deposit. It appears to be the glycerine of athamantine, of which it consists, united to valerianic acid, &c.

Ores. A term applied to the compounds of metallic elements found in nature.

Organic Analysis. *Determination of Carbon and Hydrogen.*—The object of organic analysis is to determine the weight of the elements present in an animal or vegetable

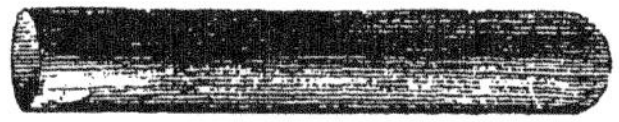

1

body. The substance purified is first deprived of moisture, by drying it in a water or oil bath. See BATH. The requisite quantity required is then weighed out in a light glass tube. It consists of carbon, hydrogen, and oxygen. To analyze it we require to convert the carbon into carbonic acid, and the hydrogen into water, by mixing and igniting it with oxide of copper, and collect the carbonic acid in caustic potash, and the water in a chloride of calcium tube. A quantity of black oxide of copper (prepared by igniting nitrate of copper, or heating copper in air), is ignited in a covered clay crucible, and while still hot it is transferred into a hard glass combustion tube of the form delineated (15 to 18 inches long) in figure 2, *a*, which is filled three-fourths with it. The organic substance is now triturated with the oxide of copper, poured from the combustion tube into a warm dry porcelain mortar. About an inch of pure oxide is allowed to remain at the closed end of the tube, and as much at the open end of the tube, free from the substance. A small air pump is then fixed to the tube, which is covered with hot sand, and the last traces of moisture thus extracted. A weighed tube, filled with chloride of calcium, is then adapted to the combustion tube by means of a cork, and at the

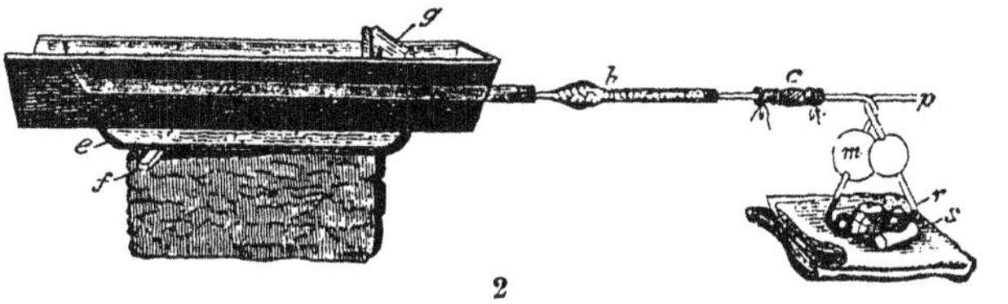

2

extremity of the chloride of calcium tube, a bulbed tube, *p*, *m*, (Liebig's potash bulbs), filled to *r* with a strong solution of caustic potash, is connected by means of a tube of caoutchouc, *c*. It is necessary before commencing the combustion to see that there is a free passage above the oxide of copper for the exit of the gas generated, and likewise that the apparatus is air-tight. This is ascertained by drawing out some air, by applying a *sucking tube*, fig. 3, at *p*, so as to make the potash fluid stand higher in the limb *m* than in *r*. If it remains permanently at a higher elevation, it is a proof that no air gains access along *a* to *m*. Heat is to be applied by means of red hot charcoal, not flaming, in small portions at a time, by means of the fender *g*, and the combustion is gradually extended along *a*, or till the whole tube has been heated, and the organic body oxidized. When bubbles of gas are no longer visible below *m*, the closed end of the combustion tube *a* is broken, a long tube, held so as to surround the broken extremity, and a current of air, drawn through the whole apparatus by means of the sucking tube at *p*. The tubes *b* and *m* are then detached, and, when cool, weighed; the increase in *b* gives the quantity of water equivalent to the hydrogen contained in the substance analyzed. This weight, divided by 9, gives the amount of hydrogen. The increase in the potash apparatus is due to carbonic acid. To obtain the carbon contained in it, we say as 2·75 : ·75 : : x : y. For the analysis of fluids, a light glass bulb is employed, drawn out into a capillary point. The bulb is heated, and the open end immersed in the fluid, and filled in the same manner as a thermometer. Fatty substances are weighed on the concave side of the half of a split tube which is slid into the combustion tube, and by heat made to mix with the oxide of copper. When the organic substance contains nitrogen, as in animal substances, it should be dried at 250° to 300°, and burned with chromate of lead instead of oxide of copper, or fused chlorate of potash may be mixed with the oxide of copper, or a current of oxygen may be passed from a gas-holder over the organic body. *See Liebig's Hand-Book of Organic Analysis*, an indispensable book to the organic analyst. *Determina-*

3

tion of Nitrogen.—When a substance contains nitrogen, a separate process must be employed for its determination. The substance (about 10 grains) is mixed with oxide of copper as before, the mixture occupying half the length of the combustion tube, fig. 4, *E*. One of the remaining fourths is filled with oxide of copper, and the rest with copper turnings, to decompose any binoxide of nitrogen formed. The gases which pass over, consisting of carbonic acid and nitrogen, are re-

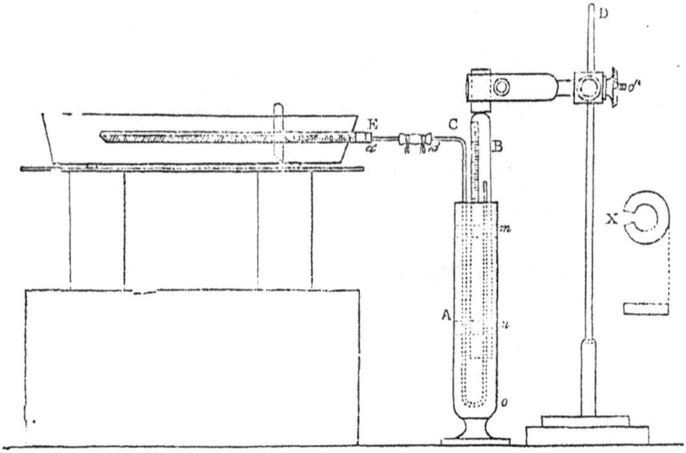

4

ceived in a graduated tube, *B*, filled with mercury. A series of these tubes (6 or 8) are thus filled and allowed to cool. They are severally transferred to a cylinder, A, filled with mercury,

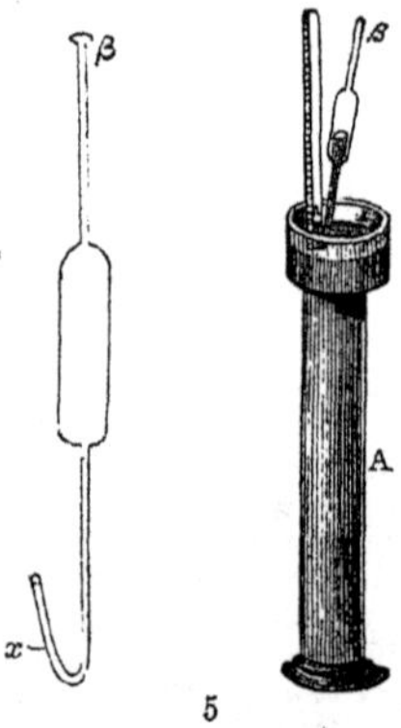

5

where, by means of a curved sucker, fig. β, a solution of caustic potash is admitted in contact with the gases, which absorbs the carbonic acid and leaves the nitrogen. To get the weight of the nitrogen, we have only to reflect that 1 volume carbonic acid corresponds to 1 atom carbon, and 1 volume nitrogen to 1 atom of this element. Hence the quantity of carbon, and the relation by volume between the gases obtained in the combustion being known, we can calculate the nitrogen from the atomic weight. According to the determination of carbon, caffeine contains 49·68 per cent.; this substance furnished nitrogen and carbonic acid in the ratio of 1 to 4; it contains, therefore, 1 atom nitrogen and 4 carbon.

Determination by Soda Lime.—The nitrogen, except in nitrates and some organic alkaloids, &c. may be estimated by mixing the nitrogenous body with 1 part caustic soda and 2 parts caustic lime in a mortar, transferring the mixture to a tube a foot long, closed at one end, fig. E, and connecting

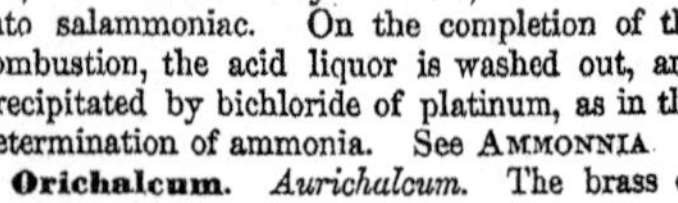

E

it with a bulbed tube containing muriatic acid. On the application of heat, as in the preceding cases, the nitrogen is changed into ammonia, and is absorbed by the acid, and converted into salammoniac. On the completion of the combustion, the acid liquor is washed out, and precipitated by bichloride of platinum, as in the determination of ammonia. See AMMONNIA.

Orichalcum. *Aurichalcum.* The brass of the ancients.

Ormolu. An alloy employed for the ornamental parts of lamps, &c. consisting of 52 zinc and 48 copper.

Orpiment. *Orpine.* Tersulphide of arsenic. See ARSENIC.

Orsellic Acid. When the Roccella tinctoria from South America is digested with milk of lime, the solution filtered, and chlorohydric acid added, a white gelatinous mass falls, which when dried dissolves in hot alcohol, and deposits white needles of *Alpha orsellic* acid, q.v.; insoluble in cold, little soluble in hot water. When this acid is mixed with water, carefully neutralized with lime or barytes, and boiled till dissolved; saturated with acid, the precipitate crystallized out of alcohol, crystals of *Alpha orsellinic* (*alphaorsellesic*) acid result, which yield with chloride of lime a transient reddish-blue. The Roccella tinctoria of the Cape of Good Hope, treated as above, yields *Beta orsellic*, and *Beta orsellinic acids*. The portion of gelatinous precipitate, insoluble in alcohol, is termed *rocelline*, $C_{38}H_{17}O_{15}$.

Orthite. Spec. grav. 3·288 to 3·63 and 3·65, H 7·. Black masses, and in rhombic prisms; powder gray, inclining to brown; fracture small, conchoidal; lustre vitreous, opaque; brittle, easily frangible. B.B. froths like zeolite, and becomes yellowish-brown, and melts into a black vesicular glass; fuses with difficulty with salt of phosphorus, leaving a silica skeleton, yellow in the oxidizing flame, disappearing on cooling; a clear glass with borax, greenish in the reducing, blood-red in oxidizing flame; decomposed but not fused with soda; gelatinizes with muriatic acid. SiO_3 36·25, CaO 4·89, Al_2O_3 14·, CeO 17·39 (LaO?), FeO 11·42, yttria 3·8, MnO 1·36, HO 8·7. Finbo, in thin veins in gneiss.

Orthoclase. *Orthose.* A synonyme of felspar.

Osmazome. An old name for the savoury portion of meat, soluble in water.

Osmelite. A synonyme of soda table spar, or Wollastonite, or pectolite.

Osmium. Os 12·375, 99. Spec. grav. 10· to 7·6. Grayish-white porous mass with metallic lustre, laminated compact mass or black powder, assuming the metallic lustre when compressed; procured by heating sesquichloride of osmium and ammonium to redness in a retort, or by passing a mixture of osmic acid and hydrogen through a tube partially ignited, or precipitating by zinc. *Separation of osmium from iridium.*—That portion of platinum ore which is insoluble in nitro-muriatic acid, consists of osmium and iridium; it is pounded first in a steel, and then in a porcelain mortar; the iron from the mortar removed by muriatic acid, mixed with its weight of nitre, and heated in a porcelain retort, first gently, and then to a full red heat as long as gas escapes; the vapour of osmic acid being passed into a tubulated receiver, and thence into a bottle containing dilute ammonia; the mass is mixed with water several times which is poured off, allowed to stand, decanted and distilled with chlorohydric acid; osmic acid passes over.

Protoxide.—OsO 13·375, ·107. Obtained by heating the *hydrate*, which is greenish-black, and formed by allowing a solution of potassio chloride of osmium to stand, when it deposits, but contains some potash; by hydrogen the metal is obtained.

Sesquioxide.—Os_2O_3. Known only in union with ammonia. Solutions of potassio sesquioxide of osmium in acids, when mixed with potash, deposit hydrous protoxide and leave in solution teroxide.—*Blue oxide*, obtained along with osmic acid by igniting in a retort containing air, seems a mixture of a proto and sesqui or of proto and binoxide. *Binoxide.*—OsO_2. Black powder, by heating potassio-bichloride of osmium with anhydrous carbonate of soda below a red heat The hydrate is obtained slowly by adding carbonate of soda to a solution of the preceding salt. The oxide and hydrate are insoluble in acids. *Teroxide, Osmous acid.*—OsO_3. Supposed to exist in the red solution on digesting a solution of potassio sesquichloride of osmium with potash.

Osmic Acid.—OsO_4. Colourless needles, subliming and fusing more easily than wax; does not redden litmus; taste acrid and sweetish; acts on the eyes and lungs, exciting expectoration; blackens organic substances; a weak acid not replacing carbonic acid; with the alkalies forms yellowish-red salts; obtained by passing oxygen over ignited osmium; by fusing osmio-iridium with nitre, and distilling with sulphuric acid.

Azo-osmic acid, Osman-osmic acid, Osmiamic acid.—OsN,OsO_4 or Os_2NO_5. A powerful acid known only in combination, decomposing carbonates and chloride of potassium. Obtained by adding caustic ammonia to a solution of osmic acid in an excess of potash, when a yellow crystalline powder separates; the acid may be obtained in solution by adding sulphuric acid to azo-osmate of barytes; the solution precipitates nitrate of silver, lemon-yellow, and crystalline. The salt of barytes consists of soluble yellow needles. The chlorides of osmium are OsCl; $OsCl_2$? $OsCl_3$? The sulphides are OsS; OsS_2, OsS_3; OsS_3; OsS_4. The sulphates are three in number, protosulphate, sulphate of blue oxide, sulphate of binoxide.

Osmium-iridium. See IRIDOSMINE.

Ossifications. A result produced in cartilaginous structures, in which, by the deposition of bone earth, the flexible cartilages assume the consistence of bone.

Osteosarcoma. (ὀστέον, bone, and σάρκος, flesh.) A species of tumour growing on animals, consisting of water 87·86, cartilage 9·85, albumen ·3, fats ·28, $CaOCO_2$ ·67, bone earth ·59, $NaOSO_3$ and $KOSO_3$ ·21, $CaOCO_2$ ·14, K and NaCl ·1.

Ostheolite. Spec. grav. 3·06. A chalky-like variety of apatite, found in Hanau in dolerite, containing 86 per cent. of phosphate of lime.

Ostranite. Spec. grav. 4·4, H 6·25. Brown rhombic prisms from the Zircon syenite of Fredericksvarn (?) B.B. infusible; fuses with soda with difficulty into a glass; insoluble in nitric acid.

Ostreocolla. A synonyme of carbonate of lime.

Otto. See ATTAR OF ROSES.

Ottrelite. Spec. grav. 4·4, H 6? Gray or greenish-black small brilliant plates; streak pale green; fracture uneven. B.B. fuses into a magnetic bead; dissolves slowly in borax, showing iron, and manganese with soda. SiO_3 43·34, Al_2O_3 24·63, FeO 16·72, MnO 8·18, HO 5·66. Ottrez, Luxemburg.

Ouwarowite. Emerald-green garnet; Bissersk, Russia.

Oxacids. Under this term are included all acids containing oxygen. They are divisible into two classes, those having a simple and those having a compound base. Thus the oxygen acids of sulphur have a simple base, sulphur, while the organic oxygen acids may be composed of two compound bodies united together. These may be termed conjugate or coupled acids. Thus tartaric acid can be resolved into oxalic and acetic acids.

Oxalates. Combinations of oxalic acid with basic oxides; they are semel, bin, and quater oxalates, the compounds being arranged in geometrical progression, and no intermediate salts being known; basic oxalates also exist as dinoxalate of lead ($2PbO,C_2O_3$) and tribasic oxalates. The alkaline oxalates and oxalates of glucina, chromium, manganese, and iron, are soluble in water, the others insoluble, or scarcely soluble. The anhydrous oxalates, when heated, leave oxides or metals, carbonic oxide and carbonic acid being evolved. The alkaline and earthy oxalates, when distilled, leave a carbonate, and give out carbonic acid, carbonic oxide, water, acetic acid, acetone and tarry matters. The oxalates may be distinguished from other salts by their yielding with hot sulphuric acid equal volumes of carbonic acid and carbonic oxide; alkaline oxalates are resolved by ferments into carbonates.

Oxalate of Lime or Mulberry calculus. See URINE.

Oxal-azo-phenylamic Acid. $C_{16}H_7N_3O_{10}$. Yellow grains, by the action of azophenylamine on oxalic acid.

Oxal-azo-phenylamide. $C_{12}H_5N_3O_8$.

Oxalhydric Acid. A synonyme of Saccharic acid.

Oxalic Acid. Acidum Sacchari (Bergman). C_2O_3 or CO,CO_2 4·5, 36; in crystals C_2O_33HO or HOC_2O_32HO 7·875, 63. Spec. grav. 1·507. This acid was discovered in 1776 by Scheele, according to Ehrhart, Hermbstadt, and Westrumb (Keir's Dict. and Elwert's Mag. for Apothecaries, 1785, p. 1, 54), although its discovery was long attributed to Bergman. It occurs as a bisalt with potash, as in Oxalis acetosella, Rumex acetosa, and also as oxalate of lime in the rhubarb, a plant also which owes its acidity to malic acid. It exists as sesquioxalate of iron in Humboldtite. Oxalic acid consists of transparent crystals, commonly as flat plates, but consisting of 6-sided irregular prisms, the primary form being an oblique 4-sided prism, with angles of 98° 30′, 63° 5′, 129° 20′, 103° 15′. The crystals, when heated a little above 212°, lose 28·57 per cent. (2 atoms) of water. The remaining 14·3 per cent. (1 atom) remains, and is not expelled till the heat be raised to 311°, when carbonic oxide, carbonic acid, and formic acids escape, and nothing remains. $12(C_2O_3HO)$ become by heat $10CO$, $12CO_2,C_2HO_3HO$ (formic acid and $10HO$). The simple hydrate of oxalic acid (HO,C_2O_3) sublimes above 212° in the form of woolly flocks. Nitric acid converts oxalic acid into carbonic acid, and sulphuric acid causes the evolution of carbonic oxide and carbonic acid. Binoxide of manganese, sesquioxides of cobalt and nickel, brown oxide of lead and chromic acid, convert oxalic acid into carbonic acid ($MnO_2\,2C_2O_3 = MnOC_2O_3, 2\,CO_2$) a reaction applied to determine the value of binoxide of manganese (q.v.). Crystals of oxalic acid dissolve in 2 parts water at 65°·7 (Bergman, 1·255), the spec. grav. of the solution being 1·0593, 8 parts of water at 60°, in 4 alcohol at 60° (Liebig); in their own weight of boiling water. 100 boiling alcohol dissolve 56 parts, cold alcohol 40 parts (Bergman). *Test.*—When pure, oxalic acid should leave no residue when dissolved in water; it should precipitate lime water and solution of sulphate of lime white; it should contain no nitric acid, and give out no NO_4 when heated. *Preparation.*—1. Oxalic acid may be prepared by heating 1 part of sugar or potato starch with 5 parts of nitric acid, spec. grav. 1·42, and 10 parts of water with the aid of gentle heat, until the evolution of gas ceases. The liquid by evaporation affords crystals, which are to be drained from the mother liquor either on paper or on a tile, until they are dry, when they are again to be dissolved in water and subjected to a second crystallization. Starch yields about the eighth part, and sugar about half its weight of acid. 2. Oxalic acid may likewise be prepared from binoxalate of potash, a salt in use in manufactures, by dissolving the salt in hot water and adding a solution of acetate of lead or of sulphide of barium. The precipitate is thrown on a filter, washed, and decomposed in the moist state with a slight excess of dilute sulphuric acid; sulphate of lead or barytes falls, which is separated by filtration, and the oxalic acid solution concentrated and crystallized.

Oxalite. A synonyme of Oxalate of iron or Humboldtine.

Oxalmethylovinide. *Oxalate of Ethyle* and *Methyle, Oxalvino-Methylide?* $C_{10}H_8O_8$.

Oxalovinic Acid. See ETHYLE, BINOXALATE OF.

Oxaluranilide. $C_{12},C_6HO_4,3NH_2H_2,O_2$. White pearly minute needles by parabanic acid and aniline.

Oxaluric Acid. $C_6\ NH_2\ O_4,\ O_4$. White acid crystalline powder, obtained by boiling parabanic acid dissolved in ammonia, and decomposing the oxalurate of ammonia formed by a mineral acid.

Oxalyle. C_2O_2. The hypothetic base of oxalic acid.

Oxamelanile. $C_{30}H_{11}N_3O_4$. Yellow crusts by chlorohydric acid on dicyanomelaniline.

Oxamethane. *Oxamate of ethyle.* $C_4NH_2H_5C_4O_6$. Pearly plates by passing dry ammonia through a tube in a retort containing dry oxalic ether.

Oxamethylane. C_2H_3,NH_2,C_4O_6. White crystalline body, by passing dry ammonia over heated oxalate of methyle (oxalic mether).

Oxamic Acid. $C_4\ NH_2\ HO_2,\ O_4$. White crystalline grains; obtained by heating binoxalate of ammonia in an oil bath till it fuses and begins to swell up. The cold aqueous solution of oxamic acid is then neutralized with ammonia, crystallized out, and decomposed with dilute sulphuric acid; it is monobasic, and forms crystalline salts.

Oxamide. $C_4NH_2O_2,O_2$. White granular powder, or crystalline plates, tasteless, odourless, neutral; in a retort incompletely fusible, partially subliming and partially decomposing into cyanogen and carbon; soluble in 10,000 cold water; converted by chlorine water into HCl, $C_2\ O_3$; by boiling with nitric acid into NO and $2CO_2$; in a metallic tube hermetically sealed and heated to 590° it yields Cy,CO_2 and NH_3. Obtained by adding ammonia to oxalic ether or by distilling oxalate of ammonia, when a portion passes into the receiver and a portion is sublimed, mixed with carbonate of ammonia. Oxamide is the type of the amides, and is oxalate of ammonia deprived of hydrogen and oxygen, in the proportion to form 2 atoms water; when heated with an alkali an oxalate is formed and ammonia evolved.

Oxamide-Oxanilide. *Oxanilamide.* $C_{16}H_8N_2O_4$. Crystals by chlorohydric acid on cyaniline.

Oxamylane. $C_{10}NH_2H_{11}C_4O_6$? Crystals by dry ammonia on dry oxalate of amyle.

Oxanilamide. $C_{12}\ (C_4\ HO_2)\ 2\ NH_2\ H_3\ O_2$. White silky flocks, by evaporating cyaniline and chlorohydric acid together.

Oxanilic Acid. $C_{12}(C_4\ HO_2)NH_2\ H_4\ O_4$. Fine acid plates, by heating aniline and an excess of oxalic acid for ten minutes.

Oxanilide. $C_{28}\ H_{12}\ N_2\ O_4$. White pearly scales, by the action of dilute acids on cyaniline.

Oxhaverite. Pale green Apophyllite from Iceland.

Oxibromides. Compounds of bromine with an oxide.

Oxichlorides. Compounds of chlorine with an oxide.

Oxidation, *Oxygenation, Oxidizement,* are terms which express the union of oxygen with another substance, and the formation of a compound body. If the product unites with acids it is termed an *oxide*, as when iron or zinc are burned in oxygen gas. If the product unites with an oxide, it is termed an *acid.* When oxygen is removed from a substance with which it is united, as for example when dry hydrogen is passed over oxide of copper at a red heat, the process is termed *deoxidation* or *deoxidizement.* The proportions in which oxygen unites with the same body are sometimes exceedingly numerous. See NOMENCLATURE.

Oxides. The oxides may be divided into *basic oxides*, which unite with acids, *acid oxides*, which neutralize basic oxides, that is, combine with them and overcome many of their properties, such as the character of the alkaline oxides to turn violets green, and themselves lose the power of reddening litmus. The *neutral oxides*, such as the binoxide of manganese (MnO_2), do not unite with either bases or acids. The term *saline oxides* has been applied to those compounds which are produced by the union of two oxides of the same metal, as the protosesquioxide of manganese ($MnOM_2\ O_3$) and lead ($PbOPb_2\ O_3$). The protoxides (MnO and PbO) may be viewed as acting the part of basic oxides, and the sesquioxides (Mn_2O_3 and Pb_2O_3) that of acids.

Oxygen. O 1, 100, 8. *Syn.* Spiritus nitro aereus (Mayow, 1672); air (Hales, 1724); dephlogisticated air (Priestley, 1774); feuerluft, or empyreal air (Scheele, 1777); oxygen, from ὀξυς, acid; and γεινομαι, I form (Lavoisier, 1787); vital air (Condorcet).

Source.—Nearly one-fifth of the bulk of the atmosphere consists of oxygen gas. Water, which constitutes the most abundant substance on the earth's surface, contains 1 part of oxygen in every $1\frac{1}{8}$ part by weight. The rocks also, which form the crust of the globe, with the animals and vegetables which subsist on it, contain oxygen in abundance.

Physical Characters.—Oxygen is a transparent, permanent, colourless gas, without taste and smell, possessing all the mechanical properties of common air. Its specific gravity 1·1111 (Thomson), 1·1026 (Berzelius and Dulong), 1·1057 (Dumas, An. Chim. 8, 201), 1·1056 (Regnault, An. Chim. 3d ser. 14, 211). Weight of 100 cubic inches at 60°, and 30 inches of pressure (100 cubic inches of air weighing 31·0117 grains), 34·1898 (Saussure and Regnault), 34·457 (Thomson), 34·289 (Dumas), 34·2138 grains (Regnault), that of air being 30·948824. Oxygen has never been condensed into a liquid, although it has been exposed to intense pressure of many atmospheres (Faraday). 100 cubic inches of water, from which the air has been expelled by boiling, dissolve 3·55 cubic inches of oxygen (Henry, Phil.

Trans. 1808). When rapidly condensed in a syringe, it evolves a heat exceeding 390°, produces a bright light, and inflames the oil of the syringe (Thenard, Döbereiner). Its refracting power, in reference to light, is small, being to that of air as ·924 to 1 (Dulong, An. Chim. 31, 166), or ·861 (Biot and Arago). Its specific heat is ·9765. When a body containing oxygen is exposed to the galvanic action, the oxygen always appears at the positive pole, and hence it is termed an electro-negative body, being the most electro-negative of any known substance.

Preparation of Oxygen Gas.—This gas is procured from various bodies, containing oxygen in a solid form. 1. *From Binoxide of Manganese* (Scheele, 1775).—The most economical substance for supplying oxygen is binoxide, or black oxide of manganese (MnO_2), a mineral which occurs in Devonshire and on the Rhine; 3 atoms of it yield 2 atoms of oxygen, and leave the red oxide (Mn_3O_4), according to the following scheme:—
3 Atoms Binoxide of Manganese.

3·447 Mn	3·447 Mn	3·447 Mn	1 1 1 1 1 1 O O O O O O
14·337 $Mn_3\,O_4$ or Red Oxide.			O_2 2 Oxygen.

3 atoms of binoxide may be also represented by the formula $Mn_3\,O_6$. When ignited 2 atoms are removed and $Mn_2\,O_4$ or $MnO\,Mn_2\,O_3$ remains, usually termed red oxide of manganese. It is always formed when the oxides of manganese are exposed to a red heat. The process admits also of a simpler statement, viz. that the binoxide loses $\frac{2}{3}$ of an atom of oxygen, so that 5·447 grains of black oxide are changed to 4·78 of red oxide (and yielding 12·3 per cent. oxygen). Commercial manganese, however, is seldom quite pure, and contains frequently carbonate of lime and other foreign matters, so that the product is inferior to those numbers. Manganite or sesquioxide of manganese ($MnO1\frac{1}{2}$ or Mn_2O_3) likewise yields a small portion of manganese. 5 grains being changed into 4·833 red oxide yield consequently only ·166 grains or 6 atoms = 30 grains afford 1 grain of oxygen or about 3 per cent. A convenient apparatus consists of an iron bottle, the mouth of which is ground so as to have a tube adapted to it. The bottle is nearly filled with binoxide of manganese in powder; a tube supplied with a joint is then luted to it by means of pipe-clay, the farther end of the tube being introduced into the lower aperture of a gas-holder; the bottle is surrounded by hot coals in a furnace, and is raised to a bright red heat. The tube should not be inserted in the gas-holder until the common air has been expelled from the bottle and tube, and until a glowing splinter of wood is re-ignited on being placed in contact with the gas issuing from the tube. When it is required that the gas should be thoroughly free from carbonic acid, an intermediate vessel filled with milk of lime should be arranged so as to wash the gas before it enters the gas-holder.

2. From *binoxide of manganese* and *sulphuric acid.*—The theory of this process is, that sulphuric acid will only combine with manganese when it contains 1 atom of oxygen; consequently the second atom, or excess of oxygen, is evolved, as appears from the diagram.

Sulphuric acid.	Black oxide of manganese.		
5 SO_3	3·447 Mn	1 O	1 O = 10·447
9·447 Mn O SO_3 Sulphate of manganese.			O = 10·447 1 Oxygen.

Common oil of vitriol ($HOSO_3$) is poured into a retort or flask, and a sufficient quantity of black oxide of manganese in powder is added, to form a paste; the apparatus being adjusted (see Gas, collection of), heat is applied to the retort and oxygen is evolved, while sulphate of manganese remains in the retort. Pure black oxide of manganese yields about 18·3 per cent. of oxygen by this process, while the sesquioxide ($MnO1\frac{1}{2} = 4{\cdot}947$) loses ·5 of an atom in 5 grains, or about 10 per cent. When the manganese contains common salt, or the sulphuric acid chlorine, that gas will contaminate the oxygen; it will be absorbed by allowing the gas to stand for some time over water, with frequent agitation.

3. *From Chlorate of Potash.*—Oxygen may be obtained in a pure state from this salt, which is a compound of chlorine, oxygen, and potassium, and occurs in commerce in the form of colourless crystalline plates. When heat is applied to this salt, the whole of the oxygen contained in it is removed, while the chlorine and potassium remain in the state of chloride of potassium ($KO\,ClO_5 = KCl$ and O_6). The chlorate of potash, in consequence of its property of yielding oxygen with facility, is much employed in the manufacture of lucifer matches. The salt is to be mixed with one-third its weight of dry black oxide of manganese, or, better still, black oxide of copper (Mitscherlich), well triturated in a mortar, and then introduced into a dry retort, by means of a funnel. The heat of a spirit-lamp or gas flame is then to be applied to the salt in the belly of the retort; the beak of the retort being placed under an inverted jar filled with water, inserted in a gas or pneumatic trough. The first bubbles of gas which pass over are not to be collected, as they consist of the common air of the retort; but as soon as the bubbles be-

gin to succeed each other rapidly, they are to be collected.

Chemical Characters.—1. Oxygen can be recognized by the peculiarity which it possesses of relighting a splinter of wood or taper, which has been made to glow, the flame being extinguished.

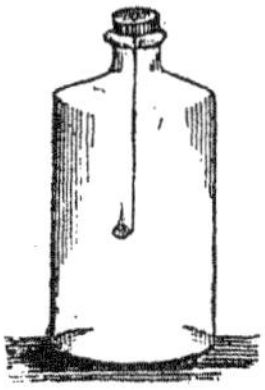

When introduced into oxygen it bursts into a flame, with an audible sound. In this case the oxygen is said to support the combustion of the wood. It unites with the carbon and hydrogen of the wood, and forms carbonic acid and water (CO_2 and HO). That carbonic acid is produced may be proved by pouring in lime water into the jar in which the experiment is made, when a white precipitate of carbonate of lime will fall. 2. That oxygen is a *supporter* of combustion, and not a *combustible* body, may be shown by inverting a jar full of this gas over a candle; the oxygen does not take fire, but the flame of the candle increases in size and brilliancy; the carbon and hydrogen of the candle unite with the oxygen, and form carbonic acid and water, as in the former case. If we repeat this experiment with a jar filled with common coal gas, we shall find that the flame of the candle will be extinguished, but that the gas will burst into flame at the mouth of the jar, and burn only where it is in contact with the oxygen of the air. Coal gas is a *combustible* body, and oxygen of the air supports its combustion. The carbon and hydrogen of the gas unite with the oxygen in the combustion and form carbonic acid and water. 3. When an iron or copper dipper, supplied with a wire, is inserted in a cork, and some powdered *charcoal* ignited in the ladle over a lamp, is immersed in oxygen, the charcoal kindles, glows with great vivacity, and even bursts into flame. Charcoal is principally composed of carbon, one of the elements of nature, and oxygen being also an element; when the two unite they form the compound, carbonic acid (CO_2). The presence of carbonic acid may be proved by means of lime water. Sulphur may also be burned in a similar manner. 4. So great is the *affinity* of oxygen for simple bodies, that iron can even be made to burn in it. Roll up a piece of iron wire spirally upon a thin tube,

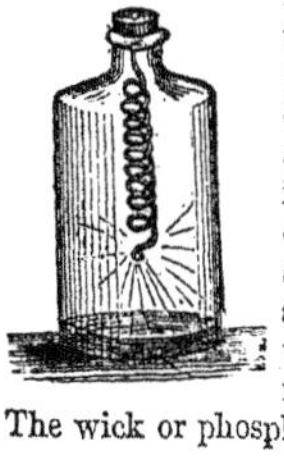

and insert one of the extremities in a cork, and to the lower end fix a loop of cotton wick dipped in melted sulphur, or a piece of phosphorus, or glowing amadou, or tinder. The oxygen should be contained in a jar, the bottom of which is strewed with sand, and covered with water, or the jar may ue open at both ends. The wick or phosphorus being ignited, the fluid is to be rapidly introduced into the oxygen, when the wire will be found to burn with great splendour (Ingenhousz, 1782, Nouv. Exp. p. 411). Oxygen may also be obtained by heating red oxide of mercury, or red lead, or bichromate of potash with sulphuric acid.

Oxygen is a term used by bleachers to designate bleaching powder.

Oxy-Hydrogen Blowpipe. A contrivance for projecting an ignited current of oxygen and hydrogen upon a body. The use of this blowpipe for chemical purposes appears to have originated at the same time (1801) with Dr. Thomson and Dr. Hare. The apparatus of Dr. Thomson is very convenient for class-room experiments. It consists of two vessels of tin japanned, one for hydrogen, double the size of that intended to hold the oxygen. These are filled with gases, as in the common gasholder. Connected with these gasholders are two tubes which terminate above in a large reservoir of water, which, by means of stop-cocks, is allowed to pass down into the gas-holders to press out the gases when they are required. The two gases, when they are to be used, escape by two tubes into a common nozzle tipped with platinum, at the point of which they are ignited. The gases being in separate vessels, there is no risk of explosion, except from a leakage in the hydrogen vessel, which would admit common air. The ignited hydrogen flame might then rush back into the mixture and explode it. Care should therefore be taken to have it perfectly airtight, and, to prevent all risk, the two tubes should be connected by Hemming's safety jet. All the illuminating and heating powers of the oxy-hydrogen blowpipe may be exhibited without much diminution of effect, by substituting common coal gas for the hydrogen. The two gases may then be kept in two separate gasholders, or mixed in the same bladder supplied with a Hemming's jet, as already described.

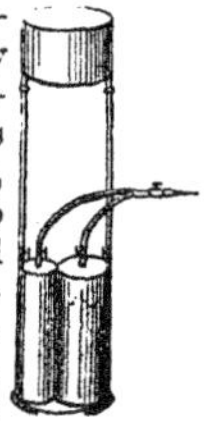

Oxylizaric Acid. $C_{15}H_5O_5$. Crystals from madder by the action of alum, in solutions of which it is soluble; soluble in hot water, alcohol, ether, alkalies, and sulphuric actd.

Oxymel. A mixture of vinegar and honey.

Oxymuriatic Acid, and **Oxymuriates.** The former name of chlorine and chlorides.

Oxyphenic Acid. A synonyme of *Pyromorintannic*, or *Phenic acid.*

Oxypicric Acid. See STYPHNIC ACID.

Oxyporphyric Acid. A synonyme of porphyric acid.

Oxyprussic Acid. A synonyme of Chlorocyanogen.

Oxysulphides. Compounds of sulphur with oxides.

Oyster. Ostrea edulis. An excellent edible

mollusc, affording pure fibrine when treated with water, alcohol, and ether.

Ozarkite. A mineral from Ozark's mountains, Arkansas. The specimens I have seen are crystalline with the colour of sphene, with a specific gravity of 2·638; H 5·5, and about 37·5 per cent. of silica; decomposed by acids.

Ozokerite. Spec. grav. ·946, ·953. F.P. 143° to 183°. B.P. 260°, 410°—570°, CH. Brown resinous waxy substance, green by transmitted light; slightly soluble in alcohol and ether; obviously a mixture. Moldavia; Urpeth, Newcastle.

Ozone. (Schönbein). The peculiar odour engendered when water is decomposed by the galvanic battery is due, according to Schönbein, to the isolation of a body, ozone. 1. It is readily prepared by introducing into large stoppered bottles, a thin layer of water, and allowing some sticks of phosphorus to stand half immersed in the water, and half in the air of the bottles, which are closed with stoppers. In twelve hours the phosphorus and water being removed, and the bottles washed out with pure water, the air of the bottles is charged with a peculiar smell, that of ozone (ὀζω, I smell). 2. It is also produced by passing electric sparks through dry oxygen gas. 3. The same odour is perceptible in dry air which has been passed from a gasometer through a tube 3½ feet long and ¼ inch in diameter, containing throughout its length sticks of phosphorus. 4. It is also formed in the decomposition of water by the galvanic battery.

Characters.—Ozonized air may be tested chemically by means of a piece of paper soaked in a paste of iodide of potassium and starch. When immersed in the ozonized air the iodine is set free, unites with the starch, and changes the paper to a blue or yellow colour, a result also of chlorine and nitrous acid. Yellow prussiate of potash immediately removes the odour from ozonized air, and is converted into red prussiate. It decomposes indigo, iodide, and bromide of potassium. In presence of water it converts iodine into iodic acid, sulphurous acid into sulphuric acid, protoxide of iron into sesquioxide. The vapour of ether when burned in common air has a similar oxidizing effect; a jet of hydrogen burned in air also oxidizes, and likewise the flame of a common candle, and different combustibles. According to Schönbein it is a peculiar substance; according to Marignac it is a modification of oxygen, or some peculiar compound of oxygen and hydrogen. According to Williamson it is a different substance, as prepared by the different processes. When furnished by the battery it is a hyperoxide of hydrogen; by the electric spark in air it is nitrous acid; by phosphorus it is hyperoxide of hydrogen. When test paper for ozone is acted on in the open air, the presence of ozone it has been inferred is indicated; chlorine and nitrous acid in the air would, however, produce similar effects.

P

Packfong. The Chinese name of Argentane.

Paco. A synonyme of Red Silver ore.

Pagodite. A synonyme of Agalmatolite.

Paints. See PIGMENTS.

Palagonite. Spec. grav. 2·43, H 5·. Silica 39·98, Al_2O_3 8·26, Fe_2O_3 17·65, CaO 8·48, MgO 4·45, KO ·43, NaO ·61, HO 18·25, matrix 1·89. *Form.* $3RO,2SiO_3+2Al_2O_3SiO_3+HO$. Frollkonugil, near Hecla, Iceland. A yellow or brownish-yellow metamorphic rock, with a resinous somewhat glassy lustre, found in the tuff of Sicily, Iceland, Galapagos. It appears to have been produced by the action of lava on limestone; and an analogous change appears to be produced wherever limestone comes in contact with pyroxenic rocks. It may be imitated by calcining together 1 part of levigated basalt with 13 of caustic lime exposed to the air. A pelagonitic sand is formed by introducing powdered basalt into a great excess of fused caustic potash, pouring water over the whole, and washing; it gelatinizes with acids (Bunsen).

Palladammonium. *Palladamine.* N,PdH_3. Microscopic 8-hedrons or brown amorphous resinous mass, not changed at 212°; obtained by precipitating a solution of chloride, bromide, or fluoride of palladium, &c. with ammonia; dissolving the precipitate in an excess, and saturating with the acid required to form the peculiar salt. From the chlorine and bromine salts the hydrogen acids precipitate the palladammonium as yellow crystalline precipitates. The base may be isolated by mixing the chlorine compound with oxide of silver, or precipitating the sulphate with hydrate of barytes; a yellow alkaline solution is obtained, which by evaporation in vacuo yields the base; when free from carbonic acid, it is quite soluble in water.

Pallad-anil-ammonium. $N,C_{12}H_7,H_3$, Pd. By adding aniline to chloride of palladium.

Pallad-ethylammonium. N,C_4H_5,H_2,Pd. Obtained by ethylamine on chloride of palladium.

Palladi-ammonium. *Palladiamine.* N,NH_4H_2 Pd. Crystalline mass; a base formed on the type of ammonium (NH_4), in which 2 atoms hydrogen are replaced by 1 palladium and 1 ammonium; obtained by treating a palladium or palladammonium salt with excess of ammonia, and evaporating; colourless crystals separate, from which the base may be isolated, as in the preceding instance.

Palladi-ethyl-ammonium. N_4H_7N,H_2, Pd. By adding ethylamine to chloride of palladammonium.

Palladiocyanogen. Cy_2Pd. A radical obtained in union with potassium, when ferro-cyanide of potassium is fused with palladium in powder.

Palladium. *Native.* Sp. grav. 12·14; 11·8, H 4·75. Steel-gray fibrous 8-hedrons with a square base? or rhombohedrons. B.B. infusible per se, but readily melts with sulphur, and leaves a globule of metal; opaque; metallic lustre; consists of palladium with some platinum and iridium. What has been described as a selenide appears to be native ore, and so does eugenesite.

Palladium. (Wollaston, 1803). Pd 6·6625, 53·3; 6·75, 54·0. Specific gravity fused 11·04, 11·3 to 11·8, hammered 11·852, 12·, 12·148; 6-sided tables as hard as platinum, or gray spongy mass; infusible in a blast furnace; obtained from the solution of palladium ore in nitro chlorohydric acid, by the addition of cyanide of mercury. The solution contains platinum, palladium, rhodium, and some iridium; cyanide of palladium falls which leaves on ignition palladium; palladium is also obtained by igniting the ammonia-chloride.

Protoxide, Semeloxide.—Pd O 7·6625, 61·3. Black mass or powder, obtained by igniting palladium, or the protonitrate of palladium, in the air. This oxide forms salts with acids, in which the hydrate is soluble. The hydrate is obtained by precipitating a solution of the protosalts of palladium with carbonate of soda.

Binoxide.—Pd O_2. Black powder by pouring caustic soda, or its carbonate in solution, on the dry potash-chloride of palladium, boiling and washing the precipitate; soluble with difficulty in acids, and forming salts, the solutions being yellow.

Protochloride.—Pd Cl. Blackish-brown or rose sublimate, obtained by evaporating the solution of palladium in nitro chlorohydric acid, or by heating sulphide in chlorine gas; decomposed by ignition; a valuable test for iodides.—*Bichloride* obtained only in solution by dissolving the binoxide in chlorohydric acid. *Iodide.*—Pd I 22·4125, 179·3 = Pd 29·73 I 70·27. Black mass by precipitating a protoxide salt with iodide of potassium, obtained in estimating iodine by chloride of palladium. *Bromide*, also a brown mass, obtained by dissolving palladium in a mixture of bromohydric and nitric acids, and evaporating. *Protonitrate* is obtained also by dissolving palladium in nitric acid containing nitrous acid. *Sulphate.*—Brown crystals, procured by dissolving the metal or the oxide in sulphuric acid. The *ammonia protochloride*, NH_3PdCl, or *red* salt, is formed in rose scales by adding a slight excess of ammonia to the protochloride of palladium. The *yellow* salt seems to possess the same composition, is in fine needles, and contains half its weight of palladium; obtained by precipitating protochloride with ammonia, redissolving the precipitate with excess, and evaporating or adding chlorohydric acid. The *diammonia protochloride*, $2NH_3$, PdCl,HO, white crystals, is procured by dissolving the red or yellow compound in ammonia and evaporating, keeping up an excess of ammonia.

Palmic Acid. $HO,C_{34}H_{32}O_5$; F.P. 113° to 114°·8. Small needles obtained by saponifying castor oil, and decomposing the soap by chlorohydric acid; the acids are then diffused through dilute nitric acid, through which a current of nitrous vapour is passed; the fat acid becomes yellow and waxy, and is purified by crystallization out of alcohol.

Palmine. F.P. 151°. $C_{37}H_{34}O_6$. White waxy or resinous body, obtained by acting on castor oil with nitrous acid; very soluble in alcohol and ether.

Palmitic Acid. $C_{32}H_{31}O_3$,HO. F.P. 140°. Pearly scales, soluble in alcohol, procured by saponifying palm oil or palmitine, or from coffee berries.

Palmitine. $C_{35}H_{33}O_4$. F.P. 118°. White crystalline body; soluble in hot ether; little soluble in hot alcohol; consisting of palmitic acid and oxide of glyceryle or lipyle.

Palmitonic Acid. $C_{31}H_{30}O_3$ HO. F.P. 125°. White granular or star-like flakes; procured by exposing palmitic acid to a temperature of 482° to 572° in air; water and carbonic acid being evolved.

Palm Oil. F.P. 90°? A deep yellow oil from the fruit of the *Elais Guineensis*, the oil palm of the coast of Guinea. It is much used in this country for making palm soap, and for this purpose requires to be bleached. This may be effected by passing steam through a worm immersed in it in the open air, or by chlorine evolved by mixing the oil with bichromate of potash and muriatic acid.

Panabase. A synonyme of gray copper ore.

Pancreatic Juice. Spec. grav. 1008·2. A limpid, viscid, alkaline fluid, secreted by the pancreas or sweet bread, resembling saliva; it contains 917 to 963 water, 36 to 15 soluble in alcohol, 15·3 to 2·8 caseine, &c. soluble in water, and 35 to 22 albumen and salts. When mixed in a glass tube with some oil or fat, the latter becomes immediately an emulsion, a fat acid and glycerine being evolved. The white appearance presented by the chyle has been ascribed to this emulsion; but it does not seem that the presence of pancreatic juice is essential to the absorption of fat by the chyliferous vessels. The pancreatic fluid converts starch into sugar.

Panification. (*Panis*, bread; *facere*, to make.) See BREAD.

Papaverine. $C_{40}H_{21}NO_8$. Needles soluble in cold and hot alcohol; insoluble in water; slightly soluble in ether; rendered blue by sulphuric acid; forms crystalline salts; obtained by treating with alcohol the brown resinous mass from crude morphine precipitated by soda from the aqueous solution of opium; the brown extract is evaporated, the residue digested with dilute acid, filtered and precipitated by ammonia.

Papin's Digester. A strong iron box, fitted with a screw, by which water may be boiled, and raised to a temperature above 400°.

Papyrine. *Woody fibre.* C 43·3, H 6·2, O 49·3. A membranous substance, obtained by steeping woody fibre in sulphuric acid for half-a-minute, and washing for a few seconds in water.

Parabanic Acid. $C_6N_2O_4 2HO$. Crystalline mass formed by heating alloxane or uric acid with an excess of nitric acid, or from the acid mother liquors of alloxane.

Paracacodyle, Oxide of. C_4AsH_6O. An oil by distilling cacodylic acid.

Parachlorocyanic Acid, with NH_3. A synonyme of Chlorocyanamide, $C_6N_3Ad_2Cl$, by ammonia on chlorocyanogen.

Paracholic Acid. See Bile.

Paracitric Acid. A synonyme of Citracic or Citraconic Acid.

Paracolumbite. In granite, containing columbic acid, in the United States?

Paracomenic Acid. $C_{12}H_4O_{10}$. Isomeric with comenic acid; in feathery crystals by the dry distillation of meconic or comenic acids.

Paracopaiba Oil. $C_{10}H_8$. Spec. grav. ·91; B.P. 485½°. A variety of Copaiba oil.

Paracyanic Acid. C_8N_4O?

Paracyanogen. $C_{12}N_6$. Black substance without smell and taste, by heating some metallic cyanides in close vessels as cyanides of mercury, silver and ferrocyanide of lead; soluble in hot sulphuric and muriatic acids, and caustic potash.

Paraellagic Acid. By heat from ellagic acid.

Paraffine. F.P. 115°·3, 110°; 105° to 115° (from coal, R.D.T.). $C_{20}H_{21}$? Spec. grav. ·87. White waxy body or silvery scales, obtained from the last portions in the distillation of coal tar, and crystallizing from ether; it is procured in large quantity in the distillation of Boghead coal at a low red heat (J. Young), and in the distillation of peat. It is probably the type of a series to which methylene, butylene, &c. belong, which are the bihydrides of the radicals of their respective acids. It has little affinity for other bodies, and hence its name.

Paraffine Oil. A clear transparent oil, lighter than water, used for machinery, apparently a mixture, consisting essentially of solid paraffine dissolved in an isomeric fluid oil; obtained from bituminous coal, such as parrot, cannel, or gas coal. The coals are broken into small pieces and introduced into a common gas iron retort supplied with a worm pipe passing through a refrigerator kept at a temperature of 55° by a stream of cold water; the retort is heated up cautiously to a low red heat, which is never pushed higher than the point required for the separation of the fluid oil; at a higher temperature gas is evolved. To purify the oil it is placed in a cistern and heated to 150°, when impurities separate from it; it is then distilled; the distilled oil mixed with 10 gallons of oil of vitriol to 100 gallons oil, and allowed to stand for twelve hours in a lead vessel; the supernatant oil is drawn off and mixed with 4 gallons caustic soda (spec. grav. 1·3) to 100 oil, and allowed to stand for eight hours in an iron vessel; the oil is run off and again distilled; a more volatile oil may now be separated by boiling the paraffine oil with water; the steam carrying over the volatile oil. It is further purified successively by sulphuric acid and chalk. When the oil is cooled to 40° or lower it deposits solid paraffine. It is a valuable oil for lubricating machinery (J. Young).

Parafumaric Acid. A synonyme of Maleic acid.

Paragenesis of Minerals. The production of minerals depending on the existence of others.

Paragonite The talc slate of St. Gothard. Without magnesia.

Paraguay Tea. *Yerva Maté.* The leaves of the Ilex Paraguaiensis, used as tea in Brazil. They contain colouring matter, two resins, tannic acid, oil, albumen, and theine or psoraleine.

Paralbumen. The albumen obtained from ovarian dropsy, precipitated imperfectly by acetic acid and boiling.

Paramaleic Acid or Fumaric acid.

Paramenispermine. A base contained in Cocculus Indicus.

Paramide. C_8HNO_4. A yellow substance like clay, obtained with euchroic acid by heating mellate of ammonia to 320°.

Paramidic Acid. $C_{24}H_5N_3O_{14}$. White needles by pouring ammonia on paramide and adding it to chlorohydric acid; it precipitates.

Paramorphine. A synonyme of Thebaine.

Paramorphism. Bodies of a similar composition, such as the naphthaline compounds, which crystallize in different systems, are paramorphous.

Paramucic Acid. $C_{12}H_{10}O_{10}, O_6$. Rectangular tables, obtained by evaporating an aqueous solution of mucic acid to dryness, dissolving in alcohol and evaporating spontaneously; its salts are more soluble than the mucates; its solutions in boiling water yield mucic acid.

Paramylene. A synonyme of Valerene.

Paramylon. $C_{12}H_{10}O_{10}$. White granules like wheat starch, insoluble in hot and cold water; unacted on by diastase, till boiled with fuming chlorohydric acid. At 392° turns brown without fusing and becomes soluble in water; not coloured by iodine; obtained from the Eugenia viridis by treating the green scum of stagnant waters with ether and alcohol, and then with boiling muriatic acid and alcohol.

Paranaphthaline. See Anthracene.

Paranicene. Spec. grav. 1·24, of vapour, 4·79; B.P. 689°. $C_{20}H_{12}$. Yellow solid in plates, with strong smell and taste; obtained by distilling chloroniceate of barytes.

Paranicine. $C_{20}H_{13}N$. Yellow substance by treating paranicene with ammonia and sulphohydric gas; forms crystalline salts.

Paranthine. A variety of Scapolite, so named from the rod-like aspect of its crystals: Arendal, Norway.

Parapectic Acid. $C_{24}H_{15}O_{21}$ 2 HO. An acid amorphous body by boiling pectic acid in water; bibasic.

Parapectine. Amorphous; by boiling pectine for some hours; it is precipitable by acetate of lead; isomeric with pectine.

Parasalicyle. $C_{28}H_{10}O_6$. 4-sided prisms, subliming; by the action of chloride of benzoyle on salicylic acid.

. **Paratartaric Acid.** A synonyme of Racemic acid.

Paratartaric Acid, Anhydrous. $C_8H_4O_{10}$. Obtained by heating racemic acid till it foams up and becomes solid.

Paratartralic Acid. $C_8H_4O_{10}$. Colourless fluid by heating racemic acid slightly above 392°; converted by water into racemic acid; saturates 1½ atom of base.

Paratartrelic Acid. $C_8H_4O_{10}$. Obtained by heating paratartralic acid for some time; unites with 1 atom base.

Paratartromethylic Acid. $C_{10}H_8O_{12}HO$. Rectangular prisms by boiling pyroxylic spirit with racemic acid.

Paratartrovinic Acid. $C_4H_6O_2,C_8H_4O_{10}$. Rhombic prisms by boiling 1 racemic acid with 4 absolute alcohol.

Paratungstic Acid. 2 $ROTnO_{12}$. A synonyme of Bitungstates.

Paravitelline. The albumen of the roe of the carp with the composition of vitelline of eggs.

Parellic Acid. $C_{21}H_8O_{10}$. Accompanies lecanoric acid in lichens and is extracted by ether; at 212° it is $C_{21}H_7O_9$.

Parenchyma. The solid matter of the organs of animals, in the liver, and in plants.

Pargasite. Dark green Pargas hornblende.

Paricine. $C_{46}H_{25}N_2O_6$. Yellow powder from China de Para.

Parietine. *Chrysophanic Acid.* When the Parmelia paretina lichen is boiled in alcohol of ·840 there deposit on cooling yellow scales, which, when redissolved in alcohol, consist of *parietine* $C_{40}H_{16}O_{14}$. The alcoholic supernatant fluid yields by spontaneous evaporation needles of *parietic acid* or *chrysophanic acid*, $C_{40}H_{16}O_{12}$ or $C_{10}H_4O_3$, which are also obtained by digesting the lichen in ammonia and neutralizing with acid. Parietine when boiled for a considerable time with alcohol becomes *oxide of parietine*, $C_{40}H_{16}O_{16}$, a bright yellow powder insoluble in alcohol. Solutions of parietine afford a very delicate test for alkalies, a fine red being produced.

Paris Blue. See BLUE, PRUSSIAN.

Parilline. See SALSEPARINE.

Parisite. Spec. grav. 4·35, H 4·5. 8($ROCO_2$), 2 CaF, RO2HO. Brownish-yellow double and 6-sided pyramids with angles of 164° 58′ and 120° 34′; streak yellowish; protoxide of cerium 23·51, lanthana and didymia and some ceria 59·44, CaO 3·17, HO 2·38, CaF 11·51. Musso valley, New Grenada. B.B. becomes brown, infusible, phosphorescent, gives a clear glass with borax, yellow when hot, colourless when cold; soluble slowly in HCl.

Parting. The operation of separating silver from gold by nitric or sulphuric acids.

Paste. The term applied to starch or flour boiled with water.

Paulite. *Labradore hypersthene.* SiO_3 46·11, MgO 25·87, FeO 12·7, MnO 5·29, CaO 5·38, Al_2O_3 4·07, HO ·48.

Peach Wood. *Brazil* or *Nicaragua Wood.* Said to be from the Casalpinia sapan; used extensively in the form of concentrated decoction as a dyestuff.

Pearl-Ash. Potashes purified by solution in water, filtration and ignition.

Pearl Mica. *Margarite* or *Diphanite.*

Pearl Sinter, or *Fiorite.* Smooth shining globular siliceous masses with a pearly lustre in cavities of volcanic tuff.

Pearl Spar. Pearly rhombohedrons of dolomite.

Pearl Stone. Spec. grav. 2·342, H 6. A gray mineral consisting of roundish balls in concentric masses; allied to obsidian or felspar, with a pearly lustre. SiO_3 70·40, Al_2O_3 11·60, Fe_2O_3 4·38, CaO 3·0, MgO 1·1, KO 5·2, HO 4·2. B.B. fuses into a frothy glass. Hungary: Cap de Gate, Spain.

Peas. (*Erbsen*, Ger.; *Pois*, Fr.) The seed of the *Pisum sativum.* Garden peas consist at 212° of nitrogen 4·42, carbon 45·12, H 6·73, O 37·92, S ·14, ash 3·18. They usually contain 13·43 water (Horsford). The ash of Giessen peas consist of KO 59·43, NaO 3·98, CaO 5·91, MgO 6·43, Fe_2O_3 1·05, PO_5 34·5, SO_3 4·91, NaCl 3·71.

Peastone. *Pisolite.* A form of calcareous spar in concentric laminæ.

Peat. *Turf.* The decaying roots and stems of plants, interspersed in some localities with peculiar resins.

Pechuran. Protoxide of Uranium.

Pectic Acid. *Maturation of fruits* (Fremy). There exists in the pulp of roots and fruits a substance insoluble in water, *pectose* quite distinct from *cellulose*, which is not rendered soluble by acids. It is converted by weak acids into *pectine*, which exists in ripe fruits. This when boiled with water precipitates acetate of lead and changes into *parapectine*, which is found in perfectly ripe fruits. By acids it changes into *metapectine*, a feeble acid, reddening slightly litmus. Along with pectine in vegetables there exist *pectase*, a ferment which transforms pectine into two gelatinous acids; *pectosic acid*, completely soluble in boiling water, but by the prolonged action of *pectase* it becomes *pectic acid*. These two acids, *pectosic* and *pectic acids*, are formed also from pectine by an alkali or an earth. When pectic acid is long boiled in water it changes into *parapectic acid*, and by still longer action of boiling water into a stronger acid, *metapectic*. When these gelatinous bodies are exposed to 390° they are converted into pyropectic acid. All of them are derived from a ternary molecule $C_8H_5O_7$, and differ only in the quantity of water which they contain. The formation of vegetable jellies is due (1) to the change of pectine into pectic acid by the action of pectase; (2) to the formation of pectosic acid by the prolonged action of pectase upon pectine; (3) to the solution of pectic acid in organic salts contained in the fruits. When we place an apple or pear in hot water, the malic and citric acids act upon the pectose and change it into *pectine*, a portion of the pectin remains in the juice imparting to it its viscidity. Further, the *pectase* acting upon the *pectine* forms *pectosic acid*, which is gelatinous on cooling.

Pectose,	?
Pectine,	$C_{64}H_{40}O_{56}$ 8 HO
Parapectine,	$C_{56}H_{40}O_{56}$ 8 HO
Metapectine,	$C_{64}H_{40}O_{56}$ 8 HO
Pectosic acid,	$C_{32}H_{20}O_{28}$ 3 HO
Pectic acid,	$C_{32}H_{20}O_{28}$ 2 HO
Parapectic,	$C_{24}H_{15}O_{21}$ 2 HO
Metapectic,	C_8 H_5 O_7 3 HO

Pectolite. See WOLLASTONITE or SODA TABLE SPAR.

Peddrazite. A gray dolomite of spec. grav. 2·613.

Peganite. Allied to wavellite, spec. grav. 2·5, H 3·5. Prismatic crusts. Al_2O_3 44·49, PO_5 30·49, HO 22·82, $CuOFe_2O_3$ 2·2.

Pegmatite. Graphic granite.

Pegmatolite. A variety of felspar, containing 78 per cent. silica.

Pegmine. The buffy coat of inflamed blood when purified from fat consists of oxidized fibrine. C 52·07, H 7·14, N 14·2, O and S 26·79.

Pelargone. $C_{17}H_{17}O$. Crystalline tables by distilling pelargonate of barytes, isomeric with the aldehyde of margaric acid.

Pelargonic Acid. $C_{18}H_{17}O_3HO$. B.P. 500°. A rancid smelling oil from the oil of Pelargonium roseum.

Pelargonic Ether. $C_4H_5O, C_{18}H_{17}O_3$. Peculiar oil, said to be the cause of the flavour of old whisky.

Pelargyle. $C_{18}H_{17}O_2$. Obtained as a colourless body in union with chlorine by distilling quintochloride of phosphorus and pelargonic acid.

Peliom. Bluish smoke-coloured Iolite in Greenland, Spain, Norway.

Pelluteine. $C_{42}H_{21}NO_7$. Brownish-yellow body by the action of air on pelosine; insoluble in ether.

Pelokonite. Spec. grav. 2·567, H 3·. Bluish-black masses; soluble in muriatic acid, with a green colour, less so in nitric acid; fracture conchoidal, streak brown, lustre vitreous, almost dull; contains iron, manganese, copper, and phosphoric acid. Tierra Amarilla and Remolinos, Chili.

Pelopium. Black powder by passing dry ammoniacal gas over chloride of pelopium.

Pelopic Acid.—PeO_3? Spec. grav. 5·495 to 6·725. Colourless powder, becoming yellow by ignition and colourless on cooling; by heating the more volatile yellow chloride from the Bavarian tantalite with water and washing. The alkaline pelopates are soluble in water, and yield a blue colour with zinc and sulphuric acid.

Chloride.—F.P. 413½°, volatile at 257°. Yellow liquid when heated; soluble in sulphuric and chlorohydric acids; decomposed by boiling water; more soluble in caustic potash and its carbonate than chloride of columbium; blackened and decomposed by sulphohydric acid, which chloride of niobium is not, at common temperatures; formed at a lower temperature than chloride of columbium.

Pelosine. *Pereirine.* $C_{36}H_{21}NO_6$. White powder, soluble in ether, from the root of the Pereira brava, by the same process as quinine; above 212° the chromate decomposes into quinoline and carbolic acid; it seems allied to bebeerine.

Pencalite. A metamorphosing dolomite.

Pennine. Spec. grav. 2·656, H 2·75. Dark green rhombohedral tabular crystals with flexible laminæ. B.B. fuses into a gray enamel; soluble slowly in boiling HCl; SiO_3, 33·39. Al_2O_3 13·24, Fe_2O_3 5·93, MgO 34·21, HO 12·8, Cr_2O_3 ·2; seems allied to chlorite.

Pennite. Spec. grav. 2·86, H 3·5. White or greenish, incrusting the surface of chrome iron ore at Texas, Pennsylvania, along with native carbonate of nickel. CO_2 44·54, CaO 20·1, MgO 27·02, FeO 7, MnO ·4, HO 5·84, Al_2O_3 ·15, NiO 1·25.

Pentaclasite. A variety of Pyroxene.

Pentathionic Acid. *Tersulpho-hyposulphuric Acid.* $S_5O_5,=SO_2,SO_3,S_3$. Obtained in union with barytes by the reaction of sulphurous acid on sulphohydric acid and saturation with carbonate of barytes. First obtained by Dr. Thomson, afterwards by Wackenroder.

Pepper. The fruit of the Piper nigrum; when dried in the sun becoming white pepper.

Peppermint Oil. Spec. grav. ·902; B.P. 365°. Yellow or greenish oil by distilling the Mentha piperita. Its stearoptene is isomeric with it ($C_{10}H_{10}O$).

Pepper Oil. $C_{20}H_{16}$. Spec grav. ·864, of vapour 4·74; B.P. 334°. Colourless; by distilling pepper.

Pepsine. $C_{48}H_{32}N_8O_{10}$? S? White precipitate, or yellow viscid mass, obtained by digesting the internal coat of the stomach of a pig, in successive portions of cold distilled water, precipitating the infusion by acetate of lead, decomposing the precipitate by sulphohydric acid, and filtering; coagulated albumen and sulphide of lead remain on the filter, while pepsine and acetic acid pass through; the filtered liquor is evaporated below 100°, and mixed with absolute alcohol; pepsine falls. This substance appears to be a modified form of albumen; in presence of acids it is said to dissolve albuminous bodies.

Per. A term prefixed to the highest oxides in 1804, by Dr. T. Thomson, to denote at that time the *thorough* oxidation of bodies. When the number of oxides became more numerous he preferred the use of Latin numerals. Thus the peroxides of iron and manganese are sesquioxide and binoxide (Fe_2O_3 and MnO_2) respectively.

Perchloric Acid. *Septachloric Acid.* Cl O_7. F.P. 113°. Oily fluid, or white crystals, of 4-sided prisms, subliming at 284°; obtained by cautiously distilling 5 perchlorate of potash and 10 sulphuric acid, diluted with 1 distilled water. When purified from sulphuric acid and chlorine, it is a clear oil like oil of vitriol; the solid is obtained by distilling with sulphuric acid the oily acid; it deposits in the neck of the retort.

Percylite. Small blue prisms, from La Sonora, Mexico, containing 2·66 lead, ·77 copper, ·84 chlorine = Pb_2Cl, Cu_2Cl, HO.

Pereirine. See PELOSINE.

Pericardial Fluid. The serosity found between the pericardium and heart was found in a child to consist of water 92·, albumen 5·5, mucus 2·, NaCl ·5.

Periclase. MgO. Spec. grav. 3·674, H 6·. Gray or green grains, cubes and 8-hedrons, consisting of MgO 92·57, FeO 6·94. B.B. infusible; soluble in nitric acid. Mount Somma, in limestone.

Pericline. A synonyme of Albite.

Peridote. A synonyme of Chrysolite.

Periodic Acid. *Septaiodic Acid, Hyperiodic Acid.* IO_7. White mass fusing by heat, or as hydrate, in oblique rhombic prisms, by heating iodate of soda with hypochlorite of soda.

Peristerine. C 69·02, H 8·74, O 22·24 = C_4H_3O. A fine red shining mass, the colouring matter of pigeon's feet and of craw fish, by alcohol; soluble in absolute ether and alcohol; insoluble in water and acetic acid; decomposed by sulphuric and nitric acids; swims on hot water in red drops; soluble in caustic potash and volatile oils; smell weak and mouldy; separated by digesting the pigeon's foot in water.

Peristerite. A beautiful iridescent variety of felspar, found at Perth, Upper Canada, by Dr. Wilson (Dr. T. Thomson).

Perlite. A synonyme of Pearlstone.

Perowskine. A synonyme of Triphylline, or trisphosphate of iron.

Perowskite. *Titaniate of Lime.* Spec. grav. 4·017, H 5·5. CaO TiO_3. Gray or black cubes, generally modified; streak gray. B.B. infusible; a clear glass with borax. Slatoust, Ural. CaO 39·2, TiO_3 58·96, FeO 2·06.

Persian Berries. From Rhamnus infectorius, yielding rhamnine and rhamneine. Used in calico printing.

Perspiration. The matters insensibly emitted from the skin of animals. According to Seguin and Lavoisier, the quantity of perspired matter varies from 9 grains to 26·25 grains per minute in a man, or, according to Cruickshanks, to 21 grains; it consists essentially of water and carbonic acid. The sensible perspiration or sweat is a transparent, colourless fluid with a saline taste, and containing lactates (?) of potash, soda, lime, and magnesia, common salt, salammoniac, traces of KCl, 'phosphates of soda and lime, and animal matter.

Persulphomolybdic Acid. MoS_4, RS.

Perthite. A fine variety of felspar from Perth, Upper Canada, with a ribbon structure of dark and light colours.

Peruvian Balsam. From the Myroxylon Peruiferum. There exists a solid and a liquid balsam. The liquid has a specific gravity of 1·14, has a bitter taste and fragrant odour when burned; soluble in alcohol.

Peruvine. $C_{18}H_{12}O_2$. Colourless oil, obtained by boiling cinnameine (the principal oil of the Peruvian balsam) with potash; it is converted into cinnamic acid and peruvine; an alcohol.

Petalite. (πεταλον, a leaf). *Berzelite.* 3 LiO 4 SiO_3, 4 (Al_2O_3 4 SiO_3). Spec. grav. 2·42, 2·45, H 6·5. Milk-white, red or greenish masses, or oblique prisms; vitreous; pearly; translucent, brittle. B.B. fuses with difficulty on the edges; forms a clear glass with borax; phosphorescent; partly decomposed by acids. SiO_3 79·212, Al_2O_3 17·225, lithia 5·761. Utön, and in United States.

Petenine. See BUTYLAMINE.

Petrolene. $C_{10}H_8$. An oil from bituminous sand at Bechelbronn.

Petroleum, *Rock Oil, Mineral Naphtha*, appears to be a solution of paraffine in different oils which have been termed naphtha $C_{14}H_{13}$, naphthene $C_{16}H_{16}$, naphthole $C_{24}H_{22}$. It occurs in Derbyshire, Persia, Birmah, &c. See NAPHTHA.

Petrosilex. Impure felspar.

Petuntze. A variety of felspar rock, containing much quartz.

Petzite. Telluride of silver.

Peucedanine. C_4H_2O. White prisms from the root of Peucedanum officinale by alcohol.

Peucyle. *Peucylene.* $C_{20}H_{16}$. A clear oil by distilling chlorohydride of terebene over lime.

Pewter. There are three kinds of pewter in common use. Plate, for plates and dishes, spec. grav. 7·248 = tin 89, Bi 2, antimony 7, copper 2; Triple, for pints and quarts, spec. grav. 7·359; and Ley, for wine measures, spec. grav. 7·963. The best pewter is said to consist of 100 tin and 17 antimony. Tin and zinc are also said to form some kinds of pewter, and likewise lead and tin. The tin, in these cases, seems to neutralize in some measure the noxious influence of the lead, as when vinegar and lemon juice are boiled for a long time in such vessels, some tin but no lead is dissolved. Ley pewter is often such an alloy, of about 3 tin and 1 lead.

Pfaffite. Sb 35·47, As 3·56, S 17·2, Pb 43·44.

Phacolite. Spec. grav. 2·162. SiO_3 47·03, Al_2O_3 21·87, CaO 10·74, KO 1·18, NaO trace, HO 21·86 (Mr. Parry, from Glenarm, Ireland). SiO_3 45·628, Al_2O_3 19·48, Fe_2O_3 ·431, NaO 1·684, CaO 13·304, MgO ·143, KO 1·314, HO 17·976 (Anderson). A mineral allied to chabasite, and sometimes confounded with Levyine.

Phæoretine. $C_{16}H_8O_9$. Yellow-brown powder, soluble in alcohol, less soluble in water and ether; from rhubarb root.

Phaiene, Sulphide of. $C_8H_5N_6S_4$. From the decomposition of persulphocyanohydric acid.

Phaiosine. *Phaiosic Acid.* Brown substance by carbonate of soda from the pericarp of laurus nobilis, after exhaustion with water, alcohol, and ether.

Pharmacanhydrite. A synonyme of Haidingerite or Diarseniate of lime.

Pharmacolite. *Sexhydrous Diarseniate of Lime.* $2CaO, AsO_5 6HO$. Spec. grav. 2·64 to 2·73, H 2·25. White or pinkish oblique rhombic prisms, with angles of 96° 46′, 90°, 117° 24′, and 121° 18′; usually consists of silky fibres arranged in stars; lustre vitreous; streak white, translucent to opaque; fracture uneven; CaO 25, AsO_5 50·54, HO 24·46. B.B. fuses into a white enamel, evolving the smell of arsenic on charcoal; soluble in acids. Wittichen, Baden; Thuringia.

Pharmacolite Magnesian. *Triarseniate of Lime and Magnesia.* Spec. grav. 2·52, H 5·5. Dirty white or honey-yellow masses; brittle. CaO 23·22, MgO 15·68, AsO_5 58·51, MnO 2·13. *Form.* $3(CaO,MgO,MnO)AsO_5$. Wermland.

Pharmacosiderite. See TRIARSENIATE OF IRON.

Phelene, Sulphide of. $C_{14}H_{14}N_{14}S_2$. From the decomposition of sulphoprussic acid.

Phenakite. Spec. grav. 2·969. Flat colourless rhomboids, with angles of 115° 25′ and 64° 35′; texture foliated, lustre vitreous; harder than quartz, not attacked by acids. B.B. unaltered *per se*; with borax fuses into a clear glass; with soda an enamel. SiO_3 55·14, glucina 44·47, Uralian mountains.

Phenamide, or Aniline.

Phenamylole. *Carbolate of Amyle.* $C_{22}H_{16}O_2$. B.P. 436°. Colourless oil by iodide of amyle on phenole.

Phene, or Benzole.

Phenetole. *Carbolate of Ethyle.* $C_{16}H_{10}O_2$. B.P. 341°. Colourless fluid, by distilling salicylate of ethyle with barytes. With nitric acid it yields nitro and dinitrophenetole.

Phenic Acid. *Oxyphenic acid, Pyromorintannic acid.* $C_{12}H_6O_2, O_2$. F.P. 233°; B.P. 467°. White shining plates similar to benzoic acid, by the distillation of catechine, morintannic acid, gum ammoniac, and peucedanine? It is metameric with hydrokinone.

Phenicine, or Sulphopurpuric acid.

Phenides. A class of bodies formed from phenylic acid ($C_{12}H_6O_2$) in the same way as ether, amide, or anilide are formed from alcohol, ammonia, or aniline. The phenides are, *benzophenide, bromophenide, dinitrobenzophenide, trinitrobenzophenide, chlorohydrophenide.*

Phenole. *Carbolic acid.* Exists in coal tar oils; is formed by passing acetic acid through an ignited porcelain tube filled with pumice; it is also formed from aniline, and by distilling urine.

Phenomethole, or Anisole.

Phenylamine, or Aniline.

Phenyle. $C_{12}H_5$. The hypothetic base of the carbolic or phenylic series. Carbolic acid is the hydrous oxide of phenyle, $C_{12}H_5OHO$.

Phenyle-ethyle-urea. $C_2H_2,C_{12}H_5,C_4H_5,N_2O_2$. Crystals by dissolving aniline in cyanate of ethyle.

Phenylic Acid. A synonyme of Carbolic acid.

Phenylomellone. $H_7NS_2O_4$. By NH_3 on chloride of sulphophenyle.

Phillipsite. *Lime Harmotome.* Spec. grav. 2 to 2·2, H 4·25. $3(CaOKO)2SiO_3, 4(Al_2O_3 2SiO_3)18HO$. White right rectangular prisms in amygdaloid; lustre vitreous; translucent, opaque. SiO_3 48·51, Al_2O_3 21·76, CaO 6·26, KO 6·33, water 17·23, BaO trace, NaO trace, Fe_2O_3 ·99. B.B. froths up before the blowpipe; soluble in HCl, and gelatinizes. Capo di Bove; Giants' Causeway.

Phillipsite. A name also given to hepatic sulphide of copper, to Christianite, and to Gismondine.

Phillyrine. Bitterish silvery plates; soluble in hot water and alcohol; nearly insoluble in ether; decomposed by NO_5 and SO_3; from the bark of Phillyrea media and latifolia.

Phlobaphene. $C_{20}H_8O_8$. Reddish-brown matter from the bark of Pinus sylvestris by exhaustion with ether and alcohol, or by extraction from the bark by an alkali.

Phlogisticated Alkali. Ferrocyanide of potassium.

Phlogiston. A hypothetic substance of the older chemists; latterly the same as hydrogen.

Phlogopite. Magnesian mica.

Phloretine. $C_{24}H_{11}O_8$ or $C_{30}H_{15}O_{10}$ or $C_{12}H_6O_4$. F.P. 356°. White, sweet plates; slightly soluble in water and ether; very soluble in alcohol; from phloridzine by sulphuric acid, which converts it into phloretine and sugar.

Phloridzeine. $C_{74}H_{45}N_3O_{12}$. Red powder by ammonia on phloridzine; with water gives a fine blue, destroyed by sulphohydric acid, but restored by oxygen.

Phloridzine. $C_{14}H_9O_9$ or $C_{32}H_{18}O_{15}$ or $C_8H_5O_4$. Spec. grav. 1·43; F.P. 320°. 4-sided bitter astringent silky prisms; soluble in boiling water and alcohol; insoluble in ether and cold water; obtained from the bark of the apple, plum, cherry, and pear trees by alcohol of 80 per cent.

Phocenic Acid. Delphinic or Valerianic acid.

Phœoretine. $C_{16}H_8O_9$. Yellow powder; soluble in alcohol; less so in water and ether; from rhubarb root.

Pholerite. Spec. grav. 2·35 to 2·57, H ·5 to 1·. White convex scales with pearly lustre; soft to the touch, and may be crushed between the fingers; disengages air bubbles under water, with which it forms a paste. B.B. infusible. SiO_3 40·75, Al_2O_3 43·886, HO 15·364. *Form.* $Al_2O_3SiO_32HO$.

Phonolite. *Clinkstone.* Spec. grav. 2·55. Felspar rock, yielding a metallic sound under the hammer.

Phorone. C_9H_7O or $C_{18}H_{14}O_2$. Spec. grav. of vapour 4·98; B.P. 406½°. Yellowish oil, lighter than water, with the odour of peppermint; soluble in sulphuric acid with a blood-red colour; obtained by distilling camphorate of lime.

Phosgene Gas. *Chlorocarbonic acid.* COCl. A colourless gas by the action of chlorine on carbonic acid, or by passing carbonic oxide through quintochloride of phosphorus.

Phosgenite. A synonyme of Chlorocarbonate of Lead.

Phosphame. P 50·4, H ·88.

Phosphamic Acid. PH_4NO_6. By the action of alkalies on nitrochloride of phosphorus; 2 atoms being converted by boiling into pyrophosphamic acid, $P_2H_5NO_{12}$, and NH_3 evolved.

Phosphamide. $PH_3N_2O_2$ or $PH_2N_2O_2$. White insoluble powder, by washing the product of the action of dry ammonia on quintochloride of phosphorus. When heated, ammonia is disengaged, and biphosphamide, PNO_2, formed.

Phosphates. Combinations of phosphoric acid with bases. There are three principal classes of phosphates, monophosphates, $ROPO_5$; diphosphates or pyrophosphates, $2RO,PO_5$; and trisphosphates or common phosphates, $3ROPO_5$.

Phosphatic Acid. P_3O_{13}. By the slow combustion of phosphorus.

Phosphides, or *Phosphurets.* Compounds of phosphorus with bases.

Phosphites. Combinations of phosphorous acid with bases. The salts are bibasic, formed on the type $2RO,PO_3,2HO$.

Phosphocerite. Spec. grav. 4·78, H 5· to 5·25. $3(CeO,LaO,DO)PO_5$. Microscopic dark red regular 8-hedrons? from the cobalt ore of Johannisberg, Sweden. It is allied to cryptolite.

Phosphorescence. The luminous appearance presented by many substances in the dark, as with stale fish, the live medusæ, or jelly fish; supposed in the latter case to be occasioned by the reproductive and motory systems from irritation. It continues when phosphorescent fish or salt water are placed in hydrogen, nitrogen, carbonic acid, and air, with a trace of ether. It is destroyed by acid and alkaline solutions, alcohol, cold, and ether, while phosphorescent sea water becomes more luminous by ammonia, alcohol, and acids. When fluor spar, which contains phosphoric acid, is heated in the dark it becomes phosphorescent.

Phosphoric Acid. *Bone Acid.* PO_5 9,72; 8·875, 71. *Source.*—This acid occurs in primary rocks as granite, mica and clay slates, and also in trap rocks. Also united with alumina, lime, magnesia, lead, copper, iron, manganese, &c. and in plants and animals.

Characters.—Sp. grav. when anhydrous 2·687, in state of glass 2·8516, in deliquescence 1·417. White flakes when anhydrous, fusing at a red heat, and subliming below a white heat (Davy). Taste sour; reaction on litmus, acid; unites with barytes, lime, and strontian; very soluble in acid, with a hissing noise; dissolving much more slowly when in union with water. After ignition the flocks are less soluble in water.

Process.—Pure anhydrous phosphoric acid may be obtained by burning phosphorus in a glass bell filled with dry atmospheric air, or oxygen, standing over mercury. White flocks are

formed which stick to the glass; are difficult to melt; not very volatile; without smell, and possessing an acrid taste.

Hydrates.—There are three hydrates of phosphoric acid, protohydrate, $HOPO_5$, or metaphosphate of water; bihydrate or diphosphate of water, $2HO,PO_5$, or pyrophosphate of water; terhydrate, trisphosphate of water, tribasic phosphate of water, $3HO,PO_5$. In uniting with bases these various hydrates form corresponding salts, the water being replaced atom for atom by the bases.

Hydrous Acid.—The most productive process for the preparation of phosphoric acid, is to separate it from the phosphate of lime or earth of bones by means of sulphuric acid; 10 parts of calcined bones are mixed with 15 to 20 times their weight of water in a porcelain or lead vessel, and 9½ parts of oil of vitriol (spec. grav. 1·85) are to be poured into the fluid. After standing for a day as much more water is to be added, and the mixture allowed to remain at rest for a similar period, being frequently stirred during the time. The liquor is then filtered through calico, or, what answers well for such filtrations, a common linen towel. The filtered fluid is then to be evaporated to the consistence of a syrup in a copper vessel, and sulphuric acid is added as long as a precipitate of sulphate of lime takes place. The whole forms into a white mass which is to be diluted with water, and again filtered through cloth. It is then evaporated to dryness. If the purest phosphoric acid is required the residue should be dissolved in alcohol in the cold, and be allowed to stand as long as phosphate of magnesia separates in crystals. The spirit is then distilled off, and the acid obtained of the consistence of a syrup.

Purity.—Phosphoric acid precipitates calomel from a solution of corrosive sublimate; it precipitates nitrate of mercury blackish; it precipitates sulphur from sulphurous acid when heat is applied; it evolves PH_3 in Marsh's apparatus, with zinc and dilute SO_3. But phosphoric acid combines with 3 different proportions of base, according to the mode in which it has been prepared, and each of the compounds thus formed has certain peculiarities.

Trisphosphoric Acid.—$3HOPO_5$. This form of salts of phosphoric acid, which is that familiarly known, may be obtained in an isolated state, by precipitating the phosphate of soda of the shops by acetate of lead, washing the precipitate with distilled water, diffusing the lead precipitate through water, and passing a current of SH through it, until the whole of the lead is separated as sulphide of lead. The liquid is then filtered, heated, and concentrated. On the escape of all the sulphohydric acid, the fluid contains phosphoric acid in solution, which is characterized by giving a yellow precipitate with nitrate of silver; when neutralized with soda it affords the common commercial phosphate of soda. A solution of this acid, when evaporated under the exhausted receiver of an air pump over sulphuric acid, crystallizes in thin deliquescent plates, or in right 4-sided prisms, or flat 6-sided prisms. When we add one atom of soda to an atom of this acid, we have a compound biphosphate formed, consisting of $NaO,2HOPO_5$. This salt is usually termed biphosphate of soda, and is obtained by adding a solution of common phosphoric acid to common phosphate of soda. If we add a second atom of soda we have the salt converted into 2NaO HO, PO_5, the common phosphate of soda formed by neutralizing phosphoric acid of bone earth with carbonate of soda. If we add an additional quantity of soda, as by treating the preceding salt with an excess of caustic soda, we obtain 3NaO, PO_5, usually termed the subphosphate of soda. Now, all of these salts afford a yellow precipitate with nitrate of silver, and the resulting silver salt is always 3AgO PO_5. The whole of these phenomena may therefore be classed as occurring to trisphosphates, or phosphates, containing 3 atoms of base. The reason why phosphoric acid in this condition is incapable of forming diphosphates has not been explained, any further than that as it is united to 2 atoms of water in this new state, these are incapable of being replaced by more than 2 of base.

Diphosphoric, or *Pyrophosphoric Acid* (Clark).—When we heat the common phosphate or trisphosphate of soda to redness, we obtain a salt which precipitates nitrate of silver white, and upon examination it is found to have undergone the following change:—

Trisphosphate before heating,........................	$PO_5+2NaO+HO$	
Diphosphate after heating,	PO_5+2NaO	
Trisphosphate of silver before heating,...............	PO_5+3AgO	yellow.
Diphosphate of silver after heating,....................	PO_5+2AgO	white.

This affords, therefore, a clear example of substitution, and is only in accordance with what we should expect from known facts. Solid SO_3 possesses some of the characters of an acid, and is incapable of combining with bases; but as soon as we add water to it, it is then in a condition to allow stronger bases to displace the water.

Protophosphoric, or *Metaphosphoric Acid*, *Glacial Phosphoric Acid*, *Phosphoric Glass.*—When the biphosphate of soda is ignited, a salt is obtained which is characterized by producing a white precipitate in solutions of albumen, and in solutions of the salts of earths and metallic oxides precipitates which are remarkable semi-fluid bodies, or soft solids, without crystallization. The acid may be isolated as in the previous instances. It is also formed by adding water to anhydrous acid. It partially volatilizes when heated with water, and entirely by ignition. All the modifications of metaphosphoric acid precipitate white of egg. On heating a concentrated solution of common phosphoric acid until it begins to volatilize, metaphosphoric acid is produced, which precipitates chloride of barium; by a rapid application of heat, an acid may be obtained which does not throw down chloride of barium. The hydrate of metaphosphoric acid, in contact with water for some days, changes into the terhydrate or common phosphoric acid. The acid obtained as concentrated as possible is a compound of 1 phosphoric acid and 1 water. The change effected on the biphosphate of soda is therefore as follows:—

Biphosphate before heating, ..	$PO_5+NaO+2HO$
Biphosphate after heating,....	PO_5+NaO
Protophosphate of lead, ...	PO_5+PbO

Phosphorous Acid.—PO_3 7·, 56; P 57·60 43·3. White voluminous flocks, easily sublimed, with a garlic odour, a peculiar acid taste; reddening moist litmus paper. It is obtained by exposing phosphorus to the air, when water is attracted, and an acid liquid formed (Steinacher). The following has been recommended:—A glass tube 10 inches long and half-an-inch wide, melted at one end so as to

leave an aperture of the diameter of a pin, and at the distance of half-an-inch or an inch from the end, the tube is to be bent at an obtuse angle; a piece of phosphorus is to be introduced near the narrow opening, and heated gradually, when it burns with a pale green flame, forming phosphorous acid, which collects in the upper part of the tube in the form of a white powder. When phosphorus is heated in highly rarefied air, according to Sir H. Davy, there are formed phosphoric acid, phosphorous acid, and oxide of phosphorus.

Hydrate of Phosphorous Acid.—$3HOPO_3$; $PO_3 = 72·46$, $HO = 27·54$. Transparent parallelipepids, changing by heat into phosphoric acid and phosphuretted hydrogen. 1. When phosphorus is heated, and its vapour made to pass through corrosive sublimate, terchloride of phosphorus is formed. When this compound is mixed with water it is converted into chlorohydric acid and phosphorous acid ($PCl_3 3HO = 3HClPO_3$). By a moderate heat the chlorohydric acid is expelled, and phosphorous acid remains. 2, It may also be formed by placing sticks of phosphorus, fig. 1, in tubes with a capillary opening, and inserting them in a funnel, fig. 2; the whole being covered with a bell jar. Phosphorous acid collects in the bottle below.

1

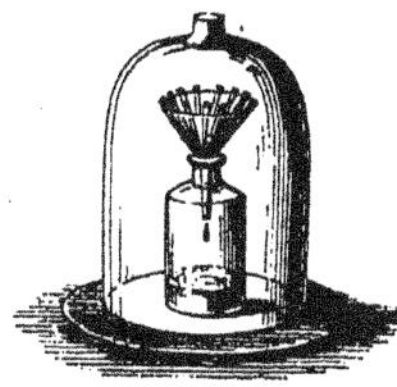
2

3. It may be procured by rapidly evaporating the product of the reaction of terchloride of phosphorus on water until the vapour of water and chlorohydric acid cease to be evolved; the residue is gently heated until a slight odour of phosphuretted hydrogen is perceptible. It is then evaporated in vacuo, and obtained in crystals. 4. It may likewise be obtained by heating phosphorus in a tube with capillary extremities, *a*, *b*.

Phosphorized Bases. Compounds of phosphorus with bases.

Phosphorite. Native phosphate of lime.

Phosphorocalcite. Hydrous pentaphosphate of copper.

Phosphorus. P, 4, 32 (Thomson, Pelouze), 3·875, 31 (Schrötter). Sp. grav. 1·748 (Thomson), 1·896 (Bockmann), 2·033 (Fourcroy), 2·089 (Böttger), of vapour 4·355 (Dumas), 4·59 (Mitscherlich, Pogg. 25, 218). Melting point 99°, 111¾° (Schrötter), 111½° (Desairs), 108° (Thomson), 112° (J. Davy), 115¼° (Heinrich). Boiling point 554° (Pelletier), 550° (Dalton), 482° (Heinrich). Specific heat = ·1740 (Regnault). Colourless, transparent when slowly cooled; with a waxy lustre after rapid cooling. According to Schönbein the vapour of phosphorus has no smell; the odour perceptible during the combustion or heating of phosphorus being due to the formation of ozone and phosphorous acid. Crystals regular 8-hedrons and 12-hedrons. In these forms it sometimes occurs by simple fusion and cooling, but when dissolved in volatile oils it separates in 8-hedrons, and from phosphide of sulphur in 12-hedrons. Melted phosphorus often cools to 40° without solidifying (Bellain, Rose). Non-conductor of electricity. In the air it yields the odour of garlic—poisonous—inflammable. When kept some time in water it becomes opaque externally, and resembles white wax in appearance and consistency. Its specific gravity is then 1·515 (Pelouze). Insoluble in water; slightly soluble in alcohol, ether, and oils.

Red Phosphorus, Amorphous Phosphorus.—White phosphorus appears to be free from oxygen, and so likewise is its red allotropic form. When phosphorus is placed in a tube, the common air displaced by dry carbonic acid, the tube heated to 212° to remove moisture, closed hermetically, and then exposed to light, the phosphorus becomes externally red in a few days at a temperature of 57°. Nitrogen and other gases, except oxygen, may be used instead of carbonic acid. When phosphorus is kept at a temperature of 464° to 482° during forty-eight or sixty hours, it separates into two layers, the upper being common phosphorus, and the lower red phosphorus. On introducing water of 140° after the phosphorus has cooled, the upper layer melts, and may be sucked up in a tube. When the red phosphorus is distilled in glass balls free from air, it is converted into fluid colourless phosphorus, which is capable of remaining many hours in a fluid state, and the longer in proportion to the frequency of distillation. The red or amorphous phosphorus is insoluble in bisulphide of carbon, in which common phosphorus is soluble. It may be further purified by caustic potash, then with water containing a few drops of nitric acid, and lastly with water. When thus prepared and dried, it is an amorphous powder, varying in colour from scarlet to carmine, and even blackish-brown. It does not alter in the air; it is insoluble in bisulphide of carbon, alcohol, ether, naphtha, protochloride of phosphorus; oil of turpentine, and, in general, all liquids which have a high boiling point, dissolve a small portion with heat. When heated in an oil bath it begins to return to the common state at 500°. It does not unite with sulphur at the fusing point of that substance or 233°, but with viscous sulphur at 446°. When it passes into the common state, light is evolved, as in the case of arsenious acid. It is soluble in hot sulphuric acid, with evolution of sulphurous acid, and in nitric acid. Hence there are five allotropic forms of phosphorus. 1. *White* phosphorus, formed by light on phosphorus kept in water. 2. Transparent phosphorus in its usual fresh state. 3. *Red* phosphorus. 4. Liquid phosphorus, produced by distilling amorphous phosphorus, sometimes remaining liquid for thirty-six days, while

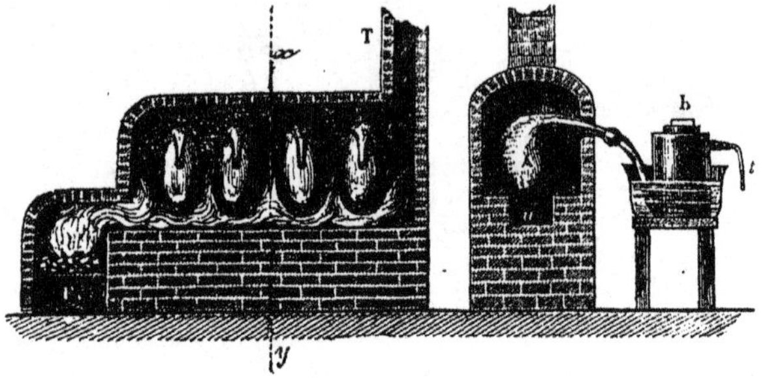

A Clay retort covered with a layer of clay; b copper receiver half filled with water; *t* tube for disengaging gases; F hearth of reverberatory furnace; *u* horizontal flue; T chimney; T, *x*, *y*, front view of the retorts.

the thermometer has been in the interval at 23°. 5. *Black phosphorus.* When heated to 158° and cooled suddenly, phosphorus becomes black, due to a molecular change in the substance.

Preparation.—1. Bones are deprived of their animal matter by being heated till they become white. Phosphate of lime remains, which is decomposed by oil of vitriol, triphosphate and sulphate of lime being formed ($8CaO3PO_5 + 5HO$ $SO_3 = 3\,CaO, PO_5, 5\,CaO\,SO_3) + 5\,HO$. The supernatant liquid is drawn off from the residual sulphate of lime, and evaporated to the consistence of a thick syrup. It is then mixed with one-fourth of finely powdered charcoal, and ignited in an iron crucible. The mixture is introduced into a porcelain retort, A. The retort is placed in a wind furnace and strongly heated, until the phosphorus ceases to drop from the extremity of the copper tube into the water. It may be found advantageous to surround the clay retort with a coating of clay. It often happens that some of the phosphorus sticks in the neck of the retort, which must be removed by means of a bent wire kept inserted into the tube of the retort. In this process the phosphate of lime ($3CaO,PO_5$), calcined with an excess of lime, is changed into neutral phosphate of lime ($2CaOPO_5$), which is not decomposed by charcoal, and into carbonic oxide and phosphorus $2(CaOPO_5) + C_3 = C_5$ $O_5, 2(CaO, PO_5 + P)$.

Purification.—Phosphorus requires to be melted under hot water, to free it from charcoal and oxide of phosphorus; to deprive it of colour it is mixed, when melted, with animal charcoal; it is also strained through chamois leather to free it from mechanical impurities. It is drawn into sticks by melting it in water, and sucking it up into glass tubes, either by aspiration or by the mouth; the finger is applied to the upper end of the tube, and the whole transferred to cold water, when it solidifies.

Applications.—When taken internally phosphorus is poisonous. In small doses of one-fourth of a grain it acts as an aphrodisiac (Leroi, Nicholson's Journ. 3, 85). Phosphorus is extensively employed in the manufacture of lucifer matches.

Hypophosphorous Acid.—PO. An acid only known in union with water or bases; obtained by placing phosphide of calcium or barium in contact with water; phosphuretted hydrogen is evolved, while phosphate and hypophosphate of lime or barytes are formed. The type of the hypophosphates is $PO2HO$ or PH_2,O_3.

Dinoxide, Yellow Oxide, Red Oxide.—P_2O. Obtained by passing oxygen into phosphorus, fused under hot water; but there is some reason to consider this red amorphous phosphorus.

Quintochloride.—PCl_5. Obtained by burning phosphorus in chlorine gas, or by passing chlorine into phosphorus as long as it takes up any gas.

Biphosphide, Solid Phosphide.—HP_2. A yellow substance, becoming red in the air; obtained by exposing phosphuretted hydrogen to the action of water.

Diphosphide.—H_2P. Colourless fluid, volatilizing and decomposing at 86°; formed by decomposing phosphide of calcium in a Wolfe's bottle with water, and condensing the resulting fluid in a U tube surrounded with a freezing mixture. Its presence in phosphuretted hydrogen in vapour is the cause of its self-inflammability.

Phosphuretted Hydrogen, Trisphosphide of Hydrogen.—H_3P 4·375, 35, H = 8·72, P = 91·28. Spec. grav. 1·185 (expt.), 1·1927 (calc.) = $\frac{1}{4}$ P = 1·0885 + $1\frac{1}{2}$H = ·1042 = 1·1927. Colourless gas; smell peculiar, resembling that of garlic; water dissolves the one-eighth of its volume; more soluble in alcohol, ether, and oil of turpentine; not a supporter of combustion, but a combustible gas; sometimes self-inflammable in contact with air, at other times non-inflammable until heated; depending on the amount of liquid phosphide diffused through the gas; has no action on litmus paper. The addition of small portions of binoxide of nitrogen to non-inflammable phosphuretted hydrogen renders it spontaneously inflammable (Graham), depending on the circumstance that the binoxide in combining with oxygen from the air to form quaternitrous acid (NO_4), produces a sufficiently high temperature (212°) to burn the non-self-inflammable gas; 1 vol. is decomposed by electric sparks into $1\frac{1}{2}$ hydrogen and a red film of phosphorus; in contact with the air, emitted bubble by bubble under water or mercury, phosphoric acid is formed in fine rings after the combustion, and oxide of phosphorus is deposited on the surface of the water. When the least inflammable gas is admitted in bubbles into a glass filled with oxygen, a beautiful combustion occurs, the only products being phosphoric acid and water. When it stands over water containing air, the gas gradually decomposes; it ignites in contact with chlorine, heat and light being evolved with explosive effect; very severe when the quantities exceed a small amount. It is made by filling a flask three-fourths with strong caustic potash, not soda, and adding some pieces of phosphorus and applying heat. The gas bubbles through water, as in the

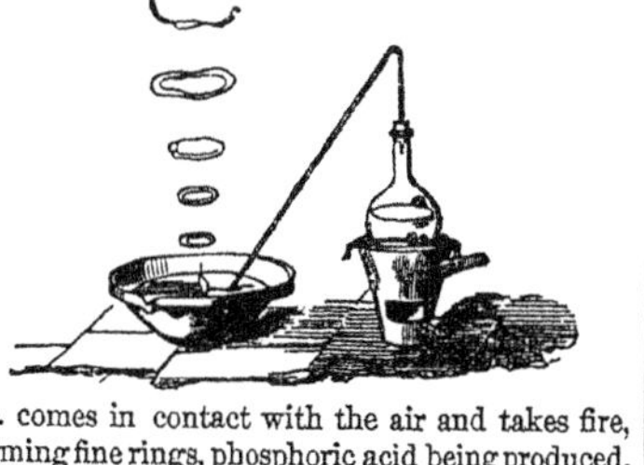

fig. comes in contact with the air and takes fire, forming fine rings, phosphoric acid being produced.

Phosphovinic Acid, or Biphosphate of Ethyle.

Photizite, or subsesquisilicate of manganese.

Photography. *Talbotype.* A mode of leaving impressions on paper by the influence of light on salts of silver.

Photolite. *Pectolite, Wollastonite,* or Soda Table Spar.

Phthalamic Acid. $C_{16}H_6NO_5HO$. By dissolving phthalic acid in alcohol, and adding ammonia.

Phthalanile. $C_{28}H_9NO_4$. Needles by fusing aniline and phthalic acid, and crystallizing from alcohol.

Phthalanilic Acid. $C_{28}H_{11}NO_6$. Thin plates by boiling the preceding with ammonia, and adding nitric acid while hot.

Phthalic Acid. See NAPHTHALIC ACID.

Phthalimide. See NAPHTHALIC ACID.

Phycic Acid. C 70·22, H 11·76, N 3·72, O 14·3. Spec. grav. ·896; F.P. 276°·8; B.P. 482°. White needles; insoluble in water; soluble in 15 boiling absolute alcohol, and in ether; by alcohol from *Protococcus vulgaris.*

Phycite. $C_4H_5O_4$, or $C_{12}H_{15}O_{12}$. F.P. $233\frac{1}{2}$°. Spec. grav. 1·59; B.P. 320°. A sugar in right rectangular prisms; fuses into a colourless fluid, giving out the odour of scorched starch; if the heat be continued it entirely disappears, showing that it is at least partially volatile; in its characters it approaches orcine; it is not altered by potash; it does not reduce sulphate of copper but with great difficulty. Nitric acid produces the same products as with sugar; it does not ferment with yeast; it has no action on polarized light; it is neutral; soluble in hot, and cold water. It is obtained from *Protococcus vulgaris* (Phyceæ); from the mother liquor of phycic acid; hence its name.

Phyllite. (φυλλον, a leaf.) Spec. grav. 2·889, H 5·75. Brownish-black thin plates, semi-metallic, splendent; opaque, sectile. SiO_3 38·4, Al_2O_3 23·68, Fe_2O_3 17·52, MgO 8·96, KO 6·8, HO 4·8. This mineral seems to be mixed with foreign matter, as another analysis gave SiO_3 23·81, Al_2O_3 44·66, Fe_2O_3 29·7, MgO 2·17, CaO 2·11.

Phylloretine. $C_{20}H_{12}$ or C_8H_5. F.P. $187\frac{1}{4}$. A resin, soluble in alcohol, from turf or fossil wood of Denmark.

Physalite. *Pyrophysalite.* A yellow-white, opaque, variety of topaz.

Piauzite. Spec. grav. 1·22, H 1·5. F.P. 599°. Brownish-black earthy resin, soluble in potash, ether, and alcohol; from brown coal at Piauze, Carniola.

Picamare. Spec. grav. 1·1; B.P. 518°. Colourless bitter oil in wood tar; insoluble in water; soluble in alcohol and ether; forms crystalline compounds with alkalies.

Pickeringite, or Magnesian Alum.

Picnite. A synonyme of topaz.

Picoline, or Odorine.

Picolite. A synonyme of Tourmaline.

Picramic Acid. $C_{12}H_4O_4 2NO_4 N,HO$. Red needles by saturating an alcoholic solution of picric acid with ammonia, and then passing sulphohydric acid through the solution; acetic acid removes ammonia; it gives a brick-red precipitate with nitrate of silver.

Picranalcime. Spec. grav. 2·257, H 5. Colourless or flesh 20-hedrons; decomposed by acids. B.B. fuses with difficulty into a white enamel. SiO_3 59·34, Al_2O_3 22·08, MgO 10·25, KO ·01, NaO ·45, HO 7·56. It is a magnesia analcime. Tuscany.

Picranisic Acid. $C_{12}H_3N_3O_{14}$. Yellow needles, slightly soluble in cold water; very soluble in hot water, alcohol, and ether; isomeric with picric acid; obtained by dilute nitric acid on the potash compound of trinitranisole and potash.

Picric Acid. *Ternitrocarbolic Acid, Welter's Bitter, Indigo Bitter, Carbazotic Acid, Nitrophenisic Acid, Chysolepic Acid.* $C_{12}H_3\,3NO_4,O_2$. Bright yellow plates or prisms; sublimable; fuses into a brownish-yellow oil; taste bitter; reaction acid; poisonous; soluble in water and hot oil of vitriol; its salts, which are crystalline, explode when rapidly heated; with lime and copperas they yield blood-red liquors, a new acid being formed; it precipitates several of the alkaloids; picrate of potash being very insoluble in alcohol, picric acid may be employed as a test for potash. Picric acid is obtained by the action of an excess of nitric acid on carbolic acid, in which 3 atoms hyponitric acid replace 3 hydrogen, likewise on salicine, indigo, gum benzoine, Botany Bay resin, silk, aloes, trinitranisole, coumarine, anilic acid.

Picrine. *Picryle.* $C_{42}H_{15}NO_4$. Colourless 8-hedrons by distilling the mass produced by the action of sulphide of ammonium on oil of bitter almonds; with nitric acid it forms *trinitropicryle.* $C_{42}H_{12}3NO_4NO_4$.

Picrite. A synonyme of Dolomite.

Picroerythrine. $C_{34}H_{24}O_{20}$, or $C_{34}H_{23}O_{20}$. Yellow crystals by neutralizing erythric acid with lime or barytes, boiling, filtering off erythrelinic acid, deposited by HCl, and allowing to stand; picroerythrine deposits.

Picrolichenine. A synonyme of Variolarine.

Picrolite. Dark green, asbestiform, serpentine.

Picromel. A name formerly given to the organic part of the bile, from its bitter taste, now disused.

Picropharmacolite. A synonyme or variety of Pharmacolite.

Picrophylle. Sp. grav. 2·75. Deep greenish-gray masses and plates; variety of serpentine. SiO_3 49·8, MgO 30·1, FeO 6·86, MnO trace, CaO ·78, Al_2O_3 1·11, HO 9·83. Sala, Sweden.

Picrosmine. (πικρὸς, bitter; ὀσμη, smell.) *Dihydrous Bisilicate of Magnesia.* Spec. grav. 2·596 to 2·68, H 2·5 to 3, $2(3MgO,2SiO_3)3HO$. Greenish-white, gray and mountain-green, or oil, leek or blackish-green rhombic 8-hedrons; streak white; dull; fracture uneven; lustre pearly. B.B. infusible, becoming black then white and harder (5); fuses with salt of phosphorus, giving a silica skeleton; becomes red with nitrate of cobalt. SiO_3 54·886, MgO 33·348, Al_2O_3 ·793, Fe_2O_3, 1·399, MnO ·42, HO 7·301. Engelsborg, near Presnitz, Bohemia.

Picro-Thomsonite. Spec. grav. 2·278, H 5. Colourless pearly masses of the rhombic system; brittle. B.B. fuses into a white enamel; decomposed by acids, leaving gelatinous silica. SiO_3 40·356, Al_2O_3 31·251, CaO 10·993, MgO 6·265, NaO, KO ·285, HO 10·70; found in Gabbro in Tuscany. It is Thomsonite in which magnesia replaces soda.

Picrotoxine. $C_{12}H_7O_5$, or $C_{10}H_6O_4$? White prisms, the bitter principle of the Hemispermum cocculus; composition uncertain.

Pictite. A synonyme of Sphene.

Pigmentum Nigrum of the eye, consists of C 58·273, H 5·973, N 13·768, O 21·986.

Pigotite. *Mudesite of Alumina.* $4Al_2O_3C_{12}H_5O_8 27HO$. Brown substance from the walls of a granite cave, Cornwall.

Pihlite. A talcy or micaceous variety from Fahlun.

Pimaric Acid. $C_{48}H_{30}O_4HO$. 4 to 6-sided prisms, from the turpentine of Pinus maritima; soluble in alcohol and ether.

Pimarone. Obtained by distilling pimaric acid at the usual pressure; contains an atom of water less than the acid.

Pimelic Acid. $C_7H_5O_3HO$. F.P. 238°. Pearly scales, soluble in hot water, ether, and alcohol; distils without change; by the action of nitric acid on wax, spermaceti, linseed oil, and stearic acid.

Pimelite. A variety of clay or halloylite with nickel. SiO_3 35, water 38·13, Al_2O_3 5·, MgO 1·25, CaO ·42, NiO 15·62, FeO 4·58.

Pimento Oil. Yellow oil smelling of cloves, heavier than water, from the covering of the fruit of Myrtus pimento.

Pinchbeck. A golden brittle alloy of zinc and copper, or brass.

Pinguite. A synonyme of chloropal.

Pinic Acid. $C_{40}H_{30}O_4$. The soluble portion in alcohol (·867) of common rosin.

Pinite. *Micarelle.* Spec. grav. 2·7575 to 2·782, H 2·25. Blackish-green, or greenish-gray, regular 6-sided prisms, usually covered with a coating of yellow ochre; lustre resinous, glimmering; slightly translucent on the edges; fracture uneven, sectile. B.B. fuses into a glass full of vesicles; with borax a transparent glass; with soda an opaque glass; with salt of phosphorus gradually decomposes, leaving a residue of silica. SiO_3 55·96, Al_2O_3 25·48, KO 7·89, NaO ·386, Fe_2O_3 5·512, MgO 3·76, HO and animal matter 1·41. Mine of Pini, Schneeberg; also in granite in Cornwall, Auvergne, Salzburg, North America.

Piotine. $2(3MgO2SiO_3)$, $Al_2O_3SiO_3 6HO$ = SiO_3 50·89, MgO 26·52, CaO ·77, Al_2O_3 9·4, Fe_2O_3 2·06, HO 10·5. A variety of soapstone.

Pipeclay. See CLAY.

Piperidine. $C_{10}H_{11}N$. Limpid colourless fluid, with the smell of ammonia; strongly alkaline, with a caustic taste; very soluble in water; its action with oxides is similar to that of ammonia; forms crystalline salts; it distils at 223°;

obtained by distilling 1 piperine with 2½ to 3 mixture of soda-lime, and retaining the distilled fluid, which passes at 223°. It unites with methyle, ethyle, amyle, forming new alkaloids. It appears to be an imide base.

Piperidine Urea. $C_{12}H_{12}N_2O_2$. White needles, by boiling sulphate of piperidine with cyanate of potash.

Piperine. $C_{70}H_{37}N_2O_{10}2HO$. F.P. 212°. Straw-coloured rhomboidal prisms, without taste and smell; soluble in alcohol and acetic acid; slightly soluble in water and ether; becomes red with sulphuric acid; by adding caustic potash to solution of pepper in alcohol, and filtering; the precipitate is crystallized out of alcohol.

Pipestone. *Scoulerite.* Spec. grav. 2·608, H 1·5. Grayish-blue mass, used by the Indians, between Nootka Sound and Columbia river, for making pipes; B.B. infusible. I found its composition SiO_3 56·6, Al_2O_3 17·42, NaO 12·8, Fe_2O_3 6·308, CaO 2·076, MgO ·29, HO 4·566. For *red pipestone* see CATLINITE.

Pisasphalte. A synonyme of Asphalt.

Pisolite. *Peastone.* A form of calc spar, composed of pea-like grains larger than oolite.

Pissophane. Sp. grav. 1·95, H 1·5. Green stalactites, or in masses; brittle. Al_2O_3 35·155, Fe_2O_3 9·738, SO_3 12·7, HO 41·69. Garnsdorf and Reichenbach, in alum slate.

Pistacite. *Pistazite*, or Ferrocalcareous Epidote.

Pistomesite. Spec. grav. 3·415. $MgOCO_2$ = 44·96, $FeOCO_2$ = 55·27. A variety of brown spar.

Pitch Blende. *Pitch Ore of Uranium, Protoxide of Uranium, Pech Uran.* Spec. grav. 6·468 to 7·, H 3·5. Grayish or iron-black, also greenish and brownish-black masses, powder or granular, and in regular 8-hedrons; fracture conchoidal, uneven; lustre imperfect, metallic. B.B. infusible; but fuses with borax into a gray scoria; soluble slowly in nitric acid. Protosesquioxide of uranium 86·5, $FeOFe_2O_3$ 2·5, PbS 6·, SiO_3 5. Seems a mixture. Accompanies ores of silver and lead at Johangeorgenstadt, Marienberg, Annaberg and Schneeberg, Saxony; and Joachimsthal and Fribus, Bohemia; Tincroft and Tollcarn mine, Cornwall.

Pitchstone. Spec. grav. 2·338 to 2·3604, H 6·5. Dark green, black, gray, and blue masses, with felspar crystals interspersed; fracture conchoidal; sometimes nearly even; lustre resinous, shining; easily frangible. B.B. easily fuses. SiO_3 63·5, Al_2O_3 12·736, CaO 4·46 FeO 3·796, NaO 6·22, vol. matter 8·. Arran, Egg, Saxony.

Pittacal. (πιττα, pitch; καλος, fine.) An indigo-coloured substance, becoming copper-coloured when rubbed, tasteless, odourless, not volatile; insoluble in water; soluble in dilute sulphuric acid, HCl, and acetic acid; not altered by light; decomposed by NO_5; it detects acids and alkalies with great delicacy; insoluble in alcohol, ether, and eupione; obtained by adding barytes water to a solution of picamare, or of oil of tar deprived of its acid; pittacal falls.

Pitticite, or Hydrous tetrarseniate of iron.

Pittizite. *Pitchy iron ore, Sulpharseniate of iron, Eisen sinter.*

Placodine. *Pentarsenide of Nickel.* (πλακωδης, tabular.) Spec. grav. 7·988 to 8·062, H 5 to 5·5. Ni_4As. Bronze-yellow tabular oblique prisms, streak black; fracture conchoidal, uneven. B.B. fuses easily, with garlic smell; with borax shows cobalt; soluble in nitric acid, with a green colour. As 39·707, S ·617, Ni 57·064, Co ·91, Cu ·862. Jungfer mine, Müsen.

Plagionite. (πλαγιος, oblique.) *Sesquisulphoantimonite of lead.* Spec. grav. 5·4, H 2·5. Blackish lead-gray rhombic 8-hedrons with axes of 1 to ·88 and ·37; fracture imperfect conchoidal, structure foliated with two cleavages, brittle. B.B. decrepitates and fuses, yielding SO_2, SbO_3 and PbO. It contains Pb 40·52, Sb 37·94, S 21·53. *Form.* 4 PbS, 3 SbS_3. Wolfsberg.

Plasma. A variety of chalcedony, with yellow and white spots.

Plasma. The liquor sanguinis of the blood, including the serum and fibrine.

Plaster of Paris. A synonyme of sulphate of lime or gypsum, from its occurrence in the Paris formation.

Plasters. *Emplastra.* External medicinal applications, having generally for their bases alkaline or metallic soaps.

Platinamine. $NHpt_2$. *Gerhardt's base.* Yellow rhomboidal prisms giving off water and ammonia when heated in a tube and leaving platinum; soluble in hot acids. Ammonia is not liberated by boiling caustic potash; unites with acids and forms salts; obtained by adding excess of ammonia to nitrate of binoxide of platinum. The precipitate is a bihydrate and loses no water at 266°. At this temperature it consists of Pt 65·66, H 3·36, N 9·7, O 21·28.

Sulphate.—$NHpt_2SO_4H_2$ or NH_3PtO_2 2 SO_3. Yellow powder with an acid taste, formed by dissolving platinamine in dilute sulphuric acid and evaporating.

Nitrate.—$NHpt_2NHO_3$, 2 HO or NH_3PtO_2 NO_5 3 HO. Yellow granular powder of rhombic tables, by boiling bimuriate of platinamine with nitrate of silver as long as chloride separates and filtering.

Diplatinamine, Gros' and Raewsky's Salts.—$N_2H_4Pt_2$ ($N_2H_4pt_2HClPtClN_2H_5$). Obtained only in combination. The hydrochlorate is formed by digesting the Platino-salammoniac (NH_3HCl $PtCl_2$ the yellow salt by precipitating salammoniac by bichloride of platinum), in strong caustic ammonia; alcohol throws down white flakes which become pale yellow by drying.

Nitrate.—$N_2H_4pt_2NHO_3$, HO or 2 NH_3PtO_2 NO_5HO. White powder by dissolving nitrate of diplatosamine in nitric acid (Sesquinitrate of diplatinamine) and boiling the resulting solid sesquinitrate with ammonia.

Bihydrochlorate of diplatinamine, or *Diammonia-bichloride of platinum.*—$N_2H_4pt_2$, 2 HCl 2 NH_3PtCl_2. Formed by passing chlorine into a boiling solution of Diammonia-protochloride of platinum ($2NH_3PtCl$) or hydrochlorate of diplatosamine.

Platinum. (Plata Silver.) Pt 12·25, 98; 12·335, 98·68. Spec. grav. *hammered*, 20·857 (Clarke), 20·98 (Borda), 21·061 (Sickingen), 21·25 (Faraday and Stodart), 21·45 (Berzelius), *wire* 21·16 (Wollaston). White metal like silver; hardness between copper and iron; very ductile, being capable of being drawn into wires of $\frac{1}{1940}$ inch in diameter; its tenacity is such that a wire of ·078 inch supports a weight of 274·31 lbs.; it may be fused by the oxy-hydrogen blowpipe; may be welded at a white heat.

Preparation.—Platinum is separated from the other metals contained in the ore by removing iron ore by the magnet. The grains are then digested for three or four days in a retort with aqua regia, the muriatic acid being diluted with an equal measure of water, and the nitric acid being common aqua fortis; the solution is poured off, and the iridium allowed to subside, and separated; 41 parts salammoniac dissolved in 205 parts of water are then added, the precipitate washed, and heated in a black lead pot; the gray powder rubbed between the hands and passed through a lawn sieve, placed in a brass mould filled with water. A circular plate of copper is next laid on the top, and pressure strongly applied in a horizontal press. The ingot is then ignited for twenty minutes, and struck on the top while hot on an anvil, and brought to a perfect ingot.

Spongy platinum is obtained by precipitating a solution of platinum in nitromuriatic acid, by salammoniac, washing with alcohol, and igniting the yellow salt.

Platinum, Black.—Specific gravity 15·78 to 22·89. Obtained by precipitating platinum from its solutions by zinc, tartaric acid, alcohol, sugar.

Protoxide, Platinoxydul.—PtO. Gray or violet powder, obtained by decomposing protochloride of platinum with a solution of potash or soda, washing and carefully heating, to remove water; it loses oxygen by ignition; forms salts with acids, which are brown and red; precipitated gradually by SH, by KI, $HgONO_5$, $KOCO_2$.

Binoxide, Platinoxyd.—PtO_2. Black powder from nitrate of binoxide by potash, or by precipitating the bichloride of platinum with SH and dissolving the sulphide in nitric acid.

Protochloride. — Pt Cl. Greenish - gray or brown powder obtained by heating the bichloride so as not to decompose the whole of it, dissolving in water and evaporating, when the protochloride remains; when ignited it leaves the metal.

Bichloride, Deutochloride, Perchloride.—$PtCl_2$. Dark red-brown mass with a radiated structure when hydrous, reddening litmus; taste astringent; colours the skin brownish-black; when heated gives off an atom of chlorine and becomes PtCl, and by ignition is reduced to the metallic state. SO_2 converts it into protochloride and the solution is not precipitated by ammonia; soluble in water, alcohol, and ether. The alcoholic solution by standing loses its power of precipitating potash salts, being converted into the inflammable chloride of platinum; obtained by dissolving platinum in nitromuriatic acid and evaporating to dryness, avoiding the disengagement of chlorine. It precipitates ammonia in the form of yellow ammonia bichloride of platinum and potash salts as yellow potassium bichloride of platinum.

Iodide.—PtI. Black powder by heating protochloride of potassium and iodide of potassium.

Bromide.—PtBr. Brown crystalline mass by dissolving platinum in nitric and bromohydric acids and evaporating.

Protosulphide.—PtS. Gray or black powder by heating sulphur and spongy platinum, or by precipitating protochloride by sulphide of ammonium.

Bisulphide.—Pt S_2. Brown, gray, or black powder by precipitating bichloride by sulphide of ammonium.

Di-Ammonia protoxide of platinum.—2 NH_3, PtO, HO. White needles or crystalline mass by adding barytes water to the protosulphate of platinum, filtering and evaporating in vacuo.

Ammonia protoxide.—NH_3PtO. Gray mass by heating the preceding a little above 230° (110° C).

Ammonia binoxide of platinum or *bihydrate of platinamine, Fulminating platinum.*—NH_3PtO_2 2 HO or $NHPt_2$ 2 HO. Brown powder exploding at 401° (205° C), formed by decomposing platinum salammoniac with a solution of caustic potash.

Chlor-ammonium bichloride of platinum, Platinum salammoniac, Ammonia bichloride of platinum, Chloroplatinate of ammonium.—NH_3HCl $PtCl_2 = NH_3$ 7·61, HCl 16·31, Pt 44·36, Cl 31·72. Lemon-yellow crystalline powder or orange regular 8-hedrons when crystallized from its solution in water; soluble in 150 cold, 80 boiling water; soluble in protochloride of tin, slightly soluble in cold caustic ammonia, very soluble when boiling, yielding a gray salt; decomposed by sulphuric acid; heated not quite to redness it leaves a greenish powder (NH_3HCl, PtCl ?); ignited it evolves nitrogen, HCl and NH_3HCl, and leaves spongy platinum; it is obtained by adding an aqueous solution of bichloride of platinum to a solution of salammoniac, and is the usual salt obtained in testing and analyzing ammoniacal salts.

Platinum Bases. When ammonia is brought in contact with certain salts of platinum, remarkable compounds are formed, which seem to be bases formed on the type of ammonia, in which platinum replaces hydrogen. Platinum again appears to unite with two different atomic weights, as *Platinosum* = 12·33 or 98·66 Pt and *Platinicum* = 6·16 or 49·3 pt, protoxide being PtO and binoxide ptO (Gerhardt). According to this view, the following bases exist:—

Reiset's 2d base,	= Platosamine	$= N\ H_2, Pt$
,, 1st base,	= Diplatosamine	$= N_2H_5, Pt$
Gerhardt's base,	Platinamine	$= N\ H,\ pt_2$
Gros' O salt,	Diplatinamine	$= N_2H_4, pt_2$

Platinum Ore. *Native platinum.* Spec. grav. mean 17·332 and 16 to 18·94, H 4·25. Steel-gray irregular grains, sometimes a little convex on one side, lustre metallic, shining, opaque, ductile; soluble in nitromuriatic acid. It consists of platinum 86·16, palladium ·35, rhodium 2·16, iridium 1·09, osmium ·97, iron 8·03, copper ·4, manganese ·1, osmium iridium 1·91. Found in Choca, Barbacoas, South America; Matto Grosso, Brazil; St. Domingo; Siberia; Uralian Mountains.

Platosamine. *Reiset's second base.* NH_2 Pt? Obtained in union with chlorohydric acid by heating bichloride of platinum to 482° (250°C.) to change it into protochloride; the solution is neutralized with carbonate of ammonia heated to the boiling point and as much ammonia added as previously; the colour from garnet becomes straw-yellow and a green substance falls on standing; if now filtered the yellow crystalline hydrochlorate separates, which decomposes at 518° (270°); it may be separated from a green substance by crystallizing from hot water.

Diplatosamine. Reiset's first base.—N_2H_5Pt. Procured in the state of *hydrochlorate* (2 NH_3Pt Cl), straw-yellow 4-sided needles, by immersing protochloride of platinum in caustic ammonia and boiling; Magnus' green salt (NH_2PtHCl) falls; the boiling is continued till the green salt disappears, the solution evaporated and the residue purified by recrystallization; decomposes at 518°, salammoniac being evolved; nitric acid converts it into two substances, one in yellow 8-hedrons and into a syrupy fluid which forms crystalline compounds with HgCl and $PtCl_2$; a garnet-coloured modification in 6-sided tables is obtained by dropping carbonate of ammonia into a boiling neutral solution of protochloride of platinum and cooling. *Chloroplatinate of diplatosamine. Magnus' green compound.* 2 NH_3, 2 ($PtCl_2$) or Pt Cl_2H, N_2H_5Pt. Green needles by adding an excess of ammonia to a solution of protochloride of platinum or by passing SO_2 through $PtCl_2$ in solution till salammoniac produces no precipitate in it; on heating the solution green needles separate.

Plattnerite. A synonyme of binoxide of lead.

Plengite. A synonyme of Anhydrite.

Pleochroism. The peculiar iridescence of certain crystals as in potash oxalate of chromium, hypersthene, Alexandrite, Antigorite, and magnesia cyanide of platinum.

Pleonaste. Black varieties of Spinel.

Pleuroclase. A synonyme of Wagnerite.

Pliniane A variety of Mispickel. As 45·46, S 20·07, Fe 34·46. St. Gothard.

Plinthite. Spec. grav. 2·342, H 2·75. Brick-red masses, texture earthy, fracture flat conchoidal, opaque, not adherent to the tongue. B.B. blackens, infusible; infusible with fluxes. It contains SiO_3 36·24, Al_2O_3 28·79, Fe_2O_3 5·08, CaO 8·86, HO 13·78, alkali, magnesia, and loss 7·25. County Antrim, Ireland.

Plombierine. A synonyme of Baregine or Glairine.

Plumbago. *Plumbagine.* A synonyme of Graphite or Black Lead.

Plumbic Acid. A synonyme of Binoxide or brown oxide of lead.

Plumbic Ochre. Native protoxide of lead.

Plumbo-Calcite. A form of rhombic calc spar containing 2·34 per cent. to 7·8 per cent. of carbonate of lead. Leadhills.

Plumboresinite. *Sexhydrous bialuminate of Lead, Plombgomme, Bleigummi.* PbO, 2 Al_2O_3, 6 HO. Spec. grav. 4·88, 6·3, 6·4, H 4·25. Yellowish and reddish-brown masses, reniform, translucent, resembling Muller's glass. B.B. decrepitates and becomes white and opaque; with borax forms a transparent glass; becomes an enamel on charcoal but does not fuse. PbO 40·14, Al_2O_3 37, HO 18·8 SO_2 ·2, CaO MnO, FeO 1·8, SiO_3 ·6. It contains 1·89 per cent. phosphoric acid probably mixed with it. Huelgoet, near Poullaouen, Brittany.

Plumbostibite. See BOULANGERITE.

Plumose Antimony. A synonyme of Feather ore of Lead.

Polianite. A hard form of pyrolusite. Platten, Bohemia.

Poliene. $C_4N_4H_4$. White powder insoluble in alcohol and ether, converted by heat into *glaucene*, C_4N_3H; a product of the decomposition of sulphocyanide of ammonium.

Polishing Slate. (*Polirschiefer*, Ger.) Spec. grav. ·6 to ·8, H 1·5. Whitish-yellow to gray, absorbing water and adhering slightly to the tongue; composed of microscopic fossil infusoria. (See ANIMALCULES.) Podosphenia nana, species of Gaillonella, Navicula and Baccillaria. The animals have a mean size of $\frac{1}{288}$ of a line; 23 millions exist in a cubic line, and a grain contains about 187 millions. Its composition is SiO_3 79, HO 14·, Fe_2O_3 4·, Al_2O_3, CaO 1·. Near coal in Bohemia, at Bilin; Planitz, Saxony; Hesse Cassel; Auvergne.

Pollen of the anthers of flowers consists in the Scotch fir (Pinus sylvestris) of cerine 2, oil trace, brownish resin 3·75, malates of KO, CaO MgO, with gum 5·, sugar 5, albumen 4·, pollenine 77·25, sulphates muriates and phosphates 3.

Pollenine. $C_{11}H_{20}O_{10}$ = C 40·, H 11·7, O 48·3 (of cedar). The principal constituent of the pollen or fructifying powder of the anthers of plants, obtained by washing the lycopodium of the shops in water, digesting in alcohol and in weak potash; it is obviously an organic body of a complex nature (John).

Pollux. Spec. grav. 2·88, H 6·25. Colourless, transparent, vitreous masses, resembling quartz; fracture conchoidal; refraction biaxial. B.B. fuses into a blebby glass, colouring the flame orange; a colourless glass with borax and salt of phosphorus; decomposed by acids. SiO_3 46·2, Al_2O_3 16·394, Fe_2O_3 ·862, KO 16·506, NaO with a trace of LiO 10·47, HO 2·321? With castor in the island of Elba.

Polyadelphite. Spec. grav. 3·767 to 3·77, H 3·25 to 4·75. Wine-yellow and greenish-yellow translucent brittle roundish grains and imperfectly foliated masses agglutinated together. B.B. blackens, infusible, a green glass with soda, black in oxidizing flame; with borax and salt of phosphorus a coloured glass. SiO_3 35·033, CaO 28·294, Fe_2O_3 28·367, MnO 5·863, Al_2O_3 2·161. Franklin, New Jersey.

Polyargite. *Rosite, Rosellane.* Spec. grav. 2·72 to 2·751, H 2·5 to 4. Red to brownish-red grains, subtransparent; fracture splintery and somewhat foliated. SiO_3 44·128, Al_2O_3 35·115, Fe_2O_3 ·961, CaO 5·547, MgO 1·428, HO 5·292, KO 6·734. Södermanland, in limestone.

Polybasite. Spec. grav. 6·214, H 2·5. Iron-black 6-sided prisms terminated by bases perpendicular to the axis of the prism; or massive; lustre metallic; opaque; fracture uneven, sectile. S 17·04, Sb 5·09, As 3·74, Ag 64·29, Cu 9·93, Fe ·06.

Polychroite. *Saffron Yellow, Crocine.* The colouring matter of saffron (Crocus orientalis); yellow or scarlet bitter substance, difficultly soluble in water, ether, and oils; very soluble in alcohol and alkalies, and precipitated by acids; obtained from the watery extract of the petals of saffron.

Polychrome. *Aesculine.* $C_8H_{4\frac{1}{2}}O_3$? Colourless bitter crystals, insoluble in water and ether; soluble in 24 alcohol, destroyed by chlorine; obtained from many plants; by decoction from the bark of chesnut, ash, quassia woods, precipitating by acetate of lead, decomposing by SH and crystallizing.

Polychromic Acid. A synonyme of Aloetic acid.

Polycrase. A synonyme of Euxenite.

Polygalic Acid. *Senegine.* $C_{22}H_{18}O_{11}$. White powder without smell, poisonous; soluble in boiling water and alcohol; insoluble in ether and oils; from the root of Polygala Senega.

Polyhalite. Spec. grav. 2·769, H 2·75. Red fibrous masses, translucent opaque, bitter, astringent. $KOSO_3$ 27·7, $MgOSO_3$ 20·03 CaO SO_3 44·47, NaCl ·19, Fe_2O_3 ·34, HO 5·95. Ischl and Aussee, in Austria. See BLOEDITE.

Polyhydrite. Spec. grav. 2·121. Brown masses, opaque with resinous lustre, a sesquisilicate of iron with 29·2 per cent. water.

Polylite. Spec. grav. 3·231, H 6·25. Black vitreous plates resembling hornblende, opaque, brittle, forming a bed ¼ inch thick in magnetic iron ore at Hoboken, New Jersey. SiO_3 40·04, FeO 34·08, MnO 6·6, Al_2O_3 9·425, CaO 11·54, HO ·398. Another analysis, SiO_3 39·644, FeO 17·305, MnO 19·284, Al_2O_3 8·012, CaO 11·54, HO ·83, MgO 2·635. B.B. infusible; with soda into a brown frit; with borax a black glass.

Polymerism is exemplified in certain bodies which differ from each other by the atom of one containing a larger number of simple atoms than that of another. Methylene C_2H_2, olefiant gas C_4H_4, amylene $C_{10}H_{10}$, are polymeric bodies.

Polymignite. Spec. grav. 4·806, H7·? Black prisms with a rectangular base with truncated edges or rhomboidal 8-hedrons; powder-brown; lustre splendent; fracture conchoidal without perceptible cleavage; opaque. B.B. infusible; with borax an iron-coloured glass; with tin a red colour; with phosphate of soda dissolves; with soda decomposed without fusion and becomes grayish-red. TiO_2 46·3, ZrO 14·4, Fe_2O_3 12·2, CaO 4·2, Mn_2O_3 2·7, Perox. Cerium 5·, Yttria 11·5. In the Zirconsyenite of Frederickvärn, Norway.

Polymorphous Bodies are those which have several crystalline forms.

Polysphoerite. A synonyme of phosphate of lead.

Polythionic Acids. Acids which contain 5 atoms oxygen united with varying quantities of sulphur; they may also be viewed as sulphuretted hyposulphuric acids.

Polytungstic Acid. 3 $ROTn_6O_{18}$. A white precipitate by HCl from the mother liquor of tungstate of ammonia after the separation of paratungstate and isotungstate of ammonia by evaporation.

Polyxene. A synonyme of native platinum.

Pompholyx. The ancient name of oxide of zinc.

Poonahlite. Spec. grav. 2·1622, H 5·25. White vitreous prismatic crystals with angles of 92° 20′, resembling mesolite. SiO_3 45·12, Al_2O_3 30·446, CaO 10·197, NaO ·657, KO trace, HO 13·386. In the Ghauts, near Poonah, Bombay.

Populine. $C_{40}H_{22}O_{16}$=Saligenine $C_{14}H_6O_4$ + Benzoic acid, $C_{14}H_8O_4$ + Starch sugar $C_{12}H_{12}O_{12}$ 4HO. White silky sweetish needles, from Populus tremula, by the same process as salicine; soluble in 70 parts boiling water, and in alcohol; when distilled yields an oil containing benzoic acid; reddened by SO_3; by acids converted into saliretine, benzoic acid, and starch sugar; boiled with barytes water, and the excess precipitated by CO_2, it forms benzoate of barytes and salicine; treated with NO_5 of 1·3, it yields needles (benzo-helicine), which appear a compound of benzoic acid, hydride of salicyle, and sugar.

Porcelain. See POTTERY.

Porcelain Clay. See CLAY.

Porcelain Jasper. Beds of clay which have been rendered semivitreous by contact with trap rocks.

Porcelain Spar. Spec. grav. 2·66, H 5·5. White square prisms like scapolite, translucent. B.B. fuses into a blebby glass; decomposed by $HCl=SiO_3$ 49·3, Al_2O_3 27·9, CaO 14·42, NaO 5·46, HO ·9.

Porphyric Acid. $C_{26}H_5N_3O_{19}$. Yellow crystalline powder, forming a red salt with carbonate of ammonia, insoluble in an excess of carbonate; with difficulty soluble in water; obtained by heating euxanthone or purrenone with cold nitric acid; when heated with hot nitric acid *oxyporphyric acid* ($C_{26}H_5N_3O_{20}$) is formed which yields a red salt with carbonate of ammonia, soluble in the carbonate.

Porphyroxine. Neutral brilliant needles, insoluble in water, soluble in alcohol and ether; becoming red by muriatic acid. From Bengal opium.

Porphyry. A rock having a basis of felspar or clay, &c. with crystals of some other mineral interspersed through it. Felspar porphyry has a felspar basis with crystals of felspar interspersed. Claystone porphyry has a clay basis with felspar crystals distributed through it. Granite is often porphyritic when large felspar crystals occur diffused through fine-grained granite.

Porter. A well-known form of fermented liquor, characterized by its dark colour, depending on the higher temperature at which the malt from which it is made is dried. This incipient charring developes a peculiarly agreeable bitter taste. The average specific gravity of porter wort is 1064·5 (Shannon); it contains 6·8 per cent. of alcohol, spec. grav. ·825 (Brande). Brown stout wort has a spec. grav. of 1062·4. The spec. grav. of the best brown stout in bottle, after some months, is 1010·6. 40 bushels of pale and 20 bushels of brown porter malt are mashed in 20 barrels water at 180°, stirred for twenty minutes, allowed to stand for three hours, and 14 barrels worts run into the underback; see BREWING; 14 barrels of water at 190° form the second mash; 12 barrels are drawn in two hours; for the third mash 8 barrels at 180° are laid on. The worts to the amount of 34 barrels are boiled for two and a-half hours, 80 lbs. Sussex hops being added. To colour and flavour it there is often added a mixture of burned sugar, liquorice, clarified sugar, and Spanish juice. The worts are now spread on the coolers and reduced to 62° of temperature. 22 barrels of worts are now *fermented* with 9 gallons of yeast; the action is continued till the head of yeast begins to slacken round the sides of the fermenting tun, when 4 lbs. flour and 2 lbs. salt being thrown over the yeast, it sinks. The worts and yeast are then mixed thoroughly and the process of *cleansing* performed in pipes or butts. See ALE and BEER. The following are the results of the experiments of my pupil, Mr. John Wright Currie, on the strength of various porters principally brewed at Perth:—

	1.	2.	3.
Common Porter,...... Spec. grav. 1014.	4·035	3·845	92·13
Brown Stout,......... Spec. grav. 1011.	4·58	3·285	92·24
Guiness' Stout,....... (1016·4).	5·27	5·	89·73
Double Brown Stout, (1013).	6·22	4·	89·78
Imperial, ... (1024).	7·09	5·5	87·41

1. per centage absolute alcohol, 2. per centage dry residue at 212°, 3. water per cent.

Portite. Spec. grav. 2·4, H ·5. White opaque, glassy, rhombic prisms. B.B. fusing into a white enamel; decomposed and gelatinizing with acids. SiO_3 58·125, Al_2O_3 27·5, CaO 1·759, MgO 4·873, NaO ·157, SO ·1, HO 7·917. It is a magnesia harmotome.

Portland Stone. A formation in the oolitic system, consisting of several beds of coarse earthy limestone. The more oolitic varieties quarried in the islands of Purbeck and Portland, are used as a building stone in London.

Potassium. K (Kalium) 5, 40 4·889, 39·2. Discovered by Sir H. Davy in 1807. Spec. grav. ·86507 (Gay Lussac and Thenard, Rech. Phys.-Chim. 1, 117). Specific heat 5·77 (Regnault). Soft and malleable at 50°, fluid at 136½°, at 32° brittle. Colour silver-white, with a metallic lustre. Sublimes in the form of green vapour when heated nearly to redness. When broken at 32° it exhibits a crystalline structure (Davy, Phil. Trans., 1808). A good conductor of heat and electricity. *Preparation.*—1. When moist pieces of caustic potash are placed upon a disc of platinum attached to the positive end of a powerful galvanic battery, and a platinum wire from the negative extremity is made to touch its upper surface, the potash is gradually decomposed, oxygen separating at the extremity of the positive pole, while globules (potassium) of a white metal like mercury appear at the side in contact with the platinum wire (Davy, ib.). 2. By heating caustic potash in contact with iron filings. 3. The best method is to prepare an iron bottle for the purpose, or to use an iron mercury bottle previously ignited to remove mercury, if any should remain. Black flux is then prepared by igniting cream of tartar in a clay crucible. The tartaric acid is decomposed ($KOHOC_8H_4O_{10}$, become $KOCO_2, C_5+4HO2CO_2$), and carbonate of potash is formed, intimately mixed with charcoal, while water and carbonic acid are evolved. It is found advantageous to triturate the black flux, formed by the ignition of cream of tartar, with an equal quantity of charcoal. This mixture is introduced into the iron bottle, which is then placed in a wind furnace, supported on bricks. To its mouth is adapted, by a screw or by grinding, a short iron tube, the other end of which passes into a receiver. This receiver consists of two copper boxes, one fitting into the other. The upper one is a thin parallelopepid, 10 inches long, 5 or 6 broad, and $1\frac{1}{6}$ thick, shut at top, and open at bottom; and is divided by a diaphragm passing to a third of the bottom. An iron wire passes through this partition, opposite the end of the iron tube, for the purpose of keeping the iron tube from being blocked up by the volatile matter in the iron bottle. The upper vessel dips into another vessel, open above and shut below, to which it is exactly adapted so as to pass to the bottom of it. A few inches of mineral naphtha are placed in it, and the air excluded by means of fat lute or putty. An aperture at the top allows gases to escape. Heat being applied to the iron bottle, the charcoal removes the oxygen from the potash, while the carbonic acid is resolved into carbonic oxide ($C_2KOCO_2 = K,3CO$). The potassium drops into the naphtha mixed with charcoal, and carbonic oxide compounds as crocoic acid, &c.

Dinoxide, Suboxide.—K_2O? Gray mass with a blue shade. When ignited becomes red; at about 80° takes fire and burns, and becomes a mixture of potash and teroxide; very fusible; obtained when potassium is heated in air, deficient in the quantity of oxygen requisite to convert it into oxide (KO), or by heating a mixture of 5 parts potassium, and 6 of potash, or protoxide of potassium.

Oxide of Potassium, Potash, Potassa, Vegetable Alkali. (*Kali*, Ger.; *Potasse*, Fr.).—KO 6·, 48·; 5·9, 47·2, K=83·05, O=16·95. Spec. grav. 2·4 (Dalton), 2·656 (Karsten). Grayish-white solid, melts a little above a red heat, and sublimes at an elevated temperature; no smell; taste strongly caustic, and corroding the tongue. Its affinity for water is very great; its solution in water changes an infusion of violets and red cabbage green, and renders reddened litmus blue; decomposed by the galvanic battery, by iron and zinc at a white heat, by chlorine at a red heat, by sulphur into sulphate of potash, and sulphide of potassium, and also similarly by phosphorus; obtained by burning potassium in dry oxygen.

Hydrate of Potash, Caustic Potash, Potential Cautery.—KOHO, 7·125; 57; 7·025, 56·2. KO =84, HO=16. Sp. grav. 2·2 (Dalton). Brittle, white substance, corrosive, and destructive of organic matter; fuses below a red heat into a clear oily fluid; volatilizing at a higher temperature; soluble in half its weight of water (Löwitz); very deliquescent; non-conductor of electricity, except when moist; it yields oxygen to the positive pole of the battery, and potassium at the negative (Davy, Phil. Trans. 1808). The potash is placed upon a disc of platinum attached to the positive pole, and a platinum wire from the negative pole is made to touch its upper surface. The amount of water in hydrate of potash is ascertained by fusing a given weight of it with fused boracic acid or silica when the water is replaced by these acids. *Process.*—1. When potassium is introduced into a tube inverted and filled with mercury, and a few drops of water are admitted to it, potash is formed, and hydrogen evolved, while the potash dissolves in the excess of the water (K. 2 HO = KO HO, H); the liquid when evaporated yields hydrate.

Solution of Hydrate of Potash in Water.—For laboratory purposes a solution of hydrate of potash, (*caustic potash, aqua potassæ, liquor potassæ*,) is made by dissolving in an iron pan 1 lb. of carbonate of potash (or bicarbonate of potash yields a still purer solution) in 10 lbs. of boiling water, adding gradually $\frac{1}{2}$ lb. of quicklime slaked and made into the consistence of a milk by water, boiling during the addition of the lime. The whole mixture is boiled for twenty minutes, withdrawn from the fire, the pan covered with a lid, and allowed to subside. The liquor is drawn off by a syphon when it has become clear. The alkaline liquor should be replaced by boiling water to wash out all potash from the lime. The liquor is then evaporated down in an iron or silver basin until it begins to deposit crystals of sulphates. It is then, when cool, to be kept in stoppered bottles. *Purification.*—Carbonate of potash very often contains silica and alumina, which, however, are removed by boiling the caustic lye during its preparation, for in the cold state lime forms a caustic alkali with carbonate of potash. The silica and alumina unite with the lime. Carbonate of potash is prevented from being reformed from the carbonate of lime by adding water, so as to preserve the original amount of fluid; that is never less than 10 lbs. water to 1 lb. carbonate of potash. When properly made, especially from bicarbonate of potash, caustic potash should not effervesce with acetic acid, and weak mineral acids. When neutralized by pure nitric acid it should give no precipitate with chloride of barium, nitrate of silver. Caustic potash usually contains some alumina, dissolved up from the lime.

Salts of Potash.—1. The salts of potash are generally soluble in water, but less soluble than those of ammonia. 2. Many of them are crystallized, but they have a less tendency to assume a regular form than salts of soda. 3. Tartaric acid precipitates salts of potash in the form of bitartrate of potash, or cream of tartar. 4. Sulphate of alumina, mixed with a salt of potash, forms octahedral crystals of alum. 5. When a solution of bichloride of platinum is poured into a concentrated solution of a salt of potash with muriatic acid, a yellow precipitate of potash chloride of platinum falls, which is converted by ignition into platinum and chloride of potassium. 6. Potash salts are not volatilized by heat: its salts, with organic acids, are converted into carbonate of potash and charcoal. 7. Salts of potash impart to the flame of the blowpipe a *violet* tinge at the moment of immersion. A mixture of borax and oxide of nickel, with a salt of potash, yields before the blowpipe a *bluish* bead.

8. Fluosilicic acid forms a semitransparent precipitate in solution of salts of potash. 9. Perchloric acid in solution forms a crystalline precipitate of perchlorate of potash with potash salts.

Chloride, Febrifuge or Digestive Salt of Sylvius, Regenerated Sea Salt (Boerhaave), *Muriate of Potash.*—9·5 or 9·325 or 74·6, K 52·63, Cl 47·37. Sp. gr. 1·836 (Kirwan), 1·9153 (Karsten), 1·945 (Kopp). Cubes and square prisms by the elongation of the cubes; taste more pungent than common salt; 1 part chloride dissolves in 2·89 water at 53¼°, in 2·87 at 56°·8, in 2·85 at 60°, and in 3·008 at 63½°; 1 water at 32° dissolves ·292 parts salt, and ·0027838 for every degree upwards; but slightly soluble in alcohol; 1 part of chloride, when dissolved in 4 parts of water, lowers the temperature 52½°, while common salt only lowers it 35½°; a mixture may be thus ascertained. From this property chloride of potassium is used in freezing mixtures. It decrepitates when heated, fuses when ignited; does not volatilize in close vessels, but volatilizes in open crucibles. Obtained by burning potassium in chlorine gas, by adding muriatic acid to carbonate of potash, but principally this salt is obtained by crystallization from kelp. (See KELP and IODINE.)

Iodide, Hydriodate of Potash.—KI 20·75 or 20·69 or 165·2. Spec. grav. 2·908 (Karsten), 3·091 (Boullay). Cubes, elongated cubes, pyramids, and 8-hedrons or square prisms. Deliquescent; alkaline reaction; 100 water dissolve 136 salt at 54½°, producing 75°·2 of cold; at 61° 141 parts; at 64½° 143 parts; at 248° 221 parts (Baup). At 54½° it dissolves in 5½ times its weight of alcohol, spec. grav. ·850, and in 39 or 40 times its weight of absolute alcohol at 56°. It is much more soluble in hot alcohol, and crystallizes out in needles on cooling. A saturated solution in water boils at 248° (Baup). Changed by nitric acid into nitre. With chlorate of potash in fusion it becomes iodate. Steam removes iodine at a red heat, and leaves an alkaline mass. Chlorine converts it into chloride, and sets iodine free. Melts below a red heat, and volatilizes in free air; undecomposed at a moderate heat. Salammoniac, when heated with it, decomposes it partially. Colours paper bleached by chlorine brown. A dilute solution is coloured brown by chlorine, which affords a delicate test; nitric acid has an analogous effect.

Process.—The method usually employed for the preparation of iodide of potassium for medical purposes is—1. To boil 3½ parts of iron wire or filings with 15¾ of iodine in 10 ounces of water, or half a pint, until the solution becomes colourless, taking care to moderate the heat, when the iodine volatilizes. 2. The solution is then filtered, and precipitated with 8¾ parts of ignited carbonate of potash at a boiling temperature (FeI,$KOCO_2$ becoming $FeOCO_2$ and KI). The solution is evaporated after filtration in vacuo, or in a dry stove. It may also be prepared by mixing 2 parts iodine, 1 caustic potash, and 6 water, evaporating to dryness, and igniting to decompose iodate. The iodide is then dissolved in water and crystallized. To decompose iodate, it has been recommended to pass sulphohydric acid through the solution. Iodide may also be formed by decomposing sulphuret of potassium by means of iodine, or by saturating caustic potash or carbonate with iodohydric acid.

Purity.—If carbonate of potash be present, acetic acid causes an effervescence, and it remains undissolved, when the iodide is taken up by alcohol. *Iodate* is detected by the brown colour which it gives when boiled with a salt of morphine. Chlorides give a precipitate with nitrate of silver, soluble in caustic ammonia, the yellow iodide being insoluble.

Cyanide, Cyanuret, Cyanodide.—KCy or KNC_2 8·139 or 65·112. White cubes and 8-hedrons; deliquescent; very soluble in water, and soluble in boiling alcohol of 60 per cent. (spec. grav. ·8952). Its aqueous solution decomposes into ammonia and formate of potash (KNC_2 and 4HO become KOC_2HO_3 and NH_3). *Process.*—If the yellow prussiate of potash, previously deprived of water, be reduced to powder and heated to redness in a close iron vessel, supplied with a small aperture for the escape of nitrogen, it is converted into cyanide of potassium and charcoal with some carbide of iron. The cyanide is to be dissolved out with cold water, and the solution evaporated to dryness in vacuo, or at a gentle heat. The white saline mass resulting is then to be boiled with alcohol of 60 per cent. when the cyanide deposits in cubes on cooling. If the experiment be made in a clay retort, the beak having adapted to it a bent tube placed under an inverted jar filled with water, the operation is at an end when the nitrogen gas ceases to be evolved. It is used to dissolve cyanide of silver in electroplating, and cyanide of gold in gilding; in medicine to form prussic acid by the extemporaneous process, and also by itself.

Sulphocyanide (Porrett, 1813).—KNC_2S_2 or $KCyS_2$ 12·139 or 97·112. Colourless channelled prisms or silky needles. *Taste* biting, resembling that of radishes, leaving a saline and cooling impression. Melts by heat into a transparent fluid, becoming crystalline on cooling; is not decomposed in close vessels, but in an open vessel it becomes sulphate of potash; deliquescent. *Process.*—Fuse together at a temperature below ignition equal parts of dry yellow prussiate and sulphur, or 2 parts yellow prussiate and 1 sulphur. Dissolve the fused mass in water, and add caustic potash, or carbonate of potash, till all the iron is thrown down, and the solution becomes, from being red, colourless; then filter and evaporate to dryness. If heated to redness, sulphide of iron would be formed, and nitrogen and sulphocarbonic acid disengaged.

Sulphide, Sulphuret.—KS 7·or 6·889 or 55·112. Red, cinnabar fibrous mass; deliquescent; tasting of sulphohydric acid; very soluble in water;

burns very slowly in the air, leaving sulphate; unites with many metallic sulphides, forming sulphur salts. It may be formed by passing hydrogen over ignited sulphate of potash, or by heating that salt with charcoal, ($KOSO_3$ and 4C becoming KS4CO). By the hydrogen process it is bright red, by the charcoal process flesh-red and crystalline. The compound formed by passing sulphohydric acid through caustic potash is a sulphohydride (KSHS) and sulphide. If the solution of the sulphide is coloured, it may be decolourized by agitating with mercury, by which is sometimes imparted a fine vermilion.

Sulphate of Potash, (Oswald Croll, 1600), *Specificum purgans Paracelsi*, *nitrum fixum*, *arcanum duplicatum*, *panacea holsatica*, *sal de duobus*, *Sal polychrestum Glaseri*, *alkali vegetabile vitriolatum* (Bergman), *vitriol of potash* (Morveau), *vitriolated tartar*. — The duke of Holstein, some time before the year 1663, purchased the preparation of this salt for 500 dollars (£75) as a remedy in fevers, stone, and scurvy. $KOSO_3$ 11· or 10·889, or 87·112; potash 54·07; sulphuric acid 45·93. Spec. grav. 2·636 (Watson's Essays, v. 67), 2·66 (Thomson), 2·66 (Kopp); 1 part soluble in 9·3 water at 65$\frac{3}{4}$°; 100 water at 32° dissolve 8·36 salt; and for every 1·8 degree above this 0·1741 (Gay Lussac). *Crystalline form*, *Dimorphous*. 1. *Right prismatic*. Primary form, a colourless right rhombic prism, the faces meeting at angles of 120° 30′. Commonly two of the opposite lateral edges are replaced by tangent planes, converting the crystals into a 6-sided prism. The edges of the base of the prism are also replaced by planes, converting the end of the prism into a pyramid, either complete or without the apex. P, M and M′ are the faces of the primary figure, *h* the face formed by replacing one of the lateral edges, and c, *e*, *e*, the pyramidal faces made by replacing the terminal edges. The measures of these faces are as follows:—M on M, 120° 30′, M on *h* 120° 45′, M on *e* 146° 22′, *h* on *c* 146° 10′, *c* on *e* 13° 12′. It also occurs in regular 6-sided prisms (Brewster, Mitscherlich, Ann. de Chim., 23, 279); Giobert, Ann. de Chim. 10, 40). 2. *Rhombohedral System*.—In preparing the salts of kelp for iodine and alum-making, the kelp is dissolved in water heated by steam; the solution is drawn off and concentrated; on cooling light is evolved in the dark, the soda separates, and a crust of sulphate of potash forms at the surface crystallized in rhombohedrons. The extreme face *o*, which replaces the point of the rhombohedron, is well developed, exactly like tabular crystals of specular iron. In most of the crystals no faces are perceived except those of the rhombohedron; but in some we observe an imperfect second rhombohedron, and sometimes the faces of the prism, with 6 sides, which depends from it. The inclination of the faces of the rhombohedron upon the face *o*, is 124°. (Discovered by John Tennent, at Campsie; and Profr. Mitscherlich, in 1842; L' Institut. 1843, 287). This light is possibly connected with a change in its crystalline form to the rhombohedral state, as occurs with arsenious acid, when it crystallizes from the opaque state in 8-hedrons; the deposit of each crystal is accompanied with a spark. Decrepitates by a red heat; converted into bisulphate by muriatic acid; fuses at a red heat; contains no soda. A solution of sulphate in hot nitric acid gives oblique prisms of spec. grav. 2·381; fusing at 302° ($2KOSO_3,HONO_5$, or $KOHO2SO_3$, $HONO_5$); and phosphoric acid yields a similar result. Taste disagreeably bitter. This salt is obtained by saturating carbonate of potash with sulphuric acid, or by saturating with the lime salt, the bisulphate of potash remaining in the retort in the preparation of nitric acid.

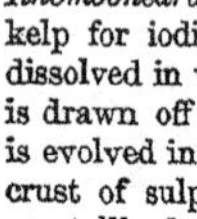

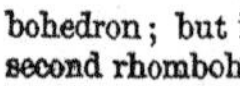

Bisulphate.—$KO,2SO_3,HO$, or $KOSO_3$ HO SO_3 17·125, or 17·025, or 136·2; potash 34·65, sulphuric acid 58·74, water 6·61. *Crystalline form*.—Thin rhomboidal plates, the plane angles of the large faces of which are 97° 20′, and 82° 40′, and the faces of the prism inclined to each other at angles of 107° 15′. Taste acid and sharp. Melts at 392° (Mitscherlich) when heated, and flows like an oil. It is obtained by fusing together 11 parts of sulphate of potash with 6$\frac{1}{3}$ parts of oil of vitriol. This salt is much used in analysis for decomposing minerals containing much alumina, as corundum, &c.

Nitrate, *Saltpetre*, *Nitre*, *Potash Nitre*.—KO NO_5 12·639, or 101·112, potash 47·05, nitric acid 52·95. Spec. grav. 1·933 (Watson's Essays, 5, 67), 2·058 (Kopp), 2·100 (Karsten). 1. In 6-sided prisms terminated by 6-sided pyramids, belonging to the right prismatic system. 2. Obtuse flattened rhombohedrons like nitrate of soda. The primary form is a rectangular 8-hedron of two 4-sided pyramids applied base to base, two of the sides being inclined to the other pyramid at an angle of 120°, the other two at 111°. Taste sharp, bitterish, and cooling; brittle. 100 water at 32° dissolve 13·32 parts, at 41° 16·72 parts, at 53° 22·23, at 64$\frac{1}{4}$° 29·31, at 76·9° 38·40, at 95$\frac{1}{4}$° 54·82, at 113$\frac{1}{3}$° 74·66, at 130$\frac{1}{2}$° 97·05, at 149·8° 125·42, at 175$\frac{1}{3}$ 169·27, at 207·8° 236·45 (Gay Lussac, Ann. of Phil. 15, 11); 1 part dissolves in 100 alcohol of spec. grav. ·878. It is more soluble in water containing common salt and nitrate of lime. When heated to redness oxygen is emitted first nearly pure, but afterwards mixed with nitrogen, forming a hyponitrite, and leaving latterly caustic potash. In this action platinum vessels are affected. Those of silver and gold resist best. When mixed with charcoal and ignited nitre *detonates*. When this experiment was made by the alchemists in a retort, the water which they obtained was called *clyssus*. Phosphorus, when struck by means of a hot hammer with nitre, detonates (Brugnatelli, Ann. de Chim. 27, 74). *Processes*.—1. Nitre occurs in the soil of various countries, as in Tirrhoot, Ceylon, Oliva, Madras, Spain, (especially in La Mancha and Arragon.) The composition of the nitre earth is as follows:—nitrate of lime 0·0, nitrate of potash, 2·4; nitrate of magnesia 0·7, water 9·4, carbonate of lime 20·5, earthy matter 60·7, animal matter, sulphate of lime, common salt, traces. In Ceylon these are washed with an equal quantity of wood ashes, filtered, and evaporated.

Theory of the Production of Nitre.—From the importance of nitre in the production of gunpowder, and its occurrence in nature, much attention has been bestowed on the theory of its formation. Stahl considering that there was only one acid in nature, supposed nitric acid to be a compound of sulphuric acid and phlogiston, produced by putrefaction. Lemery, the younger, thought that nitre previously existed in animals and vegetables, but the fact that plants only contained it when it existed in the soil, refuted this view (Bouillon). According to our present knowledge nitre may be formed, as in Ceylon, without animal matter, if ammonia be present. Ammonia consisting of NH_2H, the hydrogen is replaced by oxygen from the air NH_2H, and 3 oxygen becoming binoxide of nitrogen (NO_2), and 3 water (3HO). The binoxide by additional oxygen becomes quaternitrous acid (NO_4). This, by the action of lime or potash, becomes nitrate and nitrite of base ($2NO_4 = NO_5$, and NO_3).

Chlorate, Oxymuriate.—First obtained by Higgins, 1786 (on acetous acid 180), but first recognized by Berthollet. $KOClO_5$ 15·5, or 15·325, or 122·6. Spec. grav. 1·989 (Hassenfratz, Ann. de Chim. 28, 12), usually in small thin plates. The primitive form is an obtuse rhomboidal prism (Hauy), and belongs to the oblique prismatic system. Taste cooling, disagreeable, somewhat resembling nitre. The chlorate when heated to about 673° gives off oxygen, and if the temperature is not increased perchlorate of potash (ClO_7), and chloride of potassium remain in equal proportions, chlorite being first formed; $2ClO_5 = ClO_7ClO_3$; at a higher temperature nothing but chloride of potassium remains, the amount of oxygen lost being from 38·7 to 39·16. The presence of black oxide of manganese, or oxide of copper, and other bodies, assists the decomposition. 100 water at 32° dissolve 3·3 parts of salt, at 56° 5·6, at 59$\frac{3}{4}$° 6·03, at 76° 8·44, at 95° 12·05, at 122$\frac{1}{3}$° 18·96, at 166°·8 35·40, at 220°·6 60·24. 1 part dissolves in 120 of alcohol of 83 per cent. at 61°·8. *Explosive action.*—3 parts of chlorate, and 1 part of sulphur, triturated in a mortar, or struck on an anvil, detonate violently. Charcoal acts in a similar way, though less violently. Hence Berthollet proposed it as a substitute for nitre in gunpowder. A fatal explosion occurred in consequence at Essone, in 1788, during the trituration of the mixture, by which M. Letors and Mlle. Chevraud were killed.

Preparation.—For the following process I am indebted to Mr. James Young:—Chlorate of potash is made by forming chlorate of lime, and decomposing this by a cheap salt of potash, such as the sulphate, or chloride of potassium. To form chlorate of lime, milk of lime is warmed to 150° Fahr. and poured into a stone vessel constructed in the following manner:—It may be of any size, according to that of the stones conveniently to be had; and in practice it is found that a vessel formed of flagstones, 6 feet long × 15 inches deep × 15 inches wide inside measurement, answers well for $\frac{1}{4}$ cwt. of dry lime. This flagstone vessel is fitted with two or more inclined shelves formed also of flagstones, hollowed out on the under side so as to form an inverted shallow dish, according to the drawing shown in section, fig 1. 28 lbs. of dry caustic lime are mixed with water into a milk of lime, sufficient to fill the stone trough up to the dotted line. The chlorine passes in by the stoneware pipe *a*, escaping at the end *b*, under the inclined shaft *c*, and there spreading itself out into a thin sheet, is rapidly absorbed by the milk of lime. All unabsorbed chlorine passes through the hole *d*, to the under side of the upper shelf *e*, along which it passes, and ultimately escapes by the hole *f*; but it is only towards the end of the process, when the caustic lime is nearly all converted into chloride and chlorate, that any chlorine gas will reach the hole *f* in the upper shelf. To keep the lime in suspension during the operation, a stirrer or agitator having a pendulous motion, like a weaver's lathe, and made of wood, is fitted into the stone trough, according to the drawings shown at fig. 2 and fig. 3. Fig. 2 is a longitudinal section of the apparatus and agitator, and fig. 3 is a cross section, showing how the motion is communicated by a crank, connected with a steam engine, by which the agitator is kept constantly in motion. When the operation is completed, all the lime is dissolved, and the absorption of chlorine ceases. This is known by the clearness of the lime solution, and by the quantity of uncondensed chlorine escaping to the hole *f* in the shelf *e*. When the operation is finished, the liquor should have little or no bleaching properties; this may be tested by a protosalt

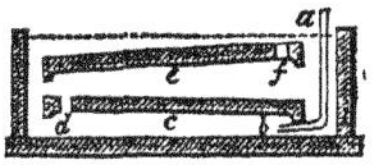

1.

2.

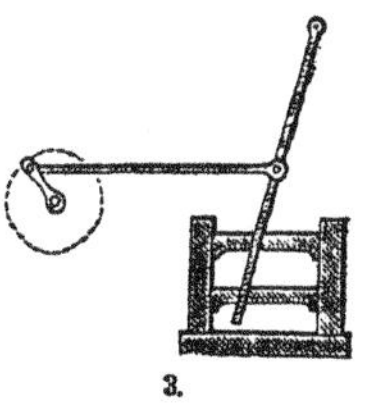

3.

of iron. The success of this process depends on the complete saturation of the lime with chlorine. As long as any free lime remains, there is little chlorate formed, but the well-known bleaching compound of chlorine and lime; on the whole lime being saturated, this changes into chlorate of lime and chloride of calcium. The quantity of chlorate in solution may be ascertained by first adding an excess of sulphuric acid, and then a measured quantity of protosulphate of iron, as in the usual way of testing bleaching powder. If sulphate of potash is used to decompose the chlorate of lime, much sulphate of lime is precipitated, which will require first to be separated before proceeding to evaporate the solution, after which the operation will be the same as when chloride of potassium is used. The solution is evaporated until it comes to 64 on Twaddel's hydrometer while hot (equal to spec. grav. 1·320); it is then run into iron vessels or coolers to crystallize, and after standing there for a week, the liquid, which is principally chloride of calcium, is run off, leaving the impure crystals of chlorate of potash in the coolers. The chloride of calcium retains some chlorate of potash in solution, but there seems no cheap method of separating it. It was tried to displace it by a cheap salt, such as common salt, but it failed. These crystals are lifted out of the coolers, and put upon a strainer made of wood or iron, pierced full of holes, to separate the solution of chloride of calcium as much as possible. They are then dissolved, and the solution evaporated in a leaden pan to 25 of Twaddel's hydrometer while hot, (equal to 1·125 spec. grav.), and run into lead vessels to crystallize. After standing a week the mother liquor is run off, and used instead of water, to mix up the lime described in the first part of this process. The crystals of chlorate of potash are redissolved, and the solution evaporated to the same specific gravity while hot (1·125), and run into leaden vessels to crystallize. In six or seven days the mother liquor is run off, and used to dissolve the impure chlorate of the first crystallization. The crystals of the chlorate of potash are lifted out of the leaden coolers, drained from the mother liquor as much as possible, and dried. These crystals still contain a trace of a chloride, but are pure enough for all manufacturing processes. If wanted quite pure they should be again dissolved in pure water, and crystallized, as described in the last operation. No oxygen is given off when the operation is properly conducted; and the solution should therefore contain 1 equivalent of chlorate of lime, and 5 equivalents of chloride of calcium. The chlorine is made as follows:—*a a*, fig. 4, is the still formed of a salt-glazed stoneware muriatic acid receiver, fitted with a false bottom, *o o*, formed of pieces of stone placed 2 inches or so above the bottom of the stoneware still, *a a*, so as to prevent manganese escaping when the muriate of manganese is run off by the plug-hole *k*, as will be described farther on. The stoneware still, *a a*, is inserted into a steam chest, *b b*, kept full of steam by the pipe *c*, communicating with a steam boiler. The glass tubes *h* and *e* are fastened into the bottle *f* by a cork covered with pitch, and the end of the bent tube *h* is secured to the still *a a* by sulphur or Roman cement—this serves for running the muriatic acid into the still *a a*, acts as a safety vessel, and prevents the escape of chlorine, as the tube *e* dips near to the bottom of the bottle *f*, while the tube *h* only goes about half way; the tube *h* should be about 2 feet long. *g* is another safety vessel made of lead, about a foot in diameter and height, placed between the still and the lime apparatus. Manganese in lumps is put into the still *a a*, until filled within 6 inches of the top; and muriatic acid, of specific gravity 1·150, is run into it by the stoneware stop-cock *d*, until it also reaches within 6 inches of the top, and steam is then let into the box *b b* by the steam pipe *c*. Muriatic acid, of 1·150 specific gravity, acts vigorously upon lump manganese at the temperature maintained in this apparatus, and chlorine will be given off in great abundance, passing by the safety vessel *g* through the lead pipe *h*, which is connected with the stoneware pipe *a b*, described in the drawing No. 1, and thence through the milk of lime. When the muriatic acid becomes saturated, (which is known by chlorine ceasing to bubble into the milk of lime solution,) the solution of muriate of manganese is run out of the still *a a* by the plug *k*; and more muriatic acid then run among the remaining lumps of manganese by the stop-cock *d*, and so on the process is continued, adding fresh manganese in lumps as required. The milk of lime saturated with chlorine as described, and now converted into chlorate of lime and chloride of calcium, is run into an iron evaporating vessel, (although the iron is slightly acted upon,) and rather more than 1 equivalent of chloride of potassium is added.

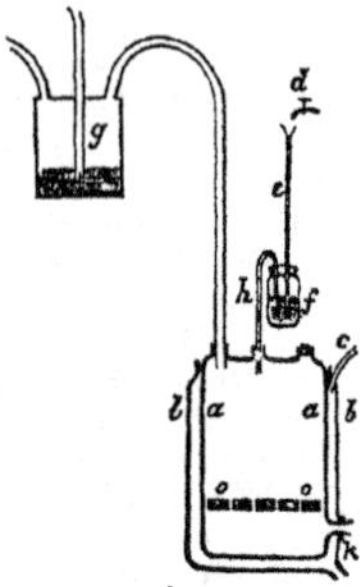

4.

Carbonate, Fixed Nitre, Salt of Tartar, Aerated vegetable Alkali.—$KO\,CO_2$ 8·75 or 8·64 or 69·12, =potash 68·57, carbonic acid 31·43; first accurately described by Bergman in 1774. White mass; spec. grav. 2·26 (Karsten); fuses at a red heat, volatile at a white heat; taste alkaline; renders litmus paper blue; obtained in the purest form by heating to redness 2 cream of tartar and 1 nitre, digesting in water, filtering, evaporating the carbonate, and igniting, or by igniting the bicarbonate of potash. When these operations are performed in a clay crucible, it is advisable to line it with a mixture of starch and gum, or even paper, to prevent the silica of the crucible from uniting with the potash. Impure cream of tartar yields likewise cyanide of potassium from the vegetable matter present.

Hydrous Carbonate.—KO CO_2 2 HO 11· or 10·9, or 87·2. Rhomboidal plates; 8-hedrons with rhombic base, the angles of the rhomb being 122° and 58; obtained by crystallization from a concentrated solution of pure carbonate of potash. A very deliquescent salt; insoluble in alcohol, but very soluble in water; soluble at 37°·4 in ·05 water, at 42°·8 0·962 water, at 54° 1·9, at 78°·8 ·747, at 158° ·49 water.

Potashes.—The potash of commerce is obtained by burning plants, as the *Salsola Kali*, and trees. The ashes which remain are digested in water in wooden vats, by which carbonate, sulphate, silicate of potash, and chloride of potassium, are dissolved. The solution is drawn off and boiled down to a brown mass, the colour proceeding from the vegetable matter of the wooden vessel. This mass is ignited in a reverberatory furnace, and is then known as *potash* of commerce—the term being derived from the *pots* in which this operation used to be performed. It is a dark mass, mixed with black spots of charcoal. *Pearl-ash.*—To convert potash into pearl-ash, the potash is dissolved in 2 parts of boiling water, filtered, allowed to stand till the sulphate crystallizes out. The clear liquid is poured off and evaporated down, and ignited in a reverberatory furnace. Chloride of potassium and sulphate may be further separated by crystallization. Silica may be removed by mixing the solution with charcoal, and allowing it to stand for twenty-four hours. Pearl-ash, so named from its lustre, is used in calico printing. *Purity.*—In purchasing it, the amount of carbonate of potash present is required to be known, together with the amounts of chloride, sulphate, and water. It might be adulterated with soda, and hence the quantity of potash must also be ascertained by means of bichloride of potassium. 50 grains are heated to 300°; the loss is water. It is then heated to low redness, the loss is organic matter. The chlorine is precipitated by nitrate of silver, from a solution of pearl-ash in nitric acid; the sulphuric acid by chloride of barium; the carbonic acid by the amount lost on the addition of sulphuric acid. (See CARBONIC ACID, and ALKALIMETRY.)

Bicarbonate.—KO,$2CO_2$ HO 12·625, or 12·525, or 100·2. Primary form, right oblique prism, the faces of which meet at angles of 103° 25. Obtained by saturating a solution of carbonate of potash with carbonic acid, evolved from carbonate of lime and chlorohydric, or sulphuric acids; to purify the carbonic acid gas, an intermediate vessel, with a small portion of water, may be used. On the small scale, a flask and bent tube, or gas bottle, may be employed. See CARBONIC ACID.

Chromate. — KOCr O_3 12·5, or 12·237; 97·896. Sp. gr. 2·6115 (Thomson). Fine yellow oblique 4-sided prisms, terminated by 4-sided pyramids; two of the opposite lateral edges of the prism are often replaced by tangent planes. M and M′ represent the primary faces of the prism meeting at angles of 107° 26′. Not altered in air. 1 grain colours 40,000 of water yellow; 1 of chromate and 20 nitre, dissolved and crystallized, yield nitre crystals of a fine yellow; 100 water at 60 dissolve 48·368 parts; soluble in all proportions in boiling water; insoluble in alcohol. When heated to 400° it does not change; ignited it becomes crimson, but becomes yellow on cooling; fused at an intense red heat it becomes permanently green oxide by losing oxygen. When any acid is added to its solution, the yellow colour is changed to orange by the removal of half its potash; 2($KOCrO_3$, and SO_3 become $KO2CrO_3$ and $KOSO_3$). It is formed by adding potash to bichromate, a salt much used in calico printing (KO $2CrO_3$ and KO becoming 2 KO CrO_3); or by fusing chrome iron ore in powder in a crucible or reverberatory furnace with nitre. It is used as a test for lead, bismuth, silver, and mercury, and to separate barytes from strontian and lime (Smith).

Bichromate.—KO $2CrO_3$ 19, 152; 18·574, 148·592. Specific gravity 1·9801 (Thomson). In fine orange crystals, of rectangular or square prisms, sometimes terminated with 4-sided pyramids. (2d Syst.) *Taste* cooling, bitter, and metallic, leaving a hot impression in the mouth not lasting so long as the chromate. At 63° 100 water dissolve 10·44 bichromate, the solution reddening vegetable blues; insoluble in alcohol. At 400° it loses 6 per cent. of hygroscopic water. At a red heat melts into a transparent red liquid; at a forge heat it is converted into a mixture of chromate and sesquioxide. The orange solution, on the addition of caustic potash or soda, becomes yellow from the formation of chromate, ($KO2CrO_3$) and KO becoming 2 ($KOCrO_3$). *Process.*—The chrome iron ore in powder is mixed with quicklime or chalk, and exposed to an intense heat in a reverberatory furnace; it is then digested in hot water, and converted into bichromate by the addition of a salt of potash. This method has been long used in this country. The amount of potash in salts of potash is determined by bichloride of potassium. A yellow salt falls, which is insoluble in alcohol. When thrown on a weighed filter, washed with alcohol, and dried at 212°, it consists of K 4·889, Cl 4·437+Pt 12·335, Cl_2 8·874=30·535. This quantity of salt, therefore, is equivalent to 5·889 potash.

Analysis.—Bichromate of potash has been long used in manufactories for testing the strength of binoxide of manganese, chloride of tin, oxides of iron, &c. See MANGANESE and TIN.

Potato. (*Pomme de Terre*, Fr.; *Kartoffel*, Ger.) The tubers of the Solanum tuberosum. They consist of water 77·5, starch 12·9, caseine 1·1, fibre 6·8, gum and sugar 1·7, ash 1·4. The ultimate composition at 212° is C 44·8, H 5·1, O 30·, N 2·3, ash 17·8. When fresh potatoes are cut up and covered with water, they ferment, a fermentation fungus appearing; the cellulose alone is effected; first the cells separate from each other, then the walls of the cells dissolve, and the starch particles fall out; twenty-four hours renders a potato soft to the depth of two lines. The same process occurs in the potato disease, where the cellulose likewise, and not the starch, is dissolved. The fluid left in contact with diseased potatoes affects sound ones. This decomposition is, therefore, not the disease itself, but the effect of it; the cause depending on the death of the plant; and as when the apices of roots die when too suddenly cooled, a sudden cold rain following a long warm winter may produce a similar condition in the potato. The ash of potatoes consists of KO 30·6, NaO 2·15, CaO 4·34, MgO 2·15, Fe_2O_3 4·21, SO_3 3·29, SiO_3 3·16, sand 11·51, NaCl 1·49, CO_2 18·86, PO_5 18·24.

Potential Cautery. The surgical name of solid hydrate of potash.

Potstone. Swedish potstone has a specific gravity of 2·88, and is a mixture of Venetian talc with a black substance in small black grains. SiO_3 49·01, MgO 30·2, FeO 11·4, Al_2O_3 6·08, HO 4·2. Sweden.

Pottery. The generic term for all kinds of clayware. When the ingredients, consisting of clay and silica, are in such proportions that they do not fuse, the fractured surface adhering to the tongue, they constitute *earthenware*. The common cups, &c. used in this country afford an illustration of this kind of ware. The proportions are 17 ball clay, 8 China clay (see CLAY), 6 flint, 2 stone, or decomposing granite. The ball clay is mixed up with water, so that a pint measure (holding 20 oz.) will weigh 24 ounces. The China clay, made up in the same way, weighs 26 ounces, and the stone 32 ounces. *Ironstone* contains an extra quantity of granite, and forms a fused mass, but is usually opaque and heavy. *China* or porcelain is made with the finer kinds of clay and bone earth. It forms a thin translucent fused body. Its composition often is 360 bone, 144 granite, 216 Rogers's China clay, 40 pitchers broken in glazed biscuit, 712 shavings of other vessels. True China from that country contains no bone earth, but is made of silica and alumina, or felspar so as to form a fusible mass. The following table exhibits the composition of different kinds of pottery, by my pupil, Mr. Roderick A. Couper:—

	1.	2.	3.	4.	5.
SiO_3, ...	39·88	40·60	72·96	71·04	68·96
Al_2O_3, ...	21·48	24·15			
FeO, ...			24·78	22·46	29·24
CaO, ...	10·06	14·22	1·04	3·82	1·60
MgO, ...		·43			
Fe_2O_3, ... } $3CaO,PO_5$, }	26·44	15·32			
NaO and loss,	2·14	5·28	1·22	2·68	

(1) English China ware. (2) Do. No. 2. (3) Berlin porcelain. (4) Fine Chinese ware. (5) Inferior Chinese ware.

The toqua, or blue-coloured ware, is coloured by cobalt, chrome, and oxide of zinc; the sage, or greenish-blue coloured ware, by nickel and cobalt; the drab by chromate of iron or nickel. The body for the cane or yellow-coloured ware is produced by a mixture of sandy clay and common red clay. The same body is also used for Rockingham ware. Common black ware is made from red clay alone. Egyptian ware is composed of ironstone, stiff clay, manganese, and red clay. Salt glazed ware is made from sandy clay, and a little sand to keep the body open, and less compact. It is glazed with common salt. *Glazes* cover the external surfaces of pottery; they are put on as a powder, which vitrifies by heat. One of the best glazes for China ware is formed of 50 borax, 50 granite, 20 flint, and 25 whiting to form a frit. 180 parts of this are mixed with 55 stone and 65 white lead. A good glaze for earthenware and ironstone consists of 18 borax, 18 flint, 9 whiting, 6 Cornish clay, well pounded, sifted, and calcined into a frit; 180 of this are mixed with 100 granite, 60 white lead, with about 1 oz. cobalt to 100 lbs. to counteract the yellow hue. The flint is brought from the chalk districts of England. The ingredients of pottery, viz. calcined flint, clay, and granite, as above—are all separately ground by large revolving granite stones, which are constantly kept moist. These ingredients are diffused in separate vessels through water; the solid particles are allowed to subside, and the supernatant liquor drawn off as it becomes clear, by a series of holes in the extremity of the wooden vessel, which are filled up with moveable pins or bungs. The clays and flint thus being finely divided, are passed through a fine sieve, and are then mixed in the appropriate proportions, in a wooden square vessel, to the proper consistence by means of water. The mixture is then run off into a long sort of bath (slip kiln), floored by heated bricks, which cause the water to evaporate, and to reduce the clay mixture, or body, as it is called, to its proper condition for placing it in the workman's hands. By means of the *potter's wheel*, it is fashioned into variously formed vessels, such as ewers, jugs, mustard pots, &c. Flat vessels, such as plates, are produced by the *whorler*, a wooden wheel, to which a rotatory motion is given. Bowls and cylindrical vessels are turned and ornamented with the *lathe*, while finer vessels, and those requiring rapid motion, are formed by the *jigger*, which requires two persons to manage it, a boy to turn a handle communicating with a wheel and strap, and an experienced man to attend to the preparation of the vessel, which is placed on a wheel similar to that of the whorler. The vessels, after being turned, are dried in a room, in the centre of which is a stove at a temperature of about 90°. Such as require ornamenting or further turning are passed through the lathe. They then receive their first *burning.* This operation is performed by packing the ware in what are technically called *seggars*, vessels resembling footpails in form, made of common fire-clay. These are packed round the kiln in circles. Those goods which require the most moderate heat are placed in the centre, while goods which are capable of standing a higher temperature occupy the sides of the kiln. The fires are then lighted, and allowed to burn for forty-eight hours. After the first heating the porcelain is painted or printed on, and glazed, by mixing the ingredients of the glaze in what is called slop, that is much like whiting and water used for whiting rooms. The vessels are rapidly dipped in succession into the glaze contained in a pail, and are then allowed to drip. The glaze completely obscures the painting or printing occupying the surface of the porcelain. But by being again heated in the gloss kiln, the glaze fuses, and forms a glass, through which ornaments can be distinctly seen. Much art is required in mixing and forming the glaze; for if too thick the figures never rise up properly, and the shading becomes too broad. The porcelain, when heated in the gloss kiln, is placed on *spurs*, triangular pieces of ware, *triangles*, and *stretchers* to separate each adjoining vessel. When taken out from the kiln, each vessel retains the marks of these bodies, which are removed by means of iron chisels, in the hands of women engaged solely in this occupation.

Poudrette. A manure used in France, consisting of solid excrement dried in the air and mixed with some copperas, gypsum, and charcoal, which have the effect of removing its odour.

Powder. See GUNPOWDER, POWDER OF ALGAROTH (Antimony).

Powder, James's. An empyrical antimonic medicine, consisting of antimonious acid 38, bone earth 62·.

Prase. Green variety of Quartz.

Praseolite. Spec. grav. 2·754, H 3·5. SiO_3 40·94, Al_2O_3 28·79, FeO 6·96, MgO 13·73, HO 7·38, MnO ·32, PbO,CuO,CaO,CoO ·5, TiO_2 ·4. Brakke, Norway. Green prisms, fusing with difficulty on the edges into a green glass.

Prasilite. Spec. grav. 2·311, H 1·. Leek-green fibres forming a bed an inch thick in amygdaloid, at Kilpatrick. SiO_3 38·55, HO 18·, MgO 15·55, CaO 2·55, Fe_2O_3 14·9, MnO 1·5, Al_2O_3 5·65, alkali 3·3?

Precipitate. A term applied to a substance obtained from a solution by the action of a reagent.

Predazzite. A bitter spar from Predazzo, Tyrol.

Prehnite. *Cape Chrysolite, Chiltonite, Koupholite, Yu.* Spec. grav. 2·9 to 2·953, H 5·. White, but generally apple, leek, jasper-green, &c.; streak white, granular and fibrous or in right rhomboidal prisms with angles of 100° and 91° 30′; lustre vitreous, semitransparent to translucent on the edges. B.B. froths and passes into a green slag; electric by heat; does not gelatinize with acids, although partially decomposed. SiO_3 43·048, Al_2O_3 23·84, CaO 26·16, HO 4·6, FeO ·64, MnO ·42, KO and NaO 1·03, 2 $CaO,SiO_3,Al_2O_3SiO_3HO$. Primary rocks, Cape of Good Hope (Col. Prehn). Amygdaloid, Kilpatrick, United States.

Princes' Metal. An alloy of 75 Cu and 25 Zn.

Proof Spirit. See ALCOHOL.

Propione. C_5H_5O. Volatile oil resembling acetone; by distilling propylate of lime or barytes.

Propionitrile. *Cyanide of Ethyle.* C_6H_5N. Volatile oil with the odour of garlic, obtained by heating propylate of ammonia with anhydrous phosphoric acid.

Propionyle. C_6H_5. Not isolated, but propylaldide, has been obtained among the products of the oxidation of albumen.

Propylamine. C_6H_9N. A base isomeric with ethyle-methyle-amine; obtained from bone oil.

Propyle. C_6H_7. The hypothetic base of butyronitrile or cyanide of propyle (C_6H_7,C_2N).

Propylene. C_6H_6. A gas obtained among the products of the decomposition of amylic alcohol yielding a series of substitution products with bromine and chlorine.

Propylic Acid. *Propionic acid, Pseudo-acetic acid, Metacetonic acid.* $C_6H_5O_3HO$. An acid resembling acetic acid, but more oily, with which it is homologous, differing from it by C_2H_2; formed by heating cyanide of ethyle with a solution of potash in alcohol and distilling the propylate with sulphuric acid.

Proteine. A term applied to a substance obtained by digesting albuminous substances in weak caustic potash and precipitating by acetic acid. It was supposed to be destitute of sulphur, but that substance can easily be detected in it by fusing it with caustic potash in a clean silver basin. The latter becomes black with sulphur. No such body as proteine has yet been obtained.

Protogine. Granite in which talc replaces mica. From Mont Blanc, it consists of SiO_3 74·25, Al_2O_3 11·58, Fe_2O_3 2·41, CaO 1·08, MgO,KO and NaO 10·01, HO ·67.

Proustite. A synonyme of Ruby Silver.

Prussian Blue. See BLUE.

Prussic Acid. See HYDROCYANIC ACID.

Prunnerite. Violet Quartz from Faroe.

Psaturose. Brittle Sulphide of Silver.

Pseudo Albite. See ANDESINE.

Pseudoerythrine. *Erythrine, Lecanorate of Ethyle.* $C_{20}H_{12}O_8$ or $C_{22}H_{13}O_9 = C_4H_5O, C_{18}H_8O_8$. Crystals by boiling lecanoric acid with alcohol; the solution after long boiling contains orcine.

Pseudomalachite. See PHOSPHOROCALCITE.

Pseudomorphine. $C_{27}H_{18}NO_{14}$. Shining scales or plates, slightly soluble in water and spirit; insoluble in alcohol and ether; soluble in alkalies; sometimes found in opium.

Pseudomorphous Form. A false form of crystal produced by one body from solution or otherwise being deposited upon another, and thus assuming the form of its crystal. Thus quartz, which belongs to the rhombohedral system, is sometimes found with the shape of fluor spar or in cubes. A large number of minerals have now been found in pseudomorphous forms. Calcareous spar as Gay Lussite, cyanite as Andalusite, steatite as hornblende, native copper as red copper ore, vitreous copper, as red silver ore, gypsum as anhydrite, &c., &c.

Pseudonepheline, or Gieseckite.

Pseudoquinine. C 76·5, H 8·1, N 10·2, O 5·2. An alkaloid found in a bark of uncertain origin.

Pseudotriplite. $3(Fe_2O_3Mn_2O_3)\ 2\ PO_5\ 2\ HO$. A mineral from Bodenmais like Triplite.

Psilomelanite. Spec. grav. 4 to 4·328, H 5 to 6. Bluish and grayish reniform and botryoidal masses, streak brownish-black, lustre imperfect, metallic, opaque, brittle. Red oxide of manganese 69·795, O 7·364, BaO 16·365, SiO_3 ·26, HO 6·216. Germany; Restormel, Cornwall; Upton Pyne, Exeter.

Psimmithium. The ancient name for White Lead.

Ptelcyle. C_6H_3. The hypothetic base obtained by acting on acetone with nitric acid.

Puchu Pat. A plant (Labiatæ) used in the East to mix with tobacco for smoking.

Puddingstone Plum-. A conglomerate of rolled stones.

Pumice. (*Bimstein*, Ger.; *Ponce*, Fr.) A vesicular species of lava which floats on water. I have obtained a similar mineral from the vitrified forts in the Highlands of Scotland, the composition of which agrees with granite or felspar which has obviously been intensely heated so as to cause fusion. Pumice consists of SiO_3 70·, Al_2O_3 16, CaO 2·5, Fe_2O_3 ·5, KO 6·5, HO 3·.

Punicine. Crystals from the root of Punica granatum by alcohol.

Purple Colours for Porcelain. *Clear Purple.* Dissolve in aqua regia 77 grains of tin turnings, evaporate in a water bath until it becomes on cooling a crystalline mass. The chloride of tin thus prepared is dissolved in distilled water and mixed with 31 grains of a solution of chloride of tin, spec. grav. 1·700; made by boiling tin filings in an excess of chlorohydric acid down to the required strength. The mixture is then diluted with 2·2 gallons distilled water; sufficient acid being present to prevent turbidity. To this liquid are added 7½ grains of gold, dissolved previously in aqua regia and evaporated to dryness on the water bath, and then dissolved in water and filtered away from the light. Add then 770 grains caustic ammonia; if no precipitate falls a few drops of sulphuric acid will produce it. It is allowed to subside, and washed by decantation five or six times. It is taken up while still moist with a silver spoon and mixed on a glass slab by means of a spatula with 310 grains of lead glass (made by fusing 2 parts minium, 1 quartzoze sand, and 1 calcined borax). It is dried in a moderately warm chamber, and then triturated with 46 grains carbonate of silver. By this process 508¼ grains of clear purple are obtained. *Deep purple.*—The clear solution of 7½ grains of gold in aqua regia is diluted with 2·2 gallons distilled water; add then 115½ grains of the prepared solution of chloride of tin above mentioned, spec. grav. 1·700. It is washed in the same way and mixed with 155 grains lead glass; it is dried and saturated with 7½ grains carbonate of silver. It yields 200 grains of deep purple.

Purple of Cassius. See GOLD.

Purple Copper. See COPPER.

Purpuric Acid. See MUREXANE.

Purpurine. A substance extracted from Garancine by Alum.

Purpurissum. Probably the Tyrian purple dye.

Purpurissum Indicum. Probably Indigo.

Purree. *Indian Yellow.* A yellow pigment from India.

Purreic Acid. *Euxanthic Acid.* $C_{20}H_9O_{11}$. Silky yellow needles soluble in boiling water, alcohol, and ether; forms precipitates with metallic salts; obtained from Indian yellow by solution in acetic acid, precipitation by acetate of lead, decomposing by SH and evaporation.

Purrenone. $C_{13}H_4O_4$. Crystalline sublimate by heating purreic acid a little above 212°. Slightly soluble in water; soluble in alcohol and ether; neutral.

Pus. The matter discharged from inflamed surfaces consists of water 88·064, cholesterine 1·046, oleine, oleic acid, chlorohydric acid 1·029, stearine ·705, liquid albumen 1·987, coagulated albumen, fibrine, 1·849, phosphate, carbonate, and sulphate of lime ·62, NaCl 4·7, $KOCO_2KO$ $SO_3,CaOSO_3$ traces.

Puschkinite. A variety of Epidote from Ural.

Putrefaction. A process of decay attended with the evolution of volatile odoriferous matter. It is considered by some to be a species of fermentation, by others to be due to the action of a vibrio. See FERMENTATION.

Putty Powder. Oxides of lead and tin; used in glass-making.

Puzzuolano. Volcanic ashes much used in mixing with mortar to produce hydraulic cement. It consists of SiO_3 44·5, Al_2O_3 15, CaO 8·8, MgO 4·7, KO 1·4, NaO 4·1, FeO and TiO_2 12, HO 9·. It is partially decomposed by HCl.

Pycnite. *Schorlous Beryl, Schorlite.* Spec.

grav. 3·5, H 7·5, SiO_3 38·43, Al_2O_3 51, HF 8·84 (or 17·09). Yellowish or reddish-white 6-sided prisms, longitudinally striated, translucent, brittle, lustre shining, resinous. B.B. infusible; with borax a transparent glass; electric by heat 2 (Al_2F_3) 4 Al_2O_3 5 SiO_3. Bohemia.

Pyine. *Mucine.* A form of albumen in pus, soluble in water and weak alcohol; does not coagulate by heat; precipitated by acetic and tannic acids and alum from an alkaline solution, by aqueous solution of iodine, and by distilled water.

Pyr. (πυρ, fire.) A term applied to many substances which have been produced by the influence of heat.

Pyrallolite. *Tersilicate of Magnesia.* Sp. grav. 2·555 to 2·594, H 3·5 to 4. White, inclining to greenish, doubly-oblique prisms, with angles of 94° 36′ and 80°; the edges being 140° 49′; lustre dull, slightly resinous, opaque, not translucent, in thin plates. B.B. becomes black, then white again, and melts on the edges into a white enamel; with borax and soda forms a transparent glass; with difficulty fuses with salt of phosphorus. SiO_3 56·62, MgO 23·38, Al_2O_3 3·38, CaO 5·58, MnO ·99, Fe_2O_3 ·99, HO 3·58, bitumen and loss 6·38. Storgord, Pargas, Finland, in limestone.

Pyrantimonite. A synonyme of Red Antimony, or Oxysulphide.

Pyrargillite. Spec. grav. 2·505, H 3·25. Black, light and shining, or bluish granular, with lustre; sometimes like a 4-sided prism; often penetrated by scales of chlorite. SiO_3 43·93, Al_2O_3 28·93, FeO 5·3, MgO 2·9, KO 1·05, NaO 1·85, HO 15·47. Allied to hydrous bucholzite.

Pyrargyrite. A synonyme of dark red Silver Ore.

Pyrene. C_5H_2, or $C_{10}H_2$, or $C_{15}H_4$. F.P. 347°. Rhomboidal microscopic plates, without taste and smell; insoluble in water; little soluble in alcohol and ether; very soluble in oil of turpentine; distils without change; charred by SO_3; obtained from coal tar; fat and resinous bodies when distilled; it occurs in the more volatile of the last matters distilled from coal tar. With nitric acid it forms *nitrite of pyrenase* $(C_{15}H_5O_4N)$.

Pyreneite. A blackish variety of ferrocalcareous Garnet.

Pyrethrine. A resin from Anthemis pyrethrum; it seems to consist of 3 sub-resins, one soluble in alcohol, insoluble in water and potash; a fixed oil, soluble in alcohol; and a yellow oil, soluble in alcohol, ether, and potash.

Pyrgom. A dark variety of Sahlite.

Pyrites. (Fire stone). A term applied to many of the metallic sulphides—as arsenical, auriferous, capillary, cellular, cockscomb, copper, hepatic, hydrous, iron, magnetic, radiated, spear, tin, variegated white iron.

Pyroacetic Spirit. A synonyme of Acetone.

Pyrobenzoline. See BENZOLONE.

Pyrocatechine. See PYROMORINTANNIC ACID.

Pyrochlore. (πυρ, fire; χλωρος, greenish-yellow.) Spec. grav. 3·802, 4·206 to 4·216, H 5. Reddish-brown small regular 8-hedrons, immersed in felspar; fresh fracture almost black; in thin splinters, translucent and brown; in larger pieces opaque; lustre resinous or vitreous; fracture conchoidal. B.B. fuses with difficulty into a blackish-brown slag; with borax into a reddish-yellow bead, becoming opaque and yellow; a grass-green glass with salt of phosphorus when cold, yellow when hot; with soda shows the green reaction of manganese. TiO_2 62·75, CaO 12·85, protoxide of uranium 6·18, CeO 6·8, MnO 2·75, FeO 2·16, SnO_2 ·61, HO 4·2. It has been found frequently to contain thorina and columbic acid. In syenite at Federicksvärn and Laurvig, Norway; Brevig; near Miask.

Pyrocitric Acid. A synonyme of Aconitic acid.

Pyrogallic Acid. $C_{12}H_6O_6$. F.P. 240°; volatile at 410°. Bitter shining scales; soluble in water, less soluble in alcohol; yields a deep blue solution with copperas, a reddish-purple with milk of lime; forms a dark solution when dissolved in caustic alkalies, absorbing oxygen with great rapidity, which renders it a valuable means of determining the amount of oxygen in air (Liebig.) The acid dissolved in water is introduced in contact with the air to be analyzed over mercury, and then caustic potash is added; immediately the absorption of the oxygen commences, and shortly terminates, the acid being converted into ulmic acid; pyrogallic acid is obtained by cautiously distilling gallic acid, when it loses 1 CO_2 and sublimes.

Pyroguaiacic Acid. See GUAIACYLE HYDRIDE.

Pyroligneous Acid. See ACETIC ACID.

Pyrolusite. See BINOXIDE OF MANGANESE.

Pyromachus. A synonyme of flint.

Pyromaric Acid. 3-sided tables, easily soluble in alcohol by distilling pimaric acid in vacuo.

Pyromeconic. $C_{10}H_3O_5HO$. Colourless prisms, becoming dark by moisture; soluble in water; unites with oxides of lead, copper, and iron, and forms definite salts; isomeric with pyromucic acid. F.P. 253°; by heating meconic acid.

Pyromeride. A bluish rock, with a fatty lustre, consisting of 88 per cent. quartz, silica, and orthoclase. Corsica; Vosges.

Pyrometer (πυρ, fire; μετρον, measure.) An instrument for measuring intense temperatures. The best seems to be Daniel's.

Pyromorintannic Acid. *Phenic, Phenylic Acid, Pyrocatechine.* $C_{12}H_6O_4$. F.P. 212°; B.P. 468°. Rectangular bitterish prisms; soluble in water, very soluble in alcohol, less so in ether; reduces salts of gold, platinum, and silver; gives a dark green precipitate with chloride of iron; obtained by distilling morintannic acid.

Pyromorphite. A synonyme of Phosphate of Lead.

Pyromucic Acid. $C_{10}H_3O_5HO$. F.P 266°. Long white shining plates; soluble in 26 cold, 4 boiling water; obtained by distilling mucic acid, which contains the elements of 1 atom pyromucic acid, 6 HO, and 2 CO_2; it forms monobasic salts.

Pyrope. See BOHEMIAN GARNET.

Pyrophorus. (πυρ, fire; φερω, I carry.) A substance which ignites in air. See HOMBERG.

Pyrophosphoric Acid. See PHOSPHORUS.

Pyrophyllite. *Fibrous Talc.* White radiated mass, with indistinct rectangular cleavage, and pearly lustre. B.B. spreads out like a fan, and increases 20 times in bulk; then is infusible; with soda forms a yellow glass, with nitrate of cobalt a blue colour. SiO_3 59·79, Al_2O_3 29·46, MgO 4·, Fe_2O_3 1·8, HO 5·62. Ural.

Pyrophysalite. See TOPAZ.

Pyroracemic Acid. Obtained by dry distillation from racemic acid, as with tartaric acid.

Pyrorthite. Specific gravity 2·19, H 2·75, Pitch-black long needles of 4-sided prisms, longitudinally streaked with a large streak in the centre, giving the prism the appearance of being divided into two; cross fracture uneven; longitudinal fracture conchoidal or splintery; lustre resinous, opaque. B.B. catches fire, and becomes white and very light; fuses into a black enamel with difficulty; with borax and salt of phosphorus into a clear glass; does not dissolve with soda; soluble in acids, leaving a black powder. SiO_3 10·43, CeO 13·92, HO and volatile matter 26·3, charcoal 30·, Al_2O_3 3·59, CaO 1·81, FeO 6·08, yttria 4·87, MnO 1·39. With gadolinite in a granite vein at Korafsberg, near Fahlun.

Pyrosclerite. Specific gravity 2·74, H 3·. Green or grayish masses, with cleavage at right angles to base, translucent, fracture uneven; lustre pearly, faint. SiO_3 37·03, MgO 31·62, Al_2O_3 13·5, FeO 3·52, Cr_2O_3 1·43, HO 11·; decomposed by HCl; allied to serpentine.

Pyrosiderite. A synonyme of hydrous sesquioxide of iron.

Pyrosmalite. Spec. grav. 3·08, H 4·25. Liver-brown, gray, or pistachio green 6-sided prisms, or tables; streak paler; fracture uneven, slightly brittle; lustre pearly. B.B. yields water and chloride of iron in a tube; on charcoal fuses to a black magnetic slag. Fuses with borax, showing the presence of iron and manganese. Soluble in strong NO_5. SiO_3 35·85, Fe_2O_3 35·48, Mn_2O_3 23·44, Al_2O_3 ·6, Cl 3·76, CaO 1·21, HO 3·6. Bjelke, Nordmark, Wermland. Named from the odour of chlorine when heated.

Pyrotartaric Acid. $HOC_5H_3O_3$. F.P. 212°; B.P. 370°. Long transparent prisms, soluble in water, alcohol, and ether; does not precipitate lime water; precipitates diacetate of lead, but not acetate or nitrates of lead; does not precipitate salts of mercury, peroxide of iron, lime, &c. Obtained by distilling tartaric acid in a retort at 374°, when a fluid passes over, which yields pyrotartaric acid in crystals by evaporation; or by distilling tartaric acid between 392° and 572°, and redistilling the distillate. It is a monobasic acid, but also forms acid salts.

Pyrotechnics or **Fireworks.** These consist of squibs and various kinds of fires, &c. *Squibs* are made of 24 mealed or finely ground powder, 4 charcoal, 1 sulphur, 3 nitre; *Chinese fire* of 16 meal powder, 2 sulphur, 5 iron filings; *Ancient fire*, of 16 meal powder, 2 charcoal; *Brilliant fire*, 16 meal powder, 4 iron filings. These are packed in paper cylinders or goose quills, and ignited. *Gold rain*, to descend from a rocket like a shower, consists of 16 nitre, 4 meal powder, 4 sulphur, 1 brass dust, 2½ sawdust, ½ glass dust.

Pyroterebic Acid. $C_{14}H_9O_7$.

Pyroterebilic Acid. $HOC_{12}H_9O_3$. Colourless oily fluid, not solid at 4°; possesses a pungent taste and a caustic reaction; soluble in water, alcohol, and ether; monobasic; formed by heating terebilic acid about 390°, when it is resolved into carbonic and pyroterebilic acids.

Pyrothonide. A brown extract, with empyreumatic taste and smell, obtained by rolling up paper or linen into a cylinder, and inflaming it in an iron vessel, which is cooled by cold water. The residue is dissolved in water, filtered, and evaporated to an extract. It contains the constituents of soot, free from nitrogenous matter.

Pyroxanthine. See EBLANINE.

Pyroxene. *Asbestus* in part, *Augite*, *Baikalite*, *Cocciolite*, *Euchysiderite*, *Fassaite*, *Pentaclasite*,

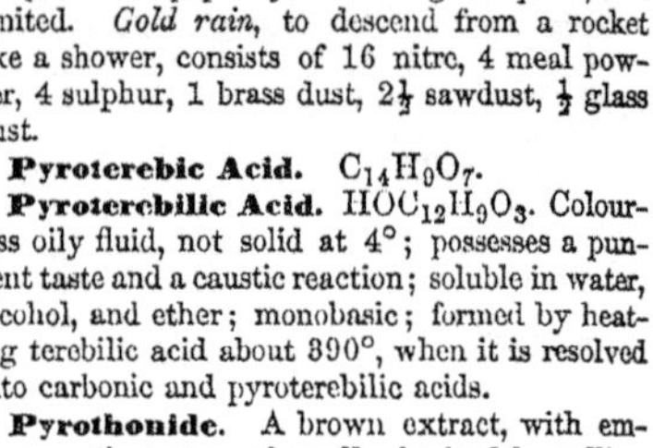

Pyrgome, *Vulcanite*. Spec. grav. 3·233 to 3·349, H 4·75. Green, brownish, or black, 4 or 6-sided oblique prisms, cleaving parallel to the faces of a rhomboidal prism; streak white; lustre vitreous inclining to resinous, sometimes translucent, but frequently opaque. B.B. fuses per se and with borax into a clear glass; sometimes showing the presence of manganese or iron. SiO_3 54·08, CaO 23·47, MgO 11·49, FeO 10·02, MnO ·61. *Form.*—3 (RO = CaO,MgO,FeO,MnO or NaO) 2 SiO_3. This mineral includes under it a number of crystalline substances varying in the nature of the bases which replace each other.

Pyroxenite. A mixture of pyroxene in plates and of felspar.

Pyroxylic Spirit. See HYDROUS OXIDE OF METHYLE.

Pyroxyline. See GUN COTTON.

Pyrrite. H 6·. Orange, subtranslucent 8-hedrons with glassy lustre. B.B. infusible; fuses into a clear glass with borax and salt of phosphorus, becoming yellowish-green with excess of flux; insoluble in HCl. Alabaschka Mursinsk in felspar cavities.

Pyrrole. An alkaloid not examined in coal tar oils.

Pyruvic Acid. $C_6H_3O_5HO$. Thick syrup, by distilling tartaric acid.

Q

Quartation. See GOLD.

Quartz. The familiar term applied to several forms of silica occurring in primary rocks. See SILICA. The chief forms of what is called quartz are the milk-white and *rose-red* varieties, from the presence of titanic or organic matter. *Smoky quartz* is found in Arran in 6-sided prisms, and a variety of it constitutes Cairngorum stone, found on that mountain. *Nectic quartz* or float stone is a vesicular form which floats on water. *Granular quartz* is a rock formed of siliceous grains. *Fibrous quartz* or *Siliceous wood* is a pseudomorphous form in which silica probably in solution has replaced the woody texture of trees. It is found in the West Indies, Burmah, Van Diemen's Land, Kerguelan's Land, Loch Neagh, &c. Hyaline quartz, hyalite, or Muller's glass, has a white or colourless aspect, and is found in grape-shaped masses or as stalactites.

Quartzite, or compact quartz, is applied to a variety consisting of SiO_3 97·75, Al_2O_3 and $Fe_2O_3 5HO$, and organic matter 1·75.

Quassite. *Quassine.* $C_{10}H_6O_3$. White bitter prisms; soluble in water (1 in 200 water) and ether; very soluble in alcohol; precipitated white by tannin and corrosive sublimate; dissolved but not decomposed by sulphuric and nitric acids; obtained from the sliced wood of the Quassia amara and excelsa by evaporating the aqueous decoction to $\frac{1}{4}$, adding dissolved lime for twenty-four hours, filtering, evaporating to dryness, dissolving and crystallizing repeatedly out of alcohol.

Quercitannic Acid. A synonyme of Tannic Acid.

Quercite. *Quercine.* White or yellow neutral bitter crystalline body from Quercus robur or oak; soluble in water and alcohol; insoluble in ether, absolute alcohol, and oil of turpentine; precipitated by salts of silver, lead, tin, and protoxide of mercury (Gerber).

Quercitrine. $C_{16}H_8O_9HO$. Crystalline pearly substance in small plates or scales, by concentrating slowly the infusion or decoction of quercitron bark (Quercus nigra). It resembles in appearance mosaic gold; very little soluble in water and ether; more so in alcohol; precipitated by barytes water, acetate of lead, chloride of tin, and acetate of copper; nitric acid renders it orange; sulphuric acid dissolves it, and water reprecipitates it; when distilled, crystals are formed in the receiver with the characters of quercitron, which is the colouring matter of the bark. It is said to be converted by boiling into needles (Quercitreine).

Quercose. *Quercite, Acorn Sugar.* $C_{12}H_{12}O_{10}$. F.P. 455°. Fine colourless prisms, losing no weight at 410°; with nitric acid it gives oxalic acid and no mucic acid; forms a conjugate acid with sulphuric acid, the lime salt of which does not crystallize; with a mixture of nitric and sulphuric acids it forms a detonating compound, *Nitro quercite*, which is white resinous, insoluble in water, and differs from nitro mannite in not crystallizing; quercose may be heated for some time with potash without producing a colour; it dissolves in barytes water and forms a compound BaO, $C_{12}H_{12}O_{10}$ 2 HO; by long boiling with sulphate of copper and caustic potash only a trace of dinoxide of copper is formed; not precipitated by diacetate of lead except on the addition of ammonia and heat; quercose does not ferment in presence of yeast or decaying cheese and water; it is obtained by alcohol from the acorn of the oak (Quercus robur).

Quicksilver. A synonyme of Mercury.

Quincite. Light carmine-red flocks forming a substance interspersed through a limestone deposit at Mohun, France, and is obtained pure by dissolving the limestone in muriatic acid; loses colour by heat; weak acids have no action on it; the strong acids imperfectly dissolve it, silica gelatinizing. SiO_3 54, MgO 19, FeO 8·, HO 17.

Quinic Acid. A synonyme of Kinic Acid.

Quinicine. A base procured by moistening disulphate of quinine with a little water and sulphuric acid on exposing it to the heat of an oil bath at 248° to 266°, for three or four hours; the whole is converted into sulphate of quinicine except some colouring matter. Quinicine like cinchonicine is almost insoluble in water, both are very soluble in strong and weak alcohol; both unite easily with carbonic acid and replace ammonia in its salts; both are precipitated from their solutions as fluid resins; both deviate the plane of polarization to the right; they are bitter and febrifuge.

Quinidine is a base which has been described as existing along with quinine in one of the new barks, but it appears to consist of two alkaloids, quinidine and cinchonidine. Quinidine is hydrous, efflorescent, isomeric with quinine, and is coloured green by chlorine and ammonia; deviates the plane of polarization to the right; cinchonidine is anhydrous, isomeric with cinchonine, exercises its rotatory power to the left, and is not rendered green by chlorine and ammonia; the salts of quinidine are easily distinguished from those of cinchonidine by their efflorescence. By heat they are converted into new isomeric bases, quinicine and cinchonicine.

Quinine. $C_{38}H_{22}N_2O_4$. F.P. 302°. C 73·4, H 7·13 (Laurent). White flocks, silky tufts or needles, intensely bitter; soluble in 200 hot water; very soluble in alcohol, from which it is deposited as a resin; more soluble in ether than cinchonine; negatively electric by rubbing; it crystallizes by spontaneous evaporation from alcohol of ·815 or by adding a small quantity of ammonia to the hot alcoholic solution and setting it aside; it is readily obtained by dissolving the disulphate in water acidulated with sulphuric acid and adding ammonia. When crystallized from alcohol it is a hydrate with 14 per cent. water, which it loses at 302°, but it recovers its water by being placed in that fluid. When distilled with strong caustic potash hydrogen is evolved and quinoline (leucoline) passes over. Quinine has been viewed as a conjugate compound of hydrous oxide of methyle and leucoline ($C_2H_4O_2$ and $C_{18}H_8N = C_{20}H_{12}O_2N$). This, however, differs from the above formula. *Pseudoquinine.*—C 76·5, H 8·1, N 10·3, O 5·2 to 4·7. Irregular prisms, tasteless, insoluble in water and ether; soluble in alcohol; neutralizes acids; becoming reddish-yellow by chlorine and ammonia; from an extract of bark of unknown origin.

Quinine, Disulphate of. $\overset{+}{Qu_2}SO_3HO$. White pearly, feathery crystals, slightly flexible and efflorescing in air; soluble in 740 water at 55°, in 30 boiling water, and 80 alcohol (·850); becomes red by heat; precipitated by tannic and gallic acids; becomes fluid by heat, and like wax. This is the commercial salt so valuable as a tonic and febrifuge. It is prepared by boiling yellow Peruvian bark in 10 times its weight of water acidulated with 12 per cent. of sulphuric or 25 per cent. of chlorohydric acid. After an hour it is strained, and repeated additions of acidulated water are boiled with the residue till the bitter taste is removed. Add milk of lime in slight excess to the acid decoction. When the lime has deposited it is thrown on a cloth; when the liquid passes through, the residue is pressed and the quinine dissolved up from it by alcohol (·847). The alcohol is distilled off and to this liquor sulphuric acid is added to form a disulphate. Oil of turpentine or pyroxylic spirit may be substituted for the alcohol. In this country it is usually made without alcohol. The colouring matter is sometimes removed in part from the bark by digesting it in a solution of carbonate of soda. The remainder of the process is as above; the colouring matter being further removed if necessary by animal charcoal. *Adulterations.*—The salt under the microscope should be entirely crystallized in needles or prisms, and no amorphous masses should appear in it which would indicate quinoidine. The presence of gypsum in this salt may be detected by combustion when the former remains; the presence of salicine may be detected by its giving a red colour with oil of vitriol; boracic acid may be detected by digesting in alcohol and burning the spirit; boracic acid will impart to it a green tinge; sugar may be fermented by yeast and the alcohol detected by bichromate of potash and sulphuric acid; margaric acid remains when the quinine is heated with water and chlorohydric acid. A beautiful test for quinine is the following:—Dissolve the disulphate in chlorine water, add a drop or two of caustic potash and allow the mixture to stand, a yellowish precipitate falls, which, especially if it possesses a greenish tinge, speedily becomes purple on the surface and gradually imparts a magnificent purple-red colour to the supernatant fluid (R.D.T.) Another test is to dissolve the disulphate in chlorine water and add to it caustic ammonia, when a deep green precipitate falls. By this test quinine may be detected in bark; boil a few grains of Peruvian bark with a drachm of water and a drop of oil of vitriol for a few seconds, add water chlorine and ammonia, when a green tinge will generally be obtained if the boiling has not been continued too long or the acid is not in excess (R.D.T.)

Sulphate.—$\overset{+}{Qu}, SO_3\ 8\ HO$. Rectangular prisms with square bases, soluble in 11 parts water at 55°, in 8 water at 72°, at 212° in its water of crystallization; more soluble in hot than in cold alcohol; obtained by adding sulphuric acid to a solution of the disulphate and evaporating.

Quinoidine. *Amorphous Quinine, Chinodine.* Isomeric with quinine, but uncrystallizable; produced in the manufacture of disulphate of quinine by the action of acids or heat; it also arises from cinchonine; insoluble in water, soluble in alcohol; forms salts which do not crystallize; procured from the mother liquors in the manufacture of quinine and cinchonine. It is said by some to answer as a substitute for quinine, as a tonic. In a variety occurring in Holland, a new base, *Beta quinine*, has been extracted to the extent of 50 to 60 per cent. by ether, in large prisms, which have the same composition as *alpha* or common quinine. *Gamma* quinine is obtained by dissolving quinine in absolute alcohol and evaporating spontaneously or in dilute ammonia.

Quinoile. *Chinoyle.* A synonyme of Kinone.

Quinoline. See LEUCOLINE.

Quinone. See KINONE.

Quioric Acid. *Knioric Acid.* An acid or bitter body, nearly insoluble in water; soluble in alcohol and ether, existing in Calasaya and Peruvian barks.

Quinovine. A synonyme of Aricine.

R

Racemic Acid. *Paratartaric Acid, Vinic Acid, Uvic Acid, Levoracemic Acid, Acid of the Vosges.* (*Traubensäure*, Ger.) (Kestner, 1822 to 1824.) $HOC_4H_2O_5$ or $2HOC_8H_4O_{10}$? Although this acid has been supposed to have been found only in the Vosges, I have been in possession for at least fifteen years of many ounces of it. It was then pointed out to me by Mr. William Blythe, of Church, Lancashire, in whose company I picked it from the sides of the tartaric acid evaporators, at the Chemical Works, in Manchester, of Mr. Edmond Thomson, at that time under Mr. Blythe's charge. Specimens have been in Dr. Thomson's Museum since 1833. Dr. T. Thomson and myself annually demonstrated its characters in our lectures, and compared it with Kestner's acid, of whch I have some fine crystals presented by himself. The existence of this acid in this country is announced in my Records of Science (vol. ii. page 97), published in 1835, but seems, in fact, to have been unknown to the French chemists.

Crystal.—Colourless doubly-oblique prisms of the 6th system which do not change in air except when warm; and lose all their water at 212° or at 302° in a dry stream of air; soluble in 5·7 cold water; its solution has no action on polarized light; begins to decompose at 383° with the evolution of a gas, 90 per cent. of which is soluble in potash. It contains when crystallized $10\frac{1}{2}$ per cent. of water. Racemic acid is distinguished from tartaric acid by the form of its crystal (tartaric acid belonging to the 5th or prismatic system) by its precipitating solution of gypsum in $\frac{1}{4}$ hour in fine needles; it precipitates more rapidly chloride of calcium and nitrate of lime, results not obtained with tartaric acid. By the action of heat racemic acid is converted into a series of acids similar to those obtained from *tartaric* acid. See PARATARTRALIC and PARATARTRELIC ACIDS. Whether racemic acid is mono or bibasic is uncertain.

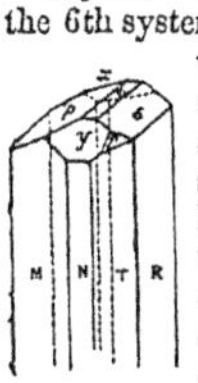

Racemic.

Radical. (*Radix*, a root.) A term applied to the root or foundation of compounds. An *acid radical* is the base of a compound which is capable of uniting with bases. A *basic radical* is the root of a compound which unites with acids. A *simple radical* is synonymous with an elementary body; while a *compound radical* acts like an elementary body or simple radical, and forms the base of other compounds. Cyanogen is (C_2N) an example of a compound radical. The name in this case is synonymous with a *Salt radical. Organic radicals* are typified by ethyle (C_4H_5), methyle (C_2H_3), &c., which are analogous in their position in the organic world to metals in the inorganic division of chemistry. They constitute the base or radical of extensive series of salts. In accordance with this definition, inorganic chemistry has sometimes been described as the chemistry of simple radicals, while organic chemistry has sometimes been characterized as that of the compound or organic radicals.

Radicle. The incipient root of the germinating seed.

Radiolite. A variety of Mesotype from Brevig, Norway.

Rain, *Earthy* and *Red*, are names which have been applied to powders which have apparently fallen from the atmosphere. The earthy rain fell in the south-east of France, and consisted of SiO_3 52·, Al_2O_3 7·5, $CaOCO_2$ 26·5, FeO 8·5, $MgOCO_2$ 2·, organic debris 3·5. The red rain fell in the kingdom of Naples, and consisted of SiO_3 33·, Al_2O_3 15·5, CaO 11·5, chromium 1·, iron 14·5, CO_2 9·, yellow resin 15·5?

Rain Water. See WATER.

Randanite. A form of gelatinous soluble silica, in fine earthy compact masses, found at, 1st, Ceyssat, near Randan, Puy de Dome; 2d, near Algiers; 3d, in England. The first consists of silica, soluble in potash 87·2, water, CO_2 and organic matter 10·, Al_2O_3 and Fe_2O_3 2·, sand ·89; the second is formed of soluble SiO_3 80·, HO 9·, insoluble silica 6·48, Al_2O_3 1·41, Fe_2O_3 ·55, CaO ·56, NaO,KO, &c., 2·; the third was found lately at Farnham, and contained 28·7 of soluble silica. Randanite consists of the casts of infusoria.

Raphilite. Spec. grav. 2·85, H 3·75. White, with a shade of bluish-green, needle-formed crystals, apparently oblique 4-sided prisms, lustre between glassy and silky; the needles or fibres are easily separable from each other, slightly flexible but easily broken. B.B. becomes opaque and white, but does not fuse; fuses with soda into a white glass, with borax into a transparent colourless glass, leaving silica, fuses imperfectly with phosphate of soda. Its composition I have found by two trials made many years ago to be SiO_3 57·06, 56·81, CaO 17·81, 14·81; Al_2O_3 2·15, 2·08; FeO 9·3, 7·08; MgO 12·6, 16·8; HO 1·08, 2·42. A former analysis gave alkali, derived from sea water. Perth, Upper Canada.

Rapidolite. A synonyme of Scapolite.

Ratofkite. A mixture of fluor spar and apatite occurring as thin plates at Ratofka, Russia.

Razoumoffskin. Snow-white, masses, opaque, passing into green; adheres strongly to the tongue. Occurs at Kosemütz along with pimelite and chrysoprase. B.B. fuses into a transparent mass, soluble in HCl, leaving the silica. Silica 50, Al_2O_3 16·88, HO 20·, NiO ·75, MgO,FeO,CaO 2·, KO 10·3.

Realgar. A synonyme of Bisulphide of Arsenic.

Red Antimony. SbO_3, 2 SbS_3. See ANTIMONY.

Red Chalk, or Reddle. A soft form of sesquioxide of iron used for drawing purposes.

Red Clay Ironstone. Sesquioxide of iron mixed with clay, &c.

Red Cobalt, or Octohedrous triarseniate of Cobalt.

Red Copper Ore, or Dinoxide of Copper.

Red Hæmatite, or Sesquioxide of Iron.

Red Iron Vitriol. Native sesquisulphate of Iron. See BOTRYOGENE.

Red Lead Ore, or Chromate of Lead.

Red Liqour. Commercial acetate of Alumina.

Red Ochre, or *Red Chalk.*

Red Silver Ore, $3AgS,SbS_3$, or Antimonial Sulphide of Silver.

Red Zinc Ore. Native oxide of zinc mixed with some protoxide of manganese.

Reddle, or Red Chalk, or Sesquioxide of iron.

Reds for Porcelain. *Rose.*—Dissolve 15½ grains of gold in aqua regia, mix its solution with a solution of 770 grains alum in 4·4 gallons water and add 23 grains solution of chloride of tin (spec. grav. 1·700); then add caustic ammonia in sufficient quantity to precipitate all the alumina. Wash the precipitate by decantation 10 times at least, throw it on a filter and dry at a gentle heat. It weighs 207⅓ grains. It is now triturated with 38½ grains carbonate of silver and 1078 grains lead glass (2 minium, 1 sand, 1 calcined borax). It is only fit for very thin layers, otherwise it separates into gold and silver. *Yellow red.*—Calcine copperas on a capsule in an open muffle, stirring with a spatula to remove the greater part of the sulphuric acid and until a sample rubbed in water on a glass slab, yields a yellow-red colour. Mix 7 parts of this with 24 lead glass (12 minium, 3 sand, 1 calcined borax) and triturate carefully. *Brownish-red* is prepared by continuing the calcination until all the acid is expelled and treating as in the preceding case. *Bluish-red* or *Pompadour.*—When copperas is still further ignited it acquires a bluish tinge, although it is not easy to hit this point. The coloured residue is mixed in equal proportions with the lead glass (5 minium, 2 sand, 2 calcined borax). *Flesh-red.*—1 red oxide of iron, 4 deep yellow, 10 lead glass (12 minium, 3 sand, 1 calcined borax), are mixed and finely triturated on a glass slab. *Venetian red* for paint.—Equal parts sulphate of iron and sulphate of lime are calcined in a reverberatory furnace until the mixture assumes a bright red colour on cooling. It is then levigated with water for twelve hours by agitation with millstones in water, then run off into cisterns and well washed to remove excess of sulphuric acid. It is then dried carefully at 145° and pulverized under an edge stone.

Refining. A term applied to the purification of various substances, as sugar, silver, gold, &c.

Refraction. A term sometimes applied to the determination of the purity of nitre.

Regulus. An impure metallic sulphide.

Reinmannite. A variety of Allophane.

Reiset, Salts of. See PLATINUM.

Rennet. A fluid prepared by digesting the salted true stomach of the calf in water for some weeks, when a portion of the fluid becomes capable of coagulating milk.

Rensselaerite A name given to a steatite which has asssumed the form of augite.

Replacement. The power which one body possesses of being substituted for another and fulfilling the same functions. See METALEPSY.

Resins. Oxides of essential oils in which the oxygen has replaced an equal amount of hydrogen. The oil or basis of these resins has the formula $C_{10}H_8$, or a multiple of it, $C_{40}H_{32}$. The simplest resin consists of $C_{40}H_{32}O$. Common rosin, obtained by distilling off oil of turpentine from turpentine balsam, has the formula $C_{40}H_{28}O_4$. Resins are divided into acid and neutral. 1. *Acid resins*, when dissolved in alcohol, redden litmus; and, if ammonia is added, they precipitate nitrate of silver with a crystalline precipitate, not redissolved by an excess of ammonia. The neutral resins are not affected in this way. Acid resins decompose carbonate of soda; they dissolve in caustic soda and potash. Their solution in spirit is not precipitated by NH_3; the precipitate produced sometimes by water is soluble in NH_3; their soaps are not precipitated by common salt. The acid resins are common rosin or colophony, dammara, gamboge, sandarach, mastic, benzoin, copal, lac, &c. When the resins are distilled they give oils of various constitution, which sometimes contain creasote, and are used to preserve timber. At higher temperatures they yield gases for lighting. 2. *Neutral resins* have their alcoholic solutions precipitated by ammonia: they do not redden litmus. To this class belong elemi, styracine, guiacum, dragon's blood. For gum resins, see GUM RESINS.

Resin Asphalt. *Retinasphalt.* $C_{21}H_{14}O_3$? Sp. grav. 1·135. F.P. 250°; quite fluid at 320°? Pale brown ochraceous yellow masses; fracture imperfectly conchoidal; it appears earthy externally; but when broken exhibits a slight resinous lustre, very soft and easily frangible; at first when dug up, elastic, flexible, but loses this property in the air; fuses when heated, smokes much, and burns with a fragrant odour; the fused mass is black and brittle. Bovey resin asphalt consists of resin 55, asphalt 41, earthy matter 3· (Hatchett), and sometimes 13 per cent. ash. Halle resin asphalt contains resin soluble in alcohol 91·, insoluble 9·. It accompanies Bovey coal; is found in vegetable earth near Helbra, Mansfield, and in brown coal near Halle.

Resineine. $C_{20}H_{15}O$. B.P. above 482°. Light coloured oil, nearly destitute of taste and smell; scarcely soluble in alcohol, not in water; very soluble in ether. Obtained by distilling common rosin or colophony.

Resineone. $C_{23}H_{18}O$. B.P. 298°·5 an oil

with a less burning taste than resinone, and less soluble in alcohol.

Resinite. *Retinite.* See Resin Asphalt.

Resinone. $C_{10}H_9O$. B.P. 172½°. White very liquid substance; taste burning; insoluble in water;' soluble in alcohol; very soluble in ether; burns like alcohol. Obtained by distilling 1 part rosin and 8 slaked lime; an ethereal oil passes over, which, when distilled in an oil bath at 320° divides into two portions; one remaining in the retort is resineine and tar; the distillate consists of resinone and resineone, which may be separated by distilling at a low temperature; the resinone being most volatile.

Respiration. The process of inspiring and expiring air in the lungs, is termed respiration. Upon it depends the production of animal heat by the union of the oxygen of the air with the carbon and hydrogen of the blood. The air which enters the lungs is oxygen and nitrogen, and that which is expired consists of oxygen, nitrogen, and carbonic acid. The following data by Vierordt are important:—In experimenting upon respiration, Vierordt uses as an expirator, a long glass flask, open below, fitted above with a stop-cock, and filled with a saturated solution of common salt. It is immersed in an exterior vessel, which is filled with a solution of the same kind, capable of containing 9200 cubic centimetres. When this is filled by 15—20 expirations, an anthrakometer is to be affixed to it, filled with common salt solution. It consists of a flask of 2442 cubic centimetres fitted to a graduated tube 1½ metre long, and 228 cubic centimetres capacity, and 2 centimetres in diameter. When the stop-cock is turned the expired air displaces the common salt solution. The stop-cocks are closed, and the tube adapted to a flask filled with caustic potash. The stop-cock is opened, and potash agitated in contact with the expired air. After some minutes the anthrakometer is shut off, the extremity plunged in water, and the stop-cock opened, and the point at which the water stands is read off. It is unnecessary to close the nostrils in expiration, as, according to Vierordt, air cannot pass from the lungs through the nostrils and mouth also, and the same occurs in inspiration.

1. *Carbonic acid at different times of the day.*

Hour.	CO2 per cent. by Vol.	Expirations per minute.	Pulsations per minute.
9	4·32	12·1	73·8
10	4·47	11·9	70·6
11	4·51	11·4	69·6
12	4·36	11·5	69·2
1	4·35	12·4	81·5
2	4·27	13·	84·4
3	4·37	12·3	82·2
4	4·21	12·2	77·8
5	4·13	11·7	76·2
6	4·12	11·6	75·2
7	4·22	11·1	74·6
11	47·33		
Mean	4·30		

2. *Influence of temperature.*

	Tempe. 47½° F.	67·° F.
Pulsations per minute,.	72·93	71·29
Expirations,..............	12·16	11·57
Vol. of expiration,......	548·	510·8
Vol. of expired air,......	6672·	6016·
— — CO_2,......	299·33	257·81
CO_2 in 100 parts,......	4·48	4·28

3. *Effect of digestion—Mean of* 18 *days.*

	Pulsation.	Expirations.	CO2 p. c.
Before 12 noon,.......	66·5	11·55	4·32
After dinner, 2 P.M.,.	82·3	12·77	4·37

4. *Effect of wine*—18 *days mean.*

	Pulsation.	Expirations.	CO2 p. c.
12 noon,...............	68·3	12·	4·33
2 P.M., after wine,...	85·8	13·22	4·20

5. *Without Wine.*

12 noon,...............	64·7	11·11	4·31
2 P.M.,.................	77·8	12·33	4·52

Quick respiration diminishes the per centage amount of carbonic acid, for while 12 expirations in a minute yield 4·34 per cent., 96 in a minute give only 2·74 carbonic acid per cent. Very slow breathing, on the contrary, increases it; 5½ expirations per minute gave 6 per cent. CO_2, or as in the following table:—

6. *Expirations per minute.*

Number.	CO2 per cent.
6	5·528
12	4·262
24	3·355
48	2·984
96	2·662

7. *Effect of deep breathing.*

	Vol. Expired.	CO2 per cent.
Common respiration,.....	591	4·69
Double in depth,.........		4·00
Common,..................	545	4·5
Three times in depth,....		3·38
Common,..................	584	4·75
Eight times in depth,....		2·78

8. *Carbonic acid in different parts of the Lungs,*

	CO2 per cent.
1st half of expiration,..................	3·72
2nd do.	5·44

9. *Effect of retaining the breath.*

Time of Retention.	Normal CO2.	CO2 in respd. Air.
20 seconds,.........	4·77	6·50
25 —	4·71	6·59
30 —	4·95	7·04
40 —	4·90	7·22
50 —	4·91	7·23
60 —	5·02	7·64

Retina. The portion of the optic nerves on which the impression of visible objects is made. It consists of water 92·9, albumen 6·25, fat ·85.

Retinalite. Spec. grav. 2·493, H 3·75. SiO_3 40·55, MgO 37·688, Al_2O_3 ·3, Fe_2O_3 ·62,

HO 20·, with some soda. Yellow resinous-looking amorphous masses; texture compact, fracture splintery, translucent. B.B. does not fuse; with borax and salt of phosphorus a clear bead. Allied to serpentine. Granville, Lower Canada (Dr. Wilson).

Retinaphtha. C_7H_4. Specific gravity ·86, of vapour 3·23. B.P. 226°. Limpid fluid; strong refractor of light; obtained by rectifying the oils from resin.

Retinite. *Resinite.* See RESIN ASPHALT.

Retinole. $C_{32}H_{16}$. Spec. grav. ·9, of vapour 7·11. B.P. 460°. Limpid oily fluid, without taste or smell, by distilling resin oils between 457° and 471°.

Retinyle. C_9H_6. Spec. grav. ·87, of vapour 4·242. B.P. 302°. Limpid and transparent fluid, not altered by light; obtained in the distillation of rosin at high temperatures.

Retinylene. $C_{18}H_{12}$. From rosin. It yields with SO_3 an acid isomeric with sulphocumenic acid ($C_{18}H_{11}S_2O_5HO$).

Retisterene. $C_{20}H_8$. F.P. 152°. A solid from rosin.

Retort. An apparatus for distilling. See GASES, PREPARATION OF.

Reussine. White 6-sided prisms, radiated, efflorescing, with saline taste. NaO SO_3 66·04, MgO SO_3 31·35, CaO SO_3 ·42, MgCl 2·19.

Reverberatory Furnace. A furnace in which the heat is applied to the body heated by a flame playing on its surface. See IRON and COPPER.

Rhacdic Acid. Dark red body; soluble in cold water, with a red colour. From red poppy flowers.

Rhaetizite. A synonyme of Cyanite.

Rhamnine. Yellowish-white crystals; bitter; soluble in water, alcohol, and ether; from Persian berries.

Rhamnoxanthine. A yellow-colouring matter by ether from the bark of the Rhamnus frangula. It has not been analyzed, but it seems to possess the characters of Parietine.

Rhaponticine. Yellow crystals; soluble in hot, insoluble in cold water, ether, and volatile oils; from rhubarb root.

Rheine. *Rhabarbarine.* Apparently the same as parietic or chrysophanic acid.

Rhenite. Hydrous Phosphate of Copper.

Rhodalite. Spec. grav. 2·, H 2·. Rose-red masses; texture earthy; apparently a congeries of rectangular prisms, mixed with earthy carbonate of lime, and apparently chabasite. B.B. infusible; forms beads with the fluxes. SiO_3 55·9, Al_2O_3 8·3, Fe_2O_3 11·4, MnO trace, CaO 1·1, MgO ·6, HO 22·. In amygdaloid, North of Ireland.

Rhodalline. A synonyme of Thiosinnamine.

Rhodalose. Sulphate of Cobalt.

Rhodeoretic Acid. $C_{42}H_{36}O_{21}$. Obtained by boiling rhodeoretine with ammonia or potash; soluble in water and alcohol, not in ether.

Rhodeoretine. *Jalapine.* $C_{42}H_{35}O_{20}$. A resin from jalap (Jalapa ipomœa) root; soluble in alcohol; insoluble in water and ether; forming a fine carmine colour with sulphuric acid.

Rhodeoretinole. $C_{42}H_{35}O_{29}$. Obtained by boiling rhodeoretine with strong acids, when it is converted into sugar and rhodeoretinole.

Rhodium. R 6·5, 52. Sp. grav. above 11·. Gray powder or silver-white brittle metal; obtained from platinum ore by digesting the ore in nitro muriatic acid, precipitating the platinum by salammoniac; the mother liquor is precipitated by metallic zinc, the black precipitate is digested in nitric acid to dissolve copper and lead; it is then dissolved in aqua regia, and common salt added; evaporated to dryness, and washed with alcohol, to dissolve out platinum and palladium oxides; the residue is a salt of rhodium, from which the metal is precipitated by zinc. *Protoxide.*—RO 7·5, 60. Black powder by heating the metal in the air. *Protosesquioxide.*—3 RO R_2O_3. By heating the metal continuously in the air. *Sesquioxide.*—R_2O_3. Black powder by fusing powdered rhodium with hydrate of potash, and digesting in chlorohydric acid. *Protochloride.*—RCl. Grayish or rose-red, formed by passing dry chlorine over heated protosulphide. *Sesquichloride.*—R_2Cl_3. Blackish-brown mass, by decomposing chlororhodiate of potassium by hydrofluosilicic acid. *Protosulphide.*—RS. Bluish-white metallic mass, by heating rhodium in sulphur. *Sesquisulphide.*—R_2S_3. Dark brown body, by precipitating chlororhodiate of sodium with sulphide of ammonium. *Chlororhodiate of potassium.*—2 KCl R_2Cl_3 2 HO. Red rectangular prisms, by fusing rhodium with chloride of potassium, and passing chlorine over it in a tube. *Sesquisulphate.*—Black powder by heating sesquisulphide with fuming NO_5, and evaporating.

Rhodizite. Spec. grav. 3·36, H 8. Gray or yellowish-white, with the form of boracite; lustre glassy, translucent. B.B. tinges flame first green and then red; fuses with borax into a glass, and also with salt of phosphorus. On red tourmalines at Mursinsk, Siberia.

Rhodizoic Acid. $C_7H_3O_{10}$, or C_7H_73HO. Colourless needles, soluble in water, alcohol, and ether; not decomposed at 212°; forms red-coloured salts; formed by passing carbonic oxide over potassium, and exposing the mass to the air, and decomposing by alcohol containing sulphuric acid.

Rhodochrome. A variety of serpentine, with a spec. grav. 2·668, H 3·5. It contains sesquioxide of chromium.

Rhodocrolite. A synonyme of Carbonate of Manganese.

Rhodoise. Arseniate of Cobalt.

Rhodonite. Silicate of Manganese.

Rhodotannic Acid. $C_{14}H_6O_7$. Amber-yellow body by alcohol and lead salt, from the leaves of Rhododendron ferrugineum.

Rhodoxanthine. $C_{14}H_7O_8$. Yellow or red substance by mineral acids on rhodotannic acid.

Rhubarb Resins. The roots of the different species of Rheum contain a variety of resins. *Aporetine*, *Phæoretine* ($C_{16}H_8O_9$), *Erythroretine* ($C_{19}H_9O_7$), and *Rheine.*

Rice. The seed of the Oryza sativa, is much used as food in tropical countries. I have found it to contain 8 per cent. of nutritive matter.

Ricinoleic Acid. HO $C_{38}H_{35}O_5$. Spec. grav. ·94. Sherry-coloured syrupy acid; miscible with alcohol and ether in all proportions; solid at 21° to 14°; yields by distillation not sebacic acid, but œnanthole and œnanthylic acid, which distinguishes it from oleic acid; obtained by saponifying castor oil.

Ricinostearic Acid. A solid acid obtained from castor oil, fusing at 158°, and corresponding with stearic or palmitic acid.

Riemannite. A synonyme of Allophane.

Rionite, or Selenide of Zinc.

Ripidolite. Spec. grav. 2·87, H 1·5. Green 6-sided prisms, or double pyramids; plates flexible. B.B. acts like chlorite, of which it is a variety. SiO_3 25·37, Al_2O_3 18·5, MgO 17·09, FeO 28·79, HO 8·96.

Rivuline. The green globules of Rivularia tubulosa, resembling in their motions those of infusoria.

Robinic Acid. Crystalline needles, from the root of the Robinia pseudoacacia, in which it exists as robinate of ammonia.

Rocellic Acid. $C_{24}H_{23}O_6$. F.P. 295°. White needles, soluble in ether, from the Rocella tinctoria.

Rocellinine. $C_{38}H_{17}O_{15}$. Neutral crystalline body, from Rocella tinctoria.

Rochelle Salt, or Potash tartrate of Soda.

Rock Cork, or Asbestus.

Rock Crystal. The crystalline form in 6-sided prisms of quartz or silica.

Rock Milk. *Agaric Mineral, Calc Tuff.* A white earthy form of carbonate of lime, deposited from water.

Rock Oil and **Tar,** or Mineral Naphtha.

Rock Salt. Native common salt.

Rock Soap. *Bole, Bergseife.* Allied to Halloylite. It consists of SiO_3 44, Al_2O_3 26·5, Fe_2O_3 8·, HO 20·5, CaO ·5. Thuringia.

Romanzovite. A variety of garnet, resembling cinnamon stone, from Kimito, Finland.

Romeite. *Romeine, Antimonite of Lime.* Hyacinth or honey-yellow 8-hedrons in minute crystals; harder than glass. SbO_4 79·31, CaO 16·67, FeO 1·2, MnO 2·16, SiO_3 ·64. B.B. melts into a black scoria; with borax a colourless glass in the reducing, violet in the oxidizing flame; insoluble in acids. St. Marcel, Piedmont.

Rosacic Acid. An acid said by Proust to exist in the red sediment of the urine.

Rosellite. Resembles cobalt bloom, except that it is in right rhombic prisms.

Rose Quartz. Quartz with a rose tinge, probably from the presence of manganese.

Roses, Oil of. See Attar.

Rosic Acid. A synonyme of Rosacic acid.

Rosin. See Colophane.

Rosin, Oils of. See Resinone, &c.

Rosite. *Rosellane. Polyargite.* Sp. gr. 2·735, H 2·5 to 4·. Red grains, foliated with a shining fracture, in limestone in Södermanland. SiO_3 44·901, Al_2O_3 34·506, Fe_2O_3 ·688, Mn_2O_3 ·191, KO 6·628, CaO 3·592, MgO 2·448, HO 6·533.

Rosolic Acid. An acid in coal tar not examined.

Rothoffite. A variety of Garnet.

Rubellane. A variety of Mica.

Rubellite. Red Tourmaline.

Ruberythric Acid. A synonyme of Xanthine?

Rubiacic Acid. $C_{31}H_7O_{15}$. Yellow flakes and powder, slightly soluble in water and alcohol, converted into rubiacine by alkalies; obtained by boiling rubiacine with sesquichloride of iron, and adding an acid.

Rubiacine. $C_{31}H_9O_{10}$, or $C_{32}H_{11}O_{10}$. Yellow resinous plates and needles, with a greenish tinge, subliming in scales; slightly soluble in boiling water; soluble in boiling carbonate of potash with a blood-red colour, and in caustic potash with a purple colour; soluble in boiling sesquichloride of iron; obtained by adding an acid to a decoction of madder roots; a dark brown deposit falls, which, when exhausted with boiling water, yields up alizarine, rubiane, pectic acid, and leaves rubiacine and a dark brown substance, which by treatment with nitric acid is destroyed, leaving the rubiacine; by boiling with sesquichloride of iron it dissolves and gives rubiacine by acids.

Rubiadine. $C_{32}H_{12}O_8$. Yellow or orange needles from alcohol, by the action of alkalies on rubiane; insoluble in boiling water; sublimes in yellow mica-like plates.

Rubiadipine. $C_{30}H_{24}O_5$. Yellowish-brown fatty body, precipitated by acetate of lead from its alcoholic solution; it most resembles in appearance and general properties rubiretine, but it is always soft; obtained from rubiane by fermentation.

Rubiafine. $C_{32}H_{13}O_9$. Yellow plates or needles by the fermenting action of erythrozyme on rubiane.

Rubiagine. $C_{33}H_{14}O_{10}$. Citron-yellow grains, insoluble in boiling water, in nitric acid; not converted into rubiacic acid by sesquichloride of iron, although it is dissolved, and thus distinguished from rubiafine; distinguished from rubianine by its insolubility in water, and from rubiadine by its not subliming; it is not precipitated by acetate of lead; formed in the action of erythrozyme on rubiane.

Rubiane. $C_{56}H_{34}O_{30}$. Hard dry amorphous

brown substance, easily soluble in water, less so in alcohol, insoluble in ether; the taste of the solutions is very bitter; converted by nitric acid into alizaric acid; decomposed by heat, and yields alizarine vapours by sublimation; it is not a colouring matter, but yields colouring substances by fermentation; by sulphuric acid it is decomposed into alizarine, rubiretine, verantine, rubianine, and sugar; prepared by making a decoction of madder, adding to the solution animal charcoal, washing the charcoal with cold water till the wash water gives no green colour with chlorohydric acid; treating the charcoal with boiling alcohol and filtering; the alcohol extract by evaporation consists of rubiane with some chlorogenine; from the latter it may be freed by frequent treatment with animal charcoal; the rubiane is finally dissolved and precipitated by acetate of lead, and the lead compound decomposed by sulphuric acid, or sulphohydric gas.

Rubianine $C_{32}H_{19}O_{15}$? Yellow silky needles by the action of sulphuric acid on rubiane.

Rubicelle. Yellow or orange-red Spinelle.

Rubichloric Acid. $C_{14}H_8O_9$. Colourless tasteless substance, soluble in water and alcohol; insoluble in ether; obtained by alcohol from the supernatant liquor after precipitating madder decoction by acetate and trisacetate of lead.

Rubinic Acid. $C_{18}H_6O_9$. A red substance, obtained in union with potash by dissolving catechine in carbonate of potash, and exposing the solution to the air without heat; excess of carbonate of potash may be removed by acetic acid.

Rubiretine. *Alpha Resin of Madder.* $C_{14}H_6O_4$? Yellow resin by the action of acids on rubiane.

Rubitannic Acid. $C_{14}H_8O_9$. The tannic acid existing in madder roots.

Rubrica. The ancient name probably for red ochre.

Ruby. A term applied to various minerals. Almandine ruby is *violet* spinell; balas ruby is *rose-red* spinell; spinell ruby is *scarlet* spinell, while oriental ruby is sapphire.

Ruby Silver. Red Silver ore.

Ruby Sulphur. Realgar or bisulphide of Arsenic.

Rue, Oil of. $C_{20}H_{20}O_2$. Spec. grav. ·837, of vapour 7·69; B.P. 444°. Yellowish or greenish oil, the aldehyde of capric acid probably; converted by nitric acid into pelargonic acid, $C_{18}H_{18}O_4$, and capric acid (rutic acid) $C_{20}H_{20}O_4$. Rue oil is formed by sulphuric acid on cod liver oil.

Rufigallic Acid. *Parellagic Acid.* $C_7H_2O_4$. Brownish crystalline body, losing water at 248°; soluble in 3500 water; changed into oxalic acid by nitric acid; into tannomelanic acid by potash; obtained by mixing 1 gallic acid with 5 strong sulphuric acid, and dropping it into water; the acid falls in flocks.

Rufimoric Acid. $C_{14}H_7O_8$. Dark red mass, easily soluble in alcohol, less so in water; very little soluble in ether, soluble in SO_3 with a red colour; obtained by acting on morintannic acid by sulphuric acid; it separates as a deposit.

Rufinc. A red substance left when phloridzine is distilled.

Rum. A spirit distilled in the West Indies from the fermented product of the sugar cane. The name is said to be derived from the last syllable of saccha*rum*, sugar.

Ruset. The yellow aqueous extract of the Berberis lycium of India; used as a substitute for bark.

Rust. The sesquioxide of iron mixed with carbonic acid and ammonia, formed on the surface of iron by exposure to moist air.

Ruthenium. Ru 6·5, 52·1. Spec. grav. 8·6. Whitish-gray porous lumps with metallic lustre; infusible by oxyhydrogen blowpipe; obtained from the osmide of iridium by heating it with common salt to low redness in a current of chlorine, the mass digested in cold water, the solution mixed with a few drops of ammonia, and heated in a porcelain capsule; sesquioxide of ruthenium and oxide of osmium fall. This precipitate is washed with nitric acid and distilled, when osmic acid passes over; the residue is fused in a silver crucible with caustic potash; the ignited mass digested in water, the solution allowed to clear and then drawn off, when sesquioxide of ruthenium separates; this is converted into the metal by passing hydrogen over it when ignited.

Protoxide.—RuO 7·5, 60·1. Gray metallic powder, reducible by hydrogen; obtained by fusing bichloride of ruthenium with carbonate of soda and digesting in water.

Sesquioxide.—Ru_2O_3 16·, 128·2. Blackish-blue powder, by heating powdered ruthenium intensely with a blowpipe; it unites with acids forming orange solutions.

Binoxide.—RuO_2 8·5, 68·1. Bluish-black powder with a green tinge; by igniting the bisulphide of ruthenium, or the bisulphate of binoxide of ruthenium; it forms yellow solutions with acids. *Chloride.*—RuCl. Black and crystalline, by passing chlorine over the metal. *Sesquichloride.*—A metallic-looking substance, by dissolving sesquioxide in muriatic acid, and evaporating. *Sesquichloride of Ruthenium and Ammonium.*—$2,NH_4Cl,Ru_2Cl_3$. Red rhombohedrons, by adding salammoniac to muriate of sesquioxide. *Sesquichloride of Ruthenium and Potassium.*—$2,KCl,Ru_2Cl_3$. Crystalline powder or orange cubes, obtained in the preparation of ruthenium. *Bichloride of Ruthenium and Chloride of Potassium.*—$KCl,RuCl_2$. Red rhombohedrons, by treating the preceding $(2\,KCl,Ru_2Cl_3)$ with aqua regia, or chlorate of potash. *Sulphides.* These appear to be a sesquisulphide and bisulphide and Ru_2S_5. *Sulphate of Binoxide.*—Yellow substance, by treating the bisulphide with nitric acid. *Rutheniate of Potash.*—An orange solution, by igniting the metal with nitre or chlorate of potash.

Rutherfordite. *Titaniate of Lime.* Spec.

grav. 5·55, H 6·5. TiO_2 58·5, CaO 10, CeO?

Rutic Acid. A synonyme of Capric acid.

Rutile. (*Rutilus*, red.) Native Titanic acid. See TITANIC ACID.

Rutine. A substance by alcohol, from the leaves of the Ruta graveolens.

Ryacolite. *Glassy Felspar.* Sp. grav. 2·576 to 2·582, H. equal to felspar. Grayish-white and oblique prisms; fracture uneven, or small imperfect conchoidal; internal lustre splendent and vitreous; transparent; optical properties different from those of adularia. SiO_3 66·6, Al_2O_3 18·5, MgO 1·, KO 8·, NaO 4·, FeO ·6. In lava.

Rye. The seed of the Secale cereale. It contains gluten (caseine?) 9·48, albumen 3·28, starch 61, sugar 3·28, gum 11·09, cellular matter 6·38. Its ash contains KO 11·43, NaO 18·89, MgO 10·57, CaO 7·05, PO_5 51·81, SO_3 ·51, SiO_3 ·69, Fe_2O_3 1·9.

S

Sabadillic Acid. See CEVADIC ACID.

Sabadilline. $C_{20}H_{13}NO_5$. White crystalline acrid base; soluble in hot water and alcohol; insoluble in ether; obtained by alcohol from the seeds of Veratrum sabadilla.

Saccharic Acid. *Oxalhydric Acid.* $C_{12}H_5O_{11}5HO$, or $C_{12}H_{10}O_{16}$. Colourless acid syrup, or colourless crystals, precipitating lime and barytes water, isomeric with mucic acid; obtained by dissolving 1 sugar or gum in 2 nitric acid diluted with 10 water with the aid of heat; the acid fluid is neutralized with chalk; neutral acetate of lead is added; a white precipitate falls which is decomposed by sulphohydric gas, and evaporated.

Saccharite. Spec. grav. 2·668. Granular variety of felspar in serpentine in Silesia, containing SiO_3 58·93, Al_2O_3 23·5, Fe_2O_3 1·27, CaO 5·67, MgO ·56, KO ·05, NaO 7·42, NiO ·39, HO 2·21.

Saccharometer. An instrument for measuring the specific gravity of saccharine solutions. That used by the officers of the Scottish inland revenue, and known under the name of Allan's saccharometer, was invented by Dr. Thos. Thomson, about the beginning of the present century.

Saccharosulphuric Acid. A substance blackening on evaporation, obtained by dissolving sugar in oil of vitriol, and saturating with chalk.

Saccholactic Acid, or Sugar of Milk in combination.

Sacchulmic Acid. $C_{30}H_{15}O_{15}$. Light brown powder, insoluble in alcohol and ether, converted into sacchulmine by long boiling with water; soluble in ammonia and the alkalies with a brown colour; obtained by boiling sugar with dilute sulphuric acid, and taking up the acid by ammonia.

Sacchulmine. A brown matter in the preceding action, insoluble in ammonia.

Safflower. See CARTHAMINE.

Saffron. The petals of the Crocus sativus, containing about half of their weight of yellow-colouring matter (Polychroite.)

Sagapenum. A gum resin in tears from Asia Minor (from Ferula Persica?) Colour yellow; taste hot and bitter, with a garlic smell; slightly soluble in water; very soluble in alcohol; yields an oil when distilled with water; the resin can be divided into two by ether. Sagapenum is used in medicine as a substitute for ammonia and asafœtida.

Sagenite. A synonyme of Titanic acid.

Sage Oil. From the Salvia officinalis; seems to be a mixture of oxides of the radical C_6H_5.

Sago. A granulated form of starch, derived, as imported from the East, from the pith of Sagus raphia and Rumphii, Phœnix farinifera, and Sagus genuina. It contains a very small amount of albuminous matter—(see ALIMENT)—and salts, which are washed out in its preparation. It is therefore by no means a nutritious substance. Sago is made on the continent in smaller grains from potato starch. See STARCH.

Sahlite. A greenish pyroxene from Sahla, Sweden.

Sal Alembroth. See ALEMBROTH.

Sal Ammoniac. See AMMONIA CHLOROHYDRIDE.

Salammoniac. See HYDROCHLORATE OF AMMONIA.

Salamstein. A synonyme of the Sapphire.

Saldanite. Native sulphate of alumina, consisting of SO_3 35·68, Al_2O_3, 14·98, HO 49·34.

Sal de Duobus. Sulphate of Potash.

Salep. *Salop.* The farinaceous portion of the tubers of the Orchis morio. It contains apparently bassorine or an analogous body, which renders it with hot water a tenacious paste.

Sal Gemme. A synonyme of Native common salt.

Salhydramide. See SALICYLIMIDE.

Salicine. $C_{26}H_{18}O_{14} = C_{12}H_{10}O_{10} + C_{14}H_8O_4$. F.P. 230° to 248°. Silky plates, scales, or needles, with a bitter taste; neutral; soluble in 5½ cold, more soluble in hot water; soluble in alcohol; insoluble in ether and oils; it has been used as a substitute for quinine, but apparently without success. When heated above its fusing point it becomes resinous and decomposes; it is not precipitated from its solutions by tannic acid, gelatine, or disacetate of lead. By *chlorine* it is converted into 3 bodies—*chlorosalicine* in the cold state, a yellow powder ($C_{26}H_{17}ClO_{14}$); at 140° a red oil, *dichlorosalicine* ($C_{26}H_{16}Cl_2O_4$), and with the further action of chlorine, *trichlorosalicine* ($C_{26}H_{15}Cl_3O_{14}$). By *nitric acid*, when strong, salicine is changed into oxalic and nitropicric acids; when dilute, anilotic, anilic acids, helicine and helicoidine are formed. By sulphuric acid, *rutiline*, a red substance, is formed, by which salicine is distinguished from quinine and its salts. Salicine is obtained by boiling willow bark with water, precipitating the decoction by acetate of lead; diffusing the precipitate after washing through water, and decomposing it by a current of sulphohydric gas, filtering, evaporating, and crystallizing out the salicine.

Salicor. The ash of the Salicornia annua, employed in France as a source of carbonate of soda, of which it often contains 15 per cent.

Salicyl-amide. $C_7H_5O_2,H_2N$.

Salicyl-benzoil-amide. $C_7H_5O_2,C_7H_5O,H_1N$. Needles insoluble in water; scarcely soluble in alcohol; very soluble in alkalies; acid reaction.

Salicyl-cumyl-amide. $C_7H_5O_2,C_{10}H_{11}O,H,N$.

Salicyle. $C_{14}H_5O_4 = Sa$. The hypothetic base of the salicyle series.

Salicyle Hydride. *Salicylous Acid, Hydride of Spiroyle.* $C_{14}H_5O_4H$. Spec. grav. 1·1731, of vapour 4·2760. B.P. 380°. Colourless oil, with an aromatic smell, burning taste, and colours the skin yellow; becomes solid at —4°; slightly soluble in water; very soluble in alcohol and ether; reddens litmus, and then whitens it; decomposed by SO_3; potash converts it into salicylic acid, with evolution of H; isomeric with benzoic acid; with bases forms salicylides. When strong ammonia is poured on it, salicylide of ammonium, $SaNH_4$, is formed, a yellow solid; with dry ammonia the compound 2 NH_3, 3 SaH is formed. With potash it forms a salicylide ($C_{14}H_5O_3KO$) yellow crystals, and bisalicylide (2 $C_{14}H_5O_3$,KO) in needles. The hydride of salicyle is obtained by distilling the flowers of meadow sweet (Spiræa ulmaria), or by distilling 1 salicine, 1 bichromate of potash, 20 water, 2½ oil of vitriol. *Salicyle Chloride* $C_7H_5O_2Cl$.

Salicylic Acid. $C_{14}H_5O_4OHO = SaO HO$. Thin prisms, subliming without decomposition; formed by heating the hydride of salicyle with potash, when H is removed, or by acting on coumarine, or gaultheria oil, with potash; distilled with barytes, it yields carbolic acid; by nitric acid it is converted into nitrosalicylic acid, and then into carbazotic acid.

Salicylimide. $C_{42}H_{18}N_2O_6$. Golden prisms, formed by adding caustic ammonia cautiously in drops to hydride of salicyle in alcohol.

Salicylous Acid. A synonyme of *Hydride of Salicyle.*

Salifiable Base. The radical of a salt.

Saligenine. $C_{14}H_8O_4$. F.P. 180°, vaporizes at 212°. Pearly rhomboids, with an unctuous feel. Soluble in hot and cold water, the solution giving an indigo colour with sesquioxide of iron; soluble in alcohol and ether; changed by heat and dilute acids into saliretine; obtained by fermenting 50 salicine with 200 water and 3 synaptase at 106° for twenty-four hours; the saligenine separates in crystals, being little soluble in water, while grape sugar remains in solution. From this it has been concluded that salicine consists of saligenine and sugar; but this is scarcely conclusive, since sugar might, with equal justice, be viewed as composed of carbonic acid and alcohol.

Saline Taste. The impression on the sense of taste by the action of a salt.

Saliretine. $C_{14}H_6O_2$. A resinous body obtained by boiling saligenine or salicine with dilute sulphuric or chlorohydric acids, or by heat at about 270°; isomeric with the essence of bitter almonds and benzoine.

Saliva. Spec. grav. 1003·8 to 1008·8. Colourless fluid secreted by the two parotid, the two submaxillary, and the two sublingual glands. The quantity of saliva secreted per day in man appears to be about 7½ ounces avoirdupois. During meals it has been found acid, at other times alkaline; on standing, white flocks (mucus) separate from saliva (5 parts from 100,000 saliva); alcohol precipitates a white substance (ptyaline) from saliva; sesquichloride of iron often gives a red tint to saliva, due to sulphocyanogen. The constituents of saliva are—

Water,	992·9	988·78	988·1
Ptyaline,	2·9	·626	1·8
Fatty acid,	—	—	·5
Albumen and soda,	—	—	1·7
Mucus, with soda,	1·4	1·64	2·6
Alkaline chlorides,	1·7	1·8	1·4
Soda,	·9	—	·5
Lactate of potash or soda,	·9	1·18	·7
Phosphate of lime,	—	·17	·6
Silica,	—	·15	—
Chloride of calcium,	—	·18	—
Sulphocyanide of potassium,	—	—	·9

Sal marine. Common salt.

Sal martis. Protosulphate of iron.

Sal mirabile. Sulphate of soda.

Sal perlatum. Phosphate of soda.

Sal prunella. Nitre melted.

Salseparine. *Smilacine, Sarsaparilline, Parilline, Parillic Acid, Parigline, Sasseparine.* $C_{18}H_{13}O_5$. Neutral crystalline body, from the root of Smilax sarsaparilla by alcohol; soluble in hot water, alcohol, and ether; with a bitter disagreeable taste.

Salt. A combination of an acid with a basic oxide, or of an acid radical with a metal of such a nature that the characters of both the acid and the base are neutralized. An oxide is not usually viewed as a salt, because it is capable of uniting with an acid; but chloride of potassium seems likewise susceptible of a similar union, if we consider the chlorochromate of potash to be composed of KCl $2CrO_3$; and there are several other known parallel cases. The theories of saline combinations are multifarious at the present time, but none of them, with the exception of the most common view, is more than empirical, since they throw no light on the mode of decomposition of the constituents of a salt in the ordinary processes of analysis. 1. According to the usual theory, nitrate of silver consists of nitric acid united to oxide of silver ($AgONO_5$). 2. According to the theory of Davy, its formula is $AgNO_6$, or nitroxide of silver. 3. According to Gerhardt's view, in order to assimilate the reactions with those which we observe among organic bodies, its formula will be NO_4O,AgO, and when chloride of sodium reacts on it we have the diagrams—

$$\begin{array}{ccc} \begin{matrix} NO_4O \\ AgO \end{matrix} & + & \begin{matrix} Na \\ Cl \end{matrix} = \\ \text{Nitrate of silver.} & & \text{Chloride of sodium.} \\ = \begin{matrix} NO_4O \\ NaO \end{matrix} & + & \begin{matrix} Ag \\ Cl \end{matrix} \\ \text{Nitrate of soda.} & & \text{Chloride of silver.} \end{array}$$

Similar changes occur when we cause chlorohydric ether (chloride of ethyle) to act on potash in alcohol, chloride of potassium and ether resulting—

$$\begin{array}{ccc} \begin{matrix} C_4H_5O \\ K\ O \end{matrix} & + & \begin{matrix} C_4H_5 \\ Cl \end{matrix} = \\ \text{Potash alcohol.} & & \text{Chlorohydric ether.} \\ = \begin{matrix} C_4H_5O \\ C_4H_5O \end{matrix} & + & \begin{matrix} K \\ Cl \end{matrix} \\ \text{Ether.} & & \text{Chloride of potassium.} \end{array}$$

Salt of Alembroth. Ammonia chloride of mercury.

Salt of Amber. Succinic acid.

Salt, Ammoniacal Fixed. Chloride of calcium.

Salt, Ammoniacal Secret. Sulphate of ammonia.

Salt, Arsenical Neutral. Arseniate of potash.

Salt of Benzoin. Benzoic acid.

Salt, Bitter Cathartic. Sulphate of magnesia.

Salt, Digestive of Sylvius. Acetate of potash.

Salt, Diuretic. Acetate of potash.

Salt, English. Sulphate of magnesia.

Salt, Epsom. Sulphate of magnesia.

Salt, Febrifuge of Sylvius. Chloride of potassium.

Salt, Fusible. Phosphate of ammonia.

Salt, Fusible of Urine. Ammonia phosphate of magnesia.

Salt, Glauber's. Sulphate of soda.

Salt, Glazer's. Sulphate of potash.

Salt of Lemery. Sulphate of potash.

Salt of Lemons. Binoxalate of potash.

Salt, Mycrocosmic, or Salt of Phosphorus. Ammonia phosphate of soda.

Salt of Paris. Sulphate of potash.

Saltpetre, Potash. Nitrate of potash.

Saltpetre, Soda. Nitrate of soda.

Salt, Polychreste. Sulphate of potash.

Salt of Saturn. Acetate of lead.

Salt, Sedative. Boracic acid.

Salt, Seidlitz. Sulphate of magnesia.

Salt of Seignette. Potash tartrate of soda.

Salt of Sorrel. Binoxalate of potash.

Salt, Spirit of. Hydrochloric acid.

Salt, Sulphureous of Stahl. Sulphite of potash.

Salt of Tartar. Carbonate of potash.

Salt, Tin. Protochloride of tin.

Salt, Vegetable. Tartrate of potash.

Salt of Vichy. Carbonate of soda.

Salt, Volatile. Carbonate of ammonia.

Salt, Wonderful. Sulphate of soda.

Sal Volatile. See CARBONATE OF AMMONIA.

Samarskite. Spec. grav. 5·45, H 5·5. A black mineral from North Carolina, containing niobic acid 54·81, U_2O_3 17·03, FeO 14·07, yttria 11·11, ceria 3·95.

Samshoo. A spirit extracted by fermentation and distillation from rice by the Chinese. I have found it to contain 26 per cent. of absolute alcohol.

Sanadoine. SiO_3 57·25, Al_2O_3 23·5, CaO 2·75, Fe 3·25, MnO ·26, NaO 8·1, HO 3·.

Sand. Particles of silica mixed with other impurities.

Sand Bath. See BATH.

Sandaraca. Realgar, or Bisulphide of Arsenic.

Sandarach. *JuniperResin.* F.P. 308°. Brown round resinous tears from the *Thuya articulata* of Barbary, resembling mastich, but crumbles when chewed. It consists of 3 resins; *a.* $C_{20}H_{31}O_5$, soluble in weak alcohol, ether, and oil of turpentine, ammonia and caustic potash; *b.* $C_{20}H_{31}O_2$, soluble in absolute alcohol and ether; *c.* $C_{20}H_{30}O_6$, soluble in alcohol of ·82°. These resins are obtained by dissolving sandarach in absolute alcohol; on adding hydrate of potash resin *c* falls, while the other two (*a* and *b*) remain in solution in union with potash; the two resins are precipitated by HCl; resin *a* is separated by weak alcohol; the white resin *b* is insoluble.

Sanders. A red dyewood from Pterocarpus santalinus. See SANTALINE.

Sandiver. (Sain de Verre). The saline scum in glass pots.

Sandyx. Probably a mixture of realgar and red chalk.

Sang Dragon. See DRAGON'S BLOOD.

Sanguinarine. $C_{37}H_{16}NO_8$. Yellow powder from Sanguinaria canadensis; insoluble in water; soluble in alcohol and ether. When dissolved in alcohol and a current of HCl passed through the solution, a beautiful blood-red solution is formed.

Sanidine. *Glassy Felspar.* SiO_3 68·18, Al_2O_3 18·33, Fe_2O_3 ·71, CaO ·51, MgO ·46, KO 7·15, NaO 4·66.

Santalic Acid. F.P. 219°. Red prisms, insoluble in water, soluble in ether; soluble in SO_3 with a red colour, from sanders wood by alcohol.

Santalic Oxide. Brown, tasteless, insoluble mass, from Sanders.

Santalide. A red, tasteless, insoluble mass.

Santalidide. Brown mass, soluble in water and hot alcohol (·912); insoluble in ether and absolute alcohol, from Sanders.

Santaline. $C_{16}H_8O_3$? Appears to consist of two or more of the preceding resins.

Santaloide. Yellow mass; soluble in water and alcohol; insoluble in ether.

Santaloidide. Brown resin, soluble in boiling alcohol of ·863, insoluble in water and ether.

Santonine. C_5H_3O. Pearly prisms; insoluble in water; soluble in alcohol, ether, and acids; from several species of Artemisia by alcohol; it unites with bases.

Santorine. A kind of tarras of volcanic origin, used as a building stone, from a Grecian island on the coast of Dalmatia. It contains 54 per cent. lime, and 22 per cent. silica.

Sap of Trees, is a fluid generally colourless, and limpid as water; that of the birch has a specific gravity of 1004 with a scarcely sweetish taste, although it contains above 1 per cent. of sugar; that of the vine contains carbonic acid, tartrate of lime, bitartrate of potash; that of the elm is composed of water and volatile matter 1027·9, acetate of potash 9·24, vegetable matter 1·06, carbonate of lime ·796. The sap ascends from the roots to the leaves, and returns by the bark.

Sap Green. A colour composed of the colouring matter of the berries of the Rhamnus catharticus and lime.

Sapan Wood. (Cæsalpinia Sapan.) *East Indian Dye Wood, False Sandal Wood, Lignum Sapan.* The species are C. coriaria and C. pulcherrima. Grown in India, Ceylon, Amboyna, and the Moluccas. The decoction under the name of Sapan liquor is used in calico printing for red colours. The specific gravity of the liquor is usually 1040, and it contains about 5897 grains of extract per imperial gallon. The composition of the wood is organic matter 987·083, carbonate and phosphate of lime 11·65, alkaline phosphate and sulphate ·85, common salt ·517 (R.D.T.)

Saponic Acid. See ESCULINE.

Saponine. $C_{52}H_{46}O_{32}$. White brittle substance, producing sneezing; froths in water, in which it dissolves; less soluble in absolute alcohol; insoluble in ether; existing in Agrostemma githago, Anagallis arvensis, Esculus hippocastanum, Saponaria officinalis, Gypsophila struthium, various species of Dianthus and Lychnis, in Quillaia saponaria, and Silene inflata.

Saponite. A synonyme of Soapstone.

Sappare. A name given by Saussure to cyanite, from an incorrect reading of a specimen marked sapphire, sent to him by a Scottish mineral dealer.

Sapphire. *Adamantine Spar, Asteria* (Pliny), *Oriental Amethyst, Corundum, Diamond Spar, Oriental Emerald, Emery, Oriental Ruby, Telesia, Oriental Topaz, Salamstein.* Al_2O_3. Sp. grav. 3·9511. Red (ruby), blue (sapphire), yellow (topaz), green (emerald), violet (amethyst), acute rhomboids; the lateral edges being frequently

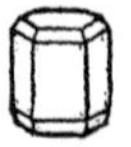

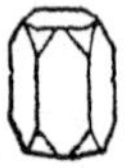

replaced by planes parallel to the perpendicular axis of the rhomboid; the common form of the corundum is a 6-sided prism; it often occurs as 12-hedrons, or 2 6-sided pyramids applied base to base; structure in plates; fracture conchoidal, uneven; lustre vitreous; transparent to opaque; refracts doubly. Sapphire consists of nearly pure alumina; but isomorphous substances sometimes exist along with it, as sesquioxide of iron. Sirian, Pegu; Saxony; Bohemia; Puy, France. The corundum is destitute of beauty, being gray, brown, or red. I have good specimens from India and China. Emery is a gray or black variety.

Sapphirine. Sp. grav. 3·450, H 7·5. Blue or green grains in plates; glassy, transparent. SiO_3 14·51, Al_2O_3 63·11, MgO 16·85, CaO ·38, Mn_2O_3 ·5, FeO 3·92, HO ·49. B.B. per se, and with fluxes, infusible. Fisknaes, Greenland.

Sarcocolline. C 57·15, H 8·34, O 34·51. Sarcocolla is in yellow oblong globules, like gum arabic, from the Penœa mucronata, a native of Persia and Arabia. When digested in ether a resin is dissolved, and absolute alcohol then takes up sarcocolline; soluble in water and alcohol, and SO_3; insoluble in ether; converted into oxalic acid by NO_5; precipitated by nutgalls, diacetate of lead (T. Thomson).

Sarcolite. *Hydrolite, Gmelenite.* Spec. grav. 2·054, H 4·. Snow-white truncated double 6-sided pyramids; translucent, frangible. B.B. increases in bulk, and becomes an enamel, but does not fuse. SiO_3 39·896, Al_2O_3 12·968, Fe_2O_3 8·27, KO 9·, HO 29·866. In the Vicentine; Glen Arm, county Antrim.

Sarcosine. (σαρξ, flesh). $C_6H_7NO_4$. F.P. a little above 212°. Colourless rhombic prisms; neutral; very soluble in water; very little soluble in alcohol, and insoluble in ether; taste slightly metallic; isomeric with lactamide, urethane, alanine, and homologous with glycocoll; colours acetate of copper deep blue like ammonia; forms needles with corrosive sublimate; forms salts with acids; obtained by adding 10 times its weight of a boiling saturated solution of hydrate of barytes to a solution of creatine, filtering, precipitating excess of barytes by carbonic acid, filtering, and crystallizing; sarcosine is obtained pure by uniting it with sulphuric acid, decomposing with carbonate of barytes, filtering, and evaporating.

Sard. Deep brownish-red calcedony.

Sardoine. A synonyme of agate.

Sarsaparilla. The root of the Smilax sarsaparilla, &c. containing oil, resin, starch, albumen, salts, &c.

Sassafrine. Brown crystalline grains, somewhat soluble in cold water and ether, acting as a weak acid; by alcohol, from the root of Laurus sassafras.

Sassafrole. *Oil of Sassafras.* Spec. grav. 1·09; B.P. 443°. Yellow volatile oil by ether, from sassafras root.

Sasseparine. A synonyme of salseparine.

Sassoline. Native boracic acid.

Satin Spar. Carbonate of lime in fibres, with a satin lustre.

Saturation. When a fluid has dissolved as much of a soluble body as it is capable of taking up, the fluid is said to be saturated; and the same expression is applied to the case where an acid is neutralized by a base, or a base by an acid.

Saucy Bark. A bark from the coast of Africa, used there as a poison.

Saussurite. *Jade, Felspath tenace.* Spec. grav. 2·801, H 7·. Gray, reddish-gray, bluish or greenish-gray, granular masses; fracture splintery, very tough, translucent on the edges. B.B. the thin edges soften, but do not fuse. SiO_3 82·168, Al_2O_3 5·072, FeO and some MnO 2·88, CaO 5·52, MgO 4·52, KO a trace. Along with diallage at the Lizard, Cornwall, also in the Verde di Corsica.

Savine, Oil of. The essence of Juniperus sabina, obtained by distilling the tops of the plant.

Savite. Spec. grav. 2·45, H 3·2. Colourless 4-sided prisms, with difficulty fusible before the blowpipe. SiO_3 49·167, Al_2O_3 19·663, MgO 13·5, NaO 10·52, KO 1·23, HO 6·575. In gabbro in Tuscany.

Saxon Blue. Sulphate of indigo.

Scammony. Spec. grav. 1·235. A cathartic gum resin from the root of Convolvulus scammonia. The Smyrna resin consists of resin 29, gum 8·, extractive 5, vegetable matter 58; from Aleppo, of resin 60, gum 3, extractive 2, vegetable debris, bark, &c. 35. It is often mixed with chalk.

Scapolite. *Arctizite, Bergmanite? Chelmsfordite, Dipyre, Meionite, Micarelle, Paranthine, Rapidolite, Spreustein? Wernerite.* Sp. gr. 2·612 to 2·65 (Meionite), 2·709 to 2·749 (Scapolite), H 4·25. White, grayish-white (Meionite), sometimes with a shade of green (Scapolite), frequently brownish-red right square prisms, terminated by a low 4-sided pyramid, with angles of 136° 22′, and 112° 5′; the surface of the prism is often streaked longitudinally; the structure is foliated; fracture imperfect, conchoidal, or uneven; lustre vitreous; from transparent to translucent on the edges; brittle. B.B. froths and fuses into an enamel or white glass. SiO_3 40·531, Al_2O_3 32·726, CaO 24,245, KO and NaO 1·812, Fe_2O_3 ·182. *Form.* 3(CaONaO) 2 SiO_3, 2Al_2O_3 SiO_3. Arendal, Norway; North America.

Scarbroite. Spec. grav. 1·48. White earthy substance, adhering to the tongue, in limestone at Scarborough. SiO_3 10·5, Al_2O_3 42·5, Fe_2O_3 ·25, HO 26·75.

Scheele's Green. Diarsenite of copper.

Scheelite. Tungstate of Lime.

Scheelitine. Tungstate of Lead.

Scheererite. Spec. grav. ·65; B.P. 197½°; F.P. 111°. C = 73, H 24 = CH_2. White or grayish fatty substance, or needles, with pearly lustre, without taste and smell, on brown coal at Uznach, Switzerland; soluble in alcohol, ether, and oils; soluble in sulphuric acid with a red colour. *Branchite.*—A variety has a spec. grav. of 1·; F.P. 167°; is colourless, and soluble in alcohol. On brown coal at Mount Vaso, in Tuscany. See also KÖNLITE.

Scheelstein. Table Spar.

Schiller Spar. *Hydrous Bisilicate of Magnesia, Diallage Metalloide, Karstine, Ottrelite.* Spec. grav. 2·668, H 2·75. Dark green foliated masses, which cleave in two directions, one of which is highly perfect and easily obtained, while the other appears only in slight traces. These cleavages are in the direction of a rhombic prism, with angles of 90° 30′, and 86° 30′; lustre metallic, pearly; colour olive-green and blackish-green, inclining to pinchbeck-brown; streak grayish-white, inclining to yellow. B.B. loses its green colour and becomes brown and metallic; the edges do not fuse, but are rounded off; gives off water in a tube with some ammonia; with borax fuses into a greenish bead from chromium; with soda infusible. SiO_3 43·9, MgO 25·856, FeO and Cr_2O_3 13·021, MnO ·535, CaO 2·642, Al_2O_3 1·28, HO 12·626. Baste, Hartz, in serpentine.

Schist, or Slate.

Schlippe's Salt. 3NaS, SbS_5 18 HO.

Schneiderite. H 3·. White opaque foliated masses, along with Humboldtite, in Tuscany. B.B. fuses to a blue enamel; decomposed by acids, leaving gelatinous silica. SiO_3 47·794, Al_2O_3 19·382, CaO 16·765, MgO 11·029, KO and NaO 1·621, HO 3·409. *Form.* 3RO2SiO_3 + $Al_2O_3SiO_3$ HO.

Schorl. Black fibrous crystals of tourmaline. *White schorl* is nepheline; *blue schorl* is disthene, vivianite, and anatase; *cruciform schorl* is staurotide; *electric schorl* is tourmaline; *octahedral schorl* is anatase; *red schorl* is rutile, rubellite; *green schorl* is epidote, amphibole; *violet schorl* is axinite.

Schorlomite. Spec. grav. 3·862 to 3·783, H 7·25. Black 6-sided prisms, with indistinct cleavage; lustre vitreous; fracture conchoidal. SiO_3 26·09, TiO_2 17·36, CaO 31·12, FeO 22·83, MgO 1·55. B.B. fuses with difficulty into a black mass, into a yellow glass with fluxes. Ozark mountains, Arkansas. *Form.* 2(3RO 2 SiO_3) 3(ROTiO_2.)

Schorlite, or *Schorlous Beryl.* A synonyme of Pycnite.

Schrotterite. Spec. grav. 1·95 to 2·05, H

3·25. Green, yellowish masses, sometimes with brown spots, resembling allophane; soluble in acids. B.B. becoming white. SiO_3 11·95, Al_2O_3 46·3, HO 36·2, Fe_2O_3 2·95, CaO 1·3, CuO ·25, SO_3 ·78 = 4 Al_2O_3,SiO_3 18 HO. Freienstein, Styria.

Schulzite. A synonyme of Geocronite.

Schutzite. Sulphate of Strontian.

Schweinfurth Green, or ARSENIO-ACETATE OF COPPER.

Scillitine. A solid substance with a caustic taste; soluble in sulphuric acid, with a purple and then black colour; dissolved and decomposed by nitric acid; obtained from the decoction of squills (Scilla maritima) by precipitating by acetate of lead; decomposing the precipitate by sulphohydric acid and evaporating.

Scleretinite. (σκληρος, hard.) Spec. grav. 1·136, H 3·, $C_{10}H_7O$ = C 76·95, H 8·95, O 10·42, ash 3·68. A fossil resin of a black colour, found on coal at Wigan; in thin pieces it has a dark reddish-brown colour, and is translucent; insoluble in water, alcohol, ether, and alkaline carbonates.

Sclerotic Coat of the Eye, containing albumen but no fibrine, consists of C 50·995, H 7·075, N 18·723, O 23·207.

Scoleca. So called from its assuming the shape of a worm, was an ancient mixture of alum, carbonate of soda, and white vinegar, from which probably resulted sulphate of soda, and acetates of soda, alumina, copper, &c.

Scolexerose. A variety of Scapolite or Labradorite (?) Pargas, Finland.

Scolezite. *Needlestone, Lime Mesotype.* Spec. grav. 2·214 to 2·27. White right rhombic prisms with angles of 91° 20′, frequently 6-sided, and often terminated by a 4-sided pyramid with angles of 116° 35′; soluble in nitric and chlorohydric acids; electric when heated; the apex of the pyramid positive, the other extremity negative; lustre vitreous; scratches glass feebly. B.B. becomes opaque, and twists up like a worm (σκωληξ), and then becomes a shining slag, and then a vesicular bead. SiO_3 46·47, Al_2O_3 25·35, CaO 14·03, NaO ·43, HO 13·63. *Form.*—CaO SiO_3, Al_2O_3 SiO_3 3HO. Iceland, Faroe, Staffa, Kilpatrick Hills.

Scolezite Anhydrous. H 6·. White crystals in scapolite, Ersby, Finland; translucent; fracture small conchoidal; lustre vitreous; difficultly fusible. B.B. behaves like scolezite. SiO_3 54·13, Al_2O_3 29·23, CaO 15·46, HO 1·07.

Scolopsite. Spec. grav. 2·53, H 5. Smoke-gray or pale reddish-white granular pieces. B.B. fuses with frothing to a greenish-white glass; decomposed by chlorohydric acid. SiO_3 44·06, Al_2O_3 17·86, Fe_2O_3 2·49, MnO ·86, CaO 15·48, MgO 2·23, NaO 11·54, KO 1·3, SO_3 4·09, trace, NaCl ·93. Kaiserthal in Breisgau.

Scoparine. $C_{21}H_{11}O_{10}$. Pale yellow amorphous mass, or small starlike crystals, without taste and smell; action neutral; little soluble in cold water, more soluble in cold alcohol; soluble in caustic and carbonated alkalies with a greenish-yellow colour; precipitated by chlorohydric and acetic acids; obtained in crystals by spontaneous evaporation from the alcoholic solution; decomposed into a resin by boiling with alkalies and acids; unsublimable; converted into picric acid by nitric acid; obtained from the aqueous decoction of broom (Cytisus scoparius) by dissolving in water and evaporating in the water bath, when chlorophyle remains undissolved.

Scoria. A slag of semi-vitrified or crystalline dross.

Scorilite. Spec. grav. 1·708, H 2·. Reddish-brown, scoriaceous-looking matter; internally dark brown, streak white; feels harsh like slag, vesicular like a cinder. B.B. infusible, but becomes whiter; fuses with soda into a green glass. SiO_3 58·02, Al_2O_3 16·78, FeO 13·328, CaO 8·62, HO 2·. Mexico.

Scorodite. See ARSENIATE OF IRON.

Scorza. A variety of Epidote. Spec. grav. 3·289. SiO_3 38·6, Al_2O_3 26·15, CaO 23·84, FeO 10·5, HO 1·3

Scotine Diagonal. See BUCKLANDITE.

Scoulerite. Mineral allied to Thomsonite, having the same composition, but an amorphous brown aspect. Ballimony, Ireland.

Sea-Foam. See MEERSCHAUM.

Sealing-Wax. Red sealing-wax is made by melting together 48 shell lac, 19 Venice turpentine, 1 balsam of Peru, 32 cinnabar. Black wax consists of 60 shell lac, 10 turpentine, 30 levigated ivory black. Yellow wax is coloured with chrome yellow, blue wax with cobalt blue, green wax with oxide of copper

Sea-Salt. Chloride of Sodium.

Sebacic Acid. $C_{20}H_{18}O_8$ = C 59·25, H 9·07. Sp. grav. when fused, 1·1317; F.P. 276°6. Crystals resembling those of benzoic acid; very little soluble in cold, very soluble in hot water; very soluble in alcohol, ether and fat bodies; sublimes unaltered; with chlorine it affords two bodies, C_{20},H_{17}Cl,O_8 and $C_{20}H_{16}Cl_2$,O_8; soluble in hot water; with nitric acid it yields succinic acid; distilled with pyroxilic spirit and sulphuric acid sebacic acid forms, dimethyle sebacic acid, $2C_2H_3O$, $C_{20}H_{16}O_6$, colourless needles, boiling at 545°, fusing at 78°. Sebamide and sebamic acid are formed by the action of ammonia on this methyle compound. Sebacic acid is formed by distilling oleine or oleic acid, and crystallizing the distillate out of water; it forms a precipitate with acetate of lead. The salts are termed sebates, and are unibasic. Sebate of ethyle has the odour of melons; becomes solid at 16°, and volatilizes above 212°.

Sebamic Acid. *Stearamic Acid.* $C_{20}H_{19}NO_6$. Granular body by adding HCl to the liquor in which sebamide has formed.

Sebamide. *Stearamide.* $C_{20}H_{20}N_2O_4$. Granular substance by digesting ammonia with stearic ether.

Secaline. See ERGOTINE.

Sedative Salt. Boracic acid.

Seidlitz Powder. An effervescing powder sold in two separate parcels, one in white papers contains 2 scruples bicarbonate of soda, 2 scruples Rochelle salt; the blue paper contains 35 grains tartaric acid. When dissolved in water they effervesce.

Seignette's Salt. *Rochelle Salt.* Potash tartrate of soda.

Selbite. A synonyme of Carbonate of Silver.

Selenaldine. An organic base, forming large crystals; insoluble in water; soluble in alcohol and ether; forms crystalline salts; obtained by passing seleniohydric acid through a solution of aldehyde-ammonia in ammonia.

Selenide of Lead and Copper. Spec. grav. 5·6 to 7·, H 2·25. Pb Se, Cu Se. Lead-coloured powder or mass. B.B. fusible, and like PbS.

Selenide of Sulphur. Orange masses like sulphur. Lipari islands, and Hawaii.

Seleniohydric Acid. HSe. Colourless gas, resembling in character and smell sulphohydric gas, precipitating metals as black selenides; obtained by acting on selenide of potassium or iron with chlorohydric acid and water.

Selenio-Mercaptane. C_4H_5Se,HSe. Colourless fœtid fluid heavier than water, burning with a blue flame, unites with HgO with a hissing noise; obtained by distilling seleniohydride of selenide of potassium (KSe, HSe, formed by passing HSe through potash) with sulphethylate of lime.

Seleniopalladite. Native palladium.

Seleniosulphide of Mercury. H 2·5. HgSe,4HgS. Lead-coloured masses, subliming by heat. San Onofre, Mexico.

Selenite. A compound of selenious acid with a base.

Selenite. Sulphate of lime.

Selenium. Se 5, 40; 4·95, 39·6. Spec. grav. 4·3. Black or brownish-black masses, or 4-sided prisms; lustre metallic; powder reddish; without taste and smell; easily fusible about 212°, and volatile, with the smell of horse-radish; non-conducter of electricity; closely resembling sulphur in its characters and affinities; obtained from some specimens of pyrites in Hungary and Germany; but has been obtained in largest quantity from a reddish precipitate in the sulphuric acid chambers near Stockholm. When dry chlorine is passed over this deposit, on the application of heat, chlorides of selenium and sulphur distil into water in the receiver; the filtered liquid, treated with sulphite of potash, deposits selenium. *Selenides* are obtained by heating together the metals and selenium.

Oxide of Selenium.—SeO. Colourless gas, with the smell of horse-radish, obtained by heating selenium in air or oxygen.

Selenious Acid.—SeO_2. White 4-sided prisms, volatilizing at the boiling point of sulphuric acid, and yielding a yellowish vapour; soluble in cold and hot water and alcohol; formed by dissolving selenium in nitro-chlorohydric acid, distilling the liquid off, and subliming in a retort. The selenites are decomposed by heat, and when heated on charcoal yield the odour of horse radish, by which they are distinguished from sulphates. The equisalts have an alkaline, the bisalts an acid reaction, isomorphous with sulphurous acid. Selenious acid is reduced to selenium when heated with SO_2.

¶ *Selenic Acid.*—SeO_3. Sp. grav. 2·61. Clear fluid, decomposed by heat into selenious acid and oxygen, evolves heat when mixed with water; dissolves zinc and iron with evolution of hydrogen; forms seleniates with bases which are isomorphous with sulphates, chromates, and manganates; obtained by acting on selenium with chlorine in water. *Seleniate of potash* is obtained by fusing selenium with potash nitre; it has the same form of crystal with sulphate of potash. *Seleniate of soda* is formed by fusing selenium with soda nitre; it is anhydrous; seleniates of barytes, strontian, lime, and lead, are insoluble. *Test.*—When selenic acid is boiled with HCl, the odour of Cl is evolved; the solution is not precipitated by SH. It gives a white precipitate with BaCl, insoluble in NO_5.

Semelin. A synonyme of Sphene.

Semen. *Spermatine.* The secretion of males. Spec. grav. heavier than water. Transparent substance like mucus; at first insoluble in water, but afterwards spontaneously fluid; in contact with water, the water becomes precipitable by gall-nuts. It contains animalculæ or spermatozoa; it is alkaline. Human semen consists of water 90, spermatine 6, phosphate of lime 3, soda 1.

Semi-Naphthalidine. $C_{10}H_5N$. Reddish-yellow prisms; insoluble in water, alcohol, and ether; by the action of ammonia and SH on nitronaphthalese.

Semiopal. Opal with less opalescence than precious opal.

Semolina. A form of wheat in which the grains are broken down.

Senegine. A synonyme of Polygalic acid.

Senna Leaves contain *Cathartine.*

Septaria. Nodules of limestone mixed with clay, forming hydraulic mortar.

Serbiane. A synonyme of Miloschine.

Sericic Acid. A synonyme of Myristic acid.

Sericite. Spec. grav. 2·897, H 1·. Greenish-white talc in the Taunus slate, consisting of SiO_3 51·83, Al_2O_3 22·21, FeO 7·5, MgO 1·38, KO 9·1, NaO 1·74, HO 5·56, PO_5 ·31, SiF 1·88.

Seroline. F.P. 97°. Minute white filaments, with pearly lustre, becoming red with sulphuric acid; soluble in ether; insoluble in alcohol of ·842, potash except after long digestion, and nitric acid; when distilled yields ammoniacal vapour, and is partly volatilized; forms no emulsion with cold water; deposited from a hot alcoholic decoction of dry serum on cooling; supposed to be a mixture.

Serpentine, Precious. *Hydrous Sesquisilicate of Magnesia, Picrolite, Ophites.* Spec. grav. 2·5 to 2·591, H 3·5, and 2·25. Leek-green, mountain-green, or dirty straw-yellow masses, or oblique 4-sided prisms; lustre resinous, inclining to pearly; translucent on the edges. B.B. infusible, but becomes brownish-red; fuses with borax into a green glass; with salt of phosphorus into a glass tinged with iron; with soda into a yellowish-brown enamel. SiO_3 43·07, MgO 40·37, CaO ·5, Al_2O_3 ·25, FeO 1·17, HO and CO_2 12·45. Sesquioxide of chromium and nickel are often associated with serpentine. *Form.* $9MgO\,4SiO_3\,6HO$? Cornwall, Ayrshire, New York, Pennsylvania. Serpentine is used for ornamental purposes from the high polish of which it is susceptible, and, being decomposible by acids, it may be used as a source of magnesia in manufactures.

Serum of Blood. Spec. grav. 1028. The amber-coloured fluid which floats over the coagulum of blood on standing. It is blood from which the fibrine and blood globules have separated: See BLOOD. It consists of water 900, albumen 86·8, chlorides of potassium and sodium 6·6, mucous extractive 4·, carbonate of soda 1·65, sulphate of potash ·35, earthy phosphates ·6 (Marcet), or water 901, albumen 81·2, extract 4·6, fats 3·4, alkaline chlorides, 5·52, alkaline, carbonate, phosphate, and sulphate 2, carbonates and phosphates of lime and magnesia ·87.

Sesquiphosphoric Acid. Phosphoric acid, which forms salts with the formula $6\,RO\,4PO_5$.

Severite. A variety of Halloylite.

Seybertite. A synonyme of Holmesite or Clintonite.

Shale. Slate containing organic matter.

Shea Butter. An oil brought from the coast of Africa, from a species of Bassia, fusing at 95° to 110°. It contains margarine, the margaric acid fusing at 142°. It is used in Africa for the purposes of butter, and when fresh is said by Mungo Park to be equal to the best English butter. (R.D.T.)

Shear Steel. Steel originally employed to make shears for dressing cloth.

Shell Lac. Lac resin fused and reduced to the state of a thin crust.

Shepardite. *Schreibersite.* Cr_2S_3?, H 4·. Brownish black prisms, in grains the size of a grain of rice, in South Carolina. B.B. emits SO_2, and fuses into a magnetic glass; with borax a yellow and green glass.

Siberite. A variety of tourmaline.

Siderite. *Siderose.* Rhombohedral carbonate of iron.

Sideritine. A synonyme of Pittizite, or Tetrarseniate of iron.

Siderochalcite. See APHANESE and OLIVENITE.

Sideroschisolite. See HYDROUS BISILICATE OF IRON and CHAMOISITE.

Sideroclepte. A synonyme of Peridote.

Sideroferrite. Metallic iron, found in a piece of fossil wood from an island in Smaland.

Siderum. Iron containing phosphorus (Bergman).

Sienite, or *Syenite.* A mixture of crystals of felspar, quartz, and hornblende.

Sienna Earth. (*Terra de Sienna*). Spec. grav. 2·1. An earth from Italy, consisting of hydrous sesquioxide of iron, with traces of arsenic, used after calcination as a pigment. In the *brown* state, at 212°, it consists of Fe_2O_3 66·, MnO 3·4, AsO_5 8·7, SiO_3 8·, HO 13·3; in the *yellow* state, Fe_2O_3 28·8, MnO 1·1, AsO_5 ·5, SiO_3 63·, HO 4·5.

Silica. *Silicic Acid, Silex.* Silicon 49·72, oxygen 50·28 (Rose); SiO 2·, 16 (Thomson); SiO_2 3·875, 31 (Gmelin); SiO_3 5·8125, 46·5 (Berzelius). Specific gravity 2·66 (Kirwan); 2·65 (Royer and Dumas, Ann. Phil. 2d, 3, 392); fine powder from minerals 2·2 (Schaffgotsch); of quartz 2·653 (ib.) Hardness 7·. White tasteless powder, feeling gritty between the teeth, soluble in water when formed by the combustion of silicon; insoluble after ignition. *Crystal;* in the form of rock crystal, 6-sided prisms, terminated by 6-sided pyramids, belonging to the rhombohedral system. When fused with several times its weight of carbonate of potash or soda, it assumes on cooling the appearance of glass, which dissolves in water. If we add chlorohydric acid to this solution to saturate the alkali and concentrate, a white jelly is produced, which, when evaporated, heated, and washed, yields pure silica. Silica fuses before the oxyhydrogen blowpipe into a clear bead, or in a powerful galvanic current, when it may be drawn into threads. It appears to volatilize with steam when exposed to a temperature above the melting point of cast iron. It fuses with carbonate of soda before the common blowpipe into a transparent colourless bead. Pure silica is soluble in a boiling solution of carbonate of soda or potash. It is hence capable of replacing carbonic acid. *Source.*—Silica occurs in crystallized and amorphous forms; 1 Crystallized, Rock crystal. Sp. grav. 2·641 to 2·69, hardness 7. Primary form a rhombohedron; usual form a 6-sided prism, terminated by a 6-sided pyramid; occurs generally in granite mountains, as in Dauphiné, Tyrol, Savoy, Brazil, from whence the crystals used for spectacles are imported. When of different colours, rock crystal receives different designations. When *violet,* it is called amethyst; *red,* rose quartz; *green,* from the presence of some epidote prase; *yellowish-green,* cantalite; *yellow,* Indian topaz, or cairngorum stone; *blue,* siderite; *smoke-gray, brown* and *black,* or *blackish-brown.* 2. *Amorphous,* or semi-crystalline. 1. *Quartz.* A white, nearly pure form of silica, crystalline. 2. *Calcedony.* A stalactitic form of silica, apparently deposited from solution in water, and occurs most usually in the cavities of amygdaloidal rocks. 3. *Carnelian* occurs of red and white colours, principally in the trap rocks of India. 4. *Onyx* is formed of alternate layers of brown and white calcedony. 5. *Heliotrope*, or *Blood stone*, is green, with blood-red spots or streaks; also abundant in India. 6. *Chrysoprase*, of an apple-green colour, due to the presence of oxide of nickel. 7. *Mocha stone*, is calcedony containing dendrites, or semblance of green mosses. 8. *Agate*, composed of alternate layers of calcedony, quartz, jasper, heliotrope, or opal. 9. *Flint*. An impure form of silica, occurring in nodules in chalk. 10. *Opal*, *wood opal*, rendered somewhat opaque by a different colour, as it were floating in the stone. Precious opal affords a fine play of colours. 11. *Jasper*. A red-coloured form of silica. 12. *Basanite*, *Lydian stone*, or *flinty slate*. A dark slate, forming the basis of basalt. 13. *Hyalite*, with a glassy lustre, containing 6 per cent. of water.

1. *Hydrate*.—$HOSiO_3$. Contains 16½ per cent. of water, obtained by precipitating at the usual temperature an alkaline silicate by an acid (Doveri). 2. $HO\ SiO_3$, formed by decomposing fluoride of silicon by water. 3. Ebelmen has obtained the terhydrate, $3\ HO\ SiO_3$, in hard and transparent masses like rock crystal, by acting on silicic ether by moist air. Spec. grav. 1·77.

Theory of Siliceous Deposits.—When we treat certain silicates with acid, the silica separates in union with water as a jelly, as in Thomsonite. In some, the silica separates as a powder as in stilbite, mesotype. In others, again, as in Labradorite, it deposits partly as a jelly, and partly as a powder; and hence we may infer that it consists of 2 minerals. A silicate of soda, soluble in water from the presence of a great excess of base, will deposit in the course of time a jelly; the latter, when washed with water, containing only traces of alkali. This throws light on the formation of some forms of silica, as carnelian and amethyst, in which traces of potash and soda have been found. In hornstone there have also been observed alkalies; and many of the forms of silica lose water by heat. Some species of opal yield ammonia when ignited, and banded opal is full of infusoria. Common quartz, when ignited and reduced to powder, is scarcely soluble in caustic potash, but powder of opal dissolves rapidly. From this fact it may be shown that chalcedony is a mixture of quartz and opal, as only a portion of it dissolves in caustic alkali. From the solubility of alkaline silicates, we must explain the circumstance that we frequently find silica in pseudomorphous crystals, that is, the silica, in gradually depositing from its solutions on other crystals, seems to have displaced the original matter and assumed its form. Thus quartz occurs in the form of rhomboids among calc spar crystals. On the other hand, brown spar occurs in the form of quartz, that is, rhomboids as 6-sided prisms. For *Analysis of Silicates*, see ANALYSIS.

Siliceous Bismuth. See SILICATE OF BISMUTH.

Siliceous Malachite. See SUBSESQUISILICATE OF COPPER.

Siliceous Manganese. See SILICATE OF MANGANESE.

Siliceous Sinter. Siliceous deposits from hot springs, of which geyserite and hyalite are species.

Siliceous Zinc. See SILICATE OF ZINC.

Silicide. *Siliciuret*, *Silicet*. A combination of silicon with a base.

Silicified Wood. The replacement of woody fibre by the deposition of silica from its solutions. It occurs in the West Indies, Birmah, Van Diemen's Land, Kerguelen's Land, Loch Neagh, Ireland, &c.

Silicite. Spec. grav. 2·666. SiO_3 54·8, Al_2O_3 28·4, CaO 12·4, FeO 4·, HO ·64. Colour white, with a shade of yellow; texture foliated; fracture small, conchoidal; lustre vitreous, resembling quartz or phenakite; hardness nearly the same as that of quartz. B.B. with soda fuses into an opaque bead; with borax into a transparent colourless bead. Coating a basaltic rock in Antrim.

Silicon. *Silicium*. Si. Sp. grav. above 1·837 after ignition. Deep brown powder, rather darker than *boron*; destitute of lustre, even when subjected to friction; stains the fingers, and adheres to glass and everything with which it comes in contact; infusible by heat, even of the blowpipe, but it becomes harder by ignition, and its properties are altered; before being strongly heated it is readily combustible in the air, and burns with a lively flame; one-third is converted into silica, while the rest is prevented from oxidizing by a crust of silica. *When ignited* it is heavier than oil of vitriol; neither burns in air nor in oxygen gas; not altered by the blowpipe, even in contact with chlorate of potash; when ignited with nitre it does not burn; fluohydric acid, or caustic potash, do not alter it even when boiling, but a mixture of fluohydric and nitric acids dissolves it easily, binoxide of nitrogen being evolved. The conditions before and after ignition have been considered *allotropic* states of silicon; but the difference may be due to the presence of hydrogen in the unignited substance, which is expelled by the heat. *Preparation*.—1. When potassium is heated in a hard glass tube, and silicofluoric (SiF_3) acid prepared from fluorspar, glass and sulphuric acid is passed over it, the metal becomes encrusted with a black layer, which at last breaks, while the potassium takes fire and burns with a red flame. When the brown scoriaceous residue is thrown into water, hydrogen is evolved, and a dark brown matter falls, insoluble in water; it is silicon mixed with fluosilicate of potash, which may be washed away, though very insoluble, by the long-continued action of hot water. The silicon is freed from the hydrogen by heating it in a covered platinum crucible just below redness. 2. Mix dry silicofluoride of potassium (heated above 212°) with three-fourths of its weight of potassium in a glass or iron tube, and heat till the potassium melts; stir the mixture intimately with an iron wire; heat over a spirit-lamp, when the potassium becomes red hot with an audible sound, and reduction occurs. The liver-brown mass (KSi,KF) is thrown into much water, when hydrogen is evolved, potash is formed, and silicon set free. If too little water is used, the caustic potash dissolves up the silicon formed. The undecomposed silicofluoride of potassium must be removed by cold water, and then by boiling with water. The silicon is then collected and dried on a filter. 3. The chloride of silicon may be decomposed by potassium.

Silicon, Chloride of. $SiCl_3$, or $SiCl = Si$ 16·71, Cl 83·29. Spec. grav. 1·53, of vapour 5·9; B.P. 138¼°. Colourless fluid, smoking in the air, and decomposing into chlorohydric and silicic acids; obtained by passing chlorine over an ignited mixture of lamp black and silica. *Silicon Bromide.*—$SiBr_3$. B.P. 300°. Colourless fluid, fuming in the air.

Silicon, Fluoride of. *Fluosilicic Acid, Silicated Fluoric Acid.* SiF_3. Specific gravity 3·57. This gas was first formed by Scheele, and was more particularly examined by Priestley. It can be readily prepared by mixing pounded glass, fluor spar, and sulphuric acid together in a retort or flask with bent tube, applying heat, and collecting the gas over mercury. The action in this process is as follows:—$3\ CaF,\ SiO_3\ 3SO_3 = SF_3,\ 3(CaOSO_3)$. When passed into water with a

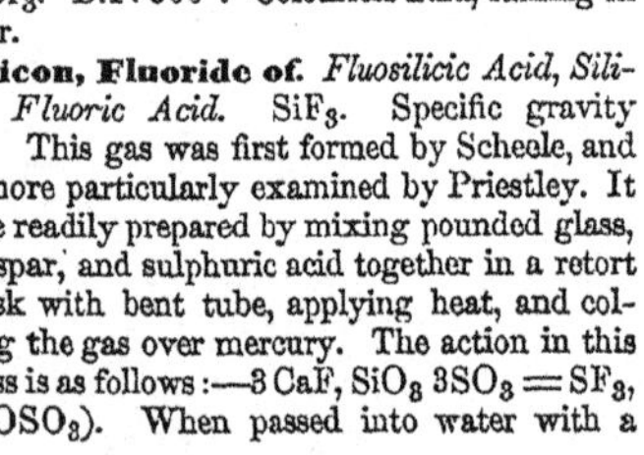

globule of mercury at the bend of the exit tube, to prevent it from being choked up with silica, the silica is deposited in a gelatinous form, and affords this substance in a very pure state. Fluosilicic acid gas is colourless, smokes in the air like hydrochloric acid gas, reddens litmus paper, and is not combustible. Its composition is as follows:—

	Theory.					Davy.
Silicon,	1	1	2·5	28 34	Silicon,	61·4
Fluorine,	3	6·75	7·125	71·66	Fluoric acid,	38 6
		7·75	9·625			

It is not altered by exposure to a red heat; potassium heated in it burns, and KF is formed with a disengagement of silicon. Fluosilicic acid unites with twice its volume of ammonia, forming a volatile salt. The water into which this gas is passed contains a distinct acid, which has been termed *Hydrofluosilicic acid.* It unites with bases, forming a series of salts.

Applications.—Fluosilicic acid is the substance which is produced in etching glass. A mixture of fluor spar and sulphuric acid is heated in a vessel incapable of being corroded, such as lead or platinum, and over this is placed a plate of glass covered with wax, except on the points intended to be etched; the hydrofluoric acid volatilizes by the heat, comes in contact with the silica of the glass, and forms fluosilicic acid, $3\ HF,\ SiO_3 = SiF_3,\ 3\ HO$. By this means we are enabled to detect fluor spar. It affords us also a method of determining the amount of alkalies in a mineral by first volatilizing the silica with which the alkalies had been in combination, and thus obtaining soluble salts by means of a mineral acid. A certain quantity of the pounded mineral is mixed with fluor spar equal to two and a-half times the amount of the silica contained in the mineral; the mixture is placed in a platinum crucible and heated with sulphuric acid, which is added in such quantity as to form a paste. So soon as the mixture ceases to give off white fumes—indicative of the expulsion of the silica—it is to be heated to redness, digested in water, and filtered; the filtered liquor is to be precipitated with oxalate of ammonia, strongly heated, to decompose all the sulphate of lime in solution, and mixed also with some caustic ammonia to precipitate alumina; or a good method is, after dissolving out the alkalies, to precipitate the sulphuric acid with chloride of barium or caustic barytes, to throw down the excess of lime and barytes by oxalate or carbonate of ammonia, and to evaporate the solution to dryness. If alumina and magnesia remain, they are separated by dissolving out the soluble salts in water and deducting the amount of the residue.

Silk. The product of the caterpillars, Phalena bombyx, and P. atlas. Silk exists in the body of the worm in a fluid state, and as it exudes hardens into a fibre, when it is termed *raw silk.* Hot water extracts from yellow silk 24·43 per cent. of albumen or caseine, and 20·66 gelatine; hot alcohol takes up, and deposits on cooling, 1·39 cerine. On evaporating the alcoholic solution ·05 colouring matter, and ·1 fatty matter and resin remains. The substance unacted on by water and alcohol is fibroine.

Sillimanite. Spec. grav. 3·238, H 6·. Dark gray long 4-sided prisms, with angles of 88° and 92°?; structure fibrous; lustre vitreous; brittle; easily frangible; translucent on the edges. B.B. infusible. SiO_3 38·13, 37·8, Al_2O_3 58·86, 59·53, Fe_2O_3 2·75, 3· (R.D.T.) Allied to, if not identical with, cyanite, fibrolite, and bucholzite (T. Thomson). Saybrook, Connecticut.

Silver. *Argentum, Luna, Diana.* (*Argent*, Fr.; *silber*, Ger.) Ag 13·5, 108·. Spec. grav. 10·481 (Fahrenheit), 10·474 (melted, Brisson, 10·51 fused), 10·5 (hammered, Muschenbroeck), 10·4812 (plates, Thomson). F.P. 1893° (Morveau), 2233° (Daniell), 1830° (Prinsep), 1832° (Pouillet). White, with a shade of yellow, cubes softer than copper, harder than gold, remarkably ductile; a single grain may be drawn out as a wire to 400 feet; a wire ·078 inch diameter supports a weight of 187·13 lbs.; becomes very brilliant when heated; boils and vaporizes above its fusing point (Claveus, Vauquelin); cooled slowly, its surface presents a crystalline appearance. *Reduction and separation.*—Silver occurs in nature in the metallic state mixed with copper, likewise as sulphide in sulphide of lead or galena, and in various other states. Silver is separated from its ores by amalgamation. This process, as practised in different countries, is described under AMALGAMATION, and the method of separating silver from lead is detailed under LEAD. *Assay of Silver.*—When silver is contained in an ore, the powdered mineral may be dissolved in strong or dilute pure nitric acid, and the silver thrown down with chlorohydric acid, and estimated as chloride of silver. In the dry assay the ore is fused with lead in a crucible, the silver is taken up by the lead, which is then placed in a cupel, ignited with a current of air directed over it; the lead is oxidized and sinks into the bone earth cupel, while the pure silver remains, and may then be weighed. For assaying and cupelling the portable furnace exhibited in the accompanying figure will be found very advantageous, especially near mines. The upright portion contains a fan, by which the fire can be urged up to a white heat. It is manufactured by Chaplin and Dixon of Glasgow. To purify silver from other metals, it may be dissolved in nitric acid precipitated by common salt, and the moist chloride decomposed by strong caustic potash solution and resin into the oxide which may be employed for making the salts, or the dry chloride may be fused with carbonate of soda and resin, and thus obtained in the metallic state; or 3 chloride and 1 resin may be heated in a crucible, gradually raised to the fusing point of silver, and a small portion of borax added; silver is speedily reduced (Mohr).

Dinoxide.—Ag_2O. Gray powder from ammoniacal solutions of oxide of silver on exposure to light, and by heating oxalate and citrate of silver.

Oxide.—AgO 14·5, 116. Grayish powder, obtained by precipitating nitrate of silver by caustic soda. Spec. grav. 7·143. It combines with ammonia, forming an explosive compound, and is dissolved by an excess. The oxide does not form a hydrate. When the salts of this oxide in solution are mixed with mercury in a glass vessel, the silver is deposited in the form of a tree, called the *Tree of Diana* (Arbor Dianæ). The experiment may be made by dissolving 15 grains of nitrate of silver in half a wine glassful of water, adding a few globules of mercury, and allowing the glass to stand at rest for some hours.

The *permanent ink* commonly sold for marking linen, contains in one bottle caustic soda, or carbonate of soda, and in the other a solution of nitrate of silver. The spot to be marked is first moistened with soda, dried, and then written on with a pen dipped in the caustic solution. The oxide is precipitated. The salts of the oxide of silver are colourless, have a metallic taste, and are poisonous, the antidote being common salt. Copper, zinc, and mercury precipitate silver from its solutions. Ammonia redissolves the oxide when precipitated.

Chloride of Silver, Horn Silver, Luna Cornea.—AgCl 17·937, 143·496. Obtained by precipitating a solution of lunar caustic ($AgONO_5$) with HCl, or common salt. It is best to precipitate silver by common salt, as the chloride is soluble in HCl. A white curdy precipitate; when moist insoluble in nitric acid; when dry, becoming purple and black. Melts at 500°, and forms a horny mass. Soluble in ammonia. Crystallizes in 8-hedrons. When moist, it may be converted into oxide by boiling with caustic soda, and the addition of a little pounded rosin. The silver may be reduced by boiling the chloride in an iron pot with water, Spec. grav. 5·55. *Iodide of Silver.*—Ag I, 29·25, 234. Yellow powder, formed by dropping iodide of potassium into nitrate of silver. *Cyanide.*—Ag Cy, 16·75. White curdy precipitate, resembling the chloride, but distinguished from it by its solubility when boiled with nitric acid. This is used for procuring hydrocyanic acid in medicine. *Phosphate of Silver.*—3 AgO PO_5. Yellow powder obtained by precipitating nitrate of silver by the common phosphate of soda. *Diphosphate* or *Pyrophosphate.*—2 AgO PO_5. White powder from the mixture of ignited phosphate of soda, and a soluble salt of silver.

Nitrate of Silver, Lunar Caustic, Lapis Infernalis.—AgO NO_5, 21·25, 170. Plates, obtained by dissolving pure silver in nitric acid, and evaporating cautiously to drive off excess of acid. Used as a reagent and caustic. For the latter purpose it is fused and poured into tubes, where it solidifies. When free from organic matter it retains its colour, but if it has touched paper it becomes dark coloured.

Alloys of Silver.—*English* silver coin contains copper; the quantity in a shilling which weighs 87·2727 grs. is 6·5455 grs. of alloy, or in 12 ounces of standard silver the copper amounts to

18 dwt. or 1 copper to 13½ silver, 1 lb. Troy yielding 66 shillings. *French* silver coin consists of 1 copper, 9 silver. *Dutch*, 1 copper to 11½ silver. *Russian*, 1 to 3½. *Hamburg*, equal parts. Silver is always *estimated* in the form of *chloride* of silver, every 18 parts of which contain nearly 13½ of metallic silver.

Silver Ores.

Native Silver.—Spec. grav. 10·338, H 4·25. White masses in strings and plates, cubes, and regular 8-hedrons, subject to tarnish, streak unaltered, shining, ductile, opaque. The composition of native silver is, silver 90·, 72, 36, copper 10, or gold 28 to 54 per cent. In primary rocks in Mexico and Peru, Cornwall.

Silver, Chloride of, Horn Silver.—Ag Cl. Spec. grav. 5·552, H 3·5. Pearl-gray masses, or cubes and needles, with shades of yellow and green, becoming brown in light; lustre resinous; translucent on the edges; fuses in the flame of a candle. B.B. almost entirely reduced on charcoal. It is also reduced when rubbed wet on zinc or iron; soluble in NH_3; insoluble in NO_5. Ag 76 to 67·75, Cl 24· to 27·32, FeO 6·, SO_3 ·5, Al_2O_3 1·75. Freiberg; Joachimsthal, Bohemia.

Silver, Iodide of.—Spec. grav. 5·504. Ag I? Yellow, with a shade of green, resinous flexible plates; translucent; in veins in steatite, in Mexico.

Silver, Bromide of.—Ag Br. Spec. grav. 5·8 to 6, H 1 to 2. Green externally, yellow internally, in concretions, seldom crystallized. Ag = 57·56, Br 42·44. B.B. melts easily; soluble in NH_3; mostly dissolved by NO_5. Mexico, Chili.

Silver, Tetra-antimonide of, Antimonial Silver Ore.—Ag_4 Sb. Spec. grav. 9·4406 to 9·8, H 3·5. Silver and tin white 6-sided prisms, with convex faces, longitudinally streaked, and amorphous; structure foliated, lustre metallic, opaque. B.B. converted into a gray smoke and a brown slag, tinging borax green, on charcoal leaves a globule of silver. Antimony 23, silver 77·. Andreasberg, Hartz; Swabia; Casalla, Spain.

Silver, Chlorobromide. See Embolite.

Silver, Arsenide of, Arsenical Silver Ore.—Fe 44·25, As 35, Ag 12·75, Sb 4·. White or blackish globular and reniform masses, with a curved lamellar structure; lustre metallic; opaque, sectile; harder than antimonide of silver. B.B. gives off arsenical fumes, leaving a globule of impure silver, surrounded by slag. Andreasberg, Hartz. Little known.

Silver Bitelluride.—Spec. grav. 8·412 to 8·465, H 2·25. Lead or steel-gray masses; structure coarse, granular; lustre metallic, splendent; malleable. B.B. melts to a black globule, with white points of silver; melts in a tube, and colours the glass yellow; in an open tube it fuses, and gives some sublimate; with salt of phosphorus forms a bead, transparent while hot, opal on cooling; a bead of pure silver with soda. Ag 62·37, Te 36·92, Fe ·37. Siberia.

Silver, Ferrosulphide of, Flexible Sulphuret of Silver.—Black tabular crystals, of right oblique angled prisms, in thin laminæ, yielding to the knife. Himmelsfürst, Saxony. See also Sternbergite.

Silver, Sulphide of, Silver Glance, Vitreous Silver.—*Glaserz*, AgS. Spec. grav. 7·196, H 2·75. Blackish lead-gray masses; cubes, regular 8-hedrons and 12-hedrons, cleaving parallel to the faces of a cube; fracture imperfect, small conchoidal; malleable, lustre metallic; opaque. B.B. fuses easily, and swells up, giving a silver globule by continuing the heat, Ag 85, S 15. Saxony and Bohemia, South America, &c.

Silver Selenide.—Spec. grav. 8, H 2·5. Black masses cleaving parallel to the faces of a cube; lustre metallic, splendent; opaque; malleable; structure foliated; streak unaltered. B.B. fuses; glows on cooling like magnetic pyrites; with soda on charcoal reduced; soluble in fuming NO_5. Ag = 65·56, Pb 4·91, Se 25·93. Hartz.

Silver, Carbonate of. Selbite.—AgO CO_2? Grayish-black masses; streak bright; lustre metallic; opaque; soft; brittle; heavy; effervesces with acids; froths when heated with borax. Ag 72·5, SbO_3 15·5, CO_2 12. Black Forest, in granite; Suabia.

Silver, Sulpho-Cupride of.—Spec. grav. 6·258. AgS $2Cu_2S$. Blackish lead-gray masses; lustre metallic, splendent; opaque; fracture flat conchoidal; brittle or sectile (?) B.B. fuses readily. S 15·782, Ag 52·272, Cu 3·478, Fe ·333. Siberia.

Silver, Antimonio-Sulphide of. Trisulphoantimoniate of Silver, Prismatic Melane Glance, Brittle Silver Glance.—Spec. grav. 6·269, H 2·25. Iron-black 6-sided prisms, terminated by faces perpendicular to the axis or 4-sided oblique prisms, having the acute edges replaced by faces, which render the prism 6-sided; fracture conchoidal, uneven; lustre metallic; opaque; sectile; streak unaltered. B.B. on charcoal gives a dark metallic bead, reducible by soda or nitre; soluble in dilute nitric acid. S 16·42, Sb 14·68, Ag 68·54, Cu ·64. Schemnitz and Freiberg.

Silver, Antimony, Sulpho-tersulphide of. See Miargyrite.

Silver Ore, Dark Red. Ruby Silver, Black Silver, Ærosite, Rothgültigerz, Subsesquisulphoantimoniate of Silver.—3AgS + SbS_3. Sp. grav. 5·8 to 5·9, H 2·25. Lead-gray, iron-black, and dark red obtuse rhomboids, with angles of 108° 30′, or 109° 56′; in 6-sided prisms, and with a 6-sided prism interposed between the two halves of the rhomboid, which constitute a triangular pyramid at either extremity; lustre metallic, adamantine; translucent to opaque; streak cochineal and aurora-red. B.B. decrepitates on charcoal, melts and emits fumes of sulphur and antimony, a globule of silver remaining. Ag 58·949, Sb 22·846, S 16·609, gangue ·299. Saxony, Bohemia, Mexico.

Silver Ore, Light Red. Subsesqui-Sulpho-arseniate of Silver, Sulpho Argento-tersulphide of Arsenic.—Spec. grav. 5·552, H 2·25. 3 AgS + As S_3. Cochineal-red rhombohedrons; streak light cochineal colour; lustre adamantine; translucent on the edges or even semitransparent; sectile. S 19·51, Sb ·69, As 15·09, Ag 64·67. The composition and crystalline form are the same as the dark red ore, with the difference that arsenic replaces antimony: Saxony and Bohemian silver mines. This, and the preceding, when boiled with sulphide of ammonium, leave sulphide of silver, while the tersulphide of arsenic is dissolved up.

Silver Ore of Bismuth.—White needles or hairs, often amorphous, opaque, tarnishing; sectile; soft. Pb 33·, Bi 27·, Ag 15·, Fe 4·3, Cu ·9, S 16·3; soluble in NO_5. B.B. fuses to a button, evolving sulphur.

Silver, Cupro-Sulphide of Antimony, and *Arsenic.* See EUKAIRITE and POLYBASITE.

Silver, Sulpho Quintosulphide of Antimony and. See XANTHOCONE.

Similor. An alloy consisting of equal parts of copper and zinc, resembling gold in colour.

Simplesite. See SCORODITE.

Sinapic Acid. $C_{22}H_{12}O_{10}$. Prisms more soluble in hot water and alcohol than when cold, forms bibasic salts; obtained by dissolving pure sinapine in potash, boiling, adding HCl, and purifying the precipitate by alcohol.

Sinapine. $C_{32}H_{26}NO_{12}$. An alkaloid, obtained in combination with sulphuric acid, by adding that acid to sulphocyanide of sinapine which is obtained from mustard powder by expressing the oil, treating the residue with alcohol, when the sulphocyanide crystallizes out in the form of crystals of quinine, having the composition $C_{34}H_{25}N_2S_2O_{10}$.

Sinapisine. A neutral fatty substance in mica-like scales, from the evaporated alcoholic tincture of black mustard; soluble in alcohol, ether, and oils; insoluble in acids and alkalies; volatile.

Sinapoline. $C_{14}H_{12}N_2O_2$. F.P. 212°. Shining plates, soluble in hot water, alcohol, and ether; formed along with sinnamine by the action of moist oxide of lead on thiosinnamine; apparently derived from 2 atoms of oil of mustard, and 6 atoms water = 1 sinapoline, 2 CO_2, and 4 SH.

Sincaline. $C_{10}H_{14}NO_2$,HO. Deliquescent colourless crystalline mass, attracting carbonic acid, precipitates metallic oxides, alumina, and sesquioxide of chromium dissolving in an excess; disengages methylamine (?) when strongly heated; formed by heating sinapine with barytes water, and allowing the sinapate of barytes to deposit; the filtered solution is acidified by SO_3, and mixed with a solution of sulphates of iron and copper; the sulphocyanide of copper is filtered off; the filtrate mixed with barytes water to precipitate excess of copper and iron, and the excess of barytes removed by CO_2. On evaporating the filtrate, carbonate of sincaline remains, which is converted into chlorohydride, and the HCl then removed by oxide of silver.

Sin-ethylamine. $C_{12}H_{10}N_2$. F.P. 212°. Bitter needles; insoluble in water; soluble in alcohol and ether, with an alkaline reaction; partially soluble in HCl; obtained by treating thiosinethylamine with moist oxide of lead.

Sinnamine. $C_8H_6N_2$. Hard 4-sided prisms, losing 9·5 per cent. water at 212°; obtained by treating thiosinnamine with oxide of lead.

Sinopis. Red chalk or reddle.

Sipeerine. An alkaloid said to exist in beeru bark.

Siricum. A mixture of sinopis and sandyx.

Sismondine. Spec. grav. 3·565. Grayish-green variety of chlorite. SiO_3 24·1, Al_2O_3 43·2, FeO 23·8, HO 7·6, TiO_2 trace. St. Marcel.

Sisserskite. The species of iridosmine consisting of $IrOs_4$.

Size. Gelatine employed for mixing with paper to destroy its porosity, and render it impervious to ink. It is usually mixed with alum to prevent it from putrifying. The best size is prepared from skins. See GLUE.

Skate Liver Oil. Spec. grav. ·928. Yellow oil by boiling the liver of the Raja clavata and batis with water; soluble in boiling ether, from which it is deposited on cooling; like cod liver oil it contains iodine.

Slag. *Scoria.* A substance produced by the union of an impurity of an ore with a flux.

Slate. A rock which splits into parallel leaves or foliæ.

Slate Spar. A slaty form of calcareous spar, with a silvery lustre.

Sloanite. Spec. grav. 2·441, H 4·5. White pearly masses, cleaving in the direction of a rhombic prism of 105°. B.B. fuses to a white enamel. SiO_3 42·187, Al_2O_3 35, CaO 8·119, MgO 2·67, NaO ·25, KO ·03, HO 12·5. Tuscany.

Smalts. *Azure, Zaffre, Cobalt Glass, Cobalt Blue.* A fine light blue colour, used as a pigment. It is essentially a potash silicate of cobalt, and is prepared by calcining cobalt ore. When roasted it is called *safflor;* the roasted ore ground up with sand and washed is called *zaffre.* Smalts are made from the roasted ore by fusing 1 part with 2 parts silica or sand, and 1 part carbonate of potash. Some arsenious acid is added to peroxidize the iron. The mixture is placed in a crucible in a furnace for five to eight hours. The upper portion of the fused mass is a light blue; the lower portion is speiss, which contains nickel, and the intermediate stratum constitutes the smalt blue. The upper portion is removed, and the cobalt blue ladled out into water, passed between rollers and washed. It is principally made in Prussia and Saxony.

Smaltine. A synonyme of Arsenide of Cobalt.

Smaragdite. A variety of Augite or Hornblende.

Smaragdochalcite. A synonyme of Dioptase.

Smaragdus. A synonyme of Beryl.

Smectite. Allied to Halloylite, from Cilly, Styria. SiO_3 51·21, Al_2O_3 12·25, Fe_2O_3 2·07, MgO 4·89, CaO 2·13, HO 27·89.

Smelite. Spec. grav. 2·168, H 2·5. Grayish white masses with a blue tinge. SiO_3 50·, Al_2O_3 32·, Fe_2O_3 2·, NaO 2·1, HO 13. Hungary.

Smilacine. A crystalline body obtained by alcohol from Smilax China.

Smithsonite. A synonyme of Hydrous dicarbonate of Zinc.

Snow, Inflammable. Said to have fallen in Russia. C 61·5, H 7·, O 31·5.

Snow Red, of Idria. A red powder found in snow, consisting of SiO_3 36·75, Al_2O_3 11·75, $CaOCO_2$ 17·5, Fe 6·25, Ti 3·75, organic matter, 24·.

Soap. (*Savon*, Fr.; *seife*, Ger.) A compound of a fatty acid with a metallic oxide. From the remarks made under OILS, it will be easily understood that the formation of a soap, that is, of an alkaline salt of stearic acid, &c. consists simply in the displacement of the oxide of glyceryle, an oily base, by an alkaline one. Soap is first mentioned by Pliny, under the term *Sapo*. He states that the Gauls invented it, and that they employed it as a detergent for the hair. He says that it was a compound of wood ashes and tallow, and that there were two kinds of it, hard and soft (spissus et liquidus). The word soap is supposed to be derived from the old German word *sepe*, a term still used in Scotland. In the second chapter of Jeremiah, verse 22, we find the words: "Though thou wash thee with nitre (carbonate of soda), and take thee much *soap*." But it is considered that the term soap here is a mistranslation. We meet in commerce with two kinds of soap commonly, *hard* and *soft* soap, as they have been well termed by Pliny. Hard soaps have a basis of soda, and are more difficult of solution in water. *Soft* soaps have a basis of potash, and readily absorb moisture from the air, which keeps them soft. It has been satisfactorily ascertained that soft soaps may be converted into hard soaps by adding to their solution in water a quantity of glauber salt or common salt. As stearic acid combines with most bases, there exists an extensive series of soaps, such as copper, lime, and iron soaps, &c. But the only soluble soaps are potash, soda, and ammoniacal soaps. The two former require to be made at a boiling temperature, and the latter at a low heat, in consequence of its volatilizing. The hard or soda soaps are usually prepared with olive oil on the continent, and in this country with tallow and other oils. The soft or potash soaps are made principally from vegetable oils, such as linseed and oil of colza, &c. In the North of France the soft soaps are usually coloured green or black. These colours are imparted either by sulphoindigotic acid or by a mixture of protosulphate of iron, nutgalls, and logwood, or by sulphate of copper. The green and black soaps are principally used for removing grease from woollen stuffs. All oils are not saponified with equal facility. Olive and almond oil are most readily saponified, then follow tallow and butter. Fish oil is difficult to convert into soap, but nut oil and linseed oil are still more refractory, and give only pasty soaps. Soda and potash soaps are very soluble in alcohol. They are also soluble in water, but in different degrees. Stearate of soda, or hard soap for example, scarcely undergoes any considerable change when brought in contact with 10 times its weight of water. The soda soaps exhibit a peculiar action with common salt. The soda soap is penetrated, or even dissolved, by a strong solution of common salt. If we bring common soap of commerce in contact with a saturated solution of common salt at the usual temperature, it swims about without apparent change; but if we heat the solution, it divides into gelatinous flocks, which collect upon the liquid, and after cooling consolidate into a firm mass, out of which the salt solution flows like water from fat. If we take out the gelatinous flocks from the hot solution, they form into an opaque firm mass, which divides between the fingers into plates. If the solution be not saturated, a division of the water takes place; the soap taking up a certain quantity of it, and the flocks divide by boiling in the solution. But if the water contain only the 400th part of common salt, none of the soap dissolves. Vauquelin observed that when a drachm of soap was dissolved in about a pint of water, the solution filtered, and mixed with a solution of common salt dissolved in 25 parts of water, on the mixture being made, a coagulation and separation of a viscid matter took place. When the quantity of common salt is sufficient, the soap is entirely decomposed. The separated coagulum is fat, insoluble in water, and when heated it floats as an oil, and on cooling crystallizes. The water in this case becomes alkaline, and if the liquid is evaporated, the common salt is found to be mixed with carbonate of soda. Other salts have the same effect, as sulphate of soda, which turns it like white of egg; salammoniac, carbonate of potash, and caustic potash act in the same way, that is, they produce a separation of a soap which is completely insoluble in alkaline liquids, and what is remarkable is, that no trace of soap remains in solution. The manufacture of soap is divisable into several important operations. 1. The preparation of the caustic lye. 2. The boiling of the soap, 3. The *cleansing* of the soap, as it is technically called. 4. The division of the soap into large pieces. 1. The lye is prepared by mixing the artificial carbonate of soda, or the best soda-ash, with lime. For the best white soap, the maker employs 13 cwt. tallow, 4 cwt. soda-ash = 2 cwt. soda. For yellow soap, 9·75 cwt. of tallow, 3·25 cwt. rosin, 4 cwt. soda-ash. These mixtures are boiled in a cast iron boiler. Sometimes, especially with oils, the boiler resembles a funnel. These have brick walls, the bottom alone being formed of thick copper, which receives the action of the fire. When the soap is to be *mottled*, it is mixed with sulphate of iron and an alumino-ferruginous soap. During the boiling, the sulphate of iron is decomposed, the soda precipitating a bluish-green hydrate composed of protoxide and peroxide, separating from the sulphuric acid. Then the sulphide of sodium, which the soda of commerce always contains, forms a black sulphide of iron, which retains the bluish tint of the hydrated oxides. This is far from being agreeable to the eye, and hence the process called mottling is adopted to throw the colour into veins. In this country this is obtained by pouring into the boiler, by means of a sort of watering-pan, a concentrated solution of crude soda which contains the sulphides. The density of the lye when passing through the pasty mass of soap accomplishes the mottling. The soap, after being properly boiled, is run off into the *frames*, boxes in which the soap is allowed to cool. When allowed to become solid, and exposed to the air, soap loses in the course of a month about 20 or 22 per cent. of its weight; hence it is not economical to buy new soap. The specifie gravity of the different kinds of soap is as follows:—White soap 1·0339, yellow 1·0340, silica 1·1368, clay 1·1408.

Windsor Soap.—This soap is made with mutton fat, or the fat or grease of bones. The finest French specimens now made contain above 30 per cent. of pork suet or olive oil. These are saponified with caustic soda. Before the boiling is finished, a mixture of essences, consisting of 6 parts essence of carvy, 1½ essence of lavender, 1½ essence of rosemary, is added to every 1000 parts of soap. The soap, when drawn off, becomes solid in twenty-four hours. In England Windsor soap is made with 9 parts of tallow and 1 part olive oil.

Transparent soaps are made by using alcohol for a solvent instead of water. This plan is principally adopted on the continent, where spirit is cheap. In this country the duty on spirits, unless modified for applications in the arts, renders this mode of manufacture impossible.

Soapstone. A synonyme of Steatite.

Soda. *Protoxide of Sodium, Mineral Alkali, Caustic Soda.*—NaO 4·, 32; 3·8717, 30·9736, 3·888, 31·. Na = 74·36, O = 25·64. Spec. grav. 2·805 (Karsten). Gray mass; non-conductor of electricity; melts in a strong red heat; corrodes organic substances. *Process.*—Obtained by igniting hydrate of soda and sodium; soluble in alcohol.

Hydrate of Soda. Hydrous Caustic Soda.—NaO HO 5·025, 40·2. NaO = 77·66, HO = 22·34. White mass; fusing point below incandescence; less volatile than hydrous potash; elevates the temperature of water during solution like hydrate of potash; distinguished from potash by its less deliquescence when exposed to the air; the soda attracting carbonic acid from the air, and becoming efflorescent. *Process.*—1. It is procured by throwing sodium on water; hydrogen is evolved, and soda dissolves; when boiling water is employed so much heat is given out that the hydrogen inflames, and, volatilizing some of the sodium, a yellow flame characteristic of soda appears. 2. When crystallized carbonate of soda is added to 5 parts of water, and ⅓ part quicklime slaked and added gradually in the form of cream of lime by admixture with water, boiled in an iron or silver pan for twenty minutes, and allowed to stand, the supernatant liquor, when drawn off by means of a syphon, is found to contain caustic soda. It may be concentrated and any deposit separated, and then evaporated to dryness, when hydrous caustic soda (NaOHO) remains.

Solution of Caustic Soda. Aqua or Liquor Sodæ, Soap-boiler's lye. When solid caustic soda or solid hydrate is exposed to the air, it attracts moisture at first, and then absorbs carbonic acid when it effloresces; 1· part of water dissolves at 64½° ·605 hydrate, or ·414 soda; at 89½° ·729 hydrate, or ·457 soda; at 131° 1· hydrate, or ·635 soda; at 176° 1·27 hydrate, or ·783 soda (Osann). When dissolved in water heat is evolved.

Soda, Salts of. 1. Soda salts are generally more soluble in water than potash salts; and they contain usually more water of crystallization than potash salts. 2. When heated they usually undergo the aqueous fusion; by continuing the heat the water is volatilized and a white powder remains. 3. No precipitate is produced by tartaric acid, or bichloride of platinum; sulphate of alumina does not form crystals of alum. 4. *Granular* antimoniate of potash dissolved in cold water produces a precipitate in soda salts of antimoniate of soda, requiring 300 water to dissolve it. 5. Before the blowpipe, soda salts impart to the flame a yellow tinge. 6. The septaiodate of potash produces a colourless precipitate in salts of soda.

Peroxide.—(Gay Lussac and Thenard, 1809); *Sesquioxide* (ib. and Davy); *Binoxide* (Millon). *Teroxide* (Gmelin.) Dirty greenish-yellow masses; fuses when heated to a high temperature; in water is reduced to soda, evolving oxygen; non-conductor of electricity; decomposed by ignition with charcoal, or tin, phosphorus, and sulphurous acids. *Process.*—By heating sodium in dry air or oxygen; by igniting caustic soda in oxygen (Recherch, Phys.-Chim. 1, 152).

Dinoxide, Suboxide.—Na_2O? Dark gray (Davy) or grayish-white mass (Gay Lussac and Thenard); decomposing water and being more combustible than sodium. It may either be a dinoxide or a mixture of sodium and soda.

Sulphate. Glauber's Salt (1658), *Sal Mira-*

bile, Thenardite (native and anhydrous), NaO SO_3 9· or 8·9, or 71·2. Sp. grav. 2·645 (Thomson). *Anhydrous Salt.*—8-hedrons, consisting of 2 or 4-sided pyramids, with a rhombic base, united together, the angles of the rhomb being 123° 43′, and 104° 18′. When ignited fuses, and on cooling concretes into a foliated brittle bitter saline mass. 100 water at 57° dissolve 10·58 parts, at 91° 50·65 parts, being a maximum (Gay Lussac). First described by Dr. T. Thomson. Obtained from Messrs. W. & J. Wilson at Hurlet, from a furnace, and by Mr. Alex. Harvey, by evaporating Glauber salt in a stove at 120°. Mitscherlich obtained it considerably under 212°. This salt is the first product in the preparation of carbonate of soda on the large scale, when chloride of sodium is converted into sulphate of soda by the action of sulphuric acid; it is impure. It may be formed by saturating carbonate of soda by sulphuric acid, and heating to redness. It occurs native as Thenardite, at Espartinas, near Madrid (Casaseca) in 4-sided prisms, and as a frequent efflorescence on bricks containing about 2 atoms water, which have been called *exanthalose* (Beaudant). With chloride of potassium and nitre yields sulphate of potash.

Decihydrated Sulphate, Common Sulphate.—$NaOSO_3$ 10 HO 20·25, 20·15, 161·2. Specific grav. 1·35 (Watson), 1·349 (Thomson). Oblique rhombic prism; the base of the prism is inclined to the lateral faces at an angle of 101° 20′, and the two contiguous lateral faces to each other at 80° 24′. (Brooke, Ann. Phil 23, 21.) The acute lateral edges are usually replaced by tangent planes, making the prism 6 or 8-sided. The faces are usually channelled longitudinally and irregular, and are easily recognized by this character, and their becoming efflorescent on exposure to the air. *Taste* bitter saline. At 60° in the air it loses all its water, about 56 per cent. (Wenzel), and likewise in vacuo over sulphuric acid; it absorbs no water in air except when the air is saturated with moisture. 100 water dissolve 48·66 salt at 64½°. Insoluble in alcohol.

Nitrate, Cubic Nitre, Chili Nitre.—$NaONO_5$ 10·75, 10·65, 85·2; NaO 37·2 per cent., NO_5 62·8. Spec. grav. 2·069 (Hassenfratz, Ann. de Chim. 28, 12). First accurately described by Margraff (Opusc. 2, 331), and repeated by Lewis (Phil. Com. 642). The *crystal* belongs to the rhombohedral system, being an obtuse rhomb. P on P′ 106° 30′, P on P″ 73° 30′ (Brooke, Ann. Phil. 2d. ser. 5, 452). *Taste* cool and sharp, somewhat more bitter than potash nitre; fuses and gives out oxygen when heated like saltpetre, nitrite of soda being formed; subsequently nitrogen and hyponitric acid fumes are disengaged. It attracts moisture from the air, and detonates less powerfully than potash nitre; it is decomposed in solution by salammoniac (Karsten), and chloride of potassium (Longchamp), which is therefore sometimes, when the price admits, used for its preparation, and sulphate of potash (Moreau). 100 water at 21° dissolve 63·1; at 32° 80 parts; at 50° 22·7; at 61° 55; at 246° 218·5; its solubility at 32° is therefore more than three times greater than at 50° (Marx). *Sources.*—1. It is procured in commerce from a large deposit in South America, in the districts of Tarapaca and Atacama, 14 leagues from Iquique, in Peru, (lat. 20° 12′,) apparently along the margin of an ancient lake; it forms a stratum 2½ feet thick, extending upwards of 150 miles. It is purified from sulphate of soda, common salt, and gypsum by repeated crystallization. As imported, it contains about 96 per cent. of nitrate, with some common salt, water, lime, and sulphuric acid. 2. It may be formed by neutralizing carbonate of soda, or caustic soda, with nitric acid. 3. By mixing nitrate of lime and sulphate of soda. It is used largely in the manufacture of sulphuric and nitric acids.

Decihydrated Carbonate, Common Carbonate.—$NaOCO_2$ 10HO 18, 17·9, 143·2. NaO 22·22, CO_2 15·27, HO 62·51. Sp. grav. 1·623 (Thomson), 1·423 (Haidinger). *Crystal.* Oblique rhombic prisms. The apex of the prism is not visible in the common crystal, but it makes angles of 108° 43′, with the faces of the prism. M M′ the two primary faces of the prism meet at angles of 76° 12′. In the following measurement P denotes the base of the primary prism; P on M or M′ 108° 43′; P on e or c′ 129° 52′; P on h 121° 20′; M on M′ 76° 12′; M on h 128° 6′; M on K 141° 54′; e on c′ 79° 44′; e on k 140° 8′ (Brooke, Ann. Phil. 2d ser. 6, 287). *Taste* similar to that of carbonate of potash, but less caustic. In the air it effloresces into a white powder, forming the protohydrated carbonate. By heat it undergoes the watery fusion, the protohydrate separating, and leaving above a liquid containing more than 10 atoms water. By a heat of 100° the protohydrate is obtained, and by 42° the quintohydrate (Schindler). In vacuo over sulphuric acid the protohydrate is formed, and, if exposed to moist air, it reabsorbs nearly its original 10 atoms (Blücher, Watson). Specific gravity of a saturated solution at 58½° 1·075; 100 water at 58° dissolve 7·74 of the anhydrous salt, or 20·64 of the crystals (Thomson), or 50 parts (Gmelin). It dissolves in rather less than its weight of boiling water, so that when dissolved in boiling water it crystallizes as the solution cools.

Manufacture of Carbonate of Soda.—Sources.—Carbonate of soda was originally prepared from the ashes of different species of Salsola, especially from the Salsola soda, from which the alkali has obtained its name. This impure soda is also called *barilla*, from its Spanish name. Most sea weeds contain carbonate of soda, and under the name of kalp or ashes of sea weeds, a large supply was formerly obtained from the Western Highlands and North of Ireland. Previous to 1822, the average price of kelp (Varec in France) for twenty-two years was £10 10s. but it sometimes rose to £22 per ton. The duty on foreign barilla being reduced in 1822, and on common salt removed in 1823, the price speedily fell to £2 and £3. In the six years, July, 1841-47, the average annual importation of kelp at Glasgow was 3232 tons (Glassford). Carbonate of soda is now obtained almost exclusively by the decomposition of common salt, although a small product is still separated in the manufacture of iodine and potash salts from kelp. Cohausen had suggested in 1717 that common salt might be decomposed by means of lime. In 1737 Duhamel obtained carbonate from it by fusing it with charcoal, dissolving the fused mass in acetic acid, and calcining the acetate. About 1758 (?) Margraff endeavoured to decompose sulphate of soda by limestone. In 1768 Hagen decomposed it by potash, and afterwards Bergman by barytes. In 1775 Scheele partially decomposed it by oxide of lead. In 1782 Morveau and Carny used felspar. In 1781 Constantine employed alum. The sulphates of lime, magnesia, ammonia, potash, bicarbonate of ammonia, iron filings, iron pyrites, alumina, steam, &c. decompose common salt. The first important step was, however, discovered in 1784 by Le Blanc and Dizé. Sulphate of soda having been formed in 1658 from common salt and sulphuric acid by Glauber, the process patented by Le Blanc in 1791 consisted in fusing sulphate of soda with carbonate of lime and charcoal. It is this important discovery which has rendered the manufactures of this country so prosperous. The process was introduced at Glasgow by Messrs. Tennant and

Fig. 1.

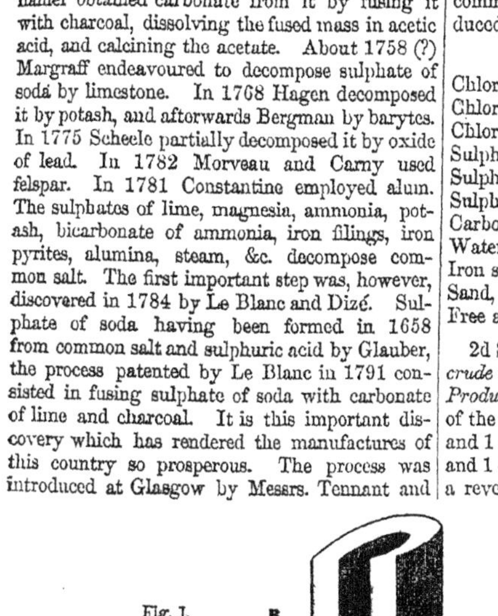

Knox in 1799. The following analyses were made by my lamented pupil Mr. John Brown in my laboratory:—

1st Stage. *Conversion of chloride of sodium into sulphate of soda by sulphuric acid.*—The salt employed in this country is derived from large brine deposits situated in the new red sandstone formation at Northwich in Cheshire. The salt, when first obtained, has a red colour, due to sesquioxide of iron. It is dissolved in water, the impurities allowed to subside, the liquor drawn off, evaporated down, the salt raked out into wicker baskets, and allowed to drain. The salt having been deposited from the sea, contains all the ingredients of sea water. 6 cwt. of the salt in fine white crystals is placed in an iron reverberatory furnace (*fig.*), and 5½ cwt. sulphuric acid (1·75 sp. gr.) is allowed to run in by a funnel at the top; heat being applied below, the chlorohydric acid evolved passes up the chimney, or it is condensed by causing it to pass through earthenware pipes to columns of moist coke, by which it becomes fluid. The sulphate of soda, when freed from the greater quantity of its muriatic acid, is pushed forward into a reverberatory furnace, where it becomes dry. It is then a white saline mass, and is drawn out. The composition of commercial salt and the sulphate of soda produced from it are as follow:—

	Commercial Salt.	Sulphate of Soda.
Chloride of sodium, ...	931·615	10·956
Chloride of potassium,.....	trace.	
Chloride of magnesium,....	1·066	
Sulphate of lime,...........	10·098	9·731
Sulphate of magnesia,.....	1·348	2·893
Sulphate of soda,...........		962·170
Carbonate of lime,.........	1·500	
Water,........................	54·373	
Iron sesquioxide,...........		2·300
Sand,..........................		3·100
Free acid,.....................		8·850

2d Stage. *Conversion of sulphate of soda into crude carbonate of soda or soda ball* (a) *substage. Production of sulphide of sodium.*—2 cwt. 2 qrs. of the sulphate of soda formed in the preceding and 1 are mixed with 2 cwts. 2½ qrs. limestone and 1 cwt. 3 qrs. coal dross, and are introduced into a reverberatory furnace and strongly heated (in France, 4 sulphate, 4 limestone, 1⅓ coal). The mixture is frequently stirred with long iron rakes, to expose a new surface to the flame. The effect of the heat and coal is to remove oxygen from the soda and sulphuric acid, and to form a sulphide of sodium, ($NaO\ SO_3$ and $C_4 = NaS\ 4(CO)$). (b) 2d substage. The heat is continued till the evolution of carbonic oxide, which burns with a blue flame, is observed to cease. The fused mass is then raked out on iron barrows or trays, and allowed to cool. In this portion of the process, for the sake of precision, we may consider that 3 atoms sulphide of sodium, in the presence of 4 atoms carbonate of lime and 1 of carbon, are converted into 3 carbonate of soda, 1 oxysulphide of calcium, and 2 carbonic oxide, ($3NaS + 4CaO\ CO_2 + C = 3NaOCO_2 + (3(CaS), CaO)\ 2CO$). The ball soda, or crude carbonate of soda, thus formed is dark coloured. Its composition is as follows:—

Soluble part.

Carbonate of soda,................	35·640
Caustic soda,........................	0·609
Aluminate of soda,................	2·350
Sulphate of soda,..................	1·160
Sulphide of sodium,..............	1·130
Chloride of sodium,..............	1·913

Insoluble part.

Ultramarine,........................	0·295
3CaS,CaO,........................	29·172
Caustic lime,........................	6·301
Sand,................................	4·285
Sulphide of iron,.......	4·917
Silicate of magnesia,..............	3·744
Carbon,..............................	7·998
Water,................................	0·700
	100·214

The ball soda, it was first observed by Mr. Brown, when heated with alcohol, gives up no caustic soda; but when water is added, it yields caustic soda and carbonate of lime.

3d Stage. *Manufacture of soda ash from ball soda.*—The ball soda is introduced into iron square pans or keaves (cuves), five or six of which are usually worked together. They are so arranged that the water heated by steam, which runs into the first pan, passes over the six in succession. From the last it runs into a large iron vessel to settle. The liquor containing carbonate of soda and other impurities is boiled down in an iron pan until it is nearly dry. It is then placed in a reverberatory furnace and oxidized, so as to convert sulphide into sulphate, and hydrate of soda into carbonate. The following are the analyses of the salt before and after furnacing:—

	Before Furnacing.	After Furnacing.
Carbonate of soda,..........	68·907	71·614
Hydrate of soda,............	14·433	11·231
Sulphate of soda,............	7·018	10·202
Sulphite of soda,.............	2·231	1·117
Hyposulphite of soda,.......	trace.	
Sulphide of sodium,.........	1·314	
Chloride of sodium,.........	3·972	3·051
Aluminate of soda,..........	1·016	0·923
Silicate of soda,..............	1·030	1·042
Insoluble matter and sand,	0·814	0·316
	100·755	99·496

The matter which remains undissolved in the hot water is termed *soda waste*, and is a great obstacle in the soda process, in consequence of its inutility for any purpose, and the difficulty of disposing of it. It consists principally of carbonate of lime, and of a compound of sulphide of calcium with lime which is insoluble in water, sulphide of calcium itself being insoluble in water. According to Dumas, Unger, and Brown, it is composed of 3 atoms sulphide of calcium united to 1 of lime.

Soda Waste.

Insoluble salts, =...............	74·741
Carbonate of lime,...............	24·220
3CaS,CaO,........................	20·363
Carbon,.............................	12·709
Silicate of magnesia,.............	5·987
Sand,................................	5·746
Sesquioxide of iron,..............	5·716
Soluble salts, = 25·382,........	—
Sulphate of lime,..................	4·281
Hyposulphite of lime,............	trace.
Bisulphide of calcium,............	3·583
Sulphide of calcium,....	8·527
Hydrate of lime,...................	5·582
Carbonate of soda,................	1·309
Water,..............................	2·100
	100·123

The soda waste gradually oxidizes; in three years it is converted into sulphate, sulphite, carbonate, and hyposulphite of lime.

4th Stage. *Carbonate of Soda Process.*—The carbonate of soda ball is lixiviated in the same way with water as for soda-ash. The liquor is pumped up into evaporating pans placed over the reverberatory furnaces, and reduced nearly to dryness. The salt is taken out in colanders, placed in a heap, and allowed to drain. The sulphide and caustic soda drain off (1). It is then furnaced (2). A finer kind of soda-ash is frequently made from (2), by dissolving in water, evaporating, and carbonating in a furnace. From this, crystallized carbonate is formed. It is dissolved in boiling water, the solution having a specific gravity of 1·250 (50° Twaddell). It is then run into a cistern, and diluted with cold water down to 1·21 (42° Twaddell). Earthy matter deposits. Bleaching powder in small quantity being now added, another deposition occurs. It passes into another pan, when it is evaporated to 1·27 (54° Twaddell); then into a cistern, and lastly into iron crystallizing pans, where it remains for eight days or less, according to the weather:—

	Carbonate of Soda.		Crystallized Carbonate of Soda.	
	1. At 212	2. Furnaced.	Hydrs.	Anhydrs.
Carbonate of soda,....	79·641	84·002	36·476	98·120
Hydrate of soda.......	2·712	1·060		
Sulphate of soda,.....	8·641	8·560	0·943	
Sulphite of soda,.....	1·238	trace		1·076
Sulphide of sodium,..	trace			
Hyposulphite of soda,	trace			
Chloride of sodium,..	4·128	3·222	0·424	0·742
Aluminate of soda,...	1·176	1·013		
Silicate of soda,......	1·234	0·984		
Insoluble matter,.....	0·972	0·716		
Water,..............			62·157	
	99·742	99·557	100·	99·938

Sesquicarbonate, *Trona*, (Sukenanear Fezzan), Urao (Mexico), (Rivero and Boussingault, Ann. Chim. 29, 116). *Natron*, *Nitrum*. 2 NaO 3 CO_2 4 HO (Phillips' Quart. Journ. 7, 296). First described by Klaproth, in 1802 (Beitrage 3, 83). *Crystal.* Doubly-oblique 4-sided prisms, with angles of about 60° and 120°. Spec. grav. 1·98 (Thomson), 2·112 (Haidinger). *Taste* alkaline, somewhat milder than common carbonate. It occurs native in considerable quantities in the province of Sukena, near Fezzan, in North Africa, and is called *Trona* by the natives. It is also found in the lakes of Debretzin in Hungary, in Columbia in the neighbourhood of Lagunilla

(Baudrimont). *Process.*—To 100 grains of simple carbonate, and 152 bicarbonate, dissolved in 4 ounces water, add 4 ounces of alcohol. Crystals of this salt separate in a few days (Winkler Repert. 48, 215). It crystallizes likewise from a solution of bicarbonate concentrated by boiling. The crystals are permanent in the air.

Bicarbonate.—$NaO2CO_2HO$ 10·625, 10·525, 84·2, NaO 37·51, CO_2 51·58, HO 10·91. *Crystal.* Oblique rectangular tables or prisms. *Taste* feebly alkaline; renders cudbear paper purple; logwood and reddened litmus paper blue, but its alkaline properties are very weak; by ignition loses about 37¾ per cent. of water and carbonic acid, by which it is reduced to the state of anhydrous carbonate. At 64° 100 water dissolve 9·33, or 1 salt dissolves in 10 water (Thomson); 1 in 13 water (V. Rose); 1 in 8 water (Berthollet). *Process.*—It may be obtained by suspending a saturated solution of common carbonate over a brewer's fermenting tun, or by exposing a solution of the carbonate in a large vessel or flask filled with carbonic acid.

Hydro-trisphosphate, — *Common Phosphate.* 2 NaO HO PO_5 24 HO 45·125, 44·85, 358·8. *Crystal.* Oblique rhombic prisms, with angles of 67° 50′. The base of the prism is inclined to the lateral faces at an angle of 166° 44′; P on M or M′ 106° 44′; P on c 129° 12′; P on g 112° 27′; P on h 121° 14′; M on M′ 67° 30′; M on h 123° 45′; M on k 146° 15′ (Brooke). Effloresces in the air between 51° and 64° (Clark); becoming white and opaque, but not losing its shape. By heat it fuses in its water of crystallization; and by ignition fuses and becomes pyrophosphate, giving then a *white* precipitate with nitrate of silver; previous to ignition it gives a *yellow* precipitate; very soluble in water, 100 water at 65° dissolve 26·77 salt; in dry air it loses more than half its water (Gay Lussac); and if exposed to the air absorbs 14 atoms. Does not affect litmus, but renders cudbear violet. At 572°, or in vacuo over sulphuric acid, it loses 24 atoms water, becoming a white mass; changing violet colours to green (2 NaO HO PO_5). When a solution is evaporated at 90°, and crystallized, the quindecihydrate, 2 NaO HO PO_5 14HO, is obtained in right rhombic prisms, meeting at angles of 98° and 92°. It does not effloresce when exposed to the air, but loses at 572° 47·6 per cent. of water, and by ignition 3·47 per cent. more. *Process.*—Obtained by adding carbonate of soda to the phosphoric acid prepared from calcined bones at 212°, till the solution has an alkaline reaction.

Bibasic Phosphate of Soda, Diphosphate of Soda, Pyrophosphate (Clark, Brewster's Journal, 7, 298).—2 NaO PO_5 10HO 27·975, 223·8; NaO 27·89, PO_5 31·90, HO 40·21. *Crystal.* Oblique prism, alkaline reaction; fuses by ignition to a glass, consisting of anhydrous diphosphate (2NaO PO_5). It regains its crystalline water when exposed to the air (10 atoms or 40 per cent.) It yields a white precipitate with nitrate of silver without liberating nitric acid ($2AgOPO_5$), and when boiled with phosphoric, sulphuric, muriatic, nitric, or acetic acids, it gives a yellow, being changed into trisphosphate. When boiled for three weeks in a flask, which it corrodes, it changes into common phosphate. *Process.* Formed by igniting common phosphate, dissolving in water, and crystallizing.

Hydro-Trisarseniate, Common Arseniate.—2 NaO HO AsO_5 24 HO 50·625, 50·3, 402·4. Sp. grav. 1·759. Oblique rhombic prisms, isomorphous with common phosphate of soda, effloresce in a dry atmosphere, and become white on the surface, but do not lose their shape. Taste cooling, resembling that of carbonate of soda; alkaline reaction; 100 water at 45° dissolve 22·268 crystals; insoluble in alcohol, but the surface of the crystals in alcohol becomes opaque from the removal of the crystalline water. Heat causes it to undergo the *aqueous fusion,* and ignition the *igneous fusion.*

Separation of Potash and Soda.—If the alkalies are in the state of chlorides in solution, evaporate to dryness in a platinum capsule, ignite and weigh. Dissolve the residue in as small a quantity of water as possible; add an excess of bichloride of platinum from a concentrated solution in water; add a considerable quantity of alcohol, and allow the whole to stand for some hours. The supernatant alcohol should be yellow coloured, indicative of an excess. The potassium bichloride of platinum is then filtered, washed with alcohol, and weighed, from which the chloride of potassium weight is obtained, and deducted from the weight of the joint chlorides; the remainder is the chloride of sodium. But the latter may be separated from the filtered liquor by evaporating away the alcohol slowly, igniting strongly in a covered capsule. During the ignition a few crystals of oxalic acid are added, which facilitates the decomposition of the sodium chloride of platinum. The chloride of sodium is then extracted with water, and the platinum left. If the aqueous solution is yellow, the salt has not been decomposed. It must be evaporated, and again strongly ignited; the residue is the chloride of sodium when free from platinum. When alkaline sulphates are present, the same process answers, but it requires much caution. The weighed sulphates are dissolved in a minimum of water, but no crystals must be seen. An excess of an aqueous solution of bichloride of platinum is now added, so as to convert both the potash and soda into double platinum salts. Some spirit and the requisite quantity of alcohol are added, and the whole allowed to stand. If a considerable excess of bichloride of platinum is not added, sulphate of soda falls from its insolubility in alcohol, which cannot be converted into the sodium chloride of platinum in its insoluble state. When the sulphate of soda is seen to precipitate, the alcohol must be poured off, and water added to dissolve the whole, and then the bichloride added again in excess. The potassium salt is then filtered, washed, dried, and weighed. To obtain the sulphate of soda, the filtered liquor is evaporated, the residue ignited with some oxalic acid, the residue exhausted with water. Chloride of sodium (not sulphate, as might be supposed,) is taken up. This is to be evaporated with sulphuric acid, ignited, and estimated as sulphate of soda. As by this process sulphate of soda may be present, if care is not taken, it is preferable to convert the sulphates into chlorides. The solution of the two sulphates (previously deprived of any free sulphuric acid by ignition) is precipitated with acetate of barytes in slight excess, the precipitate allowed to deposit in a warm place, to prevent it passing through the filter in a milky state. The filtered liquor is evaporated in a platinum capsule and ignited; the residue digested in water, and filtered. Carbonate of barytes remains, while the carbonates of the alkalies are dissolved. These are saturated with chlorohydric acid, and the chloride of potassium separated in the manner described. The acetate of barytes must be quite free from chloride of barium.

Soda Water. Liebig's apparatus of stoneware for impregnating water with carbonic acid, usually known as soda water, is represented in the accompanying figures 1 and 2. In fig. 1, B is the carbonic acid generator, the bicarbonate of soda and tartaric acid being introduced by the opening *b*, which is closed by a solid stopper by means of a bayonet joint resembling K E in fig. 2, or simply by means of a screw. The diaphragm

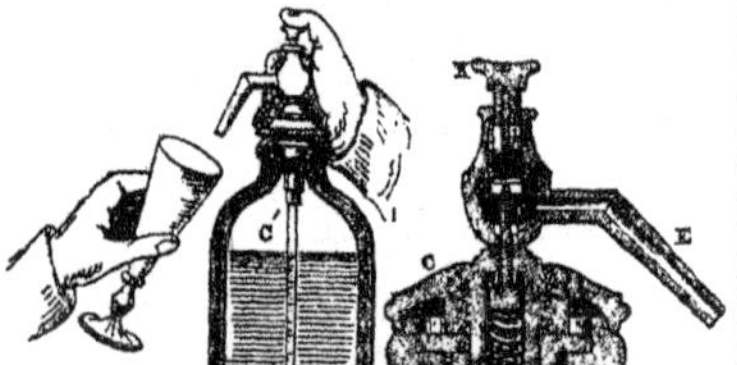

1. 2.

A is pierced with a number of capillary apertures *a*, which allow the carbonic acid to stream into the water or wine placed in the upper division of the vessel *c′ a*. The water is gradually charged with carbonic acid, which becomes thus soda water. The contrivance in fig. 2 is the spring valve, at the mouth of the bottle, for delivering the water after it is formed. When the thumb is pressed on the tin piece and piston K, the spiral spring (which retains the bottom of the piston close up on the solid piece C when the bottle is closed) is depressed, the tube F *e* E is opened. and the soda water, impelled by the gas at *c′*, is forced up the tube F, and discharged by E into a glass, as in fig. 1; K a space resting on a caoutchouc ring, G a ring of caoutchouc, B a ring of tin.

Soda Alum. Alum containing soda in place of potash.

Sodaite. A synonyme of Elæolite.

Sodalite. *Hauyne, Noseane, Spinellane.* Sp. grav. 2·295 to 2·378, H 5·75. Green, greenish-white, or snow-white masses, and rhomboidal 12-hedrons; streak white; fracture conchoidal, uneven; lustre vitreous, translucent, brittle. B.B. fuses with effervescence into a glass bead; soluble in acids. First analyzed by Dr. T. Thomson, from Greenland. SiO_3 38·40, Al_2O_3 32·04, NaO 24·47, Cl 7·3, CaO ·32. *Form.* 3 NaO, SiO_3 + 3($Al_2O_3SiO_3$) NaCl. Greenland, Vesuvius.

Soda Nitre, or Nitrate of Soda.

Soda Spodumene, or Oligoclase.

Sodium. Natrium 3, 24; 2·8717; 22·9736. Specific gravity ·97223 (Gay Lussac and Thenard), ·9348 (Davy). White metal; shade intermediate between silver and lead, crystallizes in cubes; solid and malleable at usual temperatures, so that it can be welded by pressure; still malleable at 32°, and ductile at 122°, semifluid; at 194° quite fluid; remains fixed at the melting point of plate glass (Davy); less volatile than potassium (Gay Lussac and Thenard); more volatile (Mitscherlich); decomposes water, but does not burn unless the water be boiling or be rendered mucilaginous by means of gum to diminish the motion of the metal; the hydrogen then burns with a yellow flame; good conductor of heat and electricity. Its vapour is colourless.

Process.—It may be prepared from caustic soda by electricity; by heating in an iron bottle hydrate of soda and iron filings; by igniting carbonate of soda and charcoal by Brunner's plan, as described under POTASSIUM.

Chloride, Common Salt, Kitchen Salt, Marine Salt, Rock Salt, Sea Salt, Muriate of Soda, Hydrochloride, Chlorohydrate.—NaCl 7·5, 60·; 7·308, 58·46. Sodium 40, chlorine 60 per cent. Spec. grav. 2·125 (Fahrenheit, Phil. Trans. 1724, 114), 2·257 (Mohs), 2·03 (Unger), 2·078 (Karsten), 2·15 (Kopp), spec. heat 0·214. *Taste* well known, and is what is commonly termed salt. As generally prepared it decrepitates from the presence of hygroscopic water between its plates; fuses by ignition; volatilizes at a white heat; is carried up by steam from boiling saline solutions, and likewise from the sea by winds, but not by simple evaporation; 100 water at 57¼° dissolve 36 salt; at 140, 37; at 229½°, 40·38 (Gay Lussac); at 77°, 36 parts (Kopp); at 65¾°, 38 parts (Karsten). Strong muriatic acid precipitates salt from a saturated solution; common salt is slightly soluble in dilute alcohol; 5·8 in 100 alcohol of spec. grav. ·900; 3·67 in 100 alcohol of ·872 spec. grav.; 0·5 in alcohol ·834 (Kirwan); but nearly insoluble in absolute alcohol. When heated with potassium it gives chloride of potassium and metallic sodium (Davy); with sulphuric acid it is converted into sulphate of soda and chlorohydric acid gas. When ignited with oxalic acid it yields the same gas and carbonate of soda (R.D.T.) *Quaterhydrated Chloride.*—NaCl4HO. Oblique rhombic prisms, striated, belonging to the oblique prismatic system, efflorescing in the air below 32°, deliquescing above 32°, and leaving a powder of small cubes. Obtained by exposing a saturated solution of common salt to a temperature of 14° (— 10° C.) (Löwitz), or by evaporating a dilute solution of salt at 59° on a glass plate; large 6-sided tables of the hydrate separate first, and then cubes (Ehrenberg). Sulphate of magnesia and common salt at 26°, become sulphate of soda and chloride of magnesium (Berzelius). *Sources.*—1. By heating sodium in chlorine gas perfectly pure salt is obtained. 2. Partly by spontaneous, partly by artificial evaporation salt is extracted from sea water, which contains about 3½ per cent. of it, or from salt springs as in Germany and North America. It is extracted from the earth in large quantities from deposits situated in the new red sandstone formation at Northwich, in Cheshire, in the Punjaub, &c. A large deposit of salt has recently been found at Carrickfergus. Its composition, by the analysis of my pupil, Mr. A. Buchanan, is NaCl 94·166, HO 3·433, $CaOSO_3$ 1·020, $MgOSO_3$ ·153, $CaOCO_2$ ·1, $MgOCO_2$ ·128, Fe_2O_3 ·585, SiO_3 ·42.

Iodide.—NaI 18·75, 18·65, 149·2. Cubes, and with 4 atoms of water in striated oblique rhombic prisms; it is less fusible than iodide of potassium; above 122° it crystallizes in anhydrous cubes, below this in oblique hydrous prisms; very deliquescent; 1 water dissolves 1·73 iodide; it is less volatile than the potash iodide, but more so than common salt. When fused in the air iodine escapes and also when ignited with charcoal. The hydrate effloresces in dry air, fuses when heated and leaves the anhydrous salt; the hydrate deliquesces in moist air; 1 part dissolves in 0·6 water and in alcohol. It is prepared like the potash salt by forming an iodide of iron and then decomposing this by carbonate of soda in equal atoms or by saturating iodohydric acid with soda.

Sulphide.—NaS9HO 15·125, 15·025, 120·2. Rectangular prisms (Vauquelin, Berzelius), 8-hedrons (Vauquelin, Kircher), oblique rhombic prisms (L'hermina); acid reaction; when fused gives out water and becomes anhydrous; soluble in water, less soluble in alcohol. Prepared by passing sulphohydric acid through caustic soda, and evaporating in vacuo.

Soils. The name applied to the soft materials lying on the surface of our globe on which plants grow. The soil must either be formed by the action of the atmosphere, &c. on underlying rocks, or be derived from matter transported from a distance by diluvial agencies. Soils are usually divided into sandy or light, characterized by a preponderance of sand; clay where much alumina is present; and loams, which contain much vegetable matter.

Soimonite. A synonyme of Corundum.

Solanic Acid, and **Solanine.** $C_{84}H_{63}NO_{28}$. Plumose crystals with a pearly lustre and acid taste; poisonous; insoluble in water; soluble in boiling alcohol; it forms crystalline salts; extracted from the shoots of potato and several species of solanum by digesting in SO_3, precipitating by NH_3, and crystallizing from alcohol.

Solanoleic Acid? The acid of the fluid oil of potatoes, extracted by alcohol from the solanostearic acid.

Solanostearic Acid. $C_{30}H_{30}O_4$. F.P. 122°. Crystalline plates, obtained by saponifying the fat oil from potatoes.

Solder. An alloy used for uniting metals. *Soft* solder consists of 67 lead, 33 tin, or 33 lead 67 tin; *hard* or brass solder consists of equal parts of copper and zinc; *silver* solder is formed of 66 silver, 30 copper, 4 brass; *gold* solder of 66 gold, 17 silver, 17 copper.

Solid. That condition of matter in which the particles of which it is composed are in closest contiguity.

Solution. (*Solutio*, loosening.) The combination or mixture of a solid, liquid, or gas, with a fluid; it differs from mechanical mixture, that in the latter the diffused particles are visible, and may be removed by filtration, while in a solution the dissolved particles are invisible, (as when salt is dissolved in water,) and cannot be mechanically removed.

Solvent. (*Solvens*, loosening.) The liquid or medium by which a substance is dissolved.

Somervillite. Yellow variety of Humboldilite, Vesuvius.

Sommite. A variety of Nepheline, Mount Somma.

Sooranjee. The root of a species of morinda, from India, containing a red-colouring matter. See MORINDINE.

Soot. The volatile black carbonaceous matter which rises from a coal or wood fire, and is deposited in the chimney. Water takes up *ulmine?* and deposits it on standing; the portion dissolved in the water and not deposited, but obtained by evaporation, and exhausted with alcohol, is a red substance destitute of taste (animalized matter); the portion soluble in alcohol is asboline. See ASBOLINE. Wood soot consists of ulmine 30·2, animalized matter 20·, asboline ·5, water 12·5, $CaOCO_2$ and $MgOCO_2$ 14·66, acetate of lime 5·65, $CaOSO_3$ 5·, acetate of potash 4·1, charcoal 3·85, ferrophosphate of lime 1·5, silica ·95, acetate of magnesia ·53, KCl ·36, acetate of ammonia ·2, acetate of iron *trace.* It probably also contains creasote, capnomore, &c.

Sorbic Acid. When sorbine is dissolved in potash or ammonia, and precipitated by weak chlorohydric acid, red flocks fall which, when dried at about 280°, constitute *sorbic acid*, which is isomorphous with *sorbine;* insoluble in water, alcohol, and weak acids; very soluble in alkalies. It consists of C 57·96, H = 5·51, O = 36·53. The lead salt consists of $3PbO\ C_{32}H_{18}O_{15}$. With the alkalies it forms solutions of a deep rich sepia colour, a characteristic of the acid.

Sorbine. *Sorbite.* $C_{12}H_9O_9 3HO$. Sp. grav. 1·654. C = 40·, H = 6·66, O = 53·34.—Colourless rectangular 8-hedrons. A species of sugar found in the berries of the mountain ash. When pure, exactly resembling cane sugar in sweetness, &c.; soluble in double its weight of water; slightly soluble in alcohol, from which it deposits in 8-hedral crystals; it does not ferment with yeast. It is prepared by expressing the juice of the berries, filtering, and allowing the juice to stand at rest for some months; a deposit forms; the supernatant liquor is decanted and evaporated to the consistence of a syrup, which deposits brown crystals. These are decolourized by twice treatment with animal charcoal. A solution of acetate of lead precipitates a white substance, which, dried at 212, consists of $4PbO,C_{12}H_9O_9$. It crystallizes in cubes with common salt. Weak sulphuric acid produces no change in it. Nitric acid converts it into oxalic acid; it unites with barytes and lime; it dissolves protoxide of copper, and deposits gradually red oxide; tartrates of copper and potash are also reduced. Heated between 302° and 556° it leaves a deep red residue, which is sorbic acid. Sorbine deviates to the left the plane of polarization. Sorbine differs from *inosite*, the sugar of muscle, by the latter not colouring potash, and not reducing black oxide of copper, and forming a hydrate $C_{12}H_{16}O_{16}$. It would easily be confounded with glucose by its reducing action of salts of copper.

Sordawalite. Spec. grav. 2·58, H 2·75. Gray or pitch-black masses, with a vitreous and semi-metallic lustre; opaque; fracture conchoidal; brittle; becomes reddish by exposure; streak brown. B.B. fuses into a black globule; a green glass with soda, and a slag; with borax a green glass; with salt of phosphorus leaves a silica skeleton. SiO_3 49·4, Al_2O_3 13·8, FeO 18·17, MgO 10·67, PO_5 2·68, HO 4·38. Sordawala, Finland, as a layer an inch thick on trap, resembling coal in appearance; Bodenmais, Bavaria.

Spadaite. Spec. grav. 2·5. Reddish masses, with imperfectly conchoidal fracture and pearly lustre. B.B. melts into an enamel; soluble in HCl, the silica forming a jelly. SiO_3 56, MgO 30·67, FeO ·66, Al_2O_3 ·66, HO 11·34. Capo di Bove. *Form.* 5 MgO, 4 SiO_3, 4 HO.

Spaniolitmine. $C_{18}H_7O_{16}$. Bright red substance; soluble in alcohol and ether; rarely present in litmus.

Spar. A term applied to various minerals from their lustre. *Calcareous spar.* See CARBONATE OF LIME. *Adamantine spar*, or corundum; *bitter spar*, or carbonate of magnesia and lime; *blue spar*, or lazulite; *Bologna spar*, or sulphate of barytes; *brown spar*, or ferrocarbonate of lime and magnesia; *chlorite spar*, q. v.; *cubic spar*, or anhydrite; *sparkling spar*, or felspar; *fluor spar*, or fluoride of calcium; *fossil spar*, or fluor spar; *fusible spar*, or felspar; *heavy spar*, or sulphate of barytes; *Iceland spar*, or calc spar; *pearl spar*, or dolomite; *selenite spar*, or sulphate of lime; *table spar*, or lime and soda-lime table spar.

Sparging. An operation in Scottish ale brewing. See ALE.

Sparry Iron Ore. See CARBONATE OF IRON.

Sparry Manganese, or Silicate of Manganese.

Sparteine. $C_{15}H_{13}N$. B.P. 550½° (288°C). Colourless heavy bitter oil; alkaline; becoming opaline with water, brown in air; it unites with acids; obtained by distilling the mother liquor of crude scoparine with carbonate of soda, and rectifying the distillate saturated with common salt. It forms with picric acid a crystalline salt; with $PtCl_2$ rhombic crystals.

Spear Pyrites. A synonyme of radiated pyrites.

Specific Gravity. See my SCHOOL CHEMISTRY, p. 7.

Specific Volume. See ATOMIC VOLUME.

Specular Iron. See CARBONATE OF IRON.

Speculum Metal. The specula of telescopes are made of tin 34, copper 66.

Speiss. A deposit occurring in the formation of smalts. See NICKEL and SMALTS.

Spelter. A term employed in commerce for zinc.

Spermaceti. (*Wallratt*, Ger.) Sp. grav. ·940; F.P. 112°. Crystalline pearly substance, subsiding from sperm oil; obtained from cavities in the skull of the Physeter macrocephalus, tursio, microps, orthodon, and Delphinus edentulus. This substance has been usually viewed as an ethalate of the oxide of cetyle ($C_{32}H_{33}O,C_{32}H_{31}O_3$). Heintz has found, by saponifying with an alcoholic solution of potash, that various acids may be separated, viz.; stearophanic acid ($C_{36}H_{36}O_4$), margaric acid ($C_{34}H_{34}O_4$), palmitic acid ($C_{32}H_{32}O_4$), cocic acid ($C_{26}H_{26}O_4$), myristic acid ($C_{28}H_{28}O_4$), cetic or behenic acid ($C_{30}H_{30}O_4$), ethal ($C_{32}H_{34}O_2$), stethal ($C_{36}H_{38}O_4$).

Sphærosiderite. A synonyme of Carbonate of iron.

Sphærostilbite. A variety of Stilbite.

Sphaerulite. Spec. grav. 2·416 to 2·452, H 7·25. Brown and gray masses; fracture conchoidal; translucent on the edges; opaque; brittle. B.B. nearly infusible, but becomes almost enamel at the edges. Glashütte, Schemnitz. Allied to Obsidian.

Sphene. *Brunon, Braunmekanerz, Gelbmekanerz, Greenovite, Spinellane, Spinthere.* Spec. grav. 3·468, 3·2378, 3·51. Brown, yellow, green, and gray oblique rhombic prisms, with angles of 76°·2, 93°·1, or very oblique prisms; texture foliated; cross fracture conchoidal; lustre vitreous; adamantine or resinous. B.B. fuses with difficulty on the edges into a dark-coloured glass; with soda forms an opaque glass, becoming grayish-white on cooling; with borax a yellow glass; with difficulty with salt of phosphorus. SiO_3 32·52, titanic acid 43·21, lime 24·18 = 2CaO $2SiO_3$ $3TiO_2$. In primary rocks. St. Gothard; Passau; America.

Sphenomite. H 5·5. Grayish-brown crystals on black crystals of augite; soluble in NO_5. Allied to Sphene.

Spilite. Spec. grav. 2·906. Green, violet, or black porphyry. SiO_3 54·42, Al_2O_3 20·6, FeO 9·44, MnO ·93, CaO 8·64, MgO 3·87, NaO 4·48, KO ·94, HO 1·47. Haute Saone.

Spinell. *Almandine ruby, Balass ruby, Ceylanite, Spinell ruby, Pleonaste, Rubicelle.* Spec. grav. 3·523, 3·575, H 8·. Red, blue, green, yellow, brown and black regular 8-hedrons and 12-hedrons, with rhomboidal faces and in other modifications; fracture conchoidal; lustre vitreous, sometimes splendent, sometimes dull, varies from transparent to opaque. B.B. unchanged; with borax fuses into a transparent bead. SiO_3 5·62, Al_2O_3 73·308, MgO 16·632, FeO 7·42, CaO trace = (MgO, FeO,) Al_2 O_3. The silica seems foreign.

Spinellane. A synonyme of Sphene, Sodalite, and Nosiane.

Spinell-Zinc. See GAHNITE.

Spiraea Oil. A yellow oil, obtained by distilling S. ulmaria (queen of the meadow), flowers with water; it is a hydride of salicyle.

Spirit of Wine. A term applied to rectified alcoholic fluids, with a specific gravity of ·830 to ·840.

Spirits, in dyeing; a term applied to solutions of tin in acids, as used for dyeing different colours.

Spirits, Distillation of. The term spirits is applied to various dilute forms of alcohol, all derived by fermentation from sugar existing in various plants. *Brandy* is obtained from the grape by fermentation; *gin* (Genievre, juniper) from barley, rye, and is flavoured with turpentine, junipers, &c.; *rum* (saccharum) from sugar canes; *samshoo* or *showchoo* from rice in China; *aqua ardiente* from the fermented juice of a Mexican agave; *lau* from rice in Siam; *mahwah arrack* from the flowers of Bassia butyracea in India; *tuba arrack* from palm wine in the Philippines; *koumiss* from the sugar of mare's milk; *kirschwasser* in Switzerland from cherries; *goldwasser* at Dantzic from the cerealia; *maraschino* from macarska cherry, Dalmatia; *slatkaia trava* from a sweet grass at Kamtschatka; *ywera* from the root of the Tee-root in the Sandwich islands; *whisky* (usquebagh, pronounced wiskybay) from barley and oats, in Scotland and Ireland; *muchumer* from mushrooms in Kamtschatka. For the manufacture of whisky, see that article and STILL. Dutch Geneva, or gin, is said to be made in the following manner:—112 lbs. of barley meal and 228 lbs. of rye meal are mashed with 460 gallons of water at 162°; after some time cold water is added till the strength of the wort is reduced to 45 lbs. per barrel; the whole is then put into a fermenting vat at 80°, with half-a-gallon of yeast; the temperature rises to 90°, and the process terminates in forty-eight hours. The wash and grains are placed in the still; the low wines, as usual, are distilled again, and the spirits of the second distillation are rectified, so that the Hollands pass thrice through the still; a few juniper berries and some hops are used to flavour the spirits. Rum is manufactured in the West Indies from molasses or cane juice. The term is said to be derived from the last syllable of sacchar*um*. Twelve parts of sweets are dissolved in 100 water and fermented; 14 gallons of spirits of 1 to 10 overproof are obtained from 100 gallons wash. The flavour of the rum is derived from a peculiar oil existing in the cane juice; the colour of the rum is derived from the wooden casks in which it is exported.

Spodos. An ancient mixture of oxides of zinc and copper.

Spodumene. *Triphane.* Spec. grav. 3·115, 3·17, 3·188, H 6·5. Grayish-green and greenish-white oblique prisms, with angles of 93° and 87°; lustre pearly, translucent; texture foliated; streak white. B.B. swells to a foliated reddish-yellow mass, and falling to powder on slight agitation; fusing on the outside into a glass. The name is derived from σποδος, *ashes*, from its aspect before the blowpipe. SiO_3 63·812, Al_2O_3 28·508, LiO 5·604, CaO ·728, FeO ·828, HO ·36=3(NaO, LiO), $4SiO_3$, $4(Al_2O_3\ 2SiO_3)$. Utön, Sweden; Sterzing, Tyrol; Killiney, Dublin.

Sponge. (*Schwamm*, Ger.; *eponge*, Fr.) A class of animals found in water. Sometimes they consist of siliceous spiculæ, but usually of inorganic cells. The organic matter which they contain has been represented as isomeric with fibroine of silk, with the formula $C_{39}H_{31}N_6O_{17}$ or C_{96} $H_{75}N_{13}O_{44}$.

Spunk. See AMADOU.

Squama aeris. Black oxide of copper.

Stalactite. (σταλαζω, I drop.) A pendent deposit from water covering the roofs of caves in limestone rocks and then consisting of carbonate of lime, or deposited from siliceous water, and then formed of silica.

Stalagmite. (σταλαγμος, a drop.) A broad deposit on the floors or sides of caves, consisting usually of carbonate of lime.

Stannamylium. $C_{10}H_{11}Sn$. A metallo organic alkaloid by the action of iodide of amyle on tin.

Stannates. See TIN.

Stannethylium. C_4H_5Sn. B.P. 302°. Yellow oil, obtained by acting on chloride of stannethylium by zinc. The chloride is obtained by dissolving oxide of stannethylium in dilute chlorohydric acid; the oxide is again formed by acting with metallic tin on iodide of ethyle, so as to form iodide of stannethylium, which, by the influence of alkalies, becomes oxide.

Stannic Acid. See TIN.

Stannine. A synonyme of Sulphide of tin.

Stannite. Spec. grav. 3·545, H 6·5. Yellowish-white masses; lustre nearly dull; fracture conchoidal; contains 36·5 per cent. oxide of tin, also SiO_3,Al_2O_3. B.B. infusible. Cornwall; probably a mixture.

Stannmethylium. C_2H_3Sn. Obtained by a similar process to stannethylium, by substituting iodide of methyle.

Stanzaite. A synonyme of Andalusite.

Stapeline. A bitter principle in the juice of Stapelia hirsuta, resembling aloes in taste; soluble in water, alcohol, and ether.

Staphysine. *Staphysaine.* $C_{32}H_{24}NO_4$. F.P. 392°. Yellowish solid, acrid; converted by NO_5 into a bitter resin; little soluble in water; deprived of its taste at 300° by Cl; obtained from the seeds of Delphinium staphysagria (stavesacre), in the process for delphinine, which is soluble; while staphysine is insoluble.

Starch. *Fecula, Amylum, Amiline, Amidine, Amylon.* $C_{12}H_{10}O_{10}$ = C 44·26, H 6·7, O 49·04. Spec. grav. 1·53. White granules, consisting of layers, without taste and smell; insoluble in cold water, alcohol, and ether; with water of 212° converted into a thick paste produced by the swelling out rather than solution of the granules; the granules vary from $\frac{1}{200}$th to the $\frac{18}{1000}$th of an inch in diameter; when treated with hot water and cooled it yields a blue colour with iodine in solution; when boiled with diluted acids iodine gives a purple with it, and after longer boiling no colouration (dextrine); the iodide of starch is soluble at 150°; by continuing the heat the starch is converted into sugar; nitric acid changes starch into oxalic acid; bromine forms an orange compound with starch; when heated in Papin's digester to 302°, with 10 parts of water, starch is changed into small granules, $\frac{1}{12700}$th part of an inch in diameter; and at 356° sugar is formed; solution of starch is precipitated by alcohol, disacetate of lead, lime, and tannic acid.

Starch and Water.—Water, at the usual temperature and pressure, does not dissolve starch but unites with it in different proportions. Dried between 212° and 284° in a vacuum, starch retains its equivalent of constitutional water, and forms a very hygroscopic mobile powder. It absorbs from the air 2 atoms of water, the same degree of hydration obtained by drying recently prepared starch in vacuo at 60°. When kept in store it absorbs in all 4 atoms, or 18 per cent. of water, and is then known as *dry starch.* In a damp room it absorbs 6 atoms, or 35 per cent. of water, and then the grains become adherent, and do not pass through a sieve. The largest amount of water retained by starch is in *green starch*, or 45 per cent. when it is dried on a porous substance. When in this state, and thrown on plates heated to 300°, its granules swell, and adhere together. Advantage is taken of this property to make imitations of sago, tapioca, &c. The different states of hydration of starch are as follows:—

			Starch p. c.
1. Anhydrous starch, as united with lead,...............	$C_{12}H_9O_9$		
2. Dried between 212° and 284°,............................	$HO\ C_{12}H_9O_9$		100
3. Dried at 68° in vacuum,.................................	$HO\ C_{12}H_9O_9$	2HO	90
4. Dried at 68° in air,..	$HO\ C_{12}H_9O_9$	4HO	81·82
5. In the air, saturated with moisture,.....................	$HO\ C_{12}H_9O_9$	10HO	64·39
6. Thoroughly saturated with moisture,.....................	$HO\ C_{12}H_9O_9$	15HO	54·55

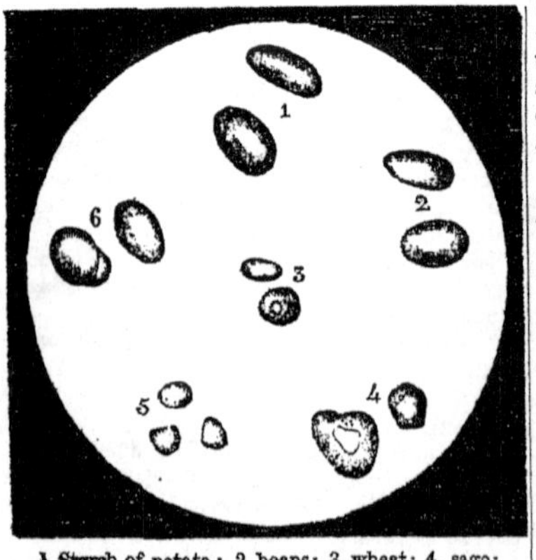

1. Starch of potato; 2. beans; 3. wheat; 4. sago; 5. maize; 6. arrow root.

Starch occurs abundantly in the seeds of wheat, oats, barley, rye, &c. and in many bulbs, as potato, &c.; *arrow root* is from the root of Maranta arundinacea; *Cassava* and *tapioca* from the root of Jatropha manihot; *sago* from Sagus rumphii and raphia; *salep* from Orchis morio, &c.

Starch may be heated to 320° without change; at 390° it becomes amber-coloured, and is partially changed into dextrine, being soluble in cold water. Starch may be heated in 12 to 15 times its weight of water to 130° without change; at 134° the younger granules begin to swell; at 160° the liquid thickens, and this increases towards the boiling point, when it becomes a consistent paste. The starch granules now occupy nearly 30 times their original volume; cold caustic soda, and potash to the extent of $\frac{2}{100}$ of the weight of starch, cause the grains to swell to 75 times their bulk. Sulphuric, chlorohydric, and nitric acids swell and dissolve starch. Starch paste under the microscope shows the granules all split; the interior layers in absorbing water swell out, and the granules have increased to 30 times their bulk. On cooling the paste to a temperature of 14°, the internal layers contract under the influence of cold water, re-enter into their envelopes, the paste loses its consistence, and the liquid acquires its former fluidity. When starch is boiled with water, and filtered through fine paper, only pure water passes through.

Preparation.—In preparing starch from potatoes, the first object is to separate the inferior potatoes from the good, the worst being used only for extracting starch. A solution of common salt is made of the spec. grav. 16° Twaddell, or 1080. The inferior potatoes swim in this liquor, while the superior sink. The former are then crushed by passing them over a cylindrical roller. The crushed matter is then thrown on a sieve, when the starch passes through. It is then placed in a series of channels, and water is run over it. It is then placed in a large cask, and further washed. It is dried in a temperature above 200°. The refuse is rinsed in a trough, along which is arranged a horizontal strainer. Most of the starch is thus removed, but much of it is not separated, and is thrown away. To make wheat starch, the flour is placed in troughs with water, and is allowed to undergo the acid fermentation, the acid dissolving the gluten. It is then filtered and dried. The soluble starch is made by introducing the dry starch into hollow revolving cylinders of iron. The cylinders are kept hot by flues from a furnace. In consequence of being pressed from behind by new additions of starch, the calcined starch makes its way to the other end of the cylinder, and discharges itself. Sago flour from Singapore is likewise extensively used for the manufacture of soluble gum. To much of the starch a blue shade is given by the addition of a small amount of smalts. Instead of salt water for ascertaining the density or levity of the potatoes, a mixture of water and clay is often employed.

Staurotide. *Grenatite, Staurolite.* Sp. grav. 3·693, H 6·75. Reddish-brown, right rhombic prisms with angles of 129° 20′, the acute edges of

a b c

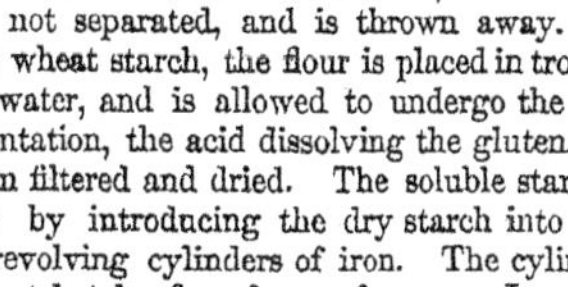

the prisms being replaced by tangent planes, making angles with the adjacent planes of the prism of 115° 20′; 2 crystals across each other in the form of a St. Andrew's cross; fracture conchoidal or uneven; lustre vitreous, inclining to resinous; translucent, frequently only on the edges; streak white. B.B. infusible, becomes dark. SiO_3 36·696, Al_2O_3 39·88, FeO 18·144, MnO 4·046, MgO ·686, H ·08. *Formula* undetermined from the tendency to replacement in its constituents. In mica slate, St. Gothard, Tyrol, Oporto, Aberdeen, United States.

Stearic Acid. (*Talgsaure*, Ger.) $C_{36}H_{36}O_4$ (Heintz), or $C_{34}H_{33}O_3HO$, or $C_{18}H_{16}O_6$ 2 HO. F.P. 167°, 156°·4 (Heintz), 154°·4 (Duffy); spec. grav., fused, ·854; solid, 1·01; solidifying point 158°. White, solid, in brilliant needles when fused and allowed to cool; pulverizable; insoluble in water; soluble in its own weight of alcohol and ether; burns with a clear flame like wax; from hot alcohol it deposits in white plates and needles; the solution reddens litmus paper; without taste and smell; by dry distillation passes over as margaric acid; heated in contact with nitric acid it becomes margaric acid; when long boiled with nitric acid it is converted into suberic and succinic acids; it is prepared by boiling mutton fat with caustic potash or soda until a clear solution can be formed in water; on adding an acid to the hot solution an oil floats on the surface, which solidifies; on being passed between porous paper and hot plates it is deprived of oleic acid, and by frequent crystallization out of alcohol and pressure it is obtained nearly pure; but there are still doubts as to its true composition. Indeed, from the difficulty of separating fatty acids, their formulæ are yet uncertain. Stearic acid candles consist of impure stearic acid, obtained by saponifying fat by means of lime. Stearic acid is the oily acid present in common white soap; it may also be obtained by heating mutton fat with half its weight of strong sulphuric acid; sulphate of oxide of glyceryle results, and impure stearic acid is set free. The stearates are neutral (MO, St) and acid (MO,HOSt); the neutral salts are soluble in 10 to 20 parts of hot water without change; the alcoholic solution of the acid salts is acid. Stearic ether, a waxy solid oil, fuses at 85° (HO,EO,St), and decomposes at 329°.

Stearine. *Stearate of Oxide of Glyceryle.* $C_{34}H_{34}O_4$. According to Duffy, mutton stearine has 3 melting points, 125°·6, 147°·5? 157°·5; solidifying point 125°·; beef stearine 123°·8, 145°·4, 152°·4; with a solidifying point of 122°·9. White pearly plates, without taste and smell; insoluble in water; soluble in $6\frac{1}{2}$ parts boiling alcohol; separating on cooling in white flocks; very soluble in boiling ether, separating on cooling, only $\frac{1}{225}$ part remaining in solution. It seems a most difficult problem to purify stearine and other fatty bodies, or to know when they are pure. Mr. Duffy obtained only $123\frac{1}{2}$ grs. stearine from 4·4 lbs. of fat; after 32 crystallizations it fused at $147\frac{1}{4}$°, its solidifying point was much below this, or $128\frac{1}{2}$°; but it fused again at 145°·4, and had passed into another isomeric form. These characters bring to mind the phenomena exhibited in the melting point of lead and its alloys. Mr. Duffy has shown that the specific gravity varies in these and other isomeric forms of stearine, being sometimes less and at other times greater than that of water. Stearine is obtained by melting mutton fat in the water bath and crystallizing out from ether, which is to be used to the extent of at least 10 times its volume. It is contained in many animal and vegetable fats and oils.

Stearoconote. An impure form of cerebric acid.

Stearolauretine. Solid fat from Laurus nobilis, separating at from 39° to 50°.

Stearolaurine. A firm fat oil, separating from the pericarp of the Laurus nobilis at 50°.

Stearole Anhydrous. $C_{68}H_{66}O_5$. F.P. 130° to 140°. Yellow brilliant, solid, lighter than water; very soluble in ether; slightly soluble in alcohol.

Stearone. $C_{68}H_{66}O_2$. F.P. 168°·8, 187°. Crystalline body, soluble in ether; by distilling 4 parts stearic acid with 1 lime. A substance, *Stearene*, $C_{28}H_{28}O$, in crystalline plates by the same process, has also been described.

Stearophanic Acid and *Stearophanine*. $C_{36}H_{36}O_4$ or $C_{35}H_{34}O_3$. F.P. 154°. By saponifying Cocculus indicus fat and spermaceti: is considered by Heintz as identical with stearic acid.

Stearoptene. The solid deposit from essential oils, frequently isomeric with the fluid oils; but is sometimes an oxide.

Steatite. *Soapstone.* Spec. grav. 2·411 to 2·396, H 1 to 1·5. Greenish-white or yellowish or variegated masses, soft at first and then soapy to the feel; texture fine earthy, feel unctuous, lustre soapy; sectile. B.B. infusible. SiO_3 43·884, MgO 24·144, Al_2O_3 9·872, HO 21·228. Cornwall, United States.

Steinheilite. See IOLITE.

Steinmannite. Spec. grav. 6·833, H 2·5. Lead-gray cubes and 8-hedrons; fracture uneven; surface smooth, lustre metallic; botryoidal, massive. B.B. decrepitates on charcoal leaving a metallic globule; it contains lead, antimony, silver, and sulphur. Przibram.

Stellite. Spec. grav. 2·612, H 3·25. Snow-white stelliform crystals; oblique 4-sided prisms; tough, resembling asbestus or nemalite, lustre silky, shining; translucent. B.B. forms a white enamel; with soda a white bead; with borax and salt of phosphorus a transparent glass. SiO_3 48·465, CaO 30·96, MgO 5·58, Al_2O_3 5·301, FeO 3·534, HO 6·108. On greenstone, Kilsyth.

Stercorite. A name applied to crystals of ammonia phosphate of soda found in guano. Spec. grav. 1·6151. $NaO,NH_3PO_5 10HO$ (Herapath).

Sternbergite. Spec. grav. 4·215, H 1 to 1·5. Pinchbeck-brown, right oblique prisms with angles of 120° 15′ or short 6-sided prisms; lustre metallic, splendent, opaque; texture foliated; laminæ flexible; very sectile. B.B. gives out SO_2 and fuses into a globule; it affords evidence of the presence of iron. Silver 33·2, iron 36, sulphur 30=4½ FeS+AgS_2. Joachimsthal, Bohemia.

Stethal, or *Hydrous Oxide of Stethyle.* $C_{36}H_{38}O_2$. A supposed constituent of spermaceti, which is then viewed as a combination of cetyle oxide and stethyle oxide with stearic, palmitic, cetic, myristic, and cocic acids.

Stibethene. $C_{12}H_{15}Sb$. The hypothetic base of the stibethyle salts (Laurent).

Stibiconise. A synonyme of antimonious acid.

Stiblite. A synonyme of dinoxide of antimony or antimonious acid.

Stib-triethyle. 3 (C_4H_5) Sb. Spec. grav. 1·3244; of vapour 7·44. B.P. 317°·3. A clear thin fluid with a disagreeable smell; solid at—20°; self-inflammable in air; difficultly soluble in water; very soluble in alcohol and ether; obtained by the action of antimonide of potassium on chloride, bromide or iodide of ethyle; by oxidation it yields oxide of stibethyle and stibethylic acid; forms salts with acids.

Stibtrimethyle. Sb, 3 (C_2H_3). Obtained by the action of iodide of methyle and antimonide of potassium. In this action occurs also *stibtetramethyle* Sb, 4 (C_2H_3) or stibmethylium.

Stilbene. $C_{28}H_{12}$. Pearly scales, obtained by distilling hydride of sulphobenzoyle ($C_{14}H_5S_2H$), a white deposit by acting on crude oil of bitter almonds in alcohol with NH_4S.

Stilbine. A synonyme of Tersulphide of Antimony.

Stilbite. *Desmine, Foliated Zeolite.* Spec. grav. 2·133 to 2·143, H 3·5. White or red slender prisms in a sheaf-like form; right prisms with rectangular bases or right rhombic prisms with angles of 101° 36′; streak white; lustre vitreous; semitransparent to translucent. B.B. an opaque vesicular bead; decomposed by acids, but does not gelatinize. SiO_3 54·805, Al_2O_3 18·205, CaO 9·83, HO 19· = $CaOSiO_3,Al_2O_3$ 3 SiO_3, 6 HO. Kilpatrick, Skye, Sweden, Nova Scotia, Kerguelen's Land, Ghauts.

Still. An apparatus for vaporizing and condensing water and volatile fluids. A still for separating spirits from water, as used in many distilleries, is represented in the following cut, where the wash descending from the reservoir into the upright column meets with steam, which rectifies it (Stein). See fig. page 469.

Stillistearic Acid. $C_{30}H_{29}O_3HO$. Spec. grav. ·818; F.P. 142°·7. Not altered at 572°; above this it distils unchanged. Pearly plates; by saponifying the oil of Stillingia sibifera or Chinese vegetable tallow. It is obviously margaric acid.

Stillolite. Allied to Opal.

Stilpnomelane. Spec. grav. 3·35, H 3·. Greenish-black foliated or radiated or compact mass; lustre glassy or pearly; streak greenish. SiO_3 46·17, FeO 35·82, Al_2O_3 5·88, MgO 2·67, KO ·75, HO 8·72. B.B. fuses at length to a black globule. Obergrund.

Stilpnosiderite. A synonyme of Protohydrous Sesquioxide of Iron.

Stinkstone. See ANTHRACOLITE.

Stolzite. A synonyme of Tungstate of Lead.

Stomach. The organ in animals destined for the digestion of food. See GASTRIC JUICE.

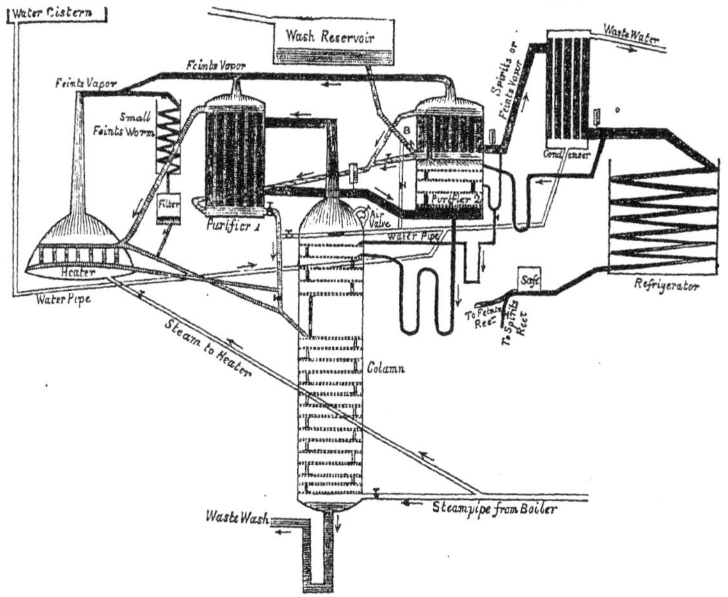

Stein's Still.

Stomoma. An ancient name for oxide of copper, formed by igniting copper.

Storax. A fragrant resin, in the state of red tears or in large reddish cakes, brittle and soft; soluble in alcohol; from the Styrax officinalis. It affords styracine, styrone, and styrole. The resin seems to consist of volatile oil, resin, styracine, and cinnamic acid.

Strahlstein. A synonyme of Tremolite or Amphibole.

Strahlzeolite. A synonyme of Stilbite.

Stralite. A synonyme of Epidote.

Stramonine. F.P. 302°. White needles without odour and taste; insoluble in water; soluble in alcohol and ether; reddened by SO_3; unacted on by NO_5; found in Datura stramonium along with daturine.

Strass. A volcanic rock from the Rhine. It consists of SiO_3 57·, Al_2O_3 16·, CaO 2·6, MgO 1·, KO 7·, NaO 1·, FeO and TiO_2 1·5, HO 9·6.

Strass, Artificial. A preparation with a fine lustre resembling diamond. SiO_3 38·8, PbO 53·, KO and NaO 8·2.

Strelite. A synonyme of Anthophyllite.

Stroganowite. A green foliated variety of cancrinite from the river Slüdänka in Dauria.

Stromeyerite. *Stromeyerine.* See SULPHO-CUPRITE OF SILVER.

Stromnite. See BARYTO STRONTIANITE.

Strontian. *Strontia, Oxide of Strontium.* SrO 6·5, 52; 6·475, 51·8. Spec. grav. 3 to 4 (Davy); 3·9321 (Karsten). Grayish-white porous mass; taste acrid, alkaline; converts vegetable blues to green; infusible by ignition, fixed; *soluble* in 160 water at 60°, forming *strontian water*; it is much more soluble in boiling water. Tinges flame of a crimson colour (Dr. Ash, 1787). The experiment may be shown by placing nitrate of strontian in the wick of a candle (Vauquelin), or by igniting alcohol, containing chloride of strontium in solution (Pelletier). *Process.*—Strontian may be obtained—1. By igniting the carbonate mixed with charcoal in a plumbago crucible (Hope). 2. By igniting the nitrate in a covered crucible. *Protohydrate*, SrOHO 7·625, 61, white powder or grayish fibrous mass, losing its water by intense heat (Smith, Phil. Mag. 9, 87); obtained by slaking strontian like lime. *Novohydrate*, SrO9HO 16·625, 133. *Crystal.*—Square prismatic system; generally as thin quadrangular plates or parallelograms; sometimes the edges are plain, but they oftener consist of two facettes meeting together like the roof of a house, and adhere together so as to form a plate one inch or more in length; sometimes cubical. Primary form, a right square prism terminated by a 4-sided truncated pyramid, the face of the prism making with the corresponding face of the pyramid an angle of 132°·12 (Brooke, Ann. Phil. 2d ser. 7, 287) *Soluble* in 51·4 water, at 60°; in 2 boiling water (Hope); 2·4 (Bucholz); loses 50 per cent. of water at 212° (Smith), without melting; fuses at a red heat, which is not the case with strontian from the nitrate (Bucholz).

Process.—Obtained on cooling from a concentrated hot solution of strontian.

Hydrous Binoxide, Peroxide. — SrO_2HO? Pearly scales, obtained by mixing binoxide of hydrogen with strontian water.

Chloride.—SrCl 10·, 79·4; 9·912, 79·296. Spec. grav. about 2·803 (Karsten). White powder or white semi-glass when ignited; fixed; non-conductor of electricity; *taste* sharp and bitter; loses chlorine and absorbs oxygen when ignited in air (Kraus, Pogg. Ann. 43, 139); *soluble* in 116·4 cold alcohol of 99·3 per cent., and in 262 boiling (Fresenius, Quant. Anal.); obtained by dissolving the carbonate of strontian in chlorohydric acid and heating to redness.

Sexhydrated Chloride.—SrCl 6 HO 16·75, 16·675, 133·4; 6-sided needles or prisms, generally with 2 faces larger than the other 4; deliquescent; *soluble* in ·66 water at 60°, in any quantity of boiling water, in 24 absolute alcohol at 60°, in 19 boiling absolute alcohol, in ·75 water at 60°, in 6 alcohol (spec. grav. 833,) (Vauquelin). The alcoholic solution burns with a fine crimson flame. By alcohol it can be separated from chloride of barium. By heat it undergoes the watery fusion, and by ignition becomes anhydrous. *Process.*—It is formed by dissolving the sulphide of strontium, prepared from the sulphate or the native carbonate in chlorohydric acid and crystallizing.

Salts of Strontian.—1. Salts of strontian are more soluble than salts of barytes, but less so than salts of lime. 2. Solutions of strontian salts are not precipitated by ammonia; precipitated by soda; by sulphate of lime only after some time. 3. Succinate of ammonia precipitates salts of barytes, but not salts of strontian. 4. Salts of strontian when heated before the blowpipe communicate to the flame a red colour, barytes rendering it green, and lime red. 5. Fluosilicic acid precipitates barytes, but not strontian. 6. Chromate of potash precipitates barytes immediately, strontian slowly. *Use.*—It may be employed in fireworks to produce a crimson flame.

Sulphate, Cœlestine.—$SrOSO_3$ 11·5 92·, SrO 56·52, SO_3 43·48. Spec. grav., artificial, 2·5883 (Karsten); native, 3·86 (Thomson); 3·953 (Breithaupt). *Crystal.* — Right rhombic prism, the faces meeting at angles of 104° and 76°, having the same form as heavy spar, from which it differs by about 2° 20′ of greater obliquity. *Tasteless. Soluble* in strong sulphuric acid and precipitated again by water, in 15,029 water at 51°·8 (Brandes and Silber), in 3600 at 60°, in 6895 at 57¼° (Fresenius), in 3840 boiling water (Hope), in 9,638 boiling water (Fresenius). Its solubility is diminished by the presence of sulphate of soda and sulphuric acid (Andrews); soluble slowly in a solution of common salt, and thus distinguished from heavy spar, and again precipitated by sulphuric acid. *Sources.*—Occurs in nature most commonly in the red sandstone formation, as at Bristol, and in North America, Paris, Sicily.

Nitrate Anhydrous.—$SrONO_5$ 13·25, 106. Spec. grav. 3·006. *Crystal.*—Transparent regular 8-hedrons or cube 8-hedrons with an adamantine lustre. *Taste* sharp and cooling; *soluble* in 1 water at 60°, in little more than 5 boiling water; insoluble in alcohol; permanent in air; deflagrates on hot coals, fuses by ignition and leaves strontian (Hope). *Process.*—First obtained by Dr. Hope, in 1791; but most completely described by Vauquelin, in 1797 (Journ. Min. Ann. vi. 7); by dissolving the carbonate in nitric acid and evaporating to dryness, dissolving in water and recrystallizing.

Nitrate Quintohydrated. — $SrONO_5$ 5 HO 18·875, 151. *Crystal.*—Oblique rhombic prisms, with angles of 103° 40′; 66° 20′. The edges are sometimes replaced by one or even two planes inclined to M and M′ respectively, at angles of 150°; the two new planes being inclined to each other at an angle of 126° (Brooke); lustre vitreous; efflorescent; by heat undergoes the watery fusion; *soluble* in 5 cold and ·5 boiling water; insoluble in absolute alcohol, a property enabling us to separate it from nitrate of lime, which is soluble in alcohol. *Process.*—It is deposited sometimes when an aqueous solution of the anhydrous salt is set aside for crystallization. *Use.*—It is employed by pyrotechnists in fireworks, the red fire being formed of nitrate of strontian well dried 40 parts, flowers of sulphur 13 parts, chlorate of potash 5 parts, tersulphide of antimony 1 part.

Carbonate, Strontianite.—$SrOCO_2$ 9·25, 74. Spec. grav. native, 3·605 (Mohs). Right prismatic system (Mitscherlich) as found native; commonly as brown and green radiated masses; artificially prepared it is a white powder, fusing at 226° Wedgewood to a clear glass (Saussure, Jour. Phys. 45, 24); *tasteless*; alkaline reaction when native (Pleischl); *soluble* in 18,045 cold water (Fresenius), in 11,862 water acidulated with chlorohydric or sulphuric acids, and in 1536 boiling water (Hope); soluble in cold solutions of salammoniac, nitrate and succinate of ammonia (Vogel). When mixed with charcoal and strongly heated in a crucible (Hope), or simply by the heat of a powerful forge it is decomposed (Gay Lussac and Thenard); volatilizes by a stream of oxygen on charcoal; fuses at the edges before the blowpipe, swelling up and becoming incandescent, giving a red colour to the flame (Berzelius). *Sources.*—It is found native, always, however, accompanied with carbonate of lime. It may be prepared by precipitating solutions of the soluble salts of strontian with alkaline carbonates.

Estimation of Strontian.—1. This earth may be estimated by adding sulphuric acid to a solution of the chloride of strontian in alcohol, and washing with alcohol. 2. It may also be precipitated by an alkaline carbonate, and estimated as carbonate of strontian. 3. Strontian may also be thrown down by oxalate of ammonia. The oxalate of strontian is, however, somewhat more soluble in water than the lime salt.

Separation of Barytes and Strontian.—1. They may be separated imperfectly by converting them into chlorides and dissolving out the chloride of strontian by alcohol. 2. The barytes may be precipitated by converting the earths into chlorides or acetates, and passing fluosilicic acid through the solution, the barytes is precipitated as silicofluoride of barium, while the corresponding salt of strontian remains in the solution, dissolved by the excess of acid. It is washed and estimated on a weighed filter. The strontian is separated from the filtered liquid by sulphuric acid. In a dilute solution the fluosilicic acid takes some time to act, the solution must therefore be allowed to stand. 3. Dilute neutral chromate of potash, free from sulphate, precipitates in very dilute solutions barytes, but not strontian. The strontian is removed from the filtered liquor by an alkaline carbonate or oxalate.

Strontianite. A synonyme of Carbonate of Strontian.

Strontium. Sr 5·5, 44·; 5·475, 43·8. White solid, much heavier than water, resembling barium, ductile, rapidly absorbing oxygen in air or water, and becoming strontian (SrO). It is procured by similar processes to those for barium.

Struvite. A name given to ammonia phosphate of magnesia (NH_3 2 MgO,PO_5, 13 HO) found at Hamburg.

Strychnic Acid. White needles, soluble in water; little soluble in alcohol; by boiling 3 strychnine, 1 chlorate of potash, a few drops of sulphuric acid, and 10 distilled water.

Strychnine. $C_{42}H_{22}N_2O_4$. Not fusible; decomposes at 600°; brilliant 8-hedrons or 4-sided prisms, terminated by 4-sided pyramids; without smell; with a bitter metallic taste; soluble in 7000 cold, 2500 hot water; its solution in water when diluted 100 times is still bitter; soluble in strong nitric acid, with a yellow or greenish-yellow colour; red when brucine is present; strong sulphuric acid colours it brownish-red, then yellow; solutions of gold and silver render its solution brownish-red, and bluish; it becomes green with the violet solution of chameleon mineral; tincture of galls renders it white; spirit of ·870 dissolves 5 per cent.; of ·934 $\frac{1}{240}$; insoluble in ether and alkalies; sulphocyanide of potassium forms white crystals with it; heated with bases it yields quinoline ($C_{18}H_8N$). Strychnine is prepared from the seeds of the Strychnos nux vomica; the bean of S. Ignatii, and wood of S. colubina. An aqueous decoction is made with the addition of dilute sulphuric acid; the decoction is precipitated with milk of lime, and the precipitate exhausted with boiling alcohol, which deposits the alkaloid in crystals on cooling; brucine being more soluble, remains in the liquid; but it may be thoroughly separated by converting both into nitrates; the strychnine salt crystallizes while the nitrate of brucine remains in solution. The *sulphate*, $SrSO_3$ 8 HO, colourless rectangular prisms with a vitreous lustre, obtained by dissolving strychnine in SO_3 and crystallizing. *Test for Strychnine.*—Moisten the powdered strychnine with a drop of strong sulphuric acid, adding a small portion of binoxide of lead or bichromate of potash, a violet colour is produced which disappears in a few minutes. Organic matter interferes with the production of the colour. The strychnine should be withdrawn from organic solutions by animal charcoal, and the latter boiled with spirit of wine and distilled, when the alkaloid remains, which is purified with potash and ether.

Stupp. A black matter mixed with mercury, obtained in the mercurial works of Idria. See Idryle.

Stylobite. A synonyme of Gehlenite.

Styphnic Acid. *Oxypicric Acid.* $C_{12}N_3H_3O_{16}$. Yellow regular 7-sided prism; partly subliming undecomposed; fusible; carried off by the vapour when boiled with water; acid reaction; soluble in water and alcohol; stains the skin yellow; taste bitter; feels gritty between the teeth; obtained by acting on asafœtida with sulphuric and nitric acids, or by acting on extract of Brazil-wood with nitric acid; it forms monobasic crystalline salts.

Styracine. $C_{24}H_{11}O_2$, or $C_{36}H_{16}O_4$. F.P. 122°. White neutral needles; insoluble in water; soluble in 3 boiling, 22 cold alcohol; in 3 ether; converted by distillation with NO_5 into cyanohydric acid, hydride of benzoyle, benzoic acid, picric acid, cinnamic acid, and resin; obtained by distilling liquid storax with caustic potash and water.

Styracone. *Styrone, Hydrous Oxide of Styryle.* $C_{18}H_{12}O_2$. B.P. 428°. An oil heavier than water; soluble in 95 cold, 30 hot water; obtained by distilling styracine with strong caustic potash; it has the same relation to hydrous cinnamic acid as alcohol to acetic acid.

Styrole. $C_{16}H_8$. B.P. 293°. An oil with an aromatic smell, evaporating at all temperatures; becoming at 399° in a close tube a solid (metastyrole), which is isomeric; it is reconverted into the oil by a stronger heat; forms substitution compounds with NO_5, Cl, and Br; it is prepared by distilling storax with water.

Styryle. $C_{18}H_9$. The hypothetic basis of the derivatives from storax.

Sub. The Latin term *under*, as applied to basic compounds.

Suberanilic Acid. $C_{28}H_{19}NO_8$. F.P. 262°. Crystalline acid from the mother liquor of suberanilide.

Suberanilide. $C_{20}H_{12}NO_2$ = Suberate of aniline, minus 2HO. F.P. 361°. Pearly scales by aniline and suberic acid.

Suberic Acid. (*Suber*, bark, Lat.; *korksaure*, Ger.) $HOC_8H_6O_3$. F.P. 130° moist, 238° dry. White crystalline powder; above its fusing point sublimes into an oil, which becomes solid in long needles; soluble in hot, and sparingly in cold water; soluble in hot alcohol, in 6 cold, in 10 hot ether; the suberates are monobasic, the alkaline salts soluble, the earthy salts insoluble; obtained by boiling nitric acid with stearic, margaric, or oleic acids, evaporating to one-half; on cooling, suberic acid separates.

Suberic Ether. $EO,C_8H_6O_3$. Spec. grav. 1003; B.P. 500°. Colourless oil, formed by passing chlorohydric acid gas through a solution of suberic acid in alcohol.

Suberone. C_8H_7O. B.P. 367°. Colourless pleasant smelling fluid, by distilling suberic acid with lime; it may be the hydride of C_8H_6O; by air or NO_5 it is converted into suberic acid.

Suberose. *Suberine.* See CORK.

Suberyle. C_8H_6O. The hypothetic radical of suberic acid.

Sublimate, Corrosive. See CHLORIDE OF MERCURY.

Sublimation. The volatilization and condensation of a solid by heat.

Subresin of Anime. The portion of animé resin insoluble in cold alcohol. It crystallizes from hot alcohol.

Substitution. See METALEPSY.

Succinamide. C_8,NH_2,N_4O_2,O_2. White needles converted by cold NO_5 into succinic acid; soluble in 220 cold, in 9 boiling water; insoluble in alcohol and ether; decomposed at 392°; obtained by precipitating succinic ether with ammonia.

Succinanile. $C_{12}(C_8H_5O_2)NH_4O_2$. F.P. 311°. Long needles, sublimable; soluble in water, HCl, NO_5, alcohol, and ether; obtained by fusing succinic acid with excess of aniline, and crystallizing out of ether.

Succinanilic Acid. $C_{12}(C_8H_5O_2)NH_2, H_4,O_4$. F.P. 314°·6. Crystalline plates, by dissolving succinanile in boiling dilute ammonia, neutralizing with NO_5, and crystallizing out of alcohol.

Succinanilide. $C_{24}(C_8H_5O_2)N,NH_2,H_9, O_2$. F.P. 428°. Fine needles, insoluble in water; soluble in alcohol and ether; formed by heating excess of aniline with succinic acid for ten minutes, and separating the succinanile by water.

Succinic Acid. (*Bernsteinsäure*, Ger.) HO, $C_8H_6O_2,O_6$. Spec. grav. 1·55; F.P. 356°; sublimed acid 320°; B.P. 455°; sublimed 467½°. Fine rhombic prisms, evaporating by heat in vapours; without smell, acid taste; obtained by distilling amber; by oxidating stearic acid or wax by NO_5; it remains in the mother liquor when suberic acid crystallizes, from which it may be separated by sublimation; also by fermentation from malic acid; soluble in 24 cold, 2 boiling water; forms monobasic and acid salts. The succinate of ammonia is used to precipitate sesquioxide of iron in separating it from manganese.

Succinimide. *Bisuccinamide.* $C_8NH_5 O_2,O_2$. White crystals by heating succinate of ammonia

Succinite, or **Colophonite,** an amber-coloured variety of garnet.

Succisterene. Probably idrialine, formed by distilling amber; a solid rendered blue by SO_3.

Sucric Acid. See SACCHARIC ACID.

Suet. A term applied to the most solid forms of fatty bodies.

Sugars. (*Saccharum*, Lat.; *zucker*, Ger.; *sucre*, Fr.) A term applied to a class of bodies which possess a sweet taste. Some of them are decomposed in contact with a ferment, while others resist this action. The following are the sugars at present known:—

Fermentable.

Cane sugar,..........	$C_{12}H_{11}O_{11}$
Glucose, including grape sugar and diabetine, ...	$C_{12}H_{14}O_{14}$
Lactine,...............	$C_{24}H_{24}O_{24}$
Ergot sugar,..........	$C_{12}H_{13}O_{13}$

Unfermentable.

Mannite,..............	$C_6H_7O_6$
Inosite,...............	$C_{12}H_{16}O_{16}$
Sorbite,...............	$C_{12}H_{12}O_{12}$
Quercite,..............	$C_{12}H_{12}O_{10}$
Phycite,...............	$C_{12}H_{15}O_{12}$
Glycyrrhizite, ...	$C_{16}H_{12}O_6$
Glycocoll,, ...	$C_4H_4O_3NHO$

Cane Sugar.—$C_{12}H_9O_9$ 2HO. Spec. grav. 1·6065 (Fahrenheit); 1·5629 (Thomson). F.P. 287½° (Proust), 356° (Peligot). Colourless 4 or 6-sided rhombic prisms, evolving light when rubbed in the dark; at 419° loses 2 atoms water, and becomes black caramel; above this, yields CO_2 and inflammable gases, acetic acid, and charcoal; 3 parts soluble in 1 cold, in all proportions in hot water, forming *syrup*, with a specific gravity of 1·086; when syrup is cooled the sugar separates in crystals; when kept long above the boiling point it loses its tendency to crystallize; the same effect is produced by the addition of oxalic, citric, and malic acids; soluble in 80 boiling alcohol, in 4 spirit (·830); sulphuric acid dissolves sugar with a brown colour; when this solution is saturated with chalk a saccharo-sulphate is obtained; treated with dilute sulphuric acid and heat, formic acid, humic acid or humus are formed; chlorohydric acid forms a blackish paste; nitric acid converts it into oxalic acid; 100 sugar yield 58 acid (Cruickshank), 67 (Thenard). Arsenic acid forms with sugar a red and then a purple colour; sugar dissolves carbonate of copper to a green fluid; hydrous oxide is soluble in sugar on the addition of an alkali; all these when boiled with an alkali yield formic acid, and deposit metallic copper, or dinoxide of copper; the last, united with a brown matter, soluble in ammonia; nitrate of silver heated with sugar yields metallic silver, and black oxide (Vogel). All these are tests of the presence of sugar; it unites with 2 atoms oxide of lead; with NaCl and 3 HO, with 1 CaO and 1 HO.

West Indian Method of Manufacture.—The juice of the cane is calculated to amount to 90 per cent. of the total weight of the plant, the remainder consisting of woody tissue and inorganic matter, and yet the amount of juice usually obtained by passing the canes through the sugar mill does not average more than 50 per cent. in the French colonies: 70 per cent. has, however, been obtained. The mill usually consists of 3 rollers, which are placed either horizontally or vertically. The hydraulic press has been tried, but not hitherto successfully. It has been proposed to dry the canes, and send them to Europe for expression. Michiel proposed to cut the canes into slices, and to act upon them with lime and water, so as to coagulate albuminous matter. Liebig suggested the coagulation of the albumen by heat previous to the extraction of the juice, so as to leave the fermenting principle in the cane, and obtain the sugar free from it; and I have excellent specimens of sugar prepared in Trinidad by this plan by Mr. Mitchell.

Defecation or Clarification of Cane Juice.—1. The juice from the mill is placed in copper or wooden cisterns lined with lead, or at once run into the clarifiers, shallow copper pans almost flat, or arched slightly upwards at bottom, capable of holding from 250 to 500 gallons, each being hung over a separate fireplace. A temperature of 140° is applied and slaked or milk of lime is added. To ascertain the amount of lime required, it is usual to fill three or four wine glasses with cane juice from the clarifiérs, and to drop into each increased quantities of lime water; that which produces the clearest solution being taken as indicating the quantity required. On adding the lime the juice is stirred, the temperature increased to near the boiling point, and until a thick scum forms on the top which cracks, and shows the liquor below. The fire is withdrawn and the juice stands till it clarifies and becomes yellow. The lime precipitates mechanical impurities, and together with the heat coagulates the albuminous matter. The lime also saturates acid formed, but still the lime may produce a change in a portion of the sugar, converting it into *glucic*, and *melassic acids*. After the action of heat and lime it gives a precipitate with diacetate of lead, and possesses a colour. Various other methods of clarifying have been tried—as by lime and filtration; (2.) by nutgalls (2 oz. to 400 galls. of juice); (3.) by sulphate of zinc, (12 oz. to 300 galls.); (4.) by alum (4 lbs. to 350 galls.); (5.) by sulphate of alumina, neutralizing with lime (1 lb. to 100 galls.); (6.) by a mixture of sulphates of alumina and zinc; (7.) by alumina and sulphate of lime, adding chalk to 2 lbs. alum till effervescence ceases, washing the precipitate, and adding it to 100 lbs. juice; (8.) by sulphuric acid; (9.) Messrs. Guyne and Young have recommended the use of acetate of lead, filtration, and then diphosphate of lime to remove excess of lead; and, lastly, (10.) Dr. Scoffern has patented the use of the diacetate with sulphurous acid to remove excess of lead. On the small scale this plan succeeds well, but on the large scale it generally leaves traces of lead; (11.) by the bark of the wild elm (Theobroma guazuma.)

Concentration of Syrups.—In the West Indies the liquor is run from the clarifiers to the coppers, where renewed defecation and evaporation are carried on by means of double steam pans to the spec. grav. 1240.

Filtration through Animal Charcoal.—The syrup thus concentrated is passed through beds of animal charcoal, arranged in a variety of ways. In Dumont's filter the charcoal is placed in a square pyramidal wooden vessel supplied with a false bottom. In Peyron's filter there is a series of copper vessels with double bottoms hermetically closed at the top, 6 feet high, and 3 in diameter.

Evaporation in Vacuo.—The syrup is concentrated by boiling in a pan or in a copper vacuum pan (Howard, 1819); and from thence passes into wooden coolers, where it becomes partially crystallized, mixed with molasses. It is then carried in buckets to the curing house, thrown into a hogshead, where it remains until the drainage of the molasses is sufficiently complete to admit of its being shipped, or the sugar may be placed in moulds for dripping.

Claying, Liquoring, or Syruping.—When fluid ceases to drop from the moulds, the means formerly employed consisted in applying to the upper surface of the sugar a mass of clay beat up with water to the consistence of a thick cream. The clay, acting as a sponge, allows it to separate from it slowly, and to percolate through the sugar, so as to replace the dark molasses. When the clay dries it is removed, and a new portion added, until the sugar is of the proper colour. Instead of claying, except in Cuba and Brazils, the sugar in the moulds is washed with a saturated solution of sugar or syrup; and washing is now conducted by introducing the moist sugar into a centrifugal machine, where the moisture is thrown off.

Sugar of Lead. See Acetate of Lead.

Sugar of Starch, or **Grapes.** *Glucose, Honey Sugar, Mushroom Sugar, Diabetic Sugar.* (*Traubenzucker*, Ger.) $C_{12}\ H_{14}\ O_{14}$. Spec. grav. 1·3861 (Guerin). Colourless square hard plates or prisms; soluble in $1\frac{1}{2}$ cold water, in all proportions in hot water; in 8 spirit; in 20 alcohol; solution is less sweet than with cane sugar; obtained from the juice of grapes also by heating starch with $\frac{1}{100}$th of oil of vitriol and 4 times its weight of water; it forms 12-

hedrons with common salt containing 75 sugar and 25 NaCl; sulphuric acid dissolves it without blackening it, converting it into saccharo-sulphuric acid, which distinguishes it from cane sugar; alkalies do not colour cane sugar, but blacken starch sugar. When placed in contact with yeast or decomposing bodies, as putrid cheese or flesh, sugar is converted into carbonic acid and alcohol, $C_{12}H_{12}O_{12}$, becoming 2 alcohol $= 2(C_4H_6O_2$ and $4\ CO_2)$.

Sulphamide. $NH_2SO_2 = AdSO_2$. A white powder, by the action of chlorosulphuric acid on NH_3.

Sulphamidic Acid. $NH_3S_3O_9$, or $5\ SO_3(SO_2NH_2)\ 3\ HO$, by the decomposition in cold water of sulphammonates.

Sulphamethylane. $C_2H_5NS_2O_6$. Crystalline body, by a current of NH_3 on neutral sulphate of methyle; it seems to be oxamethylane, in which sulphamide (NH_2SO_2) replaces oxamide, or SO_2 for C_2O_2.

Sulphammonic Acid. $8\ SO_2,NO_3\ 3\ HO$, or $7\ SO_3(SO_2NH_2)\ 4\ HO$. Obtained in union with potash, by passing a current of SO_2 into a solution of a nitrite, previously rendered alkaline; in cold water it changes into SO_3 and bisulphate of ammonia.

Sulphamylic Acid. A synonyme of sulphate of amyle and water, Ayl $OSO_3HO\ SO_3$, formed by heating fusel oil with sulphuric acid.

Sulph-amylo-sulphuric Acid. $C_{10}H_{11}S_2O_4,HO$. Probably identical with hyposulphamylic acid, which is formed by acting on sulphocyanide of amyle with nitric acid.

Sulphanisole. By the action of SO_3 on anisole.

Sulphanisolic Acid. By the action of strong SO_3 on sulphanisolide.

Sulphanisolide. $C_{14}H_7SO_4$. Silvery needles, by passing the vapour of anhydrous sulphuric acid into anisole in a freezing mixture.

Sulphantimonic Acid. Quadrosulphide of antimony when united to sulphides.

Sulphantimonic Acid. Quintosulphide of antimony when in union with sulphides.

Sulpharsenic Acid. A name applied to quintosulphide of arsenic when united to sulphides.

Sulpharsenious Acid. Tersulphide of arsenic when in union with sulphides.

Sulphasatyde. $C_{16}H_6NO_3S$. White crystalline powder by KO on sulphesatyde.

Sulphato Carbonate of Lead. See LEAD.

Sulphazic Acid. $4SO_3,NO_3,3HO$. Obtained by dissolving sulphazite of potash in an alkaline liquor, and treating it with an additional quantity of sulphurous acid; more SO_2 produces sulphazotic acid.

Sulphazidic Acid. $S_2O,NO_3\ 3HO$ or $S_2NH_2O_7HO$ or $2SO_2,NO2HO$. By the decomposition of sulphazotate of potash in water; neutral sulphazotate, bisulphate of potash, SO_2, and sulphazidate are formed.

Sulphazilic Acid. $4SO_2NO_3HO$ or S_4HNO_{12}. By treating sulphazotate of potash with binoxide of lead—metasulphazilate is also formed.

Sulphazobenzoyle Hydride. $C_{42}H_{15}S_4N_2$. Formed along with hydride of sulphobenzoyle.

Sulphazotic Acid. $5SO_2NO_33HO$ or $2SO_3,3SO_2NO_32HO$. An acid obtained by heating a sulphazotite in excess of potash with sulphurous acid.

Sulphazotous Acid. $2SO_2,NO_32HO2SO_3$. An acid in union with potash, at a particular stage of the action of sulphuric acid on an alkaline nitrite.

Sulphazous Acid. $3SO_2,NO_33HO$ or $S_3,NH_3,O_{12}3HO$. Obtained united to potash by the first action of sulphurous acid on nitrite of potash; it crystallizes out.

Sulphesatine. See SULPHESATYDE.

Sulphesatyde. $C_{16}H_6N,O_2S_2$. Greyish-yellow powder by SH on an alcoholic solution of isatine.

Sulphessale. Needles along with stilbene.

Sulphethamic Acid. $C_{16}H_{23}NO_4$. Obtained in union with NH_3 by acting on neutral sulphate of ethyle with dry NH_3.

Sulphide. A combination of sulphur with a metallic base, as sulphide of iron (FeS), bisulphide of iron (FeS_2).

Sulphindelic Acid. See INDIGO.

Sulphoacetic Acid. $C_4H_2S_2O_8,2HO3HO$. F.P. 143½°. Needles or silky fibres by anhydrous sulphuric acid on protohydrate of acetic acid. See ACETO SULPHURIC ACID.

Sulphoacetylic Acid. $C_4H_42SO_3$. Needles by the action of solid sulphuric acid on olefiant gas.

Sulphoamylolic Acid. $C_{10}H_{12}S_2O_5,HO$.

Sulphobenzide. $C_{12}H_5SO_2$. By anhydrous sulphuric acid on benzole.

Sulphobenzoic Acid. A synonyme of Hyposulphobenzoic acid.

Sulphobenzoyle Hydride. *Sulphobenzole.* $C_{14}H_5S_2,H$. White powder with the smell of garlic, by adding sulphide of ammonium to an alcoholic solution of oil of bitter almonds.

Sulphocacodylic Acid. $C_4AsH_6O_3$.

Sulphocamphic Acid. $C_{20}H_{13}S_2O_5HO$. By the action of SO_3 on camphogene.

Sulphocamphoric Acid. *Camphosulphuric Acid.* $C_9H_7O_3,SO_23HO$. F.P. 320° to 331°. 6-sided prisms; soluble in water, alcohol, and ether; insoluble in oil of turpentine and bisulphide of carbon; by anhydrous sulphuric acid on camphoric acid.

Sulphocarbamic Acid. C_2NH_2,S_3. Produced in the action of ammonia on bisulphide of carbon.

Sulphocarbamide. C_2NS_2H. From ammonia on bisulphide of carbon. Identical with hydrosulphocyanic acid.

Sulphocarbanilide. $C_{13}H_6NS = C_{12}H_6N,CS$. F.P. 284°. A crystalline substance, obtained by the decomposition of hydrosulphocyanate of aniline; slightly soluble in water, readily in alcohol and ether.

Sulphocarbolic Acid. *Sulphophenic Acid.* $C_{12}H_5O,HO+2SO_3$. Obtained by adding excess of sulphuric acid to carbolic acid, allowing it to stand for twenty-four hours, then saturating with carbonate of barytes, filtering, and evaporating till the sulphocarbolate of barytes crystallizes out. When this salt is purified by alcohol, and decomposed by dilute sulphuric acid, the sulphocarbolic acid is obtained.

Sulphocarbomethylic Acid. $C_2H_4O_2,2CS_2$. The potash salt is obtained in fine silky needles by mixing caustic potash and bisulphide of carbon with pyroxilic spirit.

Sulphocarbonic Acid. HCS_3. Reddish-brown, transparent, oily fluid, heavier than water; odour similar to that of sulphuretted hydrogen. Produced when dry sulphocarbonate of ammonia is put into dilute hydrochloric acid. Soluble in aqueous HCl or SO_3.

Sulphocetic Acid. $C_{32}H_{32}$ OHO, SO_3.

Sulphocetylic Acid. 2 SO_3 HO, $C_{32}H_{33}O$. A liquid homologous with sulphovinic acid.

Sulphocinnamic Acid. $C_{18}H_8$ O_4,SO_3.

Sulphocumenic Acid. $C_{18}H_{11}S_2O_6$.

Sulphocyanic Acid. See SULPHOCYANOHYDRIC ACID.

Sulphocyanogen. *Rhodanogen.* $C_8N_4S_8$? The radical existing in sulphocyanides.

Sulphocyanohydric Acid. *Rhodanhydric Acid.* CyS,HS. Spec. grav. 1·022. B.P. 217°. Crystallizes at 14°. Occurs in the seeds and blossoms of Cruciferæ, and in the saliva of man and sheep; by decomposing the sulphocyanide of silver by SH. A colourless, very acid fluid, possessed of a very strong odour.

Sulphocymenic Acid. $C_{40}H_{26}(SO_2)SO_3$. An oily substance by dissolving cymene (which exists in the essence of cumin) in SO_3.

Sulphodraconic Acid. $C_{24}H_{16}SO_8$. By SO_3 on the oil of Artemisia dracunculus.

Sulphoethylic Acid. A synonyme of Sulphovinic acid.

Sulpho-ethyl-sulphuric Acid. $C_4H_8O_2$ $2SO_2$. Spec. grav. 1·30. Heavy oil. By nitric acid on mercaptan.

Sulphoglyceric Acid. $C_6H_7O_5,HO2SO_3$. By the action of sulphuric acid on glycerine; readily decomposed by heat.

Sulphohelenic Acid. $C_{19}H_{25}SO_3$. Obtained by exposing helenine to the action of fuming sulphuric acid. The acid is not produced in the cold.

Sulphohydric Acid. A synonyme of sulphuretted hydrogen.

Sulphohydrokinone. *Brown.* $C_{12}H_5O_4S_2$. Formed when a solution of kinone is exposed, partially to a current of sulphuretted hydrogen. *Yellow.*—$C_{12}H_6O_4S$. By the further action of SH on kinone.

Sulphohyposulphuric Acid. $S_3O_5 = SO_2SO_3S$.

Sulphoindigotic Acid. $C_{16}H_4NO_2,S_2O_5$(?) Produced along with sulphopurpuric acid in the action of sulphuric acid on indigo. From 8 to 10 sulphuric acid to 1 of indigo are the proportions necessary.

Sulpholeic Acid. Not yet isolated. Formed when oleine is acted on by an excess of sulphuric acid.

Sulpholignic Acid. $C_8H_4SO_6$. By SO_3 on lignine.

Sulphomannic Acid. $C_6H_5O_4,2SO_3$. Obtained by pouring concentrated sulphuric acid on mannite.

Sulphomargaric Acid. A substance not yet isolated; formed when sulphuric acid in excess is added to margarine.

Sulphomellone. The supposed radical of hydrosulphomellonic acid. ($C_6N_4H_3S_4H$.) Its composition has not yet been determined.

Sulphomesitylic Acid. $C_{12}H_{10}OSO_3$. By adding to mesitylic alcohol twice its weight of concentrated sulphuric acid.

Sulphomethylic Acid. $HO,C_2H_3,2SO_3$. A crystalline substance, perfectly analogous with sulphovinic acid. Best obtained by the action of hot water on the neutral sulphate of methyle.

Sulphomethylosulphuric Acid. $C_2H_3S_2O_4HO$. By NO_5 on bisulphide of methyle.

Sulphomolybdic Acid. MoS_3. A blackish-brown powder; formed by the decomposition of molybdic acid by SH. Leaves a blackish-brown streak on porcelain or paper; slightly soluble in water.

Sulphomorphide. C_{34},H_{18},NO_5,SO_3. A white substance, becoming green when exposed to the air; obtained by the action of an excess of sulphuric acid on morphine.

Sulphonaphthaledamic Acid. $C_{20}H_7NH_2$ 2 SO_3. By the action of sulphide of ammonium on nitrosulphonaphthalic acid.

Sulphonaphthalic Acid. $C_{20}H_7(SO_2)SO_3$ By the action of concentrated sulphuric acid on naphthaline.

Sulphonaphthalide. $C_{24}H_{10}SO_2$? A crystalline powder, not fusible at 212°.

Sulphonaphthalidine-Carbamide. $C_{21}H_8NS+HS$. Colourless brilliant needles; by adding bisulphide of carbon to a solution of naphthalidine in alcohol.

Sulphonaphthaline. $C_{20}H_8SO_2$? A crystalline fusible solid.

Sulphonarcotide. $C_{46}H_{24}NO_{12}SO_3$. A dark green body, insoluble in water, soluble in alcohol; by treating narcotine with dilute sulphuric acid.

Sulphophenic Acid. A synonyme of sulphocarbolic acid.

Sulphophosphethylic Acid. C_4H_5O $2HO,PO_3,S_2$. By the action of sulphochloride of phosphorus (PS_2Cl_3) on alcohol.

Sulphophosphoric Acid. PS_2,O_3. A body very easily decomposed into HS and PO_5. It can be separated by means of an acid from the sulphophosphate of soda.

Sulphopianic Acid. $C_{20}H_9O_7S_2,HO$. An amorphous yellow powder, formed by the action of sulphuretted hydrogen on opianic acid.

Sulphoproteic Acid. $C_{21}H_{31}N_5O_{12}SO_3$.

Sulphopurpuric Acid. $C_{32}H_{10}N_2O_4 2SO_3$. A purple substance remaining when 1 part of blue indigo is left in contact with 8 parts of SO_3 for some days.

Sulphoretinylic Acid. $C_{18}H_{11}\ 2SO_2$

Sulphorufic Acid. $C_{14}\ H_8O_6,\ 2SO_3$. By the action of oil of vitriol on phloridzine.

Sulphosaccharic Acid. $C_{48}H_{40}O_{20}SO_3$. A white crystalline substance, obtained by adding $\frac{1}{2}$ part sulphuric acid in small quantities to 1 part sugar.

Sulphosalts. Salts into which sulphur enters as a constituent. These are principally sulphohydrides, bisulphocarbonates, sulphoarseniates, sulphoarsenites, bisulphoarsenites, sulphomolybdates, sulphotungstates, sulphotellurates, sulphoantimoniates. They consist of a sulphide of a metal united to another metallic sulphide.

Sulphosinapisine. C 57·92, H 7·795, N 4·94, S 9·657, O 19·688 $= C_{24}H_{22}NS_2O_7$. White crystalline bitter substance, soluble in water, alcohol, and ether; from mustard seeds.

Sulphotelluric Acid. A blackish-gray, metal-like substance, obtained by saturating a dilute aqueous solution of telluric acid with SH and allowing it to stand.

Sulphotellurous Acid. Brownish-black body by passing sulphohydric gas through the solution of a salt of telluric oxide.

Sulphovinic Acid. $C_4H_5O_2SO_3$. Sp. grav. 1·12. A colourless oily fluid, with an acrid taste, and the odour of oil of peppermint. Leaves a greasy stain on paper. See ETHYLE.

Sulphoxalenide. C_2HNS.

Sulphur. S. 2, 16. Sp. gr. 2·071, H1·5 to 2·5. F.P. 221° to 233°. Dimorphous. *1st Form.* Greenish-yellow 8-hedrons, composed of two 4-sided pyramids with a rhomboidal base (4th syst. fig.) with angles of 106° 20′ and 143° 25′. Sometimes the apices are replaced by low 4-sided pyramids. This is the form of native sulphur, and of sulphur crystallized from its solutions.

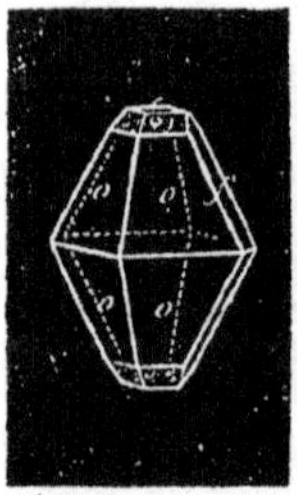

2d Form. Oblique prisms with a rhombic base (5th syst.) obtained by fusing sulphur in a clay crucible, allowing it to cool, permitting the central fluid to run off, when the cavity will be found surrounded with prisms of sulphur. Sulphur sublimes at 170°, and fuses at 218° to 233°. At 230° sulphur is a clear and limpid yellow fluid; at 320° it is more viscid; at 390° it is so thick that the crucible may be inverted without loss, and its colour is deep brown; at 750° it boils, and may be distilled; insoluble in water; slightly soluble in alcohol, ether, and oils; when fused and poured into water, it remains for some time soft, and is used in this state for taking delicate impressions; non-conductor of electricity; when held in the hand it splits with a crackling sound, and becomes negatively electric, giving out the peculiar odour of sulphur. Sulphur occurs in union with bases in nature as sulphide, and in a free state in volcanic regions. The great deposit from which the sulphur of this country is obtained is in a blue calcareous clay formation, which forms a considerable part of Sicily, occupying the central part of the south coast, extending inwards two-thirds of the island, and eastward as far as the district of Etna. It is probably a contemporaneous deposit with the gypsum beds of Paris. In hot springs sulphur is often deposited from the sulphuretted hydrogen contained in them. This gas, by the action of sulphurous acid ($SO_2\ 2\ SH = S_3\ 2\ HO$), forms water and sulphur. The sulphur ore at Muglia (according to information communicated to me by my pupil, Mr. George W. Brown, who has recently visited Sicily), is taken from the mine, broken up, and made into small heaps, when it is calcined or burned for eight days, care being taken to exclude the air by covering the fresh ore with layers of exhausted ore. It is burned

much after the manner of alum ore, but the heaps are much smaller, being perhaps fifteen to twenty feet broad. This calcination separates the sulphur from the matrix. The sulphur is then placed in clay pots capable of containing 35 pints, and being deposited in a furnace, is heated for four to five days. From this pot it distils into a similar pot, which acts as a receiver, or it is simply fused, and allowed to deposit from its impurities. From the receiving pot it is run by a tap-hole at the bottom into wooden moulds, where it is allowed to cool, and constitutes crude sulphur. It is sent to Palermo on the backs of mules, and thence to Catania in carts. In the figure, a wooden vessel filled with water is represented as placed to receive the sulphur after distillation by the tap-hole; but Mr. Brown states that this is incorrect. The mine of Muglia supplies annually 50,000 centners of sulphur. In commerce there are met with the 2d, 3d, 4th *fine*, and

4th *common* qualities. *Roll sulphur* is obtained by melting sulphur, and pouring it into wooden moulds. See fig.

Flowers of Sulphur is prepared by placing 12 or 13 cwt. of crude sulphur in a pot or furnace, C, applying heat at K, when it volatilizes and passes by the flue, D, into the chamber, A, where it condenses on the sides of the apartment; S a valve to allow heated air to escape, and to prevent common air from entering; to charge the pot another pot, M, is placed over it, heated by the waste heat of the furnace, from which sulphur is admitted by the valve, *v*, so as to prevent access of air; the sulphur at the bottom of the chamber fuses, and is drawn off by the tap-hole, O, kept heated, into moulds, to form roll sulphur. The flowers of sulphur deposited on the walls of the chamber is removed by a door at the rear of the apartment. In Saxony and Bohemia sulphur is obtained by roasting metallic sulphides in earthen or iron tubes placed 12 or 24 in a furnace, and conducting into vessels of water, where the sulphur is condensed.

Applications.—Sulphur is used to prepare sulphuric acid, to cure scabies or itch, in the form of ointment; to form a lute for leaks in pipes and steam-boilers (100 iron filings, 10 to 20 flowers of sulphur, 3 to 5 salammoniac, mixed into a paste with water). Vulcanized India rubber is formed by uniting 5 per cent. of sulphur with the fabric. Sulphur is also used in vapour to disinfect corn and seeds, and to harden graphite; it is used to prepare sulphur matches. *Lac* or *milk of sulphur* is formed by boiling sulphur with milk of lime or potash, filtering and precipitating by chlorohydric acid. What occurs in commerce often contains 50·73 sulphate of lime, 13·42 water, and 35·85 sulphur, from its being precipitated by SO_3.

Sulphuric Acid, Oil of Vitriol, Vitriolic acid, Spirit of Vitriol, Spirit of Sulphur. SO_3 5, 40.

History.—This acid was known to Geber as early as the 7th century, and was prepared by Basil Valentine from copperas in the middle of the 15th century. It is found in nature combined with bases, dissolved in well and river waters; also in veins as sulphates of barytes, strontian, lime, magnesia. In a free state in earth from Persia (Thomson), in the Rio Vinaigre (Andes), and Paramo de Ruiz. The general nature of the process consists in burning sulphur, and treating nitrate of soda with sulphuric acid in a furnace, from whence the fumes pass by a flue into large lead chambers, where they act upon each other, and are finally condensed as sulphuric acid in a layer of water deposited at the bottom of the chambers. The apparatus, therefore, for the manufacture of sulphuric acid consists of the chambers and the furnace or oven in which the materials are evolved, with the steam boiler, from which steam is injected into the chambers for the condensation of the acid. *Chambers.* — Sometimes the chambers consist of one immense oblong rectangular apartment from 70 to 120 feet long, composed of sheet lead, which is cemented at the edges, not by solder, but by fusing the lead itself either by a peculiar blowpipe or by hot lead. The chambers are raised above the floor by means of stone pillars and sleepers, to enable the bottom to be readily examined. The height of the chambers varies from 10 to 16 feet, their breadth from 14 to 20, and upwards. When the chamber is very extensive, it is divided into compartments by means of transverse divisions, which are alternately open at top and bottom, so as to allow the vapours to traverse the different compartments. Sometimes the divisions reach completely from top to bottom, but are perforated by apertures, which enable the vapours to take the course described. Generally, however, the chambers are numerous and of smaller extent, but are connected sometimes by pipes which are fixed on the exterior side of each connected chamber, or by a pipe which passes from the bottom of one chamber to the top of the adjoining chamber diame-

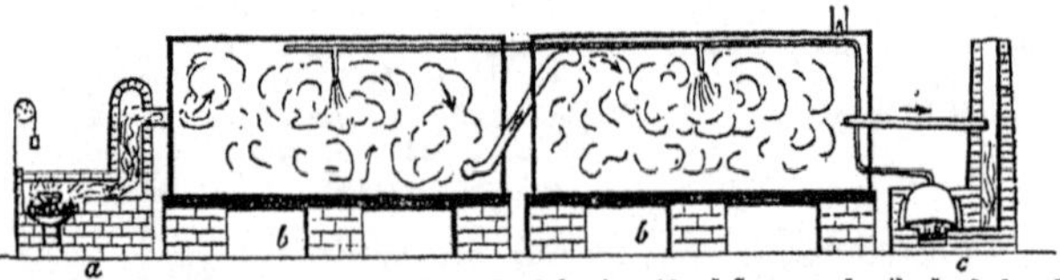

a Sulphur furnace, with cup for nitrate of soda and sulphuric acid. *b* Space under the lead chambers. *c* Steam boiler.

trically opposite, (see fig.), or by a continuation of the chamber on one side to a small extent, while the greater portion is separated by a division. The vapours traverse the various chambers or compartments of the chamber in a continuous stream, depositing the sulphuric acid in their progress, until the further extremity of the chamber is attained, where nothing remains save the nitrogen of the air, which is allowed to escape up the chimney by means of a tube leading from the upper part of the side of the chamber externally.

The Furnace, Oven, or *Burner*, *a*, consists of a brick hearth with an arched roof, supplied with a sliding door, which is raised by means of a weight. The furnace is often divided into three compartments, which may be considered as so many distinct burners, all communicating, however, with the same vent that leads into the lead chambers. The position of the furnace varies according to the pleasure of the manufacturer. Sometimes it is placed at the extremity of the chambers, at other times it is situated at the side, and at some feet distance. This is more particularly the arrangement when pyrites is roasted. A pipe carries in this case the vapours from the burner to the chamber, and is inserted in the latter midway between the top and bottom of its extremity. The furnace is floored with a thick plate of cast iron, upon which the sulphur to be burned is strewed. Upon this is placed an iron tripod or stand with three legs, which is intended to support an iron or clay saucer or vessel for holding the nitrate of soda, and sulphuric acid is added to disengage the nitric acid. The clay vessel has often no support. The quantity of nitrate of soda used amounts to about one-eighth by weight of the sulphur employed. So that in charging with 2 cwt. of sulphur, which is placed on the hearth of the burner, 28 lbs. of nitrate of soda are introduced into the basin. Upon the nitrate a sufficient quantity of sulphuric acid is poured to produce decomposition. This amounts to $6\frac{1}{8}$ parts of oil of vitriol for every $10\frac{3}{4}$ of nitrate of soda. The *steam boiler*, of common construction, is placed either at the side or extremity of the lead chamber, but in either case the steam pipe leading from it enters towards the centre of the chamber.

Mode of Conducting the Process.—It used to be the custom to cover the bottom of the leaden chambers with a layer of water mixed with sulphuric acid before commencing the process, but since the introduction of steam this practice is less generally followed. The sulphur having been spread on the floor of the furnace or burner, and the nitrate of soda and sulphuric acid being introduced into the basin, the sulphur is set on fire, and by admitting the air gradually or rapidly the combustion may be properly regulated. It was formerly the custom to light a fire under the sulphur for the first day in order to accelerate its combustion, but this method is now disused. The heat produced by the combustion of the sulphur is sufficient to cause the sulphuric acid to act upon the nitrate of soda. Vapours of sulphurous acid and nitric acid are thus disengaged, and pass from the furnace by the aperture described into the leaden chambers, where, by a reaction differently explained in consequence of the admission of steam at proper intervals, sulphuric acid is produced and deposited in solution in water at the bottom of the chambers.

Theory of the Process.—(1.) We are enabled to produce sulphuric acid on a small scale in a very simple way. For this purpose place in a gas bottle or flask, B, some copper and sulphuric acid, in another, C, copper filings and nitric acid, and in a third water. Connect these by bent tubes with a vase of glass, A, or a large receiver. The sulphurous acid should be disengaged first, and allowed to fill the receiver. The nitric acid should then be poured on the copper filings, when binoxide of nitrogen will be evolved in abundance, and combining with the oxygen of the air contained in the receiver, will be converted into hyponitric acid (NO_4). In a short space of time the sides of the glass receiver will be found to be lined with a mass of white crystals. (2.) These crystals, which are small scales, often are observed in the lead chambers whenever any sloping part occurs in the side of the chamber at some height above the floor which is covered with water. Their specific gravity is 1·831. The analysis of Dr. Thomson would make these crystals to consist of $3\,SO_2 + 2\,SO_3 + NO_5 + HO$, while that of Henry leads to the formula $5\,SO_3\,NO_3 + 5\,HO$. According to the present view, the crystalline product is only accidental, and a different action seems to precede the formation of the sulphuric acid than by the view usually entertained. Peligot considers that there are four phases in the process. 1. The sulphurous acid decomposes the nitric acid, the first being changed into sulphuric acid (SO_3), according to Dr. Thomson's view, and the second into hyponitric

SUL

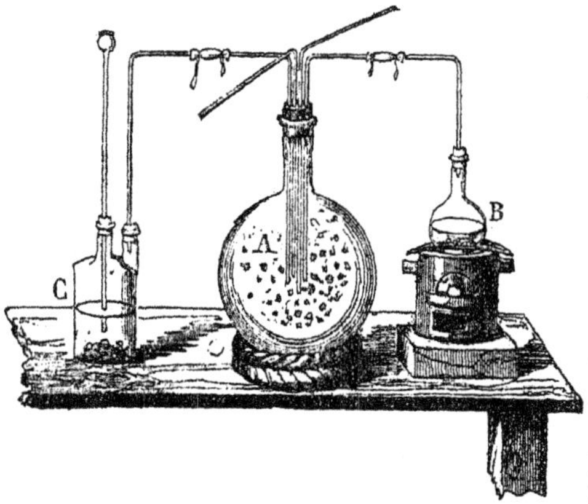

acid (NO_4). 2. Water doubles the hyponitric acid (NO_4) into nitric acid (NO_5HO) and nitrous acid (NO_3). 3. The nitrous acid (NO_3), under the action of a greater amount of water, is converted into nitric acid (NO_5) and binoxide of nitrogen (NO_2). 4. The binoxide (NO_2) is changed by the oxygen of the air into hyponitric acid (NO_4), which, by the agency of water, is transformed into nitric acid (NO_5) and nitrous acid (NO_3), according to the following formulæ:—

$$\begin{array}{llll} NO_5 + SO_2 + & HO = & SO_3 + NO_4 + HO \\ 2NO_4 + & HO = & NO_3 + NO_5 + HO \\ 3NO_3 + & HO = & 2NO_2 + NO_5 + HO \\ NO_2 + & 2O = & NO_4 \end{array}$$

$$\text{or } SO_2HO,\ NO_5 = SO_3NO_4HO$$
$$2HO\ 3NO_4 = 2(NO_5HO)NO_2$$

The principal facts in confirmation of this explanation are, that NO_2 in contact with air is always transformed into NO_4, and not into NO_3, while NO_4 is never acted on by SO_2, except under the influence of strong pressure. These experiments are in accordance with a plan adopted by some manufacturers of placing vessels in the first lead chamber filled with commercial nitric acid. They afford also a method of removing nitric acid from sulphuric acid by passing SO_2 through it. This process will not succeed, however, in effecting the object if the sulphuric acid is very concentrated, but as taken from the chambers at the gravity of 1·530 it answers perfectly, or when oil of vitriol is diluted with water the same result follows. The fact, however, shows that to avoid the presence of the nitric acid in the sulphuric acid of the chambers, it is necessary to have always an excess of sulphurous acid present. According to this explanation, the importance of the admission of steam is placed in a strong point of view, since without it the $2NO_4$ could not be converted into NO_5, that acid being incapable of existing in the absence of water. According to Mr. William Blythe, in making sulphuric acid, when the acid in the chamber reaches 1·45 spec. grav. it is impossible to make it denser without increasing the proportion of nitre, in consequence of the slowness with which acid of this strength acts in decomposing the white compound of sulphuric acid and binoxide of nitrogen formed in the process; acid of 1·5 dissolves and retains it. The crystalline compound, when dissolved in acid of 1·6, destroys the colour of sulphate of indigo. When the acid is diluted with water and concentrated, it is deprived of this nitrous compound. To obtain acid free from nitrogen compounds, so as to serve for making sulphate of indigo, Mr. Blythe recommends drawing the acid from the chamber at the spec. grav. 1·3 or 1·35. According to his views, steam is not required in the manufacture of sulphuric acid. The great objection to the view of Peligot is the difficulty of understanding how nitric acid can be generated in presence of sulphurous acid.

Concentration of the Acid.—The strength of the acid in the chambers may be raised as high as 1·53 to 1·62. The acid at this gravity, or as low as 1·3, is then drawn off from the chambers by means of a syphon into a lead tube, which conveys it as required into open pans of the same metal, where it is boiled till it attains the spec. grav. of 1·7. When of this strength it begins to act upon the lead, so that it must be drawn off into the platinum still. This is a large retort set in a cast iron pot, where the acid is boiled till it becomes of the spec. grav. 1·845 or 1·837. When it has reached this point it is allowed to run off by a worm inserted in cold water into carboys, or large globular glass bottles, packed in wicker baskets with straw. In using platinum stills, the presence of lead or chlorohydric and nitric acids would endanger the safety of the vessel, and as the expense of these is very heavy (many hundred pounds), such risks are to be carefully avoided.

Protohydrate of Sulphuric Acid or *Protosulphate of Water*, *English Sulphuric Acid*, *Oil of Vitriol.*—HO,SO_3 6·125; 49. When sulphuric acid prepared by the preceding process is concentrated as strongly as possible, it possesses the specific gravity 1·850. Sulphuric acid combines in various proportions with water. The following are the most important:—

	Sp. Grav.	
Oil of vitriol, protosulphate of water,........................	1·85	$HO\ SO_3$
Nordhausen acid, bisulphate of water,....................	1·896	$HO2SO_3$
Disulphate of water,............	1·78	$2HO\ SO_3$
Trisulphate of water, or terhydrate of SO_3,..............	1·632	$3HO\ SO_3$

Oil of vitriol is an oily viscid liquid, without colour, having a strongly acid taste, reddening vegetable blues, and immediately charring organic substances when brought in contact with them. The boiling point of acid of 1·850 is 620°, its freezing point —29°. When frozen it presents itself frequently in the form of 6-sided tabular prisms. After it has been once frozen, it again solidifies at a much lower temperature. Mr. Keir found that sulphuric acid of spec. grav. 1·78 froze at 45°, but if it contained either more or less water, it required a greater cold to freeze it. Dr. Thomson cooled it down in thermometer tubes to —36° before it congealed. When the vapour is passed through an ignited porcelain tube, 2 vols. of SO_2 and 1 vol. of oxygen are produced.

Bihydrate of Sulphuric Acid.—$2HOSO_3$ 7·25. Obtained by mixing 1 atom oil of vitriol with 1 atom water. Spec. grav. 1·78. When exposed to a temperature of 39¼° it deposits in crystals. At 390° it loses 1 atom of water, and becomes common oil of vitriol.

Terhydrate of SO_3.—Obtained by adding 1 atom of water to the bihydrate, when they undergo the maximum of contraction. When oil of vitriol is freely exposed to the air, it absorbs 15 times its weight of water.

Bisulphate of Water, Nordhausen Acid.—HO $2SO_3$ 11·125, 89. This acid is still prepared at Nordhausen in Prussia, and at Kraslitz in Bohemia. The process consists in first roasting gently the protosulphate of iron (copperas, or green vitriol), so as to remove the greater part of its water of crystallization, which amounts to about 45 per cent., the salt being composed of FeO, SO_3+7HO. The amount of water removed by the heat is about 40 per cent. One and a-half to two lbs. of the sulphate, now in the form of a reddish-yellow powder, are introduced into a stoneware retort, A, a number of which are set in pairs in a brick furnace, and connected with stone receivers, B. The heat is applied gradually at first,

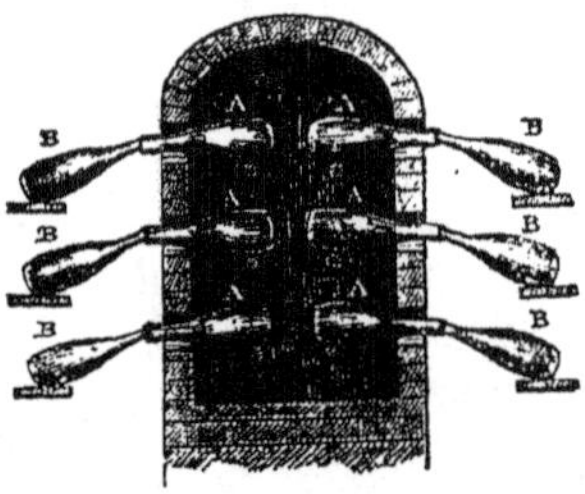

and the impure matter (spirit of vitriol) which comes over first is rejected. As soon as the white clouds of SO_3 appear, the receivers are connected with the retorts and the heat increased. In from thirty-two to thirty-six hours the process is terminated; the heat being increased to whiteness during the last six hours. The retorts are then emptied of their colcothar, but the same receivers are retained during three distillations. The acid thus prepared is brownish-yellow, often brown, from the presence of cork, straw, or organic substances; oily, possessing a specific gravity of from 1·82 to 1·896 and 1·92; it smokes in the air, giving out thick white vapours. In this state it is known under the title of the smoking or fuming acid of Nordhausen, chiefly employed to dissolve indigo for dyeing Saxon blue. When the acid is kept at a temperature under 32° it crystallizes, and the crystals are composed of $HO + 2SO_3$.

Anhydrous or *Solid Sulphuric Acid.*—SO_3 5, 40. When the fuming acid or bisulphate of water is distilled by a moderate heat in a retort, to which a receiver is adapted, and kept cool with water below 50° or ice, the strong acid comes over, and if the process is stopped before the water begins to pass over, we obtain acid in a solid form. It presents the appearance of thin white silky filaments, which are tough and difficult to cut, bearing a resemblance to asbestus, and having a waxy consistence. In the air it disappears in the form of vapour. It may be held in the hand for some time without producing any sensation; gradually, however, it absorbs water, and acts upon the skin. The solid acid liquefies when heated up to 64° (Bussy), and at 68° its spec. grav. is 1·97. The spec. grav. of its vapour theoretically is 2·777, although it was found by Mitscherlich 3·00. When allowed to fall into water, a hissing sound is produced as by a hot iron.

Sulphates.—When oil of vitriol is brought in contact with potash a case of substitution occurs by the alkali taking the place of the water, and instead of $HO+SO_3$ we have the anhydrous salt $KO+SO_3$ produced. Here then occurs, as in all cases where oxygen acids unite with bases, the intervention of oxygen in union with the base, as we have no such compound as $K+SO_3$. In this respect, therefore, there exists a remarkable anomaly in the case of the oxygen acid salts when compared with the chlorine, bromine, iodine, or fluorine acid salts, for these substances all unite with metals without the aid of oxygen. To reconcile this incongruity, it was proposed by Sir H. Davy to consider that the sulphate of water was, for example, not a sulphate of an oxide, $HO+SO_3$, but a compound of H with a new radical SO_4, sulphoxygen, so that oil of vitriol would then be a sulphoxide of H, the hydrogen and not the oxygen being the acidifying principle, since we can have no acidity without the presence of water. The compounds enumerated would then be represented by—

Oil of vitriol, H SO_4, sulphoxide of hydrogen.
Sulphate of potash, K SO_4, sulphoxide of potassium.

Applications.—Sulphuric acid is of great importance in manufactures. It is used extensively in bleaching, to remove lime, &c. from the cloths in the operation called souring. It is also employed to a great extent in the decomposition of common salt, which constitutes the first stage in the process for making soda-ash and carbonate of soda, the basis of soaps, and an important article in bleaching. It is likewise used in the preparation of NO_5 for the decomposition of nitrate of soda or potash. In medicine it is employed externally in the dilute form designated acidum sulphuricum dilutum, or dilute SO_3 acid, which is made from the commercial acid, according to the London and Edinburgh Pharmacopœia, in the following proportions:—

	London.	Edinburgh.
Sulphuric acid,	1½ fluid oz.	1 fluid oz.
Water distilled,	14½ „	13 „

A fluid drachm of the London contains about 10 grains of the strong acid, and saturates 28 grains of crystals of carbonate of soda. The dose amounts to from 10 to 40 minims. It is employed as a tonic and astringent, and as a corrective in cases of deposition of phosphate of lime in the urine. When taken in too large doses, or as a poison, the antidote is lime water and chalk, or alkaline carbonates.

Estimation.—Sulphuric acid is estimated by precipitation by means of chloride of barium. The sulphate of barytes ($BaO\ SO_3$) which falls is insoluble in nitric acid, and contains in 14½ parts 5 of SO_3.

Sulphurous Acid, Volatile Spirit of Sulphur.—SO_2, atomic weight 4, 32. Spec. grav. 2·222. Stahl first examined this acid, and termed it phlogisticated sulphuric acid. Scheele and Priestley examined it about 1774. The latter first obtained it in a pure form. It is formed by burning sulphur in atmospheric air; by heating sulphur with many metallic oxides, as with black oxide of manganese. The method of obtaining it in the purest state is to boil sulphuric acid in a retort with an equal weight of mercury or copper, and collect the gas which is disengaged over mercury. The third atom of oxygen of one atom of sulphuric acid combines with the mercury and forms oxide and sometimes suboxide of mercury, while the sulphurous acid is disengaged; a second atom of SO_3 uniting with the oxide, and producing sulphate of oxide of mercury, according to the following scheme:—$Hg\ 2SO_3 = SO_2, HgOSO_3$.

Liquid Sulphurous Acid.—By means of a freezing mixture, Bussy has shown that it may be condensed. He passes the gas through a vessel surrounded with ice to condense the aqueous vapour and dry it; it is then made to go through a tube filled with chloride of calcium. It finally enters into a receiver surrounded by a freezing mixture, consisting of 2 parts of snow and one of common salt, where it condenses into a fluid. It boils at the temperature of 14°. The cold produced by its evaporation is so great as to freeze mercury and sink the thermometer to —66°. It is a non-conductor of electricity, but is converted into a conductor by the addition of a small portion of water. De La Rive observed that solid SO_2 is formed in the vessel containing the ice in Bussy's process in the state of thin plates, which consist of $SO_2 + 10\ HO$. SO_3 absorbs SO_2 in considerable quantity. It is also absorbed by borax, according to the observation of Priestley, and in this way the separation of SO_2 and CO_2 may be effected.

The *sulphites* are formed by passing the gas into water containing bases or their carbonates in solution. They have no smell, but when soluble in water possess a sharp taste. The single sulphites of potash and soda have an alkaline, the bisulphites a neutral reaction.

Characters.—Sulphurous acid is a colourless gas possessing the mechanical properties of common air, having a specific gravity of 2·222, or being 370 times lighter than water. When acted on by a pressure of two atmospheres at 45°, it condenses into a colourless liquid, with a specific gravity of 1·42 (Faraday). The gas possesses a strong acrid and sulphurous taste; its smell is disagreeable and suffocating, and is well characterized by the smell of burning sulphur. It changes vegetable blue colours to red, and then destroys them. The petals of a rose become white in SO_2 and red in SO_3. Water absorbs 33 times its volume of this gas (Thomson), or, according to Saussure, nearly 43¾ vols., alcohol absorbing 116¾ vols. The spec. grav. of the saturated solution is 1·0513. When oxygen and SO_2 are passed through a red hot tube SO_3 is formed. Its composition is as follows:—

	Atoms.	Wt.	Vols.	Expt.	Theor.	P. C.
Sulphur,	1	2	⅙	1·1117	1·1111	50
Oxygen,	2	2	1	1·1057	1·1058	50
		4		2·2174	2·2169	100

SO_2 is rapidly decomposed by potassium and sodium. When brought in contact with peroxide of lead it receives oxygen from the peroxide, a sulphate of lead being formed. It is not decomposed by any degree of heat. In the cold state, SO_2 has no action upon any substance except the alkaline metals; but with the assistance of heat it readily acts. When passed with hydrogen through an ignited porcelain tube, water is formed and sulphur deposited, while sometimes sulphohydric acid also makes its appearance. Zinc in sulphurous acid and sulphuric acid yields sulphohydric acid. With phosphuretted hydrogen it is decomposed into water and sulphide of phosphorus. Liquid phosphorous acid forms with it, when heated, phosphoric acid and SH. Sulphurous acid decomposes chloric, bromic, and iodic acids, sulphuric acid being formed, and chlorine, bromine, or iodine. Sulphohydric acid and sulphurous acid mutually decompose each other

(SO_2, 2 SH = 3 S, 2 HO), water being formed and sulphur deposited.

Applications.—Sulphurous acid is an important agent in bleaching woollens, which do not readily lose their deteriorating colours by the action of common bleaching liquor. Blankets or other woollens to be bleached are suspended in rows upon wooden pins in the sulphur house. A pot of sulphur is then introduced and heated with hot bricks or a hot ball of iron. The apartment is then perfectly closed by cementing up the doors. The SO_2 acid fumes fill the chamber and act intimately upon the goods, which are thus decolourized in twenty-four hours or longer. The woollen goods being introduced in a moist state, the SO_2 is dissolved by the moisture, and comes in contact with the inmost fibres of the goods. It is necessary in the operation that no iron materials should touch the goods, otherwise, in consequence of the action of the SO_2, they would be discoloured. An important improvement has been made in this process. See BLEACHING.

Sulphuretted Hydrogen, Sulphohydric Acid, Hydrosulphuric Acid, Hydrothionic Acid, Hepatic Air, Sulphuret of Hydrogen.—SH 2·125, 17.

	Vol.	Sp. Gr.	Ats.	Weight.		Expt, p. c.		Calcd.
S	$\frac{1}{6}$	1·1093	1	2	16	93·855		94·11
H	1	0·0692	1	·125	1	6·145		5·89
		1·1785		2·125	17	100·	-	100·

Spec. grav. 1·1912 (Gay Lussac and Thenard, Ann. Chim. 73, 229), 1·1967 (Davy), 1·1788 (Thomson). Refracting power 2·187 (Dulong, Ann. Chim. 31, 167).

Characters.—Colourless gas with the mechanical properties of air; smell strong and fetid, resembling that of putrefying eggs; it does not support combustion but is combustible, yielding a blue flame and depositing sulphur. When electric sparks are passed through it sulphur is likewise deposited. It reddens litmus, the colour disappearing in the air. 100 vols. water absorb 108 sulphohydric acid gas at 50° (Henry), 253 (T. De Saussure), at least 300 (Gay Lussac and Thenard), 366 (T. Thomson). 100 alcohol of spec. grav. ·840 absorb 606 vols. gas (Saussure); soluble also in ether (Higgins). The aqueous solution, which may be formed by passing the gas from a flask and bent tube through water, is colourless, possesses a strong odour of the gas and a sweet nauseous taste; converts vegetable blues to red; the gas escapes when exposed to the air; when acted on by air the sulphur falls, while the hydrogen unites with the oxygen of the air (SH,O = S,HO), and some sulphuric acid is formed; chlorine, bromine, iodine, sulphurous acid ($2SHSO_2 = S_3 2HO$), selenious acid, &c. precipitate sulphur.

Preparation.—1. When sulphur is heated in hydrogen gas sulphohydric acid is formed (Scheele). 2. It is likewise produced in the decomposition of organic bodies containing sulphur. 3. For practical purposes the best mode of preparing this gas is to form a sulphide of iron (FeS) by igniting an iron bar in a smith's forge or in a furnace, and then to press against it when nearly white hot, a roll of sulphur over water, so that the sulphide formed shall drop into the water. It may also be produced by igniting sulphur and iron filings in a covered clay crucible. The sulphide in the state of rough powder is then introduced into a flask supplied with a perforated cork and bent tube. Water is poured over it and then sulphuric acid ($FeS,HO,SO_3 = SH,FeO\ SO_3$), sulphohydric acid is evolved, and sulphate of protoxide of iron (copperas or green vitriol) remains dissolved in the water of the flask. Sulphohydric acid thus obtained often contains hydrogen, in consequence of the sulphide of iron often possessing an excess of iron. Sulphohydric acid is obtained as a colourless fluid of spec. grav. ·9 by its own pressure in a strong tube, and as a solid resembling camphor by exposing the liquid to a cold of —122° and pressure by a forcing pump.

Analysis.—The gas, when heated in a tube retort (see fig. PROTOXIDE OF NITROGEN), in contact with tin, gives hydrogen and sulphide of tin without change of volume. In mineral waters the sulphur is precipitated by arsenious acid or nitrate of silver and caustic ammonia.

Applications.—It is a valuable reagent in testing and analysis. See ANALYSIS QUALITATIVE.

Sulphophenyle-benzoil-amide. $C_6H_5 SO_2,C_7H_5O,H,N$. Fine flat acid needles, little soluble in water, very soluble in alkalies, forming salts with bases.

Sulphophenyle-cumyl-benzoil-amide. $C_6H_5SO_2,C_{10}H_{11}O,C_7H_5O$. Prisms.

Sulphophenyle-dibenzoil-amide. $C_6H_5 SO_2$, 2 (C_7H_5O) N. 8-hedron with diamond lustre.

Sumac. *Young Fustic.* The dried branches of the *Rhus coriaria*, grown in Italy and south of France, used for dyeing black in consequence of the amount of tannic acid which it contains, and which gives a black with salts of iron.

Sumbul. The Persian musk root consists of water, essential oil, balsam, wax, gum, starch, &c.

Sumbulic Acid. An acid found in the root of the sumbul, approaching in character to angelic acid, but it becomes blue on adding sulphuric acid.

Sumbulolic Acid. An acid obtained by distilling balsam of sumbul with solution of potash.

Sunadine. *Sunstone.* A synonyme of potash felspar.

Sunstone. A variety of felspar with micaceous matter imbedded, giving a fine lustre.

Super. A term applied to compounds with an excess of acid, oxide, &c., as contrasted with *sub*, which is applied to excess of base.

Surinamine. A crystalline alkaloid from Geoffræa inermis.

Susanite. A synonyme of sulphato tricarbonate of lead.

Swaga. A synonyme of Borax.

Sweat. *Sensible perspiration.* A colourless or turbid fluid excreted by the skin; it contains lactates (?) of soda, potash, lime, magnesia, common salt, salammoniac, and chloride of potassium. I have found the sweat neutral in a case where about a drachm was collected.

Sweet Bay Oil. See LAUREL OIL.

Swinestone. *Stinkstone.* A variety of carbonate of limestone, from the odour which it gives out when breathed on.

Syenite. A rock first obtained at Syene, in Egypt, consisting of quartz, felspar, and hornblende, differing from granite in the replacement of mica by hornblende.

Sylvanite. A synonyme of native tellurium.

Sylvic Acid. $C_{20}H_{15}O_2$. F.P. 305½°. A resin from common rosin, soluble in absolute alcohol, ether, fixed and volatile oils.

Sylvius, Salt of. Chloride of Potassium.

Symbols. The first letters of the Latin terms for the elementary bodies are used as their brief representatives. For example O, Cl, Br, I, F, &c., denote oxygen, chlorine, bromine, iodine, and fluorine, &c. In organic chemistry, a line over a symbol denotes the substance to be an acid; while a cross over it implies that it is an alkaloid. $\bar{M}$ represents malic acid, and $\overset{+}{Qu}$ quinine. Sometimes dots over a symbol denote oxygen. Thus $\dddot{S}$ is sulphuric acid (SO_3). Commas sometimes represent sulphur, as $\overset{,,}{Fe}$ = bisulphide of iron. See ATOMIC WEIGHTS.

Symplesite. *Scorodite?* Spec. grav. 2·957, H 2·5. Indigo or green oblique rhombic prisms, pearly lustre; translucent; consists of water, arsenic, and iron; it is probably arseniate of iron.

Synaptase. $C_{20}H_{25}N_2O_{32}$? White powder, soluble in water; insoluble in alcohol and ether; precipitated by rennet coagulated at 140° heat, or by acetic acid; obtained from sweet almonds.

Synovia. Spec. grav. 1·05, (human knee) 1·099, (horse's knee) 1·027. Alkaline mucilaginous fluid secreted in the capsular ligament of joints, consisting of HO 92·8, albumen 6·4, animal matter with $NaOCO_2$ and NaCl ·6, phosphate of lime, phosphate of soda, ammoniacal salts ·15.

Synthesis. (συνθεσις, formation). Production of a compound body from its elements as contrasted with analysis or determination of the composition of a body by taking it to pieces and ascertaining the nature of the elementary bodies of which it is composed.

Syringine. The bitter principle of the Syringa vulgaris or mock orange.

T

Tabasheer. Spec. grav. 2·0824, HO 4·87, SiO_3 90·5, KO 1·1, Fe_2O_3 ·9, Al_2O_3 ·4, CaO ·1. Bluish-white masses with a pearly lustre, not unlike chalcedony, but softer; a concretion from the joints of the Indian bamboo; in water it gives out air with a crackling sound. Tabasheer from South America is said to contain 30 per cent. of alkali (Vauquelin).

Table Spar. *Schaalstein, Grammite, Wollastonite* of Hauy. Spec. grav. 2·805, 2·758, 2·895, 2·863, H 3 to 4. White with a shade of gray doubly-oblique prisms with angles of 126°, 93° 40′, 95° 15′; lustre vitreous, inclining to pearly; semitransparent; structure foliated; streak white. B.B. fuses with difficulty into a semitransparent colourless glass; with borax into a clear glass. SiO_3 52·58, CaO 44·45, FeO 1·13, MgO ·68, HO ·99, 3 CaO 2 SiO_3. It occurs in limestone, often associated with trap dikes, and I have been able to trace the conversion, in some instances, of limestone into table spar by the agency of the silica of the trap. This is well illustrated near Girvan, in Ayrshire, and near Gourock. In these localities it always contains some soda. It occurs in Ceylon, Rhode Island, &c.

Table Spar, Soda. See WOLLASTONITE.

Tabular Spar. Subsesquisilicate of lime.

Tacamahac. A resin, one species of which is yellow translucent; soluble in alcohol; acrid; is obtained from the Callophyllum inophyllum, in Madagascar; the other is a light brown, brittle resin, with an aromatic odour, partially soluble in alcohol; soluble in ether and fixed oils, and is derived from the Fagara octandra and Populus balsimifera.

Tachylite. Spec. grav. 2·52, H 6·5. Velvet-brown or black masses and plates with a vitreo-fatty lustre; opaque. SiO_3 50·22, TiO_2 1·42, Al_2O_3 17·84, FeO 10·27, CaO 8·25, MgO 3·37, NaO 5·18, KO 3·87, MnO ·4, HO and NH_3 ·5. B.B. fuses into a brown scoria, resembling obsidian. Säsebühl, near Dransfield.

Taddy, or *Toddy.* The fermented juice of a species of palm.

Tagilite. *Terhydrous Tetraphosphate of Copper* from Nischne Tagilsk. 4 CuO,PO_5 3 HO. Spec. grav. 3·5, H 3·. Green kidney-shaped masses.

Tails. A name applied to lead ore. See LEAD.

Talc Apatite. Spec. grav. 2·73. 6-sided prisms, containing CaO 37·5, MgO 7·74, PO_5 39·02, SO_3 2·1, Cl ·91, F and loss 2·23, Fe_2O_3 1·; insoluble 9·5. It is probably a mixture.

Talc Chlorite. See GLAUCONITE.

Talc Earthy. See NACRITE.

Talc Hexagonal. Spec. grav. 2·772, H 1·75. Bluish-gray, obscurely foliated, waxy 6-sided prisms; opaque; feel greasy. SiO_3

29·364, Al_2O_3 17·808, MgO 12·144, CaO 3·092, FeO 26·016, HO 11·2. Ala, Piedmont.

Talc Slate. Spec. grav. 2·877, H 2·25. White with a shade of yellow, massive and slaty; fracture flat conchoidal; opaque; sectile; lustre silky. SiO_3 57·56, MgO 27·216, CaO 7·944, Al_2O_3 1·72, FeO 4·716, HO 1·6. Novorda, Piedmont.

Talc Venetian. Spec. grav .2·697, H 1·. $4MgO3SiO_3$? Apple-green, thin flexible plates with a silvery and pearly lustre; semitransparent; very sectile. B.B. infusible. SiO_3 62·588, MgO 30·328, FeO 3·848, HO 3·4. Salzburg, Tyrol, Vermont, &c. It was used as a cosmetic at Venice.

Talcite. See NACRITE.

Tallow. The solid oil of various kinds of cattle.

Tallow Mineral. See HATCHETTINE.

Tamarinds. The fruit of the Tamarindus Indicus, is used from its containing citric acid when boiled with water as a drink in fevers, &c.

Tanacetic Acid. A crystalline body from tansy decoction by acetate of lead and SH.

Tanacetine. A bitter principle from Tanacetum vulgare.

Tancelite. Phosphate of Yttria.

Tanghicine. Transparent plates from the fruit of *Tanghinia Madagascarensis* by ether; soluble in alcohol, ether; insoluble in water; bitter; poisonous.

Tangue. Sand containing organic matter, found on the sea shore.

Tannaspidic Acid. $C_{26}H_{13}O_{10}HO$. Black substance by alcohol from fern roots.

Tannic Acid. *Tannine, Quercitannic acid.* $C_{18}H_5O_9$ 3 HO. White uncrystalline powder, soluble in water, alcohol; scarcely soluble in ether; taste astringent but not bitter; decomposes when exposed to the air in solution into gallic, ellagic, and carbonic acids, with a substance analogous to gum or sugar ($C_{18}H_8O_{12} = C_4H_4O_4$, $C_{14}H_4O_8 = 2$ gallic acid). This action takes place even in distilled water, by the influence of the fermenting principle pectase; no decomposition takes place if the tannic acid is perfectly pure, or when exposed for a long time to the heat of boiling water; it is obtained by pounding nutgalls and exhausting them in a displacement apparatus (see DISPLACEMENT), with ether saturated with water. The fluid in the lower vessel separates into two strata, the lower containing tannic acid in water, from which it separates by evaporation in vacuo; it is washed with ether; redissolved in water and evaporated in vacuo. Tannic acid precipitates gelatine and forms leather, starch, albumen, and gluten; in the animal system it is changed into gallic and pyrogallic acids, and substances resembling humus; it unites with bases and forms salts of irregular composition; its aqueous solution is precipitated by SO_3,HCl,PO_5, AsO_5,BO_3; the precipitate contains tannic acid and the acid used; the compound with SO_3, when boiled with potash, yields gallic acid.

Tannic Acids. This term is applied to a particular acid (the *quercitannic*) existing in nutgalls, which seems to be the type of a series of tannic acids, existing in various plants, which, with gallic acid, would stand as follows:—

Quercitannic,	$C_{18}H_5\ O_9$
Gallic,	$C_{14}H_6\ O_{10}$
Callutannic,	$C_{14}H_6\ O_8$
Viridic,	$C_{14}H_7\ O_8$
Rhodotannic,	$C_{14}H_6\ O_7$
Leditannic,	$C_{14}H_6\ O_6$
Rubichloric,	$C_{14}H_8\ O_9$
Caffeotannic,	$C_{14}H_8\ O_7$
Eritannic,	$C_{14}H_8\ O_7$
Tanningenic,	$C_{15}H_6\ O_6$?
Kinic,	$C_{14}H_8\ O_8$
Mimotannic,	$C_{24}H_{18}O_{13}$
Rubitannic,	$C_{15}H_5\ O_{11}$?
Aspertannic,	$C_{14}H_8\ O_8$
Humotannic,	$C_{14}H_5\ O_7$
Boheotannic,	$C_{14}H_{10}O_{12}$
Galitannic,	$C_{14}H_8\ O_{10}$
Rufimoric,	$C_{14}H_7\ O_8$
Ipecacuanhic,	$C_{14}H_8\ O_6$
Morintannic,	$C_{22}H_{11}O_{13}$
Pteritannic,	$C_{24}H_{15}O_8$
Pyromorintannic,	$C_{24}H_{12}O_8$
Kinovatannic,	$C_{14}H_8\ O_7$
Tannaspidic acid,	$C_{26}H_{13}O_{10}$
Kinovic,	$C_{12}H_9\ O_8$
Melanotannic,	$C_{14}H_4\ O_7$
Rufogallic,	$C_{14}H_4\ O_8$

Tannigenamic Acid. $C_{42}H_{20}N_3O_{23}9HO$. White fatty square tables; difficultly soluble in cold water; giving a blue with iron salts; by sulphite of ammonia on tannic acid.

Tannine. See TANNIC ACID.

Tanningenic Acid. See CATECHINE.

Tannocaffeic Acid. See CAFFEOTANNIC ACID.

Tannoxylic Acid. *Rubitannic Acid.* $C_{15}H_5O_{11}$. An amorphous substance by adding potash to a solution of tannic acid.

Tansy Oil. Spec. grav. ·952. Yellow or greenish bitter oil.

Tantalite. Spec. grav. 7·236, 7·30. Iron-black amorphous mass; lustre semi-metallic, hard, brittle; streak grayish-black. Columbic, niobic, and pelopic acids 83·2, FeO 7·2, MnO 7·4, SnO_2 ·6. Skogsböle, Kimito, Finland. See IRON TANTALITE, TORRELITE, and COLUMBITE.

Tantalite, Iron. *Ferrotantalite, Columbate of Iron.* Spec. grav. 7·936, 7·655. Black masses with a tendency to crystallization; scratches glass; powder reddish or cinnamon-brown; unacted on by acids. B.B. not altered; dissolves slowly in borax into a green glass with white particles floating; fuses with salt of phosphorus, not with soda. Columbic acid 85·85, FeO 12·97, MnO 1·61, TnO_2 ·8, CaO ·56, SiO_3 ·72. Kimito.

Tantalum. See COLUMBIUM.

Tapioca. The starch of the root of the cassava or Jatropha manihot. It contains but a small quantity of albumen, and is therefore not nutritious, but may be employed as food in small quantity, in the same way that starch would be used. See ALIMENT.

Taqua Nut. Vegetable Ivory. See IVORY, VEGETABLE.

Tar. A mixture of heavy and light oils mixed with asphalt, obtained by distilling coal and wood.

Tarandite. A dolomite consisting of CaO CO_2 54·76, $MgOCO_2$ 42·1, $FeOCO_2$ 4·19.

Taraxacine. A crystalline bitter substance, soluble in hot water, alcohol, and ether; from the root of the dandelion (Leontodon taraxacum).

Tarnowitzite. A variety of Arragoni from Russia, containing $3\frac{3}{4}$ per cent. of carbonate of lead.

Tartar, Cream of. *Bitartrate of Potash, Tartrate of Potash and Water.* See TARTRATE OF POTASH AND WATER.

Tartar Emetic. Tartrate of antimony and potash. See ANTIMONY.

Tartaric Acid. *Tartrylic Acid, Vinic Acid, Essential Salt of Tartar.* $2\,HO,C_8H_4O_{10}$. Spec. grav. 1·75. F.P. 338°. The crystals belonging to the oblique prismatic system (5th syst.) are in vertical prisms, with 4 faces M; terminated by 2 basic inclined faces; the angles M on *m* are 129° 20′, M on *r* 121° 4′, M on *w* 125° 15′; soluble in water and alcohol. When heated in a stove it becomes electric, adhering to paper; boiled with alcohol, tartaric ether is formed; potash and heat convert it into oxalate and acetate of potash ($C_8H_4O_{10}$ $2\,HO = C_4H_3O_3HO$ and $2\,(C_2O_3)\,HO$). MnO_2, and SO_3 change it into formic, carbonic acids, &c.; it precipitates lime water, but not chloride of calcium; it precipitates potash salts, bitartrate of potash falling. By the application of heat a series of acids is formed, which differ in their amount of basic water. The temperatures at which they are formed are placed beside them in the table (Fremy).

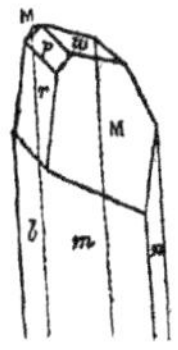

Tartaric acid.

Tartaric,	$C_8H_4O_{10}$ 2 HO,	
Tartralic,	$C_8H_4O_{10}$ $1\frac{1}{2}$ HO,	390°
Tartrelic,	$C_8H_4O_{10}$ 1	356° contd.
Anhydrous Tartaric,	$C_8H_4O_{10}$	356° contd.

According to another view (Laurent, Gerhardt), tartaric acid by heat, without loss of weight, is converted into metatartaric and isotartaric acids (see these acids), while the anhydrous acid is an isomeric form of the hydrous acid. By a more elevated temperature it is converted into an oily and crystalline acid. Solutions of tartaric acid when allowed to stand become mouldy from the formation of a mucor, and are converted into acetic and butyroacetic acid; by nitric acid they are converted into acetic, oxalic, and saccharic acids; tartaric acid is a powerful deoxidizing agent, reducing chromic, vanadic acids, &c. to the state of oxides. Tartaric acid is largely used as a discharger by calico printers. When printed on the cloth in a starch paste, it unites with the lime of bleaching liquor, when the cloth is passed through a solution of this substance and sets the chlorine free, which removes the vegetable colour on those parts where it is printed. It is also used in effervescing draughts in mixture with bicarbonate of soda; it forms double salts and retains metallic oxides in solution, so that they are not precipitated by alkalies. *Process.*—Tartaric acid is prepared by dissolving cream of tartar in water, adding milk of lime to the solution until the liquid is neutral; the tartrate of lime is filtered off and decomposed by sulphuric acid.

Anhydrous.—$C_8H_4O_{10}$. White powder, insoluble in water, alcohol, and ether; obtained by heating tartaric acid rapidly in a basin, and then in an air bath at 302; it is then washed and dried.

Tartralic Acid. $HOC_{12}H_6O_5$ or $C_8H_4O_{10}$ $1\frac{1}{2}$ HO. White uncrystalline mass by heating tartaric acid at 392° in a porcelain basin, dissolving it in water, neutralizing with chalk, decomposing by SO_3; soluble in water and alcohol; deliquescent; reconverted into tartaric acid by boiling with water.

Tartramethane. *Tartramate of Ethyle.* Formed by the action of ammonia in alcohol on tartaric ether.

Tartramic Acid. Obtained by passing dry ammonia over anhydrous tartaric acid. The salt of ammonia is soluble in water; insoluble in alcohol.

Tartramide. By the prolonged action of ammonia on tartramethane.

Tartramylic Acid. $C_{18}H_{16}O_{12}$. By the action of tartaric acid on fusel oil.

Tartrates. These salts are bibasic.

Tartrate of Ethyle. *Tartrovinic Acid.* EO, $HOC_8H_4O_{10}$. Rhombic prisms, soluble in water and alcohol by boiling alcohol and tartaric acid.

Tartrate of Potash. $KO,C_8H_4O_{10}HO$. Rhombic prisms, soluble in ·66 cold, in 248 boiling water; very deliquescent; obtained by neutralizing cream of tartar with carbonate of potash.

Tartrate of Potash and Ammonia. NH_3 $KO,C_8H_4O_{10}$ 2 HO. Rhombic soluble prisms, obtained by neutralizing cream of tartar by ammonia.

Tartrate of Potash and Soda. *Salpolychrestum, Rochelle Salt, Salt of Seignette.* Spec. grav. 1·756; F.P. 167°; B.P. 248°. KONaO $C_8H_4O_{10}$ 8 HO. Long rhombic prisms of 8, 10, or 12 sides, with their ends truncated at right angles; taste bitter; soluble in 2·62 water; obtained by neutralizing cream of tartar with carbonate of soda, and used as an aperient.

Tartrate of Potash and Water. *Bitartrate, Cream of Tartar, Tartar, Winestone.* Spec. grav.

2·4. $KOHO,C_8H_4O_{10}$ 23·525, 188·2. Irregular hard right rhombic prisms, with angles of 107° 30′, 126° 15′, 117° 2′, and 77°; soluble in 234°·6 or 240 parts cold water, in 14½ boiling water; its solubility is increased by the presence of acids; insoluble in alcohol; 100 commercial cream of tartar, by ignition, leave 40 of a mixture of 31½ carbonate of potash and 8·75 $CaOCO_2$, and CaS with some carbon; obtained by purifying the red argols of commerce by solution and crystallization; argols are deposited on the inside of some casks, in consequence of their insolubility in alcohol.

Tartrelic Acid. $C_{16}H_8O_{20}$ 2 HO or $C_8H_4O_{10}HO$. Crystalline monobasic acid by keeping tartralic acid for a considerable time at 356°; soluble in water and alcohol.

Tartromethylic Acid. *Tartrate of Methyle and Water.* $C_2H_3O,HOC_8H_4O_{10}$. White 4-sided prisms, obtained by distilling pyroxylic spirit with tartaric acid.

Tartronic Acid. $C_6H_4O_{10}$. F.P. 320°. Obtained by the spontaneous decomposition of nitro-tartaric acid in water. When heated to 356° it is converted into glycollic acid ($C_4H_4O_6$). *Glycollide.*—$C_4H_2O_4$. Insoluble body, procured from the dry distillation of tartaric acid; it is converted into glycollic acid by alkalies. *Glycolamide.*—$C_4H_5N_2O_4$. Fine crystals by heating bitartronate of ammonia, or by adding ammonia to glycollide.

Tartrovinic Acid. *Tartrate of Ethyle and Water.* $C_4H_5O,HO, C_8H_4O_{10}$. White oblique 4-sided prisms, deliquescent, heavier than water; soluble in water and alcohol; insoluble in ether; decomposed by boiling into tartaric acid and alcohol; obtained by boiling together tartaric acid and alcohol, saturating with carbonate of barytes and precipitating the barytes by sulphuric acid.

Taurilic Acid. $C_{14}H_8O_2$? Obtained among the volatile products in the distillation of cow's urine; isomeric with anisole and carbolic acid?

Taurine. $C_4NS_2H_7O_6$. Fine 6-sided prisms, terminated by 4 or 6-sided pyramids; taste sharp; neutral: soluble in 15½ parts cold water; little soluble in alcohol; gives oil and an acid liquor containing ammonia, by distillation; soluble in NO_5 without change; obtained by boiling bile with chlorohydric acid, concentrating to separate common salt, filtering, adding to the filtrate 5 times its weight of hot alcohol, when the taurine crystallizes out in radiated needles.

Tautolite. Spec. grav. 3·865, H 6·5. Allied to chrysolite, closely resembling hyalosiderite. Laacher See.

Tea. The leaves of the Thea bohea. *Tea* is made from the Tea plant in the tea districts of China, between the latitudes 25° and 31° north latitude, the best districts being between 27° and 31°. Green and Black teas are both made from the same plant. *Green Tea.*—The leaves are brought in from the plantations and spread out from one to two hours, thinly on flat bamboo trays, in order to dry off superfluous moisture. A portion is now thrown into each roasting pan, heated by a brisk wood fire; they are moved about and shaken up with both hands. They emit a crackling sound and give out vapour for five minutes; they are placed on the rolling table, made into the form of a ball with the hands; this is rolled upon the rattan work table, and compressed to remove moisture and burst the leaves. They are returned to the roasting pan, heated by a slow and steady charcoal fire and stirred by the workmen's hands. In an hour or 1½ hour they are well dried, and the colour is fixed. They are then passed through sieves of different sizes to remove dust and impurities, and divided into the different kinds known as twankay, hyson skin, hyson, young hyson, gunpowder, &c. During this process it is refired, the coarser kinds once, and the finest sorts three or four times. *Black Tea.*—The leaves brought from the trees are spread out upon large bamboo mats or trays for a night. The leaves are then tossed and beat with the hands of the workmen for a considerable time, and when they become soft and flaccid are thrown in heaps and allowed to lie for an hour or more. They then are soft and moist, and emit a fragrant smell. They are thrown into an iron pan and roasted for five minutes, as with green tea, then rolled on the rattan table; afterwards shaken out thinly on sieves and exposed to the air, out of doors, for three hours; they are put a second time into the roasting pan for three or four minutes and again rolled. The charcoal fires are now got ready; a tubular basket, narrow at the middle and wide at both ends, is placed over the fire. A sieve is dropped into this tube, covered with leaves, which are shaken on it to an inch in thickness. After five minutes they are removed from the fire and rolled a third time. They are again shaken on the sieves over the fire. Sometimes the last operation is repeated a fourth time, till the leaves have assumed their dark colour (Fortune). *Faced Green Teas* are made from black teas by means of prussian blue and sulphate of lime (Davis). They may be detected by the green aspect being confined to the surface. Hyson tea contains 2·4 per cent. theine; another kind 2·56; gunpowder 6·22; another kind 3·5. The leaves also contain much caseine.

Tears. The fluid secreted by the lachrymal glands of the eyes; alkaline, clear liquid. It consists of 96 water with mucus, common salt, soda, phosphate of lime and soda.

Teetizite. H 1·5 to 2·. Clove-brown pyramids, needles and masses; believed to be sulphated sesquioxide of iron, from Schwarzenberg, Saxony.

Teeth. The human tooth consists essentially of two portions, the *enamel*, forming a thin covering on the crown of the tooth; and the ivory, constituting the body of the tooth. There is a third body termed the *crusta petrosa* or cement, which is most conspicuously observable over the fangs. 1. The *Enamel (encaustum, adamas, substantia vitrea)* consists of earthy fibres, generally 6-sided, which

 TEE

have originally constituted delicate membranous tubes, formed of cells. The animal matter has, however, nearly disappeared, even in the temporary teeth, as is obvious from the analysis. The fibres run parallel to each other in a perpendicular direction to the surface of the ivory (Owen). 2. The *Ivory (tooth bone, dentine)* is formed of tubes, which are perpendicular to the surface of the ivory, separated from each other by intertubular matter. When a transverse section is made of the ivory, the area of the tubes can be observed by means of the microscope, filled with calcareous matter, which dissolves with effervescence on the addition of acid, and leaves the area empty (Owen). The tubes send off branches into the intertubular spaces, which are more numerous in the fangs than in the crown. The cells of the ivory are somewhat circular, $\frac{1}{3000}$th of an inch in diameter, and most abundant towards the external surface of the ivory. 3. The crusta petrosa or cement is confined to the exterior of the tooth, is thinnest on the crown, and gradually thickens towards the point of the fang. It consists of oblong cells, and contains no vascular canals. The following tables are the results of my analyses of teeth, supplied to me by the late Alex. Nasmyth, Esq. :—The substances have, with the exception of those employed for the specific gravities, been dried at the temperature of 212°. The quantity of water which the ivory naturally contains varies from 10 to 14 per cent.

Table I.—Enamel.

	1. Human Temporary.	2. Human Temporary.	3. Human Adult.	4. Hindoo Adult.	5. Elephant.	6. Hippopotamus.
Specific Gravity.	2·711	2·621	2·688			
Organic matter,............	8·218	3·76	6·160	5·573	6·80	4·102
Phosphate of lime,.........}			89·160	89·821	82·55	} 83·630
Fluoride of calcium,........}	} 90·489	85·00	·260	0·205	1·65	} 0·850
Phosphate of magnesia,...}						
Carbonate of lime,.........	0·831	}	4·010	3·578	7·65	10·620
Chloride of sodium,.........}	} 0·240	} 11·24		} 0·743	1·05	0·800
Chloride of potassium,.....}		}				
	99·778	100·00	99·59	99·920	99·70	100·000

Nos. 4 and 9 represent the composition of the teeth of the inhabitants of Bengal, living on a vegetable diet, and may be compared with the European or carnivorous tooth.

Table II.—Ivory.

	7. Human Temporary.	8. Human Adult.	9. Hindoo Adult.	10. Elephant Fine.	11. Elephant Coarse.	12. Walrus External.	13. Walrus Internal.	14. Hippopotamus.
Specific Gravity.	2·090	2·105		1·728	1·794	1·909		1·866
Organic matter,.........	31·55	28·20	26·81	45·65	44·00	39·28	38·16	33·41
Phosphate of lime,......}		60·16	66·42}	} 52·30	54·35	52·80	59·04	55·90
Phosphate of magnesia,}	} 61·23	1·66	0·62}		1·13	2·10	0·67	1·19
Fluoride of calcium,.....}								
Carbonate of lime,.......	4·71	7·04	5·63	1·35	1·33	4·38	1·02	9·14
Chloride of sodium,......}	} 2·51	2·66	0·10	0·06	0·09	0·69	0·53	0·36
Chloride of potassium,..}								
	100·00	99·72	99·58	99·36	100·90	99·25	99·42	100·00

Table III.—Comparative table of Teeth and Bones.

	15. Ox Tooth Temporary.	16. Ox Tooth Permanent.	17. Tiger Tooth.	18. Whale Rib.	19. Human Thigh Bone.
Specific Gravity.				2·102	
Organic matter,.....................	29·16	23·45	31·49	37·96	35·93
Phosphate of lime,..................}	} 67·54	74·88	63·67	{ 52·81	51·12
Fluoride of calcium,.................}				{	
Phosphate of magnesia,............	0·76	0·83	0·52	{	0·63
Carbonate of lime,...................	1·62	1·03	3·64	9·46	9·77
Alkaline chlorides,...	trace	trace	trace	0·17	0·59
	99·08	100·19	99·32	100·40	98·04

Nos. 15, 16, and 17, enable a contrast to be formed of animals living on vegetable and animal food. The teeth in these cases were used entire without separating the ivory and enamel. A portion of the difference in the amount of the animal matter in No. 16 may be attributable to the presence of a portion of enamel. The quantity of alkaline chlorides present in different kinds of teeth is always so inconsiderable, and yet constant, that there can be little hesitation in tracing their origin to the saliva with which the teeth must be always saturated, as is obvious from their absorptive power.

Table IV.—Composition of Teeth in Caries and of Tartar.

	20. Caries.	21. Caries.	22. Tartar.
Specific Gravity.	1·533		
Organic matter,.........	62·00	62·78	20·59
Phosphate of lime,...... } Fluoride of calcium,..... }	32·20	34·79	73·30
Phosphate of magnesia,	2·23		3·93
Carbonate of lime,......	2·27	1·11	0·28
Alkaline chlorides,......	1·28	0·52	0·14
	100·00	99·20	98·24

Tekoretine. $C_{10}H_9$. F.P. 113°. A fusible volatile and crystalline resin, from the fossil wood of peat, in Denmark; soluble in ether; little soluble in alcohol. C 87·17, H 12·84.

Teleoxydic Bodies are those organic substances (as pea, rape, and wheat straw), the inorganic constituents of which are in a complete state of oxidation.

Telerythrine. $C_{22}H_{10}O_{19}$. A crystalline body, produced by the action of water on amarithrine or erythrine bitter.

Telesite. *Telesie.* A synonyme of Corundum.

Tellurethyle. C_4H_5Te. Red fluid, heavier than water, with a disagreeable odour; poisonous; by distilling sulphovinate of barytes and telluride of sodium.

Telluric Bismuth. *Telluric Lead, Telluric Silver.* See these metals.

Telluric Acid. TeO_3. Orange crystalline mass obtained by fusing tellurous acid with carbonate of potash, and passing chlorine through the solution, supersaturating with NH_3 and precipitating telluric acid by chloride of barium, and decomposing by SO_3; it is also formed by fusing tellurous acid with nitre; tellurate of potash in needles is formed by boiling telluric acid with carbonate of potash and evaporating. *Chloride of Tellurium*, a black substance, is obtained by passing chlorine over tellurium. *Bichloride*, a yellow body, is formed by passing chlorine, in excess, over tellurium. *Bromide*, in black crystals, is obtained by distilling tellurium and bromine. *Iodide*, in black crystals, is likewise formed by distilling these two elements.

Tellurium. Te 8, 64. Spec. grav. 6·115, 6·1379, 6·2445, 6·343. Brown powder or acute rhombohedron, colour tin-white; volatilizes at the melting point of glass, as a yellow vapour; sublimes in a retort in shining drops or elastic needles; melts above the fusing point of lead; obtained from telluric bismuth, by fusing with carbonate of potash or soda made into a paste with olive oil; washed with water; sulphide and selenide of tellurium are precipitated from the filtrate by blowing air through it and decomposing the precipitates by chlorohydric acid; the washed tellurium is then distilled.

Tellurium, Ores of.

Tellurium, Graphic Ore of. *Schrifterz, Aurum graphicum.* Spec. grav. 5·723 to 8·28, H 1·5 to 2. Steel-gray, right rhombic prisms, with angles of 107° 44′; fracture fine-grained uneven; structure foliated; sometimes the edges are replaced by faces which obliterate the terminal face, and form a 4-sided pyramid with scalene faces. B.B. fuses into a dark gray metallic globule, which becomes brilliant and malleable; soluble in nitric acid. Te 52, Ag 11·33, Au 24·, Pb 1·5 Offenbanya, Transylvania. The name is derived from the crystals being arranged like written characters.

Tellurium, Native. Spec. grav. 5·7 to 6·1, H 2·25. Tin-white regular 6-sided prisms, the edges at the base of the prism being replaced by planes. The angles of the lateral faces are 120°; opaque; brittle; lustre metallic. B.B. melts before ignition, and on the increase of the heat, burns with a bluish-green flame, being almost entirely volatilized as a dense white vapour. It consists of Te 92·55, Fe 7·2, Au ·25. Maria Loretto mine at Facebay, near Zalathna, in Transylvania.

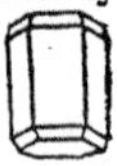

Tellurium, White or Yellow Ore of. Spec. grav. 10·678. Silver-white, inclining to brass-yellow, small right rhombic prisms, with angles of 105° 30′; fracture uneven; lustre metallic; opaque; rather brittle; soft. B.B. fuses into a metallic globule, and gives out a pungent odour. Te 44·75. Au 26·75, Ag 8·5, Pb 19·5, S ·5. Nagyag, Transylvania.

Tellurohydric Acid. *Telluretted Hydrogen.* Colourless gas, soluble in water, with the odour of sulphohydric acid; formed by acting on tellurides of potassium, iron or zinc with chlorohydric acid.

Tellurous Acid. TeO_2. White crystalline crust of 8-hedrons; becomes orange by heat, and then a dark yellow fluid, and volatilizes; obtained by burning tellurium in air, or dissolving tellurium in nitric acid and evaporating.

Tennantite. Spec. grav. 4·375 to 4·53, H 3·. Blackish-gray cubes, 8-hedrons, and rhomboidal 12-hedrons; lustre metallic; opaque; brittle. B.B. decrepitates, burning with a blue flame, emitting arsenical fumes, and fusing into a black magnetic scoria. S 28·74, As 11·84, Cu 45·32, Fe 9·26, SiO_3, 5·. *Form.*—4 Cu_2S, 4 FeS,AsS_3. Cornwall; Modum.

Tenorite. *Melaconite, Black Oxide of Copper.* CuO = C 80, O 20. Steel-gray, 6-sided prisms, tables or scales with a black streak and metallic lustre. On lava, at Vesuvius.

Tephroite. Chrysolite containing manganese.

Teratolite. *Hard Lithomarge.* Spec. grav. 2·5, H 2·75. Lavender or blue compact opaque earthy masses, with often white or red veins; B.B. infusible. SiO_3 41·66, Al_2O_3 22·85, Fe_2O_3 12·98, CaO 3·04, MgO 2·55, KO ·93, MnO 1·68, HO 14·2. Coal formation, Zwickau, Saxony.

Terbium. The hypothetic base of terbia or oxide of terbium, which has not yet been obtained in a state of purity. Its salts have a sweet and astringent taste, and an amethyst colour. The sulphate crystallizes; the crystals effloresce at 122°; the nitrate solidifies in reddish radiated crystals, which do not deliquesce. Terbia is obtained from a solution of common yttria by adding some free acid to its solution, dropping it into binoxalate of potash and stirring till the precipitate no longer dissolves; the precipitate which separates is filtered; binoxalate produces more precipitate in the filtrate; the filtrate neutralized with an alkali yields nearly pure oxalate of yttria; the first precipitates contain much erbia, mixed with some terbia and yttria, the next chiefly terbia. See YTTRIA.

Terebene. $C_{20}H_{16}$. Spec. grav. ·86; B.P. 312·8°. A colourless fluid with an odour of thyme. See COLOPHENE.

Terebenthene. Oil of Turpentine.

Terebenthene, Muriate of. (*Liquid.*) $C_{20}H_{16}HCl$. Colourless volatile fluid, insoluble in water; soluble in alcohol and ether; lighter than water.

Terebenthene, Muriate. (*Solid.*) $C_{20}H_{16}HCl$. F.P. 302°; B.P. 329. Flexible needles, insoluble in water; soluble in alcohol and ether; sublimes.

Terebenzic Acid. $C_{14}H_7O_4$. F.P. 336°. Needles resembling benzoic acid, which however consists of plates, and fuses at 248°; decomposed by hot water; by the action of nitric acid on oil of turpentine.

Terebic Acid. *Terebenic Acid.* $C_{14}H_9O_7$ HO. 4-sided prisms, not volatile; by the continued action of nitric acid on oil of turpentine.

Terebilene. *Peucyle.* $C_{20}H_{16}$. Spec. grav. ·843. Resembles camphilene, and is obtained from chlorohydride of terebene, by passing its vapours through caustic lime.

Terebilic Acid. $C_{14}H_9O_7$. F.P. 392°; by nitric acid on oil of turpentine, soluble in hot water, alcohol, and ether; above its fusing point decomposed into carbonic acid and *Pyroterebilic Acid* $C_{12}H_{10}O_4$. A colourless oily fluid.

Terechrysic Acid. $HOC_6H_3O_4$. Orange mass; deliquescent; soluble in water, alcohol, and ether; obtained by the continued action of nitric acid on oil of turpentine, along with terephthalic and terebenzic acids.

Terenite. Spec. grav. 2·53, H 2. Yellowish-white or green masses, with a pearly lustre; brittle. B.B. fuses into a white enamel; becomes blue with cobalt; allied to scapolite. Antwerp, New York.

Terephthalic Acid. $2HOC_{16}H_4O_6$. White tasteless body; insoluble in water, alcohol, and ether; changed by heating with slaked lime into CO_2 and benzole; forms crystallizable salts; isomeric with phthalic acid; obtained by the prolonged action of nitric acid on oil of turpentine.

Teretinic Acid. $C_9H_7O_5$. White crystals by heating terpenole with litharge.

Terhydrofluate of Titanic Acid. TiO_2 3 HF or TiF_3, HF is formed by the decomposition of crystallised bihydrofluate of titanic acid in water, or by dissolving titanium in hydrofluoric acid and nitric acid. *Cyanide and nitride of Titanium* consists of copper-coloured cubes found in the hearths of iron furnaces, having a specific gravity of 4·6. They were long supposed to be pure titanium.

Teropiammon. $C_{60}NH_{29}O_{26}$. White needles, sparingly soluble in any medium; becomes crimson when heated with sulphuric acid; formed by oxidizing narcotine with nitric acid of moderate strength; it is derived from 3 atoms opianic acid and 1 NH_3, with the separation of 3 HO.

Terpenole. $C_{28}H_{17}O$. Spec. grav. ·852; B.P. 334°. Colourless oily fluid, smelling of hyacinths; by dropping a few drops of a weak acid on hydrate of turpentine (Terpine).

Terpine. *Hydrate of Oil of Turpentine.* $C_{20}H_{16}$ 6 HO. F.P. 217°. Right rhombic prisms; soluble in 200 cold, in 22 hot water; more soluble in alcohol and ether; drops from oil of turpentine at low temperatures, or by nitric acid and alcohol on oil of turpentine; at 217° becomes a crystalline milky substance, boiling at 489°, and subliming in prisms.

Terra Alba. A fine white form of gypsum, used in the manufacture of confections.

Terra de Sienna. See SIENNA.

Terra Japonica, or Catechu.

Terra Sigillata. A kind of bole, formerly employed in medicine, impressed with a seal.

Tesselite. Cubic apophyllite, from Faroe, possessing, when examined optically, a tesselated or mosaic-like structure.

Test. A reagent, which, when added to another body, indicates its presence by a striking physical character. See ANALYSIS QUALITATIVE.

Tetartine. A variety of Albite.

Tetr-amyle-ammonium. N,Ayl_4. The base of the hydrate, an oil which crystallizes; formed by the action of iodide of amyle on tri-amyle-amine, or on ammonia.

Tetr-ethyle-ammonium. NE_4. The oxide of this substance is a powerful alkali, forming soaps with oils; having a caustic and bitter taste; it precipitates metallic oxides, and closely resembles potash; formed in the action of bromide of ethyle on ammonia, having the type of ammonium.

Tetraclasitc. A variety of Scapolite.

Tetra-chloro-naphthalic Acid. $C_{20}H_4Cl_4$. By chlorine on naphthalic acid.

Tetra-chloro-sulphonaphthalic Acid. $C_{20}H_3,SO_2,Cl_4SO_3$. By dissolving tetrachloronaphthalic acid in fuming SO_3.

Tetradymite. Telluric Bismuth.

Tetra-meta-phosphoric Acid. $4\ RO + 4\ PO_5$. By the action of phosphoric on oxide of lead.

Tetra-methyle-ammonium. $N,Me_4 = 4\ (C_2H_3)\ N$. The iodide is obtained by acting on iodide of methyle with ammonia, and treating the resulting iodide with oxide of silver; it is a powerful base and very caustic.

Tetraphyline. See TRIPHYLINE.

Tetrathionic Acid. $S_4O_5 = SO_2,SO_3S_2$. A synonyme of bisulpho-hyposulphuric acid; obtained by heating hyposulphite of lead with hyposulphite of soda.

Thakcetone. A brownish-yellow mass, one of the products of the action of sulphur and ammonia on acetone.

Thallite. Needle-formed epidote in Dauphiné and Isere.

Thallochlore. A green substance from the Citraria islandica.

Tharandite. A variety of dolomite containing $CaOCO_2$ 54·76, $MgOCO_2$ 42·1, FeO, MnO,CO_2 4·19.

Thebaine. F.P. 300°. *Paramorphine.* $C_{38}NH_{21}O_6$. Grains or needles; insoluble in water; soluble in alcohol and ether; taste sharp, metallic; reaction alkaline; a base contained in opium.

Theine. See CAFFEINE.

Thenardite. *Native Anhydrous Sulphate of Soda.* Spec. grav. 2·73. Rhombic 8-hedrons from Espartine near Madrid, accompanied with carbonate of soda.

Thenard's Blue. See BLUE.

Theobromine. $C_{14}H_8N_4O_4$ = caffeine $C_{16}H_{10}N_4O_4-C_2H_2$. Volatile above 482°. White crystalline powder; little soluble in hot water, less so in alcohol and ether; soluble in boiling barytes water; obtained from the seeds of the Cacao theobroma by the same process as caffeine. By means of the battery a body has been obtained from it with the formula $C_{12}N_2H_8O_{16}$, and by binoxide of lead and sulphuric acid an acid, $C_{12}N_2H_8O_{10}$, homologous with inosic acid and with alloxane.

Theory. A term derived from the Greek θεωρία, contemplation, or mode of viewing a subject. Theories are advantageous in an experimental and advancing science by classifying masses of facts which would otherwise be isolated and comparatively unintelligible. The most important theories entertained in chemistry are the *atomic theory*, *theories of affinity*, *amides*, *dimorphism*, *polymorphism*, *metamerism*, *isomorphism*, *isomerism*, *polymerism*, *metalepsy* or *substitution*, *organic radicals* or *bases*, *atomic volumes*, *types*, all of which are described under the several heads.

Theory of Amides.—The following theory has recently been announced:—Supposing that monobasic acids are formed on the type of two molecules of water, then acetic acid will be $C_4H_3HO_2O_2$. When ammonia acts on a monobasic acid, 2 molecules of hydrogen are carried to the molecules of oxygen, water is formed, and the residue, NH, replaces the oxygen. Acetamide is acetic acid, in which 2 oxygen are replaced by NH. Diacetamide is derived from anhydrous acetic acid as acetamide is from protohydrous acetic acid. The following shows the bearing of this view (Wurtz):—

Acetic Acid.	Acetamide.	Diacetamide.
$\begin{matrix}C_4H_3O_2\\H\end{matrix}\ O_2$	$\begin{matrix}C_4H_3O_2\\H\end{matrix}\ NH$	$\begin{matrix}C_4H_3O_2\\C_4H_3O_2\end{matrix}\ NH$
Oxalic Acid.	**Oxamide.**	**Oxamic Acid.**
$\begin{matrix}C_2O_2\\H\end{matrix}\ O_2$	$\begin{matrix}C_2O_2\\H\end{matrix}\ NH$	$\begin{matrix}C_2O_2\\H\end{matrix}\ NH$
$\begin{matrix}C_2O_2\\H\end{matrix}\ O_2$	$\begin{matrix}C_2O_2\\H\end{matrix}\ NH$	$\begin{matrix}C_2O_2\\H\end{matrix}\ O_2$

Theory, Radical. Binary Theory.—According to this theory, every organic body is a compound radical, or is composed of a compound radical with oxygen or electro-negative element, or with water; the radical in this theory is in place of the electro-positive element of inorganic compounds. Thus, in common salt, NaCl, the sodium, an electro-positive metal, is united with the electro-negative element, chlorine. In the chloride of methyle, MeCl, the methyle (C_2H_3) occupies the same position as the sodium in the common salt. The objection to this theory, as applied to organic compounds, is, that in cases of substitution, electro-negative replace electro-positive elements with regard to binary arrangements. Thus, by the action of chlorine on acetic acid we have chloracetic acid formed:—

Acetic acid,..................$C_4H_3O_3HO$
Chloracetic acid,...........$C_4Cl_3O_3HO$

contrary to the notions of a binary theory.

Theory of Nuclei or *Kernels* (*Noyaux Kerne*). According to this theory, organic compounds consist of kernels or nuclei, or of these nuclei united to various bodies. Nuclei consisting of carbon and hydrogen are termed primitive nuclei, as naphthaline ($C_{20}H_8$), but when the hydrogen is replaced by other elements, they are termed derivative or secondary nuclei, as chloronaphthaline ($C_{20}H_7Cl$), dichloronaphthaline ($C_{20}H_6Cl_2$), &c. The compounds capable of replacing hydrogen in the nucleus are, hyponitric acid (NO_4, often expressed by X), imidogene (NH= Im), amidogene (NH_2 or Ad), ammonia, (NH_3 or Am), arsidogene (AsH_2 or Ar), cyanogen (C_2N or Cy). Laurent arranges organic compounds in six classes:—1. *Nuclei*, which consist of a characteristic = car. and the constant

$= const.$ In methylene (C_2H_2), C_2 is the characteristic, and H_2 is the constant. The primary nucleus may be altered by the abstraction or replacement of the H_2. 2. *Protogenides.* Compounds of nuclei with 2 atoms hydrogen or oxygen, or with 2, 4, or 6 atoms of a salt radical. Carburetted hydrogen ($C_2H_2H_2$), is a protogenide of methylene (C_2H_2). 3. *Monobasic Salts* are formed by the replacement of 1 atom of hydrogen of the constant by a metal. Thus, acetic acid being C_4H_2, H_2, O_4, a metallic acetate is C_4H_2, HM, O_4. 4. *Dibasic Salts*, or compounds of a nucleus with O_6 containing 2 atoms hydrogen, replaced by a metal. Oxalic acid being C_4O_2, H_2, O_6, its metallic salts are C_4O_2, M_2, O_6. 5. *Prometallides.* Combinations of nuclei with 1 atom hydrogen; these have not been isolated, but their existence is a probable hypothesis, as cacodyle $= C_4H_3Ar + H$ ($Ar = As\,H_2$). 6. *Syndesmides* result from the union of two types of the same series (*homodesmides*), or of different series (*heterodesmides*). Thus, benzoin ($C_{28}H_{12}O_4$), produced by 2 atoms of oil of bitter almonds uniting ($C_{14}H_6O_2$), is a homodesmide; and formobenzoilic acid ($C_{16}H_8O_6$), probably composed of oil of bitter almonds with formic acid ($C_2H_2O_4$), is a heterodesmide.

Thermometer. (*θέρμη*, heat; *μέτρον*, a measure.) A graduated tube containing a fluid, the expansion of which measures increments of heat applied to it. The thermometers used in Europe are the Centigrade or Celsus thermometer, in which the freezing point is marked 0°, and the boiling point 100°. In Fahrenheit's thermometer the freezing point is at 32°, and the boiling point 212°. In Reaumur's thermometer the freezing point is 0°, and the boiling point 80°. In this country and America Fahrenheit's is universally employed; it possesses the advantage over the Centigrade of the less frequent occurrence of the minus degrees, or those under the freezing point. The Centigrade can scarcely be introduced into this country with any advantage, unless the movement were backed by the United States, since from the much greater number of observers, particularly in meteorology, in that country, where the Fahrenheit scale is always employed, the comparison with our own registers would be rendered less convenient without any corresponding advantage. The degrees of the three thermometers corresponding to each other are:—

	Fahrenheit.		Centigrade.	Reaumur.
Freezing point,	32°	=	0°	0°
Boiling point,	212°	=	100°	80°
1 degree Fahrenheit,	1°	=	1°·8	2°·25

To convert Centigrade degrees into Fahrenheit, multiply by 1·8 and add 32°. Thus 10° C $\times$ 1·8 = 18 + 32 = 50°. To convert Fahrenheit degrees into Centigrade subtract 32° and divide by 1·8. Thus 50° F — 32 = 18 ÷ 1°·8 = 10°. Reaumur's degrees are converted into Fahrenheit's by multiplying by 2·25 and adding 32. Thus 10° R $\times$ 2·25 = 22°·5 + 32 = 54°·5. To convert Reaumur's degrees into Fahrenheit subtract 32 and divide by 2·25. Thus 54°·5—32 = 22·5 ÷ 2·25 = 10° R.

Thermonatrite. Spec. grav. 1·55, H 1·25. $NaOCO_2HO$. Rectangular plates, or as an efflorescence. Found along with native sesquicarbonate of soda.

Therythrine. F.P. above 212°. Brownish-red powder or resin; insoluble in water; soluble in SO_3 and potash with deep brown colour, in bisulphide of carbon, alcohol, and acetone; by the action of ammonia and sulphur on acetone.

Thiacetonine. *Akcethine.* $C_{18}\,H_{10}\,NS_4$? Yellow lustrous rhombohedrons with an alkaline reaction; soluble in dilute acids, alcohol, ether, and acetone; little soluble in water. It is obtained by the action of sulphohydric acid and sulphur on acetone, as thialdine is formed from aldehyde.

Thialdine. $C_{12}\,H_{13}\,NS_4$. Large crystals like camphor; scarcely soluble in water; very soluble in alcohol and ether; very volatile and fusible; strong base; obtained by passing sulphohydric acid through an aqueous solution of aldehydammonia; in decomposing it possesses a highly disagreeable odour.

Thialic Oil. A synonyme of Bisulphide of Ethyle.

Thianisole. *Hydranisimide* $C_{16}H_8S_2O_2$. By SH on hydrocinnamide.

Thio. From *θεῖον*, the Greek word for sulphur, is a term applied to a number of compounds into which sulphur enters as an element.

Thiobenzoldine. $C_{14}H_6S_2$? By the action of sulphide of ammonium on an alcoholic solution of oil of bitter almonds.

Thiocinnole. $C_{18}H_8S_2$. By SH on hydrocinnamide.

Thiofurfole. $C_{10}H_4S_2O_2$. Yellow powder by SH on furfuramide.

Thiomelanic Acid. $C_{80}H_{24}O_{20}S_3$. Black mass with a shining lustre; in moist state reddening litmus; yields by distillation, water, CO_2, SO_2, SH, S, and a brown oil. Obtained by heating alcohol in contact with SO_3, or by passing the vapour of spirit into a tubulated retort containing SO_3; forms monobasic salts with bases.

Thionaphthamic Acid. $C_{20}NS_2H_8O_5HO$. An acid only obtained in union with bases; its salts are red or purple, and in scales resembling mica; on adding acids to its salts sulphuric acid and naphthaline or naphthalamine are produced; obtained by boiling an excess of a solution of sulphite of ammonia with an alcoholic solution of nitronaphthaline; naphthionic acid is also produced.

Thionaphthionic Acid. $C_{20}\,NH_8\,S_2O_5\,HO$. White powder or needles, with satiny lustre; displaces acetic acid; decomposed by NO_5, not by HCl at 392°; the salts are crystalline. By sulphite of NH_3 on nitronaphthaline.

Thionuric Acid. $C_8N, 2\,NH_2, HO_4\,2SO_2$.

White acid needles, decomposed by boiling into 2 sulphuric acid and uramile ($C_8N_3H_5O_6$). Obtained in union with ammonia by the action of SO_2 in the cold state on an aqueous solution of alloxane, or by boiling for half-an-hour alloxane with sulphite of ammonia, to which has been added an excess of carbonate of ammonia.

Thiorsauite. (*Impure Anorthite?*) Spec. grav. 2·688, H 6·. White or gray masses; transparent; fusing B.B. into a glass; soluble in HCl. SiO_3 48·36, Al_2O_3 30·59, Fe_2O_3 1·37, MgO ·97, CaO 17·16, NaO 1·13, KO ·62. Thjorsa, Hekla. Allied to Scapolite.

Thiosinnamine. *Rhodalline.* $C_8H_8N_2S_2$, or C_8N,NH_2H_4,H_2S_2. F.P. 162°·5. A bitter base in right rhombic prisms, obtained by adding ammonia to oil of mustard; not poisonous; yields by heat hydrosulphocyanic acid, an oil, and charcoal.

Thiosinn-ethylcamine. $C_{12}H_{12}N_2O_2$. A supposed base obtained in union with HCl and Pt Cl_2 by precipitating the product of the action of HCl on a mixture of ethylamine and oil of mustard.

Thomsonite. Spec. grav. 2·29 to 2·369, H 4·75. Snow-white right rectangular prisms with square bases, with angles of 135° 30′, and 90° 40′. The height of the prism is equal to about four times its breadth; structure foliated; fracture uneven; lustre vitreous, inclining to pearly; transparent to translucent; brittle. B.B. swells up like borax, becoming opaque and snow-white, but does not fuse; by ignition becomes like enamel, the edges being rounded, but without losing its shape. SiO_3 37·08, Al_2O_3 33·02, CaO 10·75, NaO 3·7, HO 13·. *Form.* 3(CaONaO) 4(SiO_3, 3($Al_2O_3 2SiO_3$) 24 HO. Lochwinnoch, Kilpatrick, Scotland; Nova Scotia. *Scoulerite*, occurring at Ballimony, Ireland, is an amorphous variety of Thomsonite. *Comptonite* is a crystallized variety.

Thorina. *Thoria.* ThO 8·5, 68; 8·449, 67·592. Th = 88·17, O = 11·83. Spec. grav. 9·402, (Berzelius), 9·366 (Damour). A fine snow-white powder, infusible before the blowpipe, becoming hard by ignition, and difficult to pulverize. *Process.*—Thorite is pulverized and digested in chlorohydric acid; it becomes yellow, evolves chlorine and dissolves, the silica gelatinizing. The whole is evaporated to dryness, and heated till the excess of chlorohydric acid is evolved, the residue pulverized and digested for twenty minutes in chlorohydric acid, water added and thrown on a filter, on which the silica is washed. Through the solution a current of sulphohydric acid is passed to remove tin and lead; it is again evaporated to dryness to remove some silica, and filtered. The solution is boiled with caustic potash, which precipitates thorina, sesquioxides of iron, and uranium and manganese, while alumina is dissolved. This precipitate being well washed, is dissolved in dilute chlorohydric or sulphuric acids. With the latter acid the liquid is evaporated; sulphate of thorina falls, soluble with difficulty in the hot solution. The mother liquor is poured off; the sulphate washed with boiling water and ignited; it leaves thorina. The mother liquor is saturated with carbonate of potash or caustic ammonia, and as much sulphate of potash added to it as it is capable of dissolving, and the precipitate washed on a filter with a saturated solution of sulphate of potash. The potash sulphate of thorina is then dissolved in cold water and precipitated by caustic ammonia or potash in the state of a hydrate. It is still yellow from the presence of manganese, from which it may be nearly purified by solution in sulphuric acid as above.

Hydrate of Thorina.—White, gelatinous, contracting much while drying; absorbs carbonic acid in the air; it ought therefore to be dried in vacuo over sulphuric acid; readily soluble in acids when moist; much less soluble when dry; insoluble in caustic alkalies; soluble in alkaline carbonates; more soluble in cold than in hot carbonate of ammonia; ammonia does not precipitate thorina from a saturated solution of carbonate of ammonia as it does zirconia; it loses water by ignition, and becomes hard and difficult to pulverize; in this state it is soluble in no acid except the sulphuric; to dissolve it, the earth reduced to a fine powder, is digested for a long time in sulphuric acid diluted with half its weight of water. It is rendered soluble in acids by calcining it with a caustic or carbonated alkali. Water added to this calcined mass forms a milky fluid.

Salts.—The hydrate is readily soluble in acids, but after ignition thorina is only soluble in sulphuric acid. The salts of thorina are colourless, and have an astringent taste; they lose their acid by ignition, are precipitated by boiling, the sulphate dissolving again on cooling. They are precipitated by *caustic ammonia, soda*, and *potash;* insoluble in excess of the reagent by which thorina is distinguished from alumina and glucina. They are precipitated by *sulphohydride of ammonia;* precipitated and redissolved by *alkaline carbonates.* They form with *sulphate of potash* a double sulphate, which is soluble in cold water, a distinguishing character from the potash sulphate of zirconia, which is almost wholly insoluble. The thorina double salt is insoluble in a saturated solution of *sulphate of potash*, which distinguishes it from yttria. Yellow prussiate of potash precipitates thorina salts, but not zirconia salts. From protoxide of cerium it is distinguished by not becoming reddish-brown, but continuing white when calcined; and by not forming a coloured bead before the blowpipe, either with borax or biphosphate of soda. Thorina salts are precipitated by oxalates and oxalic acid.

Chloride.—Th Cl? 12, 96; 11·886, 95·088. White crystalline mass, or brilliant colourless crystals, by a second sublimation. Obtained by exposing a mixture of carbonized thorina and sugar to the action of chlorine.

Thorinum. *Thorium.* Th 7·5, 60; 7·449, 59·592. A dark iron-gray heavy powder with a metallic lustre when pressed, evolving a little hydrogen when first immersed in water, from the formation of some potash; apparently malleable like aluminum; not oxidized by water even when heated; burns in the open air when ignited, and becomes thorina, a snow-white earth without any appearance of fusion; dilute sulphuric acid produces at first an effervescence with thorium, but it speedily ceases, and so slight is its action that it may be used to remove thorina from thorium. Nitric acid has little action except by long digestion. Chlorohydric acid dissolves it rapidly, especially when assisted by heat, hydrogen gas being evolved; fluohydric acid possesses but a slight action; it is not acted on by caustic alkalies. *Process.*—Chloride of thorinum is first obtained by passing a current of dry chlorine through a mixture of ignited thorina and charcoal in a porcelain tube. The chloride of thorinum is then heated in a platinum crucible or glass tube with potassium; a slight detonation occurs, and heat but no light is evolved. The residue is washed with water, thorinum remains. The double potash fluoride of thorinum may also be used along with potassium.

Thorite. Spec. grav. 4·63 to 4·8. Black masses resembling gadolinite; streak reddish-gray; powder brownish-red; easily frangible; lustre of new fracture vitreous, of old resinous; opaque; easily scratched by the knife. B.B. infusible; becomes pale red; ignited gives out FH in a tube; fuses with borax into a bead, clear when hot, opaque when cold coloured by iron; the addition of nitre, showing the presence of manganese; with carbonate of soda decomposed without fusion. ThO 57·91, SiO_3 18·98, CaO 2·58, Fe_2O_3 3·4, Mn_2O_3 2·39, MgO ·36, U_2O_3 1·61, PbO ·8, SnO ·01, HO 9·5, KO ·14, NaO ·1, Al_2 O_3 ·06, powder undecomposed 1·7 = 3 ThO,SiO_3 3 HO? In Syenite, Lövö, Brevig, Norway.

Thraulite. *Hisingerite, Hydrous-sesqui-silicate of Iron.* Sp. grav. 3·045. Black roundish nodules; powder brownish-yellow; fracture uneven, or imperfect conchoidal; splendent; opaque; brittle. B.B. edges rounded, but the lustre unaltered; after ignition attracted by the magnet. SiO_3 31·28, Fe_2O_3 43·42, FeO 5·7, HO 19·12. *Form.* 3 FeO, SiO_3, 2 Fe_2O_3, SiO_3 6 HO. Riddarhyttan, Westmanland; Bodenmais, Germany.

Thrombolite. Spec. grav. 3·39, H 3·5. Emerald or green masses with a glassy lustre; opaque. PO_5 45·22, CuO 37·68, HO 17·1 = 1½CuO,$PO_5$3HO. Retzbanya, Hungary. B.B. turns the flame blue and then green; on charcoal gives a bead of copper.

Thulite. Spec. grav. 3·1055, H 5· to 6°. Rose-red granular masses, or with a tendency to the crystalline structure of an oblique prism, with angles of 87° 30′, and 92° 30′; lustre vitreous; translucent on the edges. B.B. fuses with soda into a transparent bead, giving a violet on adding nitre, indicating manganese. SiO_3 47·9, Al_2 O_3 27·54, CaO 17·8, Fe_2O_3 2·9, HO 1·17; loss 2·69 (KO?) (R.D.T.) Souland, Tellemark, Norway. Allied to epidote.

Thumite. A variety of axinite from Thum, Saxony.

Thuringite. Spec. grav. 3·153, H 2·25. Olive-green masses with pearly lustre; soluble in HCl, forming a jelly. B.B. indications of iron; allied to hydrous yenite. SiO_3 22·41, Fe_2O_3 21·94, FeO 42·6, MgO 1·16, HO 11·89. Saalfield, Thuringia.

Thuyone. B.P. 329° to 347°. A carbohydrogen oil by the action of iodine on oil of thuya, from the Thuya occidentalis.

Thyme, Oil of. From Thymus vulgaris. Consists of 2 oils, C_{34} $H_{26}O$; B.P. 346°, and $C_{40}H_{30}O_7$; B.P. 455°; Spec. grav. of vapour 5·511; when distilled with anhydrous phosphoric acid, $C_{40}H_{26}$ (Doveri).

Thymene. $C_{20}H_{16}$. An oil isomeric with oil of turpentine, having the same density of vapour; from essence of thyme by fractional distillation.

Thymole. $C_{20}H_{14}O_2$. F.P. 111°; B.P. 446°. The stearoptene of oil of thyme; oblique prisms with a rhombic base; little soluble in water; very soluble in alcohol and ether; possesses no rotatory power on light; soluble in SO_3, forming *sulphothymic acid,* $C_{20}H_{13}S_2O_5O_2$, which is monobasic; *chlorothymole,* $C_{20}H_8Cl_6O_2$, a viscid fluid.

Tin. Sn 7·25 58. *Sources.*—Tin is found native in the state of *peroxide* or *tinstone,* in Cornwall; in the mountains separating Galicia from Portugal; in the mountains between Saxony and Bohemia; in Malacca, Mexico, and at Oruro in Upper Peru, mined for two hundred and fifty years in primary mountains. *Char.*—Spec. grav. 7·285 (Herapath, Phil. Mag. 64, 322), 7·291 and 7·299 (by hammering, Brisson), 7·178 (crystals, Miller), 7·293 (fused, ib.), 7·291 (Kupffer). *Melting Point.*—512°·6 (Morveau), 442° (Crichton, Phil. Mag. 15, 147), 446° (Kupffer), 432°·5 (G. A. Erman), 442° (Rudberg). When cooled slowly it separates as a rhomboidal prism (Pajot, Journ. de Phys. 38, 52). *Colour.*—Fine white colour like silver, and when fresh its brilliancy is very great. *Crystal.*—Square prisms, 8-sided needles, 6-sided prisms; very malleable. *Tin foil,* or leaf tin, is about $\frac{1}{1000}$th part of an inch thick, and it might be beat to half this thickness. Its ductility and tenacity are inferior. A bar of tin ¼ inch in diameter is broken by a weight of 296 lbs. (Rennie). Hence a tin wire, $\frac{1}{10}$th of an inch in diameter, would be broken by 47·36 lbs. A tin wire 0·078 inch diameter supports 34·7 lbs. (Morveau). Taste slightly disagreeable; tin emits a peculiar smell when rubbed, depending on its action on the substance with which it is rubbed. It has been obtained in crystals by passing a weak current of electricity through a solution of protochloride of tin, forming the positive pole of a plate of tin, also by melting and slowly cooling it. It boils at a white heat. It is very flexible, and emits a peculiar crackling sound when bent. Tin is not acted on by the air; but heat speedily causes it to oxidize, protoxide being first formed, and then binoxide.

Purification. — The purest kinds of tin are Malacca, Banca, and English grain tin; the least pure, common grain tin, block tin. The metals which usually contaminate it are arsenic, antimony, iron, copper, zinc, lead, bismuth. When tin is free from arsenic, it dissolves completely in chlorohydric acid; if that metal is present, it remains as an insoluble residue, which when heated yields a garlic smell. Lead, &c. may be detected by treating the tin by nitric acid, the binoxide of tin remains insoluble, while nitrate of lead, &c. dissolves. Sulphuric acid precipitates the lead, yellow prussiate of potash the iron as Prussian blue, and a plate of iron detects copper by being covered with a yellow coating. Antimony and arsenic may likewise be detected by the usual tests for these metals.

Preparation of Tin. Mode of Smelting.—Tin ore, binoxide of tin and iron, occurs in two kinds of deposits in Cornwall, in veins accompanied by other ores, and in loose grains or sand in alluvial matter in detached fragments. The tin reduced from the ore occurring in veins is most impure, and is called *common block* or *mine tin.* The purest grain tin, or *stream tin*, is procured from the alluvial ore, which is free from every impurity, except nodules of hæmatite. 1. *Grinding* or *stamping.* The ore is reduced to powder by means of stamping mills driven by steam power. The state of division is regulated by a plate of iron pierced with small holes, through which the whole passes from the stamping mill, being washed through by a rapid stream of water. 2. *Washing.*—The ground ore is then carefully washed and *dressed.* When the ore is rich there is little difficulty in washing away earths in consequence of the greater specific gravity of the tin ore, but in poorer ores this is more difficult, especially when they are mixed with copper and other ores, which are likewise heavy. 3. *Roasting for the removal of Sulphur and Arsenic.*—The ores after being made clean on the dressing floors are taken to the *burning house*, where they are placed in small reverberatory furnaces, and exposed to a moderate and regular heat. They are frequently turned over by an iron rake to expose fresh surfaces. Sulphurous acid and arsenious acid being volatilized, the sulphurous acid escapes, and the arsenious acid is condensed in long horizontal flues constructed for the purpose. 4. *Washing.*—After the ores come from the burning house, the process of dressing is completed by further washing. After this operation the ore presents the appearance of a brown moist heavy powder; the washing removes by gravitation certain other impurities, and sulphate of copper, from the solution of which the copper is gained by a plate of iron, (as recommended by Dr. T. Thomson, in 1813. Ann. Phil. 2, 352.) The ore as now prepared is sold to the smelter. 5. *Smelting*, or *Deoxidation.* (Conversion of SnO_2 into Sn.)—The furnaces

a furnace door; *b* grate; *c* door for ore; *e* grate for admitting air; *d* working door; *i* chimney with good draught.

for smelting tin are of the reverberatory kind, and hold from 12 to 16 cwt. of ore. The charge is prepared by mixing it with coal or Welsh culm and slaked lime, and heated so as to bring the whole into a state of fusion, which is kept up for seven or eight hours, when the charge is ready to draw. The lime unites with the silica and earthy matters still mixed with the ore and forms a slag, which floats on the surface of the melted metal reduced by the coal. The slag is raked off through a door, c, while the melted metal is run off by a tap-hole in the bottom of the furnace, and *laded* into moulds, so as to form plates of a moderate size, and put aside for more refinement. The slag is broken up, stamped and washed, and a quantity of tin taken from it, which is called *Prillion*, and is smelted again. 6. *Refining*, or removal of iron, copper, arsenic, tungsten, and a portion of undecomposed oxides, sulphurets, or arseniates, and of some earthy matter or slag. The plates of tin are heated in a refining furnace to a very moderate temperature, and are melted gradually.

The metal flows from the furnace into a kettle, which is kept hot by a small fire placed beneath. The more infusible substances are left in the furnace, and a further purification of the tin is obtained by agitating it for some time by an operation called *tossing*, which consists in lifting some of the metal in a ladle, and allowing it to fall back upon the fused metal from such a height as to put every part in motion. Refining is also produced by heating it with blocks of unseasoned wood. It is then skimmed, and being laded into granite or cast iron moulds, is formed into *blocks* of 3 cwt. *Grain* or *stream* tin is smelted by charcoal, and the fire is urged by bellows in a *blowing house*, being thus distinguished from *smelting houses* in which common tin is made. The furnaces consist of a cylinder of iron standing upon one end, and lined with clay. The upper end is open for receiving the fuel and ore which are thrown in alternately, and a hole at some distance from the bottom of the back of the cylinder is provided to admit the blast, and another lower down, and opposite to it, allows the metal to flow out regularly as it is reduced. The only purification it seems to require is the separation of mechanically mixed impurities. For this purpose it is laded into an iron pan or kettle, where the fusion is kept up by a gentle heat beneath, and the mass is continually agitated by plunging into it pieces of charcoal soaked in water, and retaining them at the bottom of the kettle by an iron tool. The water in the charcoal is rapidly converted into steam, and gives the mass the appearance of boiling. It is allowed to rest and skimmed, when it has a brilliant appearance. It is then laded into moulds. Grain tin is sometimes put into a different form by heating it; the blocks being so heated as to render the metal brittle, are let fall from a height, when they break into fragments resembling prisms or basaltic pillars.

Protoxide, Black oxide.—SnO 8·25, 66·. Black powder; tasteless; insoluble in water. Spec. grav. 6·66. Protoxide of tin is obtained by heating the oxalate of tin, or by dissolving tin in HCl, and precipitating the solution by carbonate of soda, washing the precipitate well, and igniting it in close vessels or in CO_2. When heated in the open air it catches fire, burns, and becomes binoxide (SnO_2). It is insoluble in NH_3; soluble in NaO; precipitated *brown* by SH; precipitated *brown* and redissolved by NH_3SH. It dissolves in SO_3, HCl, and dilute NO_5; by strong NO_5 it becomes binoxide (SnO_2).

Binoxide, Deutoxide, Peroxide, White Oxide, Stannic Acid, Tinstone.—SnO_2, 9·25; 74. Occurs in nature in yellow or black 8-hedrons, or more commonly as a 4-sided square prism. Hardness, 6·5, sp. gr. 6·55 to 6·945. Artificially it is prepared by acting on tin with NO_5, or by heating tin to whiteness in the open air, when the metal takes fire and is converted into the yellow oxide. When prepared by either of these methods, it dissolves with difficulty in acids. When prepared by precipitation with an alkaline carbonate or alkalies from the bichloride, it dissolves easily in acids. Deutoxide of tin is precipitated by NH_3, and redissolved by a great excess, also by NaO, and NH_3 SH. From copper furnaces I have artificial stannic acid in long brilliant acicular square prisms. Native tinstone contains some portion of sesquioxide of iron.

Sesquioxide.—$Sn_2\ O_3$. White gelatinous precipitate, obtained by digesting hydrous sesquioxide of iron with a solution of protochloride of tin. *Stannate of Stannous oxide.* SnO 3(SnO_2). Orange-yellow powder, formed by digesting metastannic acid in a solution of tin salts.

Metastannic Acid, Hydrous. — 5 (2 HO, SnO_2), or 10 HO, Sn_5O_{10}, consisting of SnO_2= 88·64 HO = 19·36 (Thomson, Ann. Phil. 10, 149). White friable, semitransparent, with a vitreous fracture, or as a white powder; obtained by treating tin with diluted nitric acid, washing the precipitate and drying in a dry current of air, at the common temperature. It is quite insoluble in caustic ammonia (Fremy). When dried in a vacuum it loses 5 atoms of water, and becomes 5 HO Sn_5O_{10}, containing 11·3 per cent. of water. When dried at 292°, it loses another atom of water, and becomes 4 HO Sn_5O_{10}, with 8·8 per cent. of water; at 320° it parts with another atom of water. If any of the preceding hydrates are dissolved in an alkali, and precipitated by an acid, metastannic acid insoluble in nitric acid, results, but it possesses the property of *dissolving in ammonia.* The most constant hydrate is 5 HO, $Sn_5\ O_{10}$, which, when reacted on by soda, becomes NaO $Sn_5\ O_{10}$, 4 HO. The latter salt is difficult to obtain, but the anhydrous salt (NaO, Sn_5O_{10}), is formed by acting on metastannic acid with a cold solution of soda and drying on a tile. It is white, granular, crystalline, dissolves in a great quantity of cold water, but decomposes with a slight elevation of temperature; at 140° decomposes into caustic soda and metastannic acid. Metastannic acid, according to Fremy, is represented by the formula $Sn_5\ O_{10}$. When treated with chlorohydric acid, it dissolves in a hydrous state, but produces a compound which does not possess any of the properties of bichloride of tin.

Stannic Acid, SnO_2, is prepared by decomposing bichloride of tin by an alkali, or by treating the bichloride by carbonate of lime or carbonate of barytes. It falls as a jelly, and should be washed rapidly on a filter. It is easily soluble in chlorohydric acid, and reproduces bichloride of tin; soluble in sulphuric acid even when dilute; but the solution is decomposed by boiling, and the stannic acid is completely precipitated. When precipitated, washed, and dried spontaneously in air at the usual temperature, its composition is 7 HO, 3(SnO_2), containing 21·5 per cent. of water. When this hydrate is dried in vacuo, or at 284°, it loses 5 atoms water, and becomes metastannic acid with 2 water, containing 7·4 per cent. water. In this state it does not reproduce bichloride of tin. When stannic acid is dissolved in an alkali, and precipitated by an acid, stannic acid falls, which is soluble in ammonia, while, before solution in alkali, it was *insoluble in ammonia.* It loses its solubility by drying, or by boiling for a few minutes. Bichloride of tin changes stannic acid into a yellow body resembling that afforded by metastannic acid (Sn_3O_6 SnO 3 HO). The stannates of potash and soda crystallize very easily, while the metastannates are always gelatinous and uncrystallisable (Fremy). Stannates of potash and soda are obtained by dissolving stannic acid precipitated from bichloride of tin in caustic potash or soda, or by calcining metastannate of potash or soda with an excess of alkali. That the metastannic acid has been thus converted into stannic acid is proved by precipitating by sulphuric acid. A jelly falls, which dissolves at once in chlorohydric acid, and forms bichloride of tin, which is not the case with metastannic acid; even carbonic acid precipitates hydrous stannic acid.

Stannate of Soda occurs as a dry powder in commerce. When crystallized, it consists of NaO SnO_2 9 HO, with a crystalline form resembling that of Augite. This salt is extensively used in calico printing, as a "*prepare*" for the cloth, which is dipped in a solution of it, and then passed through weak SO_3.

Protochloride of Tin, Salt of Tin—SnCl 2 HO = 13·937, 111·496. Formed by dissolving tin in HCl, with an excess of the metal until the acid is saturated. The solution is evaporated, when the protochloride separates, in long needles or 4-sided prisms. The anhydrous chloride is formed by distilling equal parts of tin and calomel; a gray body passes over, which becomes solid on cooling. The crystals are employed in dyeing, to remove oxygen, as in indigo, and also as a mordant, in calico printing. It sometimes occurs with 1 or 3 atoms water.

Bichloride, Permuriate of Tin, Fuming Liquor of Libavius.—$SnCl_2$, 16·124, 128·992, is obtained by distilling 6 parts of tin, rubbed with 1 mercury, and 33 corrosive sublimate; it is also formed by placing tin in chlorine gas. It is colourless like water. Spec. grav. 2·25; smokes in the air; 3 parts of this liquid mixed with 1 water, form *butter of tin*, a thick fluid, which dissolves in more water. It acts on oil of turpentine. It does not precipitate gold.

Sulphide of Tin—SnS 9·25, ·74. Gray brittle substance, forming, by fusing tin with sulphur.

Bisulphide, Mosaic Gold.—SnS_2 11·25. Gold-coloured substance, obtained by subliming a mixture of tin, 6 parts; mercury 3; salammoniac 3; flowers of sulphur 3½. Muriate of ammonia, cinnabar, and protochloride, are volatilized, while Mosaic gold remains.

Native Cupreous Sulphide of Tin, Tin Pyrites. Spec. grav. 4·35, H 3·25. Steel-gray, yellowish-white, or brass-yellow, masses, or in regular 6-hedrons; streak black, lustre metallic; opaque; brittle. B.B. sulphur is expelled, and a black scoria remains without a metallic button; soluble in aqua regia. S 25, Sn 34, Cu 36, Fe 2. Huel Rock, parish of St. Agnes, Cornwall.

Estimation—Tin is usually estimated by converting it into binoxide, by an excess of NO_5. For estimating the value of protochloride of tin, bichromate of potash has been long used by Mr. William Blythe, Mr. Dale, and Mr. James Young, of Manchester. I first saw it in use in the hands of Mr. Young, several years ago, and it has been in general use in the manufacturing districts for many years. It has also been used in Germany for estimating the oxides of iron when mixed. I am indebted to Mr. Blythe for the following note:—

"I find the first experiments to have been made April 1st, 1836. In these experiments it is shown that 120 grains of the crystals of protomuriate of tin are required to reduce the chromic acid in 50 grains of bichromate of potash. Previous to this date, in examining chrome ores, I reduced the chromic acid, precipitated the green oxide, and weighed it; but after this date, the protomuriate of tin was always used. I give the following from my note-book as an example.

"'July 18th, 1836.—Examination of a sample of chrome ore from St. Petersburg.

"'33 grains, carefully pounded, were fluxed with 200 grains carbonate of soda, and 40 grains of nitrate of potash. The fused mass was edulcorated till everything soluble was taken up. The yellow solution was now neutralized with nitric acid, and the chromic acid tested with protomuriate of tin. The quantity required was 72 grains, equivalent to 30 grains bichrome = 90 per cent., 47·3 of green oxide.'

"It appears from my note-book that I first began to use bichromate in the testing of manganese, February 6th, 1837. After this, in all my examinations of samples, I have merely entered the respective quantities of copperas, manganese, and bichromate used. January 19th, 1839. Sample of Swedish manganese, from Thomas Boyd, Liverpool.

"Copperas, 600 grains; manganese, 100 grains; bichrome, 20 grains.

20 Bichrome = 110·73 copperas.

600—110·73 = 489·27.

2 At. Copperas. 1 At. MnO_2.

34·75 : 5·5 489·27, : 77 per centage of black oxide."

Tincal. Said to be a native name for borax.

Tin Ore. Native binoxide of tin.

Tinplate, Sheet Tin, or **White Iron,** is made by heating sheet iron in a reverberatory furnace, dipping it in dilute muriatic acid, and reheating in the furnace. The iron is then passed through cold rollers, immersed in fluid fermenting bran, dipped in dilute sulphuric acid, and then in a bath of melted tallow. It is dipped into fused tin for an hour or two, is then removed, allowed to drain on an iron grating, and rubbed with a hemp brush. The figured appearance which it sometimes presents, is obtained by the action of NO_5.

Titanic Acid. TiO_2 5·125, 41. Specific gravity 3·9311 (artificial), 4·249 (titanite or rutile). Specific gravity 3·826 3·927 (anatase), 4·13 (R.D.T.), 4·166 (Brookite, Arkansite). It is dimorphous; being square prismatic in rutile and anatase and right prismatic in Brookite; artificially prepared, it is a white powder; becoming yellow when heated, and resuming its white colour on cooling. B.B. yields a colourless or yellow glass, with borax; acquiring a bluish colour in the reducing flame; with salt of phosphorus a colourless or yellow glass, in the oxidizing flame; by long exposure to the reducing flame on charcoal, this glass is yellowish while hot, and violet on cooling. Titanic acid occurs in two forms: *Alpha* is precipitated from solutions by ammonia, and is soluble in dilute acids, both when moist and dry, and incandesces when ignited. *Beta* is obtained by dissolving the preceding in an acid, and boiling, when beta precipitates; it is slowly soluble in oil of vitriol, &c. Titanic acid is obtained from its native minerals, by fusing them with caustic potash, or carbonate of potash, or carbonate of soda; the fused mass washed with water, and dissolved in strong muriatic acid; the solution is diluted with water and boiled, when the titanic acid falls. This process may be repeated until the iron is completely separated. Titanic acid precipitated from an alkaline solution by boiling with an acid, is the hydrate; it is a white flocculent and bulky precipitate; when precipitated by ammonia, it contains an atom of water. Titanic acid is a feeble acid in union with bases; it occurs as titaniate of iron in iserine. It is likewise a feeble base when dissolved in acids, the combination being destroyed by simple boiling. A solution of a salt of titanic acid gives with metallic salts of zinc, iron and tin, a blue colour, probably a sesquioxide of titanium

Titanic Iron. See ISERINE and ILMENITE.

Titaniferous Cerite. A black mineral from Coromandel, resembling gadolinite, containing CeO 36·5, FeO 19·8, CaO 8·, Al_2O_3 6·, HO 11·, MnO 1·2, SiO_3 19, TiO_2 8·.

Titanite. *Crispite, Gallizinite, Nigrine, Rutile, Sagenite, Titanic Acid.* Specific gravity 4·18, 4·249, 4·256, H 6·75. Brownish-red right striated square prisms, frequently terminated by a 4 or 8-sided pyramid, with angles of 132° 32′, 161° 40′, 122° 45′; fracture foliated. B.B. unaltered; forms with borax a hyacinth bead; does not fuse with salt of phosphorus; it is titanic acid with some iron. It is found in quartz in primary rocks, as at Craig Challiach and Glen Tilt, Perthshire.

Titanium. *Menakane.* Ti 3·24, 3·125, 25. Black powder, obtained by heating a mixture of potassium and anhydrous fluoride of titanium, and washing away the fluoride of potassium, or by heating in a tube the ammonia chloride of titanium, *Protoxide*, TiO, a probable oxide by heating titanic acid in a charcoal crucible. *Bichloride.* $TiCl_2$. B.P. 275°, a colourless heavy fluid with a pungent odour; obtained by passing chlorine over titanium, or over a mixture of titanic acid and charcoal at a red heat. *Bisulphide.* TiS_2. A dark green mass or bronze scales by passing bichloride and sulphohydric acid through an ignited tube.

Tiza. The native name of Boronatrocalcite.

Toadstone. A species of earthy basalt.

Tobacco. The leaves of the Nicotiona tabacum go under this name. It is a native of Virginia and Mexico, from Tobaco, a province of which country, it derives its name. The constitution of the leaves is as follows:—Nicotine ·06, nicotianine ·01, slightly bitter extract 2·87, gum, with some malate of lime 1·74, green resin ·267, albumen ·26, gluten 1·048, malic acid ·51, malate of ammonia ·12, $KOSO_3$ ·048, KCl· 063, malate and nitrate of potash ·095, phosphate of lime ·166, malate of lime ·242, silica ·088, lignine 4·969, starch trace, water 88·28. The poisonous and stimulant principle of tobacco is nicotine, which seems closely analogous in its characters to aniline. The mean composition of tobacco is—water 10·09; ashes 18·64, consisting of alkaline salts 5·15, earthy salts soluble in acids, 11·27, siliceous earth and sand 2·18. In consequence of the duty, tobacco is frequently adulterated, and principally with sugar, which is added in solution. The mode of detecting this sophistication is by fermenting the hot infusion, distilling off the spirit, and estimating its equivalent of sugar. The following table affords a view of the amount of revenue derived from tobacco and snuff:—

Account of tobacco brought into home consumption during the last ten years.

Years.	Pounds weight Tobacco Imported for Consumption. Unmanufactured. lbs.	Manufactured, and Snuff. lbs.	Duty. £
1842,......	21,788,954	172,481	3,500,560
1843,......	22,827,253	184,748	3,667,292
1844,......	24,250,010	191,361	3,896,378
1845,......	25,852,199	183,430	4,143,721
1846,......	26,557,143	192,221	4,259,750
1847,......	26,701,911	189,981	4,279,561
1848,......	26,777,399	191,246	4,432,640
1849,......	27,297,903	194,368	4,377,419
1850,.....	27,538,104	196,681	4,415,250
1851,......	27,853,253	209,588	4,470,334
1852,......	28,358,908	200,031	4,549,982

Duty on unmanufactured tobacco 3s. per lb. and 5 per cent., about 3s. 2d. per lb.

Duty on cigars, cavendish, &c. 9s. per lb. and 5 per cent.

Foreign snuff 6s. per lb. and 5 per cent.

The following table gives the composition of some kinds of common snuff as occurring in the trade:—

	1.	2.	3.	4.	5.
Water,	28·94	35·88	28·70	29·68	36·34
Soluble salts,	11·80	14·00	7·84	12·40	19·20
Insoluble salts,	7·31	12·41	10·78	9·64	9·58
Vegetable matter,	51·95	37·72	52·68	48·28	34·88

1. Thirty-sevens, 2. black rappee, 3. imperial rappee, 4. London brown, 5. brown rappee.

Tobernite, *Torbenite*. A synonyme of chalcolite, or phosphate of uranium.

Tolene, $C_{10}H_8$. Spec. grav. ·858; B.P. 338°. Volatile colourless fluid; obtained by distilling balsam of tolu with water.

Tolu, Balsam of. A reddish-brown thickish balsam, derived from incisions made in the bark of the Myroxylon toluiferum; when exposed to the air it becomes solid and brittle; the smell is fragrant; it affords, when distilled with water, benzoïc acid, tolene, and cinnameine; it is soluble in alcohol, ether, and volatile oils; it yields by destructive distillation toluole and an oily fluid with the characters of benzoic ether; it gives also benzoic and cinnamic acids. Two resins have been described as derivatives from the balsam, viz. alpha resin, $C_{36}H_{19}O_8$, F.P. 140°, a brown substance yielding toluole and benzoic acid by dry distillation; beta resin, a dull yellowish-brown tasteless resin, slightly soluble in alcohol and ether.

Toluidine. $C_{14}H_9N$, or N,Tyl,H_2. F.P. 104°; B.P. 388°. Colourless plates, slightly soluble in water; very soluble in alcohol, ether, and oils; renders fir wood and elder pith yellow. It may be represented as ammonia in which 1 atom H is replaced by toluyle; it is homologous with benzole and aniline, &c.; it is obtained by the action of ammonia in alcohol and sulphohydric acid on toluole.

Toluole. *Benzoene*. $C_{14}H_8$. Spec. grav. ·870, of vapour 3·26; B.P. 226°. A thin colourless fluid; insoluble in water; slightly soluble in alcohol; more soluble in ether; obtained from the volatile portion of the products of the distillation of balsam of tolu by fractional distillation. When heated with nitric acid it yields *nitrotoluole*, $C_{14}H_7NO_4$, and *dinitrotoluole*, $C_{14}H_6 2NO_4$; the first is analogous to nitrobenzide, and its vapour, in contact with ignited lime, is converted into aniline and carbonic acid.

Toluylic Acid. $HO,C_{16}H_7O_3$. A substance corresponding closely with benzoic acid, with which it is homologous, differing from it only by 2(CH); obtained by acting on cymole with NO_5; with nitric acid it affords nitrotoluylic acid $(C_{16}H_6NO_4)O_3HO$; diluted with lime it gives toluole (Noa).

Tombac. An alloy of 85 copper and 15 zinc.

Tombazite. A synonyme of nickel glance.

Tomosite. Silicate of manganese.

Tonka Beans. The seed of the Dipterix odorata, used to flavour snuff, from its containing coumarine.

Topaz. *Fluosilicate of Alumina, Physalite, Pyrophysalite.* Spec. grav. 3·641, 3·499, H 8·. White, yellow, green, blue, right rhombic prisms,

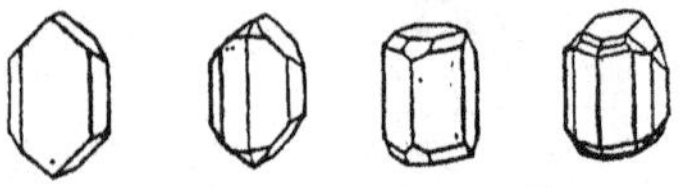

differently modified. The principal angle measures 124° 22′; the angles of the base are usually replaced by tangent planes, forming a 4-sided pyramid; structure foliated; fracture conchoidal; lustre vitreous, from splendent to glimmering; transparent to translucent on the edges. B.B. the faces of crystallization blister; with borax forms a transparent glass; crystals possessing different faces of crystallization at opposite ends, acquire different kinds of electricity when heated; by friction it acquires positive electricity. The topaz of Brazil consists of SiO_3 34·01, Al_2O_3 58·38, FH 7·79=2 AlF_3, 5 Al_2O_3,SiO_3. Ural; Altai, Brazil; Mucla, Asia Minor; Aberdeenshire; Down mountains. The term oriental topaz is applied to the yellow sapphire, and false topaz to a yellow rock crystal.

Topazolite. A synonyme of colophonite.

Tophus. A concretion described by John as derived from the joints of the arms, consisting of animal matter 73·, $CaOCO_2$ 10, 3 CaO,PO_5 17.

Topinambour. The Helianthus tuberosus, or Jerusalem artichoke. The root used as an esculent consists of water 772, sugar 148, inuline 30, cellulose? 12·2, gum 10·78, citrate of potash 10·7; peculiar body producing viscous fermentation 9·9, phosphate of lime 1·44, $KOSO_3$ 1·2, citrate of lime ·8, KO HCl ·8, $KOPO_5$ ·6, oil ·6, cerine ·3, malate of potash ·3, SiO_3 ·24, tartrate of lime ·14. The ultimate elements are C 43·02, H 5·91, O 43·56, N 1·57, ash 5·94.

Torrelite. *American Columbite*, or *Tantalite*. Spec. grav. 4·8038, H 4·25 to 5·4. Black or dark brown masses, or 4-sided oblique prisms, with angles of 157° 29′; surface iridescent; lustre imperfectly metallic, almost resinous; cross fracture granular; opaque. B.B. fuses with soda and borax into a dark red bead; with excess of soda, indications of manganese. Columbic, niobic, and pelopic acids 73·9, FeO 15·65, MnO ·8, HO ·35. Middleton, Connecticut. It was the determination of the specific gravity of this mineral, and of the other columbic acid minerals, by Dr. T. Thomson, that led H. Rose to discover niobic and pelopic acids, which he now, however, considers to be only one acid, for which the term niobic is retained.

Touchstone. A synonyme of Lydian stone, or basanite.

Touloucouna Oil. F.P. 110°. A solid buttery oil of a yellowish-red colour, from the drupas of the Carafa touloucouna in Senegal; it excites vomiting when swallowed, and is used by the natives to anoint the skin, and thus ward off insects.

Tourmaline. *Achroite*, *Aphrizite*, *Apyrite*, *Cockle*, *Daurite*, *Indicolite*, *Picolite*, *Rubellite*, *Schorl*, *Siberite*. Spec. grav. 3·, 3·06, 3·076, H 7·5 to 8. Brown, green, blue, red, white, black obtuse rhomboidal prisms (3d syst.) The figures represent 6-sided prisms of tourmaline,

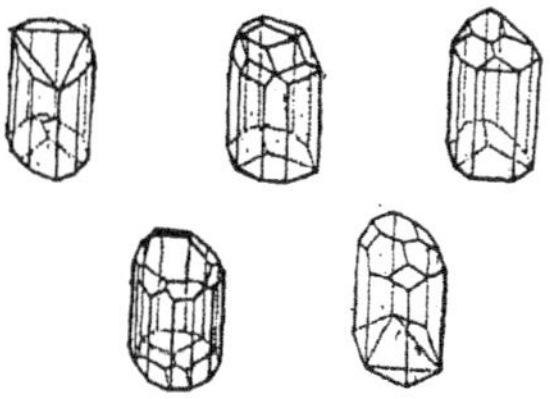

differently modified; fracture conchoidal; lustre vitreous; transparent to translucent; B.B. intumesces, but does not fuse; the black variety becomes a scoria; the transparent varieties, when heated, become electric, the termination of the prisms with most planes becoming positive, and the other end negative. Tourmalines receive different names according to their colours, &c. Aphrizite consists of small crystals from the Hartz; achroite white crystals from St. Gothard; indicolite, indigo coloured; rubellite, red variety; schorl, black varieties. They have been divided into five different groups, dependent on their composition, with the following formulæ:—

1. 3 RO, 2 SiO_3, 3 (R_2O_3 SiO_3)
2. 3 RO, 2 SiO_3, 4 (R_2O_3, SiO_3)
3. 3 RO 2 SiO_3 6 (R_2O_3, SiO_3)
4. RO SiO_3 3 (R_2O_3 SiO_3)
5. RO SiO_3 4 (R_2O_3 SiO_3)

and the composition of these groups per cent. is given in the following table:—

	1.	2.	3.	4.	5.
S. grav..	3·049	3·043	3·226	3·107	2·998
F.........	2·28	2·36	1·54	2·09	2·70
PO_5	trace	0·20	0·00	trace	0·22
SiO_3.....	38·35	38·45	35·74	38·55	41·16
BO_3	8·25	8·48	8·00	7·29	8·56
Al_2O_3...	31·32	34·56	34·40	38·40	41·83
F_2O_3....	1·27	3·31	7·61	5·13	0·00
Mn_2O_3..	0·00	0·00	0·00	0·00	0·97
FeO	0·00	0·09	8·60	2·00	0·00
MgO	14·89	9·11	1·76	0·73	0·61
CaO	1·60	0·71	0·86	1·14	0·00
NaO	1·28	2·	1·02	2·37	1·37

	1.	2.	3.	4.	5.
KO	·26	0·73	0·47	0·37	2·17
MnO.....	0·00	0·00	0·00	0·73	0·00
LiO......	0·00	0·00	0·00	1·20	0·41

1. Brown variety; 2. black, Texas; 3. black, Alabaschka; 4. green, Brazil; 5. red, Rozena (Rammelsberg),

Trachyte. (τραχυς, rough.) A volcanic rock or lava, consisting of a mass of felspar, in which glassy crystals of felspar are imbedded; it often contains also augite, hornblende, and magnetic iron. It is found on the Rhine, Auvergne, Caucasus, America.

Tragacanth. Specific gravity 1·384. A gum in hard white worm-like masses, from the Astragalus tragacantha, a thorny shrub of Candia and the Levant; becomes easily pulverizable at 115°; swells up in water, and forms a mucilage; it consists of water 11·1, ashes 2·5, arabine 53·3, bassorine and starch 33·1.

Transposition of Gases. See DIFFUSION.

Trapp. (*Trappa*, a stair, Swed.) Ancient volcanic rocks, consisting principally of hornblende, albite or felspar, augite, &c., and constituting a class which includes basalt, amygdaloid, greenstone. See GEOLOGY.

Trass. *Dukstein.* A volcanic ash or scoria, consisting of a portion decomposed and of another undecomposed by muriatic acid.

Travertine. A form of deposited carbonate of lime.

Treacle. That portion of sugar syrups which cannot be crystallized.

Tremolite. *Grammatite.* A species of amphibole. (Found in the Tremola valley.) Spec. grav. 2·9257 to 3·2, H 4·75. White, greenish, bluish, or reddish oblique rhombic prisms, with angles of 130° 15′, and 124° 30′; generally in fibrous crystals; lustre vitreous; always translucent; sometimes semitransparent; easily frangible. B.B. fuses easily into a semitransparent glass; with borax a glass; not decomposed by salt of phosphorus. SiO_3 60·1, MgO 24·31, CaO 12·73 Al_2O_3 ·42, FeO 1·, MnO ·47, FH ·83, HO ·15. Raphilite is a gray tremolite. Switzerland; Canada, United States.

Tremolite, Norwegian. Spec. grav. 3·2, H 6·. White or bluish-gray amorphous masses; lustre pearly; translucent on the edges; structure foliated. B.B. fuses with difficulty on the edges. SiO_3 63·7, CaO 27·2, MgO8·9.

Triamylamine. $C_{30}H_{33}N$. B.P. 498°. Light oil of peculiar aromatic smell and burning taste, by the action of bromide of amyle on diamylamine.

Tribromaniline. *Tribromophenylamine.* $C_{12}(H_4Br_3)N$. By treating aniline with bromine.

Tribromobenzide. $C_{12}Br_3H_3$. Silky needles, very soluble in alcohol and ether, by distilling bromobenzine with lime.

Tribromocarbolic Acid. *Tribromophenisic Acid.* $C_{12}H_3Br_3O_2$. White needles, by bromine on carbolic acid.

Tribromocodeine. $C_{36}H_{18}Br_3NO_6$.

Tribromomesitilole. $C_{18}H_9Br_3$.

Tribromosixaldid. $C_6Br_3H_3O_2$. Amber oil, heavier than water, by bromine on solution of citraconate of potash.

Tricetylamine. $C_{96}H_{99}N = 3C_{32}H_{33},N$. F.P. 102°·2. White needles, formed by the action of ammoniacal gas on iodide of cetyle.

Trichloraniline. $C_{12}(H_4Cl_3)N$. A volatile crystalline body, obtained when aniline is treated with chlorine gas.

Trichloroacetic Acid. See CHLOROACETIC ACID.

Trichlorocarbolic Acid. See CHLOROPHENISIC ACID.

Trichlorofilixic Acid. $C_{26}H_{13}Cl_3O_{10}$. Yellow powder, insoluble in water, melting in hot water; soluble in alcohol, ether, fat, and volatile oils. By chlorine on filixic acid.

Trichlorohydrokinone. $C_{12}H_3Cl_3O_4$(?) Fuses a little above 266°. Delicate colourless plates; soluble in hot water, alcohol, and ether. By excess of sulphurous acid on trichlorokinone.

Trichlorohydrokinone. *Yellow.* $C_{24}Cl_6H_4O_8$. By a deficiency of sulphurous acid on trichlorokinone.

Trichlorokinone. C_{12},HCl_3,O_4. Yellow prisms; sublimes above 266°; fuses at 320°. Obtained along with other compounds when kinic acid is distilled with a chlorine mixture.

Trichloromesitilole. $C_{18}H_9Cl_3$.

Trichloromethyl-sulphurous Acid. $C_2Cl_3HO_22SO_2$.

Trichloropteritannic Acid. $C_{24}H_{12}Cl_3O_8$. By dry chlorine on pteritannic acid.

Trichlorosulphonaphthalic Acid. $C_{20}H_4(SO_2)Cl_3,SO_3$. Crystalline substance; soluble in water and alcohol. By dissolving trichloronaphthalic acid ($C_{20}H_5Cl_3$) in hot fuming sulphuric acid, distilling with water, saturating with potash, then with acetate of lead, microscopic needles are obtained, which by decomposition by SO_3 yield this acid.

Trichlorotannaspidic Acid. $C_{26}H_{10}Cl_3O_{13}$. By chlorine on tannaspidic acid.

Trichlorovalerianic Acid. See CHLOROVALERISIC ACID.

Trichloroxamide. *Chloroacetamide.* $C_4NH_2Cl_3O_2$.

Trichloroxitannaspidic Acid. $C_{26}H_{10}Cl_3O_{14}$. By the action of muriatic acid and chlorate of potash on tannaspidic acid.

Triclasite. A synonyme of Fahlunite.

Triethylamine. $C_{12}H_{15}N$. Colourless, inflammable liquid, very alkaline; soluble in water. By the action of bromide of ethyle on diethylamine.

Triethyl-amyl-ammonium, Hydrated Oxide of. $C_{22}H_{27}NO_2$. The iodide of triethyl-amyl-ammonium is formed when triethylamine is boiled with iodide of amyle.

Triethyl-phenyl-ammonium. $3C_4H_5,C_{12}H_5,N$. By iodide of ethyle on diethylaniline.

Triethylstibin. Sb, $3(C_4H_5)$. B.P 317°. Transparent very mobile liquid, with the penetrating odour of onions. In contact with air emits dense white fumes, and frequently takes fire, burning with a white brilliant flame. Formed when iodide of ethyle is distilled with an alloy of antimony and potassium.

Trigenic Acid. Radiated prisms, little soluble in water, insoluble in alcohol; yielding quinoline by heat; obtained by the action of cyanic acid on aldehyde.

Tri-iodo-codeine. $C_{36}H_{21}NO_6I_3$. Ruby prisms by the action of iodine on codeine dissolved in alcohol; insoluble in water and ether; soluble in alcohol.

Trimetaphosphoric Acid. A modification of metaphosphoric acid, only found in the crystallization of a fused mixture of equal equivalents of phosphoric acid and soda, by allowing it to cool gradually.

Trimethylamine. C_6H_9N. Readily obtained by heating the oxide of tetramethyl-ammonium ($C_8H_{13}NO_2$). It is a gas at common temperatures, but liquefies at about 48°·2 to a mobile fluid of very powerfully alkaline reaction.

Trimethylamine-Alum. $C_6H_9NSO_3,Al_2O_33SO_3$ 24HO. 8-hedrons, obtained by mixing sulphate of trimethylamine with sulphate of alumina.

Trinitranisole. $C_{14}H_5N_3O_{14}$. F.P. 138°·2. Brilliant tables, slightly yellowish; insoluble in water; soluble in ether and hot alcohol. By heating 1 part anisic acid with 15 parts of a mixture of equal weights of sulphuric and nitric acids.

Trinitro-amarine. $C_{42}H_{15}N_5O_{12}$. White warty crystals, soluble in boiling water and alcohol. By heating trinitrohydrobenzamide with potash.

Trinitrobenzophenide. $C_{26}H_{73}(NO_4)O_4$. By heating gently an intimate mixture of picric acid ($C_{12}H_{33}(NO_4)O_2$), with chloride of benzole, till HCl ceases to be evolved, the product washed well with cold alcohol, and then crystallized from boiling alcohol. Golden-yellow rhombic plates, fusing at an elevated temperature, and deflagrating when very strongly heated.

Trinitrocarbolic Acid. See PICRIC ACID.

Trinitroethionic Acid. $C_4H_2\,3\,NO_4O$, HO?

Trinitrohydrobenzamide. $C_{42}H_{15}3NO_4N_2$. A white powder, by ammonia on hydride of nitrobenzoyle; insoluble in water, ether, and oil of turpentine.

Trinitromesitilole. $C_{18}H_9\,3(NO_4)$.

Trinitroniceic Acid. $C_{12}H_23NO_4O_3HO$?

Trinitrosalithole. A synonyme of Trinitranisole.

Trinket Gold. An alloy of gold 75, and copper 25. Spec. grav. 14·78.

Trioxymercurazotin. $3HgO(Hg_3N)2HO$. By the action of ammonia on oxide of mercury.

Tripestone. A synonyme of Anhydrite.

Triphane. A synonyme of Spodumene.

Triphyline. *Tetraphyline, Perowskine.* Sp. grav. 3·6, H 5. Greenish-gray masses, or imperfect prisms; subresinous; translucent when thin. B.B. fuses into a magnetic bead; gives a green glass with borax; soluble in acids. FeO 48·57, MnO 4·7, LiO 3·4, HO ·68, SiO_3 ·53, PO_5 41·47. Bodenmais, Bavaria.

Triplite. *Ferrophosphate of Manganese, Pitchy Iron Ore in part.* Spec. grav. 3·439 to 3·775, H 5·25. Blackish-brown resinous imperfectly crystalline masses; streak yellowish-gray; opaque, or somewhat translucent. B.B. melts into a dark scoria; a glass with borax, coloured by manganese; soluble in NO_5. PO_5 32·8, MnO 32·6, FeO 31·9, 3 CaO, PO_5 3·2 = 4(MnO FeO) PO_5. Limoges.

Tripoli. *Tripoleene.* Spec. grav. 1·86 to 2·2. Coarse dull earthy mass, rough; does not adhere to the tongue; consists of 80 to 90 silica, Al_2O_3 7·, Fe_2O_3 3·. Composed of the shields of animalcules. Tripoli, Africa; near Prague, &c. Used in polishing metals, &c.

Tris, or **Trito.** Applied to a compound which has 3 atoms of base united to 1 of acid.

Trithionic Acid. *Sulpho-hyposulphuric Acid.* $S_3O_5 = SO_2SO_3S$. A sulphur acid, obtained in aqueous solution by decomposing the potash salt. It has never been seen in the solid state.

Tritomite. Spec. grav. 4·16 to 4·66. Dark brown 4-hedron at Lamö, Norway. B.B. intumesces and becomes white; with borax a clear bead; decomposed by HCl. SiO_3 20·13, Al_2O_3 2·24, C_2O_3 40·36, LiO 15·11, YO ·46, CaO 5·15, MgO ·22, FeO 1·83, Mn, Cu, Sn, W 4·62, volatile matter 7·86.

Tritylc. *Propionic Alcohol.* C_3H_7HO. B.P. 204°·8. Spec. grav. of vapour 2·02. Colourless liquid lighter than water, pleasant odour.

Trityle-sulpho-carbonic Acid. By adding sulphide of carbon to the hydrate of trityle in potash.

Trityle-sulphuric Acid. $C_3H_7SO_3$.

Trombolite. *Terhydrous Subsesquiphosphate of Copper.* $1\frac{1}{2}$ $CuO,PO_5 3HO$.

Trona. *Nitrum.* Native sesquicarbonate of soda.

Troostite. Silicate of zinc.

Truffles. *Tuber cibarium.* An esculent fungus.

Tsantjan. A preparation of Fucus cartilaginosus, used as an esculent in the east instead of edible birds' nests.

Tschewkinite. *Tscheffkinite, Uralorthite.* A variety of Allanite. Specific gravity 4·5, by which it is separated from Allanite, the specific gravity of which is 3·6, H 4·25. Black masses, opaque; phosphoresces by heat. B.B. fuses into a black bead. It contains SiO_3 35·49, Al_2O_3 18·21, Fe_2O_3, FeO 18·03, CeO 10·85, LaO 6·54, CaO 9·25, MgO 2·06, MnO 2·37, HO 2·. Ilmen mountains.

Tuesite. Spec. grav. 2·558, H 2·5. Milk-white masses; opaque; lustre resinous, nearly dull; sectile. B.B. becomes bluish and brittle; with soda becomes an opaque mass; with borax and salt of phosphorus forms a glass. SiO_3 44·3, Al_2O_3 40·4, CaO ·755, MgO ·5, HO 13·5. Used for slate pencils. Banks of the Tweed, near Melrose. Allied to halloylite (R.D.T.)

Tufa, Calcareous. A variety of deposited carbonates of lime.

Tuff. *Tufo, Trap tuff, Earthy tuff, Granular tuff, Stone tuff.* An Italian name for a volcanic rock, consisting of agglutinated scoriæ and loose matters ejected from volcanoes. Stone tuff is reddish-brown, with orange streaks, from interspersed pumice, and forms the summit of the Tarpeian rock at Rome; it contains leucite crystals, brown mica, pyroxene, and sometimes felspar. Granular tuff is a blackish or brownish substance composed of thick grains, slightly agglutinated, intermixed with scales of leucite, pieces of augite, mica scales. The composition of volcanic tuff is SiO_3 44·5, FeO 6·5, Al_2O_3 12·, MgO ·7, KO 5·5, NaO 1·5, insoluble in acid 16·4, HO 11· (Pausilippo, Berthier).

Tuggkada Resin. A resin found in the pine in Sweden, is chewed by the country people; it has a milky fracture, and becomes soft by mastication; by boiling with water a new acid is taken up and deposited in crystals.

Tugilite. A synonyme of hydrous sesquioxide of iron.

Tumour. Excrescences on the soft tissues or bony parts of animals. A tumour of the brain consisted of water 65·, fibrine 13·23, fatty matters 1·20, mucous matter? 12·2, osmazome 8·. A tumour of the cheek was formed of fibrinous matter 18·3, soluble albumen 1·, glutine? 3·7, extractive? 2·5, fat 3·1, water 71·4.

Tungstate of Lead. See LEAD.

Tungsten. W or Tn 11·75, 94· or 95·. *Scheelium, Wolframium.* Tungstic acid was first discovered by Scheele in 1781, in a mineral termed *tung-sten*, or ponderous stone, now known to be tungstate of lime (Scheele). The Messrs. De Luyart obtained the same acid from wolfram (tungstate of iron and manganese), and reduced it to the metallic state. Tungsten occurs native in the form of tungstate of lime, a yellowish mineral found in octahedrons, consisting of (CaO TnO_3) as tungstate of lead, (PbO TnO_3) found in octahedrons, and as wolfram, of which there appear to be two species, as under:—

	I.	II.	III.
Spec. grav., ...	7·00	7·43	7·54
Tungstic acid, ...	75·33	74·66	75·85
FeO, ...	9·30	17·59	19·26
MnO, ...	15·13	5·64	4·89
Silica, ...		2·10	

The first of these species may be termed De Luyarts' wolfram, or zinnwaldite. It occurs at Zinnwald. The specimen analyzed gives the same locality as the original specimen examined by the Spanish chemists (R.D.T.) Mr. Richardson (Records of Gen. Science, i. 449, 1835) first showed the existence of at least two species, and Schaffgotsch has confirmed this view. The two species may be detected by the specific gravity; that of the De Luyarts was 6·8, and the amount of red oxide of manganese obtained 22 per cent. or 20 per cent. of protoxide. The second species may be named Cumberlandite, the original specimen being from that county. The formulæ of these two species is:—

Zinwaldite,.........3 (FeO TnO_3) $MnOTnO_3$
Cumberlandite,.....2 (FeO TnO_3) 3($MnOTnO_3$)
Montevidite,........4 (FeO TnO_3) $MnOTnO_3$

The third species, from Montevideo, contains binoxide of tungsten, according to the views of Schaffgotsch. Tungsten, in the metallic state, may be obtained by fusing wolfram in powder and carbonate of potash in a crucible; the tungstate of potash is taken up by water, the salammoniac added to the solution, and the whole evaporated to dryness and ignited in a Hessian crucible. On digesting in water, oxide of tungsten, a black powder, remains, which when ignited takes fire, and is converted into tungstic acid. The acid can then be reduced to the metallic state by passing hydrogen over it at a red heat. Tungsten is a grayish-white metal, resembling steel. Specific gravity 17·4, and therefore approaching that of gold, platinum, &c.

Oxide of Tungsten, Brown oxide.—TnO_2. When we fuse a mixture of 1 part of wolfram and 2 of carbonate of potash, dissolve the residue in water, add 1½ part salammoniac, evaporate to dryness, ignite in a Hessian crucible; on dissolving the fused mass in water, a black powder remains, which is the oxide of tungsten (Wöhler). It may be also procured by passing hydrogen over tungstic acid until the latter assumes a brown colour. It has then a metallic lustre. When tungstic acid is placed in hydrochloric acid, and a plate of zinc is brought into the solution, copper-red flocks of oxide fall. When ignited the oxide catches fire, burns like tinder, and becomes tungstic acid. The oxide thus exists in 2 forms, as fine copper-coloured plates, and as a black powder, assuming the metallic lustre by friction.

Soda Oxide of Tungsten.—NaO 2 TnO_2. A yellow salt crystallized in cubes; obtained by passing dry hydrogen over bitungstate of soda, this salt being formed by fusing tungstate of soda, and dissolving in it as much tungstic acid as it will take up. The product of the action of the hydrogen is treated with water to dissolve neutral tungstate of soda, and there remains a golden crystalline powder. It is then boiled with HCl to decompose the adhering tungstate, next with pure potash; and lastly, with pure water. It is a beautiful substance. When suspended in water, in the state of powder, it is transparent, and of a green colour; acids and alkalies do not decompose it. When heated in the air it changes colour, softens and changes partly into tungstate of soda.

Tungstic Acid. TnO_3 or WO_3 14·75, 118. A pale yellow powder, infusible, tasteless; insoluble in water; spec. grav. 6·12. Before the blowpipe it gives with borax a yellow glass; with salt of phosphorus, also a yellow bead in the oxidizing flame; blue in the reducing flame. When ignited, is not soluble in acids; in the hydrous state it is soluble in HCl and NH_3. It may be prepared by boiling wolfram powder, with a mixture of HCl and NO_5, which dissolve the iron and manganese, and leave the acid. The acid may be further purified by dissolving it in caustic ammonia and precipitating by nitric acid, or it may be dissolved by sulphohydret of ammonia and decomposed by NO_5, sulphuret of tungsten falls, which is converted by roasting into tungstic acid. It may be procured from tungstate of lime by nitric acid. Laurent represents the different salts as modifications of tungstic acid, and thus explains some singular reactions. Thus, we know that certain tungstates allow their acid to precipitate on the addition of NO_5 or HCl, while others give no precipitate. According to this view, the common tungstates, MO TnO_3, contain an insoluble acid; while $MOTn_3O_6$ and $MOTn_5O_{15}$ contain an acid which is soluble in water. All the salts, with the exception of the neutral salts, retain a certain amount of water essential to them. The following are the types:—

	Type.
1. Tungstates,........	$HOTn\ O_3$
2. Isotungstates,.....	$HOTn_2O_6$
3. Metatungstates,...	$HOTn_3O_9$ 2 HO
4. Paratungstates,...	2 $HOTn_4O_{12}$
Paratungstate of Potash,.........	(⅘KO ⅕ HO) Tn_4O_{12} 2HO
5. Homotungstates of Ammonia,.......	(⅘NH_3⅕HO) Tn_5O_{15} 8HO
6. Polytungstates,....	3 HO Tn_6O_{18}

Bichloride of Tungsten, $TnCl_2$ 20·689, is formed by heating metallic tungsten in chlorine gas, the metal catches fire and is converted in cinnabar red needles, which melt easily and volatilize in red fumes. Water turns it violet and finally red, produces oxide and hydrochloric acid. Caustic potash decomposes it, hydrogen being evolved, and the same result accompanies the action of ammonia, the brown oxide being deposited.

Terchloride, $TnCl_3$, makes its appearance in the form of yellowish-white scales, resembling boracic acid, when the brown oxide is heated in chlorine gas. It is very volatile, and rises in yellow fumes. When exposed to the air it changes into tungstic acid and hydrochloric acid. A red chloride appears along with the bichloride, in fine red needles, more volatile than the preceding chlorides; it bubbles up when thrown into water and gives out much heat. Its composition has not been determined.

Bisulphide of Tungsten.—TnS_2. Formed by passing SH over tungstic acid, ignited in a porcelain tube, is a grayish-black powder, having when rubbed a metallic lustre. The *tersulphide*, TnS_3, makes its appearance when tungstic acid is dissolved in bisulphohydride of ammonia, and the solution precipitated by NO_5; it is a black powder when dry; it dissolves in water. None of the compounds of tungsten appear to exercise any deteriorating influence on the animal system when swallowed.

Tunstate of Lime. See LIME.

Turf. *Peat.* The decomposing remains of plants. It yields when distilled illuminating gases (CH_2) and solid paraffine. In Denmark it is associated with the resins boloretine, phylloretine, tekoretine, and xyloretine. It yields a large amount of ash, and sometimes affords a considerable quantity of potash.

Turgite. A synonyme of sesquioxide of iron.

Turkey Red. See CALICO PRINTING.

Turmeric. The root of the Curcuma longa, a plant found in India, used in powder as an ingredient in curry powder; when its colour is impressed on paper, it is employed as a test for ammonia, wich renders it brown. The colouring matter is curcumine qv.

Turnerite. H above 4·. Brownish-yellow oblique rhombic prisms, the lateral faces of which are inclined at an angle of 96° 10′; lustre nearly adamantine; transparent to translucent; consists of Al_2O_3,CaO,MgO and FeO, with some silica. Mont Sorel, Dauphine.

Turnip. The bulb of species of Brassica. The fig. represents the appearance of the turnip under the microscope. The yellow turnip con-

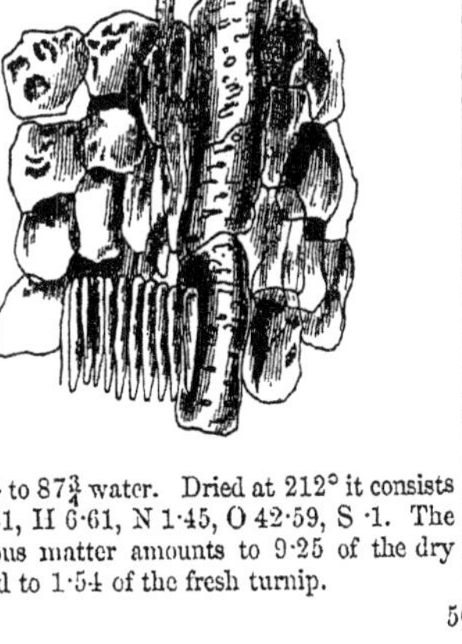

tains $83\frac{1}{4}$ to $87\frac{3}{4}$ water. Dried at 212° it consists of C 45·31, H 6·61, N 1·45, O 42·59, S ·1. The nitrogenous matter amounts to 9·25 of the dry mass, and to 1·54 of the fresh turnip.

Turnsole. The French name for Litmus.

Turpentine. A semifluid resinous body, from various species of Pinus. Common turpentine comes from the Pinus abies and sylvestris, and consists of oil of turpentine and two resins, pinic and silvic acids. *Venice Turpentine* is a light yellow fluid, extracted from Pinus larix (larch), and contains $\frac{1}{5}$ of oil of turpentine, besides succinic acid, &c. *French Turpentine* from Pinus maritima, resembling the last. *Strasburg Turpentine* is a pale yellow fluid, from Pinus picea. *Canada Balsam*, from Pinus canadensis.

Turpentine, Oil of. C_5H_4 or $C_{20}H_{16}$. Spec. grav. ·865, of vapour 4·764; B.P. 312°. Thin colourless fluid with a peculiar well-known smell; slightly soluble in water, more soluble in alcohol and ether; mixes in all proportions with *fixed* oils; obtained by distilling crude turpentine with water and rectifying. It constitutes the type of the turpentine class of oils. When dry hydrochloric acid gas is passed into it a solid artificial camphor is formed, a muriate of oil of turpentine $C_{20}H_{16}HCl$, and the mother liquor contains a fluid isomeric muriate. When the solid and fluid muriates are distilled with lime, camphylene and terebylene result. Colophene is procured by distilling oil of turpentine with strong sulphuric acid. Hydrate of oil of turpentine ($C_{20}H_{16}$ 6 HO), in prisms, is formed often in oil of turpentine bottles.

Turpeth. *Turbith Mineral.* Subsulphate of Mercury.

Turquoise. *Agaphite, Birousa, Calaite, Johnite.* Spec. grav. 2·6296 to 3·25. Is softer than quartz; bluish-green masses occurring in a conglomerate at Mount Madan, 40 miles N.W. from Nishapore, near Meshid, in Persia; lustre dull; vitreous, admits of polish. *Blue* turquoise consists of Al_2O_3 47·45, PO_5 27·34, CuO 2·02, FeO 1·1, HO 18·18, 3 CaO,PO_5 3·41, MnO ·5; the *green* is formed of Al_2O_3 50·75, PO_5 5·64, CaO 1·42, FeO 1·1, HO 18·12, phosphate of lime 18·1, SiO_3 4·26.

Tutenag. *Chinese Silver.* See PACKFONG and ARGENTANE.

Tweers. *Tuyeres.* See IRON SMELTING.

Twin, or *Compound Crystals*, are well illustrated by Staurolite, in the prismatic system, where four crystals are united on planes parallel to a plane on two terminal edges of an 8-hedron.

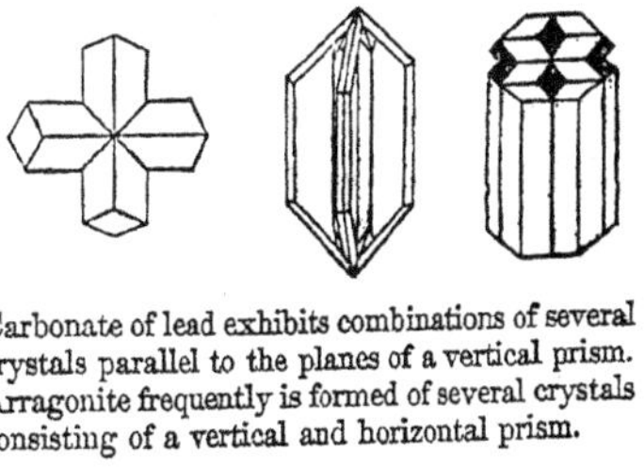

Carbonate of lead exhibits combinations of several crystals parallel to the planes of a vertical prism. Arragonite frequently is formed of several crystals consisting of a vertical and horizontal prism.

Type Metal. An alloy of 80 lead and 20 antimony; or 77 lead, 15 antimony, and 15 bismuth.

Types, The Theory of, expresses the fact that certain compounds which possess the same number of atoms in the same relative position, and the same fundamental properties, are said to belong to the same chemical type. Acetic acid, $C_4H_4O_4$, and chloracetic acid, $C_4HCl_3O_4$, are examples.

1. According to some views, *water* is the type of a series of compounds, if we represent its atoms as corresponding with its volumes H_2O. The following compounds will be written according to the annexed formula by this theory:—

Water,......................	$H + H + O$
Hydrate of potash,.......	$H + K + O$
Dry potash,................	$K + K + O$
Alcohol, ...	$C_2H_5 + H + O$
Ether,........................	$C_2H_5 + C_2H_5 + O$
Compound ethers (Williamson),........	$C_2H_5 + C\,H_3 + O$
	$C_2H_5 + C_5H_{11} + O$
	$C_2H_3 + C_5H_{11} + O$

2. *Ammonia* constitutes another type of organic radicals or bases, in which nitrogen remains constant, but the hydrogen may be replaced entirely or in part by organic bases or radicals. We have thus a series of ammonia, amide, imide, and nitrile bases, according to the amount of hydrogen replaced. ABC indicates the bases replacing the hydrogen.

Ammonia, NHHH, Amide base, N A H H
Imide base, NABH, Nitrile base, N A B C

The replacing bases may be different atoms of the same base, as in the following examples:—

Ethylamine, ...	C_4H_5,H_2,N
Diethylamine, ...	$2C_4H_5\ H,\ N$
Triethylamine,............	$3C_4H_5,\ \ N$

3. *Ammonium* is likewise the type of a series of organic radicals or bases:—

Amm. N H H H H. 1st replacd. base N H H H A
2d base, NHH A B. 3d base,.........N H A B C
4th base, NA,B C D

The examples of these substitutions will be found under organic bases.

The hydrate of oxide of ammonium may also be represented as the ammonia type united to the oxide type, thus:—

$$NHHHH,O + HO = NH_4O + HO$$

4. 2 atoms *Hydrogen* have also been regarded as a type, thus:—

Hydrogen,.....................	$H,\ H$
Hydride of methyle (marsh gas), ...	$C_2H_3,\ H$
Hydride of ethyle, ...	$C_4H_5,\ H$
Hydride of acetyle (aldehyde), ...	$C_4H_3\ O_2$
Methyle, ...	C_2H_3,C_2H_3
Ethyle, ...	C_4H_5,C_4H_5
Acetyle, ...	$C_4H_5\ O_2,C_4\ H_3O_2$
Acetylide of methyle (acetone), ...	$C_2H_3,C_4\ H_3O_2$

5. *Chlorohydric acid* is also a type.

Chlorohydric acid, ...	$H,\ Cl$
Chloride of potassium,......	$K\ \ Cl$
Chloride of methyle, ...	$C_2\,H_3,\ \ Cl$
Chloride of acetyle, ...	$C_4H_3O_2Cl$

Tyrolite. See NOVOHYDROUS PENTARSENIATE of COPPER.

Tyrosine. $C_{18}H_{11}NO_6$. Needles obtained by adding to fused caustic potash an equal weight of dry fibrine, and heating till the colour changes from brown to yellow; the matter is then dissolved in water, neutralized with acetic acid and allowed to deposit. It may also be prepared by boiling albumen or fibrine with 5 parts chlorohydric acid for six hours, or with sulphuric acid diluted with 4 parts water, evaporating the solution, the remainder dissolved in water, mixed with an excess of lime, boiled to remove ammonia, and filtered; by concentrating the solution, a mass of crystals is left, the syrupy portion of which is removed by alcohol of ·830; a mixture of crystals of leucine and tyrosine remains, which are separated by their different solubilities in water; a third crystalline body also occurs in the action of acids on fibrine, &c.; tyrosine likewise appears in the putrefaction of fibrine, along with a volatile crystalline substance; tyrosine, likewise, can be formed by the action of potash or sulphuric acid on horn; it is also found in cochineal. Tyrosine is, therefore, a product from all albuminous bodies; and is probably a proximate constituent of them; it is a base, and unites with acids. Nitrate of nitrotyrosine, $C_{18}NH_{10}NO_4O_6 + HONO_5$, consists of yellow crystals; by the action of dilute nitric acid on tyrosine.

U

Ullmannite. See ANTIMONIAL NICKEL.

Ulmaric Acid. C 24·66, H 1·95, O 15·32.

Ulmic Acid. $C_{40}H_{14}O_{12}$. A brown powder or gelatinous flocks, precipitated by an acid from the solution produced by the action of potash or soda, on mould, or soil; it becomes anhydrous at 383°; it unites with bases, and forms amorphous salts; it is also produced by the action of sulphuric acid on sugar.

Ulmine. $C_{40}H_{16}O_4$. A name given by Dr. T. Thomson, in 1804, to an exudation on the bark of the elm; afterwards ulmic acid.

Ultramarine, Blue, was formerly obtained from Hauyne or lapis lazuli, but is now prepared artificially, in consequence of the demand for it in calico printing. It is said to be successfully formed by evaporating 100 lbs. of a solution of sulphide of sodium, mixed with 25 lbs. dry china clay, and ½ lb. crystals of copperas; the dry mass is then heated in a muffle to a red heat, washed, and again heated.

Umbellic Acid. $HOC_{16}H_7O_5$. Colourless needles, a product of the oxidation of oil of anise, identical with anisic and draconic acids.

Umber. An earth composed of SiO_3 33·1, HO 10·5, Fe_2O_3 46·2, MnO 9·8.

Upas Poison. The dried juice of the Antiaris toxicaria, a tree in Java. It contains the poison *Antiarine* ($C_{14}H_{10}O_5$), colourless crystalline scales; soluble in 251 water, 70 alcohol, and 2792 ether, in 27½ hot water.

Uraconise. A synonyme of sesquioxide of uranium or ochre.

Uralite. A pseudomorphous form of Augite.

Ural Orthite. A synonyme of Tschewkinite.

Uramile. $C_8H_5N_3O_6$. Crystalline powder or feathery crystals; soluble in NH_3 and KO, oxygen being absorbed, becoming purple, and depositing murexide; boiled with oxide of mercury it yields murexide; changed by potash into uramilic acid; obtained by the action of acids on thionurate of ammonia.

Uramilic Acid. $C_{16}H_{10}N_5O_{15}$? Fine prisms by boiling uramile with potash. It has only once been obtained.

Uran, Green. *Chalcolite, Cupreophosphate of Uranium.* Spec. grav. 3·55, H 2·25. Emerald and grass-green scales. PO_5 15·57, U_2O_3 61·39, CuO 8·44, HO 15·05. B.B. fuses into a black mass, colouring the flame greenish, a green glass with borax. Gunnis Lake, near Callington; and in Corrarich, Tincroft, &c., mines, Cornwall.

Urancolumbite, or *Tantalite,* or *Niobite.* See SAMARSKITE.

Uranic Acid. Sesquioxide of Uranium.

Uran Mica. See URANITE.

Uranite. *Uran Mica, Calcareophosphate of Uranium.* Spec. grav. 3·12, H 3·25. Lemon-yellow to brown, 4-sided prisms, and in 4, 6, and 8-sided foliated tables; lustre pearly on the face of the prism; adamantine on the other faces; sectile; transparent on the edges. BaO 1·51, CaO 5·66, MgO MnO ·19, U_2O_3 59·37, PO_5 14·63, HO 14·9; stony matter 2·7, F and NH_3, no change.

Uranium. Discovered by Klaproth, in 1789, in pitch blende. The oxide may be obtained from this mineral by treating it with NO_5; the solution is evaporated to dryness, and digested in cold water; the solution filtered. Nitrate of uranium crystallizes. By calcining this nitrate we obtain the black oxide. This oxide may, by acting on it with Cl, be converted into the chloride. By treating the chloride with potassium, pure metallic uranium is separated. It is a black powder, having somewhat of a metallic lustre; very combustible; burning with a white light.

Suboxide.—U_4O_3. A brown precipitate, obtained by pouring NH_3 into the subchloride.

Protoxide. — Mistaken until lately for a metal, has been shown by Peligot to be an oxide. Obtained by decomposing the yellow oxalate by passing hydrogen over it at a red heat. It has a great tendency to combine with oxygen. When obtained from the double chloride of uranium and potassium it is in crystalline plates, having a metallic lustre.

Biprotosesquioxide, Black Oxide.—Obtained by calcining the nitrate.

Protosesquioxide, or Olive oxide. By heating the preceding oxides to redness in contact with oxygen.

Sesquioxide, Yellow Oxide.—This oxide has not yet been isolated, but exists in the yellow salts of uranium.

Suboxide, ...	U_4O_3
Black protoxide, ...	U O
Biprotosesquioxide,....	U_4O_5
Olive oxide,................................	U_3O_4
Yellow sesquioxide,.......................	U_2O_3

The protoxide, which has been hitherto considered as the metal, differs in its combinations from other oxides; the protoxide of uranium combines directly with chlorine. In 1 atom of this compound metallic radical thus combined, there are 2 atoms of protoxide of uranium, or an atom of peroxide of uranium. The base Peligot terms uranyle. One atom of uranyle unites with one of oxygen, to form the peroxide which enters into the salts of the oxide of uranyle. The following represents the series:—

$(U_2O_2)O$	Oxide of uranyle.
U_2O_2 Cl	Chloride ,,
U_2O_2 $OC_2O_3$3HO	Oxalate ,,
U_2O_2 O $NO_5$6HO	Nitrate ,,

Oxide of uranium, and uranate of potash and ammonia, have been employed to colour glass. The glasses have an orange-yellow colour, or greenish and opaline. The uranate of potash, obtained by calcining nitrate of uranium and nitrate of potash, affords a splendid orange for artists.

Potash Chloride of Uranyle.—U_2O_2Cl, KCl + 2HO crystallizes in large rhomboidal tables of a greenish-yellow colour, and is very soluble in water. The double salt of ammonia possesses a similar composition, and crystallizes in rhombohedrons. The *oxalate of uranyle,* U_2O_2O,C_2O_3 + 3HO, is obtained by pouring a saturated solution of oxalic acid into a solution of nitrate of uranyle. The double *oxalate of ammonia* may be crystallized in yellow prisms by digesting a hot solution of oxalate of ammonia on oxalate of uranyle. The double sulphates of uranyle and potash and of ammonia crystallize each with 2 atoms of water. Uranite has been employed when deprived of its matrix to colour glass. It has also been proposed to employ the uranate of potash as a pigment, prepared by the calcination of a mixture of nitrate of uranium and nitre. It possesses a rich orange colour. The salts of the sesquioxide, when not viewed as containing uranyle, have the following formulæ :—

Nitrate, ... $U_2O_3NO_56HO$
Sulphate, ... $U_2O_3SO_33HO$
Bisulphate, ... $U_2O_32SO_3HO$
Potash sulphate, ..$(U_2O_3SO_3)$ $(KOSO_3)2HO$
NH_3 carbonate, ...$(U_2O_3CO_2)$ $2(NH_3CO_2HO)$
Oxalate, ... $U_2O_3C_2O_33HO$
Potash oxalate, ...$(U_2O_3C_2O_3)$ (KOC_2O_33HO)

Uranium Ore, or *Pitch blende.*

Uranium, Sulphated Protoxide of. Emerald-green flattened prisms, with a vitreous lustre; soluble in water; found native in Elias mine near Joachimsthal.

Uranium, Sulphated Peroxide of. Sulphur-yellow botryoidal masses, friable, soiling the fingers; partly soluble in water, the remainder soluble in NO_5; found along with the preceding.

Uranochre. *Uraconise.* U_2O_3? Yellow powder on pitch blende, in Cornwall and Bohemia; soluble in acids, and precipitated yellow by NH_3. B.B. becomes orange; in the reducing flame turns green; infusible.

Uranphyllite. A synonyme of Uranite.

Urao. *Natron.* South American name of native sesquicarbonate of soda.

Urea. A substance derived from urine, but also denoting a class of bodies connected with organic radicals, forming parallel compounds to that of the urine. There exist ethyle urea, methyle urea, anilo urea, acetyle urea, amyle urea, &c.

Urea, or **Anomalous Cyanate of Ammonia.** C_2NO, NH_3HO, or $C_2O_2N_2H_4$ 7·5, 60. Specific gravity 1·35 (Prout); F.P. 248°. C 20, H 6·67, N 46·67, O 26·66. Narrow transparent 4-sided flattened prisms.

Characters.—Urea is a neutral body, as it produces no change on vegetable blues. When heated it melts; one portion sublimes without decomposing, while another part is decomposed. See Biuret. Water at 60° dissolves rather more than its own weight of urea; water at 212° dissolves any quantity of it. 4 parts of alcohol of ·816 dissolve 1 part of urea. Boiling alcohol dissolves more than its weight of urea. Urea deliquesces in moist air, but is permanent in dry air. It melts at 248° into a colourless liquid, but by a higher temperature it is decomposed into ammonia, cyanate of ammonia, and dry cyanuric acid. Alkalies cause the evolution of no ammonia from urea in the cold state, but when a solution of urea is evaporated with nitrate of silver, cyanate of silver and nitrate of ammonia are formed; and when a similar experiment is made with acetate of lead, carbonate of lead and acetate of ammonia are the result. When heated with caustic potash or soda, urea is decomposed into carbonic acid and ammonia, by taking up the elements of 2 atoms of water; nitrous acid decomposes urea into nitrogen, and carbonic acid; chlorine decomposes it into hydrochloric acid, nitrogen, and carbonic acid. The composition of urea is equivalent to 2 atoms of carbonate of ammonia, deprived of the elements of 2 atoms of water; and when it remains in contact with water it is converted into 2 atoms dry carbonate of ammonia, $C_2N_2H_4O_2 + 2HO = 2NH_42CO_2$.

History.—Urea was first described as obtained from urine under the title of *soapy* matter by Rouelle, jun. It was afterwards obtained by Cruickshanks, and by Fourcroy and Vauquelin. It is the second substance of animal origin—formic acid being the first (Döbereiner)—which has been prepared in the laboratory (Wöhler).

Preparation.—1. Urea may be obtained by mixing aqueous cyanic acid with ammonia and heating the mixture, or by spontaneous evaporation. A solution of cyanogen in water also yields urea. Cyanate of lead, heated with a solution of carbonate of ammonia, and cyanate of silver, with a solution of salammoniac, produce urea (Wöhler), or when uric acid is distilled. 2. The easiest process for obtaining urea is to evaporate urine to the consistence of a syrup with a moderate heat. Small portions are to be occasionally tested with nitric acid of 1·42 spec. grav. free from nitrous acid (NO_3). If the liquor thickens on the addition of the acid to a crystalline mass, the whole urine should be mixed with an equal measure of the acid. The crystals which result of nitrate of urea are to be purified by washing them with dilute nitric acid, and drying them on a porous tile; they are then to be dissolved in water, decolourized with freshly heated animal charcoal, and crystallized. A solution of these colourless crystals is then to be mixed with carbonate of barytes until the solution is neutral. The liquor being evaporated, the nitrate of barytes crystallizes out first, and then the urea. To obtain the urea perfectly pure, it may be dissolved and crystallized from alcohol. 3. Evaporate the urine to a syrup, dry it at 212°, dissolve up the urea by absolute alcohol, and distil off the alcohol at 212°. Dissolve the residue in water, and decolourize by animal charcoal; filter; heat to 122°, and add as much oxalic acid as it is capable of dissolving at that temperature; oxalate of urea deposits on cooling, which increases by evaporation. When the solution thickens, and has no strong acid taste, heat to 122°, and add more oxalic acid, when a new crop of oxalate is obtained. Dissolve the oxalate in boiling water, add animal charcoal, filter and evaporate. White oxalate is obtained. Dissolve these in water and add carbonate of lime, when oxalate of lime is formed and urea set free; filter and evaporate, and free the urea from an alkaline oxalate by absolute alcohol, which dissolves the urea but not the oxalate. 4. 28 parts yellow prussiate of potash (ferrocyanide of potassium), completely dried, are mixed with 14 parts of black oxide of manganese, both finely pulverized; the mixture is to be heated on an iron plate over a charcoal fire to a low red heat, where it catches fire, and gradually burns away. Its concretion is prevented by stirring it; the access of air is thus also promoted. The burned mass is, after cooling, to be treated with cold water, and the solution mixed with $20\frac{1}{2}$ parts of dry sulphate of ammonia, either as it occurs in commerce or prepared by saturating carbonate of ammonia with SO_3 and evaporating to dryness. It is proper to set aside the first strong lye from the burned prussiate; to dissolve the sulphate in the last washings, and then to mix the first washings with the rest. A copious precipitation of sulphate of potash generally takes place, from which the solution is to be poured off. It is then to be evaporated in the water bath without boiling. Crusts of $KOSO_3$ form, from which the solution is to be poured off. The last portion is to be evaporated to dryness and treated with boiling spirit of 80 to 90 per cent., which dissolves the urea. This is to be crystallized out by cooling, and diluting the spirit while the sulphate and salts remain in solution. From a lb. of prussiate 4 oz. of pure urea may be obtained. It sometimes happens that the solution containing the sulphate of potash and urea is coloured yellow by the ammonia or potash cyanide of iron, which dissolves in the spirit, and colours the urea yellow. This may be removed by the addition of some solution of sulphate of iron. After the subsidence of the Prussian blue, carbonate of ammonia is added to the solution, by which the excess of iron is precipitated, and the solution becomes clear and colourless; it is then evaporated and treated as above.

Analysis for Urea.

Lecanu's Process.—Evaporate in a water bath 900 grains of urine. Dissolve up the urea by strong alcohol; evaporate again in the water bath; add quaterhydrate of nitric acid. Place the nitrate of urea crystals on a porous tile over lime in a vacuum; next day change the tile, and allow it to remain for some days. Weigh. The nitrate contains 48·84 per cent. of urea.

Millon's.—Prepare nitrate of mercury by digesting 2604 grains of nitric acid on 1937 grains mercury. The mercury dissolves almost completely in the cold, but towards the end a gentle heat may be applied. Two volumes of water are then added to one volume of fluid. This may be preserved in stoppered bottles for months without decomposing. 40 to 45 cub. centimetres of the mercurial solution are poured into a flask, and about 300 grains of urine are added. The flask is then connected with an aspirating apparatus. The urine mixture is brought just to the boiling point, and the heat stopped, when NO_4 appears. The carbonic acid is absorbed by potash, the water being dried up by tubes filled with pumice, moistened with SO_3. The weight of the CO_2 gives the weight of the urea by multiplying by 1·731.

Bunsen's Process consists in heating urine with chloride of barium in an ammoniacal solution, in a sealed tube, at a temperature of 464°, in an oil bath. The urea being converted into carbonate of ammonia, throws down its equivalent of carbonate of barytes.

Liebig's Test for Urea.—Add a solution of the nitrate of mercury to a solution of urea, and neutralize the liberated acid by barytes or caustic soda. When, by repeated additions of salt and neutralization, the whole of the urea is separated as a white precipitate, a further quantity of nitrate gives the yellow oxide of mercury. The presence of common salt in urine presents some difficulties (see Liebig, Ann.)

Practical Applications.—Urea exists in considerable quantity in the urine of the inferior animals, and is of great value as a manure; but as it readily changes in a few hours into a volatile carbonate of ammonia, it is necessary to fix it in the form of a more permanent salt, by means either of sulphuric or hydrochloric acids. To convert $7\frac{1}{2}$ parts of urea into sulphate of ammonia, it is necessary to add $12\frac{1}{4}$ parts, or 2 atoms, of oil of vitriol, or sulphuric acid of commerce. Now, according to experiment, it appears that the urine of the cow contains 4 per cent. of urea. Hence every 100 pounds of cow's urine would require to fix its urea in the form of sulphate of ammonia, 6·53 lbs. of oil of vitriol, or about $6\frac{1}{2}$ lbs. The urine of the horse contains 0·7 per cent. of urea; every 100 lbs. of it would, therefore, only require 1·14 lbs. of oil of vitriol to fix the ammonia. To collect the urine of horses and cows, the liquid from stables, cowhouses, and dunghills should be allowed to run into tanks, properly placed for the purpose, and covered in from the external air. The acid should be placed in the tank, and the quantity to be added can be judged of by the capacity of the reservoir, and the quantity of liquid contained in it.

Nitrate of Urea.—$UrNO_5 + HO$ (Regnault). Pearly plates, difficultly soluble in NO_5, and not decomposed by boiling in the same acid. When heated up to about 220°, it decomposes, ammonia being formed. It consists of urea 48·86, NO_5 43·85, HO 7·29 (Heintz, Poggendorff, lxvi. 122). This salt was first discovered by Cruickshanks in 1797 (Rollo, Cases of the Diabetes mellitus, and Philosoph. Mag. ii. 240, 1798).

Oxalate of Urea.—Ur, Oxal. HO (Ur,C_2O_3, HO). Long slender plates. Taste cooling; when heated melts, boils, ammonia being evolved, and cyanuric acid formed.

Ureide. A compound containing the elements of a salt of urea, minus water; thus *benzoyle ureide*, $C_{50}H_{28}N_8O_8$, a white powder by the action of oil of bitter almonds on urea.

Urethane. *Carbamic Ether, Carbamate of Oxide of Ethyle.* $C_6H_7NO_4 = C_4H_5O, CO_2, CONH_2$? Crystalline body, isomeric with lactamide, sarcosine, and alanine, and homologous with glycocoll; obtained by acting on chlorocarbonic ether with ammonia, and subliming.

Urethylane. *Carbamate of Oxide of Methyle.* $C_4H_5NO_4 = C_2H_3O, C_2NH_2O_3$. Crystalline compound, formed by acting on pyroxylic spirit with chlorocarbonic acid, and mixing the product with ammonia. It is isomeric with glycocoll.

Uric Acid. *Lithic Acid.* $C_{10}H_2N_4O_4\,2HO$, or $C_5HN_2O_2,HO$. This acid was discovered by Scheele in 1776. It was termed *lithiasic acid* by Morveau before 1789, and *lithic acid* by Lavoisier in 1789; *uric* or *ouric oxide* by Dr. Pearson in 1798; and *ouric acid* by Fourcroy in 1798. It was first observed in urinary calculi, and in all urine by Scheele; in the excrement of snakes by Vauquelin; in that of silk worms by Brugnatelli; in cantharides by Robiquet. It also exists in the urine or excrement of the common fowl, and various sea birds, as exemplified in guano.

Preparation.—It may be obtained pure by boiling the excrements of serpents or birds (which consist of urate of ammonia mixed with phosphate of lime) in caustic potash or soda, filtering, and then precipitating the solution at a boiling temperature with hydrochloric acid. The precipitate is thrown on a filter, and washed with cold water. To purify it still further, it may again be dissolved in potash or soda, and boiled for fifteen minutes with hydrochloric acid. Uric acid may be procured from the urine in crystals by adding to it a few drops of nitric acid, and allowing the liquid to stand at rest for a day; the crystals of uric acid collect on the sides of the glass.

Characters.—Uric acid obtained by this process is frequently in the form of a fine white powder, and also falls in fine minute silky prisms, destitute of taste and smell, or in scales. It dissolves in 1,400 times its weight of boiling water (Dr. Henry), the solution reddening litmus paper; and it requires more than 10,000 times its weight of cold water to dissolve it. It is slightly soluble in hydrochloric acid; soluble in sulphuric acid, from which it separates on adding water. It dissolves in alkaline carbonates. Nitric acid dissolves it; equal volumes of carbonic acid and nitrogen being evolved, and alloxan, alloxantin, urea, parabanic acid, ammonia, remaining in the solution. According to Bensch, 1 part is soluble in 14 to 15,000 parts of cold, and in 1,800 to 1,900 of boiling water. Uric acid may be distinguished from other bodies by the fact, that when the nitric acid solution is mixed with ammonia and evaporated, a fine purple colour is produced. When subjected to dry distillation, the volatile products are urea, cyanic acid, cyamelide, hydrocyanic acid, and carbonate of ammonia, while there remains in the retort a coaly azotized substance. When fused with solid caustic potash, there are formed carbonate and cyanate of potash, and cyanide of potassium. When boiled with brown oxide of lead and water it is converted into allantoin, urea, and oxalic acid (Liebig and Wöhler). Sulphuric acid forms with uric acid a crystalline compound, 1 uric acid $+ 4SO_3 + 4HO$ (Fritsche). When dissolved in this acid, and water added, uric acid falls in granular crystals. Uric acid has been viewed as composed of a hypothetic *uryle*, $C_8N_2O_4$ ($4CO$, $2C_2N$,) and urea; alloxantine $=$ $Ul + O + 5HO$; alloxan $Ul + O_2$ and $4HO$.

Salts.—The bisalts contain an atom of base and an atom of water; while the neutral salts, as those of potash, soda, lime, lead, contain no water, simply an atom of base; the strontian salt contains 2 atoms HO, the barytes salt 1 atom HO (Bensch).

Urate of Lime.—The acid urate is the most soluble of the urates likely to occur in animals. It dissolves in 276 hot, 603 cold water.

Cause of Uric Acid Calculi.—Children, and particularly those of gouty and dyspeptic persons, or who inherit a tendency to urinary diseases, are liable to uric acid deposits. The presence of gravel of this description is a frequent cause of involuntary micturition at night among children. After puberty this tendency disappears. In these individuals the tendency is again apt to occur after the age of forty. It has been observed that this disease is closely connected with what is termed in common language a *scorbutic* habit of body, that is, a tendency to disease of the skin. In those predisposed to the disease, an excess of food is apt to produce an attack, and more especially food difficult of digestion—as doughy bread, and doughy dishes of all kinds. Malt liquors, strong and sweet wines, have a similar influence. The same observation applies to hard and impure waters. Exercise immediately after a meal, and what operates so as to interfere with the digestive process, is very apt to cause the deposition of uric acid. Want of exercise at the proper time is also liable to occasion this disease.

Products of the Oxidation of Uric Acid.—*Allantoin*, $C_4H_3N_2O_3$, colourless rhombic prisms, obtained by boiling 1 uric acid, 20 water; brown oxide of lead is to be added as long as its colour changes. The allantoin is obtained by filtration and evaporation; this substance occurs in the allantoic fluid of the cow. *Alloxan*, $C_8N_2H_4O_{10}$, colourless right prisms, efflorescent, obtained by dissolving uric acid in NO_5 (1·45), and allowing the mother liquid to drain off. *Alloxanic acid*, $C_8N_2H_2O_8 2HO$, radiated needles, obtained by treating alloxan with alkalies. *Leucoturic acid*, $C_6N_2H_3O_6$, white powder, obtained by evaporating a strong solution of alloxanic acid at a certain temperature. *Difluan* is found in the liquid which deposits the preceding acid. *Mesoxalic acid*, formed by heating alloxanate of barytes to the boiling point. *Mycomelic acid*, prepared by heating alloxan with an excess of ammonia. *Parabanic acid*, formed by heating alloxan or uric acid with NO_5, and evaporating. *Oxaluric acid*, obtained by heating parabanic acid with NH_3; the acid may be separated by SO_3. *Thionuric acid*, formed by causing sulphite of NH_3 to act on alloxan. *Uramile* is prepared by treating thionurate of NH_3 with HCl. *Uramilic acid*, prepared from the preceding by treating with SO_3. *Alloxantine*, $C_8N_2H_5O_{10}$, oblique 4-sided prisms, formed by adding dilute NO_5 to uric acid, and evaporating down one-third; or by passing SH through alloxan. *Dialuric acid*, obtained by passing SH through a boiling solution of alloxantine and adding NH_3; dialurate of NH_3 separates in silky crystals. *Allituric acid*, prepared by boiling alloxantin with HCl,NO_5. *Dilituric acid*, obtained from the mother liquor of the preceding body by SH. *Murexide*, $C_{12}N_5H_6O_8$, purpurate of NH_3, of Prout, 4-sided prisms, with a beautiful green metallic lustre, obtained by dissolving uric acid in dilute NO_5, and evaporating till it acquires a flesh colour; it is treated with NH_3 in slight excess; it is then diluted with half its weight of boiling water, and allowed to cool. It may also be obtained by mixing 7 grs. alloxan, 5 alloxantine, dissolved in 240 grs. water, with 80 grs. of a cold strong solution of carbonate of ammonia. *Murexan*, $C_6N_2H_4O_5$, purpuric acid of Prout, silky scales, insoluble in water; prepared by dissolving murexide in caustic potash, and adding SO_3 in excess. *Xanthic oxide*, $C_5N_2H_2O_2$, a rare calculus, dissolves in NO_5, leaving a yellow residue, which is not redissolved by NH_3; somewhat soluble in water. *Cystic oxide*, an organic base, $C_6NH_6O_4S_2$, yellowish white crystals; a rare calculus, soluble in alkalies, decomposed by heat, with evolution of SO_2, and NH_3. *Guanine*, $C_{10}H_5N_5O_2$, white crystalline powder, extracted from guano by HCl, and precipitated by alkalies; it is a base uniting with acids and forming salts. *Hyperuric acid* ($C_{10}N_4H_3O_7 2HO$) by oxidating guanine. See also LANTANURIC ACID, HIDANTOIC ACID, UROXANIC ACID, UROXILE.

Uric Oxide. See XANTHIC OXIDE.

Urinary Calculi. The deposit of granular matter in the bladder is termed gravel. When the deposits are larger, the term calculus or stone is applied to them. Calculi are divisible into two classes:—

I. *Calculi soluble in Caustic Soda or Potash.* 1. *Uric Acid, Lithic Acid.*—These have usually a yellow tint and smoothish surface, and are deposited in layers. They dissolve in soda, and afford a precipitate when an acid is added to the soda solution. They dissolve with effervescence in nitric acid, and yield a fine purple colour when ammonia is added to the acid solution (fig. 2).

2. *Urate of Ammonia* is soluble in caustic soda, with evolution of ammonia, and likewise yields the same results as uric acid. It is also soluble in alkaline carbonates (fig. 1).

3. *Cystic Oxide*, or *Cystine* (fig. 5). Pale yellow compact rounded masses; insoluble in water, alcohol, acetic, tartaric, citric acids, and bicarbonate of ammonia; soluble in HCl, NO_5, SO_3, PO_5, and C_2O_3; also in potash, soda, ammonia, lime water, bicarbonates of potash, and soda; it unites with acids, and forms crystals; before the blowpipe emits a peculiar odour. Consists of C 30·01, H 5·1, N 11·, S 25·51, O 28·38 $= C_6H_6NO_4S_2$ (fig. 5); in 6-sided scales.

4. *Xanthic Oxide, Xanthine, Uric Oxide.*—Reddish cinnamon-coloured, compact, hard, and laminated calculus; the colour being heightened by soda; soluble in boiling water with acid reaction; insoluble in bicarbonates of potash and ammonia, insoluble in chlorohydric and oxalic acids, by which it is distinguished from cystine; soluble in caustic soda, and reprecipitated by acetic acid; soluble in SO_3 with a yellow colour. The solution in nitric acid leaves by evaporation a lemon residue, partly soluble in water, to which it gives its colour; caustic potash turns the yellow residue red, while the addition of an acid destroys it. When purified by dissolving in potash and precipitating by CO_2, it consists of C 39·28, H 2·95, N 36·35, O 21·42 $= C_5H_2N_2O_2$.

II. *Calculi insoluble in caustic potash or soda.*

1. *Phosphate of Lime* or *Bone Earth.*—Pale brown masses with a smooth surface, consisting of concentric laminæ, which easily separate; soluble in nitric acid, and reprecipitated in flocks by ammonia; insoluble in acetic acid, which distinguishes it from ammonia phosphate of magnesia. B.B. becomes black, turns white, and is fusible at a very high temperature.

2. *Ammonia Phosphate of Magnesia*, or *Triple Phosphate.*—(NH_3 2 MgO, PO_5 + 10 HO) (fig. 8). Crystals and calculi. The crystals which occur

Fig. 1, Urate of ammonia (soda?); 2, uric acid; 3, yeast plant in diabetic urine; 4, oxalate of lime; 5, cystic oxide; 6, blood globules in urine; 7, pus globules in urine; 8, ammonia phosphate of magnesia, or triple phosphate.

frequently in ammoniacal or alkaline urine are shaped like the roof of a house; they are soluble in acetic acid, and are reprecipitated in stars on the addition of ammonia to the acid solution.

B.B. ammonia is disengaged, and diphosphate of magnesia remains. The larger calculi are composed of accumulations of these crystals, and often contain some phosphate of lime.

3. *Fusible Calculus*, or *mixture of Ammonia-Phosphate of Magnesia*, and *Phosphate of Lime.*—White and friable calculi resembling chalk, leaving a powder on the fingers; they separate frequently into laminæ, the interstices of which are studded with triple phosphate. B.B. they fuse into a glassy globule, in consequence of the reaction of the phosphate of magnesia on phosphate of lime. Acetic acid dissolves up the ammonia phosphate of magnesia; the solution may then be treated by caustic potash; the phosphate of lime remains, which may be dissolved by chlorohydric acid.

4. *Oxalate of Lime Crystals*, or *Mulberry Calculus* (fig. 4).—Oxalate of lime occurs very frequently in the urine in the form of 8-hedrons or dumb-bells, even when there appears no deviation from health; insoluble in acetic acid. The mulberry calculus is a brown concretion with a tuberculated surface, usually forming the nucleus of a uric acid surface; soluble in nitric and chlorohydric acids; with caustic potash it forms oxalate of potash, and lime is precipitated; with carbonate of potash, oxalate of potash in solution, and carbonate of lime, precipitated. B.B. becomes carbonate of lime, which dissolves with effervescence in acetic acid.

5. Rare calculi of carbonate of lime have also been described.

Urine. The descending aorta in its course gives off the emulgent arteries which proceed to the kidneys, and secrete the urine, a yellow or amber-coloured fluid with an aromatic odour and acid reaction. The specific gravity of urine has been differently stated. Urine when evacuated has a temperature of 94° or 96°. The following table of specific gravity was taken when the urine was emitted:—

40	Experiments	1015
18	„	1018
16	„	1017
14	„	1014
16	„	1015
15	„	1014
20	„	1016
19	„	1020
158	„	8 | 8129
		1016

I found that when the temperature fell to 60° there was a difference in the specific gravity in the 4th place to the amount of 3, so that if we add this we shall get for the mean of 158 experiments 1019 as the average specific gravity of healthy urine. These experiments were made in winter when the urine may be expected to be lighter than in summer when so much of the fluids of the body escape by the skin. This agrees almost exactly with the result of Dr. Prout's conclusions, that the average specific gravity scarcely reaches 1020. Dr. Jas. Gregory makes the specific gravity 1022½. Dr. Thomson made it 1014. My number is intermediate. Becquerel makes the mean 1017. The mean specific gravity of the urine of 163 patients in St. Thomas's Hospital, determined in my laboratory, was 1018.

Quantity of Urine.—Experiments made by me during a series of weeks gave the following results in fluid ounces:—

1st. Week. Oz.	2d.	3d.	4th.	5th.	6th.
62¾	40	42½	49½	40	52½
45¾	55½	55	57	52	47
56	45	68	52½	65	42½
36	49	48	56	36	49½
60	67	46	54	34½	35
44½	34	32	49	44½	51½
52	66	89	51½	69	36
357	356½	380½	369½	341	314

Total = 2118½ fluid ounces.

= 353 fluid ounces per week. Average of six weeks = nearly to 50¼ fluid oz. per day, or 51·4 ounces avoirdupois. According to Becquerel the average quantity in twenty-four hours is 1319 grammes=42⅓ ounces avoirdupois. According to Rayer the extremes in health are 21 and 57 ounces. The smallest quantity obtained by me in the experiments alluded to was 32 fluid ounces in a healthy person, when the thermometer was at 30°, and the fluid ingesta amounted to 53 fluid ounces; the largest amount was 68 fluid ounces, under ordinary circumstances, or 3·21 lbs. Dr. T. Thomson makes it 3,307 lbs. avoirdupois = 52·9 oz. avoirdupois. The urine is easily increased by certain articles of diet, such as coffee, if it be taken of a certain strength. The following diary shows the effect of this beverage:—

		Ingesta.	Egesta.	Sp. Grav.
5th.	9 a.m.	18 fluid oz.	fluid oz.	
	10 „	2		
	12 „		16⅓	1007
	5 p.m.	12		
	6 „	12 „ coffee	18½	1015
6th.	1 a.m.	10 „ strong coffee	28	1018
	3½ „		18	1005
	8 „		8	1017
		54 fluid oz.	89 oz.	

By a mean of forty-two days it was found that the number of times that a person passes his urine is three times daily. All these observations are made on the understanding that the person experimented on abstains from fermented liquors. Next we may turn our attention to the relation between the fluid ingesta and fluid egesta.

I have found the relation of the fluid ingesta to the egesta as follows:—

	Fluid ounces.
Fluid ingesta, ...	46
,, egesta, ...	50

or a greater amount of fluid excreted than is introduced into the system. But we must recollect that the greater portion of our food consists of water in a solid state, amply sufficient to account for this excess.

Table of Urine in Health and Disease.

	Health.	Fevers.	Anæmia.	Bright's Disease.	Diabetes.
Spec. grav....	1017	1022	1010	1012	1045
Water,.......	972·38	963·7	983·15	984·3	835·5
Urea,.........	11·58	13·1	6·4	7·8	8·3
Uric acid, hippu. acid, creatinine,	·47	1·4	·25	·46	
Animal mat.,.	8·65	14·3	6·10	6·3	11·7
Chlorine,....	0·50				
Sulphuric acid,......	0·85				
Phosphoric acid, ...	0·32				
Potash, ...	1·30	7·5	4·1	5·14	8·1
Soda,....... Lime,...... Magnesia,.	3·95				134 sugar

	Becquerel.	Thomson.
Weight of water in 24 hours,..	19801 grs.	22551 grs.
Weight of solids,..............	569 ,,	598 ,,
Total weight,..............	20370 grs.	23149 grs. or 52·9 oz. av.
Total weight, R.D.T.,.....	22487 grs. or 51·4 oz. avoird.	

Salts of Urine.—When the food has been bread or flesh, the ashes of which contain only phosphates, the urine contains the alkalies in the state of phosphates. If the food has consisted of roots, green vegetables, and fruits, the ashes of which contain as soluble salts only alkaline carbonates, the urine contains alkaline carbonates; bitartrates, tartrates, malates, and acetates appear in the urine as carbonates, and render it alkaline; and it has long been observed that after eating juicy fruits, cherries, strawberries, apples, &c., the urine becomes alkaline. The presence of alkali tends to the combustion of organic substances. Urate of potash, when given to a rabbit, was converted into oxalic acid and urea, the latter corresponding with CO_2, and the former only wanting one of oxygen to convert it into CO_2. Hence in herbivora, in consequence of the alkali, no uric acid is found in their urine. We have thus a key to the treatment of uric acid diathesis by the use of alkalies. Gallic acid, when taken without alkali, can be detected by copperas in the urine, but with alkali it is not detectable. These facts show us how much is to be gained in our treatment of disease by the judicious use of diet. The urine of a pig fed on potatoes, which is alkaline, becomes acid as soon as it gets corn or peas. When we examine the ash of the healthy blood, and also that of the urine of the same person, we find the soluble salts the same; hence we may expect to be able by analysis of the urine to reason backwards on the condition of the blood in disease. By treatment with hydrate of lime and distillation, urine affords carbolic, taurilic, and damaluric acid.

Analysis of Urine.—The water is obtained by evaporating a weighed quantity at 212°, the uric acid by adding a few drops of chlorohydric acid to a weighed quantity, and allowing it to stand for twenty-four hours. The crystals which deposit on the sides of the vessel in needles are thrown on a weighed filter, washed, and dried at 212°. The urea is estimated by evaporating a given weight of urine to the consistence of syrup, and adding an equal bulk of nitric acid; nitrate of urea falls, which is thrown on a weighed filter, and washed once with ice-cold water, dried, and weighed, or by any of the methods given under urea. Hippuric acid is obtained by evaporating to a syrup, adding chlorohydric acid, and then ether; the acid is dissolved. The acids and bases are obtained by the methods described under these heads.

Urinary Deposits.—The flocky deposits commonly observed in urine are three in number. 1. *The buffy precipitate* has been termed the precipitate of health; it is dissolved by the application of heat, and is described by some as urate of ammonia, and by others, apparently with accuracy, as urate of soda. It appears as fine cellular granules, or as star-like spiculæ (fig. 1); it is not a product of disease, but an increased quantity of a salt always existing in the urine. It is accompanied with traces of urates of lime and ammonia. It falls sometimes when there is an insufficiency of exercise. Carnivorous animals confined in cages frequently deposit it. It occurs in emphysema, in heart disease, enlargement of liver. The deposition which occurs of uric acid appears to originate from the *urate* of soda by standing. In gout we have calculi of this nature; but uric acid also occurs in the blood in Bright's disease. 2. *Red Sediment.*—When of a light tint, this deposit often appears after drinking; but when brick-red it indicates inflammatory action in some part of the system. It is believed to be urate of soda, mixed with purpurate of ammonia (murexide). 3. *Pink Sediment* occurs in chronic diseases of the viscera, in hectic and irritative fever. Besides these, there are deposited uric acid in orange diamond-like crystals, oxalate of lime in dumb-bell and 8-hedral crystals (fig. 4), &c.; ammonia phosphate of magnesia (fig. 8) in alkaline urine.

Uroerythrine. The red-colouring matter of urine.

Uroglaucine. *Cyanurine.* The blue-colouring matter of urine. Indigo has been detected in urine.

Urostealite. An urinary calculus insoluble in water; soluble in ether, hot alcohol, and in carbonate of soda and caustic soda; soluble in nitric acid, leaving a residue by evaporation, which becomes yellow with ammonia or potash; when heated gives out the smell of benzoin and shell lac (Heller).

Uroxanic Acid. $C_{10}H_9N_3O_{14}$. 4-hedra or plumose crystals; obtained by allowing a solution of uric acid in excess of potash to stand, when this acid is deposited along with urate of potash.

Uroxanthine. A yellow-colouring matter in urine of Bright's disease and in cholera.

Uroxile. $C_8H_7N_3O_8$. Yellowish mass, by heating uroxanic acid to 212°; carbonic acid and water separate.

Urrhoidine. A ruby-red matter, formed by oxidation from uroxanthine.

Uryle. $C_8N_2O_4 = 4CO,2C_2N$. The hypothetic radical of uric acid.

Urylic Acid. A synonyme of uric acid.

Usnic Acid. $C_{32}H_{18}O_{14}$. Yellowish-white needles with metallic lustre, extracted from Usnea barbata and other lichens by alcohol or by lime, and an acid and crystallization out of alcohol; by distillation it yields betaorcine

Uvarowite. See OUWAROWITE.

V

Vaccic, or *Vaccinic Acid.* $C_{20}H_{20}O_6$. Apparently a conjugate acid of capric and butyric acids, obtained on one occasion along with capric and butyric acids by saponifying butter, distilling off the volatile acids, and saturating with barytes.

Vade, or *Wad.* Binoxide of manganese.

Valencianite. A variety of Felspar.

Valentinite, or White antimony.

Valer-aceto-nitrile. $C_{26}H_{24}N_2O_6$. Spec. grav. ·79; B.P. 156°. Volatile oil; obtained by distilling the products of the oxidation of gelatine by bichromate of potash and sulphuric acid.

Valeraldehyde. *Valeral.* $C_{10}H_{10}O_2$. A fluid obtained in the oxidation of vegetable fibrine, polymeric with capric acid.

Valeramide. $C_{10}H_{11}NO_2$. Large crystals by the action of strong caustic ammonia on valerate of oxide of ethyle.

Valeranilide. $C_{22}H_{15}NO_2$. F.P. 239°. Shining rectangular plates, by the reaction of valerianic acid on aniline.

Valerene. $C_{10}H_{10}$. A volatile oil produced in the action of zinc on iodide of amyle.

Valerianic Acid. *Valeric Acid.* (*Baldriansaure,* Ger.) *Delphinic, Phocenic Acids.* $C_{10}H_9O_3HO$. Spec. grav. ·932, ·937, of vapour 3·67; B.P. 347°. Thin colourless oil, smelling of butyric and acetic acids, with an acid taste, corroding the tongue; burns like a volatile oil; converted by chlorine into chlorine acids, chlorovalerisic and chlorovalerosic acids; forms monobasic salts with water; obtained by distilling valerian root (Valeriana officinalis) and neutralizing the product with magnesia; distilling off the fluid oils, and distilling the valerianate of magnesia with sulphuric acid; also by distilling angelica root; likewise by heating oil of grain or potatoes with hydrate of potash to 392°, or by treating it with bichromate of potash and sulphuric acid till no more hydrogen is evolved; valerianate of potash remains, which is distilled with dilute SO_3; likewise from Viburnum opulus; anhydrous valerianic acid is obtained by the action of oxychloride of phosphorus on valerianate of potash.

Valerianobenzoic Acid. $C_{10}H_9O_3, C_{14}H_5O_3$. Oily fluid, heavier than water, by the action of chlorobenzoyle on valerianate of potash.

Valerine. *Valero-Glycerine.* 1. *Monovalerine.* Spec. grav. 1·100. $C_{16}H_{16}O_8$. Neutral volatile oil, changing into valeramide by NH_3. 2. *Divalerine.* $C_{26}H_{26}O_{12}$. Spec. grav. 1·059. Disagreeable smelling oil, with a bitter taste. Both of these bodies are formed by the action of valeric acid on glycerine; the first at 400°, the second at 500°.

Valerole. *Oil of Valerian.* $C_{12}H_{10}O_2$. F.P. 68°. Colourless prisms at 32°; smells in the oily state of hay; metameric with oxide of mesityle; changes in the air into valerianic acid and resin; forms a resin with NO_5; with SO_3 forms a red solution and sulphovalerolic acid; little soluble in water; easily soluble in alcohol, ether, and volatile oils; obtained along with valerianic acid and borneene in distilling valerian root with water; it is the least volatile product. It is separated by fractional distillation.

Valerone. $C_{18}H_{18}O_2$. B.P. under 212°. Colourless thin fluid, lighter than water, of agreeable ethereal smell; neutral; burns with a smoky flame; obtained by distilling valerianic acid with excess of lime, and rectifying over fresh lime.

Valeronitrile. *Cyanide of Butyle.* $C_{10}H_9N$ Spec. grav. ·81; B.P. 257°. Volatile along with valeracetonitrile in the oxidation of gelatine.

Valeryle. $C_{10}H_9$. The hypothetic radical of valeraldehyde.

Valyle. *Butyle.* C_8H_9. B.P. 226½°. Oily fluid produced in the decomposition of valerianic acid by the galvanic battery. It is a homologue of methyle, ethyle, &c.

Vanadium. V 8·5, 68. Discovered in 1830 by Sefström, and so termed from Vanadis, a Scandinavian deity. It was first observed in the iron ore of Taberg in Sweden. It occurs in the form of vanadiate of lead at Zimapan in Mexico (Del Rio, Berzelius, Wöhler, at Wanlockhead in Scotland (Johnston), in Wicklow, Ireland (R.D.T.). In 1801 Del Rio announced the existence of a new metal in the Zimapan mineral, and termed it erythronium; but Descotils having also examined it, and considering it a compound of chromic acid, Del Rio abandoned his original view of its constitution. The metal may be obtained by passing dry ammoniacal gas over the chloride of vanadium. Salammoniac is formed, which volatilizes on the application of heat, and leaves the metal. The chloride is prepared by passing chlorine gas over a mixture of oxide of vanadium and charcoal. Vanadium is a white metal resembling silver, or more strongly molybdenum. It is a good conductor of electricity, and is decidedly negative. It dissolves readily in nitric and hydrochloric acids, affording a blue solution. There are three oxides.

Protoxide of Vanadium, VO 9·5, 76·, is obtained by passing hydrogen gas over vanadic acid at a red heat, or by heating it with charcoal. It is a black semi-metallic powder resembling plumbago; and is a good conductor of electricity. It does not combine with acids. When heated in the air it takes fire, and is converted into a black fused mass (VO_2). By the action of chlorine it is converted into VCl and VO_3.

Binoxide of Vanadium, VO_2 10·5, 84, may be obtained by exposing to a white heat in an atmosphere of carbonic acid 9½ parts of protoxide of vanadium with 11½ parts of vanadic acid, or by heating vanadiate of ammonia in a retort till the gas is all evolved, or by reducing vanadic acid by sulphuretted hydrogen, oxalic acid or sugar, as in the case of chromium. The solution is precipitated with carbonate of soda. This is thrown on a filter and washed. When dried in vacuo it is the hydrate of binoxide—a gray powder. When ignited away from the air it it a black powder. It is insoluble in water, but when allowed to remain in contact with that fluid it colours it green, by undergoing a higher oxidation. It dissolves completely in acids, and forms salts. It also unites with bases and forms compounds. The salts are precipitated by carbonate of soda, and by solution of nutgalls like iron. The *sulphate of vanadium* exists as a pale blue crystalline crust, and contains 17½ per cent. of water. Few of the salts are capable of crystallizing.

Vanadic Acid, VO_3 11·5, 92, may be prepared from the vanadiate of ammonia by igniting it in an open crucible; the salt decomposes, becoming first black, then reddish-brown, and latterly a rust-coloured powder. It is destitute of taste and smell, but reddens moistened litmus paper. When ignited it fuses, and crystallizes on cooling in needles. It is a non-conductor of electricity. It is scarcely soluble in 1000 parts of water. When heated per se on charcoal before the blowpipe it gives the protoxide resembling plumbago; a green glass with ammonia, phosphate of soda, and borax. It unites with bases, and also with stronger acids; the acid salts have an orange colour; the neutral salts are colourless and yellow. The yellow salts have generally a tendency to lose their colour by heat. It combines in 2 proportions with potash. The vanadiate of ammonia is almost insoluble in a solution of salammoniac, and precipitates in the form of a white powder.

Other Oxides.—By combinations occurring between the two oxides we have various other states of oxidation produced. A *purple* oxide is procured by the action of air and moisture upon the hydrate of the binoxide. A vanadiate of the binoxide, a bivanadiate of the binoxide, and an *orange* vanadiate of binoxide exist. *Sulphide of vanadium*, VS_3, is obtained by dissolving vanadic acid in an alkaline hydrosulphuret, and precipitating by sulphuric or hydrochloric acids. *Phosphide of vanadium* is produced by igniting phosphate of vanadium.

Vanadiobronzite. A mineral identical with bronzite, in which vanadium oxide and soda replace protoxide of iron. SiO_3 49·5, Al_2O_3 5·55, CaO 18·126, MgO 14·12, FeO 3·28, VO 3·65, NaO 3·75, HO 1·77. Bracco.

Varec. See KELP.

Variegated Copper. See COPPER.

Varioline. *Variolarine, Picrolichenite.* Spec. grav. 1·176. Rhombic 8-hedrons, bitter, neutral; insoluble in cold water; slightly soluble in boiling water; very soluble in alcohol, ether, bisulphide of carbon, turpentine; soluble in SO_3; soluble in caustic potash, with a red colour; obtained by alcohol from the Variolara amara lichen.

Variolite. A variety of Felspar.

Variscite. Apple-green kidney-shaped masses, allied to turquoise. B.B. becomes white, colours the flame bluish-green; with borax and microcosmic salt forms a yellowish-green glass. Messbach, Voigtland.

Varnish. (*Firniss*, Ger.; *Vernis*, Fr.) A solution of a resin, &c. in alcohol, ether, or naphtha; which, when applied to a surface, coats it by the evaporation of the solvent. One of the most valuable is copal varnish, which is dissolved either in spirit or in turpentine.

Varvicite. See MANGANESE.

Vauquelinite. See CUPREOCHROMATE OF LEAD.

Vegetable Ivory. See IVORY.

Vegetable Tallow, Chinese. F.P. 80°. A solid oil extracted from the seeds of Stillingia sebifera, in China; contains margaric and stearic acids.

Velvet Copper. See LETTSOMITE.

Venetian Red. See REDS.

Verantine. A substance produced from madder by the action of alkalies, and also by the influence of acids on rubiane.

Veratric Acid. $C_{18}H_9O_7$. From the seeds of the *Veratrum sabadilla*. Allied to the volatile fatty acids; soluble in alcohol and water; insoluble in ether.

Veratrine. $C_{34}H_{23}NO_6$. F.P. 239°. A poisonous alkaloid existing in the root of *Hellebore* and in the seeds of *Sabadilla*. White powder, soluble in alcohol, less so in ether, and but very slightly in water.

Verd Antique. Serpentine mixed with carbonate of lime.

Verdic Acid. A reddish-brown substance; obtained by exposing verdous acid, saturated with an alkali, to the action of the air, and then decomposing by means of an acid. Verdates are green.

Verdigris. See ACETATE OF COPPER.

Verditer. See GREEN BREMER.

Verdous Acid. Yellow brittle body, extracted from the roots of the *Scabiosa succisa*, and existing also in various other families of plants. Verdites are yellow.

Verjuice. The acid liquor from unripe grapes.

Vermiculite. Hardness 1·. Spec. grav. 2·5252 (Thomson); 2·756 (Crossley). Micaceous plates, disseminated through a mealy magnesian substance; lustre soapy; feel greasy; sectile. B.B. infusible; with carbonate of soda in the reducing flame, gives a greenish; in the oxidizing flame, an amethyst-coloured glass. SiO_3 49·08, MgO 16·964, F_2O_3 16·12, Al_2O_3 7·28, HO 10·276, Mn trace (Thomson), SiO_3 35·74, MgO 27·44, Fe_2O_3 10·02, Al_2O_3 16·42, HO 10·3 (Crossley). Vermont, United States.

Vermilion. See MERCURY, SULPHIDE OF.

Vesuvian. See IDOCRASE.

Vesuvian Salt. A synonyme of sulphate of potash.

Vienna Green. Arsenio acetate of copper.

Vignite. Probably a mixture of magnetic iron, carbonate and phosphate of iron.

Villarsite. H 3, 3·5. Spec. grav. 2·975. Yellowish-green rhombic octahedrons of 139° 45′. Subtransparent; B.B. infusible; with borax a green enamel; decomposed by strong acids. SiO_3 39·40, MgO 45·33, FeO 4·3, MnO 2·86, CaO ·54, KO ·46, HO 5·8, At Traversella, in Piedmont.

Vinegar. See ACETIC ACID.

Vinester. A German term, sometimes applied to compounds of ethyle with acids.

Vinous Fermentation. See FERMENTATION.

Violan. Spec. grav. 4·233, H 5—6. Amorphous; cleavage giving a rhombic prism; lustre waxy; colour dark violet; opaque; fracture uneven to imperfectly conchoidal; B.B. fuses to a clear glass, with borax in the outer flame a brownish-yellow glass, violet-red when cold; in the inner flame a yellow glass, colourless when cold. It is a silicate of alumina, magnesia, lime, containing much FeO,NaO and MnO. With magnesian epidote, at St. Marcel, in Piedmont.

Violine. A substance analogous to *Emetine*, extracted from the *Viola odorata;* renders litmus paper green; insoluble in ether and oils; precipitated by nutgalls; acts as an emetic.

Virgineic Acid. A peculiar fatty acid, forming the principal part of an oil which exists in considerable quantity in the root of the *Polygala Senega*, a native of Virginia. This oil has a reddish-brown colour, a syrupy consistence, a bitter rancid taste, and disagreeable odour; reddens litmus.

Viridic Acid. $C_{14}H_6O_7$. By the oxidation of caffeotannic acid. Gives with BaO a bluish-green precipitate.

Viscine. *Bird Lime.* A substance exuding from the epidermis of a species of *Acacia*, and from the involucrum of the *Atractylis gummifera*, and several other plants. Soft and elastic, with a greenish or brownish colour, and adhering firmly to the fingers; insoluble in water and fixed oils; slightly soluble in alcohol, and very soluble in ether and oil of turpentine. When heated melts and burns with a white flame, giving out much smoke. *Artificial Bird Lime* is prepared from the middle bark of the *Holly*, by boiling in water till it becomes soft, allowing it to ferment for several weeks, and then pounding and washing with water.

Vitelline. C 52·264, H 7·249, N 15·061, S 1·170, Phosphorus 1·020? O 23·236. The albuminous body constituting 15·76 per cent. of yolk of egg. It is very similar to albumen, with which it was long confounded. Like albumen, it has a *soluble* and *insoluble* condition. The former is precipitated by HCl and SO_3, but not by organic or phosphoric acids. At 140° it becomes milky, and about 166° deposits flakes. It is only distinguished from soluble albumen in being coagulated by heat, without the addition of acetic acid or salts; in not being precipitated by the salts of oxide of lead or copper; and in being thrown down by ether. The other modification of vitelline possesses the same properties as the corresponding form of albumen. Vitelline seems to be one of the first stages in the metamorphosis of albumen by oxygen.

Vitriol, Blue. See SULPHATE OF COPPER.

Vitriol, Cobalt. Sulphate of Cobalt.

Vitriol, Green, or Sulphate of Iron.

Vitriol, Lead. Sulphate of Lead.

Vitriol, Ochre. See BOTRYOGEN.

Vitriol, White. A synonyme of Sulphate of Zinc.

Vivianite. *Phosphate of Iron, Mullicite, Blue Iron Earth.* 3 FeO,PO_5 + 8 HO. Spec. grav. 2·661, H 1·5 to 2. Primary form a right oblique prism; thin laminæ flexible, but not elastic; sectile; lustre pearly; almost metallic on the summit of the prism; the other faces vitreous. At first translucent, but opaque on keeping. Colour between blue and green. Becomes white when heated, and then exfoliates and melts into a black enamel, which becomes magnetic by continuance of the heat. PO_5 31·1825, FeO 41·2266, HO 27·4843. Huel kind, at St. Agnes, Cornwall.

Volatile Bases. See BASES.

Volatile Oils. See OILS.

Volborthite. Spec. grav. 3·459, 3·860, H 3·, 3·5. Green or gray tables, often aggregated together in globules; lustre pearly to vitreous; streak, clear yellowish-green, nearly yellow; translucent. VO_3 36·58, CuO 44·15, CaO 12·28, MgO ·5, MnO ·4, HO 4·62, gangue ·1 (green variety), VO_3 39·02, CuO 38·27, CaO 16·65, MgO ·92, MnO ·52, HO 5·05, gangue ·76. Friedericksrode, in Thuringia.

Volcanic Ashes. The following analyses of two portions of volcanic ashes from Barbadoes in 1812, and preserved in my museum in St. Thomas's Hospital, were made by my assistant, Mr. David Walker. They present the appearance of an organized or crystalline structure. 1st. SiO_3 58·8, Al_2O_3 19·2, Fe_2O_3 9·4, CaO 6·608, MgO·2·743, HO 1·88, NaO 1·897. 2d. SiO_3 59·36, Al_2O_3 19·40, Fe_2O_3 12·88, CaO 4·3, MgO 2·97, HO 1·24, NaO 2·85. I have in my museum a portion of the vesicular matter of which Graham's Island was composed. This island rose suddenly to the S. of Malta in 1831, in lat. 37° 11′ N. long. 12° 44′ E., and as suddenly disappeared. The specimen was taken up by A. Gilchrist, Esq. R.N., from the surface of the sea. Mr. D. Walker obtained SiO_3 48·3, Al_2O_3 and Fe_2O_3 31·, CaO 7·, MgO 2·27, NaO and KO 6·43, HO 5·.

Volcanic Glass. See OBSIDIAN.

Volcanite. *Selen Sulphur.* Resembling sulphur, but of an orange or brownish colour. B.B. on charcoal yields fumes of selenium and sulphurous acid, and burns readily. From Vulcano, one of the Lipari Islands.

Volkonskoite. Amorphous; dull; shining; bluish-green; streak bluish-green and shining; feel resinous; fracture subconchoidal; adheres slightly to the tongue; very fragile. SiO_3 37·01, CrO_3 17·93, Fe_2O_3 10·43, Al_2O_3 6·47, Mn_2O_3 1·66, MgO 1·91, HO 21·84, PbO 1·01, KO trace. A silicate of chromium. Okhansk.

Volknerite. 6 MgO, Al_2O_3 + 15 HO. Spec. grav. 2·04. Hexagonal; cleavage basal, eminent, lateral distinct; colour white; lustre pearly; feel greasy. B.B. exfoliates, and gives out light; infusible. With cobalt solution a weak rose-red. With the fluxes a clear colourless glass. Schischimsk mines, Slatoust.

Voltaite. 3 (FeO,KO) SO_3 + 2 Al_2O_3, 3 SO_3 + 12 HO. 8-hedrons; allied to iron alum; colour brown or black; soluble in water with difficulty, and decomposing. Al_2O_3 3·27, FeO 28·69, KO 5·47, SO_3 45·67, HO 15·77, gangue ·46. Solfatara, near Naples.

Voltzite. 4 ZnS + ZnO. Spec. grav. 3·66, H 4·5. Implanted spherical globules; structure, thin curved lamellar; lustre vitreous to greasy; opaque; B.B. like blende. ZnS 82·92, ZnO 15·34, Fe_2O_3 1·84, resinous matter, trace. Rosières, in Puy de Dome; and in the iron slags of Freiberg and Altenberg.

Volumes, Theory of. Soon after the discovery of the fact, by Dalton, that simple substances unite with each other in definite weights to form compound bodies, it was announced by Gay Lussac that gases unite with each other in definite volumes or measures, when they form combinations. Thus water, by electricity, is resolved analytically into 1 volume oxygen and 2 hydrogen, or ½ volume oxygen and 1 volume hydrogen; and the same proportions may be demonstrated synthetically. Thus, ½ volume oxygen and 1 volume of hydrogen, when fired by the electric spark, are resolved into 1 volume of steam or vapour, and the specific gravity of the steam will be found equivalent to half the specific gravity of oxygen = ·5528, added to the specific gravity of hydrogen = ·0692 = ·6220, the specific gravity of steam. To this important fact, the title of the theory of volumes has been given. Illustrations of its application will be found throughout the work, under the various compound gases. An example from the compounds of nitrogen and oxygen, may here suffice to exhibit the importance of the doctrine:—

	Vols. Oxygen.	Vols. Nitrogen.	Relation of Vols.
Protoxide of nitrogen	1	½	or 2 to 1
Binoxide..............	½	½	or 2 to 2
Ternitrous acid.......	1	1½	or 2 to 3
Quaternitrous acid....	½	1	or 2 to 4

When we are acquainted with the composition, by volume, of a compound gas, we can thus readily determine its specific gravity. But the specific gravity of a gas may also be obtained by multiplying the atomic weight of the compound gas by half the specific gravity of oxygen, ·5528. Thus, the atomic weight of the protoxide of nitrogen is 2·75, according to the oxygen scale; which, multiplied by ·5528, gives 1·5202 as its specific gravity, which we also obtain by adding ·9713, the specific gravity of nitrogen (equivalent to 1 volume), to half the specific gravity of oxygen, ·5528 (corresponding with half a volume) = 1·5241. This method of calculating the spec. grav. of a body from its composition has only been successfully applied to a limited extent. Various theories have been suggested for the purpose of solving this problem in regard to organic fluids.

Kopp's Theory endeavours to institute a relation between the specific gravity of a substance and its composition, and thus to determine its density. Thus alcohol may be viewed as composed of ether (C_4H_5O) and water (HO); the specific gravities of these substances, at their boiling points, have been ascertained to be water ·931, ether ·695, and alcohol ·739. When these densities are divided by the atomic weights on the hydrogen scale respectively, 9, 37, 46, there result the atomic or specific volumes 9·37, 53·24, and 62·25. If now we add the specific volumes of water and ether, 9·37 + 53·24, we obtain 62·61, the specific volume of alcohol. If we take the specific gravities of these fluids at equal numbers of degrees from their boiling points, the same correspondence will be observed. The specific volumes of carbon being 6·24, hydrogen 4·68, and oxygen 4·68, the specific volume of alcohol is 62·4. But the specific volumes of C, H, and O vary at different degrees below the boiling point. A table of these alterations being formed, Kopp endeavours to calculate the specific gravity of organic fluids at any required temperature. Thus, acetic acid ($C_4H_4O_4$) boiling at 118° C., its specific gravity at 16° C., or 102° C. below the boiling point, may be calculated. At that temperature the specific volume of C being 5·585, 4 C = 22·34; 1 spec. volume H and O = 4·19, and 4·19 × 8 = 33·52. Then 22·34 + 33·52 = 55·86 = spec. volume of acetic acid. Dividing the atomic weight of acetic acid = 60 by 55·86, we obtain 1·074 = specific gravity acetic acid at 16° C. The experimental number is 1·063, which differs much from the theoretic amount.

Schröder's Theory.—The specific volumes of C, H, and O being supposed to be equal, alcohol $C_4H_6O_2$ contains 2 double atoms of C, 3 double atoms of H, and 1 double atom of O; dividing 46, the atomic weight of alcohol, by ·9395, its specific gravity, we have 62·2 = the specific volume at the boiling point; dividing by 6 (the number of atoms), there results 10·379, = the specific volume of a double atom of C, H, and O.

Löwig's Theory.—The specific volumes of the different elements are directly related to the atomic weights. If the atomic weights of C, H, O, N, and Cl be 1, 6, 8, 14, and 36, the specific volumes will be 1, 3, 4, 7, and 9. The specific volumes are estimated as follows, by the oxygen scale:—H 44, C 132, N 308, O 176, &c. The volume of H is susceptible of condensation in organic compounds of $\frac{3}{4}$, $\frac{1}{2}$, or $\frac{1}{4}$; or even $\frac{2}{3}$ and $\frac{1}{3}$; and then, instead of 44, becomes 33, 22, or 11, &c. For example, benzine, $C_{12}H_6$, spec. grav. ·85, atomic weight = 975 ÷ ·85 = 1147 = spec. vol. By adding 12 spec. vols., C condensed to $\frac{2}{3}$ = 1056, and 6 spec. vols. H, condensed to $\frac{1}{2}$, = 132 there results 1188, = spec. vol. benzine. By dividing the atomic weight by this spec. vol. 975 ÷ 1188 = ·821 = calculated spec. grav., a considerable difference from the true gravity. According to Löwig, the specific volume of a compound is not changed by the addition of oxygen; for example, aldehyde, $C_4H_4O_2$ = atomic weight 44, spec. grav. ·79; and acetic acid, $C_4H_4O_4$, at. weight 60, spec. grav. 1·063. The spec. vol. of aldehyde is 44 ÷ ·79 = 55·7, and that of acetic acid 60 ÷ 1·063 = 56·4. It appears, therefore, that none of these theories afford very striking results.

Playfair and Joule's Theory.—According to experiment, they infer that "the atomic volumes of bodies stand in a simple multiple relation to each other. The volumes of solid bodies are multiples of a submultiple of the volume of ice (9·8), which for convenience is adopted as a standard. The force of cohesion with metals prevents the assumption of their natural volumes, which, however, appear when they are placed in a position not to yield to the solicitation of cohesion. It is probable that the differences from the natural volume produced by cohesion could be expressed by a submultiple of the volume of ice. The differences between unlike forms of dimorphous, polymorphous, and allotropic substances are expressed by a submultiple of the volume of ice." It appears that the volumes of salts are multiples of 9·8 and 11; 9·8 (the volume of ice) being composed of 1·225 × 8, and 11 of 1·225 × 9. The volume of magnesian sulphates is 22 (11 × 2); the magnesian metals have a volume of 1·225 × 3. Now, if R*v* be the volume of any radical, O*v* that of oxygen, and *Av* that of an acid, then R*v* + O*v* = 6·12 = magnesian oxide; oxygen being 1·225 × 2, and sulphur 8·57, sulphuric acid will be S*v* + O*v* × 3 = 15·921. The magnesian sulphate volume is therefore RO = 6·12, SO_3 = 15·92, $ROSO_3$ = 22·04. It is to be regretted that Messrs. Playfair and Joule have not continued their laborious and valuable researches to a conclusion, as the accuracy of the data established by them, so far as they went, appears undoubted.

Voraulite. A synonyme of *Lazulite.*

Vosgite. 3 $RO\bar{S}iO_3$ + 3 Al_2O_3 2 SiO_3. Spec. grav. 2·737, H 6·. Massive or in thin tables; lustre greasy or pearly; colour whitish, with a shade of green; translucent; fracture conchoidal; B.B. swells up and fuses to a white glass; transparent bead with borax. SiO_3 49·32, Al_2O_3 30·07, Fe_2O_3 ·7, MnO ·6, CaO 4·25, MgO 1·96, NaO 4·85, KO 4·45, HO 3·15. Ternuay, in the Vosges; Corsica.

Vulcanite. Pyroxene.

Vulpiline, or *Vulpinic Acid.* Transparent rectangular prisms, of a fine yellow colour; little soluble in cold water; very soluble in boiling water and alcohol, and also in ether. Reddens litmus; would constitute an important yellow dye. Extracted by alcohol and ether from the *Lichen vulpinus*, a native of the south of Europe.

Vulpinite. A siliceous anhydrite, called marble of Bergamo. See ANHYDRITE.

W

Wacke. A soft mineral, intermediate between basalt and clay. Colour greenish-gray; by long exposure becomes hard; no lustre. B.B. melts. In its composition bears considerable resemblance to basalt. Often contains crystals of hornblende and mica.

Wad. See MANGANESE, BINOXIDE.

Wagnerite. *Fluophosphate of Magnesia.* $3MgOPO_5 + MgF$. Spec. grav. 3·068, 2·985, H 5, 5·5. Lustre vitreous; streak white. Colour yellowish, often grayish; translucent; fracture uneven and splintery. B.B. fuses with difficulty to a dark greenish-gray glass. With borax, or salt of phosphorus, forms a colourless pearl;

with NO_5 or SO_3 gives out fumes of HF. PO_5 40·61, MgO 46·27, FeO 4·59, CaO 2·38, F 9·36. Werfen, in Saltzburg, Austria.

Walkerde. A synonyme of *Fuller's Earth.*

Walmstedtite. A variety of brown spar.

Walnut Oil. Freezing point 17°. A greenish oil, becoming pale yellow by keeping; without smell; taste agreeable; absorbs $15\frac{1}{2}$ times its vol. of oxygen, giving out carbonic acid amounting to $\frac{2}{15}$ of the oxygen; used in painting. From the fruit of Juglans regia.

Warwickite. See ENCELADITE.

Wash. See DISTILLING and BREWING.

Washingtonite. A synonyme of Ilmenite.

Water. HO 1·125, 9. Spec. grav. 1000· or 1·. Water in a pure state is colourless; transparent, and destitute of taste and smell. In large quantities it possesses a violet-blue colour by reflected, and a green colour by transmitted light, as may be shown by viewing any brilliant white object upon a white ground through a column of water $6\frac{1}{2}$ feet long contained in a tube blackened internally, or by allowing the sun light to fall through such a column of water upon a white object (Bunsen). The specific gravity of water is taken as the standard of comparison, being reckoned 1· at the temperature of 60°. Its freezing point is 32° F., 0° C. Its boiling point 212° F., 100° C. A cubic inch of water at 62°, barometer 30 inches, weighs 252·458 grains, and in a vacuum 252·722 grains. The weight of a cubic inch of water at 60° will therefore be $252\frac{1}{2}$ grs. nearly. The imperial gallon of distilled water weighs 10 lbs. or 70,000 grains, and has a capacity of 277·274 cubic inches. A cubic foot of water weighs at 62° 62·3862 lbs. avoirdupois. In France the unit of the scientific system of weights and measures is a gramme (15·438 grs.), which is equal to a cubic centimetre of water at 4°·1 C., or 39°·38 F., the point at which water has the greatest density. When the temperature of water falls to 32° an irregular prism is formed, from which others shoot out on both sides at angles of 60° and 120°. Hail crystallizes in the form of two 6-sided pyramids applied base to base (Smithson). Crystals of ice have been observed which were rhomboids resembling calcareous spar (the usual form being 6-sided prisms, terminated by 6-sided pyramids), the faces of which were inclined at angles of 60° and 120° (Dr. E. D. Clarke). The specific gravity of ice is ·920 (Thomson), 918 (Brunner, Playfair, and Joule). As the density of ice is less than that of water, it is obvious that water in freezing expands in volume. As water is capable of transmitting sounds the conclusion was drawn that it was compressible. The amount of this has been found by

Canton,....................	$\frac{44}{1000000}$	of its vol.
Perkins,....................	$\frac{45}{1000000}$	,,
Oerstedt,....................	$\frac{46}{1000000}$	,,
Colladon and Sturm,....	$\frac{48}{1000000}$	,,

and by correcting the error proceeding from the contraction of the glass this number becomes 51·3 millionths, which appears constant for 16 atmospheres. Water, as found in nature, always contains a quantity of oxygen and nitrogen gases in solution, which may be removed either by the air pump or by boiling in a retort. The air rises to the upper part of the retort. In rain water it amounts to $2\frac{1}{2}$ per cent. But the proportion of oxygen is greater than in common air, reaching from 24 to 33 per cent., while in snow water as much as $34\frac{3}{4}$ per cent. have been found (Humboldt and Gay Lussac). The air in the water of the river Clyde amounts to 3·113 per cent., and it consists of 70·9 nitrogen and 29·1 oxygen (T. Thomson). Boussingault found snow water at 16,877 feet to contain air yielding only 16 to 17 per cent. of oxygen; while Bischoff, in the Alps, found it as low as 10 per cent. The absence of fishes in the elevated Swiss lakes has been attributed to deficiency of oxygen dissolved in the water; but it is a curious fact that Dr. Thos. Thomson, junr., found large fishes in Thibet, at an elevation of 14,600 feet, in a stream near Hanlé, discharging itself into the Indus. The water had the temperature of 69°, being mixed with discharges from hot springs. These "large fish," he says, "apparently enjoying the hot water, dart about in great numbers, and in every direction." (Epist.) It has been inferred that air is necessary in water for the respiration of fishes, because, when the water is boiled, they are no longer capable of subsisting in it; but boiling will also destroy animalcules which may serve as their food. The following table presents a view of the absorption by water of various gases according to the best authorities (Dalton, Henry, and Saussure). 100 vols. of water at 60° and 30 inches pressure, absorb of the following gases:—

Fluoboric acid,..............	70,000
Chlorohydric acid gas,..	48,000 31,000
Ammonia,.................	67,000 78,000
Sulphurous acid,............	3,700
Sulphohydric acid,.........	253
Chlorine,.....................	206
Carbonic acid,...............	100
Binoxide of nitrogen,......	76
Protoxide of nitrogen,......	76
Olefiant gas,.................	12·5
Oxygen,.......................	3·7
Hydrogen,....................	1·56
Nitrogen,.....................	1·56
Carbonic oxide,.............	1·56

Hydrates.—Water unites with certain bodies and forms *hydrate* (ὕδωρ, water, Proust). See HYDRATE. The purest water is, 1. *Rain* Water, which besides a certain amount of common air in solution, contains likewise carbonate of ammonia (Saussure; W. Stark, Ann. Phil. 3, 140,

1814; Liebig, 1825); and during thunderstorms traces of nitrate of ammonia (Cavendish, Liebig). The former of these may be detected by the addition of acetate of lead. If carbonate of ammonia is present, a white precipitate of carbonate of lead falls. In distillation, the ammonia may be fixed or detained in the retort by the addition of a few drops of sulphuric acid or a few grains of alum. Rain water has been found to have an alkaline reaction (Stark, Dr. M'Lelland in Kamaon). 2. *River* water is next in purity. It generally contains common salt, sulphate of lime, carbonate of lime, and magnesia, with portions of clay diffused, or silica and alumina. In river water, besides mud, which is contained in it in a fine state of division, diatomaceous plants, in the form of a green deposit, are abundantly present. These consist principally of siliceous shields indestructible by heat. The following table gives the composition of some large rivers, per imper. gallon:—

	Seine.	Thames.	Clyde.	Isere.
Carbonate of Lime	6·17	12·76 }	5·13	7·277
.. Magnesia	2·29	1·03 }		0·175
Sulphate of Lime .	1·39	0·45	0·25	1·459
.. Magnesia	{			2·119
.. Soda	{ ·69	2·00	}	·331
... Potash	{	0·66	trace. }	
Chloride of Calcium .	{	1·75	·18	
.. Magnesium	{ ·69		.10	0·050
.. Sodium .	{		1·84	0·250
Nitrates . . .	traces.		indications	
Silica, Alumina, Fe_2O_3	·55	0·27	0·54	0·515
Organic matter . .	traces.	3·48	1·49	?
	11·78	22·40	9·53	12·476
	Pont Notre Dame.	Twickenham.	Water Works.	Grange.

Lime and magnesia are detected by adding a solution of soap; a curdy precipitate falling. River water is usually employed for manufacturing purposes, and likewise for the supply of water to towns, although such a source is highly objectionable when the sewerage of a district is run into the river above the point at which the water is taken for the purposes of food. In consequence of its condition of continual movement, river water generally contains earthy matter distributed through it, which can be partially separated by allowing the water to stand in a cistern and deposit the diffused particles. The water of rivers and streams usually contains suspended in it a certain amount of vegetable matter (diatomaceae) in addition to the earthy matter. These can be separated by filtration. See FILTRATION.

3. *Well water*.—The most impure variety of water is derived from wells. This often contains 100 grains of saline matter in the gallon. The most abundant ingredients are carbonate and sulphate of lime and common salt. The latter is always present to a greater or less extent. It has been observed that the lime of well water may supply nourishment to the bones, and it has therefore been recommended for this purpose (Boussingault) by non-medical men. On the contrary, Dr. Prout, the best authority on such subjects, strongly cautions against the use of hard waters as being dangerous, where urinary deposits, or calculi, have any tendency to form. Lime deposits in the form of oxalate of lime are exceedingly common in the urine. Distilled water is strongly advocated by Dr. Prout in many cases, and next to it soft or rain water. Margraff, in 1751, found nitre in the wells of Berlin, but none in those at some distance from the city; and afterwards Liebig found nitrates in the wells of Giessen, a town in Hesse Darmstadt, with a population of 8000 persons. I have found it in all the wells situated within the city of Glasgow, to the extent in some of 3·64 grs. in the gallon, but not in those at a distance from the sewers, or at most in mere traces. I have likewise found it abundantly in the wells in London and Liverpool, and in numerous other towns where I have examined the waters. This nitric acid is derived from the urea of urine, which it is therefore obvious gains access to city wells.

Table of Solubility of Salts in Water.

	100 Water at 60° Dissolve
Ammonia, Nitrate,	200·
Ammonia, Chlorohydride,	33·6
Ammonia, Sulphate,	50
Potassium, Chloride,	35
Potassium, Iodide,	141
Potash, Sulphate,	9·6
Potash, Nitrate,	27
Potash, Chlorate,	6
Potash, Carbonate,	121
Potash, Bicarbonate,	25
Potash, Bichromate,	9·94
Sodium, Chloride,	37
Soda, Nitrate,	43·6
Soda, Sulphate,	83·6
Soda, Carbonate,	20½
Soda, Bicarbonate,	8¾
Soda, Phosphate,	24¾
Soda, Biborate (Borax),	8⅓
Barium, Chloride,	43½
Strontium, Chloride,	151
Calcium, Chloride,	400
Lime, Sulphate,	·217
Magnesia, Sulphate,	114
Iron, Protosulphate,	70
Copper, Sulphate,	39
Silver, Nitrate,	100
Zinc, Sulphate,	102
Mercury, Chloride,	55½

The above salts are supposed to be in the usual pure forms in which they occur with water of anhydrous, or supplied with the water of crystallization.

Water, Sea.—The following table contains the constituents in 1000 grains of different sea waters:—

	English Channel.	Dead Sea. 1.	Dead Sea. 2.
Specific gravity,	1027·4	1156	1153
Chloride of Sodium,	27·059	76·50	87·192
— Potassium,	0·765	23·30	11·000
— Magnesium,	3·666	95·60	101·748
— Calcium,		22·45	22·824
— Aluminum,		0·24	3·251
Bromide of Magnesium,	0·029	2·31	2 518
Sulphate of Lime,	1·406	0·86	1·026
— Magnesia,	2·295		
Carbonate of Lime,	0·033	traces	
Ammonia,	trace		
Fixed Salts,	35·253	221·26	229·559
Water,	904·747	778·74	770·441
	1000·000	1000·00	1000·000

Water of English Channel, spec. grav. 1027·4, by Schweitzer. Dead Sea.—1. This water was taken in April, 1847, on the western shore of the lake, about a mile from the mouth of the Jordan; analyzed by my pupil Mr. R. M. Murray, spec. grav. 1156. 2. Taken about the end of April, 1853, nearly from the same place as the preceding, and analyzed by my pupil, Mr. George William Brown, specific gravity 1153.

Waters, Mineral.

1. *Saline Waters* per imperial gallon.

	Airthrie.	Cheltenham.	Leamington.
Sulphate of Magnesia,		106·08	
— Lime,	16·062	43·40	
— Soda,		144·66	323·19
Chloride of Calcium,	300·883		164·49
— Magnesium,	9·234		26·93
— Sodium,	363·825	482·20	326·16
Fixed Salts,	690·004	776·34	840·75
	R.D.T.	Brande and Parkes.	T. Thomson.

2. *Chalybeate Waters* per imperial gallon.

	Tunbridge.	Moffat.	Spa.
Chloride of Sodium,	1·5		0·768
— Calcium,	1·848		
— Magnesium,	0·348		
Free Sulphuric Acid,		5·202	
Sulphate of Soda,	1·768		0·744
Sesquisulphated Peroxide of Iron,		591·025	
Sulphate of Alumina,		112·726	
Carbonate of Lime,	0·328		0·240
— Soda,			1·716
— Magnesia,			1·260
Protoxide of Iron,	2·748		1·008
Alumina,			0·228
Manganese, Silica, &c.	0·528		1·680
	9·068	708·953	13·644
	Scudamore.	Thomson.	G. Jones.

3. *Sulphureous Waters.*

	Aix-la-Chapelle. Spec. Grav. 1034·9.	Harrowgate. Spec. Grav. 1005·1.	Moffat. Spec. Grav. 1002·55.
Chloride of Sodium,	2·63940	462·00	176·569
— Calcium,		47·20	
— Magnesium,		21·8	
Bromide of Sodium,	0·00360		
Iodide of Sodium,	0·00051		
Sulphide of Sodium,	0·00950		
Carbonate of Soda,	0·65040		

	Aix-la-Chapelle. Spec. grav. 1034·9.	Harrowgate. Spec. Grav. 1005·1.	Moffat. Spec. Grav. 1002·55.
Bicarbonate,		11·00	
Sulphate of Soda,	0·28272	14·00	16·562
— Potash,	0·15445		
— Lime,			11·579
— Magnesia,			5·474
Carbonate of Lime,	0·15851		
— Magnesia,	0·05147		
— Iron,	0·00955		
Silica,	0·06611		
Organic Matter,	0·07517		
Carbonate of Lithia,	0·00029		
— Strontia,	0·00022		
— Manganese, } Phosphate of Alumina, } Fluoride of Calcium, } Ammonia, }	trace		
Fixed Salts in 1000 parts,	=4·1090	In gal. 556·0	In gal. 210·184
Gases—Nitrogen,	669·8	10·10 cub. in.	
Carbonic Acid,	308·9	3·35	
Carburetted Hydrogen,	18·2	4·80	
Sulphuretted Hydrogen,	3·1	5·75	21·29 cub. in.
Oxygen,	0·0		
	1000·0	24·00	
	Liebig.	Hunter.	Thomson.

4. *Acidulous Waters.*

	Eger. Spec. Grav. 1004·883.	Vichi.
Sulphate of Soda,	196·154	51·57
Chloride of Sodium,	79·933	21·17
Carbonate of Soda,	47·467	320·46
— Lithia,	0·245	
— Lime,	12·936	14·91
— Strontia,	trace	
— Magnesia,	7·273	2·78
— Manganese,	0·112	
— Iron,	0·644	0·74
Phosphate of Lime, } Subphosphate of Alumina, }	0·244	
Silica,	4·473	
Fluoric Acid, } Iodine, }	traces	
Fixed Salts in gallon,	349·461	= 411·63
Carbonic Acid per gallon,	446·17 cub. in. (?)	= 132·84 cub. in.
	Berzelius.	Mossier.

Applications of Water in Medicine.—There can be but one opinion as to the beneficial and extensive influence of water, when judiciously employed, in all departments of medical practice. The effect which may be produced upon the fluids by introducing water into the stomach may be readily appreciated, when we know that it passes through the coats of the stomach with the greatest rapidity; while its influence upon the exterior of the body, in the form of baths, is understood, to a certain extent, in most diseases. Water also may be employed to communicate cold and heat to various parts of the body, as in the form of cold and hot lotions; but it has also been recently used, to some extent, in the form of local steam applications; in which condition it seems to possess a remarkably soothing influence in many skin diseases. It is necessary, in this case, to choose the diseases judiciously. If vapour of water at 212° come in contact with an ulcerated surface, the condition of the sore will probably be exasperated; whereas, the same application made to a part affected with tic, or gouty rheumatism, would be very efficacious. It would be necessary in such cases to diminish the temperature of the vapour, and perhaps to mingle it with some soothing or narcotic preparation. The diseases in which it has been found chiefly efficacious are, white swelling, synovial inflammation of the knee, chronic rheumatism, and various species of skin disease. It may be applied locally by means of artificial bags or boxes, merely covering the part affected. In conjunction with such applications, it is necessary to apply proper treatment in reference to the adjustment of any functions which may have been impaired. The great fallacy into which many persons are apt to fall, in the use of remedies, is that of one-sidedness. It is to this contemplation, of only one aspect of the picture, that the errors of hydropathy and other novelties are attributable. The importance of water to the animal system is at once apparent, from the fact, that the muscles of the body contain 79 per cent. of water, and that an adult Guanche, when dried and converted into a mummy, weighed only a pound or two. Plants consist, likewise, principally of water; the vinegar plant (Mycoderma aceti) containing 95 per cent. of this fluid.

Wavellite. *Devonite, Lasionite, Hydrous diphosphate of Alumina.* $4 Al_2O_3, 3PO_5$ 19 HO. Spec. grav. 2·337, 2·25, 2·3616, H 3·25. White, greenish, gray, yellow, brown, and black minute radiated crystals of right rhombic prisms in the form of globular masses; the angles of the crystals are 122° 15′, and 107° 26′; lustre of cleavage planes pearly to vitreous. B.B. loses its lustre, infusible; with boracic acid and iron wire yields a globule of phosphide of iron. Soluble in NO_5, SO_3; sometimes yielding fluohydric acid. PO_5 34·65, Al_2O_3 34·, Fe_2O_3 and Mn_2O_3 2·15, Water 28·75, F? In clay slate at Barnstaple, Devon; near Cork; Shiant isles; Brazil, Bohemia, Mount Vesuvius, Susquehanna.

Wax. A solid greasy substance derived from plants and animals; a kind of deoxidized oil.

Wax Andaquics. $C_{24} H_{24} O$. Spec. grav. ·917; F.P. 170½°; from the Orinoco and Amazon districts.

Wax, Bees'. Spec. grav. ·960, ·8203; F.P. 142°, 143½° (unbleached), 155° (bleached); yellow or snow-white; scarcely soluble in cold alcohol; partly soluble in 20 hot alcohol (see Cerine); ether dissolves $\frac{1}{50}$th; by alcohol it is resolved into cerine soluble, and myricine, insoluble in alcohol. Cerine consists principally of cerotic acid united to oxide of cerotyle. This acid also exists in Chinese wax, but not in bees' wax of Ceylon. Myricine is palmitate of oxide of melissyle, $C_6 N_6 O$? By bees it is produced from sugar (John Hunter).

Wax, Brazil. Spec. grav. ·980; F.P. 206°. Greenish body, soluble in alcohol, ether, and oils; from an unknown tree in Brazil.

Wax, Carnauba. $C_{30}H_{30}O_2$? F.P. 185°. From the leaves of a palm in Brazil; soluble in hot alcohol and ether.

Wax, China. $C_{108} H_{108} O_4$. F.P. 180½°. White crystalline, probably the product of the insect coccus ceriferus (?) inhabiting Rhus succidaneum; it contains cerotine, and yields by distillation cerotene and cerotic acid (Brodie).

Wax Ibucuiba. The product of the Myristica bicuiba.

Wax of Ceroxylon Andicola. $C_{35} H_{29} O_2$. F.P. above 212°. A yellow wax from the Andes; it contains resin which alcohol retains in solution.

Wax of Chamacrops Humilis. West Indies.

Wax of Cow-tree. See Galactine.

Wax of Japan. See Japan Wax.

Wax, Myrtle. See Myrtle Wax.

Wax of Ocuba. F.P. 98°; from the Myristica ocuba of Brazil and French Guiana.

Wax, Sealing. See Sealing Wax.

Wax of Sugar Cane. See Cerosine.

Websterite. See Aluminite.

Wehrlite. Spec. grav. 3·9, H 6·25. Black granular masses; lustre inclining to metallic; streak greenish; magnetic. B.B. melts with difficulty on the edges. SiO_3 34·6, Fe_2O_3 42·38, FeO 15·78, CaO 5·84, MnO_2 ·28, Al_2O_3 ·12, HO 1·. Allied to Lievrite. Zemescher district, Hungary.

Weights.

1.—Troy Weight.

Grains.	Pennyweights.	Oz.	Lb.
1	0·0416	0·00208·3	0·0001736·1
24	1	0·05	0·00416
480	20	1	0·083
5760	240	12	1

2.—Apothecaries' Weight.

Grains.	℈	ʒ	℥	Lb.
1	0·05	0·016	0·002083	0·00017361
20	1	0·33	0·0416	0·003472
60	3	1	0·125	0·010416
480	24	8	1	0·083
5760	288	96	12	1

3.—Avoirdupois Weight.

Grains.	Dr.	Oz.	Lb.	Qrs.	Cwt.	Tons.
27·34375	1	0·0625	0·00390625	0·0001395089	0·0000348772	0·00000174386
437·5	16	1	0·0625	0·002231428	0·000557857	0·0000278928
7000	256	16	1	0·35714285	0·0892857842	0·0044642857
196000	7168	448	28	1	0·25	0·0125
784000	28672	1792	112	4	1	0·05
15680000	473440	35840	2240	80	20	1

ENGLISH. TROY.	FRENCH.
Grain (24th of a pennyweight),	0·06477 grammes.
Pennyweight (20th of an ounce),	1·55456 grammes.
Ounce (16th of a pound),	31·0913 grammes.
Imperial pound troy,	0·3730956 kilogrammes.
ENGLISH. AVOIRDUPOIS.	**FRENCH.**
Drachm (16th of an ounce),	1·7712 grammes.
Ounce (16th of a pound),	28·3384 grammes.
Imperial pound avoirdupois,	0·4534148 kilogrammes.
Hundredweight (112 pounds),	50·78246 kilogrammes.
Ton (20 hundredweight),	1015·649 kilogrammes.
FRENCH.	**ENGLISH.**
Gramme,	15·438 grains troy. 0·643 pennyweights. 0·03216 ounce troy.
Kilogramme,	2·68027 pounds troy. 2·20548 lbs. avoirdupois.

Weissite. Spec. grav. 2·808, 2·826, H 1·75. Ash-gray, kidney-shaped masses in chlorite slate at Fahlun; texture foliated, perhaps rhomboidal; powder, white; lustre, pearly to waxy. B.B. becomes white, and fuses on the edges; on charcoal gives out a zinc smoke; with borax and salt of phosphorus a colourless glass; with soda, an opaque slag, finally fused into a bead. SiO_3 59·69, Al_2O_3 21·7, MgO 8·99, FeO 143, MnO ·63, KO 4·1, NaO ·68, ZnO ·3, HO and NH_3 3·2 (Wachtmeister), SiO_3 55·05, Al_2O_3 22·6, MgO 5·7, FeO 12·6, CaO 1·4, HO 2·25 (J. Tennent).

Weld or **Woad.** The stem and leaves of Reseda luteola, long used as a dye even by the ancient Britons. It contains luteoline

Wernerite. Spec. grav. 2·77, 2·712. Darkish coloured variety of scapolite, from Pargas and Ersby. SiO_3 45·1, Al_2O_3 32·76, CaO 17·84, NaO ·76, HO ·68, KO 1·04.

Wheat. The seed of the Triticum hybernum. Vienna wheat has been found composed of C 45·74, H 6·7, O 42·23, S ·23, H 3·, ash ·7, HO 13·85. The proximate constitution of flour is starch 902, fibrine 116·8, caseine 5·27, glutine 3·04, albumen 14·, gum 60·4, sugar 16·3, water 189·4. The ash is composed of KO 30·12, CaO 3·, MgO 33·12, Fe_2O_3 16·26, PO_5 48·3, SO_3 1·01, SiO_3 1·31.

Whey. The serum of milk. (See MILK.) It consists of 96·3 water, sugar 3·5, salts ·2.

Whisky. (*Usquebaugh*, Gaelic, pronounced wisky bay or ooisky bay.) A spirit distilled from the fermented infusion of malt, barley, or oats, etc. The barley is ground, the malt bruised and mixed in the proportion of 40 bushels barley, 20 bushels malt, in the mash tun; 750 wine gallons of water at 150° are run on, and thoroughly mixed or mashed by agitation for one-and-a-half hour; 500 wine gallons of water at 190° to 205° are added at intervals to keep up the heat, and the whole allowed to infuse for two hours; ¾ of the wort is drawn off from the top of the mash tun and the grains or residue washed with repeated additions of hot water. The first addition consists of 500 gallons as above. To remove all the soluble matter from the grains, about 800 gallons of boiling water are let in for the third worts. The law of Scotland required that 19 gallons of spirits (·90917) should be produced from 100 gallons of fermented wort; now it is 13 gallons. The worts are passed into the coolers, which are shallow wooden vessels, covering an extensive area, in which the worts are placed to the depth of 1, 2, or 3 inches; or the hot worts are passed through a great extent of pipe, which is immersed in a running stream of water. The first worts are let down to 70°, the second worts to 60° or 65°. To the quantity of barley and malt mentioned, when converted into wort, 27 gallons of good porter yeast are added, a portion only being added on the first day, the remainder on the second, third, and fourth days; the fermentation lasts from nine to twelve days, according to circumstances. During the first five days the fermenting tuns are left open at top, but on the sixth day they are shut up close. The wort increases 20° to 25° in temperature, generally attaining the highest heat on the fourth day. As the fermentation proceeds, the specific gravity of the wort diminishes or *attenuates*, and by the amount of *attenuation*, as indicated by the hydrometer, the distiller calculates the amount of spirit in the wash. It is usual to reckon that every 11½ lbs. of saccharine matter fermented will yield a gallon of spirits of spec. grav. ·90917 at 60°. As soon as the fermentation ceases, the wash should be distilled, otherwise acetic acid will be formed at the expense of the alcohol. A great variety of stills is now in use. The common still is most commonly used, and next to it Coffey's (see Thomson's Records of Gen. Science, vol. 3, p. 32), and Stein's (see STILL) stills are in most requisition. The wash being placed in the still, heat is applied, and the spirits being most volatile, pass over first, and are condensed by the refrigeratory in the common still. In Coffey's and Stein's still, the alcoholic fluid meets with steam introduced from the opposite end of the apparatus from a boiler, which takes the water from the spirit, and drives the spirit into a proper receiver. In the common still the distillation is continued till the liquid which flows from the worm is as heavy as water, as is ascertained by the hydrometer. The liquor remaining in the still is termed spent wash, containing saccharine matter and spirit. This is used as a feed for cattle, but by the evidence of large dairy-keepers is found to be prejudicial to animal life when used in any considerable quantity. The product of the first distillation of worts is termed *low wines*, the spec. grav. being ·978. The low wines are distilled again, a process termed doubling; the first portion which comes over is milky, and is termed *foreshot*. When the spirits run clear, they are collected separately. As soon as the specific gravity begins to rise, the product is caught in a different receiver. This third product is termed *feints*, and is mixed with the low wines, and again distilled. The foreshot contains hydrous oxide of amyle or fusel oil, a poisonous oil, which should therefore be carefully separated from all distilled fluids. The strength at which the duty is levied on whisky is from 1 to 10 above hydrometer proof, or about ·90917 specific gravity. The present duties on British spirits are in England 7s. 10d., Scotland 3s. 8d., Ireland 2s. 8d. The following table gives the number of gallons of spirits charged with duty in the three kingdoms in the three years 1850-52 (A. Young):—

	1850.	1851.	1852.
England—			
Exported	77,825	57,654	90,452
Home consumption	9,331,512	9,595,368	9,820,608
Scotland—			
Exported	209,395	194,073	227,204
Home consumption	7,122,987	6,830,710	7,172,015
Ireland—			
Exported	43,418	37,754	33,497
Home consumption	7,408,086	7,550,518	8,208,256
	24,193,223	24,266,077	25,552,082

White Antimony. See TEROXIDE OF ANTIMONY.

White Arsenic. See ARSENIOUS ACID.

White Copperas. See COQUIMBITE.

White Garnet. See LEUCITE.

White Indigo. See INDIGO.

White Iron Pyrites. See WHITE BISULPHIDE OF IRON.

White Iron Sinter. See TETRARSENIATE OF IRON.

White Lead. See CARBONATE OF LEAD.

White Precipitate. See CHLORAMIDE OF MERCURY.

White Stone, or Granular Granite. See GEOLOGY.

White Vitriol. See SULPHATE OF ZINC.

Whiting. *Common Whiting*, *Spanish White*, and *Kentish White*. London chalk is triturated in a mill with water and carefully dried, care being taken to remove sand by means of sieves. What is termed best whiting, consists of the finer particles of the chalk diffused through water; the coarser particles, or common whiting, falling to the bottom. Whiting is dried on a kiln. It is used for making putty, cleaning plate, and for various colouring purposes. *Paris white* is made by the same process, using Hull cliffstone, or compact limestone, instead of chalk, and drying it in dishes.

Wichtisite. See WICHTYNE.

Wichtyne. *Wichtisite*. Spec. grav. 3·03. Black masses or rectangular prisms (?); lustre dull; fracture angular or conchoidal. B.B. fuses into a black magnetic enamel. SiO_3 56·3, Al_2O_3 13·3, Fe_2O_3 4·, FeO 13·, CaO 6·, MgO 3·, NaO 3·5. Wichtis, Finland.

Williamsite. *Wilhelmite*, *Willemite*. See ANHYDROUS SILICATE OF ZINC.

Willuite. *Willenite*. A variety of Garnet.

Wine. The fermented juice of the grape. The spirit which wines contain, and to which they owe their stimulating properties, is derived from the fermented sugar. (See VINOUS FERMENTATION.) The odour and flavour of wines depend on certain combinations which are produced during fermentation. Old Rhenish wines contain acetic ether, many of them a very minute proportion of butyric ether, which impart to them a peculiar and agreeable odour. All wines contain œnanthic ether, upon the presence of which depends their vinous smell. These compounds are partially formed in fermentation and partly in the casks, by the influence of acids on the alcohol of the wine. Œnanthic acid appears to be produced by fermentation, at least it has not been detected in the grape. The free acids which are present in the fermenting juice, take a decided part in the formation of those aromatic substances upon which odour and flavour depend. The wines of Southern countries are produced from perfectly ripe grains; they contain tartar and no free acids, and are characterized by a different flavour from the wines of France and the Rhine (Liebig). Table I. gives the amount of absolute alcohol in wines as ascertained by Brande, and Table II. the determinations of Fontenelle:—

TABLE I.

Name of Wine.	Spec. Grav.	Absolute Alcohol per cent.
Port Wine	0·97616	19·82
Do	0·97200	23·92
Mean	0·97460	21·75
Madeira	0·97810	17·91
Do	0·97333	22·61
Sherry	0·97913	17·00
Do	0·97700	18·37
Bordeaux, Claret	0·97410	·11·95
Do	0·97092	15·11
Calcavella	0·97920	16·76
Lisbon	0·97846	17·45
Malaga	0·98000	15·98
Bucellas	0·97890	17·22
Red Madeira	0·97899	17·04
Malmsey	0·98090	15·91
Marsala	0·98190	14·31

Name of Wine.	Spec. Grav.	Absolute Alcohol per cent.
Marsala,	0·98000	15·98
Champagne (rose)	0·98608	10·46
Do.........(white)	0·98450	11·84
Burgundy	0·98300	13·34
Do.	0·98540	11·06
White Hermitage	0·97990	16·14
Red do.	0·98495	11·40
Hock	0·78290	13·31
Do.	0·98873	8·00
Vin de Grave	0·98450	11·84
Frontignac	0·98452	11·84
Cote-Roti	0·98495	11·36
Rousillon	0·98005	15·96
Cape Madeira	0·97924	16·77
Muscat	0·97913	17·00
Constantia	0·97770	18·29
Tinto	0·98399	12·32
Schiraz	0·98176	14·35
Syracuse	0·98200	14·15
Nice	0·98263	13·64
Tokay	0·98760	9·15
Raisin wine	0·97205	23·86
Drained grape wine	0·97925	16·77
Lachryma Christi	—	18·24
Currant wine	0·97696	19·03
Gooseberry wine	0·98550	10·96
Elder wine }		
Cider }	0·98760	9·14
Perry }		
Brown stout	0·99116	6·30
Ale	0·98873	8·00
Porter	—	3·89
Rum	0·93494	49·71
Hollands	0·93855	47·77
Scotch whisky	—	50·20
Irish whisky	—	49·91

TABLE II.

Wine.		Absolute Alcohol per cent.
Roussillon (Eastern Pyrenees).		
Rive-salts	18 years old	9·156
Banyulls	18 „	9·223
Collyouvre	15 „	9·080
Salces	10 „	8·580
Department of the Aude.		
Fitou and Leucaté	10 years old	8·568
Lapalme	10 „	8·790
Sejeau	8 „	8·635
Narbonne	8 „	8·379
Lezignan	10 „	8·173
Mirepeisset	10 „	8·589
Carcasonne,	8 „	7·190
Department of l'Herault.		
Nissan	9 years old	7·896
Beziers	8 „	7·728
Montagnac	10 „	8·108
Mèze	10 „	7·812
Montpellier	5 „	7·413

Wine.		Absolute Alcohol per cent.
Lunel	8 years old	7·564
Frontignac	5 „	7·098
Red Hermitage	4 „	5·848
White do.	4 „	7·056
Burgundy	4 „	6·195
Grave	3 „	5·838
Champagne (sparkling)		5·880
Do. white do.		5·145
Do. rose		4·956
Bordeaux		6·186
Toulouse		5·027

Wintergreen. See GAULTHERIC ACID.

Wiscrite. A variety of hydrous carbonate of manganese. Gonzen, Switzerland.

Withamite. Spec. grav. 3·137, 2·857, H 6·. Carmine-red and pale straw-yellow masses and minute crystals in truncated pyramids, with angles of 128° 20′ and 16° 40′; streak white. B.B. intumesces and fuses at length into a greenish-gray scoria; with salt of phosphorus with effervescence fuses into a globule containing a skeleton of silica, and becomes opaque on cooling; it corresponds in many properties with the epidote of Arendal. SiO_3 55·28, Al_2O_3 16·74, Fe_2O_3 21·13, CaO 8·13, HO 3·25. Glencoe, on the surface of a reddish trap rock.

Witherite. A synonyme of carbonate of Barytes.

Woad. *Pastel.* The blue dye formed by fermenting the leaves of the Isatis tinctoria.

Woehlerite. Sp. gr. 3·41, H 5·5. Yellow or honey-coloured tables, rectangular prisms, or grains; streak yellowish-white; transparent; subtranslucent; fracture conchoidal, splintery. B.B. fuses into a yellowish glass when strongly heated, and gives the reaction with fluxes of iron, manganese, and silica; soluble in HCl. SiO_3 30·62, CmO_3 14·47, Zr_2O_3 15·17, Fe_2O_3 2·12, MnO 1·55, CaO 26·19, MgO ·40, NaO 7·78, HO ·24 = $3Zn_2O_3,CmO_3$, $5(NaOSiO_3)$ $3CaOSiO_3$. Found with elæolite, on zircon syenite, at Brevig, Norway.

Wœlchite. Spec. grav. 5·75, H 3·. Lead-gray masses, with a tendency to rhombic cleavage; fracture uneven; fragile. S 28·602, Sb 16·647, As 6·036, Pb 29·902, Cu 17·352, Fe 1·404. Wœlch, Carinthia.

Wœrthite. Spec. grav. above 3, H 7·25. White foliated masses, with the lustre of cyanite; translucent. B.B. infusible; a colourless glass with borax; a silica skeleton in microcosmic salt; a blue with a cobalt solution. SiO_3 40·58, Al_2O_3 53·5, MgO 1·, HO 4·63; near St. Petersburg.

Wolchonskoite. See VOLKONSKOITE.

Wolfram. See TUNGSTEN.

Wolfsbergite. Spec. grav. 4·748, H 3 to 4. CuS,SbS_3. Lead-brown, gray, tabular or right rhombic prisms, with truncated edges; lustre metallic; opaque; fracture conchoidal. B.B. decrepitates; soon fuses; evolves fumes of antimony on charcoal. S 26·34, Sb 46·81, Cu 24·46, Fe 1·39, Pb ·56. Wolfsberg, Hartz.

Wollastonite. A synonyme of Table spar.

Wollastonite. *Soda Table Spar, Pectolite* Spec. grav. 2·850 to 2·876, H 5·25. White, with a shade of green, fibrous tufts, diverging from a centre; lustre inclining to silky; translucent on the edges; fracture splintery. B.B. fuses with some difficulty, and without frothing, into a white enamel; with borax into a bead, yellow while hot, colourless when cold; with salt of phosphorus forms a colourless bead with a silica skeleton, with soda an opaque bead I first analyzed this mineral in 1830, when it was named and described by Dr. T. Thomson. It seems to be the same mineral which Dr. Kennedy had analyzed from the castle rock of Edinburgh, about thirty years before, and which afterwards Karsten described from an imperfect analysis under the name of pectolite. Its constituents are—

	1.	2.	3.
SiO_3,.........	51·50	52·744	52·059
CaO,..........	32·00	31·684	32·817
NaO,..........	8·50	9·600	
MgO,..........		1·520	1·624
FeO,..........	0·50	1·200	2·682
Al_2O_3,........	0·5	·672	
HO,...........	5·	2·	

1. Edinburgh, Dr. Kennedy; 2. Kilsyth, R.D. T.; 3. Bishopton tunnel, J. C. Stevenson.

Wongshy. A red or yellow-colouring principle, the seed capsule of a Batavian species of gentian.

Woulfe's, Wolf's, Bottles. See Hydrochloric Acid.

Wood. The solid portion of trees is so termed, which consists essentially of cells (cellulose) lined or incrusted on the interior of their walls with at hard matter which yields a variety of results to chemical agents. Table I. gives the solubilities of the ligneous matter of the cells in different fluids; Table II. the ultimate analyses of woods.

Table I.

Ligneous Matter Incrusting the Cells.

Insoluble in—

Lignose,.....	Water	Alcohol	Ether	NH_3
Lignone,....	Water	Alcohol	Ether	
Lignin,......	Water		Ether	
Lignireose,..	Water			
Cellulose,....	Water	Alcohol	Ether	NH_3

Soluble in—

Lignose,....	Potash	Soda		
Lignone,....	Potash	Soda	Ammonia	
Lignin,.....	Potash	Soda	Ammonia	Alcohol
Lignireose,.	Potash	Soda	Ammonia	Alcohol, Ether

Table II.

Composition of Woods.

	C.	H.	O.	Equivalent in Charcoal.
St. Lucia,...	52·90	6·07	41·03	55·35
Ebony,......	52·89	6·00	41·15	53·75
Fir,..........	51·79	6·28	41·93	54·70
Oak,.........	50·00	6·20	43·80	53·30
Beech,......	49·25	6·40	44·65	51·40
Poplar,......	47·00	5·80	47·20	47·20
Cellulose,...	44·44	6·17	49·34	44·44

Protection of Wood from Decay.—Every one is acquainted with the fact, that when wood is buried in the earth it gradually becomes rotten or decayed. This decay may be checked or delayed by charring the exterior surface of the wood, a method frequently adopted for the preservation of stakes, rails, posts, and other preparations of wood employed in agriculture; and sailors have long known how to preserve their spars, by means of oil and tar. It was found, however, many years ago, that a more effective preservative action could be attained by the influence of corrosive sublimate; and the practice of soaking wood intended for ship-building, in tanks containing a solution of corrosive sublimate, is now pretty general. The chloride of zinc has also lately been introduced. This is a very deliquescent salt, and of a caustic nature, and may be readily detected by heating it on charcoal before the blowpipe, when it affords a sublimate of the oxide of zinc. In order to cause the complete impregnation of the wood by a solution of the salt, it is now the practice to introduce the wood into a vacuum, and to force the solution between the fibres in this situation. The view of the action of these bodies at present entertained is, that they combine with an albuminous matter in the sap, or accompanying the cellulose, and thus prevent it from producing a species of fermentation leading to that decay, so well known under the title of *dry rot.* It is therefore certain that all the metallic salts, as recommended to the Admiralty forty years ago, by Dr. T. Thomson, and all those substances which form compounds with the albuminous matter of wood, will prove equally efficacious. Various resinous and coal oils are now used. I lately examined some oils produced from resins by distilling them at a high temperature which were manufactured for this purpose, in which I detected a considerable quantity of creasote, a substance well known as capable of combining with albumen, and as a powerful antidote to decay. Raw pyroligneous acid has also been recommended as an antidote to dry rot, and from the quantity of creasote which it contains must undoubtedly be of service. Common salt has been long used for preserving both vegetable and animal substance. Tar, sulphate of iron, pyrolignate of iron, arsenious acid, Fuchs' soluble glass, &c. have all been employed with various degrees of success. To preserve wood, and also to give it variegated shades, M. Boucherie has recommended the introduction of various solutions into the current of the sap, either immediately after the tree is felled, or before this operation is performed. In the former case it is necessary to leave some tufts of leaves at the summit of the tree, and then by cutting a crucial notch at the foot, and immersing it in the solution, the whole ligneous matter becomes impregnated with the liquor. To produce hardness, a solution of pyrolignate of iron is employed. A solution of chloride of calcium possesses a good preservative action, at the same time that the tree retains its suppleness, and is rendered less combustible. The mother liquid of salt marshes is equally efficient. By introducing solutions of tannic acid, prussiate of potash after pyrolignate of iron, we obtain black and blue colours respectively. Acetate of lead, and chromate of potash, introduced in succession, produce the yellow chromate of lead colour; and by the use of pyrolignate of iron, prussiate of potash, acetate of lead, chromate of potash, one after the other—blue, green, yellow, and brown tints are formed. To insure the preservation of timber it should be felled when the least amount of sap is contained in its vessels. This period is of course during the winter. It requires then to be seasoned, that is, gradually dried spontaneously, so as to enable as much as possible of the water of the sap contained in it to evaporate. To remove the sap more completely, it has been recommended that the wood should be steeped for a fortnight in clear water, or running water, and then seasoned in the ordinary way. Wood for buildings requires two or three years' seasoning before it is adapted for its object. The wood in ships is often preserved by packing rock salt between the timbers. In this way I have known as much as 20 or 30 tons of salt stowed away in a large vessel. From careful inspection I have not found this method to produce corrosion of the ship's iron when due attention has been paid to its protection.

Wood Opal. Opal, a form of silica, which has replaced the woody matter of trees, and assumed its fibrous aspect.

Wood Tin. A fibrous variety of binoxide of tin or tin ore found in Cornwall and Brazil.

Woody Fibre. The ligneous fibres of which wood is composed.

Wool. The hairy matter of the skin of sheep is so called. It differs from cotton or linen in being a minutely jointed tube, uniting with

Wool.

colouring matters without the interposition of mordants, and in not being properly decolourized by chlorine but by sulphurous acid; its specific gravity is 1·614, that of cotton being 1·549.

Woorara. An extract obtained by digesting the Strychnos toxifera in water, and employed by the South American Indians as a poison on their arrows.

Wormwood. See ABSINTHINE.

Wormwood, Oil of. Spec. grav. ·973, of vapour 5·3. Colourless oil, pungent and bitter; isomeric with camphor from wormwood.

Wort. The saccharine aqueous infusion of grain. See BEER.

Wulfenite. A synonyme of Molybdate of lead.

Wootz. Indian steel.

X

Xanthamylate of Potash. $2\ CS_2 + C_{10}H_{11}O,KO$. Citron-white crystalline scales; greasy to the touch; soluble in water, alcohol, and ether; anhydrous. By adding CS_2 dissolved in amylic alcohol, to a solution of potash, in the same solvent.

Xantharine. $C_8H_8SO_4$.

Xanthate of Lead. $PbOC_6H_5S_4O$. Colourless crystals, by mixing bisulphide of carbon and hydrous protoxide of lead, in a solution of potash, in alcohol; insoluble in cold water, alcohol, and ether.

Xanthate of Oxide of Ammonium. $NH_4O,C_4H_5O,\ 2\ CS_2$. Lustrous crystals like urea, formed when dry ammoniacal gas is passed into an alcoholic solution of binoxisulphocarbonate of ethyl.

Xanthene. A substance obtained as sulphide in the decomposition of tersulphocyanohydric acid by heat. Formula of the sulphide, $C_8N_2\ H_2S_2$.

Xanthic Acid. A transparent colourless fluid, heavier than water, possessed of a strong odour resembling sulphurous acid, and an acid taste. Catches fire when approached by a lighted body. Obtained by adding CS_2 to a solution of potash in alcohol till the alcohol is neutralized, and then decomposing the xanthate of potash by dilute HCl or SO_3.

Xanthic Oxide. See URINARY CALCULI.

Xanthic Oil. A yellow limpid liquid, with a strong peculiar smell and sweet taste; soluble in weak alcohol, and burning with a blue flame. Obtained during the distillation of xanthate of potash.

Xanthile. $C_4H_5O_{10}$.

Xanthin. A yellow colouring matter, said by Kuhlmann to exist in madder root. It has the appearance of an extract; very soluble in water, its solution having a sweet odour; very soluble in alcohol, and very little in ether.

Xanthite. 5 ($2\ CaOSiO_3$) + 2 ($2\ Al_2O_3\ SiO_3$). Spec. grav. 3·221, H 2. Rounded grains or foliated masses, yielding by cleavage doubly-oblique prisms, whose faces are inclined to each other. Colour yellow; translucent; sometimes transparent. B.B. fuses with borax into a glass, which is yellow while hot, but colourless on cooling. SiO_3 35·092, CaO 33·08, Al_2O_3 17·428, Fe_2O_3 6·368, MnO 2·801, MgO 2·001, HO 1·68. Amity, Orange County, New York.

Xanthobetic Acid. An amorphous reddish-yellow mass, attracting moisture from the atmosphere; soluble with difficulty in ether, more readily in alcohol and water. Extracted from the root of the red-beet, *Beta vulgaris.*

Xanthocone. ($3\ AgS + AsS_5$) + 2 ($3\ AgS + AsS_3$). Spec. grav. 5, 5·2, H 2. Reniform masses; colour dull red to clove-brown; streak yellow; B.B. fuses, and AsS_3 sublimes; on charcoal gives a bead of silver. S 21·798, As 14·322, Ag 63·88. Near Freiberg.

Xanthogenamide. $C_6H_7NS_2O_2$. F.P. 96·8. Formed in the process for xanthate of ammonia, as an oil, but obtained in large pyramids by dissolving in alcohol; soluble in alcohol and ether; not easily soluble in water.

Xanthopenic Acid. A lemon-yellow crystalline powder, by the action of alkalies on opiammon, a derivative of narcotine.

Xanthophyll. Existing in seared autumn leaves, and from which they derive their yellow colour.

Xanthophyllite. Light coloured mineral resembling clintonite or holmesite, in globular and columnar masses at Slatoust, Ural; containing SiO_3 16·20, Al_2O_3 44·96, CaO 12·15, MgO 19·43, FeO 2·73, NaO ·55, volatile matter 4·33.

Xanthopicrite. A body of a greenish-yellow colour; silky lustre; destitute of smell; taste bitter and astringent; little soluble in water; soluble in alcohol; insoluble in ether. Extracted by alcohol from the bark of the *Xanthoxylon carybœum.*

Xanthoproteic Acid? $2\ C_{36}H_{27}N_4O_{12} + NO_3$. An orange-coloured powder, insoluble in water, alcohol, and ether; inodorous; combines with acids and bases; produced by nitric acid on protein.

Xanthorhamnine. $C_{23}H_{12}O_{14}$. Brown matter, obtained by boiling *Chrysorhamnine* (existing in the *Rhamnus tinctoria*, or Persian berries) in water, with free access of air; insoluble in ether; soluble in alcohol and water.

Xanthorœa. *Botany Bay Resin.* The yellow resin of the *Xanthorœa hastilis;* soluble in ether and alcohol. By the action of alkalies, cinnamic and benzoic acids are formed; and by nitric acid, picric acid results.

Xanthorthite. See ALLANITE.

Xanthosiderite. An impure sesquioxide of manganese from Ilmenau.

Xenolite. See BUCHOLZITE.

Xenotime. Phosphate of Yttria.

Xuthene, Sulphide of. $C_{10}H_6N_9S_{,3}SH$. A substance produced in the decomposition of persulphocyanohydric acid, which is insoluble in NH_3, and is thus distinguished from sulphide of phaiene.

Xylidine. $C_{16}H_{11}N$.

Xylite. See XYLITIC ACID.

Xylite. (CaO,MgO) $SiO_3 + Fe_2O_3SiO_3$ + HO. Spec. grav. 2·935, H 3. Nut-brown; opaque; asbestus like. B.B. fuses on the edges to a black mass. SiO_3 44·06, Fe_2O_3 37·84, CaO 6·58, MgO 5·42, CuO 1·36, HO 4·7. From the Urals?

Xylitic Acid. $C_{12}H_{12}O_5$. A volatile fluid from pyroxylic spirit.

Xylitic Naphtha. $C_{12}H_{12}O_3$. When an excess of potash is added to xylitic acid.

Xylitic Oil. $C_{12}H_9O$. By potash on xylitic acid.

Xylitic Resin. C_8H_6O. By the action of caustic potash in excess on xylitic acid.

Xylocryptite. See SCHEERERITE.

Xyloidin. A white insipid powder, by digesting sawdust of wood or potato starch in concentrated nitric acid, and then precipitating by water.

Xyloietin. $C_{40}H_{33}O_4$. Fossil resin found in the remains of pines, in the peat of Denmark. It is also prepared by extracting turf with alcohol, and then adding ether, when xyloietin falls; which, when crystallized, fuses at 327°.

Xylole. $C_{16}H_{10}$. A hydrocarbon homologous to benzole and toluole. Found in wood tar and coal gas naphtha.

Y

Yanolite. A synonyme of *Axinite.*

Yeast. See FERMENTATION.

Yellow, Cassel. *Hexachloride of Lead,* or PbCl, 6 PbO; obtained by fusing together 1 salammoniac and 10 litharge; used as a pigment.

Yellow Copperas. See COPIAPSITE.

Yellow Copper Ore. See COPPER PYRITES.

Yellow, English; or *Turner's*; or *Patent.* Probably a mixture of dichloride of lead and oxide of lead; prepared by mixing 1 common salt with 7 litharge in water; the precipitate is washed free from soda and ground.

Yellow Lead Ore. *Molybdate of Lead.*

Yellow Tellurium. *Aurotellurite.* See TELLURIUM, GRAPHIC ORE OF.

Yellows for Pottery. *Clear Yellow* for browns and greens; gray flux 79, antimoniate of potash 14, hydrous carbonate of zinc 7, triturate and fuse. *Jonquil Yellow.*—Pebble flux 86, mixture (of 1 tin, 3 lead) 8, calcined carbonate of soda 3, antimoniate of potash 3. *Deep Yellow* for browns and greens.—Gray flux 75, antimoniate of potash 17, hydrous carbonate of zinc 4, red oxide of iron 4. *Uranium Yellow.*—Pebble flux 75, pure oxide of uranium 25. *Chrome Orange Yellow.*—Minium 75, chromate of lead 25. *Imperial Yellow,* or inferior chrome yellow, for paints.—Add acetate of lead to water containing whiting or Paris white, diffused through it to the consistency of cream, and precipitate by bichromate of potash; dry carefully.

Yenite. See ILVAITE, SILICATE OF IRON, LIEVRITE.

Yle. The termination of organic radicals, from ὕλη, a substance from which a body is derived.

Yolk of Eggs. The yellow portion of eggs, consisting of vitelline. See EGG.

Ypadu. *Coca.* Leaves of the *Erythroxylon Coca,* an herb much valued by the Indians of Peru. See COCA.

Ypoleime. See COPPER PHOSPHATE.

Ytterbite. See GADOLINITE.

Ytterspath, or *Xenotime.*

Yttrite. See GADOLINITE.

Yttrium. Y (Gadolin, 1794). This substance is always accompanied with erbium and terbium, so that no accurate description of its characters has been published. When to a solution of a salt of yttrium, as usually prepared from yttria, is added gradually a solution of an alkali, at first there falls a precipitate, which is rendered yellow or orange by ignition, consisting of oxide of erbium, with oxides of terbium and yttrium. The second precipitate is paler yellow on ignition, consisting principally of oxide of terbium. The third precipitate remains quite white on ignition, and is yttria. The salts of erbium are red; of terbium pale red; of yttrium colourless.

Yttrocerite. *Fluate of Yttria.*

Yttrocolumbite. There are three varieties of this, the black, the yellow, the brown. *Black.*—3 (YO,CaO,FeO) CmO_3TnO_3. Indistinct 4 or 6-sided prisms and plates; spec. grav. 5·395, H 5·5. Lustre submetallic; colour iron-black; streak gray; opaque. CmO_3 57·, TnO_3 8·25, U_2O_3 ·5, YO 20·25, CaO 6·25, Fe_2O_3 3·5. *Yellow.*—Spec. grav. 5·882, H 5·, 3 YO ($CmO_3U_2O_3$). Occurs in fissures in felspar; not crystalline; lustre, resinous on the surface, vitreous in the fracture; colour yellowish-brown to greenish; streak white; opaque. CmO_3 59·5, TnO_3 1·25, U_2O_3 3·23, YO 29·90, CaO 3·29, Fe_2O_3 3·50. *Brown.*—3 (YO,CaO) CmO_3, H 4·5, 5·. Plates; lustre vitreous to resinous; colour black, with a shade of brown; streak white. Yttrocolumbite is infusible per se before the blowpipe; but decrepitates and becomes light coloured. Ytterby, Broddbo, Finbo, Sweden.

Yttroilmenite. *Samarskite, Uranotantalite.*

Yttrotantalite. A synonyme of *Yttrocolumbite.*

Yttrotitanite, or *Kilhauite.*

Z

Zaffre, or *Smalts.* See COBALT.

Zala. A synonyme of *Borax.*

Zeagonite. See GISMONDINE.

Zeasite. A synonyme of *Opal.*

Zeilanite. A synonyme of *Spinelle.*

Zeolite. The following minerals are such as have the term zeolite appended:—*Needle Zeolite,* or *Thomsonite*; *Foliated Zeolite,* or *Heulandite*; *Compact Zeolite,* or *Lehuntite*; *Crystalline,* or *Radiated, and fibrous, and mealy, and pulverulent,* or *Mesotype*; *Dodecahedral,* or *Sodalite*; *Efflorescing,* or *Laumontite*; *Hexahedral,* or *Hard Zeolite,* or *Analcime*; *Pyramidal,* or *Zeolite of d'Hellesta,* or *Apophyllite*; *Radiated, or Pearly,* or *foliated,* or *Stilbite*; *Trapezoidal,* or *Leucite*; *Feather,* or *Natrolite*; *Blue Zeolite,* or *Lapis Lazuli*; *Calcareous,* or *Stellite*; *Cubic,* or *Chabasite*; *Red,* or *Adelsforsite*; *Swedish* or *Spodumene*; *Tenacious,* or *Dysclasite.*

Zero. The Italian name for the freezing point on the Centigrade and Reaumur scales. In Fahrenheit's scale it is the temperature of mixed snow and salt.

Zeuxite. Spec. grav. 3·051, H 4·25. Small flat, interwoven, rectangular prisms; colour brown, with a light shade of green; lustre vitreous;

glistening; opaque; crystals easily broken between the fingers; loses water when heated. B.B. colour deepens, and the crystals lose their edges and become scoriaceous; infusible per se. With carbonate of soda gives bottle-green glass; with borax effervesces, and yields a very dark brown glass. SiO_3 33·48, Al_2O_3 31·848, FeO 26·01, CaO 2·456, HO 5·28. Redruth, Cornwall.

Zigueline. *Red Oxide of Copper.*

Zinc. Zn 4·0625, 32·5. Spec. grav. fused 6·862, hammered 7·215; F.P. 684° (Liebig), 752° (Schwartz), 771° (Daniell), 773° (Pelouze), 795°·75 (Person). Zinc is a metal having a bluish-white colour; crystallizes in 5-gonal 12-hedrons. It may be hammered out to a certain extent, but flies off before the hammer in small scales. Its specific gravity is about 7·0. It melts below a red heat, and when heated much above this point it burns with a very fine greenish-white flame, at the same time throwing off flakes of oxide of zinc. Zinc was unknown to the ancients as a metal, but they seem to have been acquainted with alloys of it. There is mention made of brass (an alloy of copper and zinc) in the writings of Moses; where it reads, "Out of whose hills thou mayest dig brass." But as brass is not extracted from ores (but is manufactured), it is more probable that this word (brass) means copper, or the ancients may have known an ore containing zinc and copper. The Indians seem to have some rude method of extracting zinc from the ore called "calamine" or carbonate of zinc, and zinc is used by them, in a state of purity, as a current coin. The Chinese, also, have a coin containing about 45 per cent. of zinc. These coins have a square hole in the centre, for the purpose of passing a string through and slinging them around the neck.

Ores.—The ores of zinc are various, and are brought from different localities. Blende (sulphide of zinc or black jack) is the most common variety. Calamine, hydrous, and anhydrous silicates are other kinds. The following shows an analysis of hydrous silicate of zinc:—Oxide of zinc 68·00, silica 23·60, water 8·45, copper trace. This specimen was of a bluish colour, probably due to the trace of copper; upon scratching it with a knife it left a white streak. A specimen from New Jersey, having a very fine red spotted appearance, due to the presence of oxide of manganese, gave as follows:—Oxide of zinc 66·00, protoxide of iron 25·20, oxide of manganese 3·96, silica 4·80, sulphur 1·00, lead and magnesia traces. A very poor specimen from North Wales, commonly called "blue-stone," gave—zinc 17·94, iron 14·56, copper 1·32, lead 7·88, sulphur 31·50, silica 28·80. This ore also contains about 13 oz. of silver to the ton of ore, and about ½ oz. of gold.

Smelting of the Ores.—The reduction of zinc ores at present is certainly very simple, but, nevertheless, is accompanied by an immense loss, as we shall have occasion to see as we proceed. A great quantity of the ore smelted in this country is brought from the "Isle of Man," under the name of "Black Jack." This ore, before the zinc is ready for the market, undergoes the three following processes:—1st. Calcination of raw ores, for twenty-four hours. 2d. Fusion of calcined ore. 3d. Moulding or refining the zinc.

1st *Operation—Calcination of Raw Ore.*—The raw ore, finely ground, is conveyed into a hopper or "binn," on the top of the calcining furnace, and is let down into the furnace by drawing out a slider at the bottom of the binn. These calcining furnaces are constructed with two floors, the floor of the upper serving as the roof of the lower bed. The ore first falls into the top bed, where it is spread uniformly over the bottom by iron rakes. The doors of the furnace are then shut, and the heat gradually raised and kept up for twelve hours, taking care to keep the ore occasionally stirred to expose fresh portions to the action of the air, and to prevent the particles from adhering together. When the ore has remained in this top bed for twelve hours, it is drawn through holes, in the floor of this bed, into the lower bed; and another charge of fresh ore put into the top to be treated in the same way. That in the lower bed is now spread all over the bottom, and allowed to remain with the doors open for six hours, the ore being stirred frequently with iron rakes. At the end of the six hours the doors are closed, and the heat raised gradually for other six hours; at the end of which time the ore is drawn out through holes in the sole of the furnace. The charge in the upper bed is now lowered into the lower bed, and a fresh charge again put into the top; and the whole again treated, as just described. The object of allowing the doors to remain open for the first six hours, in the lower bed, is this—When the ore is drawn from the upper to the lower bed it receives a very intense heat indeed; and were the heat to be continued at once upon the ore, there would gather a thin coating of oxide of zinc on its surface, which would prevent the free escape of the sulphur. But by merely allowing the doors of the calciner to remain open for the first six hours, this inconvenience is overcome. By this process a quantity of sulphur is got rid of and oxygen substituted, although the loss in weight occasioned is nearly 20 per cent. Calamine (or carbonate of zinc) sometimes undergoes the same operation of calcination as the sulphurets, and sometimes it is only broken up and fused as carbonate, along with calcined sulphide.

2d *Operation—Fusion of Calcined Ore.*—The fusion of the calcined ore or oxides is carried on in three different ways. First, we have the "English method," which consists in fusing the ores in large "crucibles." Second, the "Silesian method," which consists in the fusion of the ores in large muffles. Third, there is the "Vieille Montagne method," which consists in fusing the ores in long tubes or cylindrical retorts.

1st. *English Method.*—The crucibles in which this operation is conducted are made of clay, being about 3½ or 4 feet high. Six of these crucibles are arranged in a round furnace, similar at the bottom to a glass furnace, there being an opening opposite each crucible. Above the crucible, in the interior of the furnace, there is an arch, which extends over the crucibles with holes, perforated opposite each crucible for the escape of the smoke into the chimney, which is a continuation of the furnace, although the whole of the smelting operation goes on beneath the arch just mentioned. The fire is introduced through a door, F, in the centre of the furnace between the crucibles. In the centre of the bottom of the crucible there is a small aperture. There is a hole in the furnace bottom, opposite the aperture in each of the crucibles. To this hole in the furnace bottom access

C Crucibles. *d d* Doors opposite each crucible. F Fire-place. Directly under the fire-place is the ash-pit. To the right and left are culverts under the furnace through which access can be had to the receivers. *t t* Iron tubes through which the sublimed zinc passes. *r* Receivers for zinc.

can be had by a culvert under the furnace. About 3 cwt. of the calcined ore are mixed up with about ½ cwt. of coal and coke (the aperture in the bottom of the crucible being previously stopped by a wooden plug). The mixture is then introduced into the crucible by the lid on the top, and the hole closed. The whole six pots being charged in this way, the heat of the furnace is raised gradually, till it attains a bright white. The wooden plug in the crucible bottom becomes consumed, and by this time the coal which was mixed with the ore gets caked and prevents the ore from running through the orifice. At this point a long tube of sheet iron is fixed to the hole in the bottom of the furnace, through which the zinc passes and falls in the state of drops into an iron receiver, *r*. In some works it falls into water, but it has been found more economical to let it fall into a dry receiver, as when it falls into water there is a quantity of oxide of zinc formed. The space of sixty hours elapses before one charge from these crucibles is exhausted, and the metal amounts to from 30 to 35 per cent. of the ore treated. The crucible being exhausted, the iron tube from the bottom is taken away, and the refuse from the crucibles is removed by the top by working it down by an iron rod from the mouth of the crucible. The wooden plug is again put in the aperture, and the crucible filled with a new charge of calcined ore and coal, which is treated in the same way as just described.

2d. *Silesian Method.*—When the zinc ore, along with the required flux, is heated, the vapour has always a tendency to pass upwards, but in the process just described, it had to force its way down through the whole mass before it could effect an escape; and to obviate this defect large muffles have been used. These muffles are manufactured of fire-clay, and are of the following shape, having two apertures, *a* and *d*. To *a* is fitted an earthenware tube, *b*, *c*, bent in the shape of an arm, and to *d* is luted a fire tile. These muffles are arranged in rows on either side of the furnace. Each muffle is introduced into the heart of the furnace, through a small opening just large enough for its admission. This opening is then closed by an iron door, to prevent the tube, *a b*, from fracturing by the cold air. The muffles are filled half-full with the same mixture as for the crucibles by the opening *d*, and the opening again closed and luted air-tight. There is now an iron tube, *c*, joined to the tube *a*, which iron tube dips into a receptacle. The fire of the furnace being now raised, the zinc begins to volatilize and passes through the earthen tube, *a*; then by the iron tube, *c*, when it condenses and falls in drops into the receiver. These muffles only require twenty-four hours to exhaust one charge, but there is only about half the quantity of mixture put into them as into the crucibles. When the works are in full operation a greenish-white flame is to be seen at the end of each

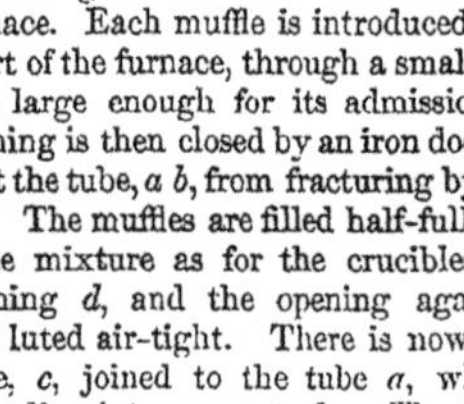

iron pipe, *c*, and white flakes of oxide of zinc flying about the house in which the operation is carried on; causing a considerable loss of zinc. The workmen employed in this process appear unhealthy, and have a very ghastly appearance, owing, no doubt, to the bad effects of the oxide of zinc, which must be breathed by them in considerable quantities. When the charge in the muffles has been exhausted, the refuse is removed, through the aperture *d*, by removing the fire tile.

3d *Operation—Refining the Zinc.*—The zinc in the form of drops, as it falls from the crucibles and muffles, is re-melted in large pots, where, when melted, it is kept covered over with charcoal to prevent its oxidation. It is then laded in small square blocks, in which state it is sent to market.

3d. *Vieille Montagne Method.*—The ores treated at Vieille Montagne are mostly carbonates, oxides, and silicates of zinc. These ores, when first brought to the works, are largely contaminated with clay, and other earthy matters, of a less specific gravity than the real ore itself, and these earthy matters are sometimes removed by a very economical process, although it occupies a considerable time. The ores are in the state of lumps allowed to remain for several months exposed to the action of the weather, when the earthy particles get washed away by the heavy rain. But, to facilitate the operation, a stream of water is sometimes allowed to run upon them. After this mechanical operation, the ores are sorted into two classes, according to their physical and chemical characters. The first class comprises those which contain little iron, and are called "White ores." The second class takes in those which contain most iron, and are known by the name of "Red ores;" the colour, of course, being due to the presence of oxide of iron. The ores, after being thus sorted, are calcined in kilns, similar in shape to those made use of for burning lime. Here the water and carbonic acid are driven away, causing a loss of about 25 per cent. After this operation, the ore is ground very fine under large rollers, and is passed through very fine sieves, when it is ready to be reduced. The reduction of the ores is conducted in large cylindrical retorts, measuring 3 feet 8 inches long, and 6 inches diameter inside. Forty-two of these retorts are arranged in one furnace, four of which are formed out of one mass of brick-work. To the mouth of each retort is fitted a cast iron conical adapter, 16 inches long; to this again is attached a wrought iron tube, tapering outwards till the bore is only half-an-inch. About 1,100 lbs. of calcined ore and 550 of coal constitute a charge for the 42 retorts, being 26 lbs. ore and 13 coal for each retort. When the charge has been introduced, the heat is gradually raised till it attains a bright white. When the retort has become very hot, carbonic oxide begins to make its escape, and burns with its characteristic blue flame at the point of the adapter of cast iron. When reduction commences to take place, a greenish-white flame issues from the tube and volumes of white fumes. At this point the conical wrought iron tube is luted on, and the fire is kept up for about two hours, when the conical tube is removed by a pair of tongs, and passed over a vessel to receive the oxide of zinc, which is rejected and fused with a future charge. One of the workmen now holds a receiver, or what they call a "Poelon," beneath the beak of the retort, while another workman rakes out into it the distilled zinc which has accumulated at the retort's mouth. This zinc is covered with a coating of oxide, which must be removed before the zinc is cast into bars. When this is completed, the cone is again luted on and another charge drawn off, in the same way, after two or three hours' firing. After the operation is finished, the residue is drawn out, which still contains a large quantity of zinc, but this must be thrown away, as it cannot be extracted from the refuse, as new compounds are formed, such as silicates. "The establishment of Vieille Montagne employs 2,646 persons, and produced 11,500 tons of zinc in 1850."

Mode of Analyzing the Ores.

I. *Cupreous Silicate of Zinc.*—The substances to be separated were silica, copper, zinc, and water.

1st. *Silica.*—25 grains of the ore finely ground were introduced into a porcelain basin, and muriatic acid poured upon it, and then placed on a hot sand bath, and allowed to evaporate to dryness, keeping it stirred with a glass rod. It was heated till it ceased to give off the smell of chlorine. It was now removed to one side and allowed to cool, and when cold a fresh portion of muriatic acid and water was added, and the whole again heated to complete the decomposition of the mineral. It is sometimes necessary to evaporate to dryness several times to complete the decomposition. When the whole was completely decomposed, a fresh portion of acid, diluted with water, was added, and when slightly heated it was filtered, and the silica collected, which was washed, dried, burned, and weighed.

2d. *Copper.*—To obtain the copper sulphuretted hydrogen was passed through the filtrate from the *silica*, which throws down sulphide of copper. When the solution was strongly saturated, it was filtered, and the sulphide washed with water, impregnated with SH. When clean washed, it was removed from the filter and dissolved in nitric acid, and the solid sulphur parted by filtration. The copper may now, if thought fit, be thrown down in the ordinary way by caustic potash or soda, but the copper in these ores has been determined by two other methods. The one is by the depth of colour, and the other by cyanide of potassium; both occupying but very little time (in fact, a few minutes), not so expensive, and at the same time quite as accurately, if not more so (if proper care be taken) as by potash and soda. One great evil in the caustic potash and soda process is the very long time it takes to free the oxide of copper from the alkalies, and oxide of copper is slightly soluble in caustic potash.

Colour Process.—If it is wished to obtain the per centage of copper by the colour process, the mode of procedure is this—First of all, a standard must be made by dissolving one grain of pure copper in an alkalimeter, then diluting with water and adding ammonia. The alkalimeter is now filled up to the 100 with water, and the whole must now be poured into a tube, which must be corked air-tight, and placed in a tube frame on the window. This forms a standard of 1 per cent.; to form one of $\frac{1}{2}$ per cent., $\frac{1}{2}$ a grain of copper is treated in the same way, and so on. To determine the quantity of copper in the substance under examination in this way, the sulphide, after being dissolved in NO_5, and the solid sulphur separated, ammonia was added to the solution, which gave a slight blue colour, showing only a trace of copper. But had there been a sufficient amount of copper to determine, part of the ammoniacal solution would have been poured into a tube of exactly the same thickness and colour as those containing the standards, and would have been compared with the standards. If it had not been exactly the same colour as either of the standards, it would have been poured back into the beaker and diluted with water until it was precisely the same colour. The whole solution would now have been measured out with the alkalimeter, and the precise quantity noted; and if the colour agreed with the 1 per cent. standard, the quantity of solution obtained is the per centage of copper in the ore.

Cyanide of Potassium (KCy) *Process.*—To determine the per centage of copper in a substance by this process, a standard solution of cyanide of potassium is first made, by dissolving a portion of cyanide of potassium in water, in a bottle; a grain of pure copper is now dissolved in nitric acid, then diluted with water, and ammonia added. An alkalimeter is now filled with the solution of KCy, and is poured into the ammoniacal solution of copper (keeping it constantly agitated), till the whole of the blue colour disappears. A note is now made of the number of spaces of the measure it has taken of the KCy to decolourize the grain of copper. This forms the standard. It is always advisable to make two or three experiments of this kind before the standard is decided upon. The standard having been found, we may now proceed to take the per centage of copper in the substance under examination. We proceed precisely in the same way as that described in the colour process, but instead of taking the depth of colour, we pour in from an alkalimeter as much KCy as will decolourize the whole, taking great care to keep the copper solution in a state of motion while pouring in the KCy, so that it may diffuse itself through it. When the whole of the colour has been removed, the quantity of KCy employed is noted; then we say, If so many spaces of the alkalimeter decolourize 1 grain of copper, how many grains will the number of spaces or 100ths required to decolourize the substances treated be, supposing the quantity of the substances treated to be 100 grains? For example:—Supposing the standard of the KCy to be 30·, or that it took 30 spaces or 100ths of the alkalimeter to decolourize 1 grain of copper. Then supposing it took 200 spaces of the alkalimeter of KCy to decolourize the quantity of copper obtained from 100 grains of the mineral treated, we would state the question thus:—If 30 spaces decolourize 1 grain, how many grains will 200 spaces decolourize? or, as 30 : 1 : : 200 : 6·666. By this method, very accurate results may be obtained, if the process be carefully carried on; but if the KCy stands long dissolved, it is apt to decompose and lose its strength; it is, therefore, requisite to test the strength of the solution at intervals. Another thing to be attended to in this process is, that some of the other metals, such as zinc, nickel, cobalt, &c. interfere, and when present give erroneous results; and it is always necessary when any of these metals are present, to throw down the copper by SH, and then dissolve the sulphuret, and proceed as before. This process is one of Mr. Parkes's, of Pembrey Copper Works, South Wales.

3d. *Zinc.*—The solution which ran through from the sulphide of copper, was heated till the odour of SH was gone. It was then boiled in a flask with carbonate of soda, which threw down carbonate of zinc. After boiling for some time, the carbonate of zinc was collected on a filter, washed with hot water, then dried and burned, by which operation it was decomposed, and converted into oxide, when it was weighed.

4th. *Water* was determined from the loss of weight by heating. For the preceding details I am much indebted to my pupil, Mr. James Napier, jun.

Dinoxide.—Zn_2O. Blackish-gray powder, obtained by calcining the oxalates, and forms on the surface of zinc when exposed to the air.

Oxide.—Zn O 5·0625. Zinc tarnishes in the air by uniting with a small portion of oxygen. Water is decomposed by the hot metal. When zinc is strongly heated in an open clay crucible it takes fire, burns with a white flame, and gives out abundant white flocks of oxide (*philosophic wool*). It may also be prepared by precipitating any of the salts of zinc, as the sulphate, by an alkaline carbonate, washing and igniting the white carbonate; oxide of zinc remains, which is yellow when hot, white when cold. It is tasteless, soluble in ammonia and caustic alkalies, and can thus be separated from iron.

Binoxide.—$Zn\,O_2$. White powder, obtained

by mixing hydrous oxide of zinc with binoxide of hydrogen.

Chloride of Zinc.—Zn Cl 8·5. Gray, solid, deliquescent; formed by dissolving zinc in hydrochloric acid, and igniting gently. It may be sublimed, when it crystallizes in needles. It is used as a caustic, also to preserve wood and cordage, and as a disinfecting agent.

Sulphate of Zinc.—Zn O SO_3 7 HO.—Prisms obtained in the preparation of hydrogen by acting on zinc with dilute sulphuric acid; used in the proportion of 1, 2, or 3 grains to an ounce of water, as an eye water, in inflammation: it acts also as an emetic. It forms with sulphate of potash a double salt; the potash salt replacing one atom of water, Zn O SO_3, KO SO_3 6 HO. Magnesia, manganese, &c. form similar double salts.

Carbonate of Zinc, Zn O CO_2, *Zinc Spar*, *Calamine*, the common ore of zinc, a yellowish solid. The precipitated carbonate consists of a mixture of carbonate and hydrate of the oxide (2 Zn O CO_2, 3 Zn O HO). A plate of zinc precipitates lead from its solution. *Galvanized iron* is obtained by plunging iron into melted zinc. An alloy is formed on the surface, which prevents oxidization. Zinc is usually estimated by boiling with carbonate of soda. To separate it from nickel it is necessary to precipitate both metals by an alkaline carbonate, to dissolve the carbonate in acetic acid, and pass a current of sulphuretted hydrogen through the solution; Zn S is precipitated, while the nickel remains in solution.

Zinc Ores.

Zinc Bloom. Dicarbonate of zinc.

Zinc, Sulphide of. *Blende, Black Jack.* Spec. grav. 4—4·2, H 3·5—4. Lustre adamantine, resinous; colour yellow, black, red, green, brown; streak white, reddish-brown, transparent. Fracture conchoidal; brittle; botryoidal. Structure columnar, also massive, compact. B.B. infusible per se and with borax. On charcoal vapours of zinc are evolved. Soluble in NO_5. S 32·75, Zn 62·62, Fe 2·2, Cd 1·78. Found in England, America, Sweden, and other countries.

Oxysulphide. *Volzite.* Spec. grav. 3·66, H 4·5. ZnS 81, ZnO 15·, FeO 1·8, organic matter 2·2 = 4½ ZnS, ZnO. Yellowish-red, interspersed with brown bands; hemispherical tubercles coating substances in a vein at Rosiers, in Puy de Dome; opaque, or feebly translucent; lustre pearly on the natural coats, but vitreous or resinous in other directions.

Selenide. Spec. grav. 5·56. Opaque gray mass with the metallic lustre. B.B. burns with a fine violet-coloured flame, exhaling a strong smell resembling that of rotten cabbage; heated in a retort, selenium, mercury, and some sulphur sublime, while oxide of zinc remains. Soluble in acids. Se 49, Zn 24·, Hg 19·, S 1·5, CuO 6·.

Zinc-Amyle. *Zincamylium.* $C_{10}H_{11}$ Zn. Formed in the same way as zinc-ethyle and zinc-methyle, by substituting iodide of amyle. Colourless transparent fluid.

Zinc, Anhydrous Carbonate of. *Calamine in part.* Spec. grav. 4·334, H 5·. ZnO CO_2. White, gray, green, or brown, massive or in obtuse rhomboids, with angles of 107° 40′, in acute rhomboids and 4-sided tables; foliated; fracture imperfect, conchoidal; lustre vitreous, inclining to pearly; semitransparent to translucent; brittle. B.B. infusible, becomes opaque; soluble in acids with effervescence; becomes negatively electric by friction. CO_2 34·8, ZnO 65·2. Found with galena and blende. *Common Calamine* is sulphate of barytes, bole, and chalk.

Zinc, Anhydrous Silicate of. *Williamsite, Wilhelmite, Willemite, Hebetine, Troostite.* Spec. grav. 3·935, H 5·. Yellow or brownish amorphous masses, or right angled prisms; fracture uneven, granular; lustre resinous; brittle; crystals varying from transparent to translucent. B.B. action the same as the hydrous silicate. Switzerland. ZnO 68·77, SiO_3 26·97, Fe_2O_3 1·48, Al_2O_3, with traces of ZnO and Fe_2O_3 1·44, HO 1·25.

Zinc, Arseniate of. *Köttigite.* Spec. grav. 3·1, H 2·5. Red or peach masses, or crystalline prisms; streak reddish-white, translucent. ZnO 30·52, AsO_5? 37·17, CaO 6·91, NiO 2·, HO 23·4. B.B. melts into a pearl, evolving arsenical fumes; soluble in acids. Schneeberg.

Zinc, Ferrosulphide of. See MARMATITE.

Zinc, Hydrous Dicarbonate of. *Zinc Bloom, Zinconise.* Spec. grav. 3·589, H 2·25, 2ZnO CO_2 2HO = ZnO 69·36, CO_2 13·52, HO 15·1. White, grayish, or yellowish-white masses, resembling chalk, encrusting other minerals; dull; opaque; absorbs water. B.B. becomes yellow, but resumes its colour on cooling; in reducing flame entirely dissipated, oxide of zinc being deposited on the charcoal; a colourless glass with borax and microcosmic salt; infusible with soda. Found with the preceding.

Zinc, Hydrous Silicate of. *Calamine in part, Electric Calamine.* Spec. grav. 3·434, 3·379, H 5·. White, blue, green, yellow, or brown masses, or stalactites, or right rhombic prisms, with angles of 102° 35′; foliated; lustre vitreous; brittle; transparent to translucent; streak white. B.B. decrepitates and gives out a green light; a glass with borax and microcosmic salt; infusible with soda; electric by heat; gelatinizes in NO_5 or HCl. SiO_3 25·, ZnO 68·3, HO 4·4. In limestone in Carinthia; lead mines in England and Scotland. Its blue colour seems due to copper. See ZINC.

Zinc, Phosphate of. See HOPEITE.

Zinc, Red. *Manganesian Oxide of Zinc.*—Sp. grav. 5·432, ZnO 88·, Mn_2O_3 12·. Bright-red, with a shade of yellow, foliated grains, cleaving in the direction of a regular 6-sided prism; streak orange-yellow; cross fracture conchoidal; brittle;

lustre adamantine; translucent; in thin plates; transparent; becomes dull in the air, and is covered with a white coating. B.B. infusible; gives with borax a yellow glass; soluble in NO_5.

Zinc, Sulphate of. Spec. grav. 2·036, H 2·25. White, or peach-blossom, or violet-blue masses, or stalactitic, or in 4-sided prisms, terminated by 4-sided pyramids; brittle; lustre vitreous, transparent to translucent; taste astringent, metallic, and nauseous; soluble in water. B.B. froths and covers the charcoal with white flocks. SO_3 22·7, ZnO 27·5, Mn_2O_3 ·5, HO 49·3 (Klaproth). Rammelsberg, Hartz; Hungary; Sweden; Flintshire.

For other Zinc Minerals, see AURICHALCITE, BURATITE.

Zincenite. Spec. grav. 5·303, H 3 to 3·25. 6-sided prisms, terminated by low 6-sided pyramids; lustre strongly metallic; opaque. B.B. on charcoal decrepitates and fuses; with $NaOCO_2$ a globule of lead. S 22·58, Pb 31·84, Cu ·42, Sb 44·39 = PbS, SbS_3.

Zinc-ethyle. *Zinc-ethylium.* C_4H_5Zn. Analogous to zinc-methyle, but less volatile; by substituting iodide of ethyl for iodide of methyle; it is prepared in a similar way to zinc-methyle.

Zinc Glass. *Silicate of Zinc.*

Zincite. *Red Oxide of Zinc.*

Zinc-methyle. *Zincmethylium.* C_2H_3 Zn. Colourless penetrating fluid, with a disagreeable odour; vapour poisonous; decomposing water, like potassium. Burns in contact with air, with a greenish-blue flame, affording dense clouds of oxide of zinc. By acting on iodide of methyle, by means of zinc, a white residue remains; which, on distillation, yields zinc-methyle.

Zinconise. A synonyme of *Zinc Bloom.*

Zinc Spar. *Carbonate of Zinc.*

Zinc Vitriol. *Sulphate of Zinc.*

Zinnober. *Cinnabar.*

Zinnstein. *Binoxide of Tin.*

Zinnwaldite. *Lepidolite.*

Zippeite. See URANOCHRE.

Zircon. *Hyacinth, Jargon, Silicate of Zirconia.* Spec. grav. 4·681, 4·721, 4·505, H 6·5 to 7·5. Zr_2O_3, SiO_3 = zirconia 66·58, silica 33·40. Red, brown, yellow, gray, green, white 8-hedrons, with a square base with angles of 123° 20′ and 84° 20′; the angles at the base are often replaced by planes, which convert the crystal into a 4-sided prism terminated by a 4-sided pyramid with a square base (*see fig.*); fracture concoidal; streak white; lustre more or less adamantine; transparent to translucent; brittle. B.B. loses its colour, retaining its transparency; infusible; with soda and microcosmic salt infusible; a transparent glass with borax. Expailly in Auvergne, Ceylon, United States, Carinthia.

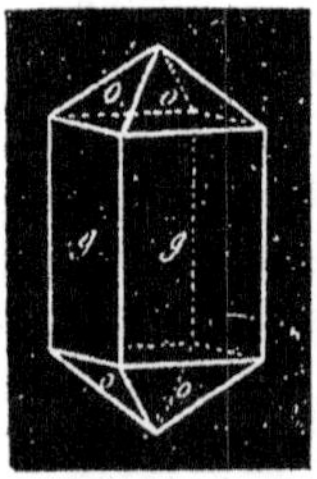

Zirconia. ZrO or Zr_2O_3 3·75, 30; 38, 30·4. Spec. grav. 4·30. Zr 73·68, O 26·32. Fine white powder, feeling somewhat harsh between the fingers, without taste and smell, infusible when pure; when fusible it contains potash, and then resembles porcelain; strikes sparks with steel, scratches glass, and is insoluble in acids, excepting strong sulphuric acid; in the blowpipe flame it evolves a fine light; insoluble in caustic alkalies, soluble in alkaline carbonates. According to Svanberg, what is usually called zirconia, consists of a mixture of various new earths, which are sesquioxides; their atomic weights vary from 9·375 to 13·2. In one of them, *noria,* (from Nore, the old name of Norway,) the oxalate and chloride are not so soluble in acids as the others; the sulphate of noria with excess of acid is peculiar and more crystallizable than that of the others. *Process.*—The zircon or silicate of zirconia is reduced to a fine powder in the diamond mortar, and afterwards in an agate mortar, and fused with three parts of solid caustic potash or soda, in a silver crucible or in a platinum crucible with carbonate of soda, the mineral being introduced into the centre of the powder and mixed with the hydrate of soda, to prevent the platinum from being injured. Bisulphate of potash, or better still, bisulphate of soda, answers well for decomposing this mineral, or carbonate of soda may be alone employed. The fused mass is digested in dilute chlorohydric acid, the silica separated in the usual manner, filtered and washed. The chlorides of zirconium and iron in solution are then precipitated by caustic potash or soda, when hydrous zirconia contaminated with some sesquioxide of iron falls. To remove the iron the washed precipitate is boiled with a solution of oxalic acid, the oxalate of iron dissolves, while the oxalate of zirconia remains undissolved. The oxalate of zirconia, when ignited, leaves zirconia (Dubois and Silveira, Ann. Chim. 14, 110). The iron may also be dissolved up by suphurous acid (Berthier). If the impure zirconia be dissolved in chlorohydric acid the solution evaporated to a paste, the paste washed with strong chlorohyric acid, till the washing gives no black precipitate with sulphohydride of ammonia, the iron dissolves, and the chloride of zirconium remains. (Chevreul, ib. 13, 245.)

Hydrate of Zirconia.—2 ZrO, HO 8·625, 69; 8·725, 69·8, ZrO=87·11, HO=12·89. White, bulky, semigelatinous precipitate, readily soluble in acids when moist, but very slowly soluble when dry; it contracts when dry into a gummy mass, which is translucent with a conchoidal fracture. The moist precipitate by alkaline carbonates is soluble in excess of the reagent; but moist zirconia is insoluble in carbonates of soda and potash, though slightly so in carbonate of ammonia. When ignited it begins to glow like a live coal at incipient redness.

Potash Fluoride of Zirconium.—$K_2Zr_3F_5$= 2KF 3ZrF 29·9, 239·1. Granular crystals, yielding no water by ignition nor undergoing change; slightly soluble in cold, very soluble in boiling water, formed by dropping fluohydride of zirconia into an excess of fluohydride of potash. Another salt, KF 2 Zr F 17·5, 140·1, with similar characters, is prepared by dropping fluohydride of potash into an excess of fluohydride of zirconia (Berzelius).

Chloride.—White fixed matter, or sublimate, formed by heating zirconium in chlorine, or by passing chlorine over ignited zirconia and charcoal; it absorbs ammonia at the usual temperature (Wöhler, Pog. Ann. 48, 94). *Hydrous Chloride.*—Small colourless silky needles, becoming opaque at 122°, losing half their acid, and their water of hydration; ignition removes the whole of the acid; soluble in water and alcohol; very slightly soluble in strong chlorohydric acid; obtained by dissolving the choride in water, or zirconia in chlorohydric acid and evaporating. A *dichlorohydride* (2 ZrO,HCl), a gummy mass, is formed by evaporating the preceding salt to dryness at 140° (Berzelius).

Sulphate.—$ZrOSO_3$ 8·80, 70·4. Crystalline mass, slowly soluble in cold, quickly in hot water; loses its acid by ignition; alcohol precipitates a mixture of sulphate and bisulphate, from which the sulphate is removed by washing with alcohol; obtained by dissolving hydrous zirconia in an excess of sulphuric acid, evaporating to dryness, and heating just below redness. An acid solution of the salt yields hydrous crystals, from which the excess of acid may be removed by alcohol. The *disulphate.*—2 ZrO,SO_3. Soluble in water, but decomposed by much water into sulphate and trisulphate; loses its acid at a white heat; prepared by saturating the preceding salt with hydrous zirconia. *Trisulphate.*—3 ZrO,SO_3 16·4, 131·2. White flakes, soluble in chlorohydric acid; insoluble in water; formed by adding alcohol to a solution of the sulphate, and washing the sulphate and trisulphate precipitated, with alcohol or water till the trisulphate remains pure (Berzelius).

Sulphite.—White powder, insoluble in water; soluble in excess of sulphurous acid, from which it precipitates by boiling.

Ammonia Sulphate.—Soluble in hot and cold water, and acids; obtained by mixing strong solutions of sulphate of ammonia and sulphate of zirconia.

Potash Sulphate of Zirconia.—$KOSO_3$ 6 ZrO SO_3? Crystalline, slightly soluble in water, the solution being again precipitated by sulphate of potash; when washed with a small quantity of water it is soluble in acids, but insoluble when thoroughly washed after ignition, during which it loses water and acid; it is insoluble in boiling sulphuric acid.

Zirconite. A synonyme of Zircon.

Zirconium. Zr 2·75, 22; 2·8, 22·4. Black powder resembling charcoal, converted by burnishing into thin graphite-like shining scales; non-conductor of electricity; if pure it is not altered when ignited to the fusing point of glass in an atmosphere of hydrogen or in vacuo; but when hydrous zirconia is mixed with it, the metal becomes incandescent by absorbing the oxygen of the water. When heated in the open air it takes fire before it becomes red hot; and becomes white zirconia; when mixed with hydrate it explodes; when mixed with chlorate of potash and struck, it takes fire and burns without detonating; in fused nitre it does not burn below redness; when mixed with carbonate of potash, it burns at the expense of the carbonic acid, some light being evolved; it burns also in fused borax, at the expense of the water of hydration, and likewise in alkaline hydrates; it is insoluble in sulphuric, chlorohydric, nitric acids, and in aqua regia, but is quite soluble in fluohydric acid, in a mixture of nitric and fluohydric acids; insoluble in caustic potash and soda. *Process.*—The potassio fluoride of zirconium is heated to remove water of hydration; it is mixed in alternate layers with potassium in an iron tube ¼ inch wide, 1¼ inch long, contained in a covered platinum crucible; it is heated just sufficiently to melt the potassium, and stirred with an iron wire; it is then heated to low ignition, when the potassium deprives the salt of its oxygen silently; when cold, the residue is thrown into cold water; the fluoride of potassium dissolves with slight evolution of hydrogen; it is washed with cold water to remove hydrate of zirconia, it is digested in equal parts of water and chlorohydric acid, of the temperature of about 112°, for six hours. The zirconium is thrown on a filter and washed with water containing salammoniac and then with alcohol. It may be also obtained minutely by the galvanic battery, (in 4 sided tables) by acting on a mixed solution of chlorides of zirconium and iron.

Zœestine. *Celestine. Sulphate of strontian.*

Zoiodin. Violet plates, inodorous, tasteless, insoluble in water. Obtained when the *glairin* of sulphuretted water, containing no other matter in admixture, is collected and allowed to drain; the water which comes through deposits zoiodin.

Zoisite. *Lime Epidote.* Spec. grav. 3·32—3·3207, H 6·25. Oblique rhombic prisms, with angles of 116° 30′. Colour gray, yellowish, greenish, or bluish-gray, or brown. Structure foliated; lustre pearly, shining, translucent. B.B. swells and fuses on the edges; with borax diaphanous glass. SiO_3 39·3, Al_2O_3 29·448, CaO 22·956, FeO with trace of MnO 6·48, HO 1·36. Carinthia. Allied to Epidote.

Zomidine. A constituent of the extract of flesh. Not precipitated by tannin or HgCl, but by neutral and basic acetates of lead. Considered by Berzelius to give the savour to boiled and roasted meat.

Zootinsalz. *Nitrate of Soda.*

Zurlite. A synonyme of *Humboldtilite.*

Zwieselite. *Phosphate of Iron and Manganese.* Spec. grav. 3·97, H 5. Hexagonal; lustre greasy; colour clove-brown; streak grayish-white. Fracture uneven, or imperfect conchoidal. B.B. decrepitates, and melts to a bluish-black magnetic glass. PO_5 30·33, FeO 41·42, MnO 23·25, F 6. Zwiesel, Bavaria.

Zygadite. Allied to *Castor.* Spec. grav. 2·511. Broad rhombic prism. Lustre glassy; colour red and yellowish-white. Feebly transparent. Contains silica, alumina, and lithia, without water.

SUPPLEMENT.

Acetidine. $C_{10}H_9O_7$. Spec. grav. 1·184. A fluid by acetic acid on glycerine at 392° to 527°.

Acetine. $C_{10}H_{10}O_8$. Spec. grav. 1·2. Neutral ethereal fluid by the action of acetic acid on glycerine at 212°.

Acetonic Acid. $HOC_8H_7O_5$. Prisms with an acid taste, homologous with lactic acid; soluble in water, alcohol, and ether; fused with hydrate of potash it yields acetone; decomposed by sulphuric acid. With zinc forms a salt with 2 atoms water, similar to lactate of zinc, but which is little soluble in water, and insoluble in alcohol and ether.

Acetonine. $C_{18}H_{18}N_2$. Colourless fluid base with alkaline reaction, a smell of urine, and a peculiar taste; soluble in water, alcohol, and ether; obtained by allowing acetone and ammonia to evaporate spontaneously, or by heating at 212° in a sealed tube acetone saturated with ammonia.

Acids. Opposite Permanganic *substitute* Mn_2O_7.

Aethalone. $C_{62}H_{62}O_2$. Pearly plates by distilling aethalic acid with hydrate of lime.

Alban. F.P. 320°. White resin by alcohol from gutta percha.

Alcoometry. Page 24, 2d col., line 21 from bottom, *substitute* yields no green colour and no odour of aldehyde.

Alphene, Sulphide of. $C_{10}H_{10}N_{10},S_2$. White body, little soluble in water; fusing by heat; produced in the decomposition of sulphocyanide of ammonium.

Alum Shale. My pupil, Mr. John Wilson, junr., has found some of the Hurlet shale to consist of SiO_3 48·28, Al_2O_3 26·99, FeO 3·67, CaO 2·38, HO 2·02, Fe 5·19 (FeS_2 11·13), S 5·94, C 4·02, H 1·07.

Amylo Urethane. For carbonate, read carbamate.

Anilides. A class of bodies, the first of which was derived from aniline, produced by the separation of oxygen and hydrogen in the same proportion as in water from the salts of aniline; they are similar to amides.

Arachic Acid. $HO,C_{40}H_{39}O_3 = C$ 76·84, H 12·96. F.P. 167°. Small shining pearly plates, soluble in cold and hot alcohol and ether; obtained by saponification from ground nut oil (Arachis hypogæa). It agrees in its characters with stearic acid.

Atomic Weights. Page 82, Palladium, *substitute* 666·25.

Azophenylamine. $H_2, C_{12}H_5NO_4$, N,N. Red needles by NH_4S on dinitraniline.

Azorite. *Columbate of Lime?* White 8-hedrons, translucent or opaque, with a vitreous fracture. B.B. infusible; with borax slowly forms a transparent globule, sometimes with a green tint; with microcosmic salt a pale green bead. Azores, on albite, tourmaline, and pyrölite.

Bases, Organic, add the following:—

Aconitin,	$C_{60}H_{27}NO_{14}$
Ammelide,	$C_{12}H_9N_9O_6$, or $C_6H_4N_4O_4$
Amylamine urea,	$C_{12}H_{14}N_2O_2$
Aniline urea,	$C_{14}H_8N_2O_2$
Benzidine,	$C_{12}H_6N$
Biethylaniline,	$C_{20}H_{15}N$
Biethylotoluidin,	$C_{22}H_{17}N$ (?)
Bichlorocinchonin,	$C_{38}H_{20}Cl_2N_2O_2$
Bicyanocodein,	$C_{40}H_{21}N_3O_6$
Binitranisidine,	$C_{14}H_7N_3O_{10}$
Biplatinamin,	$H_8PtN_2O_4$ (?)
Biplatosamin,	H_6PtN_2O
Bromocinchonin,	$C_{38}H_{20}BrN_2O_2$
Bromocodein,	$C_{36}H_{20}BrNO_6$
Butyrothialdin,	$C_{24}H_{25}NO_4$ (?)
Carbothialdin,	$C_5H_5NS_2$
Chlornicin,	$C_{10}H_6ClN$
Chlorostrychnine,	$C_{42}H_{21}ClN_2O_4$
Cumidine,	$C_{18}H_{13}N$
Cyanoæthin,	$C_{18}H_{15}N_3$
Diphenine,	$C_{12}H_6N_2$ / $C_{24}H_{12}N_4$
Ethylamin urea,	$C_6H_8N_2O_2$
Ethylobiplatosamin,	C_8H_{14},Pt,N_2O
Ethyl-chloranilin,	$C_{16}H_{10}ClN$
Fabine,	$C_{10}H_6N$ (?)
Flavine,	$C_{13}H_6NO$ / $C_{26}H_{12}N_2O_2$
Harmaline,	$C_{27}H_{14}N_2O_2$ / $C_{28}H_{14}N_2O_2$ (?)
Harmine,	$C_{27}H_{12}N_2O_2$ / $C_{28}H_{12}N_2O_2$
Mercuramine,	$H_3N,4HgO,2HO$
Methyl-ethyl-diethyl-ammonum (hydrate),	$C_{20}H_{24}NO\,HO$

2 N

Methylamin-urea, $C_4H_6N_2O_2$
Methylamylanilin, $C_{24}H_{19}N$
Methyl-biplatosamin, $C_4H_{10}PtN_2O$
Metoluidin, $C_{30}H_{17}N_3$

Bdellium. *Substitute* Heudelotia for Hendolotia.

Benzo-helicine. $C_{40}H_{20}O_{16}$. Needles by NO_5 of 1·30 on populin; containing benzoic acid hydride of salicyle and sugar.

Benzoycine. $C_{20}H_{12}O_8$. Spec. grav. 1·228. By benzoic acid on glycerine at 392° to 518°.

Bismuth Teroxide. For soluble, read insoluble in caustic soda.

Butic Acid. $C_{40}H_{40}O_4$? An acid said to exist as butine in butter.

Butyrine. There are three butyrines. Monobutyrine, $C_{14}H_{14}O_8$, spec. grav. 1·088, by excess of glycerine on butyric acid at 392°; dibutyrine, $C_{22}H_{22}O_{12}$, spec. grav. 1·081, at 392° to 500°; and butyridine, $C_{14}H_{13}O_7$, spec. grav. 1·084, oily fluid by 4 butyric acid on 1 glycerine at 392°.

Butyrolimnodic Acid. $C_{32}H_{32}O_4$. Silky crystals by saponification from bog butter

Caffeone. Brown liquid oil; slightly soluble in water; heavier than water; soluble in ether; obtained by distilling infusion of coffee.

Camphorine. By glycerine on camphoric acid.

Carbanilidic Acid, or *Benzamic Acid.*

Carbothiacetonine. $C_{20}H_{18}N_2S_4$. Yellow crystals formed by allowing acetone, carbide of sulphur, and ammonia to stand in contact with each other. It is formed by the replacement of 3 atoms acetone, by 2 atoms carbosulphide of ammonium; it is therefore the sulphohydride of a sulphur base.

Castelnaudite. *Hydrous phosphate of Yttria.* H 4·5. Grayish-white 4-sided prisms, terminated with 4-sided pyramids, with angles of 96° 35′, and 124° 23′ 30″. B.B. infusible; gives a colourless opaque bead with borax and microcosmic salt; soluble in sulphuric acid.

Chiviatite. Spec. grav. 6·92. $2(PbS,Cu_2S)$ $3Bi_2S_3$, or S 18·, Bi 60·95, Pb 16·73, Cu 2·42, Fe 1·02, Ag trace, insol. mat. ·59. Lead-gray mineral with metallic lustre, cleaving in the direction of a plane to which the second is inclined at 153°, and the third at 133°. B.B. behaves like needle ore of bismuth. Civiato, Peru.

Chlorhydrine. $C_6H_7ClO_4$. Sp. grav. 1·31. Neutral oil by heating glycerine with HCl gas at 212° for thirty-six hours.

Citranilide. $C_{48}H_{23}N_3O_3$. Colourless truncated prisms = tribasic citrate of aniline deprived of 6 atoms water; by the action of citric acid on aniline.

Citrene. $C_{20}H_{16}$. Spec. grav. ·847, of vapour 4·73; B.P. 325°. Obtained by passing HCl gas into oil of lemons and distilling the camphor ($C_{20}H_{16}HCl$) with water, and then passing it through a tube filled with lime at 320°; has not the property of deviating the rays of polarized light to the right.

Citrilene. *Citronile.* Spec. grav. ·88, of vapour 5·08; B.P. 331° to 347°. Volatile oil by decomposing the liquid camphor of oil of lemons.

Citrobianile. $C_{36}H_{16}N_2O_8$. 6-sided plates = bibasic citrate of aniline, less 6 atoms water; in the formation of citranilide.

Citrobianilic Acid. $C_{36}H_{18}N_2O_{10}$. Silky needles = citrobianile + 2 HO; by adding ammonia to citrobianile.

Citromonanilic Acid. $C_{24}H_{11}NO_{10}$ = equicitrate of aniline — 4 HO; by heating equicitrate of aniline between 284° and 302°.

Copper, p. 184. For column under, "after calcining," *substitute* copper 16·42, sulphur 4·56, Fe_2O_3 40·, silica 38·.

Cyanoxysulphide. A synonyme of Sulphocyanogen.

Elaoptene *Oleoptene.* Fluid part of essential oils.

Erythroleic Acid. $C_{26}H_{22}O_8$. Purple fluid; soluble in ether and alcohol, but almost insoluble in water; soluble in alkalies; obtained by evaporating the ether solution from Orchil.

Ethers. A number of compounds have been obtained by replacement, formed on the type of water. Ethyl-œnanthyle ether = C_2H_5,C_7H_{15},O; Methyl-œnanthyl ether, $C_2H_3,C_7H_{15}O$; Amyl-œnanthyle ether, C_5H_{11},C_7H_{15},O, by treating œnanthylic alcohol with iodides of ethyle, methyle, and amyle.

Ethyle-toluidine. By the action of iodide of ethyle in a closed tube on toluidine four bases have been obtained, three of the ammonia, and the fourth of the ammonium type.

Fluavile. F.P. 212°. Yellow resin from gutta percha by alcohol.

Fluosilicanilide. $C_{48}H_{33}F_{11}Si_4O_6$. White body by terfluoride of silicon on aniline.

Geoceraine. $C_{56}H_{56}O_4$. Wax by distilling brown coal.

Geoceric Acid. $C_{56}H_{56}O_4$. Gelatinous mass by alcohol from brown coal.

Geocerinone. $C_{110}H_{110}O_2$. 6-sided tables from the distilled matter from yellowish brown coal.

Geomyricine. $C_{68}H_{68}O_4$. Hair crystals by alcohol from brown coal.

Georetic Acid. $C_{24}H_{21}O_7$. Needles by ether and alcohol, from the wax and resins of brown coal from Gerstewitz.

Heterologous. Dissimilar bodies which result the one from the other. The formic, acetic, and propionic series are examples.

Hidrotic Acid. $C_{10}H_8NO_{13}$. Syrupy fluid, obtained from sweat by alcohol, in which it is soluble.

Hipparaffine. $C_{16}H_8NO_2$. F.P. 392°. Prismatic needles, little soluble in boiling water; neutral; soluble in alcohol; very soluble in ether; not decomposed by boiling HCl and SO_3; decomposed by NO_5; slightly sublimable; formed along with benzamide by treating hippuric acid with binoxide of lead.

Homologous. Bodies are said to be homologous when their composition differs by a constant quantity or multiple of it. Thus the formula of methyle differs from that of ethyle by C_2H_2; and a series of homologues exist of which methyle may be viewed as the simplest term. The fatty acids enumerated under that title are homologues. The organic radicals, their alcohols and derived acids, are examples of homologous bodies.

Hydrargyromethilium and **Methide,** instead of Hydrargyromethylnine and Methose.

Iridium and **Ammonia.** *New bases.* The ammoniacal protochloride of iridium, yields with NO_5 the base $IrClN_2H_6O$, corresponding with the platinum base of Gros, and with SO_3 the base $IrONH_3$. Another base is $IrON_2H_6$.

Iron, page 305, 2d column, line 23, introduce, instead of the "carbon is," "the carbonaceous iron is."

Isologous. Similar bodies are so termed which do not present the same relation in their differences as in homologous bodies. Acetic and benzoic acids are isologous.

Leucene, Sulphide of. $C_6H_5N_5S_2$. Grayish product by heating persulphocyanohydric acid to 437°.

Leucopetrin. $C_{50}H_{42}O_3$. Needles from the brown coal of Gerstewitz by ether and alcohol.

Liliacine, *substitute* for **Lilaicne.**

Manganate and **Permangate** of potash. *Manganic Acid.* Mn O_8 6·5. Formed by fusing equal parts of black oxide and saltpetre or caustic potash; a black mass results, which becomes in water green, blue, purple, and reddish-brown, turning into the red oxide. This substance is termed the chameleon mineral, in consequence of the changes of colour depending on the absorption of oxygen. *Hypermanganic acid,* Mn_2O_7 14·, prepared by heating $3\frac{1}{2}$ parts chlorate of potash, 4 parts black oxide of manganese, and 5 parts dry caustic potash dissolved in water. (*Gregory*). The whole is evaporated to dryness in a porcelain crucible, and exposed to a low red heat, but not fused. The powder is taken out, digested in water, the clear liquor poured off and evaporated; black crystals of hypermanganate of potash crystallize out. The solution of these crystals in water is finely red, but cannot be concentrated, in consequence of the tendency of the acid to give out oxygen. When passed through filtering paper it decomposes.

Margarine. $C_{40}H_{40}O_8$. F.P. 132°·8. Formed at ordinary temperatures by margaric acid in glycerine. Tetramargarine is obtained by heating monomargarine at 212° with excess of margaric acid.

Niobium. It has been recently ascertained by H. Rose that niobium and pelopium are identical; he proposes to retain the name of niobium.

Octyle. A synonyme of Hydrous oxide of capryle.

Oleine. $C_{42}H_{40}O_8$. Spec. grav. ·947; F.P. 59°. *Dioleine.* $C_{78}H_{74}O_{12}$. Spec. grav. ·921; F.P. 59°, by oleic acid on glycerine.

Organoleptism. The effect produced on the organs of sense by bodies, as touch, taste, smell, &c.

Othyle. A hypothetic radical. The othyle theory represents acetic acid as containing the radical C_2H_3O (derived from ethyle C_2H_5 by the substitution of O for h_2). It connects the formulæ of the acetic series with that of water hh,O, by writing them in the following manner:—

Formic Acid.	Acetic Acids.
$\begin{matrix} C\,h\,O \\ h \end{matrix}\ O$	$\begin{matrix} C_2\,h_3\,O \\ h \end{matrix}\ O$

Palmitine. $C_{38}H_{38}O_8$. F.P. 136°. Dipalmitine, $C_{70}H_{70}O_4$, F.P. 138°. Tetrapalmitine, F.P. 140°, are obtained by the action of palmitic acid on glycerine.

Phalene, Sulphide of. $C_{12}H_{12}N_{12}S_2$. A body produced in the decomposition by heat of sulphocyanide of ammonium.

Pigments. See blue, black, green, red, yellow, &c. colours.

Poliene. $C_4H_4N_4$. A substance formed in the decomposition of sulphocyanide of ammonium.

Pyropine. C 53·33—53·5, H 7·52—7·66, N 15, O 23·63, ash ·52. A beautiful ruby-coloured substance, formed in the interior of the tusk of an elephant; insoluble in water, but becoming soft in it, and sometimes smelling of glue; the ash is red.

Salicyle-methyle, Benzoate of. $C_7H_4(CH_3)O_2 + C_7H_5O,O$.

Salicyle-methyle, Cuminate of. $C_7H_4(CH_3)O_2 + C_{10}H_{11}O,O$.

Salicyle-methyle, Hydrate of. $C_7H_4(CH_3)O_2HO$.

Sebine. $C_{32}H_{30}O_{10}$. Crystals by glycerine on sebacic acid at 392°.

Stearine. By heating equal parts stearic acid and glycerine at 500° for twenty-four hours, monostearine in needles results $= C_{42}H_{42}O_8$; equal parts heated to 212° for a hundred and fourteen hours, give distearine in plates; by heating natural stearine with excess of glycerine at

392° for twenty-two hours, tetrastearine is formed ($C_{150}H_{146}O_{16}$).

Tartronic Acid. $C_6H_4O_{10}$. F.P. 320°. Obtained by the spontaneous decomposition of nitrotartaric acid in water. When heated to 356° it is converted into glycolic acid ($C_4H_4O_6$). *Glycollide.*—$C_4H_2O_4$. Insoluble body proceeding from the dry distillation of tartronic acid; it is converted into glycolic acid by alkalies. *Glycolamide.*—$C_4H_5N_2O_4$. Fine crystals by heating bitartronate of ammonia, or by adding ammonia to glycollide.

Theory of Acids and Bases, &c. According to Gerhardt the greater number of organic compounds at present known are derived from a limited number of types in inorganic chemistry, viz. water, chlorohydric acid, ammonia, &c. The derivatives of these types form a series, the properties of each differing, but being in progression. The acids are placed at one extremity, which may be termed the *negative* end, the bases at the other or positive end, while the neutral bodies placed in the intermediate space constitute the transition point of the series. Thus methyle, ethyle, &c. when replacing hydrogen in the above types, produce positive or basic derivatives; while benzoyle, acetyle, cumyle, give origin to negative derivatives or bodies more or less analogous to acids. The following is an illustration of water as a type, viewing its formula as H_2O:—

Water.	Acetic Acid Hydrous.	Do. Anhydrous.
$\left.\begin{matrix}H\\H\end{matrix}\right\}O$	$\left.\begin{matrix}C_2H_3O\\H\end{matrix}\right\}O$	$\left.\begin{matrix}C_2H_3O\\C_2H_3O\end{matrix}\right\}O$

Train Oil. From the Balena misticetus and rostrata. Spec. grav. ·927; B.P. 600°. Colour brown. 100 alcohol dissolve 122 parts.

Xanthoxyline. C 61·09, H 6·45, N ? White oblique prisms from Japanese pepper (Xanthoxylum piperitum) by alcohol.

GLASGOW: PRINTED BY BELL AND BAIN.

Printed in Dunstable, United Kingdom

91220875R00312